Cambridge Handbook of Engineering Education Research

The *Cambridge Handbook of Engineering Education Research* is the critical reference source for the growing field of engineering education research, featuring the work of world experts writing to define and inform this emerging field. Since a landmark issue of the *Journal of Engineering Education* in 2005, in which senior scholars argued for a stronger theoretical and empirically driven agenda, engineering education has quickly emerged as a research-driven field with an increasing quality and quantity of both theoretical and empirical work that draws on many social science disciplines, disciplinary engineering knowledge, and computing. The *Handbook* draws extensively on contemporary research within the engineering education community and allied fields such as the learning sciences. The *Handbook* is organized into six parts and thirty-five chapters.

Aditya Johri (Ph.D., Stanford University) is an associate professor of Engineering Education, Computer Science (courtesy), Industrial and Systems Engineering (courtesy), and Science and Technology in Society (courtesy) at Virginia Tech. His current projects focus on situated engineering learning, shifts in engineering work practices due to globalization and technology diffusion, the role of engineers and engineering in development, and learning analytics. He is a past recipient of the U.S. National Science Foundation Early Career Award. Starting January 2014, he will join George Mason University as an associate professor in the Department of Applied Information Technology and director of a research center focused on engineering education and learning analytics research.

Barbara M. Olds is Professor Emerita of Liberal Arts and International Studies at the Colorado School of Mines, where she most recently served as Associate Provost for Educational Innovation. She is also Senior Advisor in the Directorate for Education and Human Resources at the U.S. National Science Foundation (NSF). She served as the Director for the Division of Research, Evaluation, and Communication at NSF from 2003 to 2006. Olds is a Fellow of the American Society for Engineering Education, for which she has served in a number of capacities, including a term on the Board of Directors and as Chair of the International Advisory Committee. She was also a member of the Advisory Committee for the NSF's Office of International Science and Engineering and chaired the Committee of Visitors for that program in 2008. She was a Fulbright lecturer/researcher in Sweden in 1999.

Cambridge Handbook of Engineering Education Research

Edited by Aditya Johri, Virginia Tech, and
Barbara M. Olds, Colorado School of Mines/National Science Foundation

The *Cambridge Handbook of Engineering Education Research* (*CHEER*) will be an important reference source for the growing field of engineering education research (EER). EER has become an increasingly important field internationally, as evidenced by the growing prestige and subscriber base of its key journal, the *Journal of Engineering Education (JEE)*, the founding of several Ph.D.-granting engineering education departments at prestigious institutions, and the growth of an international community of engineering education researchers who hold global meetings and have a variety of publication venues. Despite the tremendous growth of the field, there is currently no book that provides an overview of EER. Thus we believe *CHEER* will fill an important gap internationally in the EER field and will be used as a textbook for graduate courses, a reference book by engineering faculty in disciplinary engineering areas, and a resource by policy makers, K–12 engineering curriculum designers, informal science educators, and others.

Cambridge Handbook of Engineering Education Research

Edited by

Aditya Johri

Virginia Polytechnic Institute and State University

Barbara M. Olds

Colorado School of Mines and National Science Foundation

CAMBRIDGE
UNIVERSITY PRESS

CAMBRIDGE
UNIVERSITY PRESS

32 Avenue of the Americas, New York, NY 10013-2473, USA

Cambridge University Press is part of the University of Cambridge.

It furthers the University's mission by disseminating knowledge in the pursuit of education, learning, and research at the highest international levels of excellence.

www.cambridge.org
Information on this title: www.cambridge.org/9781107014107

© Cambridge University Press 2014

First published 2014

Printed in the United States of America

A catalog record for this publication is available from the British Library.

Library of Congress Cataloging in Publication Data
Cambridge handbook of engineering education research / edited by Aditya Johri, Virginia Polytechnic Institute and State University, Barbara M. Olds, National Science Foundation/Colorado School of Mines.
   pages cm
Includes bibliographical references and index.
ISBN 978-1-107-01410-7 (hardback)
1. Engineering – Study and teaching – Handbooks, manuals, etc.   I. Johri, Aditya, 1976–   II. Olds, Barbara M.
III. Title: Handbook of engineering education research.
T65.C345   2013
620.0071–dc23   2013012156

ISBN 978-1-107-01410-7 Hardback

We dedicate this volume to our esteemed colleagues
Dr. Kamyar Haghighi and Dr. David Jonassen.

Dr. Haghighi was founding head of the School of Engineering Education (2004–10) at Purdue University. He was a national force for engineering education reform based on research and scholarship, and under his leadership, the School of Engineering Education (ENE) at Purdue launched the world's first Ph.D. program in engineering education; attracted a critical mass of nationally and internationally recognized faculty; created INSPIRE, the Institute for P–12 Engineering Research and Learning (the first such institute to combine research on childhood learning of engineering with outreach to teachers); saw the ABET accreditation of the undergraduate Multidisciplinary Engineering Program; and transformed the first-year engineering program with a design-focused curriculum – aligned with the "Purdue Engineer of 2020" paradigm – that integrates seamlessly with the new Ideas to Innovation (i2i) Learning Laboratory. In 2009, the American Society for Engineering Education honored Dr. Haghighi with the Chester F. Carlson Award, which recognizes an individual innovator in engineering education who, by motivation and ability to reach beyond the accepted traditions, has made a significant contribution to the profession.

Dr. David H. Jonassen, who has contributed a chapter to this volume, passed away in December 2012, at his home in Columbia, Missouri, after living life fully, the last two years with advanced-stage lung cancer. David had been a professor of learning technologies and educational psychology at the University of Missouri since 2000. In 2010, he was named curators' professor – the University of Missouri's highest honor for world-renowned scholars. Previously, he held professorships at Penn State University; the University of Colorado, Denver; and the University of North Carolina, Greensboro. Over his nearly 40-year academic career, he wrote 37 books, 182 journal articles, and 67 book chapters, as well as numerous other types of publications. He made 400 presentations in the United States and 28 other countries. He also was an invited visiting scholar in countries such as Australia, Austria, The Netherlands, Norway, Singapore, and the United Kingdom. His work has attracted more than $12 million in external funding from sources such as the National Science Foundation, the U.S. Department of Education, the Australian Research Council, NATO, and the European Union. Dave was the recipient of more than 40 scholarly awards, including 19 awards for outstanding publications and books, the 2001 Presidential Service Award, and the Lifetime Achievement Award for Excellence in Research and Theory, both from the Association for Educational Communications and Technology.

Contents

Editors

ADITYA JOHRI (Ph.D., Stanford University) is an associate professor of Engineering Education, Computer Science (courtesy), Industrial and Systems Engineering (courtesy), and Science and Technology in Society (courtesy) at Virginia Tech. His current projects focus on situated engineering learning, shifts in engineering work practices due to globalization and technology diffusion, the role of engineers and engineering in development, and learning analytics. He is a past recipient of the U.S. National Science Foundation Early Career Award. Starting January 2014, he will join George Mason University as an associate professor in the Department of Applied Information Technology and director of a research center focused on engineering education and learning analytics research. E-mail: ajohri@vt.edu, ajohri@gmail.com.

BARBARA M. OLDS is Professor Emerita of Liberal Arts and International Studies at the Colorado School of Mines, where she most recently served as Associate Provost for Educational Innovation. She is also Senior Advisor in the Directorate for Education and Human Resources at the U.S. National Science Foundation (NSF). She served as the Director for the Division of Research, Evaluation, and Communication at NSF from 2003 to 2006. Olds is a Fellow of the American Society for Engineering Education, where she has served in a number of capacities, including a term on the Board of Directors and as Chair of the International Advisory Committee. She was also a member of the Advisory Committee for the NSF's Office of International Science and Engineering and chaired the Committee of Visitors for that program in 2008. She was a Fulbright lecturer/researcher in Sweden in 1999. E-mail: bolds@mines.edu.

Contributors

Advisory Board

R. KEITH SAWYER, Washington University

JACK R. LOHMANN, Georgia Tech

KARL A. SMITH, Purdue University

WENDY C. NEWSTETTER, Georgia Tech

Contributing Authors

ANTHONY LISING ANTONIO is Associate Professor of Education at Stanford University. His research focuses on stratification and postsecondary access, racial diversity and its impact on students and institutions, student friendship networks, and student development. In 2004, he received the Promising Scholar/Early Career Award from the Association for the Study of Higher Education. His current research projects include studies of engineering education and college counseling programs in schools. E-mail: aantonio@stanford.edu.

CYNTHIA J. ATMAN is the founding director of the Center for Engineering Learning & Teaching (CELT), a professor in Human Centered Design & Engineering, and the inaugural holder of the Mitchell T. & Lella Blanche Bowie Endowed Chair at the University of Washington. She

also directed the NSF-funded Center for the Advancement of Engineering Education (CAEE). Her research centers on engineering design learning with a focus on issues of context in design. E-mail: atman@uw.edu.

CAROLINE A. BAILLIE is the Chair in Engineering Education at The University of Western Australia. Caroline has been involved in engineering education for many years and has extensive experience in threshold concepts. She is currently directing a major national and international project to review potential thresholds in first-year engineering courses globally. Caroline coedited the most recent book on threshold concepts, *Threshold Concepts and Transformational Learning*, containing papers presented at the conference she hosted in Ontario in 2009. E-mail: caroline.baillie@uwa.edu.au.

BROCK E. BARRY is an associate professor in the Department of Civil & Mechanical Engineering at the U.S. Military Academy, West Point. He is licensed as a professional engineer and accumulated ten years of practice experience prior to entering academics. Dr. Barry is a graduate of Purdue University's Engineering Education Ph.D. program and also holds a master's degree from the University of Colorado at Boulder and a bachelor's degree from Rochester

Institute of Technology. E-mail: Brock.Barry@ usma.edu.

ANGELA R. BIELEFELDT is a professor in the Department of Civil, Environmental, and Architectural Engineering at the University of Colorado at Boulder. She currently serves as the Associate Chair for Undergraduate Education for this department. She holds a B.S. in Civil Engineering from Iowa State University and M.S. and Ph.D. degrees in Civil Engineering from the University of Washington. She is also a licensed P.E. Her engineering educational research focuses around sustainability, service-learning, first-year students, and gender issues. E-mail: Angela.Bielefeldt@colorado.edu.

JIM L. BORGFORD-PARNELL is the assistant director and instructional consultant for the Center for Engineering Learning & Teaching at the University of Washington. He taught design drawing, design theory, research methods, educational theory, and adult and higher education pedagogy courses for more than twenty-five years. Jim has been involved in instructional development for more than ten years and currently does both research and instructional development in engineering education. E-mail: bparnell@u .washington.edu.

MAURA BORREGO is an associate professor in the Department of Engineering Education at Virginia Tech, serving as a Program Director at the National Science Foundation. She is an associate editor for the *Journal of Engineering Education*. Dr. Borrego's engineering education research awards include PECASE, CAREER, and two outstanding publication awards from the American Educational Research Association. She holds a Ph.D. in Materials Science and Engineering from Stanford University. E-mail: mborrego@vt.edu.

REBECCA BRENT is President of Education Designs, Inc., a consulting firm in Cary, North Carolina. Her areas of expertise are faculty development in engineering and the sciences, evaluation of educational programs at both precollege and college levels, and classroom uses of instructional technology. Dr. Brent received her B.A. in Music Education from Millsaps College in 1978, her M.Ed. from Mississippi State University in 1981, and her Ed.D. from Auburn University in 1988. From 1997 to 2003, she directed the NSF-sponsored SUCCEED Coalition faculty development program, and she currently coordinates faculty development activities for the North Carolina State University College of Engineering. E-mail: rbrent@mindspring.com.

SEAN P. BROPHY is an associate professor in the School of Engineering Education at Purdue University. Dr. Brophy is a mechanical engineer with experience in aerospace and automobile control systems applications. His advanced degrees are in computer science, education, and human development. His areas of research include adaptive expertise, reasoning with mathematics and models, conceptual change using simulations, technology-supported learning environments, and designing assessment for learning. E-mail: sbrophy@purdue.edu.

SHANE BROWN is an associate professor in the School of Civil and Construction Engineering at Oregon State University. He earned a B.S. and Ph.D. in Civil Engineering from the Oregon State University and an M.S. in Environmental Engineering from University of California, Davis. He is a recipient of the National Science Foundation CAREER Award. His research interests are in conceptual change and situated cognition, particularly an understanding of how science and engineering concepts are used in the engineering workplace. E-mail: shane.brown@oregonstate.edu.

SAMANTHA R. BRUNHAVER is a Ph.D. candidate in Mechanical Engineering at Stanford University. Her research focuses on early career engineering pathways and the effect of sociocognitive factors on career persistence. Samantha is a Stanford Diversifying Academia Recruiting Excellence (DARE) Fellow. She holds a B.S. in Mechanical Engineering from Northeastern University and an M.S. in Mechanical Engineering from Stanford. E-mail: sbrunhaver@stanford.edu.

MONICA E. CARDELLA is an associate professor in the School of Engineering Education at Purdue University. She holds a B.Sc. (Mathematics) from the University of Puget Sound and an M.S. and Ph.D. (Industrial Engineering) from the University of Washington. Her NSF CAREER Award examines the interplay between design thinking and mathematical thinking. She is also the Director of Informal Learning Environments Research for the Institute for P–12 Engineering Research and Learning (INSPIRE). E-mail: cardella@purdue.edu.

JENNIFER M. CASE is a professor in the Department of Chemical Engineering at the University of Cape Town, with a special responsibility for academic development. She teaches in the undergraduate chemical engineering program, and her research on the student experience of learning has been widely published. She is the founding president of the South African Society for Engineering Education (SASEE). She currently acts as Assistant Dean for Academic Development in the Faculty of Engineering and the Built Environment. E-mail: jenni.case@uct.ac.za.

HELEN L. CHEN is a researcher in the Department of Mechanical Engineering and the director of ePortfolio initiatives in the Office of the Registrar at Stanford University. Her research interests include academic and professional persistence in engineering education and the use of ePortfolios for teaching, learning, and assessment. She and her colleagues Tracy Penny-Light and John Ittelson are the authors of *Documenting Learning with ePortfolios: A Guide for College Instructors* (2011). E-mail: hlchen@stanford.edu.

JOHN C. CHEN is Professor of Mechanical Engineering at California Polytechnic State University. His research interests include thermal methods for nondestructive evaluation and testing of advanced engineering materials, student autonomy and lifelong learning, and active learning techniques. He is an associate editor for *Advances in Engineering Education*, an online journal of the American Society for Engineering Education. E-mail: jchen24@calpoly.edu.

CLAIRE DAVIS studied at the University of Cambridge, leaving with a first class honours degree in Natural Sciences and a Ph.D. in Metallurgy. She then joined the University of Birmingham, where she is now a Professor in Ferrous Metallurgy within the College of Engineering and Physical Sciences. Her interests are in microstructure-property relationships in metals, covering topics such as solidification and segregation, rolling, and microstructure development, predominantly in steel. She is a National Teaching Fellow and has held an ExxonMobil Excellence in Teaching award and a Birmingham University teaching fellowship. She is particularly interested in case study learning, student engagement, and public engagement with science and technology. E-mail: c.l.davis@bham.ac.uk.

ERIK DE GRAAFF is a psychologist specializing in higher education. After working on medical education at Limburg University in Maastricht and on engineering education at Delft University of Technology he was appointed as a professor at Aalborg University in 2011. He has published more than a hundred papers and presented more than fifty keynotes and invited lectures. Since 2008 he has been editor-in-chief of the *European Journal of Engineering Education* (EJEE). E-mail: degraaff@plan.aau.dk.

LOUIS V. DIBELLO is Associate Director of the Learning Sciences Research Institute and Research Professor at the University of Illinois at Chicago. His research interests include measurement and psychometrics with a focus on diagnostic classification modeling. After working in research within the testing industry for twelve years, he currently is collaborating on research projects focused on a variety of classroom assessments, including embedded assessments in elementary school mathematics curricula and the Physics Diagnoser facet-based assessments of Jim Minstrell and colleagues. E-mail: ldibello@uic.edu.

GARY LEE DOWNEY is Alumni Distinguished Professor of Science and Technology Studies at Virginia Tech. A mechanical engineer (Lehigh) and cultural anthropologist (Chicago), he works to build the field of engineering studies. His *The Machine in Me* critically examined engineers' claims of jurisdiction over technology, and *Cyborgs and Citadels* theorized critical participation in engineering as "hiring in." He edits the *Engineering Studies* journal, the Engineering Studies series, and the Global Engineering series. E-mail: downeyg@vt.edu.

OZGUR ERIS is an Associate Professor of Design Theory and Methodology at Delft University of Technology. His research interests include design thinking and theory, design informatics, and engineering education. He received a B.S. from the University of Washington and an M.S. and a Ph.D. in Mechanical Engineering with a focus on design from Stanford University. He is the author of *Effective Inquiry for Innovative Engineering Design* (2004) and has been the recipient of ASEE's William Elgin Wickenden award. E-mail: O.Eris@tudelft.nl.

RICHARD M. FELDER is the Hoechst Celanese Professor Emeritus of Chemical Engineering at North Carolina State University. He is coauthor of *Elementary Principles of Chemical Processes* (2008), an introductory chemical engineering text now in its third edition. He has contributed more than 200 publications to the fields of science and engineering education and chemical process engineering, and he writes "Random Thoughts," a column on educational methods and issues for the quarterly journal *Chemical Engineering Education*. E-mail: rmfelder@mindspring.com.

NORMAN L. FORTENBERRY is the executive director of the American Society for Engineering Education, an international society of individual, institutional, and corporate members. Previously, Fortenberry served as the founding Director of the Center for the Advancement of Scholarship on Engineering Education at the National Academy of Engineering. He served in various executive roles at the National Science Foundation. Fortenberry has also served as executive director of the National Consortium for Graduate Degrees for Minorities in Engineering and Science, Inc., and as a faculty member in the department of mechanical engineering at Florida A&M University – Florida State University College of Engineering. Dr. Fortenberry was awarded the S.B., S.M., and Sc.D. degrees (all in mechanical engineering) by the Massachusetts Institute of Technology. E-mail: N.Fortenberry@asee.org.

JEFFREY E. FROYD is a research professor in the Center for Teaching Excellence and Director of Academic Development in the Texas Engineering Experiment Station. He served as Project Director for the Foundation Coalition, an NSF Engineering Education Coalition and helped create the Integrated, First-Year Curriculum in Science, Engineering and Mathematics at Rose-Hulman Institute of Technology. His current interests are learning and faculty development. E-mail: froyd@tamu.edu.

SHANNON K. GILMARTIN is Consulting Associate Professor in Mechanical Engineering at Stanford University and Director of SKG Analysis. Her research focuses on education and workforce development in science and engineering fields. Shannon received her B.A. in American Studies at Stanford University and her M.A. and Ph.D. in Education at UCLA. E-mail: sgilmartin@skganalysis.com.

ELIZABETH GODFREY is an independent consultant in engineering education, working with universities in Australia and New Zealand offering research support and project management. Her research predominantly relates to understanding engineering education culture and its interaction with gender, and transformational change. More than twenty years of involvement with the University of Auckland and the Australasian Association for Engineering Education has also resulted in research collaborations on topics such as attrition, academic support, and pathways and access for women and indigenous students. E-mail: lizgodfrey@gmail.com.

JOSEPH R. HERKERT is Lincoln Associate Professor of Ethics and Technology in the School of Letters and Sciences and the Consortium for Science, Policy & Outcomes at Arizona State University. He has taught engineering ethics and related courses for twenty-five years. His work on engineering ethics has appeared in engineering, law, social science, and applied ethics journals. Dr. Herkert is the past editor of *IEEE Technology & Society* and an associate editor of *Engineering Studies*. E-mail: joseph.herkert@asu.edu.

GEOFFREY L. HERMAN is a visiting assistant professor with the Illinois Foundry for Innovation in Engineering Education at the University of Illinois at Urbana-Champaign. He holds a Ph.D. in Electrical and Computer Engineering from the University of Illinois and was a postdoctoral researcher in the School of Engineering Education at Purdue University. He currently develops theories about conceptual change and effects education reform by promoting students' intrinsic motivation to learn in traditional courses. E-mail: glherman@illinois.edu.

JOHN HEYWOOD. After R&D work in the radio communications industry he began teaching. In 1960 his first research in technical education was published. In addition to several policy-inspired investigations he has authored some sixty papers and articles in the field of engineering and technology education. He is a Professorial Fellow Emeritus of Trinity College Dublin; a Life Senior MIEEE; a Fellow of ASEE, and a Fellow of the Chartered Management Institute. E-mail: heywoodj@eircom.net.

BRENT K. JESIEK is an assistant professor in the Schools of Engineering Education and Electrical and Computer Engineering at Purdue University. He is also an Associate Director of Purdue's Global Engineering Program and leads the Global Engineering Education Collaboratory research group. Dr. Jesiek draws on expertise from engineering, computing, and the social sciences to advance understanding of geographic, disciplinary, and historical variations in engineering education and professional practice. E-mail: bjesiek@purdue.edu.

DAVID H. JONASSEN (1947–2012) was Curators' Professor of Learning Technologies and Educational Psychology at the University of Missouri. He had previously taught at Pennsylvania State University, University of Colorado, University of Twente, the University of North Carolina at Greensboro, and Syracuse University. During his academic career he published thirty-five books and hundreds of articles, papers, and reports. His research focused on the cognition of problem solving, culminating in *Learning to Solve Problems: A Handbook for Designing Problem-Solving Learning Environments* (2010).

ANTHONY E. KELLY is Professor of Educational Psychology at George Mason University. His teaching and research interests include research, methodology, and assessment. Twice theme editor for the *Educational Researcher* journal, he will edit special issues of *Irish Education Studies* and *Theory Into Practice*. He was program director at the U.S. National Science Foundation and special advisor on research to the U.S. Department of Education's Office of Educational Technology. He was a 2009 Fulbright Fellow. E-mail: akelly1@gmu.edu.

ANETTE KOLMOS is Professor in Engineering Education and PBL and Chairholder for UNESCO in Problem Based Learning, Aalborg University. She was the president of SEFI 2009–2011 (European Society for Engineering Education), Founding Chair of the SEFI-working group on EER, and associate editor for the *European Journal of Engineering Education*. She is a member of several organizations and committees within EER, national government bodies, and committees in the EU. E-mail: ak@plan.aau.dk.

JANET L. KOLODNER is Regents' Professor in Interactive Computing at the Georgia Institute of Technology, co-founder and former Executive Officer of the International Society of the Learning Sciences (ISLS, http://www.isls.org), and Founding and Emerita Editor-in-Chief of *The Journal of the Learning Sciences*. Her Learning by Design project is described at http://www.cc.gatech.edu/projects/lbd/home.html. Project-Based Inquiry Science, her middle school science curriculum, is described at http://www.its-about-time.com/pbis/pbis.html. She is currently on loan to the National Science Foundation, managing the new program called Cyberlearning: Transforming Education. E-mail: jlk@cc.gatech.edu or jkolodne@nsf.gov.

LISA R. LATTUCA is a professor of higher education in the University of Michigan's Center for the Study of Higher and Postsecondary Education. For the past decade she has been studying engineering education with collaborators from the fields of engineering and education. These studies rely on both quantitative and qualitative research methods to identify curricular, instructional, and organizational conditions that foster the development of key engineering skills, including design, problem solving, and contextual and interdisciplinary competencies. E-mail: llatt@umich.edu.

JON A. LEYDENS is an associate professor in the Division of Liberal Arts and International Studies at the Colorado School of Mines, where he has been since 1997. His research and teaching interests include communication and social justice. Dr. Leydens is a coauthor of *Engineering and Sustainable Community Development* (2010). He recently served as guest editor for an engineering communication theme issue of *Engineering Studies* and designed a course on Rhetoric, Energy, and Public Policy. E-mail: jleydens@mines.edu.

GARY LICHTENSTEIN is the principal of Quality Evaluation Designs, a firm specializing in education research, evaluation, and policy. His intellectual interests include engineering education, mixed-methods research, and community-based research. He has conducted research and evaluation for K–12 schools, higher education institutions, and nonprofit and government organizations nationwide. He has researched multiple segments of the STEM pathway on projects with the National Academy of Engineering, the Center for the Advancement of Engineering Education, and the Carnegie Foundation for the Advancement of Teaching. E-mail: gary@QualityEvaluationDesigns.com.

GREGORY LIGHT is the director of the Searle Center for Teaching Excellence and an associate professor in the School of Education and Social Policy at Northwestern University. His research focuses on the theory and practice of learning and teaching in higher and professional education. He is the author of *Learning and Teaching in Higher Education: The Reflective Professional* (2001, 2009) and the forthcoming *Making Scientists: Six Principles for Effective College Teaching*. E-mail: g-light@northwestern.edu.

EUAN D. LINDSAY is the Dean of the School of Engineering & Built Environment at Central Queensland University. His research interests focus on remote laboratory classes and the way in which the combination of physical separation and technology-mediated interfaces impacts on students' learning outcomes. He was the 2010 President of the Australasian Association for Engineering Education and is a Fellow of the U.K. Higher Education Academy. E-mail: e.lindsay@cqu.edu.au.

THOMAS A. LITZINGER is director of the Leonhard Center for the Enhancement of Engineering Education and a professor of Mechanical Engineering at Penn State University. He has been involved in engineering education research and reform efforts for more than twenty years. He also conducts research on the effects of fuel composition on emissions from gas turbines and internal combustion engines. He is a Fellow of the ASEE and ASME. E-mail: tal2@psu.edu.

JACK R. LOHMANN is professor emeritus at Georgia Institute of Technology where he served as vice provost for Faculty and Academic Development and professor of Industrial and Systems Engineering. His principal responsibilities included faculty development and promotion; the initiation, development, and accreditation of Georgia Tech's academic programs; and serving as the president's liaison to the Commission on Colleges of the Southern Association of Colleges and Schools (COC/SACS) and the National Collegiate Athletic Association. He was also a member of the Board of Trustees of the COC/SACS. Dr. Lohmann has held appointments at the University of Michigan, the University of Southern California, l'Ecole Centrale Paris, and the National Science Foundation. Dr. Lohmann served as the editor of the *Journal of Engineering Education* and is a Fellow of the Institute of Industrial Engineers, the American Society of Engineering Education, and the European Society for Engineering Education. E-mail: jack.lohmann@carnegie.gatech.edu.

SUSAN M. LORD is Professor and Coordinator of Electrical Engineering at the University of San Diego. Her teaching and research interests include optoelectronics, student persistence, and lifelong learning. She was guest coeditor of the special issue of the *International Journal of Engineering Education* on Applications of Engineering Education Research. She and colleagues won the 2011 best paper awards for the *Journal of Engineering Education* and the *IEEE Transactions on Education*. E-mail: slord@sandiego.edu.

KRISHNA MADHAVAN is an assistant professor of Engineering Education at Purdue University. He is also the Education Director for the NSF-funded Network for Computational Nanotechnology (http://www.nanohub.org). His work focuses on large-scale data analysis and interactive visualization for personalizing learning and understanding academic impact. He also focuses on the design of instrumentation for large-scale engineering and science cyberenvironments. He won a CAREER Award for work on personalizing learning with engineering cyberenvironments. E-mail: cm@purdue.edu.

THERESA A. MALDONADO is director of the Division of Engineering Education and Centers of the Directorate for Engineering at the National Science Foundation. Prior to joining NSF in 2011, she served in several leadership positions in the Texas A&M University System, including as Associate Vice Chancellor for Research and as Interim Vice President for Research. She was founding director of the Energy Engineering Institute, and she is professor of electrical and computer engineering. E-mail: tmaldona@nsf.gov.

SALLY A. MALE is a researcher at The University of Western Australia. She is qualified in electrical engineering, which she has taught at The University of Western Australia and Curtin University. Sally's Ph.D. is on competencies required by engineering graduates. Her research interests include engineering education and women in engineering. Sally has served on several Western Australian committees and the National Women in Engineering Committee of Engineers Australia. E-mail: sally.male@uwa.edu.au.

JANET MCDONNELL is Professor of Design Studies at Central Saint Martins College of Arts and Design, University of the Arts London, where she is Director of Research. She holds a B.Sc. (Eng.), became a chartered electrical engineer in 1982, and has an M.Sc. and a Ph.D. in Computer Science. Her research focus is design practice in natural settings motivated by interests in supporting reflective practitioners, enabling user engagement in design, and understanding creative design collaboration. She is the editor of the *International Journal of CoDesign*. E-mail: j.mcdonnell@csm.arts.ac.uk.

ANN F. MCKENNA is Chair and Associate Professor of Engineering in the College of Technology and Innovation at Arizona State University. Prior to joining ASU she served as a program officer at the National Science Foundation in the Division of Undergraduate Education and was on the faculty at Northwestern University. Dr. McKenna received her B.S. and M.S. degrees in Mechanical Engineering from Drexel University and Ph.D. from the University of California at Berkeley. E-mail: ann.mckenna@asu.edu.

LISA D. MCNAIR is an Associate Professor of Engineering Education at Virginia Tech, where she also serves as Assistant Department Head of Graduate Education and co-director of the VT

Engineering Communication Center (VTECC). She received her Ph.D. in Linguistics from the University of Chicago and a B.A. in English from the University of Georgia. Her research interests include interdisciplinary collaboration, design education, communication studies, identity theory, and reflective practice. E-mail: lmcnair@vt.edu.

DEVLIN MONTFORT is an assistant professor in the School of Chemical, Biological and Environmental Engineering at Oregon State University. His research interests include the theoretical, methodological, and philosophical peculiarities of conceptual change and personal epistemology in the context of engineering. He holds a Ph.D. in Civil Engineering from Washington State University. E-mail: devlin.montfort@oregonstate.edu.

BARBARA M. MOSKAL is a professor in Applied Mathematics and Statistics and the Director of the Trefny Institute of Educational Innovation at the Colorado School of Mines. She is a senior editor for the *Journal of Engineering Education*. Her research interests include educational outreach, gender issues in engineering, and educational assessment. She has been involved in educational research in engineering for more than fifteen years. E-mail: bmoskal@mines.edu.

NANCY J. NERSESSIAN is Regents' Professor of Cognitive Science at the Georgia Institute of Technology. Her research focuses on the creative research practices of scientists and engineers, especially how model-based reasoning leads to fundamentally novel insights and how interdisciplinary engineering sciences research laboratories foster and sustain creative practices and learning. She is a Fellow of the Cognitive Science Society and AAAS. Her book, *Creating Scientific Concepts* (2008), received the inaugural Patrick Suppes Award in Philosophy from the American Philosophical Society. E-mail: nancyn@cc.gatech .edu; www.cc.gatech.edu/~nersessian.

WENDY C. NEWSTETTER is the Director of Educational Research and Innovation in the College of Engineering at Georgia Institute of Technology as well as the Director of Learning Sciences Research in the Coulter Department of Biomedical Engineering. Her research interests include interdisciplinary cognition and learning, model-based approaches to reasoning and problem-solving, and the design of rich learning environments that support and nurture complex, real-world problem solving. She is also interested in developing new models for faculty develop-

ment. She is a senior associate editor and special issues editor for the *Journal of Engineering Education*. E-mail: wendy.newstetter@coe.gatech.edu.

KEVIN O'CONNOR is assistant professor of educational psychology at University of Colorado, Boulder. His scholarship focuses on human action, communication, and learning as socioculturally organized phenomena. One major strand of research has explored the varied trajectories taken by students as they attempt to enter professional disciplines such as engineering, and focuses on the dilemmas encountered by students as they move through these institutionalized trajectories. Another strand of research has explored community organizing efforts that aim to construct new trajectories into valued futures for youth, especially those of nondominant communities. He is coeditor of a 2010 National Society for the Study of Education Yearbook, *Learning Research as a Human Science*. Other work has appeared in *Linguistics and Education*; *Mind, Culture, and Activity*; *Anthropology & Education Quarterly*, the *Encyclopedia of Cognitive Science*; and the *Journal of Engineering Education*. His teaching interests include developmental psychology; sociocultural theories of communication, learning, and identity; and discourse analysis. E-mail: Kevin.OConnor@colorado.edu.

MARIE C. PARETTI is an associate professor in the Department of Engineering Education at Virginia Tech, the co-director of the Virginia Tech Engineering Communication Center, and the former director of the Engineering Communication Program for Materials Science and Engineering and Engineering Science and Mechanics at VT. She holds degrees in both chemical engineering and English, and her research interests include communication pedagogies in engineering classrooms as well as communication practices in engineering workplaces. E-mail: mparetti@vt.edu.

KURT PATERSON currently serves as Head of the Department of Engineering at James Madison University. Dr. Paterson is an award-winning educator, author, mentor, and photographer, as well as a noted workshop designer and public speaker on community engagement in engineering. He leads several national initiatives, recently launching ASEE's newest division, Community Engagement in Engineering Education. He is PI on research projects assessing the impacts of community engagement on students, faculty, and communities around the world. E-mail: paterskg@jmu.edu.

ALICE L. PAWLEY is an associate professor in the School of Engineering Education at Purdue University and an affiliate faculty in the Women's Studies Program and Division of Environmental and Ecological Engineering. She holds a B.Eng. (Chemical) from McGill University and an M.S. and a Ph.D. (Industrial) from the University of Wisconsin-Madison. Her NSF CAREER Award uses feminist theory and methods to understand underrepresentation in engineering. She received a PECASE award in 2012. E-mail: apawley@purdue.edu.

JAMES W. PELLEGRINO is Liberal Arts and Sciences Distinguished Professor and Distinguished Professor of Education at the University of Illinois at Chicago and Co-director of UIC's interdisciplinary Learning Sciences Research Institute. His research and development interests focus on children's and adult's thinking and learning and the implications of cognitive research and theory for assessment and instructional practice. He has chaired several National Academy of Sciences study committees, is a past member of the Board on Testing and Assessment of the National Research Council, and is a lifetime member of the National Academy of Education. E-mail: pellegjw@uic.edu.

MICHAEL J. PRINCE is a Professor of Chemical Engineering at Bucknell University and co-director of the National Effective Teaching Institute. He is the author of several education-related papers for engineering faculty and has given more than 100 faculty development workshops to local, national, and international audiences. His current research examines a number of topical issues within engineering education as well as how to increase the use of research-supported instructional strategies by engineering faculty. E-mail: prince@bucknell.edu.

TERI REED is assistant vice chancellor of academic affairs for engineering and assistant dean of academic affairs for the Dwight Look College of Engineering, and associate professor in the Harold Vance Department of Petroleum Engineering at Texas A&M. She received her B.S. in petroleum engineering from the University of Oklahoma and spent seven years in the petroleum industry, during which time she earned her M.B.A. She subsequently received her Ph.D. in industrial engineering from Arizona State University. Professor Reed is a Fellow of the American Society for Engineering Education and member of the Institute of Electronics and Electrical Engineers and the Institute of Industrial Engineers. E-mail: terireed@tamu.edu.

DONNA RILEY is Associate Professor of Engineering at Smith College, the first U.S. women's college to establish an engineering program. In 2005 she received an NSF CAREER Award for critical and feminist pedagogies. She holds a B.S.E. in Chemical Engineering from Princeton and a Ph.D. in Engineering and Public Policy from Carnegie Mellon. Her books include *Engineering and Social Justice* (2008) and *Engineering Thermodynamics and 21st Century Energy Problems* (2011). E-mail: driley@smith.edu.

WOLFF-MICHAEL ROTH is Lansdowne Professor of Applied Cognitive Science at the University of Victoria, British Columbia. He studies knowing and learning in school and non-school settings, focusing on science, mathematics, technology, and language. His recent books include *Possibility: At the Limits of the Constructivist Metaphor* (2011) and *Geometry as Objective Science in Elementary School Classrooms: Mathematics in the Flesh* (2011). He is a Fellow of the American Association for the Advancement of Science and of the American Educational Research Association. E-mail: wolffmichael.roth@gmail.com.

SHERI D. SHEPPARD is Professor of Mechanical Engineering at Stanford University. Besides teaching design-related classes, she conducts research on applied finite element analysis and on how people become engineers. She is lead author of *Educating Engineers: Designing for the Future of the Field* (2008) and is currently Associate Vice Provost of Graduate Education. Before Stanford, she worked in the automotive industry. Sheri's graduate work was done at the University of Michigan. E-mail: sheppard@stanford.edu.

AMY E. SLATON is a professor of history at Drexel University. She holds a Ph.D. in the History and Sociology of Science from the University of Pennsylvania. She is the author of *Race, Rigor and Selectivity in U.S. Engineering: The History of an Occupational Color Line* (2010) and other books, and she produces the blog STEMequity.com, centered on equity in technical education and workforce issues. E-mail: slatonae@drexel.edu.

KARL A. SMITH is Cooperative Learning Professor of Engineering Education, School of Engineering Education, at Purdue University; and Morse-Alumni Distinguished Teaching Professor and Emeritus Professor of Civil

Engineering at the University of Minnesota. Karl has been actively involved in engineering education research and practice for more than thirty years and has worked with thousands of faculty all over the world on pedagogies of engagement, especially cooperative learning, problem-based learning, and constructive controversy. E-mail: ksmith@umn.edu.

REED STEVENS is professor of Learning Sciences in the School of Education and Social Policy at Northwestern University. His research examines and compares cognitive activity in a range of settings including classrooms, workplaces, and science museums. On the basis of this comparative work, he is exploring new ways to conceptualize cognition and organize learning environments. Dr. Stevens' specific interests include how mathematical activity contributes to various settings and how technology mediates thinking and learning. His multidisciplinary research draws on cognitive science, interactionist traditions, and the social studies of science and technology. He also designs curriculum, activities, and technologies, including Video Traces software that allows people to collect digital video clips and annotate them with talk or gestures. E-mail: reed-stevens@northwestern.edu.

RUTH A. STREVELER is an associate professor in the School of Engineering Education at Purdue University. She earned a B.A. in Biology from Indiana University-Bloomington, an M.S. in Zoology from the Ohio State University, and a Ph.D. in Educational Psychology from the University of Hawai'i at Mānoa. Her primary research interests are investigating students' understanding of difficult concepts in engineering science and helping engineering faculty conduct high-quality research in engineering education. E-mail: streveler@purdue.edu.

SCOTT A. STRONG is a teaching associate professor in the Department of Applied Mathematics and Statistics at the Colorado School of Mines who has been trained in the Cornell Office of Research Evaluation System's Evaluation Protocol, which he has used to evaluate local research experience for undergraduate and engineering education programs. E-mail: sstrong@mines.edu.

MARILLA D. SVINICKI currently is a Full Professor in Educational Psychology at the University of Texas at Austin. Prior to that she served as the University of Texas Faculty Development Director for thirty years. She has served as the President of the POD Network, an international organization for professionals in faculty development. She is currently the chair of the Faculty Teaching Development and Evaluation Special Interest Group of the American Educational Research Association. Her expertise is in the application of educational psychology principles to teaching and learning. She has published in the areas of faculty development and teaching expertise, including serving as the Editor-in-Chief of *New Directions for Teaching and Learning* (2006, No. 108), writing a monthly column on teaching for the National Teaching/Learning Forum, and editing *Teaching Tips: Strategies, Research and Theory for College and University Teachers* (2013).

CHRISTOPHER SWAN is an associate professor of Civil and Environmental Engineering (CEE) at Tufts University with additional appointments in the Jonathan M. Tisch College of Citizenship and Public Service and the Center for Engineering Education and Outreach. He has also served as chair of Tufts CEE department (from 2002 to 2007). Dr. Swan's current research interests in engineering education focus on project-based learning and service-based pedagogies. E-mail: chris.swan@tufts.edu.

KAREN L. TONSO is Associate Professor (Education) at Wayne State University, following engineering for fifteen years in the petroleum industry. She studies structures of learning settings using anthropological lenses, focusing on social justice. Her books include *On the Outskirts of Engineering and Women's Science* (2007). Awards include recognition for "Student engineers and engineer identity" in CSSE and the Betty Vetter Award for Research from WEPAN in 2009. She was PI for WSU's NSF ADVANCE-PAID grant. E-mail: ag7246@wayne.edu.

KRISTEN BETHKE WENDELL is Assistant Professor of Elementary Science Education at the University of Massachusetts Boston. Her teaching and research interests include pre-service teacher education in engineering and the integration of engineering design into children's science, reading, and writing experiences. She was a graduate policy Fellow at the National Academy of Engineering and received her B.S.E. from Princeton, her M.S. in Aeronautics and Astronautics from MIT, and her Ph.D. in Science Education from Tufts University. E-mail: kristen.wendell@umb.edu.

AMAN YADAV is an Associate Professor in Educational Psychology with a courtesy appointment in Computer Science at Purdue University. In addition to a Ph.D. in Educational Psychology and Educational Technology, he holds a bachelor and a master's degree in Electrical Engineering. Dr. Yadav's teaching and research focus on problem-based learning, technology, cognition, and computational thinking. His work has been published in a number of leading journals, including *Journal of Engineering Education, International Journal of Engineering Education*, and *Communications of the ACM*. E-mail: amanyadav@purdue.edu.

Foreword

Norman L. Fortenberry

Executive Director, American Society of Engineering Education (ASEE)

Although isolated individual engineering faculty members had pursued engineering education research (EER) for as long as long as there has been formalized instruction in engineering, EER has made noteworthy progress as a critically important discipline, particularly within the United States, during the last fifteen years. In the United States much of this change was driven or supported by the shift, in 1996, by the engineering accreditation agency ABET to an outcomes focus. This change mandated measurement of student learning outcomes, and the need to assess student learning outcomes fostered a demand for the research findings of EER as well as broader faculty interest in the field. I recall being involved in early EER efforts when as a National Science Foundation (NSF) program officer I managed the award to Richard Felder of North Carolina State University for the first National Science Foundation grant (in 1991) to examine student learning styles. I later had the privilege of being one of the three program officers who recommended the 1993 grants to Cynthia J. Atman (University of Washington) and Martin Ramirez (then of Johns Hopkins) for the first NSF National Young Investigator awards to be made in engineering education research.

Since those first prominent NSF awards, the field has matured in terms of its topical focus and the quality of the research. Of particular note are the contributions and the intellectual integration of the work from the global community of researchers. The first International Conference on Research in Engineering Education (now called the Research in Engineering Education Symposium – REES) was held in 2007 in Honolulu, Hawaii, with fewer than 25% of the participants from outside North America. The most recent REES meeting in Madrid still had a strong minority of North American representatives but drew the majority of participants from throughout the global community of EER scholars.

As with most new fields, EER has drawn extensively on lessons learned from a variety of other fields. It has drawn extensively from the social sciences, educational research, and the cognitive and learning sciences to create a unique synthesis with its own expanding scholarly base. Although EER originally focused primarily on students and their learning, it rapidly expanded to encompass virtually all aspects of formal and informal learning systems including:

- Learners, instructors, administrators
- Teaching, learning, and assessment systems
- Curricula, laboratories, and instructional technologies
- Goal and objective systems for students, instructors, administrators, and external stakeholders
- Constraints (economic, social, political, etc.) inherent in the learning system

This volume reflects the richness and comprehensiveness of the emerging field of EER and will be of tremendous utility to students as well as those seeking an introductory overview. The editors have carefully selected authors who are leading, globally recognized scholars writing in their areas of expertise.

This volume comes at a particularly opportune time as the broad engineering education community seeks to define better the required elements of a twenty-first-century engineering education in an increasingly integrated global community where the solutions to a growing array of social, political, and economic challenges require access to engineering knowledge. The traditional image of the solitary engineer is no longer valid. Successful engineering research, innovation, and practice require collaboration with and integration of professionals from other fields. The engineering education system faces stresses from larger changes within education systems overall, especially political pressures to reduce costs, and stresses on engineering faculty to be increasingly effective in teaching, research, and service. There is a tremendous need for a robust research base to inform future practice in this unfamiliar environment. This volume serves as an important tool for accomplishing this goal.

Acknowledgments

We would like to acknowledge, first and foremost, the contributions of our fellow scholars. Without their hard work and efforts this volume would not exist. We also thank Peter Gordon and Sarika Narula at Cambridge University Press for their help in bringing this volume together. Peter worked tirelessly with the authors since their first meeting at the annual conference of ASEE 2010 to bring this volume to fruition. R. Keith Sawyer from Washington University was of great assistance during the proposal preparation and served on our advisory board. We also want to thank our other advisory board members, Jack R. Lohmann, Karl A. Smith, and Wendy C. Newstetter. Thanks also to Norman L. Fortenberry, the Executive Director of ASEE, who helped us gain permission to reprint material from ASEE publications and for writing the foreword for the *Handbook*. Chapters 3, 21, and 27 have previously appeared in the *Journal of Engineering Education*.

Aditya Johri would like to thank Barbara M. Olds for agreeing to work on this volume and for her mentorship throughout the process. Aditya expresses his sincere gratitude to Sue Kemnitzer and Alan Cheville for their intellectual guidance and support. He would also like to thank the U.S. National Science Foundation that partially supported his work on this project through Awards EEC-0954034 and EEC-0935143. Any opinions, findings, and conclusions or recommendations expressed in this material are those of the author(s) and do not necessarily reflect the views of the National Science Foundation.

Barbara Olds would like to thank Aditya Johri for his intellectual leadership of the project and for his tireless work to bring the volume to fruition.

In addition to the authors, the following individuals graciously agreed to review the chapters in this volume: Erin Crede, David B. Knight, Lee Martin, Holly Matusovich, Jacob Moore, Jay Pembridge, Olga Pierrakos, Ken Stanton, Hon Jie Teo, James Trevelyan, and Jennifer Turns.

Introduction

Aditya Johri and Barbara M. Olds

The Cambridge Handbook of Engineering Education Research (CHEER) is an important reference source for the growing field of engineering education research (EER). EER has become an increasingly important field internationally, as evidenced by the growing prestige and subscriber base of its key journal, the *Journal of Engineering Education (JEE)*, the founding of several Ph.D.-granting engineering education departments at prestigious institutions, and the growth of an international community of engineering education researchers who hold global meetings and have a variety of publication venues. Despite the growth of the field, there is currently no book that provides an overview of EER.[1] Thus we proposed CHEER with the belief that it will fill an important gap internationally in the EER field and will be used as a textbook for graduate courses, a reference book by engineering faculty in disciplinary engineering areas, and a resource by policymakers, K–12 engineering curriculum designers, informal science educators, and others.

Engineering education research draws on many social science disciplines in addition to disciplinary engineering knowledge and computing. Research in engineering education has traditionally focused on reporting of classroom interventions and generally lacked definition as a discipline until the late1990s and early 2000s. Since a landmark issue of *JEE* in 2005, which included papers by senior scholars in the field who argued for a stronger theoretical and empirically driven agenda for the field, engineering education has quickly emerged as a research driven field and subsequently has seen a substantial increase in both the quality and quantity of theoretical and empirical work. Research in engineering education is focused primarily on formal settings but work on informal learning in settings such as museums and after school programs is starting to appear. Chapters in this volume draw extensively on contemporary research in the learning sciences to include how technology affects learners and learning environments, and the role of social context in learning.

This volume contains thirty-five chapters organized into six sections authored by seventy-three scholars. We have selected these themes based on the research agenda developed for engineering education through a series of interdisciplinary colloquia funded by the U.S. National Science Foundation and published in the *Journal of Engineering Education* in October 2006.[2] We have modified the titles of the themes to make them fit better with a handbook but the intent remains the same. We have also added a theme to the five originally proposed. The first chapter, which is not part of any of the sections, provides a historical overview of engineering education research. The first section, "Engineering Thinking and Knowing," contains six chapters that focus on "research on what constitutes engineering thinking and knowledge within social contexts now and into the future." Part 2, "Engineering Learning Mechanisms and Approaches," contains six chapters and looks at "research on engineering learners' developing knowledge and competencies in context." Part 3, "Pathways into Diversity and Inclusiveness," explores "research on how diverse human talents contribute solutions to the social and global challenges and relevance of our profession" and consists of five chapters. In Part 4, "Engineering Education and Institutional Practices," five chapters highlight "research on the instructional culture, institutional infrastructure, and epistemology of engineering educators." Part 5, "Research Methods and Assessment," contains six chapters that focus on "research on, and the development of, assessment methods, instruments, and metrics to inform engineering education practice and learning." Finally, Part 6 "Cross-Cutting Issues and Perspectives," contains six chapters that address themes/topics that have emerged within engineering education research and

have been pursued by a critical mass of scholars.

The authors of the handbook chapters represent the who's who of the engineering education research community and come from all corners of the world. Our goal in producing this handbook has been to publish an easily accessible volume that will be widely used by researchers in the field of engineering education but will also support the needs of students, engineering faculty, and policymakers.[3]

Footnotes

1. The only current text that comprehensively addresses some issues relevant to EER is John Heywood's 2006 publication, *Engineering Education: Research and Development in Curriculum and Instruction*, which provides a synopsis of nearly 2,000 articles related to engineering education published since 1960. However, unlike this volume, Heywood's book provides only an overview to the field and its focus is not on recent theoretical and empirical developments in the social and learning sciences related to engineering education.

2. "The Research Agenda for the New Discipline of Engineering Education," *Journal of Engineering Education*, 94(4), pp. 259–261 (2006). DOI: 10.1002/j.2168-9830.2006.tb00900.x

3. Many authors refer to work that has appeared in the ASEE and FIE conferences and proceedings from those conference are available online at the following links:

ASEE Conference Proceedings Search: http://www.asee.org/search/proceedings

FIE IEEE Explore Digital Library: http://ieeexplore.ieee.org/xpl/conhome.jsp?punumber=1000297

CHAPTER 1

Chronological and Ontological Development of Engineering Education as a Field of Scientific Inquiry

Jeffrey E. Froyd and Jack R. Lohmann

Introduction

Engineering education as an area of interest for curriculum development and pedagogical innovation emerged in the United States in the period around 1890 to 1910 with the founding of the Society for the Promotion of Engineering Education (SPEE) in 1893 (American Society for Engineering Education, n.d.). Founding dates for a few other engineering education associations may provide some indication of when interest in engineering education emerged across the world: Internationale Gesellschaft für Ingenieurpadagogik (IGIP, 1972); Société Européenne pour la Formation des Ingénieurs (SEFI, 1973); and Australasian Association of Engineering Education (AAEE, 1989). Other associations interested in engineering education include Associação Brasileira de Educação em Engenharia (ABENGE), Asociación Nacional de Facultades y Escuelas de Ingenieria (ANFEI), International Association for Continuing Engineering Education (IACEE), Korean Society for Engineering Education (KSEE), Latin American and Caribbean Consortium

of Engineering Institutions (LACCEI), and Mühendislik Dekanlari Konseyi (MDK). Given the date of the founding of the SPEE and historical data available on the society and its growth, in relation to similar information about other engineering education associations or societies, the authors have elected to use the chronology of events in the United States as the principal framework to describe the evolution of engineering education as a field of scientific inquiry with references to similar events internationally.

A transition, which is not nearly complete, to an interdisciplinary, more scholarly field of scientific inquiry into engineering education is occurring nearly 100 years later (Borrego & Bernhard, 2011; Continental, 2006; Haghighi, 2005; Jesiek, Newswander, & Borrego, 2009; Lohmann, 2005). Contextual factors, which are too numerous to describe exhaustively, have and will influence evolution of the field of engineering education research; however, the authors would like to draw attention to four important factors. First, although engineering is taught at K–12, undergraduate, and graduate levels, professional licensure currently requires a baccalaureate degree in engineering. Therefore, undergraduate education has been the primary avenue through which engineers enter the profession, and the literature in engineering education has focused predominantly on undergraduate education. As a result, research questions in undergraduate engineering education have tended to dominate attention of researchers; however, this is changing. Second, unlike mathematics and science education in K–12, K–12 engineering education has traditionally been lacking. As a result, research in K–12 engineering education has been minimal. However, the situation is changing. "Although K–12 engineering education has received little attention from most Americans, including educators and policy makers, it has slowly been making its way into U.S. K–12 classrooms. Today, several dozen different engineering programs and curricula are offered in school districts around the country, and thousands of teachers have attended professional development sessions to teach engineering-related coursework. In the past 15 years, several million K–12 students have experienced some formal engineering education" (Committee on K–12 Engineering Education; Linda Katehi, 2009, p. 1). As a result of increasing interest in engineering education in K–12, research questions associated with this focus are growing in importance. However, a large percentage of engineering faculty members, who traditionally have been viewed as primary stakeholders in findings from engineering education research, may not take much interest in findings from engineering education research in K–12. Third, research in education and the learning sciences can make significant contributions as researchers in any disciplinary-based educational field address their complex research questions (Froyd, Wankat, & Smith, 2012; Johri & Olds, 2011; also see Chapters 2 by Newstetter & Svinicki and 29 by Pellegrino, DiBello, & Brophy in this volume). However, much of the scholarly literature in education and the learning sciences has focused on precollege education, an area that traditionally has attracted less

attention from most engineering faculty members (Johri & Olds, 2011), because of their focus on undergraduate education. It will take time and energy for familiarity and interest of engineering practitioners at the undergraduate level in research and learning sciences to reach a level that it begins to influence practice. Fourth, engineering educators, in general, receive little or no formal preparation for their instructional duties during their doctoral training or later as faculty. As a result, for most engineering faculty members, lack of familiarity with the education and learning sciences literature, reliance on familiar research methodologies that were often ill suited for educational studies, and complacency with accepting student satisfaction surveys as indicators of efficacy of course changes generated "a rich tradition of educational innovation, but until the 1980s assessment of innovation was typically of the 'We tried it and liked it and so did the students' variety" (Wankat, Felder, Smith, & Oreovicz, 2002, p. 217). Changing practice in engineering education so that faculty members apply findings in engineering education research, education research, and research in the learning sciences to their practice in engineering classrooms is a major challenge for engineering education practice and research (Jamieson & Lohmann, 2009, 2012).

Catalysts, including major National Science Foundation (NSF) funding for educational research and development beginning in the late 1980s and emergence of the outcomes-based ABET Engineering Criteria led to significant publications in engineering education research in the 1990s. In the last twenty years, engineering education research has begun to emerge as an interdisciplinary research field seeking its own theoretical foundations from a rich array of research traditions in the cognitive sciences; learning sciences; education; and educational research in physics, chemistry, and other scientific disciplines.

The remainder of the chapter is divided into two parts. First, we provide a brief chronology of the development of engineering education as a field of study. We then describe the ontological transformation of the field into engineering education research using criteria for defining the field of science education research (Fensham, 2004). A brief conclusion projects the near future of the field.

The Chronological Evolution of U.S. Engineering Education as a Field of Scientific Inquiry

The first engineering program in the United States, civil engineering, was established at the United States Military Academy, which was founded in 1802 to reduce the nation's dependence on foreign engineers and artillerists in times of war (United States Military Academy, 2010). Other parts of the world also began engineering programs during the 1800s and especially the latter half of the century (Continental, 2006). Nonetheless, higher education was largely inaccessible to many Americans until the passage of the Morrill Act[1] in 1862 (Lightcap, 2010), which accelerated the nation's growth throughout the last half of the century fueled by such engineering efforts as the transcontinental railroad, electric power, the telegraph and telephone, and steam and internal combustion engines. Mechanical, electrical, and chemical engineering emerged as distinct disciplines toward the end of the nineteenth century and near the beginning of the twentieth century (Grayson, 1993). Other engineering disciplines, for example, industrial, biomedical, environmental, petroleum, mining, and nuclear, emerged during the twentieth century.

For the first half of the twentieth century, U.S. engineering and engineering education was characterized by its practical arts (Seely, 1999; also see Chapter 7 by Stevens, Johri, & O'Connor in this volume). This focus changed abruptly when the world observed the power of science and its applications during World War II (Seely, 1999). When coupled with creation of NSF in 1950

(National Science Foundation [NSF], 2010), and several other programs within existing federal agencies, federal funding largely transformed the American higher education system into research-based institutions of higher learning, especially in science and engineering. Engineering education shifted from hands-on, practicum-oriented curricula to ones that emphasized mathematical and scientific foundations (Grayson, 1993; Seely, 1999). The shift was codified when ASEE issued its landmark study commonly called the Grinter Report in 1955 (American Society for Engineering Education, 1994). It outlined more research-oriented and science-based curricula, from which initial transitions to more design-oriented curricula are recent occurrences (Froyd et al., 2012).

The first engineering society, the American Society of Civil Engineers, was established in 1852 (American Society of Civil Engineers, 2010) and the first engineering education society, the Society for the Promotion of Engineering Education (SPEE), was founded in 1893 (Reynolds & Seely, 1993) and is now known as the American Society for Engineering Education (ASEE). As mentioned in the introduction, the growth of similar engineering education societies appears to have occurred mostly after the Second World War (*Journal of Engineering Education*, 2010). SPEE established the first periodical "devoted to technical education" in 1910, called the *Bulletin* (American Society for Engineering Education, 1910), which nearly a century later evolved into the discipline-based (engineering) education research journal *Journal of Engineering Education* (*Journal of Engineering Education*, 2010; Lohmann, 2003).

In 1986, the National Science Board issued an overdue wake-up call about the state of U.S. engineering, mathematics, and science education (National Science Board, 1986). Its report provided a number of recommendations and made clear that one among them played a critical role: "The recommendations of this report make renewed demands on the academic community – especially that its best *scholarship* be applied to the manifold activities needed to strengthen undergraduate science, engineering, and mathematics education in the United States" (National Science Board, 1986, p. 1, emphasis added). It was instrumental in reviving the NSF's role to "initiate and support science and engineering education programs at all levels and in all the various fields of science and engineering" (NSF, 2006, p. 5). The report was also among those that sparked a vigorous national dialogue on the role of scholarship in improving the quality of U.S. higher education. For example, the highly influential 1990 report, "Scholarship Reconsidered: Priorities of the Professoriate," by Ernest Boyer of the Carnegie Foundation, offered a new taxonomy and terminology to describe academia's multifaceted forms of scholarship (Boyer, 1990). In engineering, introduction of EC2000 by ABET in the 1990s was a major driver to improve the quality of engineering education (ABET, 1995; Prados, 2005). Its outcomes-focused, evidence-based cycle of observation, evaluation, and improvement characterized many aspects of a scholarly approach to educational innovation.

Dialogue and decisions made in the 1990s paved the way for engineering education to become a field of scientific inquiry as it became increasingly clear that the intuition-based approaches of the past were not producing the quantity and quality of engineering talent needed to address society's challenges (Continental, 2006; National Academy of Engineering [NAE], 2004; National Research Council [NRC], 2005; NSF, 1992). More scholarly and systematic approaches based on the learning sciences were needed (Gabriele, 2005; Haghighi, 2005; NRC, 2000, 2002); concurrently, research on engineering science should contribute to the development of the learning sciences (Johri, 2010; Shulman, 2005), especially in areas closely linked to engineering, such as design. Consequently, embryonic and globally diverse communities began to emerge and collaborate such that by the mid-2000s engineering education as a

scientific field of inquiry (research) had passed the "tipping point" both within the United States and elsewhere (Borrego & Bernhard, 2011; Jesiek, Borrego, & Beddoes, 2010). Integrating and expanding these communities was a major point of discussion in a recent NSF-funded ASEE study, Creating a Culture for Scholarly and Systematic Innovation in Engineering Education (Jamieson & Lohmann, 2009, 2012).

For a more detailed chronological description of the development of engineering education and engineering education, the authors (together with the support of others in the engineering education research community [please see Acknowledgments]) have compiled a timeline in Appendix 1.1. In the next section, we describe the current state of engineering education research, much of it having been created within the last decade or so. Figure 1.1 presents a picture of the largest authorship network within engineering education research and shows a core group that is linked to several other groups and nodes on peripheries.

An Ontological Description of the State of Engineering Education Research

A chronological description of a field of research uses time and temporal ordering as its organizational framework. An alternative organizational framework describes entities and relationships among the entities, that is, a conceptualization (Genesereth & Nilsson, 1987). To describe a conceptualization of the state of engineering education research requires an ontology, that is, a specification for a conceptualization (Gruber, 1993). An ontology for evaluating maturation of fields of disciplinary-based education research has been formulated with three categories of criteria: structural, research, and outcome, as summarized in Table 1.1 (Fensham, 2004). We believe this framework is appropriate for organizing and critiquing the evolution and maturity of engineering education research.

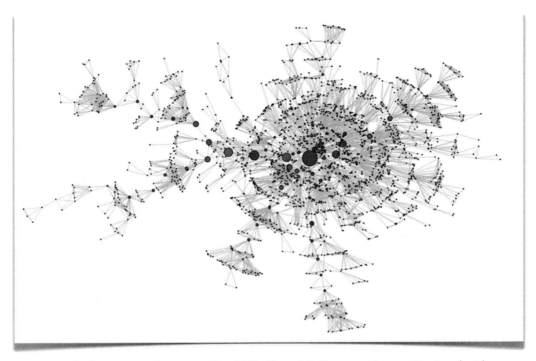

Figure 1.1. The largest co-author network in EER. (From Madhavan et al., 2011. Reprinted with permission.)

Table 1.1. Fensham's (2004) Criteria for Defining the Field of Science Education Research

Category	Criteria	Exemplars of Criteria
Structural	Academic Recognition	Full faculty appointments in the area of research
	Research Journals	Successful journals for reporting quality research
	Professional Associations	Healthy national and international professional associations
	Research Conferences	Regular conferences for the direct exchange of research that enable researchers to meet in person
Research	Scientific Knowledge	Knowledge of science content required to conduct the research
	Asking Questions	Asking distinctive research questions not addressed by other fields
	Conceptual and Theoretical Development	Theoretical models with predictive or explanatory power
	Research Methodologies	Invention, development, or at least adaptation of methodologies, techniques, or instruments
	Progression	Researchers are informed by previous studies and build on or deepen understanding
	Model Publications	Publications that other researchers hold up as models of conduct and presentation of research studies in the field
	Seminal Publications	Publications recognized as important or definitive because they marked new directions or provided new insights
Outcome	Implications for Practice	Outcomes from research that are applications to the practice of science education

Structural Criteria

1. *Academic Recognition:* Academic recognition examines extent to which scholars in the field are recognized by their institutions. One metric for recognition is establishment of organizational units for scholarship in the discipline, that is, centers for engineering education research. In Europe, a "specific goal of the Bologna declaration is to promote mobility amongst engineering students in Europe. As a consequence, universities will have to engage in an international competition to attract students. This results in a growing interest for improvement and innovation in engineering education. All over Europe "Centres of Expertise on Learning and Teaching" are being established or, in case of older existing institutes are re-installed. The position of a centre of this kind within the university organisation varies as well as tasks and responsibilities. Some establishments are divided into a research group and a teacher-training and consultant division" (Hawwash, 2007, p. 30). In the United States, there are about twenty centers involved in engineering education research of which most were established in the last decade (Center for the Advancement of Scholarship in Engineering Education, 2010). Departments of Engineering Education were established at Purdue and Virginia Tech, and were the first to provide tenured positions in engineering education. Later, Utah State University established a Department of Engineering Education and Clemson University established a Department of Engineering and Science Education.

2. *Research Journals:* The field has one journal focused exclusively on research, the *Journal of Engineering Education* (JEE), and five whose missions encompass research: *Engineering Studies, European Journal of Engineering Education* (EJEE), *International Journal of Engineering Education* (IJEE), *Engineering Education,* and *Chemical Engineering Education* (Borrego & Bernhard, 2011).¹ Two are listed on Thomson-Reuters citation indices (IJEE and JEE) and three are ranked by the Australian Research Council (EJEE, IJEE, and JEE).

3. *Professional Associations and Research Conferences:* There are many international engineering education societies including a federation of such societies (International Federation of Engineering Education Societies, 2010). The dominant ones are ASEE, the Australasian Association for Engineering Education (AAEE), and the Société Européenne pour la Formation des Ingénieurs (SEFI). Annual conferences focus on curriculum development; however, increasingly some host engineering education research tracks and some have groups whose focus is engineering education research, notably AAEE, ASEE, and SEFI. An independent research symposium Research on Engineering Education Symposium (REES) was established in 2007 to facilitate a periodic global gathering of researchers in the field (Research in Engineering Education Network, 2010).

4. *Funding and Honors:* The authors believe there are two additional structural criteria of importance to engineering. Peer-reviewed extramural support has been a critical to U.S. engineering research since World War II. Educational initiatives, however, have been supported mostly within university budgets. In the late 1980s, the NSF established programs for curriculum development and pedagogical innovation whose support mirrored their technical research counterparts, and a number of programs are now available for discipline-based education research. Awards and honors for teaching are ubiquitous but recognitions for engineering education research are nearly nonexistent. Two publication awards include ASEE's Wickenden Award for the best paper published annually in JEE and the Outstanding Research Publication Award by Division I (Education in the Professions) of the American Educational Research Association.

Research Criteria

1. *Scientific Knowledge and Asking Questions:* The NSF-funded Engineering Education Research Colloquies held in 2004–2005 were among the more notable efforts to begin to frame a scientific basis for thinking about the research challenges in the field of engineering education (The Steering Committee of the National Engineering Education Research Colloquies, 2006a, 2006b). They produced a taxonomy organized around "five priority research areas (Engineering Epistemologies, Engineering Learning Mechanisms, Engineering Learning Systems, Engineering Diversity and Inclusiveness, and Engineering Assessment)" that merge disciplinary engineering and learning sciences knowledge. Other efforts have recently emerged in the European community (Borrego & Bernhard, 2011; European and Global Engineering Education Network, 2010). Although the global community has not reached consensus on a taxonomy, it clearly feels a pressing need for such and is working to develop it (Borrego & Bernhard, 2011).

2. *Conceptual and Theoretical Development and Research Methodologies:* These two areas form the intellectual core of any disciplinary-based educational research field. Currently, conceptual and theoretical frameworks and research methodologies in engineering education research show considerable similarity to those of educational research in general, a condition that reveals its lack of maturity. Like other educational research fields, one foundation is research in the learning sciences, with its vast literature base and different theoretical frameworks (Greeno, Collins, & Resnick, 1996). At present, theoretical frameworks for research in engineering education do not distinguish themselves itself from frameworks for educational research in general, which tend to emphasize individual learning. Research in the cognitive sciences, for example, brain physiology, might contribute to a theoretical framework; however, constructing bridges from functions of individual or small groups of neurons to complex engineering concepts and processes would be a formidable task (Johri & Olds, 2011). Also, because engineering faculty members teach as collections or organizations of individuals, a potential contributor to future theoretical frameworks may be organizational change (Weick & Quinn, 1999).

Similar statements can also be made about applicable research methodologies, that is, engineering education research does not have a distinctive set of research methodologies. Engineering faculty members who apply engineering education research have backgrounds that condition them to understand quantitative research methodologies more easily than qualitative or mixed methodologies. As a result, efforts have been made to educate a large segment of the audience for engineering education research about the nature and value of the latter two sets of methodologies (Borrego, Douglas, & Amelink, 2009), but further progress is required.

3. *Progression, Models, and Seminal Publications:* Strobel, Evangelou, Streveler, and Smith (2008) think the first doctoral thesis on engineering education was published in 1929, and additional theses appeared occasionally up to about 1980. However, for the years between 1980 and 1989, they found five to eleven theses published every year; thereafter, thesis production increased markedly, and several widely cited articles on research in engineering education were published (Atman, Chimka, Bursic, & Nachtmann, 1999; Besterfield-Sacre, Atman, & Shuman, 1997; Felder, Felder, & Dietz, 1998) in the 1990s. These papers laid foundations for (i) further understanding of how students learn the engineering design process and how verbal protocol analysis methodologies can support the research (Atman & Bursic, 1998; Atman et al., 1999); (ii) rigorous assessment and adoption of cooperative learning (and later, other innovations) in engineering (Felder

et al., 1998; Haller, Gallagher, Weldon, & Felder, 2000); and (iii) the importance of and instruments for understanding engineering student attitudes and the roles they play in retention and learning (Besterfield-Sacre et al., 1997). In the first decade of the new millennium, significant publications in engineering education research have become too numerous to mention in this short review.

Outcome Criteria (Implications for Practice)

One key set of criteria in evaluating maturity of any research field are its influences on practice. Examining one metric related to the criteria was a survey of engineering department chairs about the extent to which seven innovations in engineering education had been adopted in engineering departments (Borrego, Froyd, & Hall, 2010). Each of the innovations was well supported by research demonstrating its efficacy. Survey results showed that engineering department chairs were aware of the innovations, but adoption of the innovations lagged well behind awareness. These findings in engineering echo similar findings in physics education (Dancy & Henderson, 2012).

Anticipating these findings, in 2006, ASEE launched a major initiative in engineering education community to persuade members of the synergistic and complementary roles played by innovation and research, beginning with the ASEE Year of Dialogue. Culmination of this initiative was publication of two ASEE reports: *Creating a Culture for Scholarly and Systematic Innovation in Engineering Education* (Jamieson & Lohmann, 2009) and *Innovation with Impact* (Jamieson & Lohmann, 2012). However, the fact that such an initiative was required is indicative of a culture in which most engineering education practitioners are content to continue to focus on innovations and less concerned about theoretical foundations that might catalyze innovations or methodologies with which the efficacy of the innovations might be evaluated.

Other factors besides a focus on innovations contribute to the lack of influence of engineering education research on practice in engineering classrooms. Research in physics education suggests that researchers expect that their curricular innovations will be adopted by faculty members "with minimal changes, while faculty expect researchers to work with them to incorporate research-based knowledge and materials into their *unique* instructional situations" (Henderson & Dancy, 2008, p. 79, emphasis added). For example, a study of adoption of research-based instructional strategies by chemical engineering faculty members showed that the primary faculty concern was classroom time that might be required to implement the instructional strategy (Prince, Borrego, Cutler, Henderson, & Froyd, 2013), but efficacy with respect to student learning is often a primary focus when evaluating an instructional strategy. Other factors that influence adoption lie outside of the control of an individual faculty member. These include student attitudes toward school (Henderson & Dancy, 2007); expectations of content coverage (J. L. Cooper, MacGregor, Smith, & Robinson, 2000; M. M. Cooper, 1995; Henderson & Dancy, 2007), which may be linked to classroom time; time required to prepare a lecture period (Henderson & Dancy, 2007; Prince et al., 2013); departmental norms (Henderson & Dancy, 2007); student resistance (J. L. Cooper et al., 2000; Henderson & Dancy, 2007); class size and room layout (M. M. Cooper, 1995; Henderson & Dancy, 2007); and constraints imposed by how class periods are scheduled (Henderson & Dancy, 2007).

In addition to the aforementioned factors, numerous articles have suggested that adoption of innovations from disciplinary-based educational research, educational research, and research in the learning sciences is hindered by institutional reward systems that value research far more than they value teaching (Cuban, 1999; Diamond, 1993; Handelsman et al., 2004). An often-repeated rationale for emphasis on research is that quality in research and teaching are

Figure 1.2. A case study of frontiers in education conference – ten years (Madhavan et al., 2010. Reprinted with permission.)

correlated (Fairweather, 2002; Hattie & Marsh, 1996; Prince, Felder, & Brent, 2007). However, scholarly inquiries into this key assertion have not found support for it (Fairweather, 2002; Hattie & Marsh, 1996). Thorough inquiries into questions such as the following are beyond the scope of this chapter, but might be possibilities for future research:

1. What is the extent of compatibility between university reward systems that placed greater emphasis on teaching and short-term (five to ten years) progress toward achievement of institutional missions?

2. How might faculty evaluation systems be modified to address more balanced emphases on teaching and teaching (Arreola, 1995)?

3. What external incentives might promote changes in faculty evaluation and reward systems?

There are many possible ways of examining these issues, and one possible solution is the use of cyberinfrastructure to analyze patterns of topics and authorship over time. Madhavan et al. (2010) adopted such an approach to re-create visually the

growth of one important conference associated with engineering education research – Frontiers in Education. An image is presented in Figure 1.2 and the movie is available online: http://www.youtube.com/watch?v=Oqd6vpjzqBI.

Conclusions: The Path Forward

Engineering education research has become an established field within the last decade, although its recognition and acceptance within the broader engineering community remains a challenge. It has established the critical physical infrastructure, for example, centers, departments, journals, conferences, and funding, necessary for it to now devote increasing attention to its intellectual growth, for example, conceptual and theoretical development, research methodologies, and progression. We foresee two major developments in the next decade: (1) major national or regional efforts to better integrate engineering education research into engineering programs, such as ASEE's effort *Creating a Culture for Scholarly and Systematic Innovation in Engineering Education* and the European effort, European and Global Engineering Education Network

(EUGENE), and (2) increasing collaboration (and occasional tensions) among the growing global communities of engineering education researchers as the field continues to mature.

Forecasts for development of the scholarly, interdisciplinary field of engineering education research are often, perhaps inevitably, intertwined with conversations about reform, improvement, and change in engineering education practice. In response to concerns that engineering education has never and will never change, inquiry into the history of engineering education of engineering shows that substantive changes have occurred and are occurring (Froyd et al., 2012; Seely, 1999). Studies have also demonstrated periodic reform in science and mathematics education at the K–12 level (Tyack & Cuban, 1995). However, pace of the changes may not satisfy many stakeholders with aspirations for major changes by 2020 (NAE, 2004).

Acknowledgments

Alone, the two authors could only provide an incomplete portrait of the development of engineering education and engineering education research. Other people have contributed to provide a more complete picture and we would like to acknowledge their valuable contributions: Cynthia J. Atman, Richard M. Felder, Larry J. Shuman, and Karl A. Smith.

Footnotes

1. The Morrill Act (often called the Land Grant Act) gave each U.S. senator and representative 30,000 acres of land, which was to be used to provide for colleges in each of the states. The colleges were to educate citizens in agriculture, home economics, mechanical arts (i.e., engineering), and other professions practical for the times.

2. There are other engineering education journals but their primary focus is curriculum development and pedagogical innovation.

References

ABET. (1995). Vision for change: A summary report of the ABET/NSF/Industry workshops, 95-VIS. Baltimore, MD: ABET, Inc.

American Society of Civil Engineers. (2010). History of ASCE. Retrieved from http://www.asce.org/PPLContent.aspx?id=2147485181.

American Society for Engineering Education. (1910). *Bulletin of the Society for the Promotion of Engineering Education*, Vol. XVIII, No. 1. Ithaca, NY: Society for the Promotion of Engineering Education.

American Society for Engineering Education. (1994). Report on evaluation of engineering education (reprint of the 1955 report). *Journal of Engineering Education*, 83(1), 74–94.

American Society for Engineering Education. (n.d.). American Society for Engineering Education: Our history. Retrieved from http://www.asee.org/about-us/the-organization/our-history.

Arreola, R. A. (1995). *Developing a comprehensive faculty evaluation system: A handbook for college faculty and administrators on designing and operating a comprehensive faculty evaluation system.* Bolton, MA: Anker.

Atman, C. J., & Bursic, K. M. (1998). Verbal protocol analysis as a method to document engineering student design processes. *Journal of Engineering Education*, 87(2), 121–132.

Atman, C. J., Chimka, J. R., Bursic, K. M., & Nachtmann, H. L. (1999). A comparison of freshman and senior engineering design processes. *Design Studies*, 20(2), 131–152.

Besterfield-Sacre, M. E., Atman, C. J., & Shuman, L. J. (1997). Characteristics of freshman engineering students: Models for determining student attrition in engineering. *Journal of Engineering Education*, 86(2), 139–149.

Borrego, M., & Bernhard, J. (2011). The emergence of engineering education research as an internationally-connected field of inquiry. *Journal of Engineering Education*, 100(1), 14–47.

Borrego, M., Douglas, E. P., & Amelink, C. T. (2009). Quantitative, qualitative, and mixed research methods in engineering education. *Journal of Engineering Education*, 98(1), 53–66.

Borrego, M., Froyd, J. E., & Hall, T. S. (2010). Diffusion of engineering education innovations: A survey of awareness and adoption rates in U.S. engineering departments. *Journal of Engineering Education*, 99(3), 185–207.

Boyer, E. L. (1990). *Scholarship reconsidered: Priorities of the professoriate*. San Francisco, CA: Jossey-Bass.

Center for the Advancement of Scholarship in Engineering Education. (2010). Engineering education centers roundtable. Retrieved from http://www.nae.edu/cms/11785.aspx.

Committee on K–12 Engineering Education; Linda Katehi, Greg Pearson, and Michael Feder, & National Academy of Engineering and National Research Council (Eds.). (2009). *Engineering in K–12 education: Understanding the status and improving the prospects*. Washington, DC: The National Academies Press.

Continental. (2006). *In search of global engineering excellence: Educating the next generation of engineers for the global workplace*. Hanover, Germany: Continental, AG.

Cooper, J. L., MacGregor, J., Smith, K. A., & Robinson, P. (2000). Implementing small-group instruction: Insights from successful practitioners. *New Directions in Teaching and Learning*, 81, 64–76.

Cooper, M. M. (1995). Cooperative learning: An approach for large enrollment courses. *Journal of Chemical Education*, 72(2), 162–164.

Cuban, L. (1999). *How scholars trumped teachers: Change without reform in university curriculum, teaching, and research, 1890–1990*. New York, NY: Teachers College Press.

Dancy, M. H., & Henderson, C. (2012). Pedagogical practices and instructional change of physics faculty. *American Journal of Physics*, 78(10), 1056–1063.

Diamond, R. M. (1993). Changing priorities and the faculty reward system. *New Directions for Higher Education*, 81, 5–12.

European and Global Engineering Education Network. (2010). Home page. Retrieved from http://www.eugene.unifi.it/.

Fairweather, J. S. (2002). The mythologies of faculty productivity: Implications for institutional policy and decision making. *The Journal of Higher Education*, 73(1), 26–48.

Felder, R. M., Felder, G. N., & Dietz, J. E. (1998). A longitudinal study of engineering student performance and retention. V. Comparisons with traditionally-taught students. *Journal of Engineering Education*, 87(4), 469–480.

Fensham, P. J. (2004). *Defining an identity: The evolution of science education as a field of research*. New York, NY: Springer.

Froyd, J. E., Wankat, P. C., & Smith, K. A. (2012). Five major shifts in 100 years of engineering education. *Proceedings of the IEEE*, 100(13), 1344–1360.

Gabriele, G. A. (2005). Advancing engineering education in a flattened world. *Journal of Engineering Education*, 94(3), 285–286.

Genesereth, M. R., & Nilsson, N. J. (1987). *Logical foundations of artificial intelligence*. Los Altos, CA: Morgan Kaufmann.

Grayson, L. P. (1993). *The making of an engineer: An illustrated history of engineering education in the United States and Canada*. New York, NY: John Wiley & Sons.

Greeno, J. G., Collins, A. M., & Resnick, L. B. (1996). Cognition and learning. In D. Berliner & R. Calfee (Eds.), *Handbook of educational psychology*. New York, NY: Macmillan.

Gruber, T. R. (1993). A translation approach to portable ontologies. *Knowledge Acquisition*, 5(2), 199–220.

Haghighi, K. (2005). Quiet no longer: Birth of a new discipline. *Journal of Engineering Education*, 94(4), 351–353.

Haller, C. R., Gallagher, V. J., Weldon, T. L., & Felder, R. M. (2000). Dynamics of peer interaction in cooperative learning workgroups. *Journal of Engineering Education*, 89(3), 285–293.

Handelsman, J., Ebert-May, D., Beichner, R., Bruns, P., Chang, A., DeHaan, R., . . . Wood, W. B. (2004). Scientific teaching. *Science*, 304(5670), 521–522.

Hattie, J., & Marsh, H. W. (1996). The relationship between research and teaching: A meta-analysis. *Review of Educational Research*, 66(4), 507–542.

Hawwash, K. (2007). Promotion of pedagogical abilities of engineering lecturers. In C. Borri, C., & F. Maffioli (Eds.), *Re-engineering engineering education in Europe* (pp. 17–44). Florence, Italy: Firenze University Press.

Henderson, C., & Dancy, M. H. (2007). Barriers to the use of research-based instructional strategies: The influence of both individual and situational characteristics. *Physical Review Special Topics – Physics Education Research*, 3(2), 020102-1–020102-14.

Henderson, C., & Dancy, M. H. (2008). Physics faculty and educational researchers: Divergent expectations as barriers to the diffusion of innovations. *American Journal of Physics*, 76(1), 79–91.

International Federation of Engineering Education Societies. (2010). Home page. Retrieved from http://www.ifees.net.

Jamieson, L. H., & Lohmann, J. R. (2009). *Creating a culture for scholarly and systematic innovation in engineering education, Phase 1.* Washington, DC: American Society for Engineering Education.

Jamieson, L. H., & Lohmann, J. R. (2012). *Innovation with impact: Creating a culture for scholarly and systematic innovation in engineering education.* Washington, DC: American Society for Engineering.

Jesiek, B. K., Borrego, M., & Beddoes, K. (2010). Advancing the global capacity for engineering education research (AGCEER): Relating research to practice, policy, and industry. *Journal of Engineering Education, 99*(2), 107–199.

Jesiek, B. K., Newswander, L. K., & Borrego, M. (2009). Engineering education research: Discipline, community, or field? *Journal of Engineering Education, 98*(1), 39–52.

Johri, A. (2010). Creating theoretical insights in engineering education. *Journal of Engineering Education, 99*(3), 183–184.

Johri, A., & Olds, B. M. (2011). Situated engineering learning: Bridging engineering education research and the learning sciences. *Journal of Engineering Education, 100*(1), 151–185.

Journal of Engineering Education. (2010). *Journal of Engineering Education,* home page. Retrieved from http://www.jee.org.

Lightcap, B. (2010). Morrill Act. Retrieved from http://www.nd.edu/~rbarger/www7/morrill.html.

Lohmann, J. R. (2003). Mission, measures, and ManuscriptCentral™. *Journal of Engineering Education, 92*(1), 1.

Lohmann, J. R. (2005). Building a community of scholars: The role of the *Journal of Engineering Education* as a research journal. *Journal of Engineering Education, 94*(1), 1–6.

Madhavan, K., Xian, H., Johri, A., Vorvoreanu, M., Jesiek, B. K., Wankat, P. C. (2011). Understanding the engineering education research problem space using interactive knowledge networks. In *Proceedings of 2011 Annual Conference and Exposition of the American Society of Engineering Education,* Vancouver, BC, Canada.

Madhavan, K. P. C., Xian, H., Vorvoreanu, M., Johri, A., Jesiek, B., Wang, A., & Wankat, P. (2010). The FIE Story 1991 to 2009. Invited video presentation featured at the Frontiers in Education Conference 2010, October 2010. The video is available online at: http://www.youtube.com/watch?v=bKA4zJc3bsA.

National Academy of Engineering (NAE). (2004). *The engineer of 2020.* Washington, DC: The National Academies Press.

National Research Council (NRC). (2000). *How people learn: Brain, mind, experience, and school* (expanded ed.). Washington, DC: The National Academies Press.

National Research Council (NRC). (2002). *Scientific research in education.* Washington, DC: The National Academies Press.

National Research Council (NRC). (2005). *Rising above the gathering storm: Energizing and employing America for a brighter economic future.* Washington, DC: The National Academies Press.

National Science Board. (1986). *Undergraduate science, mathematics and engineering education,* NSB 86010. Washington, DC: National Science Foundation.

National Science Foundation (NSF). (1992). *America's academic future: A report of the Presidential Young Investigator colloquium on U.S. engineering, mathematics, and science education for the year 2010 and beyond,* NSF 91–150. Washington, DC: National Science Foundation.

National Science Foundation (NSF). (2006). Approved Minutes, Open Session, 393rd Meeting, National Science Board. Retrieved from http://www.nsf.gov/nsb/meetings/2006/0809/minutes.pdf.

Prados, J. W. (2005). Quality assurance of engineering education through accreditation: The impact of Engineering Criteria 2000 and its global influence. *Journal of Engineering Education, 94*(1), 165–184.

Prince, M. J., Borrego, M., Cutler, S., Henderson, C., & Froyd, J. E. (2013). Use of research-based instructional strategies in core chemical engineering courses. *Chemical Engineering Education, 47*(1), 27–37.

Prince, M. J., Felder, R. M., & Brent, R. (2007). Does faculty research improve undergraduate teaching? An analysis of existing and potential synergies. *Journal of Engineering Education, 96*(4), 283–294.

Research in Engineering Education Network. (2010). Research in Engineering Education Network, home page. Retrieved from http://grou.ps/reen/home.

Reynolds, T. S., & Seely, B. E. (1993). Striving for balance: A hundred years of the American Society for Engineering Education. *Journal of Engineering Education*, 82(3), 136–151.

Seely, B. E. (1999). The other re-engineering of engineering education, 1900–1965. *Journal of Engineering Education*, 88(3), 285–294.

Shulman, L. S. (2005). If not now, when? The timeliness of scholarship of the education of engineers. *Journal of Engineering Education*, 94(1), 11–12.

Strobel, J., Evangelou, D., Streveler, R. A., & Smith, K. A. (2008). *The many homes of engineering education research: Historical analysis of PhD dissertations*. Paper presented at the Research in Engineering Education Symposium, Davos, Switzerland. Retrieved from http://www.engconfintl.org/8axabstracts/Session%204A/rees08_submission_16.pdf.

The Steering Committee of the National Engineering Education Research Colloquies. (2006a). Special report: The National Engineering Education Research Colloquies. *Journal of Engineering Education*, 95(4), 257–258.

The Steering Committee of the National Engineering Education Research Colloquies. (2006b). Special report: The research agenda for the new discipline of engineering education. *Journal of Engineering Education*, 95(4), 259–261.

Tyack, D., & Cuban, L. (1995). *Tinkering toward utopia: A century of public school reform*. Cambridge, MA: Harvard University Press.

United States Military Academy. (2010). A brief history of West Point. Retrieved from http://www.westpoint.edu/wphistory/SitePages/Home.aspx.

Wankat, P. C., Felder, R. M., Smith, K. A., & Oreovicz, F. S. (2002). The scholarship of teaching and learning in engineering. In M. T. Huber & S. Morreale (Eds.), *Disciplinary styles in the scholarship of teaching and learning: Exploring common ground* (pp. 217–237). Washington, DC: American Association for Higher Education/Carnegie Foundation for the Advancement of Teaching.

Weick, K. E., & Quinn, R. E. (1999). Organizational change and development. *Annual Review of Psychology*, 50, 361–386.

Appendix 1.1. Timeline of Events in Engineering Education and Engineering Education Research

Year	Events	Reports	ABET	Papers	NSF
1973				Stice paper: Hereford, S. M., & Stice, J. E. (1973). A course in college teaching in engineering and the physical sciences. Paper presented at the ASEE Annual Conference & Exposition.	
1976				Stice papers: (1) Stice, J. E. (1976). A first step toward improved teaching. *Engineering Education*, 66(5), 394–398; (2) The what, why, and how of faculty development, or who, me?"	
1981	Cooperative learning is introduced at the Frontiers in Education Conference, one of the two major conferences on engineering education in the United States.			First in a series of papers on cooperative learning published in *Engineering Education*, Smith, K. A., Johnson, D. W., & Johnson, R. T. (1981). Structuring learning goals to meet the goals of engineering education. *Engineering Education*, 72(3), 221–226.	
1986					NSB releases the Neal Report calling for more scholarship in engineering, science, and mathematics education.

1987

Stice publications: (1) Stice, J. E. (1987). Using Kolb's learning cycle to improve student learning. *Engineering Education, 77,* 291–296; (2) Stice, J. E. (Ed.). (1987). *Developing critical thinking and problem-solving abilities* (Vol. 30). San Francisco, CA: Jossey-Bass.

1988

NSF launches first program for curriculum development; NSF funds grant that leads to development of the E4 program at Drexel University (E4 program was the foundation for the Gateway Engineering Education Coalition funded in 1992); NSF funds grant that leads to development of the Engineering Core Curriculum at Texas A&M University the Engineering Core Curriculum was a pillar in the formation of the Foundation Coalition, a NSF Engineering Education Coalition funded in 1993).

(continued)

17

Appendix 1.1 (*continued*)

Year	Events	Reports	ABET	Papers	NSF
1989					NSF funds grant that leads to development of the Integrated, First-Year Curriculum in Science, Engineering and Mathematics (IFYCSEM) (IFYCSEM was a pillar in the formation of the Foundation Coalition, a NSF Engineering Education Coalition funded in 1993).
1990	ECSEL and Synthesis Engineering Education Coalitions started. EXCEL and Synthesis were engineering education Coalitions. NSF invested about $30 million in each Coalition to catalyze systemic improvement in engineering education; Leonhard Center for Enhancement of Engineering Education was established at Pennsylvania State University; Presidential Young Investigator (PYI) Colloquium held and report published.	America's Academic Future: A Report of the Presidential Young Investigator Colloquium on U.S. Engineering, Mathematics, and Science Education for the Year 2010 and Beyond			

(continued)

1991	Cooperative Learning Publications	Johnson, Johnson & Smith, Cooperative Learning ASHE-ERIC Research Report	First edition published: Johnson, D. W., Johnson, R. T., & Smith, K. A. (1998). *Active learning: Cooperation in the college classroom* (2nd ed.). Edina, MN: Interaction Book Company.
1992	SUCCEED and Gateway Engineering Education Coalitions started.		"ABET President John Prados challenged the Board of Directors to consider radical revisions in accreditation philosophy, criteria, and procedures" (Prados, J. W., Peterson, G. D., & Lattuca, L. R. (2005). Quality assurance of engineering education through accreditation: The impact of Engineering Criteria 2000 and its global influence. *Journal of Engineering Education*, 94(1), 165–184, p. 168)

Year	Events	Reports	ABET	Papers	NSF
1993	Foundation Coalition (Engineering Education Coalition) started.			Part I of longitudinal study of cooperative learning in chemical engineering: Felder, R. M., Forrest, K. D., Baker-Ward, L., Dietz, E. J., & Moh, P. H. (1993). A longitudinal study of engineering student performance and retention. I. Success and failure in the introductory course. *Journal of Engineering Education*, 82(1), 15–21.	NSF funds the first Presidential Young Investigator (PYI) awards in engineering education (PYI was the precursor to CAREER program).
1994	Greenfield Coalition (Engineering Education Coalition) started.	Report: Engineering Education for a Changing World. Report by the Engineering Deans Council and the Business Roundtable of the American Society for Engineering Education	ABET holds three consensus-building workshops for the Accreditation Process Review Committee. Workshops involve more than 125 participants from academia, industry, and government.	Part II of longitudinal study of cooperative learning in chemical engineering: Felder, R. M., Mohr, P. H., Dietz, E. J., & Baker-Ward, L. (1994). A longitudinal study of engineering student performance and retention. II. Rural/urban student differences. *Journal of Engineering Education*, 83(3), 209–217.	
1995	Frontiers in Education Conference (FIE): The FIE Conference held in		ABET Board of Directors approved the publication of new	Part III and IV of longitudinal study of cooperative learning in	

(continued)

Year			
	Atlanta was the first of a new format for the conference that attracted a relatively large crowd and made a substantial profit, enabling Education Research and Methods (ERM) Division of ASEE to begin to take on a number of creative activities.	criteria for evaluating engineering programs – Engineering Criteria 2000 (EC2000) for public comment.	chemical engineering: (1) Felder, R. M., Felder, G. N., Mauney, M., Charles Hamrin, J., & Dietz, E. J. (1995). A longitudinal study of engineering student performance and retention. III. Gender differences in student performance and attitudes. *Journal of Engineering Education, 84(2)*, 151–163; (2) Felder, R. M. (1995). A longitudinal study of engineering student performance and retention. IV. Instructional methods. *Journal of Engineering Education, 84(4)*, 361–367.
1996		Pilot evaluations conducted using Engineering Criteria 2000 at five diverse institutions.	USEME (precursor to DUE) formed.
1997	Frontiers in Education (FIE) Conference: The FIE Conference held in Pittsburgh introduced the New Faculty Fellows program and published the outstanding papers from the Conference in the Journal of Engineering Education.		

Appendix 1.1 *(continued)*

Year	Events	Reports	ABET	Papers	NSF
1998	Center for Engineering Learning and Teaching (CELT) at University of Washington established.			Part V of longitudinal study of cooperative learning in chemical engineering: Felder, R. M., Felder, G. N., & Dietz, E. J. (1998). A longitudinal study of engineering student performance and retention. V. Comparisons with traditionally-taught students. *Journal of Engineering Education*, 87(4), 469–480.	
1999	National Research Council (NRC) Board on Engineering Education moved to National Academy of Engineering (NAE) and renamed to Committee on Engineering Education (CEE).			Atman et al publish first study comparing performance with respect to engineering design. Paper compared engineering design performance of first-year and senior engineering students: Atman, C. J., Chimka, J. R., Bursic, K. M., & Nachtmann, H. L. (1999). A comparison of freshman and senior engineering design processes. *Design Studies*, 20(2), 131–152.	NSF funds the first Engineering Research Center (VaNTH) with a focus on engineering education; NSF initiates its Action Agenda program to facilitate adaptation of innovations in engineering and science education
2000	Center for Engineering Education was established at the Colorado School of Mines.				

Year			
2001	The Engineer of 2020 Project was started by the Committee on Engineering Education of the NAE.	Outcomes-based criteria (formerly referred to as Engineering Criteria 2000) become the only criteria used for accrediting engineering degree programs.	
2002	Center for the Advancement of Scholarship in Engineering was established at the NAE as one of the initiatives of the Committee on Engineering Education.		
2003	National Academy of Engineering (NAE) established the Bernard M. Gordon Prize for Innovation in Engineering and Technology Education ($500,000 award); *Journal of Engineering Education* focuses exclusively on research in engineering education.		NSF funds the first Center on Teaching and Learning (CLT) on engineering education, Center for Advancement of Engineering Education (CAEE).
2004	Purdue University and Virginia Tech each create a Department of Engineering Education.	Report: Engineer of 2020	

(*continued*)

Appendix 1.1 (continued)

Year	Events	Reports	ABET	Papers	NSF
2005	Engineering Education Research Colloquies (EERC). A series of four colloquies designed to spur discussion on the future of engineering education. The first of these colloquies occurred in September 2005. Invited scholars known for their experience and expertise in the field begin preparing a roadmap for engineering education and an engineering education research agenda; JEE publishes first special issue, the Art and Science of Engineering Education Research, whose articles are the most cited since the launch of JEE as a research journal.			Special Issue of *Journal of Engineering Education*; John Heywood book "Engineering Education: Research and Development in Curriculum and Instruction."	
2006	*Advances in Engineering Education* (AEE) launched.				Not sure of the exact year but about this time the EEC/ENG division started creating EER programs.

(continued)

2007 The first conference on engineering education research, International Conference on Research in Engineering Education (ICREE) launched with meeting in Honolulu; Global Colloquium in Engineering Education (GCEE) hosts its first track on engineering education; JEE initiates a new monthly column in Prism, "JEE Selects: Research in Practice."

2008 ASEE launches *Advances in Engineering Education* (AEE) as a repository for successful applications to complement JEE's focus on research; JEE begins forming international partnerships, ten created by 2010; Research in Engineering Education Symposium (REES), successor to ICREE, holds its first meeting in Davos, Switzerland.

Appendix 1.1 *(continued)*

Year	Events	Reports	ABET	Papers	NSF
2009	REES holds its second meeting in Palm Cove, Queensland, Australia.	Creating a Culture for Scholarly and Systematic Innovation in Engineering Education (CCSSIEE), Phase 1 Report			
2010		Creating a Culture for Scholarly and Systematic Innovation in Engineering Education (CCSSIEE), Phase 2 Report			
2011	JEE celebrates centennial issue.				

Part 1

ENGINEERING THINKING AND KNOWING

CHAPTER 2

Learning Theories for Engineering Education Practice

Wendy C. Newstetter and Marilla D. Svinicki

Introduction

In his book *Discussion of the Method*, Billy V. Koen characterizes the *engineering method* as "the strategy for causing the best change in a poorly understood situation within the available resources" (Koen, 2003, p. 7). This characterization could easily apply to the *instructional method*: the strategy is to effect the best change in a learner or in the relations between the learner and a larger community of practice in an imperfectly understood situation with available resources. The worlds of engineering and education are not so far apart after all. Whereas engineers seek to change the material world, instructors seek to change students who inhabit that world, both for the better. "Better," for the engineer, can mean less expensive, faster, or more durable materials and processes while for the instructor "better" can mean a more skilled, knowledgeable, or expert person who is essentially changed as a result of participating in a new community of practice. In both cases, "best" is accomplished through careful, informed design. And just

as physical principles, that is, engineering fundamentals, inform engineering design, learning theory should inform instructional design. Thus, we make the case that designing learning environments without learning theory is comparable to designing a bridge without mechanical laws and principles. In both cases, the goal is unlikely to be accomplished; the learner fails to change in desired ways and the bridge collapses.

To address a common scenario in post-secondary education where faculty are not instructional designers with a toolkit of learning fundamentals but rather accomplished disciplinary experts, we offer a primer on theories of cognition and learning as they relate to engineering education. As with other theoretical disciplines, advances in our understanding of cognition and learning have proceeded through a series of Kuhn-like paradigm shifts. These shifts have occurred as psychology, linguistics, philosophy, anthropology, artificial intelligence, and recently neuroscience have engaged the same two questions: *What is the nature of knowing? What is the nature of learning?*

These questions are bound together, for conceptualizations of knowing have significant implications for conceptualizations of learning and for the design of learning environments. As the conceptualization of knowledge changes, so should the interventions enacted in the educational setting.

Over the years, different conceptualizations of knowing and learning have taken on oppositional tones, with one theory vying for ascendancy over others. Thus, designs for what should happen in the classroom can shift dramatically. As an example in K–12 education, differing conceptualizations of the requisite knowledge needed to read have produced the phonics and the whole language approaches, each articulating a different notion of instruction associated with it. How reading gets taught in classrooms across the nation is dependent on the ascendant conceptualization. More recently, many have come to see the ecumenical value in blending and merging conceptual frameworks in the quest to design, test, and redesign optimal learning environments. Our goal here is to help engineering educators both to understand their current classroom activities in terms of learning theory and have the tools to develop new designs for "best change" in their students. In addition, because classroom instruction most often derives from one's conception of how students learn, a second goal is to assist engineering faculty in being more reflective about their own (implicit) theory of learning. A more ambitious objective is to provide engineering faculty and researchers with tools for thinking about, identifying, and designing educational research studies. As a field, engineering education is young and the number of experienced researchers investigating the complex learning questions associated with becoming an engineer is still sparse. But this means it is an exciting time; a great deal is yet to be discovered. In this chapter we aim to take a first step in suggesting those areas where careful research can make important and powerful contributions to the field. Later chapters more fully address specific areas where research

has been undertaken and where it needs to occur.

As with scientific and engineering research, thoughtful, well-designed investigations are informed by theory or, to use a better term, *conceptual frameworks*. Unlike the physical sciences, where established laws regularly guide researchers, the learning and cognitive sciences have few firm laws that have been validated by scientific processes of repeatability or reproducibility. The minute one puts a variety of learners together and offers instruction, the learning outcomes can be highly variable for many reasons. Learning scientists and educational psychologists want to understand those reasons better to minimize variability, but few expect to discover immutable laws of learning. Therefore, conceptual frameworks developed by the community of cognitive and learning scientists govern well-designed investigations of learning. These frameworks have a certain resemblance to theory but without the well-established rules or laws found in the physical sciences. Rather a conceptual framework is *a set of principles addressing the conditions and processes – social and environmental – that support or impede cognition and learning.* Well-designed educational studies are firmly grounded in and informed by these conceptual models and thus they are essential tools for engineering education researchers.

Creating a primer for a variety of engineering educators wishing to conduct educational research is a challenging task. It is tantamount to compressing a century of research and theory building into a finite number of pages. To maximize usefulness, we have embraced the tenet that people can understand best when an explanation is followed by a concrete example. We start each section with an overview of the particular learning paradigm in regard to its tenets of knowing and learning. We then exemplify that theory in an engineering classroom scenario that is most suited to that particular theory of knowing and learning. We end each section with the current status of the theory.

Figure 2.1. The behaviorist focus-S-R chain.

The Behaviorist Framework

Initially proposed in 1913 (Watson, 1913) and developed through the 1970s (Guthrie, 1935; Skinner, 1938, 1950, 1971; Tolman, 1932), behaviorism was the most dedicated attempt to apply the practices of what many psychologists considered to be "science" to the subject of human behavior. Rejecting the internally focused theories built on individual introspection, such as psychoanalysis, Watson and others wanted psychology to be based on the view that the understanding of behavior needed to be derived from empirical observation (Harzem, 2004). The behaviorists advocated explaining, predicting, and controlling behavior by dealing strictly with observable behavior in the observable environment without the need to consider internal mental processes. In engineering terms, this would be like looking at the inputs and outputs to a system while black boxing the system itself, as represented in Figure 2.1.

In the behaviorist framework, "knowing" consists of long chains of stimulus (S)–response (R) pairs that have been associated with past events and their consequences often enough to form a connection (Greeno, Collins, & Resnick, 1996; Johri & Olds, 2011). If the consequences had been positive (reinforcement), those chains had a high probability of being triggered when the initial stimulus was encountered again. Observing that increase in the response's probability was believed to be more verifiable, and therefore behaviorists believed that they had the appropriate paradigm to study human behavior scientifically.

If "knowing" is the established chain of stimulus–response pairings, then "learning" is the creation of those stimulus–response connections through exposure, repetition, and consequences (Thorndike, 1931). "We may define learning as a change in probability of response but we must also specify the conditions under which it comes about" (Skinner, 1950, p. 200). Instruction, therefore, is arranging for the sequence of "expose, response, consequate and repeat" until the behavior is learned. For example, to learn vocabulary, students can be given a list of terms, definitions, and examples to practice by pairing the term (stimulus) and its meaning (response) both in isolation and in the context of the course material. Correct use is rewarded by approval of the instructor or with points on a test (the consequence). Repeated pairings, as with flash cards, eventually result in the students' ability to respond with the definition or example in class or on a test. All the parts of the process are observable and verifiable. A difficulty with the behavioral procedure is that it takes a long time and much trial and error. To deal with this problem, behaviorists proposed another procedure called "shaping." In shaping, the terminal response is broken into smaller steps, beginning with a behavior that is already a part of the learner's repertoire. When the stimulus is presented, if the learner exhibits any part of the terminal behavior, it is reinforced. Once that step has been mastered, the criterion for reinforcement increases closer and closer to the terminal behavior until the terminal behavior itself is displayed. After that point, only the final form of the behavior is reinforced. Think of this as starting at the student's current level and proceeding to the final behavior gradually. The fundamentals of the behaviorist view of learning are exemplified by the following design principles.

Design Principles for Behaviorist Instructional

Instructional objectives: A key to behavioral design is a good understanding of the end goal of the instruction (Gagne, 1965). In behaviorist instruction, the instructor focuses on the instructional objective, and assessments are based directly on it. Mastery of the learning is defined as performance of the terminal, observable, measurable behavior. Creating objectives also makes instruction more standardized and less dependent on the vagaries of the situation. An assessment can be performance on a test, completion of a lab procedure, submission of a written paper, or any activity that provides an observable outcome. Instructional objectives have become an integral part of education (Gronlund, 2000).

Task analysis: This is the practice of breaking down the behavior to be learned into its logically sequenced component parts as in shaping (Aronson & Briggs, 1983). The same principle is used in backward design except in reverse. Instruction begins with the last step and works backward toward the first step (Wiggins & McTighe, 1995).

Shaping in small steps: Behavior-based instruction includes the idea of starting the process at the learner's current level of performance (Gagne, 1968). Most of the time learning has to proceed in gradual steps, as in shaping. The task analysis described earlier delineates the steps, and the instructional objective (also described earlier) indicates the terminal behavior. The concept of shaping encourages instructors to monitor ongoing learning and intervene with feedback and reinforcement. This self-paced learning system is fully grounded in the behaviorist theory and it works fairly well (Kulik, Kulik, & Cohen, 1979; Kulik, Cohen, & Eberling, 1980a; Kulik, Kulik, & Cohen, 1980b; Kulik, Kulik, & Bangert-Downs, 1990).

Observable stimulus–response association: Instruction often consists of exposing the learner to the stimulus associated with the response desired and then providing positive reinforcement if he or she makes the appropriate response. For example, in some classrooms, students are shown an example of a problem solution followed by single-answer questions (stimuli) to which they must respond correctly using an audience response system (observable responding by clickers). If the majority of the class chooses the correct answer, the instructor moves on; if not, the instructor goes back and repeats until the students can respond correctly (consequences).

Mastery and self-pacing: Behaviorism requires having a learner continue to work on an objective until mastery. This allows the student's abilities and efforts to determine progress rather than requiring each student to keep up with the class as a whole (Bloom, 1984).

Reinforcement: In behaviorism, learning requires reinforcement. Therefore, there are frequent samplings of behavior throughout the learning process so that the learner can be rewarded for progress.

The Behaviorist Classroom: An Example

Dr. Holland uses a behaviorist approach in one aspect of his class by assigning a self-paced, computer-assisted module to help students attain mastery of unit conversions in a step-by-step process. This course section has one module with clearly articulated learning objectives that match the kinds of questions that are asked at the end of the module. The students log on to complete a "prior knowledge" quiz to assess their knowledge of unit conversions. Based on the outcome of this pre-quiz, the computer determines what the student has already mastered and what requires more review. Each student is then tracked to the first section in which he or she showed deficits. Each section of the module takes the student through the main types of unit conversions interspersed with frequent questions to answer. If the student answers correctly, he or she moves on to the next section. If not, he or she receives feedback about what the problem might be and is routed back to the section to try again. If the student is still having problems after three tries, a teaching assistant is signaled and comes over to offer individual help. Students cannot proceed to the next section until they have successfully mastered the previous section. When the student reaches the end of the module, there is a post-test covering approximately the same material as the pre-test to confirm if the student has mastered the objectives. Students can work at their own pace unless they fall behind a predetermined optimal schedule of module completion, at which point the instructor sends a reminder to keep

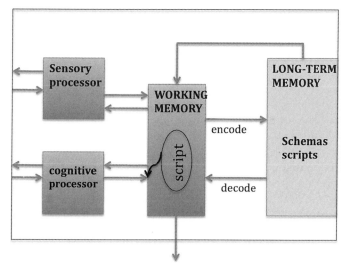

Figure 2.2. The cognitivist focus-internal units, relationships, and mechanism.

up the pace. Dr. Holland also sends congratulatory notes to those who are making good progress.

Current Status of the Framework and Implications for Research in Engineering Education

As noted earlier, behaviorism was most prominent during the late 1960s and early 1970s but has declined as a framework for instructional design and educational research as cognitivism has risen in popularity. However, it is still prominent as a behavior management system in schools (e.g., teachers controlling behavior through points or praise) and in work with special populations (e.g., students with learning disabilities or behavior control problems). The biggest change in the theory has been accepting the concept of thinking as a behavior subject to the same learning principles as observable behavior. Many therapies, for example, are based on shaping the way the client interprets events, which leads to changes in behavior, which can then be reinforced. Cognitive–Behavioral therapy is the best example of this process (Foreyt & Rathjen, 1978; Meichenbaum, 1977).

The Cognitivist Framework

A cognitive revolution was brewing in the 1950s, even as behaviorism maintained its position as the primary research-based theory of learning. George Miller, a prominent figure in experimental psychology, points to a 1956 conference that assembled psychologists, linguists, communications experts, and computer scientists as the birth of cognitive science (Gardiner, 1985). At the same time, a possible model for understanding the workings of the mind had come onto the scene – the computer. A central processing unit, input–output configurations, and the manipulation of symbols toward higher-level functions all seemed analogous to how the mind might operate, fostering a new wave of research by those interested in artificial intelligence that required discovering what might be inside the black box. Thus, in the shift from behaviorism to cognitivism, the importance of the internal workings, or, in more engineering terms, the possible mechanisms in a multiunit system – the mind – became the focus of investigations of knowing and learning, as depicted in Figure 2.2.

One main reason for this shift was that observing a learner's behavior alone did not give very accurate explanations or predictions of what a learner would or could do (Sperry, 1993; Wittrock, 2010 [original 1974]). For example, when presented with a list of things to remember, learners reorganize information to make it easier (Bousfield, 1953). They often use strategies such as mnemonics (e.g., ROY G. BIV for the color spectrum) to make information more manageable or meaningful and structure things that are difficult to remember (Foer, 2011). In another example, two people seeing the same stimulus will "see" it differently, which is the basis for the poor record of eyewitness testimony (Deffenbacher, 1991). Something more had to be happening in the brain to explain all the discrepancies.

Another contrast between behaviorism and cognitivism was the time needed to "shape" a behavior. In reality, one can learn a new behavior quickly through watching or being told what to do rather than through step-by-step shaping. To understand behavior, psychologists shifted to trying to understand how a learner was thinking or interpreting the situation. This was the basis for the rise of cognitivism (Sperry, 1993). All versions of cognitive theory state that "knowing" consists of having mental models that have been created and stored in the learner's long-term memory as a function of interacting with the environment. Referred to as concepts, models, sometimes as schemata or schema, "an active organization of past reactions, or of past experiences" (Bartlett, 1932, p. 201), these connections represent the stored experiences of that individual organized in clusters within a network structure across the brain (Gertner & Stevens, 1983). These models were deemed essential to reasoning and problem solving in any expert practice from playing chess to working on high-energy physics. Unlike novices who are either lacking or have poorly developed models, experts deploy these models to analyze, design, interpret, diagnose, and predict – all higher-order cognitive processes.

If "knowing" consists of having these organized models, then "learning" is the process of creating those models. The

processes follow a general sequence such as the following:

The learner is constantly monitoring and responding to the environment through his senses and his behavior. He notices incoming information, especially when things change, and moves the information into working memory, a temporary storage system. New information gets compared to already stored models and is either assimilated into the existing models, used to create new models, or modify old ones. Those models are then stored in long-term memory. Things that are stored in a structured or richly interconnected manner can be retrieved to guide behavior when a similar situation is encountered in the future. Long-term memory is continuously under revision as the learner is constantly responding to the changing world. The learner also engages in self-monitoring and organization through a process called "metacognition." Metacognition should be thought of as self-regulatory executive functioning that keeps the process flowing smoothly and the changes consistent with reality. (Atkinson & Schiffrin, 1968)

In engineering education, students invariably bring preexisting conceptions (mental models) of force developed through personal observations and stored in long-term memory. But these prior experiences, in school and out, may have led to an incorrect or naïve understanding of force (a misconception). Aware of the variety of misconceptions students bring, the savvy instructor performs a demonstration of force, but first asks the students to predict what will happen based on prior knowledge (the preexisting models stored in their long-term memory). This brings those old models into working memory (activates student prior knowledge) where they can be compared with the actual results. Then the instructor performs the demonstration. The students pay attention to what is happening, which brings the current situation into working memory, where it can be compared with the predictions based on their preexisting models. If their prediction was accurate, their preexisting model is confirmed and strengthened. If not, the students then must find a way to modify their prior models to accommodate the new information from the demonstration. This particular example of how the cognitive model deals with misconceptions is the subject of Chapter 3, but it is a good description of what happens even when there are no misconceptions (inaccurate mental models).

Current Subtheories of the Cognitive Framework

There are many versions of cognitive theory as applied to instruction. Each one that follows is typical of a different type of instruction derived from cognitive theory.

Information-processing theory: This version of cognitive theory assumes that what is in memory is a fairly accurate representation of the environment. The learner processes information and creates an accurate representation of what is presented in instruction (Mayer, 1996). It is often compared to the workings of a computer. Information processing theory is most applicable to learning facts, theories, and the overall structure of a particular field. It allows the instructor the maximum amount of control over the organization of the learning and the final structure of the content stored in long-term memory. Lectures, complemented by some individual hands-on work, are probably the best examples of information processing designs for instruction. However, they ignore the key idea that it is the learner who needs to be actively engaged in deep processing, and not just listening to the instructor.

Social cognitive theory (observational learning): A second version of cognitive theory is social cognitive theory, particularly *observational learning*. This theory emerged to explain why a great deal of learning seemed to come from observing the behavior of others rather than shaping. Its main proponent, Albert Bandura, proposed that the basis for learning from observed behavior was the creation of a "mental model." Bandura emphasized the importance of interaction between the learner and other people. This is why it is called "social" "cognitive" theory: it depends on interacting with others (social) and involves a mental model (cognitive) (Bandura, 1986).

Social cognitive theory can be used to explain skill learning (e.g., watching an instructor operate a piece of machinery) including intellectual skills (e.g., watching an instructor solve a math problem). Instruction derived from this theory would best be illustrated by demonstration with an emphasis on noting the key features of the skill, followed by practice of the skill by the learners with feedback from an instructor (Bandura, 1989).

Constructivism: This variant of cognitivism is the most prevalent one in the learning literature at this point (Tobias, 2010). Constructivism adheres to the mechanisms of creating and storing mental models, but with the learner in control (Bartlett, 1932; Dewey, 1916; Piaget, 1973; von Glasersfeld, 1989). In constructivism the mental model learned is a "construction" by the learner. As a result each person has a slightly different model that is a combination of all of his or her past experiences and his interpretations of the current situation (Smith, Sheppard, Johnson, & Johnson, 2005).

Instruction based on constructivism has learners exploring their own understanding as in inquiry learning. In inquiry learning, proponents assert that students should be put into authentic environments and allowed to explore. Problem-based learning strategies often take this form (Prince & Felder, 2006). A more structured version is guided inquiry, in which the learning is "guided" by instructions or features of the environment (Cordray, Harris, & Klein, 2009; Mayer, 2004). As mentioned, there are many variants of cognitive theory and many subtheories that attempt to explain different parts of learning (Johri & Olds, 2011). What follows is an attempt to extract the common elements in the cognitive paradigm.

Design Principles for Cognitivist Instruction

A focus on emphasizing the key features of the concepts to be learned: The first step in any instruction designed according to cognitive theory involves ensuring that the learner focuses on the key features of the concept being learned so that those key features can be used to make connections with the learner's prior knowledge. For instance, in our example, the instructional designer would make sure that the key features of force are emphasized in presentations, demonstrations, or activities.

Taking advantage of prior knowledge and experience of the learner: Because the goal of learning in cognitive theory is to make connections between new information and information already in long-term memory, instruction should start with what the learner already knows. In the example, what the learners already know about force in a naïve way or have experienced in the real world would be "activated" by making them predict what will happen. Most of the time the goal is to connect new knowledge with prior knowledge. Sometimes, however, prior knowledge is incorrect or not well formulated, in which case instruction should expose those misconceptions. Research on student misconceptions has been a particular focus for learning research in the physics education community. Hestenes and colleagues (Hestenes, Wells, & Swackhamer, 1996; Kintsch, 1988) developed the most extensive taxonomy of misconceptions in mechanics and an instrument for assessing those misconceptions. In engineering education, concept inventories designed to measure student misconceptions have been developed in statics (von Glasersfeld, 1989), heat transfer and thermodynamics (Piaget, 1980), and material science (Piaget, 1973), to name a few. But for DiSessa (1988, 1993), prior knowledge is not a set of stable, stored knowledge structures, but rather a set of more abstract, fundamental cognitive structures he termed P-Prims or phenomenological primitives. Examples of P-Prims include proximity and intensity, dying away, resistance and interference – all abstract concepts, not misconceptions – that DiSessa proposed students use as building blocks rather than impediments. These two alternative views of student prior knowledge have served as conceptual frameworks for a significant number of studies in science, technology, engineering, and mathematics (STEM) fields.

Aiming for deep processing of information (learning with understanding) rather than passive dependence on surface features: The general idea of deep processing is that learners should understand the structure of information to be learned, such as the main ideas and how they relate to one another and to sub-ideas that might derive from them (Chi, Feltovich, & Glaser, 1981; Perkins & Unger, 1999). Information in long-term memory entails very complex networks of associations. The more organized that network, the easier it is to remember and use. This organized structure has been proposed as one of the bases for the differences between novices and experts in a field (Alexander, 2003; Lajoie, 2003; Litzinger, Wise, & Lee, 2011; Streveler, Litzinger, Miller, & Steif, 2008). Understanding the network allows information to be stored more easily and retrieved more readily, and even speculated about when the details are not known (Bransford et al., 2006). This is the value of concept inventories; they catalog the knowledge and structure that represents the field (Streveler et al., 2008).

Involving the learner actively in selecting, organizing and integrating new information: It is critical that the learner be actively involved in creating the structure through deep engagement with content (Prince 2004; Smith et al., 2005). Only the learner can make the meaningful connections to his prior knowledge that result in clear models to govern use and further learning.

Developing metacognitive knowledge that allows students to control their own learning: We accept the fact that eventually all students must become independent learners who can set and monitor their own goals and processes (Bransford et al., 2006; Litzinger et al., 2005).

The Cognitivist Classroom: An Example
Professor Anderson is teaching a mass and energy balances course. The course goal is for students to develop problem-solving strategies

that use the laws of conservation of mass and energy. In this section of the course, they are focusing on the general energy balance equation and how to use it to solve problems. Although the complete equation, written in full without an understanding of its organization, is quite complex, Professor Anderson points out to the students that the equation can really be considered composed of just two parts: part 1 is the energy that resides within the system, and Part 2 is the energy that crosses the system boundary, either into or out of the system.

The equation can be broken down further into smaller parts, each of which is introduced in an organized way to help the students systematically build their knowledge. To explain the different parts of the energy balance equation, the professor chooses to use an example that would be familiar to all of the students – the pouring and drinking of a hot cup of coffee. Using a familiar example activates their prior knowledge, even if it is only personal experience. Dr. Anderson wants students to build on their existing mental models, correcting misconceptions or elaborating their model. By using the same situation to explain each part of the energy balance equation, Dr. Anderson makes it easier for students to understand the structural underpinnings of each type of energy and how they are related to one another.

Before each class period he posts skeletal notes interspersed with problems that the class will work through in class. The notes highlight the key in an organized format so students understand how the parts fit together. In class, Dr. Anderson keeps pointing to the key ideas in the outline.

In class when students reach a sample problem in the notes, he has them work the problem by themselves and then uses an audience response system (clickers) to report their results. Working through problems is an opportunity to engage students in deep processing of the material. If the entire class gets the correct answer, Dr. Anderson acknowledges their success and gives them a moment to write a note about their own solutions, right or wrong. By having students summarize their solution strategies, including any errors they made, Dr. Anderson increases the depth of processing of

the solution. If a majority of the class did not get the correct answer, Dr. Anderson can tell from their answer choices which misconceptions they still have or calculation errors they most likely made. He has them compare answers with one another and explain why they chose what they did, going through their problem set-up and calculations. Then he goes through the three incorrect choices to illustrate common errors in problems of this type. This kind of attempted self-explanation followed by the deeper processing of the problems illustrated by the instructor helps the students see where they went wrong much more readily than just having the instructor explain the problem.

At week's end, the class is devoted to students doing more involved and extensive problems embedding the week's work in real situations where this particular problem might be encountered. This is done in small groups, each group having a different version of the problem requiring a slightly different solution strategy. The group must also explain their solution and how their problem fits into the overall picture of mass and energy balances. They are encouraged to draw a concept map or flow chart to show how their problem is related to the main theory.

Current Status of the Framework

The constructivist version of cognitive theory is the dominant interpretation used in instructional design. It has made the largest difference in how instruction is designed by situating the learning in the learner rather than in the instruction.

The Situated Framework

While cognitive scientists were accounting for internal processes by which representations of information are constructed, stored, retrieved, and modified, social scientists from anthropology, sociology, and social psychology were conducting *interactional studies* (Star, 1996) of how individuals and groups interact with, coordinate, and use the resources in the external environment to accomplish meaningful objectives. This interest in "knowledge as distributed among people and their environment, including the objects, artifacts, tools, books and the communities they are part of" (Greeno et al., 1996, p. 20) is one feature of what has come to be known as the socio-cultural or *situative* perspective. Another equally important feature is a focus on identity formation accomplished through changing participation opportunities and choices in a community of practice (Lave & Wenger, 1991; Wenger, 1998). In contrast to the laboratory studies used by behaviorists and cognitivists, situationists wanted to understand "cognition in the wild" or how everyday folk make use of the material and social environment to solve problems and accomplish goals successfully (Brown, Collins, & Duguid, 1989; Hutchins, 1993, 1995; Lave, 1988; Suchman, 1987).

Knowing in this paradigm is attributed to both individuals and to groups. Individual knowing is the ability to participate successfully in the regular activities – *practices* – of a specific community. Lave and Wenger describe learning as a movement from peripheral forms of participation in a community of practice to full participation facilitated by apprenticeship opportunities to observe and then practice activities (Lave & Wenger, 1991). Thus, learning is signaled by changes in how the learner is able to participate more fully and effectively in an already existing community. And this learning is helped along by mentoring and apprenticing of newcomers by fuller participants in the community of practice. Consequently, in this framework learning is not generalized but always constrained by what a community values and adheres to as the members of the community work through everyday problems.

Knowing is also "attunements" to the affordances and the constraints of particular activity systems, tools, and practices as resources for accomplishing goals. This requires understanding the meanings and functions associated with situations, tools, and interactions (Barwise & Perry, 1983; Greeno & Middle School Mathematics Through Applications Project

Group, 1998). Knowing in the situative paradigm also involves meaning making such that the activities one participates in have both cultural and personal significance. At the same time, groups also "know" collectively how to carry out cooperative activities like navigating a nuclear submarine into port (Hutchins, 1995). Learning in the situative perspective is therefore a process of becoming a fuller participant in a community of practice as depicted in Figure 2.3.

Variations on the Situated Perspective

DISTRIBUTED COGNITION

In this perspective, knowledge and cognition are not "protected from the external world" (Newell, Rosenbloom, & Laird, 1989) but rather cognition is the mind working in conjunction with the tools available in the environment; cognition is distributed across the person attempting to accomplish a task and the material/informational resources in a given cultural environment. "Culture is not any collection of things, whether tangible or abstract. Rather, it is a process. It is a human cognitive process that takes place both inside and outside the minds of people. It is the process in which our everyday cultural practices are enacted" (Hutchins, 1995, p. 354). More to the point, in the distributed cognitive perspective the boundary between the internal (mind) and the external (environment) is rejected. People can take on complex problem-solving behaviors because they can bring together and leverage both internal and external representational structures. "'Cognition' observed in everyday practice is distributed – stretched over, not divided among – mind, body, activity and culturally organized settings (which include other actors)" (Lave, 1988, p. 1). When an engineer works on a problem, he or she will not just sit there "thinking" and miraculously arrive at a solution. Rather, he or she will shuttle between the cultural tools of diagrams (free body, for example) and symbols (math) on a piece of paper and the principles and laws, prior experiences, mental models, and so forth, thus harnessing a complete cognitive system that

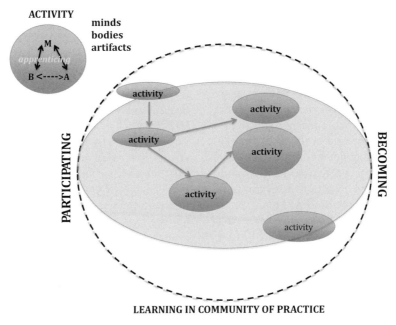

Figure 2.3. The situative focus. Learning as mind–body–artifact interaction in community of practice.

integrates the internal and external. Often a distributed cognitive system can be a group of people. A team that works together over a sustained period of time develops strategies for distributing and using information to accomplish a larger task. In this case, the system comprises the collective capabilities of the team and the material/social environment in which they work.

In a community or culture, the learner experiences new and changing forms of participation that facilitate the development and organization of internal knowledge representations. By adaptively responding to both repetitive and changed external circumstances, the learner is continually reorganizing and building knowledge and skill. Effective learning environments support the learner in developing an ability to integrate the external environmental structures and internal knowledge in problem solving.

NEO-VYGOTSKIAN SOCIOCULTURAL APPROACHES

A number of related approaches derive from the work of the Russian psychologist Lev Vygotsky (1962, 1978) and later extended by Leont'ev (1978), Luria (1976), and Wertsch

(1998). Like distributed cognition, sociocultural theory generally is concerned with the intersection and the integration of the material/social/cultural environment and the mind. Vygotsky (1981) proposed that learning happens simultaneously on two planes – the social plane between people ("interpsychological") and the personal plane "intrapsychological." What exists as communal knowledge develops into individual understanding through engaging in community-valued activities and forming societal relations. This integration of the communal and personal happens through participation in *activity systems* (Engestrom, 1987). Essentially, an activity system comprises an individual person intent on achieving a personally motivated goal by using environment-specific resources that support the process. The means (mediation) by which the subject is enabled to act toward achieving a goal is of central concern, for every time the form of mediation changes it impacts both the subject and the intended goal. In learning terms, every time the form of mediation in a classroom changes (e.g., lecture, clickers, work sheets), both the students and the desired goal will be impacted.

Another major concept is the zone of proximal development (ZPD) (Vygotsky, 1962, 1978). This is the gap between what a learner can accomplish on his or her own and what he or she can attain in collaboration with someone more expert. More commonly referred to as *scaffolding* (Wood, Bruner, & Ross, 1976), a learner and expert working collaboratively in this zone enables the learner to carry out a task that he or she could not manage on his or her own, and build competence eventually to complete such a task alone (Mercer & Fisher, 1993). A final concern of activity theory is identity formation. By participating in personally meaningful activities, the learner tries on and builds an identity that enacts the concerns and goals that are both community and individually valued (Roth & Lee, 2007).

Guiding Principles for Situationist Instructional Design

Creating opportunities for learners to actively participate in the social and material practices of a target community: Because the goal of instruction is to make it possible for learners progressively to participate more fully in the practices of a given community, classroom activities simulate the valued and repeated practices of a given community. Learning occurs when students interact with tools, people, and the physical world to develop an understanding of the affordances and constraints of culturally condoned tools and artifacts.

Encouraging students to try on the identity of members in the community of practice: By working with the tools, interacting with community members, and practicing community approaches to working through problems, a learner's notion of self as a member of a community of practice is constructed and validated.

Creating a learning environment that allows learners to chart their own learning path: Learning outcomes in a situated approach are not specified for each student in advance. Rather the project or problem used to engage and embed the learning is rich and complex enough to support

the exploration of multiple, self-determined learning pathways.

The Situative Classroom: An Example

Dr. Monica Carson attended a research conference and heard a talk about a new potential strategy for stopping the spread of HIV. The idea was to engineer infectious pseudoviruses that could potentially out-compete real HIV viruses for binding sites on cells and halt the spread of HIV. Dr. Carson thought this strategy could be the basis for a project for her biomedical engineering students. She brought the paper back and explained that the approach was entirely theoretical and not proven. She said they had an opportunity to "test" the hypothesis by using mathematical and computational approaches. Students formed teams and started doing research on the topic. Each member of the team decided how he or she could contribute to the project. As a team, they talked with experts who could be useful. They read journal articles to understand how modeling is used to test hypotheses; they checked out books that explained the pathophysiology of the HIV virus. Dr. Carson, her grad students, and post docs met with the teams, helping them by facilitating discussions and offering suggestions for resources they could use. Dr. Carson gave a talk on the centrality of modeling to engineering work practices and offered some mathematical approaches that could be useful. Another faculty member was invited to talk about his research and how he uses modeling as a tool for generating certain kinds of knowledge. The teams each developed a modeling strategy based on their work. The student groups made presentations on their models to faculty members who gave them feedback. They used the feedback to improve their models and then each group wrote a journal article on their model. One was later published in the Journal of Young Investigators.

Current Status of the Framework

This approach to instructional design has rarely been adopted into mainstream engineering educational settings. Community-based and -driven activities tend to be seen as extracurricular, add-on enrichment opportunities. However, with the growing interest in service- and community-based learning in engineering education, some programs are moving toward this model (Oakes, Duffy, Jacobius, Lord, Schultz, & Smith, 2002; Tsang & Van Haneghan, 2001). In addition, co-opting, internships, design competitions, and often capstone design courses are places where a situated approach to learning can be found. Advocates for online gaming environments are also experimenting with ways to have learners join online communities of practice as a first learning step.

Bridging the Cognitive–Situative Divide

Although we have kept these conceptual frameworks separate in the discussion so far, in actuality, many (Greeno & van de Sande, 2007; Kintsch, 1988; Piaget, 1980; von Glasersfeld, 1989) argue for an ecumenical approach to supporting learning that brings the situative and cognitivist approaches together in instructional design. This blending strategy also seeks to develop deep conceptual understanding but not through traditional lecturing or constrained problem solving in a classroom. Rather, conceptual development will occur when students participate in authentic community-sanctioned activities that offer opportunities for individual exploration and identity development. As with the other frameworks, we offer a classroom example.

The Blended Classroom: An Example

Professor Anderson is teaching a mass and energy balances course. A problem he deals with every day is that his coffee gets cold before he can finish drinking it, so generally he drinks only half the cup and throws the rest away. He wants his students to design a cup with the help of industrial design students that is lightweight, durable, low cost, and stylish but most importantly will keep his coffee warm for an hour. He assigns this challenge to the class on the second day after he explains on the first how the class will be structured – around solving real problems. Mixed teams of engineers and industrial designers form and start doing research on this problem. They find out what is already on the market, do testing with various kinds of cups, and investigate the properties of materials. Dr. Anderson, who is cognizant of where the teams will be at particular times, gives just-in-time lectures on engineering fundamentals relevant to the problem. He shows them how to model the problem and do calculations. He shows a video on an IDEO team at work so they can get a sense of designers in industry. Teams use the content of these lectures in the developing designs. Dr. Anderson acts as a facilitator-consultant to the teams during class meetings. Teams generate design proposals and during a class session explain/defend them while receiving feedback from Professor Anderson and an industrial design professional he has invited to the session. Teams revisit their designs based on the feedback and move forward to sketches and prototypes of their designs. Varied representational formats that leverage both engineering and design are encouraged. At the end of the project, teams make a presentation to a team of marketing students from the management school. Professor Anderson assesses the teams using engineering analysis of their proposed designs.

Translating Learning Theory into Classroom Practices and Research

As a way to make these theories more relevant and concrete for classroom practitioners, we end this chapter with a comprehensive table that translates tenets from each theory into specific classroom practices along six dimensions: *prototypical instruction, teacher/student roles, content outcomes, class activities, assessment,* and *purpose of grades* (Table 2.1). This translation is intended to illustrate how a change in what the student is expected to know implies a particular type of learning; in turn, the type of learning and knowing should be reflected in the

Table 2.1. An Analysis of Critical Instructional Components under Different Theoretical Orientations

Instructional Component	Behaviorism	Cognitivism	Situativity
Prototypical Instruction	Self-paced, mastery-based design (e.g., Keller plan, computer-assisted instruction) *Although many people have the impression that lecture/objective testing is a behaviorist model, lecture lacks the key features of self-pacing, small steps with immediate feedback, mastery at each step.*	1. Lectures emphasizing structural understanding of content (information processing subtype) 2. Demonstrations of skills by model (social cognitive subtype) 3. Guided inquiry such as experiential learning (constructivist subtype)	1. Learning in practice settings (e.g. service learning, apprenticeships, project work in real work settings) or in simulated practice settings. 2. Simulations of authentic problems (e.g. computer-based simulations)
Teacher/Student Roles	*Teacher role* – Define clear objectives in observable terms; do task analysis; set up mastery criteria; lay out instruction in routines to match objectives; evaluate mastery level, provide reinforcement and tutoring. *Student role* – Proceed through instructional steps outlined by instructor, mastery at each level.	*Teacher role* – Analyze and specify learning outcomes; provide necessary background and instructional materials; incorporate opportunities for learners to construct understanding; provide feedback during learning and evaluate final level achieved. *Student role* – Actively draw on own background, making connections, interpretations, their own mental models; test and revise on the basis of feedback	*Teacher role* – Member of community of practice or a stand-in for that community; identifier of learning settings; facilitator of learning processes; co-learner; model of expert processes: evaluator of final level *Student role* – Increasingly skilled actor in the community of practice, starting with legitimate peripheral participation and advancing to higher levels through increasingly central roles in community
Content Outcomes	Observable performance of behavior as specified in instructional objectives.	Demonstrate use of mental models in new situations.	Ability to function in the community of practice; understanding the constraints and affordances of that situation
Class Activities	Stepped presentation of content or skills according to task analysis with frequent overt practice followed by positive or negative feedback	Presentations or activities that require learners to construct and demonstrate understanding through problem solving, active practice of skills, and feedback.	Participation in authentic learning environments in which students solve real problems typical of the community of practice

Instructional Component	Behaviorism	Cognitivism	Situativity
Assessment	Tests with items or tasks matched to the instructional objectives	Test items or activities requiring deeper processing of application of models to complex situations.	Participation in practices of the community often demonstrated through self-reflections or portfolios of work
Purpose of Grades	Reinforcement or punishment; document mastery for purposes of advancing	Feedback on learners' model constructions; documentation of level of achievement reached	Formative feedback during learning; documentation of level of achievement reached at the end of the process, usually in the format of a written description

design of the classroom experience and in its intent. Too often in engineering classrooms, the instructional activities required of the students are not aligned with the kind of knowledge those activities are intended to foster. For example, if the intent is to develop complex problem-solving skills using fundamental laws and axioms, simple repetition of the type found in mastery of the multiplication tables will fail. Yet too often, instructors and students alike see homework sets as repetition rather than as conceptual model building. Engineering faculty often label student homework strategies as "plug and chug" in which students mindlessly map a homework problem onto a textbook example without deep processing. However, students possibly assume this practice because the faculty failed in the classroom to promote reflection and deep processing. We are hopeful that with Table 2.1, coupled with the historical account and the diagrammatic representations of each theory, engineering faculty will be able to both reflect on their own practices and design more optimal learning environments for best change.

At the same time, engineering education research must always situate itself in appropriate conceptual frameworks. We offer

Table 2.1 as a possible roadmap for the many significant directions that research in engineering learning can and needs to take, directions that are discussed in much greater detail in the following chapters. One could imagine using each box to generate a set of research questions and derivative studies. Take the first entry in the second column addressing prototypical instruction informed by behaviorism – *self-paced, mastery-based design*. A number of research questions easily come to mind: Which areas of engineering content are best suited to computer-mediated adaptive instruction? Is mastery learning the optimal way to acquire engineering fundamentals? If not, why not? Do personal response systems, which require a quick response, really develop sustained, transferable knowledge? Mastery learning assumes that smaller learning steps eventually lead to more complex understanding and application. Is this true for engineering problem solving? Does engineering education have the smaller steps right? For example, in mechanics it is assumed that statics needs to precede dynamics. Has this sequence been tested and validated? Or is it just how it has always been done, commonly accepted practice? Does using a textbook to check answers (feedback) foster bad habits such as plug and chug?

Let's take a box in a very different region of this table – teacher and student roles in a situative learning environment. A number of research questions and studies could easily be generated focusing on the teacher as *facilitator of learning processes*. First questions might be: What are the most effective facilitation moves? The least effective? How should/does facilitation change when the classroom activity changes or as the students become more experienced/knowledgeable? How much should the facilitator intervene or stay out of the way? Is there a taxonomy of facilitation moves that can be identified and described? Are there content areas in which facilitation is not the best approach but direct instruction is required? What is the best way to train teachers to be facilitators? Can a computer be a facilitator?

Of course, each of these questions implies a different research strategy, a different method, some quantitative and some qualitative. It should be clear, though, that engineering researchers need to always be cognizant of the context they are investigating and the origins or underpinnings of that context from a theoretical perspective and use conceptual frameworks as the basis for rigorous and well-designed and executed investigations.

References

Alexander, P. (2003). The development of expertise: The journey from acclimation to proficiency. *Educational Researcher, 32*, 10–14.

Aronson, D., & Briggs, L. (1983). Contributions of Gagne and Briggs to a prescriptive model of instruction. In C. Reigeluth (Ed.), *Instructional design theories and models: An overview of their current status* (pp. 75–100). Hillsdale, NJ: Lawrence Erlbaum.

Atkinson, R. C., & Schiffrin, R. M. (1968). Human memory: A proposed system and its control processes. In K.W. Spence & J. T. Spence (Eds.), *The psychology of learning and memory*. Advances in research and theory, Vol. 2. New York, NY: Academic Press.

Bandura, A. (1986). *Social foundations of thought and action: A social cognitive theory*. Englewood Cliffs, NJ: Prentice Hall.

Bandura, A. (1989). Human agency in social cognitive theory. *American Psychologist, 44*, 1175–1184.

Bartlett, F. C. (1932). *Remembering: A study in experimental and social psychology*. New York, NY: Macmillan.

Barwise, J., & Perry, J. (1983). *Situations and attitudes*. Cambridge, MA: MIT Press.

Bloom, B. S. (1984). The 2 sigma problem: The search for methods of instruction as effective as one-to-one tutoring. *Educational Researcher, 13*(6), 4–16

Bousfield, W. A. (1953). The occurrence of clustering in the recall of randomly arranged associates. *Journal of General Psychology, 49*, 229–240.

Bransford, J., Stevens, R., Schwartz, D., Meltzoff, A., Pea, R., Roschelle, J., . . . & Sabelli, N., et al. (2006). Learning theories and education: Toward a decade of synergy. In P. Alexander & P. Winne (Eds.), *Handbook of educational psychology* (pp. 209–244). Mahwah, NJ: Lawrence Erlbaum.

Brown, J. S., Collins, A., & Duguid, P. (1989). Situated cognition and the culture of learning. *Educational Researcher, 18*(1), 32–42.

Chi, M. T. H., Feltovich, P. J., & Glaser, R. (1981). Categorization and representation of physics problems by experts and novices. *Cognitive Science, 5*(2), 121–152.

Cordray, D. S., Harris, T. R., & Klein, S. (2009). A research synthesis of the effectiveness, replicability, and generality of the VaNTH challenge-based instructional modules in bioengineering. *Journal of Engineering Education, 98*(4), 335–348.

Deffenbacher, K. A. (1991). A maturing of research on the behaviour of eyewitnesses. *Applied Cognitive Psychology, 5*(5), 377–402.

Dewey, J. (1916). *Democracy and education*. New York, NY: The Free Press.

DiSessa, A. (Ed.). (1988). *Knowledge in pieces*. Hillsdale, NJ: Lawrence Erlbaum.

DiSessa, A. (1993). Toward an epistemology of physics. *Cognition and Instruction, 10*, 105–122.

Engestrom, Y. (1987). *Learning by expanding: An activity-theoretical approach to developmental research*. Helsinki, Finland: Orienta-Konsultit.

Foer, J. (2011). *Moonwalking with Einstein: The art and science of remembering everything*. New York, NY: Penguin Press.

Foreyt, J. P., & Rathjen, D. P. (1978). *Cognitive behavior therapy: Research and application*. New York, NY: Plenum Press.

Gagne, R. M. (1965). *The conditions of learning*. New York, NY: Holt, Reinhart & Winston.

Gagne, R. M. (1968). Learning hierarchies. *Educational Psychologist*, 6, 1–9.

Gardiner, H. (1985). *The mind's new science: A history of the cognitive revolution*. New York, NY: Basic Books.

Gertner, D., & Stevens, A. L. (Eds.). (1983). *Mental models*. Hillsdale, NJ: Lawrence Erlbaum.

Greeno, J. G., Collins, A. M., & Resnick, L. B. (1996). Cognition and learning. In R. Calfee & D. Berliner (Eds.), *Handbook of educational psychology* (pp. 15–47). New York, NY: Macmillan

Greeno, J. G., & Middle School Mathematics Through Applications Project Group. (1998). The situativity of knowing, learning and research. *American Psychologist*, 53, 5–26.

Greeno, J. G., & van de Sande, C. (2007). Perspectival understanding of conceptions and conceptual growth in interaction. *Educational Psychologist*, 42(1), 9–23.

Gronlund, N. E. (2000). *How to write and use instructional objectives*. Upper Saddle River, NJ: Merrill-Prentice Hall.

Guthrie, E. R. (1935). *The psychology of learning*. New York, NY: Harper.

Harzem, P. (2004). Behaviorism for new psychology: What was wrong with behaviorism and what is wrong with it now. *Behavior and Philosophy*, 32, 5–12.

Hestenes, D., Wells, M., & Swackhamer, G. (1996). Force concept inventory. *The Physics Teacher*, 30, 141–158.

Hutchins, E. (1993). Learning to navigate. In S. Chaiklin & J. Lave (Eds.), *Understanding practice: Perspectives on activity and context* (pp. 35–63). Cambridge: Cambridge University Press

Hutchins, E. (1995). *Cognition in the wild*. Cambridge, MA: MIT Press.

Johri, A., & Olds, B. (2011). Situated engineering learning: Bridging engineering education research and the learning sciences. *Journal of Engineering Education*, 100(1), 151–185.

Kintsch, W. (1988). The role of knowledge in discourse comprehension: A construction-integration model. *Psychological Review*, 95, 163–182.

Koen, B. V. (2003). *Discussion of the method*. New York, NY: Oxford University Press.

Kulik, C. C., Kulik, J. A., & Bangert-Downs, R. I. (1990). Effectiveness of mastery learning programs: A meta-analysis. *Review of Educational Research*, 60, 265–299.

Kulik, J. A., Cohen, P. A., Eberling, B. J. (1980a). Effectiveness of programmed instruction in higher education: A meta-analysis of findings. *Educational Evaluation and Policy Analysis*, 2(6), 51–64.

Kulik, J. A., Kulik, C. C., & Cohen, P. A. (1979). A meta-analysis of outcome studies of Keller's Personalized System of Instruction. *American Psychologist*, 34, 307–318.

Kulik, J. A., Kulik, C. C., & Cohen, P. A. (1980b). Effectiveness of computer-based college teaching: A meta-analysis of findings. *Review of Educational Research*, 50, 525–544.

Lajoie, S. (2003). Transitions and trajectories for studies of expertise. *Educational Researcher*, 32, 21–25.

Lave, J. (1988). *Cognition in practice*. Cambridge: Cambridge University Press.

Lave, J., & Wenger, E. (1991). *Situated learning: Legitimate peripheral participation*. Cambridge: Cambridge University Press.

Leont'ev, A. N. (1978). *Activity, consciousness, and personality*. Englewood Cliffs, NJ: Prentice-Hall.

Litzinger, T. A. Latuca, L., Hadgraft, R., & Newstetter, W. (2011). Engineering education and the development of expertise. *Journal of Engineering Education*, 100(1), 123–150.

Litzinger, T. A., Wise, J. C., & Lee, S. A. (2005). Self-directed learning readiness among engineering undergraduate students. *Journal of Engineering Education*, 94(2), 215–221.

Luria, A. R. (1976). *Cognitive development: Its cultural and social foundations*. Cambridge, MA: Harvard University Press.

Mayer, R. E. (1996). Learners as information processors: Legacies and limitations of educational psychology's second metaphor. *Educational Psychologist*, 59(1), 151–161.

Mayer, R. E. (2004). Should there be a three-strikes rule against pure discovery learning? The case for guided methods of instruction. *American Psychologist*, 59(1), 14–19.

Meichenbaum, D. H. (1977). *Cognitive-behavior modification*. New York, NY: Plenum Press.

Mercer, N., & Fisher, E. (1993). How do teachers help children to learn? An analysis of teachers' interventions in computer-based activities. *Learning and instruction, 2*, 339–355.

Newell, A., Rosenbloom, P. S., & Laird, J. E. (1989). Symbolic architectures for cognition. In M. Posner (Ed.), *Foundations of cognitive science* (pp. 93–132). Cambridge, MA: MIT Press.

Oakes, W., Jacobius, T., Linos, P., Lord, S., Schultz, W.W., & Smith, A. (2002). Service learning in engineering. In *Proceedings of the ASEE/IEEE Frontiers in Education Conference*, Boston, MA (pp. F3A1–F3A6).

Perkins, D., & Unger, C. (1999). Teaching and learning for understanding. In C. Reigeluth (Ed.), *Instructional-design theories and models: A new paradigm of instructional theory*, Vol. II. Mahwah, NJ: Lawrence Erlbaum.

Piaget, J. (1973). *To understand is to invent: The future of education*. New York, NY: Grossman.

Piaget, J. (1980). The psychogenesis of knowledge and its epistemological significance. In M. Piattelli-Palmarini (Ed.), *Language and learning*. Cambridge, MA: Harvard University Press.

Prince, M. J. (2004). Does active learning work? A review of the research. *Journal of Engineering Education, 95*(3), 223–231.

Prince, M. J., & Felder, R. M. (2006). Inductive teaching and learning methods: Definitions, comparison, and research bases. *Journal of Engineering Education, 97*, 123–138.

Roth, W., & Lee, Y. (2007). Vygotsky's neglected legacy: Cultural-historical activity theory. *Review of Educational Research, 77*(2), 186–232.

Skinner, B. F. (1938). *The behavior of organisms: An experimental analysis*. New York, NY: D. Appleton-Century.

Skinner, B. F. (1950). Classics in the history of psychology. *Psychological Review, 57*, 193–216.

Skinner, B. F. (1971). *Beyond freedom and dignity*. Indianapolis, IN: Hackett .

Smith, K. A., Sheppard, S., Johnson, D., & Johnson, R. (2005). Pedagogies of engagement: Classroom-based practices. *Journal of Engineering Education, 94*(1), 87–101.

Sperry, R. W. (1993). The impact and promise of the cognitive revolution. *American Psychologist, 48*(8), 878–885.

Star, S. L. (1996). Working together: Symbolic interactionism, activity theory and information sustems. In Y. Engstrom & D. Middleton (Eds.), *Cognition and communication at work*. Cambridge: Cambridge University Press.

Streveler, R. A., Litzinger, T. A., Miller, R. L., & Steif, P. (2008). Learning conceptual knowledge in the engineering sciences: Overview and future research directions. *Journal of Engineering Education, 98*(3), 279–294.

Suchman, L. (1987). *Plans and situated action*. New York, NY: Cambridge University Press.

Thorndike, E. L. (1931). *Human learning*. New York, NY: The Century Co.

Tobias, S. (2010). Generative learning theory, paradigm shifts, and constructivism in educational psychology: A tribute to Merl Wittrock. *Educational Psychologist, 45*(1), 51–54.

Tolman, E. C. (1932). *Purposive behavior in animals and men*. New York, NY: The Century Co.

Tsang, E., & Van Haneghan, J. (2001). A report on service learning and engineering design. *International Journal of Engineering Education, 17*(1), 30–39.

von Glasersfeld, E. (1989). Cognition, construction of knowledge, and teaching. *Synthese, 80*, 121–140.

Vygotsky, L. S. (1962). *Thought and language*. Cambridge, MA: MIT Press.

Vygotsky, L. S. (1978). *Mind in society: The development of higher psychological processes*. Cambridge, MA: Harvard University Press.

Watson, J. B. (1913). Psychology as the behaviorist views it. *Psychological Review, 20*, 158–177.

Wenger, E. (1998). *Communities of practice: Learning, meaning and identity*. Cambridge: Cambridge University Press.

Wertsch, J. V. (1998). *Mind as action*. New York, NY: Oxford University Press.

Wiggins, G., & McTighe, J. (1995). *Understanding by design*. Upper Saddle River, NJ: Merrill-Prentice Hall.

Wittrock, M. (2010, original 1974). Learning as a generative process. *Educational Psychologist, 45*(1), 40–45.

Wood, D., Bruner, J., & Ross, G. (1976). The role of tutoring in problem solving. *Journal of Child Psychology and Psychiatry, 17*, 89–100.

CHAPTER 3

Situative Frameworks for Engineering Learning Research

Aditya Johri, Barbara M. Olds, and Kevin O'Connor

Introduction

There is increased concern with developing a better understanding of how people learn engineering, as prior efforts to improve engineering education have often followed an ad hoc trajectory. The field lacks a systematic understanding of how engineering learning occurs and there is a paucity of knowledge on which to draw (Felder, Sheppard, & Smith, 2005; Chapter 1 by Froyd & Lohmann, this volume). To help redress this situation, in this chapter we review scholarship on learning with the aim of building a framework that can guide future research on engineering learning. Specifically, we hope to make the case for a framework that focuses on situativity and learning in engineering settings. This chapter complements other chapters in this volume including Chapters 2 by Newstetter and Svinicki; 4 by

This chapter is a substantially revised version of Johri & Olds (2011), which appeared in the *Journal of Engineering Education* (Vol. 100, Issue 1, pp. 151–185). We are grateful to the American Society of Engineering Education (ASEE) for permission to reproduce portions of the text.

Roth; and 5 by Streveler, Brown, Herman, and Monfort that also focus on learning.

An Introduction to Learning

During the past couple of centuries, scholars from a wide range of disciplines including philosophy, psychology, anthropology, and sociology have spent considerable time trying to answer questions related to learning, such as: How does cognitive development take place? How do we grow from a child with rudimentary abilities and knowledge into a highly skillful adult? How are humans able to engage in highly complex activities? Some scholars whose work has had a major influence on research on learning include Lev Vygotsky (1962, 1978), Jean Piaget (1952, 1964), John Dewey (1896, 1934), Harold Garfinkel (1967), William James (1890/1950), George Herbert Mead (1934), Gregory Bateson (1978), Michel Polanyi (1967), and Jerome Bruner (1990, 1960). Core ideas of these scholars adopted by learning researchers in their intellectual and methodological trajectory include

Vygotsky's cultural historical theory, Piaget's genetic epistemology, Dewey's transactional account, James's pragmatism and realism, Polanyi's tacit knowledge, and Garfinkel's ethnomethodology. These ideas have not only shaped theoretical development of the field of learning but have also influenced the design of learning environments including our schools and curricula. Many central ideas that we take for granted in educational practice, such as the progression of child development through specific stages and the value of group work and collaborative learning, can be traced back to these influential scholars.

Greeno, Collins, and Resnick (1996) provide a useful classification of research on learning that occurred over the twentieth century and divide the work into three broad areas: behaviorist, cognitive, and situative. Chapter 2 by Newstetter and Svinicki provides a comparative analysis of these three perspectives. In this chapter we draw primarily on the situative perspective and discuss its implications for research on engineering learning. The situative perspective views knowledge "as distributed among people and their environments, including objects, artifacts, tools, books, and the communities of which they are a part" (Greeno et al., 1996, p. 17) and learning is conceptualized as meaningful participation in a community of practice. There is an understanding that "the constraints and affordances of social practices and of the material and technological systems of environments" (Greeno et al., 1996, p. 17) shape learning significantly. The situative movement differs significantly from prior approaches such as the behaviorist and cognitive perspectives in its emphasis on the role of the environment on an individual's conception of knowing and how he or she learns – knowledge is not something that an individual possesses or stores in the brain but is present in all that he or she does. Clancey (1997) succinctly summarizes the situative perspective and how it differs from the cognitive perspective when he argues, "The idea that knowledge is a possession of an individual is as limited as the idea that culture is going to the opera. Culture is

pervasive; we are participating in a culture and shaping it by everything we do. Knowledge is pervasive in all our capabilities to participate in our society; it is not merely beliefs and theories describing what we do" (p. 271).

The Situative Perspective on Learning

As introduced in the preceding section, one significant change in research on learning over the past couple of decades is a move toward examining learning as a situated activity. The situative perspective is broad and owes a debt to many scholars and ideas. Its seeds can be traced back to the work of Dewey (1934). This perspective has been referred to as situated cognition (Brown, Collins, & Duguid, 1989; Greeno, 1989; O'Connor & Glenberg, 2003), cognition in practice (Lave, 1988), situated learning (Lave & Wenger, 1991), situated action (Suchman, 1987), sociocultural psychology (Rogoff, 1990; Wertsch, 1993), activity theory (Engeström, 1987), and distributed cognition (Hutchins, 1995). Greeno (2006) refers to the perspective as *situative* and/or *situativity*, as opposed to situated learning, to prevent the misconception that only some action, cognition, or learning is situated. He argues, as do others (Lave, 1988; Suchman, 1987), that *all* action, cognition, and learning are situated, whether in informal settings or formal school settings. The situative perspective views human knowledge as arising dynamically, being constructed and/or reinterpreted, within a specific social context (Clancey, 2009). Furthermore, knowledge is socially reproduced and learning occurs through participation in meaningful activities that are part of a community of practice (Lave & Wenger, 1991). This participation is mutually constituted through, and is a reflection of, our thinking and literacy skills (Gee, 1997). According to Sawyer and Greeno (2009) the core commitment of the situative perspective is "to analyze performance and transformation of activity systems that usually comprise multiple people and a variety of technological arti-facts" (p. 348). In other words, a central aim of the situated perspective is to understand learning as situated in a complex web of social organization rather than as a shift in mental structures of a learner.[1]

Recently, many scholars have argued for research that bridges the cognitive and sit-uative perspectives on learning (Greeno & van de Sande, 2007; Vosniadou, 2007). These efforts are driven by a desire to overcome what some see as a dichotomy between approaches that see learning in terms of acquisition of knowledge by individuals and those that see learning in terms of participa-tion in forms of social organization (Sfard, 1998). Cognitivists ask, if all learning is situ-ated – and participatory – then how do we account for transfer (Bransford & Schwartz, 1999)? One way in which situativists answer this criticism is by arguing that we apply what we know in a new activity based on fea-tures common to that activity and previous activities and by reframing the learning con-texts (Engle, 2006).[2] Still, this answer does not satisfy all conditions of transfer – such as use of knowledge in a novel domain or sit-uation – and remains a critical challenge for the situative perspective. Arguing that both acquisition and participation metaphors can provide useful guidance for research, some scholars have proposed a third metaphor, "knowledge creation," as a way to provide a better overall framework – consisting of all three metaphors – to advance our under-standing of learning (Paavola, Lipponen, & Hakkarainen, 2004).

Another thorny issue between the pro-ponents and opponents of the situative per-spective is the question of whether indi-viduals learn or learning is a characteristic of an activity system (Salomon & Perkins, 1998). Although situativity scholars argue that situativity does not preclude learn-ing from occurring at an individual level, empirical studies of situated activity have primarily dealt with a group-level anal-ysis. These debates show the continuing importance of cognitivist and situativist approaches but they also highlight the devel-opment of the situative perspective along

two slightly different trajectories – sociocog-nitive and sociocultural. Whereas research in the sociocognitive tradition follows the cognitive tradition in its focus on the indi-vidual in his/her immediate surroundings, sociocultural scholars look at participation of learners in broader communities.

We believe that a deeper understanding of the situative perspective can provide valu-able lessons for engineering educators, par-ticularly in their efforts to develop theoret-ical insights into engineering learning. To facilitate this process, we next outline and discuss three analytical aspects of situative learning. First, we look at the importance of *action* as the primary analytic focus of situ-ative approaches. Second, we examine the role of *mediation* in the conduct and devel-opment of action. Third, we explore the ideas of *participation and identity* in relation to situativity. Finally, we discuss some cur-rent critiques of situativity. After discussing each concept in detail, we explore their sig-nificance for engineering learning and how they can help inform future research.

Primacy of Action and Interaction

One of the primary aims of situative ap-proaches to cognition and learning has been a critique of dominant paradigms of knowl-edge, learning, and schooling, and, most par-ticularly, of the views of cognitivism (e.g., Anderson, Reder, & Simon, 1996, 1997; Vera & Simon, 1993). A major difference between cognitivism and situated perspectives lies in their respective views of the relation-ship between mind and action. All situ-ative theories begin from the assumption that primary analytical emphasis should be placed on human action or activity[3]; they thus reject in principle the cognitivist notion that cognition and learning are phenomena that can be considered as analytically sepa-rate from and prior to action (O'Connor & Glenberg, 2003). Cognitivism assumes that individuals and the world are fundamen-tally separate and that the world has a single, objective, and knowable character. Knowledge about the world, in the form of stable mental representations located in individual minds, is taken to underlie and enable action or behavior in concrete contexts, which are assumed to have a determinate character apart from human activity and interpretation. In this view, the power of knowledge depends on its degree of abstractness and generality – the more abstract and general, or "decontextualized," knowledge is, the more contexts in which it will allow for action, or "intelligent behavior" (Vera & Simon, 1993). Learning, from this perspective, involves an individual's movement away from the concrete, situated, and purportedly faulty and inefficient forms of thought taken to characterize everyday life, and toward the acquisition of abstract, general, and universally applicable conceptual knowledge. Learning is best brought about by separating learners from the complexities of everyday experience and providing them with instruction designed to allow them to acquire explicit decontextualized concepts that can be transferred to and applied at other times and in other places.

Situative approaches have challenged these cognitivist assumptions about the relationship between mind and action, including the presumed separation between individuals and an objective and stable world, and about the nature of knowledge and learning. Theories of situated learning start not with a given world, but with a world in process, and "contexts" do not completely predetermine human action and interpretation. Instead, human agents flexibly *contextualize* (Duranti & Goodwin, 1992; Lave, 1993; McDermott, 1999; Miller & Goodnow, 1995; O'Connor, 2003) their ongoing activity. That is, through their activity, people construe, and thereby constitute, the context of that activity as a context of a certain type, involving participants of a certain type. Activity is partially structured through the use of material and semiotic resources that have evolved within and are associated with particular practices. This is what makes activity recognizable and reproducible across occasions. The meaning of activity is not determined by the use of resources associated with particular practices, however, because there are always contingencies on any particular occasion

that ensure that a given instance of activity will be unique (Wertsch, 1998). The constitution of contexts out of the flow of activity, then, requires ongoing reflexive judgments as to the type of context that participants are engaged in, and as to the adequacy and appropriateness of any particular contribution to a context of that type (Lave & Wenger, 1991). This is an inherently evaluative process in which all participants in activity position themselves with regard to one another, the ongoing activity, and broader forms of social organization. Of course, if contexts are indeterminate and constituted through activity, then knowledge cannot be understood as inhering in decontextualized and stable mental representations of a given, unchanging, and objective world. Instead, knowledge is seen as located in ever-changing forms of situated, embodied activity in and with the world (Lave, 1988; Scribner, 1997a), and as continually negotiated by participants in activity. This view removes knowledge from the heads of individuals and locates it in broader forms of social organization. In this view, learning is not limited to didactic activities in formal environments and occurs across diverse settings and in and through interaction among people and objects (Lave & Wenger, 1999; Scribner & Cole, 1973). Rather, schools are taken to offer atypical educational experiences that are divorced from daily life experiences (Scribner & Cole, 1973); they are inauthentic (Brown et al., 1989) and consist of decontextualized knowledge (Donaldson, 1978).

Mediation in Social and Material Context

The focus on action has led researchers associated with the situative perspective to examine how people engage in a wide variety of practical tasks, from the seemingly mundane activity of what Lave (1988) ironically called "just plain folks" (grocery shoppers, cooks, bartenders, street vendors, factory workers, and the like) to those engaged in activities taken to be the height of rationality, such as science, mathematics, and engineering (e.g., Goodwin, 1994; Latour, 1987, 1999). A central focus of this work has been on the ways in which action is mediated by artifacts, both material (such as laboratory equipment and calculating devices) and semiotic (such as language, mathematical notation, and graphical representation).

The significance of this can be seen in the foundational work of Sylvia Scribner on problem solving in the activity of dairy workers. Scribner (1997b) challenged a number of ideas promoted by cognitivist models of problem solving, including that problems are given to the problem-solver in a complete form; that problems of the same logical class will be solved by the same sequence of operations on all occasions; and that learning involves increasing independence from the concrete particularities of a context. Scribner's ethnographic and experimental work showed instead that what appeared, in theory, to be formally identical problems were, in practice, flexibly formulated and solved by the problem solver according to the contingencies of the environment, including the mediating artifacts that were available. She showed, for instance, that the dairy workers reformulated abstract computational problems into problems that depended on the concrete physical array of the product they are working with, such as the cartons being loaded and the crates into which they were being placed. She thus argued, against cognitivist views of problem solving, that these mediating artifacts play a "constitutive role" in cognition. She argued that, "skilled practical thinking incorporates features of the task environment (people, things, information) into the problem-solving system. It is as valid to describe the environment as part of the problem-solving system as it is to observe that problem-solving occurs 'in' the environment" (1997b, p. 329).

The general point for situativity is that all activity is mediated by artifacts. Even material artifacts, as Scribner's work shows, not only have a physical dimension but also allow us to represent symbols and manipulate those symbols (Hutchins, 1995, 1993; Norman, 1993; Pea, 1993b; Perkins, 1993).

When the dairy workers transformed abstract computational problems into problems that were tied to the physical environment in which they were situated, they did so by using materials (e.g., milk crates) designed for noncognitive purposes (e.g., holding bottles) to mediate their formulation and solution of a cognitive task. Thus, from a situative perspective, representations are not abstract, disembodied, and detached from the world, as they are for cognitivism; rather, they arise in the course of situated activity and get their meaning from their use. As Dourish (2001) puts it, "representations are as representations do."

In summary, mediation is central to the situative perspective. Representations have been a strong focus of research within the learning sciences and have played an especially strong role in, for example, our understanding of science and mathematics practices (Danish & Enyedy, 2006; Greeno & Hall, 1997; Hall, 1996; Lee & Sherin, 2006). Mediation and representational abilities have been shown to be central in learning science and mathematics, especially for expertise development (Pea, 1993b).

Participation and Identity

Research on mediation points to another central aspect of situative perspectives, which is that action is located within broader systems of organization. Not only do people modify their environment by creating mediating artifacts, as Scribner's work shows; these artifacts can, as Hutchins (1995) argues, transform subsequent possibilities for action by embodying "partial solutions to frequently encountered problems" (Hutchins, 1995, p. 374) of an individual or group. That is, once introduced into activity, forms of mediation that have been useful in solving problems or resolving dilemmas (Lave, 1988) can be preserved and passed on to subsequent generations. The participation by a person or group of people in practices making use of such a preserved mediating artifact situates its user within the tradition that developed, passed on, and continues to maintain that artifact.

Meaningful participation in practices is a central concept within the situative perspective. A great deal of research, much of it building on the foundational ideas of Vygotsky (1978, 1987), has focused on how people develop into these traditions. According to this perspective, it is through situated engagement in motivated action (Goodwin, 2000), using tools and in interaction with others, that we learn some of our most essential skills. For instance, a child first acquires language through its use with parents. Clancey (1997) has conceptualized an activity as a *participation framework*, that is, "an encompassing fabric of ways of interacting that shapes what people do" (p. 266). So, through participating in broader social systems, known variously as communities of practice (Lave & Wenger, 1991; Wenger, 1998), activity systems (Engeström, 1987), figured worlds (Holland, Lachicotte, Skinner, & Cain, 1998), actor networks (Latour, 1999; Nespor, 1994), or Discourses (Gee, 1992), newcomers move along a trajectory from "legitimate peripheral participation" (Lave & Wenger, 1991), in which the authentic practices of a community are within their "horizon of observation" (Hutchins, 1995; Stevens, O'Connor, Garrison, Jocuns, & Amos, 2008), but in which they have less than full responsibility for their performance, toward full participation in the practices of the community (Paretti, 2008). According to this view, all actions and activities are guided toward a larger goal and learning is about understanding that larger goal and aligning actions with it. In this view, then, learning takes place not through transmission of abstract knowledge, but through engagement in the "knowledgeable skills" that are realized in the everyday activities of a community; that is, people become good at the practices that they routinely participate in, gaining understanding of how to engage successfully under varying conditions by flexibly adapting their performance to the contingencies of particular occasions.

It is important to note, however, that learning, in this view, is understood as more than mastery of the "knowledgeable skills"

of a community. This is because, as Lave and Wenger argue:

> *Activities, tasks, functions, and understandings do not exist in isolation; they are part of broader systems of relations in which they have meaning. These systems of relations arise out of and are reproduced and developed within social communities, which are in part systems of relations among persons. The person is defined by as well as defines these relations. Learning thus implies becoming a different person with respect to the possibilities enabled by these systems of relations. To ignore this aspect of learning is to overlook the fact that learning involves the construction of identities. (Lave & Wenger, 1991, p. 53)*

Consequently, then, as one engages in the practices of a community, she is not simply becoming adept at carrying out those practices; she is also becoming identifiable as a certain kind of person within the community (Holland et al. 1998). It is important to note here that identities are not determined by "the possibilities enabled by [the] systems of relations" of a community; rather, participants actively identify themselves and others in terms of those possibilities, in the process both reproducing and transforming the community.

The Situative Perspective and Engineering Learning

We believe that the situative perspective offers many useful avenues for research on engineering learning given three distinguishing characteristics of engineering learning: use of representations, alignment with professional practices, and the emphasis on design. The first element of engineering that is central to engineering learning and practice is the use of *representations*. Like many other practitioners, engineers are surrounded by tools, and the purpose of many of these tools is to lead to representations that can help guide the work of engineers. Graphs, charts, and visuals are all examples of representations that engineers use on a regular basis. As a matter of fact, scholars have suggested that engineering can be seen as a discipline that teaches students how to convert one form of representation into another (McCracken & Newstetter, 2001). For instance, many problems that engineering students work on, and that a practitioner faces, are expressed as text that needs to be converted into another symbol system, often visual. A free body diagram is a prime example of such a conversion (also see Chapter 4 by Roth in this volume). Increasingly, the role of producing representations is being played by digital technology, leading to an era of production and exchange of representations that is unprecedented in human history. The use of tools is also leading to collaboration among engineers – aided by representations – that is swiftly but decisively reinventing engineering cognition (Pea, 1985) and practice, akin to the change brought about by the first wave of information technology use in manufacturing (Zuboff, 1989) and now being fostered by digital environments and large-scale cyberinfrastructure. The role of technological tools, particularly digital tools, is extremely under-theorized in engineering education and a perspective of representational mediation can prove useful to develop a deeper understanding of technology use and design. The potential exists for changing not only how we teach and learn, but also our research practices themselves.

A second critical aspect of engineering and engineering learning is its close association with *professional practice*. A majority of engineering students pursues the major to be able to work as engineers. Therefore, an inherently large aspect of their training is learning to become a part of the community of practice of professional engineers. This includes developing an identity as an engineer – in numerous ways and forms. Professional practice is also collaborative in nature and therefore learning to work as part of groups and teams is essential to engineering learning. Of course, work settings have played a crucial role in informing early work in the field (Scribner, 1997a, 1997b; Wenger, 1998), and this is one aspect of engineering learning that we believe has been studied extensively by scholars in aligned disciplines such as technical

communication (Winsor, 1996), science education (Lemke, 1997), technology studies (Bucciarelli, 1994), architecture (Schön, 1983), and also learning sciences (Hall & Stevens, 1995; Stevens, 2000). Lemke (1997), for instance, provides a comprehensive discussion of the disjuncture between school practices and professional practices and the effort that must be made to integrate students' school and professional trajectories. The discussion has particular importance for engineering learning given the professional and applied nature of our practices. Yet, the disjuncture between school-engineering and work-engineering remains intact and significant efforts are needed to bridge this gap (Stevens, Johri, & O'Connor, Chapter 7, this volume). Engineers in the workplace often say that even technical skills are easier to learn on the job as compared to formal training. They often complain that very little of what they learn in school is of any use to them. There is an issue of situated learning and transfer. Recent work has started to capture this tension (Stevens et al., 2008). There has to be effort that links research with practice on an ongoing basis as the environment for work and learning changes rapidly.

A final element of engineering learning that is unique to it compared to mathematics and science learning is *design*. Engineers are by definition designers. Engineering design thinking and learning are central to the development of an engineer (Dym, Agogino, Eris, Frey, & Leifer, 2005). Yet, design has its own unique ways of developing cognitive and situated skill requirements. It requires skills with materials, ability to work collaboratively, and the ability to become part of a community of practice. Models of teaching design therefore differ from teaching engineering science-based content to students. Design is also a useful metaphor to think about engineering learning research that can lead to innovations. Design-based research (a descendent of work on design experiments) is a useful paradigm that can be adopted by engineering learning researchers and highlights that ideas from core engineering disciplines can be used to improve engineering learning – if applied with the right understanding of context (see Chapter 25 by Kelly in this volume).

Corresponding to the three characteristics of engineering we have outlined earlier, we believe that the three elements of situated learning we have identified can make a significant contribution to furthering our understanding of engineering learning in unique ways, similar to its role in biomedical engineering education (Harris, Bransford, & Brophy, 2002). These connections have already started to manifest themselves in recent articles in engineering education publications. For instance, Paretti (2008) uses the situated learning approach, in conjunction with activity theory, to examine communication practices in a capstone design class. Pierrakos, Beam, Constantz, Johri, and Anderson (2009) use emerging literature on situated identities to investigate the different pathways and experiences of students who persist with engineering versus those who switch out of engineering. Similar examination of identity has also been done by Stevens et al. (2008). Gill, Sharp, Mills, and Franzway (2008) build on anthropological investigations of the workplace and the communities of practice literature to draw attention to gender issues in professional engineering settings. They find that the positive self-image women had in school – in relation to engineering – was not maintained in the workplace given a lack of women role models in immediately higher up positions in the office hierarchy. Johri (2011) introduces the concept of "sociomaterial bricolage" to capture the essence of engineering work practices on global teams. Software engineers in his study make do with whatever resources are available to them to develop work practices that span geographic dispersion, which, he argues, is the essence of most engineering work. In relation to digital technology, Johri and Lohani (2011) draw on guided participation and communities of practice frameworks to examine the role of representations in large engineering classes. They argue that a content-transfer model of technology use undercuts the benefits that are available with digital

technology – particularly pen-based technology – to engage students in mutual construction of engineering representations. These studies illustrate welcome progress but we believe that enormous potential still exists to make significant theoretical contributions (Johri, 2010), and with that aim in mind we have outlined some potential research ideas in Table 3.1.

Critical Approaches to Situativity

The emergence over the past couple of decades of situated theories of cognition and learning have profoundly shaped the learning sciences, to the point that situative approaches should now be viewed not as a small critical movement in opposition to cognitivism but as a full-fledged approach to studying learning that stands alongside cognitivism. This development has been remarkable. At the same time, virtually since the onset of the rise of situative theories, there have been voices arising from within the situative approach that have challenged some of the ways in which that approach has developed and been used. Indeed, Jean Lave, perhaps the most influential thinker in the development of situative theory, has herself recently critiqued Lave and Wenger's *Situated Learning* (1991) and some of the uses to which that work has been put. Lave (2008) suggests that "[m]any who use the concept of 'communities of practice' now seem ignorant of the original intent (and its limitations), and simply assimilate it into conventional theory" (2008, p. 283). In what follows, we outline several of the internal critiques of situative theory, suggesting that these might be useful in sharpening the conceptualization and uses of situativity within engineering education research.

In her critical review, Lave (2008) points out that Lave and Wenger's original position "was specifically not intended as a normative or prescriptive model for what to do differently or how to create better classrooms or businesses" (2008, p. 283; cf. Lave & Wenger, 1991, pp. 40–1). Lave and Wenger wrote:

Table 3.1. Elements of Situativity and Implications for Engineering Learning Research

	Relation to Engineering	Implications for Engineering Learning Research
Activities and Interactions	Engineering work is usually project based, accomplished by teams, and is highly collaborative.	Empirical studies of team work and collaboration Empirical studies of role of interaction in engineering practice (peer learning, informal learning)
Mediation in Social and Material Context	Engineering is highly dependent on mediating artifacts, including representations, and uses significant kinds and amounts of physical materials as well.	Empirical studies of role of representations Empirical studies of mediation by tools used in learning and practice Empirical studies of differences between the use of representations and materials in engineering design and engineering science and their relationship
Participation and Identity	Engineers have a strong community of practice – which often varies across disciplines – especially when they practice. Engineering identity is distinct entity.	Engineering community formation Engineering identity formation and differences in school versus work identity Situated identities and conflict between identities Open organizing and other emergent forms of practices based on technology use

legitimate peripheral participation is not itself an educational form, much less a pedagogical strategy or a teaching technique. It is an analytical viewpoint on learning, a way of understanding learning. . . . Undoubtedly, the analytical perspective of legitimate peripheral participation could – we hope that it will – inform educational endeavors by shedding a new light on learning processes, and by drawing attention to key aspects of learning experience that may be overlooked. But this is very different from attributing a prescriptive value to the concept of legitimate peripheral participation and from proposing ways of "implementing" or "operationalizing" it for educational purposes. (Lave & Wenger, 1991, pp. 40–1)

Two points concern us here with respect to the above. First, it is important to note that Lave and Wenger clearly did not oppose the use of their work to "inform educational endeavors," and our view is that certain implementations of their ideas have certainly been quite useful in "shedding a new light on learning processes, and . . . drawing attention to key aspects of learning experience that may be overlooked." Situated learning theory has undoubtedly pushed both its advocates and those whom it critiqued to understand learning in different, we think deeper and more complex, ways (e.g., Anderson, Greeno, Reder, & Simon, 2000; Sfard, 1998).

Second, though, we also agree with Lave's (2008) assessment that in some ways, the uses of situated learning theory do indeed "assimilate it into conventional theory," and that this assimilation threatens to rob the approach of some of its critical intent. It is important to note that we do not mean to imply that attempts to create "cognitive apprenticeships" (Brown et al., 1989) or "practice fields" (Barab & Duffy, 2000) are not a useful and legitimate aspect of the situativity perspective. Rather, our aim is to show how different ways of understanding situated learning open up different analytic strategies in the study of engineering education. These different strategies are at least potentially complementary, though we would hasten to add that we think it is always a good idea to consider critically the effects, intended and unintended, of efforts to change pedagogical practices.

It is informative to examine the analytic strategy adopted by Lave and Wenger. These authors considered apprenticeships in a variety of communities of practice (e.g., midwives, Alcoholics Anonymous groups, naval navigation teams, tailor shops, supermarket meat departments) to show that learning, understood as increasing access to positively valued participation in a community, bears no necessary relationship to formal educational objectives or structures. This was part of the effort of theorists of situated cognition and learning to argue against dominant ways of understanding learning and its relationship to schooling, including those of cognitivism. To accomplish this, Lave and Wenger adopted a strategic focus on communities of practice with certain characteristics. They examined well-established communities with clear boundaries (Nespor, 1994; O'Connor, 2001, 2003), such that each community could be taken to be enclosed and unchanging (Lave, 2008). These communities were homogeneous (Lave, 2008), in the sense that differences among members were treated largely in terms of their relative advancement toward full participation in the community, rather than, for example, in terms of the ways that different participants had different histories before arriving at the periphery of the given community of practice, how they were simultaneously located in other communities, and the like (Lave, 2008; Nespor, 1994; O'Connor, 2001, 2003). Furthermore, Lave and Wenger's communities were, for the most part, explicitly benign,[4] in the sense that it was possible, and expected, that all or nearly all apprentices would move toward full participation within that community (Lave, 1996; O'Connor, 2001, 2003). This focus was strategic. Lave and Wenger started with communities of practice that were arranged so as routinely to produce positive outcomes for virtually all participants

and examined how learning was organized in these communities. The observed absence of explicit transmission of abstract knowledge, together with the successful learning of apprentices in these communities, provided important evidence against cognitivist accounts of how successful learning happens.

This strategy had unintended consequences, however, and likely played into another of Lave's critiques of Lave and Wenger. Lave (2008) suggests that "*Situated Learning* has all too often been read as painting a view of social life as closed, harmonious, and homogeneous, so that participants are 'members'" (p. 288). This reading, in our view, is one of the primary sources of what Lave (2008) described as the assimilation of situated learning theory, at least as Lave and Wenger had conceptualized the perspective, into conventional theory.

But Lave and Wenger (1991) had made it quite clear that these should not be taken to be general characteristics of communities of practice, and in subsequent work they and others have attempted to move beyond this limited focus toward increased attention to the heterogeneity of social practice and the resultant tension and conflict that characterize all social practice (e.g., Engestrom & Cole, 1997; Kirshner & Whitson, 1998; Lave, 1993, 1996, 2008; Lemke, 1997; O'Connor, 2001, 2003; O'Connor & Allen, 2010; Penuel & O'Connor, 2010). This work has called for a move away from analysis of participation in communities and practices that are assumed to be stable, bounded, and benign and toward a focus on the interconnections among practices and the tensions that arise as a result of this heterogeneity. Lave (1993), for example, called for attention to "interrelations among local practices," suggesting that "local practices must inevitably take part in constituting each other, through their structural interconnections, their intertwined activities, their common participants, and more" (Lave, 1993, p. 22).

This more complex approach to understanding what we earlier called "contextualization" problematizes some of the ways in which situative research has been conducted, especially work that has involved attempts to design learning contexts around the ideas of Lave and Wenger and others. A basic strategy for such "prescriptive" (Lave, 2008; O'Connor, 2001) work has been to retain some aspects of traditional approaches to school learning while at the same time transforming learning contexts in fundamental ways. Recognizing the social value of mastery of such "knowledge domains" as science, mathematics, and engineering, these researchers have kept the cognitivist emphasis on these and other traditional school subjects. But prescriptive approaches have diverged from cognitivism by arguing for the inherent social and material situatedness of knowledge and learning. For example, Greeno et al. (1997) point out that a major goal of their work is to "create environments in which students can learn to participate in practices of productive inquiry and use of concepts and principles that are characteristic of subject matter disciplines" (Greeno et al., 1997, p. 99). However, rather than being understood as a body of abstract knowledge, these disciplines are understood as communities of practice with a range of what Lave (1991) calls "knowledgeable skills." Researchers and educators working within a prescriptive approach to situated learning explicitly model these knowledgeable skills, and use these models to design "structures of participation" (Greeno et al., 1997) for placement in classrooms and other learning contexts. So, through participation in practices modeled on those in particular target communities or disciplines, students serve as apprentices in the social practices associated with those communities, a process that is intended to result in "improved participation" in those practices and communities (Greeno et al., 1997).

From the point of view, though, of "analytical" (Lave, 2008) or "critical" (O'Connor, 2001) approaches to situated learning, work within the prescriptive approach might be seen as paying insufficient attention to contextualization processes. The problem here is pointed to in Brown and colleagues' (1989) influential article on situated cognition and learning. These authors, in critiquing traditional approaches to schooling, including those of cognitivism, argued:

> School activity too often tends to be hybrid, implicitly framed by one culture, but explicitly attributed to another. Classroom activity very much takes place within the culture of schools, although it is attributed to the culture of readers, writers, mathematicians, historians, economists, geographers, and so forth. Many of the activities students undertake are simply not the activities of practitioners and would not make sense or be endorsed by the cultures to which they are attributed. This hybrid activity, furthermore, limits students' access to the important structuring and supporting cues that arise from the context. (Brown et al., 1989, p. 34)

This argument was a powerful and generative critique of cognitivism, to be sure. However, it also can be seen as the basis for a critique of the work of at least some who take a prescriptive approach to situated learning. The strategic focus of early critical research on benign, stable, and bounded communities has often been adopted by prescriptive researchers as a general strategy, through their emphasis on highly stabilized disciplines. This prescriptive work has moreover tended to privilege "official" understandings of learning contexts by using a particular model of practice based in idealized understandings of disciplines as the basis for understanding the meaning of participation and for assessing learning or "improved participation" (Greeno et al., 1997). This strategy, however, backgrounds some of the subtle ways in which participants in activity draw on heterogeneous resources, both "official" and "unofficial," as they negotiate the meaning of the context, their ongoing activity, and their own emerging identities. As a result, the kinds of learning environments developed within prescriptive approaches offer promising sites for adopting a critical perspective on "the relationships between local practices that contextualize the ways people act together, both in and across contexts" (Lave, 1993). Insofar as these learning environments involve an attempt to reproduce in schools conditions that will allow students to participate in the practices of some "target" context, such as professional communities of practice, outside of schools, they are inherently heterogeneous contexts. Critical analyses of these sites examine more fully the various ways in which participants orient themselves to these contexts and negotiate the meaning of their participation, and such analyses also follow the consequences of these negotiations beyond the immediate event.

An example of such an analytical or critical approach to studying situated learning within engineering education can be found in O'Connor (2001, 2003), who studied participation in an undergraduate engineering project involving students from several universities with distinct institutional cultures and of widely varying status. This project involved the use of communication technologies such as videoconferencing to establish a "virtual organization" made up of students from each of the schools. Because such technologies allow for the possibility of connecting historically separated institutions, they offer transformed possibilities for the negotiation of social identities and relationships. O'Connor showed that, as it developed, this project became shaped by preexisting relationships among its participating institutions in ways that were unanticipated by the project's developers. For example, participants' "official" project roles, such as project manager or control system designer, and their "unofficial" institutional roles as students at particular schools existed in tension with one another in participants' interactions throughout the project, and in the course of constructing knowledge, participants also constructed identities that were consistent with historically evolved status differences among the schools. In this way, participation in the project served as a site not only for the negotiation of knowledge, but also for the negotiation of who the participants were, both with respect to each other and within their discipline more broadly. These included diminished status for some students in the project, and the reimagination of their possible trajectories within the discipline. While the

project could certainly have been viewed in terms of increasing mastery of some of the knowledgeable skills of engineering, such as planning, design, analysis, and communication, such a focus pursued apart from more critical understanding of the reproduction of status differences would have disregarded some central tensions and dilemmas that faced students participating in the project and disregarded the ways in which participation in designed communities of practice is not only not benign, but can also actively serve to marginalize the participation of some students (cf. Hodges, 1998; Eisenhart & Finkel, 1998; and Tonso, 2006 for related analyses pertaining to science, technology, engineering, and mathematics [STEM]disciplines).

A second example of a critical approach to situated learning in engineering education is found in Stevens et al. (2008). These authors conducted an ethnographic study of learning across the four years of students' undergraduate programs. However, rather than adopting the common approach in situativity theory of looking for trajectories of increasing knowledgeability and identification with respect to specific forms of practice, this study focused more broadly on the complex processes by which students become, or do not become, engineers. A central strategy of this work was to look across three "dimensions" of learning. The first dimension, accountable disciplinary knowledge, is the one most traditionally associated with the concept of learning. But instead of seeing learning as the acquisition of a unified and stable body of knowledge, Stevens et al. viewed disciplinary knowledge as at least potentially both less unified and more variable in terms of what practitioners know and do, and sought to identify what *counts as* disciplinary knowledge in different situations. The second dimension, identification, concerned how a person both identifies with engineering and is identified by others as an engineer. The third dimension, navigation, focused on how engineers-in-the-making moved through and constructed various pathways, both personal and institutional, both official and unofficial, as they progressed through their undergraduate careers.

The insights gained through this three-dimensional framework were developed partly through case studies. A comparison of two such case studies focused on two students at one school who were interested in the same prestigious and competitive engineering major. One, Simon, gained admission to that major after his second year; the other, Jill, switched majors at that same point without applying to an engineering department. Simon and Jill were similar in important ways, including that both had roughly identical GPAs (which were lower than the average of students admitted to their desired major), and that both were wavering in their respective identifications with engineering as a major. There were important differences, though. Jill relied on an official navigational route into the major and focused her efforts on her GPA. When she foundered there, she saw few other options for pursuing an engineering major and became progressively less identified with the field, eventually electing to leave. Simon, meanwhile, with the help of a family friend who was a professor in his desired engineering department, had obtained a job at a departmental testing facility. In large part as a result of his experience in this position, he was able to make a case for admittance to the major, despite his lower-than-average GPA. Both students went on to be quite successful through their final two years, Simon as an engineering major and Jill as a business major. Stevens et al. used these cases to argue against a view of learning that is overly focused either on individual knowledge or on individual motivation. They argued that Jill *could* be viewed as having insufficient knowledge or motivation to succeed in engineering; Simon's success *could* be seen as the unfolding of the more or less straightforward trajectory of an intelligent and motivated student who followed his interests until he ultimately became recognized for his strengths. But Stevens et al. claimed that, although these understandings are possible, they are misleading in that they background

the contingency of the pathways and the extensive organizing work, by Simon and Jill along with others, that shaped their careers at every point along the way. This study shows the value of attention to the multiple and complex "dilemmas of becoming" experienced by students over an extended time period, and how those dilemmas are shaped by and resolved within a tension-laden institutional framework. Such work points to a way beyond the focus by some prescriptive situativity theorists on the progressive mastery of specific disciplinary knowledgeable skills, at least in the absence of the collective work that is done to make these noticeable and consequential within always contingent pathways, and especially within not necessarily benign institutions that do not allow for all to become equally successful (cf. Varenne & McDermott, 1998).

A third and final example of a critical approach to learning within engineering education is taken from Eisenhart's recent work on the FREE Project (Female Recruits Explore Engineering; e.g., Eisenhart, 2011). This project was designed to allow for mostly minority, nonprivileged, high-achieving girls to participate in an engineering-focused after-school program. At the start of the project during their sophomore year in high school, and despite strong performance in math and science, fewer than 20% of these girls were considering studying engineering in college. Among the central research interests was to see the extent to which these girls, once exposed to engineering as a discipline and potential profession, subsequently pursued engineering. During their time in the after-school club, the girls attended engineering career fairs on college campuses, visited engineering labs, were introduced to practicing engineers, undertook guided Internet explorations of engineering, and talked with researchers and advisors about the pluses and minuses of engineering careers. They also undertook their own small-scale engineering projects, guided by engineering consultants who helped with specific projects; these design projects included a playground for disabled children, clothing accented with light-emitting diode (LED) displays, solar-powered jewelry, and adjustable high-heeled shoes for women. Among Eisenhart's key findings were that, although the girls knew almost nothing about engineering when the FREE program began, it was easy to spark their interest, and that high school was not too late to get them interested. The girls participated eagerly, actively, and successfully in the various technical and nontechnical aspects of the program. By the end of the program, many began to think about majoring in engineering in college. To use Lave and Wenger's (1991) terms, these girls had been participating on the periphery of the "community of practice" of engineering, and were learning both in the sense of gaining some of the knowledgeable skills of the community and also envisioning a trajectory for themselves that would lead to fuller participation.

Several barriers effectively cut off this trajectory for many of the girls, however. Some of these barriers were economic, including the high cost of attending college. This was made even more difficult for some of the girls by a second barrier, their immigration status; legislation in their home state prevented undocumented students from attending state schools at in-state rates. Other challenges related to a lack of social and cultural capital (Foor, Walden, & Trytten, 2007) among the girls, their families, and their schools. There was, for example, a lack of well-communicated information about colleges, costs, and scholarships, and about differences between high school graduation requirements and college entrance requirements. Many did not know how to navigate online college or scholarship applications. Most parents and older siblings did not have their own experiences – with either college entrance requirements or financial aid – to help. The uncertainty introduced by these challenges introduced still others. These included challenges involving the social positioning in U.S. society of women of color, for whom success in school is tied deeply to their ability to counter negative images of themselves and their group (Hurtado, 1996). Their identities as

students, as respectable persons, and as valued members of their families (Lopez, 2003) were at stake in the choices that they made about attending college and the majors they would pursue. In the face of these barriers and the related fear and uncertainty, the FREE girls scaled back college expectations and plans. Engineering, a "hard major," became especially risky, and most decided that it was too much of a risk. Eisenhart (2011) writes that:

> ... it does not seem that failure to pursue engineering can be attributed to an engineering education intervention program that "did not work." It cannot be attributed to a lack of interest in engineering. It cannot be attributed to discrimination, stereotyping, or harassment in engineering.... The outcome was due primarily to social, political and economic barriers that interfered with access to college, undermined academic confidence, and threatened important identities that had been nurtured in and were dependent on school.... The girls responded to this unfamiliar and threatening context by protecting their identities as they headed to college: They went, but they did so in ways that limited their exposure to risks that threatened their status as good students, respectable people, and worthy family members. This practically eliminated engineering from serious consideration. (p. 20)

Eisenhart's study complicates the prescriptive idea that learning involves the creation of "environments in which students can learn to participate in practices of productive inquiry and use of concepts and principles that are characteristic of subject matter disciplines" (Greeno et al., 1997, p. 99). The FREE program certainly was successful in allowing its members to participate in this way. But Eisenhart's work also points to a major limitation of prescriptive work based on Lave and Wenger's apprenticeships, specifically, that trajectories beyond the learning context are rarely examined. This was not the case for Lave and Wenger, whose apprentices were almost literally taken into their futures by community "old-timers," who actively organized the movement of their apprentices along a trajectory toward a valued place in the community. Eisenhart's work points to the importance of attending to the organizing of trajectories in complex, often non-benign contemporary conditions.

Eisenhart's work resonates with recent arguments that, if learning is conceived as it is in critical situative approaches in terms of access to valued participation within the complex systems of relations that characterize contemporary society (cf. Sawyer & Greeno, 2009, p. 348), then research on learning must centrally involve attention to processes of the organizing of processes through which people move along trajectories into their futures (O'Connor & Allen, 2010; Penuel & O'Connor, 2010), including the conditions in which people become recognized (Gee, 1999; Taylor, 1992), or not, as valued participants in social worlds.

Conclusion

Complexity of human social and engineered life has never been higher. This significantly affects our ability to make sense of the world around us and to act intelligently. This change is reflected in the professional lives of engineers where they have to work with novel technologies, with a diversity of people around the world, as part of highly interdisciplinary teams, and on projects that are complex both in scale and expertise. What engineers need is adaptive expertise that allows them to be both innovative and efficient at what they do (Bransford, 2007; also see Chapter 12 by McKenna in this volume). From an educator's viewpoint, this is the world we inherit and need to prepare our students to enter. Therefore, it is imperative that we reflect on the skill development we need to facilitate and design implementable learning environments for our students. This forms the core agenda for the emerging discipline of engineering education (Haghigi, 2005). In this chapter we have laid out a broad theoretical agenda and an analytical framework to serve as a toolkit for understanding and designing engineering learning to make consistent progress toward this goal of improving the education of engineers.

Footnotes

1. In addition to the scholars cited above, readers can refer to the following edited volumes for in-depth work on this topic: *Perspectives on Socially Shared Cognition* (Resnick, Levine, & Teasley), *Handbook of Situated Cognition* (Robbins & Aydede, 2009), *Mind, Culture & Activity* (Cole, Engeström, & Vasquez, 1997), *Everyday Cognition* (Rogoff & Lave, 1984), *Perspectives on Activity Theory* (Engeström, Miettinen, & Punamaki, 1999), *Communication and Cognition at Work* (Engeström & Middleton, 1998), *Sociocultural Studies of Mind* (Wertsch, Río, & Alvarez [Eds.], 1995), *Distributed Cognitions* (Salomon, 1993), and *Situated Cognition: Social, Semiotic, and Psychological Perspectives* (Kirshner & Whitson, 1997).

2. Another response would be to reject the idea of "application" and to look instead at the ratification or recognition of participation as appropriate on a given occasion (e.g., Stevens et al., 2008).

3. Different approaches within the broader umbrella of situativity theory (e.g., ethnomethodology, mediated action theory, cultural-historical activity theory) understand action or activity in different ways. It is not our aim to differentiate these different perspectives here. It is sufficient for our purposes to point out that each of these approaches takes action, interaction, or activity in some form to be central and primary in the analysis of cognition and learning.

4. In practice, most contexts are not this benign, including in engineering education. The literature is full of studies that document the problem engineering education has typically faced in retaining students (see Chapter 16 by Lichtenstein, Chen, Smith, and Maldonado in this volume).

References

ABET (2011). *Criteria for accrediting engineering programs – Program outcomes and assessment.* Baltimore, MD: Accreditation Board for Engineering and Technology.

Anderson, J. R., Greeno, J. G., Reder, L. M., & Simon, H. A. (2000). Perspectives on learning, thinking, and activity. *Educational Researcher*, 29, 11–13.

Anderson, J. R., Reder, L., & Simon, H. A. (1996). Situated learning and education. *Educational Researcher*, 25, 5–11.

Anderson, J. R., Reder, L., & Simon, H. A. (1997). Situative and cognitive perspectives: Form versus substance. *Educational Researcher*, 26, 18–21.

Barab, S. A., & Duffy, T. (2000). From practice fields to communities of practice. In D. Jonassen & S. M. Land (Eds.), *Theoretical foundations of learning environments* (pp. 25–56). Mahwah, NJ: Lawrence Erlbaum.

Bateson, G. (1978). *Steps to an ecology of mind*. New York, NY: Ballantine.

Becker, H. S. (1972). A school is a lousy place to learn anything in. *American Behavioral Scientist*, 16, 85–105.

Bereiter, C., & Scardamalia, M. (1989). Intentional learning as a goal of instruction. In L. B. Resnick (Ed.), *Knowing, learning, and instruction: Essays in honor of Robert Glaser* (pp. 361–392). Hillsdale, NJ: Lawrence Erlbaum.

Berger, P. L., & Luckmann, T. (1966). *The social construction of reality: A treatise in the sociology of knowledge*. Garden City, NY: Anchor Books.

Bransford, J. (2007). Preparing people for rapidly changing environments. *Journal of Engineering Education*, 96(1), 1–3.

Bransford, J. D., Brown, A. L., & Cocking, R. R. (2000). *How people learn: Brain, mind, experience, and school* (expanded edition). Washington, DC: The National Academies Press.

Bransford, J. D., & Schwartz, D. (1999). Rethinking transfer: A simple proposal with multiple implications. In A. Iran-Nejad & P. D. Pearson (Eds.), *Review of research in education* (Vol. 24, pp. 61–100). Washington, DC: American Educational Research Association.

Brown, J. S., Collins, A., & Duguid, P. (1989). Situated cognition and the culture of learning. *Educational Researcher*, 18, 32–42.

Bruner, J. S. (1960). *The process of education*. Cambridge, MA: Harvard University Press.

Bruner, J. S. (1990). *Acts of meaning*. Cambridge, MA: Harvard University Press.

Bucciarelli, L. L. (1994). *Designing engineers*. Cambridge, MA: MIT Press.

Clancey, W. J. (1997). *Situated cognition: On human knowledge and computer representations*. Cambridge: Cambridge University Press.

Clancey, W. J. (2009). Scientific antecedents of situated cognition. In P. Robbins & M. Aydede (Eds.), *The Cambridge handbook of situated cognition* (pp. 11–34). New York, NY: Cambridge University Press.

Cole, M. (1996). *Cultural psychology: A once and future discipline*. Cambridge, MA: Harvard University Press.

Cole, M., Engeström, Y., & Vasquez, O (1997). Introduction. In M. Cole, Y. Engeström, & O. Vasquez (Eds.), *Mind, culture and activity* (pp. 1–21). Cambridge: Cambridge University Press.

Danish, J., & Enyedy, N. (2006). Unpacking the mediation of invented representations. In *Proceedings of the Seventh International Conference on the Learning Sciences*, Indiana University, Bloomington, IN (pp. 113–119).

Dewey, J. (1934). *Art as experience*. New York, NY: Minton, Balch.

Dewey, J. (1896). The reflex arc concept in psychology. *Psychological Review*, 3(4), 357–370.

Donaldson, M. (1978). *Children's minds*. Glasgow: Fontana.

Dourish, P. (2001). *Where the action is: The foundations of embodied interaction*. Cambridge, MA: MIT Press.

Duranti, A., & Goodwin, C. (1992). *Rethinking context: Language as an interactive phenomenon*. New York, NY: Cambridge University Press.

Dym, C. L., Agogino, A. M., Eris, O., Frey, D. D., & Leifer, L. J. (2005). Engineering design thinking, teaching, and learning. *Journal of Engineering Education*, 94(1), 103–120.

Eckert, P. (1989). *Jocks and burnouts: Social categories and identity in the high school*. New York, NY: Teachers College Press.

Eisenhart, M. A. (2011, April). "We can't get there from here:" The meaning and context of girls' engagement with STEM. Division G invited lecture, American Educational Research Association, New Orleans, LA.

Eisenhart, M. A., & Finkel, E. (1998). *Women's science: Learning and succeeding from the margins*. Chicago, IL: University of Chicago Press.

Engeström, Y. (1987). *Learning by expanding: An activity-theoretical approach to developmental research*. Helsinki: Orienta-Konsultit Oy.

Engeström, Y., & Cole, M. (1997). Situated cognition in search of an agenda. In D. Kirshner &

J. A. Whitson (Eds.), *Situated cognition: Social, semiotic, and psychological perspectives* (pp. 301–309). Mahwah, NJ: Lawrence Erlbaum.

Engle, R. A. (2006). Framing interactions to foster generative learning: A situative explanation of transfer in a community of learners classroom. *Journal of the Learning Sciences*, 15(4), 451–498.

Felder, R. M., Shepard, S. D., & Smith, K. A. (2005). A new journal for a field in transition. *Journal of Engineering Education*, 94(1), 7–10.

Foor, C. E., Walden, S. E., & Trytten, D. A. (2007). "I wish that I belonged more in this whole engineering group:" Achieving individual diversity. *Journal of Engineering Education*, 96, 103–115.

Garfinkel, H. (1967). *Studies in ethnomethodology.* Englewood Cliffs, NJ: Prentice Hall.

Gee, J. P. (1992). *The social mind: Language, ideology and social practice.* New York, NY: Bergen and Garvey.

Gee, J. P. (1997). Thinking, learning and reading: The situated sociocultural mind. In D. Kirshner & J. A. Whitson (Eds.), *Situated cognition: Social, semiotic and psychological perspectives* (pp. 37–55). Mahwah, NJ: Lawrence Erlbaum.

Gee, J. P. (1999). *An introduction to discourse analysis: Theory and method.* London: Routledge.

Gee, J. P. (2003). *What video games have to teach us about learning and literacy.* New York, NY: Palgrave.

Gill, J., Sharp, R., Mills, J., & Franzway, S. (2008). I *still* wanna be an engineer! Women, education and the engineering profession. *European Journal of Engineering Education*, 33(4), 391–402.

Goodwin, C. (1994). Professional vision. *American Anthropologist*, 96, 606–633.

Goodwin, C. (2000). Action and embodiment within situated human interaction. *Journal of Pragmatics*, 32, 1489–1522.

Greeno, J. (2006). *Learning in activity.* In Sawyer, K. (Ed). *Cambridge handbook of learning sciences* (pp. 79–96). New York, NY: Cambridge University Press.

Greeno, J. G. (1989). A perspective on thinking. *American Psychologist*, 44(2), 134–141.

Greeno, J. G. & The Middle School Mathematics Through Applications Project Group (1997). Theories and practices of thinking and learning to think. *American Journal of Education*, 106, 85–126.

Greeno, J., Collins, A., & Resnick, L. (1996). Cognition and learning. In R. Calfee & D. Berliner (Eds.), *Handbook of educational psychology* (pp. 15–46). New York, NY: Macmillan.

Greeno, J., & Hall, R. (1997). Practicing representation learning with and about representational forms. *Phi Delta Kappa*, 78(5), 361–367.

Greeno, J., & van de Sande, C. (2007). Perspectival understanding of conceptions and conceptual growth in interaction. *Educational Psychologist*, 42(1), 9–23.

Greeno, J. G. (1997). On claims that answer the wrong question. *Educational Researcher*, 26(1), 5–17.

Haghigi, K. (2005, October). Quiet no longer: Birth of a new discipline. *Journal of Engineering Education*, 351–353.

Hall, R. (1996). Representation as shared activity: Situated cognition and Dewey's Cartography of Experience. *The Journal of the Learning Sciences*, 5(3), 209–238.

Hall, R., & Stevens, R. (1995). Making space: A comparison of mathematical work at school and in professional design practice. In S. L. Star (Ed.), *Cultures of computing* (pp. 118–145). London: Basil Blackwell.

Harris, T. R., Bransford, J. D., & Brophy, S. P. (2002). Roles for learning sciences and learning technologies in biomedical engineering education: A review of recent advances. *Annual Review of Biomedical Engineering*, 4, 29–48.

Hodges, D. C. (1998). Participation as disidentification with/in a community of practice. *Mind, Culture, and Activity*, 5, 272–290.

Holland, D., Lachicotte, W., Skinner, D., & Cain, C. (1998). *Identity and agency in cultural worlds.* Cambridge, MA: Harvard University Press.

Hurtado, A. (1996). *The color of privilege: Three blasphemies on race and feminism.* Ann Arbor, MI: University of Michigan Press.

Hutchins, E. (1993). Learning to navigate. In S. Chaiklin & J. Lave (Eds.), *Understanding practice: Perspectives on activity and context* (pp. 35–63). Cambridge: Cambridge University Press.

Hutchins, E. (1995). *Cognition in the wild.* Cambridge, MA: MIT Press.

James, W. (1890/1950). *The principles of psychology*, Vols. I and II. New York, NY: Dover Publications.

Johri, A. (2010, July). Guest editorial: Creating theoretical insights in engineering Education. *Journal of Engineering Education*, 183–184.

Johri, A. (2011). Sociomaterial Bricolage: The creation of location-spanning work practices by global software developers. *Information and Software Technology, 53*(9), 955–968.

Johri, A., & Lohani, V. (2011). A framework for improving engineering representational literacy through the use of pen-based computing. *International Journal of Engineering Education, 27*(5), 958–967.

Kirshner, D., & Whitson, J. A. (1997). *Introduction*. In D. Kirshner & J. A. Whitson (Eds.), *Situated cognition: Social, semiotic, and psychological perspectives*. Mahwah, NJ: Lawrence Erlbaum.

Kirshner, D., & Whitson, J. A. (1998). Obstacles to understanding cognition as situated. *Educational Researcher, 27*, 22–28.

Latour, B. (1987). *Science in action: How to follow scientists and engineers through society*. Cambridge, MA: Harvard University Press.

Latour, B. (1999). *Pandora's hope: Essays on the reality of science studies*. Cambridge, MA: Harvard University Press.

Lave, J. (1988). *Cognition in practice*. Cambridge, UK: Cambridge University Press.

Lave, J. (1991). Situated learning in communities of practice. In L. B. Resnick, J. M. Levine, & S. D. Teasley (Eds.), *Perspectives on socially shared cognition* (pp. 63–82). Washington, DC: American Psychological Association.

Lave, J. (1993). The practice of learning. In S. Chaiklin, & J. Lave (Eds.), *Understanding practice: Perspectives on activity and context* (pp. 3–34). New York, NY: Cambridge University Press.

Lave, J. (1996). Teaching, as learning, in practice. *Mind, Culture, and Activity, 3*, 149–164.

Lave, J. (2008). Situated learning and changing practice. In A. Amin & J. Roberts (Eds.), *Community, economic creativity, and organization* (pp. 283–296). New York: Oxford University Press.

Lave, J., & Wenger, E. (1991). *Situated learning: Legitimate peripheral participation*. Cambridge: Cambridge University Press.

Lee, V. R., & Sherin, B. (2006, June). Beyond transparency: How students make representations meaningful. In *Proceedings of the 7th International Conference on Learning Sciences*, Indiana University, Bloomington, IN (pp. 397–340).

Lemke, J. L. (1997). Cognition, context, and learning: A social semiotic perspective. In D.

Kirshner & J. A. Whitson (Eds.), *Situated cognition: Social, semiotic and psychological perspectives* (pp. 37–55). Mahwah, NJ: Lawrence Erlbaum.

Leont'ev, A. N. (1978). *Activity, consciousness, and personality*. Englewood Cliffs, NJ: Prentice-Hall.

Lewin, K. (1948). *Resolving social conflicts: Selected papers on group dynamics*. New York, NY: Harper.

Lopez, N. (2003). *Hopeful girls, troubled boys: Race and gender disparity in urban education*. New York, NY: Routledge/Falmer.

Luria, A. R. (1976). *Cognitive development: Its cultural and social foundations*. Cambridge, MA: Harvard University Press.

McCracken, W. M., & Newstetter, W. C. (2001). Text to diagram to symbol: Representational transformations in problem-solving. In *31st Annual Frontiers in Education Conference, 2001* (Vol. 2, pp. F2G–13). Piscataway, NJ: IEEE.

McDermott, R. (1999). Culture is not an environment of the mind. *Journal of the Learning Sciences, 8*(1), 157–169.

Mead, G. H. (1934). *Mind, self, and society: From the standpoint of a social behaviorist*. Chicago, IL: University of Chicago Press.

Meltzoff, A. N., Kuhl, P. K., Movellan, J., & Sejnowski, T. J. (2009). Foundations for a new science of learning. *Science, 325*, 284–288.

Miller, P. J., & Goodnow, J. J. (1995). Cultural practices: Toward an integration of culture and development. In J. J. Goodnow, P. J. Miller, & F. Kessel (Eds.), *Cultural practices as contexts for development* (pp. 5–15). San Francisco, CA: Jossey-Bass.

National Research Council (2001). *Knowing what students know: The science and design of educational assessment*. Washington, DC: The National Academies Press.

National Research Council (2005). *How people learn: Brain, mind, experience, and school* (expanded edition). Washington, DC: The National Academies Press.

Nespor, J. (1994). *Knowledge in practice: Space, time and curriculum in undergraduate physics and management*. London: Routledge.

Newell, A. (1989). Putting it all together. In D. Klahr & K. Kovosky (Eds.), *Complex information processing: The impact of Herbert A. Simon* (pp. 399–440). Hillsdale, NJ: Lawrence Erlbaum.

Newell, A., & Simon, H. A. (1972). *Human problem solving*. Englewood Cliffs, NJ: Prentice Hall.

Norman, D. A. (1993). *Things that make us smart: Defending human attributes in the age of the machine*. New York, NY: Addison-Wesley.

O'Connor, K. (2001). Contextualization and the negotiation of social identities in a geographically distributed situated learning project. *Linguistics and Education*, 12, 285–308.

O'Connor, K. (2003). Communicative practice, cultural production, and situated learning: Constructing and contesting identities of expertise in a heterogeneous learning context. In S. Wortham & B. Rymes (Eds.), *Linguistic anthropology of education* (pp. 63–91). London: Praeger.

O'Connor, K., & Allen, A. (2010). Learning as the organizing of social futures. In W. R. Penuel & K. O'Connor (Eds.), *Yearbook of the National Society for the Study of Education*, 108, 1: Learning research as a human science (pp. 160–175).

O'Connor, K., & Glenberg, A. M. (2003). Situated cognition. In L. Nadel (Ed.), *Encyclopedia of cognitive science*. London: Nature Publishing Group.

Paavola, S., Lipponen, L., & Hakkarainen, K. (2004). Models of innovative knowledge communities and three metaphors of learning. *Review of Educational Research*, 7(4), 557–576.

Paretti, M. C. (2008, October). Teaching communication in capstone design: The role of the instructor in situated learning. *Journal of Engineering Education*, 491–503.

Pea, R. D. (1985). Beyond amplification: Using computers to reorganize human mental functioning. *Educational Psychologist*, 20, 167–182.

Pea, R. D. (1993a). Practices of distributed intelligence and designs for education. In G. Salomon (Ed.), *Distributed cognitions: Psychological and educational considerations* (pp. 47–87). New York, NY: Cambridge University Press.

Pea, R. D. (1993b). Learning scientific concepts through material and social activities: Conversational analysis meets conceptual change. *Educational Psychologist*, 28(3), 265–277.

Penuel, W. R., & O'Connor, K. (2010). Learning research as a human science: Old wine in new bottles? In W. R. Penuel & K. O'Connor (Eds.), *National Society for the Study of Education*, 108, 1, 268–283.

Perkins, D. N. (1993). Person-plus: A distributed view of thinking and learning. In G. Salomon (Ed.), *Distributed cognitions. Psychological and educational considerations* (pp. 88–110). New York, NY: Cambridge University Press.

Piaget, J. (1952). *The origins of intelligence in children*. Trans. M. Cook. New York, NY: International Universities Press.

Piaget, J. (1964). Development and learning. In R. E. Ripple & V. N. Rockcastle (Eds.), *Piaget rediscovered*. Ithaca, NY: Cornell University Press.

Piaget, J. (1970). Genetic epistemology. New York, NY: W. W. Norton.

Piaget, J. (1972). Intellectual evolution from adolescence to adulthood. *Human Development*, 15(1), 1–12.

Pierrakos, O., Beam, T. K., Constantz, J., Johri, A., & Anderson, R. (2009). On the development of a professional identity: Engineering persisters vs engineering switchers. In *Proceedings of 39th ASEE/IEEE Frontiers in Education Conference*, San Antonio, TX (pp. M4F-1–M4F-6).

Polanyi, M. (1967). *The tacit dimension*. New York, NY: Anchor Books.

Resnick, L. (1987). Learning in school and out. *Educational Researcher*, 16(9), 13–20.

Rogoff, B. (1990). *Apprenticeship in thinking: Cognitive development in social context*. New York, NY: Oxford University Press.

Rogoff, B. (1995). Observing sociocultural activity on three planes: Participatory appropriation, guided participation, and apprenticeship. In J. Wertsch, P. del Río, & A. Alvarez (Eds.), *Sociocultural studies of mind*. New York, NY: Cambridge University Press.

Rogoff, B. (2003). *The cultural nature of human development*. New York, NY: Oxford University Press.

Rogoff, B., & Lave, J. (1984). *Everyday cognition: Its development in social context*. Cambridge, MA: Harvard University Press.

Salomon, G. (Ed.). (1992). *Distributed cognitions: Psychological and educational considerations*. New York, NY: Cambridge University Press, pp. 47–87.

Salomon, G., & Perkins, D. N. (1998). Individual and social aspects of learning. In P. D. Pearson & A. Iran-Nejad (Eds.), *Review of Research in Education*, 23, 1–24.

Sawyer, R. K., & Greeno, J. G. (2009). Situativity and learning. In *The Cambridge handbook of situated cognition* (pp. 347–367). New York, NY: Cambridge University Press.

Schauble, L., Leinhardt, G., & Martin, L. (1997). A framework for organizing a cumulative research agenda in informal learning contexts. *Journal of Museum Education, 22*(2–3), 3–7.

Schön, D. (1983). *The reflective practitioner. How professionals think in action*. London: Temple Smith.

Scribner, S. (1997a). Knowledge at work. In E. Tobach, R. J. Falmagne, M. B. Parlee, L. M. W. Martin & A. S. Kapelman (Eds.), *Mind & social practice: Selected writings of Sylvia Scribner* (pp. 308–318). Cambridge: Cambridge University Press.

Scribner, S. (1997b). Studying working intelligence. In E. Tobach, R. J. Falmagne, M. B. Parlee, L. M. W. Martin, & A. S. Kapelman (Eds.), *Mind and social practice: Selected writings of Sylvia Scribner* (pp. 338–366). Cambridge: Cambridge University Press.

Scribner, S., & Cole, M. (1973). Cognitive consequences of formal and informal education. *Science, 182*(4112), 553–559.

Sfard, A. (1998, March). On two metaphors for learning and the dangers of choosing just one. *Educational Researcher*, 4–13.

Stevens, R. (2000). Divisions of labor in school and in the workplace: Comparing computer and paper-supported activities across settings. *The Journal of the Learning Sciences, 9*(4), 373–401.

Stevens, R., O'Connor, K., Garrison, L., Jocuns, A., & Amos, D. (2008). Becoming an engineer: Toward a three dimensional view of engineering learning. *Journal of Engineering Education, 97*, 355–368.

Suchman, L. (1987). *Plans and situated action*. New York, NY: Cambridge University Press.

Taylor, C. (1992). *Multiculturalism and "the politics of recognition."* Princeton, NJ: Princeton University Press.

Tonso, K. L. (2006). Teams that work: Campus culture, engineer identity, and social interactions. *Journal of Engineering Education, 95*, 25–37.

Varenne, H., & McDermott, R. P. (1998). *Successful failure: The school America builds*. Boulder, CO: Westview Press.

Vera, A., & Simon, H. A. (1993). Situated action: A symbolic interpretation. *Cognitive Science, 17*, 7–42.

Vosniadou, S. (2007). The cognitive-situative divide and the problem of conceptual change. *Educational Psychologist, 42*(1), 55–66.

Vygotsky, L. S. (1962). *Thought and language*. Cambridge, MA: MIT Press.

Vygotsky, L. S. (1978). *Mind in society: The development of higher psychological processes*. Cambridge, MA: Harvard University Press.

Vygotsky, L. S. (1997). *The collected works by L. S. Vygotsky* (6 vols.). New York, NY: Plenum Press.

Wenger, E. (1998). *Communities of practice: Learning, meaning and identity*. Cambridge: Cambridge University Press.

Wertsch, J. V. (1998). *Mind as action*. New York, NY: Oxford University Press.

Wertsch, J. V. (1993). *Voices of the mind*. Cambridge, MA: Harvard University Press.

Winsor, D. A. (1996). *Writing like an engineer: A rhetorical education*. Mahwah, NJ: Lawrence Erlbaum.

Zuboff, S. (1989). *In the age of the smart machine: The future of work and power*. New York, NY: Basic Books.

CHAPTER 4

The Social Nature of Representational Engineering Knowledge

Wolff-Michael Roth

What we call *"descriptions"* are instruments for particular uses/applications. Think of a machine-drawing, a cross-section, an elevation with measures, that an engineer has before him. Thinking of a description as a word-picture of the facts has something misleading about it: One tends to think only of pictures, as they hang on our walls; they appear simply to portray how a thing looks like, what it is like.

(Wittgenstein, 1953/1997, p. 99, my translation)

Introduction

The purpose of this chapter is to articulate a perspective on the nature of representation in engineering that has been developed on the basis of ethnographic and sociological studies across science and technology. It is a sociocultural and cultural-historical perspective that has some decided advantages over the "cognitive approach" for the teaching of engineering, an approach that has an exclusive focus on what goes on in the mind and hidden from view.

Consistent with the social-psychological diction that all higher cognitive functions are societal relations that come to shape those who participate in them (Vygotsky, 1989), this chapter focuses on the social dimensions of representations because these are the origin of anything that we may attribute to the mind. But because these social dimensions *are* what we subsequently attribute to mind, the perspective developed here also is a cognitive one. However, rather than speculating about hidden mental processes, this approach allows us to study psychological functions in the very public arena where they originate. We may express this fundamental fact in the following aphorism: *engineering representations are in the mind because they are integral to the societal relations engineers entertain.* Among those who study representations-in-use, the term "inscription" tends to be employed. Inscriptions include diagrams, photographs, formulas, and tables, that is, anything other than language that features in scientific research and communication. In this chapter, I move from the term *representation* to *inscription*, because the latter allows us to eschew the frequent confusion between "internal" and "external" representations. I present some of the advantages for engineering education that come with this way of thinking about representational engineering knowledge.

Ours is a visual culture – the adage "a picture is worth a thousand words" is but one indication of the truth of this phenomenon, even though, as the quotation suggests, there is something misleading about equating words and images. The visual nature of knowledge is especially evident in the natural sciences and engineering, where many forms of representations of nature other than text abound. In fact, the sciences and engineering would not exist in their present-day aspects if it were not for visual representations (Edgerton, 1985); engineering practices, whether oriented toward the design of general application or specific scientific applications, are essentially predicated by the relations between image and logic (Galison, 1997). External representations (inscriptions), especially in this age of computer-aided drafting (CAD), are the predominant means of going about designing and simulating the production of future artifacts (e.g., Christensen & Schunn, 2009), doing reverse engineering (Mengoni, Germani, & Mandorli, 2007), and the teaching of engineering practices (Ibrahim & Rahimian, 2010). Visual aspects embodied in external representations epitomize engineering vision as process: both as a way of seeing the world and of envisioning alternatives (Bucciarelli, 2004). Thus, specifically the development of figurative representations during the Renaissance and the subsequent Industrial Revolution enabled the taking off of all formal, scientifically managed cultural practices (Foucault, 1975). Visual representations make possible the objectivity of the sciences in the face of the inherently subjective nature that comes with the individual execution, enactment, and application of discipline-specific knowledge (Husserl, 1939). These representations allow knowledge to be "handed down" and associated practices to be reproduced at will in and by future generations of scientists and engineers. That is, engineers do

visualize and use gesture in their communication so that representation may be said to be both internal and external: representations generally and mathematical representations specifically lie at and constitute *the interface of body and culture* (Roth, 2009).

Historically the development of a more scientific approach to some specific practice coincides with the adoption of, and focus on, forms of representations that formalize the existing, experience-based practical knowledge. The full split between disciplines, however, sometimes had to await more modern times, such as that between scientists, architects, engineers, and master builders (Garrison, 1999). For example, a historical study of the emergence of architecture shows that prior to the construction of Gothic cathedrals, much of the knowledge required existed in the embodied skills of the master builders (Turnbull, 1993). They worked with plumb line, hammer, and templates to produce wonderful buildings many of which have lasted to the present day through wars and punishing climates. The success of any such construction depended on the actual presence of these master artisans, who, with these very simple tools, did all of the design work. But with the increasing size of the cathedrals, the structural knowledge embodied in the craft knowledge came to its limit. A split occurred in the profession leading to master craftsmen heading the work at the construction site, on the one hand, and to architects concerned with design and structure, on the other hand. Work with diagrams, which can be sent to the different work sites, means the presence of the architect at the construction location no longer is required. His or her ideas could be made present again, represented, and stored (archived) independent of the individual using them.

By the time of the Renaissance, the use of visual representations became more widespread and allowed, as apparent in the drawings of Leonardo da Vinci, more formal investigations in proportions, structures, and design of artifacts (Kemp, 2007). His drawings on paper, the epitome of external representations (i.e., inscriptions), served as a primary tool for analyzing complex relations in a variety of contexts that today would be attributed to design and engineering (technology). Thus, among Leonardo's famous drawings are those that pertain to flight generally and the study of wings more specifically. This split of thinking practices from material practices has paid off:

> There are fewer boiler explosions than formerly, now that the thickness of the walls no longer is determined by feeling but by making such-and-such calculations instead. Or since each calculation done by one engineer got checked by another. (Wittgenstein, 1958/1997, p. 134)

The very fact that engineers work with external representations, mathematical or otherwise, allows checking by others whose practices, public as those of a third or fourth engineer, also can be checked for consistency with those practices of engineering more generally.

Survey

As a way of exemplifying the pervasiveness of visual representations (inscriptions) in engineering – anything other than text, generally including numbered equations – I conducted a survey of engineering journals from the eighteen fields listed in Thomson ISI Web of Science with the term "engineering" in the name (e.g., agricultural engineering, engineering: aerospace, engineering: electrical, or metallurgy & metallurgic engineering). For each of these fields, I produced a list sorted according to the impact factor for 2010. I randomly selected one of the top five journals for inclusion in the survey (as long as my library had electronic access to the journal) and then randomly selected one article from the most recent of the 2011 issues available to me. I used a classification scheme developed for the analysis of visual representations in scientific journals and science textbooks (Roth, Bowen, & McGinn, 1999) to establish their frequencies in different categories. The total number of pages includes the reference section and abstract (but excludes any table of contents).

The 18 randomly chosen journal articles amounted to 327 pages of text, ranging from a minimum of 5.25 to a maximum of 93.5 with a mean of 18.2 pages per article (SD = 20.4). In these pages, there are a total of 855 labeled visual representations (or numbered equations) yielding a mean of 2.61 per page. The unweighted mean, which gives equal weight to the mean inscriptions of short and long papers, amounts to 3.07 inscriptions per page (SD = 1.65). This is higher than the 1.46 mean number of inscriptions per page reported for the study of ecology journals (Roth et al., 1999) and more similar to the 3.95 reported for *Physics Review Letters* or the 5.4 inscriptions per page in *Science* (Lemke, 1998). By far the largest frequency of representations fell to the category of equations (Fig. 4.1). This number is heavily biased by one very long paper ($N = 93.5$ pages) with a total of 125 equations averaging 2.3 equations per page; 7 articles had no equations at all but 5 articles had between 15 and 47. However, this trend is similar to that observed, for example, in a selection of ecology journals in which the mean number of equations per page ranged from about 0.02 to 0.60 (Roth et al., 1999). A comparative study in the field of physics shows that the frequency of equations is much higher and equivalent to that in the extreme paper in this review; thus, in the journal *Physics Review Letters*, the average number of equations per page is 2.7 (Lemke, 1998). In mathematics-intensive engineering journals, there are 3–4 equations per page, whereas the survey of technical articles in *Science* reveals 1.9 equations per page.

In the engineering journals I surveyed, there is a large number of scatter plots and line graphs compared to the remaining types of inscriptions (Fig. 4.1). The frequency of the photographs, micrographs, and drawings hides the fact that micrographs often come in plates with several images side by side or as inserts. This frequency is much higher than that reported for the ecology journals (Roth et al., 1999).

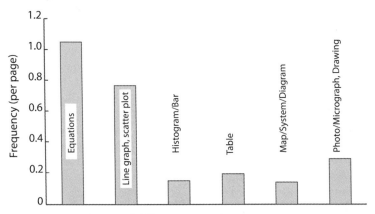

Figure 4.1. Frequency of different types of inscriptions per page in eighteen randomly chosen articles from 18 high-impact engineering journals.

This brief overview shows how important inscriptions are, even though the precise number and type of these may vary across journals and articles and their focus on empirical results and theoretical developments. For a comparison, I randomly pulled an issue of *Science, Technology, and Human Values* off my bookshelves, a journal in which authors would write *about* knowledge construction in engineering and technology. I found 4 representations in 190 pages amounting to 0.02 representations/page. Similarly, I pulled 2 issues of the *Journal of the Learning Sciences*, a journal in which authors might write about learning through engineering design activities in middle and high schools. Here, I counted a total of 49 representations (28 drawings, photographs, and diagrams and 21 tables) in a total of 322 pages, which amounts to 0.15 representation/page. These numbers are similar to those that have been reported for the *Journal of Research in Science Teaching* (Roth et al., 1999). That is, there is an order of magnitude of difference between the per page frequency of visual representation in engineering and those social science journals reporting about the psychological and social aspects of learning in the areas of science, technology, engineering, and mathematics. This may even point to an essential gap between social science and (systems) engineering, formality and informality, and prescription and negotiation (McCarthy,

2000). We may conclude that representations constitute a large part of what counts as knowing in engineering. At the same time, these representations do not go by themselves, but rather they come with a large amount of language-based presentation.

From Representations to Inscriptions

The term *representation* literally means "making something present again" when it is in fact absent. Without this ability to make present something that was or will be present again no cognition of the kind that distinguishes humans from other forms of life would be possible: there would be no memory of past events and no anticipation of and planning for the future (Husserl, 1980). There is some confusion around the term, however, because representations may be thought of as something in the mind or as a text or visual encoding in some medium (e.g., stone, tablets, paper, or monitor). To allow making clear and unambiguous distinctions between the two, the term *inscription* has been proposed: it denotes all those *external* representations that are not text – including graphs, tables, lists, photographs, micrographs, drawings, diagrams, spreadsheets, and equations (Latour, 1987).

Inscriptions are characterized by different levels of abstractness along a single dimension that has the real world as one extreme and language as the other (Latour, 1993; Roth, Bowen, & Masciotra, 2002). The resulting quasi-continuum forms, within a particular research project, a cascade of related inscriptions. Such cascades come about as researchers – middle school students (Roth & Bowen, 1994) or research scientists (Latour, 1993; Roth & Bowen, 1999b) – interact with the natural world in some way to generate data (points), which then are transformed into graphs and equations. On the other hand, a cascade may come about as design engineers go from some rough and barely articulated idea – for example, building a device that accomplishes a particular task – through design drawings, prototypes, and established applications (Roth, 1996a). That is, the work of generating knowledge in the sciences and engineering tends to move from the real world in the direction of the production of graphs, equations, and text (concepts, theories), whereas in engineering and technology, designs tend to move from the state of general ideas through various inscriptional phases (e.g., drawings, diagrams) to working prototypes to the final product in the world (Fig. 4.2). In the example depicted, the purpose of the aerospace engineering project is the production of the satellites depicted in the photograph with software that allows two of these to fly in tandem. From the initially vague idea, the researchers use mathematics to develop the computer code that will control the flight formation. The more abstract an inscription, the more it tends to be able to summarize and compact information. Thus, it is evident that the presentation of a scatterplot is a more economical presentation than a table of values; the means with standard deviations (error) or regression equations are more economical (e.g., in terms of space on a page of printed paper) than tables or graphs.

The very advantage of abstraction in terms of generalization also comes with a price: it leaves out the contextual particulars that any *practical* engineering application has to confront – there are many examples in which something that worked on paper either does not work in the world or that collapses under certain environmental conditions, as the infamous Tacoma Bridge or the nuclear reactors of Fukushima in Japan show. Sometimes, engineers get an application to work a first time but then find themselves unable to replicate it on a second attempt (Sørensen & Levold, 1992). At each stage of going from a more abstract inscription to a more concrete inscription, additional contingencies enter on the trajectory from engineering design to full-scale application (e.g., Bucciarelli, 1994). Theoretically, there is a gap *in kind* – in technical terms, an "ontological" gap – between any two inscriptions. (In mathematics, the relation between the domain and the co-domain of a function constitutes a *mapping*, not a self-identity.) This is especially the case when two neighboring or consecutive inscriptions compared are of different kind, such as a line graph with slope 1 through the origin and the function $f(x) = x$. This gap exists even though competent practitioners see a perfect equivalence between graph and function. This equivalence is in fact hidden in and embodied by their practice (i.e., in the mapping process), but it does not exist for the novice, whose training consists precisely in the process of *making* the relevant connections. That there are differences between naturalistic drawings and photographs of the same natural objects has been shown, for example, in studies of field guides, where, interestingly enough, the former provide greater classificatory help to new ornithologists than the latter (Law & Lynch, 1990). Because there is no natural continuity but an ontological gap between any two inscriptions, *work* is required to bridge any two inscriptions said to belong to each other. The example from ornithology also allows us to understand why and how opportunities arise for teaching "representations" by focusing on the social aspects of inscriptions rather than on supposed (hidden) mental processes.

Properties of Inscriptions

Research conducted in a variety of contexts including engineering bureaus has

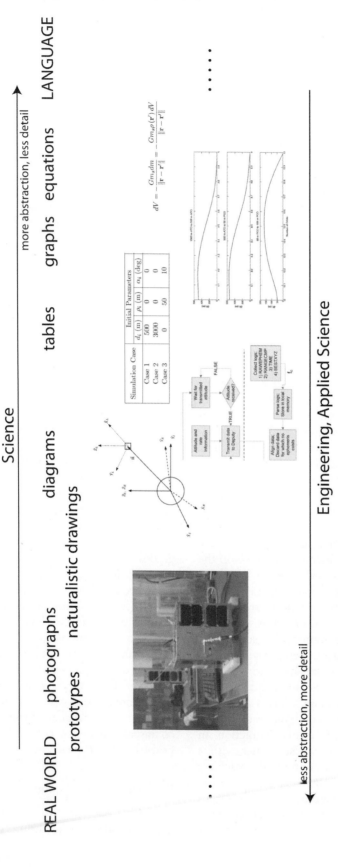

Figure 4.2. Inscriptions can be classified along a scale of abstractness, which ranges from the real world, on the one end, to conceptual language, on the other end. Any two related inscriptions are separated by a gap. (Inscriptions used with permission of R. E. Zee & N. H. Roth, UTIAS Space Flight Laboratory.)

identified at least eight characteristics common to inscriptions. First, because inscriptions are expressed externally in material form, they are mobile: inscriptions can be sent easily using mail, courier, facsimile, or computer networks (e.g., Star & Griesemer, 1989). Second, in the displacement process, inscriptions are *immutable*: they do not change their properties or internal relations during transport (e.g., Latour, 1993). Photographs, diagrams, or formulas sent via e-mail from one end of the country to another – including those used in this chapter – remain the same, though their signification may be different in the new context. Digital images can be made available, without alterations, at any location where there is electronic access (e.g., Lynch & Edgerton, 1988). Third, inscriptions are easily incorporated into different contexts, including this chapter. They contribute to the multiplication of signification, which arises from the presence of different forms of inscriptions (Bastide, 1985; Lemke, 1998). Fourth, inscriptions are easily rescaled to produce larger or smaller images without changing their internal relations (Latour, 1987). A photograph, such as the one of the CanX-6 satellite (Fig. 4.2), may appear in larger format in a report and then appear in miniature format in a chapter of representation in engineering. Fifth, inscriptions are easily combined and superimposed (Star, 1995). A grid can be drawn on top of a photograph of the bubbles on the interface between two media, thereby allowing precise location of the measurement sites; or diagrams of atoms at aluminum and fluorinated carbon interfaces may appear in scatterplots of data, allowing the coordination of measurement and structure information. This leads to the heterogeneous and composite nature of layered inscriptions (e.g., Fig. 4.2). Sixth, inscriptions are reproducible at low cost – from economic, cognitive, and temporal perspectives. Thus, the lecture notes for an engineering course are readily distributed using electronic means almost instantly to all students of the course. Seventh, inscriptions can be merged with geometrical information. Thus, the imposition of a grid over the photograph of the bubbles on a material interface provides a territory with geometrically delineable shapes wherein the object under study can be seen to move through time and across space (Lynch, 1990; Suchman, 1993). Eighth, inscriptions are easily translated into other inscriptions that are again translated, forming the aforementioned cascades of inscriptions (e.g., Fig. 4.2).

Conscription Device

The focus on inscriptions also allows us to understand and theorize the collaborative nature of the work of engineers (e.g., Suchman, 1990); this collaborative context of the work also constitutes an important site for the enculturation of new engineers (e.g., Ibrahim & Rahimian, 2010). Thus, rather than having to rely on guesses what the participants in a design session have in their minds, participants to the session can gather around inscriptions and discuss or argue alternative understandings and perspectives (e.g., Ehn, 1992). In this manner, inscriptions are understood to function as *conscription devices* (Henderson, 1991): they enlist and gather collaborators, focus their attention to a joint object, and thereby organize their joint work. The inscriptions become the focal artifacts that allow engineers to hold each other accountable for their expressions and thoughts (Suchman, 1993). There is no difference then whether we look at the designing activities of children, of high school or university students, or at practical engineers (Roth, 1998). Take the following example from an upper elementary classroom where students were learning a variety of scientific concepts through engineering design projects. The fragment from the videotape followed a tug of war in which, mediated by a block and tackle, the teacher won the competition against twenty sixth- and seventh-grade students. The discussion concerned the design of devices that give a mechanical advantage. There had already been a number of students on the chalkboard providing alternative designs and arguments for why the students lost and for how to redesign the situation.

Episode 1

```
01 Teacher:  and where do YOU pull.
02           (1.58)  (Several students chuckle.)
03 Shamir:   and there is anOTHer bANister
04           (1.05)  (Several students chuckle.)
05           here
06           (0.61)
07           and we pull
08           (0.81)  (Writes "class," Fig. 4.3a)
09           here.
10           (2.71)  (Several Ss clap.)
11 Student:  oh yeah.
12 Shamir:   <<p>you khh is>  (Gestures as in Fig. 4.3b,c)
13 Student:  uhhh.
14           (0.57)
15 Aslam:    but then mister doctor roth has nothing to PULL at
16 Shamir:   thats just it.
```

In this situation, the participants to the design situation – students and teacher – hold each other accountable in and through the design drawings that emerge on the common representation medium, including the force arrows and magnitudes that they layer upon the design itself. In fact, the group has decided to use the chalkboard because they no longer understood each other and there was a conceptual confusion as long as they were just talking (Roth, 1996b). In the end, the chalkboard was littered with alternative designs, some of which were discarded as viable options including the one that Shamir was proposing. Aslam realizes (turn 13) that in the design Shamir has proposed, the class is pulling against the wall and the teacher (Roth) does not have to pull to hold his own.

The presence of the pictorial inscription of the design provides opportunities (affordances) to the manner in which speakers articulate pertinent issues – even if, and especially when, the participants to the setting represent different communities and different levels of understanding (Ehn & Kyng, 1991). For example, Shamir points (Fig. 4.3a) with his hand and fingers (technically, he uses a *deictic* gesture) to a particular location in the design, thereby increasing the definiteness of the topic of talk and decreasing the amount of interpretive flexibility that his listeners can attribute to what he is saying. Perhaps even more interesting, Shamir uses a hand gesture that moves along one of the "ropes" in the pulley configuration he designed (Fig. 4.3b, c). The movement of this gesture is in the direction of the "pull," that is, the force that acts in the system. The movement therefore bears an *iconic* relation to the direction of the force. We may say that he *enacts* the (mental, embodied) schema that goes with and underlies the gestural movement (e.g., Núñez, 2009). As a reified whole, the gesture is an iconic signifier that stands for the rope (the signified). The presence or creation of pictorial inscription, and the gestures it affords, contribute to the emergence of culture and engineering language (Roth & Lawless, 2002).

As conscription devices, inscriptions provide an (external) focus of attention to interaction participants. Here they may function as topic (e.g., when Shamir talks *about* the design) and as ground of the discussion (e.g., when Shamir expresses himself with a gesture the signification of which comes from its iconic relation to the figure). Inscriptions therefore serve a double function: they represent the content of the issue at hand, and they serve as the *con*text that is required for the constitution of the sense of the text.

Figure 4.3. One of a number of students has come to the chalkboard as part of a discussion about the affordances of a block and tackle in a particular design situation. He labels a diagram to show where certain forces act **(a)** and uses gestures **(b, c)** to show the direction of forces and where in the design this act.

Boundary Objects

The preceding example also exhibits another important concept: inscriptions function as *boundary objects* that coordinate people who have qualitatively different understandings of the situation (Star, 1989). In the example it is the difference between an everyday understanding of the forces in a system brought by the children and the understanding of a physicist with a graduate degree. A typical example of a boundary object exists in the design drawings of an airplane used in a manufacturing company (Henderson, 1991). The same set of engineering design drawings is used in different departments of an airplane manufacturing company including the shop floor, engineering design studio, accounting department, inventory management office, and structural engineering bureau. That is, boundary objects allow the sharing of knowledge across knowledge boundaries (Barrett & Oborn, 2010). But this does not mean that inscriptions are used in the same manner when they cross from one disciplinary practice into another: they are used differently, which is to say that the associated practices also differ in the different departments. Boundary objects, therefore, allow the management of institutional and disciplinary differences that interdisciplinary work brings with it (Akkerman & Bakker, 2011). They constitute a normative force that brings different (disciplinary) practices into alignment (Star, 2010). This is especially true for those instances in which scientific and engineering consultants work with their clients with very different knowledge profiles and disciplinary practices (Hall, Wieckert, & Wright, 2006). Those individuals with competencies in more than one domain, who appropriately use inscriptions within two or more very different technological contexts, obtain wizard or guru status (Star, 1995). Sometimes they are referred to as *boundary spanners*: individuals who serve as the intermediaries between developers and users of open source software (Barcellini, Détienne, & Burkhardt, 2009).

Those who envision some interdisciplinary collaboration involving designers or engineers but who do not yet have a shared language may also produce boundary objects (Broberg & Hermund, 2007). Thus, in the design of new workplaces, the system engineers spent three weeks with the future users; they jointly produced inscriptions and built mock-ups (Ehn & Kyng, 1991). In the course of their collaborative work, the engineers and the future users established a common language anchored in and on the inscriptions and artifacts that they were building together. As a result, they were able to overcome the gaps that naturally exist between the languages of their respective communities: jointly produced inscriptions supported the building of bridges. Their function is to articulate and coordinate the very different practices of engineering and user communities. That is, in such studies, the participants come to see the points of view of the members from other communities and practices, which therefore highlights the reflective and reflexive nature of inscriptions when they function as boundary objects. Thus, as one study in the field of process control shows, boundary crossing and change in perspectives occur when boundary "objects facilitate communication between different activity systems by making explicit the knowledge and assumptions mobilized in the interpretation of the object" (Hoyles, Bakker, Kent, & Noss, 2007, p. 335).

Cartography of Inscriptions: A Framework for Producing, Reading, and Teaching Inscriptions

Inscriptions do not come alone but are integrated into what is generally referred to as the main text. It is certainly the earlier cultural-historical origin of language that makes text the dominant form to establish signification generally and the signification of inscriptions more specifically. Based on several studies of scholarly, educational, and popular scientific texts I established what we might call a *cartography of inscriptions* (Pozzer & Roth, 2003; W.-M. Roth, 2010; Roth, Bowen, & McGinn, 1999). These studies show that there a multiple ways in which different parts and kinds of text and inscriptions come to be related in explicit ways and by means of explicit markers. To understand an engineering text, students and other newcomers to the field need to be able to understand this cartography (even if in an imminent way) or they would be lost in the reading much in the same way that a hiker or a tourist might be lost without a map of the novel terrain they enter. I work out some of the relations in the following example from a master's thesis in aerospace engineering entitled *Navigation and Control Design for*

the CanX-4/-5 Satellite Formation Flying Mission (N. H. Roth, 2010).

As readers know from their experience, orientations for reading a particular piece – and, with it, for reading the inscriptions it comes with – is provided by the title of a piece of work (book, journal article). This title organizes the reading of the text such that reading will approach the thesis as an example of an engineering genre rather than another literary genre (poem, scientific novel). I use the term reading rather than reader, because it is in any case an anonymous culturally enabled reading practice that comes to be enacted in any particular situation in which an individual engages with a text. This text in turn has motivated the title. The title already orients readers toward specific parts of the main text (Fig. 4.4). We therefore understand "target attitudes" or "predicted and actual attitude targets" in this aerospace engineering context rather differently than if the context had prepared us for reading an experimental psychological text or in an educational text on teaching method, in which "target attitudes" have very different significations (attitudes toward something that are to be studied or attitudes toward something that need to be changed). Individual readers do not even have to think whether they are reading an engineering text, an online news feature reporting a new engineering feat, or a poem. This is so because the cultural forms of reading that we embody have devices built in that orient the process differently as a function of the particulars of the text at hand (e.g., Livingston, 1995; W.-M. Roth, 2010).

The main text and graphs are linked in multiple ways. For example, the phrase "shown in Fig. 5.3" directs readers to look for a figure with the number "5.3." This is important, for there could be multiple figures and the one appearing on the same page as the index may not be the one indexed by the representation "Fig. 5.3" – in the present instance, the figure in fact appears on the following page. The main text also provides a description of what the reader is to look for and how to read "Fig. 5.3." For example, the text describes: "each line denotes

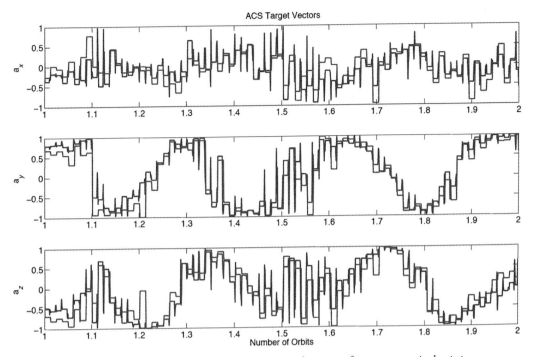

Figure 4.4. Main text, Cartesian line graph, and associated caption from a master's thesis in aerospace studies and the associated caption. (From N. H. Roth, 2010, pp. 61, 62. Used with permission.)

the normalized component of the commended attitude vector in the GCI frame." The main text further describes what would be ideally found ("the blue lines would be straight or mildly kinked") – thereby comparing what is presented to an absent norm. It also describes what each jump represents. The text then orients competent reading to spikes, which, when it moves to the caption, it finds again: "Attitude spikes resulting from the prediction scheme" (Fig. 4.4). This repeated description also constitutes an instruction for the reading to orient itself toward the spikes rather than toward the other, minor differences between the red and blue curves. That is, whereas the spikes may not be immediately salient in the reading, the repeated description/instruction orients the process in such a manner. This difference turns out to become especially salient when the reader comes to the point where the red lines are described as denoting "a spike-free series of targets."

The caption is not a superfluous phenomenon that we could get away with leaving out. This is so because any inscrip-

tion potentially can be read in multiple ways – even highly educated and successful individuals from the same discipline read graphs from undergraduate textbooks in different, often incongruent ways (Roth & Bowen, 2003); and professors may talk about inscriptions during lectures that are clearly incorrect (Roth & Bowen, 1999a). Thus, it is precisely in the contrast of the caption opening "Attitude spikes resulting from prediction scheme" and the closing sentence "The red lines denote a spike-free series of targets" that readers can find the pointer to the spikes as the phenomenon of interest. The caption then has at least three important functions. First, it *describes* precisely those features that are relevant to the overall argument that the authors make in the main text. Scientists tend to do a tremendous amount of work so that their readers find in an inscription precisely what the authors intend for them to find (Bastide, 1985). Second, and at the same token, the caption is *an instruction* that guides reading what to look for in the inscription and how to go about finding it. Third, by looking at the visual aspects, competent reading finds the spikes. That is, the figure thereby *authenticates* what the text is describing.

The caption and inscription thereby relate back to the main text in multiple ways. First, they validate, in a visual form, what the main text claims: variations, which the author attributes to "the variation due to real-time updates of the commanded attitude." The caption marker "Figure 5.3" allows readers concerned with these inscriptions first to find the relevant information in the main text. Main text and caption therefore "collaborate" to establish the salience of a particular perceptual feature, and this feature, once discovered by the reading process, validates (reinforces) in a different manner what the text has directed it to look for.

Advantages of a Social Psychological/Pragmatic Approach

> For us, to speak about the external *process* means to speak of the social. Any higher psychological function was external; this means that it was social; before becoming a function, it was the social relation between people.
> (*Vygotsky*, 1989, p. 56)

When discussing knowledge, psychologists tend to focus on mental representations. The problem is that no researcher has direct access to mental representation, what it might look like, and how it might be used. A practice approach to scientific representation therefore is especially useful in teaching and learning at all levels of the formal and informal educational processes (Roth & McGinn, 1998). The introductory quotation to this section provides us with help here, as it suggests that any higher psychological function initially is external; it is not just found in external form, such as, for example, a diagram that engineers might also generate – in less elaborate form – in their mind. Indeed, the quotation suggests that the higher function *is* a social *relation* first. It is therefore not surprising to find that a text on how to follow engineers around in their work suggests examining "first the

many ways through which instructions are gathered, combined, tied together and send back... before attributing any special quality to the mind or method" of the engineers (Latour, 1987, p. 258). Only when there is something left unexplained do we have to draw on hidden variables (cognitive factors) to explain the work of engineers.

This position that moved the discussion from internal representation to inscriptions has shocked quite a number of scientists and engineers when initially formulated. But it actually provides us with some advantages for thinking about what we might want to teach in engineering and how we might go about it. This is so because if anything we wish an engineer to do and think is external first, then we do not have to theorize how we get something into his or her mind. All we need to do is allow future engineers to encounter inscriptions and engage in relations where they are used; and we need to assist those on a trajectory into an engineering field to become competent. The real training in engineering is at the advanced levels of graduate work, which is precisely what has been reported for science (Traweek, 1988). It is quite evident that up until the undergraduate levels and even into graduate school, newcomers are left on their own, to study and memorize engineering inscriptions and practices rather than experiencing them first hand at the elbow of someone who already is competent in their use or in groups of engineers *using* inscriptions. *Use* is important, as it rather than some mental "meaning" is the pertinent and defining aspect of a representation/word (Wittgenstein, 1953/1997). A special orientation to the reasoned use of inscriptions may foster learning environments in which students' competencies to produce, analyze, and interpret increase. Thus, one study showed that eighth-grade students – who designed and conducted investigations and reported them for the purpose of convincingly reporting their findings to their audiences – outperformed, at statistically significant levels, pre-service science teachers who had already obtained either a bachelor's or master's degree in science (Roth, McGinn,

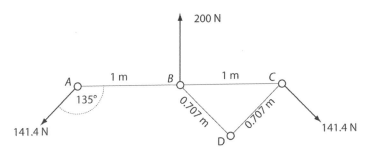

Figure 4.5. Example of static structure with forces in a mechanical engineering course.

& Bowen, 1998). The eighth-grade students used more abstract inscriptions – including plots, line graphs, and statistics – than the future teachers who, less than a year after the study, took positions of teaching in middle and high schools. Detailed quantitative and qualitative analyses of the inscription use among these eighth-graders exhibited the tremendous inscription-related competencies in contexts that push them to mathematize their experience (Roth, 1996c; Roth & Bowen, 1994).

A very simple example of focusing on the external aspects of learning engineering is through the teaching by examples. Thus, an engineering lecture on forces in structures might present two equations, one for forces (1) and another for the moments (2):

$$\underline{F} = F_x\underline{i} + F_y\underline{j} + F_z\underline{k} \qquad (1)$$

$$M_i = F_i d_i \qquad (2)$$

The lecturer might present a third equation for a static two-dimensional system that is in equilibrium:

$$\sum_{i=1}^{n} F_{x,i} = 0; \quad \sum_{i=1}^{n} F_{y,i} = 0; \quad \sum_{i=1}^{n} M_i = 0 \qquad (3)$$

Rather than letting the students do problems on their own, the lecturer might actually "apply" the equations to a real, concrete situation. Thus, he or she would exhibit *what an engineer does with these equations given a concrete example*. Such an approach has been successful for teaching mathematics at the undergraduate level, for it makes explicit what expert practitioners do

and how they think (Schoenfeld, 1985). For example, a configuration of some forces around specific points in a structure may be given (Fig. 4.5). The lecturer then takes the students through the example in a step-by-step fashion. This is important because the whole point is the *showing*, the relation that is thereby established between the lecturing engineer and his or her students – who should be encouraged to ask questions because this furthers the level and intensity of their interaction. It is precisely in the sequentially ordered interaction that the relation exists of which L. S. Vygotsky speaks in the introductory quotation. The demonstration involves further inscriptions, which exist in shared space and really precisely because of their materiality (talk alone would be more ephemeral).

The lecturer might point to the symmetry of the system, which therefore constitutes an equilibrium condition. For point *B*, we can therefore apply the first of the three sums in (3):

$$\sum_{i=1}^{3} F_{x,i} = 141.4\cos(45°)N$$

$$+ 0N - 141.4\cos(45°)N = 0 \qquad (4)$$

Because we now think in terms of *relation* involving lecturer and students, the lecturer must not leave *implicit* anything that appears to go without saying, because it is precisely the objective presence in public of everything that we later want to be the higher psychological function that counts. Thus, sometimes lecturers might leave out the fact that the three "i's" refer to the *A*, *B*, and C points where there are forces. Thus, $F_{x,1}$ is the force at A along the x-axis. It also has to be made explicit that there is a negative sign in the third force, which indicates that it points in the opposite direction.

This simple example is provided to highlight the fact that inscriptions do not have *inherent* "meaning" that students can somehow pick up or make in their minds. From the social psychological perspective articulated here, every higher psychological function first exists in the form of a concrete relation in material life. From a pragmatic perspective in L. Wittgenstein's language philosophy, *use* is what matters; and use is experienced in concrete relations rather than existing in ephemeral "mental frameworks" and "mental constructions." One set of lecture notes that I saw makes explicit the fact that some textbooks leave implicit certain ideas by discussing examples of *"Why didn't the book just say that...?"* (Ronney, 2010, p. 24). That is, the nature of engineering knowledge, which relies heavily on inscriptions (visual representation), makes it ideal for teaching and learning purposes.

Conclusion

Visual representation (inscription) not only is a faddish sign of our times but also is a constitutive feature of engineering knowledge. This essentially external aspect of engineering knowledge constitutes an advantage both for teaching engineering and for doing research in engineering education. It allows teachers and researchers to focus on the use and transformations of inscriptions. This makes it possible for us to follow engineers around at their work rather than having to try and enter the realm of their private thoughts. Inscriptions allow us to study engineering in action, whereby the point is to do so before knowledge is blackboxed into the common sense of engineering culture or in following the controversies during which established knowledge and norms come to be challenged. Engineering education research therefore is much better situ-

ated to capitalize on the possibilities arising from the social perspective on representation than other fields, where knowing is relegated to the mental sphere alone. The nature of engineering knowledge of being continually present in inscriptions provides us with affordances that other fields where the prevalence of inscriptions is much lower – social sciences, humanities – just do not have.

References

Akkerman, S. F., & Bakker, A. (2011). Boundary crossing and boundary objects. *Review of Educational Research*, 81, 132–169.

Barrett, M., & Oborn, E. (2010). Boundary object use in cross-cultural software development teams. *Human Relations*, 63, 1199–1221.

Bastide, F. (1985). Iconographie des textes scientifiques: Principes d'analyse. *Culture Technique*, 14, 132–151.

Barcellini, F., Détienne, F., & Burkhardt, J. (2009). Participation in online interaction spaces: Design-use mediation in an open source software community. *International Journal of Industrial Ergonomics*, 39, 533–540.

Broberg, O., & Hermund, I. (2007). The OHS consultant as a facilitator of learning in workplace design processes: Four explorative case studies of current practice. *International Journal of Industrial Ergonomics*, 37, 810–816.

Bucciarelli, L. L. (1994). *Designing engineers.* Cambridge, MA: MIT Press.

Christensen, B. T., & Schunn, C. D. (2009). The role and impact of mental simulation in design. *Applied Cognitive Psychology*, 23, 327–344.

Edgerton, S. (1985). The renaissance development of the scientific illustration. In J. Shirley & D. Hoeniger (Eds.), *Science and the arts in the Renaissance* (pp. 168–197). Washington, DC: Folger Shakespeare Library.

Ehn, P. (1992). Scandinavian design: On participation and skill. In P. S. Adler & T. A. Winograd (Eds.), *Usability: Turning technologies into tools* (pp. 96–132). New York, NY: Oxford University Press.

Ehn, P., & Kyng, M. (1991). Cardboard computers: Mocking-it-up or hands-on the future. In J. Greenbaum & M. Kyng (Eds.), *Design at work:*

Cooperative design of computer systems (pp. 169–195). Hillsdale, NJ: Lawrence Erlbaum.

Foucault, M. (1975). *Surveiller et punir: Naissance de la prison*. Paris: Gallimard.

Galison, P. (1997). *Image and logic: A material culture of microphysics*. Chicago, IL: University of Chicago Press.

Garrison, E. G. (1999). *A history of engineering and technology: Artful methods*. Boca Raton, FL: CRC Press.

Hall, R., Stevens, R., & Torralba, A. (2002). Disrupting the representational infrastructure in conversations across disciplines. *Mind, Culture, and Activity*, 9, 179–210.

Hall, R., Wieckert, K., & Wright, K. (2006, April). *Learning, teaching, and generalizing statistical concepts as statisticians consult across client domains*. Paper presented at the 2006 annual meeting of the American Educational Research Association, San Francisco, CA.

Henderson, K. (1991). Flexible sketches and inflexible databases: Visual communication, conscription devices, and boundary objects in design engineering. *Science, Technology, & Human Values*, 16, 448–473.

Hoyles, C., Bakker, A., Kent, P., & Noss, R. (2007). Attributing meanings to representations of data: The case of statistical process control. *Mathematical Thinking and Learning*, 9, 331–360.

Husserl, E. (1939). Die Frage nach dem Ursprung der Geometrie als intentional-historisches Problem. *Revue Internationale de Philosophie*, 1, 203–225.

Husserl, E. (1980). *Vorlesungen zur Phänomenologie des inneren Zeitbewußtseins*. Tübingen, Germany: Niemeyer.

Ibrahim, R., & Rahimian, F. P. (2010). Comparison of CAD and manual sketching tools for teaching architectural design. *Automation in Construction*, 19, 978–987.

Kemp, M. (2007). *Leonardo da Vinci: The marvelous works of nature and man*. Oxford: Oxford University Press.

Latour, B. (1987). *Science in action: How to follow scientists and engineers through society*. Milton Keynes, U.K.: Open University Press.

Latour, B. (1993). *La clef de Berlin et autres leçons d'un amateur de sciences* [The key to Berlin and other lessons of a science lover]. Paris: Éditions la Découverte.

Law, J., & Lynch, M. (1990). Lists, field guides, and the descriptive organization of seeing: Birdwatching as an exemplary observational activity. In M. Lynch & S. Woolgar (Eds.), *Representation in scientific practice* (pp. 267–299). Cambridge, MA: MIT Press.

Lemke, J. L. (1998). Multiplying meaning: Visual and verbal semiotics in scientific text. In J. R. Martin & R. Veel (Eds.), *Reading science* (pp. 87–113). London: Routledge.

Livingston, E. (1995). *An anthropology of reading*. Bloomington, IN: Indiana University Press.

Lynch, M. (1990). The externalized retina: Selection and mathematization in the visual documentation of objects in the life sciences. In M. Lynch & S. Woolgar (Eds.), *Representation in scientific practice* (pp. 153–186). Cambridge, MA: MIT Press.

Lynch, M., & Edgerton, S. Y. (1988). Aesthetics and digital image processing: Representational craft in contemporary astronomy. In G. Fyfe & J. Law (Eds.), *Picturing power: Visual depiction and social relations* (pp. 184–220). London: Routledge.

McCarthy, J. (2000). The paradox of understanding work for design. *International Journal of Human-Computer Studies*, 53, 197–219.

Mengoni, M., Germani, M., & Mandorli, F. (2007). Reverse engineering of aesthetic products: Use of hand-made sketches for the design intent formalization. *Journal of Engineering Design*, 18, 413–435.

Núñez, R. (2009). Gesture, inscriptions, and abstraction: The embodied nature of mathematics or why mathematics education shouldn't leave the math untouched. In W.-M. Roth (Ed.), *Mathematical representations at the interface of body and culture* (pp. 309–328). Charlotte, NC: Information Age Publishing.

Pozzer, L. L., & Roth, W.-M. (2003). Prevalence, function, and structure of photographs in high school biology textbooks. *Journal of Research in Science Teaching*, 40, 1089–1114.

Ronney, P. D. (2010). *Basics of mechanical engineering: Integrating science, technology, and common sense*. Retrieved from http://ronney.usc.edu/AME101F10/AME101-F10-LectureNotes.pdf

Roth, N. H. (2010). *Navigation and control design for the CanX-4/-5 satellite formation flying mission*. Unpublished master's thesis, Institute for Aerospace Studies, University of Toronto. Retrieved from https://tspace.library.utoronto.ca/bitstream/1807/25908/3/Roth_Niels_H_201011_MASc_thesis.pdf

Roth, W.-M. (1996a). Art and artifact of children's designing: A situated cognition perspective. *The Journal of the Learning Sciences*, 5, 129–166.

Roth, W.-M. (1996b). Thinking with hands, eyes, and signs: Multimodal science talk in a grade 6/7 unit on simple machines. *Interactive Learning Environments*, 4, 170–187.

Roth, W.-M. (1996c). Where is the context in contextual word problems? Mathematical practices and products in Grade 8 students' answers to story problems. *Cognition and Instruction*, 14, 487–527.

Roth, W.-M. (1998). *Designing communities*. Dordrecht, The Netherlands: Kluwer Academic.

Roth, W.-M. (Ed.). (2009). *Mathematical representation at the interface of body and culture*. Charlotte, NC: Information Age Publishing.

Roth, W.-M (2010). An anthropology of reading science texts in online media. *Semiotica*, 182, 409–442.

Roth, W.-M., & Bowen, G. M. (1994). Mathematization of experience in a grade 8 open-inquiry environment: An introduction to the representational practices of science. *Journal of Research in Science Teaching*, 31, 293–318.

Roth, W.-M., & Bowen, G. M. (1999a). Complexities of graphical representations during lectures: A phenomenological approach. *Learning and Instruction*, 9, 235–255.

Roth, W.-M., & Bowen, G. M. (1999b). Digitizing lizards or the topology of vision in ecological fieldwork. *Social Studies of Science*, 29, 719–764.

Roth, W.-M., & Bowen, G. M. (2003). When are graphs ten thousand words worth? An expert/expert study. *Cognition and Instruction*, 21, 429–473.

Roth, W.-M., Bowen, G. M., & Masciotra, D. (2002). From thing to sign and 'natural object': Toward a genetic phenomenology of graph interpretation. *Science, Technology, & Human Values*, 27, 327–356.

Roth, W.-M., Bowen, G. M., & McGinn, M. K. (1999). Differences in graph-related practices between high school biology textbooks and scientific ecology journals. *Journal of Research in Science Teaching*, 36, 977–1019.

Roth, W.-M., & Lawless, D. (2002). Science, culture, and the emergence of language. *Science Education*, 86, 368–385.

Roth, W.-M., & McGinn, M. K. (1998). Inscriptions: A social practice approach to "representations." *Review of Educational Research*, 68, 35–59.

Roth, W.-M., McGinn, M. K., & Bowen, G. M. (1998). How prepared are preservice teachers to teach scientific inquiry? Levels of performance in scientific representation practices. *Journal of Science Teacher Education*, 9, 25–48.

Schoenfeld, A. (1985). *Mathematical problem solving*. Orlando, FL: Academic Press.

Sørensen, K. H., & Levold, N. (1992). Tacit networks, heterogeneous engineers, and embodied technology. *Science, Technology, & Human Values*, 17, 13–35.

Star, S. L. (1989). The structure of ill-structured solutions: Boundary objects and heterogeneous distributed problem solving. In L. Gasser & M. N. Huhns (Eds.), *Distributed artificial intelligence* (Vol. 2, pp. 37–54). London: Pitman.

Star, S. L. (1995). The politics of formal representations: Wizards, gurus, and organizational complexity. In S. L. Star (ed.), *Ecologies of knowledge: Work and politics in science and technology* (pp. 88–118). Albany: State University of New York Press.

Star, S. L. (2010). This is not a boundary object: Reflections on the origin of a concept. *Science, Technology, & Human Values*, 35, 601–617.

Suchman, L. A. (1990). Representing practice in cognitive science. In M. Lynch & S. Woolgar (Eds.), *Representation in scientific practice* (pp. 301–321). Cambridge, MA: MIT Press.

Suchman, L. A. (1993). Technologies of accountability: Of lizards and aeroplanes. In G. Button (Ed.), *Technology in working order: Studies of work, interaction, and technology* (pp. 113–126). London and New York: Routledge.

Traweek, S. (1988). *Beamtimes and lifetimes: The world of high energy physicists*. Cambridge, MA: MIT Press.

Turnbull, D. (1993). The ad hoc collective work of building gothic cathedrals with templates, string, and geometry. *Science, Technology, & Human Values*, 18, 315–340.

Vygotsky, L. S. (1989). Concrete human psychology. *Soviet Psychology*, 27(2), 53–77.

Wittgenstein, L. (1997). *Philosophische Untersuchungen / Philosophical investigations* (2nd ed.). Oxford: Blackwell. (First published in 1953)

CHAPTER 5

Conceptual Change and Misconceptions in Engineering Education

Curriculum, Measurement, and Theory-Focused Approaches

Ruth A. Streveler, Shane Brown, Geoffrey L. Herman, and Devlin Montfort

Introduction

Recent research has shown that many students continue to understand phenomena in simplified or unproductive ways, even after those understandings are directly contradicted in educational settings (Hake, 1998; Miller et al., 2006). In the context of engineering education, many engineering graduates still do not understand the foundational concepts of solid and fluid mechanics, physics, thermodynamics, digital logic, or other fields. The study of conceptual change and misconceptions is one attempt to understand and address this issue.

Because this field of study is fractious and diverse, we briefly establish some shared vocabulary and understanding of the fundamental processes underlying conceptual change and misconceptions. The following section introduces three primary theories of conceptual change: curriculum, measurement, and theory-focused efforts in engineering education. The chapter concludes with a brief summary and discussion of future directions for research.

We must define conceptual understanding somewhat carefully for our terminology to be useful across the various theoretical frameworks discussed in this chapter. An individual's *conceptual understanding* of a topic is the collection of his or her concepts, beliefs, and mental models, where the following definitions apply:

- *Concepts* are pieces or clusters of knowledge, for example, "force," "mass," "causation," and "acceleration."
- *Beliefs* are propositional relationships between concepts, for example, "a force on a mass causes acceleration."
- *Mental models* are groups of meaningfully related beliefs and concepts that allow people to explain phenomena and make predictions; for example, an expert dynamics instructor would use her mental model of Newtonian physics to predict an object's motion.

Conceptual change is the process of altering one's conceptual understanding. A central research question in conceptual change theory is why some misconceptions are more difficult to change than others. We define a *misconception* as any aspect of an individual's conceptual understanding that resists conceptual change and contributes to an incorrect, naïve, or unproductive conceptual understanding. In our usage, a misconception is a specific type of *preconception*, which is one's conceptual understanding of a topic developed before a time of interest. Note that students' first formal education on a topic is the most common "time of interest" used to differentiate preconceptions from conceptual understanding.

With these defined terms, we commit to a more rigorous definition of the problem facing engineering education. We argue that engineering students' low conceptual understanding of fundamental engineering concepts is due to misconceptions limiting or preventing conceptual change, so that their preconceptions built on limited and uncritically observed life experience or previous coursework endure, even in the face of contradictory and more productive means of describing and understanding. We turn to the work of conceptual change theorists to look for more detailed explanations of how and why conceptual change is sometimes so difficult.

Theories of Conceptual Change

Despite the emphasis on theoretical differences between contemporary conceptual change researchers (e.g., see diSessa, 2008; Mayer, 2002), the various approaches to conceptual change research are based on a largely shared historical development. At the core, all theories of conceptual change share the assumption that an individuals' conceptual understanding can have a strong effect on their learning. This effect is sometimes summarized in terms of potential responses to new information, with the two general kinds of responses being to either change one's understanding, or to change the new information (Carey, 1985; Chinn & Brewer, 1993). In Piaget's work (1970) and much that followed, conceptual change is

taken to be the rare case in which people choose to change their existing system of knowledge to accommodate new information, as opposed to altering the new information to better fit with existing beliefs. In other words, what an individual learns is at least partially controlled by what they already know.

Theorists disagree, however, as to how existing knowledge is structured, and how it affects conceptual change. Three prominent theories of conceptual change are built on opposing descriptions of conceptual understanding as either (1) a hierarchical categorization of knowledge, (2) a weak grouping of experiences and perceptions, or (3) a form of social behavior.

Conceptual Understanding as Categories

Some cognitive theorists describe conceptual understanding as a hierarchy of categories and subcategories. The category "engineering," for example, might contain the subcategory "civil engineering," which might contain the sub-category "geotechnical engineering," and so on. Concepts, beliefs, and mental models are organized within these categories and take on properties associated with them (all of geotechnical engineering shares some properties with other kinds of civil engineering, for example).

Chi and Roscoe (2002) and Chi, Roscoe, Slotta, Roy, and Chase (2012) argue that there is one fundamentally difficult categorization that accounts for many misconceptions. Drawing on examples from human physiology, thermodynamics, and electricity, Chi (2005) found that students often fail to categorize "emergent processes" appropriately. An emergent process is one in which an observable macro-scale pattern emerges from chaotic, unpredictable microscopic phenomena. A commonly cited example involves cream diffusing in coffee. The observable phenomenon is that the cream distributes itself in the coffee relatively quickly, and a kind of steady-state is reached wherein the cream and coffee are evenly mixed. The true cause of this event (the distribution of cream in coffee), however, is the continuous and ongoing molecular interactions that, on average, result in observable diffusion. Chi posits that without properly forming and applying the category of emergent process, people tend to treat every process as intentionally instigated by an agent. We say that the cream "wants to spread out," for example, and believe the process to be completed once the cream has achieved this goal.

Vosniadou, Vamvakoussi, and Skopeliti (1994, 2008) similarly write that most apparently difficult conceptual changes are those that require students to make changes at a very high level in the hierarchy. In this approach, students' conceptual understandings are all contextualized, based on what general domain they relate to. Each domain (e.g., the physical and social worlds) is characterized based on a "framework theory," which defines the basic assumptions about what is possible and expected within that domain. For example, it is acceptable to discuss a nonspecific car in a physics problem, because it is assumed that processes in the physical world will be the same regardless of the make of the car. But it does not fit within the framework theory of the social world to be similarly nonspecific (i.e., to talk about a nonspecific car when filing a police report about a stolen vehicle). Vosniadou argues that synthetic frameworks (Vosniadou, 2007; Vosniadou & Brewer, 1992) arise when learners inappropriately conflate what they learned in the classroom with their everyday experiences. For example, children may create a synthetic framework that combines instruction about a round earth with experiences of a flat earth to create a model that is pancake shaped. These synthetic frameworks interfere with related learning, problem-solving, and reasoning tasks.

Conceptual Understanding as Component Pieces

Some theorists directly oppose the assumption that conceptual understanding holds to a consistent and hierarchical structure.

These theorists argue instead that knowledge exists primarily in pieces, and it is not until a relatively high level of expertise is reached in a content area that consistent and repeatable relationships between those pieces are established (diSessa, 2008; Minstrell, 2001). More specifically, diSessa argues that knowledge is largely made up of atomistic intuitions he calls "phenomenological primitives." These small pieces of knowledge are called phenomenological because they are based on individuals' specific life experiences. They are primitives, because they exist without interpretation. They are raw, in a sense. For diSessa, conceptual change consists of processing these and other primitives, learning to relate them, and then learning when to use those relationships (diSessa, 2002, 2004).

For example, life experience inspires a basic intuition in most people that objects are motionless unless something is pushing them. This intuition is a strong phenomenological primitive, because it is reinforced often in daily life as we throw our keys on the table or sweep dust repeatedly to move it across the floor. This belief is directly contradicted by the First Law of Newtonian mechanics. To believe the First Law of Newtonian mechanics, students must undergo conceptual change. This change process requires them to somehow relate their phenomenological primitive to the First Law. To achieve change, students must first put their intuitions and the First Law in compatible forms. For example, students may not even notice the contradiction, because an intuition and an explicitly stated, scientific law are such drastically different things. Second, after identifying the contradiction, the student must fit his or her experience into the broader context of Newtonian physics, perhaps by noting the unaccounted-for forces that would cause their keys and dust to stop moving. Finally, the student must also learn when to apply his or her newly constructed understanding. There is a point during conceptual change in which students might have to catch themselves relying on naïve relationships and consciously replace them with newly developed ones.

Conceptual Understanding as Social Behavior

Whereas diSessa and Chi differ about the cognitive qualities of students' conceptual understandings, some researchers reject the entire idea of focusing solely on cognitive processes. Säljö (1999), for example, argues that instead of attempting to describe students' conceptual understanding and conceptual change – what they are supposedly *thinking* – we should focus on their use of discursive resources – what they are actually *saying*. This argument is, in part, based on questioning the common assumption that what people say in interviews is an accurate and meaningful revelation of their conceptual understanding (Halldén, Haglund, & Strömdahl, et al., 2007).

This theory of conceptual change goes beyond simply pointing out a potential flaw in a common methodology, however. Säljö and similar theorists argue that conceptual change is actually the process of learning how to communicate more complex ideas more effectively. Although other theorists might say that a student incorrectly believes that forces cause velocity, for example, theorists such as Säljö would instead say that the student lacks the discursive tools to make the distinction between velocity and acceleration. Conceptual change, then, is a process of language acquisition and enculturation (Brown, Collins, & Duguid, 1989; Ivarsson, Schoultz, & Säljö, 2002). This emphasis on language and culture subtly alters conceptual change from an internal process of moving from incorrect to correct ways of thinking, to a more social process of learning the values and distinctions that are important in a certain group, and learning to speak like that group. This line of reasoning has substantial parallels to the theory that learning is the act of becoming a part of a community of practice (Wenger, 1998), and that knowledge does not exist as a consistent entity in the mind but is a negotiation between the mind and the context in which knowledge is applied (Robbins & Aydede, 2009). This distinction has major implications for best pedagogical practice.

Research on Conceptual Change in Engineering Education

We have divided the existing research on conceptual change in engineering education into three branches, based on the apparent goals and emphases of published work: (1) curriculum-focused research, (2) measurement-focused research, and (3) theory-focused research. We recognize that the categorization of conceptual change research into three branches is artificial, and that there is overlap between branches. These divisions represent our synthesis of the field, and as such are colored by our approaches to studying conceptual understanding and misconceptions.

Each of the three categories is discussed in terms of the characteristic theoretical framework, methods, and findings. Our approach in this section is to present each category with examples that are drawn from established and sustained research programs. The researchers cited in this section have reviewed the draft and have agreed with the characterizations of their research programs.

We distinguish the three research divisions by their predominant research questions. Curriculum-focused conceptual change research focuses on asking action research questions such as, "What can the engineering educator do to address students' misconceptions?" or, "How do students respond to conceptual change interventions?" In curriculum-focused conceptual change research, the researcher often plays a dual role of instructor and researcher.

Measurement-focused conceptual change research focuses on asking how we can quickly and effectively measure a person's conceptual frameworks. Consequently, this branch of research asks questions like, "How reliable is X measure of conceptual change?" or, "What exactly does instrument Y measure?" Efforts in measurement-focused conceptual change research have centered on the development and testing of concept inventories. Concept inventories are often used as measures of conceptual change, and more indirectly, as measures of pedagogical practices that may promote conceptual change (Hake, 1998).

Theory-focused conceptual change research is often more open-ended, and investigates the existence, importance, and interrelatedness of implied cognitive phenomena by asking questions like, "How do students relate concepts A and B in contexts X and Y?" As is the case with other kinds of theoretical research, the ultimate goal of this kind of research is to inform the design of applications that are useful to society. Theory-focused conceptual change research hopes to ultimately inform the design of learning environments and pedagogical methods that will promote conceptual change in students.

Curriculum-Focused Research

There is a rich body of research that investigates ways to improve students' conceptual understanding and address their misconceptions. Although improving pedagogy is a long-term goal for many researchers in conceptual change, this branch of the field is distinguished by a practical emphasis on how conceptual understanding is observable and important in the classroom. Unlike the other two branches of research discussed here, which attempt to draw insights from "offline" studies and then apply the findings to the classroom, this branch of research derives insights about conceptual change from the classroom and tests these insights in the classroom. The work of Steif in statics and Krause in materials science provides clear and extensive examples of this branch of research.

APPLIED THEORIES OF CONCEPTUAL CHANGE

Krause uses a definition of conceptual knowledge that is adopted from Norman (1983) and Gilbert (1995). Krause and Tasooji (2007) write:

> A conceptual framework is comprised of mental models, *which are transformed representations of real-world systems or phenomena called* modeled target systems or phenomena. *As such,* mental models *are defined as simplified, conceptual representations that*

are personalized interpretations of modeled target systems or phenomena *in the world around us. . . . An individual communicates his/her* mental models *with some form of external representation which are* expressed models. *(p. 4)*

From this definition, Krause proposes that conceptual change can be tracked and measured by comparing students' expressed models (Krause & Tasooji, 2007).

Krause's conceptual change research is largely based on a pedagogically driven framework derived by Taber that was intended to facilitate the repair of misconceptions (Krause, Kelly, Corkins, Tasooji, & Purzer, 2009; Purzer, Krause, & Kelly, 2009; Taber 2001, 2004). Taber classifies misconceptions according to how they impede learning. He specifies two general types: *null impediment* refers to misconceptions that impede learning because students lack information to construct the scientifically accepted conception. *Substantive impediment* refers to faulty concept models that students have constructed.

Steif's work is closely tied to the pedagogical importance of students' experiences, as is evidenced by his significant effort in cataloging the preconceptions students bring to their respective fields (Steif & Hansen, 2006), as well as in developing learning modules that are specifically designed to help students visualize or otherwise perceive important concepts (Steif, 2002a, 2002b; Steif & Dollar, 2003; Steif & Gallagher, 2004). This emphasis on the students' experiences as a potential avenue to addressing misconceptions is similar to diSessa's theory of conceptual change in the ways it explicitly includes perceptions and observations as pieces of conceptual understanding. Steif also carefully distinguishes between concepts that can be expressed as statements of belief and skills that are specific tasks. These distinctions are used in Steif's work as potential sources of errors (Steif, Lobue, Kara, & Fay, 2010).

METHODS

Curriculum-focused research in conceptual change is typically conducted with a combination of broad sampling of students (surveys, focus groups, or semistructured interviews as opposed to case studies) combined with the researchers' own extensive experience with learning and teaching in their fields. These data-collection methods are often directed at the development of assessment tools (discussed in more detail in the following section) or typologies of student difficulties. These tools are then applied in the classroom to assess students' conceptual understanding or to develop specific interventions or learning modules.

To guide his curriculum-focused research, Krause applied his experience as an instructor to create the Materials Concept Inventory (MCI) (Krause, Decker, & Griffin, 2003a). Additional conceptual change data were collected through "intuition quizzes," in which students were asked to express their mental model before instruction (Krause, Decker, Niska, & Alford, 2003b). This preliminary work was supplemented through the use of focus groups that elicited students' misconceptions (Krause, Tasooji, & Griffin, 2004). Students were asked to solve paired questions that covered both *what* students thought would happen in a situation and *why* students thought the situation would happen. Purzer et al. (2009) conducted a mixed-method study (a mixture of think-aloud interviews, MCI data, and a literature survey) to classify the origin and robustness of various materials misconceptions. K–12 misconceptions literature was examined to determine when students developed their misconceptions. In later studies, Krause used multimodal conceptual assessment tools (MMCAT) that were administered to students in a materials course twice per week. The MMCAT elicited students' mental models by asking them to provide sketches or descriptions, answer MCI questions, and answer other conceptual questions. The MMCAT were used to assess the effectiveness of instruction in creating conceptual change.

Similar to Krause's approach, Steif combined his extensive experience teaching statics with a research-based approach to examining student difficulties to develop

lists and clusters of common student difficulties (Steif & Gallagher, 2004). This effort ultimately led to the development of the Concept Assessment Tool for Statics (CATS), formerly known as the Statics CI, one of the most recognized CIs in engineering education (Steif & Dantzler, 2005). Steif's work following the CI development largely focused on relating CI performance to other classroom performance measures (Steif et al., 2005; Steif & Hansen, 2006) to provide additional validity evidence for CATS. One study used interview techniques to elicit students' reasoning about difficult concepts (Newcomer & Steif, 2007). More recent work uses the CATS to measure conceptual change to develop pedagogical and curricular approaches to improve student learning.

FINDINGS

From his initial studies, Krause observed that students' misconceptions develop from poor observations and reasoning ("I have never seen Ni gas, so it must not exist"), informal learning experiences ("I heard in a PBS special that . . . "), and improper prior instruction ("My fifth grade teacher [taught this misconception]") (Krause et al., 2004). Later studies revealed that different misconceptions were created through different causes (Purzer et al., 2009). For example, misconceptions related to the nature of crystalline structure and unit cells were caused by null deficiency impediments, but misconceptions related to material processing were attributed to substantive misinterpretive and experiential impediments. Purzer et al. (2009) classified several misconceptions and hypothesized that substantive pedagogic impediments exist through all levels of formal education, and are the most difficult to overcome.

In the spirit of this branch of research, Krause and his colleagues immediately applied these findings to the classroom. Krause, Kelly, Kurpius-Robinson, and Baker (2010) and Kelly, Krause, and Baker (2010) used these classifications to determine whether different pedagogies better remedy the different classes of misconceptions. Through MMCAT, Kelly et al. (2010) found

Table 5.1. Summary of the Skills and Concepts Underlying Common Student Errors in Statics

Common Student Error	Underlying Concepts (C)	Underlying Skill (S)
Leaving a force off the free body diagram	C1: Forces act between bodies. C4: Equilibrium conditions are imposed on a body.	S1: Discern separate parts of an assembly where each connects the others.
Ignoring a couple that could act between two bodies, or falsely presuming its presence	C2: Combinations and/or distributions of forces acting on a body are statically equivalent to a force and couple. C3: Conditions of contact between bodies or types of bodies imply simplification of forces.	S2: Discern the surfaces of contact between connected parts and/or the relative motions that are permitted between two connected parts.
Failure to take advantage of the options of treating a collection of parts as a single body, dismembering a system into individual parts, or dividing a part into two	C4: Equilibrium conditions are imposed on a body.	S3: Group separate parts of an assembly in various ways and discern external parts that contact a chosen group.

Adapted from Steif (2004).

that different modes of assessment revealed different conceptual deficiencies. For example, students could verbally describe atomic bonding, but they could not produce accurate sketches of these same atomic bonds. They also found that some pedagogies remedied students' misconceptions more effectively than others. For example, they ranked the following five pedagogies from most effective to least effective: (1) team discussions with hands-on activities and concept sketching; (2) team discussions with contextualized concept mini-lectures and activities; (3) team discussions, contextualized concept lectures and activities, plus pre-/post- topic assessments and daily reflections; (4) lecture with some discussions; and (5) lecture only with no team discussions or activities.

In a paper characterizing the concepts that define statics, particularly as it is distinct from Newtonian physics, Steif (2004) defined four concept clusters and four skills that underlie the learning of statics. He further listed common student errors with examples and evidence of how students express those errors. Steif also related each type of common error to the conceptual lapses and missing skills that related to those student errors. Three examples of

common student errors (Steif identified eleven) are included in Table 5.1. Recently, work has begun to examine the conceptual similarities that might underlie students' common errors, as well as investigating how those commonalities might be utilized to improve classroom instruction (Santiago Román, Streveler, Steif, & DiBello, 2010). In particular, Steif notes that emphasizing the role of bodies in statics (i.e., the actual objects that are in physical contact) seems to improve the development of conceptual understanding (Steif et al., 2010).

Measurement-Focused Research

Conceptual change research in this category is concerned with the development and use of assessment instruments, often concept inventories, as a means to measure students' conceptual knowledge. Concept inventories continue to be developed in many areas of engineering; however, cataloging this area is beyond the scope of this chapter, and several summaries are available elsewhere (see Santiago Román, 2009). Rather than providing a list of concept inventories, we provide an in-depth discussion of one concept inventory, the Thermal and Transport

Science Concept Inventory (TTCI), because it was developed based on established practices in assessment (Streveler et al., 2011).

APPLIED THEORIES OF CONCEPTUAL CHANGE

The goal of the TTCI is to measure student understanding of particularly difficult concepts in fluid mechanics, thermodynamics, and heat transfer. Owing to its status as an assessment instrument, the TTCI was developed using best practices in assessment design, most notably the framework called the "Assessment Triangle" (Pellegrino, Chudowsky, & Glaser, 2001).

The Assessment Triangle is composed of three elements or "corners": the cognition corner, the observation corner, and the interpretation corner. The cognition corner, the starting point of assessment development, requires the researcher to explicitly commit to formal or informal theories about how people learn in the target discipline – in the case of the TTCI, thermal and transport sciences.

Thus, the cognition corner identifies the chosen learning theory as the foundation for developing any assessment. In the case of the TTCI, Chi's theories about emergent processes (Chi 2005; Chi et al., 2012) were chosen for the cognition corner, because she addressed why students might have difficulty learning about topics such as equilibrium and diffusion, two important concepts in the thermal sciences.

The observation corner then requires the researcher to create assessment tasks that make the learning observable. The final corner of the assessment triangle, the interpretation corner, describes the interpretation system one will use to make sense of the assessment tasks. The interpretation corner is therefore related to the kind of analyses that are appropriate, given the assessment tasks and the kind of learning that is taking place.

METHODS

In the TTCI, two approaches were used to investigate the cognition corner in thermal and transport science. In addition to using Chi's theories as a guide, a Delphi survey (Linstone & Turoff, 2002) of experienced engineering educators was used to identify what concepts in thermodynamics, fluid mechanics, and heat transfer were both important and difficult for students to understand (Streveler et al., 2011). The ten most important and difficult concepts identified by the Delphi survey were then used as the basis for creating the TTCI. After identifying the concepts of interest, the questions themselves (the observation corner) were developed by following commonly used practices (Downing, 2006):

1. Following suggestions from Chi, open-ended questions were developed for each concept of interest to elicit students' thinking about *WHY* a phenomenon was observed, not *WHAT* was happening.

2. Students were interviewed with the open-ended questions to elicit responses that became the basis of the incorrect alternative answers, or *distractors*.

3. Questions were revised through focus groups (usability testing), expert feedback (validation), and repeated administration of beta versions of the TTCI.

4. Classical Test Theory was used to calculate reliability of the TTCI. Item Response Theory was also used to determine item difficulty and item discrimination. (See Streveler et al., 2011 to follow one example question through this revision process.) The current version of the TTCI is available in a Web-based form [http://www.thermalinventory.com].

FINDINGS

Three major findings related to student learning have emerged. First, the three conceptual distinctions that are the most difficult for students to understand are (1) heat versus temperature, (2) temperature versus energy, and (3) equilibrium versus steady-state. Second, results suggest that a significant number of advanced engineering students, perhaps as many as 30%, may still misunderstand these three conceptual distinctions (Miller et al., 2006). Finally, additional research by Miller and colleagues suggests that instruction that provides students with a way to think about the emergent properties of a phenomenon – such as heat – may help students understand these concepts more completely (Miller, Streveler, Yang, & Santiago Román, 2011).

Theory-focused Research

The theory-focused branch of conceptual change research asks questions that both draw from and contribute to existing cross-disciplinary efforts. Consequently, while the curriculum branch primarily develops findings targeted to engineering education practitioners, and the measurement branch links theoreticians and practitioners, the results of the theory-focused branch of research may also be of interest to cognitive scientists, educational psychologists, and other conceptual change theorists. This branch attempts to move from identifying specific instances of misconceptions and their remedies toward establishing, confirming, or denying theoretical frameworks that allow findings to be transferred between classrooms and content areas. The work of Brown and Montfort in mechanics of materials and that of Herman, Loui, and Zilles in digital logic are examples of this branch of research.

APPLIED THEORIES OF CONCEPTUAL CHANGE

The research in this branch tends to draw from multiple theories of conceptual change in efforts to develop new syntheses. Recent efforts to build new theories of conceptual change have been shaped by the increasingly prominent concerns of theorists about the potentially context-specific and situated nature of conceptual understanding as it develops (Limón, 2002) and is measured (Halldén et al., 2007; Ivarsson et al., 2002) and applied (Sinatra, 2002). In this use, a context is a real-world situation in which an individual makes use of their conceptual understanding (e.g., being asked a question by a relative, taking an exam, providing expert testimony).

METHODS

In response to these concerns, research in this branch typically involves one-on-one, semistructured interviews with students (Brown & Lewis, 2010; Herman, Loui, Kaczmarczyk, & Zilles, 2012), which present concepts in multiple contexts. Contexts are loosely defined as written, verbal, pictorial, and social representations of a concept. Analyses then explicitly investigate the consistency of student reasoning and apparent understanding across contexts. For example, in Figure 5.1, students are presented with the same task twice with slightly different contexts: one presentation that asks for a Boolean expression and one that asks for a truth table.

Work in this research branch depends on qualitative methods. Because there was sparse documentation of students' misconceptions in digital logic, Herman, Loui, and Zilles (2011) used a grounded-theory-based approach. In their interviews, students were asked to solve digital logic problems in multiple contexts; for example, canonical analysis problems versus unconventional comparison tasks, or the same problem presented multiple times in different contexts (see Figure 5.1; Herman, Loui et al., 2011, 2012; Herman, Zilles et al., 2011, 2012).

In another example of theory-focused conceptual change research, Brown and Montfort have investigated student understanding of stresses and strains in two central topics in mechanics of materials: axially loaded members and bending beams. Interview protocols with multiple contexts, including physical demonstrations (Brown, Montfort, & Findley, 2007), ranking tasks (Montfort, Brown, & Pollock, 2009), open-ended problems (Brown & Lewis, 2010), and traditional textbook-style problems (Brown, Montfort, & Hildreth, 2008), were used. Students were presented with several different contexts, such as physical demonstrations, photographs of research procedures, ranking tasks, open-ended discussions, and computational tasks (Brown & Lewis, 2010). Figure 5.2 below shows some of the various contexts referred to in these interviews.

Alex, Beth, and Chris want to order a single, large pizza that they all will like to eat.
Use the variables, p = 1 when the pizza has pepperoni; s = 1 when the pizza has sausage; o = 1 when the pizza has olives

Alex will eat pizzas with olives if and only if the pizza also has pepperoni.
Beth will eat only pizzas that have pepperoni without sausage.
Chris will eat only pizzas that have exactly two ingredients.

Presentation (1): Write a single Boolean expression that specifies what pizzas they can order.

Presentation (2): Fill out the truth table below to show what pizzas the group can order

p	s	o	
0	0	0	
0	0	1	
0	1	0	
0	1	1	
1	0	0	
1	0	1	
1	1	0	
1	1	1	

Figure 5.1. Two presentations of digital logic tasks that used slightly different contexts.

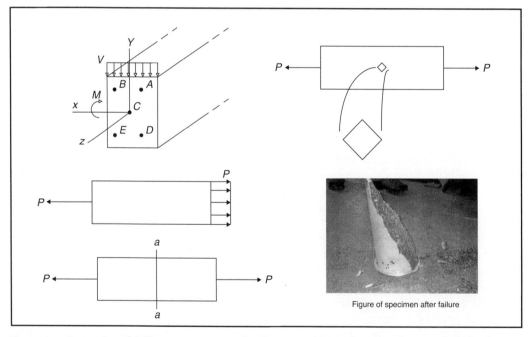

Figure 5.2. Examples of different contexts used in Brown and Montfort's Mechanics of Materials interviews.

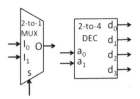

Figure 5.3. Typical representations of multiplexers and decoders in digital logic courses.

FINDINGS

The findings that arise from this branch of research have two main components. First, researchers are developing in-depth descriptions of students' conceptual understanding in a variety of topics. Brown and Montfort have found, for example, that students often cannot accurately describe the stresses that develop under simple loadings, and that they are particularly inclined to ignore shear stresses, or stresses acting in directions other than the direction of loading (Brown et al., 2007, 2008). Despite students' considerable abilities to calculate or even define various types of stresses, students ranging from sophomores to graduate students were still unable to predict stresses reliably and accurately, even under the simplest loadings (Montfort et al., 2009). Students expressed inconsistent or incorrect beliefs about these fundamental concepts, even when they demonstrated a deeper understanding of more complex concepts (Brown & Lewis, 2010).

Herman, Loui et al. (2011) have found more evidence for the importance of perception in conceptual change when students incorrectly conflate the concepts of "multiplexers" and "decoders." Although multiplexers and decoders have different behaviors and purposes in a circuit, students often develop faulty associations between the two components, and conceive of them as structural or functional opposites. Herman et al. believe that students develop this misconception because the two concepts are perceptually similar: boxes with arrows pointing in and out (see Figure 5.3).

Herman et al. have also found that despite considerable ability to perform calculations, students struggle with interpreting numbers that are represented in bases other than base 10 (also called "decimal"). When asked to compare two numbers in different bases, students often used faulty overgeneralizations to perform the comparison (Herman, Zilles et al., 2011). For example, sophomores in computer engineering falsely claimed that $(1)_2$ was smaller than $(1)_{10}$ because "numbers in base 2 are always less than numbers in base 10."

Herman and colleagues have also documented the context dependence of students' conceptual knowledge, which seems to shift when the contexts within the interview change. For example, the term and concept of "state" was particularly troublesome for students. In digital logic, state is the binary encoded information stored in the circuit's storage elements ("flip-flops") at any given time. However, over the course of an interview, students used an average of four different conceptions of state. Not only did students use different conceptions, but they also used contradictory, mutually exclusive conceptions (e.g., inputs and outputs in one context, and then flip-flops in a different context; Herman, Zilles et al., 2012). Furthermore, when interpreting an "if A then B" logic condition in a card game context, only 20% of students interpreted the statement correctly, but in a drinking bar context, more than 90% of students interpreted the statement correctly. Similarly, when students were simply asked to derive a Boolean expression of the "pepperoni without sausage" specification in Figure 5.1, most students derived the same incorrect expression. But when they were asked to derive a Boolean truth table first, most students derived the correct expression. The conclusion is that different contexts present different perceptual cues that affect how students access their conceptual knowledge (Herman, Loui et al., 2012).

Herman et al. propose a theory of conceptual change that focuses on the associations that people develop among words, experiences, images, and contexts. For example, students' misconceptions about multiplexers and decoders described earlier are created from the faulty associations between images (similarity of representation) and experiences (proximity of instruction; Herman, Loui et al., 2011). Based on these findings, Herman et al. suggest that although students' knowledge is context dependent, it is not altogether fragmented. Rather, conceptual change is created through the promotion and decay of associations between ideas. They further suggest that conceptual change can be promoted by helping students become aware of the associations that they possess for different concepts. In this case, instruction might effectively create conceptual change by explicitly making students aware of their implicit associations (Herman, Loui et al., 2012). This suggestion is similar to the broader recommendation of Chi, in which she calls for the development of curricular materials specifically intended to help students create a conceptual category for emergent processes (Slotta & Chi, 2006).

Discussion

By dividing existing research in conceptual change in engineering education into three groups, we do not intend to imply that there is no overlap between different research approaches in this area. There are strengths and weaknesses to each approach, and we would argue for the value of more synthesized efforts.

Curriculum-focused research has the benefit of generating products that directly affect students involved in the research program. The clarity of purpose and admirable focus on students lead to innovative ways to translate understanding of conceptual change more efficiently into improved education. An important contribution of Steif's findings is the correlation of students' scores on CATS to their course performance (Steif & Hansen, 2006) and problem-solving ability (Steif, Dollar, & Dantzler, 2005), therefore linking concept inventory scores with

classroom performance. This highlights the validity of CI's in representing other forms of understanding and performance in the classroom and renders results more applicable to curricular and pedagogical improvements.

Measurement-focused research generates metrics that can serve as motivation and catalysts for broad-scale changes in engineering education. For example, the comparison of Force Concept Inventory (FCI) learning gains versus pedagogical approach made famous by Hake (1998) has been used to promote more interactive approaches in physics education. The use of conceptually focused questions championed by Mazur (1997) was catalyzed by his comparison of his Harvard students' FCI scores with other populations of students that found that the Harvard students did no better than students at other institutions.

Theory-focused research benefits from drawing on established fields to craft research that is carefully situated to inform and support research goals that extend beyond any one classroom or even educational model. The context dependence of conceptual understanding is important, for example, because it allows conceptual change researchers to move beyond cataloging lists of student difficulties into perhaps understanding the processes underlying those difficulties, and thereby predicting and preventing future problems.

Fundamentally, all conceptual change research in engineering education works toward the same goal of improving engineering education by increasing student understanding. Each of the three categories identified in this chapter form an essential component of achieving that larger goal. The theories define and guide the measurements and curricular interventions, while providing constructs that allow them to be assessed critically, and potentially improved. By situating curricular innovations within a particular theoretical framework, researchers add to the potential transferability of our successes, and perhaps more importantly, provide some means of salvaging progress from failures. Theory without curricular intervention, however, would never reach the students. The art and science of designing and implementing curricula form a very real part of fostering conceptual change and should inform theory-focused research. Similarly, theory and curriculum-focused research is somewhat blind without reliable, valid measurements. If curricula form the connection between developing theory and affecting students, measurement completes the circuit by providing a flow of information back to the development of theory and curricula.

Finally, it is important to note that, despite differences in rhetoric and approach, the common goal of improving engineering education has likely already led to some implicit synthesis among the three categories. The curricular efforts of Steif and Krause, for example, have strong implications for theory that have not yet been teased out. Krause's investigations into the sources of students' misconceptions suggest a new tactic in investigating the structure, context sensitivity, and even definition of misconceptions in conceptual change research. Steif's findings that students' conceptual understanding improved when they were encouraged to emphasize the role of bodies in their explanations clarifies and supports potential lines of research investigating the role of language in forming and changing conceptual understanding (Ivarsson et al., 2002; Sinatra & Pintrich, 2003), as well as research searching for cognitive causes for conceptual difficulties (Chi, 2005). Likewise, Herman is predominantly theory-focused (e.g., Herman, Loui et al., 2011), but his research provided the foundation for the creation of a digital logic concept inventory (Herman & Loui, 2011). And Miller et al. (2011) began the design of their measurement-focused research by considering theories about how people learn in thermal and transport science.

One particularly exciting line of research is the development of "ontology training" (Slotta & Chi, 2006) to help students undergo conceptual change in thermal and transport sciences (Miller et al., 2011). This work is fundamentally curricular and theoretical, as well. It is explicitly and powerfully based on Chi's theoretical approach to conceptual change, which argues that students often struggle to learn science concepts because the category of emergent process is difficult to identify and create (Chi et al., 2012). Curricular interventions have been designed, implemented, and assessed using the previously mentioned measurements that were designed as part of the same effort. Work in this area is ongoing and has the potential benefit of being immediately applicable to new contexts owing to its integration of theory, curriculum, and measurement.

Conclusion

As we strengthen the link between the three branches of conceptual change research, we can find some common threads that should inform and direct future research efforts. First, all three branches of research have revealed that students' conceptual knowledge is sensitive to changes in contexts and students' prior knowledge. Students' ability to use conceptual knowledge depends on perceptual contexts such as visual diagrams or verbal cues, or it can depend on changes in social contexts. Second, there is often a blurry line between procedural and conceptual knowledge. Both Steif and Herman document how concepts are deeply embedded within their associated procedures. So it becomes unclear whether students' difficulties are conceptual or procedural in nature.

We propose three lines of future research. First, we will need to investigate how these changes in context affect how students' concepts change, and how students and experts can successfully switch between contexts. Second, we will need to investigate how social and motivational factors influence conceptual change and create "hot" cognition. Third, we will need to investigate whether there is indeed a difference between procedural and conceptual knowledge, and what precisely distinguishes these two types of knowledge.

Perceptual Contextualization of Conceptual Change

In a seminal study, Chi et al. (1981) demonstrated that novices and experts categorize physics problems differently: novices categorize based on physical features of the problems (incline planes, pulleys, etc.), whereas experts categorize based on conceptual features (conservation of energy, momentum, etc.). Although it is known that students rely more on perceptual features of a problem, it is not fully known how these perceptual cues affect students' formation and use of conceptual knowledge.

For example, in mechanics of materials, students often describe shear forces and stresses as vertical. This association of "shear" (concept) and "vertical" (perception) may be created because students are first introduced to the concept of shear forces when they learn about shear and bending moment diagrams, where the shear force is applied vertically on a horizontal beam (Montfort et al., 2012). Consequently, when students discuss shear forces, stresses, and strains, they often refer to them as vertical forces, stresses, and strains. When students transfer this conceptual knowledge to axially loaded members, where the member is often still drawn horizontally, some students are able to reason well about the shear strain, while using the nonexpert language of vertical forces and vertical shears. Would these same students' conceptual change about shear stress be hampered or hindered if the axially loaded members were always drawn vertically? Future research should examine how much the diagram conventions in various disciplines help or hinder conceptual change.

We propose that conceptual change interviews should use more isomorphic problems that share the same solution strategy and require the same conceptual knowledge but present different perceptual cues. These problems could clarify the degree to which students' access of conceptual knowledge is based on the specific surface features of a problem.

Future research should also investigate the effects of the proximity or sequencing of events on students' conceptual knowledge. For example, students often imitate the language of the interviewer. In some interviews, a student may never use a technical term until he or she *hears* the interviewer use the word, or until the student encounters a context that prompts his or her use of the term. After hearing the term, the student is more likely to use the term throughout the rest of the interview. Similarly, students' use of conceptual knowledge can also be affected by which problems they recently solved. For example, in the Boolean logic interviews of Herman, Loui et al. (2012), students often solved a problem with a methodology that they used to solve a recently completed problem. If these students were asked to solve an isomorphic problem later in the interview, they often used different strategies and reasoning to solve the problem.

Social Contextualization of Conceptual Change

Future conceptual change research should examine the effects of different social contexts on students' conceptual knowledge. These different environments can be social contexts of practice, or they can be emotional contexts.

Students need to be able to access and use their conceptual knowledge in a variety of social contexts of practice, including exams, homework, and office hours. Students' use of their conceptual knowledge may not be the same in these different contexts. In high-pressure, time-sensitive contexts such as exams, students may attempt to use a memorized procedure rather than conceptual knowledge to solve problems. In contrast, during office hours or in collaborative study sessions, students may more readily engage in reflection and the formation of conceptual knowledge. Interviews could experiment with comparing the quality of students' conceptual knowledge in pressured situations versus more reflective situations.

While most conceptual change research has focused on comparing students' (novices') conceptual knowledge with faculty's (experts') conceptual knowledge, little research has been done to understand how other groups of experts, such as practicing engineers, use their conceptual knowledge. Future research should examine the conceptual change of practicing engineers, and how they can apply their conceptual knowledge to standard concept questions.

Students' conceptual change can also be moderated by "hot," or "emotional," processing of information. For example, if a student is asked to solve a problem multiple times during an interview, he or she begins to question the interviewer's intentions and change his or her conceptual knowledge more often (right to wrong or wrong to right). The students' sense of doubt and uncertainty may motivate conceptual change. For another example, researchers have shown that students are less likely to change their conceptions when they have an emotional aversion to the scientific reasoning about a concept, such as evolution or the demotion of Pluto to a dwarf planet (Broughton, Sinatra, & Nussbaum, 2013; Sinatra & Broughton, 2011). Future research should examine how students' personal emotional contexts affect their conceptual change. Studies in this line of research might conduct pre-/post-intervention assessments of students' attitudes toward the topic, in addition to measuring students' conceptual change from the intervention.

Finally, the use of peer instruction has become a favored teaching technique for promoting conceptual change. However, because the nature and structure of conceptual knowledge is still being debated, the precise mechanism by which peer instruction promotes conceptual change has not been clearly articulated. It may be because the students must articulate their thoughts. Or is it because students have a lowered emotional resistance, and more trust in their peers? Is the promotion of conceptual change through peer instruction cognitive, emotional, social, or all of the above?

Distinguishing the Development of Procedural Knowledge and Conceptual Change

Engineers' work with models and abstractions of concepts is significant to the description of the field, and may be one of the key epistemological features distinguishing engineering from science (Pirtle, 2010; Vincenti, 1990). In many of these models and abstractions, the concepts cannot be fully understood apart from the procedures of the model. On one level, this is a methodological concern in that procedural familiarity may be hard to distinguish from increased conceptual understanding (Steif, 2004). If students become more adept at constructing free-body diagrams over time, for example, it may be difficult to determine whether this facility is due to conceptual change or simply to practice. Differentiating procedural knowledge from conceptual knowledge is also important, in that it may help to identify the knowledge and ways of thinking that are most important to include in the curriculum. A shared understanding of the differences between procedural and conceptual knowledge would facilitate discussions of the role of undergraduate education in preparing engineers.

Acknowledgments

The authors thank the U.S. National Science Foundation (NSF) for supporting the work described in this chapter through grants 0127806, 0227558, 0550169, 0618589, 0837749, and 0918531. Any opinions, findings, and conclusions or recommendations expressed in this material are those of the authors and do not necessarily reflect the views of the NSF. The authors also thank Paul Steif and Stephen Krause for reading sections of this draft and the anonymous reviewers whose careful critique strengthened the chapter.

References

Broughton, S. H., Sinatra, G. M., & Nussbaum, E. M. (2013). "Pluto has been a planet my whole life!" Emotions, attitudes, and conceptual change in elementary students learning about Pluto's reclassification. *Research in Science Education, 43*(2), 529–550.

Brown, J. S., Collins, A., & Duguid, P. (1989). Situated cognition and the culture of learning. *Educational Researcher, 18*(1), 32–42.

Brown, S., & Lewis, D. (2010). *Student understanding of normal and shear stress and deformations in axially loaded members.* Paper presented at American Society for Engineering Education Annual Conference & Exposition, Louisville, KY.

Brown, S., Montfort, D., & Findley, K. (2007). *Student understanding of states of stress in mechanics of materials.* Paper presented at American Society for Engineering Education Annual Conference & Exposition, Honolulu, HI.

Brown, S., Montfort, D., & Hildreth, K. (2008). An investigation of student understanding of shear and bending moment diagrams. In W. Aung, J. Mecsi, J. Moscinski, I. Rouse, & P. Willmot (Eds.), *Innovations 2008: World innovations in engineering education and research* (pp. 81–101). Redding, CT: Begell House.

Carey, S. (1985). *Conceptual change in childhood.* Cambridge, MA: MIT Press.

Chi, M. T. H. (2005). Commonsense conceptions of emergent processes: Why some misconceptions are robust. *The Journal of the Learning Sciences, 14*(2), 161–199.

Chi, M. T. H. (2009). Active-constructive-interactive: A conceptual framework for differentiating learning activities. *Topics in Cognitive Science, 1*(1), 73–105.

Chi, M. T. H., & Roscoe, R. D. (2002). The processes and challenges of conceptual change. In M. Limón & L. Mason. (Eds.), *Reconsidering conceptual change: Issues in theory and practice* (pp. 3–27). Dordrecht, The Netherlands: Kluwer Academic.

Chi, M. T. H., Roscoe, R. D., Slotta, J. D., Roy, M., & Chase, C. C. (2012). Misconceived causal explanations for emergent processes. *Cognitive Science, 36*(1), 1–61.

Chinn, C. A., & Brewer, W. F. (1993). The role of anomalous data in knowledge acquisition: A theoretical framework and implications for science instruction. *Review of Educational Research, 63*(1), 1–49.

diSessa, A. A. (2002). Why "conceptual ecology" is a good idea. In M. Limón & L. Mason (Eds.),

Reconsidering conceptual change: Issues in theory and practice (pp. 29–60). Dordrecht, The Netherlands: Kluwer Academic.

diSessa, A. A. (2004). Coherence versus fragmentation in the development of the concept of force. *Cognitive Science, 28*, 843–900.

diSessa, A. A. (2008). A bird's eye view of 'pieces' vs 'coherence' controversy. In S. Vosniadou (Ed.), *Handbook of conceptual change research* (pp. 35–60). Mahwah, NJ: Lawrence Erlbaum.

Downing, S. M. (2006). Twelve steps for effective test development. In S. M. Downing & T. M. Haladyna (Eds.), *Handbook of test development* (pp. 3–26). Mahwah, NJ: Lawrence Erlbaum.

Gilbert, J. (1995). *The role of models and modeling in some narratives in science learning.* Paper presented at Annual Meeting of the American Educational Research Association, San Francisco, CA.

Hake, R. R. (1998). Interactive-engagement versus traditional methods: A six-thousand-student survey of mechanics test data for introductory physics courses. *American Journal of Physics, 66*(1), 64–74.

Halldén, O., Haglund, L., & Strömdahl, H. (2007). Conceptions and contexts: On the interpretation of interview and observational data. *Educational Psychologist, 42*(1), 25–40.

Herman, G. L., & Loui, M. C. (2011). *Administering the Digital Logic Concept Inventory at multiple institutions.* Paper presented at American Society for Engineering Education Annual Conference & Exposition, Vancouver, BC, Canada.

Herman, G. L., Loui, M. C., Kaczmarczyk, L., & Zilles, C. (2012). Discovering the what and why of students' difficulties in Boolean logic. *ACM Transactions on Computing Education, 12*(1), 1–28.

Herman, G. L., Loui, M. C., & Zilles, C. (2011). Students' misconceptions about medium-scale integrated circuits. *IEEE Transactions in Education, 54* (4), 637–645.

Herman, G. L., Zilles, C., & Loui, M. C. (2012). Flip-flops in students' conceptions of state. *IEEE Transactions in Education, 55* (1), 88–98.

Herman, G. L., Zilles, C., & Loui, M. C. (2011). How do students misunderstand number representations? *Computer Science Education, 23* (3), 289–312.

Ivarsson, J., Schoultz, J., & Säljö, R. (2002). Map reading versus mind reading: Revisiting children's understanding of the shape of the earth.

In M. Limón & L. Mason (Eds.), *Reconsidering conceptual change: Issues in theory and practice* (pp. 77–99). Dordrecht, The Netherlands: Kluwer Academic.

Kaczmarczyk, L., Petrick, E., East, J. P., & Herman, G. L. (2010). *Identifying student misconceptions of programming.* Paper presented at Forty-First ACM Technical Symposium on Computer Science Education, Milwaukee, WI.

Kelly, J., Krause, S., & Baker, D. (2010). *A prepost topic assessment tool for uncovering misconceptions and assessing their repair and conceptual change.* Paper presented at ASEE/IEEE Frontiers in Education Conference, Washington, DC.

Krause, S., Decker, J. C., & Griffin, R. (2003a). *Using a materials concept inventory to assess conceptual gain in introductory materials engineering courses.* Paper presented at ASEE/IEEE Frontiers in Education Conference, Boulder, CO.

Krause, S., Decker, J. C., Niska, J., & Alford, T. (2003b). *Identifying student misconceptions in introductory materials engineering classes.* Paper presented at American Society for Engineering Education Annual Conference & Exposition, Nashville, TN.

Krause, S., Kelly, J., Corkins, J., Tasooji, A., & Purzer, S. (2009). *Using students' previous experience and prior knowledge to facilitate conceptual change in an introductory materials course.* Paper presented at ASEE/IEEE Frontiers in Education Conference, San Antonio, TX.

Krause, S., Kelly, J., Kurpius-Robinson, S., & Baker, D. (2010). *Effect of pedagogy on conceptual change in repairing misconceptions of differing origins in an introductory materials course.* Paper presented at American Society for Engineering Education Annual Conference & Exposition, Louisville, KY.

Krause, S., & Tasooji, A. (2007). *Diagnosing students' misconceptions on solubility and saturation for understanding of phase diagrams.* Paper presented at American Society for Engineering Education Annual Conference & Exposition, Honolulu, HI.

Krause, S., Tasooji, A., & Griffin, R. (2004). *Origins of misconceptions in a materials concept inventory from student focus groups.* Paper presented at American Society for Engineering Education Annual Conference & Exposition, Salt Lake City, UT.

Limón, M. (2002). Conceptual change in history. In M. Limón & L. Mason (Eds.) *Reconsidering conceptual change: Issues in theory and practice.* (pp. 259–289). Dordrecth, The Netherlands: Kluwer Academic.

Linstone, H. A., & Turoff, M. (Eds.). (2002). *The delphi method: Techniques and applications.* Reading, MA: Addison-Wesley.

Mayer, R. (2002). Understanding conceptual change: A commentary. In M. Limón & L. Mason (Eds.), *Reconsidering conceptual change: Issues in theory and practice.* Dordrecht, The Netherlands: Kluwer Academic.

Mazur, E. (1997). *Peer instruction.* Upper Saddle River, NJ: Prentice Hall Series in Educational Innovation.

Miller, R. L., Streveler, R. A., Olds, B. M., Chi, M. T. H., Nelson, M. A., & Geist, M. R. (2006). *Misconceptions about rate processes: Preliminary evidence for the importance of emergent schemas in thermal and transport sciences.* Paper presented at American Society for Engineering Education Annual Conference and Exposition, Chicago, IL.

Miller, R. L., Streveler, R. A., Yang, D., & Santiago Román, A. I. (2011). Identifying and repairing misconceptions in thermal and transport science: Concept inventories and schema training studies. *Chemical Engineering Education*, 45(3), 203–210.

Minstrell, J. (2001). Facets of students' thinking: Designing to cross the gap from research to standards-based practice. In K. Crowley, C. D. Schunn, & T. Okada (Eds.), *Designing for science: Implications for professional, instructional and everyday science* (pp. 415–443). Mahwah, NJ: Lawrence Erlbaum.

Montfort, D., Brown, S., & Pollock, D. (2009). An investigation of students' conceptual understanding in related sophomore to graduate-level engineering and mechanics courses. *Journal of Engineering Education*, 98(2), 111–129.

Montfort, D., Herman, G. L., Streveler, R., & Brown, S. (2012). *Assessing the application of three theories of conceptual change to interdisciplinary data sets.* Paper presented at ASEE/IEEE Frontiers in Education Conference, Seattle, WA.

Newcomer, J. L., & Steif, P. S. (2007). Gaining insight into student thinking from their explanations of a concept question. In *Proceedings of the 1st International Conference on Research in Engineering Education*, Honolulu, HI (pp. 1–8).

Norman, D. (1983). Some observations on mental models. In D. Genter & A. Stevens (Eds.),

Mental models (pp. 7–14). Hillsdale, NJ: Lawrence Erlbaum.

Pellegrino, J., Chudowsky, N., & Glaser, R. (Eds.). (2001). *Knowing what students know: The science and design of educational assessment*. Washington, DC: The National Academies Press.

Piaget, J. (1970). *Science of education and the psychology of the child*. New York, NY: Orion.

Pirtle, Z. (2010). How the models of engineering tell the truth. In I. van de Poel & D. E. Goldberg (Eds.), *Philosophy and engineering: An emerging agenda* (pp. 95–108). New York: Springer.

Purzer, S., Krause, S., & Kelly, J. (2009). *What lies beneath the materials science and engineering misconceptions of undergraduate students?* Paper presented at American Society for Engineering Education Annual Conference & Exposition, Austin, TX.

Robbins, P., & Aydede, M. (Eds.). (2009). *The Cambridge handbook of situated cognition*. New York, NY: Cambridge University Press.

Säljö, R. (1999). Concepts, cognition and discourse: From mental structures to discursive tools. In W. Schnotz, S. Vosniadou, & M. Carretero (Eds.), *New perspectives on conceptual change* (pp. 81–90). Oxford: Pergamon Press.

Santiago Román, A. (2009). *Fitting cognitive diagnostic assessment to the concept assessment tool for statics*. Unpublished doctoral dissertation, Purdue University.

Santiago Román, A. I., Streveler, R. A., Steif, P., & DiBello, L. (2010). *The development of a Q-matrix for the concept assessment tool for statics*. Paper presented at American Society for Engineering Education Annual Conference & Exposition, Louisville, KY.

Sinatra, G. M. (2002). Motivational, social, and contextual aspects of conceptual change: A commentary. In M. Limón & L. Mason (Eds.), *Reconsidering conceptual change: Issues in theory and practice* (pp. 77–99). Dordrecht, The Netherlands: Kluwer Academic.

Sinatra, G. M., & Broughton, S. H. (2011). Bridging reading comprehension and conceptual change in science education: The promise of refutation text. *Reading Research Quarterly*, 46(4), 374–393.

Sinatra, G. M., & Pintrich, P. R. (Eds.). (2003). *Intentional conceptual change*. Mahwah, NJ: Lawrence Erlbaum.

Slotta, J., & Chi, M. T. H. (2006). Helping students understand challenging topics in science through ontology training. *Cognition and Instruction*, 24(2), 261–289.

Steif, P. S. (2002a). *Courseware for problem solving in mechanics of materials*. Paper presented at American Society for Engineering Education Annual Conference & Exposition, Montreal, ON, Canada.

Steif, P. S. (2002b). *Enriching statics instruction with physical objects*. Paper presented at American Society for Engineering Education Annual Conference & Exposition, Montreal, ON, Canada.

Steif, P. S. (2004). *An articulation of the concepts and skills which underlie engineering statics*. Paper presented at ASEE/IEEE Frontiers in Education Conference, Savannah, GA.

Steif, P. S., & Dantzler, J. A. (2005). A statics concept inventory: Development and psychometric analysis. *Journal of Engineering Education*, 94(4), 363–371.

Steif, P. S., & Dollar, A. (2003). *A new approach to teaching and learning statics*. Paper presented at American Society for Engineering Education Annual Conference & Exposition, Nashville, TN.

Steif, P. S., Dollar, A., & Dantzler, J. (2005). *Results from a statics concept inventory and their relationship to other measures of performance in statics*. Paper presented at ASEE/IEEE Frontiers in Education Conference, Indianapolis, IN.

Steif, P. S., & Gallagher, E. (2004). *Use of simplified FEA to enhance visualization in mechanics*. Paper presented at American Society for Engineering Education Annual Conference & Exposition, Salt Lake City, UT.

Steif, P. S., & Hansen, M. (2006). Comparisons between performances in a statics concept inventory and course examinations. *International Journal of Engineering Education*, 22(5), 1070–1077.

Steif, P. S., Lobue, J. M., Kara, L. B., & Fay, A. L. (2010). Improving problem solving performance by inducing talk about salient problem features. *Journal of Engineering Education*, 99(2), 135–142.

Streveler, R. A., Miller, R. L., Santiago Román, A. I., Nelson, M. A., Geist, M. R., & Olds, B. M. (2011). Using the "assessment triangle" as a framework for developing concept inventories: A case study using the thermal and transport concept inventory. *International Journal of Engineering Education*, 27 (5), 1–17.

Taber, K. S. (2001). The mismatch between assumed prior knowledge and the learner's conceptions: A typology of learning impediments. *Educational Studies*, 27(2), 159–171.

Taber, K. S. (2004). Learning quanta; Barriers to simulating transitions in student understanding of orbital ideas. *Science Education*, 89, 94–116.

Vincenti, W. (1990). *What engineers know and how they know it: Analytical studies from aeronautical history*. Baltimore, MD: The Johns Hopkins University Press.

Vosniadou, S. (1994). Capturing and modeling the process of conceptual change. *Learning and Instruction*, 4, 45–69.

Vosniadou, S. (2007). The conceptual change approach and its reframing. In S. Vosniadou, A. Baltas, & Z. Vamvakoussi (Eds.), *Reframing the conceptual change approach in learning and instruction* (pp. 1–15). Amsterdam: Elsevier.

Vosniadou, S., & Brewer, W. F. (1992). Mental models of the earth: A study of conceptual change in childhood. *Cognitive Psychology*, 24, 535–585.

Vosniadou, S., Vamvakoussi, Z., & Skopeliti, I. (2008). The framework theory approach to the problem of conceptual change. In S. Vosniadou (Ed.), *International handbook of research on conceptual change* (pp. 3–34). New York: Routledge.

Wenger, E. (1998). *Communities of practice: Learning, meaning, and identity*. New York, NY: Cambridge University Press.

CHAPTER 6

Engineers as Problem Solvers

David H. Jonassen

Engineers as Problem Solvers

Evers, Rush, and Berdrow (1998) identify numerous disconnects between skills acquired in college and those required of the workplace. Among the most important skills that ABET Inc., the primary engineering accreditation institution in the United States, has identified for the preparation of engineers are the abilities to identify, formulate, and solve workplace engineering problems and to function on multidisciplinary teams. Learning to solve workplace problems is an essential learning outcome for any engineering graduate. Every engineer is hired, retained, and rewarded for his or her ability to solve problems. However, engineering graduates are ill prepared to solve complex, workplace problems (Jonassen, Strobel, & Lee, 2006).

The research and recommendation in this chapter are a synthesis of fifteen years of research on problem solving, culminating in the book *Learning to Solve Problems: A Handbook for Designing Problem-Solving Learning Environments* (2010), Routledge.

Problem solving from a cognitive perspective has been the primary focus of my research for the past decade and a half. My theory differs from traditional theories of problem solving in that I argue there are different kinds of problems that vary between contexts. The kinds of problems that practicing engineers solve are different from the problems that most undergraduate science, technology, engineering, and mathematics (STEM) students learn to solve. In most undergraduate classes, students learn to solve textbook problems that are constrained and well structured, with known solution paths and convergent answers (capstone courses are an exception). Workplace problems, on the other hand, tend to be ill structured and unpredictable because they possess conflicting goals, multiple solution methods, non-engineering success standards, non-engineering constraints, unanticipated problems, distributed knowledge, and collaborative activity systems (Jonassen et al., 2006). Learning to solve classroom problems does not effectively prepare engineering graduates to solve workplace

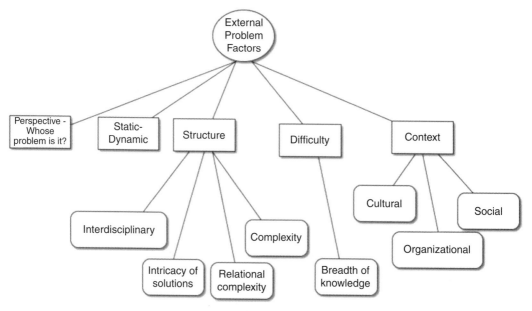

Figure 6.1. Differences in problems.

problems. To prepare engineering graduates, it is necessary to articulate the differences between educational problems and workplace problems. To do that, I first describe how problems vary.

How Does Problem Solving Vary?

I have argued that there are different kinds of problems, so learning to solve different kinds of problems requires different kinds of learning activities. Traditional models of problem solving (Bransford & Stein, 1984; Ploetzner, Fehse, Kneser, & Spada, 1999) define problem solving as phases that include problem identification, information searching, solution identification, solution evaluation, and monitoring. Phase models of problem solving tend to treat all problems the same. However, as illustrated in Figure 6.1, problems vary in several ways. Note that problem solving varies also based on numerous personal (internal factors) characteristics of the problem solver, which are beyond the scope of this chapter.

Structuredness of Problems

The most descriptive dimension of problem solving is structuredness. Problems vary along a continuum from well structured to ill structured (Arlin, 1989; Jonassen, 1997; Newell & Simon, 1972). It is important to emphasize that the level of structuredness is not a dichotomous attribute of the problem. Rather, it is better represented as a continuum from very well-structured to very ill-structured.

The most commonly encountered problems, especially in formal educational contexts, are well-structured problems. Typically found at the end of textbook chapters, well-structured problems present all of the information needed to solve the problems in the problem representation; they require the application of a limited number of regular and circumscribed rules and principles that are organized in a predictive and prescriptive way; and they have knowable, comprehensible solutions where the relationships between decision choices and all problem states are known or are probabilistic (Wood, 1983). These problems have also been referred to as transformation problems (Greeno, 1980) that consist of a well-defined initial state, a known goal state, and a constrained set of logical operators.

Ill-structured problems, on the other hand, are the kinds of problems that are encountered in everyday life and work, so they are typically emergent and not self-contained. Because they are not constrained by the content domains being studied in classrooms, their solutions are not predictable or convergent. Ill-structured problems usually require the integration of several content domains, that is, they are usually interdisciplinary in nature. Workplace engineering problems, for example, are ill structured because they possess conflicting goals, multiple solution methods, non-engineering success standards, non-engineering constraints, unanticipated problems, distributed knowledge, collaborative activity systems, and multiple forms of problem representation (Jonassen et al., 2006); they possess multiple solutions, solution paths, or no solutions at all (Kitchner, 1983); they possess multiple criteria for evaluating solutions, so there is uncertainty about which concepts, rules, and principles are necessary for the solution and how they are organized; and they often require learners to make judgments and express personal opinions or beliefs about the problem. One or more of the problem elements in ill-structured problems are unknown or not known with any degree of confidence (Wood, 1983), that is, they possess a degree of intransparency (the more we do not know about the problem). For example, predicting weather is considered an extremely difficult task because it contains a great number of variables about which forecasters are uncertain. Ill-structured problems contain higher levels of intransparency.

Along the continuum from well-structured to ill-structured problems are several different kinds of problem that are described later. For example, story (word) problems tend to be the most well-structured, whereas decision making and troubleshooting are more ill structured, with complex planning and design problems being the most ill structured. An underlying assumption of instructional course design is that different learning outcomes call on different sets of learning skills. For example, learning concepts requires different skills than learning rules. Analogously, learning to solve different kinds of problems calls on different sets of skills. For example, learning to solve story problems calls on different skills than learning to solve design problems. Traditionally, information processing theories aver that "in general, the processes used to solve ill-structured problems are the same as those used to solve well structured problems" (Simon, 1978, p. 287). However, more recent research in situated and everyday problem solving makes clear distinctions between thinking required to solve well-structured problems and ill-structured problems. Allaire and Marsiske (2002) found that measures that predict well-structured problem solving could not predict the quality of solutions to ill-structured, everyday problems. "Unlike formal problem solving, practical problem solving cannot be understood solely in terms of problem structure and mental representations" (Scribner, 1986, p. 28), but rather include aspects outside the problem space, such as environmental information or goals of the problem solver. Hong, Jonassen, and McGee (2003) found that solving ill-structured problems in an astronomy simulation called on different skills than well-structured problems, including metacognition and argumentation. Argumentation is a social and communicative activity that is an essential form of reasoning in solving ill-structured, everyday problems (Chapman, 1997). Jonassen and Kwon (2001) showed that communication patterns in teams differed when solving well-structured and ill-structured problems. Finally, groups of students solving ill-structured economics problems produced more extensive arguments than when solving well-structured problems because of the importance of generating and supporting alternative solutions when solving ill-structured problems (Cho & Jonassen, 2003). Clearly more research is needed to substantiate these differences, yet it appears that well-structured and ill-structured problem-solving methods engage different intellectual skills.

Complexity (Difficulty) of Problems

There exist many descriptions of problem complexity or difficulty. At the base level, problem complexity is a function of external factors, such as the number of issues, functions, or variables involved in the problem; the number of interactions among those issues, functions, or variables; and the predictability of the behavior of those issues, functions, or variables. According to Meacham and Emont (1989), problems vary in terms of complexity (how clearly the problem's initial and goal states are identified). Wood (1985) suggested that there are three kinds of problem complexity: component complexity, coordinative complexity, and dynamic complexity. Component complexity describes the number of distinct acts required to solve the problems along with the diversity of kinds of information needed to perform these acts. Jonassen and Hung (2008) described this attribute as intricacy of problem solutions. Coordinative complexity described the variety of relationships among problem-solving acts, otherwise referred to as relational complexity (Jonassen & Hung, 2008). Halford, Wilson, and Phillips (1998) described relational complexity as the number of relations that need to be processed in parallel during a problem-solving process. The more complex the relations in a problem, the more cognitive processing load is required during problem solving. Dynamic complexity describes changes in those relationships over time. With workplace problems, it is often difficult to determine what the problem is because problems change in light of new developments (Roth & McGinn, 1997). That is, many problems are dynamic, because their conditions or contexts change over time, altering the nature of the problem. Jonassen and Hung (2008) also identified the breadth of knowledge required to solve the problem as a factor in problem complexity. For example, designing a football stadium

equipped with a retractable roof is much more complex than designing a simple aluminum warehouse because it involves much more advanced architecture, structural engineering, civil engineering, and environmental and other related knowledge.

There is a significant correlation between complexity and structuredness. Ill-structured problems are more difficult to solve, in part because they tend to be more complex. The more complex that problems are, the more difficulty students have to choose the best solution method (Jacobs, Dolmans, Wolfhagen, & Scherpbier, 2003). Also, problem solvers represent complex problems in different ways that lead to different kinds of solutions (Voss, Wolfe, Lawrence, & Engle, 1991). Novices over-rely on quantitative representations (equations) without understanding the nature of the problem.

Although complexity and structuredness invariably overlap, complexity is more concerned with how many components are represented implicitly or explicitly in the problem, how those components interact, and how consistently they behave. Complexity has direct implications for working memory requirements. The more complex a problem, the greater the cognitive load that must be distributed across problem elements. Given the limitations of working memory (Sweller, 1988), complex problems can become very demanding. Most well-structured problems, such as textbook math and science problems, are not very complex. They involve a constrained set of factors or variables. Although ill-structured problems tend to be more complex, well-structured problems can be extremely complex and ill-structured problems fairly simple. For example, applying a multistage computer algorithm that performs a number of routine tests on a user's design is a complex but well-structured problem and a low drag body shape for an experimental vehicle is an ill-structured but simple problem.

An important research agenda for engineering educators is the ideal levels of structuredness and complexity for undergraduate engineering students. Problem-solving instruction most often begins with well-structured textbook problems that are usually taught via worked examples. How to support learning to solve workplace problems is an important issue. Perhaps the most common approach is to scaffold the development of problem-solving skills by beginning with well-structured textbook problems and progressing to more complex and ill-structured problems. Not surprisingly, Reisslein, Sullivan, and Reisslein (2007) showed that faster transitioning from worked examples to independent problem solving in electrical engineering favored the high prior knowledge students, while lower prior knowledge students benefitted from slower transitioning. This approach is used at Olin College, which delivers the majority of their engineering curricula via problem-based learning. Students spend their first year solving simpler, predetermined problems and engaging in hands-on activities. Problems of increasing complexity are integrated throughout the curriculum. Students spend their fourth year solving an authentic engineering problem that is sponsored by an established engineering client.

Role of Problem Context

The structuredness of problems is significantly related to the context in which the problem is encountered. That is, well-structured problems tend to be encountered more often in formal educational contexts that emphasize correctness of solutions. Ill-structured problems tend to be more embedded in and defined by everyday or workplace contexts, making them more subject to belief systems that are engendered by social, cultural, and organizational drivers in the context (Jonassen, 2000; Meacham & Emont, 1989; Smith, 1991). Workplace engineering problems are made more ill structured by the context that often creates unanticipated problems, conflicting goals, and uncertain solution methods (Jonassen et al, 2006).

As illustrated in Figure 6.1, any situation or context may be constrained by cultural (history, belief systems, customs), sociopolitical (power relations), or organizational factors (administration, budgets, etc.). Workplace problem-solving methods are ill structured because they present multiple belief systems embedded in the problem-solving context.

What Kinds of Problems Do Engineers Solve?

The kinds of workplace problems that engineers solve everyday are unlike the kinds of problems that engineering students most commonly learn to solve, with the exception of capstone courses (also see Chapter 7, this volume). Jonassen (2000) articulated eleven different kinds of problems that vary largely in the degree of structuredness (well-structured to ill-structured). Those problem types include:

- Algorithms (algorithmic procedures), including mathematical operations
- Story problems, shallow contexts supporting plug-and-chug solution methods
- Rule using/rule induction problems, how to adapt a process or deduce how something is supposed to work
- Decision-making, choosing among limited alternatives
- Troubleshooting, identifying fault state and repairing system
- Diagnosis-solution problems, troubleshooting causes and deciding among alternative solutions
- Strategic performance problems, real-time, high-pressured, and high-stakes decision making
- Policy problems, complex issues determined by multiple perspectives
- Planning problems, project management and planning of systems
- Design problems, typically the most complex and ill structured of all problems
- Dilemmas, which frequently have no acceptable solution

Workplace engineering problems share a number of attributes. Jonassen et al. (2006)

identified eleven attributes of workplace engineering problems. Those attributes include:

- Workplace problems are ill structured; they have no obviously correct solution.
- Ill-structured problems are often composed of numerous well-structured problems that do possess clear methods and success standards.
- Workplace engineering problems often have multiple, often conflicting goals.
- Workplace engineering problems have many varied solution strategies and methods.
- Success is rarely measured entirely by engineering standards; most often it is under budget and on time.
- Most problem constraints are non-engineering.
- Problem-solving knowledge is distributed among team members.
- Most problems require extensive collaboration; engineers seldom solve problems individually.
- Engineers rely primarily on experiential knowledge, which is what engineering students lack.
- Engineering problems often encounter unanticipated problems, making the initial problem that much more complex and dynamic.
- Engineers use multiple forms of problem representation, rarely using equations.

The three most common kinds of problems that practicing engineers solve include decision making, troubleshooting, and design.

Decision-making Problems. Decision-making problems require individuals or social groups to decide which solution, issue, or course of action to pursue. Choice-type decisions include decisions such as:

- Which material do we select for constructing this component? Are the material properties acceptable for this component?
- Which contractors do we use for supplying this component?

- Which chemical should we use for this reaction?

Acceptance decisions might include:

- Should we admit a specific individual to our professional association?

Engineers are often charged with making evaluation decisions, such as:

- How much effort should we expend on developing this new product?
- How much should we charge for providing this new service?

Though decision problems typically have a limited number of viable solutions, the number of factors to be considered in deciding among those solution options can be very complex.

Decision making often represents the core processes in more complex and ill-structured problems (Jonassen, 2012). Decision making is a critical component within more complex problems such as diagnosis, negotiation, design, situation assessment, and command and control (Means, Salas, Crandall, & Jacobs, 1993). Although some problems require only decision making, other more ill-structured and complex problems entail sequential or iterative decision-making processes. Imagine a planning problem in which you were planning a conference for a few thousand delegates (e.g., ASEE). How many decisions would you have to make before, during, and even after the convention? As described later, design problem solving can also be dissociated into multiple decisions. Decisions are central to virtually every kind of human cognitive processing and problem solving (Jonassen, 2012); that is, most kinds of complex, ill-structured problems can be deconstructed into various decisions.

The most commonly described, normative approaches to decision-making include rational choice and cost–benefit models of decision making that emphasize comparing, contrasting, and weighing the values associated with alternate solutions. Using techniques such as decision matrices, decision makers assign quantitative values to solution options and then select the option that has the highest value. Interestingly, decision makers often fail to follow the prescriptions of these normative models (Hogart & Kunreuther, 2005; Kahneman, Slovic, & Tversky, 1982). Rational choice methods of decision making are based on the concept of unbounded rationality, which does not describe how real people think. People are not consistently rational in their thinking, not even engineers.

Decision making is often less rational than we believe. Rather than basing decisions on probabilistic calculations, people often base decisions on explanations about decision options. Why? Decisions are rarely content neutral or domain neutral. People typically know a great deal about decision options and harbor knowledge and beliefs about them. Rettinger and Hastie (2001) argue that the more important the decision, the more elaborate will be the narrative processing. Important decisions elicit elaborate story construction, especially with personal experience. Jonassen et al. (2006) found that prior knowledge and experience play major roles in engineering problem solving. Decision makers, especially in complex policy problems, often construct story-based scenarios that describe possible outcomes when making decisions, rather than rationally analyzing alternatives. Companies considering major operating changes construct scenarios that predict the outcomes of different options. That is, decision makers often base their decisions on hypotheticals that are connected by causally related elements and events. For a more detailed analysis of decision-making processes, see Jonassen (2012).

Troubleshooting Problems. Many engineers are regularly engaged in troubleshooting problems, such as

- Debugging software programs
- Discovering why a material failed
- Determining why a process produces spurious results

Maintaining automobiles, aircraft, or any complex system requires troubleshooting

(diagnosis). When a system is not functioning properly, its symptoms have to be diagnosed and matched with the user's knowledge of various fault states (Jonassen & Hung, 2006). Troubleshooters use symptoms to generate and test hypotheses about different fault states.

The most common method for teaching troubleshooting is a procedural demonstration of a troubleshooting process (here's how to troubleshoot). While students receiving procedural demonstrations perform more accurately (Swezey, Perez, & Allen, 1988), they were less able to transfer their skills to similar tasks (Morris & Rouse, 1985). Learners receiving procedural instruction are less able to transfer troubleshooting skills because their mental models (a.k.a. problem schemas) for the system being diagnosed are deficient. In addition to troubleshooting procedures, learners must construct multiple kinds of knowledge about the system, including:

- Domain knowledge, including the general theories and principles on which the system or device was designed.
- System or device knowledge, an understanding of "(1) the structure of the system, (2) the function of the components within the system, and (3) the behavior of those components as they interact with other components in the system" (Johnson & Satchwell, 1993, p. 80), and the flow control within the system (Zeitz & Spoehr, 1989).
- Strategic knowledge, knowing when and why to perform certain tests plays an essential role in troubleshooting by reducing the problem space, isolating the potential faults, and testing and evaluating hypotheses and solutions (Johnson, Flesher, & Chung, 1995).
- Experience and recall of historical information is the most frequent strategy for failure diagnosis (Konradt, 1995).

Troubleshooting (diagnosis) among skilled practitioners is less of a diagnostic process than it is a matter of pattern recognition.

Experienced practitioners recognize a pattern of conditions or symptoms and base their diagnosis on it. Experienced physicians, for example, fire "illness scripts" when diagnosing medical problems.

Jonassen and Hung (2006) recommend troubleshooting instruction be focused around a complex models of the system being troubleshot, including

- Pictorial layer, containing pictures of the device or system as it exists
- Topographic layer, illustrating the components of the system, their locations, and their interconnections
- State layer, conveying normal states or values for each component and a symptom overlay conveying symptoms associated with each component malfunction
- Functional layer, illustrating and describing the flows of entities through the system based on causal effects
- Strategic layer, rule-based representations of alternative decisions regarding the states described on the state layer, including diagnostic heuristics that support fault finding
- Action layer, including descriptions of procedures for conducting various tests or operations (a job aid or just-in-time instruction).

Needless to say, developing such a sophisticated level of system understanding requires extensive learning and experience. These layers, however, could constitute the structure of a learning system designed to support troubleshooting in which students troubleshoot problems that are presented in a simulator that requires learners first to select a troubleshooting action; then select a fault hypothesis; predict the probability that the hypothesis he or she has chosen is actually the fault; and, based on the results of that action, interpret whether the hypothesis is consistent with results. In support of troubleshooting efforts, a case library of similar engineering experiences is provided (see Prior Experiences, later in the chapter). The important principle is that problems solving cannot be learned from demonstrations alone. Experience, especially failure, is essential to learning how to solve just about any kind of problem. This model represents a cognitive apprenticeship in engineering troubleshooting (Brown, Collins, & Duguid, 1989).

Design Problems. Probably the most common kind of problem regularly solved by engineers is design. Mechanical engineers design equipment, parts, and systems. Biological engineers design medical treatments and related equipment. Electrical engineers design electronic devices, circuits, and systems. Civil engineers design bridges, infrastructure, and transportation systems. Design is an extremely common activity among engineers. Whether an electronic circuit, a mechanical part, or a new manufacturing system, design requires applying a great deal of domain knowledge with a lot of strategic knowledge resulting in an original design. Design problems are typically the most complex and ill structured of all problems. Despite the apparent goal of finding an optimal solution within determined constraints, design problems usually have vaguely defined or unclear goals with unstated constraints. They possess multiple solutions, with multiple solution paths. Perhaps the most difficult aspect of design problems is that they possess multiple criteria for evaluating solutions, and these criteria are often unknown.

Engineering design is most often conceived and conveyed as a process. For example, the engineering design process includes the following phases according to Dym and Little (2004):

1. In the *problem definition*, from the client statement, clarify objectives, establish user requirements, identify constraints, and establish functions of product by providing a list of attributes.
2. In the *conceptual design* phase, establish design specifications and generate alternatives.

3. In the *preliminary design*, create a model of design and test and evaluate the conceptual design by creating morphological charts or decision matrices.

4. During the *detailed design*, refine and optimize the chosen design.

5. For the *final design*, document and communicate the fabrication specifications and the justifications for the final design.

Such models of design assume that design is a predictable process that will result in an optimal solution. This is a questionable assumption for a few reasons. First, the goal of most designers is not an optimal solution. Because design problems usually have vaguely defined or unclear goals, multiple criteria for evaluating solutions, and many unstated constraints that must be discovered during the design process, we can rarely know what an optimal solution is. During the problem definition stage, engineers analyze constraints in order to develop goals that represent an optimal solution. However, rather than optimizing a solution, designers most often seek to satisfice (Simon, 1955), a strategy that attempts to meet criteria for adequacy, rather than identifying an optimal solution. Second, designers seldom perform all of the activities defined by normative design processes. Although experienced designers do use engineering design processes to some degree, they also rely extensively on their prior experiences to recommend solutions and solution processes.

Design is an iterative process of decision making and model building. "The principal role of the designer . . . is to make decisions. Decisions help to bridge the gaps between idea and reality. . . . , decisions serve as markers to identify the progression of the design from initiation to implementation to termination" (Marston & Mistree, 1997, p. 1). Most design decisions are based on multiple constraints and constraint operations in the design space, not an agreed upon set of rules and heuristics. The design process consists totally of reasoning about constraints in order to determine parameter values (Brown & Chandrasekaran, 1989). Gross (1986) introduced the idea of design as constraint exploration. Constraints are the formal and informal "rules, requirements, conventions, and principles that define the context of learning" (p. 10). Designing is a process of exploring and expressing constraints that include operations such as describing and structuring constraints and objectives, exploring fixes, resolving conflicts, and comparing alternatives (Gross, Ervin, Anderson, & Fleisher, 1988). Jonassen (2008) described design as iterative cycles of decision-making based on constraints, beliefs, and biases. With each decision, the degrees of freedom in the design problem space are reduced, until identifying the engineering solution that satisfices. Therefore, it is important to engage engineering students in design activities in which they construct arguments (theoretical, technical) to support their design decisions.

What Kinds of Problems Do Engineering Students Learn to Solve?

In engineering classrooms, students most commonly learn to solve story (word) problems. Story problems typically present a set of variables embedded within a shallow story context. Story problems are normally solved by identifying key values in the short scenario, selecting the appropriate equation, applying the equation to generate a quantitative answer, and hopefully checking their responses (Sherrill, 1983). Despite our intentions, learners usually employ a tactical, problem avoidance strategy to solving story problems:

- Search for key words.
- Select equation based on key words.
- Translate relationships about unknowns into equations.
- Solve equations to find the value of the unknowns.

Referred to as plug-and-chug, this approach may yield verifiably correct answers but usually results in the absence of conceptual understanding of the concepts and principles represented in the problem (Catrambone & Holyoak, 1989; Gick & Holyoak, 1983; Ross, 1987, 1989). Learners too often fail to recall or reuse examples appropriately because their retrieval is based on a comparison of the surface features of the examples with the target problem, not their structural features, whereas experienced problem solvers represent problems in terms of their principles, emphasizing conceptual understanding. Successful problem solving requires the construction a conceptual model of the problem and the application of solution plans that are based on those models. It is the quality of their conceptual models that most influences the ease and adequacy with which the problem can be solved. Those conceptual models are mental representations of the pattern of information that is represented in the problem (Riley & Greeno, 1988). Each kind of problem (e.g., work-energy, kinematics, rotational dynamics in physics) requires a conceptual model that describes meaning for each of the entities in the problem, the causal relationships among those entities, as well as the equations needed to solve the problem. When students try to understand a problem in only one way, especially when that way conveys no conceptual information about the problem, students do not understand the underlying systems they are working in. So, it is necessary to help learners to construct a qualitative model of the problem as well as a quantitative. Qualitative models both constrain and facilitate the construction of quantitative representations (Ploetzner & Spada, 1998). Ploetzner et al. (1999) showed that when solving physics problems, qualitative problem representations are necessary prerequisites to learning quantitative representations. Qualitative representation is a missing link in novice problem solving.

Even if students learn to construct rich conceptual models for each kind of problem they are solving, it is doubtful that those models or skills will transfer to complex and ill-structured workplace problems. Dunkle, Schraw, and Bendixen (1995) concluded that performance in solving well-defined problems is independent of performance on ill-defined tasks, with ill-defined problems engaging a different set of epistemic beliefs.

Preparing Engineering Students for Workplace Problem Solving

Because engineering students learn to solve problems that are unlikely to transfer to workplace problem solving, engineering educators must adopt new pedagogies if they are committed to enabling their graduates to become effective engineers. The obvious pedagogy that needs to be adopted is problem-based learning (PBL, see Chapter 8 by Kolmos and Erik De Graaff, this volume). PBL is an instructional methodology. The primary goal of PBL is to enhance learning by requiring learners to solve problems. PBL is:

- Problem-focused, in which learners begin learning by focusing on an authentic, ill-structured problem. The content and skills to be learned are organized around problems, rather than as a hierarchical list of topics, so there is a reciprocal relationship between knowledge and the problem.
- Student-centered, because faculty cannot dictate learning.
- Self-directed, where students individually and collaboratively assume responsibility for generating learning issues and processes through self-assessment and peer assessment and access their own learning materials. Required assignments are rarely made.
- Self-reflective, where learners monitor their understanding and learn to adjust strategies for learning (Hung, Jonassen, & Liu, 2008).

PBL has become more prominent in engineering education. One of the major differences in PBL implementations is the scope of the problem-based program, which is a measure of an institution's commitment to problem-based learning. At the one end of the implementation continuum, numerous engineering educators have described the development and implementation of course modules. A number of PBL implementations have sought to convert entire semester-long courses into PBL. For example, Henry, Jonassen, Winholtz, and Khanna (2010) implemented a PBL version of a materials science course in mechanical engineering. This course requires students to solve seven authentic problems throughout the semester. Such course-level implementations present multiple difficulties for students, including an apparent lack of topical structure in the course, problems in learning to collaborate in groups, and the lack of clear direction in how to study. They perceived a serious disconnect between the problems and the exams. Course-level implementation of PBL may not prove to be effective because changing expectations of students who have rehearsed strategies for studying for and succeeding in traditional courses is very demanding.

Capstone engineering courses are often implemented as a form of PBL. Capstone courses attempt to apply what students have learned in their first three plus years of study to an authentic workplace problem. Most engineering programs include capstone courses, which are generally perceived as important and effective.

A few engineering programs have implemented PBL throughout their curriculum. PBL has been implemented as a partial strategy for Mechanical Engineering and Biomedical Engineering at Technische Universiteit Eindhoven (Perrenet, Bouhuijs, & Smits 2000). The oldest curricular implementation of PBL in engineering is offered by Aalborg University in Denmark. Faculty members meet with industry partners to solicit everyday problems, which they redact for the students. Based on those problems, the students determine the project's content and structure with approval of the faculty supervisors. A significant benefit of this method is the very high retention rates among students.

Impediments to Workplace Problem Solving in Engineering Education

Innovation in engineering education, as it is in all disciplines, is a vexing change problem. Changing pedagogical strategies presents multiple challenges and risks to faculty members. Engineering faculty members have learned and practiced teaching content, principles, and theories. Most faculty members argue that they would be remiss if they did not expose students to ideas that they may use in the future. Most faculty members also believe that students need to learn the content well before they can apply it.

The implication of curricular reform poses an important question: Should engineering students learn content about engineering or should they learn to solve engineering problems? Clearly they learn to do both; however, learning by solving problems results in fundamentally different ways of knowing and working. Engineering students think like students, not engineers, challenging instructors to clarify and simply the content they are distributing. If students are to learn to think like engineers, they must be challenged to solve authentic, complex problems.

Supporting Learning How to Solve Workplace Problems

Learning how to solve workplace problems in engineering classrooms represents challenges to students and faculty alike. The next section describes some of the components that constitute problem-based learning environments (PBLEs; Jonassen, 2011).

Problem to Solve

The focus of any problem-based learning is the problem that students are required to solve. All learning emerges from understanding and learning to solve the problem. The problem to be solved should be engaging, but should also address the curricular issues required by the curriculum. PBLEs most often present the problem to be solved as a story, embedding relevant information in the story context (Jonassen, 2011).

Designers of problem-solving learning environments assume that the problem to be solved should be complex and authentic. There are two broad conceptions of authenticity: preauthentication and emergent authenticity. Preauthentication refers to analyzing activity systems and attempting to simulate an authentic problem in a learning environment that students solve. Preauthentication is what Barab and Duffy (2000) refer to as a practice field, in which students can practice learning how to function in a disciplinary field by solving what engineering faculty members deem authentic problems. Emergent problems occur during practice within a disciplinary field, where problems are embedded in authentic settings, allowing students to learn a skill by engaging in the activities germane to that field (Barab, Squire, & Dueber, 2000). Service learning initiatives normally engage students in solving emergent problems. PBLEs usually engage students in solving preauthenticated problems, which requires that the designer work with experienced practitioners to articulate the most authentic contexts.

To support learning how to solve authentic, workplace problems, the following pedagogical supports may be included in any PBLE.

Case Studies

After presenting the problem to solve, exposing students to similar case studies may be useful (also see Chapter 9 by David and Yadav, this volume). In case studies, students study an account (usually narrative) of a problem that was previously experienced. Frequently guided by questions, students analyze the situation and processes and evaluate the methods and solutions. This analysis is usually ex post facto. In most case studies, students are not responsible for solving the problems, only for analyzing how others solved the problems and engaging in what-if (counterfactual) thinking. The goals of the case study method are to embed learning in authentic contexts that requires students to apply knowledge rather than acquire it. Case studies are examples of ill-structured problems that may be used to help students understand more complex and ill-structured problems. That is, students can analogically compare case studies with complex and ill-structured problems to solve in order to construct problem schemas and consider alterative perspective and solutions.

Prior Experiences

Another way of supporting problem solving is by analogy to prior experiences. Prior experiences are stories about how similar problems were solved. When a new problem is encountered, most humans attempt to retrieve cases of previously solved problems from memory in order to reuse the old case. If the solution suggested from the previous case does not work, then the old case must be revised (Jonassen & Hernandez-Serrano, 2002). When either solution is confirmed, the learned case is retained for later use. Case-based reasoning is based on a theory of memory in which episodic or experiential memories, in the form of scripts (Schank & Abelson, 1977), are encoded in memory and retrieved and reused when needed (Kolodner, 1993; Schank, 1990).

To help students solve the encountered problem, provide similar problems (perhaps case studies) in an online library of annotated problem cases. The students may reuse the advice from that problem or adapt it. To determine the most relevant and similar stories in the library, it is necessary to index them (Jonassen & Hernandez-Serrano, 2002) to make them accessible to learners when they encounter a problem. Those indexes may identify common contextual elements, solutions tried, expectations violated, or lessons learned. The indexes answer the question, Why was that prior experience remembered when this problem was encountered? Students who retrieved and reviewed cases as prior experiences while solving complex problems outperformed students who reviewed expository help in lieu of the stories on tests assessing problem-solving skills, such as reminding, identifying and recognizing the problem, identifying and explaining failure, selecting solutions, adapting solutions, explaining success or alternate strategies, and identifying needed information (Hernandez-Serrano & Jonassen, 2003). The ability to argue for a problem solution is enhanced by retrieving and reviewing failure cases rather than best practices (Tawfik & Jonassen, 2013).

Modeling Problems in Multiple Ways

"Scientific practice involves the construction, validation, and application of scientific models, so science instruction should be designed to engage students in making and using models" (Hestenes, 1996, p. 1). Mental models are enhanced and confirmed by the construction of external models. Those models may be quantitative (equations) or qualitative. Students in engineering often provide only quantitative models of problems by solving equations. That is an important part of the engineering culture. However, if they do not construct qualitative models, such as causal maps or concept maps, they often fail to understand the problem or its lessons. Both are essential to understanding and solving problems. Several types of modeling tools, including databases, concept maps, expert systems, systems dynamics tools, and graphic tools may be used to construct external models (Jonassen, 2006). While students are analyzing problems, they should be constructing models of the components and relationships in the problem. Those models will help students to hypothesize and confirm solutions to the problem (Jonassen, 2011).

Conclusion

Engineers are hired, retained, and rewarded for solving problems, problems that are unlike the well-structured problems that they learn to solve in most engineering education programs. Workplace problems are ill structured because they possess conflicting goals, multiple solution methods, non-engineering success standards, non-engineering constraints, unanticipated problems, distributed knowledge, collaborative activity systems, and multiple forms of problem representation. Engineers regularly solve combinations of decision-making problems, troubleshooting problems, and most commonly design problems. To help engineering students learn to solve workplace problems in order to become more functional engineers, some form of problem-based learning must be implemented. However, several impediments to implementing PBL programs exist. To support students solving workplace problems, case studies, case libraries, and modeling tools are especially effective.

References

Allaire, J. C., & Marsiske, M. (2002). Well-defined and ill-defined measures of everyday cognition: Relationship to older adults' intellectual ability and functional status. *Psychology and Aging, 17* (1), 101–115.

Arlin, P. K. (1989). The problem of the problem. In J. D. Sinnott (Ed.), *Everyday problem solving: Theory and applications* (pp. 229–237). New York, NY: Praeger.

Barab, S. A., & Duffy, T. (2000). From practice fields to communities of practice. *Theoretical Foundations of Learning Environments, 1*(1), 25–55.

Barab, S. A., Squire, K. D., & Dueber, W. (2000). A co-evolutionary model for supporting the emergence of authenticity. *Educational Technology Research and Development, 48*(2), 37–62.

Bransford, J., & Stein, B. S. (1984). *The IDEAL problem solver: A guide for improving thinking, learning, and creativity*. New York, NY: W. H. Freeman.

Brown, D. C., & Chandrasekaran, B. (1989). *Design problem solving: Knowledge structures and control strategies*. San Mateo, CA: Morgan Kaufman.

Brown, J. S., Collins, A., & Duguid, P. (1989). Situated cognition and the culture of learning. *Educational Researcher, 18*(1), 32–42.

Catrambone, R., & Holyoak, K. J. (1989). Overcoming contextual limitations on problem solving transfer. *Journal of Experimental Psychology: Learning, Memory, and Cognition, 15*(6), 1147–1156.

Chapman, O. (1997). Metaphors in the teaching of mathematical problem solving. *Educational Studies in Mathematics, 32*(3), 201–228.

Cho, K. L., & Jonassen, D. H. (2003). The effects of argumentation scaffolds on argumentation and problem solving. *Educational Technology: Research & Development, 50*(3), 5–22.

Dunkle, M. E., Schraw, G., & Bendixen, L. D. (1995, April). *Cognitive processes in well-defined and ill-defined problem solving.* Paper presented at the Annual Meeting of the American Educational Research Association, San Francisco, CA.

Dym, C. L., & Little, P. (2004). *Engineering design: A project-based introduction.* New York, NY: John Wiley & Sons.

Evers, F. T., Rush, J. C., & Berdrow, I. (1998). *The bases of competence: Skills for lifelong learning and employability.* San Francisco, CA: Jossey-Bass Publishers.

Gick, M. L., & Holyoak, K. J. (1983). Schema induction and analogical transfer. *Cognitive Psychology, 15*, 1–38.

Greeno, J. (1980). Trends in the theory of knowledge fro problem solving. In D. T. Tuma & F. Reif (Eds.), *Problem solving and education: Issues in teaching and research* (pp. 9–23). Hillsdale, NJ: Lawrence Erlbaum.

Gross, M. D. (1986). *Design as exploring constraints.* Doctoral dissertation, Massachusetts Institute of Technology.

Gross, M. D., Ervin, S. M., Anderson, J. A., & Fleisher, A. (1988). Constraints: Knowledge representation in design. *Design Studies, 9*(3), 133–143.

Halford, G. S., Wilson, W. H., & Phillips, S. (1998). Processing capacity defined by relational complexity: Implications for comparative, developmental, and cognitive psychology. *Behavioral & Brain Science, 21*, 803–864.

Hegarty, M., Mayer, R. E., & Monk, C. A. (1995). Comprehension of arithmetic word problems: A comparison of successful and unsuccessful problem solvers. *Journal of Educational Psychology, 87*(1), 18–32.

Henry, H., Jonassen, D. H., Winholtz, R. A., & Khana, S. K. (2010, November). Introducing problem-based learning in a materials science course in the undergraduate engineering curriculum. In *Proceedings ASME 44434*, Vol. 6: Engineering Education and Professional Development (pp. 395–403). doi: 10.1115/IMECE2010-39049.

Hernandez-Serrano, J., & Jonassen, D. H. (2003). The effects of case libraries on problem solving. *Journal of Computer-Assisted Learning, 19*, 103–114.

Hestenes, D. (1997, March). Modeling methodology for physics teachers. *AIP Conference Proceedings, 399*, 935–958.

Hogarth, R. M., & Kunreuther, H. (1995). Decision-making under ignorance: Arguing with Yourself. *Journal of Risk and Uncertainty, 10*, 15–36.

Hong, N. S., Jonassen, D. H., & McGee, S. (2003). Predictors of well-structured and ill-structured problem solving in an astronomy simulation. *Journal of Research in Science Teaching, 40*(1), 6–33.

Hung, W., Jonassen, D. H., & Liu, R. (2008). Problem-based learning. In J. M. Spector, J. G. van Merrienboer, M. D., Merrill, & M. Driscoll (Eds.), *Handbook of research on educational communications and technology* (3rd ed., pp. 485–506). Mahwah, NJ: Lawrence Erlbaum.

Jacobs, A. E. J. P., Dolmans, D. H. J. M., Wolfhagen, I. H. A. P., & Scherpbier, A. J. J. A. (2003). Validation of a short questionnaire to assess the degree of complexity and structuredness of PBL problems. *Medical Education, 37*(11), 1001–1007.

Johnson, S. D., Flesher, J. W., & Chung, S. -P., (1995, December). *Understanding troubleshooting styles to improve training methods.* Paper presented at the Annual Meeting of the American Vocational Association, Denver, CO. (ERIC Document Reproduction Service No. ED 389 948).

Johnson, S. D., & Satchwell, R. E. (1993). The effect of functional flow diagrams on apprentice aircraft mechanics' technical system understanding. *Performance Improvement Quarterly, 6*(4), 73–91.

Jonassen, D. H. (1997). Instructional design model for well-structured and ill-structured problem-solving learning outcomes. *Educational Technology: Research and Development, 45*(1), 65–95.

Jonassen, D. H. (2000). Toward a design theory of problem solving. *Educational Technology: Research & Development, 48*(4), 63–85.

Jonassen, D. H. (2006). *Modeling with technology: Mindtools for conceptual change.* Upper Saddle River, NJ: Prentice-Hall.

Jonassen, D. H. (2008). Instructional design as a design problem solving: An iterative process. *Educational Technology, 48*(3), 21–26.

Jonassen, D. H. (2011). *Learning to solve problems: A handbook for designing problem-solving learning environments.* New York, NY: Routledge.

Jonassen, D. H. (2012). *Designing for decision making. Educational Technology: Research and Development, 60,* 341–359.

Jonassen, D. H., & Hernandez-Serrano, J. (2002). Case-based reasoning and instructional design: Using stories to support problem solving. *Educational Technology: Research and Development, 50*(2), 65–77.

Jonassen, D. H., & Hung, W. (2006). Learning to troubleshoot: A new theory-based design architecture. *Educational Psychology Review, 18,* 77–114.

Jonassen, D. H., & Hung, W. (2008). All problems are not equal: Implications for PBL. *Interdisciplinary Journal of Problem-Based Learning, 2*(2), 6–28.

Jonassen, D. H., & Kwon, H. I. (2001). Communication patterns in computer-mediated vs. face-to-face group problem solving. *Educational Technology: Research and Development, 49*(1), 35–52.

Jonassen, D. H., Strobel, J., & Lee, C. B. (2006). Everyday problem solving in engineering: Lessons for engineering educators. *Journal of Engineering Education, 95*(2), 1–14.

Kahneman, D., Slovic, P., & Tversky, A. (1982). *Judgment under uncertainty: Heuristic and biases.* Cambridge: Cambridge University Press.

Kitchner, K. S. (1983). Cognition, metacognition, and epistemic cognition: A three-level model of cognitive processing. *Human Development, 26,* 222–232.

Kolodner, J. (1993). *Case-based reasoning.* New York, NY: Morgan Kaufman.

Konradt, U. (1995). Strategies of failure diagnosis in computer-controlled manufacturing systems. *International Journal of Human Computer Studies, 43,* 503–521.

Lave, J. (1988). *Cognition in practice: Mind, mathematics and culture in everyday life.* Cambridge: Cambridge University Press.

Marston, M., & Mistree, F. (1997, April). *A decision based foundation for systems design: A conceptual exposition.* Paper presented at Decision-Based Workshop, Orlando, FL. Retrieved from from:http://dbd.eng.buffalo.edu/pdf/CIRP.10.97.PDF

Meacham, J. A., & Emont, N. C. (1989). The interpersonal basis of everyday problem solving. In J. D. Sinnott (Ed.), *Everyday problem solving: Theory and applications* (pp. 7–23). New York, NY: Praeger.

Means, B., Salas, E., Crandall, B., & Jacobs, T. O. (1993). Training decision makers for the real world. In G. A. Klein, J. Orasanu, R. Calderwood, & C. E. Zsambok (Eds.), *Decision making in action: Models and methods* (pp. 306–326). Norwood, NJ: Ablex.

Morris, N. M., & Rouse, W. B. (1985). Review and evaluation of empirical research in troubleshooting. *Human Factors, 27*(5), 503–530.

Newell, A., & Simon, H. (1972). *Human problem solving.* Englewood Cliffs, NJ: Prentice Hall.

Perrenet, J. C., Bouhuijs, P. A. J., & Smits, J. G. M. M. (2000). The suitability of problem-based learning for engineering education: Theory and practice. *Teaching in Higher Education, 5*(3), 345–358.

Ploetzner, R., Fehse, E., Kneser, C., & Spada, H. (1999). Learning to relate qualitative and quantitative problem representations in a model-based setting for collaborative problem solving. *Journal of the Learning Sciences, 8*(2), 177–214.

Ploetzner, R., & Spada, H. (1998). Constructing quantitative problem representations on the basis of qualitative reasoning. *Interactive Learning Environments, 5,* 95–107.

Reisslein, J., Sullivan, H., & Reisslein, M. (2007). Learner achievement and attitudes under different paces of transitioning to independent problem solving. *Journal of Engineering Education, 96*(1), 45–56.

Rettinger, D. A., & Hastie, R. (2001). Content effects on decision making. *Organizational Behavior and Human Decision Process, 85*(2), 336–359.

Riley, M. S., & Greeno, J. G. (1988). Developmental analysis of understanding language about quantities and of solving problems. *Cognition and instruction, 5*(1), 49–101.

Rogoff, B., & Lave, J. (1984). *Everyday cognition: Its development in social context.* Cambridge, MA: Harvard University Press.

Rohlfing, K. J., Rehm, M., & Goecke, K. U. (2003). Situatedness: The interplay between context(s) and situation. *Journal of Cognition and Culture, 3*(2), 132–156.

Ross, B. H. (1987). This is like that: The use of earlier problems and the separation of similarity effects. *Journal of Experimental Psychology: Learning, Memory, and Cognition, 13,* 456–468.

Ross, B. H. (1989). Distinguishing types of superficial similarities: Different effects on the access and use of earlier problems. *Journal of Experimental Psychology: Learning, Memory, and Cognition, 15,* 629–639.

Roth, W. M., & McGinn, M. K. (1997). Toward a new perspective on problem solving. *Canadian Journal of Education, 22*(1), 18–32.

Schank, R. C. (1990). *Tell me a story: Narrative and intelligence.* Evanston, IL: Northwestern University Press.

Schank, R., & Abelson, R. (1977). *Scripts, plans, goals, and understanding: An inquiry into human knowledge structures.* Hillsdale, NJ: Lawrence Erlbaum.

Scribner, S. (1986). Thinking in action: Some characteristics of practical thought. In Sternberg, R. & Wagner, R. (Eds.), *Practical intelligence: Nature and origins of competence in the everyday world* (pp. 13–30). Cambridge: Cambridge University Press.

Sherrill, J. M. (1983). Solving textbook mathematical problems. *Alberta Journal of Educational Research, 29,* 140–152.

Simon, D. P. (1978). Information processing theory of human problem solving. In D. Estes (Ed.), *Handbook of learning and cognitive process* (pp. 287–302). Hillsdale, NJ: Lawrence Erlbaum.

Simon, H. A. (1955). A behavioral model of rational choice. *Quarterly Journal of Economics, 69,* 99–118.

Smith, M. U. (1991). A view from biology. In M. U. Smith (Ed.), *Toward a unified theory of problem solving* (pp. 1–20). Hillsdale, NJ: Lawrence Erlbaum.

Sweller, J. (1988). Cognitive load during problem solving: Effects on learning. *Cognitive Science, 12,* 257–285.

Swezey, R. W., Perez, R., & Allen, J. (1988). Effects of instructional delivery system and training parameter manipulations on electromechanical performance. *Human Factors, 30*(6), 751–762.

Tawfik, A., & Jonassen, D. (2013). The effects of successful versus failure-based cases on argumentation while solving decision-making problems. *Educational Technology Research and Development, 61*(3), 385–406.

Tessmer, M., & Wedman, J. F. (1990). A layers-of-necessity instructional development model. *Educational Technology: Research and Development, 38*(2), 77–85.

Voss, J. F., Wolfe, C. R., Lawrence, J. A., & Engle, R. A. (1991). From representation to decision: An analysis of problem solving in international relations. In R. J. Sternberg & P. A. Frensch (Eds.), *Complex problem solving* (pp. 119–157). Hillsdale, NJ: Lawrence Erlbaum.

Wood, P. K. (1983). Inquiring systems and problem structure: Implications for cognitive development. *Human Development, 26,* 249–265.

Wood, P. K. (1985). *A statistical examination of necessary but not sufficient antecedents of problem solving behavior.* Doctoral dissertation, University of Minnesota.

Zeitz, C. M., & Spoehr, K. T. (1989). Knowledge organization and the acquisition of procedural expertise. *Applied Cognitive Psychology, 3*(4), 313–336.

CHAPTER 7

Professional Engineering Work

Reed Stevens, Aditya Johri, and Kevin O'Connor

Introduction

The focus of our chapter is on current research-based understandings of professional engineering work. We argue for the relevance of these understandings to engineering education. We will also argue, as others have as well (Barley, 2004; Trevelyan, 2007, 2010; Vinck, 2003), that research on professional engineering work is too sparse. Therefore a good part of this chapter is oriented in a programmatic, agenda setting direction.

From the perspective of engineering education, the sparseness of research on professional engineering work is puzzling for a number of reasons. First, engineering education is often reorganized against the backdrop of claims about what professional engineering work is now or will be in the future. Without trustworthy and specific representations of engineering work practice and of the dispositions, skills, and identity orientations of professional engineers, how are engineering educators to know whether engineering education is preparing

engineering students to be successful, creative, or impactful engineers? A prominent consensus report from the National Academy of Engineering highlights a "disconnect between engineers in practice and engineers in academe" (National Academy of Engineering [NAE], 2005, pp. 20–21). The report stated that "the great majority of engineering faculty, for example, have no industry experience. Industry representatives point to this disconnect as the reason that engineering students are not adequately prepared, in their view, to enter today's workforce" (National Academy of Engineering [NAE], 2005, pp. 20–21). It is important that a focus on "preparation" of future engineers not be tied to an agenda that solely emphasizes what professional engineering "needs" and economic competitiveness. It also is possible to organize an engineering educational system to prepare recent graduates to be change agents and participants in new social movements within engineering work practice. However, in either case, concrete images of engineering work are critical resources for rethinking engineering education and making empirically based assessments of progress.

The lack of concrete and trustworthy images of professional engineering quite naturally extends to engineering students as well. Students often have vague images of professional engineering work, and the images they do have are strongly colored by the experiences in their educational careers that allowed them to navigate into and through engineering education – that being exceptional past performance in textbook, problem set, and test-based mathematics and science courses. As a result, students often ignore, discount, or simply do not see images of engineering that emphasize its nontechnical, noncalculative sides and its non-individual aspects (Stevens, O'Connor, & Garrison, 2005; Stevens, O'Connor, Garrison, Jocuns, & Amos, 2008). The idea that engineering work can be creative, collaborative, and oriented toward agendas of social good (not just financial gain) are aspirational positions that students sometimes adopt, but for those students for whom these are non-negotiable core values and interests, the absence of direct images of engineering that support these values can be decisive for whether even high-achieving students stay in or leave engineering (Stevens et al., 2008). So, more concrete images of engineering work can be an important resource for students themselves, as they can for institutions of engineering education broadly.

With regard to images of engineering, it is worth noting here that engineering – unlike other professions such as teaching, medicine, law, and even natural science – is not widely represented within popular cultural media. One can easily bring to mind television shows, films, and novels that depict teachers, lawyers, doctors, and even natural scientists. How easily can readers bring to mind similar representations of engineers and their work on film or television? The recent visibility of a character on the popular television show *The Big Bang Theory* is an exception that seems to confirm the rule. And the engineer, Howard Wolowitz, M. Eng., routinely endures status-based teasing from his friends

who have Ph.D.s and degrees in higher status scientific fields.

A third clear reason to have detailed research-based images of professional engineering work is that even if extant studies were sufficient in their capacity to represent engineering as it is (which we argue they are not), the images of these studies would need continual updating because engineering is properly and widely understood to be a rapidly changing form of work, under the forces of globalization; offshoring; and new technologies of communication, design, and production. These multiple reasons argue strongly that it is important to establish what we already know and what we still need to learn about professional engineering work.

What We Know About Professional Engineering Work

In this section, we review and synthesize *empirical* research on professional engineering work, drawn largely from field studies, but also from laboratory studies and surveys that include professional engineers as subjects. Every method has its strengths and weaknesses, but field studies are the only type of research that can tell us what engineering work is like *in context*. Field studies are typically conducted in workplace settings, and although the duration of fieldwork varies across studies as do the types of data captured (e.g., fieldnotes, video-recordings of engineering work, semistructured interviews), most field studies have a broadly ethnographic goal: to describe adequately the specific qualities of work practices, to understand and represent the meaning of those work practices for the people being studied, and to understand engineering work as constitutive of unique forms of work culture and social organization.

Among the social scientists who have taken an active research interest in engineering, a number have highlighted the puzzling dearth of empirical descriptions of professional engineering work (Downey and Lucena, 2004; Trevelyan, 2010; Vinck, 2003).

The academic field that probably has given professional engineering work the most extended attention has been Science and Technology Studies (STS). Bruno Latour, a leading scholar in STS, wrote an early synthetic book titled *Science in Action: How to Follow Scientists and Engineers Through Society* (1987). Despite this early programmatic announcement that engineers would be followed along with scientists, STS has mostly forgotten to follow the engineers (Downey, 1989), at least in comparison to the attention devoted to higher status natural scientists. As Downey and Lucena put it, "In research in science and technology studies (STS), engineers often make cameo appearances but rarely do they get lead roles" (Downey & Lucena, 2004, p. 395).

When STS studies have followed engineers' work, the resulting accounts diverge sharply from normative images of what historian Rosalind Williams calls "the ideology of engineering" (Williams, 2002), that is, the view that engineers have a distinct technical domain of knowledge that they can apply rationally and in a more-or-less linear manner to the solution of technical problems. Under this ideology, the social and technical do not mix. In strong and vivid contrast, STS studies have established that engineering work involves "complexity, ambiguity, and contradictions" (Hughes, cited in Williams) and that the social and technical are almost inextricably tied up together in any engineering project, at least in any project that is realized successfully. STS scholar John Law (1987) gave a name to this alternative image of engineering work; he called it "heterogeneous engineering."

Law's idea of heterogeneous engineering revolves around the imagery of engineers as "system builders" (Law, 1987, p. 112) in which any stabilized system they contribute to building is composed of heterogeneous elements that are both human and technological. Law's own case study of heterogeneous engineering is historical and concerns Portuguese colonial expansion in the late 1400s. Law also references Hughes' study of Edison through which Law makes the key point succinctly:

Edison's problem (his reverse salient) was simultaneously (italics added) economic (how to supply electric lighting at a price that would compete with gas), political (how to persuade politicians to permit the development of a power system), technical (how to minimize the cost of transmitting power by shortening lines, reducing current, and increasing voltage), and scientific (how to find a high-resistance incandescent bulb filament). (Law, 1987, p. 112)

A contemporary case study that offers a "canonical example of heterogeneous engineering" (Suchman, 2000, p. 314) involved a major bridge-building project. The study's author, Lucfy Suchman, notes that the design and technical practices that engineers view as "the real work" of engineering did take place in this project; however, her analysis shows the work of "sensemaking, persuasion and accountability" (p. 315) are as consequential for the realization of the bridge project (cf. Trevelyan, 2010). These practices are equally important parts of engineering work because of the vast number of heterogeneous actors (see Table 7.1) – human and nonhuman, small and large – that must be assembled and maintained into a stable network for a project to be realized. The critical conceptual point that undergirds the heterogeneous engineering perspective – as well as the broader perspective of actor-network theory (cf. Latour, 2005) – is that the commonsense dichotomy between the "technical" and "social" is unnecessary and in fact misleading when trying to understand how projects are realized. A recalcitrant code reviewer or problematic environmental impact statement can threaten a project as easily as can a tensile strength limitation. Put in the idiom of Actor Network Theory, to successfully realize an engineering project, nature and material forces will resist and these resistances must be overcome; the same can be said of humans and their institutions; they resist and must be overcome (e.g., persuaded, adequately paid, made silent, etc.). As Suchman writes in conclusion, "the construct of *heterogeneous engineering* is meant to underscore the extent

Table 7.1. Partial Enumeration of Relevant Actors

Federal/State	County/Region	City	Department	Other
agencies				
Federal Highway Administration (FHWA)	Two county Board of Supervisors	Two cities on north and south shores	Department Headquarters	Delta smelt
Governor	Conservation and Development Committee	Southtown Improvement Association	District	Harvest mouse
State Transport Improvement Program (TIP)	Metropolitan Transportation Committee (MTC)	Mayor of Northtown	Toll bridges	Hazardous waste
Environmental Impact Statement (EIS)	Regional Transportation Plan (RTP)	Home-owners	Structures	C&H Sugar
Federal Emergency Management Agency (FEMA)			Design	Railroad
State Historic Preservation Office (SHPO)			Bridge Replacement Project	Rights-of-way
Fish and Wildlife				Utilities
Coast Guard				
Army Corps				

Reprinted from Suchman (2000, p. 317). This represents a *partial* list of human and nonhuman elements that had to be organized and stabilized for the successful realization of the bridge-building project.

to which the work of technology construction is, to a significant degree, also the work of organizing" (Suchman, 2000, p. 324). And it is organizing of *both* the physical and the human world.

If heterogeneous engineering represents a broad conceptualization of professional engineering work, what does the work itself look like – the day-to-day practices of engineers? Some of the earliest fieldwork about professional engineering practice was conducted by Bucciarelli (1988, 1994). Bucciarelli studied "the design process" within two engineering firms, using participant observation techniques. Consistent with a general ethnographic stance, Bucciarelli did not first stipulate a definition of design process and then collect data that aligned with that definition; instead, he constructed his account of the design process on the basis of how the members of the cultural groups he

studied (i.e., the professionals at the two firms) defined and enacted design. Bucciarelli's study clearly establishes that "[engineering] design is a social process" (Bucciarelli, 1988, p. 161) not in some trivial sense that it involves people working together but rather that "[design] only exists in a collective sense" (p. 161), that it can only be seen as a process that is distributed across different sub-communities, which in turn requires social and technical coordination to bring different parts of a project's work together. Bucciarelli introduces the concept of an "object world" to identify the firms' "different worlds of technical specializations, with their own dialects, systems of symbols, metaphors and models, and craft sensitivities" (Bucciarelli, 1988, p. 162). For example, whereas the electrical engineer's object world is filled with voltage potentials and involves sketching objects like diodes, the mechanical engineer's object world is populated by beams and steel and requires an understanding of metal machining process. The manager's object world is inhabited by schedules, milestones, and critical paths. Bucciarelli highlighted that these different object worlds must be brought into some kind of coordination often involving negotiation among the inhabitants of different object-worlds making a point similar to that made by Suchman's later study, that "organizational effort *is* part of designing" (Bucciarelli, 1988, p. 162).

Another early study of engineering design by Henderson (1991, 1999) described the "visual culture" of engineering design, a culture in which sketching is the way that engineers think and communicate and in which sketches are objects through which organizational actions are frequently coordinated and negotiated. Engineering drawings and sketches are shown to be "devices that socially organize the workers, the work process, and the concepts workers manipulate in engineering design" (Henderson, 1991, p. 452). Henderson's study shows that engineers gathered around sketches, talked and revised their ideas with sketches and drawings at the center of their activity. She also uses the ubiquity and centrality of sketching and drawing (as actions) and these sketches and drawings (as objects) to offer a critique of a then dominant ideology that paper was soon to be a thing of the past in engineering, to be replaced by computer-aided design (CAD). Because Henderson showed the centrality of sketching and sketches and because CAD systems of that time period rigidly specified drawing practices and drawing forms, Henderson argued that CAD lacked the requisite *flexibility* needed to support the collaborative work practices of engineering design.[1]

In another field study of civil engineers designing roadways, Hall and Stevens (1995) found that engineers worked with both CAD and paper interfaces, each having their own affordances within the work practices and accountabilities of engineering work. In general, however, this study concurred, as did a later study of architectural design

by Stevens (2000), with Henderson's finding that the work of envisioning, exploring, and revising design alternatives was a paper-based practice because of paper's "flexibility" as a mobile, collaborative, and expressive medium. It remains an open question, one that has seemingly not been addressed in a field study, whether the idea of designing within the computer environment is still more of an ideological fiction rather than a routine fact of work practice. It seems plausible that the current generation of engineering designers, having grown up as so-called "digital natives," may have substantially shifted the balance from being engineers producing and iterating design ideas on paper to doing so more fully within CAD environments, which of course have themselves become more flexible and friendly to sketching practices with touch sensitive tables, better GUIs, and bigger screens.

Drawing on the broader thematic interests of STS, these studies of professional engineering focused attention on the importance of documenting engineers' *representational practices* (cf. Lynch & Woolgar, 1990; Greeno & Hall, 1997; Vinck, 2003) – how people use representations to make sense, solve problems, and to persuade and communicate with others. One fine-grained analysis of the representational practices of engineering work can be found in Stevens and Hall's development of the concept of "disciplined perception" (1998, cf. Stevens, 1999). *Disciplined perception* refers to the learned ways that people in a discipline see and interpret their focal phenomena – through their tools and representations. In focusing on these discipline-specific practices, disciplined perception is a concept consistent with Henderson's focus on the *visual cultures* of engineering as well as Bucciarelli's focus on the distinct culturally constituted *object-worlds* of engineering.

Using the methods of interaction analysis (Goodwin & Heritage, 1990; Jordan & Henderson, 1995) to analyze the moment-to-moment unfolding of civil engineering project work, Stevens and Hall's account of engineers' disciplined perception also provided a way to understand how disciplined perception *develops* in practitioners. The account analyzed interactions in which intersubjective gaps (i.e., different ways of seeing what representations were saying) between collaborating engineers working together provided occasions for a more experienced engineer to *discipline* the perception of a relative newcomer. Stevens and Hall's account thereby articulated the interactional mechanisms through which an engineer might gradually "learn to see" as an engineer, through proximal apprenticeship with more knowledgeable others in the context of daily work. This account can also be tied to the idea of heterogeneous engineering, in that the disciplined perception of engineers involves reading a range of heterogeneous interests and constraints directly from representations. For example, in this study the authors recount how the engineers provide a coordinated reading across plans, sections, and elevations to recover a rationale for a consequential decision to exceed code-allowable grade on a stretch of proposed roadway, which in turn could avoid the financial cost and potential slowdown of their project that would be set in motion by a damaging environmental impact statement. In the context of single stretches of interaction, the technical elements (such as "cut" and "fill" and "allowable grade") are intermixed with social issues (such as satisfying budgets and environmental concerns of some project stakeholders). This account also resonates with Suchman's account, which highlighted that engineers' work is often about making persuasive arguments to secure project interests, with arguments assembled via embodied performances with visual representations.

Comparing the Work of Engineering Education and the Work of Professional Engineering Practice

Another line of research on engineering workplaces, also relatively sparse, is work that compares the work practices of undergraduate education to the work practices of professionals on the job. This comparative research is often conducted with an eye toward possible programmatic implications for engineering education. Comparative research on problem solving in the undergraduate curriculum and in professional engineering work suggests that the types of problems that are solved and the processes of problem solving in these different contexts differ in both substance and structure (Jonassen et al., 2006; Stevens, Garrison, & Satwicz, 2007; Stevens et al., 2008). Engineering problems found in school – particularly in coursework apart from senior capstone experiences – are organized to develop facility in solving "well-structured" problems (Jonassen et al., 2006), with clearly stated goals and knowable, correct solutions attainable through application of a small, finite set of rules and principles. This recalls a venerable distinction from Rittel and Webber (1973) regarding "tame" (i.e., well-structured) vs. "wicked" problems, with wicked problems being the norm for professionals. The well-structured problems that engineering students learn to solve tend to be aimed toward advancing students' individual mastery of concepts of engineering science (Korte, Sheppard, & Jordan, 2008; Stevens et al., 2008).

Trevelyan and Tilli (2007) conducted an extensive review of different literatures on engineering work, including, among others, the technical literature on engineering research and development, competency standards of professional organizations within engineering, engineering education literature, engineering management literature, and ethnographic studies of engineering work. They concluded that most of these literatures provide an inadequate picture of engineering work. They diagnosed a prevalence of prescriptive claims about engineering work based on personal experience or anecdotal evidence, an undue privileging of design engineering over and above other engineering practices, a neglect of tacit aspects of work, and a tendency by respondents to limit what they view as "real engineering" to a narrow range of technical aspects of their work (cf. Faulkner, 2008 and Suchman, 2000).

These studies, along with others, point to the promise of qualitative and ethnographic research for broadening and deepening our understanding of the work practices of engineers. There are also other research traditions that have productively informed an understanding of professional engineering work. Atman et al. (2007), working within the tradition of expert-novice studies in cognitive science, conducted a comparative analysis of the problem-solving strategies of students and professional engineers. These authors showed that, in laboratory-based simulations of engineering problem solving, experts display not only more extensive engineering science knowledge, but also more awareness of and judgment regarding other aspects of problem solving. This kind of detailed analysis of problem solving can be valuable in demonstrating that differences between experts and novices can be understood in terms of differences in organization of knowledge and cognitive strategies. Jonassen, Strobel, and Lee (2006) used different methods to make related points about problem solving in engineering work. Based on interview accounts of engineers' past problem-solving experiences, these authors argued that workplace problems are "ill-structured" or "wicked" (Jonassen et al., 2006; Rittel & Webber, 1973), most often involving vaguely defined goals and inviting no clear solution or solution path. Workplace problems were shaped from the outset by nontechnical constraints as befits a general image of heterogeneous engineering, such that success was rarely measured solely by technical or scientific standards but also included such considerations such as timeliness, budget, and customer satisfaction. Workplace problems foreground the importance of communication of technical concepts to others, rather than individual mastery or understanding of these concepts (Korte et al., 2008). And they require distributed expertise to solve them (Stevens

et al., 2007). As a result, judgment is a critical aspect of problem solving by engineering professionals (Eraut, 2000; Stevens, 1999; Stevens et al., 2007; Vinck, 2003). Such ill-structured problems stand in contrast to the well-structured problems found in school, which are characterized by clearly stated goals and by knowable, correct solutions attainable through application of a small, finite set of rules and principles. And these tend to be practiced and evaluated through the standard forms of problem sets, quizzes, and exams (Stevens et al., 2008), forms that have no natural home in the problem solving practices of professional engineers. Thus, both expert–novice task analysis and interview studies of simulated or reported work practices echo the finding of ethnographic field studies and caution against a view of engineering work as consisting primarily in "technical rationality" or what Williams called "the ideology of engineering."

Trevelyan (2007, 2010) conducted a qualitative study that involved ethnographic interviews supplemented by field observations with Australian and Pakistani engineers. This study is noteworthy for its delineation of ten categories of engineering practice, including several that tend to be neglected by more prescriptive and normative typologies. One major category of engineering practice identified by Trevelyan is what he terms "technical coordination," that is, "working with and influencing other people so they conscientiously perform some necessary work in accordance with a mutually agreed schedule" (2007, p. 191). At least two aspects of technical coordination are important to understanding engineering work. First, technical coordination takes place outside of formal lines of authority. That is, coordination practices are not simply the province of managers; rather, all engineers seem to engage regularly in these social processes. Second, like Atman et al. (2007) and others, and against commonsense views among engineers that "the social" and "the technical" are separable (Faulkner, 2008, coordination is a hybrid of social and technical aspects of work, and therefore social interaction is at the core of accomplished engineering practice.

This body of work that compares the knowledge practices of engineering students and engineering professionals is suggestive and important, but it is also sparse and limited in how it allows us to understand connections and disconnections between engineering education and professional engineering work. Interview studies that focus on accounts of past problem-solving experiences (e.g., Jonassen et al., 2006) have been valuable in pointing to some major areas of difference between schools and workplaces. However, interviews are less than fully sensitive to the broad range of knowledge practices, especially its tacit dimensions (Eraut, 2000; Polanyi, 1966), which are largely inaccessible to reflective awareness. Laboratory studies (e.g., Atman et al., 2007) have been valuable in demonstrating that differences between experts and novices can be understood in terms of differences in the organization of knowledge and cognitive strategies. However, laboratory-based expert–novice studies are limited in their ability to represent the locally situated aspects of problem solving, including characteristic ways in which particular workplaces (or school settings) organize teamwork; access to and communication of information; and the distributed, material, and embodied properties of cognitive activities (Hall & Stevens, 1995; Hutchins, 1995; Stevens & Hall, 1998). Ethnographic studies in undergraduate engineering education and in engineering workplaces (O'Connor, 2001, 2003; Stevens & Hall, 1998; Stevens et al., 2005, 2008) have been conducted to capture these aspects of engineering practices (e.g., its embodied, material, and distributed qualities) that are easily missed in other styles of research. However, although they offer suggestive accounts about how engineering education knowledge might relate to, or fail to relate to, workplace knowledge (O'Connor et al., 2007; Stevens et al., 2005, 2008), these studies have not examined directly the specific *learning processes* of engineers making the transition from school to work. Simply put, too little is known about how the practices of undergraduate education are applied and adapted in the workplace and equally little is know about what knowledge practices from one's engineering education experience have little or no clear use at work. Lines of research that look directly at these transitions from school to work are much needed.

The Identity Dimension

Thus far we have focused our attention largely on characterizations of engineering work. In this section, we look at another important dimension of professional engineering – identity. If we are interested in a full understanding of professional engineering, we must attend not just to what people learn and know but also to who they *are* and what is their place in the world among their consociates *as engineers*, both within their local professional networks and within social life more broadly. Personal, social, and disciplinary identities intersect in complex ways among professional engineers. We understand identity formation as a two-sided process in which persons identify with certain groups (e.g., engineers) and forms of activity (e.g., engineering) and are in turn identified with certain groups and forms of activity by others. "Identities" have been argued theoretically to result from a complex, nondeterministic stabilization of these two dialectically related processes (Skinner, Valsiner, & Holland, 2001).

A few key studies bear directly on identity issues among professional engineers. In a largely historical analysis, Downey and Lucena (2004) highlight the shaping influence of "codes" to which engineers responded at a particular time and in a particular place. Tracing a set of codes through Western Europe into the U.S. context, they argue that the historical trend has been for engineering identities to be increasingly shaped by the view that "progress" in engineering is tied to participation in and affiliation with large industrial corporations. They argue that, especially in the period from the Second World War to the current day, American engineers came to see themselves in terms of corporate metrics of progress, namely the mass production of low-cost consumer goods and corporate profit. In general, they claim that "[American] engineers have made embracing private industry a patterned feature of their identity" (Downey & Lucena, 2004, p. 411), something they note is not necessarily characteristic of other professionals in law, medicine, and the clergy. Downey and Lucena's perspective resonates with a more forceful claim in David Noble's *America by Design* that American engineers in the postwar period became a "domesticated breed" (Noble, 1977, p. 322 who "in reality served only the dominant class in society" (Noble, 1977, p. 324).

Research by Faulkner (2008) points directly to some identity issues for engineers who must navigate a dominant *technicist* ideology of engineering (i.e., what we have been referring to as technical rationality) and the reality of professional practice as *heterogeneous engineering*. Despite the fact that *all* engineers do heterogeneous work and most recognize this work (though not using the term "heterogeneous"), there is a tendency among engineers to define "real" engineering in terms of the technical, "nuts and bolts," scientific and mathematical labor, and to locate the social aspects of heterogeneous engineering outside of "real" engineering (cf. Trevelyan, 2010). Faulkner suggests that identifying with these features of engineering work allows engineers to maintain a unique identity of technical competence amidst interdisciplinary collaborations with people both within their firms (e.g., managers) and outside them (e.g., architects). This may be an implicit response to what historian of technology Rosalind Williams calls the "expansive disintegration" (Williams, 2002) of engineering as a distinct and bounded form of professional knowledge and competence. If, as Williams argues, society is increasingly coming to perceive that nearly everything involves engineering, then nothing is distinctly engineering.

Faulkner's analysis suggests further that the distinction between the technicist and the heterogeneous registers is a gendered one, with the "nuts and bolts" identity being one that men can more comfortably perform to reconcile their dual identities as both men and engineers. Faulkner's study supports a point from a related pair of studies she cites (Robinson, 1992; Robinson & McIlwee, 1991) that "found that men engineers often engage in 'ritualistic displays of hands-on technical competence' *even when the job does not require this competence*. Women engineers do not generally participate in this 'engineering culture,' as they call it, and can lose out in career terms as a result" (Faulkner, 2008).

We also can compare engineering student identities and professional engineering identities, as we compared knowledge practices among students and professionals. Existing empirical work is suggestive but incomplete about how identity formation processes in engineering student culture might shape transitions to engineering workplaces. There is a plausible tension in the way a student and a professional might understand herself if most students' educations are based predominantly on coming to understand engineering as a form of technical rationality. Such an understanding of engineering could result in both direct and indirect tensions in understandings of one's work as an early career professional. Directly, new engineers who identify strongly enough with a model of technical rationality are likely to struggle to understand themselves as engineers if they perceive a dilution of "pure" engineering work by what they perceive as "nonengineering" work in professional practice (Eisenhart & Finkel, 1998; Faulkner, 2008; Korte et al., 2008). A similar dilution is experienced by architecture students who come to see "designer" to be their dominant identity during their time in design school, only to find that as practicing architects (especially early career) one does everything *but* design and that design often "hangs in the balance," meaning that it is readily pushed out by other aspects of architectural project work, such as negotiating and coordinating with clients and contractors, managing construction, and ensuring that design drawings meet state and federal codes (Cuff, 1992; cf. Stevens, 1999). A possible indirect effect of the curricular model of technical rationality is that it may prevent students from developing other imagined futures as engineers; engineers do varied kinds of work and play varied roles in their professional work lives but this diversity of experience is hardly visible in undergraduate education (Eisenhart & Finkel, 1998; Foor, Walden, & Trytten, 2008; Steering Committee of the National Engineering Education Research Colloquies, 2006). Whether the tensions are direct or indirect, what is obscured for students are the identity elements of heterogeneous engineering practice that engineering students typically do not see or learn to value as central – those related to communication, coordination, organizing, and persuasion amidst people *and* technical practices and objects.

Two examples from ethnographic case studies are suggestive of possible tensions across the boundary of engineering education and early career professional work (O'Connor et al., 2007; Stevens et al., 2008). In a first example, an engineering student whose identity was heavily invested in his sense of himself as a solver of well-structured problems in the technological rationalist mode (i.e., textbook math and science problems that bracket out all the nontechnical aspects of engineering problems) found himself quite confused and even angry when *late* in his engineering educational career he first encountered a substantive version of engineering as heterogeneous and collaborative. When he first participated in capstone design projects, and into the early months of his first new position as an engineer, this young man reported being rather at sea, because the mathematics puzzle–solving skills that were so central to how he saw his worth as a would-be engineer suddenly had little practical value in the formal and informal evaluations of his new workplace community. A second example from these case studies involves a young

woman who opted out of her engineering major into a communications-related field during the middle stages of her undergraduate educational career. This was a student who was technically proficient, earned solid grades, and was quite adept socially. Her experience in engineering education was soured, however, because she came to see engineering as a field with little room for how she understood herself, as a collaborative "people person." Based on her engineering education, she decided engineering itself would be too individualistic and competitive and did not feel that she belonged. Ironically, the very aspects of her identity that caused her to opt out of engineering might have made her a very valuable and unique contributor to an engineering firm, where these sorts of interpersonal skills seem to be in high demand, especially when they are combined with technical competence (Bucciarelli & Kuhn, 1997), which this young woman was clearly on the road to developing. Taken together, these case studies point to a range of possible dilemmas and complex transitions of personal and disciplinary identification, as engineering students become engineering professionals. This is a topic of significant importance for future research.

The Changing Character of Engineering Work

In this section, we draw on current studies of engineering work to address some of the ostensibly major changes engineering work is undergoing. We say ostensibly because, as is often the case, rhetoric and ideology may run ahead of demonstrable empirical evidence. In particular we consider the following four issues as possible candidates for major sources of change in engineering work: the role of new technologies involved in engineering work, globalization, new kinds of engineering problems, and changes to engineering work because of changes to the contemporary cultures of young people who are entering the profession.

New Technologies in Engineering Work

Broad conceptualizations of how to manage engineering projects such as concurrent engineering and product life cycle management put a premium both on information technology systems and on practices of computer-supported collaborative work. Clearly, engineers are making increased use of computational technologies that allow them to model and convert physical artifacts in digital forms (Boland et al., 2007; Yoo et al., 2010). Some commentators have argued that an almost complete elimination of manual labor in engineering is on the near horizon, suggesting that it will be fully displaced by symbolic labor at the interface. According to Zussman, "engineering practice today is characterized by a near total absence of that physical, hands-on labor that is a central attribute of craft work. Engineers manipulate symbols that refer to physical objects, mostly equipment and products, but they do not manipulate those objects themselves" (p. 77). According to this view, there is a clear division of labor in which human mechanical labor and craftwork is the purview of machinists, mechanics, technicians, and automated machines. That the lingua franca of engineering work is increasingly realized in "digital form" probably cannot be doubted as a general historical direction. However, lacking systematic empirical studies of actual engineering work, we should be cautious in subscribing to this view, because it has an ideological dimension. Similar ideological perspectives about technology, in particular CAD-CAM in the 1990s, far outran the empirical realities of work practice of the time in which paper-based representations remained central (Downey, 1992, 1995; Hall & Stevens, 1995; Henderson, 1991). Field studies of engineering work practice seem much needed in this area to understand how technological change within engineering work is changing the work itself and to understand how these changes are differentially affecting different participants in engineering projects and concerns (e.g., newcomers vs. old timers). For example, in a study of roadway design in civil engineering described earlier (Hall & Stevens, 1995; Stevens & Hall, 1998), the authors related a story told by one of the project engineers they were studying. The engineer said that at a recent meeting of all the firm's engineers, the president of the firm announced that if you had not learned to use CAD in the next couple of years, "there would not be a place for you" at the firm. The engineer then recounted that he had heard from some older engineers that they were choosing to retire early rather than retool in this way. Just as paper survived as a critical medium for work in the purported age of the "paperless office" (Sellen, 2003), manual labor too may be thriving in engineering work, if we see engineering work broadly. Again, this is the sort of question that could be answered substantively with field studies of contemporary engineering work.

Globalization and Offshoring

Major shifts in the use of information technology have combined with significant changes in the global economy over recent decades to increase the globalization of engineering work practice. Boeing's approach to their new "dreamliner" represents one promise, and perhaps a cautionary tale, about global distributions of engineering work. As has been widely reported, the ability to use networked technologies allows engineers to bypass the traditional boundaries of the workday to move projects along "24/7" and to "offshore" significant parts of engineering project work. This "follow-the-sun" model was initiated by General Electric's initiatives in the early 1990s and involves setting up coordinated teams across the globe so that work can be handled by the team where the sun was up, and then work can be handed to the next team starting their day in a different part of the globe. For instance, a typical globally distributed team might have workers in Japan, Singapore, India, France, and the United States. Again, ideology may outrun reality with respect to these globally distributed "teams." Surely, an opportunity for provocative and important empirical field research exists in this area, both for an investigation of the work practices themselves as well as for how cultural differences are understood and managed (Downey et al., 2006; cf. O'Connor, 2001, 2003 for analyses of distributed work processes in engineering education), in what are undoubtedly conditions of dramatically asymmetric power relations across global sites.

A recent article by Will-Zocholl and Schmiede (2011) draws on interviews with engineers and managers to characterize some the changes in the automobile industry that have been driven by the dual forces of new information technologies and increased globalization. Based on an analysis of these interviews, the authors argue that the automotive engineering work has been significantly reorganized. Among the changes they point to are significant offshoring of "standardized" engineering work such as simulation processes and calculations. The authors also argue that "the core of engineers' work, design, is becoming increasingly marginal . . . Other tasks, such as communicating, coordinating, and traveling, as well as administrative duties, are becoming more important" (Will-Zocholl & Schmiede, 2011, p. 13). A third point the authors make relates directly to the issue we just discussed – whether automotive engineering is becoming "dematerialized" (to use a term from Williams, 2002, p. 48) – the idea that manual aspects of engineering are disappearing. Although these authors indicate that some aspects of the work are dematerialized in the form of computer-based models, they imply that understanding materiality remains important because at some point the two- and three-dimensional images must be translated back into three-dimensional, moving objects. Finally, this study casts some light on potential existential or identity effects of these changes; they highlight that the engineers feel both less autonomous and more insecure (about their jobs and their futures) than they did in the past because

of these changes toward globalization. In fact they may have reason for this insecurity, as recent articles argue that almost 40% of work, even high-status "knowledge work" of engineering, can potentially be offshored. This puts the number of potentially outsourceable jobs at 20 million (Blinder, 2006; Smith & Rivkin, 2008). It would seem valuable to build on this article to conduct field studies that capture directly through observations or recordings (rather than through retrospective reports as were the data source in this study) the character of globalization-induced changes to work and the shifting meanings those changes have for engineers.

New Problems for Engineers to Solve

If engineers have been tied to the corporation and its broad goals of profit via the mass production and sale of low-cost consumer goods during the post–Second World War period, the role of the engineer may arguably change in the near future and with it the qualities of professional engineering work. Engineering is clearly implicated in solving some of the planet's biggest problems, including sustainable energy, climate change, and famine. These are problems that call for a full-scale recognition of heterogeneous engineering and its artful practice, because these problems can be ameliorated only through the organizing, maintaining, and adapting of complex, large-scale *sociotechnical* systems. Whether or not David Noble's view that engineers were domesticated servants for the ruling class was a provocation made from exaggeration, it is clear that *if* engineers of the near future are to contribute substantially to solving these kinds of global problems, they will need to work for constituencies other than large multinational industrial corporations. This is because these often are the very organizations that are seeking to impede progress in solving these problems, at least if it threatens their bottom line and short-term growth outlook, which often it does (cf. Hess, Breyman, Campbell, & Martin, 2007). A study of engineering students' beliefs about engineering, conducted across four very different universities, suggests that although social good is a loosely held aspiration for engineering students, the reason they most convincingly give for pursuing engineering, and enduring its difficulties, is to make a good living and have a comfortable material lifestyle, presumably within a corporate engineering setting (Stevens et al., 2008). This study suggests that engineering education has not yet strongly registered a shift to a different image of engineering work and that the identification with corporate participation and financial reward described by Downey and Lucena (2004) and by Noble (1977) is a dominant ethos 'on the ground' among contemporary engineering students.[2]

Changes in the Population of Young People Who Become Engineers

There was a time when engineers were not trained, as is currently the nearly universal norm, through accredited university engineering degree programs. Auyang argues that from the Renaissance until about World War I, "engineers and their predecessors came mostly from working families, toiled with their hands, relied more on their thinking and experience than on schooling, and were obliged to deliver products on demand" (p. 114). During this period, apprenticeship and on-the-job training were the norm and the current gap between the academy and paid engineering work, if it existed, would be narrow compared to the contemporary situation. Gradually, engineering became professionalized and the training of engineers through accredited programs became the norm. In the United States, ABET certifies engineering programs, a process that began in 1932 with ABET's predecessor organization, EPCD. In the postwar war period, an image of engineering competence was built out of a dual commitment to an engineering science perspective and the image of the engineer of the technological rationalist. During this period, engineers were seen as a distinct category of people and engineering as a distinct category of work. However, in the current era,

engineering seems to be undergoing a social-historical transformation that Williams calls "expansive disintegration":

> There is no "end to engineering" in the sense that it is disappearing. If anything, engineering-like activities are expanding. What is disappearing is engineering as a coherent and independent profession this is defined by well-understood relationships with industrial and other social organizations, with the material world, and with guiding principles such as functionality. Engineering is "ending" only in the sense that nature is ending: as a distinct and separate realm. Engineering emerged in a world in which its mission was the control of non-human nature and in which that mission was defined by strong institutional authorities. Now it exists in a hybrid world in which there is no longer a clear boundary between autonomous non-human nature and human-generated processes. Institutional authorities are also losing their boundaries and their autonomy. (Williams, 2002, p. 31)

With these changes and the growing ubiquity of engineering-like activities across society, the kinds of young people who aspire to and matriculate into engineering degree programs is bound to change. Williams reports a number of changes over the recent decades at MIT, seeing what she calls "a new breed of engineering student" (Williams, 2002, p. 58). This new breed includes greater ethnic and gender diversity, greater international diversity, more upper middle class young people, more young people of urban rather than rural communities, and more young people who see engineering from the very beginning of the higher education careers as a route to entrepreneurship, technology innovation, and management. These are reports from an elite university for engineering students, so it is an open question whether these demographic shifts are similar across the wider spectrum of accredited engineering colleges and university programs.

It also seems as if the current generation of young people who have "grown up digital" (Brown, 2000) are likely to bring with them a very different set of interests, assumptions, and capacities from just a generation before them. And with the rapid changes in social media and digital, mobile technologies, this set of interests and capacities is likely to remain a moving target for some time. These incoming students will have been weaned on mobile devices, instantaneous Web searching, and games – both single player and massively multiplayer – on a dizzying array of platforms. And the consumer objects that they most aspire to own (e.g., iPods, tablets, gaming systems) are fairly clear hybrids of engineering and aesthetics, suggesting that the aesthetic dimensions of engineering may become more prominent. The current movement to educate for STEAM rather than just STEM (i.e., including the Arts in Science, Technology, Engineering, and Mathematics) is suggestive in this regard. These young people also have and will continue to grow up in a culture that makes heroes of technology innovators who accumulate vast sums of money, people like Steve Jobs, Bill Gates, Mark Zuckerberg, Serge Brin, and Larry Page.[3]

Conclusion

In this chapter, we have described research on professional engineering work as it relates, or fails to relate, to engineering education. We have advanced an image of professional engineering work as heterogeneous. It is our assumption that this image, at least among many engineering educators, may appear foreign and may not align well with what often counts as engineering within the academy. If this assumption is true, it seems worth trying to understand why. One likely reason has to do with a gap we have mentioned so far, but have not brought into focus. This is the gap between the work of engineering professionals outside of the academy and the work of engineering faculty within the academy. Here we are in speculative waters, because there is not a body of research that undertakes this comparison empirically. But some differences seem clear and relevant. Like that of most academics (ourselves included),

the work of engineering faculty involves teaching and research, as well as many activities that maintain the going concerns of their workplaces, which are universities. Engineering research is of course a form of engineering work, but its accountabilities are clearly different from the work practices of engineering professionals outside of academia who are involved in realizing engineering projects. In short, there is a clear gap between what engineering faculty do in their work and what most of their students *will* do in theirs.

What we see as an appropriate response to this gap is by no means to argue for the disruption of the disciplinary research practices of engineering faculty. What we do see as an appropriate response is to infuse engineering education with new, more diverse, research-based images of professional engineering work. These images are as much for faculty – who do different kinds of work – as they are for the students. A fertile sub-field of research dedicated to studying professional engineering work and connecting it – both practically and conceptually – to engineering education is much needed. A thriving subfield of this kind will address *questions* that we believe all stakeholders share about the continuities and discontinuities between the work practices and identities of engineering students and professional engineers.

It is important also to state that specific directionalities for change are not assumed in advocating this program of research. It is an easy assumption to make that advocating this program of research implies advocacy for reorganizing engineering education to mirror professional work more closely. We are by no means making this assumption, though it is among the possibilities we see. To understand why this does not necessarily follow from the fact of a comparative program of research on engineering education and professional engineering work, consider the following analogy. Each of the authors of this chapter identifies with the interdisciplinary field of the learning sciences. The learning sciences as a field has, like engineering, both "basic" and "applied" elements. Basic elements involve studying how people learn in a wide range of contexts and applied elements involve trying to design tools, materials, and environments to improve learning and often to organize it new ways. Most in the learning sciences are emphatically not advocates for the ways that learning is organized in K–12 schools, but most would agree that without an understanding of how school-based learning is organized, attempts to disrupt, change, or reorganize K–12 practices will be difficult if not fruitless (O'Connor & Allen, 2010; Penuel & O'Connor, 2010). By analogy, engineering educators may see similar problems with professional engineering work and want to see it take very different forms than it currently takes. We are arguing that only by *understanding* the organization of professional engineering work and its effects on people, on nature, and on society are those efforts to change professional engineering work from the outside likely to bear fruit.

At the same time, we do see possible directions for change running from research-based understandings of professional engineering work into engineering education. Specifically, the idea of heterogeneous engineering is a potentially productive disruption for engineering education, which arguably remains the bastion for the technical rationalist view of engineering. This is not true everywhere but in many places. So engineering education could itself become more heterogeneous. This recognition would lead to a rebalancing of engineering education's portfolio of learning opportunities, influences, and requirements to make it clear that the human, organizing aspects of engineering are as important to engineering as its technical aspects, moving toward what Stevens has sketched as a *sociotechnical engineering education* (in Adams et al., 2011, cf. Trevelyan, 2010). Achieving this rebalancing will almost certainly require diversifying engineering departments themselves to include people who emphasize the social aspects of engineering, so that a future generation of students may come to see the resistances of a local code review agency or policymaker equally as relevant to the successful realization of engineering projects as the resistances of a conductive material.

We also see possible directions for change running from engineering education into professional engineering practice, changes that would themselves deserve study in the comparative research program we have outlined here. In this chapter, we have discussed research that has argued that engineering education is currently tilted toward an image of success in engineering that emphasizes technical rationality, technical innovation, business entrepreneurship, and participation in the corporate profit-making enterprise. We have shared research that shows that across a wide range of contemporary engineering students, many have vague but compelling aspirations that engineering *could be* a force for "social good" but most are clear that they are participating in engineering education first and foremost for the perceived financial security and comfortable material existence it promises. And few seem to see those values of engineering for "social good" enacted in their curriculum or in their broader educational experiences. But this could change and seems to be changing in a handful of respected engineering education programs. Some new initiatives are flying under the banners of "humanitarian engineering" and "social entrepreneurship" (Wikipedia entries) and others, such as *Design for America*, are framed around interdisciplinary efforts to design for solving "wicked problems" in society broadly and for underserved communities. These initiatives are very compatible with the image of heterogeneous engineering we have foregrounded here and go further to enact values aligned with an "ethic of care" (Riley, Pawley, Tucker, & Catalano, 2009) that is in some, if not significant, tension with goals for engineering education that primarily stress novel technical innovation and financial gain. Should future generations of engineering students have substantive experiences and identification with this image of engineering – how it can help solve pressing social problems through heterogeneous engineering based in an ethic of care as much as in an ethic of financial gain – newcomers to engineering firms could be part of a historical process of transformation in professional engineering work. Such changes would in turn invite new research about how student identification with engineering changes, who chooses to go into engineering, and how this alternative image reshapes broader images *and effects* of engineering in society.

In this conclusion we have sketched just a few ways that engineering education and professional engineering work *could* influence each other in the coming years and challenged easy assumptions about how they *should* influence each other. These images are informed by our own values and, of course, the actual relations between engineering education and professional engineering work may evolve very differently. Regardless of whether these directions of influence take hold, we hope to have argued that studies of professional engineering work are not an elective for engineering education research, but required coursework.

Footnotes

1. For another critique of CAD's insertion in engineering practice, and education, from a different perspective, see Downey's (1992) "CAD-CAM saves the nation?"

2. This study did not seek to identify the source of this ethos among engineering students, but this is a topic of some interest because this strongly held rationale for pursuing engineering seems unlikely to come directly from engineering faculty, who often do espouse different values and themselves do not typically practice engineering in corporate contexts.

3. Noticeable in assembling a list of this kind is the absence of female technology icons.

References

Adams, R., Evangelou, D., English, L., De Figueireo, A. D., Mousoulides, N., Pawley, A. L., . . . Wilson, D. M. (2011). Multiple perspectives on engaging future engineers. *Journal of Engineering Education: Centennial Special Issue,* 100(1), 48–88.

Anderson, K. J. B., Courter, S., McGlamery, T. Nathans-Kelley, T., & Nicometo, C. (2010). Understanding engineering work and identity: A cross-case analysis of engineers within six firms. *Engineering Studies,* 2(3), 153–174.

Aneesh, A. (2006). *Virtual migration: The programming of globalization.* Durham, NC: Duke University Press.

Atman, C. J., Adams, R. S., Cardella, M. E., Turns, J., Mosborg, S., & Saleem, J. (2007). Engineering design processes: A comparison of students and expert practitioners. *Journal of Engineering Education,* 96(4), 359–379.

Auyang, S. Y. (2004). *Engineering: An endless frontier.* Cambridge, MA: Harvard University Press.

Barley, S. R. (2004). What we know (and mostly don't know) about technical work. In S. Ackroyd, R. Batt, P. Thompson, & P. Tolbert (Eds.), *The Oxford handbook of work and organization.* Oxford: Oxford University Press.

Blinder, A. S. (2006, March/April). Offshoring: The next Industrial Revolution? *Foreign Affairs,* 113–128.

Boland, R. J., Jr., Lyytinen, K., & Yoo, Y. (2007, July/August). Wakes of innovation in project networks: The case of digital 3-D representations in architecture, engineering, and construction. *Organization Science,* 18, 631–647.

Brown, J. S. (2000, March/April). Growing up digital: How the Web changes work, education, and the ways people learn. *Change.*

Bucciarelli, L. (1988). An ethnographic perspective on engineering design. *Design Studies,* 9(3), 59–168.

Bucciarelli, L. L. (1994). *Designing engineers.* Cambridge, MA: MIT Press.

Bucciarelli, L. L., & Kuhn, S. (1997). Engineering education and engineering practice: Improving the fit. In R. Barley & J. E. Orr (Eds.), *Between craft and science: Technical work in U.S. settings.* London: Cornell University Press.

Cuff, D. (1992). *Architecture: The story of practice.* Cambridge, MA: MIT Press.

Downey, G. L. (1992). CAD/CAM saves the nation?: Toward an anthropology of technology. *Knowledge and Society,* 9, 143–168.

Downey, G. L., & Lucena, J. C. (2004). Knowledge and professional identity in engineering. *History and Technology,* 20(4), 393–420.

Downey, G. L., Lucena, J. C., Moskal, B. M., Parkhurst, R., Bigley, T., Hays, C., . . . Nichols-Belo, A. (2006). The globally competent engineer: Working effectively with people who define problems differently. *Journal of Engineering Education*, 95(2), 107–122.

Eisenhart, M. A., & Finkel, E. (1998). *Women's science: Learning and succeeding from the margins*. Chicago, IL: University of Chicago Press.

Eraut, M. (2000). Non-formal learning and tacit knowledge in professional work. *British Journal of Educational Psychology*, 70, 113–136.

Faulkner, W. (2008). Nuts and bolts and people. *Social Studies of Science*, 37(3), 331–356.

Faulkner, W. (2009). Doing gender in engineering workplace cultures: Observations from the field. *Engineering Studies*, 1(1), 3–18.

Foor, C. E., Walden, S. E., & Trytten, D. A. (2008). "I wish that I belonged more in the whole engineering group:" Achieving individual diversity. *Journal of Engineering Education*, 97(2), 103–115.

Friesen, M. (2011). Immigrants' integration and career development in the professional engineering workplace in the context of social and cultural capital. *Engineering Studies*, 3(2), 79–100.

Gainsburg, J., Rodriguez-Lluesma, C., & Bailey, D. E. (2010). A 'knowledge profile' of an engineering occupation: Temporal patterns in the use of engineering knowledge. *Engineering Studies*, 2(3), 197–220.

Goodwin, C., & Heritage, J. (1990). Conversation analysis. *Annual Review of* Anthropology, 19, 283–307.

Greeno, J., & Hall, R. (1997). Practicing representation: Learning with and about representational forms. *Phi Delta Kappan*, 78, 361–367.

Hall, R., & Stevens, R. (1995). Making space: A comparison of mathematical work at school and in professional design practice. In S. L. Star (Ed.), *Cultures of computing* (pp. 118–145). London: Basil Blackwell.

Henderson, K. (1991). Flexible sketches and inflexible data bases: Visual communication, conscription devices, and boundary objects in design engineering. *Science, Technology, and Human Values*, 16(4), 448–473.

Henderson, K. (1999). *On line and on paper: Visual representations, visual culture and computer graphics in design engineering*. Cambridge, MA: MIT Press.

Hess, D. J., Breyman, S., Campbell, N., & Martin, B. (2007). Science, technology, and social movements. In E. J. Hackett, O. Amsterdamska, M. Lynch, & J. Wajcman (Eds.), *The handbook of science and technology studies* (3rd ed.). Cambridge, MA: MIT Press.

Holland, D. C., Lachicotte, W., Skinner, D., & Cain, C. (1998). *Identity and agency in cultural worlds*. Cambridge, MA: Harvard University Press.

Hutchins, E. (1995). *Cognition in the wild*. Cambridge, MA: MIT Press.

Johri, A. (2010). Open organizing: Designing sustainable work practices for the engineering workforce. *International Journal of Engineering Education*, 26(2), 278–286.

Johri, A. (2011). Sociomaterial Bricolage: The creation of location-spanning work practices by global software developers. *Information and Software Technology*, 53(9), 955–968.

Jonassen, D., Strobel, J., & Lee, C. B. (2006). Everyday problem solving in engineering: Lessons for engineering educators. *Journal of Engineering Education*, 95(2), 139–151.

Jordan, B., & Henderson, A. (1995). Interaction analysis: Foundations and practice. *The Journal of the Learning Sciences*, 4(1), 39–103.

Korte, R., Sheppard, S., & Jordan, W. (2008). *A qualitative study of the early work experiences of recent graduates in engineering*. Paper presented at the American Society for Engineering Education, Pittsburgh, PA.

Latour, B. (1987). *Science in action: How to follow scientists and engineers through society*. Cambridge, MA: Harvard University Press.

Latour, B. (2005). *Reassembling the social: An introduction to Actor-Network-Theory*. Oxford: Oxford University Press.

Law, J. (1987). Technology and heterogeneous engineering: The case of the Portuguese expansion. In W. Bijker, T. Hughes, & T. Pinch (Eds.), *The social construction of technological systems* (pp. 111–134). Cambridge, MA: MIT Press.

Lynch, M., & Woolgar, S. (Eds.). (1990). *Representation in scientific practice*. Cambridge, MA: MIT Press.

National Academy of Engineering (NAE) (2005). *Educating the engineer of 2020: Adapting engineering education to the new century*. National Academies Press, Washington D.C.

Nespor, J. (1994). *Knowledge in motion: Space, time and curriculum in undergraduate physics and management*. London: Routledge.

Noble, D. (1977). *America by design*. New York, NY: Alfred A. Knopf.

O'Connor, K. (2001). Contextualization and the negotiation of social identities in a geographically distributed situated learning project. *Linguistics and Education*, 12, 285–308.

O'Connor, K. (2003). Communicative practice, cultural production, and situated learning: Constructing and contesting identities of expertise in a heterogeneous learning context. In S. Wortham & B. Rymes (Eds.), *Linguistic anthropology of education* (pp. 63–91). London: Praeger.

O'Connor, K., & Allen, A. (2010). Learning as the organizing of social futures. *NSSE Yearbook*, 109(1), 160–175.

O'Connor, K., Bailey, T., Garrison, L., Jones, M., Lichtenstein, G., Loshbaugh, H.,... Stevens, R. (2007). Sponsorship: Engineering's tacit gatekeeper. In *Proceedings of the 2007 ASEE Annual Conference and Exposition*, Honolulu, HI.

Orr, J. (1996). *Talking about machines: An ethnography of a modern job*. Ithaca, NY: Cornell University Press.

Penuel, W., & O'Connor, K. (2010). Learning research as a human science: Old wine in new bottles? *National Society for the Study of Education*, 109(1), 268–283.

Polanyi, M. (1966). *The tacit dimension*. New York, NY: Doubleday.

Radcliffe, D. F. (2006). Shaping the discipline of engineering education. *Journal of Engineering Education*, 95(4), 263–264.

Riley, D., Pawley, A., Tucker, J., & Catalano, G. D. (2009). Feminisms in engineering education. *NWSA*, 21(2), 21–40.

Rittel, H. W. J., & Webber, M. W. (1973). Dilemmas in a general theory of planning. *Policy Sciences*, 4(2), 155–169.

Robinson, J. G., & McIlwee, J. S. (1991). Men, women, and the culture of engineering. *Sociological Quarterly*, 32(3), 403–421.

Sellen, A. J., & Harper, R. (2003). *The myth of the paperless office*. Cambridge, MA: The MIT Press.

Skinner, D., Valsiner, J., & Holland, D. (2001). Discerning the dialogical self: A theoretical and methodological examination of a Nepali adolescent's narrative. *Forum: Qualitative Social Research*, North America. Retrieved from http://www.qualitative-research.net/index.php/fqs/article/view/913/1994

Smith, T., & Rivkin, J. (2008). A replication tudy of Alan Blinder's "How many US jobs might be offshorable?" *Harvard Business School Strategy Unit Working Paper* (08-104).

Steering Committee of the National Engineering Education Research Colloquies. (2006). Special report: The research agenda for the new discipline of engineering education. *Journal of Engineering Education*, 95(4), 259–261.

Stevens, R. R. (1999). *Disciplined perception: Comparing the development of embodied mathematical practices in school and at work*. Doctoral dissertation. University of California, Berkeley, 1999. *Dissertation Abstracts International*, 60, 06A.

Stevens, R. (2000). Divisions of labor in school and in the workplace: Comparing computer and paper-supported activities across settings. *Journal of the Learning Sciences*, 9(4), 373–401.

Stevens, R., & Hall, R. (1998). Disciplined perception: Learning to see in technoscience. In M. Lampert & M. L. Blunk (Eds.), (*Talking mathematics in school: Studies of teaching and learning* (pp. 107–149). Cambridge: Cambridge University Press.

Stevens, R., Amos, D. Jocuns, A. & Garrison, L. (2007). Engineering as lifestyle and a meritocracy of difficulty: Two pervasive beliefs among engineering students and their possible effects. In *Proceedings of the 2007 American Society for Engineering Education Annual Conference*, Honolulu, HI.

Stevens, R., Garrison, L., & Satwicz, T. (2007). *"Mathematics" as a shifting object across the early career of an engineer*. Paper presented at Invited Symposium for Education and the Professions Division, AERA, Chicago, IL.

Stevens, R., O'Connor, K., & Garrison, L. (2005). Engineering student identities in the navigation of the undergraduate curriculum. In *Proceedings of the 2005 Association of the Society of Engineering Education Annual Conference*, Portland, OR.

Stevens, R., O'Connor, K., Garrison, L., Jocuns, A., & Amos, D. (2008). Becoming an engineer: Toward a three dimensional view of engineering learning. *Journal of Engineering Education*, 97(3), 355–368.

Suchman, L. (2000). Embodied practices of engineering work. *Mind, Culture and Activity*, 7(1–2), 4–18.

Trevelyan, J. P. (2007). Technical coordination in engineering practice. *Journal of Engineering Education*, 96(3), 191–204.

Trevelyan, J. (2010). Reconstructing engineering from practice. *Engineering Studies* 2(3), 175–196.

Trevelyan, J. P., & Tilli, S. (2007). Published research on engineering work. *Journal of Professional Issues in Engineering Education and Practice*, 133(4), 300–307.

Vinck, D. (Ed.) (2003). *Everyday engineering: An ethnography of design and innovation*. Cambridge, MA: MIT Press.

Williams, R. (2002). *Retooling: A historian confronts technological change*. Cambridge, MA: MIT Press.

Will-Zocholl, M., & Schmiede, R. (2011). Engineers' work on the move: Challenges in automobile engineering in a globalized world. *Engineering Studies* 3(2), 101–121.

Yates, J. (1989). *Control through communication: The rise of system in American management*. Baltimore, MD: The John Hopkins University Press.

Yoo, Y., Henfridsson, O., & Lyytinen, K. (2010). Research commentary – the new organizing logic of digital innovation: An agenda for information systems research. *Information Systems Research*, 21(4), 724–735.

Youngjin, Y., Lyytinen, K. J., Boland, R. J., & Berente, N. (2010). *The next wave of digital innovation: Opportunities and challenges*. A Report on the Research Workshop Digital Challenges in Innovation Research. Retrieved from http://papers.ssrn.com/sol3/papers.cfm?abstract_id=1622170

Zuboff, S. (1988). *In the age of the smart machine: The future of work and power*. New York, NY: Basic Books.

Zussman, R. (1985). *Mechanics of the middle class: Work and politics among American engineers*. Berkeley, CA: University of California Press.

Part 2

ENGINEERING LEARNING MECHANISMS AND APPROACHES

CHAPTER 8

Problem-Based and Project-Based Learning in Engineering Education

Merging Models

Anette Kolmos and Erik de Graaff

Introduction

In the practice of engineering education, there is a wide variety of implementations of problem-based or project-based learning (PBL). In this chapter we aim to explain the relationships between different types of problem-based and project-based learning to help teachers and educational managers make innovative choices and provide benchmarks for educational researchers. We present a combined understanding of problem- and project-based learning, the theoretical and historical background, and the different models of PBL that can capture the existing practices, ranging from small- to large-scale practice, from classroom teaching to institutional models, and from single-subject to interdisciplinary and complex knowledge construction.

It is well known that one-way dissemination of knowledge by means of lectures is not very effective in achieving learning (van der Vleuten, 1997). In higher education concepts such as "self-directed-learning," "case-based learning," "inquiry based learning," "experiential learning," "service learning," "project-based service learning," "active learning," CDIO (Conceive, Design, Implement, and Operate), "project-based learning," and "problem-based learning" were introduced in the decades after the Second World War. All these new learning concepts come under the umbrella of learner-centered or student-centered learning models. Problem-based and project-based learning, both known as PBL, originate from the reform universities, and the new educational models, established between 1965 and 1975. In problem-based learning, problems form the starting point for students' learning emphasizing a self-directed learning process in teams. The educational model problem-based learning was introduced at curriculum scale at the medical faculty of McMaster University, Canada, followed by Maastricht University in the Netherlands and many others. Project-based learning shares the aspect of students working on problems in teams, but with the added component that they have to submit a project report completed collaboratively by the project team. The problem- and project-based/project organized model adopted at

Aalborg University and Roskilde University, Denmark, was inspired by the critical pedagogy in Europe after the student revolts of the 1960s. At Aalborg University both models of PBL were eventually combined in problem-based project organized learning, which was practiced at all faculties – the Faculty of Engineering and Science being the largest. This combined approach is the central point of reference for this chapter, as the pedagogical development in engineering education indicates that both educational practices are successful in their own way and the abbreviation PBL is here defined as including both practices.

A curriculum can be regarded as a social construction depending on culture, national regulations, institutional policies, and academic staff. During the last forty to fifty years, many changes have occurred, especially in engineering education, and universities all over the world have developed diverse PBL models and practices (Beddoes, Kacey, Jesiek, Brent, & Borrego, 2010). Several notable PBL implementations are to be found all over the world. In U.S. engineering education, in general, there is a trend that the capstone projects are carried out as a team-based and problem-based project (McKenzie, Trevisan, Davis, & Beyerlein, 2004), but at a more comprehensive level, there is the P5BL model at Stanford University (Fruchter & Lewis, 2003) and the work with projects at Olin College – although they do not claim any relation to any global community (Sommerville et al., 2005). In Mexico, Monterey Tech's has developed diverse PBL models at different campuses (Lopez-Islas, 2001). In Brazil, University of São Paolo's implemented change of programs (Ulisses, Valéria, & Homero, 2009). In Australia the University of Queensland (Crosthwaite, Cameron, Lant, & Lister, 2006), Central Queensland University (Joergensen & Howard, 2005), and Victoria University (Ozansoy & Stojcevski, 2009) have many years of project work practiced. In Asia, University of Technology and University Tun Hussein Onn, both in Malaysia, have course modules with PBL (Yusof, Tasir, Harun, & Helmi, 2005);

Singapore Polytechnique is running a special PBL version known as 'one problem per day' (O'Grady & Alvis, 2002). In India, there is an increasing use of various PBL models (Shinde & Kolmos, 2011). In Europe, the Bologna process, with a focus on student-centered learning and learning outcomes, has created increasing interest in PBL, and there are similar trends all over the world (Barrett & Moore, 2011), including in many European universities such as Aveiro University, Portugal (Graaff & Kolmos, 2007); Université Catholique de Louvain, Belgium (Galand & Frenay, 2006; Galand, Raucent, & Frenay, 2010); and universities in the United Kingdom, for example, Coventry, Newcastle, Sheffield, and Manchester (Graham, 2009).

However, mentioning specific universities is unadvisable, as there is no full overview of all the diverse PBL practices, ranging from large-scale implementation at an institutional level to small-scale implementation in a class room or a laboratory. Nevertheless, this indicates that the PBL practice is becoming more diverse and the definition of problem- and project-based learning more unclear. The concept of PBL is used for both small team-based projects, working on cases more or less defined in terms of learning content by one lecture, and for large mega-projects such as construction of a satellite or a racing car, requiring multidisciplinary content.

Need for New Engineering Competences

As early as the 1960s, there was a formulated need for more qualified academics and engineers, leading to an expansion of higher education capacity and the establishment of many new institutions both general universities and specifically engineering colleges (de Graaff & Sjoer, 2006). The relation between universities and society changed. Gibbons et al. (1994) note that the relation between science and society changed radically from an approach characterized by "science speaking" to society to "society speaking to science" (Gibbons et al., 1994; Nowotny, Scott, & Gibbons, 2001). Recently, several reports on new demands for engineering knowledge, skills, and competences followed the same reasoning, that the universities have to pay much greater attention to real-life problems and to societal needs, in order, especially, to address employability agenda including collaboration with companies (National Academy of Engineering, 2004; Royal Academy of Engineering, 2007). Industry requires graduates who are able to participate in engineering project organizations and to collaborate (Chinowsky, 2011; Kolmos & Holgaard, 2010). At the same time, the accreditation bodies have defined transferable or professional skills as an important part of the curriculum (ABET, EURACE), which indicates that there is a growing awareness of not only the learning of isolated discipline skills and competences, but also an integrated curriculum approach. This emphasizes learning as an integrated and exemplary act that goes beyond technical knowledge and skills. With new requirements for complex and intercultural competences, there will be a need for establishing mega-projects across disciplines and cultures and, therefore, a constant development of the PBL practices.

The requirement for complex competences will increase with the need for more green engineers (Jamison, 2001). This will further necessitate the creation of new interdisciplinary techno-science programs and, for students to learn to analyze relevant problems leading to new innovations, the societal needs and social structures, and contextual issues must be brought into engineering education together with collaborative competences (Hyldgaard Christensen, Delabousse, & Meganck, 2009; Jamison, Hylgaard Christensen & Botin, 2011). Bucciarelli (1994) has pointed out that, combined with technical knowledge, there is an ongoing negotiation of understanding of the definitions of concepts in a collaborative design process. Because engineers have a global workplace, education needs to address these new types of requirements. Therefore, students should already, have experienced some

of these activities during their education. It is expected that the ability to handle complexity and intercultural collaboration will be an embedded engineering capacity in the future, and innovation is based on collaborative knowledge construction (Sawyer, 2008).

The signal from society clearly indicates a need for engineers with relevant knowledge, skills, and competences. The challenge for universities is to meet these demands. Strobel and van Barneveld (2009) point out that the main reason many engineering universities implement PBL is that previous graduates were ill prepared for the labor market and the need for new competences. In the report from the Bernhard M. Gordon MIT Engineering Leadership program, there is a strong focus on project-oriented programs as a variable for leadership competences (Graham, Crawley, Mendelsohn, 2009). Studies on employers' needs for engineering competences very often come out with results on more project experiences (Kolmos & Holgaard, 2010) and that there is a need for engineering students to know more about engineering practice and business models and to act reflectively as practitioners (Schön, 1987). Another driver identified by Strobel and van Barneveld (2009) is the institutional interest in increased retention rate among students.

A Historical Perspective on PBL

The new universities – also known as reform universities – that were founded during the 1960s and 1970s were a response to the societal needs and had new pedagogical approaches and educational models. These universities were based on three ideas: (1) a need for new knowledge and skills in the labor market, (2) the study programs were too fragmented and without relation to the outside world, and (3) there was a need for more democracy and student influence (Carter, Eriksen, Horst, & Troelsen, 2003; Nielsen & Webb, 1999). These reform universities proved that new teaching and learning models were applicable and that the labor market welcomed the graduates.

Prominent among the educational models that were used was problem-based learning. The first place to apply PBL principles as the core for a new curriculum in medicine was McMaster University in (Neufeld & Barrows, 1974; Woods, 1994). Maastricht in the Netherlands; Linköping University, Sweden; and Newcastle in Australia followed the example set by McMaster, establishing PBL curricula in their medical schools. Around the same time, a problem-oriented and project-organized model was developed in Europe. This model was introduced at Aalborg University and Roskilde University, Denmark, and in the first years at Bremen University, Germany.

The new pedagogy that grew out of the reform universities has a history, although there is no complete agreement on where these ideas came from. The Danish reform universities were based primarily on German theories such as experiential learning formulated as an emancipatory pedagogy for the working class by Negt and Kluge (Illeris, 1976). However, in reviewing the literature, one can find that American researchers from the Chicago group such as Dewey and Kilpatrick formulated ideas about project work and democracy years earlier (Dewey, 1938; Kilpatrick, 1918; Kilpatrick, Bagley, Bonser, Hosic, & Hatch, 1921). Also, Brazilian influence dating back to Paulo Freire may be identified (Illeris, 1976). In the period of the reform universities, these authors were very much regarded as the champions of a new form of education for democracy and more equal rights in society. The reform universities share many experiences in establishing new teaching and learning practices; however, there is also an individual history related to each of them.

McMaster followed by Maastricht organized a curriculum with thematic blocks, in which the year is divided into a series of blocks of approximately six weeks. Each period focuses on a particular theme, such as a group of complaints or organ systems. A block book presents a series of cases, which the students are to analyze. The case is the problem that is to trigger the student's learning process. A problem in PBL does not need to be solved; it just serves to initiate the learning process (Norman, 1988). The students work together in self-directed study groups, discussing and analyzing specially prepared case descriptions. They organize the analyses of the cases using a procedure called the Seven Jump (Schmidt & Moust, 2000). The objective is to define learning goals, which are studied individually. The results of this individual study are reported and reviewed during the next group session. Just as in real-life practice, there is an integration of subject disciplines. The typical study group (eight to twelve students) meets one or two times a week. In the study group, each student presents his or her work, which is discussed, and the group discusses who should continue with what tasks. Research has shown the effectiveness of the self-directed learning process (Yew & Schmidt, 2008). The role of the teacher who attends the meetings is primarily to facilitate the learning process, in other words, to facilitate the group's collaboration and internal communication.

The Aalborg and Roskilde models are problem based by definition, as every student project starts with a problem. Each semester groups of five to eight students work on a problem, which they normally define by themselves within a thematic framework and, after analyzing and solving the problem, they write a project report. Normally, the projects run for a semester (half year) and the project work allows them to develop competencies in project management and collaboration. The more a problem replicates real life, the more students experience it as motivating. Working on a project can be seen as a way of organizing the learning process, but it also gives competences in how to start, carry out, and finalize a project, which is needed by companies (Kjaersdam & Enemark, 1994; Kolmos, Fink, & Krogh, 2004; Kolmos & Holgaard, 2010; Olsen & Pedersen, 2005).

The fact that PBL grew out of practices that were institutionalized has contributed to the success and dissemination of this pedagogy. The new educational models were

Table 8.1. Original Learning Principles at the PBL Universities

McMaster and Maastricht Universities *Problem-based learning*	*Aalborg and Roskilde Universities* *Problem-based and project organized learning*
• Problems form the focus and stimulus for learning. • Problems are the vehicle for development of problem-solving skills. • New information is acquired through self-directed learning. • Student-centered • Small student groups • Teachers are facilitators/guides • (Barrows, 1996)	• Problem orientation • Interdisciplinary • Exemplary learning • Participant-directed • Teams or group work • (Illeris, 1976)

institutionalized with a new approach, and the practice did not just cease if one academic staff member moved from the institution. It was possible to build up a sustainable community by recruiting academic staff who were positive about these new ideas and to evaluate and experiment with the practices. At first there was much skepticism toward the new graduates, as the labor market did not know what to expect of them. However, after about ten years, the attitudes toward the new universities became much more positive. During the 1970s and the 1980s, the reform universities proved their value by getting very good responses from the companies that employed the graduates.

Definitions of PBL

Returning to the roots of PBL, Barrows (1996) defines six characteristics that are all related to the learning process, whereas Illeris (1976) defines five characteristics that are related to both the learning process and the implications for the social and the content dimension of learning (see Table 8.1).

An important difference between the two aforementioned PBL approaches can be found in the content dimension. The Danish tradition emphasizes the interdisciplinary and exemplary aspect. The exemplar approach, as stipulated by Klafki (2001), is an important principle for the selection of content which should be of current and future significance, as well as in accordance with

the overall learning outcomes of the curriculum. The exemplar learning process is characterized by building on prior knowledge, discovery learning and being both meaningful and challenging to students. This is an essential principle with respect to the management of the learning process, both in terms of what to learn (identifying problems that are often real problems) and how to decide on the relevance to the curriculum.

PBL problems can be small or big, authentic or scholastic, practical or theoretical, etc. There are many combinations of small problems and big projects, and there is a need for a more theoretical definition of PBL that allows for variation in practice.

Many of these definitions suggest that there is a substantial difference between problem-based and project-based learning. In particular, there are differences in the way the curriculum is organized. For instance, Prince and Felder define problem-based and project-based learning as follows: "Problem-based learning (PBL) begins when students are confronted with an open-ended, ill-structured, authentic (real-world) problem and work in teams to identify learning needs and develop a viable solution, with instructors acting as facilitators rather than primary sources of information" (Prince & Felder, 2006, p. 128). In comparison, "Project-based learning begins with an assignment to carry out one or more tasks that lead to the production of a final product – a design, a model, a device or a computer simulation. The culmination of the project is normally a

Problem-based learning	Project-based learning
Process	Product and outcome
Focus on the problem	Focus on problem solving
Students work out learning needs	Lectures
Facilitation	Supervision
Can be from the beginning	Often in the end of degree
Learning cross disciplines is a necessity	Can bring together taught subjects

Figure 8.1. Differences between problem-based and project-based learning according to Savin-Baden. (Supervision is here understood as advice – U.S. term.)

written and/or oral report summarizing the procedure used to produce the product and presenting the outcome" (Prince & Felder, 2006, p. 130).

According to this interpretation, a project is a narrowly formulated task, and problem-based learning is an open-ended and ill-defined approach. Savery (2006) and Savin-Baden (2003, 2007) are very much in accordance with the definitions in the preceding text, describing the difference between the two models as shown in Figure 8.1. In their view "problem-based" stands for the process-oriented approach, whereas projects involve a much more instrumental and product-oriented approach.

Several other researchers propose their own definitions. Capraro and Slough define project-based learning as: "A well-defined outcome and ill-defined task. PBL for the purposes here is the use of a project that often results in the emergence of various learning outcomes in addition to the ones anticipated" (Capraro & Slough, 2009, p. 5). Algreen Ussing and Fruensgaard define a project as a complex, unique and situated task that cannot be repeated and will always involve an open approach (Algreen-Ussing & Fruensgaard, 1990). These definitions of projects and project-based learning are different from the definitions proposed by Prince and Felder, Savery, and Savin-Baden, as they stipulate that a project will always

start with a problem and that this problem, by definition, will always be unique. Such an interpretation of project-based learning clearly indicates that project-based learning cannot exist without a problem-oriented approach (Kolmos, 1996), and this approach constitutes the basis for the understanding of PBL in this chapter.

De Graaff characterizes the relationship between problem-based learning (Maastricht model) and problem-based and project-organized learning (Aalborg model) in an input–output model designating five dimensions (Graaff, 1995).

Input:	The first dimension concerns the kind of stimulus given to the students. It varies from discipline knowledge as available in textbooks to problems from professional practice.
Situation:	The second dimension presents different possibilities in the environment, ranging from the well-organized classroom to the real workshop.
Qualification teachers:	The third dimension relates to the qualifications of the teachers. Do you need primarily teachers with thorough didactic training, or

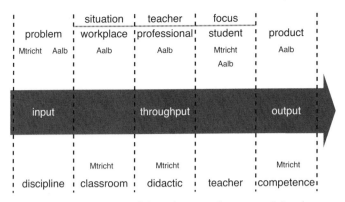

Figure 8.2. Dimensions of the educational process. (Mtricht = Maastricht University and Aalb = Aalborg University.)

Orientation: do you cherish professional experience? The fourth dimension refers to the orientation of the educational process. This ranges from teacher-centered (the teacher is the focal point) to student-centered (independent, self-directed learning).

Output: The fifth dimension concerns the type of result that is most valued. Is there a concrete product that can be shown to the world? Or is there just the invisible gain of knowledge or skills?

The extremes of the dimensions represent the two most ancient forms of teaching: "on the job training" and "frontal classroom teaching." As can be seen in Figure 8.2, both the Maastricht PBL model and the project-based model lean toward the "on the job" training, the latter even more so. However, this model does not aim to pin down educational models. Rather, it aims to be a tool to identify a particular position in the development of a curriculum to distinguish it from the ultimate objective.

PBL as a Set of Learning Principles

History and the dissemination of PBL indicate that the concept of PBL can no longer be defined only at the concrete curriculum level, because there are so many different practices. PBL has to be defined at a more abstract level, as a philosophy and set of learning principles (Graaff & Kolmos, 2003, 2007). Owing to contextual conditions (cultural, national, institutional), specific curriculum models cannot be transferred directly between countries, as any PBL practice is a social construction and what might work in one cultural or institutional setting might not work in another. The element of cultural practice is more dominant in a student-centered learning model compared with a more instructional model based on textbook approaches. Thus, a more abstract notion of PBL is needed to capture diversity in the concrete curriculum and classroom approach cross culture and subjects.

PBL is a very comprehensive system of organizing the content in new ways and students' collaborative learning, enabling them to achieve diverse sets of knowledge, skills, and competences. As already stated, there is an increasing demand for competences that goes beyond the technical field and reaches into process competences and integrative personal competences such as collaborative and creative knowledge processes. This means that education can no longer address only the cognitive part; it is necessary to think in broader terms.

Wilkerson and Gijselaers (1996) connected PBL with educational theory and formulated three basic learning theory principles: (1) learning is a constructive and not a receptive process; (2) metacognition affects learning; and (3) social and contextual factors influence learning. These three principles indicate that in order to understand and to develop a PBL curriculum, cognitive learning principles are not sufficient.

Barnett and Coate (2004) define the curriculum as knowing, acting, and being. First, this views the curriculum as a space for learning processes. Second, the curriculum should encompass knowing, acting, and being. Knowing and acting cover the normal notion of a curriculum, whereas being addresses citizenship, social responsibility, values, and personal and social identity. The knowing, acting, and being are interconnected, and the identity students develop in a PBL curriculum is both as an individual and a collaborative learner, whereby individuals are in a constant dialogue with each other about the content of the project (knowing and acting).

This broader approach to the understanding of curriculum and competences implies a broader understanding of learning. Knud Illeris defines learning as "any process that in living organisms leads to permanent capacity change and which is not solely due to biological maturation or ageing" (Illeris, 2007, p. 3). This is a very broad definition; however, the intention is to overcome more limited definitions such as learning as outcome of learning processes or learning as interaction. Illeris (2003, 2007) has developed a more holistic learning approach based on both cognitive and affective learning as one dimension and individual and social learning as another dimension.

Klafki (2001) argues that the exemplary learning process should connect to the learner's cognitive and affective stage, but also that there need to be challenges creating tension to trigger the learner to develop knowledge, skills, or competences. Furthermore, the classroom allows the learner to go beyond teaching to discover learning.

The PBL learning principles are based on this holistic view of curriculum and learning encompassing both the individual and the social, cognitive, and affective. By analyzing the specific PBL models and PBL prac-tices, combined with theoretical reflections, three main clusters of learning principles emerge: learning, social, and contents. The three learning principles are interconnected in several ways, but each of them also has its own domain.

The **learning approach** concerns the learning process of working with *problems*, which involves identification, analysis, and solution. It can be real-life/authentic, practical or purely theoretical problems. Many authors claim PBL as a means of working with authentic problems; however, this depends on the entire combination of the different learning principles and the overall objective for the learning process. The problem forms a starting point for the learning process, but it is also a reference point for the learning process, as the problem indicates the purpose of the learning process. This means that students can orient their reading toward this particular problem to gain a deeper understanding.

Ownership of the problem is also an important principle. If students define the problems by themselves, it might be as motivating to work with a theoretical problem as with an authentic or real-life problem. However, if the problem is given by the teacher, it could be made more important if related to real-life problems.

The learning principles also encompass the organization of the learning process, which can be *case-based* or *project-based*. Case-based means that the students are working with predefined cases containing either well-defined problems or ill-defined problems. Project-based contains a unique task that involves more complex and situated problem analyses and problem-solving strategies. Many researchers, from the Danish project work tradition in particular, have indicated that working with problems can be understood as a *research* process. The phases in a research process are identical to many of the formulated problem learning phases for analyzing cases and working on projects.

The **social approach** covers team-based learning and requires interaction between the individual and the group. The *team learning* aspect underpins the learning process as a social act, in which learning takes place through dialogue and communication. Furthermore, the students are not only learning from each other, but they also learn to share knowledge and organize the process of collaborative learning and collaborative knowledge construction.

The social approach also covers the concept of *participant-directed learning*, which indicates a collective ownership of the learning process and, especially, the identification of the problem. This is typically called self-directed learning. However, as it takes place in a collaborative setting, it is the participants who together negotiate the direction for learning.

The **contents approach** covers selection of knowledge and skills. PBL involves *interdisciplinary learning* across traditional subject-related boundaries. As soon as students are working with real-life problems, it is necessary to use knowledge from different disciplines or at least acquire the knowledge of the limitations of the problem analysis and problem-solving proposals.

If students have freedom to choose projects within a given theme, *exemplary practice* becomes an important tool to ensure that learning outcome is exemplary to the overall objectives. Thus, the interaction between students and supervisors is important. Furthermore, in the process of analyzing and solving problems, students apply theories, which enhances the understanding of the relation between *theory* and *practice*. This type of learning process is identical to a research process and training core research methodologies such as formulating a problem/research question, analyzing a problem by literature review, specifying and defining the final problem/research question, theoretical and methodological considerations, problem solving and data analysis, and conclusion.

PBL Models in Engineering Education

A review of the literature on PBL in engineering education clearly shows that most often projects are used – either these are more discipline based (task based) or problem based. Beddoes et al. (2010) have reviewed international engineering education research on PBL. They emphasize that most of the literature review on effectiveness is within subject fields other than engineering education. However, there are plenty of articles from all over the world reporting and assessing PBL engineering practices within existing courses.

For both PBL models, the problem is the start of the learning process. However, the seven jumps procedure used at Maastricht University focuses on the analysis of the problem, whereas the project phases also encompass the problem-solving part and the overview of the methodology used in the problem analysis and the problem-solving phases (see Figure 8.3).

In the engineering field, it is both the problem analysis and the problem solving phases that are important learning phases (Powel & Weenk, 2003). Engineers need to know what the problems are, to formulate the requirements, and to solve the problem by development of relevant technological solutions. Furthermore, it is important that these learning processes are team based in order to acquire the knowledge sharing within a smaller team as well as globally, and that the collaboration is oriented toward both process and product so that engineers learn the competence of collaborative knowledge construction.

Model for the Implementation of PBL

There is a need for a more concrete model for the analysis, construction, and implementation processes in the curriculum. To meet the implementation needs, a PBL curriculum model has been developed that is based on the PBL learning principles and on the curriculum theories of alignment and social construction (Kolmos, Graaff, & Du, 2009; Savin-Baden, 2003, 2007; Savin-Baden & Wilkie, 2004). The PBL curriculum model is linked to the PBL learning principles, but the seven elements have been identified as objectives and outcomes, types of

	Seven jumps (Gijselaers, 1996)	Project phases (Algreen Ussing, 1995)
Problem analysis	1. Clarify terms and concepts not readily comprehensible 2. Define the problem 3. Analyse the problem and offer tentative explanations 4. Draw up an inventory of explanations 5. Formulate learning objectives 6. Collect further information through private study 7. Synthesise the new information and evaluate and test it against the original problem. Reflect on and consolidate learning	1. Initiating problem (the trigger for the problem – what starts it out) 2. Problem analysis (analysis of the problem – for whom, what and why) 3. Definition and formulation of problem (specification requirement)
Problem solving		4. Problem solving methodologies (overview of possible solutions and assessment of impact) 5. Demarcation (argumentation for the choice of solution) 6. Solving the problem (carry out the solution – construction/design/ further analysis) 7. Implementation (prototype and sometimes real systems) 8. Evaluation and reflection (Impact, effect and efficiency of solution)

Figure 8.3. Differences between the seven jumps and project phases.

problems and projects, students' learning, progression and size, academic staff and facilitation, space and organization, and, finally, assessment and evaluation, see Figure 8.4. These elements were identified on the basis of a theory on relationship models and principles of alignment (Biggs, 2003; Hiim & Hippe, 1993). All components are elementary in a basic PBL curriculum and must be aligned to a certain degree. Depending on subject areas and cultural and national requirements, there might be more elements to consider and numerous dimensions for each element.

The principle of alignment is an underlying assumption for this PBL curriculum model. If there is a change in one element, it will affect all of the other elements as well. However, each of the elements can be interpreted in different ways, and bearing in mind that in practice there is a huge difference in the definitions of problem-based learning and project-based learning, there are different approaches to PBL.

There are basically two poles in the interpretation and implementation of the curriculum elements: a teacher-controlled approach on the one side, and an innovative and learner-centered approach on the other. Even if PBL is a student-centered learning methodology, a teacher-controlled approach does exist, in which the teacher to a certain degree knows the problem, the methodology, and the results. In comparison, the innovative and learner-centered approach implies that the teacher might not know the problem beforehand. He or she knows the variety of methodologies but not the solution.

In each of the curriculum elements, there are several dimensions as illustrated in Figure 8.5. Between the two approaches, there are mixed modes, and most of the PBL practice is defined as some kind of a mixed mode. For example, concerning the assessment system, there are many ways of practicing assessment to support the learning objectives such as peer assessment, formative assessment, and so forth. Therefore, the point of formulating these poles is to create awareness of the choices that have to be made in the implementation process of PBL regardless of whether this is at a single course level or a system level. Between

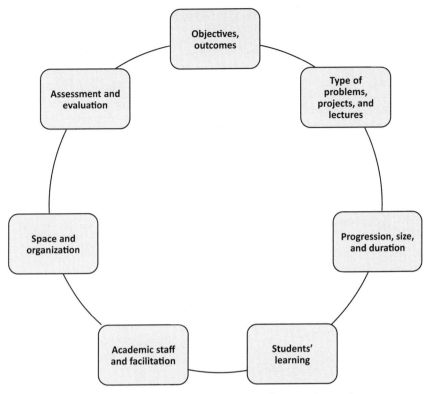

Figure 8.4. PBL curriculum model (Kolmos et al., 2009).

Curriculum element	Discipline and teacher centered approach	Innovative and learner centered approach
Outcomes	Traditional discipline objectives	PBL and interdisciplinary methodological objectives
Type of problems and projects	Well-defined problems	Ill-defined problems
	Disciplined projects	Problem projects
	Lectures determine the project	Lectures to support the project
Progression, size and duration	No visible progression	Visible and clear progression
	Minor part of the curriculum	Major part of course/curriculum
Students' learning	No supporting courses	Supporting courses
	Acquisition of knowledge	Construction of knowledge
	Collaboration for individual learning	Collaboration for innovation
Academic staff and facilitation	No training	Training courses
	Teacher-controlled supervision	Facilitator/ process guide
Space and organization	Administrative support to courses	Administrative support to PBL curriculum
	Lecture rooms	Group rooms facilitate teamwork
Assessment and evaluation	Individual assessment	Group assessment
	Summative course evaluation	Formative evaluation

Figure 8.5. Dimensions of PBL curriculum elements (Kolmos et al., 2009).

the poles or extremes, there are plenty of variation possibilities.

Outcomes

In PBL there is a need for definition of learning objectives and learning outcomes as reference points for students' projects. The learning outcomes might be formulated within disciplines and across disciplines. However, for a PBL curriculum, it is important that the formulations are not too tight and narrow and have more characteristics of methodological objectives. This will allow variation in the students' projects. Most PBL curricula and courses also formulate process skills such as collaboration, project management, communication, and so forth. These skills are embedded in the learning process. Nevertheless, the learning of process skills will be enhanced by reflecting, experimenting and conceptualizing the experienced learning process (Kolmos et al., 2004).

Types of Problems, Projects, and Coherence with the Courses

The type of problem that students should work with is related to the learning objectives and outcomes. If there are very narrow learning outcomes, the problem will also be narrow. If the learning outcomes are formulated in broader terms, it will allow for more variation. Sockalingam and Schmidt (2011) distinguish between features and functions of problems, where the features refer to the design of the problem, and the function refers to the desired learning outcomes.

The projects of a curriculum should vary; for some semesters, it will be necessary to have discipline projects with well-defined problems, and for other semesters, there might be more open learning outcomes and open innovation projects.

There are also lectures in the PBL curriculum, and the relation between lectures and projects – as well as the role of projects – has to be considered, as either applying the knowledge taught and/or creating new knowledge.

Progression, Size, and Duration

How does PBL progress during the curriculum? Does it move from more narrowly disciplined projects toward more open innovation projects? The same questions apply as to where in the system smaller projects or cases would be of greater benefit and where it would be more efficient to have more comprehensive projects. Savin-Baden (2003) has developed a variation of models for progression in blended PBL curricula ranging from full integration to smaller pockets of PBL.

Students' Learning

Students do not know how to collaborate or to run projects. There are various collaboration modes and project management systems, and it will be an advantage for the learning of these process skills to adopt a more experimental and reflective learning approach. Reflection is an important part of developing competences in handling the process. Studies indicate that reflection methods, such as portfolios with individual and team-based reflections, have an impact on students' approaches to the learning process (Kolmos et al., 2004; Turns, Cuddihy, & Guan, 2010).

Academic Staff and Facilitation

Many PBL universities have mandatory training or offer training that introduces PBL in the system, as well as already established systems for new staff. Most often the content of the training concerns the facilitator/adviser role, which is a very difficult teacher role that requires knowing when to give direct answers to students' questions or to encourage them to be independent in searching and learning processes.

Space and Organization

Space is an issue in the facilitation of students' team learning. ICT can be a type of solution, but there will still be a need for physical space. Distribution and recalculation of resources between traditional lecturing and facilitation of students' learning have to be considered.

Assessment and Evaluation

According to Gibbs (1999), assessment is one of the core drivers for students' learning. Therefore, the type of assessment system used is important for the success of PBL. There is a tremendous body of research on formative assessment, self-assessment, peer assessment, and summative assessment methods. Assessment in a PBL learning environment can be aligned with the learning outcomes and the learning methods. Formative assessment is very much in alignment with the principles of PBL, because it supports awareness of the learning process, and the summative assessment practice should judge both the individual performance as well as the group performance. If assessment has to focus on both learning and performance, this involves an oral assessment combined with types of written work such as portfolios, projects, and so forth (Brown & Glasner, 1999). The assessment practice is very often a group based assessment with individual judgment and grading of an individual's performance (Holgaard & Kolmos, 2009).

Alignment among all the curriculum elements is important, and this PBL curriculum model is aimed to create awareness of different options. The model can be used for both a course level and the entire curriculum.

Successes and Failures of PBL

Reviewing the successes and failures of PBL is quite an undertaking. If a valid study is to be made, it should provide a contextual background for all the studies, as the specific PBL practice varies from situation to situation. There is also the fact that failures are not reported, and neither are the institutions that have implemented PBL at full scale and that have had a halfway drawback to traditional learning. However, across all the different practices and subject areas, the review indicates the following.

One of the significant results from research on PBL in all types and variations is that students' motivation for learning increases (Galand et al., 2010; Schmidt & Moust, 2000; Thomas, 2000; Walker & Leary, 2009); students are increasingly curious, asking more questions, and using more time to study. This is a significant result from research on PBL across subjects and cultures; for instance, according to Yadav, Subedi, Lundeberg, & Buntung (2011), the PBL approach has led to greater motivation for learning. However, studies also indicate that the PBL curriculum might be increasing students' stress, as they are using more time for study and, as a result, can become de-motivated (Bédard et al., 2010). Studies on gender and engineering indicate that women are motivated by knowing the context; they analyze problems and the contextual aspect is embedded in the problem approach. Indeed, many PBL students are not learning isolated disciplines, but working on more comprehensive problems and relations from real life (Ihsen & Gebauer, 2009).

Students' development of skills and competences seems to be significant (Van Barneveld & Ströbel, 2009; Dochy, Segers, Van den Bossche, & Gijbels, 2003; Schmidt & Moust, 2000; Strobel & van Barneveld, 2009). Dochy et al. (2003) have made a review of the literature from the 1990s on the evaluations of long-term effects of using PBL. They concluded that PBL results in an improvement of the development of transferable skills such as process skills. The impact on knowledge acquisition is missing or not significant. However, PBL students do not acquire less knowledge compared with students educated the traditional way. Studies document that there is a significant improvement of process skills. Galand and Frenay (2006) have conducted an empirical study of a transformation process at a particular institution. Their main conclusion was the same: students obtain process competence. Crosthwaite, Cameron, Lant, & Lister (2006) documents that the students' own perception of the achievement of skills had been significantly improved by PBL learning.

Studies on PBL can present results of improved retention rate or decreased drop-out rate (Burch, Sikakana, Yeld, Seggie, & Schmidt, 2007), and PBL students seem to get higher grades (Graham, 2012).

Schmidt and Moust (2000) have reviewed existing literature and concluded that PBL seems to have an effect on long-term retention of knowledge such as remembering and understanding various concepts. Other studies are based on employers' responses to education, and Danish research shows that employers are very satisfied with candidates from the PBL institutions and that these candidates are easy to integrate (Krogh & Rasmussen, 2004). In 2004 and 2008, a survey was conducted by Danish Society for Engineers (Ingeniøren, 2004, 2008) in which the companies rated Aalborg University as the most innovative university. The response from the companies in general was very positive toward the PBL universities (Kolmos & Holgaard, 2010).

Much of the early PBL literature concerns conceptualization of the new emerging practices such as conceptual understandings of the facilitator's role, for example, the development of the tutor role, defining types of projects and problems, collaboration in teams, and so forth (Biggs, 2003; Kolmos, 1996; Savin-Baden, 2003; Savin-Baden & Wilkie, 2004). This is a natural development of researching a new practice, as there is first a need for building up a language before comparing the effectiveness.

As noted previously, not many failures have been reported, nor is there literature on failures in the change process. There are institutions that started out with high ambitions of changing the entire curriculum and ended up with almost no change at all. Other institutions have implemented an advanced PBL curriculum, but with change of management they changed the curriculum again with more traditional teaching and learning methodologies. However, this knowledge is based on the authors' experiences and was never scientifically reported.

At any rate, there have been several critical voices; during the 80's Danish criticism pointed out that PBL was relying too much on experience and reflection of experience, resulting in novices teaching novices in the teams. Recently, in similar criticism, Kirschner, Sweller, and Clark (2006) emphasized strong instructional guidance rather than knowledge construction in teams with minimal guidance. There has been heavy debate surrounding these issues in the journal *Educational Psychologist*. Our response would be that it is very seldom to have a PBL curriculum without some level of instruction. For most PBL curricular and courses, there will be a mixed mode and the PBL curricula are normally a mix between lectures and students' collaborative activities.

Barnett (1994) formulates further criticism of PBL as being too instrumental, focusing so much on authentic problems, and asserts that learning therefore becomes too concerned with problems without gaining academic overview. Another very commonly formulated criticism is that PBL does not cover as much material as conventional lecture-based courses. However, this is a problem not only in PBL practices. Generally, there is more and more scientific knowledge and the curriculum cannot cover everything, but must be selective. The importance is to be aware of the criteria for the selection of content and that the learned outcomes are exemplary of the overall objectives.

Future Directions

Most of the PBL implementation in engineering education is at the single course level, as this is the easiest way to change practice for academic staff. From a strategic point of view, it is very important that academic staff experiment with their teaching and try out new models. The traditional system is characterized by parallel and rather independent courses most often without coordination across the disciplines, so for enthusiastic teachers it might be difficult to initiate a more comprehensive PBL approach across the disciplines.

For the students, however, a more comprehensive and interdisciplinary approach would be an advantage. With no coordination, none in the system knows where the students have learned problem analysis, project management, and collaboration and whether the learning is at a stage upon which it is possible to build. Sometimes students experience PBL in several parallel courses. This might create overload, as research indicates that they are using more time for study. A coordinated PBL approach is preferable across single courses (Shepard et al., 2008).

CDIO (Crawley, 2001; Crawley, Malmquist, Ostlund, & Brodeur, 2007) is one way to build up a more complex and coherent structure at the system level, where there is an overview of where students learn certain process skills as an integrated part of existing courses.

The most efficient way to utilize the PBL approach is implementation at the institutional level – either the entire institution, faculty, or department level (Fullan 2001; Kolmos, 2002; Scott, 2003; Thomas, 2000). This type of learning approach is supported by the whole system, for example, physical facilities, training of academic staff, distributing resources, development of new culture, and so forth. Systemic change requires both top-down and bottom-up decisions in the system – and it is not a change that can be implemented overnight. It is a long, serious process of change from the traditional paradigm of learning to a new paradigm of collective, cognitive learning with the aim of achieving interdisciplinary knowledge for analyzing and solving problems.

Kotter (1995) stresses the importance of urgency and the creation of visions. External reasons are most often the motivation for institutional change. However, external factors such as cuts in resources or demands for new skills might create urgency for institutional change, but not for creating visions. Without visions, there is a risk of short-sighted change strategies. Knoster indicates that the condition for a successful change is that all of the following elements are in place: vision, consensus in the organization, skills for new teaching and learning methods, incentives for academic staff, resources (as the change process will be resource intensive), and an action plan for what to do and how to do it (in Thousand & Villa, 1995). If all of these elements are in place, there might be a conceptual change among academic staff and change in the curriculum. However, if any of the elements are ignored, it might be even more difficult to establish systemic change.

As Litzinger, Lattuca, Hadgraft, and Newstetter (2011) conclude in their article, there is a missing alignment between the goal of educating engineers with a series of new requirements and the existing engineering curricula. It might be easier for institutions to use an adding strategy instead of changing teaching and learning methodologies at the systemic level. However, the adding strategy, in which new courses and activities are added to the established system, might partly contribute to an overload in the curriculum and to a theoretical and/or disintegrated learning of the required skills and competences. Educational change is a challenge, but it is a constant condition for educating engineers of today and tomorrow and it is important to find a balance between top-down strategies and bottom-up strategies in facilitating sustainable change toward more student-centered learning.

PBL Communities

Finally, there is much inspiration to be found in participating in the established international communities. The Pan-American Network for Problem-based Learning runs international biennial conferences (http://www.udel.edu/pan-pbl/). In the Asian community, the International PBL Symposium, organized by Republic Polytechnic, Singapore, runs International Symposia every second year (http://www.rp.sg/symposium/2009/). There are many national networks like Facilitate – a PBL network (www.facilitate.ie), which is an Irish Network, and the Finnish Network on Problem-Based Learning in Higher Education ProBell (http://www.uta.fi/tiedekunnat/kasv/eduta/probell/index.html). All of these PBL communities are rooted in the McMaster and Maastricht traditions of problem-based learning.

The only community within PBL that focuses on engineering education is the UNESCO Chair in Problem Based Learning in Engineering Education (UCPBL), Aalborg University, Denmark (www.ucpbl.net). The UCPBL is running research symposia every second year, and the activities are based on philosophy and learning principles across different PBL practices, which derive from educational research and practice. This community has also declared its commitment to researching the effectiveness of PBL and strategies for changing PBL in diverse cultural and institutional settings. The evidence of the effect of PBL is an important argument for change.

There are similar communities for engineering education such as the ALE (Active Learning in Engineering) community encompassing a wide range of learning philosophies and active learning methodologies (http://ale2011.ing.uchile.cl/). The CDIO community is a very structured community with formulated standards. The CDIO community also encompasses PBL at the curriculum level; while the CDIO standards address the institutional system level covering curricula, quality assurance procedures, and training of academic staff, the PBL address mainly the curriculum (http://www.cdio.org/).

Finally, there are also interdisciplinary journals in the field: the *Interdisciplinary Journal of Problem-Based Learning* (http://www.edci.purdue.edu/ijpbl/) and the newly established *Journal of PBL in Higher Education* (http://www.pbl.aau.dk/Journal+of+Problem+Based+Learning+in+Higher+Education/).

References

Algreen-Ussing, H., & Fruensgaard, N. O. (1990). *Metode i projektarbejde.* Aalborg, Denmark: Aalborg University Press.

Barnett, R. (1994). *The limits of competence – Knowledge, higher education and society.* Buckingham, U.K.: Open University Press.

Barnett, R., & Coate, K. (2004). *Engaging the curriculum: Higher education and society.* Buckingham, U.K.: Open University Press.

Barnett, R., & Napoli, R. D. (2008). *Changing identities in higher education, voicing perspective.* New York, NY: Routledge/Taylor & Francis.

Barrett, T., & Moore, S. (2011). *New approaches to problem-based learning – Revitalising your practise in higher education.* New York, NY: Routledge/Taylor & Francis.

Barrows, H. S. (1996). Problem-based learning in medicine and beyond: A brief overview. In L. Wilkerson & W. H. Gijselaers (Eds.), *Bringing problem-based learning to higher education: Theory and practice* (pp. 3–12). New Directions for Teaching and Learning No. 68. San Francisco, CA: Jossey-Bass.

Bédard, D., Lison, C., Dalle, D., & Boutin, N. (2010). Predictors of student's engagement and persistence in an innovative PBL curriculum: Application for engineering education. *International Journal of Engineering Education, 26* (3), 511–522.

Beddoes, K. D., Jesiek, B. K., & Borrego, M. (2010). Identifying opportunities for collaborations in international engineering education research on problem- and project-based learning. *Interdisciplinary Journal of Problem-based Learning, 4*(2), Article 3. Retrieved from http://docs.lib.purdue.edu/ijpbl/vol4/iss2/3

Biggs, J. (2003). *Teaching for quality learning at university – What the student does.* Buckingham, U.K.: Open University Press.

Brown, S., and Glasner, A. (1999). *Assessment Matters in Higher Education.* UK:The Society for Research into Higher Education and Open University Press

Bucciarelli, L. L. (1994). *Designing engineers.* Cambridge, MA: MIT Press.

Burch, V. C., Sikakana, C. N., Yeld, N., Seggie, J. L., & Schmidt, H. G. (2007). Performance of academically at-risk medical students in a problem-based learning programme: A preliminary report.*Advances in Health Science Education: Theory and Practice*, 12(3), 345–358.

Capraro, R. M., & Slough, S. W. (Eds.) (2009). *Project-based learning – An integrated science, technology, engineering, and mathematics (STEM) approach*. Rotterdam: Sense.

Carter, J., Eriksen, K., Horst, S., & Troelsen, R. (2003). *If reform is the answer – What are the questions?* Copenhagen: Center for Naturfagenes Didaktik, Københavns Universitet.

Hyldgaard Christensen, S., Delabousse, B., & Meganck, M. (eds). (2009). *Engineering in context*. Aarhus : Systime Academic.

Chinowsky, P. (2011). Engineering project organization: Defining a line of inquiry and a path forward. *The Engineering Project Organization Journal*, 27(3), 170–178.

Conway, J., & Williams, A. (1999). *Themes and variations in PBL*. Callaghan, NSW: Australian Problem Based Learning Network.

Crawley, E. F. (2001). *The CDIO syllabus: A statement of goals for undergraduate engineering education*. MIT CDIO Report no. 1, 2001. Retrieved from http://www.cdio.org

Crawley, E., Malmquist, J., Ostlund, S., & Brodeur, D. (2007). *Rethinking engineering education – The CDIO approach*. New York, NY: Springer Science +Business Media.

Crosthwaite, C., Cameron, I., Lant, P., & Litster, J. (2006). Balancing curriculum processes and content in a project centred curriculum – In pursuit of graduate attributes. In *Education for Chemical Engineers*. Trans IChemE, Part D, no. 1, pp. 39–48. Retrieved from http://www.icheme.org/ECEsamplepaper.pdf

Dewey, J. (1938). *Experience and education*. New York, NY: Collier and Kappa Delta Pi.

Dochy, F., Segers, M., Van den Bossche, P., & Gijbels, D. (2003). Effects of problem-based learning: A meta-analysis. *Learning and Instruction*, 13, 553–568.

Fruchter, R., & Lewis, S. (2003). Mentoring models in support of P5BL in architecture/ engineering /construction global teamwork. *The International Journal of Engineering Education*, 19(5), 663–671.

Fullan, M. (2001). *The new meaning of educational change* (3rd ed.). New York, NY: Teachers College Press.

Galand, B., & Frenay, M. (Eds.) (2006). *Problem and project-based learning in higher education: Impact, issues, and challenges*. Louvain-la-Neuve, Belgium: Presses Universitaires de Louvain.

Galand, B., Raucent. B., & Frenay, M. (2010). Engineering students-self-regulation, study strategies, and motivational believes in traditional and problem-based curricula. *International Journal of Engineering Education*, 26(3), 523–534.

Gibbs, G. (1999). Using assessment strategically to change the way students learn. In S. Brown & A. Glasner (Eds.), *Assessment matters in higher education*. London: The Society for Research into Higher Education and Buckingham, U.K.: Open University Press.

Gibbons, M., Limoges, S., Nowotny, H., Schwartzman, S., Scott, P., & Trow, M. (1994). *The new production of knowledge. The dynamics of science and research in contemporary societies*. London: SAGE.

Gijselaer, W. H. (1996). Connecting problem-based practices with educational theory. In L. Wilkerson & W. H. Gijselaers (Eds.), *Bringing problem-based learning to higher education: Theory and practice*. San Francisco, CA: Jossey-Bass.

Graaff, E. de (1995). Models of problem-based learning. Paper presented at the Fourth World Conference on Engineering Education, Saint Paul, MN . In E. Rex Krueger & F. A. Kulacki (Eds.), *Restructuring engineering education for meeting world needs* (Vols. 1–4). Technology Based Engineering Education Consortium, The William C. Norris Institute.

Graaff, E. de (2010). Research on engineering education: The development of a field of applied research. In *Proceedings of the International Conference on Engineering Education ICEE*, Gliwice, Poland.

Graaff, E. de, & Kolmos, A. (2003). Characteristics of problem-based learning, *International Journal of Engineering Education*, 19(5), 657–662.

Graaff, E. de, & Kolmos, A. (2007). *Management of change implementation of problem-based and project-based learning in engineering*. Rotterdam, The Netherlands: Sense.

Graaff, E. de, & Sjoer, E (2006). Positioning educational consultancy and research in engineering education. In *Proceedings of the 34th Annual SEFI Conference*, Uppsala: Uppsala University (pp. 98–101).

Graham, R. (2009). *Approaches to engineering project-based learning*, White Paper. Retrieved from http://web.mit.edu/gordonelp/ukpjblwh itepaper.pdf

Graham, R. (2012). *Achieving excellence in engineering education: the ingredients of successful change*. London: The Royal Academy of Engineering.

Graham, R., Crawley, E., & Mendelsohn, B. R. (2009). *Engineering leadership education: A snapshot review of international good practice*. White Paper, Bernard M. Gordon-MIT Engineering Leadership Programme.

Hiim, H., & Hippe, E. (1993). *Learning through experience,understanding and action*, (Læring gjennom oplevelse, forståelse og handling: En studiebok i didaktikk.) Oslo: University Publishing.

Holgaard, J. E., & Kolmos, A. (2009). Group or individual assessment in engineering, science and health education: Strengths and weaknesses. In X. Du, E. de Graaff, & A. Kolmos (Eds.), *Research on PBL practice in engineering education* (pp. 57–69). Rotterdam, The Netherlands: Sense.

Ihsen, S., & Gebauer, S. (2009). Diversity issues in the engineering curriculum. *European Journal of Engineering Education*, 34(5), 419–424.

Illeris, K. (1976). *Problemorientering og deltagerstyring*: Oplæg til en alternative didaktik. (*Problem orientation and participation: Draft for an alternative didactic*.) Copenhagen: Munksgaard.

Illeris, K. (2003). Towards A contemporary and comprehensive theory of learning.*International Journal of Lifelong Education*, 22(4), 396–406.

Illeris, K. (2007). *How we learn – Learning and non-learning in school and beyond*. New York, NY: Routledge.

Ingeniøren. (2004, March 3). *Aalborg – Kandidaters Tekniske Viden er i Top*.

Ingeniøren. (2008, May 30). *Virksomheder dumper nye ingeniøruniversiteter*.

Jamison, A. (2001). The *making of green knowledge: Environmental politics and cultural transformation*. Cambridge: Cambridge University Press.

Jamison, A., Hylgaard Christensen, S. & Botin, L. (2011). *A hybrid imagination. Science and technology in cultural perspective*. Golden, CO: Morgan & Claypool.

Jarvis, P. (1992). *Paradoxes of learning: On becoming an individual in society*. San Francisco, CA: Jossey-Bass.

Jorgensen, D., & Howard, P. (2005). *Ten years in the making – a unique program in engineering*. Paper presented at WACE 2005.

Kilpatrick, W. H. (1918). The project method. *Teachers College Record*, 19, 319–335.

Kilpatrick, W. H. (1925). *Foundations of method*. New York, NY: Macmillan.

Kilpatrick, W. H., Bagley, W. C., Bonser, F. G., Hosic, J. F., & Hatch, R. W. (1921). Dangers and difficulties of the project method and how to overcome them. *Teachers College Record*, 22, 283–321.

Kirschner, P. A., Sweller, J., & Clark, R. E. (2006). Why minimal guidance during instruction does not work: An analysis of the failure of constructivist, discovery, problem-based, experiential, and inquiry-based teaching. *Educational Psychologist*, 41(2), 75–86.

Kjaersdam, F., & Enemark, S (1994). *The Aalborg experiment – Project innovation in university education*. Aalborg, Denmark: Aalborg University Press.

Klafki, W. (2001). *Dannelsesteori og didaktik: Nye studier* (Theories of Bildung and didactics: New studies). Århus, Denmark: Forlaget Klim.

Kolb, D. (1984). *Experiential learning – Experience as the source of learning and development*. Englewood Cliffs, NJ: Prentice Hall.

Kolmos, A. (1996). Reflections on project work and problem-based learning. *European Journal of Engineering Education*, 21(2), 141–148.

Kolmos, A. (2002). Facilitating change to a problem-based model. *The International Journal for Academic Development*, 7(1), 63–74.

Kolmos, A., de Graaff, E., & Du, X. (2009). Diversity of PBL: PBL learning principles and models. In X. Du, E. de Graaff, & A. Kolmos (Eds.), *Research on PBL practice in engineering education* (pp. 9–21). Rotterdam: Sense.

Kolmos, A., Fink, F., & Krogh, L. (2004). *The Aalborg PBL model*. Aalborg, Denmark: Aalborg University Press.

Kolmos, A., & Holgaard, J. (2010). Responses to problem based and project organised learning from industry. *International Journal of Engineering Education*, 26(3), 573–583.

Kotter, J. B. (1995, March–April). Leading change: Why transformation efforts fail. *Harvard Business Review*, 59–67.

Krogh, L., & Rasmussen, J. G. (2004). Employability and problem-based learning in project-organized settings at the University. In A.

Kolmos, F. K. Flink, & L. Krogh (Eds.), *The Aalborg PBL Model: Progress, diversity and challenges* (pp. 37–56). Aalborg, Denmark: Aalborg University Press.

Litzinger, T. A., Lattuca, L. R., Hadgraft, R. G., & Newstetter, W. C. (2011). Engineering education and the development of expertise. *Journal of Engineering Education, 100*(1), 123–150.

Lopez-Islas, J. R. (2001). *Collaborative learning at Monterrey Tech-Virtual University.* Paper presented at the Invited Symposium on Web-based Learning Environments to Support Learning at a Distance: Design and Evaluation, Asilomar, Pacific Grove, CA.

McKenzie, L. J., Trevisan, M. S., Davis, D. C., & Beyerlein, S. W. (2004). Capstone design courses and assessment: A national study. In *Proceedings of the 2004 American Society of Engineering Education Annual Conference & Exposition, American Society for Engineering Education.*

National Academy of Engineering. (2004). *The engineer of 2020 – Visions of engineering in the new century.* Washington, DC: The National Academies Press.

Neufeld, V. R., & Barrows, H. S. (1974). The McMaster philosophy: An approach to medical education. *Journal of Medical Education, 49,* 1040–1050.

Nielsen, Jørgen Lerche., & Webb, Thomas W. (1999). *Project work at the new reform university of Roskilde – Different interpretations?* In H. Salling Olesen & Højgaard Jensen, J. (Eds.), *Project studies – A late modern university reform.* Roskilde, Denmark: Roskilde University Press.

Norman, G. R. (1988). Problem-solving skills, solving problems and problem-based learning. *Medical Education, 22,* 279–286.

Nowotny, H., Scott, P., & Gibbons, M. (2001). *Re-thinking science: Knowledge and the public in age of uncertainty.* Cambridge, U.K.: Polity Press.

O'Grady, G., & Alvis, W. A. M. (2002). *One day one problem: PBL at Republic Polytechnic.* Paper presented at 4th Asia-Pacific Conference on PBL, Hatyai, Thailand.

Olsen, P. B., & Pedersen, K. (2005). *Problem-oriented project work – A workbook.* Roskilde, Denmark: Roskilde University Press.

Ozansoy, C., & Stojcevski, A. (2009). *Problem-based learning in electrical and electronic engineering education.* In X. Du, E. de Graaff, & A. Kolmos (Eds.), *PBL – Diversity in research questions and methodologies.* Rotterdam: Sense.

Powel, P., & Weenk, W. (2003). *Project-led engineering education.* Utrecht, The Netherlands: Lemma.

Prince, M. J., & Felder, R. M. (2006). Inductive teaching and learning methods: Definitions, comparisons, and research bases. *Journal of Engineering Education, 95*(2), 123–138.

Royal Academy of Engineering. (2007). *Educating engineers for 21st century,* Retrieved from http://www.raeng.org.uk/news/release/pdf/Educating_Engineers.pdf

Savery, J. R. (2006). Overview of problem-based learning: Definitions and distinctions. *Interdisciplinary Journal of Problem-Based Learning, 1*(1), 9–20.

Savin-Baden, M. (2003). *Facilitation problem-based learning.* London: Society for Research into Higher Education and Buckingham, U.K.: Open University Press.

Savin-Baden, M. (2007). Challenging models and perspectives of problem-ased learning. In E. de Graaff & A. Kolmos (Eds.), *Management of change: Implementation of problem-based and project-based learning in engineering* (pp. 9–30). Rotterdam, The Netherlands: Sense.

Savin-Baden, M., & Wilkie, K. (Eds.). (2004). *Challenging research in problem-based learning.* London: Society for Research into Higher Education and Buckingham, U.K.: Open University Press.

Sawyer, K. (2008). *Group genius – The creative power of collaboration.* New York, NY: Basic Books.

Schmidt, H. G., & Moust, J. H. C. (2000). Factors affecting small-group tutorial learning: A review of research. In D. H. Evensen & C. E. Hmelo (Eds.), *Problem-based learning: A research perspective on learning interactions.* Mahwah, NJ: Lawrence Erlbaum.

Schön, D. A. (1987). *Educating the reflective practitioner.* San Francisco, CA: Jossey-Bass.

Scott, G. (2003). Effective change management in higher education. *EDUCAUSE Review, 38*(6), 64–80. Retrieved from http://net.educause.edu/ir/library/pdf/erm0363.pdf

Sheppard, S. D., Macatangay, K., Colby, A., & Sullivan, W. M. (2008). *Educating engineers: Designing for the future of the field.* San Francisco, CA: Jossey-Bass.

Shinde, V., & Kolmos, A. (2011). Problem based learning in Indian engineering education:

Drivers and challenges. In *Proceedings of Wireless VITAE 2011: 2nd International Conference on Wireless Communication, Vehicular Technology, Information & Theory and Aerospace & Electronic System Technology* (pp. 179–184). Bangalore, India: IEEE Press.

Sockalingam, N., & Schmidt, H. G. (2011). Characteristics of problems for problem-based learning: The students' perspective. *Interdisciplinary Journal of Problem-Based Learning*, 5(1), 6–33.

Somerville, M., Anderson, D., Berbeco, H., Bourne, J. R., Crisman, J., Dabby, D., Zastavker, Y. (2005). The Olin Curriculum: Thinking toward the future. *IEEE Transactions on Education*, 48(1), 198–205.

Strobel, J., & van Barneveld, A. (2009). When is PBL more effective? A meta-synthesis of meta-analyses comparing PBL to conventional classrooms. *Interdisciplinary Journal of Problem-Based Learning*, 3(1), 44–58.

Thomas, J. W. (2000). *A review of research on project-based learning*. San Rafael, CA: Autodesk Foundation.

Thousand, J. S., & Villa, R. A. (1995). Managing complex change towards inclusive schooling. In R. A. Villa & J. S. Thousand (Eds.), *Creating an inclusive school*. Alexandria, VA: Association for Supervision and Curriculum Development (ASCD).

Turns, J., Cuddihy, E., & Guan, Z. (2010). I thought this was going to be a waste of time: How portfolio construction can support student learning from project-based experiences. *Interdisciplinary Journal of Problem-Based Learning*, 4(2). Retrieved from http://docs.lib.purdue.edu/cgi/viewcontent.cgi?article=1125%26;context=ijpbl

Ulisses, A., Valéria A. A., & Homero, F. F. (2009). Ensino de Sensoriamento Remoto através da Aprendizagem Baseada em Problemas ePor Projetos: Uma proposta metodológica. In *Anais XIV Simpósio Brasileiro de Senso-riamento Remoto*, Natal, Brazil (pp. 2365–2371).

van Barneveld, A., & Ströbel, J. (2009). Problem-based learning: Effectiveness, drivers and implementation challenges. In X. Du, E. de Graaff, & A. Kolmos (Eds.), *Research on PBL practice in engineering education*. Rotterdam: Sense.

Vleuten, C. P. M. van der (1997). De intuïtie voorbij [Beyond intuition] *Tijdschrift voor Hoger Onderwijs*, 15(1), 34–46.

Walker, A., & Leary, H. (2009). A problem-based learning meta analysis: Differences across problem types, implementation types, disciplines, and assessment levels. *Interdisciplinary Journal of Problem-Based Learning*, 3(1), Article 3. Retrieved from http://docs.lib.purdue.edu/ijpbl/vol3/iss1/3

Wenger, E. (1998). *Communities of practice*. Cambridge: Cambridge University Press.

Wilkerson, L., & Gijselaers, W. H. (Eds.). (1996). *Bringing problem-based learning to higher education: Theory and practice*. San Francisco, CA: Jossey-Bass.

Woods, D. (1994). *How to gain most from problem-based learning*. Hamilton, Ontario: McMaster University.

Yadav, A., Subedi, D., Lundeberg, M. A., & Bunting, C. F. (2011). Problem-based learning. *Electrical Engineering – Journal of Engineering Education*, 100(2), 253–280.

Yew, E. H. J., & Schmidt, H. G. (2008). Evidence for constructive, self-regulatory, and collaborative processes in problem-based learning. *Advances in Health Science Education*, 14(2), 251–273.

Yusof, K., Tasir, Z., Harun, J., & Helmi, S. A. (2005). Promoting problem-based learning (PBL) in engineering courses at the Universiti Teknologi Malaysia. *Global Journal of Engineering Education* 9(2) pp. 175-184. Retrieved from http://www.wiete.com.au/journals/GJEE/Publish/vol9no2/Yusof.pdf

CHAPTER 9

Case Studies in Engineering

Claire Davis and Aman Yadav

What Are Case Studies?

What are "cases" or "case studies"? Cases are narratives that present real-life scenarios/problems and allow students to experience how professionals address problems encountered in the field. Cases have three main elements: (1) they are based on real-life events or realistic situations that allow students to experience problems they are not likely to encounter first-hand; (2) they present both contextual and technical information that is based on careful research and study; (3) they may present no clear-cut solutions to allow students to develop multiple perspectives (Merseth, 1994). Hence, cases mimic real-world scenarios that engage students in solving authentic problems and experiment within the safe confines of a classroom (Demarco, Hayward, & Lynch, 2002).

The use of case studies has a long and effective history in business, law, and medical fields to teach students the complexities and ill-structured nature of those disciplines (Herreid, 2007; Mayo, 2004). Case-based

teaching dates back to 1870, when Christopher Langdell, a law professor, argued that using actual legal cases was the best way to study law (Garvin, 2003). Langdell stated that using cases would develop students' understanding "via induction from a review of those appellate court decisions in which the principles first took tangible form" (Garvin, 2003). Case studies also formed the instructional method at the Harvard Business School in 1908 (Merseth, 1991). However, unlike in law schools, case study implementation was slow in business schools owing to the lack of ready-made cases and faculty expertise in teaching with cases (Merseth, 1991). The use of cases became predominant at Harvard Business School after Wallace B. Donha became the Dean in 1919 and supported faculty to gain experience in teaching with cases. He also created the Bureau of Business Research to put together high-quality teaching materials including cases (Merseth, 1991). Case-based instruction has also been implemented in medical education via problem-based learning where students learn in cooperative groups by studying the records of an actual patient (Williams, 1992). See Chapter 11 by Atman, Borgford-Parnell, McDonnell, Eris, and Cardella in this volume for a discussion of the history of problem-based learning in medical education at McMaster University. Within science, technology, engineering, and mathematics (STEM) disciplines, James B. Conant at Harvard University was the first science educator in the late 1940s to teach an entire course using case studies via the lecture format (Conant & Van Deventer, 1951).

Within engineering, cases have been used since the 1950s with the first uses in chemical and civil engineering (Vesper, 1964). So what is an "engineering case"? Fuchs (1974) stated that an engineering case is a "written account of an engineering job as it was actually done, or of an engineering problem as it was actually encountered... [and] it involves not only quantitative relations amenable to computations, but other more subtle factors such as the interaction of people, the malevolence of inanimate objects, and the pressures of time and resources under which engineers work." Kardos and Smith added that engineering cases are meant to enhance students' learning about engineering principles and practices by including the background and complexities of an engineering event. It is important to note that cases present the information by minimizing the bias to allow students to come to their own conclusion (Kardos & Smith, 1979).

Cases can serve as *microcontexts* "that focus on a specific subset of a larger problem or domain" or as *macrocontexts* that "enable the exploration of a problem space for extended periods of time from many perspectives" (Merseth, 1994). This suggests that cases can vary in length from those that last one class period to ones that last weeks. Short cases typically do not require any student work outside of the class while the long cases may include outside student work. Along the same lines, Harris (2003) suggested that there are three types of cases – micro cases, macro cases, and exemplary cases. Micro cases present dilemmas that individual engineers face in their daily lives

and decisions, whereas macro cases involve societal issues that have the potential to impact the larger community. Exemplary cases, on the other hand, present situations that require individual(s) to take commendable actions to solve the problems (Harris, 2004).

Vesper (1964) mentioned that Fuchs suggested four additional types of cases – Critical Instance Case, Case Problem, State-of-the-Art Case, and Case History. The Critical Instance Case provides students with background information on a situation that requires them to propose a suitable course of action and say what is expected from it. The Case Problem, on the other hand, simulates a situation in which the student is given a problem along with data, some of which may not be relevant to the case. This case stops short of presenting a decision or course of action, instead allowing students to practice skills of analyzing, making decisions, and planning a course of action based on their judgment of various possible outcomes (Merseth, 1991). The third type of case, State-of-the-Art Case, familiarizes students with a "portion of the frontier in an art with the suggestion that he indicate a way to advance it further" (Vesper, 1964). Finally, Case History presents students with an actual engineering project including various phases, problems encountered, solutions, and results.

Newey (1987) proposed three major types of cases used at The Open University: descriptive/analytical cases, problem-solving/synthetical cases, and problem identification cases. Descriptive/analytical cases present products or systems that are considered technological breakthroughs to make students aware of how things work and inherent complexity in these systems. Students also learn to apply analytical techniques and assess the technological needs of the systems. The problem-solving/synthetical case is similar to the Case Problem proposed by Fuchs, but expanded by proposing six types of problem-solving case: convergent/worked, convergent/aided, convergent/unaided, open-ended/worked, open ended/aided, and open ended/unaided. A case is "convergent" when there is a unique solution to the problem whereas an "open-ended" case signifies that there are a number of satisfactory outcomes. In a "worked" case a student follows the teacher's approach to solving the problem, whereas in an "unaided" case a student solves the problem independently. An "aided" case presents a problem in which the teacher plays some role in helping students arrive at the solution. In a problem identification case, a specific situation or activity is considered with the aim of identifying any problems it may have (e.g., manufacturing industry, energy generation).

Jonassen (2011) also provided a detailed discussion of how cases can be used for a variety of pedagogical purposes. For example, cases could be used as problems to solve where students work to solve authentic tasks, generate a solution, and justify it. Cases could be worked examples of well-structured problems, which provide the process for how the problem may be solved. Jonassen argued that the use of worked examples is to help students "construct schemas for the worked examples that may then be generalized or transferred to new problems" (Jonassen, 2011). Jonassen also suggested that cases could be used as analogies to allow instructors to connect new ideas to students' prior knowledge. Furthermore, the cases can be used as prior experiences in the form of stories, which would allow students to experience complexity of the workplace and learn vicariously.

In all these descriptions the common factor is that the case study presents information on real-life problems; however, in some instances the students are either led through the data and guided to the outcomes seen, or in others they are presented with the data and allowed to explore the case study to come to their own conclusions. In sum, cases are stories with messages that educate (Herreid, 2007).

It is important to note there are some similarities and some significant differences between case study learning and problem-based learning. Burgess (2004) suggested that cases differ from problem-based

learning in that they are an account of current/past activity and contain information that reflects the perspectives of a previous problem solver or case author. Problem-based learning, on the other hand, presents real-world problems to students who require new skills or knowledge to solve the problem. Therefore the students need to discover what is required to solve the problem and build a case (Burgess, 2004). In spite of these differences, Burgess suggested that cases could be developed that provide students with authentic experience of solving real-world problems and promote flexible, adaptive learning needed for the workplace. These similarities and differences are summarized in Table 9.1 and problem-based learning is discussed in detail in Chapter 11.

Why Use Case Studies?

As outlined in the report by the National Academy of Engineering, "The Engineer of 2020: Visions of Engineering in the New Century" (National Academy of Engineering, 2004), there is general recognition that an engineer of the future will face challenges that require much broader skill sets than those honed in typical engineering coursework that primarily emphasizes the technical facets of the field. Engineers operating in the future will require interdisciplinary approaches to their work that are characteristic of complex problems that link science, technology, and social systems (Klein, 1996, 2004). Such work environments will be inherently ill structured and complex owing to conflicting goals, unanticipated problems, multiple solution methods, nontechnical success standards, and unavoidable constraints (Jonnasen, Strobel, & Lee, 2006).

One of the ways of situating students' learning in authentic and meaningful contexts and allowing them to experience the complexities of the engineering discipline is through apprenticeship opportunities. However, this task is easier said than done given the challenges of finding and placing students in authentic environments. The Cognition and Technology Group at

Table 9.1. Differences and Similarities Between Case Study Learning (similar in structure to project-based learning [PrjBL][1]) and Problem-Based Learning (PBL)

Case Study Learning	Problem-Based Learning
Predominantly task orientated with activity often set by tutor. Tutor supervises. Students are required to produce a solution or strategy to solve the problem. Could involve considering a worked example. May include supporting lectures that equip students to undertake activity; otherwise students are expected to draw on knowledge from previous lectures. Can be embedded in lectures.	Problem usually provided by tutor but what and how they learn is defined by students. Tutor facilitates. Solving the problem may be part of the process but the focus is on problem management, not on a clear and bounded solution. Lectures not usually used on the basis that students are expected to define the required knowledge needed to solve the problem.

Vanderbilt (CGTV) argued for "anchored instruction" as a way to "recreate some of the advantages of apprenticeship training in formal education settings" (Merseth, 1994). The goal of anchored instruction is to overcome the inert knowledge problem by creating environments that enable students and teachers to experience the problems and opportunities experts in the field face and the tools they use to tackle the problems. Furthermore, anchored instruction models are how experts change their own thinking based on new concepts and theories, rather than looking at the new information as facts or mechanical procedures that need to be memorized (CGTV). Brown, Collins, and Duguid (1989) also argued for learning to be situated in authentic contexts to allow students to develop useable, robust knowledge. Specifically, Brown and colleagues argued that conceptual knowledge should be thought of as a set of tools, which can be fully developed only by actively using them in the context. The use of cases is one such approach that anchors students' knowledge, that is, "when, where, and why to use the knowledge they are learning," and require them to integrate multiple sources of information in an authentic context (Bransford, Brown, & Cocking, 2000). Cases would provide students with opportunities to use the domain's conceptual tools in an authentic activity and wrestle with real-world problems, which allows them to understand

how professionals (i.e., engineers, mathematicians, etc.) look at the world and solve emergent problems (Brown et al., 1989). Thus, the use of cases is rooted in a cognitive and sociocultural constructivist perspective that emphasizes students taking an active role in learning through social interaction with peers using problem-related and collaborative practices (Bilica, 2004; Mayo, 2002, 2004).

Cases provide students with a multifaceted educational experience that will help them develop the *abilities, knowledge,* and personal and professional *qualities* that define an individual who can master technical challenges. Specifically, cases bring in "outside reality into the classroom" and sensitize students to the sort of experiences they are likely to encounter as engineers (Fuchs, 1974). Cases allow students to become familiar with the process of engineering and how engineers formulate problems, deal with incomplete data, and develop alternate solutions. Hence, cases are ideally suited for engineering students to prepare them to work in an environment that is continuously changing and complex.

Research on Cases in Engineering

Some of the earliest research evaluating the use of cases in engineering was conducted by Vesper and Adams (1969). They used student evaluations of teaching methods, including cases, from the University of Santa Clara and Stanford University to assess the impact of cases. The authors reported that students found cases to be valuable in allowing them to see what engineers do and how they work. Students also found cases to enhance their "skills of spotting key facts among less relevant data, identifying and defining practical problems, and foreseeing consequences of alternative actions."

More recently a number of researchers have examined students' perceptions of the impact of cases on their learning and engagement. For example, Raju and Sankar (1999) used a sixteen-item questionnaire to evaluate the effectiveness of cases. The authors found that students thought cases were useful, clear, attractive, and challenging. Specifically, students felt that cases were relevant and interesting, brought realism to the classroom, and allowed them to transfer theory to practice. The authors also reported that students thought cases exposed them to nontechnical issues and experience and to work with limited information. Garg and Varma (2007) also examined student perceptions of cases in a software engineering course. One hundred and eighty-six students completed a survey at the end of the course to evaluate various learning activities that corresponded with lectures and cases. They found that students reported cases to be significantly more helpful than lectures in allowing them to understand course concepts, apply software engineering skills, think critically, and improve their communication ability. These results showcase that cases allow instructors to engage students in the course content and make the course content relevant. However, the majority of this research has focused on student perceptions of the cases rather than on assessing actual learning outcomes. Lundeberg and Yadav (2006a) argued that it is important to go beyond student perceptions and examine whether cases are beneficial in terms of promoting deep understanding, problem-solving, and enabling the transfer of ideas to new contexts. Lundeberg and Yadav (2006b) also suggested using sophisticated

qualitative methods, such as conducting "focus-group interviews with the students to add additional insight into how and what students were learning from the case situations" (p. 411).

Martinez-Mones et al. (2005) used student focus group interview data and open-ended questions to examine the influence of cases in a project-based course on students' learning and collaboration in the course. Specifically, students worked in pairs to help a customer (i.e., a case) purchase a computing system for his business. The instructors used five different cases (i.e., customers) from different fields and each group worked with only one customer. To encourage collaboration, instructors used a collaborative tool (Basic Support for Cooperative Work [BSCW]) and required pairs to share information and compare their solutions with other groups working with the same case as well as groups working with different cases. The authors reported that students achieved deep and broad learning as highlighted by their ability to generalize their learning from cases experienced in class to a mini-case on the final exam. However, it must be noted that the final exam mini-case resembled one of the cases used in the class, which some students worked on during the semester and other students, who did not work on that case, at least had opportunities to discuss via BSCW. This raises an important confounding variable and limits the researchers' ability to attribute students' learning to the use of case studies. The focus group interview and survey results suggest that students' felt cases developed their collaboration skills.

In another study, Yadav, Shaver, and Meckl (2010) compared the influence of cases versus traditional lectures on students' learning in a mechanical engineering course. The cases were based on thermal and fluid mechanics content and were introduced using a counter-balanced design in two sections of the same systems modeling course. The counterbalanced design was used to remove any bias for the topic (thermal vs. fluid mechanics) and instructional method (cases vs. traditional lectures). Data were collected in pre–post test format and end of the semester surveys. The authors found that even though students had positive attitudes toward cases, they did not have any significant impact on the students' understanding of the course concepts compared to the traditional lecture approach. The authors hypothesized that the new implementation of the cases, student resistance to, and lack of experience of, new approaches, and the lack of sensitive measures might explain the results. They suggested that carefully constructed measurement tools need to be utilized to gauge true differences. The issue of assessment and evaluation is discussed further in the sections below.

Typical Case Study Considerations

Case studies can take numerous different styles (as described by Fuchs (1974), Merseth (1994), and Newey (1987)) and formats (e.g., in terms of length, content, assessment). However, there are several key factors that should be included, or at least considered, when teaching with case studies. Some of these aspects will be common to other teaching methods, such as PrjBL and PBL. A summary of the key topics is provided in this section, with a more in-depth discussion of the areas common to PrjBL, PBL, and case studies provided in Appendix 9.1.

Introduction and Key Skills

Case studies often involve activities such as independent research, group work, data analysis, presentations (e.g., poster, oral, reports, etc.). It is important to recognize that students may not already possess these key skills (self-directed learning, teamwork, analytical, and communication skills) at the level required to participate fully in the case study. Gibbs (1995) suggested that students should have a trial run when taking part in, or carrying out, new activities before the first time marks are awarded. In addition it should be indicated to the students that the case studies are about developing these key skills as well as the

academic knowledge of the case study topic. It is essential that the students be given very clear instructions at the start of the case study of what is expected of them during the more student-centered activity (Gibbs, 1995). Another strategy is to invite students from previous semesters, who learned via case studies, to share their perspectives about the case teaching method, for example, via peer tutoring (see Appendix 9.2 for further discussion).

Selecting the Case Study Topic

Obviously the main considerations when selecting the topic for the case study are to ensure that it allows the key theoretical concepts from the course curriculum to be covered and that it is an area that has sufficient resources for the students to use. The second of these considerations will often be determined by the way in which the case study is developed, that is, whether the case study follows a lecturer's research interests, is delivered by an industrialist, and so forth. Students have suggested that they are more motivated by case study topics that they can relate to (Wilcock & Davis, 2006). For example, a case in which an engineering failure is being considered an event that is more recent and/or higher profile is favored by students. This is echoed by Mustoe and Croft (1999), who emphasized the importance of implementing case studies based on modern and recent technologies and applications. Another consideration is the amount of information available on a topic, for example, for high-profile topics there can be an enormous amount of detail available (e.g., on the Internet), which can be daunting for some students. For these circumstances guidance and resources for the students about core issues surrounding the case to be considered is important.

Supporting Lecture(s)/Background Information

Case studies can be used to teach basic concepts and lecturers can emphasize important points and procedures in a way that students can enjoy learning them (Henderson, Bellman, & Furman, 1983). However, if a case study is being used to develop the students' understanding of a new topic containing complex or numerous concepts, then a supporting lecture(s) may be required to ensure that all the students understand the key areas that will be included. Some lecturers may prefer to give the students the essential information via a set of notes or references to relevant literature or on a support website. A key differentiation between problem-based learning and case study (or project-based learning) approaches is that the required information to tackle the case study is generally provided (either directly in a lecture, or indirectly through support materials) by the lecturer (Savin-Baden, 2007).

Coping with Ambiguity and Complexity

Cases introduce the ambiguity and complexity commonly seen in real-world problems, especially if there is no one correct solution to the problem. Students learning from cases must learn to monitor their own learning and thinking (Gallucci, 2006). This can lead to frustration among students, which might lead to student resistance to a case-based learning approach (Yadav et al., 2010). One approach to addressing student resistance is to be transparent about the goals of case-based learning in a safe environment and allow students to engage and reflect on the issues presented in the case through case discussions (Gallucci, 2006). When students compare and contrast their claims with peers in the class, it allows them to uncover holes in their own thinking and see value in learning from cases (Galluci, 2006). In addition, instructors can ask students from previous classes who have gone through case-based learning to share their experiences about coming to grips with ambiguity of learning from cases.

Supporting Independent Research

Case studies are a good vehicle for encouraging students to carry out independent research, that is, outside of the lecture/tutorial environment, and this can be useful for promoting active learning and self-regulated learning. In addition, case studies require resource investigation, encouraging students to utilize a number of different sources, that is, Internet, library, laboratory results, and contacting experts in industry. It is important to explain to students that critical analysis of resources is required to avoid a surface approach to the case study. Therefore, detailed instructions should be given to the students as to what is expected in terms of their independent research (e.g., minimum number of journal papers to be reviewed, examples of depth of analysis required), particularly at the early stages of their undergraduate career. Students may need assistance in determining which sources of information are most appropriate to consider. Students may also need guidance on how to use the Internet to gather information, for example, on how to determine validity of the information source (Goett and Foote, 2000). Goett and colleagues state that it is important that students should understand the difference between citing a source and plagiarizing it; how copyright law applies to the resources they wish to use; and that Web pages will differ in their accuracy, currency, completeness, and authority. These issues could be addressed either through briefings, Web pages, or notes.

Facilitating Group Activities

Where students are expected to work in a group, care is needed to ensure that all members of the group contribute. Students should be encouraged to separate the work required into sub-tasks that can be allocated to different group members. It may be appropriate for the lecturer to carry out this task division for the students if it is their first experience of carrying out this type of activity. For longer case studies, that is, those carried out over several weeks, the lecturer may wish to consider having formal meetings with the student groups to ensure progress is being made (Wilcock, Jenkins,

& Davis, 2002). An alternative approach would be for the groups to have to keep a record of meetings they hold which would then be submitted as part of the assessment procedure. Peer tutoring could also be used to provide guidance to the group on time management as well as providing technical assistance (Kieran & O'Neill, 2009).

Fostering Laboratory Skills

Case studies can be designed to involve laboratory work. It may be possible to use existing laboratory exercises in a course within the case study, perhaps by changing the way the laboratory exercise is presented. Hands-on activities can add extra dimensions to case studies and can give students a real 'feel' for the area they are studying. For example, experimental work was incorporated into a suite of case studies at the University of Birmingham (U.K.) and the feedback obtained during the evaluation stages was very positive: "It (the experimental work) was an additional source of interest and an extra dimension to case study work." In this instance the experimental work consisted of laboratory classes (hardness, microstructure, and failure analysis) to examine the materials used in bicycles (Davis & Wilcock, 2005). The students covered the required module learning outcome of materials characterization but in the context of the case study topic. This was a modification of previous laboratory classes.

Methods for Developing Case Studies

There are a number of ways to develop new case studies. The following list covers the main methods, provides some examples of the case studies generated, and discusses their advantages and disadvantages. A complete list of where to find more information on specific case studies (not just those mentioned in the text that follows) is provided in Appendix 9.2.

Developing a Case Study to Replace or Supplement More Traditional Teaching on the Same Topic

When developing a new degree course there is considerable scope for introducing new content and new approaches to teaching. However, most degree programs are well established and the existing courses may be taught in a traditional style (lectures and laboratory classes). In such courses, case studies can be introduced to replace traditional teaching methods for the same subject content. Although the lecture format is useful for teaching large blocks of information and theory, research suggests that students must do more than just listen and receive knowledge (Chickering & Gamson, 1987). By using case studies, or other student-centered activities, students can engage in active learning and study skills can be developed. In terms of Bloom's taxonomy, which ranks orders of learning in a pyramid with knowledge at the bottom rising through comprehension, application, analysis, synthesis, and evaluation at the top, case studies and active learning pedagogies allow students to move up the pyramid and take part in analytic, synthetic, and evaluative work (Conway, 2001).

When teaching an existing course by implementing case studies instead of lectures, it is important to evaluate whether the students receive the same breadth and depth of learning. Often case studies can result in good depth of learning related to the case study topic; however, the breadth may be reduced because less content may be covered (Yadav et al., 2010). There can also be problems with ensuring that the replacement case study equates to the same amount of student effort as the previous teaching activities.

Materials selection (choosing the correct material for an application) is a topic taught in many engineering courses, where case studies can be used to complement traditional teaching. The approach to materials selection can be taught in a traditional manner with several short case studies being used to support and extend the learning.

The approach adopted by Ashby is an excellent example of how case studies can be used to enhance the students learning and his book contains many worked case studies of materials selection and design (Ashby, 2011). Some case studies are also available on the Internet (see Case Studies in Materials Selection in Appendix 9.2). The Cambridge Engineering Selector (CES) software for materials selection (CES EduPack™) can be used to support the Ashby selection case studies and is now used in more than 800 universities and colleges worldwide (GrantaDesign, 2012).

Developing a Case Study Based on the Research Interests of Staff

A lecturer's in-depth knowledge and interest in the topic will add to the students' case study experience; in addition, resources to support the case study will generally be available, making this approach time-efficient to develop. Examples of this approach include both case studies based on academic research activities (e.g., Dr. Nigel Mills from the University of Birmingham (U.K.)) and industrial contract work conducted by the lecturer (e.g., Professor Neil James from the University of Plymouth (U.K.)); see "Case studies on polymer foams used in sport and the health service" and "Industrial failure-analysis case studies" respectively in Appendix 9.2 for more detail. There are some national databases of case studies developed by lecturers, for example, the National Center for Case Study Teaching in Science (NCCSTS) in the United States and the Learning and Teaching Support Network (LTSN) for Engineering in the United Kingdom (see Appendix 9.2 for details).

Requesting Students to Develop Case Studies Based on Personal Interests

Involving students in the writing of case studies can be a valuable experience and can benefit students in terms of the development of communication skills and problem-solving skills. For example, Smith (1992) suggests that students can write their own case studies and he gave an account of two courses in which students were required to study and discuss a number of published cases and then write up a case study based on a real industrial project. In each course, students worked with a project engineer ("clinical professor") to learn about a project and how it had been accomplished. Many students on engineering courses undertake industrial placements and this may provide good opportunities for developing case studies.

Dr. Irene Turner from the University of Bath (U.K.) produced a portfolio of case studies using undergraduate and postgraduate students and recent graduates to generate ideas and content (I. Turner, personal communication, 2003, 2011). It was found that undergraduate students did not fully engage in developing the cases and only submitted inferior cases, probably because there was no academic credit associated with their assignment (a financial incentive was offered). However, contributions from recent graduates and postgraduates were of higher quality and were used in the undergraduate courses and as examples during university recruitment events.

Therefore using students to develop case studies can be a good learning experience; however, the resulting output may require significant modification before it can be used as a teaching resource to ensure it has sufficient depth and breadth.

Inviting/Involving External Lecturers (e.g., from Industry) to Contribute to a Case Study

Involving external sources can add new dimensions to the learning activity. Raju and Sankar developed a case study based on an actual problem at a steam power plant, which involved considerable financial and ethical decisions (Raju & Sankar, 1999). In designing the case study, they worked closely with the engineer and manager involved in solving the problem to produce the written text. The engineer and manager from the plant also attended the classes where the case studies were discussed and were requested to make changes to ensure authenticity and accuracy. The feedback received for the case study was extremely favorable and students urged the authors to create more such case studies.

Dr. James Busfield at Queen Mary University of London (U.K.) also used a similar approach and designed seven industrial case studies, each delivered by a representative from industry. The main motivation behind this was to raise awareness of materials design and the role of a materials engineer in industry and to expose students to role models in industry. In each case, the industrialist provides written support materials for the students, presents the case studies to the students, assesses how the students dealt with the case study, and gives feedback on completion; sometimes ongoing support via e-mail is also provided. The case study topics are generally new each year, with many companies being involved repeatedly. This approach has been running for almost 10 years with about 250 students per year, which suggests this method is sustainable. However, the original format of three occasions when industrialists provide on-site input had to be reduced to two days owing to logistical and scheduling issues (J. Busfield, personal communication, 2011). Student responses to the case studies have been very positive, as suggested by this student quote: 'Course gives an insight into my potential future role in UK industry'.

The approach of using external experts works best if they are linked to a University to ensure that both the technical content and teaching approach are appropriate. A good example of how this can be achieved is the Engineering for Sustainable Development scheme at the Royal Academy of Engineering in the United Kingdom. The scheme was set up in 1998 as it was felt that convincing case studies in this area can be developed only by leading-edge industrial practitioners working as visiting professors in unison with experienced teachers. Examples of the case studies generated include those in civil engineering (Jubilee River), chemical and manufacturing engineering (laundry cleaning products), product design in electrical and electronic engineering (mobile phones), and civil engineering and building (Mossley Mill regeneration in Northern Ireland); further details are given in Appendix 9.2.

Assessment Methods

Case studies typically build on both course content (i.e., knowledge) and key skills; hence careful consideration is required on how to assess these different areas. Similar difficulties exist with assessing problem-based learning and project-based learning. Reeves and Laffey (1999) argued that assessing student performance from problem-based learning approaches is complicated given the open-ended nature of the task(s) involved. This also raises students' resistance to active learning approaches owing to their discomfort with assessment of learning. Yadav, et al. (2011) suggested that instructors interested in using active learning pedagogies should, at least initially, focus on the process of learning rather than grades. Once the students are comfortable with the new pedagogies, instructors can use more innovative assessment methods, such as Mazur's (1998) paired-problem test (more on this in the next section).

Problems can also arise where significant amounts of group work are involved. Stronger students might feel that group marking is disadvantageous to them because their grades are being pulled down by their team-mates (Serpil Acar, 2004). Group-based case studies often require students to produce one or more outputs between them (generally a report and/or presentation/poster). Learning to collaborate is a useful skill and to produce a group output is an important part of this. To meet the requirements of a university course, individual grades also have to be awarded to contribute to the final student classification; hence for any group work a means of assessing the individual should be considered. Peer assessment is now a common approach used in higher education, and there is much support for involving students in the assessment process; however, there are also associated problems. Swanson, Case, and van der Vleuten (1991) questioned the validity of peer assessment, suggesting that the process is not truly reliable. Feedback from students has presented mixed views on this type of assessment (Wilcock et al., 2002). Although the majority of students recognized its benefits, many felt that there were problems with its use. Some of these issues were overcome by adjusting the level of contribution of the peer assessment to the overall grade and ensuring opportunities were available for students to comment confidentially on the peer assessment grades (M. Strangwood, University of Birmingham, personal communication, 2003). An overview of the different approaches to assessing an individual's contribution to the group can be found in Booklets 9 and 12 of the LTSN Generic Centre Assessment Booklet Series (LTSN Generic Centre, 2011) and in the guide by Orsmond (2004).

Disadvantages, Advantages, and Common Pitfalls

As with the introduction of most new teaching styles into a course, there are several areas that can cause problems when adopting case study learning. A description of how case studies have been introduced into a degree course and some of the specific issues identified can be found in Davis and Wilcock's work (Davis & Wilcock, 2005). Newey (1987) also describes the advantages and disadvantages he encountered with case study teaching at the Open University. A review of the main issues identified is given in the sections that follow.

Added Workload for Students and Staff

It has been found that when traditional lecture-based teaching is replaced by case studies, students can spend more time working toward the case study than they would have been in the original mode of learning. This is often the case with project work, and students will sometimes spend up to three times longer on this type of learning than they would on conventionally taught modules (Gibbs, 1995). In addition there is increased workload on the staff to develop case study content. This is true when introducing problem- and project-based learning content as well, which is a disincentive to staff to change traditional teaching approaches. However "this (added) effort is largely rewarded by the enthusiasm and dedication of students" (Lima, Carvalho, Flores, & Van Hattum-Janssen, 2007, p. 345).

Ensuring that Students Are Engaged with the Case Study Subject Materials

Newey found although many students may be very interested in the chosen case study, some may not; the reason for this will vary from boredom to antipathy (Newey, 1987). An important lesson from this is not to base a course on only one case study but to use a variety. Carrying out an evaluation after the introduction of new case studies is important to determine if there has been good student engagement with the case.

Simplification of Case Study

Real-world situations are very complex, and generally some degree of simplification may be required to translate the example selected into a format that can be taught effectively. Although this may not affect the learning outcomes, care is needed to ensure that the case study does not appear contrived or that critical factors are omitted.

Explanation of Case Study Requirements

The use of case study teaching at the University of Birmingham, U.K. (Davis & Wilcock, 2004) has shown that students often want and need more details on what is expected from them in case study (or project-/problem-based) learning, for example, the level of independent research, and more specifically, sufficient information on how to write reports, give presentations, and design and present posters, as also reported by Gibbs (1995). This is particularly important at the start of the course because for many students this may be a very different form of learning than what they are used to.

Depth of Learning

It is possible that during research-based case studies students can derive all of their information from the Internet. Although this is a valuable resource, it can often result in only surface learning where the students copy or paraphrase content available on the Internet. One way of addressing this problem is to provide references to textbooks and journal papers that the students are expected to consult. Including a laboratory component to the case is also a useful way of achieving more in-depth study. Ensuring that there is progression of learning skills development (e.g., analysis to synthesis) when using a series of case studies is important, rather than too much repetition of the same skills. However, Newey (1987) pointed out that some essential engineering skills can be obtained only by repeated practice, most significantly that of numerical problem solving. He suggested that it is difficult to achieve this using a solely case study approach to a topic and that some digression into classwork drills would be necessary.

Case Study Grade Allocation

Students will often spend a disproportionate amount of time on project-related work in comparison to traditionally taught courses and there needs to be careful consideration of how many credits/grades should be allocated to the case study. It is important that students feel that they receive enough credit for the effort expected. One of the main reasons for introducing case studies is to increase students' motivation and enjoyment of the topic, but if they feel that they are not being rewarded for the work it may lead them to be disillusioned with the learning process. A useful way of determining appropriate assessment weighting is

to examine the proportion of student effort hours it involves (Gibbs, 1995).

Evaluation – Research on Cases: Where Do We Go?

Even though case studies have been used since the early 1950s within engineering education, there is limited research on the impact of cases on student learning outcomes. The research, so far, has mainly focused on student attitudes toward cases and how students think cases improve their learning and engagement in the classroom (Yadav et al., 2010). Even though perceptions provide valuable insight into student attitude toward the content and engagement in the classroom, they are not a good proxy for student learning. Prior research in education and psychology has suggested that students are not good judges of their own learning and tend of overestimate their own learning (Dunlosky & Lipko, 2007; Glenberg, Wilkinson, & Epstein, 1982). Hence, it is important to examine the impact of cases beyond student perceptions and assess their learning from case studies. In addition to examining student learning, future research on the impact of cases should include how cases should be taught, at what level should cases be taught, and how much background knowledge students need to effectively learn from cases. The following section discusses avenues for research on cases and proposes some research designs to evaluate cases.

Berliner (2002) argued that education research is the "hardest science of them all" because of the contextual factors at play; thus, there is rarely a perfect study and researchers need to make tradeoffs continually. For example, to conduct an experimental research design, a researcher needs to assign students randomly to a control (e.g., lecture) classroom and experimental (case) classroom all else being equal (e.g., both classes taught by the same instructor). However, given the confines of course schedules and availability, more often than not a true experimental research design is not possible.

A typical scenario is an instructor teaching a course and trying out case studies for some of the topics. Single-subject research designs (AB, ABA, ABAB) offer a possible solution to evaluate the impact of cases in such a setting when there is only one class available (Yadav & Barry, 2009). Within the three single-subject designs, ABAB is the most robust in allowing a researcher to attribute changes to the experimental condition. Specifically, ABAB design would include four phases that alternate between control and experimental condition across four different topics while assessing student outcomes during each phase. In this design, participants act as their own control, which also allows for more participants in each condition and is an important consideration when you have a small sample size (see Yadav, Subedi, Lundeberg, & Bunting [2011] for an example of an ABAB design).

In a scenario when two classes are available, but are taught by two different instructors, a counter-balanced research design could be used to remove any instructor effect. In this design, the control and experimental would be counter-balanced across two topics and the two instructors. Yadav et al. (2010) used this design to examine the impact of cases on student learning in a mechanical engineering course as compared to a traditional lecture across two topics (thermal systems and fluid mechanics). Instructor A used the case method for the thermal systems while instructor B used the traditional lecture and vice versa for the fluid mechanics. The outcomes of this study have been discussed earlier.

In addition to an effective research design, researchers also need robust assessment that measures students' learning outcomes (Lundeberg & Yadav, 2006a, 2006b). Previous research has suggested that case-based teaching is similar to traditional pedagogical approaches in improving students' factual knowledge; however, they are significantly better at improving students' critical thinking and problem-solving skills (Antepohl & Herzig, 1999; Dochy, Segers, Van den Bossche, & Gijbels, 2003). Hence,

it is important to use carefully constructed learning measures that go beyond assessing students' factual knowledge and assess their conceptual understanding, problem-solving skills, and ability to transfer their learning to novel situations. It is important that conditions of learning should match the assessment used to measure learning outcomes. Jonassen suggested ways to measure students' problem schemas, problem-solving performance, cognitive skills required to solve problems, and ability to defend their solutions (Jonassen, 2011). An approach suggested by Mazur is *problem-paired testing*, which involves using two types of questions to assess what students are actually learning – one traditional and one conceptual question (Mazur, 1998). Mazur found that students do very well on traditional problems, but very poorly on simple qualitative problems that deal with the same underlying concept. Researchers could use Mazur's approach by assessing students' conceptual understanding through open-ended questions. More specifically, each traditional question that examines students' ability to solve a problem could be followed up with a qualitative question that asks them to explain what's going on in the problem (see Yadav et al., 2011 for an example of such tests). In summary, researchers should use appropriate assessment tools to meet the goals of case-based instruction.

Conclusions

Case studies can be developed using a variety of approaches, from employing industrial experts to write content to basing the case on a lecturers research interests. The common factor is that the case study should be based on real-world scenarios that allow students to consider problems that they otherwise would not experience.

Case study learning can enhance a student's motivation and enjoyment of engineering subjects. The disadvantages and problems that can be encountered when using this approach have been highlighted, along with a discussion on methods for evaluation. A common theme throughout has been the need to ensure students are given appropriate training and direction on the expectations of the new learning approach so that the desired learning outcomes are achieved, and the need for appropriate assessment methods.

Appendix 9.1. Key Skills in Case Study–Based Teaching, PrjBL, and PBL

Case study–based teaching, PrjBL, and PBL are useful methods to develop transferable skills. Key skills that can be embedded into these forms of teaching include group working, peer tutoring, and communication skills. Discussion on these topics, as they relate particularly to case studies, is provided in the sections that follow.

Group Working

The ability to work well in a group is a valuable skill and one that can be developed during a student's time at university. Group learning can be used to promote active learning (Bonwell & Eison, 1991) and aid the development of communication, leadership, organization and problem-solving skills (Butcher et al., 1995); it also has clear vocational relevance. Although most students recognize and acknowledge the benefits of group working, many are concerned with conflicts and uneven workload within groups (Davis & Wilcock, 2003). Feedback on group work has shown that this can present a particular problem for some students, comments include:

> I don't like working as part of a team because there are always lazy people who don't do any work and if you don't want that to effect your own mark you end up doing everything. I work well in a team and am quite a good organiser, but tend to do too much of the work.

Many students entering their first year of university will have had little or no experience of group working and this may heighten the problems that can arise in a group situation. Hence an initial session on group working skills before running case studies has been found to be beneficial (Davis & Wilcock, 2003). This session could include a discussion of group dynamics, group roles, group meetings, and a short group exercise. Student feedback from a group working skill session linked to a case study included, "I found the session really useful and it was interesting to look at strategies used in group working" and "I got to know the group better and we eventually worked better as a team."

Some lecturers have found that carefully selecting group membership to ensure smooth group operation can overcome the potential problems experienced by students during group work. There are various ways in which group membership can be determined. Allowing students to select their own groups can go some way to alleviate problems with group dynamics; however, this can often result in the grouping of high-ability students and low-ability students, which may not be useful if you want peers to support each other (Brown, 1996). Random selection is often seen as the fairer mode of forming groups and can be more representative of industrial and professional work where team members are often selected on the required skills for the project rather than friendships (Gibbs, 1995). However, groups may encounter difficulties and conflicts, particularly if they have little experience in working with others. Another alternative is to form groups based on learning styles or abilities. This can be useful if the instructor wants to build on students' prior experiences or skills, but it can be difficult to organize because it relies on the instructor and students having a good knowledge of the way in which individuals work and interact with each other (Brown, 1996). Borges, Galvão Dias, and Falcão e Cunha (2009) reported on their evaluation of two group formation methods including self-selection and using a student questionnaire and then an algorithm aiming to achieve maximum diversity within groups and homogeneity among groups. The authors found that when the new group–formation method was used, students were forced to adapt to new circumstances and an additional effort was needed to achieve the goals proposed by the project. They reported that although teachers see the additional challenge as an opportunity to induce the acquisition of new skills, students fear that the required extra effort will have a negative impact on their final grade. However, the results showed that, contrary to students' beliefs, the new group–formation method did not negatively affect the final outcomes.

Communication Skills

Case studies often form part of a course work requirement in which the students have to present their work in a variety of formats, these include, oral presentations, articles, posters, and reports. The ability to communicate effectively is strongly emphasized by employers (Horton, 2001) and it is important that such skills should be addressed during university study. By incorporating a variety of modes of communication (i.e., reports, essays, articles, presentations, posters) students will be well prepared for the range of skills required in the workplace. Coordination is required between all the lecturers on a course to ensure that any particular communication technique is not used excessively, for example, students only producing posters rather than giving oral presentations. Examples of good and bad practice should be available for students, and feedback on performance should be provided. Asking students to self-assess and peer assess presentations (posters and oral) in a formative manner can provide added insight.

Peer Tutoring

Peer tutoring and peer-assisted learning are powerful tools for increasing motivation, enhancing learning experience, and improving performance. Both approaches can be used successfully in engineering (Kieran, 2009) within traditional teaching

and case study/problem- and project-based learning. As case studies are a vehicle for students to identify learning activities and to conduct self-directed study, it is also possible to include students from different courses and levels in the same class (Wilcock & Davis, 2006), thereby including either formal or informal peer tutoring. When using peer tutoring during case studies, it is important that sufficient consideration is given to the level of lecturer assistance offered. During the use of peer tutoring/teaching (third year students managing a first-year design exercise within civil engineering, see http://www.ebl.bham.ac .uk/bham/case9.shtml), it was found that there was a need to "identify staff who are able to take an appropriate hands-off approach and allow the third year students to learn by their mistakes.... By contrast if the approach is too 'hands-off' students can become frustrated by continued failure."

Appendix 9.2. Examples of Case Studies in Engineering

The details given here are not exhaustive, and are correct at the time of writing.

Details of case studies can be found by searching the resource listings at the following websites:

The National Center for Case Study Teaching in Science (NCCSTS) website contains 416 case studies from across science and engineering subjects; however, the number of resources currently available for engineering at an undergraduate level is limited. There is a searchable database that allows resources to be located by subject, educational level, and learning style. See http://sciencecases.lib.buffalo.edu/cs/ Answers to some of the case study questions are obtainable from the center and are released to lecturers.

Learning and Teaching Support Network (LTSN) for Engineering website. The Engineering LTSN is a one of the twenty-four subject centers that form the subject network of the Higher Education Academy. The Engineering Subject Centre delivers subject-based support to promote quality learning and teaching. There is a resource database (searchable by keyword) that contains approximately 1,000 resources. Specific engineering case studies are available that contain content and suggested teaching approaches. Unfortunately using case study as the keyword returns more than 300 resources that include the LTSN case studies on learning approaches, not just specific engineering case study content. A couple of examples of specific case studies are described below from this site. See http://www.engsc.ac.uk/.

UK Centre for Materials Education (UKCME) is a subject center of the Higher Education Academy and is based at the University of Liverpool. See http://www.materials.ac.uk/ for general details. A direct link to developed case study resources is http://www.materials.ac.uk/teachingdev/resource.asp, which contains about five different specific case studies and other learning resources.

Collaborative Open Resource Environment – for Materials. The CORE-Materials repository contains 1,586 open educational resources (OERs) in Materials Science and Engineering, available under a range of Creative Commons licenses. There are forty-eight resources listed as case studies including industrial corrosion and failure analyses and lecture notes on materials selection and design of components. Also on the site are videos, images, and data sets that can be used to support learning activities. See http://core.materials.ac.uk/.

Details and links for some specific case studies are given below:

Case studies on engineering for sustainable development can be found at in *Engineering for Sustainable Development: Guiding Principles*, published by The Royal Academy of Engineering, London, in 2005. The guide contains details of seven case studies that were developed by visiting professor at U.K. universities in the areas of civil, mechanical, and electrical engineering. Sufficient details and further references are provided to allow the case studies to be used within a course.

http://www.raeng.org.uk/events/pdf/Engineering_for_Sustainable_Development.pdf

Case studies on polymer foams used in sport and the health service. The case study topics are flexible foams that cushion falls (e.g., sport crash mats, etc.), closed-cell polystyrene and polypropylene foams for packaging and foam cushions for wheelchairs. Detailed content is provided on a dedicated website developed by Dr Nigel Mills from the University of Birmingham (www.foamstudies.bham.ac.uk). Each topic on the website contains background information, videos, design programs, laboratory experiments, links to other websites, and interactive tests.

Industrial failure-analysis case studies. Case studies include wire rope failure; undercarriage leg failure; aircraft towbar failure; hail damage; insulator caps; and a fractography resource. The examples are chosen and presented so as to guide learners through the analytical steps and thought processes used in solving such problems. Detailed information is available for students and lecturers at http://www.tech.plym.ac.uk/sme/Interactive_Resources/tutorials/FailureAnalysis/index.html.

Materials science case studies used in materials-related degree courses. Information on the case study topic, assessment methods, and links to resources are provided at http://www.birmingham.ac.uk/schools/metallurgy-materials/cases/index.aspx. The website has been designed to act as a support mechanism for students taking part in the case studies, but also as a means of demonstrating to other institutions how the case studies operate. Support resources on group working and assessment are also provided. The case studies included cover biomedical implants, sports equipment, fuel cells, chocolate and engineering failures.

Case studies in materials selection. Case studies of materials selection for six different applications (cycle helmets, seals, table legs, heat exchanger tube, knife edges, and aperture grills for cathode ray tubes). The case studies include an introduction, details of the design requirements and mathematical equations needed, and material selection charts and details of how these can be used for the specific applications.

http://www.grantadesign.com/resources/materials/casestudies/casestudies.htm

Forensic Materials Engineering: Case Studies. This book contains a range of case studies from both the United States and the United Kingdom in the field of forensic engineering. The information included covers not only the technical detail, but also the wider background including human error and the legal situation. The authors are Peter Rhys Lewis, Ken Reynolds, and Colin Gagg. The book was published by CRC Press in 2003.

Corrosion and failure analysis. This website (http://corrosionlab.com/Failure-Analysis-Studies) contains details of failure investigations, often linked to corrosion or fatigue, in engineering components. For each example a description of the background to the component is given, along with results of the investigation and discussion of the findings. More than forty investigations are reported.

Engineering ethics case studies. This Web page (http://www.engsc.ac.uk/ethics/case-studies) presents twelve engineering ethics case studies. The case studies cover a variety of topics including safety, intellectual property, sustainability, and professionalism within a range of engineering disciplines. The case studies are designed for one- or two-hour sessions with small group discussions. Each case study sets out a scenario, presents questions for discussion, provides notes, and so forth. Downloadable notes, including additional references, are available from the website.

Pumplight: A case study in engineering entrepreneurship. The case study illustrates elements of entrepreneurship in engineering and traces the development of an initial idea for a "Pumplight" for a bicycle through to the production of a prototype product. It provides ideas for use as a teaching resource and links to the matrix of generic teaching materials that are available on the engineering subject center website. See http://www.engsc.ac.uk/downloads/Entre/pumplight.pdf.

Space shuttle disasters. Case study material on ethics, materials engineering, and data presentation. A great deal of information is available on the *Challenger* and *Columbia* failures that allows case studies to be developed. The book by Diane Vaughan, *The Challenger Launch Decision: Risky Technology, Culture, and Deviance at NASA*, provides an excellent reference to the complex issues surrounding the *Challenger* disaster. The NASA website contains excellent support material, including technical data and images, that can be used.

Footnote

1. Project-based learning in engineering generally involves the design and production of a product or answer to a specific (often "real world") problem and therefore has more in common with case study learning.

References

Antepohl, W., & Herzig, S. (1999). Problem-based learning versus lecture-based learning in a course of basic pharmacology: A controlled, randomized study. *Medical Education, 33*(2), 106–113.

Ashby M. F. (2011). *Materials selection in mechanical design* (4th ed.), Burlington, MA: Butterworth-Heinemann.

Berliner, D. C. (2002). Educational research: The hardest science of all. *Educational Researcher, 31*(8), 18–20.

Bilica, K. (2004). Lessons from experts: Improving college science instruction through case teaching. *School Science and Mathematics, 104*(6), 273–278.

Bonwell, C. C., & Eison, J. A. (1991). Active learning: Creating excitement in the classroom. *ASHE-ERIC Higher Education Report No. 1.* Washington, DC: The George Washington University, School of Education and Human Development.

Borges, J., Galvão Dias, T., & Falcão e Cunha, J. (2009). A new group-formation method for student projects. *European Journal of Engineering Education*, 34(6), 573–585.

Bransford, J., Brown, A., & Cocking, R. (2000). *How people learn: Brain, mind, and experience & school*. Washington, DC: The National Academies Press.

Brown, S. (1996). The art of teaching small groups 1. *New Academic*, 5(3), 3–5.

Brown, J. S., Collins, A., & Duguid, P. (1989). Situated cognition and the culture of learning. *Educational Researcher*, 18(1), 32–42.

Burgess, K. L. (2004). Is your case a problem? *Journal of STEM Education: Innovations and Research*, 1/2, 42–44.

Butcher, A. C., Stefani, L. A., & Tario, V. N. (1995). Analysis of peer-, self- and staff-assessment in group project work, *Assessment in Higher Education*, 2(2), 165–185.

Chickering, A. W., & Gamson, Z. F. (1987). Seven principles for good practice. *American Association for Higher Education Bulletin*, 39(7), 3–7.

Conant, J. B., & Van Deventer, W. C. (1951). The growth of the experimental sciences: An experiment in general education. *Science Education*, 35(2), 112–114.

Conway, P. (2001). Using cases and activity learning with undergraduate economic classes. *The House Journal of the European Case Clearing House*, 26, 18–19.

Davis, C. L., & Wilcock, E. (2003). Thematic booklets – Case studies. UK Centre for Materials Education. Retrieved from http://www.materials.ac.uk/guides/casestudies.asp

Davis, C. L., & Wilcock, E. (2004). Case studies in engineering. In C. Baillie & I. Moore (Eds.), *Effective learning and teaching in engineering* (pp. 51–71). New York, NY: Routledge.

Davis, C., & Wilcock, E. (2005). Developing, implementing and evaluating case studies in materials science. *European Engineering Education Journal*, 30(1), 59–69.

Demarco, R., Hayward, L., & Lynch, M. (2002). Nursing students' experiences with and strategic approaches to case-based instruction: A replication and comparison study between two disciplines. *Journal of Nursing Education*, 41(4), 165–174.

Dochy, F., Segers, M., Van den Bossche, P., & Gijbels, D. (2003). Effects of problem-based learning: A meta-analysis. *Learning & Instruction*, 13(5), 533–568.

Dunlosky, J., & Lipko, A. R. (2007). Meta-comprehension: A brief history and how to improve its accuracy. *Current Directions in Psychological Science*, 16(4), 228–232.

Fuchs, H. O. (1974). On kindling flames with cases. *Engineering Education*, 64(6), 412–415.

Gallucci, K. (2006). Learning concepts with cases. *Journal of College Science Teaching*, 36(2), 16–20.

Garg, K., & Varma, V. (2007). *A study on the effectiveness of case study approach in software engineering education*. Paper presented at the 20th Conference on Software Engineering Education and Training, Dublin, Ireland.

Garvin, D. A. (2003). Making the case: Professional education for the world of practice. *Harvard Magazine*, 106(1), 56–65.

Gibbs, G. (1995). *Assessing student centred courses*. Oxford: The Oxford Centre for Staff Development.

Glenberg, A. M., Wilkinson, A. C., & Epstein, W. (1982). The illusion of knowing: Failure in the self-assessment of comprehension. *Memory & Cognition*, 10(6), 597–602.

Goett, J. A., & Foote, K. E. (2000). Cultivating student research and study skills in Web-based learning environments. *Journal of Geography in Higher Education*, 24(1), 92–99.

GrantaDesign. Retrieved from: http://www.grantadesign.com/education/

Harris, C. E. (2003). Methodologies for case studies in engineering ethics. In *Emerging technologies and ethical issues in engineering: Papers from a workshop* (pp. 79–93). Washington, DC: The National Academies Press.

Henderson, J. M., Bellman, L. E., & Furman, B. J. (1983, January). A case for teaching engineering with cases. *Engineering Education*, 288–292.

Herreid, C. F. (2007). *Start with a story: The case study teaching method of teaching college science*. Arlington, VA: National Science Teachers Association Press.

Horton, G. (2001). The need for professional skills training in engineering programs. In *First Baltic Sea workshop on education in mechatronics*. Kiel, Germany: Fachhochschule Kiel.

Jonassen, D. H. (2011). *Learning to solve problems: A handbook for designing problem-solving learning environments*. New York, NY: Routledge.

Jonnasen, D., Strobel, J., & Lee, C. B. (2006). Everyday problem solving in engineering: Lessons for engineering educators. *Journal of Engineering Education*, 95(2), 1–14.

Kardos, G., & Smith, C. O. (1979). On writing engineering cases. In *Proceedings of ASEE National Conference on Engineering Case Studies*.

Kieran, P., & O'Neill, G. (2009). Peer-assisted tutoring in a chemical engineering curriculum: Tutee and tutor experiences. *Australasian Journal of Peer Learning*, 2(1), 40–67.

Klein, J. T. (1996). *Crossing boundaries: Knowledge, disciplinarities, and interdisciplinarities*. Charlottesville: University Press of Virginia.

Klein, J. T. (2004). Prospects for transdisciplinarity. *Futures*, 36, 515–526.

Lima Rui, M., Carvalho, D., Flores, M., & Van Hattum-Janssen, N. (2007). A case study on project led education in engineering: Students' and teachers' perceptions. *European Journal of Engineering Education*, 32(3), 337–347.

LTSN Generic Centre. (2001). Assessment series. Retrieved from http://www.bioscience.heacademy.ac.uk/resources/themes/assessment.aspx

Booklet 5: *A briefing on key skills in higher education* by Roger Murphy.

Booklet 7: *A briefing on key concepts* by Peter Knight.

Booklet 9: *A briefing on self, peer and group assessment* by Phil Race.

Booklet 12: *A briefing on assessment of large groups* by Chris Rust.

Lundeberg, M. A., & Yadav, A. (2006a). Assessment of case study teaching: Where do we go from here? Part 2. *Journal of College Science Teaching*, 35(6), 8–13.

Lundeberg, M. A., & Yadav, A. (2006b). Assessment of case study teaching: Where do we go from here? Part I. *Journal of College Science Teaching*, 35(5), 10–13.

Martinez-Mones, A., Gomez-Sanchez, E., Dimitriadis, Y. A., Jorrin-Abellan, I. M., Rubia-Avi, B., & Vega-Gorgojo, G. (2005). Multiple case studies to enhance project-based learning in a computer architecture course. *IEEE Transactions on Education*, 48(3), 482–489.

Mayo, J. A. (2002). Case-based instruction: A technique for increasing conceptual application in introductory psychology. *Journal of Constructivist Psychology*, 15, 65–74.

Mayo, J. A. (2004). Using case-based instruction to bridge the gap between theory and practice in psychology of adjustment. *Journal of Constructivist Psychology*, 17, 137–146.

Mazur, E. (1998). *Moving the mountain: Impediments to change*. Paper presented at the Third Annual NISE Forum, Washington, DC.

Merseth, K. K. (1991). The early history of case-based instruction: Insights for teacher education today. *Journal of Teacher Education*, 42(4), 243–249.

Merseth, K. K. (1994). Cases, case methods, and the professional development of educators. *ERIC Digest*. Retrieved from http://eric.ed.gov/ERICWebPortal/detail?accno=ED401272

Mustoe, L. R., & Croft, A. C. (1999). Motivating engineering students by using modern case studies. *European Journal of Engineering Education*, 15(6), 469–476.

National Academy of Engineering. (2004). *The engineer of 2020: Visions of engineering in the new century*. Washington, DC: The National Academies Press.

Newey, C. (1987). A case study approach to the teaching of materials. *European Journal of Engineering Education*, 12(1), 59.

Orsmond, P. (2004). Self- and peer-assessment – Guidance on practice in the biosciences. In *Teaching bioscience enhancing learning series 2004*. Retrieved from http://www.bioscience.heacademy.ac.uk/ftp/TeachingGuides/fulltext.pdf

Raju, P. K., & Sankar, C. S. (1999). Teaching real-world issues through case studies. *Journal of Engineering Education*, 88(4), 501–508.

Reeves, T., & Laffey, J. (1999). Design, assessment and evaluation of a problem based learning environment in undergraduate engineering. *Higher Education Research and Development*, 18(2), 219–232.

Savin-Baden, M. (2007). Challenging models and perspectives of problem-based learning. In E. de Graaff & A. Kolmos (Eds.), *Management of change: Implementation of problem-based and project-based learning in engineering* (pp. 9–30). Rotterdam, The Netherlands: Sense.

Serpil Acar, B. (2004). Analysis of an assessment method for problem-based learning, *European Journal of Engineering Education*, 29(2), 231–240.

Smith, C. O. (1992). Student written engineering cases. *International Journal of Engineering Education*, 8(6), 442–445.

Swanson, D., Case, S., & van der Vleuten, C. P. M. (1991). Strategies for student assessment. In D. J. Boud & G. Felletti (Eds.), *The challenge of problem-based learning* (pp. 260–274). London: Kogan Page.

Vesper, K. H. (1964). On the use of case studies for teaching engineering. *Journal of Engineering Education*, 55(2), 56–57.

Vesper, K. H., & Adams, J. L. (1969). Evaluating learning from the case method. *Engineering Education*, 60(2), 104–106.

Wilcock, E., & Davis, C. (2006). Group working and peer tutoring in case studies. *British Journal of Engineering Education*, 5(1), 3–10.

Wilcock, E., Jenkins, M., & Davis, C. (2002). *A study of good practice in group learning in Sports Materials Science using case studies*. Paper presented at the 2nd Annual UK & USA Conference on the Scholarship of Teaching and Learning (SoTL), TUC Centre, Holborn, London.

Williams, S. M. (1992). Putting case-based instruction into context: Examples from legal and medical education. *The Journal of Learning Sciences*, 2(4), 367–427.

Yadav, A., & Barry, B. E. (2009). Using case-based instruction to increase ethical understanding in engineering: What do we know? What do we need? *International Journal of Engineering Education*, 25(1), 138–143.

Yadav, A., Shaver, G. M., & Meckl, P. (2010). Lessons learned: Implementing the case teaching method in a mechanical engineering course. *Journal of Engineering Education*, 99(1), 55–69.

Yadav, A., Subedi, D., Lundeberg, M. A., & Bunting, C. F. (2011). Problem-based learning: Influence on students' learning in an electrical engineering course. *Journal of Engineering Education*, 100(2), 253–280.

CHAPTER 10

Curriculum Design in the Middle Years

Susan M. Lord and John C. Chen

Introduction

This chapter aims to call attention to the "middle years" of engineering study, which are often overlooked from a research perspective. There have been tremendous efforts over the past two decades focused on the first year and final year of engineering education. Although there have been efforts targeted at the middle years, which we summarize in this chapter, we argue that there is a need for more attention and research-driven innovations for these formative years for the emerging engineer.

The second and third years of engineering study are times when students are focused on their specific engineering disciplines, beginning with the foundational engineering sciences and leading to discipline-specific core courses. Our goal is to highlight creative and successful efforts in these years. We hope this can inspire other instructors to use these examples as models for adaptation to their own disciplines and courses. We focus on several major engineering disciplines, but deliberately leave out computer science because at many institutions

this department may not reside in the engineering school. Furthermore, we do not review the myriad literature related to distance and online learning, service-learning, or cooperative education. Our intended audience includes new engineering faculty, all engineering faculty interested in curriculum reform, and engineering education researchers, including graduate students pursuing this field of study. Finally, given our space limitations, we admittedly take a narrow view in focusing primarily on work in the United States.

Although the term "sophomore slump" has been around for decades (Hunter et al., 2010), research on the "sophomore slump" in engineering is in its infancy (Holloway, Reed-Rhoads, & Groll, 2010, 2011). Students undergo tremendous changes during their time in college. As engineering educators, we are concerned mainly with the engineering skills and knowledge they gain, but the courses we teach are also an important part of the students' identity development as emerging engineers and their motivation to persist in their studies. A study using the Academic Pathways of People Learning

Engineering Survey (APPLES) instrument (Sheppard et al., 2010) found major differences between how first-year students and seniors experience their education. Senior students reported being significantly less satisfied with their instructors, less involved with their engineering courses, and less satisfied with college overall. What is happening to our students during the middle years, and what can we do with our course and curriculum design to improve their experience and help them in their development as emerging engineers?

The motivation for change in the first-year engineering experience has primarily been to enhance student retention and motivation by giving them a sense of what engineering is and what engineering design entails rather than being driven by a curricular need to teach specific content. In the final year, the capstone design imparts a synthesis experience, when students apply much of their prior learning to solving a complex, real-world design problem. Students also practice associated professional skills such as communications and teamwork and, in most programs, gain exposure to ethics and the global and societal impact of engineering. Again, rather than being driven by a need to help students learn specific content, the motivation for these capstone design courses is to help bridge the gap between school and the workplace and thus prepare them for engineering practice. Kotys-Schwartz, Knight, and Pawlas (2010) investigated this "bookend curriculum" with design at the beginning and end, in particular asking if this approach is effective for gaining skills.

Why are the middle years largely ignored? What is different about the curriculum in these years and what are and should be the goals for bridging the beginning and end of an undergraduate program? What challenges do faculty face in bringing about change in the middle years?

The middle years have largely remained unchanged for decades, except for some new courses being added and, arguably, the infusion of active-learning teaching techniques. They are composed of individual, mostly disconnected courses that impart discipline-specific content – content that is needed to become an engineer and to do complex designs, though relatively little design is included in the courses. The purpose of the middle years, it seems, is to build the students' "toolbox" of knowledge and skills and to lay the foundation for doing complex design in the final year. Many of us who teach these courses in the middle years may have become comfortable with the compartmentalized nature of these courses. Reluctance to change may be rooted in fears of having insufficient knowledge to integrate material across courses or insufficient time to incorporate innovative pedagogies in critical foundational courses that serve as prerequisites for other courses in the discipline. The stakes might seem higher for making changes in these content-focused courses than in the design courses at the first and senior years. We argue that faculty should rethink the goals for the middle years and find creative solutions to problems that continue to linger in engineering education, such as attrition in the sophomore year and low motivation. Engineering educators should find ways to motivate students to learn, provide relevant context and supportive environments for this learning, and to prepare them better to take on the challenges in the senior year and beyond. Research on learning and human behavior could provide useful guidance here, and engineering faculty would do well to learn from this rich literature and apply key ideas to the design of curriculum in the middle years, not just the first and senior design courses.

The remainder of this chapter begins with a brief review of research results that have informed engineering curricular design and guidelines for such design and then presents examples of curriculum design at the individual course level and integrated curriculum design with an emphasis on the second and third years. The impetus for many of these changes, namely the major funding source for implementing the changes and the accreditation body, are also described.

We conclude with a research agenda for this area.

Research Bases for Curricular Design

Calls for curriculum design in engineering education based on theoretical frameworks or sound educational research have been around for more than three decades (Stice, 1976). The adoption of outcomes-based assessment in engineering around the globe including the United States, Australia, and Europe brought a larger emphasis on learning objectives (or learning outcomes) and assessment processes tied to these. Some researchers viewed this shift as an opportunity to approach course planning and learning assessment more purposefully (Safoutin et al., 2000) and to approach instruction systematically, considering teaching, learning, and assessment as an interrelated triad (Pellegrino, 2006). This integrated and thoughtful approach has been described as design, something critically important and familiar in engineering. Duderstadt (2008) asserts that faculty members in the twenty-first century must "set aside their roles as teachers and instead become designers of learning experiences, processes, and environments." Streveler, Smith, and Pilotte (2011) encourage engineering educators to make a more conscious effort to align content, pedagogy, and assessment of outcomes for curricula, framing this as an "engineering design approach." We argue that this approach should be used for all years of the engineering curriculum and not reserved for the "special" design courses for first-year and senior students.

A large body of research exists on learning theories, models, and frameworks. Some of this is summarized in Chapter 2 by Newstetter & Svinicki in this volume. Much of this literature is unfamiliar to engineering educators but some work has been done in bridging this gap. In the next section, we describe the research results that have had the most impact on engineering education. We also introduce learning theory that is becoming prevalent in engineering education. Certainly, there is considerable work to be done on more widely integrating curriculum design with robust theories.

Theories, Frameworks, Models, and Concepts

The first educational research results to be appreciated by engineering educators are Bloom's taxonomy of educational objectives (1956) and Kolb's learning cycle including "Why?" "What?" "How?" and "What if?" (1984) which are nicely summarized in Harb and Terry (1993). Because the goal of engineering education is to help students become effective contributors to the profession or "become engineers" with the majority going into industry rather than academia, curricula are often tied to the profession. Design is critical in engineering and thus there have been numerous calls for integrating design into the engineering curriculum. Although much of this has focused on the first year and the senior capstone, there have also been attempts at design across the curriculum, which has included efforts in the middle years. The work of Bloom and Kolb has implications for teaching design. The cognitive skills needed for design fall in the highest levels of Bloom's taxonomy including synthesis, analysis, and evaluation. Incorporating design throughout the curriculum "is a natural consequence of completing the [Kolb] learning cycle for every concept which we teach" (Harb & Terry, 1993, p. 3). However, too often in engineering education, instructors teach at the lower levels of Bloom's taxonomy but assess at the higher levels (Stice, 1976) or do not consider all aspects of the Kolb learning cycle. This mismatch results in frustration and poor learning outcomes particularly for design. Incorporating open-ended problems is effective for addressing the Bloom and Kolb frameworks and there have been many examples proposed of how to incorporate these in engineering classrooms (Felder, 1987).

Engineering educators have also considered another aspect of the purposeful design of curriculum by focusing on the learners themselves. Recognition of different learning styles of students and choosing appropriate pedagogies to improve student learning has been an important topic in engineering education research. Several learning style models have been used including Kolb's (1984) and Felder and Silverman's (1988). Note that these early ideas about learning styles and Bloom's taxonomy have been critiqued by learning theorists and should be used with caution today. Nonetheless, they have had an important impact on helping instructors to reflect on their pedagogy and on catalyzing the earliest calls for a change from traditional lectures.

There has been an explosion of robust research on how people learn over the last thirty years so that "we know a considerable amount about the cognitive processes that underlie expert performances and about strategies for helping people increase their expertise" (Bransford, Vye, & Bateman, 2004, p. 165). (More details on expert–novice differences can be found in Chapter 12 by McKenna in this volume.) Pellegrino (2006) highlights three important principles about learning and understanding that instructors should consider:

1. Students come to the classroom with preconceptions about how the world works which include beliefs and prior knowledge acquired through various experiences.

2. To develop competence in an area of inquiry, students must (a) have a deep foundation of factual knowledge, (b) understand facts and ideas in the context of a conceptual framework, and (c) organize knowledge in ways that facilitate retrieval and application.

3. A "metacognitive" approach to instruction can help students learn to take control of their own learning by defining learning goals and monitoring their progress in achieving them.

Both teachers and learners benefit from solid conceptual frameworks. These principles also emphasize the developmental nature of learning, which is important to bear in mind for the engineering curricula in the "middle years."

Key aspects of modern cognitive theory are summarized in the landmark National Academy of Sciences' report *How People Learn* (HPL; National Research Council [NRC], 2000). The framework presented in HPL is organized around four overlapping aspects or lenses through which we can view the learning environment: knowledge centered, learner centered, community centered, and assessment centered. Bransford, Vye, and Bateman (2004) provide a nice discussion of each of these lenses. Each is relevant for engineering educators but, not surprisingly, the most effort to date has gone into the knowledge-centered arena. Here, the need is to emphasize connected knowledge in a discipline to help students move from novice to expert including prioritizing into categories such as "enduring ideas of the discipline" to "important things to know and be able to do" to "ideas worth mentioning" (Wiggins & McTighe, 1997). Key aspects of learner-centered classrooms include recognition of "expert blind spots" and misconceptions of novice learners. (Misconceptions are discussed in more detail in Chapter 5 by Streveler, Brown, Herman, & Monfort in this volume.) Bransford et al. briefly describe other components of being learner-centered as "honoring students' backgrounds and cultural values and finding special strengths that each may have that allow him or her to connect to information being taught in the classroom" (2004, p. 169). This echoes the ideas of different learning styles investigated earlier. This aspect of valuing the student as a person also has particularly important consequences in engineering, which is problematically lacking in diversity.

The lens of community is based on Vygotsky's theory that learning is culturally mediated and occurs in a social environment (1978). Thus it is important to make the classroom a safe space for learning. This is related to ideas of a community of practice (Lave & Wenger, 1991), learning communities, and a constructivist approach to learning. Interestingly, such approaches have been shown to have benefits that go beyond the social and include higher levels of complex problem-solving ability (Bransford, Vye, & Bateman, 2004), which is particularly important for engineering. These ideas are also important for learning communities, integrated curricula, and broadening participation in engineering, which we describe later in this chapter.

The assessment lens has perhaps been too narrowly focused in engineering on traditional testing and satisfying accreditation requirements. Researchers argue for improvement in this arena, particularly formative assessment, aimed at improving performance and understanding before final summative assessment for grading or accreditation (NRC, 2001). In addition to the many in-class assessment techniques that have become commonplace over the past twenty years (Angelo & Cross, 1993; Felder, 1995), a popular formative assessment recently adopted by engineering educators is classroom response systems or "clickers" (Bruff, 2007; Crews, Ducate, Rathel, Heid, & Bishoff, 2011). Bransford et al. (2004) argue that educators need to rethink "traditional approaches to instruction to make them higher quality" and that this should extend beyond individual activities to entire courses and curricula. Chapter 29 by Pellegrino, DiBello, & Brophy in this volume focuses on assessing learning.

Curricular Design Guidelines

There have been several publications linking theory to effective engineering curricular design and providing specific guidelines and recommendations for engineering educators (see Engel & Giddens (2011) for an overview). For example, Froyd (2008) identifies eight "promising practices" for undergraduate science, technology, engineering, and mathematics (STEM) education that overlap with the four lenses of *How People Learn* (NRC, 2000). Froyd evaluates each of these practices against standards for ease of implementation and effectiveness for student learning. Ratings against these standards ranged from Fair to Strong. Of those practices listed, those that appear most promising are the following:

- Best (Strong for implementation and student performance):
  ○ Organizing Students in Small Groups
  ○ Designing In-class Activities to Actively Engage Students
- Next best (Strong for implementation and Good for student performance):
  ○ Preparing a Set of Learning Outcomes
  ○ Providing Students Feedback through Systematic Formative Assessment
- Third best (Good to Strong for Implementation and Good for student performance)
  ○ Scenario-based Content Organization – This includes techniques such as problem-based learning (PBL), inquiry-based learning, service learning, and model-eliciting activities (MEAs).

Note that more information on PBL is available in Chapter 8 by Kolmos, De Graaff, & Nyborg in this volume. Overviews of approaches and evidence for effectiveness for a variety of pedagogies are provided in Prince and Felder (2006, 2007).

Sheppard, Macatangay, Colby, and Sullivan (2009) criticize the traditional dominant curricular model:

> . . . which might be best described as building blocks or linear components, with its attendant deductive teaching strategies, structured problems, demonstrations, and assessments of student learning does not reflect what the significant and compelling body of research on learning suggests about how students learn and develop and how experts are formed. (p. 6)

They encourage engineering educators to teach for professional practice and propose four principles to redesign engineering education:

Principle 1: Provide a professional spine.
Principle 2: Teach key concepts for use and connection.
Principle 3: Integrate identity, knowledge, and skills through approximations to practice.

Principle 4:　Place engineering in the world: encourage students to draw connections.

These guidelines provide useful frameworks that are understandable for engineering educators who are not experts in educational research.

Curriculum Design at the Course Level

In this section, we focus on research-based curriculum design for the middle years at the course level, where individuals, groups or a new idea have resulted in course innovations.

Motivations for Change: Resources and Requirements

Realistically, significant impactful change in most arenas requires strong motivation in the form of resources ("carrots") or requirements ("sticks"). In engineering education in the United States, perhaps the most powerful "carrot" has been funding from the National Science Foundation (NSF) and the most powerful "stick" being the outcomes-based accreditation requirements of ABET (formerly known as the Accreditation Board for Engineering and Technology). From 1990 to 2005, the majority of NSF's funding for engineering education supported several Engineering Education Coalitions (EECs) with the far-reaching goal of effecting change in undergraduate engineering education (Froyd, 2005). The eight groups of forty member schools represented a broad cross-section of engineering programs across the United States. Today, after the funding for the coalitions has ended, many of the changes brought about by them have persisted. Much of this information is accessible via the NEEDS (National Engineering Education Delivery System) website (NEEDS, 2012).

A significant legacy of the NSF EECs was a shift to more integration of teaching, learning, and assessment. In her analysis of the publication patterns of four of the coalitions (Foundation, SUCCEED, ECSEL, and Gateway), Borrego (2007) showed that although 21% focused on first year students and only 2% on sophomores, 45% focused on "other undergrad" which included some efforts in the second and third years. Most (68%) of the publications focused on course and curricular changes including instructional technologies, student teams, and integrated curricula. Except for integrated curricula, most efforts were by a single faculty or small groups, which seems to be the preferred model for engineering faculty although not perhaps the most effective for effecting change (Lattuca, 2006). Seventy-four percent of the contributions were classified as "experiences" of the authors. Only 6% were classified as research with an increase in this category over time. Overall, only "4 percent of the publications mentioned theory from the literature, described experiments designed with control groups, or reported on statistical data analysis." Thus the infusion of theory into engineering education has been slow but the EECs provided an opportunity for engineering faculty to work with faculty in other disciplines such as education and the social sciences to bring more rigor to their education research including relying more on theories and frameworks. Borrego, Froyd, and Hall (2010) describe the diffusion of engineering education innovations highlighting the contributions of the NSF coalitions in this process.

Another significant event that has promoted change in engineering education in the United States is the adoption by ABET in 1996, of new standards for engineering program accreditation called *Engineering Criteria 2000* (EC2000) (ABET, 2010). In adopting EC2000, ABET (formerly known as the Accreditation Board for Engineering and Technology) changed its accreditation philosophy from one that was directive (stipulating what courses must be in the curriculum, for example) to one that was based on measures of student achievement of eleven outcomes. In addition, several of these outcomes may be considered to be "professional skills," such as teamwork, communications, and lifelong learning, which were never previously emphasized for accreditation. Accreditation is critical for most engineering programs in the United States and requires significant involvement by faculty to demonstrate student achievement and a continuous improvement process for curriculum development. Thus most faculty must participate to some degree, even those who may not be focused on pedagogy.

The significant change in not only how programs are accredited but also what outcomes are assessed brought about major changes in engineering curricula and cultural changes among engineering faculty. In a large study of accredited institutions, researchers found several significant impacts (Lattuca, Terenzini, & Volkwein, 2006; Volkwein, Lattuca, Terenzini, Strauss, & Sukhbaatar, 2004), including:

- EC2000's emphasis of professional skills resulted in faculty members providing students more opportunities to practice open-ended design, teamwork, and communications.
- Faculty members also reported an increase in the use of modern engineering tools and active-learning methods, such as group work, case studies, and projects.

Examples of Course Design in the Middle Years

There have been some examples of curriculum design for specific courses in the middle years in engineering based on theory or frameworks but this is still not the norm. The examples highlighted here are intended to provide readers with a sense of the range of what has been done and suggest future possibilities. Thus examples are intentionally drawn from different engineering disciplines including the largest ones, electrical and mechanical, and the ones with the reputation for being the most innovative in terms of pedagogy, chemical and industrial.

CONCEPT INVENTORIES

The concept inventory (CI) is the most prominent application of learning theory to have gained acceptance in engineering education where it is primarily used in second- and third-year courses. CIs are based on the need to identify and repair misconceptions to promote learning as described in modern learning theories (NRC, 2000). A typical CI contains multiple-choice questions that test students' understanding of concepts and requires little or no calculation, which minimizes the students' tendency to use rules and formulas. Furthermore, incorrect solutions, called "distracters," are devised so that application of common misconceptions will lead to their selection. Use of CIs is complementary to traditional assessments of student learning as typified by calculation-based mid-term or final examinations. Taken together, these measures provide guidance for professors on how best to modify a course to meet educational objectives. Engineering-related inventories have been developed for the core foundational engineering science courses taken during the middle years including statics, dynamics, mechanics of materials, thermodynamics, fluid mechanics, heat transfer, circuits, electronics, systems and signals, material science, and statistics. Reed-Rhoads and Imbrie (2008) provide a thorough review of the history, development, use, and current status of concept inventories in engineering education. Most recently, a website (www.cihub.org) is under development that will serve as a virtual community for the growing numbers of concept inventory developers, researchers, and users.

CHEMICAL ENGINEERING

This discipline was arguably the earliest in adopting changes in teaching and learning throughout the curriculum owing to the influence and leadership of James Stice (University of Texas at Austin), Phillip Wankat (Purdue University), and Richard Felder (North Carolina State University). Their ideas and methods reached beyond chemical engineering and have influenced nearly all engineering disciplines. Stice published widely on teaching engineering, and was a pioneer in promoting such now-familiar concepts as learning objectives and Bloom's taxonomy. Wankat's classic book (co-authored with F. Oreovicz), *Teaching Engineering*, is freely available on the Internet (http://engineering.purdue.edu/ChE/AboutUs/Publications/TeachingEng/index.html). Felder's many contributions to engineering and science teaching are archived at his website (http://www4.ncsu.edu/unity/lockers/users/f/felder/public/RMF.html).

There has been heightened interest in the past decade in helping students develop deep conceptual learning of engineering concepts. It is believed that such learning is critical for both durability and transfer of the learned knowledge (NRC, 2000). Concomitant with this idea is that students often develop misconceptions, which often persist even after successful completion of a course. Prince, Vigeant, and Nottis (2009, 2010) have developed tools to uncover these misconceptions in thermodynamics and heat transfer, two core courses in the middle years of chemical and mechanical engineering curricula. Furthermore, they have developed a set of inquiry-based activities – short tabletop experiences, computer simulations, or take-home exercises – that repair such misconceptions. They demonstrate through experiments that the misconceptions can be corrected and that students also demonstrate improved application of the learned concepts in new but related contexts (transfer).

Another educational movement that has moved into engineering education in the middle years (as well as the first and final years) is liberative pedagogies (Riley, 2003, 2011) based on Paulo Freire's pedagogy of the oppressed (2006) which directly considers politics in education and aims to empower students. These pedagogies represent a radical departure from traditional pedagogy familiar to engineering educators and hold promise for creating a more inclusive context for learning while addressing critically important learning outcomes such as critical thinking, reflection, ethics, social context, and contemporary issues – outcomes that stakeholders and accreditation bodies recognize as important but are difficult to address in most traditional classrooms. Riley

provides concrete examples of application in a thermodynamics course.

Using a study design based on constructivist learning theory and the variation theory of learning, Fraser, Pillay, Tjatindi, and Case (2007) developed computer simulations to address student difficulties with three concepts in fluid mechanics (again, as with thermodynamics and heat transfer, a core, middle-years course shared with several other engineering disciplines). They showed through measurements using a concept inventory that students improved their understanding significantly in two of the three areas addressed. Although literally hundreds of studies have employed computer simulations to help students learn engineering concepts, this paper represents one of the few that used a learning theory as its foundation and redesigned the course based on that.

INDUSTRIAL ENGINEERING

Industrial engineering has been described as having a unique culture within engineering (Brawner, Camacho, Lord, Long, & Ohland, 2012; Foor & Walden, 2009). Perhaps this culture contributes to the willingness of some faculty in this discipline to adopt innovative ideas from outside of engineering. Model Eliciting Activities (MEAs) present student teams with an open-ended, real-world, client driven problem and have been shown to improve conceptual learning, teamwork, and problem-solving skills (Lesh, Hoover, Hole, Kelly, & Post, 2000). These activities arose from work in mathematics and were first used in engineering education in first-year courses. However, they are now being incorporated into courses in the middle years. Yildirim, Shuman, and Besterfield-Sacre (2010) describe the theoretical basis for MEAs and their benefits in engineering education. They successfully incorporated MEAs into a variety of industrial engineering courses including statistics courses at the sophomore and junior levels for students from many engineering disciplines. They propose strategies to help engineering educators effectively use this novel pedagogy.

Concept maps are graphical representations of the relationships among concepts in a domain. These knowledge representations are based in cognitive and constructivist theory and have been used to demonstrate the differences between novices and experts (Moon, Hoffman, Novak, & Cañas, 2011). Daley, Lovell, Perez, and Stern (2011) provide an overview of the use of concept maps in engineering education focusing on product design. Turns, Atman, and Adams (2000) describe the use of concept maps in engineering education as an effective assessment at the course and program level. The courses examined included a sophomore level product dissection course and human factors and statistics courses within industrial engineering. They used concept maps to assess domain-level expertise and students' concepts of their discipline and the engineering profession. These concepts are linked to identity development, which is particularly important for the middle years of engineering education.

ELECTRICAL ENGINEERING

One of the largest disciplines, electrical engineering (EE), has a reputation for intensive content-filled classes and a reluctance to innovate. However, research-based innovations have been demonstrated in core electrical engineering courses in the middle years. For example, drawing on constructivist and knowledge transformation frameworks, Lord (2009) developed guidelines for writing to communicate experiences in engineering. She provided examples of implementing these guidelines in the standard required EE classes of circuits for sophomores and analog electronics for juniors.

Design contests as an alternative to traditional lecture and laboratory approaches have been used in electrical engineering courses in the middle years. Hussmann and Jensen (2007) provide a nice overview of the use of such design contests and describe their own work with fourth and sixth semester students in Germany. The research bases for these contests come from a desire to emulate the professional engineering environment, which is related to the community lens of *How People Learn*, and literature on approaches to effective problem solving such as Polya (2004). The particular importance of having design experiences in the junior year to prepare students for the capstone design is emphasized by Gregson and Little (1999). They used design contests in a junior analog electronics course to increase motivation and expose students to open-ended design although they do not explicitly mention a theoretical basis.

Structured on a PBL-based approach, Cheville and Bunting (2011) describe in detail the experiences of an electrical engineering department in enacting curricular change funded by an NSF Department Level Reform (DLR) grant. They focused their *Engineering Students for the 21st Century* (ES21C) program on a subset of courses within the curriculum. Arguing that "to *become* engineers students need to continually practice *being* engineers (p. 6), " these efforts were aimed at helping students develop as engineers and thus included work in the middle years. As the authors state (p. 4):

> Engineering Students for the 21st Century *seeks to make programs development-based rather than knowledge-based by creating a set of classes focused on developing the broad set of skills students need to understand problems in depth. To develop both knowledge and skills, students are walked through the process of solving the problem while learning concepts needed to understand the problem. The project posits that it is of greater importance to develop the skills needed to solve in-depth problems than to try to cover a large breadth of content in electrical engineering.*

They present case studies of four required EE classes including two in the middle years: circuits for sophomores and semiconductor physics for sophomores and juniors. These are nice examples of thoughtful research-based curricular design of standard core classes. They also discuss the challenges for faculty, particularly engineering faculty not familiar with educational research terminology, and struggles to create cultural change.

MECHANICAL ENGINEERING

Mechanical engineering (ME) today is the most popular engineering major and has the largest enrollment in the United States. Similar to electrical engineering, it has an arguably deserved reputation for its lack of curriculum innovation and slowness to change. There are signs that this is changing, however, given the increasing number of journal publications in this field. Interestingly, it appears that the most recent ME innovations in the middle years are focused on the engineering science courses. These courses (statics, dynamics, mechanics of materials, thermodynamics, fluid mechanics) are common core courses for several engineering majors, and thus improving them has a large impact on many students.

In a study based in the mechanics course of statics, Chen and colleagues (2008, 2010) assessed the value of rapid and elaborated feedback on students' learning. They showed that such feedback, whether provided via letter-coded flashcards or clickers, was significantly better than no feedback at all, even when other active-learning techniques were employed. The authors believe that the peer discourse that constituted the feedback (and used in conjunction with either the flashcards or clickers) was responsible for the enhanced learning. The research design was based on the learning framework described in *How People Learn* (NRC, 2000) and well-established best practices in undergraduate teaching (Chickering & Gamson, 1987; Chickering & Ehrmann, 1996). Two other recent examples of innovations in statics illustrate the attention that has been given to promoting problem solving in engineering education and the acceptance of peer discourse and peer instruction as a routine, in-class activity (Steif, Lobue, & Kara, 2010; Litzinger et al., 2010).

In a junior-level systems modeling course, Yadav, Shaver, and Meckl (2010) implemented a case-study teaching method. Case-based instruction is widely used in other professional studies such as medicine and business, and is known to promote students' critical thinking and problem-solving skills, conceptual change, and motivation (Dochy, Segers, VandenBossche, & Gijbels, 2003; Gallucci, 2007; Yadav & Koehler, 2007; also see Chapter 9 by Davis and Yadav in this volume). Yadav, Shaver, and Meckl taught students using both traditional lecture and case-based method and compared their conceptual understanding and attitudes toward the material. Their results showed that students' conceptual understanding was no different using either pedagogy, but the students found the case-based approach to be more interesting, realistic, and relevant.

Curriculum Design with a Holistic Perspective

In this section, we highlight some programs that underwent a curriculum design based on research and known best practices. This summary is not meant to be exhaustive but to present illustrative examples of what can be done. Some programs were in existence and underwent reforms, while others were new programs and had the advantage of starting with a blank slate. Although our focus is on the middle years of engineering study, as readers will see, these program designs were "holistic" in that the entire four-year curriculum was considered as a whole and thus, the middle years cannot be described in isolation from the first and senior years.

Beginning in 1988, Drexel University began to reform its lower division curriculum (first and sophomore years) across all engineering departments, and the resulting changes are described by Quinn (1993). Elements of this new program, called "E4" (for An Enhanced Educational Experience for Engineering Students), were later disseminated to other schools through the Gateway Coalition. The impetus for change came from external sources, including the National Science Board, ABET, and American Society for Engineering Education (ASEE), as well as internally, as Drexel recognized that its students needed new skills but that the already packed curriculum could not accommodate more courses. Instead, it relied heavily on a report issued by ASEE (1987) and undertook a holistic restructuring of the preparation of the emerging engineer's educational needs and settled on four common subject matters that all engineering students should experience. They are: The Fundamentals of Engineering, The Mathematical and Scientific Foundations of Engineering, The Engineering Laboratory, and The Personal and Professional Enrichment Program. E4's elements either replaced or integrated 37 courses in the traditional curriculum and focused on fundamental skills and knowledge of the profession, experimental methods, and the professional skills of personal communications and lifelong learning. Consistent with the time, there was also a heavy emphasis on integrating computer use in the learning process. Quinn (1993) reported that evaluations of the program were positive, including improved retention rates, superior performance test results, better communications skills, and high levels of satisfaction from both faculty and students.

The Sooner City project represents a curriculum reform by the School of Civil Engineering and Environmental Science at the University of Oklahoma (Kolar, Muraleetharan, Mooney, & Vieux, 2000; Kolar, Sabatini, & Muraleetharan, 2009). It was implemented in fall 1998 and elements of it are still in existence today. In brief, Sooner City is a four-year-long design project that students in the major undertake to apply what they learn in the classroom to design a city, including such tasks as designing the sewer and water infrastructure and steel buildings. It "unifies the curriculum and allows material learned in early courses to carry forward." The framework for the project is that of Fink (2003), whose work was directed at the course design level. Sooner City's creators extended Fink's concepts to the entire curriculum (Kolar et al., 2009). Fink (2003) identified six kinds of significant learning experiences in a taxonomy, all of which are incorporated into Sooner City: foundational knowledge, learning how to learn, application, integration, human dimension, and caring. Sooner City proved to be an extremely successful curriculum, both for student achievement and faculty engagement (Kolar et al., 2009).

Industrialist and electrical engineer Henry M. Rowan's substantial gift to Glassboro State College resulted in the founding of the College of Engineering at (the renamed) Rowan University in 1994. The College formed a national board of advisors and, led by the founding dean and department chairs, designed the curricula for the four engineering disciplines (chemical, civil and environmental, electrical and computer, and mechanical engineering). Although these curricula follow mostly a traditional path, there is one major strand of innovation, the Engineering Clinics. The Clinics were based on the Engineering Clinic model developed by Harvey Mudd College in the 1960s (Bright & Phillips, 1999). While the Mudd Clinic's design rationale was not described explicitly in terms of learning theories or frameworks, the elements that were included are clearly recognizable today to be consistent with several educational frameworks. These include group work on projects, acquaintance with the profession, and design. Rowan's Clinics built on these elements and added three additional ones: vertical and horizontal connections, project-based learning, and integration of communications. The vertical and horizontal connections recognized that learning engineering through authentic, project-based, team-based design requires continuous practice in a multidisciplinary environment. The Clinics, therefore, are a common eight-semester-long course sequence that all engineering students take during their four-year study (vertical connection), and they are multidisciplinary (horizontal connection across all majors in the same year of study). In the first-year and sophomore Clinics, multidisciplinary faculty teams (including communications faculty) develop the projects that student teams tackle (Marchese, Ramachandran, Hesketh, & Schmalzel, 2003; Newell, Marchese, Ramachandran,

Sukumaran, & Harvey, 1999; Ramachandran & Marchese, 2002), while the final two years of Clinics rely on projects sponsored by industry or technical research (Dahm, Newell, Harvey, & Newell, 2009; Marchese, Schmalzel, Mandayam, & Chen, 2001). The teams are always formed based on the technical needs of a project, so the disciplinary make-up of each team varies. The Rowan Clinics evolve continuously based on faculty interest and industry's needs, and in the past have incorporated entrepreneurialism (Marchese et al., 2001) and service learning (Mehta & Sukumaran, 2007). The Engineering Clinics exhibit many positive outcomes for student learning, and are especially promising for supporting women students (Hartman & Hartman, 2006; Kadlowec et al., 2007).

With funding from an NSF Departmental Level Reform Grant entitled "Triple Bottom Line Awareness in Design" (Chen et al., 2009), the Materials Engineering Department at California Polytechnic State University (Cal Poly) did significant and innovative holistic reform throughout its curriculum. This effort included project-based learning in each of the four years of the curriculum guided by their belief that

> Only by integrating project-based learning experiences throughout the undergraduate curriculum will we give students the opportunity to develop a mastery of the fundamentals of science, engineering and mathematics along with providing them with the contextual environment for developing the skills necessary to practice engineering such as project management, teamwork and effective communication. (Savage, Chen, & Vanasupa, 2007, p. 2)

This echoes the knowledge, learner, and community aspects of HPL. Their assessment was based on learning objectives and explicitly aimed to reach some of the higher levels of Bloom's taxonomy. The Sophomore Experience was "Designing for Sustainability" and the Junior Experience was "A Systems Approach to Engineering." Each of these experiences involved multiple project-based learning activities and is

described in Savage et al. (2007) and Chen et al. (2009).

In 1997 the F.W. Olin Foundation established the Franklin W. Olin College of Engineering (Olin College) with the broad mission of creating an engineering school for the twenty-first century. With this blank slate, the Olin curriculum was created based on learning theory and frameworks, best practices at existing schools, and the input and experiences of 30 "Olin Partners," recent high school graduates who agreed to volunteer for a year to co-develop and test the new curriculum (Somerville et al., 2005). The broad curricular features include (1) integration of and coordination between courses, which gain efficiency in the curriculum and also make it interdisciplinary, hands-on, and motivational; (2) opportunities to gain and practice professional skills such as entrepreneurialism, teaming, communications, and lifelong learning; (3) an emphasis on all four years of the students' education, with design incorporated throughout the curriculum and near-equal resources devoted to the first two years of study; (4) a commitment to "educate the whole person," not just the engineering side; and (5) flexibility, with accountability, for students to design their own curriculum. The resulting curriculum consists of three components: the foundation, which emphasizes the fundamentals; the specialization, where students gain in-depth knowledge of their chosen field; and realization, in which students synthesize their learning and apply it to problems to emulate professional practice.

The final case of holistic curriculum design that we wish to highlight is integrated engineering curricula. Froyd and Ohland (2005) define integrated curricula to be that in which (1) faculty members from multiple disciplines collaborate to develop and implement the curriculum, (2) assessment data are reported on specific outcomes that result from the curriculum, and (3) students enroll in courses from different disciplines or in a course that combines multiple disciplines. Integrated curricula have solid foundations based on learning theory that are nicely summarized by Froyd and Ohland (2005). Furthermore, the authors cite studies indicating that integrated curricula should appeal more than traditional curricula to underrepresented groups in engineering. The available data on integrated curricula support these claims (Froyd & Ohland, 2005). Despite these benefits, however, integrated curricula have not become widespread in engineering education with the exception of a handful of first-year integrated programs and just two known sophomore-level programs, at Texas A&M and Rose-Hulman Institute of Technology, in which multiple engineering science courses were integrated. Again, interested readers are referred to the excellent paper by Froyd and Ohland (2005) for a review of these programs and the associated assessment data.

Summary and Action Agenda

Throughout the engineering education community, there is widespread recognition that there are problems with the current state of the field. More curricular design that is based on well-established theory and frameworks can help to improve engineering education. In this chapter, we have highlighted some educational innovations that are research based but clearly much more work needs to be done, especially, we argue, for the middle years of engineering study. Engineering instructors need to carefully consider their pedagogy and the learning environment, as well as their content. Most have no training in education, however, so it is understandable that it has been difficult. Also, the typical incentive system in academia is much more focused on research than teaching. Challenges for wider incorporation go beyond the need for evidence of effectiveness or uniqueness of engineering or specific engineering disciplines (Duderstadt, 2008; Fairweather, 2008). These challenges need to be addressed to help students be successful. Fairweather (2008) nicely summarizes the key assumptions that underlie most approaches to educational reform.

1. Empirical evidence of successful learning outcomes is a precondition for STEM faculty involvement in pedagogical reform.
2. Pedagogical effectiveness varies by academic discipline.
3. Instructional role can be addressed independently from other aspects of the faculty position, particularly research, and from the larger institutional context.

Then he argues that the first one has been repeatedly done and disputes the second saying "careful reviews of the substantial literature on college teaching and learning suggest that the pedagogical strategies most effective in enhancing student learning outcomes *are not discipline dependent* (Pascarella & Terenzini, 2005)" (Fairweather, 2008, p. 4). The barriers to adoption in STEM are not due to lack of evidence but are more complicated and tied to the academic institutional culture. Thus, he emphasizes the importance of pedagogical reforms that go beyond single instructors and the need for cultural change in academia, which is a daunting task.

Hopeful Signs

The examples presented in this chapter show promise of improving the situation for student learning in the second and third years of engineering curricula. There are other hopeful signs of rising interest in high-quality engineering education tied to research. Ph.D. degrees in engineering education are now offered throughout the world. The American Society for Engineering Education (ASEE) *Journal of Engineering Education* (JEE) adopted a more rigorous approach in 2003. The engineering education community enthusiastically responded to a 2010 special issue of the *International Journal of Engineering Education* on Applications of Engineering Education Research. This issue included a variety of examples of how engineering educators are bringing theory to practice including the knowledge centered aspects of developing engineering competencies (Lord & Finelli, 2010a) and the learner- and community-centered lenses of building engineering communities (Lord & Finelli, 2010b).

Learner-/Community-Centered Aspects Tied to Diversity

The HPL framework of effective learning environments being knowledge centered, learner centered, community centered, and assessment centered has great potential. The learner and community aspects of this suggest critical directions for research and curriculum design. If engineering educators were to truly embrace the ideas encompassed in the learner- and community-centered aspects of effective learning environments and adopt an intersectional lens considering factors such as race and gender, this could promote powerful change. Given the persistent low numbers of women of all races and African American, Latino, and Native American students in engineering, engineering educators should take steps to learn more about how to facilitate learning for a diverse group of students. Research has been done on these topics including liberative and feminist pedagogy, stereotype threat, and the "chilly climate," but this has not generally been incorporated into engineering education. Bridging the gulf that too often exists between those with expertise in engineering, education, gender, and ethnic studies is critical to making true progress in these areas. Including gender and race in designing engineering curricula should not be superfluous but an integral part of the design equation. Recent research on identifying the culture of engineering (Godfrey & Parker, 2010; see Chapter 22 by Godfrey in this volume) may help to dismantle aspects of the culture that do not facilitate building community. Despite the stereotype of engineering education being impersonal, in a study of engineering educators who attended the *Frontiers in Education (FIE)* conference in 2006, researchers found that "building a sense of community in the classroom" was one of the top three teaching practices for participants (Lord & Camacho, 2007). So there is promise for improvement in the culture particularly if innovations can be spread beyond the progressive community that attends FIE.

Research Agenda

We propose the following agenda for future research focused on the middle years of engineering education. They echo those of other broader research agendas and call for incorporating theory into engineering education (Fortenberry, 2006; Johri, 2010; Lohmann & Jamieson, 2009).

- Research focused specifically on sophomore year and junior year in engineering. This could build on and contribute to the multidisciplinary discourse on the "sophomore slump".
- Engineering educators should be more purposeful and thoughtful in designing educational experiences for the second and third year by integrating content, outcomes, assessment and pedagogy. Think of it as design. Bring in engineering expertise.
- Make the learner and community an integral part of teaching process beyond the first year. Address diversity as part of the equation, not as an afterthought. Learn from decades of research on gender and race.

References

ABET. (2010). *Criteria for accrediting engineering programs*. Baltimore, MD: ABET, Inc.

American Society for Engineering Education (ASEE). (1987). *A national action agenda for engineering education*. Report of the ASEE Task Force on a National Action Agenda for Engineering Education. Washington, DC: ASEE.

Angelo, T. A., & Cross, K. P. (1993). *Classroom assessment techniques: A handbook for college teachers* (2nd ed.). San Francisco, CA: Jossey-Bass.

Bloom, B. S. (Ed.) (1956). *Taxonomy of educational objectives: Handbook I: Cognitive domain*. New York, NY: Longmans, Green.

Borrego, M. (2007). Development of engineering education as a rigorous discipline: A study of the publication patterns of four coalitions. *Journal of Engineering Education, 96*(1), 5–18.

Borrego, M., Froyd, J. E., & Hall, T. S. (2010). Diffusion of engineering education innovations: A survey of awareness and adoption rates in U.S. engineering departments. *Journal of Engineering Education, 99*(3), 185–207.

Bransford, J., Vye, N., & Bateman, H. (2004). *Creating high quality learning environments: Guidelines from research on how people learn.* Washington, DC: The National Academies Press.

Brawner, C. E., Camacho, M. M., Lord, S. M., Long, R. A., & Ohland, M. W. (2012). Women in industrial engineering: Stereotypes, persistence, and perspectives. *Journal of Engineering Education, 101*(2), 228–318.

Bright, A., & Phillips, J. R. (1999). The Harvey Mudd engineering clinic past, present, future. *Journal of Engineering Education, 88*(2), 189–194.

Bruff, D. (2007). Clickers: A classroom innovation. *NEA Higher Education ADVOCATE.* Retrieved from http://www.hunter.cuny.edu/~shp/centers/hpec/docs/clickers_and_classroom_dynamics.pdf

Chen, J. C., Kadlowec, J. A., & Whittinghill, D. C. (2008). Using handheld computers for instantaneous feedback to enhance student learning and promote interaction. *International Journal of Engineering Education, 24*(3), 616–624.

Chen, J. C., Whittinghill, D. C., & Kadlowec, J. A. (2010). Classes that click: Fast, rich feedback to enhance student learning and satisfaction. *Journal of Engineering Education, 99*(2), 159–168.

Chen, K. C, Vanasupa, K., London, B., Harding, T., Savage, R., Hughes, W. & Stolk, J. (2009). Creating a project-based curriculum in materials engineering. *Journal of Materials Education*

Cheville, R. A., & C. Bunting, C. (2011, Summer). Engineering students for the 21st century: Student development through the curriculum. *Advances in Engineering Education,* 1–37.

Chickering, A. W., & Ehrmann, S. C. (1996). Implementing the seven principles: Technology as lever. *AAHE Bulletin, 49*(2), 3–6.

Chickering, A. W., & Gamson, Z. F. (1987). Seven principles for good practice in undergraduate education. *AAHE Bulletin, 39*(7), 3–7.

Crews, T. B., Ducate, L., Rathel, J. M., Heid, K., & Bishoff, S. T. (2011). Clickers in the classroom: Transforming students into active learners. *Research Bulletin 9,* Boulder, CO: EDUCAUSE Center for Applied Research.

Retrieved from http://www.educause.edu/ecar

Dahm, K., Newell, J. A., Harvey, R., & Newell, H. (2009). The impact of structured writing and developing awareness of learning preferences on the performance and attitudes of engineering teams. *Advances in Engineering Education, 1,* 1–17.

Daley, B., Lovell, M. R., Perez, R. A., & Stern, N. E. (2011). Using concept maps within the product design process in engineering: A case study. In B. M. Moon, R. R. Hoffman, J. Novak, & A. Canas (Eds.), *Applied concept mapping: capturing, analyzing, and organizing knowledge* (pp. 229–252). Boca Raton, FL: CRC Press.

Dochy, F., Segers, M., VandenBossche, P., & Gijbels, D. (2003). Effects of problem-based learning: A meta-analysis. *Learning & Instruction, 13*(5), 533–568.

Duderstadt, J. J. (2008). *Engineering for a changing world: A roadmap to the future of engineering practice, research, and education.* The Millennium Project, The University of Michigan. Retrieved from http://milproj.dc.umich.edu/

Engel, R. S., & Giddens, D. P. (2011). From our reading list to yours: A summary of key reports. *Journal of Engineering Education, 100*(2), 220–224.

Fairweather, J. (2008). Linking evidence and promising practices in science, technology, engineering, and mathematics (STEM) undergraduate education: A status report for the National Academics National Research Council Board of Science Education. Retrieved from http://www7.nationalacademies.org/bose/Fairweather_CommissionedPaper.pdf

Felder, R. M. (1987). On creating creative engineers. *Engineering Education, 77,* 222.

Felder, R. M. (1995). A longitudinal study of engineering student performance and retention. IV. Instructional methods and student responses to them. *Journal of Engineering Education, 84*(4), 361–367.

Felder, R. M., & Silverman, L. K. (1988). Learning and teaching styles in engineering education. *Engineering Education, 78,* 674.

Fink, L. D. (2003). *Creating significant learning experiences: An integrated approach to designing college courses.* San Francisco, CA: Jossey-Bass.

Foor, C. E., & Walden, S. E. (2009). "Imaginary engineering" or "re-imagined engineering": Negotiating gendered identities in the borderland of a college of engineering. *NWSA Journal*, 21(2), 41–64.

Fortenberry, N. L. (2006). An extensive agenda for engineering education research. *Journal of Engineering Education*, 95(1), 3–5.

Fraser, D. M., Pillay, R., Tjatindi, R. L., & Case, J. M. (2007). Enhancing the learning of fluid mechanics using computer simulations. *Journal of Engineering Education*, 96(4), 381–388.

Freire, P. (2006). *Pedagogy of the oppressed, 30th Anniversary* ed. New York, NY: Continuum.

Froyd, J. E. (2005). The engineering education coalitions program. In National Academy of Engineering (Ed.), *Educating the engineer of 2020: Adapting engineering education to the new century* (Appendix A, pp. 82–97). Washington, DC: The National Academies Press.

Froyd, J. E. (2008). *White paper on promising practices in undergraduate STEM education.* Commissioned paper presented at NRC Workshop on Evidence on Promising Practices in Undergraduate Science, Technology, Engineering, and Mathematics (STEM) Education. Retrieved from http://www7.nationalacademies.org/bose/Froyd_Promising_Practices_CommissionedPaper.pdf

Froyd, J. E., & Ohland, M. W. (2005). Integrated engineering curricula. *Journal of Engineering Education*, 94(1), 147–164.

Gallucci, K. (2007). *The case method of instruction, conceptual change and student attitude.* Doctoral dissertation, Science Education, North Carolina State University.

Godfrey, E. G., & Parker, L. (2010). Mapping the cultural landscape in engineering education. *Journal of Engineering Education*, 99(1), 5–22.

Gregson, P. H., & Little. T. A. (1999). Using contests to teach design to EE juniors. *IEEE Transactions on Education*, 42(3), 229–232.

Harb, R. E., & Terry, J. N. (1993). Kolb, Bloom, creativity, and engineering design. In *Proceedings of the 1993 ASEE Annual Conference*, Urbana, IL.

Hartman, H., & Hartman, M. (2006). Leaving engineering: Lessons from Rowan University's College of Engineering. *Journal of Engineering Education*, 95(1), 49–61.

Holloway, B. M., Reed-Rhoads, T., & Groll, L. M. (2010). Defining the "sophomore slump" within the discipline of engineering. In *Proceedings of the Global Colloquium on Engineering Education*, Singapore.

Holloway, B. M., Reed-Rhoads, T., & Groll, L. (2011). Women as the miner's canary in undergraduate engineering education. In *Proceedings of the ASEE Annual Conference*, Vancouver, BC, Canada.

Hunter, M. S., Tobolowsky, B. F., & Gardner, J. N. (Eds.) (2010). *Helping sophomores succeed: Understanding and improving the second-year experience.* San Francisco, CA: Jossey-Bass.

Hussmann, S., & Jensen, D. (2007). Crazy Car Race Contest: Multicourse design curricula in embedded system design. *IEEE Transactions on Education*, 50(1), 61–67.

Johri, A. (2010). Creating theoretical insights in engineering education. *Journal of Engineering Education*, 99(3), 183–184.

Kadlowec, J., Bhatia, K., Chandrupatla, T., Chen, J., Constans, E., Hartman, H., ... Zhang, H. (2007). Design integrated in the mechanical engineering curriculum: Benefits and assessment of the engineering clinics. *Transactions of the ASME: Journal of Mechanical Design: Special Edition on Design Education*, 129(7), 682–691.

Kolar, R. L., Muraleetharan, K. K., Mooney, M. A., & Vieux, B. E. (2000). Sooner City – Design across the curriculum. *Journal of Engineering Education*, 89(1), 79–87.

Kolar, R. L., Sabatini, D. A., & Muraleetharan, K. K. (2009). Sooner City: Reflections on a curriculum reform project. *New Directions for Teaching and Learning*, 2009(119), 89–95.

Kolb, D. (1984). *Experiential learning: Experience as the source of learning and development.* Englewood Cliffs, NJ: Prentice-Hall.

Kotys-Schwartz, D., Knight, D., & Pawlas, G. (2010). First-year and capstone design projects: Is the bookend curriculum approach effective for skill gain? In *Proceedings of the 2010 ASEE Annual Conference*, Louisville, KY.

Lattuca, L. R. (2006). Learning to change: A study of NSF-funded planning grants for educational innovation. In *Proceedings of the 2006 American Educational Research Association (AERA) Annual Meeting*, San Francisco, CA.

Lattuca, L. R., Terenzini, P. T., & Volkwein, J. F. (2006). *Engineering change: A study of the impact of EC2000.* Baltimore, MD: ABET, Inc.

Lave, J., & Wenger, J. (1991). *Situated learning: Legitimate peripheral participation.* Cambridge: Cambridge University Press.

Lesh, R., Hoover, M., Hole, B., Kelly, A., & Post, T. (2000). Principles for developing thought-revealing activities for students and teachers. In A. Kelly & R. Lesh (Eds.), *The handbook of research design in mathematics and science education* (pp. 591–646). Mahwah, NJ: Lawrence Erlbaum.

Litzinger, T. A., VanMeter, P., Firetto, C. M., Passmore, L. J., Masters, C. B., Turns, S. R. . . . Zappe, S. E. (2010). A cognitive study of problem solving in statics. *Journal of Engineering Education,* 99(2), 337–353.

Lohmann, J. R., & Jamieson, L. H. (2009). Creating a culture for scholarly and systematic innovation in engineering education: Ensuring U.S. engineering has the right people with the right talent for a global society, Washington, DC: ASEE. Retrieved from http://www.asee.org/about-us/the-organization/advisory-committees/CCSSIE/CCSSIEE_Phase1Report_June2009.pdf

Lord, S. M. (2009). Integrating effective 'Writing to Communicate' experiences in engineering courses: Guidelines and examples. *International Journal of Engineering Education,* 25(1), 196–204.

Lord, S. M., & Camacho, M. M. (2007). Effective teaching practices: Preliminary analysis of engineering educators. In *Proceedings of the 2007 Frontiers in Education Conference,* Milwaukee, WI.

Lord, S. M., & Finelli, C. J. (2010a). Guest editorial for Special issue on applications of engineering education research – Part 1. Developing engineering competencies. *International Journal of Engineering Education,* 26(4), 746–747.

Lord, S. M., & Finelli, C. J. (2010b). Guest editorial for special issue on applications of engineering education eesearch – Part 2. Building engineering communities. *International Journal of Engineering Education,* 26(5), 1031.

Marchese, A. J., Ramachandran, R. P., Hesketh, R. P., & Schmalzel, J. L. (2003). The competitive assessment laboratory: Introducing engineering design via consumer product benchmarking. *IEEE Transactions on Education,* 46(1), 197–205.

Marchese, A. J., Schmalzel, J. L., Mandayam, S. A., & Chen, J. C. (2001). A venture capital fund for undergraduate engineering students at Rowan University. *Journal of Engineering Education,* 90(4), 589–596.

Mehta, Y., & Sukumaran, B. (2007). Integrating service learning in engineering clinics. *International Journal for Service Learning in Engineering,* 2(1), 32–43.

Moon, B. M., Hoffman, R. R., Novak, J., & Cañas, A. (Eds.). (2011). *Applied concept mapping: Capturing, analyzing, and organizing knowledge.* Boca Raton, FL: CRC Press.

National Engineering Education Delivery System (NEEDS). http://www.needs.org/needs/?path=/public/community/nsf/index.jhtml%26; (Accessed August 13, 2012). Also, more information can be found through publications archived by the coalitions.

National Research Council (NRC). (2000). *How people learn: Brain, mind, experience, and school, Expanded edition.* Committee on Developments in the Science of Learning. J. D. Bransford, A. L. Brown, & R. R. Cocking (Eds.), with additional material from the Committee on Learning Research and Educational Practice. Commission on Behavioral and Social Sciences and Education. Washington, DC: The National Academies Press. Retrieved from http://www.nap.edu/catalog.php?record_id=9853

National Research Council (NRC). (2001). *Knowing what students know: The science and design of education assessment.* Committee on the Foundations of Assessment. J. W. Pellegrino, N. Chudowsky, & R. Glaser (Eds.), Board on Testing and Assessment, Center for Education. Division of Behavioral and Social Sciences and Education. Washington, DC: The National Academies Press.

Newell, J. A., Marchese, A. J, Ramachandran, R. P., Sukumaran, B., & Harvey, R. (1999). Multidisciplinary design and communication: A pedagogical vision. *International Journal of Engineering Education,* 15(5), 376–382.

Pascarella, E., & Terenzini, P. (2005). *How college affects students: A third decade of research.* San Francisco, CA: Jossey-Bass.

Pellegrino, J. W. (2006). *Rethinking and redesigning curriculum, instruction and assessment: What contemporary research and theory suggests.* Paper commissioned by the National Center on Education and the Economy for the New Commission on the Skills of the American Workforce.

Polya, G. (2004). *How to solve it: A new aspect of mathematical method.* Princeton, NJ: Princeton University Press.

Prince, M. J., & Felder, R. M. (2006). Inductive teaching and learning methods: Definitions, comparisons, and research bases. *Journal of Engineering Education, 95*(2), 123–138.

Prince, M. J., & Felder, R. M. (2007). The many faces of inductive teaching and learning. *Journal of College Science Teaching, 36*(5), 533–568.

Prince, M. J., Vigeant, M., & Nottis, K. (2009). A preliminary study on the effectiveness of inquiry-based activities for addressing misconceptions of undergraduate engineering students. *Education for Chemical Engineers, 4*(2), 29–41.

Prince, M., Vigeant, M., & Nottis, K. (2010). Assessing misconceptions of undergraduate engineering students in the thermal sciences. *International Journal of Engineering Education, 26*(4), 880–890.

Quinn, R. G. (1993). Drexel's E4 Program: A different professional experience for engineering students and faculty. *Journal of Engineering Education, 82*(4), 196–202.

Ramachandran, R. P., & Marchese, A. J. (2002). Integration of multidisciplinary design and technical communication: An inexorable link. *International Journal of Engineering Education, 18*(1), 32–38.

Reed-Rhoads, T., & Imbrie, P. K. (2008). *Concept inventories in engineering education.* National Academies Board on Science Education, Evidence on Promising Practices in Undergraduate Science, Technology, Engineering, and Mathematics (STEM) Education Commissioned Papers. Retrieved from http://www7.nationalacademies.org/bose/PP_Commissioned_Papers.html

Riley, D. (2003). Employing liberative pedagogies in engineering education. *Journal of Women and Minorities in Science and Engineering, 9*(2), 137–158.

Riley, D. (2011). *Engineering thermodynamics and 21st century energy problems: A textbook companion for student engagement.* San Rafael, CA: Morgan & Claypool.

Safoutin, M. J., Atman, C. J., Adams, R., Rutar, T., Kramlich, J. C., & Fridley, J. L. (2000). A design attribute framework for course planning and learning assessment. *IEEE Transactions on Education, 43*(2), 188–199.

Savage, R., Chen, K. C., & Vanasupa, L. (2007). Integrating project-based learning throughout the undergraduate engineering curriculum. *Journal of STEM Education, 8*, 1–13.

Sheppard, S., Gilmartin, S., Chen, H. L., Donaldson, K., Lichtenstein, G., Eris, Ö ... Toye, G. (2010). *Exploring the engineering student experience: Findings from the Academic Pathways of People Learning Engineering Survey (APPLES).* Technical Report CAEE-TR-10-01. Seattle, WA: Center for the Advancement for Engineering Education.

Sheppard, S., Macatangay, K., Colby, A., & Sullivan, W. (2009). *Educating engineers: Designing for the future of the field.* San Francisco, CA: Jossey-Bass.

Somerville, M., Anderson, D., Berbeco, H., Bourne, J. R., Crisman, J., Dabby, D., ... Zastavker, Y. (2005). The Olin Curriculum: Thinking toward the future. *IEEE Transactions on Education, 48*(1), 198–205.

Steif, P. S., Lobue, J. M., & Kara, L. B. (2010). Improving problem solving performance by inducing talk about salient problem features. *Journal of Engineering Education, 99*(4), 135–142.

Stice, J. E. (1976). A first step toward improved teaching. *Engineering Education, 66*, 394.

Streveler, R. A., Smith, K. A., & Pilotte, M. K. (2011). Workshop – Aligning content, assessment, and pedagogy in the design of engineering courses. In *Proceedings of the 2011 Frontiers in Education Conference*, Rapid City, SD.

Turns, J., Atman, C. J., & Adams, R. (2000). Concept maps for engineering education: A cognitively motivated tool supporting varied assessment functions. *IEEE Transactions on Education, 43*(2), 164–173.

Volkwein, J. F., Lattuca, L. R., Terenzini, P. T., Strauss, L. C., & Sukhbaatar, J. (2004). Engineering change: A study of the impact of EC2000. *International Journal of Engineering Education, 20*(3), 318–328.

Vygotsky, L. S. (1978). *Mind in society: The development of higher psychological processes.* Cambridge, MA: Harvard University Press.

Wiggins, G., & McTighe, J. (1997). *Understanding by design.* Alexandria, VA: Association for Supervision and Curriculum Development.

Yadav, A., & Koehler, M. J. (2007). The role of epistemological beliefs in preservice teachers' interpretation of video cases of early-grade literacy instruction. *Journal of Technology and Teacher Education, 15*(3), 335–361.

Yadav, A., Shaver, G. M., & Meckl, P. (2010). Lessons learned: Implementing the case teaching method in a mechanical engineering course. *Journal of Engineering Education*, 99(1), 55–69.

Yildirim, T. P., Shuman, L., & Besterfield-Sacre, M. (2010). Model-eliciting activities: Assessing engineering student problem solving and skill integration processes. *International Journal of Engineering Education*, 26(4), 831–845.

CHAPTER 11

Engineering Design Education

Research, Practice, and Examples that Link the Two

Cynthia J. Atman, Ozgur Eris, Janet McDonnell, Monica E. Cardella, and Jim L. Borgford-Parnell

Introduction

Designing is a key component of professional practice in many fields of human endeavor (e.g., architecture, engineering, industrial design, art, and literature). For engineers, designing integrates engineering knowledge, skill, and vision in the pursuit of innovations to solve problems and enable modern life.

With this understanding, engineering educators have, for several decades, been infusing their programs with design curricula and pedagogical experiences in order to enhance the design competencies of engineering graduates. Paralleling the development of these curricula and experiences, a growing body of research has been providing a scholarly basis for engineering design education.

The goal of this chapter is to acquaint readers with engineering design education research and practice. To situate engineering design education in the larger context, we first present a brief history of research on design processes across several fields and then move to a more specific description of research on engineering design processes. We then focus on research that investigates effective ways to teach and assess the design process and review curricular structures and pedagogies that are commonly used in undergraduate engineering programs.

We build on our overview of research and practice related to engineering design education with two worked examples of moving research into practice in the classroom, following the Scholarship on Teaching and Learning model suggested by Boyer (1990). The first example illustrates how laboratory-based research findings on engineering design processes can be brought into the classroom. The second example illustrates how educators can execute an observe–analyze–intervene cycle within the context of a project-based engineering design course. These illustrations demonstrate the efficacy of two different models for leveraging research to improve student learning.

A Brief History of Research on the Design Process

Research into design (design, designing, design processes, design expertise, design thinking, and so on) dates back about fifty years and embraces many fields of design, including engineering. In a landmark collection of papers just under halfway through this period, *Developments in Design Methodology*, the editor, Nigel Cross, organized the contributions in terms of a movement through stages, the first three of which he characterized as *prescription* of an ideal design process, *description* of the intrinsic nature of design problems, and *observation* of the reality of design activity (Cross, 1984). This characterization also usefully identified shifts in the objects of research, with findings and issues raised in each stage influencing studies that followed, and each new emphasis leading to a reinterpretation of the value and meaning of earlier work. In the era of *prescription*, the design methods movement (e.g., Alexander, 1964; Jones, 1970) looked to improve design through the development of systematic methods to encourage better attention to user needs and a broader range of contextual factors. A desire to manage design activity and to relate design methods systematically to parts of the design process demanded models of the design process that prescribed what should take place as design occurs. Dubberly's extensive compendium (2004) includes a selection of such design process models dating back to the early 1960s.

The failure of the design methods movement to impact practice substantially (among other reasons) led to the focus of effort shifting toward *description* – investigations of the nature of design problems. From this era, we have the legacy of understanding better that typical design problems are poorly specified; goals are vague; what is relevant (e.g., the set of constraints) is not completely knowable in advance; that designs are always amenable to improvement; and that these characteristics are *inherent* to design. Seminal writings

include Simon (1996, first published in 1969), which observed that design problems are, at minimum, ill defined, and Rittel and Webber (1973), which offered the term *"wicked problems"* to encapsulate the characteristics of design problems (see Jonassen's discussion of ill-structured problems in Chapter 5, this volume). This characterization implies that design problems have to be both *set* and *solved* by designers. The setting part, comprising the mysterious *intuitive* elements of designing, was later described by Jones (1991).

It is not surprising, therefore, that researchers' attention turned to studying designers themselves – the stage of *observation*. In such work, the focus is on studying how it is, given the complexity of designing, that a designer is able to design anything at all. The object of research is the designer, and we have learned a great deal from observation of individual designers in experimental and other settings. For example, we have learned that designers are solution oriented rather than problem focused (Lawson, 1980); that they employ a range of strategies to impose order on a design situation and to generate solution ideas (Akin, 1986; Darke, 1979; Rowe, 1987); that they can maintain mutually incompatible sets of beliefs (e.g., to support deferral of decisions, and the pursuit of parallel possibilities); they move between breadth and depth of the problem and solution (Akin, 1986); and they complete tasks opportunistically when such behavior results in efficiency gains (Visser, 1992). These and other findings about design expertise, based on observational and psychological studies and conversations with designers, are compiled in *Designerly Ways of Knowing* (Cross, 2006).

Cross' three stages are useful for indicating the different foci of attention in design research, but it would be misleading to imply that any of the objects of study in *prescription*, *description*, or *observation* research have ever been abandoned in favor of the next one, or that empirical studies of naturally occurring design activity started only in the 1980s. For example, Marples (1961) was studying engineering design decision making in the later 1950s and the ideas, findings, and shortcomings of attempts to prescribe still fuel design research agendas. Lifecycle models of the design process not only remain in use, but are also constantly refined through research and evaluation. Highly systematic approaches to design based on detailed descriptions of stages and steps are routine in engineering design (probably the most well-known example being the work of Pahl and Beitz, which was first published in German in 1977 and was recently released in 2007). A collection of studies published as *Analyzing Design Activity* (Cross, Christiaans, & Dorst, 1996), which included studies of small teams designing, provides evidence that models to describe as well as prescribe aspects of design processes continue to serve research purposes.

Over at least the last twenty years, some design researchers have taken a particular interest in studying design as it takes place in natural settings and in paying attention to the social process of design. It has always been recognized that a great deal of design does not take place in the mind of a single designer and that many stakeholders beyond design professionals have a formative influence on design processes. In the early days of design research, the prescriptive models were intended to manage a task largely viewed as an operational, objective challenge. This perspective derives, in turn, from an operational research agenda and attempts to apply it to large, complicated tasks of all kinds. Following on, perhaps as a natural development, from studies of individual designers, came studies of design teams, design collaborations, and multidisciplinary and other stakeholder engagement in design.

Research of this kind has opened up many more aspects of design to inspection. Two studies of note in this area focused on engineering design. A landmark ethnographic study by Bucciarelli (1994), *Designing Engineers*, was able to claim to reveal significant mismatches between still current idealized notions of design as an instrumental process and design as a reality. In Bucciarelli's case, the reality revealed was one in which engineering design is essentially a situated, social process of coming to an agreement about an engineering design decision through negotiating uncertainty and ambiguity rather than (solely) a process of providing factual information. Minneman (1991) arrived at similar conclusions based on field studies by stating that engineering design "can be seen as not something shaped by an externally imposed context, but rather as attending to and creating a recognizable order in the ongoing social interaction" (p. iv).

The study of design as a social process has broadened the scope and reach of design research into many new areas. These include design team interactions; the roles and uses of objects and the functions of gestures in design and design communication; and the influences of design setting variations on outcomes, such as the effects of the nature and form of design briefs on the course of the designing. It also covers how and what kind of shared representations are developed during a design process and what purposes they serve. It extends to what types of language, roles, and structures support or impede design; the effect of evaluation of others' contributions to design and of the evolving design itself on the course of events; and the roles played by uses of analogy during designing. Examples of work in each of these areas and others can be found in the collection of studies on both architectural and engineering design published as *About: Designing Analysing Design Meetings* (McDonnell & Lloyd, 2009).

Engineering Design Process Research

Engineering design process research is a domain specific case of the research on design processes described in the previous section. It aims to advance our understanding of engineering practice, increase product development performance, and inform engineering education. Some studies focus on specific parts of the engineering design process whereas others are more concerned with synthesizing a unified understanding of it. In many respects, what differentiates engineering design research from design research as a field is the role of scientific and technical knowledge in engineering. Traditionally, generation and application of scientific knowledge has been seen as the fundamental driver of engineering design practice (Carvalho, Dong, & Maton, 2009). However, as outlined in the previous section, that framing has been revisited.

Such considerations have led to the development of multiple viewpoints, which characterize and advocate systematic approaches to practicing engineering design; the term "process" is being interpreted broadly. Therefore, the scope of the research has become more encompassing than the application of scientific principles during certain parts of the design process.

Engineering design is increasingly seen as an interdisciplinary activity that accounts for the entire product lifecycle in which designers, interacting with stakeholders, identify opportunities; frame goals; generate and test solutions; and plan for the manufacturing, marketing, and servicing of products. It is also important to note that the object of engineering design, the nature of engineered products, has evolved to include services and experiences in addition to physical artifacts. This development is a manifestation of a research implication (mentioned earlier) wherein it is important that engineers are involved in both the *setting* and the *solving* of design problems.

Different viewpoints on engineering design can be grouped into three overarching categories which treat engineering design practice as a:

1. Work Process: Characterizes the types of tasks that need to be undertaken and the relationships between them. Work flow, project decomposition and planning, and resource management are on the foreground of such considerations (Dym & Little, 2009; Hubka, 1982; Otto & Wood, 2001; Pahl & Beitz, 1996; Roozenburgh & Eekels, 1995; Ullman, 1992;

Ulrich & Eppinger, 1995). This viewpoint often suggests normative approaches for executing the identified tasks. The approaches can be based on understandings emerging from the other process categories outlined later, but might lack conclusive validation.

2. Cognitive Process: Characterizes the cognitive mechanisms that enable such activities. Information acquisition, processing, creation, and sharing are on the foreground of such considerations. Special attention has been paid to theories of creativity, decision-making, expertise, visualization, and informatics in the contexts of individual designers and design teams (Atman et al., 2007; Badke-Schaub, Goldschmidt, & Meijer, 2010; Ball, Evans, & Dennis, 1994; Baya & Leifer, 1996; Cross, 2006; Dorst & Cross, 2001; Eris, 2004; Finke, Ward, & Smith, 1992; Gero, 1996; Goldschmidt, 2003; Hatamura, 2006; Hazelrigg, 1998; Lawson, 1997; McDonnell, 1997; Pahl, 1997; Schön, 1983; Suh, 1990; Visser, 2006).

3. Social Process: Characterizes the role of social interactions and context in engineering design. There are two key distinct drivers for such a consideration: the ultimate goal of engineering design is to add value to the lives of human beings, and thus, society; and the observation that most modern engineering design takes place within teams (due to increasing product complexity and globalization as a function of the increasing affordances of modern communication technology). Team dynamics, collaboration (including distributed work), communication, and human values are some of the main issues driving such considerations (Brereton, Cannon, Mabogunje, & Leifer, 1996; Bucciarelli, 1994; Cross & Cross, 1996; Le Dantec, 2010; Lloyd, 2000; McDonnell & Lloyd, 2009; Minneman, 1991).

Clearly, these categories are not necessarily orthogonal given the complexity of design activity, but the differentiation has proven to be meaningful in seeding discussion within the field.

From a topical perspective, engineering design process research also aims to advance the domain knowledge within engineering disciplines. More specifically, the generation of new domain-specific scientific and technical knowledge results in the development of new heuristics for application. Moreover, the broadening of the definition of engineering design has meant that areas that have previously not intersected sharply with engineering in an academic context such as biology and entrepreneurship, as well as emerging areas such as sustainability, are gaining significant attention.

Finally, new technologies are constantly being developed based on the findings of the types of research outlined earlier in order to aid designers and design teams in practicing design. Those technologies manifest themselves in a variety of ways, including design knowledge sharing and visualization systems; design decision support systems; and product simulation, realization, and testing tools.

Engineering Design Education Research

Engineering design education research aims to develop effective pedagogical approaches to teach design to engineering students. As such, it explores what and how engineering students learn. It also explores students' experiences with and attitudes toward engineering design, as well as educators' knowledge, attitudes, and practices related to engineering design education. In addition, students' experiences with specific areas, such as sustainable design and human-centered design, have also become areas of inquiry. Finally, new approaches for assessing students' engineering design knowledge and practices have also received attention.

Recent Research

In 2005 Dym and colleagues wrote an extensive review article titled "Engineering Design Thinking, Teaching and Learning" that included close to 200 references (Dym, Agogino, Eris, Frey, & Leifer, 2005). Here we recap some key observations from that article and present findings from work that was published since 2005. In their article, Dym et al. argued that the complex nature of design practice makes it challenging to teach and learn. In addition, engineering programs have historically had difficulty with integrating it in their curricula. They illustrated the multidimensionality of design practice by discussing some of the key elements of design thinking in an engineering context: thinking at the scale of systems, making estimates, conducting experiments, managing ambiguity through convergent-divergent inquiry, decision making under uncertainty, communicating in diverse languages (text, graphics, shape grammars, and mathematical models), and functioning as a part of an interdisciplinary team with teammates from demographically diverse backgrounds.

Pedagogically, Dym et al. concluded that the more recent focus on experiential learning as a framework for engineering design education has been successful and articulated potential reasons for that success. The application of project-based learning (PBL) at various points of the engineering curriculum was examined, including PBL in a global (distributed) context. For instance, PBL-oriented design courses taught during the first year in college (referred to as cornerstone design courses) serve a unique role not only by engaging students in design thinking but also by motivating them to continue into higher level engineering science courses, and more broadly, engineering as a discipline. On the other hand, capstone PBL experiences allow students to integrate knowledge they gain in different disciplines and learn to function across disciplinary boundaries. Dym et al. identified key research studies that support the use of PBL for design education, but also suggested further research on PBL and PBL in a global context as a key opportunity for future design education research. Since the publication of Dym et al., researchers have continued the research on PBL for design education. For example, Du, de Graaff, and Kolmos (2009) have investigated the factors that maximize the potential learning outcomes with PBL.

In addition, researchers have continued to explore the themes identified in the Dym et al. article, including research on thinking at the scale of systems (Hadgraft, Carew, Sardine, Blundell, & Blundell, 2008); the complexity of design which includes issues of social and environmental sustainability (Baillie, Feinblatt, Thamae, & Berrington, 2010; Mann, Dall'Alba, & Radcliffe, 2007; Zoltowski, Oakes, & Cardella, 2012); engineering students' uses of estimates and mathematical modeling in design (Carberry & McKenna, 2011; Cardella, 2006); overcoming fixation through convergent–divergent inquiry (Hatchuel, Le Masson, & Weil, 2011); learning the discourse of design (Atman, Kilgore, & McKenna, 2008); and functioning as a part of an interdisciplinary team with teammates from demographically diverse backgrounds (Hirsch & McKenna, 2008; Kilgore, Atman, Yasuhara, Barker, & Morozov, 2007; Laeser, Moskal, Knecht, & Lasich, 2003; Okudan & Mohammed, 2006). Some of these themes are also discussed in relation to the framework of Adaptive Expertise in Chapter 12 by McKenna in this book.

Recently, researchers have also begun to theorize different learning trajectories for engineering design education. Adams began to lay a foundation for the research community's understanding of design learning trajectories as she characterized different possible shapes of learning trajectories (Adams, Turns, & Atman, 2003). Her work is developed further in a more recent paper (Crismond & Adams, 2012) that combines a synthesis of empirical findings and theoretical perspectives. Researchers have also used phenomenography as an approach to characterizing different "spaces" which can be understood as stages in a learning trajectory. Daly's study of designers (Daly, 2008) identified six major ways that the designers in the study experienced the phenomenon of design: as (1) evidence-based decision-making, (2) organized translation, (3) personal synthesis, (4) intentional progression, (5) directed creative exploration, and (6) freedom. Daly's work suggests that design

educators can help students examine their understanding of what it means to develop design skills and what it is to practice design.

Atman and colleagues also studied practitioners, comparing their design processes with the processes of first-year and graduating engineering students (Atman, Chimka, Bursic, & Nachtmann, 1999; Atman et al., 2007). The analysis demonstrates that as expertise increases, there are differences in several design process measures. Specifically, compared to first year, seniors (1) have higher quality designs, (2) scope the problem more effectively by considering more categories of information, (3) make more transitions among design steps, (4) iterate more effectively (Adams et al., 2003), and (5) demonstrate a more comprehensive design process by including decision and communication activities. In addition, compared to students, experts (1) spend more time solving the problems in all design stages, (2) exhibit a "cascade" pattern of transitions, and (3) scope the problem more effectively by gathering more information (explicitly) and covering more categories of information. These results point to design skill acquisition that could be augmented with direct instruction.

Recent research has also explored how students learn to respond to some of the complexities associated with design, specifically within the contexts of sustainable design and human-centered design. Mann identified five categories which represent five ways of experiencing sustainable design: (1) solution finding, (2) reductionist problem solving, (3), holistic problem solving, (4) social network problem solving, and (5) a way of life. A key implication of this research is the need for a shift away from an instructor-transfers-knowledge model to a developing-a-professional-way-of-being model, which incorporates both engineering skills development, such as engineering design, and a way of experiencing practice (Mann et al., 2007). Finally, Zoltowski interviewed university students who had "designed something for others" (Zoltowski, 2010). The findings from her study enable us to look at a learning "outcome space" for human-centered design (HCD). In this "outcome space" an empathic design process (where stakeholders are consistently engaged for feedback and input) represents a comprehensive understanding of HCD. This is in contrast to a design process of occasionally seeking input or feedback from prospective users or involving the user only in usability studies (Zoltowski et al., 2012). Zoltowski's work has been used as a basis for designing assessment instruments to measure students' understanding of human-centered design (Zoltowski, Cardella, & Oakes, 2011b).

A Focus on Assessment

One of the key areas of emphasis in design education research since the publication of the Dym et al. (2005) article is the development of assessment tools for engineering design education. Many of these assessment tools are focused on measuring students' understanding of the design process as well as their skill in executing the process. Bailey and colleagues' approached assessing students' understanding of the design process by developing an instrument that asks students to critique another designer's process as depicted in a Gantt chart (Bailey & Szabo, 2006). This instrument has been adapted to measure primary and secondary teachers' and students' design knowledge, based on their evaluation of a given process provided in the form of a simplified Gantt chart (Hsu, Cardella, Purzer, & Diaz, 2010). McMartin, McKenna, and Youssefi (2000) used scenario assignments to assess students' knowledge of engineering practices.

To characterize students' and practitioners' conceptions of engineering design, Mosborg et al. (2005) augmented a list of design activities from prior research (Newstetter & McCracken, 2001) and produced a survey question that has since been used in multiple studies (e.g., Adams & Fralick, 2010; Atman et al., 2008). The survey question asks respondents to identify the design activities that are more/less important. These researchers observed similar changes in engineering undergraduates' conceptions

of design during the course of their studies. Adams and Fralick (2010) also asked students to explain why they believe a design activity is (or is not) important. Purzer and Hilpert's approach is to administer repeated surveys over the course of the academic term (as students are engaged in a term-long design project), asking students to identify the aspects of design that are most challenging at that point in time (2011). In another approach that combined assessment and curriculum development, the Bloom's taxonomy of educational objectives (Bloom, Engelhart, Furst, Hill, & Krathwohl, 1956) was instantiated for engineering design learning objectives. The resulting detailed description of various elements of design was used to both teach and assess design learning (Safoutin, Atman, & Adams, 2000a; Safoutin et al., 2000b).

Other researchers have focused on specific aspects of engineering design. Davis and colleagues have developed formative assessments that prompt students to articulate and self-assess effectiveness of their problem scoping, concept generation, and solution realization design activities toward producing a design solution that meets varied stakeholder needs (Davis, Gerlick, Trevisan, & Brown, 2011). Melton and colleagues focus specifically on assessing students' human-centeredness in their design process (Melton, Cardella, Oakes, & Zoltowski, 2012).

Although many of these tools have been developed initially to support research on engineering design education (to measure changes in students' understanding of design and design abilities based on interventions), these assessment tools have also been used to improve engineering design education by providing a means for educators to measure students' knowledge and skills, and provide feedback to the learners.

A Sampling of Methods

Design education research is differentiated from design research in general not by the methods that are used but rather by the research questions being asked, the participants being studied, the research settings, and the body of literature being informed. Many types of qualitative and quantitative research methods are used to study engineering design education. A more in-depth description of a broad set of methods is included in Chapter 24 by Lattuca and Litzinger and Chapter 25 by Kelly in this volume. In this section, we briefly describe and present examples of some of the research methods that have recently been used in engineering design education.

Current research includes both experimental as well as naturalistic approaches with differential emphasis on products as well as processes. Where process is a focus, the emphasis includes both process from an –etic perspective (how the researcher describes the designer's process) and an –emic perspectives (how the designer describes the process, such as is emphasized in phenomenography). In terms of data collection, observations, surveys, interviews, and the collection an analysis of design products and by-products (i.e., sketches, journals) all have precedent. Another source of variation is in terms of the design problems, with both carefully crafted problems for laboratory studies and naturally occurring problems being considered. Below are additional notes on specific methodological choices with additional pointers into the research literature.

1. Protocol analysis: Participants are asked to respond to a design brief and engage in design tasks. The resulting activities are often recorded, which constitute data. The outcomes of the activities can also constitute data, and be analyzed independently or in conjunction with the activity records. Activity records can be in the form of audio or visual recordings, or any other type of information that can be collected that may be the result of participation actions, such as touchscreen interactions or mouse clicks.

Researchers conduct protocol analysis to observe and identify patterns in the behavior of participants while they design as individuals or in groups. Analysis variables associated with the activity can be verbal (discourse) or nonverbal (interacting with an artifact, gesturing, sketching, etc.). When studying individual designers, speech is often induced with think-aloud rules to help participants externalize their mental processes. When studying teams, the natural dialog of a team can be used as a basis (such as Nefcy, Gummer, & Koretsky, 2012; Purzer, Chen, & Yaday, 2010; Purzer & Hilpert, 2011; Roberts et al., 2007a).

By analyzing patterns, researchers make inferences about the information handling behaviors, collaboration mechanisms, and thought processes of participants, such as allocating time across different design activities (Atman & Bursic, 1998; Atman et al., 1999); making design process moves between function, behavior and structure variables (Gero, 1990; Kan & Gero, 2009; Williams, Gero, Lee, & Paretti, 2011), interacting with different representations (Cardella, Atman, & Adams, 2006), or using design schema (Ball, 1990; Stacey, Eckert, & Earl, 2009).

Protocol analysis has also been used to measure participants' understanding of the design process (Atman et al., 1999, 2007), ability to consider the broader context of the design task (Kilgore et al., 2007; Kilgore, Jocuns, Yasuhara, & Atman, 2010), or understanding of human-centered design (Melton et al., 2012).

Finally, protocol studies can also involve dialogs between researchers and participants during which participants are prompted to describe a planned design process or to critique another designer's process.

2. Interviews: Students and instructors are interviewed about their design teaching and learning experiences in and out of the classroom. Engineering design education interviews are often open-ended or semistructured (Marra, Palmer, & Litzinger, 2000). Recently, researchers have conducted phenomenographic interviews (Marton & Booth, 1997) to identify and compare different ways in which people understand and experience design. Phenomenography has been used to characterize designers' experiences with design at a broad level (Daly, Mann, & Adams, 2008) as well as specific

variants of design, such as sustainable design (Mann et al., 2007) and human-centered design (Zoltowski, Cardella, & Oakes, 2011a).

3. Surveys: Surveys can capture a large number of responses about design teaching and learning experiences in and out of the classroom. For example, Hadgraft and colleagues surveyed students to learn about their experiences with learning systems thinking (Hadgraft et al., 2008). Surveys have also been used to capture affective dimensions of design education, such as respondents' beliefs about which design activities are the most/least important (Adams & Fralick, 2010; Newstetter & McCracken, 2001) as well students' impressions of the most challenging aspects of design at different points in a design project (Purzer & Hilpert, 2011) or students' design related self-efficacy (Carberry & McKenna, 2011; Purzer, 2011).

4. Ethnographies/field studies: Researchers have also conducted direct in situ observations of students and instructors engaged in design teaching and learning. These studies range from conducting a few observations "in the field" (e.g., Green & Smrcek, 2006) to full ethnographies that reveal patterns in students' experiences with design education (Downey & Lucena, 2003; Garzca, Palou, Lopez-Malo, & Garibay, 2009; Nersessian, Newstetter, Kurz-Milcke, & Davies, 2002; Newstetter, 1998).

5. Analysis of learning artifacts: The educational setting also affords research methods that are more natural to the educational context than to that of professional practice. For example, researchers can analyze students' project reports and design journals (Sobek, 2002) or portfolios (Eliot & Turns, 2011) to examine students' design knowledge and practices. McDonnell and her colleagues (McDonnell, Lloyd, & Valkenburg, 2004), and Jones and her colleagues (Jones, Cardella, & Purzer, 2012) have also asked students to create video narratives of their design projects to prompt students' reflection on their design practices. These video narratives likewise provided rich "natural" data.

Just as there are many different methodologies available for educators to use to conduct research on engineering design education, and a variety of assessment instruments to measure students' design skills and knowledge, there are a variety of curriculum models and pedagogical approaches that design educators can use in creating engineering design learning experiences. These pedagogical approaches are described in the next section.

Engineering Design Education

Our rapidly evolving understanding of the engineering design process, which was discussed at the beginning of the chapter, presents a formidable challenge to engineering educators striving to ensure that students not only become technically and scientifically knowledgeable but also are prepared to engage competently in the complexities of design processes. The time criticality of this challenge has been articulated by industry, engineering forums, and accrediting agencies such as the Accreditation Board for Engineering and Technology (ABET).

The "dense" nature of engineering education programs, with little room for additional courses or course content, constitutes another dimension of this challenge; educators endeavor to include appropriate design learning experiences for their students within the existing closely packed curricular frameworks. Regardless, the momentum toward more and better design learning has steadily increased, and the capstone design course and PBL pedagogy, highlighted by Dym et al. (2005), have become standards in engineering design education.

Design in Engineering Curriculum

The design content and learning experiences that are built into typical engineering programs, as well as the learning outcomes criteria by which these programs are assessed, demonstrates that design is now generally accepted as a core body of engineering knowledge. However, the positioning of design experiences in the curricular sequence and the pedagogical mechanisms through which engineering design knowledge is acquired varies widely. During the last two decades, engineering design education has gravitated toward three basic curricular models:

1. The stand-alone capstone design experience: A synthesis design course (or two-course sequence) that often culminates an engineering undergraduate degree. Many capstone courses feature real design problems supplied by clients from industry (Dutson, Todd, Magleby, & Sorenson, 1997; Dym et al., 2005; Todd, Magleby, Sorenson, Swan, & Anthony, 1995). When educational programs collaborate with industry, they are often obliged to feature their more advanced (senior) students, which tends to delay design experiences to latter terms in a program. This contributes to the popularity of the capstone design model. Moreover, because engineering design has traditionally been considered to be mostly concerned with the application of scientific and technical knowledge, it is understandable why those types of knowledge are emphasized early on in engineering education and design experiences are provided late in a learning sequence (Sheppard, Macatangay, Colby, & Sullivan, 2009). Several authors have characterized current trends in capstone courses within the United States with regard to pedagogical approaches (e.g., Howe & Wilbarger, 2005; Pembridge & Paretti, 2010) and assessment practices (McKenzie, Trevisan, Davis, & Beyerlein, 2004).

2. The cornerstone-capstone model: Essentially bookends an undergraduate program, with little in the way of design experiences in between. Cornerstone design courses, as noted by Dym et al. (2005), serve a variety of purposes, from introducing students to conceptual design to helping them to become comfortable with teamwork. Heywood (2005) posited that the first-year (cornerstone) design course is used primarily in the United States to interest and retain students in engineering programs (p. 310). As conceptions of engineering design processes have broadened to encompass more of what engineers do in practice (e.g., problem framing, resource identification, communication, interdisciplinary interactions), and engineering educators have come to understand the complexity of design thinking (Dym et al., 2005; McKenna, Colgate, Carr, & Olson, 2006) required by competent designers, the single capstone course model and the cornerstone/capstone models are being recognized as insufficient (Jonassen, Strobel, & Lee, 2006) and are slowly giving way to more integrated approaches.

3. The integrated or sequential design approach: Programs that distribute design courses and coursework through the curriculum often include cornerstone and capstone design experiences. In his overview of engineering education, Heywood (2005) suggested that examples of sequential approaches to teaching design can be found as far back as the University of Reading (U.K.) program in the 1960s. The current trend is a gradual movement toward the more integrated models. The wide variety of ways that design is integrated in curriculum was illustrated by a Neeley, Sheppard, and Leifer (2006) study of courses that were designated as design in five different Mechanical Engineering programs. Those researchers developed a four-part classification: (a) *Design as Experience,* in which "Students typically work on open-ended projects in teams" (p. 7). This activity generally takes place in the capstone design course; (b) *Design as the Technical* in which students learn "practical knowledge and skills" (p. 8) (e.g., computer-aided design [CAD]); (c) *Design as Analysis* which are courses that "provide knowledge necessary for design, yet there is little explicit design content or activity. They appear to adhere to the notion that, in all engineering content, there must be *some* design"

(p. 9); and (d) *"Design as Pedagogy"* in which "design activity is used to promote more complete understanding of the subject matter and is secondary to this subject matter" (p. 10).

There are several recent and successful examples of such integrated and holistic curriculum redesigns: the influential Aalborg University curriculum model that provided a foundation for project-based learning (PBL) pedagogies (Luxhoj & Hansen, 1996); the civil engineering program at Monash University in Australia (Mills & Treagust, 2003); the Franklin W. Olin College curriculum that holds design learning as a core driver for the engineering education it offers (Somerville et al., 2005); the Conceive, Design, Implement, and Operate (CDIO) curriculum initiative that has found international adoption (Crawley, Malmquist, Ostlund, & Brodeur, 2007); and the forty-year-old "WPI Plan" curriculum at Worcester Polytechnic Institute (Mello, 2000).

Co-Curricular and Elective Approaches

Revamping entrenched engineering curricula is difficult, and it is therefore understandable why the single capstone course model is so common. Another approach is to promote extracurricular, or co-curricular design experiences that can have limited impact on curriculum structures but a positive impact on developing students' design skills. For example, when students described significant experiences "designing for others" in Zoltowski's article (2010), they included co-curricular projects such as Engineers Without Borders and Engineers for World Health, as well as undergraduate research, internships, and other work experiences. The students also described community service projects, projects completed while in high school, scouting projects, the design of athletic practices, the design of projects for competitions (e.g., Rube Goldberg, FIRST) and hobbies such as jewelry making. Similarly, in a quantitative study, Palmer, Terenzini, McKenna, Harper, & Merson (2011) found that several co-curricular experiences

had a positive influence on students' contextual competence (i.e., their ability to consider the context of the design). Those experiences included being active in an engineering-related nonprofessional organization related to women or minority students (such as the National Society of Black Engineers [NSBE] or Women in Science & Engineering [WISE]) or other non-engineering clubs and activities, or participating in humanitarian engineering projects (such as Engineers Without Borders) or other non-engineering service work. In addition to the co-curricular approaches there are also successful elective course models that include the Institute for Design Engineering and Applications (IDEA) model at Northwestern University, in which students can choose to earn "a certificate in engineering design alongside their bachelor's degree" (McKenna et al., 2006), and the Engineering Projects in Community Service (EPICS) program at Purdue University, where students choose to enroll in a design course in addition to the classes or to fulfill requirements for their program (such as technical electives). In the EPICS model, they then participate in vertically integrated (i.e., first-year students working alongside juniors and seniors) and multidisciplinary teams to meet real community needs (Coyle, Jamieson, & Oakes, 2005).

Studios

The studio setting is becoming more prevalent in engineering design education, although there is no explicit agreement on the implications of the term (Little & Cardenas, 2001). The basic notion of the "design studio" originated within the arts not only as an education medium but also as a primary environment for practice. It can be argued that there are parallels within the engineering tradition in the form of "laboratories."

In either disciplinary context, the term *studio* often refers to the physical environment *and* its affordances. The main affordances of the studio environment can be considered according to the framework we used to categorize engineering design practice earlier in the chapter; an ideal design studio enables designers to better structure and manage their workflow, augments their cognition, and facilitates social interaction. These affordances can be particularly relevant when the nature of the work that is carried out is open-ended and interdisciplinary. Therefore there is a natural coupling between the studio environment and PBL.

In design education, studio-based learning shifts the focus of the interaction to "work done by students" from general principles that may be discussed in lectures (Wilson & Jennings, 2000). This is not to say that studio-based education is not concerned with general principles, but rather that general principles are visited in the context of ongoing student work during localized interactions between teachers and students, and even between students and students. This type of interaction also redefines the role of the teacher from being a source of codified knowledge to acting as a coach because students take more responsibility of their learning goals and exercise a higher degree of autonomy over their learning processes in the studio.

Design Curriculum Emphasis

Design curriculum is also being used to convey and instill particular approaches, attitudes, and competencies as they gain relevance in the engineering field. Design experiences have been used to develop students' "professional" skills such as team-working and project management skills (Hyman, Khanna, Lin, & Borgford-Parnell, 2011), communication skills (Paretti, 2008), and professional ethics (Catalano, 2004; Dutson et al., 1997). Recently, programs have emphasized sustainable design as well as human-centered design. Educators have taken many approaches to sustainable design, with some emphasizing environmental, social, and financial sustainability equally, and others emphasizing environmental sustainability (Lau, 2010; Vanasupa et al., 2010). Sustainable design extends the product life-cycle such that the designer considers not only the process of moving from conceptual design to product deployment, but also how the product might be reused or recycled at the end of the product life-cycle (Ehrenfeld, 2008; Gerber, McKenna, Hirsch, & Yarnoff, 2010). In addition, it has been argued that there is a paradigm shift occurring in design from "technology-centered design" to "human-centered design" (Krippendorff, 2006). Human-centered approaches to design contribute to innovations in engineering design (Brown, 2008) and have been shown to increase productivity, improve quality, reduce errors, improve acceptance of new products, and reduce development costs (Maguire, 2001). In both cases, a leading emphasis in design education is to consider sustainable design and human-centered design as "normal" design (Cardella, Hoffman, Ohland, & Pawley, 2010).

Learning Engineering Design

The sophisticated abilities characterizing good design thinking that were identified by Dym et al. (2005), mentioned earlier, or what McKenna et al. (2006) called "unique knowledge and skills, or habits of mind," (p. 1) are developed and reinforced within learning systems. The integrated curricular structures discussed previously are systems that can enable the type of learning described by Ambrose, Bridges, DiPietro, Lovett, and Norman (2010), wherein students "develop a set of key component skills, practice them to the point where they can be combined fluently and used with a fair degree of automaticity, and know when and where to apply them appropriately" (p. 95). In their influential book *Educating Engineers*, Sheppard et al. proposed that if engineering programs based their curricula on current research from the learning sciences and best practices from other professional education, design experiences would be seeded throughout (Sheppard et al., 2009).

Project-based learning (PBL) has emerged as the most common and the most well-substantiated pedagogy applied in capstone (and other) design experience courses. PBL (discussed in Chapter 7 by

Stevens, Johri, & O'Connor in this volume) is a sophisticated learning model that is well supported by significant learning theory and best practices, such as experiential learning (Kolb, 1984), cooperative learning (Johnson, Johnson, & Smith, 1998; Smith, Sheppard, Johnson, & Johnson, 2005), team-based learning (Michaelsen, Knight, & Fink, 2004), cognitive apprenticeship (Collins, 2006), reflection in action (Schon, 1983), and situated cognition (Lave & Wenger, 1991). Project-based learning is common in engineering education and is often considered a form of problem-based learning. Problem-oriented PBL activities generally begin with a case or story containing a challenging problem. Students (usually in teams) are expected to identify the problem and devise a plan for solving it (Hmelo-Silver, 2004). The plan is often the expected end result of the activity. This is an inductive learning method that, among other learning goals, helps to motivate students and foster their self-directedness and metacognitive skills. Project-oriented PBL differs from other PBL pedagogies by incorporating an engineering project, which makes the learning experience more like an authentic engineering task (Mills & Treagust, 2003). In an earlier study, Barron et al. suggested that pairing project-oriented and problem-oriented activities was advantageous, given that an early problem-oriented activity could help student teams develop a shared knowledge base and prepare them to then undertake a project (Barron et al., 1998). This pairing of problem and project is the basis for Aalborg's project-based curriculum, described as "strongly problem-oriented" (Luxhoj & Hansen, 1996, p. 184).

In engineering education, PBL is often framed by a design project and incorporates important design activities, such as gathering and evaluating information, framing problems, generating and testing solutions, making and communicating decisions, and iterating. These basic design activities provide a vehicle for students to experience complete learning cycles, as described by Kolb's experiential learning theory (1984). The act of designing in a PBL learning context also offers an opportunity for students to achieve the full range of learning outcomes proposed by Barrows (Barrows, 1986) as cited in Du et al. (2009), which are: "(1) structure knowledge for use in authentic contexts, (2) develop critical thinking, reasoning, and problem solving skills, (3) develop self-directed learning skills, and (4) increase motivation for learning" (p. 38). Research has shown that to maximize the potential learning outcomes with PBL "a foundational knowledge base, prior knowledge on which to build problems, was critical" (Du et al., 2009). Therefore, the more integrated curricular approaches are better able to support the effectiveness of PBL in the development of competent engineering designers.

Capstone design courses and PBL pedagogies are familiar topics in recent engineering design education scholarship (Agrawal, Borgford-Parnell, & Atman, 2011), with educators describing both the potential benefits and common deficiencies in how they are utilized in engineering education. As conceptions of engineering design and engineering practice increasingly overlap, there is even more expectation of engineering design educators to (1) focus on current topics such as sustainable design and human-centered design; (2) promote a broader range of learning outcomes, such as the ability to communicate professionally, to work in interdisciplinary teams, to be entrepreneurial, and to be reflective-in-action; and (3) experience more "real life" aspects of design processes and contexts, such as iteration, using rapid-prototyping or virtual prototyping tools, or working in distributed teams. Some of these same topics drive the engineering design education research described previously. Because practicing engineers deal with "wicked" problems, there is great need for educational experiences that provide students more authentic contexts and call for the use and development of a range of sophisticated design knowledge and skill. PBL as an overall pedagogical strategy is flexible enough to provide opportunities for these important outcomes and experiences. However, those opportunities may be lost or diminished without additional integration of design courses and coursework throughout the engineering curriculum.

Research Informing Practice: Two Examples from Engineering Classrooms

In this section we present two examples of studies focused on improving engineering design learning (Boyer, 1990). Our primary intent is to illustrate pathways by which engineering design education research can be linked to classroom practices and support learning. The cases we present illustrate two different potential mechanisms for conducting such linkages. In the first case, experimental findings from lab-based research on design were translated into a classroom intervention and tested in multiple classrooms. In contrast, the second case treats the classroom as a learning laboratory in which the phenomenon that needed attention was identified and acted on primarily within the context of the classroom. The cases also highlight some of the complexities of teaching and learning engineering design that were discussed in previous sections.

Example One: Learning About the Design Process

This first example illustrates a translation of research from a large, laboratory-based research program to a classroom setting. Specifically, results from an in-depth verbal protocol study of the design processes employed by participants with three levels of expertise were used as the basis of an in-class exercise to teach junior and senior engineering students about design processes. The research, described briefly in the section on engineering design education research, demonstrates how the engineering design processes employed by practicing professionals in the study are more sophisticated than the processes of senior engineering students who were studied, which in turn are more sophisticated than the processes

of first-year engineering students who were studied. The metrics that were used to determine "sophistication" include amount of time, number of problem scoping efforts, as well as completeness of the process and complexity as measured by effectiveness of iterations (Adams et al., 2003; Atman et al., 1999, 2007).

The design processes revealed in these lab-based studies can be displayed as time-lines, which provide effective visual representations to convey the differences that were documented across the three groups. To demonstrate, Figure 11.1 presents time-lines for three first-year and three senior students. These are representative timelines from the students in each group who scored low, average, and high on the quality of the final product of the design task. In each timeline, a mark or block on a horizontal line indicates time spent on one of eight design activities that are defined in the figure.

The researchers found that these representations quickly conveyed the results of their research and resonated with both industry and academic colleagues. Colleagues from several engineering departments at the University of Washington (UW) invited the researchers to use these research results to teach students about design processes. The resulting classroom activity was designed by personnel from the Center for Engineering Learning & Teaching (CELT) as a tangible outcome of their research to practice mission (Atman, Turns, Borgford-Parnell, & Yasuhara, 2012). The exercise took less than an hour and included a brief description of the research methods and the data used to create the design timelines, and then the presentation of the timelines and questions that appear in Figure 11.1. The researchers presented this exercise to teams of upper-class engineering students in three engineering departments at the University of Washington. Responses to this activity in a senior-level design course in Materials Science and Engineering and a junior-level engineering analysis class with a design project in Aeronautical and Astronautical Engineering are described in

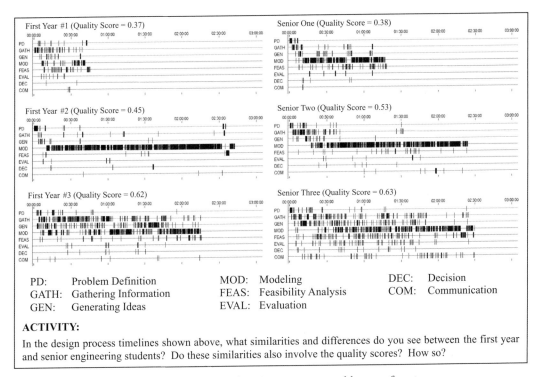

ACTIVITY:

In the design process timelines shown above, what similarities and differences do you see between the first year and senior engineering students? Do these similarities also involve the quality scores? How so?

Figure 11.1. Timeline activity handout. Timelines represent typical low-performing, average-performing, and high-performing first year and senior engineering students. Timelines originally presented in Atman (1999). (Reprinted with permission.)

detail elsewhere (Borgford-Parnell, Deibel, & Atman, 2010).

To summarize, the task was effective in helping students generate insights about the design process. Written responses that individual students had to the task were collected and coded using a bottom-up constant-comparative analysis method, and nine categories of responses emerged. The categories and representative quotes from the students are presented in Figure 11.2. Their insights often aligned with the researchers' findings, and they invariably compelled engaging classroom discussions.

This example offers an illustration of how research findings can directly be utilized to effect classroom learning. The opportunity for these classroom interventions came about through interactions of the faculty who taught these engineering courses with the scholars at CELT who conducted the intensive lab-based research. These conversations are designed to occur in the CELT model where research on engineering education informs practice in engineering classrooms through faculty development activities. Ideas for research then, in turn, come from interaction with engineering faculty, completing a feedback loop (Atman et al., 2012).

The study results presented in this example have also been used to support teaching and learning interventions beyond the one presented here. For example, the timeline representations have been used to help graphic design students develop models of their own design processes (Molhave, McDonnell, & Atman, 2011). In addition, the research itself is currently being extended to understand design processes of teams of engineering students (Purzer, 2011; Roberts et al., 2007b; Yasar-Purzer, Henderson, McKay, Roberts, & de Pennington, 2008) and to characterize how high school students approach engineering design (Becker, Mentzer, & Park, 2012; Mentzer, Becker, & Park, 2011; Park, Pieper, Mentzer, & Becker, 2010).

Breadth of Design Activity Equates with Higher Quality Design Scores	• "The people who spent their time on multiple categories generally scored higher." (Senior) • "Activity spread correlated better to quality than class standing." (Junior)
Seniors Engage in More of the Design Activities than Freshmen	• "Freshmen students spend more time on the first half of the process, especially on modeling, while senior students spend more time more evenly on every process." (Junior). • "Freshmen don't spend as much time with the follow-up portion of the design process." (Junior)
Seniors Spend More Time Scoping the Problem	• "The freshmen did not plan out the problem in the beginning nor take as long to plan it out as the seniors did." (Senior) • "Seniors' concentration was more spread out and they spent more time in the beginning on PD and GATH before starting to model. Similarly, the more spread out the concentrations was, the better the quality." (Senior)
More Time Problem Scoping Correlates with Higher Design Quality	• "Those who did more planning, gathering of information, etc., received a higher quality score." (Senior) • "Less time on problem definition – less score." (Senior)
Seniors Do More Designing Alternative Solutions and Project Realization Activities	• "It seems as if the seniors performed more evaluation than the freshmen. The amount of work for each section (gathering, feasibility, etc.) were more spread out over time." (Junior) • "Senior students appear to have generally more segments of feasibility, evaluation, decision, and communication." (Senior)
Time Spent Gathering Information Equates to Design Quality	• "Students with higher quality scores do a lot more gathering." (Junior) • "A lot of varieties in the information gathering step – all proceed at the beginning but some (mostly seniors) would gather info along the process." (Senior)
Everyone Spends Most Design Time in Modeling	• "Modeling seems to be the dominant characteristic among freshmen and seniors." (Junior) • "Avoid getting stuck in the modeling phase. Continually gather information & check to be sure you are working to make your goals." (Senior)
Iteration Is Tied to Design Quality	• "Those who constantly looked back to gather info, put it together, then made sure to properly evaluate it got much higher quality scores." (Senior) • "More dynamic interplay between modeling and secondary processes for seniors, lots of back and forth." (Junior)
A Good Design Process Has a Shape	• Sketch on diagram labeled "Ideal Project Envelope" 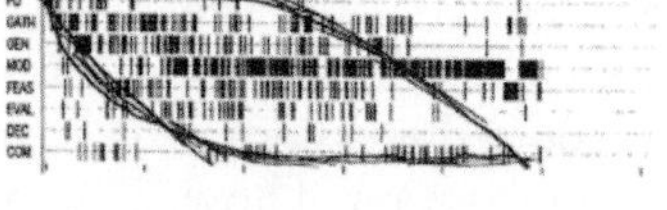

Figure 11.2. Student insights from classroom activity.

Example Two: Balancing Learning Orientation with Task Orientation in a PBL Environment

In contrast to the first example, the second example illustrates a rapid research-intervention cycle within the context of a design studio and involves using the previously described research as a backdrop for formulating research instruments appropriate for the classroom context. Moreover, this example illustrates how qualitative observations performed by instructors in a first-year engineering design studio instigated a relatively rapid research and intervention cycle that led to positive change in learning outcomes of undergraduate engineering students. The instructors acted on preliminary observations in order to frame a research question, formulate a research framework, develop an instrument, perform measurements in the studio, and intervene on the basis of those measurements. This example has been addressed in greater detail elsewhere (Linder, Somerville, Eris, & Tatar, 2010) and is summarized in the text that follows.

The key observation was that teams working on a design project in a PBL context, when overly focused on project outcomes, delegated responsibility and tasks to their members in a way that maximized the potential utility of existing and/or perceived competencies of individuals. Broadly speaking, although that type of task appropriation might be desirable in professional practice, it is highly problematic in a design studio for several reasons.

First, it depicts learning and acquiring different competencies as a highly "costly" endeavor, which a design team cannot afford if it is to maximize its project performance. If the team is to "pull it off," team members need to put the team's performance before their personal learning goals. From a learning orientation, such an approach is problematic because it limits the learning "horizon" of the individual student. It can also prematurely influence students to gravitate toward their perceived strengths and to stay away from their perceived weaknesses. That would lead to a one-dimensional growth pattern, which is not congruent with the multidimensionality of design practice as identified by Dym et al. (2005) and discussed earlier in this chapter.

Second, it can bring up cultural stereotypes and other perceptions – in terms of what competencies individuals possess – that are false. Unlike professionals, first-year undergraduate students in a design studio do not have formal credentials that speak to their competencies, and when facing the "unknown," too much can be read into limited prior experiences.

As reported in Linder et al. (2010), the research team identified literature that pointed toward a relevant conceptual distinction between learning and goal performance orientation; in the context of individuals (Dweck & Leggett, 1988), as well as teams (DeShon, Kozlowski, Schmidt, Milner, & Wiechmann, 2004). In the design studio, DeShon et al. observed that the emphasis on goal performance that the student design teams were operating under manifested itself in the form of a clear gender-correlated task division. Specifically, male students seemed to be more engaged in generating digital product models and prototypes, whereas female students seemed to be more engaged in understanding user needs and communicating project outcomes. Based on these insights, the research team developed a survey instrument that measured students' self-reported contribution to various tasks within the project and administered it after the project was complete as an exploratory tool; the results confirmed their qualitative observations.

As a response, in the next edition of the course, the researchers intervened by sharing the first set of survey results with the students and facilitated a class-wide reflection and discussion session around them. They also introduced a new assignment to the course, which asked students to identify their learning goals based on the reflection session. These interventions were carried out at the beginning of the project. The survey instrument was administered again upon the completion of the project in an offering of the course after the interventions were put in place. Post-intervention responses were then compared to the pre-intervention responses. Figure 11.3 shows the differences in means of responses between genders post-intervention as well as the statistically significant shifts when those responses were compared to the pre-intervention responses. There were significant increases in the contribution of female students to building 3D sketch models, modeling using CAD, and building prototypes, and of male students to preparing presentations. The shifts suggested a

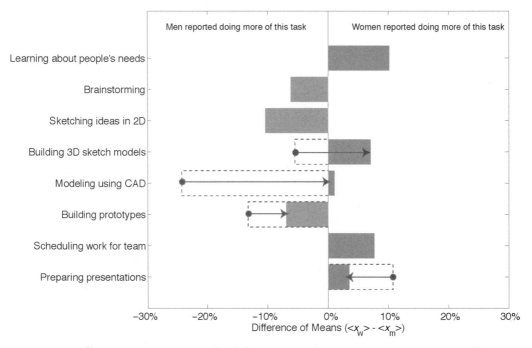

Figure 11.3. Differences in mean activity levels between genders post intervention. Statistically significant post- and pre-intervention differences are shown for the variables that exhibited shifts (as indicated by the start and end points of the arrows). The dashed bars indicate pre-intervention levels for those variables. Bars on the right are associated with tasks women reported doing more of, and bars on the left with tasks men reported doing more of.

more balanced distribution across project tasks, which was confirmed by the qualitative in-studio observations conducted by the instructors.

Two conclusions can be drawn from this case within the scope of the discussion in this chapter. First, engineering educators can and should take the initiative to formulate and implement interventions based on what they are observing in the classroom that are grounded in research, which might include developing and deploying new assessment instruments. Second, although we argued that new pedagogical approaches have bridged the gap that has historically existed between professional engineering design practice and engineering design learning, such an overlap is not necessarily entirely desirable as learning situations have unique and different goals. Educators should exercise caution when mapping the two worlds.

Conclusion

The goal of this chapter has been to acquaint readers with engineering design education research and practice. Through the content of the chapter – a brief review of the history of design research, presentation of research associated with design practice and design learning, enumeration of prominent pedagogies for design education, and two examples that connect design research and design education – we have aimed to provide a starting point for readers wanting to leverage or contribute to this body of research. For the reader who is interested in leveraging the work, the information provided helps the reader know what themes to expect from the literature but also what research methods are being used, what venues the research is published in, and what might be possible in terms of connecting research to educational practice.

There is clearly opportunity to not just leverage the work but also to contribute to it. For readers interested in contributing to a body of research related to engineering design education, one path is to extend scholarship described in this chapter, such as through studies that attempt to replicate findings or explore ideas with new populations. It is also possible to look across the work and think about broader opportunities. For example, given the growth in the research base, there is an increasing opportunity to synthesize across existing studies. There is also opportunity to understand better how design research can interact with design education.

This chapter and the work described in it represent a backdrop for thinking about "what next?" What seems clear, however, is that an emphasis on design and design education is no longer "what is next" for engineering education but rather "what is now." If we want to educate engineering students who can contribute to the grand challenges of the twenty-first century, they need to be prepared as designers. The work represented in this chapter not only suggests momentum in this area, but also provides a jumping off point for next steps.

Acknowledgments

We would like to acknowledge the funders for some of the work including NSF grants 9358516 and 0943242, and the Center for Engineering Learning & Teaching at the University of Washington. We thank the anonymous reviewers of the chapter who gave us very useful guidance. Finally, we acknowledge colleagues who collaborated with us on various parts of this work, including Robin Adams, Anukrati Agrawal, Katherine Deibel, Clive Dym, Paolo Feraboli, Sally Fincher, Brian Flinn, Vipin Kumar, Benjamin Linder, Annegrete Molhave, Marie Paretti, Julie Provenson, Mark Somerville, Nick Tatar, Jennifer Turns, and Ken Yasuhara.

References

Accreditation Board for Engineering Technology (ABET). Criteria for Accrediting Engineering Programs, 2013–2014. Retrieved from http://www.abet.org/DisplayTemplates/DocsHandbook.aspx?id=3149

Adams, R. S., & Fralick, B. (2010). *Work in progress – A conceptions of design instrument as an assessment tool.* Paper presented at the Annual ASEE/IEEE Frontiers in Education Conference, Washington, DC.

Adams, R. S., Turns, J., & Atman, C. J. (2003). Educating effective engineering designers: The role of reflective practice. *Design Studies, 24*(3), 275–294.

Agrawal, A., Borgford-Parnell, J., & Atman, C. J. (2011). Citations on engineering design education (2003–2011) *CELT-11–11.* Seattle: University of Washington (contact celtad@uw.edu).

Akin, Ö. (1986). *Psychology of architectural design.* London: Pion.

Alexander, C. (1964). *Notes on the synthesis of form.* Cambridge, MA: Harvard University Press.

Ambrose, S. A., Bridges, M. W., DiPietro, M., Lovett, M. C., & Norman, M. K. (2010). *How learning works.* San Francisco, CA: Jossey-Bass.

Atman, C. J., Adams, R. S., Cardella, M. E., Turns, J., Mosborg, S., & Saleem, J. (2007). Engineering design processes: A comparison of students and expert practitioners. *Journal of Engineering Education, 96*(4), 359–380.

Atman, C. J., & Bursic, K. M. (1998). Verbal protocol analysis as a method to document engineering student design processes. *Journal of Engineering Education, 87*(2), 121–132.

Atman, C. J., Chimka, J. R., Bursic, K. M., & Nachtmann, H. L. (1999). A comparison of freshman and senior engineering design processes. *Design Studies, 20*(2), 131–152.

Atman, C. J., Kilgore, D. E., & McKenna, A. (2008, July). Characterizing design learning: A mixed-methods study of engineering designers' use of language. *Journal of Engineering Education,* 309–326.

Atman, C. J., Turns, J. E., Borgford-Parnell, J., & Yasuhara, K. (2012). The CELT model: Linking engineering education research and practice at the Center for Engineering Learning & Teaching *CELT Technical Report CELT 12–01.* Seattle, WA: University of Washington (contact celtad@uw.edu).

Badke-Schaub, P., Goldschmidt, G., & Meijer, M. (2010). How does cognitive conflict in design teams support the development of creative ideas? *Creativity and Innovation Management*, 19(2), 119–133.

Bailey, R., & Szabo, Z. (2006). Assessing engineering design process knowledge. *International Journal of Engineering Education*, 22(3), 508.

Baillie, C., Feinblatt, E., Thamae, T., & Berrington, E. (2010). *Needs and feasibility: A guide for engineers in community projects – The case of waste for life*. San Rafael, CA: Morgan and Claypool.

Ball, L. (1990). *Cognitive processes in engineering design*. Plymouth, U.K.: Polytechnic South West.

Ball, L., Evans, J., & Dennis, I. (1994). Cognitive processes in engineering design: A longitudinal study. *Ergonomics*, 37, 1753–1786.

Barron, B. J. S., Schwartz, D. L., Vye, N. J., Moore, A., Petrosino, A., Zech, L., & Bansford, J. D. (1998). Doing with understanding: Lessons from research on problem- and project-based learning. *Journal of the Learning Sciences*, 7(3–4), 271–311.

Barrows, H. S. (1986). A taxonomy of problem-based learning methods. *Medical Education*, 20(6), 481–486.

Baya, V., & Leifer, L. (1996). Understanding information management in conceptual design. In N. Cross, H. Christiaans, & K. Dorst (Eds.), *Analyzing design activity* (pp. 151–168). Chichester, U.K.: John Wiley & Sons.

Becker, K., Mentzer, N., & Park, K. (2012). *High school student engineering design thinking and performance*. Paper presented at the American Society for Engineering Education conference, San Antonio, TX.

Bloom, B. S., Engelhart, M. D., Furst, E. J., Hill, W. H., & Krathwohl, D. R. (1956). *Taxonomy of educational objectives: The classification of educational goals. Handbook I: Cognitive domain*. New York, NY: Longmans, Green.

Borgford-Parnell, J., Deibel, K., & Atman, C. J. (2010). From engineering design research to engineering pedagogy: Bringing research results directly to the students. *International Journal of Engineering Education*, 26(4), 748–759.

Boyer, E. (1990). *Scholarship reconsidered*. Princeton, NJ: Carnegie Endowment for the Advancement of Teaching.

Brereton, M., Cannon, D., Mabogunje, A., & Leifer, L. (1996). Collaboration in design teams: How social interaction shapes a product. In N. Cross, H. Christiaans, & K. Dorst (Eds.), *Analyzing design activity* (pp. 319–342). Chichester, U.K.: John Wiley & Sons.

Brown, T. (2008). Design thinking. *Harvard Business Review*, 86(6), 84–92.

Bucciarelli, L. L. (1994). *Designing engineers*. Cambridge, MA: MIT Press.

Carberry, A. R., & McKenna, A. F. (2011). *Analyzing engineering student conceptions of modeling in design*. Paper presented at the Frontiers in Education Conference, Rapid City, SD.

Cardella, M. E. (2006). *Engineering mathematics: An investigation of students' mathematical thinking from a cognitive engineering perspective*. Ph.D. dissertation, University of Washington.

Cardella, M. E., Atman, C. J., & Adams, R. S. (2006). Mapping between design activities and external representations for engineering student designers. *Design Studies*, 27(1), 5–24.

Cardella, M. E., Hoffman, S. R., Ohland, M. W., & Pawley, A. L. (2010). Sustaining sustainable design through "normalized sustainability" in a first-year engineering course. *International Journal of Engineering Education*, 26(2), 366–377.

Carvalho, L., Dong, A., & Maton, K. (2009). Legitimating design: A sociology of knowledge account of the field. *Design Studies*, 30(5), 483–502.

Catalano, G. D. (2004). Senior capstone design and ethics: A bridge to the professional world. *Science and Engineering Ethics*, 10, 409–415.

Collins, A. (2006). *Cognitive apprenticeship*. In S. R. K. (Ed.), *The Cambridge handbook of the learning sciences* (pp. 47–60). Cambridge: Cambridge University Press.

Coyle, E. J., Jamieson, L. H., & Oakes, W. C. (2005). EPICS: Engineering projects in community service. *International Journal of Engineering Education*, 21(1), 139–150.

Crawley, E. F., Malmquist, J., Ostlund, S., & Brodeur, D. (2007). *Rethinking engineering education: The CDIO approach*. New York, NY: Springer.

Crismond, D., & Adams, R. S. (2012). The informed design teaching and learning matrix. *Journal of Engineering Education*, 101(4), 738–797.

Cross, N. (1984). *Developments in design methodology*. Chichester, U.K.: John Wiley & Sons.

Cross, N. (2006). *Designerly ways of knowing.* London: Springer.

Cross, N., Christiaans, H., & Dorst, K. (1996). *Analysing design activity.* Chichester, U.K.: John Wiley & Sons.

Cross, N., & Cross, A. (1996). Observations of teamwork and social processes in design. In N. Cross, H. Christiaans, & K. Dorst (Eds.), *Analyzing design activity* (pp. 291–318). Chichester, U.K.: John Wiley & Sons.

Daly, S. (2008). *Design across disciplines,* Ph.D. dissertation, Purdue University.

Daly, S., Mann, L., & Adams, R. S. (2008). *A new direction for engineering education research: Unique phenomenographic results that impact big picture understandings.* Paper presented at the Australasian Association for Engineering Education, Yeppoon, Queensland, Australia.

Darke, J. (1979). The primary generator and the design process. *Design Studies,* 1(1), 36–44.

Davis, D. C., Gerlick, R., Trevisan, M. S., & Brown, S. A. (2011). *Establishing interrater agreement for TIDEE's teamwork and professional development assessments.* Paper presented at the 2011 American Society for Engineering Education Annual Conference & Exposition, Vancouver, BC, Canada.

DeShon, R. P., Kozlowski, S. W. J., Schmidt, A. M., Milner, K. R., & Wiechmann, D. (2004). A multiple-goal, multilevel model of feedback effects on the regulation of individual and team performance. *Journal of Applied Psychology,* 89(6), 1035–1056.

Dorst, K., & Cross, N. (2001). Creativity in the design process: Co-evolution of problem-solution. *Design Studies,* 22(5), 425–437.

Downey, G., & Lucena, J. (2003). When students resist: Ethnography of a senior design experience in engineering education. *International Journal of Engineering Education,* 19(1), 168–176.

Du, X., de Graaff, E., & Kolmos, A. (Eds.). (2009). *Research on PBL practice in engineering education.* Boston, MA: Sense.

Dubberly, H. (2004). How do you design? A compendium of models. San Francisco, CA. Retrieved from http://www .dubberly.com/wp-content/uploads/2008/06/ ddo_designprocess.pdf

Dutson, A. J., Todd, R. H., Magleby, S. P., & Sorenson, C. D. (1997, January). A review of literature on teaching engineering design through project-oriented capstone courses. *Journal of Engineering Education,* 17–28.

Dweck, C. S., & Leggett, E. L. (1988). A social cognitive approach to motivation and personality. *Psychological Review,* 95(2), 256–273.

Dym, C. L., Agogino, A. M., Eris, O., Frey, D. D., & Leifer, L. (2005). Engineering design thinking, teaching, and learning. *Journal of Engineering Education,* 94(1), 103–120.

Dym, C. L., & Little, L. (2009). *Engineering design: A project-based introduction* (2nd ed.). New York, NY: John Wiley & Sons.

Ehrenfeld, J. R. (2008). *Sustainability by design.* New Haven, CT: Yale University Press.

Eliot, M., & Turns, J. (2011). Constructing professional portfolios: Sense-making and professional identity development for engineering undergraduates. *Journal of Engineering Education,* 100(4), 630–654.

Eris, O. (2004). *Effective inquiry for innovative engineering design.* Dordrecht, The Netherlands: Kluwer Academic.

Finke, R. A., Ward, T. B., & Smith, S. M. (1992). *Creative cognition: Theory, research, and applications.* Cambridge, MA: MIT Press.

Garzca, L., Palou, E., Lopez-Malo, A., & Garibay, J. M. (2009). *Ethnography of a first-year design experience in the introduction to engineering design course.* Paper presented at the Frontiers in Education Conference, San Antonio, TX.

Gerber, E., McKenna, A., Hirsch, P., & Yarnoff, C. (2010). Learning to waste and wasting to learn? How to use cradle to cradle principles to improve the teaching of design. *International Journal of Engineering Education,* 28(2), 314–323.

Gero, J. (1990). Design prototypes: A knowledge representation schema for design. *AI Magazine,* 11, 26–36.

Gero, J. S. (1996). Creativity, emergence and evolution in design. *Knowledge-Based Systems,* 9(7), 435–448.

Goldschmidt, G. (2003). Cognitive economy in design reasoning. In U. Lindemann (Ed.), *Human behavior in design* (pp. 53–62). New York, NY: Springer.

Green, G., & Smrcek, L. (2006). On the developing role of physical models in engineering design education. *European Journal of Engineering Education,* 31(2), 191–200.

Hadgraft, R. G., Carew, A. L., Sardine, A., Blundell, T., & Blundell, D. (2008). *Teaching and assessing systems thinking in engineering.* Paper presented at the Research in Engineering Education Symposium, Davos, Switzerland.

Hatamura, Y. (2006). *Decision-making in engineering design: Theory and practice.* London: Springer.

Hatchuel, A., Le Masson, P., & Weil, B. (2011). Teaching innovative design reasoning: How C-K theory can help to overcome fixation effect. *Artificial Intelligence for Engineering Design, Analysis and Manufacturing, 25*(1), 77–92.

Hazelrigg, G. A. (1998). A framework for decision-based engineering design. *Journal of Mechanical Design, 120*(4), 653–658.

Heywood, J. (2005). *Engineering education: Research and development in curriculum and instruction.* Hoboken, NJ: John Wiley & Sons.

Hirsch, P. L., & McKenna, A. F. (2008). Using reflections to promote teamwork understanding in engineering design education. *International Journal of Engineering Education, 22*(2), 377–385.

Hmelo-Silver, C. E. (2004). Problem-based learning: What and how do students learn? *Educational Psychology Review, 16*(3), 235–266.

Howe, S., & Wilbarger, J. (2005). 2005 *National survey of engineering capstone design courses.* Paper presented at the Proceedings of the 2006 American Engineering Education Annual Conference and Exposition, Chicago, IL.

Hsu, M. C., Cardella, M., Purzer, Ş., & Diaz, N. M. (2010). *Assessing elementary teachers' perceptions of engineering and familiarity with design, engineering and technology: Implications on teacher professional development.* Paper presented at the 2010 American Society for Engineering Education Annual Conference & Exposition, Louisville, KY.

Hubka, V. (1982). *Principles of engineering design* (W. E. Eder, Trans.). Guildford, U.K.: Butterworth Scientific.

Hyman, B., Khanna, S., Lin, Y., & Borgford-Parnell, J. (2011). *A case study of using capstone design as basis for curriculum-wide project based learning.* Paper presented at the ASME 2011 International Mechanical Engineering Congress & Exhibition, Denver, CO.

Johnson, D. W., Johnson, R. T., & Smith, K. A. (1998). *Active learning: Cooperation in the college classroom.* Edna, MN: Interaction Book Company.

Jonassen, D., Strobel, J., & Lee, C. B. (2006). Everyday problem solving in engineering: Lessons for engineering educators. *Journal of Engineering Education, 95*(2), 139–151.

Jones, J. C. (1970). *Design methods: Seeds of human futures.* Chichester, U.K.: John Wiley & Sons.

Jones, J. C. (1991). *Designing designing.* London: Architecture Design and Technology Press.

Jones, T. R., Cardella, M. E., & Purzer, Ş. (2012). *Using and comparing paper and media to improve student reflection in science and design courses.* Paper presented at the National Association of Research on Science Teaching Annual International Conference, Indianapolis, IN.

Kan, J., & Gero, J. (2009). Using the FBS ontology to capture semantic design information in design protocol studies. In J. McDonell & P. Lloyd (Eds.), *About: Designing – analysing design meetings* (pp. 213–229). London: Taylor & Francis.

Kilgore, D. E., Atman, C. J., Yasuhara, K., Barker, T., & Morozov, A. (2007, October). Considering context: A study of first-year engineering students. *Journal of Engineering Education,* 321–334.

Kilgore, D. E., Jocuns, A., Yasuhara, K., & Atman, C. J. (2010). From beginning to end: How engineering students think and talk about sustainability across the life cycle. *International Journal of Engineering Education, 26*(2), 305–313.

Kolb, D. A. (1984). *Experiential learning: Experience as the source of learning and development.* Englewood Cliffs, NJ: Prentice-Hall.

Krippendorff, K. (2006). *The semantic turn: A new foundation for design.* Boca Raton, FL: CRC Taylor & Francis.

Laeser, M., Moskal, B. M., Knecht, R., & Lasich, D. (2003). Engineering design: Examining the impact of gender and the team's gender composition. *Journal of Engineering Education, 92*(1), 49–56.

Lau, A. (2010). Sustainable design: A new paradigm for engineering education. *International Journal of Engineering Education, 26*(2), 252–259.

Lave, J., & Wenger, E. (1991). *Situated learning: Legitimate peripheral participation.* Cambridge: Cambridge University Press.

Lawson, B. (1980). *How designers think.* London & Westfield, NJ: Architectural Press; Eastview Editions.

Lawson, B. (1997). *How designers think: The design process demystified* (completely rev. 3rd ed.). Oxford and Boston: Architectural Press.

Le Dantec, C. A. (2010). Situating design as social creation and cultural cognition. *CoDesign*, 6(4), 207–224.

Linder, B., Somerville, M., Eris, O., & Tatar, N. (2010). *Taking one for the team: Goal orientation and gender-correlated task division*. Paper presented at the ASEE/IEEE Frontiers in Education Conference, Arlington, VA.

Little, P., & Cardenas, M. (2001). Use of "studio" methods in the introductory engineering design curriculum. *Journal of Engineering Education*, 90(3), 309–318.

Lloyd, P. (2000). Storytelling and the development of discourse in the engineering design process. *Design Studies*, 21(4), 357–373.

Luxhoj, J. T., & Hansen, P. H. K. (1996). Engineering curriculum reform at Aalborg University. *Journal of Engineering Education*, 85(3), 183–186.

Maguire, M. (2001). Methods to support human-centered design. *International Journal of Human-Computer Interaction*, 55(4), 587–634.

Mann, L., Dall'Alba, G., & Radcliffe, D. (2007). *Using phenomenography to investigate different ways of experiencing sustainable design*. Paper presented at the American Society for Engineering Education Annual Conference & Exposition, Honolulu, HI.

Marples, D. L. (1961). The decisions of engineering design. *IRE Transactions on Engineering Management*, EM-8(2), 55–71.

Marra, R. M., Palmer, B., & Litzinger, T. A. (2000). The effects of a first-year engineering design course on student intellectual development as measured by the Perry Scheme. *Journal of Engineering Education*, 89(1), 39–46.

Marton, F., & Booth, S. (1997). *Learning and awareness*. Mahwah, NJ: Lawrence Erlbaum.

McDonnell, J. (1997). Descriptive models for interpreting design. *Design Studies*, 18(4), 457–473.

McDonnell, J., & Lloyd, P. (2009). *About designing: Analysing design meetings*. London: Taylor & Francis.

McDonnell, J., Lloyd, P., & Valkenburg, R. C. (2004). Developing design expertise through the construction of video stories. *Design Studies*, 25, 509–525.

McKenna, A. F., Colgate, J. E., Carr, S. H., & Olson, G. B. (2006). IDEA: Formalizing the foundation for an engineering design education. *International Journal of Engineering Education*, 22(3), 671–678.

McKenzie, L. J., Trevisan, M. S., Davis, D. C., & Beyerlein S. W. (2004). *Capstone design courses and assessment: A national study*. In *Proceedings of the 2004 American Society for Engineering Education Annual Conference and Exposition*, Salt Lake City, UT.

McMartin, F., McKenna, A., & Youssefi, K. (2000). Scenario assignments as assessment tools for undergraduate engineering education. *IEEE Transactions on Education*, 43(2), 111–119.

Mello, N. A. (2000). *How can universities provide global perspective for engineers? One institution's solution*. Paper presented at the American Society for Engineering Education Conference, St. Louis, MO.

Melton, R. B., Cardella, M. E., Oakes, W. C., & Zoltowski, C. B. (2012). *Development of a design task to assess students' understanding of human-centered design*. Paper presented at the 42nd Annual ASEE/IEEE Frontiers in Education Conference, Seattle, WA.

Mentzer, N., Becker, K., & Park, K. (2011). *High school students as novice designers*. Paper presented at the American Society for Engineering Education, Vancouver, BC, Canada.

Michaelsen, L. K., Knight, A. B., & Fink, L. D. (2004). *Team-based learning*. Sterling, VA: Stylus.

Mills, J. E., & Treagust, D. F. (2003). Engineering education – is problem-based or project-based learning the answer? *Australasian Journal of Engineering Education*. Retrieved from http://www.aaee.com.au/journal/2003/mills_treagust03.pdf

Minneman, S. L. (1991). *The social construction of a technical reality: Empirical studies of group engineering design practice*. Ph.D. dissertation, Stanford University.

Molhave, A., McDonnell, J., & Atman, C. J. (2011). Seeing and hearing: Elucidating the design process – A workshop at Central Saint Matins College of Art and Design *CELT Technical Report* (Vol. CELT – 11–03): University of Washington (contact celtad@uw.edu).

Mosborg, S., Adams, R. S., Kim, R., Atman, C. J., Turns, J., & Cardella, M. (2005). *Conceptions of the engineering design process: An expert study of advanced practicing professionals.* Paper presented at the American Society of Engineering Education Conference, Portland, OR.

Neeley, W. L., Sheppard, S. D., & Leifer, L. G. (2006). *Design is design is design (or is it?) What we say vs. what we do in engineering education.* Paper presented at the American Society for Engineering Education Annual Conference, Chicago, IL.

Nefcy, E. J., Gummer, E. S., & Koretsky, M. D. (2012). *Characterization of student modeling in an industrially situated virtual laboratory.* Paper presented at the 2012 American Society for Engineering Education, Annual Conference and Exposition, San Antonio, TX.

Nersessian, N. J., Newstetter, W. C., Kurz-Milcke, E., & Davies, J. (2002). *A mixed-method approach to studying distributed cognition in evolving environments.* Paper presented at the International Conference on Learning Sciences, Seattle, WA.

Newstetter, W. C. (1998). Of green monkeys and failed affordances: A case study of a mechanical engineering design course. *Research in Engineering Design,* 10(2), 118–128.

Newstetter, W. C., & McCracken, W. M. (2001). Novice conceptions of design: Implications for the design of learning environments. In *Design learning and knowing: Cognition in design education* (pp. 63–78). New York, NY: Elsevier.

Okudan, G. E., & Mohammed, S. (2006). Task gender orientation perceptions by novice designers: Implications for engineering design research, teaching and practice. *Design Studies,* 27(6), 723–740.

Otto, N. K., & Wood, L. K. (2001). *Product design: Techniques in reverse engineering and new product development.* Upper Saddle River, NJ: Prentice-Hall.

Pahl, G. (1997). *How and why collaboration with cognitive psychologists began.* Paper presented at the Designers: The Key to Successful Product Development, Darmstadt Symposium, Darmstadt, Germany.

Pahl, G., & Beitz, W. (1996). *Engineering design: A systematic approach.* London: Springer-Verlag.

Palmer, B., Terenzini, P. T., McKenna, A. F., Harper, B. J., & Merson, D. (2011). *Design in context: Where do the engineers of 2020 learn this skill?* Paper presented at the American Society of Engineering Education Annual Conference, Vancouver, BC, Canada.

Paretti, M. (2008). Teaching communication in capstone design: The role of the instructor in situated learning. *Journal of Engineering Education,* 97(4), 491–503.

Park, K., Pieper, J., Mentzer, N., & Becker, K. (2010). *Exploring engineering design knowing and thinking as an innovation in STEM learning.* Paper presented at the P12 Engineering and Design Education Research Summit, Seaside, OR.

Pembridge, J., & Paretti, M. (2010). *The current state of capstone design pedagogy.* Paper presented at the Proceedings of the 2010 American Society for Engineering Education Annual Conference and Exposition, Louisville, KY.

Purzer, Ş., Chen, J., & Yaday, A. (2010). *Does context matter? Engineering students' approaches to global vs local problems.* Paper presented at the Frontiers in Education Conference, Arlington, VA.

Purzer, Ş., & Hilpert, J. C. (2011). *Special session: Cognitive processes critical for ill-defined problem solving: Linking theory, research, and classroom implications.* Paper presented at the 41st ASEE/IEEE Frontiers in Education Conference, Rapid City, SD.

Purzer, Ş. Y. (2011). The relationship between team discourse, self-efficacy, and individual achievement: A sequential mixed-methods study. *Journal of Engineering Education,* 100(4), 655–679.

Rittel, H. W. J., & Webber, M. M. (1973). Dilemmas in a general theory of planning. *Policy Sciences,* 4(2), 155–169.

Roberts, C., Yasar, S., Morrell, D., Henderson, M., Danielson, S., & Cooke, J. (2007a). *A pilot study of engineering design teams using protocol analysis.* Paper presented at the American Society for Engineering Education Conference, Honolulu, HI.

Roberts, C., Yasar, S., Morrell, D., Henderson, M., Danielson, S., & Cooke, N. (2007b). *A pilot study of engineering design teams using protocol analysis.* Paper presented at the Proceedings of the 2007 ASEE Annual Conference, Honolulu, HI.

Roozenburgh, N., & Eekels, J. (1995). *Product design: Fundamentals and methods.* New York, NY: John Wiley & Sons.

Rowe, P. G. (1987). *Design thinking.* Cambridge, MA: MIT Press.

Safoutin, M. J., Atman, C. J., & Adams, R. S. (2000a). The design attribute framework *CELT*

Technical Report 99–01. Seattle, WA: University of Washington (contact celtad@uw.edu).

Safoutin, M. J., Atman, C. J., Adams, R. S., Rutat, T., Kramlich, J. C., & Firidley, J. L. (2000b). A design attribute framework for course planning and learning assessment. *IEEE Transactions on Education*, 43(2), 188–199.

Schön, D. A. (1983). *The reflective practitioner: How professionals think in action.* New York, NY: Basic Books.

Sheppard, S., Macatangay, K., Colby, A., & Sullivan, W. M. (2009). *Educating engineers: Designing for the future of the field.* San Francisco, CA: Jossey-Bass.

Simon, H. A. (1996). *The sciences of the artificial* (3rd ed). Cambridge, MA: The MIT Press.

Smith, K. A., Sheppard, S. D., Johnson, D. W., & Johnson, R. T. (2005). Pedagogies of engagement: Classroom-based practices. *Journal of Engineering Education*, 94(1), 87–101.

Sobek, D., II. (2002). *Preliminary findings from coding student design journals.* Paper presented at the American Society for Engineering Education Conference, Montreal, Canada.

Somerville, M., Anderson, D., Berbeco, H., Bourne, J. R., Crisman, J., Dabby, D., & Zastavker, Y. (2005). The Olin College curriculum: Thinking toward the future. *IEEE Transactions on Education*, 48(1), 198–205.

Stacey, M., Eckert, C., & Earl, C. (2009). From Ronchamp by sledge: On the pragmatics of object references. In J. McDonell & P. Lloyd (Eds.), *About: Designing – analysing design meetings* (pp. 361–379). London: Taylor & Francis.

Suh, N. P. (1990). *The principles of design.* Oxford: Oxford University Press.

Todd, R. H., Magleby, S. P., Sorenson, C. D., Swan, B. R., & Anthony, D. K. (1995, April). A survey of capstone engineering courses in North America. *Journal of Engineering Education*, 165–174.

Ullman, G. D. (1992). *The mechanical design process.* New York, NY: McGraw-Hill.

Ulrich, K. T., & Eppinger, S. D. (1995). *Product design and development.* New York, NY: McGraw-Hill.

Vanasupa, L., Burton, R., Stolk, J., Zimmerman, J. B., Leifer, L., & Anastas, P. T. (2010). The systemic correlation between mental modes and sustainable design: Implications for engineering educators. *International Journal of Engineering Education*, 26(2), 428–450.

Visser, W. (1992). Designers activities examined at 3 levels – Organization, strategies and problem-solving processes. *Knowledge-Based Systems*, 5(1), 92–104.

Visser, W. (2006). *The cognitive artifacts of designing.* Mahwah, NJ: Lawrence Erlbaum.

Williams, C. B., Gero, J., Lee, Y., & Paretti, M. (2011). *Exploring the effect of design education on the design cognition of mechanical engineering students.* Paper presented at the ASME 2011 International Design Engineering Technical Conference & Computers and Information in Engineering Conference (IDETC/CIE 2011), Washington, DC.

Wilson, J. M., & Jennings, W. C. (2000). *Studio courses: How Information technology is changing the way we teach, on campus and off.* Paper presented at the IEEE Conference.

Yasar-Purzer, Ş., Henderson, M., McKay, A., Roberts, C., & de Pennington, A. (2008). *Comparing the design problem solving processes of product design and engineering student teams in the US and UK.* Paper presented at the American Society for Engineering Education Conference, Pittsburgh, PA.

Zoltowski, C. B. (2010). *Students' ways of experiencing human-centered design.* Ph.D. dissertation, Purdue University.

Zoltowski, C. B., Cardella, M., & Oakes, W. C. (2011b). *The development of assessment tools using phenomenography.* Paper presented at the Research in Engineering Education Symposium, Madrid, Spain.

Zoltowski, C. B., Cardella, M. E., & Oakes, W. C. (2011a). *Phenomenographic study of human-centered design: Educational implications.* Paper presented at the 2011 American Society for Engineering Education, Annual Conference & Exposition, Vancouver, BC, Canada.

Zoltowski, C. B., Oakes, W. C., & Cardella, M. E. (2012). Students' ways of experiencing human-centered design. *Journal of Engineering Education*, 101(1), 28–59.

CHAPTER 12

Adaptive Expertise and Knowledge Fluency in Design and Innovation

Ann F. McKenna

Introduction

This chapter presents an overview of the adaptive expertise framework and describes how this framework holds promise for exploring knowledge fluency in the context of engineering design and innovation. Adaptive expertise is linked to research on knowledge transfer and the development of expertise. Each of these theoretical approaches explores how one recognizes when knowledge learned in one setting may apply in a novel setting, and the mechanisms for transfer of knowledge when challenged by an unfamiliar or ambiguous situation.

The adaptive expertise framework is directly applicable in the context of design and innovation given the emphasis on how one develops adaptiveness in learning, and how to apply knowledge fluidly. The process of design and innovation involves developing a solution where one does not yet exist. From this perspective, every design situation is novel, embeds ambiguity, and has no one correct answer. The very nature of design requires one to recognize how prior knowledge might apply under new circumstances. Furthermore, research suggests that the nature of knowledge one employs in the process of design innovation is nuanced, complex, and spans the cognitive, metacognitve, and affective domains.

The chapter provides an overview of how adaptive expertise and the concept of innovation relate to the literature on knowledge transfer, and the mechanisms for scaffolding knowledge transfer in the context of adaptive expertise. The framework of adaptive expertise has been applied in a variety of educational contexts both at the pre-college and postsecondary levels, and in studies of practicing professionals. These studies of students, instructors, and professionals represent a variety of lenses for exploring adaptive expertise from the perspective of instructional and professional practices, assessment approaches, and developing different subject matter expertise. The chapter situates the research on adaptive expertise specifically in the context of engineering design and innovation and explores how one might interpret the framework in this context. Finally, the chapter poses

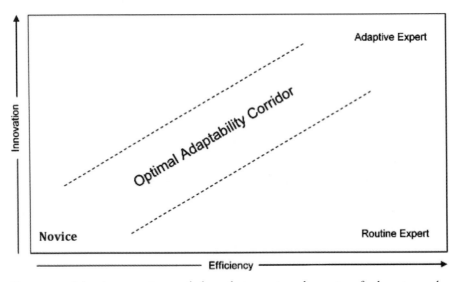

Figure 12.1. Adaptive expertise as a balance between two dimensions for learning and assessment: Efficiency and innovation.

further areas of investigation that could provide new insights into characterizing adaptive expertise as it relates to engineering design.

Adaptive Expertise: An Overview

The concept of adaptive expertise was introduced to extend our understanding of the meaning of expertise, as well as a way of thinking about how to prepare learners to respond flexibly to new learning situations. Hatano and Inagaki (1986) contrasted two types of expertise: routine and adaptive. They claim, "routine experts are outstanding in speed, accuracy, and automaticity of performance but lack flexibility and adaptability to new problems" (p. 266). Hatano and Oura (2003) explained that the majority of studies on expertise "have shown that experts, who have had many years of problem-solving experiences in a given domain, can solve familiar types of problems quickly and accurately, but often fail to go beyond procedural efficiency" (p. 28). In contrast, adaptive experts can go beyond procedural efficiency and "can be characterized by their flexibility, innovative, and creative competencies within the domain" (p. 28).

Schwartz, Bransford, and Sears (2005) proposed that adaptive expertise emerges from a balance between efficient use of knowledge and the innovation skills associated with accessing prior knowledge as well as generating new ideas and new knowledge. The framework defines adaptive expertise on two axes representing efficiency and innovation (see Figure 12.1). The efficiency scale indicates an individual's competence to apply domain-specific knowledge and skills fluently to complete activities he or she has significant experience performing. Individuals' ability to *replicate* their performance with increased accuracy and speed improves as they gain more experience. Novices can accurately perform on only a small set of problems, so they are on the lower end of the scale. Experts have a wider range of experience and therefore solve a larger class of problems quickly and accurately.

The innovation scale is presented in contrast to the emphasis on speed, accuracy, and replication of the efficiency dimension. Innovation refers to achieving learning in a challenging context or to generating knowledge to achieve a novel and appropriate goal. New knowledge can improve on old ideas or identify completely new directions for approaching one's goal. Therefore, part of the innovation axis relates to inquiry and self-regulating skills necessary to identify and comprehend a problem, identify what additional knowledge is necessary, and generate ideas and leverage existing knowledge to facilitate recognition of relevant information. The innovation scale relates to the skills used to recognize and *apply* the knowledge along the efficiency scale to support learning in a challenging context or to generate new knowledge in novel contexts. There has been some discussion as to how to characterize one who is high on the innovation scale but low in efficiency. Schwartz et al. (2005) have suggested the term "frustrated novice." The idea is that one may be generative with his or her knowledge; however, the knowledge innovation can be based on flawed interpretations or wrong assumptions due to limited domain-specific knowledge of the context, hence leading to "frustration."

Figure 12.1 includes an "optimal adaptability corridor" (OAC). As described by Schwartz et al. (2005), the function of the OAC is to ensure that innovation and efficiency develop together. One purpose of the OAC is to remind educators of the importance of these two dimensions and as a framework for gauging or developing instructional experiences. However, there are possibly many different trajectories one might take to navigate to the goal of reaching adaptive expertise.

Many studies have been conducted to examine differences between experts and novices. For example, the literature reports that experts (1) solve problems faster and more accurately than novices (e.g., Larkin, McDermott, Simon, & Simon, 1980), (2) have well-organized knowledge structures (e.g., Bruer, 1993), and (3) notice meaningful patterns of information and can flexibly retrieve important aspects of their knowledge with little attentional effort (Bransford, Brown, & Cocking, 2000). Expert–novice studies have led educators to a more formal understanding of the differences in problem-solving approaches and cognitive abilities between experts and novices in a variety of domains (e.g., Chi, Feltovich, & Glaser, 1981; Patel & Groen, 1991; Scardamalia & Bereiter, 1991).

The concept of adaptive expertise builds on the expert–novice research literature by providing a new perspective on the nature of knowledge that is transferred among different learning activities that allow for solutions to new problems. Transfer studies often produce disappointing results because, as Bransford and Schwartz (1999) argue, they occur in settings that do not allow for testing new ideas and revising as necessary, and transfer studies too narrowly focus on measuring knowledge used in a replicative way. Schwartz et al. (2005) state: "for many new situations, people do not have sufficient memories, schemas, or procedures to solve a problem, but they do have interpretations that shape how they begin to make sense of the situation" (p. 9). How one interprets new situations and how one frames a problem "has major effects on subsequent thinking and cognitive processing" (p. 9). The focus on transfer and more specifically on the type of knowledge that gets transferred into *and out of* situations sets the stage for research on the concept of adaptive expertise.

Innovation in the Context of Knowledge Transfer and Adaptive Expertise

Of central importance to understanding the process of knowledge innovation are transfer mechanisms (e.g., Anderson, Corbett, Koedinger, & Pelletier, 1995; De Corte, 1999; Marini & Genereux, 1995; Reed, 1993). Detterman and Sternberg (1993) defined transfer as the degree to which previously acquired knowledge will be applied in a new situation. However, as pointed out by Schwartz, Varma, and Martin (2008), the idea of simply replicating a behavior in a new environment is quite different from innovating a completely new conceptual structure based on prior knowledge. The recognized need for a broader understanding of transfer has caused some researchers to propose new ways in which transfer might be defined (e.g., Barnett & Ceci, 2002; Broudy, 1977; Butterfield & Nelson, 1991) and methodologically approached (e.g., Fortus, Krajcik, Dershimer, Marx, & Mamlok-Naaman, 2005; Mayer, 1999; Novick, 1988). For example, some have suggested studying students' abilities to achieve new learning, guided by previously acquired knowledge, as a mechanism for understanding transfer for innovation (Bransford & Schwartz, 1999; Chi et al., 1981; Greeno, Smith, & Moore, 1993; Wineburg, 1998).

Bransford and Schwartz (1999) proposed focusing on students' ability to pick up on and learn from the key components required for adaptation to and performance in an information-rich environment. They point out:

> When organizations hire new employees, they do not expect them to have learned everything they need for successful adaptation. They want people who can learn, and they expect them to make use of resources (e.g., texts, computer programs, colleagues) to facilitate this learning. The better prepared they are for future learning, the greater the transfer (in terms of speed and/or quality of new learning). (p. 68)

This view leads to what Bransford and Shwartz (1999) refer to as "preparation for future learning" (PFL). Those who advocate examining transfer from this perspective recognize the importance assessment plays in documenting learning and recommend that great care be placed on crafting assessments to align with desired outcomes. A consequence of redesigning assessment to emphasize PFL places increased emphasis on creating environments in which students become better equipped to recognize when their knowledge applies, what they might need to learn more about to apply their knowledge correctly, and where they can find the information they need to perform in their present environment. This is particularly relevant in the domain of engineering, where design problems by their nature are ambiguous and complex, and almost always require knowledge integration from a range of sources, disciplines, and perspectives. Preparing students to recognize when different knowledge applies, in the appropriate context, is at the core of preparing students for future success.

Understanding how educators might promote both the efficient use of disciplinary knowledge (i.e., traditionally defined transfer) as well as the ability to use that knowledge to learn about and learn in new situations (i.e., PFL view of transfer) is the basis for the adaptive expertise framework.

Studies that have approached transfer and learning from the perspective of the adaptive expertise framework have routinely suggested that students can best be prepared for future knowledge transfer by early opportunities to be inventive (Schwartz, Chang, & Martin, 2008; Schwartz & Martin, 2004). For example, in a study on teaching statistics to ninth-grade students, researchers tested whether or not initial attempts by students to "invent" statistical relationships from provided data would better prepare them for learning from a lecture on related ideas (Schwartz & Martin, 2004). The study found students who had taken part in the "Inventing to Prepare for Learning (IPL)" activities were better prepared to learn from the lecture and subsequently performed better when tested on the statistical concepts. Within the domain of engineering, PFL-like activities have also been demonstrated to promote the development of characteristics common among adaptive experts (e.g., Pandy, Petrosino, Austin, & Barr, 2004; Rayne, Martin, Brophy, & Diller, 2006; Walker, Corday, King, & Brophy, 2006).

Others have applied adaptive expertise to a range of activities such as analyzing historical texts (Wineburg, 1998), avionic device troubleshooting (Gott, Hall, Pokorny, Dibble, & Glaser, 1992), the practice of teaching (Crawford, 2007), and business expertise in the context of restaurant management (Barnett & Koslowski, 2002). Gott et al. (1992) make a distinction of *intentional* transfer; that is, the need for transfer is apparent to learners. This distinction has implications for how one might interpret particular behaviors when observing individuals performing a task. Specifically, observing individuals as they engage in intentional transfer suggests that one is actually attempting transfer. The focus is not on whether the attempt is being made, but rather, *how* one is attempting to do so. Crawford (2007) suggests that in her observations of teachers' practices "adaptiveness is not an intrinsic trait but a set of cognitive and self-regulative skills and abilities, as well as habits of mind and dispositions" (p. 7). Each of these studies provides additional insights on how one might frame studies of adaptive expertise in different contexts.

Despite having a more narrow focus, the traditional transfer literature is still quite useful to understanding why people use and modify previously learned concepts in some contexts but not in others (Schwartz et al., 2008b). Transfer research has addressed key questions such as (1) how individuals use prior knowledge to make innovative new knowledge (Chi & Ohlsson, 2005; Salomon & Perkins, 1989); (2) how new concepts emerge from prior learning (Carraher & Schliemann, 2002; Chi & Ohlsson, 2005; diSessa & Wagner, 2005; Schwartz et al., 2008b); (3) why knowledge is transferred in some cases, but not in others (Bransford et al., 2000; Chi & Ohlsson, 2005; Di Vesta & Peverly, 1984; Kimball & Holyoak, 2000; Mayer, 1974); and (4) why individuals sometimes transfer prior learning to support change, and in other instances transfer routines that prevent change (Chi, 2006; Hatano & Oura, 2003; Salomon & Perkins, 1989). Answers to these questions, with ties to transfer research, and by extension adaptive expertise, are all important to developing an understanding of how to support individuals in recognizing how and when prior knowledge applies in a new setting.

Some have made distinctions between two different types of transfer: similarity transfer and dynamic transfer. Similarity transfer occurs when people use prior knowledge to help solve a novel situation. One type of similarity transfer involves recognition of similar surface features between two problems (often referred to as near transfer or surface feature recognition), while another type involves the recognition

of more structural or relational similarities (also known as far transfer or deep feature recognition).

Studies of analogical transfer, one form of similarity transfer, suggest it may be particularly useful in helping learners achieve innovative solutions (Gick & Holyoak, 1983; Reed, Ernst, & Banerji, 1974). Analogical transfer occurs when an analogy based on a well-understood concept is used to propose an explanation for a new situation. Students who are able to construct analogies on their own are said to have achieved spontaneous transfer while analogies provided by instructors to assist student learning have been called prompted transfer (Schwartz et al., 2008b). Both of these are useful; however, their distinction is important to the discussion of designing interventions that can help students achieve knowledge innovation.

Dynamic transfer occurs when environments or contexts help people coordinate component abilities to create a novel concept. Instead of statically transferring knowledge from one task to a new situation, dynamic transfer requires iterative interactions between the learner and the learning situation. Numerous studies have documented the existence of transfer as a dynamic process (e.g., Lobato, 2006). This can involve transforming the transfer task so that it can be considered from a new perspective (Cobb & Bowers, 1999; Greeno et al., 1993), resituating current conceptual knowledge (Carraher & Schliemann, 2002), or revisiting prior knowledge in light of a novel problem (Rebello et al., 2005; Schwartz & Martin, 2004).

Based on these definitions, several likely paths to knowledge innovation have been proposed. Schwartz and colleagues (2008b) refer to fault-driven transfer, in which individuals encounter a difficulty (or fault), which then drives the search for an innovation process to address the fault At this point they may fail to adapt their prior knowledge to fit the new situation and, unable to move forward, they will disengage from the problem-solving process. Alternatively, on encountering this obstacle, one might also create an innovative analogy to account for the new situation, achieving a far transfer leap toward innovation.

Innovation might also be achieved through dynamic transfer whereby a number of interactions with the problem-solving environment may lead one to transfer several pieces of prior knowledge to the new situation. Through these interactions with and manipulations of the environment, a coordination of previously acquired concepts is eventually constructed that deliver one to an innovation. This route to innovation presents a dip in productivity during interaction with the environment, although it generally proves less challenging than coming up with the necessary innovative leap required to follow the path of fault-driven transfer.

Scaffolding of Knowledge Innovation

Descriptions of the dynamic transfer process emphasize the importance of interaction between the learner and the problem-solving environment. This interaction requires that the learner be learning from and about the problem while continually reflecting on, and possibly reshaping, prior knowledge and experiences. Literature on scaffolding the learning process describes scaffolding as a support that " . . . not only assists learners in accomplishing tasks, but also enables them to learn from the experience" and emphasizes its usefulness in educational settings where " . . . it is important to stress the dual aspects of both (a) accomplishing the tasks and (b) learning from one's efforts, that is, improving one's performance on the future tasks in the process" (Reiser, 2004, p. 275). Similarly, Hogan, Nastasi, and Pressley (1999) point out that scaffolding involves both the support of learners as they develop solutions, as well as stimulation to engage actively in the learning process. This similar focus on learning from problem-solving experiences suggests the use of scaffolding in efforts to promote the use of transfer for innovation by students.

Traditionally, scaffolding has been used to describe the process by which a teacher,

mentor, or more knowledgeable person helps a learner achieve a task that would otherwise be out of reach (Vygotsky, 1978; Wood, Bruner, & Ross, 1976). Scaffolding may take many different forms including reductions in degrees of freedom, accentuation of relevant task features, or modeling solutions (Wood et al., 1976). Although the metaphor of scaffolding traditionally has been interactions between learners and more knowledgeable peers or adults, it has been extended to include the idea that representations and physical artifacts – including, for example, paper artifacts (Davis & Miyake, 2004; Hall, Kibler, Wegner, & Truxaw, 1989) and computer software (Brush & Saye, 2001; Davis & Linn, 2000; Linn, 2000) – can also act as scaffolds. As learners become advanced, the scaffolding is slowly removed, a process known as fading (Collins, Brown, & Neuman, 1989).

There is a need for more study of the less traditional forms of scaffolding because the traditional view of scaffolding has been extended beyond learner–mentor interactions (Davis & Miyake, 2004). Reiser (2004) has defined two mechanisms of scaffolding, structuring and problematizing. Structuring guides the learner through key components of the problem, supports planning and carrying out of the solution, and prompts recognition of important considerations. In contrast, problematizing shapes performance and understanding in terms of key disciplinary knowledge and strategies; forces engagement with complex solutions, making the problem-solving process a more productive learning opportunity; and provokes learners to devote resources to issues the might not otherwise address. Problematizing is important because it creates cognitive conflict (Webb, 1996) in the mind of the learner.

Relating to engineering design in particular, others have found that embedding specific details in the problem-solving statement can serve as a type of scaffolding that supports the dynamic transfer process (Hutchison & McKenna, 2008b). In the context of an engineering design problem detailed information can be included in the form of a design specification or constraint, that is, how quickly, and to what temperature one must cool a device. Such design requirements can help learners recognize what knowledge may apply in the situation and helps them frame more detailed searches for information.

Innovation in the Context of Engineering Design

Up to this point, the chapter has focused on knowledge innovation. There is a relationship between *knowledge* innovation and the notion of *design* innovation as typically used in the engineering or product design communities. In the design and engineering communities, design innovation often refers to the tangible product, process, or system that gets produced as a result of identifying a need and working through a process to develop a unique solution to that need. However, in addition to generating a tangible product, one generates knowledge in the process of design. This section situates innovation from an engineering design perspective and makes the link between *knowledge* innovation and *design* innovation.

One succinct attempt at making the link is to suggest that knowledge innovation occurs as one engages in the process of design innovation. That is, knowledge innovation is a requirement for developing design innovations; therefore, by virtue of being a design innovator, one is also an adaptive expert. To elucidate this point, it is helpful to provide perspective on what defines engineering as a field of practice and on the nature of design and design problem solving. Several influential educators have asserted that design is the core of engineering (Dym et al., 2005; Simon, 1996) and that experiential learning is the basis for developing meaningful and robust understanding (Dewey, 1938; Kolb, 1984). As Simon (1996) states, "design, so construed, is the core of all professional training . . . schools of engineering are centrally concerned with the process of design" (p. 111). Design requires unique knowledge and skills, or habits of mind,

common to all engineering disciplines, and it is these skills and habits of mind that distinguish engineering as a profession.

Similarly, Dym and colleagues (2005) assert that "engineering design is a systematic, intelligent process in which designers generate, evaluate, and specify concepts for devices, systems, or processes whose form and function achieve clients' objectives or users' needs while satisfying a specified set of constraints" (p. 104). Central to the engineering education community primarily within the United States is Accreditation Board for Engineering and Technology (ABET), which establishes the criteria by which engineering schools are evaluated to receive accreditation. ABET defines design as "the process of devising a system, component, or process to meet desired needs. It is a decision-making process (often iterative), in which the basic sciences, mathematics, and the engineering sciences are applied to convert resources optimally to meet these stated needs" (ABET, 2000).

In addition to defining engineering as a field of practice, and recognizing the central role of design within the field, it is also useful to characterize the nature of design and design problem solving. Design represents a particular class of problem solving that is distinguished by ambiguity, the existence of multiple solutions (as well as multiple problem representations), and a lack of procedural and declarative rules (Cross, 2000; Goel & Pirolli, 1992; Jonassen, Strobel, & Beng Lee, 2006). Design is situated in real contexts, involves social processes (Bucciarelli, 1996), and involves people with different perspectives (designers, non-designers, users, etc.) from different disciplines (within and outside of engineering) working together to solve complex technological problems that address societal needs.

The nature of problem solving in the context of design is innately open ended and ambiguous. That is, there is always more than one solution, and developing the solution requires a diverse knowledge base inclusive of competencies related to mathematical and scientific principles, as well as a sophisticated appreciation of environmental, societal, economical, and other contextual factors as the factors relate to the product or process under consideration. In this way the domain of engineering, in particular how learners engage in the process of developing design solutions, provides a rich opportunity to explore how innovation in learning can be facilitated and how learners begin to make the transition from knowing to applying fundamental theories fluidly. A curricular environment that will produce productive, successful problem solvers in this domain most certainly must include a developed understanding of how to promote innovation and transfer among learners.

Knowledge Fluency in Design: Operationalizing the Efficiency and Innovation Dimensions

One of the challenges in applying the adaptive expertise framework for instructional or research purposes is the lack of specificity for what might characterize "efficiency" and "innovation," as illustrated in Figure 12.1. As one example, the lack of specificity raises challenges with regard to developing metrics such that one might measure adaptive expertise, or even be able to recognize it when it occurs. It is difficult to develop instruments to detect the phenomena or to know how to structure experiences in order to develop adaptive expertise.

The efficiency dimension suggests that there is a knowledge basis that one operates on in the context of knowledge innovation, thereby indicating the development of adaptive expertise is not domain or context independent. Even to be a routine expert one must have a fluency in applying known schemas or procedures to familiar problems. The domain-specific nature of adaptive expertise is evidenced by the many domain-specific studies cited throughout the chapter. That is, one develops adaptive expertise in the context of learning statistics, analyzing historical texts, or developing effective teaching practices. This raises a question regarding the nature of knowledge that is invoked as one engages

in "adaptiveness," specifically, within a particular domain or context (e.g., Martin, Petrosino, Rivale, & Diller, 2007; Martin, Rayne, Kemp, Hart, & Diller, 2005; Verschaffel et al., 2009), and how one might parse and account for knowledge (or efficiency) that could be cognitive, metacognitive, motivational, or otherwise.

The context of engineering design is inherently non-routine, requires process knowledge as well as knowledge of fundamentals, and is often performed in teams, thereby embedding social and interpersonal skills and dispositions. Several studies have aimed to provide more specificity with regard to what may constitute the dimensions of efficiency and innovation in the context of engineering design and problem solving. For example, Pandy et al. (2004) used the How People Learn (HPL) framework (Bransford et al., 2000) and STAR-Legacy cycle (Schwartz et al., 1999) to develop instructional materials that focus on the development of adaptive expertise in biomechanics. To quantify adaptive expertise they introduced a weighted formula that includes factual knowledge, conceptual knowledge, and transfer. The rubric used for scoring responses included specific problem-solving techniques such as writing equations, drawing a diagram, and correctly identifying variables. Their findings as defined by the rubric indicated that HPL approaches to instruction increased students' conceptual knowledge and ability to transfer knowledge.

Walker et al. (2006) investigated the concept of adaptive expertise in the context of an introductory engineering science course and a senior design course in biomedical engineering. They used a design scenario approach (McMartin, McKenna, & Youseffi, 2000) to evaluate students' responses to an open-ended problem. Based on students' responses they evaluated the quality of strategies, the quality of students' questions, and confidence. Furthermore, they categorized the quality of strategies as the efficiency dimension of adaptive expertise and the quality of students' questions as the innovation dimension. Their findings suggest that fourth-year students devised more efficient and innovative solutions than first-year students and over time all students became more confident in their approach.

Fisher and Peterson (2001) suggested four constructs to define adaptive expertise: multiple perspectives, metacognition, goals and beliefs, and epistemology. Based on these constructs they created a forty-two–item survey, and conducted pilot testing with biomedical engineering students and faculty. Based on their pilot data they calculated an average adaptive expertise score and found increasing levels from first-year students to seniors to faculty. Although this study acknowledges that more research is necessary to validate the survey, this work is useful in its attempt to present a more precise definition of adaptive expertise.

Others have conducted related expert–novice studies within the context of design. Atman, Chimka, Bursic, and Nachtmann (1999) conducted verbal protocol analysis of freshman and senior engineering students as they solved a playground design problem. They found many similarities between how freshmen and senior students approached the design problem and how they allocated time on specific activities as well as overall time on task. However, they also found that seniors developed a higher quality product, suggesting that the end result is not just attributable to time on task but to how one spends his or her time during the process. In particular, they identified several design process activities that led to higher quality results: gathering information, considering multiple alternatives, transitioning between steps throughout the process, setting up the problem, and progressing through final steps.

Cross (2004) identifies specific behaviors that distinguish the nature of expert performance in design. In his overview of several protocol studies he synthesized that expert designers appear to spend substantial time and attention on defining the problem. He refers to this as "ill-behaved" problem solving but recognizes this as appropriate behavior because successful design depends on adequate problem scoping, and a directed approach to gathering information. Cross also explains that expert designers are solution focused and take effort to frame the problem appropriately.

In an ethnographic study of expert product design teams, Eris and Leifer (2003) found similar but distinct expert design behaviors. Results from this study shed additional insights into how design expertise functions as a result of team interaction. In particular, Eris and Leifer found that product development knowledge emerges out of the combined interaction of those involved in the process. Tasks and procedures are interpreted and contextualized, and this can vary depending on situation, setting, and resources available.

Popovic (2004) presents a network-type model that suggests the level of expertise in product design plays an important role in problem representation, that is, visual representations manifested by sketching. Using a knowledge connection model, Popovic suggests that experts and novices differ in how they organize knowledge, the amount of information they use, how they access domain-specific knowledge, how they apply domain-specific goal strategies. Although studies such as those of Popovic, Cross and Atman et al. did not focus on adaptive expertise per se, their findings are consistent with the theory and provide helpful insights as to the nature of expertise in design.

McKenna (2007) has applied the adaptive expertise framework to explore how design process knowledge may transfer between design tasks. She found that students who have participated in user-focused engineering design courses develop a level of fluency (significant gains on pre-/post- measures) in design process skills such as generating alternatives, building mock-ups, getting feedback from users and other stakeholders, and revising based on feedback and analysis. Her findings suggest that fluency in design process knowledge is a component of adaptive expertise, and that elements of design process are concepts that endure such that they are recognized and transferred across different design contexts.

Further analyses showed that students transitioned from taking a solution-focused approach to a more process-focused approach, and equally important, appropriated a "language" of design to communicate effectively with their team members and stakeholders in describing a solution path to the task (Atman, Kilgore, & McKenna, 2008). Consistent with Vygotsky's (1978) emphasis on social interaction and the role of speech as a tool, these findings support the notion that language mediates the way students know and learn design. The language of engineering design presented in any educational setting can empower students to shift their focus from solution to process. Dialogue and use of design-specific language provide the mechanism to allow students to make connections between the activities involved in design. This is more dynamic than just knowing engineering design as a fixed set of steps with few or no interactions in between and little to no interactions among people and the environment. How one introduces and appropriates the language of engineering design can affect how one perceives of and engages in the process.

In addition to focusing on design process, several studies have focused on exploring more discipline-specific types of knowledge that may be used in the process of design. That is, disciplinary knowledge is required to perform detailed design activities such as setting up and running simulations; conducting appropriate experimental procedures; or modeling the system for descriptive, predictive, or optimization purposes.

McKenna, Linsenmeier, and Glucksberg (2008) introduced the notion of "computational adaptive expertise" (CADEX) as a way of specifically focusing on more disciplinary aspects of engineering knowledge and as a way to supplement prior work focused on design process skills and knowledge. One motivation for the emphasis on CADEX is the practical reality that a major portion of an engineering curriculum focuses on developing analytical and computational knowledge. Yet, as several studies have suggested, students often struggle with applying or transferring computational knowledge in the context of design. Engineering courses often focus on reductive thinking and "require students to learn in unconnected pieces, separate courses whose relationship to each other and to the engineering process are not explained until later in a baccalaureate education, if ever" (Bordogna, Fromm, & Ernst, 1993, p. 5). This structure of a curriculum works against the development of adaptive expertise.

One aspect of CADEX is the specific role of modeling in the process of design and innovation. McKenna and Carberry (2012) found that the primary conception for a method of modeling in the design process, from both a student and faculty perspective, is to build some type of physical representation and, moreover, that abstract representations such as mathematical or theoretical/conceptual models are mentioned and used less often. These findings indicate that students develop a fluency in the descriptive nature and purposes of modeling in design but less so on the predictive utility of modeling (Starfield, Smith, & Bleloch, 1994).

These results highlight an important element of developing adaptive expertise in the context of design. In particular, at some point in the engineering design process one will need to use specific tools (e.g., solid modeling or circuit layout software), theories (e.g., knowledge of how fluid flows, how mechanisms interact, or how energy is transferred or converted, etc.), or more abstract analyses (e.g., mathematical modeling of systems) to develop realistic, functional, and fully vetted design solutions. Abstract modeling is pervasive in an engineering curriculum; students manipulate mathematical equations, sketch diagrams that demonstrate interactions, and implicitly make assumptions about behaviors of systems regardless of engineering discipline. However, reminiscent of the issues discussed regarding transfer, there is a disconnect when how curricula are structured to learn techniques, tools, and methods for modeling in one setting are not coupled with the adaptive expertise that is necessary to recognize how they might be usefully applied in a novel setting.

Additional studies focused on CADEX have suggested that there are some targeted activities that can lead to enhanced ability of mathematical modeling in the context of engineering design. Cole, Linsenmeier, Molina, Glucksberg, and McKenna (2011) guided students through a structured sequence of reflective mathematical modeling activities within a biomedical engineering capstone design course and found that this led to adaptive modeling behaviors such as stating assumptions better, creating appropriate equations more frequently, and more often identifying the direction of the mismatch between model outputs and the physical situation.

Summary and Future Areas of Research

This chapter presented an overview of the adaptive expertise framework, how adaptive expertise and the concept of innovation relate to the literature on knowledge transfer, and how the framework of adaptive expertise can be useful to study the processes of engineering design and innovation. Several examples were presented that showed the variety of studies that have applied the framework of adaptive expertise in a range of contexts.

While these studies have shown promising findings with respect to developing aspects of adaptive expertise within engineering and other contexts, and specifying the nature of efficiency and innovation, there are several areas for further research. For example, some have suggested that on average it takes approximately ten years of deliberate practice, along with the accumulation of experience to develop internationally recognized levels of expertise (Ericsson & Lehmann, 1996; Ericsson, Charnes, Feltovich, & Hoffman, 2006; Ericsson, Prietula, & Cokely, 2007). Acknowledging the time scale involved in the development of expertise, the studies described in this chapter focus mostly on relatively brief snap shots in time. The literature would benefit from studies that examine adaptive expertise from a more longitudinal perspective. Examining adaptive expertise over a more extended period of time could yield useful insights into the types of developmental trajectories that can put one on the path to becoming an adaptive expert.

Several studies cited throughout have alluded to self-regulation, habits of mind, and affective factors such as dispositions and motivation in regard to developing adaptive expertise. The influence of beliefs, motivation, and self-efficacy (e.g., Carberry, Lee, & Ohland, 2010) on learning has been well documented in the literature but these factors have been studied less so as they relate to adaptive expertise. Several new research directions could be conducted that examine the role of personal characteristics within the efficiency and innovation dimensions of adaptive expertise.

It is also important to note that engineering design activity almost always involves working in teams. Moreover, in addition to working with colleagues, design process requires interaction with others outside the immediate team such as users, vendors, experts, clients, and a variety of other stakeholders. The embedded social nature of design begs the questions about how adaptive expertise might account for group processes, in contrast to just focusing on an individual's path to adaptive expertise.

One line of research might focus on how the adaptive expertise framework provides any utility for capturing the social nature of learning, and its relationship to helping one develop adaptive expertise. Related to this idea, another area of research could involve investigating how the adaptive expertise framework might be applied to groups, organizations, or collections of individuals such that the unit of analysis is the group, not the individual. Changing the unit of analysis raises very interesting lines of research for defining and measuring aspects of innovation and efficiency. This also raises the argument that there might be a phenomenon of "group adaptive expertise" separate from individual adaptive expertise.

Several colleagues have raised the question about whether the adaptive expertise framework, with just two orthogonal axes are sufficient for capturing all that might be involved with developing expertise. This is an open and ongoing discussion and is particularly relevant in the context of accounting for affective factors, and applying the framework to study group processes.

Finally, much of the current work on adaptive expertise has taken place in educational settings. The settings have been somewhat comprehensive, that is inclusive of both pre-college and higher education environments across different subject matter. However, by virtue of the context these studies mostly involve students, teachers, and faculty. As Cross (1998) cautions, many studies of designer behavior are based on students or designers of "relatively modest talents" (p. 141), limiting our understanding to a narrow band of expertise. Cross makes a cogent point, and we duly note that in the context of engineering design, additional studies that include participants from a variety of professional settings including industry, government, and non-profit organizations could yield important insights into the nature of adaptive expertise, and the knowledge, skills, and habits of mind that are essential to the process.

Continued research into the development of adaptive expertise holds promise for helping educators design learning and assessment experiences that can put learners on the path to adaptive expertise. While there may be ongoing debate about what constitutes efficiency and innovation, or if in fact there is a difference between adaptive vs. traditional studies of expertise, there is consensus around the importance of helping individuals recognize and use knowledge in novel situations. This is the core of both *knowledge* innovation and *design* innovation. There is a growing group of researchers who are investigating adaptive expertise in a variety of settings and disciplines. It is exciting to imagine the potential of what we might collectively uncover as we continue down this path of discovery.

Acknowledgments

The author would like to acknowledge Adam Carberry and Taylor Martin for their thoughtful review of earlier drafts of this chapter. Their input helped to refine the chapter for clarity and substance. This material is partly based upon work supported by the National Science Foundation under Grant Nos. 0648316 and 1118659. Any opinions, findings, and conclusions or recommendations expressed in this material are those of the author and do not necessarily reflect the views of the National Science Foundation.

References

Accreditation Board for Engineering Technology (ABET). (2000). Engineering Criteria 2000. Retrieved from http://www.abet.org, 2000

Anderson, J. R., Corbett, A. T., Koedinger, K., & Pelletier, R. (1995). Cognitive tutors: Lessons learned. *The Journal of the Learning Sciences, 4,* 167–207.

Atman, C. J., Chimka, J. R., Bursic, K. M., & Nachtmann, H. L. (1999). A comparison of freshman and senior engineering design processes. *Design Studies,* 20(2), 131–152.

Atman, C. J., Kilgore, D., and McKenna, A. F. (2008). Characterizing design learning: A mixed-methods study of engineering designers' use of language. *Journal of Engineering Education,* 97(3), 309–326.

Barnett, S., & Ceci, S. J. (2002). When and where do we apply what we learn? A taxonomy for far transfer. *Psychological Bulletin,* 128, 612–637.

Barnett, S., & Koslowski, B (2002). Adaptive expertise: Effects of type of experience and the level of theoretical understanding it generates. *Thinking & Reasoning,* 8(4), 237–267.

Bordogna, J., Fromm, E., & Ernst, E. W. (1993). Engineering education: Innovation through integration. *Journal of Engineering Education,* 82(1), 3–8.

Bransford, J. D., Brown, A. L., & Cocking, R. R. (Eds.) (2000). *How people learn: Brain, mind, experience, and school.* Washington, DC: The National Academies Press.

Bransford, J. D., & Schwartz, D. L. (1999). Rethinking transfer: A simple proposal with multiple implications. *Review of Research in Education, 24,* 61–100.

Broudy, H. S. (1977). Types of knowledge and purposes of education. In R. C. Anderson, R. J. Spiro, & W. E. Montague (Eds.), *Schooling and the acquisition of knowledge* (pp. 1–17). Hillsdale, NJ: Lawrence Erlbaum.

Bruer, J. T. (1993). *Schools for thought: A science of learning in the classroom.* Cambridge MA: MIT Press.

Brush, T., & Saye, J. (2001). The use of embedded scaffolds with hypermedia-supported student-centered learning. *Journal of Educational Multimedia & Hypermedia, 10,* 333–356.

Bucciarelli, L. L. (1996). *Designing engineers.* Cambridge, MA: MIT Press.

Bucciarelli, L. L. (2003). Designing and learning: A disjunction in contexts, *Design Studies, 24*(3), 295–311.

Butterfield, E. C., & Nelson, G. D. (1991). Promoting positive transfer of different types, *Cognition and Instruction, 8*(1), 69–102.

Carberry, A. R., Lee, H-S., & Ohland, M. W. (2010). Measuring engineering design self-efficacy. *Journal of Engineering Education, 99*(1), 71–79.

Carraher, D., & Schliemann, A. D. (2002). The transfer dilemma. *The Journal of the Learning Sciences, 11*(1), 1–24.

Chi, M. T. H. (2006). Two approaches to the study of experts' characteristics. In K. A. Ericsson, N. Charness, P. J. Feltovich, & R. R. Hoffman (Eds.), *The Cambridge handbook of expertise and expert performance* (pp. 21–30). New York, NY: Cambridge University Press.

Chi, M, Feltovich, P., & Glaser, P. (1981). Categorization and representation of physics problems by experts and novices. *Cognitive Science, 5*(2), 121–152.

Chi, M. T. H., & Ohlsson, S. (2005). Complex declarative learning. In K. J. Holyoak & R. G. Morrison (Eds.), *Cambridge handbook of thinking and reasoning.* New York, NY: Cambridge University Press.

Cobb, P., & Bowers, J. (1999). Cognitive and situated learning perspectives in theory and practice. *Educational Researcher, 28*(2), 4–15.

Cole, J. Y., Linsenmeier, R. A., Molina, E., Glucksberg, M. R., & McKenna, A. F. (2011). Assessing engineering students' mathematical modeling abilities in capstone design. In *Proceedings of the American Society for Engineering Education (ASEE) Annual Conference,* Vancouver, BC, Canada.

Collins, A., Brown, J. S., & Newman, S. E. (1989). Cognitive apprenticeship: Teaching the crafts of reading, writing, and mathematics. In L. B. Resnick (Ed.), *Knowing, learning, and instruction* (pp. 453–494). Hillsdale, NJ: Lawrence Erlbaum.

Crawford, V. M. (2007). Adaptive expertise as knowledge building in science teachers' problem solving. In *Proceedings of the Second European Cognitive Science Society* (EuroCogSci07), Delphi, Greece.

Cross, N. (1998). Expertise in engineering design. *Research in Engineering Design, 10*(3), 141–149.

Cross, N. (2000). *Engineering design methods: Strategies for product design.* Chichester, U.K.: John Wiley & Sons.

Cross, N. (2004). Expertise in design: An overview. *Design Studies, 25*(5), 427–441.

Davis, E. A., & Linn, M. C. (2000). Scaffolding students' knowledge integration: Prompts for reflection in KIE. *International Journal of Science Education, 22,* 819–837.

Davis, E. A., & Miyake, N. (2004). Explorations of scaffolding in complex classroom systems. *The Journal of the Learning Sciences, 13*(3), 265–272.

De Corte, E. (1999). On the road to transfer. *International Journal of Educational Research, 31,* 555–559.

Detterman, D. K., & Sternberg, R. J. (1993). The case for the prosecution: Transfer as epiphenomenon. In D. K. Detterman & R. J. Sternberg (Eds.), *Transfer on trial: Intelligence, cognition, and instruction.* Norwood, NJ: Ablex.

Dewey, J. (1938). *Experience and education.* New York, NY: Touchstone.

diSessa, A. A., & Wagner, J. F. (2005). What coordination has to say about transfer. In J. P. Mestre (Ed.), *Transfer of learning from a modern multidisciplinary perspective.* Greenwich, CT: Information Age.

Di Vesta, F. J., & Peverly, S. T. (1984). The effects of encoding variability, processing activity, and rule-example sequence on the transfer of conceptual rules. *Journal of Educational Psychology, 76,* 108–119.

Dym, C. L., Agogino, A. M., Eris, O., Frey, D. D., & Leifer, Larry (2005). Engineering design thinking, teaching, and learning. *Journal of Engineering Education*, 94(1), 103–120.

Ericsson, K. A., & Lehmann, A. C. (1996). Expert and exceptional performance: Evidence of maximal adaptation to task. *Annual Review of Psychology*, 47, 273–305.

Ericsson, A. K., Charness, N., Feltovich, P., & Hoffman, R. R. (2006). *Cambridge handbook on expertise and expert performance*. Cambridge: Cambridge University Press.

Ericsson, A. K., Prietula, M. J., & Cokely, E. T. (2007, July). The making of an expert. *Harvard Business Review*, 114–121.

Ericsson, A. K., & Simon, H. A. (1984). *Protocol analysis: Verbal reports as data*. Cambridge, MA: MIT Press.

Eris, O., & Leifer, L. (2003). Facilitating product development knowledge acquisition: Interaction between the expert and the team. *International Journal of Engineering Education*, 19(1), 142–152.

Fisher, F. F., & Peterson, P. (2001). A tool to measure adaptive expertise in biomedical engineering students. In *Proceedings of the 2001 American Society for Engineering Education Annual Conference*, Albuquerque, NM.

Fortus, D., Krajcik, J., Dershimer, R. C., Marx, R. W., & Mamlok-Naaman, R. (2005). Design-based science and real-world problem-solving. *International Journal of Science Education*, 27, 855–879.

Gick, M. L., & Holyoak, K. J. (1983). Schema induction and analogical transfer. *Cognitive Psychology*, 15, 1–38.

Goel, V., & Pirolli, P. (1992). The structure of design problem spaces. *Cognitive Science*, 16, 395–429.

Gott, S., Hall, P., Pokorny, A., Dibble, E., & Glaser, R. (1992). A naturalistic study of transfer: Adaptive expertise in technical domains. In D. Detterman & R. Sternberg (Eds.), *Transfer on trial: Intelligence, cognition, and instruction* (pp. 258–288). Norwood, NJ: Ablex.

Greeno, J., Smith, D. R., & Moore, J. L. (1993). Transfer of situated learning. In D. K. Detterman & R. J. Sternberg (Eds.), *Transfer on trial: Intelligence, cognition, and instruction*. Norwood, NJ: Ablex.

Hall, R., Kibler, D., Wegner, E., & Truxaw, C. (1989). Exploring the episodic structure of algebra story problem solving. *Cognition and Instruction*, 6(3), 223–283

Hatano, G., & Inagaki, K. (1986). Two courses of expertise. In H. Stevenson, H. Azuma, & K. Hakuta (Eds.), *Child development and education in Japan*, (pp. 262–272). New York, NY: W. H. Freeman.

Hatano, G., & Oura, Y. (2003). Commentary: Reconceptualizing school learning using insight from expertise research. *Educational Researcher*, 32(8), 26–29.

Hogan, K., Nastasi, B. K., & Pressley, M. (1999). Discourse patterns and collaborative scientific reasoning in peer and teacher-guided discussions. *Cognition and Instruction*, 17, 379–432.

Hutchison, M. A., & McKenna, A. F. (2008a). *The influence of problem structure on students' use of innovative learning strategies*. Paper presented at the Research in Engineering Education Symposium (REES), Davos, Switzerland.

Hutchison, M. A., & McKenna, A. F. (2008b). Promoting innovative design. In *Proceedings of the American Society for Engineering Education (ASEE) Annual Conference*, ASEE 2008, Pittsburgh, PA.

Jonassen, D. H. (2000). Toward a design theory of problem solving. *Educational Technology Research and Development*, 48(4), 63–85.

Jonassen, D., Strobel, J., & Beng Lee, C. (2006). Everyday problem solving in engineering: Lessons for engineering education. *Journal of Engineering Education*, 95(20), 139–151.

Kimball, D. R., & Holyoak, K. J. (2000). Transfer and expertise. In E. Tulving & F. I. M. Craik (Eds.), *The Oxford handbook of memory* (pp. 109–122). New York, NY: Oxford University Press.

Kolb, D. A. (1984). *Experiential learning: Experience as the source of learning and development*. Englewood Cliffs, NJ: Prentice-Hall.

Larkin, J., McDermott, J., Simon, D. P., & Simon, H. A. (1980). Expert and novice performance in solving physics problems. *Science*, 208(4450), 1335–1342.

Linn, M. C. (2000). Designing the knowledge integration environment. *International Journal of Science Education*, 22, 781–796.

Lobato, J. (2006). Alternative perspectives on the transfer of learning: History, issues, and challenges for future research. The *Journal of the Learning Sciences*, 15(4), 431–449.

Marini, A., & Genereux, R. (1995). The challenge of teaching for transfer. In A. McKeough, J. Lupart, & A. Marini (Eds.), *Teaching for transfer: Fostering generalization in learning* (pp. 1–20). Mahwah, NJ: Lawrence Erlbaum.

Martin, T., Petrosino, A. J., Rivale, S., & Diller, K. (2007). The development of adaptive expertise in biotransport. *New Directions in Teaching and Learning*, 108, 35–49.

Martin, T., Rayne, K, Kemp, N. J., Hart, J., & Diller, K. R. (2005). Teaching for adaptive expertise in biomedical engineering ethics. *Science and Engineering Ethics*, 11(2), 257–276.

Martin, L., & Schwartz, D. L. (2009). Prospective adaptation in the use of external representations. *Cognition and Instruction*, 27(4), 370–400.

Mayer, R. E. (1974). Acquisition processes and resilience under varying testing conditions for structurally different problem-solving processes. *Journal of Educational Psychology*, 66, 644–656.

Mayer, R. E. (1999). Multimedia aids to problem-solving transfer. *International Journal of Educational Research*, 31, 611–623.

McKenna, A. F. (2007). An investigation of adaptive expertise and transfer of design process knowledge. *ASME Journal of Mechanical Design*, 129(7), 730–734.

McKenna, A. F., & Carberry, A. (2012). Characterizing the role of modeling in innovation. *International Journal of Engineering Education*, 28(2), 263–269.

McKenna, A. F., Linsenmeier, R., & Glucksberg, M. (2008). Characterizing computational adaptive expertise. In *Proceedings of the American Society for Engineering Education (ASEE) Annual Conference*, ASEE 2008, Pittsburgh, PA.

McMartin, F., McKenna, A., & Youseffi, K. (2000). Scenario assignments as assessment tools for undergraduate engineering education. *IEEE Transactions on Education*, 43(2), 111–119.

Novick, L. R. (1988). Analogical transfer, problem similarity, and expertise. *Journal of Experimental Psychology: Learning, Memory, and Cognition*, 14(3), 510–520.

Pandy, M. G., Petrosino, A. J., Austin, B. A., & Barr, R. E. (2004). Assessing adaptive expertise in undergraduate biomechanics. *Journal of Engineering Education*, 93, 211–222.

Patel, V. L., & Groen, G. J. (1991). The general and specific nature of medical expertise. In K. A. Ericsson & J. Smith (Eds.), *Toward a general theory of expertise: Prospects and limitations* (pp. 93–125). New York, NY: Cambridge University Press.

Popovic, V. (2004). Expertise development in product design – strategic and domain-specific knowledge connections. *Design Studies*, 25(5), 527–545.

Rayne, K., Martin, T., Brophy, S., & Diller, K. R. (2006). A cross-sectional examination of the development of adaptive expertise in bioengineering ethics. *Journal of Engineering Education*, 95(2), 165–173.

Rebello, N. S., Zollman, D., Albaugh, A. R., Engelhardt, P. B., Gray, K. E., Hrepic, Z., & Itza-Ortiz, S. F. (2005). Dynamic transfer: A perspective from physics educaiton research. In J. Mestre (Ed.), *Transfer of learning from a modern multidisciplinary perspective* (pp. 217–250). Greenwich, CT: Information Age.

Reed, S. K. (1993). A schema-based theory of transfer. In D. K. Detterman & R. J. Sternberg (Eds.), *Transfer on trial: Intelligence, cognition, and instruction* (pp. 39–67). Norwood, NJ: Ablex.

Reed, S. K., Ernst, G. W., & Banerji, R. B. (1974). The role of analogy in transfer between similar problem states. *Cognitive Psychology*, 6, 436–450.

Reiser, B. J. (2004). Scaffolding complex learning: The mechanisms of structuring and problematizing student work. *The Journal of the Learning Sciences*, 13(3), 273–304.

Salomon, G., & Perkins, D. N. (1989). Rocky roads to transfer: Rethinking mechanisms of a neglected phenomenon. *Educational Psychologist*, 24(2), 113–142.

Scardamalia, M., & Bereiter, C. (1991). Literate expertise. In K. A. Ericsson and J. Smith (Eds.), *Toward a general theory of expertise: Prospects and limitations*. New York, NY: Cambridge University Press.

Schwartz, D. L. (1993). The construction and analogical transfer of symbolic visualizations. *Journal of Research in Science Teaching*, 30, 1309–1325.

Schwartz, D. L., Bransford, J. D., & Sears, D. (2005). Efficiency and innovation in transfer. In J. Mestre (Ed.), *Transfer of learning: Research and perspectives* (pp. 1–51). Greenwich, CT: Information Age.

Schwartz, D. L., Chang, J., & Martin, L. (2008a). Instrumentation and innovation in design experiments: Taking the turn toward efficiency. In A. E. Kelly, R. Lesh, & J. Y. Baek (Eds.), *Handbook of innovative design research methods in science, technology, engineering, mathematics (STEM) education.* New York, NY: Routledge.

Schwartz, D. L., Lin, X., Brophy, S., & Bransford, J. D. (1999). Toward the development of flexibly adaptive instructional designs. In C. M. Reigeluth (Ed.), *Instructional design theories and models* (pp. 183–213). Mahwah, NJ: Lawrence Erlbaum.

Schwartz, D. L., & Martin, T. (2004). Inventing to prepare for future learning: The hidden efficiency of encouraging original student production in statistics instruction. *Cognition and Instruction, 22*(2), 129–184.

Schwartz, D. L., Varma, S., & Martin, L. (2008b). Dynamic transfer and innovation. In S. Vosniadou (Ed.), *Handbook of research on conceptual change.* Mahwah, NJ: Lawrence Erlbaum.

Sherin, B., Reiser, B. J., & Edelson, D. (2004). Scaffolding analysis: Extending the scaffolding metaphor to learning artifacts. *The Journal of the Learning Sciences, 13*(3), 387–421.

Simon, H. A. (1996). *The sciences of the artificial.* Cambridge, MA: MIT Press.

Starfield, A. M., Smith, K. A., & Bleloch, A. L. 1994. *How to model it: Problem solving for the computer age.* Edina, MN: Burgess.

Verschaffel, L. Luwel, K., Torbeyns, J., & Van Dooren, W. (2009). Conceptualizing, investigating, and enhancing adaptive expertise in elementary mathematics education. *European Journal of Psychology of Education, 24*(3), 335–359.

Vygotsky, L. S. (1978). *Mind and society: The development of higher psychological processes.* Cambridge, MA: Harvard University Press.

Walker, J. M. T., Cordray, D. S., King, P. H., & Brophy, S. P. (2006). Design scenarios as an assessment of adaptive expertise. *International Journal of Engineering Education, 22*(3), 645–651.

Webb, N. M. (1996). Group processes in the classroom. In D. Berliner & R. Calfee (Eds.), *Handbook of educational psychology.* New York, NY: Macmillan.

Wineburg, S. S. (1998). Reading Abraham Lincoln: An expert/expert study in the interpretation of historical texts. *Cognitive Science, 22,* 319–346.

Wood, D., Bruner, J. S., & Ross, G. (1976). The role of tutoring in problem solving. *Journal of Child Psychology and Psychiatry and Allied Disciplines, 17,* 89–100.

CHAPTER 13

Learning Disciplinary Ideas and Practices Through Engineering Design

Kristen Bethke Wendell and Janet L. Kolodner

Introduction

When pre-college students are given opportunities to engage in engineering, and those opportunities are carefully structured to include particular affordances, much more than designing and building can take place. Engineering education researchers who focus on elementary and secondary education are finding that engineering design can help create an environment for the learning of ideas and practices in a variety of academic disciplines. In this chapter, we put forth a framework, case-based reasoning, that suggests ways of carefully conceiving engineering design activities to support disciplinary learning. We highlight K–12 engineering education research that exemplifies the structures and practices suggested by the case-based reasoning framework. These research studies inform our understanding of how students learn disciplinary ideas and practices through engineering design at the pre-college level.

For a first glimpse into this work, picture the following scene in a third-grade science classroom. Two students are consulting with their teacher about the final assignment of their musical instrument engineering unit. The goal is a novel instrument that can play three different pitches. After a brief discussion to help the students see the strengths and weaknesses of their initial plan, the teacher sends them on their way to revise their ideas. As they return to brainstorming, an intense debate transpires:

Teacher:	Back to the planning stages. I think you have some ideas. [Students walk back to desks.]
Adam:	Sean, what should we do, man?
Sean:	I don't know!
Adam:	Oh, I know! We can make a triangle, and we can put connector pegs on it, and we can make elastics, on the connector pegs. One small, one medium, and one –
Sean:	Yeah, good idea!
Adam:	Long.
Sean:	Yeah! Make a box, and
Adam:	No, wait, we can do a triangle, triangle –
Sean:	Box, box –
Adam:	Triangle.
Sean:	A box. Box.

Adam: Dude, a triangle is more better, cuz we have –

Sean: Box! Cuz we have more sides on a box.

Adam: So how are we gonna make one shorter, one bigger, and one littler? See? A triangle's more better!

Sean: I know! Let me show you how. Look. It's gonna be like this, look. [Grabs a pencil and starts drawing on Adam's journal page]. It's gonna be a box. We're gonna put connector pegs, then this is the middle. Middle, small. Just do a box, just do a box, and put a bunch of connector pegs. Just put connector pegs everywhere.

Adam: [Starts drawing on his journal page]

In this episode, Sean and Adam are ostensibly engaged in engineering design. But more than that, they are beginning to reason about scientific ideas and to participate in practices of science. In this chapter, by *ideas*, we mean pieces of knowledge that are important for engaging in the problems of the discipline. By *practices*, we mean the ways of "doing" science, or mathematics, or any other discipline. Disciplinary practices are employed by experts in the discipline, and they require the coordination of both knowledge and skill (National Research Council [NRC], 2012).

For Adam and Sean's instrument to have three different pitches, Adam suggests that they use three different elastics – one small, one medium, and one long. Here he is considering the scientific idea that there is a consistent, predictable relationship between the physical characteristics of an object and the pitch of the sound it makes. For their instrument to support those different elastics, Adam and Sean express conflicting proposals about the shape of the instrument's frame. Sean says a box is better because it has "more sides" for the elastics, while Adam envisions a triangle, making it easier to have "one shorter, one bigger, and one littler." Here they are beginning to participate in practices of scientific argumentation: making claims, attempting to add evidence and reasoning to convince each other, and drawing graphical representations to communicate their ideas.

This chapter's central thesis, which we illustrate with selected research, is that engineering design can provide a deep, rich set of feedback for students to incorporate into their understandings and abilities. Consequently, engineering design in the pre-college classroom is well-suited for integration with ideas and practices from other disciplines. Moreover, the framework of case-based reasoning suggests how to conceive of K–12 engineering design challenges so that students can learn disciplinary ideas and practices while engaging in engineering design. Sean and Adam's box-versus-triangle debate is the beginning of the kind of K–12 experience we are considering here.

We begin the remainder of the chapter with a discussion of the affordances of engineering design for learning. Then, we explain case-based reasoning, the theoretical principle behind our claim that students can learn ideas and practices through design activity. We go on to provide examples of learning science through design from two particular efforts in our own work based on case-based reasoning. In the first example, middle school students explore force and motion through vehicle design, and in the second, elementary school students explore sound through musical instrument design. Following these examples, we acknowledge some challenges of learning disciplinary ideas and practices through design, and we propose some possible solutions. Here we draw on the work of other researchers who have integrated engineering into disciplinary learning. Finally, we explain why the potential benefits of learning disciplinary content through engineering design inspire us to overcome these challenges.

Our Theoretical Grounding: Case-Based Reasoning and Engineering Design

Case-based reasoning (CBR; Kolodner, 1993; Schank, 1982, 1999) originated in the artificial intelligence research of Roger Schank and his team (e.g., Schank & Abelson, 1977). CBR claims that our minds are filled with thousands of "cases" – some the stories of specific events and some more generalized, script-like stories about common sequences of everyday events – and that intelligent behavior is based on a person's ability to identify the important features of a new situation and retrieve the best-matching case from this large repertoire.

CBR explains how people apply their previous experiences to solve problems in new situations. For instance, special experiences at restaurants would be available to guide future reasoning at restaurants when appropriate – for example, the time the clumsy waitress spilled soup on you might be remembered the next time a waitress seems clumsy, and you might consider not ordering soup or being careful to move away when the waitress brings soup; the time you had excellent Chinese food in a vegetarian restaurant may be remembered the next time you are in that same restaurant or another vegetarian restaurant, and you might use that memory to decide to order a Chinese dish. In the general case of going to a restaurant without any novel characteristics, the restaurant script itself would allow you to navigate your way through the situation.

By making computer programs that carried out CBR, Schank's group learned more about scripts, memory, and problem-solving. They eventually used case-based reasoning programs to try to model the reasoning in novice learners – to learn how a reasoner might go from having very little knowledge about a task or domain to becoming expert in its vocabulary and practices. CELIA (Cases and Explanations in Learning; an Integrated Approach), for example, used cases to model the troubleshooting and learning of an apprentice mechanic (Redmond, 1992). These studies all contributed to a case-based theory of learning (Kolodner, 1993, 1997; Schank, 1982, 1999):

1. Learning will happen best in contexts of trying to achieve goals of interest. When someone is interested in achieving a particular outcome, he or she takes the initiative to develop the new skills or construct the new ideas that will contribute to success.

2. To learn well from their experiences, learners need to interpret their experiences so as to make them into well-articulated cases in their memories. The better learners connect their goals to their reasoning about achieving the goals, the more useful the case will be for later reasoning.

3. Experience applying cases from memory allows further learning. Failures at application and failures of expectations tell the learner that more needs to be learned and provide an opportunity to reinterpret an old experience. The more opportunities learners have to apply their cases, the better they are able to debug their interpretations and add to their knowledge and capabilities.

4. Learners can learn from the cases of others as well as from their own. For instance, if a friend tells you about her very unpleasant experience having hot soup spilled on her lap by a waitress, you might decide not to order soup in a busy restaurant with many waiters and waitresses buzzing around.

5. Learners can best learn from their mistakes and expectation failures if they get immediate feedback so that they have a way to recognize their errors and expectation failures, and if they can explain why they happened and what they should have done differently.

In summary, case-based reasoning suggests that the best learning experiences are those that afford clear feedback in a timely way and where learners have a chance to develop their ideas further based on that feedback and apply what they are learning in another iteration. We can see this in the experience of Sean and Adam, the third-grade students who opened this chapter. Prior to beginning their final challenge, Sean and Adam had experienced a sequence of smaller engineering design activities that they referred back to (sometimes explicitly but more often implicitly) as they worked on their final design. For instance, they had solved the problem of designing a miniature drum, they had created a functioning maraca, and they had determined how to change the pitch of the strings on a miniature guitar. With each of these smaller design problems, they had been asked to persevere until their artifact was functional, and they had been encouraged to use writing, drawing, and oral language to express their ideas about how each artifact worked. They had been given multiple opportunities to explain how the visible characteristics of musical instruments influence their audible characteristics. When they came to designing their final musical instrument, they referred back to those experiences for ideas and to critique those ideas. They were not simply coming up with ideas willy-nilly; they were using ideas (and the scientific concepts associated with them) that they had studied previously and applying them in this new situation, in the process identifying what else they needed to understand to know how to modify those ideas for their new product.

Moreover, there is an excellent match between the processes and activities involved in engineering design and what CBR tells us about sequences of activities that promote learning. The engineering cycle of designing, building, and testing artifacts provides the kind of feedback CBR tells us is important for learning. Constructing working physical objects gives students the motivation to learn and the opportunities to discover what they need to learn, to use disciplinary ideas and practices, and to test their conceptions and discover the gaps in their knowledge and capabilities. Design challenges also provide opportunities for engaging in and learning complex cognitive, social, and communication skills.[1]

Engineering design has many powerful affordances for promoting both idea development and learning of disciplinary practices:

- Design challenges focus learning and provide opportunities for learners to make their mental models concrete.

- Students' construction failures are opportunities for testing and revising newly developing conceptions. The functioning (or nonfunctioning) of a physical artifact offers students feedback on their ideas.

- Designing a working artifact naturally involves iterations in design; if done well, each can contribute to iterative refinement in understanding of concepts and gradual learning of skills and practices.

- Doing and reflecting, aimed at helping students turn their experiences into accessible, reusable cases, can be easily interleaved with each other. Students' desire to achieve a design challenge successfully provides a natural motivation for discussing the rationale behind their own design decisions, for wanting to hear about the designs and rationales of others, for identifying what else they need to learn, and for wanting to learn the ideas that will allow them to come up with better solutions.

- Designing affords learning of communication, representation, decision making, and collaboration skills – designers must show their design ideas to others and sell them.

- Designing in the particular way that engineers design gives students a chance to take on the identity of a practitioner in a discipline and to reflect on the connections between the discipline of engineering and the discipline of science (or math, or any other discipline brought to bear during a design challenge).

Looking back again at third-grade students Sean and Adam, we can see some of these affordances. They were truly personally invested in building an instrument that they would be able to play in their classroom "band." Thus they were motivated to learn what they needed to learn to succeed. The fact that they had to create one single instrument between the two of them forced them to express their ideas about how best to construct the instrument, and when they did not agree, they had an opportunity to find scientific evidence to justify their design ideas. As their work on the final design challenge went on, their slight dissatisfaction with each successive prototype prompted them to test their ideas about how physical characteristics of instruments influence pitch.

Two Examples from Our Work

To illustrate further how case-based reasoning and learning through engineering design go hand in hand, we turn to two examples from work by our respective research groups at Georgia Tech and at the Center for Engineering Education and Outreach at Tufts University. The first example is set at the middle school level and comes from Georgia Tech's LEARNING BY DESIGN™ program. The second example takes us to the elementary school level with the CEEO's *Science Through LEGO™ Engineering* project.

Before diving into the examples, a few notes may be helpful for readers not familiar with engineering in pre-college education. One question that may surface is how engineering-based units fit into the overall K–12 curriculum. Although these learning experiences do intentionally involve students in engineering design, they were also carefully designed to meet standards from the *National Science Education Standards* (NRC, 1996) and to serve as complete science curriculum units rather than as science unit extensions. The curriculum developers for both programs began with a set of science learning objectives and then identified engineering challenges with affordances for meeting those objectives. Interestingly, because engineering practices and core ideas are included prominently in the new national *Framework for K–12 Science Education* (NRC, 2012) and *Next Generation Science Standards* (Achieve, Inc., 2013), the engineering-based units profiled below now meet additional standards that are likely to be adopted by the majority of states. Readers may also wonder how teachers become prepared for teaching through engineering

design. The LEARNING BY DESIGN™ and *Science Through LEGO™ Engineering* programs involved extended professional development for in-service teachers, and the written curriculum materials were designed to address questions and concerns that typically arise in the course of teaching an engineering-based unit for the first time. However, the issue of supporting teachers in bringing engineering to the classroom is a major one, and we return to it at the end of the chapter. Finally, in the following sections, some readers may see, and question, an implicit assumption of student engagement – an expectation that students are automatically "hooked" or motivated by engineering design challenges. For learning through design to succeed, it is important that students work through to the completion of the design challenge, and thus some degree of motivation is required. This is where the importance of building and testing, rather than designing only on paper, comes in. Of course not every student has the same level of excitement about every engineering challenge. But we have tried to support all students by providing appropriate materials, time, and scaffolding to create a functioning artifact, *and* by designing challenges to be developmentally appropriate, culturally relevant, and not strongly associated with a single gender. We have found that when given the opportunity to work on these challenges, nearly all students show a desire to construct an artifact that works.[2] Constructing working artifacts is at the heart of the two examples presented in the next section.

Middle School Example

LEARNING BY DESIGN™ (LBD; Kolodner, 2006; Kolodner et al., 2003) is a project-based inquiry approach in which students attempt to achieve design challenges, and learning cycles and activity structures promote the kinds of reflection on experience needed to learn productively from those experiences. Figure 13.1 shows how LBD's activities are sequenced. Activities in the

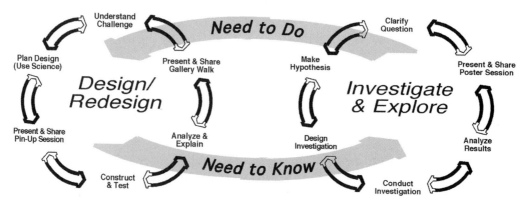

Figure 13.1. LEARNING BY DESIGN's cycles.

design/redesign cycle (on the left) are those needed to achieve a design challenge successfully. Because successful achievement of a challenge often requires investigation, an investigative cycle (on the right) is a natural part of LBD. Activity begins at the top of the design/redesign cycle, and when students discover a need to learn something new, they engage in investigation. Results of investigations, in turn, provide content for application to the design in progress. Individual activities in each cycle are designed to move learners toward successful achievement of a challenge, and they integrate a variety of science, design, collaboration, and communication practices.

Enactment of LBD's cycles of activities involves participation in a variety of carefully designed and repeated activity structures and sequences (Table 13.1). These classroom activity structures sequences are designed so that they allow success at carrying out the tasks in the cycles in Figure 13.1 at the same time that they provide practice at scientific reasoning and use of newly learned science ideas.

An LBD Scenario

The most studied LBD unit is called *Vehicles in Motion*. In this unit, middle school students learn about forces and motion in the context of designing a small "hybrid" vehicle that can travel straight and far and navigate a set of hills. The unit has four modules. In the first, students design the axle, wheel, and bearings system of a coaster car to make it go straight and far. Through this process, students learn about two forces (friction and gravity) and some of their interactions, speed and velocity, and some simple measuring techniques. In the second, students design a propulsion system, made from balloons and straws, and optimize it to make their car go as far as possible, continuing to focus on interacting forces, and beginning to think about rate of change. In the third module, students focus on a more powerful propulsion system, one that uses rubber bands, which can also take the car over a hill. And in the fourth module, students pull it all together to create a hybrid propulsion system.

Wrapped around all of these activities is the Grand Challenge and the questions its achievement raises. Their Grand Challenge is to design the hybrid car that can go over several hills and as far and straight as possible after. Before the coaster car module formally begins, students launch into the overall unit by exploring the workings of several toy cars (an activity called "messing about"), some of which perform better than others, and, as a class, they generate questions about why the different cars perform differently; the effects of wheel size, surface, and inclines on car performance; and how to design a good propulsion system. The teacher then leads students to the first of the modules – the coaster car challenge – where they address the questions they asked about making a car go straight and far. When finished with that challenge, they consider the Grand Challenge again and what else they need to

Table 13.1. A Selection of LBD's Scripted Activity Structures (classroom scripts)

Function(s) in Cycle	LBD Activity Structure	Type and Venue	Description
Design an investigation	**Design an experiment**	Action: small group	Given a question to investigate (in the form of discovering the effect of a variable), design an experiment in which variables are controlled well, with appropriate number of trails, etc.
Analyze results; analyze and explain, present and share	**Creating and refining design rules of thumb**	Action, discourse: small group	Identify trends in data and behaviors of devices; connect scientific explanations so as to know when the trends apply (small groups suggest new rules of thumb and the need for changes in existing ones).
Present and share (investigate cycle)	**Poster session**	Discourse: Present & Share: whole class	Present procedures, results, and analysis of investigations for peer review; followed by rules of thumb.
Plan design	**Plan design**	Action: small group	Choose and integrate design components to achieve the design challenge, basing choices on evidence.
Present and share (design/redesign)	**Pin-up session**	Discourse: Present & Share: whole class	Present design ideas and design decisions and their justifications for peer review, followed by plan design or by construction and test of design.
Construct and test	**Test design**	Action: small group	Run trials of constructed device, gathering data about behavior, attempt to explain, followed by gallery walk.
Present and share (design/redesign)	**Gallery walk**	Discourse: Present & Share: whole class	Present design experiences and explain design's behavior for peer review and advice, followed by whiteboarding and rules of thumb.

learn, generating additional questions about propulsion, and the teacher introduces them to the Balloon Car Challenge, the unit's second module. We begin the scenario here. The cycle of activities is the same in every LBD module, and the cycle of each unit is as described for *Vehicles in Motion*.

UNDERSTANDING THE CHALLENGE

In the Balloon Car Challenge, students are challenged to design and build a propulsion system from balloons and straws that can propel a coaster car as far as possible. They begin by exploring ("messing about") in small groups to identify what might affect a balloon engine's behavior. Then they report what they observed to the class and work as a class to identify investigations they might do to learn more about those effects. During this public session, they volunteer what they have observed (e.g., "It seems like a wider straw makes the car go farther,"), argue

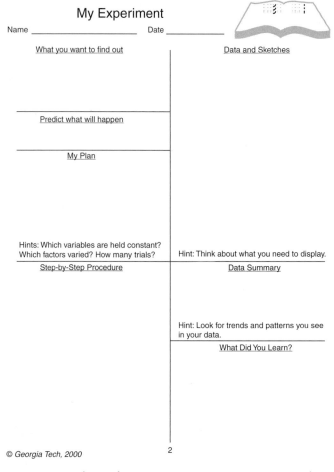

Figure 13.2. A design diary page: "My Experiment." Notice that it prompts learners for some of the important issues they need to discuss and/or plan for (see Puntambekar & Kolodner, 2005, for further details).

about what they saw and how to interpret it (e.g., "I don't think we can compare across those cars because they didn't go exactly the same distance off the ramp."), try to explain what they observed (e.g., "I think the wider straw makes it go farther because more air comes out of it, and that must mean more force."), and identify variables whose effects they want to know about conclusively (e.g., length of straw, extra engines, amount of air in the balloon). The teacher helps them turn their initial questions into questions that can be answered through well-controlled experiments (e.g., what effect does the size of a balloon have on the distance the car will travel?), and each group is assigned one question to investigate.

INVESTIGATE AND EXPLORE

As they are designing their experiments, students use a "Design Diary" page (see Figure 13.2) to prompt them on what to pay attention to in designing and running an experiment and collecting data.

Students spend a day or two designing and running their experiments and analyzing their data, and at the beginning of the following day, each group prepares a poster to present. Posters show their experimental design, data, and interpretations, and, if they can do it, a piece of advice for the class in the form of a rule of thumb. Each group presents to the class in a "poster session." Because students need each other's investigative results to be successful balloon car designers, they listen intently and query each other about the gathering and interpretation of data (much as in a professional poster session). This provides an opportunity to discuss the ins and outs of designing and running experiments. When some groups' results are not trustworthy yet, the class decides they should redo their experiments, and the cycle of activities just described is repeated.

When the class agrees that results of most groups are believable, the teacher helps students abstract over the full set of experiments and results to notice commonalities and extract out "design rules of thumb," for example, "By using double-walled balloon engines, the car goes farther because a larger force is acting on the car." To learn the explanations behind these phenomena, students read about the science content involved, and the teacher might perform demonstrations that provide evidence for the explanation. Students then revisit the rule of thumb, producing a more informed and complete statement, for example, "By using double-walled balloon engines, the car goes farther because a larger force is acting on the air inside, so then an equally large force from the air acts on the car."

BACK TO THE DESIGN CHALLENGE:
DESIGN PLANNING

With investigation complete, activity returns to the design/redesign cycle, and each group uses the results of the class's investigations to plan its balloon car design. Each group prepares another poster, this time presenting their design ideas along with the evidence that justifies each decision and their predictions about how it will perform. They present to their peers in a "pin-up session." Then the class discusses the ideas everyone has presented, followed by discussion of the practices and skills students have just engaged in.

CONSTRUCT AND TEST; ANALYZE AND
EXPLAIN; GALLERY WALK

Students modify their designs based on what they've discussed in class and heard from their peers, and then they construct and test their first balloon-powered engine. They use another design diary page or other planning tool here, this time with prompts helping them to keep track of their predictions, the data they are collecting as they test, whether their predictions are met, and if not, explanations of why not.

None of their balloon cars work exactly as predicted, sometimes because of construction problems and sometimes because of incomplete understanding of scientific principles and the results of experiments. After working in small groups to try to explain their results, the class engages in a "gallery walk," with each group's presentation focused on what happened when its design was constructed and tested, why it worked the way it did, and what to do next so that it will perform better. Gallery walks are followed by classroom discussion summarizing the set of experiences presented, and design rules of thumb are revisited and revised, with a focus on better explaining why each one works. Discussion also focuses on the explanations students made and what good explaining entails.

REDESIGN: ITERATION AND FINISHING UP

In traditional classrooms, after solutions have been generated and discussed, the class moves on to its next topic or project. But in LBD, as CBR suggests, learners are given the opportunity to try again, often several times. Students revise their designs, based on explanations their peers have helped them develop and their developing scientific conceptions. They construct and test their new designs and present results to the class for discussion, iterating toward better solutions and better science understanding in parallel.

The entire balloon car module takes ten to twelve 45-minute class periods. At the end, the class holds a final gallery walk and a competition, and they compare and contrast across designs to understand better the scientific principles they are learning, going back to the rules of thumb yet again to revise and explain them better. They finish up, as well, by discussing their collaboration experience, their design process, their use of evidence, and so on. After all of this group work, each student writes up and hands in a project report – including a summary of the reasoning behind his or her group's final design and what they've learned about collaboration, design, use of evidence, and so on.

Assessing the Success of LBD

Field tests have been carried out in over a dozen classrooms and compared knowledge and capabilities of students participating in LBD environments to students in matched comparison classes (with matched teachers). The assessments addressed two areas: (1) assessing content learning by comparing change from pre-to post-curriculum on written, mostly multiple-choice exams and (2) assessing students' use of science practices as they occur during data-gathering and analysis activities and during experimental design activities. Results show that LBD students consistently learn science content as well or better than comparison students. However, the data that are most encouraging – given the science education community's recent emphasis on scientific practices (NRC, 2012) – come from the analysis of students' experimental design and data-gathering activities. This data set shows large, consistent differences between all LBD classes and their comparisons. While they are engaging in science activities, LBD students greatly outperform comparison students in their abilities to design experiments, plan for data gathering, and collaborate. Indeed, some mixed-achievement LBD classes outperform comparison honors students on these measures (Gray, Camp, Holbrook, Fasse, & Kolodner, 2001; Kolodner & Gray, 2002).

Elementary School Example

Inspired by the success of LBD, researchers at the Center for Engineering Education and Outreach (CEEO) at Tufts University embarked on a research program to explore the relationship between science learning and engineering design activity in the upper elementary grades. Adapting the LBD approach for younger students, they developed four curriculum units that use engineering design problems and LEGO™ artifact construction as the basis for elementary-school science learning.[3] The curriculum is called *Science Through LEGO™ Engineering,* and its units address the science domains of animal adaptations, the properties of materials and objects, simple machines, and sound (Wendell & Lee, 2010; Wendell & Rogers, in press).

Each *Science Through LEGO™ Engineering* unit builds on the LBD approach by interweaving design/redesign with investigation and exploration in the context of a grand engineering design challenge (see Figure 13.3). As suggested by case-based reasoning and LBD, each unit begins with the unveiling of a challenge that will captivate students' interest and give them a goal to achieve. The first experience in each unit focuses on specifying the *grand engineering design challenge* and *the big science question.* Students write about and discuss what they already know that will help them solve the problem and answer the question. And they identify what they still need to learn. Then, to prepare students for success on that grand design challenge, the unit leads them through several smaller challenges that incorporate LEGO™ engineering activities. In these smaller challenges, students move back and forth from investigating scientific questions and constructing/reconstructing artifacts to perform particular functions. Although the students' work on the grand design challenge takes them through only one overarching engineering design cycle, within that cycle students are iterating on physical artifacts throughout the smaller challenges and investigations. These smaller challenges serve as cases that students can use later on when they are reasoning about how to solve the grand design challenge.

An *Engineer's Journal* guides students to make these cases well articulated by prompting them to draw and write about their science ideas and their design experiences. It provides introductory open-response questions, building and observation instructions, data recording prompts,

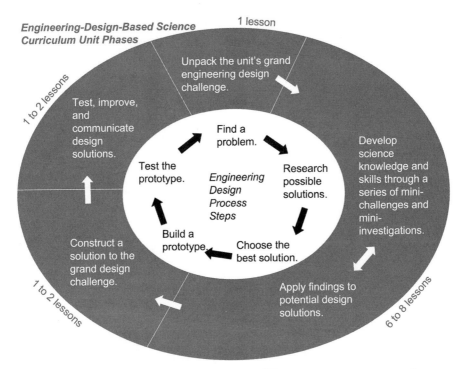

Figure 13.3. Cycle for the *Science Through LEGO*TM *Engineering* units compared to a simplified model of an engineering design process. (Adapted from the eight-step process model in Massachusetts Department of Education, 2006.)

and reflection questions. The prompts and questions ask for writing, drawing, and numerical inscriptions, and each is an opportunity for students to record their emerging ideas related to the unit's science domain.

A Science Through LEGO™ Engineering Unit

One *Science Through LEGO*TM *Engineering* unit is the third-grade unit *Design a Musical Instrument: The Science of Sound*. In the opening session, students learn that their engineering design challenge is to create a new musical instrument that can play at least three different notes and contribute to a classroom band. Over the next six sessions, students build a series of artifacts and investigate them to explore how sounds are produced, transmitted, and varied across different sound producers. Using LEGO™ construction kit elements and craft materials, they create a miniature drum, pan pipe, guitar, and maraca. Students explore

the structural design of these instruments, observe how they look and sound when played, and consider how visible characteristics are related to sound characteristics. For example, how does the size of an object influence the pitch of the sound it makes? Throughout the unit, students are encouraged to consider how these relationships can inform their design of a new musical instrument. In the unit's two concluding sessions, students design, construct, and demonstrate musical instruments of their own invention (see Figure 13.4). They also demonstrate what they have learned about the unit's big science question: *How are sounds made?*

The teacher's guide for the *Design a Musical Instrument* unit is intended both to guide the teacher's facilitation of the learning and to support growth in the teacher's science and pedagogical content knowledge. It suggests that teachers and students follow a similar sequence of events in each individual session. This begins with introducing the scientific question to be investigated. Next,

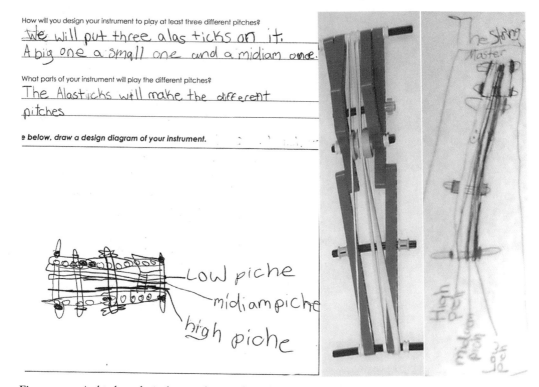

Figure 13.4. A third-grader's design plans and corresponding three-pitch musical instrument from the curriculum's overarching engineering design challenge.

students work independently for five minutes to respond to a brief brainstorming prompt – called an *exploration question* – related to that goal. For the majority of the session, students work in dyads to complete the challenge and investigation of the day. Each session concludes with a whole-class discussion to make sense of the challenge and investigation and return to the scientific question for the day. This sequence supports students in forming well-articulated cases that will be useful for later reasoning about the final design challenge.

Investigating Student Learning

A variety of approaches were used to investigate student learning within the *Science Through LEGO*[TM] *Engineering* units. For instance, using paper-and-pencil tests of science content knowledge, researchers conducted a quasi-experimental study comparing performance of *Science Through LEGO*[TM] *Engineering* students to that of

students using their district's status quo curricula. The experimental and comparison cohorts were students of the same teachers but over the course of two consecutive years. Each teacher taught with his or her typical science curriculum in the first year, and then in the second year taught the same science content but with the engineering-based unit. The two student cohorts were matched for demographic characteristics, and they completed the same multiple-choice and open-response items assessing the science learning objectives. These items were chosen from a bank of test items created and tested in a validation pilot study by a research team including an educational psychologist, a science education researcher, and engineering education researchers. (Items were created by the research team rather than chosen from national or state standardized science tests because many of the research participants were third graders, and the national and state science tests are written for fourth and fifth graders, respectively.)

Stated very briefly, the finding of the quasi-experimental study was that the increase in science content scores from pre- to post-test were greater for the engineering-design–based students than for the comparison students (Wendell & Rogers, in press).

Another study focused on students' scientific reasoning and investigated whether students using the *Design a Musical Instrument* curriculum were reasoning deeply about physical causes and effects related to sound. The goal of this study was to look for changes in reasoning about mechanisms. When students reason about mechanisms, they describe a physical process that links the cause of a phenomenon to its effect (Hammer, 2004). Before and after the *Design a Musical Instrument* unit, thirty-one third- and fourth-grade students participated in twenty-minute clinical interviews. In each interview, students answered questions about sound production, sound transmission, and pitch as they interacted with three separate sound-producing devices: a xylophone, a set of strings and hooks attached to wooden rectangular boards, and a series of rubber bands suspended between screws. Using Russ, Scherr, Hammer, and Mikeska's (2008) framework for identifying reasoning about mechanisms in students' oral discourse, the *Science Through LEGO*™ *Engineering* team analyzed the third- and fourth-graders' interview responses for indicators of reasoning about mechanisms. In accordance with Russ and colleagues' analytical framework, students' interview responses were characterized as featuring reasoning about mechanisms if they identified the entities involved in the phenomenon and (1) identified their activities, (2) identified their properties, (3) identified their organization, or (4) chained backward or forward in time (see Wendell, 2011, for further details).

Before the curriculum, the students' responses had strong evidence of reasoning about mechanisms 26% of the time when discussing sound production, 7% of the time when discussing sound transmission, and 8% of the time when discussing pitch. By contrast, after completing the *Design a Musical Instrument* unit, there was strong evidence

for reasoning about mechanisms 69% of the time when discussing sound production, 46% of the time when discussing sound transmission, and 40% of the time when discussing pitch. Statistical testing (McNemar's test for paired binomial samples) revealed that for each topic, the increase in reasoning about mechanisms was significant.

Third-grader Rena is an example of a student who shifted toward reasoning about mechanisms when responding to questions about pitch. In her pre-interview, Rena noticed only the relationship between size and pitch for all three of the instruments. She observed that repeatedly, "The small ones are always the highest ones." Her discussion of pitch stopped there. By post-interview, however, Rena was attempting to explain the process by which the smallest xylophone bar generated the highest pitch: "I think because the red one's shorter... Because they don't have that much, like, um, room to travel, so they have um, so we, it gets to our ears quickly. And it vibrates the fastest." Despite conflating vibration speed with propagation speed, Rena accurately reasoned that small objects offer sound less "room to travel," and that this fast traveling process underlies their higher pitches. At least five features of her explanation can be characterized as indicators of reasoning about mechanisms (Russ et al., 2008). First, Rena pointed out the *target phenomenon* of smaller components producing higher pitches. She also identified the shortest xylophone bar as the crucial *entity* in this phenomenon. She then focused on a *property of that entity* – its short length – and reasoned about how that property would influence the *entity's activity* – it would vibrate fast. She also *chained* two activities: the bar had to "travel" before the sound could "get to our ears."

The students' reasoning about the mechanisms of sound is an example of their participation in the scientific practice of constructing explanations. This is one of the eight key scientific practices emphasized in the new national *Framework for K-12 Science Education*, (NRC, 2012), which serves as a guide for the *Next Generation Science*

Standards. The principles of CBR and LBD are strongly aligned with the new framework's focus on students' participation in practices, idea development over recall of facts, and depth of experience over breadth.

Challenges of Promoting Learning Through Design

The *Design a Musical Instrument* example illustrates how CBR and LBD lay the groundwork for successfully using engineering design to promote ideas and practices. Yet there are challenges to learning disciplinary ideas and practices through engineering design. Here we consider how strategies used by other learning-through-design researchers address some of these challenges.

Iteration Toward Understanding, Not Just Toward Good Design

One challenge is to help teachers and students focus on not just iterating toward a good design solution but rather on iteration toward understanding. Teachers and students need to find opportunities to grapple with what they need to do next and move in a direction that will afford the best iterative refinement of student conceptions. The LBD approach addresses this challenge through its repeated activity structures and sequences, which are intended to provide practice at scientific reasoning and use of newly learned science ideas. Experiences with both LBD and *Science Through LEGO*™ *Engineering* show that when learners engage in these repeated kinds of activity sequences, they become habit and the children begin to reflect on their design experiences in ways that promote learning. But to make that happen, LBD had to integrate a set of iterative practices around design rules of thumb and *Science Through LEGO*™ *Engineering* had to integrate engineer's journals that scaffold the sequence of learning experiences. Without activities that explicitly ask teachers and students to reflect on

the science behind their design successes and failures, it doesn't happen naturally.

Silk, Schunn, and Cary (2009) similarly realized that students in a design-based science experience would do more scientific reasoning if they had ample opportunities to stop and reflect about their design experiences. Their approach, in the *Design for Science* curriculum, was to embed scaffolding throughout to make the engineering design processes explicit. For example, when students are designing an alarm system, they explicitly write the requirements for their alarm system and consider a variety of alternative options for meeting each requirement. They also break down the alarm system into several subsystems, each of which focuses on particular science concepts. They analyze the connections among the subsystems and consider how each contributes to the function of the entire system. Silk and colleagues looked at the domain-general science reasoning of the Design for Science students compared to those using another reform-based curriculum, Model Assisted Reasoning in Science, and to those using a traditional textbook curriculum. They assessed reasoning with six multiple-choice items on controlling variables and drawing conclusions from data about the relationships among variables (Lawson, 1978). They found significant increases by all three groups of students on these items, but the gains were largest for the Design for Science students. This finding is intriguing because from the students' perspective, the electrical alarm unit focused on engineering design, not on scientific reasoning.

For their research into teaching mathematics through robot design, Silk, Higashi, Shoop, and Schunn (2010) confirmed another important principle for promoting iteration toward understanding rather than just toward design solutions. They found that it is important to require students to explain their designs to an external audience, or "client." In their *Robot Synchronized Dancing* curriculum for middle school students, Silk and his colleagues asked students to produce a toolkit for synchronizing any robots on any dance routine, rather than to create just one routine in which several robots moved together (a challenge that could be accomplished by guess-and-check). Furthermore, Silk and colleagues insisted that students communicate their design ideas to a "choreographer" client. Students had to explain their synchronizing toolkit to a choreographer who might need to sync an infinite number of dancing robots to an infinite number of dance routines. This approach required students to represent mathematical models of the relationships among their dancing robots and enabled them to develop their skills and knowledge of proportional reasoning.

Young Students and Planning

A second challenge is that young students often do not engage in the planning phase of a design process (Rogers & Wallace, 2000; Welch, 1999), even though taking the time to plan can improve both the design product and the learning outcomes (Fortus, Dershimer, Krajcik, Marx, & Mamlok-Naaman, 2004; Song & Agogino, 2004). This challenge exists because if students are given enough time and materials, they can often use trial and error methods, rather than advance planning, to complete many engineering design problems. Therefore in the pre-college classroom, teachers need to find ways to make planning obviously useful. For example, they can set constraints on the time students will have to build, the number of times they can visit the materials bin, or the amount of material they can use. Alternatively, they can guide students in how to plan, and allow students to discover how the planning benefits them. Portsmore (2010) explored the issue of students' planning by carrying out two slightly different versions of her *Goldilocks Engineering* curriculum for first-grade students. One version included direct guidance on methods for planning, including drawing, talking, and making lists of needed materials. The other version allowed students to plan their designs but did not provide guidance on how to do so. Portsmore found that the students in the "planning" classroom were more

likely to request planning sheets for design challenges at the end of the unit, and to add labels to their drawings. They were also more likely to decode a design problem's implicit requirements, such as the need for Baby Bear's bed to be as long as his body, and to persist with their initial ideas for solving a design problem. These findings suggest both that young students' use of planning can be increased through specific supports for planning, and that young students' design processes can be facilitated by planning.

Educators of middle and high school students might also consider helping students become better planners by using a particular kind of design challenge called a model eliciting activity (Diefes-Dux, Moore, Zawojewki, Imbrie, & Follman, 2004; Lesh, Hoover, Hole, Kelly, & Post, 2000). For example, the *Robot Synchronized Dancing* curriculum created by Silk and colleagues (2010) was inspired by the model eliciting activities approach. Model eliciting activities are real-world, client-driven problems with multiple possible solutions but structured requirements for students to model the problem mathematically, document their solution method, and self-assess its effectiveness. As an engineering education tool, model eliciting activities "are not so much about product as they are about process, the development of higher order understandings that lead to solutions" (Diefes-Dux et al., 2004, p. 3). In a model eliciting activity created by Diefes-Dux and colleagues (2004) for a college freshman engineering course, students were required to develop a method to measure the "nano roughness" of a material given microscopic images. To complete the assignment, the students had to describe their method mathematically, apply it to different samples and evaluate its effectiveness, and write a memo to a company explaining how to use the method and what other information would be needed. Orienting the students' attention toward process over product, these requirements illustrate the six principles of model eliciting activities: *Model Construction* (students construct an explicit model of a mathematically significant situation), *Reality* (activity is posed in a realistic context), *Self-Assessment* (students test and revise their models on their own), *Model Documentation* (students document their models), *Construct Share-Ability and Re-Usability* (models can be shared with others and modified for other situations), and *Effective Prototype* (models are as simple as possible but still mathematically significant and scientifically meaningful) (Diefes-Dux et al., 2004).

A New Paradigm for Teachers

For most teachers, learning disciplinary ideas and practices through engineering design is a new, unfamiliar approach. It can be difficult to facilitate learning experiences where there are always multiple "right answers" and multiple paths to arrive at each answer. For in-service teachers, week-long professional development sessions in the summer have been found to build their confidence in this way of teaching (Capobianco, Diefes-Dux, & Mena, 2011; Kendall & Wendell, 2012). Pre-service teachers can be prepared to include engineering in their future classrooms by exploring engineering in their science methods courses. However, to teach disciplinary content through engineering design, teachers' content knowledge needs to be as strong as their engineering preparation. After a class session in which each student has created a different design construction, focusing conversation on a specific disciplinary idea requires the teacher to quickly apply his or her own content knowledge to the unpredictable creations of students, and then to transition students from physically manipulating materials to cognitively operating on ideas. For instance, when helping students explain how their musical instruments create sounds with different characteristics, teachers need their own firm understanding of how physical variables affect pitch and volume. To address this challenge of design-based learning, curriculum materials need to provide support for teachers to develop their content understanding as well as strategies for how to apply it to the particular engineering activities their students will be completing.

Hynes (2009) discovered that middle-school teachers who were new to teaching engineering, but not new to teaching, relied on experiences from hobbies and previous jobs to support their students' engineering. The teachers used examples from their own lives to illustrate the practices of engineering in which they wanted their students to engage. For example, one teacher who had previously worked in food services connected the engineering design process to his experiences designing kitchen layouts. Hynes also found that teachers benefit from sustained opportunities to work hands-on with the exact construction or software materials that their students will be using. Hynes' results also suggested that teachers with more years of teaching experience in general made more connections between math and science ideas and their new engineering activities. This may mean that when a school begins an effort to have students learn disciplinary content through engineering design, more experienced teachers could serve as coaches to assist novice teachers in making connections from engineering to other disciplines.

Another strategy for easing teachers' foray into learning through design experiences is to shift much of the burden of facilitation and assessment to students' themselves. For example, Kafai's Learning Science By Design program (Kafai & Muir Welsh, 2007) gave fourth and fifth grade students the necessary structure and opportunities to assess their own and their peers' designs in progress. Their design task was to create instructional software that would help third graders in their school learn about human physiology. Each student team had to generate its own question about the human body, find the answer to the question, and communicate that answer through software simulations that third graders would understand. The third graders visited two times to review the software simulations and ask questions of the fourth and fifth graders. The fourth and fifth graders also made presentations to each other and provided feedback. The teachers arranged the logistics of these reviews and presentations, and they provided guiding questions to focus students on four key components of assessment (understandability, science content representation, science content quantity, and simulation quality). However, it was the *students'* responsibility to offer constructive criticism that would move their own and their peers' work forward. In the end, students and researchers gave the final projects similar scores on the 20-point scale (5 points possible for each of the four assessment components). The average peer-assessment score was 12.16, and the average score given by researchers was 12.91. There was a significant and strongly positive correlation ($R = 0.85$) between the peer-assessment and researcher-given scores (Kafai & Muir Welsh, 2007). This suggests that students understood the criteria of understandability, science content representation, science content quantity, and simulation quality. They not only learned the physiology ideas explained in the software projects themselves, but they also learned practices of communicating science ideas to a less knowledgeable audience.

Time Limitations

Finally, learning through engineering design can be difficult simply because of the limited available time within the pre-college school schedule, which is often broken up into short 30-minute to 60-minute blocks. Many design-based units require substantially longer sessions than that because learning experiences that weave back and forth between design endeavors and disciplinary explorations can take longer than those focused on only one kind of activity. For example, the teachers who piloted the *Design a Musical Instrument* curriculum found that sometimes they had to allow students to make final touches to their musical instruments during free-choice period, or to complete their engineering journal work during language arts block.

Another way to deal with this challenge is to shift the goal of curricular planning away from maximum "coverage" and toward deep idea development and learning of

practices, as suggested by the new *Framework for K–12 Science Education* (National Research Council, 2012). If students are given the opportunity to experience learning through engineering design multiple times throughout the school year, they will be able to learn practices and develop ideas even if each specific design challenge is relatively brief or broken up into short blocks. The LBD approach and others based on it, such as Project-Based Inquiry Science (It's About Time, 2011; Kolodner et al., 2008), are most powerful when students have the chance to experience their cycles again and again throughout the school year.

The challenge of time constraints can also be addressed by intentionally integrating learning experiences in different domains together into one longer time block. Researchers from the CEEO have recently initiated a new *Integrating Engineering and Literacy* program that integrates engineering with language arts learning in grades K through 5 (McCormick & Hynes, 2012). Using the texts that they are already reading as part of the language arts curriculum, teachers and students discuss the problems faced by the characters and identify which problems could be solved by engineering design. For instance, in the book *The Mouse and the Motorcycle* (Cleary, 1990), the main character, Ralph, gets stuck in a trash can and could hypothetically work his way out of this predicament by engineering an escape device out of pieces of trash. Each student pair then chooses one problem to define more narrowly and solve by planning, prototyping, and testing. Finally, they engage in meaningful writing by composing letters about their solutions or proposing new endings to the story. During an *Integrating Engineering and Literacy* experience, teachers are addressing learning standards for both language arts and engineering, and they are also finding that science and mathematics ideas are being revisited. In a pilot study (McCormick & Hynes, 2012), two fourth-grade students decided to tackle an obstructed-view problem faced by the characters in a story. To solve the problem, the fourth graders wanted to design and build a periscope. This idea required them to reconsider their understanding of mirrors and light and angles, important topics in the science and mathematics curricula. Thus in *Integrating Engineering and Literacy*, students are able to engage in multiple academic disciplines in one long block, which requires no more time than a sequence of shorter blocks each dedicated to a single discipline.

Concluding Remarks

Despite the challenges described in this chapter, the potential benefits of learning experiences based in engineering design make worthwhile the effort associated with implementing such experiences. The power of the design-based approach stems from its ability to create opportunities for three kinds of activities: deep reasoning about ideas, sustained experimentation with physical or virtual materials, and authentic planning and construction of artifacts to meet specific requirements. In the research efforts profiled in this chapter, the affordances of engineering design are evident as pre-college students contribute to lively whole-class discussions, produce sophisticated drawing and text to express their disciplinary ideas, and enthusiastically collaborate with other students to bring design ideas to fruition.

As demonstrated by the research discussed in this chapter, engineering design can provide a deep, rich set of feedback for students to incorporate into their understanding and abilities. As a result, engineering design in the pre-college classroom is well-suited for integration with ideas and practices from other disciplines.

Of course, any curriculum approach on its own cannot do it all. Teacher knowledge, perspectives, and practices are all important factors in the success of students engaged in learning through engineering design. In studies of LBD modules, the degree to which students connected science to their designs seemed to depend on the extent to which teachers modeled science talk and the degree to which teachers facilitated students' participation in science practices

during pin-up sessions and gallery walks (Kolodner et al., 2003). When teachers thought about learning as iterative refinement of ideas (not just refinement of designs), students engaged enthusiastically with developing ideas, participating in practices, *and* constructing engineering solutions. Determining effective and supportive teacher professional development remains a major challenge in broadly disseminating any approach based on CBR and LBD.

Acknowledgments

This material is partially based on work supported by the National Science Foundation (DRL-0633952, ESI-9553583, and ESI-9818828), the McDonnell Foundation, and the BellSouth Foundation. The opinions, findings, and recommendations expressed in this material are those of the authors. We wish to thank all the colleagues, graduate and undergraduate students, and teachers who have contributed to curriculum and assessment development. Contributors for *Science Through LEGO*[TM] *Engineering* include Kathleen Connolly, Christopher Wright, Linda Jarvin, Chris Rogers, and Merredith Portsmore at Tufts University and Mike Barnett and Ismail Marulcu at Boston College. We also wish to thank the K–12 students and teachers whose enthusiastic participation has made this work possible. Advice from Aditya Johri and anonymous reviewers was very helpful in focusing this chapter. Thanks to all.

Footnotes

1. By engineering design, we refer to the full range of activities that a professional engineer engages in to solve a design problem fully – understanding the problem and the environment in which its solution must function well (Atman et al., 2007); generating ideas (Cross, 2004); learning new concepts necessary for its solution (through a variety of means, ranging from asking an expert to reading to carrying out an investigation); building models and testing them (Hazelrigg, 1999), analyzing

(Petroski, 1996), rethinking, and revising; and going back to any of the previous steps to move forward, repeating until a solution is found (Massachusetts Department of Education, 2006; National Research Council, 2012).

2. In addition, the *Science Through LEGO*™ *Engineering* team administered an attitudinal survey to 254 third- and fourth-graders at the end of engineering-design–based science units and found that 94% of the students agreed or strongly agreed with the statement "I like using LEGOs to learn science," and 93% of students agreed or strongly agreed with the statement "I liked using engineering to learn science" (Wendell & Rogers, in press).

3. Other previous teaching experiments, including those of Roth (1996, 2001); Penner, Giles, Lehrer, and Schauble (1997); Penner, Lehrer, and Schauble (1998); and Sadler, Coyle, and Schwartz (2000), also influenced the curriculum development work.

References

Achieve, Inc. (2013). *Next generation science standards.* Retrieved from http://www.nextgen science.org/next-generation-science-standards/

Atman, C. J., Adams, R. S., Cardella, M. E., Mosborg, S., Turns, J., & Saleem, J. (2007). Engineering design processes: A comparison of students and expert practitioners. *Journal of Engineering Education, 96*(4), 359–379.

Capobianco, B. M., Diefes-Dux, H. A., & Mena, I. B. (2011). Elementary school teachers' attempts at integrating engineering design: Transformation or assimilation? In *Proceedings of the 118th American Society for Engineering Education Annual Conference & Exposition.* Vancouver, BC, Canada.

Cleary, B. (1990). *The mouse and the motorcycle.* New York, NY: Harper Collins.

Cross, N. (2004). Expertise in design: An overview. *Design Studies, 25*(5), 427–441.

Diefes-Dux, H. A., Moore, T., Zawojewski, J., Imbrie, P. K., & Follman, D. (2004). A framework for posing open-ended engineering problems: Model-eliciting activities. In *Proceedings, 34th ASEE/IEEE Frontiers in Education Conference,* Savannah, GA (pp. F1A-3–F1A-8). Piscataway, NJ: IEEE.

Fortus, D., Dershimer, R. C., Krajcik, J. S., Marx, R. W., & Mamlok-Naaman, R. (2004). Design-based science and student learning.

Journal of Research in Science Teaching, 41(10), 1081–1110.

Gray, J., Camp, P., Holbrook, J., Fasse, B., & Kolodner, J. L. (2001). Science talk as a way to assess student transfer and learning: Implications for formative assessment. Retrieved from http://www.cc.gatech.edu/projects/lbd/ pubtopic.html.

Hammer, D. (2004). The variability of student reasoning, lecture 1: Case studies of children's inquiries. In E. Redish & M. Vicentini (Eds.), *Proceedings of the Enrico Fermi Summer School, Course CLVI* (pp. 279–299). Bologna, Italy: Italian Physical Society.

Hazelrigg, G. A. (1999). On the role and use of mathematical models in engineering design. *Journal of Mechanical Design, 121*(9), 336–341.

Hynes, M. M. (2009). *Teaching middle-school engineering: An investigation of teachers' subject matter and pedagogical content knowledge* (Unpublished doctoral dissertation). Tufts University, Medford, MA.

It's About Time. (2011). Project-Based Inquiry Science: PBIS™. Retrieved from http://www .its-about-time.com/pbis/pbis.html

Kafai, Y. B., & Muir Welsh, K. A. (2007). Evaluating students' multimedia science design projects in the elementary classroom. In R. Pintó & D. Couso (Eds.), *Contributions from science education research* (pp. 435–449). New York, NY: Springer.

Kendall, A. L. M., & Wendell, K. B. (2012). Understanding the beliefs and perceptions of teachers who choose to implement engineering-based science instruction. In *Proceedings of the 119th American Society for Engineering Education Annual Conference & Exposition,* San Antonio, TX.

Kolodner, J. L. (1993). *Case based reasoning.* San Mateo, CA: Morgan Kaufmann.

Kolodner, J. L. (1997). Educational implications of analogy: A view from case-based reasoning. *American Psychologist, 52*(1), 57–66.

Kolodner, J. L. (2006). Case-based reasoning. In K. L. Sawyer (Ed.), *The Cambridge handbook of the learning sciences* (pp. 225–242). Cambridge: Cambridge University Press.

Kolodner, J. L., Camp, P. J., Crismond, D., Fasse, B., Gray, J., Holbrook, J., Puntambekar, S., & Ryan, M. (2003). Problem-based learning meets case-based reasoning in the middle-school science classroom: Putting Learning by Design™

into practice. *Journal of the Learning Sciences*, 12(4), 495–547.

Kolodner, J. L., & Gray, J. (2002). Understanding the affordances of ritualized activity structures for project-based classrooms. In P. Bell, R. Stevens, & T. Satwicz (Eds.), *Keeping learning complex: International conference of the learning sciences* (ICLS) (pp. 221–228). Mahwah, NJ: Lawrence Erlbaum.

Kolodner, J. L., Starr, M. L., Edelson, D., Hug, B., Kanter, D., Krajcik, J., . . . Zahm, B. (2008). Implementing what we know about learning in a middle-school curriculum for widespread dissemination: The Project-Based Inquiry Science (PBIS) story. In *Proceedings of the International Conference of the Learning Sciences*, Utrecht, Netherlands (Vol. 3, pp. 274–281). International Society of the Learning Sciences.

Lawson, A. E. (1978). The development and validation of a classroom test of formal reasoning. *Journal of Research in Science Teaching*, 15(1), 11–24.

Lesh, R., Hoover, M., Hole, B., Kelly, A., & Post, T. (2000). Principles for developing thought-revealing activities for students and teachers. In *Handbook of research design in mathematics and science education* (pp. 591–645). Mahwah, NJ: Lawrence Erlbaum.

Massachusetts Department of Education. (2006). *Massachusetts science and technology/engineering curriculum framework*. Malden, MA: Author.

McCormick, M., & Hynes, M. M. (2012). Engineering in a fictional world: Early findings from integrating engineering and literacy. In *Proceedings of the 119th American Society for Engineering Education Annual Conference & Exposition*, San Antonio, TX.

National Research Council (NRC). (1996). *National science education standards*. Washington, DC: The National Academies Press.

National Research Council (NRC). (2012). *A framework for K-12 science education: Practices, crosscutting concepts, and core ideas*. Washington, DC: National Academies Press.

Penner, D., Giles, N. D., Lehrer, R., & Schauble, L. (1997). Building functional models: Designing an elbow. *Journal of Research in Science Teaching*, 34(2), 125–143.

Penner, D. E., Lehrer, R., & Schauble, L. (1998). From physical models to biomechanics: A design-based modeling approach. *Journal of the Learning Sciences*, 7(3/4), 429–449.

Petroski, H. (1996). *Engineering by design: How engineers get from thought to thing*. Cambridge, MA: Harvard University Press.

Portsmore, M. D. (2010). *Exploring how experience with planning impacts first grade students' planning and solutions to engineering design problems* (Unpublished doctoral dissertation). Tufts University, Medford, MA.

Puntambekar, S., & Kolodner, J. L. (2005). Toward implementing distributed scaffolding: Helping students learn from design. *Journal of Research in Science Teaching*, 42(2), 185–217.

Redmond, M. (1992). *Learning by observing and understanding expert problem solving*. (Unpublished doctoral thesis). College of Computing, Georgia Institute of Technology, Atlanta, GA.

Rogers, G., & Wallace, J. (2000). The wheels of the bus: Children designing in an early years classroom. *Research in Science & Technology Education*, 18(1), 127–135.

Roth, W.-M. (1996). Art and artifact of children's designing: A situated cognition perspective. *Journal of the Learning Sciences*, 5(2), 129–166.

Roth, W.-M. (2001). Learning science through technological design. *Journal of Research in Science Teaching*, 38(7), 768–790.

Russ, R. S., Scherr, R. E., Hammer, D., & Mikeska, J. (2008). Recognizing mechanistic reasoning in student scientific inquiry: A framework for discourse analysis developed from philosophy of science. *Science Education*, 92, 499–525.

Sadler, P. M., Coyle, H. P., & Schwartz, M. (2000). Engineering competitions in the middle school classroom: Key elements in developing effective design challenges. *Journal of the Learning Sciences*, 9(3), 299–327.

Schank, R. C. (1982). *Dynamic memory*. Cambridge: Cambridge University Press.

Schank, R. C. (1999). *Dynamic memory revisited*. Cambridge: Cambridge University Press.

Schank, R. C., & Abelson, R. L. (1977). *Scripts, plans, goals, and understanding*. Mahwah, NJ: Lawrence Erlbaum.

Silk, E. M., Higashi, R., Shoop, R., & Schunn, C. D. (2010, December–January). Designing technology activities that teach mathematics. *The Technology Teacher*, 21–27.

Silk, E. M., Schunn, C. D., & Cary, M. S. (2009). The impact of an engineering design curriculum on science reasoning in an urban setting. *Journal of Science Education & Technology*, 18, 209–223.

Song, S., & Agogino, A. M. (2004). *Insights in designers' sketching activities in new product design teams. Proceedings of the 16th International Conference on Design Theory and Methodology*, Salt Lake City, UT (Vol. 3a, pp. 351–360). American Society of Mechanical Engineers.

Welch, M. (1999). Analyzing the tacit strategies of novice designers. *Research in Science & Technology Education*, 17(1), 19–34.

Wendell, K. B. (2011). *Science through engineering in elementary school: Comparing three enactments of an engineering-design-based curriculum on the science of sound* (Doctoral dissertation). Retrieved from ProQuest Dissertations & Theses Database (3445103).

Wendell, K. B., & Lee, H.-S. (2010). Elementary students' learning of materials science practices through instruction based on engineering design tasks. *Journal of Science Education and Technology*, 19(6), 580–601.

Wendell, K. B. & Rogers, C. (in press). Engineering-design-based science, science content performance, and science attitudes in elementary school. *Journal of Engineering Education*.

Part 3

PATHWAYS INTO DIVERSITY AND INCLUSIVENESS

CHAPTER 14

Engineering Identity

Karen L. Tonso

Identity is a concept that figuratively combines the intimate or personal world with the collective space of cultural forms and social relations.

(Holland, Lachicotte, Skinner, & Cain, 1998, p. 4)

In everyday speech referring to someone as an engineer signals their having an engineering identity – marking that person as belonging to a group of people who practice engineering. In the popular imagination, engineers mistakenly tend to be considered (at least in the United States) as socially inept sorts who are fascinated with gadgets and fixing things, more practical than scientists, and somehow brainier than technicians. As such, though "engineer" provides vernacular shorthand for everyday conversation, research on engineering identity unpacks nuanced ways of being engineers, as well as delineates how being an engineer relates to doing engineering. Engineers' identification with their profession can be critical for persistence, both as a student and then as a professional (see Chapter 16 by Lichtenstein, Chen, Smith, &

Maldonado, this volume). Studies show that a lack of identification with, and by, engineering often motivates students to migrate out of engineering into other majors and can be a barrier for students in other majors to move into engineering (Seymour & Hewitt, 1997; Tonso, 2006a; see also Chapter 15 by Sheppard, Gilmartin, Brunhaver, & Antonio, this volume). In fact, identity and learning prove interconnected, as delineated by several scholars in the situated learning tradition from cultural anthropology (Greeno, 2006; Johri & Olds, 2011; Lave & Wenger, 1991; Tonso, 1997; Wenger, 1998). Within this tradition, learning is itself conceptualized as a change in identity that comes with participation. As learners move from novices to mature practitioners, they likewise shift from peripheral participation to fuller participation; they undergo identity transformations and their identifying with a community of practice becomes a central part of the learning that takes place. According to this perspective, students learn as they engage in meaningful activities, involved not only in the learning of content, but also in learning how to put content learning into perspective, gain conceptual understanding, and become part of a community. But, though self-identification grounds an emerging engineering identity, as Tonso (1999, 2006a) demonstrated, being identified by the community as belonging also proves crucial, and lack of identification can interfere with belonging for some with well-regarded competencies.

The centrality of identity in engineering education raises several important questions. What kind of experiences – institutional, organizational, group, and team – should educators provide students to encourage identifying with engineering? In what ways might educators anticipate dilemmas impeding students' identifying with engineering and how might learning opportunities use diversity advantageously? How might educators ensure identification of a wide range of engineering identities, those supported and nurtured (identified as belonging) within engineering? This chapter on engineering identity centers on two questions: In what ways has identity been conceptualized in the social sciences, and how have engineering education researchers understood engineering identity?

Social Science Conceptions of Identity

Identity has been theorized and conceptualized variously in social science thought ranging from developmental psychology, to social psychology, sociology, and cultural anthropology. Historically, psychological perspectives gave prominence to individuals versus community, with social psychology concerned with the influences of context on self and cultural anthropology situating selves in community. Where psychological perspectives saw autonomous actors, social psychologists saw persons structured (framed or otherwise produced) by social circumstances, and cultural anthropologists understood persons as both framed by and constituting cultural settings. Conversations about engineering identity tend to take up persons acting, reacting, and resisting in context, and thus tend to depend on social psychology and cultural anthropological perspectives.

Brubaker and Cooper argued that identity was "both a *category of practice* and a *category of analysis*" (2000, pp. 4–5, italics mine). As a category of practice, identity meant that persons acted in context to make sense of themselves and of events, to situate themselves relative to others within their everyday lives. And, considerable engineering education research follows this sense of identity. As a category of analysis, identity signaled that group membership provided a coalescing force for collective action. Here, researchers used such affiliations as a basis for analysis, understanding that considerable diversity within a particular group existed. Thus, following Brubaker and Cooper, this chapter divides conceptions of engineering identity into two realms: (1) engineer as collective affiliation and (2) engineer as individual, whether as autonomous actor or as agent in social interactions imbued with local culture and societal forces.

When taken as a collective formed for action or as a community affiliation, identity provided ways for individuals to coalesce around claims about group membership (e.g., ethnicity), such as the Hispanic community, or for political action, such as advocating gay rights or breast cancer awareness (after Brubaker & Cooper, 2000). As such, these groups make possible considering how members of a group survive difficult circumstances when social support networks are blind to their group's needs and interests (see, e.g., Carol Stack's 1974 study of a low-income African American urban community), as well as allow groups persistently left out of mainstream conversations a way to advocate for their group's interests, such as the emergence of a broad-based African American community during the civil rights struggle. Here, the term "culture" implies national identity, ethnic identity, or racial identity, and focuses on the shared goals or histories of members of the group. But collective terms do not otherwise imply uniformity across group members. A presumption of individual sameness within collectives could be interpreted as stereotyping individual group members, or as essentializing the group. To account for within-group group diversity, and to understand how diversity and collectivity were co-constructed, sociologists and cultural anthropologists began to consider how identity emerged through everyday social interactions in particular social and cultural contexts. Here, cultural anthropologists took the word "culture" to imply that a group with historical persistence shared a set of understandings (or meanings) about their world, and these understandings provided ways to make sense of events and to locate oneself in events. Feminists, critical race scholars, and others expanded on this interpretation of culture to locate the tensions within a particular culture and to uncover multiple meanings in a particular setting. (For further details, see Eisenhart's 2001 discussion of culture.)

Early conceptions of individual identity provided ways to consider claims about selfhood. Many of these conceptions emanated from psychological theorists, such as from Erik Erikson's stage theory of development (1968) and others from sociological theories of the self, such as G. H. Mead's social psychology (2001). Erikson's theory understood that social circumstances contributed to individual development, but such an identity development seemed a one-way process with social context impacting individual, but not vice versa. Mead considered the "I," or mental sense of oneself, as emerging from self-conversation (thinking) with the "Me," or social sense of oneself (2001, originally early twentieth century thought). Here, identity theorists considered individuals autonomous actors, and presentations of self the substance of identity.

Scholars within social psychology advanced other perspectives on identity. The social identity perspective (Oakes, 1987; Tajfel & Turner, 1979; Turner, Hogg, Oakes, Reicher, & Wetherell, 1987) argued that groups (and their norms) exert a regulatory influence on their members' identities and their behaviors (see Haslam, 2004, for a review). Here, individuals hold a personal identity as well as a range of social identities, or aspects of the self derived from memberships in social groups. Others in social psychology advanced identity theory or identity role theory (Stryker & Burke, 2000), a self-categorization of oneself as "an occupant of a role" (Stets & Burke, 2000, p. 225). In role-based identity, uniqueness, rather than uniformity, became emphasized by those who shared the same role (Stets & Burke, 2000), contrasting with social identity theory, which highlights uniformity among group members. Finally, Gee (2001) defined identity as "being recognized as a certain type of person," having four aspects that interconnected to form an individual's identity: (1) nature identity coming via uncontrollable forces, (2) institutional identity arising from being granted a position in an institution, (3) discourse identity a trait recognized in the self and by others through speech and action, and (4) affinity identity being a member of a group and engaging in the practices of that group. Thus, social psychology conceptions view identity as primarily about an individual, though accounting for the ways a context within which one is formed and one's group affiliations impinge on subjectivities.

Among cultural anthropologists, identity theories considered how individuals come to be considered members of a group, to learn identities in a community, and locate oneself in a particular cultural milieu. Here, "culture" in contemporary anthropological thought implies historical persistence and shared meanings about the nature of the world, about ways to make sense of events, and about their salience in particular settings. Some education researchers documented the ways that school settings *reproduced* the social order. Connell, Ashenden, Kessler, and Dowsett's (1982) study of school life in Australia illustrated how middle class schools reinforced Aussie culture's middle-class and working-class social divide, and how individuals developed identities consistent with group life. Connell and colleagues' study typifies identity theorizing that grew out of social *reproduction* theory. However, some sociologists and cultural anthropologists argued that considerably more interaction occurred between an individual and their social settings, arguing that individuals were not simply passive recipients of prescribed outcomes, and this led to new conceptions about identity.

Ethnographies of schooling contributed to understanding situated identities. Paul Willis's (1977) study of British middle-class and working-class young men, in a high school modeled after middle-class ways of life, illustrated the myriad ways that working-class lads actively resisted being made into middle-class graduates. However, their resistance contributed to remaining in a working class where a shrinking pool of jobs diminished students' economic futures. Students' resistance contributed to their social-class subordination in British life. Subsequent anthropological studies of peer-group life in U.S. high schools suggested social institutions neither produced uniformity across graduates nor pigeonholed students into existing social locations, because actors (students) made decisions that not only brought in affiliations from outside a particular institution, but also selectively chose what to make of various aspects of social institutions. In a suburban Midwestern U.S. school, Eckert (1989) found that middle-class Jocks affiliated with the school's "corporate" culture, but working-class Burnouts did not. The school's structuring of life framed students' decisions, but did not guarantee them, even though the school offered opportunities that asked some students to forego affiliations. Foley (1990) reported that, in a U.S. border town with ethnic diversity, student identities such as, in student vernacular, a jock, vato, band fag, cheerleader, kicker, good girl, or farm boy signaled social standing, ethnicity, and gender. Here, some students moved across identity borders, but ethnic boundaries proved more obdurate. Thus, cultural *production* theory inferred that "... people are not seen simply as living enactments of core cultural themes," (Holland, Lachicotte, Skinner, & Cain, 1998, p. 31), "... [but] persons develop through and around the cultural forms *by which they are identified,*

and identify themselves, in the context of their affiliation or disaffiliation with those associated with those forms and practices," via a process best termed "*codevelopment* – the linked development of people, cultural forms, and social positions in particular historical worlds" (p. 33, emphasis added).

Lines of thought from social psychology (summarized in Turner & Oakes, 1986), sociology (summarized in Brubaker & Cooper, 2000), and cultural anthropology (summarized in Holland et al., 1998, and in Levinson, Foley, & Holland, 1996) thus situate self in world. These sociocultural approaches highlight "becoming" and "belonging," instead of simply "being." Here, individuals are not isolated from their social and cultural circumstances, but viewed as actors in a continuously produced sociocultural world; participants who simultaneously shape their world and are being shaped by it. "[T]his vision emphasizes that identities are improvised – in the flow of activity within specific social situations – from the cultural resources at hand" (Holland et al., 1998, p. 4). Thus, conceptions of identity moved from being considered the product of individual *autonomy*, to social *reproduction* processes, to social *agency* in social contexts, to being considered persistently in flux via sociocultural *production* processes.

Ultimately, identity theorists began to consider how group affiliations (claims about race, ethnicity, gender, and sexuality, in particular) made a difference in identity productions in social institutions. By delineating difficulties, and resistance, when the associations of individuals differed from those of a social institution, researchers unpacked the masculinity of science, technology, engineering, and mathematics (STEM) areas of study (e.g., Carlone & Johnson, 2007; Eisenhart & Finkel, 1998; Seymour & Hewitt, 1997; Tonso, 1996, 1999; Traweek, 1988), the whiteness of STEM fields in higher education (e.g., Carlone & Johnson, 2007; Foor, Walden, & Trytten, 2007; Johnson, Brown, Carlone, Azita, & Cuevas, 2011; Malone & Barabino, 2009; Seymour & Hewitt, 1997), the heteronormativity of higher education (e.g., O'Connor,

1998; Talburt, 2000; Tonso, 2007), and the real power that accrued from affiliating with an institution's preferred way of life (e.g., Fordham, 1996; Tonso, 2007). Johnson (2001) wrote about women of color in undergraduate and graduate science majors who were variously constrained by science presumptions of value neutrality, with many "self-select[ing] out of that path" (p. 452), while other students challenged these presumptions in their research, "without renouncing their core values – altruism, the importance of relationships, wanting to be judged as an individual, pride in ability and achievement" (p. 452). Identity also involved being thought to belong, or *recognized* as belonging (Carlone & Johnson, 2007; Tonso, 1999, 2006a). Thus, sociocultural conceptions of identity provided increasingly complex understandings of interactions between and among individuals, the affiliations from which they come, and the institutions wherein they produce identities and are identified as belonging by such institutions.

Engineering identity research spans the social science conceptions of identity. Using illustrative studies, I synthesize engineering identity research findings. Rather than employing all the perspectives introduced in the preceding text, I follow Brubaker and Cooper's parsing into two notions of identity. I begin with engineering identity as collective identity, then turn to studies that focus on individuals, and then to studies of individuals situated in social settings.

Engineering Identity as Collective Identity

Engineering identity centered in collective identity provides a way to think globally about variations in engineers, and as Downey and Lucena argue, ways to think about how to prepare engineers for work globally. Historically, and owing to different national circumstances impinging on the evolution of engineering and engineering education, considerable cross-national variation existed in engineering identity

(Downey & Lucena, 2004). For instance, France gave prominence to mathematical theory, and thought it "should precede and guide both experimental research and design" (p. 4), while the United States elevated "experimental research and empirical practice" as the guide for design work (p. 5). Comparing British and American engineering concerned how engineering drawings ("dimensional plans") were used. For the British, dimensional plans were used "as a sophisticated instrument for the design of heavy machinery," and in America "as a production-control instrument, to subordinate work and thus shift the balance of power over production from workers to engineer-managers" (quoting Brown, p. 5). As these and other examples illustrated, engineering identity differed nationally and was socioculturally produced, not only being influence by national political and engineering education differences, but also influenced by the various ways these nations practiced capitalism.

Downey and Lucena also highlighted how engineering identities became diverse within national communities (Lucena, Downey, Jesiek, & Elber, 2008). Since at least the late twentieth century, multinational companies expect engineers not only to work with engineers from other countries, but also to situate their work in complex international circumstances (Lucena et al., 2008). Thus, with the landscape of engineering shifting globally, and with engineers needing to reposition themselves relative to these shifts, Downey and Lucena (2004) "offer[ed] the metaphor 'code-switching,'"[1] to name "the process through which engineers build legitimacy for themselves and their knowledge simultaneously in professional and popular terms" (p. 4). They described an Egyptian engineer switching between competing roles (codes), as an engineer and as a father, for example, or as an American-trained versus Egyptian-born and -practicing engineer. Here, caught between competing perceptions or imaginings about one's actions in the world, engineers made choices about who they were and could become and became diverse.

Downey and Lucena (2005; Lucena et al., 2008) also suggested ways to promote these global engineering identities, building on engineering education reforms from the late twentieth century that incorporated design engineering (courses intended to mimic everyday engineering work typical of engineering careers by adding teamwork and teaching communications skills) (e.g., Downey & Lucena, 2003; Dutson, Todd, Magleby, & Sorenson, 1997; Tonso, 1996). Downey and Lucena developed "Engineering Cultures," a course examining how "what counts as an engineer and engineering knowledge has varied over time and from place to place," where students consider themselves "global engineers" who recognize and value diverse perspectives (from http://www.engcultures.sts.vt.edu/overview.html). Their efforts raised the bar for bringing engineering identity into question and for opening conversations about how engineers from one set of circumstances could contribute to appropriate – culturally aware and humane – engineering solutions in unfamiliar places, where different goals and traditions exist.

Further, interested in changes occurring in engineering education globally and how these impacted institutional notions of engineering identity (Downey, Hegg, & Lucena, 1993), Lucena and his colleagues (2008) performed an historical ethnography of emerging change, informed by fieldnotes and interviews from (international) engineering education professional meetings (dating from the late 1990s to the early 2000s). Change occurred in the United States, Europe, and Latin America at different rates. Change was fastest in the United States, with U.S. organizations promoting a global presence. For instance, the *Journal of Engineering Education* (the central journal of the American Society of Engineering Education) transformed itself into the outlet for *research* about engineering education, and the Accreditation Board for Engineering and Technology (ABET) scaled up from a United States to a global role to reinvent itself as an international accreditation board. EC 2000, the culmination

of conversations about diversifying engineering expertise (accepted in the United States in 1997), encompassed not only mathematical, scientific, and engineering principles, but also other key areas: designing and conducting experiments; analyzing and interpreting data; matching a design to a particular context; working on multidisciplinary teams; problem identification, formulation, and solution; communicating effectively; understanding the impact of engineering solutions in light of contemporary issues; and using tools needed for engineering practice (ABET, http://www.abet.org/eac-current-criteria/). Here, a novel notion of engineer seemed likely to flow from diverse competencies.

Change came more slowly elsewhere. European reform centered engineering-competency conversations in "a new *regional identity* in terms of continental mobility and economic competitiveness" (p. 434, emphasis added). This represented an inversion of past practice with engineering identity now leading engineering competencies, instead of flowing from them, and to encompass both educational qualifications and job classification. Elsewhere, Latin American conversations about engineering competencies stalled over two competing notions of engineering identity (p. 434). Thus, at the macro level of geopolitical entities and regions, engineering competencies and identities proved to be inextricably interwoven, with each side simultaneously shaping, and being shaped by, the other. Here, engineering identity did not simply result, in an uncomplicated way, from engineering students moving through coursework and developing requisite engineering competencies, but engineering identity influenced changes in engineering competencies. Furthermore, global engineers became diverse practitioners, with diversifying competencies, an engineering identity termed *heterogeneous* engineers (MacKenzie, 1996).

Thus, engineering identity in this perspective produced complex understandings about the ways that nationality, global economic demands, and social and professional institutions set the stage for individual engineers' identity productions. These macrolevel discussions proved central to thinking about ways to ensure that engineering became more malleable as a profession, and thus more responsive to rapid changes in global circumstances. On the other end of the spectrum, engineering identity conversations took up both developmental psychology perspectives on identity and sociocultural perspectives.

Engineering Identity Conceived from Developmental Psychology Perspectives

Those researchers whose work incorporates an approach grounded in an individual's fit with engineering tend to theorize identity as a set of variables or factors common to, or needed by, engineering students to succeed in engineering education. For instance, consider researchers who considered personality traits and attributes needed to be a successful engineering student (Loui, 2005; Meyers, Ohland, Pawley, Siliman, & Smith, 2012; Pierrakos, Beam, Constantz, Johri, & Anderson, 2009). Meyers and colleagues studied key factors of students' identity related to self-identification as engineering students self-identified with engineering. Using survey data guided by a stage theory approach to engineering identity development (after Erikson, 1968), they found most respondents (more than 85%) agreed with statements also found to be important for becoming an adult (citing Arnett's 2000 study): being able to make competent decisions, work with others, accept responsibilities, speak using accurate technical terminology, complete their degree, and make moral decisions (p. 125). Loui (2005) studied students from engineering ethics classes and found "four [similar] kinds of characteristics of ideal professional engineers" (p. 385):

- Technical competence: technical knowledge, problem-solving skills, creativity
- Interpersonal skills: communication skills, effective teamwork
- Work ethic: conscientiousness, diligence, persistence
- Moral standards: honesty, integrity

Meyers et al. (2012) found that referring to students as engineers, understanding both the centrality of the first-year experience to engineering identity development and the qualitative differences between the first year and subsequent years, and appreciating the links between persistence and identifying as an engineer encouraged self-identification with engineering. Pierrakos and her colleagues (2009) took up Meyers and co-workers' final point, using interviews of eight freshmen engineering students, and comparing the responses of the four who persisted to the four who switched to focus on students' perceptions about engineering and their identifying with engineering. Persisters and switchers differed in ways that signaled persisters identified more with engineering than switchers.

These studies illustrate research about students' perceptions of what makes an engineer, but leave several important matters unstudied. For instance, such studies overlook calls for engineering education to produce global engineers and to broaden the participation of underrepresented groups by changing the culture of engineering. To take up sociocultural production of engineering identity among engineering students, researchers moved into campus and peer-group life to examine what engineering students did, how this related to their being recognized as engineers, and what notions of engineering identity emerged.

Engineering Identity Conceived from Sociocultural Perspectives

Engineering identity conceptions from sociocultural perspectives range from conversations about professional identity (e.g., Eliot & Turns, 2011; Anderson, Courter, McGlamery, Nathans-Kelly, & Nicometo,

2010), to technical/social dualisms (e.g., Faulkner, 2007), to identity productions in complex sites of practice (coproducing racial/ethnic, social class, gender, and sexuality affiliations) (e.g., Foor et al., 2007; Pawley & Tonso, 2011; Stevens, O'Connor, Garrison, Jocuns, & Amos, 2008; Tonso, 1996, 1999; 2006a, 2006b, 2007).

Professional Identity

Conversations about professional identity document "a negotiation between the social expectations related to a specific professional role and the needs, wants, and aptitude of the individual engaging in that role" (Eliot & Turns, 2011, p. 47). Internships, co-opportunities, and group work promoted a professional engineering identity (Eliot, Turns, & Xu, 2008), as did durable productions, such as portfolios (Turns, Sattler, Eliot, Kilgore, & Mobrand, 2012).

Anderson et al. (2010), in a cross-case analysis of engineers from six engineering firms (grounded in interview data), found that: "Engineering identity is a complex equation that factors in problem solving, teamwork, learning, and personal contributions" (p. 170). Communication, especially during team meetings, emerged as central to engineering work, though engineers wished for less time in meetings and more time doing "hands-on" engineering. A tension existed between business constraints and quality engineering, with engineers worried that business constraints might detract from problem solutions. And, these engineers generally gave lower priority to serving the public good, preferring "solving problems well – for themselves, for their team, for their organization, and for their client" (p. 170).

Studies focused on professional identity located important competencies underpinning engineering professional identity, and these mapped onto accreditation standards such as ABET criteria. However, these studies encompassed a presumption that exhibiting these engineering competencies results in having an engineering identity. Other sociocultural researchers, especially those with a critical perspective, found otherwise.

Straddling a Technical/Social Dualism

Some sociocultural researchers wondered about the ways that societal structures and individual affiliations were implicated in engineering identity productions. Focused on interactions among engineers, Wendy Faulkner (2007) unpacked engineers' identities in a variety of engineering work settings. She found competing explanations of engineer self – "technicist" and "heterogeneous" engineers. On the one hand, as Faulkner's friend intimated, different jobs in different engineering sectors are "all engineering really – all nuts and bolts, . . . well, nuts and bolts and people" (p. 331). The "nuts and bolts" implied the technicist engineer, which he shifted to heterogeneous engineer by adding "and people." Engineering identity encompassed having competencies in "technical" areas, but also having competencies that her engineer participants considered "social." But engineers in her studies proved uncomfortable straddling "a deep technical/social dualism at the heart of engineers' identities as engineers" (p. 332).

She found that gender stereotypes – technical's association with masculinity, and social's association with femininity – underpinned engineers' discomfort with a heterogeneous engineering identity. In fact, in everyday interactions as engineers, men valorized their "nuts and bolts" expertise, whereas a woman engineer was miscast as underprepared because of a lack of "nuts and bolts" experiences, though these had little to do with men's or women's engineering work. Plus, men downplayed the social skills central in their engineering. Faulkner wrote: "As my ethnography of building design engineering demonstrates, *heterogeneous engineering requires heterogeneous genders* – in the sense that it requires various mixes of stereotypically masculine and feminine strengths" (p. 351). Thus, Faulkner illustrated how societal stereotypes about gender constrained engineering identity production for both men and women, with men denying their social skills, but women's technical skills being questioned. Stereotypes also interfered with practicing the kinds of engineering envisioned for engineering education.

Identity Productions in Engineering Education

Also, engineering identity studies encompassed how engineering campuses frame engineering identity productions (Stevens et al., 2008; Tonso, 1999, 2006a, 2007). In longitudinal ethnographic studies of student design teams engaged in performing engineering for government and industry clients, engineering identity emerged as a complex interplay among campus cultural knowledge, sociocultural identity performances, and campus recognition routines (Tonso, 1997, 2006a).[2] Tonso's ethnographic mapping of a complex identity terrain illustrated how campus practices for noticing some kinds of students ahead of others set up two ways to belong: among the social and academic *over-achievers* who affiliated with technicist engineering, or among *nerds*, who demonstrated heterogeneous engineering. (See also Walther, Kellam, Sochacka, & Radcliffe, 2011, for an example of the complexity of engineering competency terrain.) Ultimately, technicist engineers and academic-science engineering were given prominence on campus (via the curriculum, teaching, grading, notice from faculty and administrators, and award distribution). By the time they were seniors, over-achiever students accrued considerable power, which they deployed in teamwork to advance their version of problem solutions (Tonso, 2006b). Unfortunately, in design teams, these over-achiever students demonstrated little in the way of engineering competence (either the technicist proficiencies presumed to underpin their academic success, or the design, social, and communication skills associated with heterogeneous engineering). Nerds, wary of powerful colleagues, seldom openly challenged them and worked behind the scenes. Thus, men nerds demonstrated diminished standing relative to men over-achiever colleagues, even though nerds'

engineering competence proved far superior to over-achievers'. However, other social group status arrangements occurred.

In addition to learning these intended curricula of the campus, other learning occurred that influenced identity productions, and these signaled other engineering education norms that contributed to unequal relations of power associated with societal group affiliations. First, some researchers documented how being of one sex or the other had implications for engineering identity development (e.g., Chu, 2008; Foor et al., 2007; Hacker, 1989; McIlwee & Robinson, 1992; Pawley & Tonso, 2011), while others studied how becoming an engineer produced a gendered order as a part of engineering identity, especially documenting how becoming an engineer preferred some forms of masculinity over others, and most masculinities over femininities (Faulkner, 2007; Lohan & Faulkner, 2004; Tonso, 1999, 2006a, 2007). Chu's (2008) study of women engineers and their on-campus experiences described considerable gender inequity. Here, women engineering students used a range of approaches on campus: some felt coerced and left, some embraced normative practices, and some creatively resisted circumstances they found troubling (echoing Johnson's college science majors, 2001; and Seymour & Hewitt's STEM majors, 1997). Tonso (1997, 1999) found that contra a wide variation in masculinity across engineering identities, only one identity term signaled a respectable female-marked engineering identity among *social* over-achievers (echoing Faulkner's technical/social gender stereotypes). Other female-marked terms (again among *social* over-achievers) signaled disreputable femininities. Over time students learned how to deploy gendering cultural knowledge, to use it to make sense of themselves and others in gender-stereotyping ways. Women engineers became invisible as engineers, but hyper-visible as women (1999). Campus incidents (faculty comments, administration pronouncements, student assumptions) routinely characterized women as unusual, which signaled engineering's normative masculinity (Tonso, 1999), and cast women as interlopers, with classroom climates often quite chilly (Tonso, 1996, 1999), echoing a long history of women being thought to invade engineering (Bix, 2000; Pawley & Tonso, 2011). Masculinities were ranked as well, with men who demonstrated the best kinds of social skills falling into disfavor. For instance, one man was miscast as gay because of his social-organization prowess, communication skills, and helpful demeanor, suggesting how identity contestations became gendered and implied social status (Tonso, 2006a).

Second, though considerable identity research in science education (K–12) and among college science majors takes up issues of social class as an analytic identity affiliation (e.g., Calabrese Barton, 1998), others demonstrate how social standing becomes produced as a part of engineering education (Tonso, 2007). Also, science education researchers study racial and ethnic identity (reviewed in Carlone & Johnson, 2007), but few in engineering or engineering education explicitly do so, and almost always consider race/ethnicity an analytic identity affiliation. An analysis of media representations of women (1930–1970) found only white, middle-class depictions of women engineers (Pawley & Tonso, 2011), and U.S. engineering's middle-class affiliations have been reported elsewhere (Oldenziel, 1999). Foor and colleagues (2007) present a watershed case study of a student engineer, "Inez," who clearly found her campus difficult, saying: "I wish that I belonged more in this whole engineering group" (p. 103). A multiminority woman, Inez came from lower socioeconomic circumstances than those of her classmates. Rather than her gender or her racial and ethnic affiliations underpinning her difficulties forming an engineering identity, in fact her social class most differentiated her from her classmates. "Her response [to why women do not pursue engineering] focused exclusively on socioeconomic circumstances which mark her as an outsider" (p. 107). This outsider status was reinforced on campus when faculty encouraged those few students who

participated in internships to share their experiences with the whole class, thinking retellings would benefit nonparticipants. However, "this practice contributes to a faculty-student dynamic of favoritism and exclusion and a student-student dynamic of elitism and 'us' versus 'them'" (p. 110). Inez, and others, countered outsider status, which she did with her "inspirational story of perseverance, creativity, motivation, and dedication – the essential qualities of a good engineer" (p. 106). But, low retention suggests not all students shifted to the margins of engineering will persist.

Finally, in rare instances, engineering identity findings reveal a heterosexist social order in which members of the lesbian, gay, bisexual, or transgendered (LGBT) community may not be welcome. In Pawley and Tonso's (2011) analysis of media representations of women engineers between 1930 and 1970, heterosexual assumptions overwhelmed representations. Women engineers were considered to be "pre-married" or dating material, especially after World War II. Likewise, comments about heterosexual social relations routinely occurred in team meetings, but students spoke nothing about homosexuality, though in interviews a few student engineers expressed outrage at the lack of respect given gay men on campus and an "out" lesbian hid her sexuality on campus because of the campus climate (Tonso, 2007). Such actions and inactions suggest how students learn to appear heterosexual to become engineers.

Thus, as students moved through engineering education, they participated in complex sociocultural productions underpinning engineering identity. When studied, *normative* (hegemonic) engineering identity encoded technicist expertise (contra calls for more heterogeneity) and masculinities (in spite of women's growing participation), as well as heterosexuality, middle-class values, and whiteness. But these norms could not constrain engineering identity productions, because engineers created selves that did not align ideologically. Engineering students with affiliations outside engineering's norms faced complex choices about how they practiced engineering, about what kinds of engineer they became, and about how they were identified as engineers. As the cases of Inez (Foor et al., 2007), nerd engineers (men and women) (Tonso, 2007), and heterogeneous engineers (Faulkner, 2007) illustrate, some engineering students created spaces to inhabit meaningful and productive engineer identities. However, other students (Tonso's Marianne and Pamela, for instance, 2007) exhibited heterogeneous engineering consistent with that of men students, but failed to be recognized as engineers by teammates, by the campus, and by on-campus industry recruiters. As Eisenhart noted: "Institutional identities [such as engineering identity] take form in relation to others both inside and outside the institutions, thereby creating situations that contain both hegemonic and counter-hegemonic possibilities" (1996, p. 183).

Beginning in the last decade of the twentieth century and continuing today, engineering education research developed increasingly complex ways to conceptualize engineering identity as both collective identity and individual identity mediated, to varying degrees, by social circumstances. Here, scholars in the sociocultural perspectives delineated engineering identity's coproduction with engineering education; that is, engineer was shaped by engineering's practices, as engineers simultaneously shape engineering through their creative engineering practices. Tonso (2007) argued that evolving engineering practice exemplified situated learning processes for coproducing engineering identities, genders, heterogeneous expertise, and differences in power. Such engineering identity conceptions grounded in critical anthropology perspectives suggest deep taken-for-granted grooves for becoming engineers, as well as established routines for identifying engineers as belonging. These frame engineering identity productions, and in particular influence how individuals navigate engineering realms in complex ways. Such navigations are mediated, both as they play out during social interactions in engineering work and in student course- and teamwork, and as institutions and workplaces make sense of individuals. Ultimately, some students' navigations between campus practices and personal dreams, hopes, and desires, impinge on their identifying as an engineer, which can move them (sometimes in novel ways) toward becoming engineers, or present challenges that interfere with their efforts to belong. In fact, some even fall outside the scope of identification routines, even though they considered themselves heterogeneous engineers (as did their colleagues), and though they demonstrated heterogeneous engineering (diverse competencies consistent with ABET's criteria). Finally, claims about engineering identity are ultimately grounded in identity politics in which contestations occur about what an engineer might be. Here, some sorts of engineers can be valued ahead of others, which is not only influenced by the weight of past practices, but also contributes to future visions for engineering identity norms. Thus, the term "engineering identity" signals the active production of engineer selves imbued with diverse engineering competencies, a process that moves some engineers through time and space, connecting them with other engineers and with engineering, both engineering's history and its future.

Identity Matters

As suggested by this chapter, identity matters. Engineering education provides a crucible for becoming engineers – activities, historically salient understandings about engineers and engineering, and routines for recognition as engineers – all of which frame how students navigate educational opportunities and, for some, become engineers thought to belong. And since for engineering identity, engineering education matters, the final issue of this chapter concerns next steps: What future directions for research might be considered? How might engineering educators make use of engineering identity research?

First, as to engineering identity research in engineering education, owing in large measure to the speed of change in social science conceptions of identity, interdisciplinary research programs at these frontiers seem to promise the greatest contribution to furthering understandings of engineering identity. As this review illustrates, there are important aspects of identity to be studied both as a category of analysis (group conceptions) and as a category of practice (everyday actions in context). Downey and Lucena's research suggests the importance of following change movements in engineering education, while ethnographic researchers (Chu, 2008; Faulkner, 2007; Foor et al., 2007; Stevens et al., 2008; Tonso, 1997, 2006a) suggest ways to follow processes of becoming an engineer. Far more of these studies are needed to broaden the scope of what is known about important processes, especially learning, and to deepen understandings about engineering education. Such research will entail securing funding for research that at present is difficult to find. Also, more insights are needed about the nature of the changes to engineering competencies, and how they fare in engineering-identity politics (such as in the technicist/heterogeneous, and over-achiever/nerd hierarchies). Instead of narrowing learning conversations to disciplinary knowledge (the *intended* curriculum), research about the ways *unintended* curricula proceed and their impact on engineering identity, about how reforms in engineering are taken up in identity productions, and about how vestigial structures (all too often related to racial, ethic, gender, and sexuality divisions) persist in the face of considerable progress toward substantive change in engineering education. In the main, more research needs to be undertaken that makes explicit the implicit understandings that faculty, students, and engineers hold about engineering and to consider the implications of these taken-for-granted understandings for students' developing engineering identities.

Second, as to engineering educators, research illustrates that engineering education and engineering identity are two sides of the same coin that mutually shape one another. Unlike the unidirectionality of cause-and-effect statements, engineering identity emerges via a process where mutual shaping occurs. Here, neither side can guarantee the other, but as the process of becoming an engineer during the college years moves through time, both sides influence change in the other. On the one hand, students mold themselves into beings recognizable as engineers, and in time as engineers who constitute what counts in engineering education they mold educational processes. This is apparent in reforms in engineering education: with industry's push to add design coursework in the 1980s and to expand beyond mathematical, scientific, and engineering principles and competencies to incorporate a wider range of competencies in late 1990s; with possibilities of a global engineer and international discussions about engineering competencies. And, faculty can be mindful of engineering's history and ensure that engineering education provides all students with meaningful ways to think of themselves as engineers, and that faculty find novel ways to recognize diverse competencies. With changes to engineering in a global economy, more diversity of expertise seems warranted, rather than less. By being alert to the ways in which engineering has made it difficult for some to become engineers and to be recognized as engineers, and by disrupting notions that this is the way engineering should be now, engineering faculty in ground-breaking, as well as in small, seemingly inconsequential ways, can signal the centrality of diversity to engineering's future. Also, faculty can become even more aware of potential for identity categories of analysis to be used as stereotypes, to act as if every group member has the same needs, interests, and desires. If, as Nasir, Rosebery, Warren, and Lee argue, "diversity is a pedagogical asset, rather than a problem to be solved" (2006, p. 498), then faculty can learn about students' expertise, interests, and aspirations and find ways to make these come to life in every course. Finally, opportunities to become engineers should not be so narrow that they offer students choices that entangle becoming an engineer with giving up some of who students are; instead students deserve opportunities for producing engineering identities that recognize diversity, are worth wanting, and lack opportunity costs associated with engineering identity.

Acknowledgments

I express my gratitude to Aditya Johri for his review of an early draft of this chapter, especially his helpful advice about identity conceptions in social psychology, as well as to Jrene Rahm for her continuing friendship and conversations about identity conceptions.

Footnotes

1. "Code-switching" emerged initially from linguistics studies in multilingual settings, and referred to speakers frequently using two or more languages as they spoke, to switching back and forth between languages. (See Benson 2001 for a discussion of these roots.) In more contemporary times, code-switching has come to imply moving among other symbolic terrains.

2. Stevens et al., 2008, reviewed additional research in this area, and suggested a similar process model of becoming an engineer, a three-dimensional version of engineering identity grounded in disciplinary knowledge, navigation, and identification.

References

Anderson, K. J. B., Courter, S. S., McGlamery, T., Nathans-Kelly, T. M., & Nicometo, C. G. (2010). Understanding engineering work and identity: A cross-case analysis of engineers within six firms. *Engineering Studies*, 2(3), 153–174.

Benson, E. J. (2001). The neglected early history of codeswitching research in the United States. *Language & Communication*, 21(1), 23–36.

Bix, A. S. (2000) "Engineeresses invade campus": Four decades of debate over technical coeducation. *IEEE Technology and Society*, 19(1), 20–26.

Brubaker, R., & Cooper, F. (2000). Beyond 'identity'. *Theory and Society*, 29(1), 1–47.

Calabrese Barton, A. (1998). Teaching science with homeless children: Pedagogy, representation, and identity. *Journal of Research in Science Teaching*, 35(4), 379–394.

Carlone, H. B., & Johnson, A. (2007). Understanding the science experiences of successful women of color: Science identity as an analytic lens. *Journal of Research in Science Teaching*, 44(8), 1187–1218.

Chu, H. (2006). *Being a female engineer: Identity construction and resistance of women in engineering schools.* (Order No. 3231625, Texas A&M University). *ProQuest Dissertations and Theses*, 253 pp.

Connell, R. W., Ashenden, D., Kessler, S., & Dowsett, G. W. (1982). *Making the difference: Schools, families and social division.* Sydney: Allen & Unwin.

Downey, G. L., Hegg, S., & Lucena, J. C. (1993). *Weeded out: Critical reflection in engineering education.* Paper presented at the 1993 American Anthropological Association (AAA) Annual Meeting, Washington, DC.

Downey, G. L., & Lucena, J. C. (2003). When students resist: Ethnography of a senior design experience in engineering. *International Journal of Engineering Education*, 19(1), 168–176.

Downey, G. L., & Lucena, J. C. (2004). Knowledge and professional identity in engineering. *History and Technology*, 20(4), 393–420.

Downey, G. L., & Lucena, J. C. (2005). National identities in multinational worlds: Engineers and 'engineering cultures. *International Journal for Continuing Engineering Education and Lifelong Learning*, 15(3/4), 252–260.

Dutson, A. J., Todd, R. H., Magleby, S. P., & Sorenson, C. D. (1997). A review of literature on teaching engineering design through project-oriented capstone courses. *Journal of Engineering Education*, 86(1), 17–18.

Eckert, P. (1989). *Jocks and Burnouts: Social categories and identity in the high school.* New York, NY: Teachers College Press.

Eisenhart, M. (1996). The production of biologists at school and work: Making scientists, conservationists, or flowery bone-heads? In B. A. Levinson, D. E. Foley, & D. C. Holland (Eds.), *The cultural production of the educated person: Critical ethnographies of schooling and local practice* (pp. 169–185). Albany: State University of New York Press.

Eisenhart, M. (2001). Changing conceptions of culture and ethnographic methodology: Recent thematic shifts and their implications for research on teaching. In V. Richardson (Ed.) *Handbook of research on teaching* (4th ed., pp. 209–225). Washington, DC: American Educational Research Association.

Eisenhart, M., & Finkel, E., with Behm, L., Lawrence, N., & Tonso, K. (1998). *Women's science: Learning and succeeding from the margins.* Chicago, IL: University of Chicago Press.

Eliot, M., & Turns, J. (2011). Constructing professional portfolios: Sense-making and professional identity development for engineering undergraduates. *Journal of Engineering Education,* 100(4), 630–654.

Eliot, M., Turns, J., & Xu, K. (2008). *Engineering students' extrinsic and intrinsic strategies for the construction of professional identity.* Paper presented at the Research in Engineering Education Symposium, Davos, Switzerland. Retrieved from http://www.engconfintl.org/

Erikson, E. H. (1968). *Identity, youth, and crisis* (1st ed.). New York, NY: W.W. Norton.

Faulkner, W. (2007). "Nuts and bolts and people": Gender-troubled engineering identities. *Social Studies of Science,* 37(3), 331–356.

Foley, D. (1990). *Learning capitalist culture: Deep in the heat of Tejas.* Philadelphia, PA: University of Pennsylvania Press.

Foor, C. E., Walden, S. E., & Trytten, D. A. (2007). "I wish that I belonged more in this whole engineering group": Achieving individual diversity. *Journal of Engineering Education,* 96(2), 103–115.

Fordham, S. (1996). *Blacked out: Dilemmas of race, identity, and success at Capital High.* Chicago, IL: University of Chicago Press.

Gee, J. P. (2001). Identity as an analytic lens for research in education. *Review of Research in Education,* 25, 99–125.

Greeno, J. (2006). Learning in activity. In K. Sawyer (Ed.), *The Cambridge handbook of the learning sciences* (pp. 79–96). Cambridge: Cambridge University Press.

Hacker, S. (1989). *Pleasure, power, and technology: Some tales of gender, engineering, and the cooperative workplace.* Boston, MA: Unwin Hyman.

Haslam, S. A. (2004). *Psychology in organizations: The social identity approach* (2nd ed.). London: SAGE and Cambridge: Cambridge University Press.

Holland, D. C., Lachicotte, W., Jr., Skinner, D., & Cain, C. (1998). *Identity and agency in cultural worlds.* Cambridge, MA: Harvard University Press.

Johnson, A. C. (2001). *Women, race, and science: The academic experiences of twenty women of color with a passion for science.* (Order No. 3005063, University of Colorado at Boulder). *ProQuest Dissertations and Theses,* 485 pp.

Johnson, A. C., Brown, J., Carlone, H., Azita, K., & Cuevas, A. K. (2011). Authoring identity amidst the treacherous terrain of science: A multiracial feminist examination of the journeys of three women of color in science. *Journal of Research in Science Teaching,* 48(4), 339–366.

Johri, A., & Olds, B. (2011). Situated engineering learning: Bridging engineering education research and the learning sciences. *Journal of Engineering Education,* 100(1), 151–185.

Lave, J., & Wenger, E. (1991). *Situated learning: Legitimate peripheral participation.* Cambridge: Cambridge University Press.

Levinson, B. A., Foley, D. E., & Holland, D. C. (Eds.). (1996). *The cultural production of the educated person: Critical ethnographies of schooling and local practice.* Albany: State University of New York Press.

Lohan, M., & Faulkner, W. (2004). Masculinities and technologies: Some introductory remarks. *Men and Masculinities,* 6(4), 319–329.

Loui, M. C. (2005). Ethics and the development of professional identities of engineering students, *Journal of Engineering Education,* 94(4), 383–390.

Lucena, J., Downey, G., Jesiek, B., & Elber, S. (2008). Competencies beyond countries: The re-organization of engineering education in the United States, Europe, and Latin America. *Journal of Engineering Education,* 97(4), 433–447.

MacKenzie, D. (1996). *Knowing machines: Essays on technical change.* Cambridge, MA: MIT Press.

Malone, K. R., & Barabino, G. (2009). Narrations of race in STEM research settings: Identity formation and its discontents. *Science Education,* 93(3), 485–510.

McIlwee, J. S., & Robinson, J. G. (1992). *Women in engineering: Gender, power, and workplace culture*. Albany, NY: State University of New York Press.

Mead, G. H. (Edited by M. J. Deegan) (2001). *Essays in social psychology*. New Brunswick, NJ: Transaction.

Meyers, K. L., Ohland, M. W., Pawley, A. L., Siliman, S. E., & Smith, K. A. (2012). Factors relating to engineering identity. *Global Journal of Engineering Education*, 14(1), 119–131.

Nasir, N. S., Rosebery, A. S., Warren, B., & Lee, C. D. (2006). Learning as a cultural process: Achieving equity through diversity. In K. Sawyer (Ed.), *The Cambridge handbook of the learning sciences* (pp. 489–505). Cambridge: Cambridge University Press.

Oakes, P. J. (1987). The salience of social categories. In J. C. Turner, M. A. Hoog, P. J. Oakes, S. D. Reicher, & M. S. Wetherell (Eds.), *Rediscovering the social group: A self-categorisation theory* (pp. 117–141). Oxford: Blackwell.

O'Connor, A. D. (1998). *The cultural logic of gender in college: Heterosexism, homophobia and sexism in campus peer groups*. (Order No. 9838392, University of Colorado at Boulder). *ProQuest Dissertations and Theses*, 387 pp.

Oldenziel, R. (1999). *Making technology masculine: Men, women and modern machines in America, 1870–1945*. Amsterdam: Amsterdam University Press.

Pawley, A., & Tonso, K. L. (2011). "Monsters of unnaturalness": Making women engineers' identities via newspapers and magazines (1930–1970), *Journal of the Society of Women Engineers – 60th Anniversary Edition*, 60, 60–75.

Pierrakos, O., Beam, T. K., Constantz, J., Johri, A., & Anderson, R. (2009). *On the development of a professional identity: Engineering persisters vs. engineering switchers*. Presentation at the 39th ASEE/IEEE Frontiers in Education Conference, San Antonio, TX.

Seymour, E., & Hewitt, N. M. (1997). *Talking about leaving: Why undergraduates leave the sciences*. Boulder, CO: Westview Press.

Stack, C. (1974). *All our kin: Strategies for survival in a black community*. NY: Harper & Row.

Stets, J. E., & Burke, P. J. (2000). Identity theory and social identity theory. *Social Psychology Quarterly*, 63, 224–237.

Stevens, R., O'Connor, K., Garrison, L., Jocuns, A., & Amos, D. M. (2008). Becoming an engineer: Toward a three dimensional view of engineering learning. *Journal of Engineering Education*, 97(3), 355–368.

Stryker, S., & Burke, P. J. (2000). The past, present, and future of an identity theory. *Social Psychological Quarterly*, 63, 284–297.

Tajfel, H., & Turner, J. C. (1979). An integrative theory of intergroup conflict. In W. G. Austin & S. Worchel (Eds.), *The social psychology of intergroup relations* (pp. 33–47). San Anselmo, CA: Brooks/Cole.

Talburt, S. (2000). *Subject to identity: Knowledge, sexuality, and the academic practices in higher education*. Albany, NY: State University of New York Press.

Tonso, K. L. (1996). The impact of cultural norms on women. *Journal of Engineering Education*, 85(3), 217–225.

Tonso, K. L. (1997). *Constructing engineers through practice: Gendered features of learning and identity development*. (Order No. 9800565, University of Colorado at Boulder). *ProQuest Dissertations and Theses*, 456 pp.

Tonso, K. L. (1999). Engineering gender – gendering engineering: A cultural model for belonging. *Journal of Women and Minorities in Science and Engineering*, 5(4), 365–404.

Tonso, K. L. (2006a). Student engineers and engineering identity: Campus engineer identities as figured world. *Cultural Studies of Science Education*, 1(2), 1–35.

Tonso, K. L. (2006b). Teams that work: Campus culture, engineering identity, and social interactions. *Journal of Engineering Education*, 95(1), 25–37.

Tonso, K. L. (2007). *On the outskirts of engineering: Learning identity, gender, and power via engineering practice*. Rotterdam, The Netherlands: Sense.

Traweek, S. (1988). *Beamtimes and lifetimes: The world of high energy physics*. Cambridge, MA: Harvard University Press.

Turner, J. C., Hogg, M. A., Oakes, P. J., Reicher, S. D., & Wetherell, M. S. (1987). *Rediscovering the social group: A self-categorization theory*. Cambridge, MA: Basil Blackwell.

Turner, J. C., & Oakes, P. J. (1986). The significance of the social identity concept for social psychology with reference to individualism, interactionism and social influence. *British Journal of Social Psychology*, 25(3), 237–252.

Turns, J., Sattler, B., Eliot, M., Kilgore, D., & Mobrand, K. (2012). Preparedness portfolios and portfolio studios. *International Journal of ePortfolio*, 2(1), 1–13.

Walther, J., Kellam, N., Sochacka, N., & Radcliffe, D. (2011). Engineering competence? An interpretive investigation of engineering students' professional formation. *Journal of Engineering Education*, 100(4), 703–740.

Wenger, E. (1998). *Communities of practice: Learning, meaning, and identity*. New York, NY: Cambridge University Press.

Willis, P. (1977). *Learning to labor: How working class kids get working class jobs*. New York, NY: Columbia University Press.

CHAPTER 15

Studying the Career Pathways
of Engineers

An Illustration with Two Data Sets

Sheri D. Sheppard, Anthony Lising Antonio, Samantha R. Brunhaver, and Shannon K. Gilmartin

Introduction

While calls for the strengthening of U.S. education once again surface in the name of global competitiveness, a primary issue facing engineering education is retention in the profession. As Lowell and Salzman (2007) have argued, the demand for engineers and scientists remains strong and the overall production of engineers and scientists appears more than adequate. The troubling trend over the last two decades, however, is that the highest performing students and graduates are leaving science and engineering pathways at higher rates than are their lower performing peers (Lowell, Salzman, Bernstein, & Henderson, 2009). This finding is significant for engineering education as it identifies an important direction for research in this area. Based on their study of pathways through and beyond college, Lowell et al. (2009) conclude that "students are not leaving STEM pathways because of lack of preparation or ability" and that research efforts should turn to "factors other than educational preparation or student ability in this compositional shift to lower-performing students in the STEM pipeline" (p. iii).

Our understanding of the aforementioned shift is limited even while the study of engineering career pathways began as early as the late 1970s with the work of LeBold, Bond, and Thomas (1977) on black engineers at Purdue University. Although the literature on engineering education and the profession has proliferated since that time, relatively few studies have looked carefully at the career decisions of engineering graduates. For instance, much of the work on engineering career pathways simply accounts for the numbers of engineers at different points in the pathway to quantify attrition points and rates (e.g., Bradburn, Nevill, Forrest, Cataldi, & Perry, 2006; Choy, Bradburn, & Carroll, 2008; Forrest Cataldi et al., 2011; Frehill, 2007a; Reese, 2003; Regets, 2006) and provides little information on differential pathways or the factors which influence these pathways. More recent work investigates aspects of early career engineers that reflect a focus beyond educational preparation and training and academic and technical ability (e.g., Fouad & Singh, 2011; Ro, 2011), but a thorough review reveals a collection of data sets and studies that remain incomplete for comprehensively understanding the early career pathways of engineers.

In this chapter, we characterize our current understanding with a review of this recent work and then present analyses of two data sets as an illustration of the kind of future research that is needed in the field. The first data set, drawn from the Academic Pathways of People Learning Engineering Survey (APPLES), focuses on engineering students' post-graduation career plans, while the second, drawn from the National Survey of Recent College Graduates (NSRCG), contains engineering graduates' actual career decisions. Together, they provide a deeper understanding of early career engineering pathways and highlight data and analysis needs for the next generation of studies on retention in the engineering profession.

Previous Studies of Engineering Career Pathways

Table 15.1 summarizes our review of previous research on engineering career pathways. For the purpose of our review, we focused on quantitative studies, studies that included engineering bachelor's recipients, and studies conducted in the United States within the past twenty years. We classified the resulting twelve studies along four dimensions: the focus of study, the scope of the sample, the study design, and the scope of the analysis.

Focus of Study

Studies of engineering pathways tend to be focused on one of two discrete measures: student career plans or post-graduation career paths. Just three of the published studies we reviewed analyze future plans and they are all recent. Eris et al. (2010) examined students' intentions to practice, conduct research in, or teach engineering for at least three years after graduation.

Not surprisingly, they found that the intentions of students who eventually completed engineering degrees increased in college, while the intentions of those who switched to other majors lagged behind. In another study, Amelink and Creamer (2010) identified correlations between engineering students' undergraduate experiences and their plans to work in engineering careers ten years later. Men were more likely to have engineering plans than were women, and both were more likely to have engineering plans if they reported satisfaction with various faculty- and peer-related factors, including the number of female role models in the department. Finally, Ro (2011) examined engineering students' post-graduation plans inside and outside of engineering and found that men had higher odds of pursuing engineering work and graduate school. In addition, students whose programs emphasized technical and professional skills and who had high self-confidence in their engineering and design skills were more likely to have engineering plans, whereas students whose programs emphasized professional values and who had high self-confidence in their contextual competencies were more likely to have non-engineering plans.

Student reports of career plans provide considerable knowledge about future pathways; in fact, some researchers have suggested that senior plans are the best proxy for actual career choice or graduate school enrollment (Astin, 1993; Pascarella & Terenzini, 2005). Nevertheless, these measures are imperfect because some students change their minds before graduation or develop new interests after initially joining the workforce as an engineer.

Other work has addressed this issue through their investigations of engineering alumni. Several of these studies focus on gender-related issues, looking at the employment patterns of men and/or women (Fouad & Singh, 2011; Frehill, 2007a, 2007b, 2007c, 2008; Robst, 2007). Reese (2003) examined recent civil engineering graduates, concluding that most are employed within private consulting firms, change jobs infrequently, and do not hold additional degrees.

Table 15.1. Studies of Engineering Career Pathways Since 1992

Study	Focus of Study	Data Source	Sample[1]	Sample Scope	Study Design	Scope of Analysis	Career Constructs
Amelink & Creamer, 2010	Student career plans	The Engineering Student Survey	1,629 freshman through senior engineering students in 2004	Nine institutions	Cross-sectional	Inferential analyses	Plans for engineering work
Eris et al., 2010	Student career plans	Persistence in Engineering (PIE) Survey	160 freshman students who entered college in 2003 tracked for four years	Four institutions	Longitudinal	Inferential analyses	Plans for engineering work
Ro, 2011	Student career plans	Prototype to Production (P2P) survey	5,239 sophomore through seniors engineering students in summer/ spring 2009	Thirty-one institutions	Cross-sectional	Advanced inferential analyses	Plans for engineering work and/or graduate school and for non-engineering work
Bradburn, Nevill, Forrest Cataldi, & Perry, 2006	Post-graduation career paths	Baccalaureate and Beyond 1992/1993 Cohort	9,000 bachelor's recipients earning degrees from 1992–93, 6.4% of whom earned engineering degrees (weighted; unweighted *n* not reported)	National	Longitudinal	Descriptive analyses	Labor force status Salary Additional education Career decision factors

(*continued*)

Table 15.1 (continued)

Study	Focus of Study	Data Source	Sample	Sample Scope	Study Design	Scope of Analysis	Career Constructs
Choy, Bradburn, & Carroll, 2008	Post-graduation career paths	Baccalaureate and Beyond 1992/1993 Cohort	9,000 bachelor's recipients earning degrees from 1992–93, 6.4% of whom earned engineering degrees (weighted; unweighted n not reported)	National	Longitudinal	Descriptive analyses	Labor force status Salary Occupational field Career decision factors
Forrest Cataldi, et al., 2011	Post-graduation career paths	Baccalaureate and Beyond 2008 Cohort	16,000 bachelor's recipients earning degrees from 2007 to 2008, including an unknown number of engineering degrees (weighted; unweighted n not reported)	National	Cross-sectional	Descriptive analyses	Labor force status Salary Additional education
Fouad & Singh, 2011	Post-graduation career paths	Project on Women Engineers' Retention (POWER)	3,700 women who earned bachelor's degrees in engineering surveyed in 2009	Thirty institutions	Cross-sectional	Advanced inferential analyses	Labor force status Occupational field Career decision factors
Frehill, 2007a	Post-graduation career paths	2003 National Survey of College Graduates (NSCG)	15,223 men and women who earned engineering degrees between 1980 and 2002	National	Cross-sectional	Descriptive analyses	Labor force status Occupational field Primary work activity Organizational sector Additional education

Frehill, 2007b, 2007c, 2008	Post-graduation career paths	Society of Women Engineers (SWE) National Survey about Engineers	5,027 men and women who earned bachelor's degrees in engineering from 1985 to 2005	Twenty-five institutions	Cross-sectional	Descriptive analyses	Labor force status Occupational field Career decision factors
Reese, 2003	Post-graduation career paths	American Society of Civil Engineers (ASCE) Young Members Study	1,165 early career civil engineers ages twenty-two to thirty-five in fall 2001	One professional society	Cross-sectional	Descriptive analyses	Years worked Positions held Organizations worked for Salary Additional education Career decision factors
Regets, 2006	Post-graduation career paths	2003 Science and Engineering Statistical Database surveys	Engineering alumni ten-plus years post-graduation in 2004 (n not reported)	National	Cross-sectional	Descriptive analyses	Primary work activity Relatedness of occupation to degree Additional education
Robst, 2007	Post-graduation career paths	1993 National Survey of College Graduates (NSCG)	124,063 men and women earning bachelor's degrees before 1993; 0.029% of women and 0.26% of men earned engineering degrees	National	Cross-sectional	Descriptive analyses	Relatedness of occupation to degree

Three studies have looked at engineering graduates of both genders and from different majors in the aggregate (Bradburn et al., 2006; Choy et al., 2008; Regets, 2006). Bradburn et al., for example, reported that nearly 60% of all engineering graduates from 1992–3 were still employed in engineering, architecture, or computer science and 30% had gone on to pursue additional education ten years after graduation. Although helpful for understanding the pathway while in the workforce, these studies are unable to link pre-labor force factors such as college experiences to post-graduation employment and/or graduate school enrollment.

Sample Scope

Research on the career trajectories of engineers has been undertaken with either large, national samples or smaller, more limited data sets. The National Science Foundation (NSF) offers the former with its Scientists and Engineers Statistical (SESTAT) Data System. SESTAT integrates data from three NSF-sponsored surveys administered every two to three years: the National Survey of College Graduates (NSCG), the National Survey of Recent College Graduates (NSRCG), and the Survey of Doctoral Recipients (SDR).

The SESTAT surveys capture a range of information on the employment and education of engineering graduates. An advantage of SESTAT is the ability to compare engineering alumni with graduates from other fields of study. For example, Regets (2006) analyzed the 2003 SESTAT surveys and found that engineering bachelor degree recipients were less likely than those from other science and science-and-engineering–related fields to pursue additional education; however, they were among the most likely to report that their job was related to their undergraduate education. The size of SESTAT data sets also facilitates analysis of gender differences. Robst (2007), for example, analyzed 1993 NSCG data and found that 16% of women engineering graduates perceived a mismatch between their college major and their job, compared to 11%

of men. Furthermore, women reporting a high degree of mismatch often cited family reasons as the main cause for the mismatch while men cited a change in career interests. Frehill (2007a) conducted a similar analysis and found that women were less likely than men to be employed in engineering at all points in their careers, more likely to earn master's degrees, and less likely to earn doctorates.

A few researchers have used the nationally representative Baccalaureate and Beyond (B&B) data set available from the National Center for Education Statistics. The B&B data set is a longitudinal panel study that explores post-baccalaureate education and work experiences. Several researchers have used it to examine engineering graduates' labor force status, salaries, and post-bachelor's degrees (Bradburn et al., 2006; Choy et al., 2008; Forrest Cataldi et al., 2011). For example, Bradburn et al. found that, relative to graduates in other majors, engineering graduates were among the most likely to be employed full time, to earn high salaries, and to view their undergraduate degrees as very important to their work and careers. As government-sponsored databases, the B&B and SESTAT provide relatively easy access to data on engineering graduates. A drawback to these national samples is the lack of engineering-specific depth due to their broad coverage of many fields.

Researchers have utilized smaller data sets to provide greater depth into engineering careers, including reasons for leaving the profession. For example, Fouad and Singh (2011) and Frehill (2007c, 2008) provide a more nuanced understanding of women engineers' career decisions. Similar to studies with men engineers (e.g., Robst, 2007), both found that women often leave because they lose interest in engineering or develop a new interest in another field. They further found, however, that women also leave due to a dislike of the work culture or a perceived lack of advancement opportunities. Surveys that are administered to targeted samples can allow for in-depth analyses of specific engineering disciplines (e.g., mechanical or chemical engineering) as well. In her investigation of the effects of major on engineering students' plans, for example, Ro (2011) discovered that, compared to mechanical engineering, those who major in general engineering have greater odds of pursuing non-engineering careers.

Study Design

Three of the studies reviewed employed longitudinal designs (Bradburn et al., 2006; Choy et al., 2008; Eris et al., 2010). Longitudinal studies are preferable because they allow researchers to measure directly change in career aspirations and attainment. They can also help identify potential contributors to change. For example, Choy et al. used B&B data to illustrate the migration of 1992–93 engineering graduates to business, management, and computer science in the ten years after graduation. Unfortunately, the B&B data do not allow an exact reason to be determined, although Choy et al. suggested career advancement as a factor.

Most research on engineering career pathways has been cross-sectional. Cross-sectional surveys are less expensive and simpler to administer. Data from cross-sectional surveys are also available sooner to reflect the contemporary status of engineers. They cannot measure explicit change, of course, and are therefore limited in their use to address questions regarding the impact of academic and work experiences or institutional or programmatic effects over time. Some studies compare data from multiple cohorts, for example, sophomores through seniors (Ro, 2011) or engineering graduates at different points in their careers (Frehill, 2008). In the latter study, Frehill compared engineering graduates from 1985–2000 and found that women from earlier cohorts were more likely to have left engineering for time- and family-related reasons than were women from more recent ones, reflecting an increase in responsibilities outside of work later in life. Care must be taken, however, in how differences (or "changes") between cohorts are interpreted because each cohort is subject to a unique set of unmeasured experiences and contexts.

Scope of Analysis

With little exception, most studies of engineering alumni career paths have been descriptive, reporting the proportions of graduates who have remained in the labor force or engineering profession (Bradburn et al., 2006; Choy et al., 2008; Forrest Cataldi et al., 2011; Frehill, 2007a, 2007b, 2007c, 2008; Reese, 2003; Regets, 2006; Robst, 2007). Studies of students' career plans, on the other hand, tend to feature inferential statistics. For example, Amelink and Creamer (2010) used simple correlations to identify significant relationships between students' undergraduate experiences and future plans, and Eris et al. (2010) used analysis of variance to study differences between students who persisted in engineering majors and those who did not.

Ro (2011) and Fouad and Singh (2011) adopted more advanced statistical techniques. Ro used multinomial logistic regression to examine the effects of personal and programmatic factors on student career intentions and found that pre-college characteristics were less important than were college experiences and engineering abilities. In their study of women engineers, Fouad and Singh also used regression techniques and found that women who were strongly committed to engineering were more likely to feel well-supported by the organization and to have high confidence in their abilities to perform engineering, manage their multiple life roles, and navigate their way at work. These studies illustrate the clear benefit of multivariate techniques in producing a more sophisticated understanding of engineering pathways.

Additional Factors

The studies in Table 15.1 differ in other ways. For example, some focus on particular subgroups (Fouad & Singh, 2011; Reese, 2003) whereas others compare several groups, for example, women and men (Frehill 2007a,

2007b, 2007c, 2008; Robst, 2007) or different disciplines (Ro, 2011). The period in which a population is studied and the types of questions asked are also of importance, especially when early career engineers are of interest. Studies of engineering pathways will have limited generalizability to early career engineers the more distal the sample is from their graduation. Similarly, there may be different implications for engineering students' intentions to practice engineering three years post-graduation (Eris et al., 2010; Ro, 2011) as opposed to ten years (Amelink & Creamer, 2010). Researchers interested in studying engineers' career goals and actions should consider these issues when designing their studies.

Two Data Sets, Two Explorations of Early Career Pathways

Our review in the previous section suggests that research on engineering pathways has been somewhat constrained by the data available, the scope of analysis, and the particular focus of individual studies. In this section, we introduce two complementary data sets, which, when analyzed in tandem, can provide a more thorough exploration of early career pathways of engineers. The breadth, depth, and data collection period of these data sets – they represent some of the most recent available data on engineering pathways in the United States – make them ideal candidates for investigation.

The first data set is drawn from the Academic Pathways of People Learning Engineering Survey (APPLES). APPLES data focus on the collegiate aspect of the early career pathway, capturing undergraduates' experiences with engineering education, orientations toward engineering work, and current post-graduation career plans. A cross-sectional data set, the strength of APPLES is its rich data on the undergraduate experience that may contribute to the formation of students' post-graduation plans. A diverse range of engineering schools participated in APPLES, providing a means to examine

institutional and programmatic characteristics that may be associated with students' plans. APPLES is not nationally representative, however, and actual post-graduation destinations and experiences are not within the scope of the data set.

The second data set is drawn from the National Survey of Recent College Graduates (NSRCG). The NSRCG is administered by the NSF as part of the SESTAT system and collects information from individuals who recently obtained bachelor's or master's degrees in science, engineering, or health (SEH) fields. The NSRCG focuses on the post-graduation portion of the early career pathway, allowing researchers to examine what recent engineering graduates actually do with their degrees. For our work here, the key features of the NSRCG are its large, nationally representative sample and focused set of questions on respondents' occupational field and degrees pursuing and earned since their first engineering bachelor's degree.

The Academic Pathways of People Learning Engineering Survey

APPLES is one of the research tools developed and used by the NSF-funded Academic Pathways Study (APS), one of several projects undertaken by The Center for the Advancement of Engineering Education (CAEE).[2] It is the third and last in a series of surveys designed as part of APS. The instrument comprises fifty questions across the following domains:

- Motivation to pursue engineering studies
- Confidence in engineering-related skills and perceived importance of these skills in professional engineering work
- Academic experiences in engineering and non-engineering classes, as well as co- and extracurricular involvement in engineering and non-engineering activities (e.g., engineering research, internships and co-ops, clubs and organizations)
- Self-assessed knowledge of the engineering profession

- Perceptions of knowledge of key engineering and design skills
- Satisfaction with college, level of stress, and financial concerns
- Major field of study
- Socioeconomic background and other demographic characteristics (citizenship, gender, age, race/ethnicity, etc.)
- Intention to persist in the engineering profession, and specific plans after graduation

Each of these domains of questions was developed specifically for the study of engineering academic and career pathways, building on earlier APS instrumentation and findings of engineering students (see Atman et al., 2010). Most relevant to this chapter, they include student estimates of their likelihood of working in an engineering job, working in a non-engineering job, going to graduate school in an engineering discipline, and going to graduate school in a non-engineering discipline after graduation. Numerous additional measures are available as potentially important correlates and predictors of plans. For example, students' post-graduation plans can be influenced by motivation to pursue engineering studies, involvement in extracurricular activities, specific engineering major, as well as citizenship, race, and gender.

APPLES was administered to more than 4,000 undergraduates across twenty-one U.S. institutions in the spring of 2008. The average survey response rate was 14% (individual school response rates varied from 49% at a small institution to 5% at a medium-large institution) (Donaldson, Chen, Toye, Clark, & Sheppard, 2008). Ninety-two percent of the 4,266 respondents self-reported as engineering majors, including 1,753 first-year and sophomore students and 2,143 juniors, seniors, and fifth-year students (the balance did not mark their cohort status).

Women are overrepresented among APPLES junior and senior respondents as compared with the U.S. population of students who earned an engineering degree in 2008 (NSF, 2011). In terms of field,

mechanical engineering majors are slightly overrepresented, and electrical engineering majors are slightly underrepresented, compared with 2008 engineering degree-earners (NSF, 2011). The APPLES data are not weighted due to incomplete population data at participating schools. Accordingly, researchers using the APPLES data set have tended to conduct all analyses separately by gender and, at times, major, to flag significant between-group differences and provide appropriate qualifications to inferences using aggregate data.

The institutional sample for APPLES is a resource for researchers. The twenty-one institutions are geographically diverse, representing seventeen states across major U.S. regions, and structurally diverse in terms of size, control (public vs. private), academic selectivity, and Carnegie classification. Like the student sample, the institutional sample is not nationally representative. However, it does provide some flexibility for researchers who are interested in studying how various institutional contexts are differentially associated with the career plans of engineering undergraduates as well as how they may vary with student characteristics and college experiences. We provide an example of this type of inquiry in the next section.

Sheppard et al. (2010) elaborate on the technical details of APPLES sampling and data. One important aspect to emphasize is that, similar to other studies on post-graduation career plans, the APPLES data set is cross-sectional, capturing students' plans and perceptions at only one point in time. However, the APPLES data set is distinctive due to its inclusion of a large number of variables focused on studying engineering and non-engineering career pathways and its diverse, multi-institutional sample.

For this chapter, we have augmented the original APPLES data set with three additional sources of data in order to conduct more comprehensive analyses of engineering pathways. We obtained median annual wage data by engineering field and U.S. state from the Bureau of Labor Statistics (BLS), National Cross-Industry Estimates,

May 2007 (see BLS, 2011). We also collected institutional characteristics such as selectivity, size, and aggregate student body characteristics from the Integrated Postsecondary Education Data System (IPEDS; see National Center for Education Statistics [NCES], 2011) and the American Society for Engineering Education's (ASEE) Engineering and Engineering Technology College Profiles (see ASEE, 2011).[3]

The National Survey of Recent College Graduates

The National Survey of Recent College Graduates (NSCRG) is sponsored by the NSF's Division of Science Resources Studies (SRS) and is one of three SRS surveys of graduates and professionals in SEH fields in the United States. The NSRCG has been conducted every two to three years since 1974. For the purpose of this chapter, data from the 2006 administration of the NSRCG are used. Although individuals earning bachelor's and master's degrees in the target years of 2002–3, 2003–4, and 2004–5 were included in the 2006 administration, our sample consists of only those who earned bachelor's degrees in engineering. Details about the administration can be found in the 2006 NSRCG technical report (NSF, 2006).

The NSRCG instrument collects a wide range of information about graduates' additional education since completing their "first bachelor's degree," as well as details about their current employment. The survey comprises 68 questions across the following domains:

- Undergraduate background (e.g., GPA, major field of study)
- Additional degrees earned or currently pursuing (e.g., degree type, degree field, financial support for study, reasons for pursuing study)
- Current employment status ("working for pay or profit" [yes/no] at the time of the survey)
- Characteristics of current and principal job (e.g., title, job category, type of

employer, work activities, relationship to highest degree, salary)
- Additional work experiences (e.g., workshops, training)
- Demographic characteristics (gender, race/ethnicity, partnership status, number of children, etc.)

These measures allow researchers to construct major markers of graduates' education and employment pathways in the roughly two years after college. (Throughout this chapter, we use "two years out of college" to denote the average span of time from graduation to survey completion for NSRCG respondents.) Not every detail about these years is captured – for instance, it is possible only to identify characteristics of respondents' current jobs (or, for those not currently working, their most recent jobs), rather than characteristics of each position that the respondent may have held in this time period. The data do show, however, if these graduates are employed in an engineering occupation, in a science or science-and-engineering-related occupation, or in a non–science-and-engineering occupation two years out of college, and if they have earned second bachelor's degrees, master's degrees, or professional/doctoral degrees in an engineering field, in a science or science-and-engineering–related field, and/or in a non–science-and-engineering–related field. Like APPLES, other measures in the NSRCG can serve as important covariates of these education and employment outcomes, such as undergraduate major, gender, and reasons for currently pursuing an additional degree.

The NSRCG data set is large and designed to be nationally representative. The 2006 administration includes 18,000 individuals who earned degrees in SEH fields from 295 institutions during the target years. A sample weight is included to adjust for unequal selection probabilities, nonresponse at the institutional and graduate level, and de-duplication of graduates sampled more than once. The weighted sample includes more than 1.98 million individuals

Table 15.2. A Comparison of APPLES and NSRCG Data Sets

	APPLES 2008	NSRCG 2006
Sample	2,143 junior and senior engineering students 0–1 years from graduation	183,070 engineering bachelor's recipients 1–3 years post-graduation (weighted)
Focus of study	Student career plans	Post-graduation career paths
Sample scope	Twenty-one institutions	National
Study design	Cross-sectional	Cross-sectional
Career constructs	Plans for engineering and/or non-engineering work and/or graduate school	Occupational field, relatedness of occupation to degree, additional education, career-related decisions

from 300 institutions; 1.57 million received bachelor's degrees between 2002 and 2005, including 183,070 graduates in engineering. Public-use data are available online through SESTAT, and restricted-use data can be licensed from the NSF.

Comparing Data Sets

In the remainder of this chapter, we illustrate how analysis of the two data sets introduced in this section can broaden understanding of early career pathways. Although these data sets are complementary, it is important to keep in mind the differences between these data sets beyond the fact that one encompasses current undergraduate students (APPLES) and the other, individuals roughly two years out of their undergraduate programs (NSRCG). APPLES was administered to undergraduate students in 2008, whereas NSRCG was administered to college graduates two years earlier in 2006. APPLES is designed for engineering students only, and although not nationally representative, samples across engineering programs at twenty-one universities; its measures are focused on affective and behavioral dimensions of the college experience and intention to persist in engineering work or study after graduation. NSRCG is a national data set, weighted to represent the entire population of U.S. SEH graduates during the coverage period; its measures are focused on education and employment status, with a few highlighting the more affective dimensions of career choices. These distinctions,

summarized in Table 15.2, are important not only to qualify, but also leverage in comparative work.

Pathways and Predictors from APPLES – *Students Looking Forward*

We begin our comparative analysis with a look at APPLES. APPLES captures the undergraduate portion of the early career pathway. In this section, we present an analysis that illustrates its potential for studying post-graduation plans. Specifically, we use APPLES data to explore (1) variation in students' post-graduation plans for engineering and non-engineering career options and (2) how individual and contextual factors may differentiate these plans. Findings help to address the question of whether and why engineering students either choose to pursue an engineering career or envision different types of opportunities ahead.

The APPLES Sample

For these analyses, we examine a subset of the APPLES data set, the 2,143 respondents who indicated that they were "juniors," "seniors," or "fifth-year seniors or more." This sample was selected to examine plans among students within approximately one year of graduation. Respondents in each academic cohort were similar on key demographic measures and measures of professional plans, thus allowing us to aggregate cohorts and consider predictors of plans

Table 15.3. Junior and Senior Engineering Majors: Major Field of Study by Gender and URM Status (APPLES, $n = 2{,}143$)

Major	Percentage	Percentage Who Are:			
		URM Women	Non-URM Women	URM Men	Non-URM Men
Aerospace engineering	4.6	1.0	12.2	9.2	77.6
BioX engineering	5.6	10.1	37.0	5.9	47.1
Chemical engineering	5.8	9.6	44.0	6.4	40.0
Civil and environmental engineering	14.1	9.3	29.5	14.2	47.0
Computer science/engineering	10.7	7.4	12.7	21.8	58.1
Electrical engineering	13.8	7.1	12.2	19.7	61.0
Industrial engineering	7.8	11.4	40.1	8.4	40.1
Mechanical engineering	27.3	3.4	16.6	14.2	65.8
Other engineering	10.5	7.1	38.4	8.0	46.4

across a larger student group. We augmented the data set with institutional data from three additional sources as described earlier.

The distribution of students across nine self-reported fields of study is shown in Table 15.3. Note again that mechanical engineers are overrepresented and electrical engineers are underrepresented in this data set, relative to the U.S. population of engineering degree-earners. The distributions of respondents by gender and underrepresented racial/ethnic minority (URM) status are also shown because of relevance of these measures to post-graduation plans indicated in earlier studies of APPLES data (Sheppard et al., 2010). The highest percentages of URM men across fields are in computer science/engineering and electrical engineering. The highest percentages of URM women are in bio-x and industrial engineering. In addition, nearly nine of ten

APPLES respondents are U.S citizens (87%), 5% identified as permanent residents, and 8% marked "Other" for their citizenship status.[4]

Student Plans

Using the APPLES data, we can develop a picture of the post-graduation plans of current engineering undergraduates in the sample institutions. The survey asked students to mark the likelihood of each of four post-graduation plans: pursuing an engineering job, non-engineering job, engineering graduate school, and non-engineering graduate school. The distribution of the junior and senior respondents for each option is shown in Table 15.4. Just over 80% of students report plans for working as a practicing engineer. In contrast, 26% have plans for working in a non-engineering position. Graduate school plans are mixed; more than 40%

Table 15.4. Junior and Senior Engineering Majors: Post-graduation Plans (APPLES, $n = 2{,}143$)

	Percentage Marking:		
	Definitely/ Probably Not	Maybe	Definitely/ Probably Yes
Engineering job	9.4	9.5	81.1
Non-engineering job	44.5	30.0	25.5
Engineering graduate school	28.2	28.9	42.9
Non-engineering graduate school	45.7	26.0	28.3

Table 15.5. Junior and Senior Engineering Majors: Post-graduation Plans by Gender (APPLES, women $n = 661$, men $n = 1,482$)

		Percentage Marking:			
		Definitely/Probably Not	Maybe	Definitely/Probably Yes	
Engineering job	Women	11.6	10.9	77.5	*
	Men	8.5	8.9	82.7	
Non-engineering job	Women	39.9	30.3	29.8	**
	Men	46.6	29.9	23.5	
Engineering graduate school	Women	26.9	27.8	45.2	NS
	Men	28.7	29.4	41.9	
Non-engineering graduate school	Women	39.2	30.7	30.2	***
	Men	48.6	23.9	27.5	

Chi-square tests of association. ***$p < .001$, **$p < .01$, *$p < .05$. NS = not significant.

of students indicate plans for attending a graduate program in engineering and about a third have graduate plans in other fields. Comparisons of post-graduation plans by gender are presented in Table 15.5. Women are considering non-engineering graduate school and non-engineering jobs at higher rates than are men, whereas men are more likely to have engineering job plans. Women and men are considering engineering graduate school at comparable rates.

We also applied a more stringent definition of post-graduation plans by creating three mutually exclusive groups based on the same survey responses:

- *"Engineering focused students"* are those who marked probably/definitely yes to one or both engineering options (job, graduate school), and probably/definitely no to both non-engineering options (job, graduate school).

- *"Non-engineering focused students"* are those who marked probably/definitely yes to one or both non-engineering options, and probably/definitely no to both engineering options.

- *"All other plans"* include all other possible response combinations to the four post-graduation survey items.

Under this more strict definition, fewer than 30% of students appear to be on engineering-only pathways (see Table 15.6). Interestingly, most engineering students – 65.4% – remain open to both engineering and non-engineering pathways. We note two possible meanings for this large proportion: (1) students conceive of their careers as spanning several fields or (2) students are unsure about what they will do in the future and are keeping both options open. There is one gender difference: men are more engineering

Table 15.6. Junior and Senior Engineering Majors: Post-graduation Plans Using Three Mutually Exclusive Categories (APPLES, $n = 2,143$)

	Percentage Who Are:			
	All Students	Women	Men	
Engineering focused	28.1	24.6	29.9	*
Non-engineering focused	6.5	7.9	5.9	NS
All other plans	65.4	67.5	64.2	NS

Independent sample t-tests for women vs. men. ***$p < .001$, **$p < .01$, *$p < .05$. NS = not significant.
Women $n = 661$; men $n = 1,482$.

focused in their career plans than are women.

Factors Influencing Engineering Focused Plans

The analytical value of the APPLES data we highlight here is that we can model post-graduation plans and begin to develop an understanding of what differentiates engineering focused students, non-engineering focused students, and students with mixed or uncertain plans before entering the workforce or graduate school. Below we use multinomial logistic regression to analyze the mutually exclusive categories in Table 15.6, and estimate how various individual and contextual factors distinguish engineering focused pathways from non-engineering focused and "all other" pathways. In these analyses, "contextual factors" include educational experiences and environments, labor market characteristics, and institutional characteristics.

Results of the model differentiating engineering focused pathways and non-engineering focused pathways are presented in Table 15.7. Owing to the nested nature of the data set (2,143 respondents across 21 institutions), we conducted these analyses using hierarchical linear modeling (HLM) techniques (see Gilmartin, Antonio, Brunhaver, Chen, & Sheppard [forthcoming] for additional methodological details, including coding schemes for independent variables and reliability statistics for constructs). The first three panels in the table present the log-odds coefficients for individual, academic, and market factors ("Level 1" variables in the HLM model) and the last panel contains the log-odds coefficients for institutional factors ("Level 2" variables in the HLM model). Note that when exponentiated, the coefficients represent the multiplicative increase in the odds of being engineering focused relative to being non-engineering focused for a one unit increase in the independent variable. For clarity, only the statistically significant institution-level variables are displayed.

INDIVIDUAL FACTORS

In the first panel of the Table 15.7 (7A), we see that two motivational factors are important – intrinsic psychological motivation and financial motivation. Intrinsic psychological motivation is a measure of an individual's positive orientation to engineering; its positive coefficient suggests that individuals who are excited about and intellectually stimulated by engineering are more likely to plan on staying in engineering. In our sample, a one standard deviation increase in psychological motivation increases the odds of being engineering focused versus non-engineering focused by a factor of 2.7 (exp(0.98)). Similarly, students who are motivated by the perceived financial rewards of engineering work tend to be more engineering focused; a one standard deviation increase in financial motivation increases those odds over being non-engineering focused by a factor of 1.6. One factor was identified as a negative predictor of being engineering focused. APPLES respondents who reported being more confident in their interpersonal and professional skills are less likely to have engineering focused pathways after graduation, or conversely, are more likely to have non-engineering focused pathways.

CONTEXTUAL FACTORS

In looking at the professional plans of engineering juniors and seniors, we next examined the role of three sets of contextual variables: educational experiences and environments (Table 15.7, 7B), the labor market for new engineers (7C), and institutional characteristics (7D). We defined educational experiences and environments broadly to include engagement with courses and teachers, extracurricular activities, and internships and co-ops. We also considered majors within engineering (e.g., mechanical, electrical, industrial) as potentially distinguishing environments, as each represents the collective values, approaches, and history of a particular professional community.

Table 15.7. Junior and Senior Engineering Majors: Predictors of Engineering Focused Plans (vs. non-engineering focused plans) (APPLES, $n = 2,143$)

	Log-odds Coefficient	Significance
7A. Level-1 Conditional Model: Individual Cognitive and Demographic Characteristics		
Financial motivation (standardized)	0.49	***
Intrinsic psychological motivation (standardized)	0.98	***
Professional/interpersonal confidence (standardized)	−0.68	***
URM women [1]	0.95	NS
Perceived family income (standardized)	−0.07	NS
GPA (standardized)	−0.08	NS
7B. Level-1 Conditional Model: Students' Educational Experiences and Environments		
Exposure to engineering work (standardized)	0.38	**
Involvement in engineering classes (standardized)	0.39	**
Participation in non-engineering activities (standardized)	−0.46	**
Aerospace engineering [2]	0.43	NS
BioX engineering [2]	−1.44	**
Chemical engineering [2]	−1.19	*
Civil/Environmental engineering [2]	1.31	**
Computer science/engineering [2]	−0.67	NS
Electrical engineering [2]	−0.17	NS
Industrial engineering [2]	−0.78	NS
Other engineering [2]	−0.32	NS
7C. Level-1 Conditional Model: Students' Expected Engineering Salary		
Expected salary given major field and state (using median salaries from BLS occupation data, 2007) (standardized)	0.34	*
7D. Level-2 Conditional Models: Institutional Characteristics [3]		
SAT composite (math 75th percentile + reading 75th percentile) (standardized)	−1.34	***
Financial aid: Federal grants (standardized)	2.13	**
Percentage of undergraduates who are part-time students (standardized)	1.07	*
Percentage of total students who are graduate students (standardized)	−0.93	*
Private institution [4]	−2.37	**

[1] Reference group: All other students.
[2] Reference group: Mechanical engineering.
[3] Owing to the small number of institutions in the APPLES sample, institutional characteristics were tested in separate Level-2 models. Level-2 coefficients reflect models holding all Level-1 variables constant. Only statistically significant Level-2 predictors are shown here. Additional details about the Level-2 models are available on request.
[4] Reference group: Public institution.
***$p < .001$, **$p < .01$, *$p < .05$. NS = not significant.

Students who have engineering focused plans are more involved in their engineering classes (e.g., attend class more, turn in assignments more consistently) and have had greater exposure to engineering through engineering co-ops and internships. These findings are consistent with their higher intrinsic psychological motivation.

Although we do not know precisely how classroom engagement and intrinsic motivation interact (if intrinsic interest fosters engagement, or vice versa, or both), we note they did appear as statistically significant, independent of one another in our models.

Specific engineering majors also exhibit differences. Relative to students in mechanical engineering, civil/environmental engineering majors are more likely to be engineering focused (3.7 times greater odds), while bio-x and chemical engineering students are more likely to be non-engineering focused in their future plans (4.2 and 3.3 times greater odds).

There is a weak but positive relationship between engineering focused plans and our measure of the labor market for this study, "expected salary" given a student's major and location of study (Table 15.7, 7C). This is consistent with the finding that students who see the financial rewards of engineering are more likely to have engineering rather than non-engineering focused plans.

Institutional factors (7D) further differentiate students' post-graduation plans. Although we examined seventeen different institutional characteristics, here we discuss only those found to be statistically significant ($p < .05$). Relative to non-engineering focused students, engineering focused students tend to be enrolled at institutions that have student populations with lower SAT scores, higher rates of receiving federal financial aid, and higher rates of part-time attendance. Students attending public institutions in our sample are much more likely to be engineering focused than are their counterparts at private institutions (10.7 times greater odds). Graduate student enrollment is associated with having non-engineering post-graduation plans. The broad range of institutions in the APPLES sample ensures some variability in these institutional characteristics, but it is important to keep in mind that the sample is not inclusive of all institution types and features. Findings, therefore, should be interpreted cautiously.

Differentiating "All Other Plans"

As noted previously, 65% of students are not focused on an engineering-only or non-engineering-only pathway, whether due to uncertainty about the future or to interests spanning multiple fields and options. Our independent variables better differentiate engineering focused students from non-

engineering focused students than they do engineering focused students from students with this wide range of "unsettled" or cross-field plans. However, there are six statistically significant predictors in this model (see Gilmartin et al., forthcoming, for full analysis). Students who have stronger psychological motivation to study engineering, students who are highly involved in their engineering classes, students with higher grade point averages, and students with lower professional/interpersonal confidence are more likely to be engineering focused than to fall into this "all other plans" group. Civil/environmental engineering majors are less likely to have "other plans," relative to the reference major, mechanical engineering, while industrial engineering majors are more likely to be classified in this group.

Summary from APPLES Data

Analyses of APPLES data provide a rich picture of the potential pathways of engineering students after graduation and the individual and contextual factors that may be important in differentiating them. First, they suggest that a majority of engineering graduates are not focused on an engineering-only professional pathway. Many may conceive of their careers as combining engineering and non-engineering. We do not know whether such students envision their careers as starting out as engineers, then being promoted into more supervisory roles within engineering industries, or as working in several disparate fields over a lifetime. These data do not give us insight into the types of work that students are considering in each of these broad "engineering" and "non-engineering" categories. It is apparent, however, that students focused on an engineering-only career are motivated differently and engaged in their engineering studies more intensively than are their peers who are looking beyond engineering after college. Engineering focused students also seem to attend particular kinds of institutions. Together, these findings indicate that career pathways begin early, as students self-select into certain kinds of institutions, and that college experiences themselves may influence the development of students' professional plans.

These findings can be particularly valuable to engineering programs and universities in assessing how well they are meeting their mission, and for students to recognize that their engineering degree equips them for a variety of careers. However, to look only at plans fails to address whether students realize those plans. This is where the NSRCG comes in, in two ways: (1) by providing a picture of what was actually realized and (2) by illuminating a variety of unforeseen factors that affect plans, from the intrinsic (e.g., change in professional interest) to the extrinsic (e.g., family, job not available, working conditions).

Pathways and Important Factors from NSRCG – *Graduates Now and Looking Back*

Relative to APPLES, the NSRCG data contain the actual career outcomes of graduates who obtained a bachelor's degree in engineering approximately two years prior. Keeping in mind the analysis of career plans of engineering students, in the following subsections we illustrate how NSRCG data allow us to peer into the early career to examine more closely engineering and non-engineering pathways. Specifically, we examine current occupations, alignment of occupations with engineering degree, and additional education since graduation.

The NSRCG Sample

The NSRCG samples recent graduates with science, engineering, or health-related (SEH) degrees and then weights this sample to the total U.S. population of recent SEH graduates. For our analysis, we selected the subsample of respondents receiving a bachelor's degree in engineering between 2003 and 2005. The NSRCG data set reports participants' first bachelor's engineering degree,

their highest degree, and their most recent degree. Engineering majors are grouped into five categories: chemical engineering, civil engineering, electrical, electronics, and communications engineering (referred to in this chapter as just electrical engineering), mechanical engineering, and other engineering. In cases in which the respondent's major could not be determined (e.g., because the engineering degree they earned between 2003 and 2005 was not their first bachelor's degree nor their most recent degree), we added them to the "other engineering" category. The final weighted sample totaled 183,070 engineers.

The composition of the weighted sample by major is shown in in Table 15.8. Compared to the APPLES sample, there are higher proportions of chemical and electrical engineering majors and lower proportions of civil, mechanical, and all other engineering majors.[5] The weighted NSRCG sample also contains a higher proportion of U.S. citizens (93%) relative to the APPLES sample and a lower percentage of permanent residents (3%). Another 3% were temporary resident visa holders. A breakdown by gender and race/ethnicity is also shown.[6] URM and non-URM men are more likely to be in civil, electrical, and mechanical engineering fields as compared with chemical and "other" engineering fields; the trend for non-URM and URM women is reversed.

Occupational Choice

With the NSRCG data, we can delineate the occupations that engineering graduates choose roughly two years after earning their bachelor's degrees. We assigned each respondent to one of three analytic groups based on the job code they selected on the survey, if they were employed. These groups included: employed in engineering, employed in the sciences or in a science-and-engineering–related occupation, or employed in a non-science-and-engineering occupation. If the respondent was not employed but pursuing a degree, s/he was assigned to the "student only" category. Respondents who were

Table 15.8. Early Career Graduates with Engineering Bachelor's Degrees: Major Field of Study by Gender and URM Status (2006 NSRCG, $n = 183,070$)

Major	Percentage	Percentage Who Are:			
		URM Women	Non-URM Women	URM Men	Non-URM Men
Chemical engineering	7.6	7.3	29.9	8.9	53.9
Civil engineering	12.8	5.0	17.2	12.1	65.8
Electrical engineering	33.4	3.9	12.3	15.6	68.2
Mechanical engineering	22.0	2.6	10.3	10.8	76.3
Other engineering	24.1	6.4	25.1	10.0	58.5

neither employed nor pursuing a degree were assigned to the "unemployed" category. Table 15.9 displays the percentage of weighted respondents in each of these five categories.

Some 60% of early career engineering graduates were employed in engineering jobs at the time of the survey in April 2006. This percentage is lower than the 80% of students with plans to pursue engineering options in APPLES but on par with the 60% of engineering graduates working in engineering-related fields ten years out in the 1992–3 Baccalaureate and Beyond Cohort (Bradburn et al., 2006). Furthermore, the NSRCG data suggest that some engineering students are ultimately attracted to other technical fields outside of engineering. Eighteen percent of the weighted respondents identified themselves as working in a science or science and engineering (S&E) related field; these jobs spanned the computer and mathematical sciences (66%), the physical sciences (6%), the life sciences (5%), the social sciences (1%), and other fields (23%). Therefore nearly 80% of engineering graduates were employed in S&E.

Of the remaining 23% not employed in S&E jobs, nearly two thirds were employed in non-S&E jobs and the other third were enrolled as students. Of those employed in non-S&E jobs, more than half (57%) reported that their work was closely or somewhat related to their engineering degree. We also note that less than 3% of the engineering graduates from 2003–5 reported being unemployed – almost half the national unemployment rate in April 2006 (4.7%, based on the Bureau of Labor Statistics, May 2006; BLS, 2011).

Lastly, we call attention to gender differentials in employment patterns. In line with our APPLES results, men were more likely than women to have chosen

Table 15.9. Early Career Graduates with Engineering Bachelor's Degrees: Post-graduation Employment Status (2006 NSRCG, $n = 183,070$)

Employment Status	Percentage Marking:			
	All Graduates	Women	Men	
Employed in engineering	60.1	54.9	61.6	***
Employed in a science or S&E-related field	17.5	17.3	17.6	NS
Employed in non-S&E	13.8	16.8	13.0	***
Student only	6.1	7.7	5.6	***
Unemployed	2.5	3.3	2.3	***

Independent sample t-tests for women vs. men. $***p < .001$, $**p < .01$, $*p < .05$. NS = not significant.
Women $n = 39,431$; men $n = 143,639$.

Table 15.10. Early Career Graduates with Engineering Bachelor's Degrees: Occupational Alignment with Engineering Degree (2006 NSRCG, $n = 167,410$)

| Major | Total | Engineering Occupation | | Non-engineering Occupation | |
		Occupation in Same Engineering Degree Field as Degree	Occupation in Different Engineering Field than Degree	Occupation in a Science or S&E-related Field	Occupation in a Non-S&E Field
Chemical engineering	71.8	44.4	27.4	14.5	13.7
Civil engineering	88.0	79.6	8.3	3.4	8.7
Electrical engineering	48.8	40.2	8.6	37.6	13.5
Mechanical engineering	79.1	57.7	21.3	7.7	13.2
Other engineering	62.2	44.7	17.5	14.4	23.5

engineering occupations, whereas women were more likely to be employed in non-S&E. Women were also more likely to be full-time students or not in the labor force.

Occupational Alignment

Just as student post-graduation plans vary by major in APPLES, so too do the employment patterns of early career engineering graduates in NSCRG. Table 15.10 shows the alignment between the weighted respondents' occupation and engineering bachelor's degree.[7] The majority (71–88%) of civil, chemical, and mechanical engineering degree recipients reported that their occupation was in engineering. However, occupational placement within their "home" fields varied. Eighty percent of employed civil engineering graduates reported working in the same field as their undergraduate major, compared to 58% of employed mechanical engineering and 44% of employed chemical engineering graduates. These results are consistent with the variation of post-graduation plans by major in the APPLES data set.

Electrical engineering is an interesting case, where we see greater balance between close engineering alignment (40%) and working in another S&E field (38%). Further analysis shows that nearly nine in ten electrical engineering majors employed in another S&E field are working in math-

and computer science-focused positions. Because math and computer science programing skills are central to the curriculum in electrical engineering, it is feasible that many electrical engineering majors are able to find "natural" alignment across a broader range of fields than are other majors.

Among "other engineering" graduates, almost one in four are heading into non-S&E work, the largest proportion among the major groups. This result is not surprising, given that the category includes industrial engineering majors (whose studies may include business-related topics) as well as students majoring in biomedical fields.

The NSRCG data also shed light on the roughly 9% of employed engineering graduates who reported that their work was "outside of the field of their highest degree", regardless of their current occupation (i.e., engineering, S&E, or non-S&E). As shown in Table 15.11, the state of the job market seems to be an important factor, as one in four of these graduates reported not being able to find a job aligned with their highest degree. Other reasons for pursuing unrelated careers varied significantly by gender. Women's primary reasons were a change in their professional or career interests or unfavorable working conditions. For men, pay and promotion opportunities appear prominent. These results are consistent with those found in Frehill (2007a, 2008).

Table 15.11. Early Career Graduates with Engineering Bachelor's Degrees: Primary Reason for Working in Occupation Unrelated to Highest Degree (2006 NSRCG, $n = 15,579$)

Reason	Percentage Marking:			
	All Graduates	Women	Men	
Job in highest degree field not available	26.0	26.9	25.6	NS
Change in professional/career interests	21.0	24.0	19.7	***
Pay, promotion opportunities	19.3	11.5	22.6	***
Working conditions	16.9	22.8	14.4	***
Other reason	8.6	6.1	9.6	***
Job location	6.4	6.7	6.3	NS
Family	1.8	1.9	1.8	NS

Independent sample t-tests for women vs. men. ***$p < .001$, **$p < .01$, *$p < .05$. NS = not significant.
Women $n = 4,675$; men $n = 10,904$.

Additional Education

In addition to or in lieu of employment, some engineering graduates pursue further education in their early career. Of the 183,070 individuals who graduated with engineering bachelor's degrees in 2003–5, 95% did not obtain an additional degree (see Table 15.12). Just under 4% of this cohort of graduates have gone on to obtain a master's degree in engineering, which is far fewer than the percentage of U.S.-educated undergraduate engineering students getting advanced engineering degrees that has been reported historically (15%, based on Regets, 2006). This difference is likely due in part to the narrow window (1–3 years) captured by the NSRCG. In addition, of the 50,978 weighted respondents who reported taking courses at the time of the survey, nearly 80% were also working, suggesting longer times to completion.

To form a more complete picture of degree attainment, we look at the 50,978 weighted respondents who are still pursuing education. Table 15.13 shows that some 57% of these individuals were working on advanced engineering degrees (22,437 at the MS level and 6,486 at the Ph.D. level). These numbers, combined with those who had already earned their advanced engineering degree, suggest that at least 18% will accomplish this milestone in their lifetimes. Still, this proportion falls short of the 40% of engineering students with plans to attend engineering graduate school in APPLES, suggesting that many may abandon these plans after either encountering new opportunities or gaining a better understanding of what is needed to succeed in engineering.

Significantly, nearly one in three graduates are working on a degree outside of engineering, and they pursue these degrees for reasons very different than those

Table 15.12. Early Career Graduates with Engineering Bachelor's Degrees: Highest Degree Earned (2006 NSRCG, $n = 183,070$)

Highest Degree	Percentage With:		
	Engineering	Science or S&E-Related	Non-S&E
Bachelor's	94.6	0.4	0.1
Master's	3.7	0.6	0.5
Professional	0.0	0.1	0.0

Table 15.13. Early Career Graduates with Engineering Bachelor's Degrees: Highest Degree Currently Pursuing (2006 NSRCG, $n = 50,978$)

Degree Pursuing	Percentage Marking:			
	No Specific Field	Engineering	Science or S&E-Related	Non-S&E
No specific type	4.7			
Bachelor's		7.3	1.4	0.3
Master's		44.0	7.3	8.7
Doctorate		12.7	2.4	0.1
Professional		0.0	6.2	3.8
Other		0.0	0.5	0.7

pursuing engineering degrees (see Table 15.14). Perhaps not surprisingly, more graduates enroll in non-engineering programs than in engineering programs to change fields, obtain special certification, and broaden their knowledge base before embarking on a career, and more graduates enroll in engineering programs to acquire new skills and advancement opportunities.

Summary from NSRCG Data

Our analyses of NSRCG data reveal that the majority of engineering graduates (60%) work in engineering early in their careers. Among the balance of graduates who are employed, roughly half are working in science or S&E-related fields while the other half work outside of S&E.

Alignment between graduates' occupation and engineering degrees varies considerably by major. Whereas four in five civil engineering majors are working in their discipline, electrical engineers are more likely to be employed outside of engineering altogether. Although electrical engineering graduates may be equipped with a broad set of skills that makes crossing into other fields easier, the data indicate that many engineering graduates in unrelated occupations could not find work in their home fields, suggesting that the job market has a powerful influence on graduates' career decision making.

Although fewer than 4% of the weighted respondents have earned advanced degrees in engineering, the data suggest that this proportion will rise to approximately 18% in the near future. An additional 30% of graduates

Table 15.14. Early Career Graduates with Engineering Bachelor's Degrees: Reasons for Taking Additional Coursework (2006 NSRCG, $n = 50,978$)

Reason	Percentage Marking:			
	All Graduates Taking Classes	Enrolled in Engineering	Enrolled in Non-engineering	
Acquire skills	81.2	89.1	69.3	***
Advancement	76.7	80.6	73.6	***
Further education before career	70.9	71.8	74.6	***
Personal interest	47.5	50.9	42.8	***
Other reason	46.5	48.8	39.3	***
Change field	20.8	10.2	39.0	***
Licensing	17.1	11.2	28.9	***

Independent sample t-tests for engineering vs. non-engineering enrollment groups. $***p < .001$, $**p < .01$, $*p < .05$. NS = not significant; n enrolled in engineering = 32,674; n enrolled in non-engineering = 15,930.

Table 15.15. Engineering and Non-engineering Plans and Choices

	Percentage Among:	
Career Plans/Choices	Early Career Professionals (1–3 years since graduating)	Engineering Juniors and Seniors (0–1 years from graduating)
Engineering only	60.1	28.1
Non-engineering only	31.3	6.5
All other plans		65.4
Dataset	NSRCG	APPLES

currently taking courses are seeking degrees in science and other non-engineering fields, their primary motivation for which includes changing fields or seeking special certification.

Taken together, the APPLES and NSRCG show a professional picture of engineering graduates that is still evolving. In other words, they show that engineering students' and recent graduates' career plans are in flux, moving between and combining in various ways engineering and non-engineering options. That is part of the power of looking at two data sets that represent two different time points in engineers' career trajectories. Both show that either picture alone is not enough. The next section illustrates the power of considering findings from both data sets in concert.

Learning from Both APPLES and NSRCG

The two survey-based data sets explored in this chapter give us different perspectives on the pathways of engineering graduates. APPLES, which queried engineering juniors and seniors from across the country, gives us a glimpse into the plans of engineering students and what factors may be influencing those plans. In contrast, NSRCG, which queried engineering graduates some two years after graduation, lets us see what engineering graduates actually do vocationally and some of the reasons for their career choices. Taken together, these two data sets provide additional insights.

Early Career Planning

Table 15.15 repeats the engineering and non-engineering employment patterns of recent graduates and the plans of junior and seniors still in college. Greater proportions of individuals are working in engineering (60%) and non-engineering (31%) fields than might be indicated by students with sharply focused plans. This indicates that as students graduate and make initial career decisions, many with mixed or uncertain plans are beginning their careers in engineering fields and a somewhat smaller proportion choose careers outside of engineering. Some of the reasons for these choices are suggested by NSRCG data – availability of jobs, changes in professional interests, or pay and promotion opportunities. Although we cannot clearly associate specific reasons with engineering and non-engineering initial career choice, it is apparent that many students keep their options open and eventually do pursue a variety of careers with an engineering degree. This finding supports other work showing that, for many students, an engineering degree does not necessarily presuppose an engineering job after college (Lichtenstein et al., 2009). Some students may even be planning with career flexibility in mind.

Additional Education

Both data sets indicate engineers' interest in additional education. Among APPLES respondents, 43% expressed plans for graduate education in engineering and 28% in non-engineering programs (with some 8%

planning on both). In NSRCG, some graduates have realized these school plans; nearly 4% have already received a graduate degree in engineering, and about 1% has received one in a non-engineering field. Another 28% either combine work and school or are full-time students. The top four reasons given for seeking additional formal education are acquiring new skills (88% said this was a reason), career advancement (77%), further education before career (71%), and personal interest (47%). Only one in five pursued additional education to change fields.

Because the fraction of engineering graduates in NSRCG that report obtaining a graduate degree or currently working toward one is only about a third, it falls short of the more than two-thirds of students who reported plans for seeking graduate degrees in APPLES. Given these data, it is clear that some of those not enrolled in school some two years after graduation will enroll at some point in the future, and a sizeable number of those with student plans for additional engineering or non-engineering education will not realize these plans. We are only beginning to capture fully the dynamic trajectory of the early career pathway of engineers.

The Importance of Gender and URM Status

The APPLES and NSRCG data also suggest that gender and URM status influence the decisions of engineering students to enter certain majors and careers. In both data sets, men were found to be more focused on engineering careers than were women. In both data sets, URM men were more likely to major in electrical engineering than in other fields. In APPLES, URM women were more likely to major in bio-x and industrial engineering; it is possible that the higher rates of URM women in these fields mirror the high rates of URM women majoring in "other engineering" fields in NSRCG. Future work is needed to understand the factors that influence the early career trajectories of URM men and women in engineering.

The Importance of Environmental and Contextual Factors

Finally, both data sets point to specific environmental and contextual factors that can shape early career pathways. While in college, one's major as well as several student experiences seem to contribute to the construction of particular pathways for engineering majors. Where one attends college appears relevant as well. Once in the workforce, working conditions, job location, and family contexts add to this construction, and major continues to distinguish career paths. Examining the two data sets together reveals that some environmental factors may be important to students and early career professionals alike (e.g., the culture of particular engineering majors and fields), whereas others may be emergent, becoming important as students mature to professional status (e.g., changes in career or professional interests, job market concerns). For instance, the culture of and curriculum in civil engineering programs may reinforce a strong school-to-work link, but variable job markets may give even the most loyal engineers impetus to look beyond their field in seeking employment. Again, our analyses illustrate dynamism and flux among the factors at play during the early career pathway.

Next Steps and Further Lines of Inquiry

This chapter explored the use of surveys to examine early career engineering pathways. Complementary analyses of APPLES and NSRCG data suggest that "pathways" are a useful way to study the school-to-work transition. It recognizes that there are a number of different pathways to take, and that any one person's pathway is an emergent journey. The data also reveal pathways still in flux during the early career period spanning the completion of college and the first years post-college. Even as seniors embark on their next steps, the career plans of many remain open, subject to influence by various personal and contextual factors. Of these factors, major field of study and institutional characteristics seem to be especially important. Career motivations and interests are important as well, as strong engineering interests were associated with plans to pursue engineering-only options after graduation.

In addition to contributing new understanding to the college–career transition, our analyses illustrate where knowledge gaps still remain. Because APPLES and the NSRCG were designed to address different research questions, they provide disparate information about their respective populations. APPLES contains data on psychological measures such as motivation and self-confidence, undergraduate experiences (e.g., research, internship/co-op), and because of its sampling design, institutional characteristics as well. As a forward-looking survey, APPLES is able to capture a broad range of student plans, but the data set does not give visibility into the rationale for these plans, only the characteristics of students pursuing them. Alternatively, the NSRCG data set looks only at recent graduates' employment and education patterns; it does not have comprehensive measures of affective traits and outcomes of early career professionals (such as self-confidence), nor does it allow for exploration of the effects of undergraduate experiences on decision-making.

A significant limitation of both data sets is that what is characterized as engineering work remains undefined. APPLES asked students about their intentions to pursue engineering and non-engineering careers but students were left to define those categories for themselves. The NSRCG survey asks respondents about "alignment" with their engineering degree but suffers the same lack of definition.

Furthermore, the APPLES and NSRCG data sets are cross-sectional and represent snapshots of two different samples at two different points in time. As such, they provide limited information about the ways that students operationalize their plans into career decisions. Significantly, there may be times between senior year and 1–3 years post-graduation when the career pathway is particularly fluid.

These limitations clearly warrant a longitudinal study of engineering students from the time they are juniors to roughly five years after obtaining their degrees. Such a study could be conducted in three phases: The first phase would query students about their future plans, the reasons for these plans, and their perceptions and expectations of work. The second phase (two years out) might focus on the now-graduates' next steps and how they relate to their initial career experiences. The last phase (five years out) might measure graduates' success at building and settling into a career, and how this success impacts their long-term career commitment. The results of our analyses further suggest investigating more deeply the ways that disciplines and institutions help to shape pathways within and beyond college. As demonstrated by Ohland et al. (2011), institutional differences matter in the retention of undergraduates in engineering programs. Given the results of our APPLES analyses, it seems reasonable then that institutional differences would be a key differentiator with respect to post-graduation retention as well.

Models for such a longitudinal study abound. The Baccalaureate and Beyond studies discussed in this chapter are just one example of studies that have tracked the same participants (albeit college graduates only) over time. Smaller, mixed methods studies have also emerged as a way to build a rich and robust picture of engineering pathways. A follow on to the Academic Pathways Study, the NSF-funded Engineering Pathways Study (EPS) (Sheppard, Matusovich, Atman, Streveler, & Miller, 2011) used a sequential, exploratory[8] design to consider career pathways and persistence among former APS (student) participants. Interviews conducted during EPS were combined with APS data to examine engineering students' career goals and actions longitudinally (Winters, 2012). Findings from the interviews of these alumni were also used to inform the development of a survey of engineering alumni (Brunhaver, Gilmartin, Chen, & Sheppard, 2012; Chen et al., 2012).

A logical next step for smaller, mixed methods studies (such as EPS) is to provide foundational understanding for a larger study similar to Seymour and Hewitt's (1997) extensive longitudinal work on science, math, and engineering majors. Just as Seymour and Hewitt discovered reasons why undergraduates leave science majors that have little to do with performance, a similar study on engineering graduates could help to explain the observations of Lowell et al. (2009) discussed previously.

Lastly, we recognize the potential of this work to not only leverage existing pathways studies as resources, but also to contribute new knowledge and ideas toward their improvement in the future. For example, research agencies such as the NSF and the NCES could extend their suite of surveys to traverse both sides of the college–career divide. They could also incorporate more psychological and affective measures to help contribute to a more refined understanding of early engineering careers.

Much effort has been invested in better understanding what factors help to attract and retain a diverse student body to engineering – this is certainly an important part of the engineering pathway (see Chapters 19 by Litzinger and Lattuca and 16 by Lichtenstein, Chen, Smith, & Maldonado in this volume). This chapter illustrates the need to further explore the next part of the pathway, namely from school into the workforce. Only then can we be better understand (and therefore support) young talent entering into the engineering workforce.

Acknowledgments

This research was supported by grants from the National Science Foundation (Grant No. ESI-0227558) and Sloan Foundation, as well as a Stanford Graduate Fellowship. Any opinions, findings, conclusions, or recommendations expressed in this publication are those of the authors and do not necessarily reflect those of the NSF or Sloan.

Footnotes

1. The sample sizes for Bradburn et al. (2006), Choy et al. (2008); Forrest Cataldi et al. (2011), and Fouad & Singh (2011) are approximate, as shown in the original references.

2. The other two major projects of CAEE were the Scholarship on Teaching Engineering and the Institute for Scholarship on Engineering Education. CAEE's five partner institutions are Colorado School of Mines, Howard University, Stanford University, University of Minnesota, and University of Washington, the lead institution. The final CAEE report (Atman et al., 2010) includes details about the design of APS, as well as an overview of project findings. See http://www.engr.washington.edu/caee/.

3. For measures of selectivity (i.e., SAT percentile scores) at a small number of schools, 2007–8 IPEDS data were used in the absence of such data for 2006–7; for the student-to-faculty ratio at each institution, 2008–9 IPEDS data were used in the absence of such data for 2006–7.

4. "Bio-x" is so titled to account for the range of bio-engineering programs across institutions, e.g., bioengineering, biomedical engineering, etc. The majority of institutions in the APPLES sample house civil and environmental engineering in one department, so these majors were aggregated; the aggregate major includes 280 respondents marking "civil" and 22 respondents marking "environmental." Computer science/engineering includes only those students marking on the survey "computer science/engineering (in engineering)" (as distinct from a computer science program outside of engineering). "Other engineering" includes the following categories: agricultural engineering, construction engineering, engineering math and physics, engineering operations research and business, general engineering, materials and metallurgical engineering, nuclear engineering, ocean engineering, and other engineering (these categories, along with bio-x programs, were identified based on students' write-in responses to an "other engineering major" open-ended field on the survey; all other categories were listed on the

instrument). Two APPLES institutions offer a "general engineering" degree only; all respondents at these two schools are classified as "other engineering" majors. Mechanical engineering is the reference group for all regression models in this section.

5. The list of engineering majors in APPLES is more extensive than that in NSRCG owing to differences in the way data were collected and made available. In the NSRCG data set, "other engineering" includes aerospace, bio-x, computer, and industrial engineering majors.

6. For these analyses, graduates classified as URM include those marking Hispanic and/or one of the following racial backgrounds: American Indian/Alaska Native, black/African American, Native Hawaiian or other Pacific Islander.

7. Includes only weighted respondents who were employed at the time of the survey (April 2006).

8. Qualitative interviews, followed by a quantitative survey.

References

Amelink, C., & Creamer, E. G. (2010). Gender differences in elements of the undergraduate experience that influence satisfaction with the engineering major and the intent to pursue engineering as a career. *Journal of Engineering Education*, 99(1), 81–92.

American Society for Engineering Education. (2011). Engineering and engineering technology college profiles for 2006. Retrieved from http://profiles.asee.org/

Astin, A. W. (1993). *What matters in college? Four critical years revisited*. San Francisco: Jossey-Bass.

Atman, C. J., Sheppard, S. D., Turns, J., Adams, R. S., Fleming, L. N., Stevens, R., . . . Lund, D. (2010). *Enabling student success: The final report for the Center for the Advancement of Engineering Education*. San Rafael, CA: Morgan and Claypool.

Bradburn, E. M., Nevill, S., Forrest Cataldi, E., & Perry, K. (2006). *Where are they now? A description of 1992–93 bachelor's degree recipients 10 years later (NCES 2007–159)*. Washington, DC: The U. S. Department of Education, National Center for Education Statistics.

Brunhaver, S., Gilmartin, S., Chen, H., & Sheppard, S. (2012). *Correlates of post-graduation*

paths among early career engineering graduates. Paper presented at the Association for the Study of Higher Education Annual Conference, Las Vegas, NV.

Bureau of Labor Statistics. (2006). *The employment situation: April 2006*. Washington, DC: United States Department of Labor.

Bureau of Labor Statistics. (2011). National cross-industry estimates, May 2007. Retrieved from http://www.bls.gov/oes/2007/may/oes_dl.htm#2007

Chen, H. L., Grau, M. M., Brunhaver, S. R., Gilmartin, S. K., Sheppard, S. D., & Warner, M. (2012). Designing the Pathways of Engineering Alumni Research Survey (PEARS). In *Proceedings of the American Society for Engineering Education Annual Conference*, San Antonio, TX.

Choy, S. P., Bradburn, E. M., & Carroll, C. D. (2008). Ten years after college: Comparing the employment experiences of 1992–93 bachelor's degree recipients with academic and career-oriented majors (NCES 2008–155). Washington, DC: The U. S. Department of Education, National Center for Education Statistics.

Donaldson, K. M., Chen, H. L., Toye, G., Clark, M., & Sheppard, S. D. (2008). *Scaling up: Taking the Academic Pathways of People Learning Engineering Survey (APPLES) national*. Paper presented at the American Association for Engineering Education/Frontiers in Education Conference, Saratoga Springs, NY.

Eris, O., Chachra, D., Chen, H., Sheppard, S., Ludlow, L., Rosca, C., . . . Toye, G. (2010). Outcomes of a longitudinal administration of the Persistence in Engineering Survey. *Journal of Engineering Education*, 99(4), 371–395.

Forrest Cataldi, E., Green, C., Henke, R., Lew, T., Woo, J., Shepherd, B., . . . & Socha, T. (2011). 2008–09 Baccalaureate and Beyond Longitudinal Study (B&B:08/09): First look. Washington, DC: The U. S. Department of Education, National Center for Education Statistics.

Fouad, N. A., & Singh, R. (2011). *Stemming the tide: Why women leave engineering*. Milwaukee, WI: The University of Wisconsin-Milwaukee.

Frehill, L. (2007a). What do women do with engineering degrees? In *Proceedings of the 2007 Women in Engineering Programs and Advocates Network (WEPAN) Conference*, Lake Buena Vista, FL.

Frehill, L. (2007b). The Society of Women Engineers National Survey about Engineering: Are women more or less likely than men to be retained in engineering after college? *SWE Magazine*, 53(4), 22–25.

Frehill, L. (2007c). The Society of Women Engineers National Survey about Engineering: Is the engineering workplace 'warming' for women? *SWE Magazine*, 53(5), 16–20.

Frehill, L. (2008). The Society of Women Engineers National Survey about Engineering: Why do women leave the engineering work force? *SWE Magazine*, 54(1), 24–26.

Gilmartin, S. K., Antonio, A. L., Brunhaver, S. R., Chen, H. L., & Sheppard, S. D. (Forthcoming). Career plans of undergraduate engineering students: Characteristics and contexts. In R. Freeman & H. Salzman (Eds.), *U. S. engineering in the global economy*. Cambridge, MA: National Bureau of Economic Research.

LeBold, W., Bond, A., & Thomas, M. (1977). *Recruitment and retention of Black Americans in engineering at Purdue: A follow-up study of Black engineering graduates*. West Lafayette, IN: Purdue University.

Lichtenstein, G., Loshbaugh, H. G., Claar, B., Chen, H. L., Jackson, K., & Sheppard, S. D. (2009). An engineering major does not (necessarily) an engineer make: Career decision-making among undergraduate engineering majors. *Journal of Engineering Education*, 98(2), 227–234.

Lowell, B. L., & Salzman, H. (2007). *Into the eye of the storm: Assessing the evidence on science and engineering education, quality, and workforce demand*. Washington, DC: Urban Institute.

Lowell, B. L., Salzman, H., Bernstein, H., & Henderson, E. (2009). *Steady as she goes? Three generations of students through the science and engineering pipeline*. Paper presented at the Annual Meetings of the Association for Public Policy Analysis and Management, Washington, DC.

National Center for Education Statistics. (2011). Integrated postsecondary education data system, 2006. Retrieved from http://nces.ed.gov/ipeds/datacenter/Default.aspx

National Science Foundation (NSF), Division of Science Resources Statistics. (2011). Women, minorities, and persons with disabilities in science and engineering: 2011. Special Report NSF 11–309. Arlington, VA. Retrieved from http://www.nsf.gov/statistics/wmpd/.

National Science Foundation (NSF), National Center for Science and Engineering Statistics. (2006). NSRCG PUBLIC 2006 data file. Retrieved from http://sestat.nsf.gov/datadownload/

Ohland, M. W., Brawner, C. E., Layton, R. A., Long, R. A., Lord, S. M., & Wasburn, M. H. (2011). Race, gender, and measures of success in engineering education. *Journal of Engineering Education*, 100(2), 225–252.

Pascarella, E. T., & Terenzini, P. T. (2005). *How college affects students. A third decade of research*. San Francisco, CA: Jossey-Bass.

Reese, C. (2003). Employment history survey of ASCE's younger members. *Leadership and Management in Engineering*, 3(1), 33–53.

Regets, M. C. (2006). *What do people do after earning a science and engineering bachelor's degree? Info Brief 06–234*. Washington, DC: The National Science Foundation.

Ro, H. K. (2011). *An investigation of engineering students' post-graduation plans inside or outside of engineering* (Doctoral dissertation). University Park, PA: The Pennsylvania State University. Retrieved from http://etda.libraries.psu.edu/paper/12289/7999.

Robst, J. (2007). Education and job match: The relatedness of college major and work. *Economics of Education Review*, 26(4), 397–407.

Seymour, E., & Hewitt, N. M. (1997). *Talking about leaving: Why undergraduates leave the sciences*. Boulder, CO: Westview Press.

Sheppard, S., Gilmartin, S., Chen, H. L., Donaldson, K., Lichtenstein, G., Eris, O., . . . Toye, G. (2010). *Exploring the engineering student experience: Findings from the Academic Pathways of People Learning Engineering Survey (APPLES). (TR-10–01)*. Seattle, WA: Center for the Advancement of Engineering Education.

Sheppard, S., Matusovich, H. M., Atman, C., Streveler, R. A., & Miller, R. L. (2011). *Work in progress – Engineering Pathways Study: The college–career transition*. Paper presented at the American Association for Engineering Education/Frontiers in Education Conference, Rapid City, SD.

Winters, K. E. (2012). *Career goals and actions of early career engineering graduates* (Doctoral dissertation). Blacksburg, VA: Virginia Polytechnic Institute and State University. Retrieved from http://scholar.lib.vt.edu/theses/available/etd-03292012-111315/.

CHAPTER 16

Retention and Persistence of Women and Minorities Along the Engineering Pathway in the United States

Gary Lichtenstein, Helen L. Chen, Karl A. Smith, and Theresa A. Maldonado

Introduction

Countries around the world rely on the contributions of engineers to support national interests and maintain economic competitiveness. In the United States, government and industry leaders have long regarded engineers and other members of the science, technology, engineering, and mathematics (STEM) workforce as vital to the nation's economy and security. It is hardly surprising, then, that issues surrounding student retention and persistence in engineering degree programs and the engineering workforce are of special interest to engineering educators.

Since the 1970s, federal policy and funding have specifically focused on attracting and retaining women and minorities in science and engineering fields. Yet progress has been halting. In one comprehensive study, the United States ranked 30th of 35 countries in the proportion of female Ph.D.s in engineering, manufacturing, and construction, and 24th of 30 with respect to growth in the proportion of female Ph.D.s in these sectors (European Commission, 2009, p. 51).[1]

In this chapter, we examine the influence of U.S. federal policy on engineering education over the past forty years, with special attention to the impact of efforts to increase the number of women and minorities in the STEM workforce.

The National Science Foundation Seeks a STEM Workforce

In the history of engineering education in the United States, few events have proven more decisive than the creation of the National Science Foundation (NSF). The agency was formed in 1950 "to promote the progress of science; to advance the national health, prosperity, and welfare; to secure the national defense...." (National Science Foundation Act of 1950, Public Law 81-507). One of the agency's core missions is to "cultivate a world-class, broadly inclusive science and engineering workforce" to ensure that the nation produces sufficient numbers of scientists and engineers to keep the United States at the forefront of scientific discovery and technological innovation (http://www.nsf.gov/about/).

Because NSF allocates significant funding to these efforts, attracting talented individuals to careers in STEM fields has become a central theme in STEM education research. NSF administers a total of $1.2 billion annually, or 34%, of all federal dollars ($3.4 billion) spent on STEM education. This investment is the largest expenditure for this purpose by any federal agency. Moreover, NSF is the only agency that houses programs devoted solely to engineering education. Nearly all these funds are for programs and initiatives designed to promote STEM education and workforce development; the annual investment dedicated to engineering education research is only $14 million, or less than 0.005% of the total federal STEM education annual investment (Committee on STEM Education, 2011).

1970–1985: *Expanding the STEM Workforce by Targeting Women and Minorities*

For decades, efforts to create a talented STEM workforce focused on the population stereotypically associated with science and engineering. The average scientist was "a 30-year old [White] male from the middle-Atlantic states (usually New York) with a Ph.D., ten years or less experience in the field, was employed in either academia or in industry, and more likely to know German than Spanish" (Lucena, 2005, p. 51; NSF, 1964). During the 1970s, however, this focus began to change. As concerns about threats from other countries receded, concerns about domestic issues, from poverty to environmental degradation, came to the fore. Leaders of organizations and academic institutions that predominantly served minorities argued to Congress that problems afflicting inner-city communities could best be addressed by applied scientists who came from and were knowledgeable about those communities (Lucena, 2005).[2] Also during this period, the second wave of the feminist movement was expanding traditional conceptualizations of science and the role

of women in scientific fields (Espinosa, 2011; Keller, 1985).

Both the civil rights and the women's movements impacted federal policymaking. In one notable example, Janet Welsh Brown, head of the American Association for the Advancement of Science (AAAS) Office of Opportunities in Science and president of the Federation of Organizations for Professional Women, testified before Congress, calling for greater efforts by NSF to broaden participation of women and minorities in the sciences (U.S. Congress, House Committee on Appropriations, 1976). Brown was also a co-author, with Shirley Malcom and Paula Hall, of *The Double Bind: The Price of Being a Minority Woman in Science*, a report still referenced today for its analysis of the challenges women – and women of color in particular – face in overcoming gender and ethnic bias in the pursuit of science and engineering careers (Malcom, Hall, & Brown, 1976). Armed with statistics showing the dearth of minority and female scientists, Brown and other leaders were powerful advocates for policies and funding to increase the numbers of minorities and women in STEM (Lucena, 2005).

Such efforts resulted in the Science and Technology Equal Opportunity Act of 1980 (PL96-516).[3] Enacted thirty years to the day after the first NSF executive board meeting, the Act made promoting scientific and engineering talent among women and minorities a federal priority:

> *The Congress finds that it is in the national interest to promote the full use of human resources in science and technology and to insure the full development and use of the scientific talent and technical skills of men and women, equally, of all ethnic, racial, and economic backgrounds.*[4]

By the mid-1980s, economic competition from Japan, combined with a recession in the United States, renewed concerns about identifying and training a workforce to preserve U.S. global technological superiority. The Committee on the Education and Utilization of the Engineer, formed by the National Research Council (NRC), set about forecasting the availability of a sufficiently trained science and engineering workforce (NRC, 1986). In 1986, the Committee, whose members were all engineers, created a mathematical "flow model" based on the balance equation from traditional energy and material balances.[5] The reports presenting this model were accompanied by an elaborate schematic that looked rather like a diagram of a complex computer circuit. The language of "flow," "input," and "output" that accompanied the figure led very naturally to a vivid and powerful *pipeline* metaphor, which drove policy and research for the next twenty years. Indeed, the pipeline metaphor persists today, in spite of attempts to supplant it with a more expansive *pathways* metaphor that accommodates multiple points of entry, exit, and reentry (Adelman, 2006; Atman et al., 2008; Sheppard, Macatangay, Colby, & Sullivan, 2008).

The NRC Committee's mathematical model in 1986 projected a 5% decrease over the following decade in the college enrollment of White male 22-year-olds, the primary population that until then had earned bachelor's degrees in science and engineering. At the same time, analysts noted increases in the number of bachelor's degrees awarded to minorities and women during the previous five years and speculated that women and minorities could be relied on to fill the gap left by the sagging numbers of White males. From then until now, researchers and policy analysts looking for ways to bolster the STEM workforce have emphasized the need to encourage greater participation among women and underrepresented minorities (Hartline & Poston, 2009; Nelson & Rogers, 2007; Satcher, 2001).

1986–2000: *Retention and Persistence Emerge as Obstacles for the STEM Workforce*

Even as the U.S. government established an imperative to increase the numbers of women and minorities in science and engineering, several research articles and policy reports from the mid-1980s to the early 1990s documented low retention rates for all students in science, engineering, and math, especially among minorities and women. One federal report found that 50% of undergraduates entering science and engineering programs did not complete degrees in those fields, and only 30% of those who did complete degrees went on to graduate school (U.S. Congress, Office of Technology Assessment, 1989). Other studies reported disproportionately low interest and participation in STEM fields among women and minorities. Minorities accounted for less than 6% of undergraduate engineering degrees awarded in 1984. In 1986, only 2.7% of female undergraduates reported an intention to major in engineering, compared to 17.8% of males (see Malcom, George, & Van Horne, 1996). If women and minorities were to fill the need for talented individuals in the STEM workforce, their participation in STEM majors, and their rates of completion of STEM degrees, would have to increase.

The seminal work to come out of this research was – and still is – *Talking About Leaving*, by Elaine Seymour and Nancy Hewitt (Seymour & Hewitt, 1997). The authors relied primarily on qualitative data (interviews) to explore patterns of persistence and switching among more than 345 undergraduates from seven diverse colleges and universities. Seventy-three percent of the students in the sample were White, and 27% were people of color (African American, Native American, Hispanic, Japanese, Chinese, Laotian, Pakistani, Cambodian, Korean, and East Indian). In both groups, just over half of the students were women.

Talking About Leaving provided a vivid picture of the challenges faced by many who pursued STEM majors. The data showed that those who left science, math, and engineering (SME) and those who did not had similar capabilities, as indicated by grade point averages and SAT scores. Differences between switchers and persisters had to do with perceptions of and attitudes toward the culture and climate of science and engineering classrooms and majors. Seymour and Hewitt found that faculty teaching

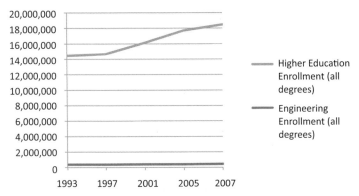

Figure 16.1. Snapshot 1: Engineering degree enrollment (all degrees) compared to higher education enrollment (all degrees), 1993–2007. (*Source*: National Science Board (2010). Science and Engineering Indicators – 2010. Appendix tables 2–4 & 2–15. Arlington, VA: National Science Foundation (NSB 10–01). Retrieved from http://www.nsf.gov/statistics/seind10/c/cs1.htm.)

styles and other environmental factors affected students differentially, ultimately causing many women and students of color to migrate out of SME programs and into programs they perceived to be more compatible with their goals and learning styles.

Seymour and Hewitt's work had two effects that continue to shape today's research into STEM retention and persistence. First, *Talking About Leaving* married concerns about recruitment of minorities and women into STEM with the issue of undergraduate retention. Second, Seymour and Hewitt argued that improving minority and female retention would require changes in classroom instruction and institutional policies in higher education. These findings prompted a fundamental shift from forty years of prior research and policy on the STEM workforce. Whereas previous efforts were aimed at identifying talented individuals and recruiting them into STEM fields, Seymour and Hewitt focused attention on retaining those who demonstrated both the interest and capacity to succeed. Notably, the kinds of changes advocated in *Talking About Leaving* in order to sustain motivation and increase the retention and persistence of minorities and women in STEM were eventually recognized as advantageous not just for those groups, but for *all* students pursuing STEM majors (Kyle, 1997).

STEM Retention and Persistence Today

Policy and research documents continue to argue that the United States needs a diverse STEM workforce to address a broad range of domestic and international challenges (Committee on Prospering in the Global Economy of the 21st Century, 2007; Committee on Underrepresented Groups and the Expansion of the Science and Engineering Workforce Pipeline et al., 2011; Hartline & Poston, 2009). However, while enrollment in science fields has increased relative to higher education enrollment overall, engineering enrollment as a proportion of higher education enrollment has declined (see Figure 16.1).

Although the proportions of women and underrepresented minorities enrolled in undergraduate programs generally have increased since the 1980s, minimal progress has been made in recruiting and retaining students, and especially women and minorities, into engineering programs (see Figure 16.2).

At all postsecondary academic levels and in the workforce, women are underrepresented in engineering and in many scientific fields, with the exception of biological sciences (Committee on Underrepresented Groups and the Expansion of the Science and Engineering Workforce Pipeline et al.,

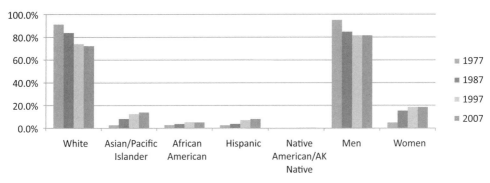

Figure 16.2. Snapshot 2: Engineering bachelor's degrees awarded by race and gender, 1977–2007. (*Source*: 1977–1997 data: National Science Foundation (2002). *Science and Engineering Indicators – 2002*. Tables 2-16 & 2-17. Arlington, VA: National Science Foundation, 2002 (NSB-02–1). Retrieved March 31, 2012 from http://www.nsf.gov/statistics/seind02/pdf_v2.htm#c2. 2007 data: National Science Foundation (2010). *Science and Engineering Indicators – 2010*. Tables 2-12 & 2-13 (appendix). Arlington, VA: National Science Foundation (NSB 10–01). Retrieved March 31, 2012 from http://www.nsf.gov/statistics/seind10/c/cs1.htm.)

2011; Ong, Wright, Espinosa, & Orfield, 2011). In fact, women are underrepresented in two ways. First, women who are not White or Asian are significantly underrepresented in higher education relative to their proportion in the general population. Second, women of all races are underrepresented in engineering (17.9%) relative to the proportion of total undergraduate degrees they earn (57.5%) – see Figure 16.3.

Women and minorities continue to face the same obstacles first reported fifty years ago, yet the issues are better documented now than they were in the past. Furthermore, research has revealed strategies that successfully promote the retention and persistence of all populations, including women and minorities, in STEM majors and careers. The remainder of this chapter provides an overview of the literature, followed

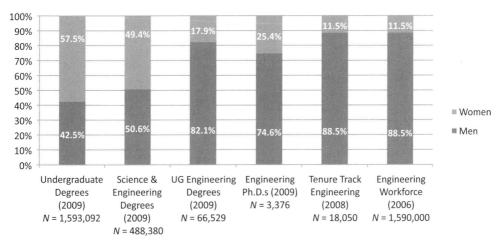

Figure 16.3. Engineering attainment at career junctures by gender. (*Sources*: National Science Foundation (2011). *Women, Minorities, and Persons with Disabilities in Science and Engineering: 2011*. Special Report NSF11–309. Division of Science Resources Statistics. Tables 5-3, 7-2, 9-26, 97. Arlington, VA. Retrieved January 31, 2012 from http://www.nsf.gov/statistics/wmpd/2013/tables .cfm.)

by a discussion of effective approaches to developing a strong, diverse STEM workforce.

Literature Review

Understanding Persistence and Retention

Before reviewing the relevant research, we provide some groundwork for discussing *persistence* and *retention*. Much of the research in these areas looks at students in engineering and in the sciences collectively. Those studies that disaggregate data show varying profiles of retention and persistence across STEM fields. Interestingly, engineering as a whole has higher overall retention than other science and technology fields, but less participation by women. There are also dramatic differences across engineering disciplines. Having said that, we believe that findings across engineering and the sciences differ more in degree than in kind, and we would be overlooking important lessons if we focused exclusively on studies of retention and persistence in engineering. Therefore, the body of literature reviewed in this chapter includes a judicious combination of studies dealing with engineering generally, with engineering subfields, and with engineering and science disciplines more broadly.

In the discussion that follows, *retention* refers to remaining in STEM during each phase of one's education or career (e.g., completing an engineering degree). *Persistence* refers to making the transition from one career juncture to another (e.g., progressing from an undergraduate STEM program into a graduate STEM program). Lowell, Salzman, Bernstein, and Henderson (2009) propose a similar framework that examines retention and persistence during the periods from high school to STEM undergraduate degree (five years after high school), completion of STEM undergraduate degree to first job (three years after college), and completion of STEM degree to employment in STEM occupation at mid-career (ten years after college completion).

It is with some hesitation, however, that we adopt this distinction between retention and persistence, for we recognize that doing so risks fragmenting the research literature. For example, we have used the word "persistence" in the titles of previous articles, referring to what we are now calling "retention." But we believe that *not* making the distinction fragments the research in a more troublesome way, because using the terms "retention" and "persistence" interchangeably glosses over significant challenges faced by women and people of color navigating STEM pathways.

Persistence Along the STEM Pathway

The landscape of research into persistence in engineering and the sciences is extremely nuanced. Findings vary by gender as well as by race and ethnicity (Johnson, 2011). Yet in spite of apparent contradictions at a superficial level, careful observation of specific populations and subpopulations yields consistent and revealing trends.

Persistence from High School to Two- and Four-Year Colleges and Universities

Several studies have shown that young women and minority high school students and undergraduates are as likely as young White men to express an interest in or intention to pursue an undergraduate STEM degree (Hurtado, Eagan, & Chang, 2010; Riegle-Crumb & King, 2010; Staniec, 2004; Varma & Hahn, 2007). There is also evidence that young women are no less qualified than young men to pursue STEM careers. Students who complete college degrees in engineering and the sciences have usually taken higher levels of high school math and/or scored well on the Advanced Placement math exam compared to their peers. The K–12 gender gap on these important indicators, significant in the past, has closed over the past two decades, with females performing comparably to males on most measures of math and science achievement (AAAS, 2010; Freeman, 2004; Peter & Horn, 2005). On the other hand, although girls enroll in high school biology and environmental science in greater numbers than boys, they enroll less often in calculus, physics, chemistry, and computer science (Hill, Corbett, & St. Rose, 2010, p. 6).

Unfortunately, the racial achievement gap in math and science has not closed. On the 2007 National Assessment of Educational Progress (NAEP), the scores of minority high school students continued to lag behind those of Whites and Asians (Committee on Equal Opportunities in Science & Engineering, 2009). In addition, African American and Hispanic youth have less access to Advanced Placement courses that can give students an edge in STEM. From 1990 to 2009, 41.2% of Asian/Pacific Islanders and 16.0% of White students took Advanced Placement/International Baccalaureate calculus classes, compared to 6.5% of African American and 9.4% of Hispanic students (NSF, 2012, Table Apx 1-8). Nevertheless, as shown later, the proportion of engineering degrees earned by minority males is comparable to, or slightly higher than, the proportion of undergraduate degrees they earn.

After (or during) high school, many students take courses at two-year community colleges. This pathway is more likely to include members of populations underrepresented in engineering and the sciences (Malcom & Malcom, 2011; NSF, 2011). Tsapogas (2004) found that between 45% and 51% of African Americans, Latinos, and Native Americans who completed four-year degrees in these fields first attended community colleges.

Research has shown that the transition from a two-year to a four-year institution is a critical juncture along the STEM pathway. Many students switch from STEM to non-STEM majors or leave higher education altogether. The evidence suggests that these outcomes have less to do with academic preparation than with the shift from a supportive to a competitive academic environment, lack of effective advising, and feelings of isolation (Johnson & Sheppard, 2004; Packard, Gagnon, LaBelle, Jeffers, & Lynn, 2011; Reyes, 2011; Townsend, 2007; Valenzuela, 2006).

Representation by Race and Gender at the Undergraduate Level

When we disaggregate the data by gender and race, we find that women make up a smaller proportion of students earning engineering degrees than of students earning bachelor's degrees overall. The differential between men's and women's degree attainment in engineering is greatest among Whites, but the pattern holds for all racial and ethnic groups. In 2009, 1.3% of women who earned bachelor's degrees completed a B.S. in engineering, compared to 8.2% of men. Majority men (White and Asian) completed undergraduate engineering degrees at six times the rate of majority women, and minority men (African American, Hispanic, and Native American) completed such degrees at five times the rate of minority women. Figures 16.4 and 16.5 show the percentages of all B.S. degrees, B.S. degrees in engineering and science, and B.S. degrees in engineering earned by women and men, respectively.

The proportion of undergraduate engineering degrees earned by minority men is comparable to the proportion of bachelor's degrees earned by minorities overall; yet, like minority women, they are extremely underrepresented in higher education relative to their proportions in the general population. For example, U.S. Census Bureau (2008) figures show that Hispanic men comprised approximately 15% of twenty- to twenty-four-year-olds in the U.S. and that African American men comprised approximately 5% of twenty- to twenty-four-year-olds, yet Hispanic and African American men each earned less than 3.5% of all bachelor's degrees in 2009 (also see National Research Council and National Academy of Engineering, 2012).[6]

Postgraduate and Workforce Pathways

NSF is especially interested in producing highly qualified, domestically prepared, science and engineering Ph.D.s (Committee on Prospering in the Global Economy of the 21st Century, 2007). Yet doctoral engineering

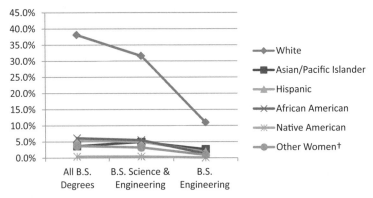

Figure 16.4. Undergraduate degrees, by race, earned by women, shown as a proportion of total degrees earned by all men and women (2009). (*Source*: National Science Foundation (2011). *Women, Minorities, and Persons with Disabilities in Science and Engineering: 2011.* Special Report NSF11–309. Division of Science Resources Statistics. Tables 5-3 & 5-4. Arlington, VA. Retrieved January 31, 2012 from http://www.nsf.gov/statistics/wmpd/.) †Includes those who reported mixed race and "Unknown."

degrees conferred on U.S. citizens declined 21% in 2009 compared to 2000 (National Science Foundation, 2011).

Beyond the undergraduate level, White and Asian men comprise an increasing proportion of engineering graduate students and professionals, while the proportions of women and minorities have dropped sharply. Although engineering doctorates awarded to women and minorities have increased over the past thirty years, overall numbers remain low (Cosentino de Cohen & Deterding, 2009). For example, Malcom and Malcom (2011)

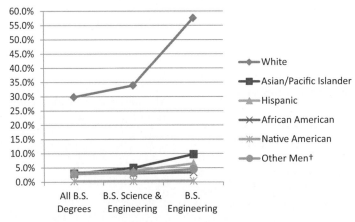

Figure 16.5. Undergraduate degrees, by race, earned by men, shown as a proportion of total degrees earned by all men and women (2009). (*Source*: National Science Foundation (2011). *Women, Minorities, and Persons with Disabilities in Science and Engineering: 2011.* Special Report NSF11–309. Division of Science Resources Statistics. Tables 5-3 & 5-4. Arlington, VA. Retrieved January 31, 2012 from http://www.nsf.gov/statistics/wmpd/.) †Includes those who reported mixed race and "Unknown."

Table 16.1. The Engineering Pathway by Race and Gender from Ph.D. to Tenure-track Positions and the Workforce[a]

Race	Engineering Ph.D.s (2009)[a] (%) N = 3,376		Tenure Track Engineering Positions (2008)[a] (%) N = 18,050		Engineering Workforce (2006)[a] (%) N (est.) = 1,590,000	
	Female	Male	Female	Male	Female	Male
White	15.6	50.6	6.1	59.8	7.5	68.7
Asian	4.1	10.9	3.3	21.1	2.3	12.5
Hispanic	1.1	3.4	0.3	2.8	0.9	4.7
African American	1.3	2.8	0.3	2.8	0.7	2.5
Native American/ Alaskan Native	0.2	0.4	0.6	0.6	0.1	0.2
Other[b]	3.1	6.5	1.2	1.4	±[c]	±[c]
Total	100		100		100	

[a] National Science Foundation, 2011, updated January, 2012. Tables 7-2, 9-26, 9-7. Each column reflects most recent data available. Comparisons across columns are for illustrative purposes only, though changes between 2008 and 2009 were negligible.

[b] Includes those who reported mixed race and "Unknown." In 2009, NSF began tracking numbers of Hawaiians and Pacific Islanders in the Tenure Track category. Those numbers (0.6% for men and women) were placed in the "Other" categories to enable comparison across years.

[c] "Other" not reported.

report that in 1975, no underrepresented minority women earned engineering doctorates (p. 167). In 2009, eighty-eight engineering Ph.D.s were awarded to underrepresented minority women (see Table 16.1).

Table 16.1 shows the representation of men and women, by race, at different points along the graduate and postgraduate STEM pathway. In 2009, of 3,376 engineering Ph.D.s awarded nationally to U.S. citizens, 804 (25.4%) were earned by women; 11% of those women (approximately 2.6% of total doctorate recipients) were underrepresented minorities. African American, Hispanic, and Native American men, combined, earned 223 (6.6%) of the engineering Ph.D.s conferred in that year (NSF, 2011).[7] Similarly, very few women and minorities take the pathway to tenure-track faculty positions in engineering. According to NSF's estimate, there were 18,050 such positions at four-year institutions in 2008. Of these, 80.9% were held by White (59.8%) and Asian (21.1%) men. African American and Hispanic men each held 2.8% of these positions, while White and Asian women held 6.1% and 3.3%,

respectively. There were too few African American, Hispanic, and Native American women to be reported (NSF, 2011).[8]

The research shows one further difference in the outcomes between men and women. For the most part, the proportion of male science and engineering Ph.D.s is similar to the proportion of male faculty members and STEM workforce participants. (Asian men are an exception: their representation among tenure-track faculty is nearly twice their representation among doctorate holders.) On the other hand, the proportion of female faculty members is half or a third the proportion of females among all doctorate holders (see Table 16.1).

Even though enrollment rates for women in doctoral engineering programs are extremely low, their completion rates are as high as or higher than the rates for men. Nevertheless, relatively few women who earn the doctorate continue on the engineering pathway (AAAS, 2010; Committee on Women in Science and Engineering, 1994; Hosoi & Canetto, 2011). Women of all races combined comprise less than

12% of all tenure-track engineering faculty members and workforce employees. Yet studies confirm that women who do pursue tenure-track positions fare well in the interview and hiring process, and perform comparably to men on indicators of success, even achieving tenure at higher rates than men do (Committee on Gender Differences in Careers of Science, Engineering, and Mathematics Faculty, 2010).

In fact, studies suggest that gender is not a strong predictor of doctorate completion, tenure-track position seeking, or participation in the workforce (Hosoi & Canetto, 2011; Preston, 2004, 2006). Long (2001) found that marital status and childrearing responsibilities were the strongest predictors of science and engineering labor force participation, for both men and women, but differentially. Although single men and women were found to participate in the workforce at similar rates, being married and having young children predicted *increased* rates of engineering employment for men and *decreased* rates of engineering employment for women.

Among women who choose the engineering pathway, we see retention rates at each level of education that are comparable to and sometimes higher than those of men. However, we also see dramatic rates of decline in persistence across each juncture along the engineering pathway (Chubin, May, & Babco, 2005). Women who have succeeded at one level of preparation in engineering often opt out of further engineering study or out of the profession (Fouad, Fitzpatrick, & Liu, 2011).

Retention Within Each Phase Along the STEM Pathway

Recent research indicates that students in undergraduate engineering programs at four-year institutions have retention rates similar to or higher than those of students in other majors (George-Jackson, 2011; Lord et al., 2009; McCormick, Lichtenstein, Chen, Puma, & Sheppard, manuscript submitted for publication; Ohland et al., 2008).

At all academic levels, underrepresented minorities and women remain in engineering programs at rates comparable to those of White and Asian men (Cosentino de Cohen & Deterding, 2009; Hosoi & Canetto, 2011).

These findings, although pervasive, are typically based on longitudinal studies whose data sources cannot capture students if they move from one institution to another, or if they leave higher education altogether. Some research suggests that minority students who transfer from a two-year to a four-year institution struggle disproportionately when they encounter a competitive environment, and that they experience feelings of social and cultural isolation when they are no longer surrounded by peers of the same race or ethnicity (Crawford & Macleod, 1990; Hagedorn, Maxwell, Cypers, Hye Sun, & Lester, 2007; Russell & Atwater, 2005; Townsend, 2007).

Moreover, retention is a negative outcome when students remain on pathways that do not lead to degree completion. Chen and Weko (2009) found that engineering had the highest proportion of students (19.3%) persisting in their major without earning a degree after six years. Hurtado et al. (2010) looked at degree completion rates among 201,588 students in 326 four-year institutions. The authors found that minorities and non-minorities have comparable rates of retention in STEM majors, but minorities in STEM have significantly lower rates of degree completion – the five-year degree completion rates for Asians (42%) and Whites (33%) are nearly twice those of Latinos (22%), African Americans (18%), and Native Americans (19%).

Yet discerning completion rates can be complicated. Ohland et al. (2011) found notable differences in the patterns of undergraduate degree completion by race, based on institutional differences. Although some factors were consistent by race, analyses that did not factor in institutional policies and program structures distorted completion rates.

Retention rates are comparable among STEM students regardless of ethnicity, and retention in degree programs among women tends to be higher than that of White and Asian men. But the conditions under which women and minorities study differ from those experienced by majority men, and these conditions can affect their completion rates. Faced with unsupportive institutional policies and negative classroom environments, women and underrepresented minorities consistently report experiences of isolation, self-doubt, and questioning about continuing in engineering programs and post-degree work settings. This phenomenon is described throughout the research literature as *chilly climate* (Hall & Sandler, 1982, 1984; Johnson & Lucero, 2003; Morris & Daniel, 2008; Sandler, Silverberg, & Hall, 1996).

Chilly climate results from several conditions that have been consistently documented in the research literature for decades, including negative interpersonal relations, subtle and overt denigration of skills, attribution of attainment to affirmative action policies, avoidance of eye contact, favoritism toward male and majority students, sexual harassment, and, in the workplace, a dearth of opportunities to advance, failure to be recognized for contributions, and wage disparities (CEOSE, 2009; Crawford & MacLeod, 1990; Daempfle, 2003; Miner-Rubino & Cortina, 2004; Malcom & Malcom, 2011; National Academy of Sciences, 2007; Seymour & Hewitt, 1997; Wyer, 2003). Brown (2000) found chilly climate to be a more challenging issue for women and people of color in academia than lack of financial support, recruitment practices, or composition of faculty. Chilly climate is rarely reported by White and Asian men, who comprise the majority of engineering students and workforce employees. But it is commonly mentioned in interviews with minorities and women, and it has been cited as a common cause of attrition from all levels of engineering education and from the workforce.

Other barriers also impede the progress of women and underrepresented minorities in engineering, both during and after degree completion. For undergraduates, these may include an absence of role models, limited interaction with faculty, and a lack of effective advising (Johnson & Sheppard, 2004; Seymour & Hewitt, 1997; Tsui, 2010). For graduate students and early-career faculty, a lack of role models and mentors is an especially critical factor (Davidson & Foster-Johnson, 2001; Office of Scientific and Engineering Personnel, 1994; Nelson & Rogers, 2007; Reybold & Alamia, 2008). In the workplace, common barriers include lack of mentoring and lack of support in balancing work/family tensions (Fouad et al., 2011; Miner-Rubino & Cortina, 2004; Preston, 2004, 2006).

Lack of financial aid is also cited as a barrier to minorities' attainment of STEM degrees (Chubin et al., 2005; Clewell & Tinto, 1999). Studies consistently show that financial aid is a predictor of college retention among minorities (Carter, 2006; Georges, 1999; Lee, 1991). Recent research finds that financial aid increasingly takes the form of loans as opposed to grants and scholarships, and increasingly fails to keep pace with the rising costs of higher education (Lee & Clery, 2004; NRC, 2007). Among the obstacles to degree completion, lack of financial aid is cited less often than chilly climate, lack of faculty contact and mentors, and work/family tensions. Even so, in an era of high unemployment and housing market declines, all of which disproportionately affect minorities and women, cutbacks in grant and scholarship programs could have an increasingly significant impact on retention in engineering specifically and in STEM more generally (Joint Economic Committee, 2011; Kochhar, 2011; Kochhar, Fry, & Taylor, 2011).

Family circumstances are commonly cited by women and minorities as influencing their retention in STEM. Ong et al. (2011) conclude, "Family and community support is perhaps the most salient and influential factor that women of color identify as encouraging to their completion of a STEM degree" (p. 186). Fouad et al. (2011) conducted a qualitative study of female engineers, fourteen of whom were currently practicing and eleven of whom had left the field. The authors identified five domains in which these women encountered challenges, including *support and/or barriers in work and family*. Women who had left engineering cited among their reasons the need to care for their children, in addition to issues specific to the workplace environment, including lack of movement into management roles and a dislike of engineering tasks or environment. Women who remained in engineering also mentioned family responsibilities as challenges. Valenzuela (2006) examined the impact of family support on Latina transfer students. Russell and Atwater (2005) documented the importance of family support to African American undergraduates at a predominantly White university. Bellisari (1991) found that Asian and African American women's choice of and retention in science majors were strongly influenced by family expectations. Moller-Wong and Eide (1997) found that marital status (being married) was a high-risk factor related to retention in the engineering major.

Trenor, Yu, Waight, Zerda, and Sha (2008), in their mixed-methods study of ethnically diverse female engineering students, found that motivations for pursuing engineering varied for different ethnic groups. Li, Swaminathan, and Tang (2009) reviewed internal and external factors that predict success in engineering, including community and college influence (external characteristics), cognitive and affective characteristics (internal), and demographic variables. Their findings highlight a need for more sophisticated measurement approaches that can model the interaction effects among these various factors on educational outcomes. Besterfield-Sacre, Atman, and Shuman (1998) and Besterfield, Moreno, Shuman, and Atman (2001) found that African American and Hispanic students exhibit more positive impressions of engineers than do White students or members of other ethnic minorities (e.g., Asian-Pacific Islanders). Yet the self-efficacy of African American and Hispanic students

was found to decrease during the freshman year (Besterfield-Sacre et al., 1998, 2001). Why is this so? The evidence is overwhelming that, for both minorities and women, environmental factors play a decisive role in limiting retention and persistence in engineering.

Promoting Retention and Persistence in STEM: What the Research Tells Us

Several studies and policy documents indicate that increased diversity in the classroom and workplace enhances scholarship, achievement, and productivity in general (AAAS, 2010; Gurin, Dey, Hurtado, & Gurin, 2002; Page, 2008; Satcher, 2001). Although it has proven difficult to raise persistence rates of women and underrepresented minorities in engineering and the sciences, successes have been documented. Furthermore, research has identified strategies that consistently promote recruitment and persistence in K–12 and higher education institutions and workplaces. Concerted attempts to improve academic and workplace climates and to boost recruitment of minorities and women have yielded encouraging results (see Tinto, 1993; Hill, Corbett, & St. Rose, 2010). The research suggests that effective policies and practices respond to a broad range of student needs and benefit all students, not just women and underrepresented minorities. In this section, we begin with classroom strategies and then consider strategies implemented at the institutional level.

Research-Based Strategies for Improving Classroom Climate

Learner-centered pedagogical approaches have proven effective in engaging diverse student populations. Amelink and Creamer (2010) highlight the impact of peer interactions on academic and professional persistence. Not surprisingly, the extent to which women students feel respected as classmates can influence their assessment of their own engineering abilities and their emerging identities as prospective engineers. Seeking a better understanding of engineering student achievement and career plans, Jones, Paretti, Hein, and Knott (2010) differentiated between *expectancy-related constructs* (i.e., engineering self-efficacy and expectations to succeed in engineering) and *value-related constructs* (i.e., engineering identity and beliefs about the importance and usefulness of engineering). Findings from a survey of 363 first-year engineering students showed that expectancy-related constructs tended to predict achievement and that value-related constructs predicted career plans in engineering for both men and women.

In addition to peer interactions, faculty–student interactions play a significant role in forming engineering identities and subsequent career decision making, leading to greater student satisfaction and undergraduate persistence in engineering (Lattuca, Terenzini, & Volkwein, 2006; Pascarella & Terenzini, 2005). Faculty–student interactions both inside and outside the classroom have been shown to influence the career goals of female students (Amelink & Creamer, 2010; Bernold, 2007; Braxton, Milem, & Sullivan, 2000; Eskandari et al., 2007; Springer, Stanne, & Donovan, 1997; Terenzini, Cabrera, Colbeck, Parente, & Bjorklund, 2001). Particularly for minority students, positive interactions with engineering faculty role models can have a significant influence on students' decisions to pursue graduate study in engineering (May & Chubin, 2003; Tinto, 1975). Positive faculty–student interactions contribute to greater student retention and academic success by enhancing engineering students' confidence in their problem-solving, engineering design, and professional and interpersonal skills (Bjorklund, Parente, & Sathianathan, 2004; Chen, Lattuca, & Hamilton, 2008). Chen et al. (2008) emphasize that faculty promote student engagement in engineering by helping students understand the importance of developing professional and interpersonal skills, providing undergraduate research experiences, and encouraging academic achievement.

In recent years, *pedagogies of engagement*, including cooperative learning, problem-based and project-based learning, case-based learning, and service learning, have garnered increasing interest by engineering faculty (Smith, 2011; Smith, Sheppard, Johnson, & Johnson, 2005; Vaughn & Seifer, 2004). Pedagogies of engagement, which are characterized by small groups of students working together to solve meaningful problems, have been shown to promote academic learning and social integration, both of which are critical for retention of all students in college. McKeachie, Pintrich, Yi-Guang, and Smith (1986) found that learning how to engage in critical thinking depends on student participation in class, teacher encouragement, and student-to-student interaction. Tinto (2002, p. 2) identifies five essential conditions for student learning and retention in college: (1) being held to high expectations, (2) receiving academic and social support, (3) receiving feedback about performance, (4) academic and social integration, and (5) relevant learning. Pedagogies of engagement contribute to establishing these conditions.

Cooperative learning is probably the most thoroughly researched pedagogy of engagement. There are currently more than 400 articles on cooperative learning in STEM disciplines, including several meta-analyses (Johnson, Johnson, & Smith, 2006; Smith, Sheppard, Johnson, & Johnson, 2005). First introduced to the engineering education community more than thirty years ago (Smith, Johnson, & Johnson, 1981), cooperative learning has demonstrated its efficacy in promoting student learning and a positive classroom environment. Research has focused primarily on student achievement, but there are studies related to student persistence and retention as well.

The conceptual framework underlying cooperative learning (Deutsch 1949, 1962) addresses issues pertinent to the educational experiences of women and underrepresented minorities, who often report finding traditional classroom environments "chilly," "aggressive," and inimical to diverse learning styles (Crawford & McLeod, 1990).

Balancing challenging material with support for learning has been identified as an important factor in (1) promoting high-level performance among all students and (2) improving retention and persistence among women and underrepresented minorities alienated by competitive classroom structures (Edmonson, 2008; Pelz & Andrews, 1966). A 1997 meta-analysis of cooperative learning in introductory-level STEM classes showed moderate to high mean effect sizes for achievement, persistence, and attitudes: 0.51, 0.46, and 0.55, respectively (Springer, Stanne, & Donovan, 1999).

Johnson et al. (2006) found through their meta-analysis of twenty-four studies that cooperative efforts promoted greater liking among students than did competition (effect size = 0.68) or independent work (effect size = 0.55). College students learning cooperatively perceived more academic and personal support from peers and instructors than did students working competitively (effect size = 0.60) or individualistically (effect size = 0.51). Extensive research indicates that the benefits of cooperation are felt among all students, regardless of race and culture, language, social class, ability, and gender (Johnson & Johnson, 1989). Cooperative learning is one example of an evidence-based practice within the family of pedagogies of engagement that promotes learning and retention among all students, and warms chilly classroom environments.

Warming Chilly Climates in Higher Education and Workplaces

There is a growing body of evidence concerning successful efforts on the institutional level to recruit and retain women and minorities, both in academia and in the STEM workforce. The National Academy of Engineering's Center for the Advancement of Scholarship in Engineering Education (CASEE) includes on its website a list of research-based programs that have proven effective in increasing student retention and persistence. Although these programs focus on women, the criteria for inclusion on the list include demonstrated efficacy for multiple demographic groups (National Academy of Engineering, 2012).

The research literature consistently refers to several strategies that have effectively promoted retention and persistence among STEM undergraduates, graduate students, faculty, and members of the workforce. These strategies include providing hands-on undergraduate and graduate research experiences, aligning new programs with successful existing ones, extending the "personal touch" with plenty of mentoring and attention to individual needs, providing incentives to laboratory researchers to broaden participation by women and underrepresented minorities, and having personnel dedicated to tracking students in academic programs (CEOSE, 2009; French, Immekus, & Oakes, 2005).

"High-impact practices" such as first-year seminars, capstone projects, and learning communities benefit students from diverse backgrounds, increasing rates of retention and engagement and supporting student persistence (Kuh, 2008; Kuh et al., 2005; Pascarella & Terenzini, 2005; Terenzini et al., 2001). These practices promote student interactions with faculty and peers around meaningful questions and issues, thereby deepening students' investments in these activities as well as their commitment to the academic program and the institution as a whole. Internships that provide candidates with work experience in industry can be important for building professional networks that increase affiliation with engineering and thereby encourage persistence (Ong et al., 2011).

Muraskin, Lee, Wilner, and Swail (2004) compared undergraduate institutions with higher and lower completion rates for low-income students. Those with Higher Graduate Rates (HGR) tended to attract more students who had recently finished high school and who were better prepared academically. HGR institutions had more full-time faculty, lower student/faculty ratios, and greater financial subsidies than did Low Graduate Rate (LGR) institutions. HGR colleges and universities shared several institutional policies and practices, including small class sizes, careful academic planning for students, special programs for students at academic risk, programs to promote a sense of belonging to the campus community, extensive academic support services, accessible faculty, modest selectivity, and initiatives aimed at promoting retention and high graduation rates.

Fox, Sonnert, and Nikiforova (2009) conducted a mixed-methods study examining forty-nine programs designed to support women undergraduates in the sciences and engineering. The authors began with a regression analysis, using numbers of women graduating from these programs between 1984 and 2001 as a dependent variable. Then they interviewed the directors of the ten most successful and the ten least successful programs. The authors found that leaders of successful programs had a well-elaborated understanding of issues related to recruitment and retention within their institutions, promoted a range of strategies, actively linked students with resources, and integrated their services with other university programs.

Tsui (2010) conducted 110 interviews and 25 focus groups with mechanical engineering faculty, staff, and senior undergraduates at six universities nationally. She found that access to women- or minority-focused organizations (e.g., Society of Women Engineers, National Society of Black Engineers, Hispanic Engineering & Science Organization) helped dispel students' feelings of isolation. Participation in other student organizations, especially those with a humanitarian or community focus, as well as participation in regional and national conferences, helped women and minorities affiliate with the engineering profession and find mentors and role models. Students who enrolled early in an engineering department demonstrated higher degree commitment and sense of belonging than did students who spent the first two years taking non-engineering courses from non-engineering departments.

In a comprehensive synthesis of literature on undergraduate and graduate women of color in STEM fields, Ong et al. (2011) showed that role models and mentors are especially important to the success of women, particularly women of color, in doctoral programs. Mentoring by faculty, the authors wrote, "was rare but incredibly valuable" (p. 193). The authors found that mentors need not be of the same gender or ethnicity as those they mentor. Brown (2000) found that "few minority women had true mentors while in graduate school, but those who did reported exceptional relationships and experiences" (p. 259). Mentors can play important roles in students' decisions to attend graduate school, choose a particular doctoral program, or remain in their programs.

Fouad et al. (2011) surveyed more than 3,700 women who graduated with engineering degrees. They analyzed factors that predicted women's decisions to remain in the engineering workforce. Perceptions of workplace climate, the presence or absence of support from colleagues and supervisors, and the availability of opportunities for advancement influenced career decision making. Women were more likely to remain when company policies and practices supported the challenges of balancing multiple life roles. Yet women who left engineering did not attribute their decisions to their childrearing responsibilities. Women who left engineering jobs tended to leave the engineering profession entirely.

Summary and Future Research

Summary

In spite of a federal policy imperative to expand the engineering workforce, the proportion of college and university students pursuing engineering degrees continues to decline (President's Council on Jobs and Competitiveness, 2011). And in spite of a policy agenda targeted at boosting participation of women and underrepresented minorities in the engineering workforce, progress has been slow.

Research over the past several decades has consistently identified factors that reduce participation of women and underrepresented minorities within and between phases of the engineering pathway, including competitive classroom pedagogies, inadequate advising, lack of mentors and role models, discrimination and bias, limited access to financial aid, and lack of attention to and accommodation of family issues. Several of these factors have been associated with *chilly climate* (Hall & Sandler, 1982), an inhospitable environment that has been documented in the postsecondary classrooms, higher education workplaces, and business work environments.

On the other hand, research has also identified policies and practices that overcome these obstacles. Strategies include pedagogies of engagement at the classroom level and policies and practices within academic institutions and work environments that (1) promote students' affiliation with their department and the institution overall, and (2) help students and employees balance academic and professional advancement with family responsibilities. These pedagogies, policies, and practices benefit all engineering candidates and working engineers, not just women and underrepresented minorities.

Areas for Further Research

Understanding the interplay of factors contributing to recruitment, retention, and persistence in engineering, especially among women and underrepresented minorities, requires the application of a range of academic disciplines, including psychology, education, economics, and sociology as well as engineering. Engineering educators, who hail from a variety of disciplines, are well situated to tackle this critical issue.

Research and policy documents related to the *supply* of engineers are extensive. But the *demand* for engineers has received little attention (see, however, Carnevale, Smith, & Melton, 2011). The field would benefit from studies of the need for domestically trained engineers, including what kinds of

engineers are in greatest demand across various industries.

In preparing this review, the authors found more literature pertaining directly to women than to minorities, although there is crossover in many articles. The field would benefit from more research on underrepresented minorities, including African Americans, Hispanics, Native Americans, and Asian subgroups consisting largely of first- or second-generation immigrants (see Ngo, 2006). Because participation patterns vary significantly between men and women within these groups, studies should disaggregate by gender.

We need a refined understanding of the circumstances that influence women and underrepresented minorities at critical decision making junctures – as they decide whether to study engineering in the first place, and then whether to continue along each segment of the engineering pathway. Why do so few women and minority undergraduates choose to major in engineering? Why do we lose so many women between junctures? How do family circumstances, including marital status, childrearing responsibilities, and cultural expectations, affect decisions to participate in engineering at every step along the way? There are several studies of students who chose engineering and the factors that influenced their persistence (see, e.g., Fouad et al., 2011; Lichtenstein, Loshbaugh, Claar, Bailey, & Sheppard, 2007; Lichtenstein et al., 2009; Russell & Atwater, 2005; Valenzuela, 2006). Yet qualitative research concerning those who did not choose engineering is scant.

Higher education would benefit from studies tracing the differential effects of institutional policies (e.g., transfer policies, structure of majors, workload) on various student populations (Lichtenstein, McCormick, Puma, & Sheppard, 2010). Longitudinal student studies, both quantitative and qualitative, that can reach across institutional boundaries will be necessary to further our understanding of how women and underrepresented minorities negotiate the engineering pathway, especially because many of these students transfer from two-year to four-year institutions. Greater attention could be paid to this pathway (see National Research Council and National Academy of Engineering, 2012).

To date, much of the research into retention and persistence has been correlational (Johnson, 2011). Studies that draw causal relationships between specific policies, practices, and conditions and how those impact participation in STEM fields will be essential for identifying strategies that can increase the representation of women and minorities in these fields.

In this chapter, we have reviewed strategies that have been proven to promote retention and persistence of all students in engineering, and of women and underrepresented minorities in particular. But how do we get colleges and universities to adopt these strategies? The incentive structures and institutional cultures of higher education are notoriously intractable and carry with them the tensile memory of steel. Colleges and universities have struggled to translate research findings into sustainable policies, programs, and practices. Yet, in many cases where faculty and administrators – individually and/or collectively – were sufficiently committed to reform, these obstacles have been overcome (see, e.g., O'Meara, 2012). An emerging literature on the diffusion of innovation in engineering education is directly relevant to issues of retention and persistence (see Chapter 19 by Litzinger & Lattuca, this volume). Lessons from this literature hold promise for supporting all students on STEM pathways, as well as addressing issues of retention and persistence that for decades have limited the participation of women and underrepresented minorities in engineering.

Acknowledgments

The authors are grateful to Arthur Evenchik for his exceptional editorial support, and to Dr. Sheri Sheppard, Dr. Stephanie Adams, and CHEER reviewers for their thoughtful suggestions. We are also grateful to Ms. Susan Olmsted, National Science Foundation reference librarian, for her invaluable assistance.

Footnotes

1. Data were not available concerning ethnic minority employment in these countries.

2. In the 1970s, "minorities" referred to non-Whites generally. Among minorities, African Americans were better organized than other ethnic groups, largely because of the civil rights movement, and therefore African American voices were heard more often in testimony to Congress and in other political arenas. Yet "minorities" also included Hispanics, Native Americans, and Asians. By the 1990s, the term "underrepresented minorities" emerged. This category typically excludes Japanese and Chinese Americans, who are overrepresented in higher education in comparison to their numbers in the general population. The terms "minority" and "underrepresented minority" refer to U.S. citizens and do not include international students.

3. Science and Engineering Equal Opportunities Act, Section 32(b), Part B of P.L. 96–516, 94 Stat. 3010, as amended by P.L. 99–159.

4. The Science and Technology Equal Opportunity Act requires NSF to submit a report to Congress every two years detailing its activities and progress in expanding the talent pool in science and engineering among minorities and women (in recent years, the charge has expanded to include people with disabilities). The standing committee charged with this task, the Committee on Equal Opportunities in Science & Engineering (CEOSE), was chaired by Dr. Theresa Maldonado, one of this chapter's co-authors, in 2009–2010.

5. Interestingly, all of the science and engineering workforce forecasts since the 1950s have calculated only the supply side, not the demand side of the equation.

6. See U.S. Census Bureau, Age & Sex in the United States, 2008: http://www.census.gov/population/www/socdemo/age/age_sex_2008.html).

7. The proportion of foreign students earning doctoral degrees in engineering has increased. In 2009, foreign nationals accounted for 55% of all engineering Ph.D.s conferred in the United

States, up from 49.9% in 2000 (Brown, 2000; Gibbons, 2009). Statistics for foreign nationals are not included in this chapter.

8. NSF does not include counts when estimates are less than 50 because a count might compromise confidentiality. The relevant data table, Table 9-26, shows science and engineering doctorate holders employed in universities and four-year colleges, classified by field, sex, race/ethnicity, and tenure status. Six cells within engineering contained "D"s, indicating a count smaller than 50. For current purposes, "D"s were treated as 50, overestimating the number of tenured and tenure-track positions and the numbers of faculty members who hold them.

References

Adelman, C. (2006). *The toolbox revisited: Paths to degree completion from high school through college.* Washington, DC: U.S. Department of Education.

Amelink, C. T., & Creamer, E. G. (2010). Gender differences in elements of the undergraduate experience that influence satisfaction with the engineering major and the intent to pursue engineering as a career. *Journal of Engineering Education,* 99(1), 81–92.

American Association for the Advancement of Science (AAAS). (2010). *Handbook on diversity and the law: Navigating a complex landscape to foster greater faculty and student diversity in higher education.* Washington, DC: Author.

Atman, C. J., Sheppard, S. D., Fleming, L., Miller, R., Smith, K., Stevens, R., . . . Lund, D. (2008, June). Moving from pipeline thinking to understanding pathways: Findings from the Academic Pathways Study of engineering undergraduates (AC 2008-1307). In *Proceedings of the American Society for Engineering Education Annual Conference,* Pittsburgh, PA. Washington, DC: American Society for Engineering Education (ASEE).

Bellisari, A. (1991). Cultural influences on the science career choices of women. *Ohio Journal of Science,* 91(3), 129–133.

Bernold, L. E. (2007). Preparedness of engineering freshman to inquiry-based learning. *Journal of Professional Issues in Engineering Education and Practice,* 133(7), 99–106.

Besterfield-Sacre, M. E., Atman, C. J., & Shuman, L. J. (1998). Engineering student attitudes assessment. *Journal of Engineering Education,* 87(2), 133–141.

Besterfield-Sacre, M. E., Moreno, M., Shuman, L. J., & Atman, C. J. (2001). Gender and ethnicity differences in freshmen engineering attitudes: A cross-institutional study. *Journal of Engineering Education,* 90(4), 477–489.

Bjorklund, S. A., Parente, J. M., & Sathianathan, D. (2004). Effects of faculty interaction and feedback on gains in student skills. *Journal of Engineering Education,* 93(2), 153–160.

Braxton, J. M., Milem, J. F., & Sullivan, A. S. (2000). The influence of active learning on the college student departure process: Toward a revision of Tinto's Theory. *Journal of Higher Education,* 71(5), 569–590.

Brown, S. V. (2000). The preparation of minorities for academic careers in science and engineering: How well are we doing? In G. Campbell, R. Denes, & C. Morrison (Eds.), *Access denied: Race, ethnicity, and the scientific enterprise* (pp. 239–260). New York, NY: Oxford University Press.

Carnevale, A., Smith, N., & Melton, M. (2011). STEM webinar. Presented at Georgetown Public Policy Institute, Center for Education and the Workforce. Washington, D.C.: Georgetown University. August 14, 2012 from http://cew.georgetown.edu/stem/

Carter, D. F. (2006). Key issues in the persistence of underrepresented minority students. *New Directions for Institutional Research,* 130, 33–46.

Chen, H. L., Lattuca, L. R., & Hamilton, E. R. (2008). Conceptualizing Engagement: Contributions of Faculty to Student Engagement in Engineering. *Journal of Engineering Education,* 97, 339–353.

Chen, X., & Weko, T. (2009). *Students who study science, technology, engineering, and mathematics (STEM) in postsecondary education* (NCES 2009–161). Washington, DC: U.S. Department of Education, National Center for Education Statistics (NCES).

Chubin, D., May, G. S., & Babco, E. L. (2005). Diversifying the engineering workforce. *Journal of Engineering Education,* 94(1), 73–86.

Clewell, B., & Tinto, V. (1999). *A comparative study of the impact of differing forms of financial aid on the persistence of minority and majority doctoral students*. Syracuse, NY: Syracuse University.

Committee on Equal Opportunities in Science & Engineering (CEOSE) 2007–2008. (2009). *Biennial report to Congress: Broadening participation in America's STEM workforce*. Arlington, VA: National Science Foundation.

Committee on Gender Differences in Careers of Science, Engineering, and Mathematics Faculty. (2010). *Gender differences at critical transitions in the careers of science, engineering, and mathematics faculty*. Washington, DC: The National Academies Press.

Committee on Prospering in the Global Economy of the 21st Century. (2007). *Rising above the gathering storm: Energizing and employing America for a brighter economic future*. Washington, DC: The National Academies Press. Retrieved from http://www.nap.edu/catalog/11463.html/

Committee on STEM Education, National Science and Technology Council. (2011). *The Federal Science, Technology, Engineering, and Mathematics (STEM) Education Portfolio*. A Report from the Federal Inventory of STEM Education Fast-Track Committee. Retrieved from http://www.whitehouse.gov/sites/default/files/microsites/ostp/costem__federal_stem_education_portfolio_report.pdf

Committee on Underrepresented Groups and the Expansion of the Science and Engineering Workforce Pipeline; Committee on Science, Engineering, and Public Policy; Policy and Global Affairs; National Academy of Sciences, National Academy of Engineering, and Institute of Medicine. (2011). *Expanding underrepresented minority participation: America's science and technology talent at the crossroads*. Washington, DC: The National Academies Press.

Committee on Women in Science and Engineering (CWSE). (1994). *Women scientists and engineers employed in industry: Why so few?* National Research Council. Washington, DC: The National Academies Press.

Cosentino de Cohen, C., & Deterding, N. (2009). Widening the net: National estimates of gender disparities in engineering. *Journal of Engineering Education, 98*(3), 211–226.

Crawford, M., & MacLeod, M. (1990). Gender in the college classroom: An assessment of the "chilly climate" for women. *Sex Roles, 23*(3/4), 101–122. Retrieved from http://www.springerlink.com/content/u77g4hk6177551vj/fulltext.pdf

Daempfle, P. (2003). An analysis of the high attrition rates among first year college science, math, and engineering majors. *Journal of College Student Retention: Research, Theory and Practice, 5*(1), 37–52.

Davidson, M. N., & Foster-Johnson, L. (2001). Mentoring in the preparation of graduate researchers of color. *Review of Educational Research, 71*(4), 549–574.

Deutsch, M. (1949). A theory of cooperation and competition. *Human Relations, 2*, 129–152.

Deutsch, M. (1962). Cooperation and trust: Some theoretical notes. In M. R. Jones (Ed.), *Nebraska symposium on motivation* (pp. 275–319). Lincoln, NE: University of Nebraska Press.

Edmonson, A. C. (2008). The competitive imperative of learning. *Harvard Business Review, 86*(7/8), 60–67.

Eskandari, H., Sala-Diakanda, S., Furterer, S., Rabelo, L., Crumpton-Young, L., & Williams, K. (2007). Enhancing the undergraduate Industrial Engineering curriculum: Defining desired characteristics and emerging topics. *Education Training, 49*(1), 45–55.

Espinosa, L. (2011). Pipelines and pathways: Women of color in undergraduate STEM majors and the college experiences that contribute to persistence. *Harvard Educational Review, 81*(2), 209–240.

European Commission. (2009). *She figures 2009 – Statistics and indicators on gender equality in science*. Commission for Science and Research. Luxembourg: Publications Office of the European Union. Retrieved from http://ec.europa.eu/research/science-society/document_library/pdf_06/she_figures_2009_en.pdf

Fouad, N., Fitzpatrick, M., & Liu, J. P. (2011). Persistence of women in engineering careers: A qualitative study of current and former female engineers. *Journal of Women and Minorities in Science and Engineering, 17*(1), 69–96.

Fox, M. F., Sonnert, G., & Nikiforova, I. (2009). Successful programs for undergraduate women in science and engineering: "Adapting" versus "adopting" the institutional environment. *Research in Higher Education, 50*(4), 333–353.

Freeman, C. E. (2004). *Trends in educational equity of girls & women: 2004* (U.S. Department of Education, National Center for Education Statistics 2005–016). Washington, DC: U.S. Government Printing Office.

French, B. F., Immekus, J. C., & Oakes, W. C. (2005). An examination of indicators of engineering students' success and persistence. *Journal of Engineering Education*, 94(4), 419–425.

George-Jackson, C. (2011). Stem switching: Examining departures of undergraduate women in STEM fields. *Journal of Women and Minorities in Science and Engineering*, 17(2), 149–171.

Georges, A. (1999). Keeping what we've got: The impact of financial aid on minority retention in engineering. *NACME Research Letter*, 9(1), 1–19.

Gibbons, M. (2009). Engineering by the numbers. Washington, DC: American Society of Engineering Education. Retrieved from http://www.asee.org/papers-and-publications/publications/college-profiles/2009-profile-engineering-statistics.pdf

Gurin, P., Dey, E. L., Hurtado, S., & Gurin, G. (2002). Diversity and higher education: Theory and impact on educational outcomes. *Harvard Educational Review*, 72(3), 330–336.

Hagedorn, L. S., Maxwell, W., Cypers, S., Hye Sun, M., & Lester, J. (2007). Course shopping in urban community colleges: An analysis of student drop and add activities. *Journal of Higher Education*, 78(4), 464–485.

Hall, R. M., & Sandler, B. R. (1982). *The classroom climate: A chilly one for women*. Washington, DC: Association of American Colleges.

Hall, R. M., & Sandler, B. R. (1984). *Out of the classroom: A chilly campus climate for women?* Washington, DC: Association of American Colleges.

Hartline, B. K., & Poston, M. (2009). The mandate for broadening participation: Developing the best minds and solutions. In M. Boyd & J. L. Wesemann (Eds.), *Broadening participation in undergraduate research: Fostering excellence and enhancing the impact* (pp. 13–21). Washington, DC: Council on Undergraduate Research.

Hill, C., Corbett, C., & St. Rose, A. (2010). *Why so few? Women in Science, Technology, Engineering, and Mathematics*. Washington, DC: American Association of University Women (AAUW).

Hosoi, S. A., & Canetto, S. S. (2011). Women in graduate engineering: Is differential dropout a factor in their underrepresentation among engineering doctorates? *Journal of Women and Minorities in Science and Engineering*, 17(1), 11–27.

Hurtado, S., Eagan, K., & Chang, M. (2010, January). *Degrees of success: Bachelor's degree completion rates among initial STEM majors* (HERI Research Brief). Los Angeles: Higher Education Research Institute at UCLA.

Johnson, A. (2011). Accomplishments and challenges for women in STEM: Implications for future research and programs. *Journal of Women and Minorities in Science and Engineering*, 17(1), 5–10.

Johnson, D. W., & Johnson, R. (1989). *Cooperation and competition: Theory and research*. Edina, MN: Interaction Book.

Johnson, D. W., Johnson, R. T., & Smith, K. A. (2006). *Active learning: Cooperation in the college classroom* (3rd ed.). Edina, MN: Interaction Book.

Johnson, M. J., & Sheppard, S. D. (2004). Relationships between engineering student and faculty demographics and stakeholders working to affect change. *Journal of Engineering Education*, 92(2), 137–151.

Johnson, S. D., & Lucero, C. (2003). Transforming the academic workplace: Socializing under-represented minorities into faculty life. In M. A. Fox (Ed.), *Pan organizational summit on the U.S. science and engineering workforce* (pp. 138–144). Washington, DC: The National Academies Press.

Joint Economic Committee. (2011). *Assessing the impact of the great recession on income and poverty across states: Highlights from the Census Bureau's release of data from the 2010 American Community Survey*. Washington, DC: United States Congress. Retrieved from http://www.jec.senate.gov/public/?a=Files.Serve&File_id=bb857721-f48a-44cc-8214-508dbd24cc7

Jones, B. D., Paretti, M. C., Hein, S. F., & Knott, T. W. (2010). An analysis of motivation constructs with first-year engineering students: Relationships among expectancies, values, achievement, and career plans. *Journal of Engineering Education*, 99(4), 319–336.

Keller, E. F. (1985). *Reflections on gender and science*. New Haven, CT: Yale University Press.

Kochhar, R. (2011, July 6). *In two years of economic recovery, women lost jobs, men found them*. Washington, DC: The Research Center. Retrieved from http://pewresearch.org/pubs/2049/unemployment-jobs-gender-recession-economic-recovery

Kochhar, R., Fry, R., & Taylor, P. (2011, July 26). *Wealth gaps rise to record highs* between whites, blacks, Hispanics. Washington, DC: Pew Research Center. Retrieved from http://www.pewsocialtrends.org/2011/07/26/wealth-gaps-rise-to-record-highs-between-whites-blacks-hispanics/10/

Kuh, G. D. (2008). High-impact educational practices: What they are, who has access to them, and why they matter. Washington, DC: Association of American Colleges and Universities.

Kuh, G. D., Jinzie, J., Schuh, J. H., & Whitt, E. J. (2005). Student success in college: Creating conditions that matter. San Francisco, CA: Jossey-Bass.

Kyle, W., Jr. (1997). The imperative to improve undergraduate education in science, mathematics, engineering, and technology. Journal of Research in Science Teaching, 34(6), 547–549.

Lattuca, L. R., Terenzini, P. T., & Volkwein, J. F. (2006). Engineering change: A study of the impact of EC2000. Final Report. Baltimore, MD: ABET.

Lee, C. (1991). Achieving diversity: Issues in the recruitment and retention of underrepresented racial/ethnic students in higher education: A review of the literature. Alexandria, VA: National Association of College Admission Counselors.

Lee, J., & Clery, S. (2004). Key trends in higher education. Bethesda, MD: JBL Associates. Retrieved from http://www.aft.org/pdfs/highered/academic/june04/Lee.qxp.pdf

Li, Q., Swaminathan, H., & Tang, J. (2009). Development of a classification system for engineering student characteristics affecting college enrollment and retention. Journal of Engineering Education, 98(4), 361–376.

Lichtenstein, G., Loshbaugh, H. G., Claar, B., Chen, H., Jackson, K., & Sheppard, S. D. (2009). An engineering major does not (necessarily) an engineer make: Career decision-making among undergraduate engineering majors. Journal of Engineering Education, 98(3), 227–234.

Lichtenstein, G., Loshbaugh, H. G., Claar, B., Bailey, B., & Sheppard, S. D. (2007, June). Should I stay or should I go? Engineering students' persistence is based on little experience or data. ASEE09-122209 In Proceedings of the American Society for Engineering Education Annual Conference, Honolulu, Hawaii. Washington, DC: The American Society for Engineering Education (ASEE).

Lichtenstein, G., McCormick, A., Puma, J., & Sheppard, S. (2010). Comparing undergraduate experience of engineers to all other majors: Significant differences are programmatic. Journal of Engineering Education, 99(4), 305–317.

Long, J. S. (2001). From scarcity to visibility: Gender differences in the careers of doctoral scientists and engineers. Washington, DC: The National Academies Press.

Lord, S., Camacho, M., Layton, R., Long, R., Ohland, M., & Wasburn, M. (2009). Who's persisting in engineering? A comparative analysis of female and male Asian, Black, Hispanic, Native American, and White students. Journal of Women and Minorities in Science and Engineering, 15(2), 167–190.

Lowell, B. L., Salzman, H., Bernstein, H., & Henderson, E. (2009, October). Steady as she goes? Three generations of students through the science and engineering pipeline. Paper presented at the Annual Meetings of the Association for Public Policy Analysis and Management, Washington, DC. Retrieved from http://policy.rutgers.edu/faculty/salzman/SteadyAsSheGoes.pdf

Lucena, J. (2005). Defending the nation: U.S. policymaking to create scientists and engineers from Sputnik to the 'War Against Terrorism.' Lanham, MD: University Press of America.

Malcom, S., George, Y. S., & Van Horne, V. V. (Eds). (1996). The effect of the changing policy climate on science, mathematics, and engineering diversity. Washington, DC: American Association for the Advancement of Science.

Malcom, S., Hall, P., & Brown, J. (1976). The double bind: The price of being a minority woman in science. Office of Opportunities in Science (Report 76-R-3). Washington, DC: American Association for the Advancement of Science (AAAS).

Malcom, L., & Malcom, S. (2011, Summer). The double bind: The next generation. Harvard Educational Review, 81(2), 162–171.

May, G. S., & Chubin, D. E. (2003, January). A retrospective on undergraduate engineering success for underrepresented minority students. Journal of Engineering Education, 92(1), 1–13.

McCormick, A., Lichtenstein, G., Chen, H., Puma, J., & Sheppard, S. (2013). Is the supply of engineering majors a problem of many losses or few gains? Persistence and migration of engineering majors compared to other fields. Manuscript submitted for publication.

McKeachie, W., Pintrich, P., Yi-Guang, L., & Smith, D. (1986). *Teaching and learning in the college classroom: A review of the research literature*. Ann Arbor, MI: The Regents of the University of Michigan.

Miner-Rubino, K., & Cortina, L. M. (2004). Working in a context of hostility toward women: Implications for employees' well-being. *Journal of Occupational Health Psychology*, 9(2), 107–122.

Moller-Wong, C., & Eide, A. (1997). An engineering student retention study. *Journal of Engineering Education*, 86(1), 7–15.

Morris, L. K., & Daniel, L. G. (2008). Perceptions of a chilly climate: Differences in traditional and non-traditional majors for women. *Research in Higher Education*, 49(3), 256–273.

Muraskin, L., Lee, J., Wilner, A., & Swail, W. S. (2004). *Raising the graduation rates of low-income college students*. Washington, DC: Pell Institute for the Study of Opportunity in Higher Education. Retrieved from http://www.luminafoundation.org/publications/PellDec2004.pdf

National Academy of Engineering. (2012, April 2). EEES Research-Informed practices for student recruitment, retention, and success. Retrieved from http://www.nae.edu/activities/20676/21702/26338/35823/eeeshome/research-informedpractices.aspx

National Academy of Sciences. (2007). *Beyond bias and barriers: Fulfilling the potential of women in academic science and engineering*. Washington, DC: The National Academies Press.

National Research Council (NRC). (1986). *Engineering infrastructure diagramming and modeling*. Washington, DC: The National Academies Press.

National Research Council (NRC). (2007). *Beyond bias and barriers: Fulfilling the potential of women in academic science and engineering*. Washington, DC: The National Academies Press.

National Research Council and National Academy of Engineering. (2012). *Community colleges in the evolving STEM education landscape: Summary of a summit*. In S. Olson & J. B. Labov (Rapporteurs), Planning Committee on Evolving Relationships and Dynamics Between Two- and Four-Year Colleges, and

Universities. Board on Higher Education and Workforce, Division on Policy and Global Affairs. Board on Life Sciences, Division on Earth and Life Studies. Board on Science Education, Teacher Advisory Council, Division of Behavioral and Social Sciences and Education. Engineering Education Program Office, National Academy of Engineering. Washington, DC: The National Academies Press.

National Science Foundation (NSF). (1964). *Annual report*. Washington, DC: National Science Foundation.

National Science Foundation (NSF). (2012). NSB science and engineering indicators report. Retrieved from http://www.nsf.gov/statistics/seind12/append/c1/at01-08.pdf

National Science Foundation (NSF). (2006). NSB science and engineering indicators report. Retrieved from http://www.nsf.gov/statistics/seind06/

National Science Foundation (NSF). (2011). Women, minorities, and persons with disabilities in science and engineering: 2011 (Special Report NSF11-309). Arlington, VA: Division of Science Resources Statistics. Retrieved from http://www.nsf.gov/statistics/wmpd/

National Science Foundation Act of 1950, Public Law 81-507. Retrieved from http://www.nsf.gov/about/history/legislation.pdf

Nelson, D. J., & Rogers, D. C. (2007). *A national analysis of minorities in science and engineering faculties at research universities*. Norman, OK: University of Oklahoma.

Ngo, B. (2006). Learning from the margins: The education of Southeast and South Asian Americans in context. *Race, Ethnicity, and Education*, 9(1), 51–65.

Office of Scientific and Engineering Personnel. (1994). *Reshaping the graduate education of scientists and engineers*. Washington, DC: The National Academies Press.

Office of Scientific and Engineering Personnel, National Research Council. (1994). *Women scientists and engineers employed in industry: Why so few?* Washington, DC: National Academies Press. Retrieved from http://www.nap.edu/openbook.php?isbn=0309049911

Ohland, M. W., Brawner, C. E., Camacho, M. M., Layton, R. A., Long, R. A., Lord, S. M., & Wasburn, M. H. (2011). Race, gender, and measures of success in engineering education.

Journal of Engineering Education, 100(2), 225–252.

Ohland, M. W., Sheppard, S. D., Lichtenstein, G., Eris, O., Chachra, D., & Layton, R. (2008). Persistence, engagement, and migration in engineering programs. *Journal of Engineering Education*, 97(3), 259–278.

O'Meara, K. (2012). *Because I can: Exploring faculty civic agency.* Kettering Foundation Working Paper (2012-1). Dayton, OH: Kettering Foundation.

Ong, M., Wright, C., Espinosa, L. L., & Orfield, G. (2011, Summer). Inside the double bind: A synthesis of empirical research on undergraduate and graduate women of color in science, technology, engineering, and mathematics. *Harvard Educational Review*, 81(2), 172–209.

Packard, B. W., Gagnon, J., LaBelle, O., Jeffers, K., & Lynn, E. (2011). Women's experiences in the STEM community college transfer pathway. *Journal of Women and Minorities in Science and Engineering*, 17(2), 129–147.

Page, S. E. (2008). *The difference: How the power of diversity creates better groups, firms, schools and societies.* Princeton, NJ: Princeton University Press.

Pascarella, E. T., & Terenzini, P. T. (2005). *How college affects students.* San Francisco, CA: Jossey-Bass.

Pelz, D., & Andrews, F. (1966). *Scientists in organizations: Productive climates for research and development.* Ann Arbor, MI: Institute for Social Research, University of Michigan.

Peter, K., & Horn, L. (2005). *Gender differences in participation and completion of undergraduate education and how they have changed over time* (U.S. Department of Education, National Center for Education Statistics 2005–169). Washington, DC: U.S. Government Printing Office.

President's Council on Jobs and Competitiveness. (2011). Jobs Council solutions: Graduate 10,000 more engineers. Retrieved from http://files.jobs-council.com/jobscouncil/files/2011/09/Jobs-Council-10K-Engineers-Fact-Sheet.pdf

Preston, A. (2004). *Leaving science: Occupational exit from scientific careers.* New York, NY: Russell Sage Foundation.

Preston, A. (2006). Women leaving science. Retrieved from http://www4.gsb.columbia.edu/rt/null?&exclusive=filemgr.download&file_id=29107&rtcontentdisposition=filename% 3DAnnePreston_Columbia_conference_paper .pdf

Reybold, L. E., & Alamia, J. (2008). Academic transitions in education: A developmental perspective on women's faculty experiences. *Journal of Career Development*, 35(3), 107–128.

Reyes, M. E. (2011). Unique challenges for women of color in STEM transferring from community colleges to universities. *Harvard Educational Review*, 81(2), 241–262.

Riegle-Crumb, C., & King, B. (2010). Questioning a white male advantage in STEM: Examining disparities in college major. *Educational Researcher*, 39(9), 656–664.

Russell, M., & Atwater, M. M. (2005). Traveling the road to success: A discourse on persistence throughout the science pipeline with African American students at a predominantly white institution. *Journal of Research in Science Teaching*, 42(6), 691–715.

Sandler, B. R., Silverberg, L. A., & Hall, R. M. (1996). *The chilly classroom climate: A guide to improve the education of women.* Washington, DC: National Association for Women in Education.

Satcher, D. (2001). Our commitment to eliminate racial and ethnic health disparities. *Yale Journal of Health Policy Law and Ethics*, 1(1), 1–14.

Seymour, E., & Hewitt, N. (1997). *Talking about leaving: Why undergraduates leave the sciences.* Boulder, CO: Westview Press.

Sheppard, S. D., Macatangay, K., Colby, A., & Sullivan, W. M. (2008). *Educating engineers: Designing for the future of the field.* San Francisco, CA: Jossey-Bass.

Smith, K. A. (2011). Cooperative learning: Lessons and insights from thirty years of championing a research-based innovative practice. FIE-CL-1240-10 In *Proceedings of the 41st ASEE/IEEE Frontiers in Education Conference*, Rapid City, SD.

Smith, K. A., Johnson, D. W., & Johnson, R. T. (1981). Structuring learning goals to meet the goals of engineering education. *Engineering Education*, 72(3), 221–226.

Smith, K. A., Sheppard, S. D., Johnson, D. W., & Johnson, R. T. (2005). Pedagogies of engagement: Classroom-based practices (Cooperative learning and problem based learning). *Journal of Engineering Education*, 94(1), 87–102.

Springer, L., Stanne, M. E., & Donovan, S. S. (1997). *Effects of small-group learning on undergraduates in science, mathematics, engineering, and technology: A meta-analysis.* Madison, WI: National Institute for Science Education.

Springer, L., Stanne, M. E., & Donovan, S. S. (1999). Effects of small group learning on undergraduates in science, mathematics, engineering, and technology: A meta-analysis. *Review of Educational Research, 69*(1), 21–51.

Staniec, J. F. (2004). The effects of race, sex, and expected returns on the choice of college major. *Eastern Economic Journal, 30*(4), 549–562.

Terenzini, P. T., Cabrera, A. F., Colbeck, C. L., Parente, J., & Bjorklund, S. (2001). Collaborative learning vs. lecture/discussion: Students' reported learning gains. *Journal of Engineering Education, 90*(1), 123–130.

Tinto, V. (1975). Dropout from higher education: A theoretical synthesis of recent research. *Review of Educational Research, 45*(1), 89–125.

Tinto, V. (1993). *Leaving college: Rethinking the causes and cures of student attrition* (2nd ed.). Chicago, IL: University of Chicago Press.

Tinto, V. (2002). *Taking student retention seriously: Rethinking the first year of college.* Speech delivered at the annual meeting of the American Association of Collegiate Registrars and Admission Officers, Minneapolis, MN.

Townsend, B. K. (2007). Interpreting the influence of community college attendance upon baccalaureate attainment. *Community College Review, 35*(2), 128–136.

Trenor, J. M., Yu, S. L., Waight, C. L., Zerda, K. S., & Sha, T. L. (2008). The relations of ethnicity to female engineering students' educational experiences and college and career plans in an ethnically diverse learning environment. *Journal of Engineering Education, 97*(4), 449–465.

Tsapogas, J. (2004). *The role of community colleges in the education of recent science and engineering graduates* (InfoBrief NSF 04–315). Arlington, VA: National Science Foundation, Direc-

torate for Social, Behavioral, and Economic Sciences.

Tsui, L. (2010). Overcoming barriers: Engineering program environments that support women. *Journal of Women and Minorities in Science and Engineering, 16*(2), 137–160.

U.S. Census Bureau. (2008). *Age & sex in the United States, 2008.* Retrieved from http://www.census.gov/population/age/. www/socdemo/age/age_sex_2008.html

U.S. Congress, House Committee on Appropriations. (1976). Appropriations for *HUD and independent agencies, 1977,* 94th Congress, 2nd Session.

U.S. Congress, Office of Technology Assessment. (1988, June). *Educating scientists and engineers: Grade school to grad school* (OTA-SET-377). Washington, DC: U.S. Government Printing Office.

U.S. Congress, Office of Technology Assessment. (1989, March). *Higher education for science and engineering – A background paper* (OTA-BP-SET-52). Washington, DC: U.S. Government Printing Office.

Valenzuela, Y. (2006). Mi fuerza/My strength: The academic and personal experiences of Chicana/Latina transfer students in math and science. (Doctoral Dissertation). Retrieved from ProQuest Dissertation and Theses database. (Publication No. AAI3243278).

Varma, R., & Hahn, H. (2007). Gender differences in students' experiences in computing education in the United States. *International Journal of Engineering Education, 23*(2), 361–367.

Vaughn, R. L, & Seifer S. (2004). *Quick guide: Service-learning in engineering education.* National Service Learning Clearinghouse.

Wyer, M. (2003). Intending to stay: Images of scientists, attitudes toward women, and gender as influences on persistence among science and engineering majors. *Journal of Women and Minorities in Science and Engineering, 9*(1), 1–16.

CHAPTER 17

Social Justice and Inclusion

Women and Minorities in Engineering

Donna Riley, Amy E. Slaton, and Alice L. Pawley

Introduction

Evelynn Hammonds (1994) made a field-shaping move in her essay "Black (W)holes and the geometry of Black female sexuality (More Gender Trouble: Feminism Meets Queer Theory)." Her piece was part of a special issue on queer theory in the journal *differences*, the second that journal had produced. She describes her experience of picking up the journal's first issue on queer theory, as well as the *Gay and Lesbian Studies Reader*, looking for articles that reflected in some way her experience. She notes that even when writers of color were included, "the text displays the consistently exclusionary practices of lesbian and gay studies in general. In my reading, the canonical terms and categories of the field . . . are stripped of context in the works of those theorizing about these very categories, identities, and subject positions. Each of these terms is defined with white as the normative state of existence" (127–128). She goes on to note the lack of reflexivity among queer theorists, who did not examine intersectional categories of difference in their own work.

The theorists' engagement with inequity, although not necessarily insincere, nonetheless stopped short of questioning the structural conditions in which the theorists themselves gained intellectual credence and other forms of cultural privilege.

What can the field of engineering education learn from this critical moment in queer theory? We seek to emulate Hammonds's process in this chapter, which explores how the engineering education community tends to take up issues of diversity. In engineering, efforts to expand participation in the field have hewed to concerns that are well intentioned, but like the queer theorists' approach, somewhat lacking in reflexivity and self-limiting in their impacts. After years of struggle, so far engineering has come to include *women and [racial and ethnic] minorities* to a certain extent (albeit treated as two separate categories, largely ignoring the intersection). Queerness, class, nationality, disability, age, and other forms of difference are for the most part not seen as requiring address. The focus has been on expanding the roster of who is present, using only those categories and methods familiar from decades of educational policy and practice. That is, we customarily count women and people of color, and measure gains or losses in numeric representation of those groups, as programs to end discrimination in education, recruiting, hiring, and promotion have been put into practice.

This familiar approach to diversity cannot maximize the inclusiveness of engineering occupations. First, entire categories of practitioner identity and experience remain under-considered: the underrepresentation of persons identifying as queer, disabled, elderly, or poor in engineering has historically garnered little explicit attention, even from those most concerned with diversity. (Assistive technology developers and disability services for science, technology, engineering, and mathematics [STEM] students have made significant contributions, but institutional commitments to these interventions remain few.) In failing to consider the nature of discrimination faced by individuals of these backgrounds and identities, not only do we place those particular individuals at a disadvantage, but we also fail to probe with depth and complexity exactly how exclusionary social patterns perpetuate in our culture.

Second, and crucially, *counting* white and nonwhite, male and female persons present in engineering (the crudeness of those rubrics aside) has been a self-limiting reformist exercise. Reflexivity is too easily foreclosed in such quantitative excursions. We have seen that even when marginalized voices are included, there are problems with *how* this inclusion is structured: tokenism and compartmentalization often configure institutional diversity programs.

What is more, literature in engineering, engineering education, and engineering education research regularly relegates concerns of marginalized groups to specialized publications, or to a chapter or section where those who are concerned may read about it, but those with certain kinds of privilege need not encounter it. Many works on diversity pose little challenge to accepted thought structures in engineering (aspiring to *solve the problem* of underrepresentation or *repair*

the leaky pipeline without deeply interrogating those structures).

The lens of "social justice" emerging in engineering education research could shift conversations about numeric representation to far more incisive discussions on power. Such a shift demands not least a systematic consideration of the *content* accepted as constituting engineering in schools or workplaces. With power relations made visible in this way, incidences of discrimination that have customarily gone unanalyzed may gain attention.

We seek to critique the "normative state of existence" (to use Hammonds' words) in engineering education and we argue that social justice issues should actually pervade *every other* consideration of engineering education research; however this volume or the field as a whole might compartmentalize these concerns. To illustrate how social justice and inclusion (going well beyond sexism and racism to engage heterosexism, ableism, ageism, globalizing labor forces, and other forms of injustice) could, and we argue should, be considered throughout engineering education, we structure this chapter as the volume is structured: we consider social justice and diversity in theoretical and epistemological issues, classroom practice, institutional practice, and research methods and assessment. Developing a routine sensibility around difference, power, and privilege that informs our approach to the discipline as a whole is one way in which other fields have found an opening to move forward.

Part of this sensibility is enhanced reflexivity about critical practice itself. For example, this is a handbook chapter, the format of which is intended to summarize the state of the research in a way that allows readers to quickly grasp accepted knowledge. However, critical research necessitates our recognition that the chapter's authors have made choices about what to include or not, that the context of the chapter's writing is embedded in an historical and geographical space, and that the actual reader may or may not be who the authors have imagined. As a result, we anticipate that readers will find the tone of this chapter different from that of others in this volume. We engage this topic not through a direct report of a series of positions researchers (including ourselves) have taken, but in order systematically to critique those positions, to assert that there may be other ways of considering the topic at hand. Our aim is to encourage ongoing skepticism about approaches to social justice and inclusion in engineering education, to ensure that privileged expertise regarding these matters (including our own) is consistently one of the types of power under address.

We include ourselves as educated, employed individuals among those functioning from a position of power, and see our position as authors as a privileged condition. We do not presume our analysis to be free from the problematics of privilege – indeed, we acknowledge that our context within U.S. academic networks and culture functions effectively to sanction a limited awareness of parallel conversations about social justice and engineering education outside of the United States. However, with these considerations we hope more deeply to interrogate the ways in which discussions of diversity in engineering have fallen short of inclusivity despite decades of (spoken) commitments to such reform.

Theory and Epistemology

Within the Western academy, engineering education research draws primarily from two research traditions: basic research, which focuses on how we understand the world around us and how it functions; and applied research, where we try to use what we have learned to improve the human condition in varied (and often contradictory) ways. However, there is a third research tradition well developed in the humanities, social sciences, and educational research, small but growing in engineering education: critical research (Sullivan & Porter, 1997). This tradition examines the products, processes, and social relations generated through various research enterprises with

an eye to studying power, prompting those producing such work to reflect about the broader social consequences of their practice. Common to critical research across disciplines are questions about gender, race, and class in particular systems of interest. In this section we discuss critical theories in engineering and engineering education, considering what perspectives on power can offer our theoretical and epistemological foundations in engineering education research. We then illustrate this thinking applied to innovation discourses that theoretically inform engineering education research.

Critical Research Expanding Theory, Challenging Epistemic Assumptions

We are at a time of resurgence of interest in the relationship between engineering education and philosophy. One arm of this resurgence is focused on applying classical Western philosophy to the study of engineering and engineering education, including engineering practice, pedagogy, and curricular content (Bucciarelli, 2003; Grimson, 2007; Heywood, 2008a, 2008b; Heywood, Smith, & McGrann, 2007). This group brings to the study of engineering education historical insights informed by Western philosophies of knowing and some classic studies of "what engineers know and how they know it" (Vincenti, 1993).

The other arm of this resurgence has developed through a connection with critical theory, including feminist, antiracist, humanist, and social justice movements, and seems to have roughly aligned itself under the conceptual umbrella of *engineering, social justice, and peace*. This arm reflects on the ad hoc and performed philosophies of engineering practitioners, engineering educators, and engineering students, as well as the subject of, content of, and participants in engineering education and practice. Applying critiques developed through science and technology studies, the history of science and technology, women's studies, critical race theory, labor history, and curriculum and instruction, these scholars prompt us to consider engineering and technology with reflexivity. They provoke scrutiny on power relations embodied through our choices and practices as engineers, on the notions of privilege, structure, and reproduction of each, to ask ourselves collectively and individually both who "we" define "ourselves" as, and whom we exclude from such performable definitions (Baillie & Catalano, 2009; Catalano, 2006; Riley, 2008a).

At the center of social justice theory applied in engineering education is a consideration of how the profession serves structures of power, most notably militarism and corporate power in national and multinational settings. Under neoliberalism, engineers serve governments that are increasingly withdrawing from social services, labor protection, and environmental regulation in favor of "free market" solutions that increase income inequality worldwide (Goldberg & Traiman, 2001; Steger, 2009). Engineering education exists by and large within these received frameworks and, despite its own vulnerability to the deskilling, destabilizing regimes of globalizing neoliberal labor arrangements, rarely includes an awareness of these structures of power, let alone approaches to challenging them (Head, 2003).

Although both the classical philosophical and the critical engineering education research directions make use of philosophy as a process of inquiry about humanity's condition, each deploys it differently. The former tends to situate engineering ways of knowing as objective, disaggregating complex sociotechnical systems into cleanly differentiated types of work or practice, or social and technical inquiries. The latter explores a critical reflection on the former, including asking what ways of knowing are thereby excluded from "traditional engineering," and how the language we use to construct the subject of engineering inquiry also limits the tools available for such inquiry.

Some of this critical research approaches issues of inclusion implicitly through questions about who decides what constitutes "engineering" and how this value-laden content is reproduced through the training of new engineers, or about who feels the consequences of such decisions. Historians Amy Slaton (2001) and Amy Bix (2002) and social scientists Lisa Frehill (2004) and Karen Tonso (2007), as well as engineering education researcher Alice L. Pawley (2007, 2009, 2012) have argued that *who* has constructed engineering as a discipline has consequences for the content, which in turn determines who becomes an engineer. Specifically, the content and practice of engineering have been organized around historically masculinized and racialized discursive spaces in part because white men have been the people making such discipline-defining decisions, and those decisions in turn meet the interests and expectations of white men searching for their future careers.

Other critical research focuses explicitly on the topic of *women and (ethnic and racial) minorities*, including Juan Lucena's (2000) historical discussion of the making of the grouping label "women and minorities" as a new category of policy interest (Bowker & Star, 1999). In this work, Lucena notes that the focus on white women and "ethnic minorities" is not a natural consequence of intellectual curiosity, nor of self-righteous outrage arising from injustice. Instead, he notes how the category related to the rhetoric around engineering as well-suited to meet the nation's defense and economic needs during the Cold War, which then situated arguments for increased gender and race diversity on a platform of national necessity and global competitiveness.

Other very recent work considers social categories beyond gender and race. Erin Cech and Tom Waidzunas (2011) expand demographic categories to include lesbian, gay, and bisexual engineers, while George Catalano and Maggie Howell (2006) include nonhuman animals as engineering clients. Student economic background is being taken up in studies of students who enter technical fields through sub-baccalaureate programs at vocational schools or community colleges, including Lisa McLoughlin's challenge to the stigmatizing rubric, "non-traditional" students (McLoughlin, 2010; Slaton, 2010a). Martin (2011) uses social capital as a theoretical framework for understanding engineering career choice, while Orr, Ohland, and colleagues (Ohland, Orr, Lundy-Wagner, Veenstra, & Long, 2012; Orr, Ramirez, & Ohland, 2011) examine the effect of socioeconomic status on student educational outcomes in large national data sets. Anthropologist Ajantha Subramanian has begun to address the role of caste in patterns of engineering educational attainment in India (Subramanian, A. (2011). Gifted: Knowledge and value in Indian technical education. Unpublished manuscript).

On a theoretical and epistemic level, this critical research inherently questions the boundaries of engineering practice (where knowledge and expertise constitute those of authoritative engineers, or "merely" those of students, technicians, or the lay population), thus revealing how hierarchies of prestige and credentialing map onto social identities.

Uncovering Theoretical and Epistemic Assumptions of Innovation Discourses

Critical research contrasts sharply against current theoretical trends around "innovation." References to "innovation" are pervasive in engineering education research, including National Science Foundation funding programs such as IEECI (Innovations in Engineering Education, Curriculum, and Infrastructure); university centers such as Princeton's Keller Center for Innovation in Engineering Education, OSU's Engineering Education Innovation Center; the Stevens Center for Innovation in Engineering and Science Education; and the International Network of Engineering Education and Research (iNEER) conference: Innovations 2010.

"Innovation" is ideologically linked not only to foundational capitalist theory (competition creates innovation), but also to the financialization trend of the past decade that brought on the recent global financial crisis (Dallyn, 2011).

More than a buzzword, "innovation" has come to inform how we theorize the process of engineering education research; in their ASEE white paper, Jamieson and Lohman (2009) model the research process as an "innovation cycle." The title of this report, "Creating a culture for scholarly and systematic innovation in engineering education: Ensuring U.S. engineering has the right people with the right talent for a global society" may be a vague reference to diversity, but the authors never take a position on who the "right people" are. The report uses more corporate-centered language in advocating that "Engineering education must move from its customer-supplier model of delivering engineering instruction to a collaborative-distributed model of shared investment in the formation of engineers" (p. 14). Clearly the notion of students as customers is problematic (Svensson & Wood, 2007), but the notion of students as investors (or shareholders) seems equally so in foreclosing reflexivity about the aims and consequences of engineering as currently practiced. By contrast, Svensson and Wood offer a model of community citizenship to prescribe better a collaborative relationship between students and universities.

The epistemological leanings of "Creating a Culture" are quite positivist, going so far as to rename the discipline of education "learning science." In this regard the report typifies the materialist stance of much engineering education research: large quantitative data sets or electroencephalogram outputs are more convincing than what we learn through relationships with individuals. This sort of positivism, however, may recapitulate the stress on technical novelty, intellectual closure, and productivity that make innovation a central trope of engineering professions; that is, the premise that human problems can be solved through the discovery and application of new technical knowledge. Many preferred methodologies in engineering education research reflect these same biases toward reductionism, solving problems, innovation, and positivist ways of knowing, to the detriment of nuanced, inclusive, and reflexive inquiry that might better support social justice.

Classroom Practices

Pedagogy

Trends in engineering education have recently included active and cooperative learning, problem-based learning, and service learning. Advocates of each approach claim the potential to improve participation of women and minorities in engineering, among other learning benefits (Coyle, Jamieson, & Oakes, 2005; Du & Kolmos, 2009; Felder, Felder, Mauney, Hamrin, & Dietz, 1995).

However, as feminist science scholar Maralee Mayberry (1998) has argued, many of these active learning paradigms simply train students in reproducing conventional practices of scientific inquiry, and ironically create scientists without science's purported culture of questioning. Alternative pedagogies that question the fundamental assumptions about the practice of science and engineering have been taken up by some members of the engineering education research community in a series of special sessions and workshops at the Frontiers in Education Conference (Eschenbach, Cashman, Waller, & Lord, 2005; Pawley, Riley, Harding, Lord, & Finelli, 2009; Riley, Catalano, Pawley, & Tucker, 2007; Tucker, Pawley, Riley, & Catalano, 2008; Waller, 2005; Waller, Riley, Cashman, Eschenbach, & Lord, 2006). These scholars draw on critical, antiracist, feminist, and decolonizing pedagogical traditions (Darder, Baltodano, & Torres, 2008; Freire, 1970; hooks, 1994). These pedagogies' explicit attention to race, class, gender, and transnational power relations hold the greatest promise for addressing inclusion and social justice in engineering education. However, engineering education research places greater emphasis on active learning strategies that may be collaborative or learner-centered, but do not

consciously take up power considerations (Bransford, Brown, & Cocking, 2000).

In *problem-based learning*, approaches range from teacher- to learner-controlled (Chapter 8 by Kolmos & deGraaf in this volume). The focus is on solving a problem in the context of completing a particular project. Fenwick and Parsons (1997) critique the use of problem-based learning in professional preparation, arguing that:

> The competent professional is cast as the heroic problem-solver, bringing salvation to the passive others in the "problematic" situation, incarnating human desire for mastery and control based on normative ideals for the world. Problem-based professional practice is seen as supporting the professional role as the rightful epistemic authority, thus perpetuating a class of professional elite who dominate social order and knowledge. (p. 8)

Although this scathing analysis is perhaps too sweeping in that it does not leave room for conceptions of problem-based learning that practice an awareness of power, for engineers in particular it gets to the heart of how we define professional competence and delimit the contributions of those deemed inexpert.

Service learning – a term preferred by engineers over alternative framings such as community-based learning or participatory action research – has a particular set of power relations embedded in its language. Students serve a community and learn from the experience. But some have critically examined the model's assumptions and motivations, asking: Who is being served? Who is learning what from whom? (Strand, Marullo, Cutforth, Stoeker, & Donohue, 2003; Wade, 2000; Ward & Wolf-Wendell, 2000). These critics point out that many aspects of the service learning structure are vulnerable to replicating unjust power relations, increasing dependency of communities, and diverting precious resources from community groups toward often relatively privileged students. By placing students in heroic roles in communities both close to home and abroad, service learning teaches a particular set of power relations around engineering expertise (Lucena, Leydens, & Schneider, 2010; Nieusma & Riley 2010; Riley & Bloomgarden, 2006).

Some scholars have identified points of tension in service learning for students who do not fit a presumed norm of race and class privilege. Henry (2005) problematized the binary of "privileged server" and "under-privileged served" in her work on first-generation college students' involvement in service learning. Swaminathan (2005) found African American and Latino students encountered negative perceptions of service learning where it was conflated with court-ordered community service; by contrast, white students tended to receive accolades for their involvement.

As bell hooks argues, "No education is politically neutral" (hooks, 1994, p. 37); every pedagogy imparts a particular set of power relations in the classroom. Although no pedagogy unto itself can eliminate or equalize power relations, critical pedagogies consciously confront power relations. Although these pedagogies have many techniques and strategies in common with active, cooperative, problem-based, and service learning, their concern with power should lead to reflexive action for change in the classroom and in society. Without attention to power relations and social justice outcomes, active learning approaches are entirely compatible with the current state of affairs in engineering education in which race, class, sexuality, disability, and other axes of diversity are rarely mentioned. Although engineering educators may think they are simply not teaching about diversity while using mainstream active learning approaches, we believe this silence is itself a particular kind of diversity curriculum that engineers absorb and replicate.

Content

Although there is an extensive discussion in engineering education research circles regarding the *practice and performance* of teaching through discussions of *pedagogy*, there is less of a discussion about the *content* of engineering education itself. Nonetheless, the engineering curriculum embeds messages about the focus, scope, and values of engineering practice; there is little systematic consideration of how these content choices may be discouraging many minority groups (including women, people of color, people with disabilities, LGBT people, and more) from entering or persisting through the engineering curriculum. In fact, the practice of critical pedagogies discussed in the previous section demands a reconsideration of the canon in engineering along precisely these lines. After considering aspects of social justice and inclusion in the overt curriculum, we discuss the notion of the hidden curriculum in engineering education and its implications.

OVERT CURRICULUM

Diversity is rarely mentioned explicitly in engineering classrooms, although sometimes it comes up in design decisions and teamwork (e.g., Wulf, 1998). The conventional wisdom is that diverse teams represent a broader range of perspectives and therefore produce a greater variety of potential solutions, leading to better outcomes that benefit the bottom line in a globalized economy. Neoliberal ideologies of economic expansion are evident here; the whole world represents clientele for Western corporations, and the celebration of multiple identities in the corporate workforce supports the ability to move with ease through foreign markets (Lippit, 2004; National Science Foundation, 2007). "Diversity," as Avery Gordon (1995) has so clearly shown, becomes a means by which personal experiences of one's gender, ethnicity, age, disability, queerness, or other and melded identities may be claimed by a student or employee without threatening the historic privileges of dominant groups.

What would happen if we could speak about diversity in ways that bring power and privilege to the fore? Although this objective is still new to engineering classrooms, experience suggests that difficult and important conversations can ensue (Bothwell & McGuire, 2006; Pawley, Pfund, Lauffer, & Handelsman, 2006; Riley & Sciarra,

2006). Replacing the curriculum of silence (or of ahistorical, depoliticized celebrations of diversity) with a curriculum of dismantling privilege can create a conversation around understanding and addressing exclusion and social injustice in engineering.

Catalano, Byrne, Nieusma, and Riley (2008) have developed new course modules for engineering and social justice instruction. Jens Kabo's (2010) study of students' development of social justice thinking in three courses suggests that explicit instruction is necessary for students to work effectively with this threshold concept. Alternative approaches to engineering ethics education are discussed elsewhere in this volume (Chapter 33 by Barry & Herkert), as are approaches toward internationalizing the curriculum (Chapter 32 by Johri & Jesiek). Riley (2008b) and Riley and Claris (2006, 2010) have experimented with teaching critiques of engineering from science and technology studies alongside traditional content. Catalano (2006) has revised the engineering design process toward social justice, while Pantazidou and Nair (1999) have incorporated a feminist ethic of care in engineering design.

The design curriculum is a particularly productive site for reimagining engineering content. Universal design and design for accessibility, as well as assistive technology, offer a direct connection to inclusion and social justice, and are beginning to be incorporated in some engineers' educations (Erlandson, Enderle, & Winters, 2007; National Industries for Severely Handicapped [NISH], 2011). Some assistive technologies are themselves implemented in educational settings, increasing access to engineering for people with disabilities (Burgstahler, 1994). There is also increasing interest in human-centered design within engineering education (Oehlberg & Agogino, 2011; Zoltowski, Oakes, & Cardella, 2012).

Such moves are promising in two ways. First, if the notion of "disability" is created by society's failure to accommodate people's impairments (Davis, 1996), then engineering has a role to play in facilitating access to all those settings and activities configured by engineering. However, it is significant that such inclusive efforts do not necessarily alter foundational structures of privilege and access. For example, the creation of assistive technologies for the operation of private automobiles may serve a smaller number of end users than would the redesign of public transport systems. As is always the case, inclusion is not tantamount to equity.

Second, we should consider the relationship between designer and user as itself one that involves power. Some types of human-centered design, particularly participatory design, have the potential to shift the balance of power toward users. However, it is absolutely essential that those who implement human-centered approaches to design more fully understand the differences between various approaches and the political commitments of each in relation to power, inclusion, and social justice. Nieusma (2004) provides a helpful explication of different alternative approaches to design, with their advantages and drawbacks, for those concerned about design for marginalized groups and design to counter a wide range of social problems. He warns in particular that participatory design frameworks, developed to include voices of those traditionally excluded from design process, are unfortunately being collapsed into the term "user-centered design," in which users may be consulted about their views, but are not necessarily brought in to the process in ways that counter existing inequalities between designers and users. This is analogous to the distinction between learner-centered and critical pedagogies; attention to power makes all the difference.

HIDDEN CURRICULUM

As we reimagine content, however, we must not forget that the classroom of the twenty-first century follows historical patterns in supporting larger social organizations, including the production of citizens for whom difference is nonproblematic: not meaningless, but not requiring address. That is, education plays a role in naturalizing for future engineers the differentials in experience accorded by the culture on the basis of one's gender, race, sexual identity, age, and other factors (Apple, 1996, 2000; Bourdieu, 1990). As an example, "team building" involves consensus-forming around approved curricular topics, and not around, say, political resistance or minority consciousness, helping to inculcate "correct worker attitude" (Agostinone-Wilson, 2006). The absence of explicit discussions of identity, power, and understandings of difference is unquestionably constitutive of rigorous engineering (Slaton, 2010b).

Therefore a key question situated in the critical research domain for engineering education researchers is, What is the "hidden curriculum" (Jackson, 1968; Snyder, 1971) of engineering? What underlying values are taught to students in the form of ostensibly value-free content? The concept of the hidden curriculum is tightly bound to educational theories of social reproduction, in which social hierarchies are reproduced through educational systems. In some accounts, schools replicate socioeconomic hierarchies in order to produce disciplined workers for a socially stratified labor force that serves industrial capitalism (Bowles & Gintis, 1976), while in others, public education systems are constructed around cultural class-based norms of behavior, value, and content (Bourdieu, 1986). As a classic example, Jean Anyon (1980, 1981) has described how the math curriculum (as well as other curricula) in the late 1970s varied across U.S. public elementary schools that served different classes of students, including a working class, middle class, an affluent professional class, and an executive elite managerial class. Although students were held to the same public education standards of the time, the mechanisms of the teaching of this content also taught students about their presumed and inevitable lot in life – preparing menial workers, secretaries and civil servants, artists and academics, and business leaders. These studies shaped the field of education, making political economy an essential consideration in educational policy and analysis (Anyon, 2011). The hidden curriculum

remains a current topic for education research in the twenty-first century, emerging in fields including medicine (Hafferty, & Castellani, 2009), business (Blasco, 2012, and physics education (Redish, 2010).

Critical researchers in engineering education are making a parallel argument on the content and practice of teaching engineering students. Pawley (2009) has argued how the reinforcement of engineering identities around "solving problems," "making things," and "applying math and science" systematically excises the historical problems and experiences of women, people of color, poor U.S. communities, and communities in the global South from the attention and concern of engineers and engineering. Gary Downey (2005) has argued that, by positioning themselves as "problem solvers," engineers sacrifice the right to structure the focus and scope of their work. Instead, he recommends that engineers become "problem definers," an identity characteristic with both more professional prestige and responsibility. This would necessarily challenge any number of institutional strategies currently in use in university engineering departments, vital to garnering resources from industrial, governmental, and military audiences who believe they have already defined their "problems" (and as Scheman [1993] explicates, problem framing activities are heavily laden with power) and need ready solutions (in the form of engineered technologies or trained personnel).

Institutional Practices

What institutional practices support engineering education research? In the section of this volume dedicated to this topic, authors take up important questions related to the dissemination and implementation of research findings among engineering educators as a vehicle for institutional change (Chapter 19 by Litzinger & Lattuca; Chapter 20 by Baillie & Male; Chapter 21 by Felder, Brent, & Prince; Chapter 22 by Godfrey; and Chapter 23 by Borrego & Streveler). If we ask specifically what current institutional practices support inclusion, we might immediately look to Minority Engineering Programs (MEPs), Women in Engineering Programs (WEPs), and K–12 programs designed to attract diverse students to engineering. However, social justice approaches might further consider culture (as Godfrey does in her chapter) as well as the practices of larger social institutions such as higher education or K–12 education nationally, or economic institutions such as multinational corporations operating in a global economy. Broad social structures of opportunity both sustain, and have been sustained by, researchers' relatively shallow inquiries into power relations at all levels of institutional function.

Minority Engineering Programs and Women in Engineering Programs

In the decades since the conclusion of the nation's strongest involvement with civil rights reform, roughly from 1980 onward, institutions that educate engineers in America or oversee that education have shown sharply defined commitments to altering patterns of gender and ethnic underrepresentation. Many MEPs and WEPs have been instituted in U.S. universities; some of these initiatives are now thirty years old and the majority are staffed by energetic and remarkably committed personnel. Yet year after year, major reports on the matter issued by academic, governmental, and nonprofit agencies express disappointment in the limited degree to which historic patterns of underrepresentation have been reversed since the major legal changes of the 1960s and 1970s (Slaton, 2010b). We briefly outline here some reasons for this pattern, but all may be introduced with the general notion that expectations for gender and racial equity in American engineering have remained diluted since shortly after their most ardent expression in the civil rights era.

Engineering is not alone in this discriminatory habit: the prevailing social and economic climate post-Reagan has discouraged the pursuit of authentic gender and ethnic

diversity by universities, funding organizations, and accrediting bodies in virtually every highly skilled occupation. Neoliberal sensibilities not only subordinate gender, ethnic, and other forms of diversity within knowledge-making institutions to agendas of global capitalism, but also cast those earlier eras of heightened minority consciousness as innocuous episodes of "self-empowerment" or "freedom," as Cotten Seiler explains, disguising the subsequent "retrenchment of various forms of privilege" (Seiler, 2009). More immediately damaging to the prospects of marginalized groups, conservative economic projections in recent years have identified a "skills gap" in the United States workforce at middle levels of industrial labor where higher education is not required (although of course, all such divisions of skill and labor are themselves arbitrary). This logic has produced the associated point that "too many" people are pursuing higher education. High dropout rates for college enrollees of low socioeconomic standing are invoked to bolster this point (Harvard Graduate School of Education, 2011). These seemingly pragmatic diagnoses of American productivity problems justify continuing the exclusion of groups historically marginalized from highly skilled occupations: those who cannot afford secondary education of the quality that prepares one for success in college, a group in which women and minority Americans figure disproportionately. Anti-affirmative action activism, present in the country since the early 1990s, finds corroboration in these supposedly objective economic prescriptions. Legal challenges to gender, and more frequently, race-based educational programs occur often enough to discourage some schools from pursuing such targeted interventions; or at the very least, to make the absence of such interventions seem acceptable to many observers (Slaton, 2010b).

In this landscape, economic, social, and political ideologies are co-terminus. In some sense, it tells us little to distinguish a tax cut that undermines funding for public minority programming from arguments that claim we (in the United States in 2012) have an African American president so no longer need racial protections. Engineering educators and policymakers who authentically wish to expand opportunities for the underrepresented face obstacles ranging from diminished public funding for K–12 STEM programming to retrenchments in the post-secondary minority education programs that emerged in many colleges in the 1970s and early 1980s. But without taking away from the tremendous impediments such systemic conditions present, we should also note that STEM education and accreditation have in some ways proposed and designed only limited reforms over the years. Well-intentioned as they may be, some inclusive programs stop short of radically reimagining opportunity structures in the United States.

For one thing, standards of intellectual merit in STEM fields, routinely invoked to justify a female or minority individual's inclusion in a recruitment or support program (i.e., the project of "finding" the nation's "missing" talent pools), remain largely unanalyzed, politically. Even where the inequitable and misleading nature of standardized math and science testing are understood, ideas of talent and student potential are left unquestioned (McLoughlin, 2009). Important efforts to redress racial discrimination, for example, may seek out "gifted youngsters" for special services and thus empower the students' engagement with engineering, but what counts as "gifted" (and indeed, what counts as an age or life stage at which one may reasonably be offered services) remains narrow. The power relations that inhere in American higher education in general are hidden when standards are not examined.

But because technical fields are commonly presumed to function on the basis of knowledge that is not only esoteric but also entirely objective and empirical, the very notion that power may reside in engineering's intellectual features is hard to address. One notable exception is MIT's recent self-study on faculty diversity (Massachusetts Institute of Technology, 2010), for which

1. Pre-college summer bridge programs in math and science
2. Mentoring programs with mentors of color
3. Undergraduate research experiences
4. Tutoring, where students can "catch up" or "keep up"
5. Career counseling and awareness
6. Learning centers
7. Professional or academic skills workshops and seminars
8. Academic advising to help students access cultural capital
9. Financial support
10. Reforming curriculum and teaching practices

Figure 17.1. 10 types of interventions designed to increase the number of underrepresented students of color in STEM disciplines (Tsui, 2007).

researchers brought into question the whole idea that science and education function as meritocratic institutions.

Conventions of Inclusive Programming

The activities that constitute MEPs and WEPs often proceed from a set of premises about underrepresentation that denies the political nature of gender and race relations in the United States. Figure 17.1 summarizes types of approaches common in MEPs. Remedial coursework, tutoring, financial aid, and social support all have established roles in the correction of educational and occupational discrimination. However, since their inception in American education during the early 1960s, such programs have emphasized explanations for student success and failure that focus on individual capacity to achieve, or self-efficacy, to the detriment of contextual explanations. Even as programming for the first nine interventions in Figure 17.1 may occur at the institutional level, each intervention seeks change at the level of individual students. The tenth *looks* different in that it claims institutional scope, but in reality it addresses teaching attitudes, practice and design, not the structure of the university itself. Similarly, the AAUW (formerly The American Association for University Women)'s recent summary of critical research on gender in STEM disciplines describes six common types of research, most of which center on student "mindsets" or "self-efficacy" (Hill, Corbett, & St. Rose, 2010). The function of the individual in ameliorating discrimination is paramount, putting social structures in a secondary causal role.

Even where prevailing social conditions (including sexism and racism) are acknowledged, intervention on behalf of the underrepresented often includes the cultivation of a set of attitudes and behaviors by which an individual supposedly may adapt his or her conduct to that of the "mainstream" setting he or she seeks to enter. Conversations about bringing economically disadvantaged groups into technical disciplines and jobs commonly invoke the need for such behavioral adjustments, including corrections to habits of dress, speech, and promptness, or to that all-encompassing and apparently timeless social arbiter, manners. That some features of individual conduct may derive from cultural or community affinities is seen to be disturbing, or to interfere with effectual institutional participation (in school or workplace). Mainstream culture is the unmarked category, and old "deficit models" of minority economic marginality, gain new life in these assimilationist, Foucauldian operations.

On another level, programs directed toward increasing the representation of women and minorities in postsecondary engineering curricula appear by definition to correct exclusion, but are often given only a marginal status within a university or federal agency. Services directed at all or "typical" students define normalcy; the white, male student himself resides in an unmarked category. Centers or programs expressly aimed at "othered" populations are cast as add-ons or accommodations in the eyes of those who are not directly involved with them. With such marginal status, their advocates are unable to garner deep institutional

support for these programs. Certainly the problem of rewards for those within technical disciplines who choose to work with issues of inclusion and diversity is significant: faculty tenure, promotion, and research funding are customarily put at risk by such commitments, which are seen to run counter to the prosecution of serious, rigorous technical teaching and labor (Massachusetts Institute of Technology, 2010; Slaton, 2010b).

Research Methods and Assessment

Having discussed social justice in the context of educational practice and content and in the structural components of educational institutions, we now turn our critical gaze to research and assessment practices. First we take up research methods, discussing existing approaches to studying gender and race and proposing alternative mechanisms that allow for the study of power relations and that align with the goals of action research to aim for social justice. Second, we consider assessment methods that are appropriate to the alternative theories, epistemologies, classroom and institutional practices, and research methods we propose.

Research Methods and Methodologies

How do efforts for inclusion in engineering achieve or fall short of criticality through their choice of investigative process? Empiricism does not guarantee an address (or redress) of power and privilege, and in fact, some highly rigorous and precise investigative methodologies foreclose the revelation of power relations by virtue of their apparent objectivity. The impact of research resides in choices regarding levels and units of analysis, choices between quantitative and qualitative methods, choices of representational conventions, and above all, researchers' ideas about what actors, characteristics, experiences, or policies require address. Sociologists of knowledge understand how form and content, question and answer, shape one another in academic and scientific research, and that dynamic may frame our own critical inquiry into the social features of engineering education research.

Research that focuses on understanding the underrepresentation of women and people of color in engineering disciplines in the United States often succumbs to several methodological problems: (1) aggregating across different social categories to maintain statistical power, (2) naturalizing the categories used to represent gender and race (in addition to the exclusion of other categories), and (3) overlooking studies of institutional structure and larger power relations.

AGGREGATING ACROSS SOCIAL CATEGORIES MASKS
IMPORTANT RESEARCH QUESTIONS

As hinted at throughout this chapter so far, white women and people of color remain significantly underrepresented in engineering undergraduate education in the United States compared both to college students of all disciplines and to the general population; the same is true for faculty members in engineering. To focus on undergraduate students as an example, women earned 17.8% of all bachelor's degrees awarded in engineering in 2009, a decrease from the previous year for the ninth year in a row (Gibbons, 2010). African American students earned 4.6% and Hispanic students 6.6% of engineering bachelor's degrees in 2009 (Gibbons, 2010). The percentages of each nonwhite racial group in engineering are significantly smaller than those representing each group enrolled in university. However, these dismal numbers mask the experience of women of color who become invisible through the aggregation across race of the gender numbers, and the aggregation across gender of the race numbers. Importantly, although the numbers of women in each underrepresented racial group are startlingly small when compared to the total numbers of all engineering students, the proportion of women engineering students in several nonwhite racial groups is actually higher than the proportion within engineering of either white women or women aggregated across races. In 2010, 25%

of African American engineering students, 17% of Hispanic engineering students, and almost 23% of Native American engineering students were women, compared to 16% of white engineering students and 18% of all engineering students (Gibbons, 2010).

However, researchers regularly aggregate all women together, regardless of race, and all members of an ethnic group together regardless of gender. In part this is because the small numbers of white women and people of color in engineering programs mean either that statistical analyses produce nonsignificant results, or that researchers must aggregate across gender or race in order to make statistically significant conclusions. Specifically, the small numbers of women of color often limit the power of calculating any interaction terms between main effects. The result that only main effects can be reported to any statistical significance can erroneously be interpreted to mean that gender and race are independent of each other – that women of color have the problems of white women simply "added" to the problems of men of color – whereas the consensus of the extensive literature from womanists and feminists of color and work on intersectionality demonstrate this is not the case (Anzaldúa, 1987; Crenshaw, 1989; Hill Collins, 1998). Existing research demonstrates that women of color's experiences of education are different from either of the other two groups (see Berry & Mizelle, 2006; Rockquemore & Laszloffy, 2008 as examples). Little of this focus on intersectionality has migrated into engineering education research, with two notable exceptions: (1) Cynthia Foor and colleagues' case study of "Inez" articulates the complexity of a multiracial, first-generation-college woman's experience in an undergraduate engineering program (Foor, Walden, & Trytten, 2007); and (2) Donna Riley's research on liberative pedagogies in engineering education makes use of intersectional and critical theories of gender and race (e.g., Riley, 2003). Engineering education researchers must develop studies that explore ways gender and race interact, as well as considering additional critical characteristics of identity, such as class, sexuality, age, ability status, and others.

To understand people's educational experiences at the intersection of many social categories, we must use methods specifically designed to examine the experiences of small numbers of people; after all, *small numbers of white women and people of color in engineering is what we have*. Qualitative methods that have at their heart the deep exploration of small numbers of stories are well suited to this sort of examination. Foor and colleagues' (2007) "ethnography of one" discussed previously eschewed criticism that generalizable results cannot be drawn from the study of a single person. Slaton's (2010b) historical analysis of engineering education through three cases of engineering education reform showed how racial underrepresentation persists in educational institutions despite calls for institutional reform. Phenomenographic variation theory and threshold concept theory were combined to understand how students come to view engineering through a social justice lens (Kabo, 2010). Feminist analyses of the publication patterns of engineering education journals – one content analysis exploring *Journal of Engineering Education* (JEE), *International Journal of Engineering Education* (IJEE), and *European Journal of Engineering Education* (EJEE) (Beddoes & Borrego, 2011) and one using domain analysis exploring JEE exclusively (Nelson & Pawley, 2010) – demonstrated the limited research frameworks being applied to the study of gender in engineering education.

CATEGORIES OF RACE AND GENDER ARE NOT AS SIMPLE AS THEY SEEM

Engineering education researchers need to be aware of varying definitions of the very concepts of gender and race. "Gender" is the complex categorization system that defines social characteristics as "male" or "female," by which we organize much of the world (Connell, 1987). Historically these categories have been linked to biological differences between males and females, but social scientists have shown that they are socially constructed, vary across time and place,

and are not as static, dichotomous, or independent as previously assumed (Kimmel & Aronson, 2008). Once thought to be biological, race has similarly emerged from a history of complicated social relations organized in and around immigration, nation, labor, class, and gender, with racial groups changing significantly over time (Jacobson, 1999; Omi & Winant, 1994). Engineering education researchers should become more familiar with the development of, for example, the commonly used categories of Black, Hispanic, Native American, or White, and how racial categories are different from ethnic ones (Omi & Winant, 1994), and make explicit the processes of identification and self-identification from which their studies necessarily derive.

MICRO-LEVEL ANALYSIS MUST BE DONE IN THE COMPANY OF MESO- AND MACRO-LEVEL STUDIES

Although the number of research projects that use qualitative methods to investigate the experiences of underrepresented students and faculty in engineering has increased dramatically over the last decade, their main focus often remains on remediation efforts that promote change at the *micro* level of individual students and faculty. However, understanding gender and race involves understanding the construction, maintenance, and resistance of *macro* social structures (such as "U.S. education" or "community" or "engineering"), which researchers and scholars still do not understand very well. How individuals interact with these macro structures is via *meso* institutions: specific organizations themselves (DiMaggio, 1991) such as school districts, university systems, or ABET.

In the 1980s and 1990s, feminist sociologists and others exploring gender and racial access issues started theorizing that women's low participation in positions of institutional, economic, or political power was neither simply an effect of women's missocialization, nor due to men's individual sexism. Instead, these scholars argued that the institutions in which women sought inclusion are themselves gendered, raced, and classed in ways that promote the success of people participating in an exclusive set of social relations: middle-to-upper class white men (Acker, 1992, 1998).

There is exceptionally little research in engineering education investigating gender or race in these ways, and even if there were a robust research effort here, we are still neglecting the investigation of age, class, sexuality, global power relations, and so on, both intersectionally with gender and race and on their own terms.

In summary, to move toward a more socially just research enterprise, engineering education research as a community must begin to embrace the kind of research that (1) explores how gender and race interact with each other and with other social categories; (2) treats gender and race as complex constructs rather than biological inevitabilities; (3) allows us to "learn from small numbers" (Pawley, 2013); and (4) seeks to understand how the very *structure of our institutions* maintains relations of social inequality.

Assessment

Because engineering education research toward social justice and inclusion differs from mainstream research at the levels of theory and epistemology, pedagogy and classroom practice, institutional location and practices, and methodology and method, it is not surprising that assessment methods will need to follow suit. Riley and Claris (2008) have developed an approach based on liberative pedagogies in which students participate actively in devising and performing assessment, and assessment is integrated with learning and reflection activities. These methods are necessarily quite different from more common approaches based on large numbers, empirical epistemologies, and statistically based methods of analysis (see, e.g., Chapter 29 by Pellegrino, DiBello, & Brophy in this volume). Assessment for social justice and inclusion might draw on process-oriented and qualitative approaches (see Chapter 27 by Case & Light in this volume), incorporate emotion

through reflection, and involve students as subjective authorities.

In the bigger picture we might ask about the purposes and goals of assessment, including overarching assessment structures, particularly ABET. Although the Criteria 2000 were generally applauded by engineering education reformers for moving away from bean counting, assessment systems still leave much to be desired when it comes to diversity and social justice (Riley, 2012; Slaton, 2012).

In terms of content, reforms in the 2000 criteria created an opening for a shift in emphasis away from purely technical material. Although social context, ethics, working in teams, and communication might each imply a need to understand diversity and work across lines of difference, such a requirement is never explicit. Substantively, diversity doesn't count. Social justice is even further afield, likely considered to be a priority that is too ideologically driven for engineering education, even as serving military and corporate ends is taken as politically neutral.

Looking more deeply one might interrogate the notion of outcomes-based education that completely devalues and ignores learning processes. For all of the research emphasis on approaches to learning that promote diversity, ABET only considers what abilities students have at the end of their education, not how they acquired them. Capper and Jamison (1993) detail the brutalities of outcomes-based education, revealing its structural-functionalist assumptions (in which the aim of educational institutions is to socialize students to perform as part of a larger social order). "Success means mastering what people other than the student deem important and performing the mastered material in schools and society as they are currently structured" (p. 439). They note that a focus on mastery means that when students fail, it is their fault, not the fault of educational structures, the curriculum, or pedagogies employed.

Assessment then needs to be entirely reframed. As long as the reformist impulses of universities are grounded in maintaining existing decision-making structures (both within the university and the accreditation system), or fail to acknowledge those structures, as is certainly the norm in the United States at least, such reframing seems unlikely. Historians, sociologists, and progressive educators have long understood that the goals of any educational system are not only arbitrary, but predicated on the interests of those most influential in their formation. If the correction of race-, gender-, age-, disability-, or any other identity-based discrimination is truly our aim, why not *begin* by bringing in members of underrepresented groups as a foundational step in formulating assessment practices, rather than awaiting their arrival following pedagogical or curricular reforms in which they have had little voice?

Conclusion

Inclusion and social justice in engineering education research requires nothing less than a pervasive consideration of difference, power, and privilege throughout our research practices including the theoretical, epistemic, methodological, pedagogical, and the practical. Engineering has not historically been on the cutting edge of inclusion and social justice, but can therefore benefit from prior research in this area from a variety of other disciplines. Here we have reviewed only a small sampling of this literature with the intent of showing how its integration in engineering education research can fundamentally and holistically transform our approach, realigning our work toward goals of inclusion and social justice.

Specifically, we as members of the engineering education research community recommend shifting our theoretical base to incorporate feminist theories, critical race theories, and other philosophical approaches that help us to make serious considerations of diversity, inclusion, and social justice an integral part of the fabric of rigorous engineering education research. In addition, we need to elevate ways of knowing traditionally excluded and devalued in engineering. These can in turn help us to develop methods that enable us to both frame and answer research questions situated around power relations, and that enable us to expand categories in our research to consider groups not widely included to date. Finally, we need to develop a culture of critical reflection on our own practice as engineering education researchers and as engineering educators, as our pedagogies and practice need to interrogate systems of power operating in engineering practice, engineering education, and engineering education research. In this critical and reflexive practice, we collectively can begin to work for comprehensive structural change.

Acknowledgments

A few short passages in this chapter have been included in conference proceedings from the 2013 American Society for Engineering Education National Conference and Exposition (Pawley, 2013); used with permission of ASEE. Some of this material is based upon work supported by the National Science Foundation under Grant No. 1055900. Any opinions, findings, and conclusions or recommendations expressed in this material are those of the author(s) and do not necessarily reflect the views of the National Science Foundation.

References

Acker, J. (1992). From sex roles to gendered institutions. *Contemporary Sociology, 21*(5), 565–569.

Acker, J. (1998). Hierarchies, jobs, bodies: A theory of gendered organizations. In K. A. Myers, B. J. Risman, & C. D. Anderson, (Eds.). *Feminist foundations: Towards transforming sociology* (pp. 299–317). Thousand Oaks, CA: SAGE.

Agostinone-Wilson, F. (2006). Downsized discourse: Classroom management, neoliberalism, and the shaping of correct workplace attitude. *Journal for Critical Education Policy Studies, 4*(2).

Anyon, J. (1980). Social class and the hidden curriculum of work. *Journal of Education, 162*(1), 67–92.

Anyon, J. (1981). Social class and school knowledge. *Curriculum Inquiry, 11*(1), 3–42.

Anyon, J. (2011). *Marx and education.* New York, NY: Routledge.

Anzaldúa, G. (1987). *Borderlands/La frontera: The new Mestiza.* San Francisco, CA: aunt lute books.

Apple, M. (1996). *Cultural politics and education.* New York, NY: Teachers College Press.

Apple, M. (2000). *Official knowledge: Democratic education in a conservative age.* New York, NY: Routledge.

Baillie, C., & Catalano, G. D. (2009). *Engineering and society: Working toward social justice.* San Rafael, CA: Morgan and Claypool.

Beddoes, K., & Borrego, M. (2011). Feminist theory in three engineering education journals. *Journal of Engineering Education, 100*(2), 281–303.

Berry, T. R., & Mizelle, M. D. (Eds.). (2006). *From oppression to grace: women of color and their dilemmas within the academy.* Sterling, VA: Stylus Publishing.

Bix, A. S. (2002). Equipped for life: Gendered technical training and consumerism in home economics, 1920–1980. *Technology and Culture, 43*(4), 728–754.

Blasco, M. (2012). Aligning the hidden curriculum of management education with PRME: An inquiry-based framework. *Journal of Management Education, 36,* 364–388.

Bothwell, M., & McGuire, J. (2006). Difference, power and discrimination in engineering education. In J. Xing, J. Li, L. Roper, & S. M. Shaw (Eds.), *Teaching for change: The difference, power, and discrimination model.* Lanham, MD: Rowman & Littlefield.

Bourdieu, P. (1986). The forms of capital. In J. E. Richardson (Ed.) & R. Nice (Trans.), *Handbook of theory of research for the sociology of education* (pp. 241–258). Westport, CT: Greenwood Press.

Bourdieu, P. (1990). *Reproduction in education, society and culture.* Thousand Oaks, CA: SAGE.

Bowker, G. C., & Star, S. L. (1999). *Sorting things out: Classification and its consequences.* Inside Technology. Cambridge, MA: MIT Press.

Bowles, S., & Gintis, H. (1976). *Schooling in capitalist America: Educational reform and the contradictions of economic life.* New York, NY: Basic Books.

Bransford, J. D., Brown, A. L., & Cocking, R. R. (2000). *How people learn: Brain, mind, experience, and school.* Washington, DC: The National Academies Press.

Bucciarelli, L. L. (2003). *Engineering philosophy.* Delft, The Netherlands: DUP Satellite Press.

Burgstahler, S. (1994). Increasing the representation of people with disabilities in science, engineering, and mathematics. *Journal of Information Technology and Disability, 24*(4).

Campbell, M. L., & Gregor, F. M. (2004). *Mapping social relations: A primer in doing institutional ethnography.* Walnut Creek, CA: AltaMira Press.

Capper, C. A., & Jamison, M. T. (1993). Outcomes-based education reexamined: From structural functionalism to poststructuralism. *Educational Policy, 7*(4), 427–446.

Catalano, G. D. (2006). *Engineering ethics: Peace, justice, and the earth.* San Rafael, CA: Morgan and Claypool.

Catalano, G. D., Baillie, C., Byrne, C., Nieusma, D., & Riley, D. (2008). Increasing awareness of issues of poverty, environmental degradation and war within the engineering classroom: A course modules approach. In *Proceedings of the 38th ASEE/IEEE Frontiers in Education Conference,* Saratoga Springs, NY. Piscataway, NJ: IEEE.

Catalano, G. D., & Howell, M. (2006). Work in progress – Using wolves to teach engineering design. In *Proceedings of the 36th ASEE/IEEE Frontiers in Education Conference,* San Diego, CA. Piscataway, NJ: IEEE.

Cech, E. A., & Waidzunas, T. (2011). Navigating the heteronormativity of engineering: The experiences of lesbian, gay, and bisexual students. *Engineering Studies, 3*(1), 1–24.

Connell, R. W. (1987). *Gender and power: Society, the person and sexual politics.* Stanford, CA: Stanford University Press.

Coyle, E. J., Jamieson, L. H., & Oakes, W. C. (2005). EPICS: Engineering Projects in Community Service. *International Journal of Engineering Education, 21*(1), 1–12.

Crenshaw, K. (1989). Demarginalizing the intersection of race and sex: A black feminist critique of antidiscrimination doctrine, feminist theory, and antiracist politics. *University of Chicago Legal Forum, 1989,* 139–168.

Dallyn, S. (2011). Innovation and financialisation: Unpicking a close association. In Proceedings of the 7th International Critical

Management Studies Conference, Naples, Italy. Retrieved from http://www.organizzazione.unina.it/cms7/proceedings/proceedings_stream_24/Dallyn.pdf

Darder, A., Baltodano, M. P., & Torres, R. D. (Eds.). (2008). *The critical pedagogy reader* (2nd ed.). New York, NY: Routledge.

Davis, K. (1996). The social model of disability: Setting the terms of a new debate. Derbyshire Coalition of Disabled People. With an Introduction by Grant Carson, Glasgow Centre for Inclusive Living. Retrieved from http://www.gcil.org.uk/FileAccess.aspx?id=59

DiMaggio, P. (1991). The micro-macro dilemma in organizational research: Implications of role-system theory. In J. Huber (Ed.), *Macro-micro linkages in sociology* (pp. 76–98). Newbury Park, CA: SAGE.

Downey, G. L. (2005). Keynote lecture: Are engineers losing control of technology? From "problem solving" to "problem definition and solution" in engineering education. *Chemical Engineering Research and Design*, 83(A8), 1–12.

Downey, G. L., & Lucena, J. C. (1997). Engineering selves: Hiring in to a Contested Field of Education. In G. L. Downey & J. Dumit, (Eds.), (*Cyborgs and citadels: Anthropological interventions in emerging sciences and technologies* (pp. 117–141). Santa Fe, NM: The SAR Press.

Du, X., & Kolmos, A. (2009). Increasing the diversity of engineering education: A gender analysis in a PBL context. *European Journal of Engineering Education*, 34(5), 425–437.

Erlandson, R., Enderle, J., & Winters, J. M. (2007). Educating engineers in universal and accessible design. In J. Winters & M. Story (Eds.), *Medical instrumentation: Accessibility and usability considerations* (pp. 123–146). Boca Raton, FL: CRC Press.

Eschenbach, E., Cashman, E., Waller, A. A., & Lord, S. (2005). Incorporating feminist pedagogy into the engineering learning experience. In *Proceedings of the 35th ASEE/IEEE Frontiers in Education Conference*, Indianapolis, IN. Piscataway, NJ: IEEE.

Felder, R., Felder, G., Mauney, M., Hamrin, C., & Dietz, J. (1995). A longitudinal study of engineering student performance and retention. III. Gender differences in student performance and attitudes. *Journal of Engineering Education*, 84(2), 151–163.

Fenwick, T. J., & Parsons, J. (1997). *A critical investigation of the problems with problem-based learning*. Research Report 143, United States Department of Education. ERIC ED 409272. Retrieved from http://www.eric.ed.gov/ERICWebPortal/detail?accno=ED409272

Foor, C. E., Walden, S. E., & Trytten, D. A. (2007). "I wish that I belonged more in this whole engineering group:" Achieving individual diversity. *Journal of Engineering Education*, 96(2), 103–115.

Frehill, L. M. (2004). The gendered construction of the engineering profession in the United States, 1893–1920. *Men and Masculinities*, 6(4), 383–403.

Freire, P. (1970). *Pedagogy of the oppressed* (M. Bergman Ramos, Trans.). New York, NY: Seabury Press.

Gibbons, M. T. (2010). *Engineering by the numbers*. 2010. Retrieved from http://asee.org/publications/profiles/upload/2009ProfileEngOverview.pdf

Giroux, H. A., & Purpel, D. E. (1983). *The hidden curriculum and moral education: Deception or discovery?* Berkeley, CA: McCutchan.

Goldberg, M., & Traiman, S. (2001). Why business backs education standards. *Brookings Papers on Education Policy*, 2001, 75–129.

Gordon, A. (1995). The work of corporate culture: Diversity management. *Social Text*, 44, 3–30.

Grimson, W. (2007). The philosophical nature of engineering: A characterisation of engineering using the language and activities of philosophy. In *Proceedings of the American Society for Engineering Education Annual Conference & Exposition*, Honolulu, HI. Washington, DC: ASEE.

Hafferty, F. W., & Castellani, B. (2009). The hidden curriculum: A theory of medical education. In C. Brosnan & B. S. Turner (Eds.), *Handbook of medical education* (pp. 15–35). London: Routledge.

Hammonds, E. (1994). Black (w)holes and the geometry of black female sexuality. *differences: A Journal of Feminist Cultural Studies*, 6 (2+3), 126–145.

Harvard Graduate School of Education. (2011). *Pathways to prosperity: Meeting the challenges of preparing young Americans for the 21st century*. Cambridge, MA: Harvard Graduate School of Education.

Head, S. (2003). *The new ruthless economy: Work and power in the digital age*. New York, NY: Oxford.

Henry, S. E. (2005). "I can never turn my back on that:" Liminality and the impact of class on service-learning experience. In D. W. Butin (Ed.), *Service-learning in higher education: Critical issues and directions* New York, NY: Palgrave Macmillan.

Heywood, J. (2008a). Philosophy and engineering education: A review of certain developments in the field. In *Proceedings of the 38th ASEE/IEEE Frontiers in Education Conference*, Saratoga Springs, NY. Piscataway, NJ: IEEE.

Heywood, J. (2008b). Philosophy, engineering education and the curriculum. In *Proceedings of the American Society for Engineering Education Annual Conference & Exposition*, Pittsburgh, PA. Washington, DC: ASEE.

Heywood, J., Smith, K. A., & McGrann, R. (2007). Special session: Can philosophy of engineering education improve the practice of engineering education? In *Proceedings of the 37th ASEE/IEEE Frontiers in Education Conference*, Milwaukee, WI. Piscataway, NJ: IEEE.

Hill, C., Corbett, C., & St. Rose, A. (2010). *Why so few? Women in science, technology, engineering and mathematics*. Washington, DC: AAUW.

Hill Collins, P. (1998). The social construction of black feminist thought. In K. A. Myers, B. J. Risman, & C. D. Anderson (Eds.), (*Feminist foundations: Towards transforming sociology* (pp. 371–396). Thousand Oaks, CA: SAGE.

hooks, b. (1994). *Teaching to transgress: Education as the practice of freedom*. New York, NY: Routledge.

Jackson, P. W. (1968). *Life in classrooms*. New York, NY: Holt, Rinehart, and Winston.

Jacobson, M. F. (1999). *Whiteness of a different color: European immigrants and the alchemy of race*. Cambridge, MA: Harvard University Press.

Jamieson, L., & Lohman, J. (2009). *Creating a culture for scholarly and systematic innovation in engineering education: Ensuring U.S. engineering has the right people with the right talent for a global society*. Washington, DC: American Society for Engineering Education. Retrieved from http://www.asee.org/about-us/the-organization/advisory-committees/CCSSIE/CCSSIEE_Phase1Report_June2009.pdf

Kabo, J. D. (2010). *Seeing through the lens of social justice: A threshold for engineering* (Doctoral dissertation). Department of Chemical Engineering, Queens University. Retrieved from http://qspace.library.queensu.ca/handle/1974/5521

Kimmel, M. S., & Aronson, A. (Eds.). (2008). *The gendered society reader*. New York, NY: Oxford University Press.

Lippit, V. D. (2004). Class struggles and the reinvention of American capitalism in the second half of the twentieth century. *Review of Radical Political Economics, 36*, 336–343.

Lucena, J. C. (2000). Making women and minorities in science and engineering for national purposes in the United States. *Journal of Women and Minorities in Science and Engineering, 6*(1), 1–31.

Lucena, J., Leydens, J., & Schneider, J. (2010). *Engineering and sustainable community development*. San Rafael, CA: Morgan and Claypool.

Martin Trenor, J. (2011). Influence of social capital on under-represented engineering students' academic and career decisions. In *Proceedings of the American Society of Engineering Education Annual Conference and Exposition*, Vancouver, BC, Canada. Washington, DC: ASEE.

Massachusetts Institute of Technology. (2010). Report on the Initiative for Faculty Race and Diversity. Cambridge, MA: Massachusetts Institute of Technology. Retrieved from http://web.mit.edu/provost/raceinitative.

Mayberry, M. (1998). Reproductive and resistant pedagogies: The comparative roles of collaborative learning and feminist pedagogy in science education. *Journal of Research in Science Teaching, 35*(4), 443–459.

McLoughlin, L. A. (2009). Success, recruitment, and retention of academically elite women students without STEM backgrounds in U.S. undergraduate engineering education. *Engineering Studies, 1*(2), 151–168.

McLoughlin, L. A. (2010). *Becoming an engineer at a community college*. Paper presented at Engineering, Social Justice and Peace Sixth Annual International Conference, London.

National Industries for Severely Handicapped (NISH). (2011). AbilityOne Network Design Challenge. Retrieved from http://a1design challenge.org

National Science Foundation. (2007). *Moving forward to improve engineering education*. Washington, DC: National Science Foundation.

Nelson, L., & Pawley, A. L. (2010). Using the emergent methodology of domain analysis to answer complex research questions. In *Proceedings of the American Society for Engineering Education Annual Conference and Exposition*, Louisville, KY. Washington, DC: ASEE.

Nieusma, D. (2004). Alternative design scholarship: Working toward appropriate design. *Design Issues*, 20(3), 13–24.

Nieusma, D., & Riley, D. (2010). Designs on development: Engineering, globalization, and social justice. *Engineering Studies*, 2(1), 29–59.

Oehlberg, L., & Agogino, A. M. (2011). Undergraduate conceptions of the engineering design process: Assessing the impact of a human-centered design course. In *Proceedings of the American Society of Engineering Education Annual Conference and Exposition*, Vancouver, BC, Canada. Washington, DC: ASEE.

Ohland, M. W., Orr, M. K., Lundy-Wagner, V., Veenstra, C. P., & Long, R. A. (2012). Viewing access and persistence in engineering through a socioeconomic lens. In C. Baillie, A. L. Pawley, & D. Riley (Eds.), *Engineering and social justice: In the university and beyond* (pp. 157–180). West Lafayette, IN: Purdue University Press.

Omi, M., & Winant, H. (1994). *Racial formation in the United States: From the 1960s to the 1990s*. New York, NY: Routledge.

Orr, M. K., Ramirez, N., & Ohland, M. W. (2011). Socioeconomic trends in engineering: Enrollment, persistence, and academic achievement. In *Proceedings of the American Society of Engineering Education Annual Conference and Exposition*, Vancouver, BC, Canada. Washington, DC: ASEE.

Pantazidou, M., & Nair, I. (1999). Ethic of care: Guiding principles for engineering teaching and practice. *Journal of Engineering Education*. 88(2), 205–221.

Pawley, A. L. (2007). *Drawing the line: Academic engineers negotiating the boundaries of engineering* (Doctoral dissertation). Department of Industrial Engineering, University of Wisconsin–Madison.

Pawley, A. L. (2009). Universalized narratives: Patterns in how faculty describe "engineering." *Journal of Engineering Education*, 98(3), 309–319.

Pawley, A. L. (2012). What counts as 'engineering'?: Towards a Redefinition. In C. Baillie, A. L. Pawley, & D. Riley (Eds.), *Engineering and social justice: In the university and beyond*

(pp. 59–85). West Lafayette, IN: Purdue University Press.

Pawley, A. L. (2013). "Learning from small numbers" of underrepresented students' stories: Discussing a method to learn about institutional structure through narrative. In *Proceedings of the American Society of Engineering Education Annual Conference and Exposition*, Atlanta, GA. Washington, DC: ASEE.

Pawley, A. L., Pfund, C., Lauffer, S. M., & Handelsman, J. (2006). A case study of "Diversity in the College Classroom:" A course to improve the next generation of faculty. In *Proceedings of the American Society for Engineering Education Annual Conference and Exposition*, Chicago, IL. Washington, DC: ASEE.

Pawley, A. L., Riley, D., Harding, T., Lord, S. M., & Finelli, C. J. (2009). Special session – From active learning to liberative pedagogies: Alternative teaching philosophies in Cset education. In *Proceedings of the 39th ASEE/IEEE Frontiers in Education Conference*. Retrieved from http://fie-conference.org/fie2009/papers/1231.pdf

Redish, E. F. (2010). *Introducing students to the culture of physics: Explicating elements of the hidden curriculum*. Plenary, Physics Education Research Conference. Retrieved from http://arxiv.org/ftp/arxiv/papers/1008/1008.0578.pdf

Riley, D. (2003). Employing liberative pedagogies in engineering education. *Journal of Women and Minorities in Science and Engineering*, 9(2), 137–158.

Riley, D. (2008a). *Engineering and social justice*. San Rafael, CA: Morgan and Claypool.

Riley, D. (2008b). Ethics in context, ethics in action: Getting beyond the individual professional in engineering ethics education. In *Proceedings of the American Society for Engineering Education Annual Conference and Exposition*, Pittsburgh, PA. Washington, DC: ASEE.

Riley, D. (2012). Aiding and ABETing: The bankruptcy of outcomes-based education as a change strategy. In *Proceedings of the American Society for Engineering Education Annual Conference and Exposition*, San Antonio, TX. Washington, DC: ASEE.

Riley, D., & Bloomgarden, A. (2006). Learning and service in engineering and global development. *International Journal for Service Learning in Engineering*, 1(2), 48–59.

Riley, D. M., Catalano, G. D., Pawley, A. L., & Tucker, J. (2007). Special session – Reimagining engineering education: Feminist visions for transforming the field. In *Proceedings of the 37th ASEE/IEEE Frontiers in Education Conference*, Milwaukee, WI. Piscataway, NJ: IEEE. Retrieved from http://fie-conference .org/fie2007/papers/1540.pdf

Riley, D., & Claris, L. (2006). Power/Knowledge: Using Foucault to promote critical understandings of content and pedagogy in engineering thermodynamics. In *Proceedings of the American Society for Engineering Education Annual Conference and Exposition*, Chicago, IL. Washington, DC: ASEE.

Riley, D., & Claris, L. (2008). Developing and assessing students' ability to engage in lifelong learning. *International Journal of Engineering Education*, 24(5), 906–916.

Riley, D., & Claris, L. (2010). This engineering which is not one: Encountering Irigaray in heat and mass transfer. In *Proceedings of the 40th ASEE/IEEE Frontiers in Education Conference*, Washington, DC. Piscataway, NJ: IEEE. Retrieved from http://fie-conference.org/ fie2010/papers/1456.pdf

Riley, D. & Sciarra, G.L. (2006). "You're all a bunch of fucking feminists!" Addressing the perceived conflict between gender and professional identities using the Montreal Massacre. In *Proceedings of the 38th ASEE/IEEE Frontiers in Education Conference*, Chicago, IL. Washington, DC: ASEE. Retrieved from http://fie-conference.org/fie2006/papers/1485.pdf.

Rittel, H. W. J., & Webber, M. M. (1973). Dilemmas in a general theory of planning. *Policy Sciences*, 4, 155–169.

Rockquemore, K. A., & Laszloffy, T. (2008). *The black academic's guide to winning tenure – Without losing your soul*. Boulder, CO: Lynne Rienner.

Scheman, N. (1993). *Engenderings: Constructions of knowledge, authority, and privilege*. New York, NY: Routledge.

Seiler, C. (2009). Putting the market in its place. *American Quarterly*, 61, 943–953.

Slaton, A. E. (2001). *Reinforced concrete and the modernization of American building*, 1900–1930. Baltimore: Johns Hopkins University Press.

Slaton, A. E. (2010a). Ambiguous reform: Technical workforce planning and ideologies of class and race in 1960s Chicago. *Engineering Studies*, 2, 5–20.

Slaton, A. E. (2010b). *Race, rigor, and selectivity in U.S. engineering: The history of an occupational color line*. Cambridge, MA: Harvard University Press.

Slaton, A. E. (2012). The tyranny of outcomes: The social origins and impacts of educational standards in American engineering. In *Proceedings of the American Society for Engineering Education Annual Conference and Exposition*, San Antonio, TX.Washington, DC: ASEE.

Snyder, B. R. (1971). *The hidden curriculum*. New York, NY: Alfred A. Knopf.

Steger, M. (2009). *Globalization: A very short introduction*. New York, NY: Oxford University Press.

Strand, K., Marullo, S., Cutforth, N., Stoeker, R., & Donohue, P. (2003). *Community-based research and higher education*. San Francisco, CA: Jossey-Bass.

Subramaniam, B., Ginorio, A. B., & Yee, S. J. (1999). Feminism, Women's Studies, and Engineering: Opportunities and Challenges. *Journal of Women and Minorities in Science and Engineering*, 5, 311–322.

Sullivan, P., & Porter, J. E. (1997). *Opening spaces: Writing technologies and critical research*. Westport, CT: Ablex.

Svensson, G., & Wood, G. (2007). Are university students really customers? When illusion may lead to delusion for all! *International Journal of Educational Management*, 21(1), 17–28.

Swaminathan, R. (2005). "Whose school is it anyway?" Student voices in an urban classroom. In D. W. Butin (Ed.), *Service-learning in higher education: Critical issues and directions* (pp. 25–44). New York, NY: Palgrave Macmillan.

Tonso, K. (2007). *On the outskirts of engineering: Learning identity, gender and power via engineering practice*. Rotterdam, The Netherlands: Sense.

Tsui, L. (2007). Effective strategies to increase diversity in STEM fields: A review of the research literature. *Journal of Negro Education*, 76(4), 555–581.

Tucker, J., Pawley, A. L., Riley, D., & Catalano, G. D. (2008). New engineering stories: How feminist thinking can impact engineering ethics and practice. In *Proceedings of the 38th ASEE/IEEE Frontiers in Education Conference* Saratoga Springs, NY. Piscataway, NJ: IEEE. Retrieved from http://fie-conference .org/fie2008/papers/1788.pdf

Vincenti, W. G. (1993). *What engineers know and how they know it: Analytical studies from aeronautical history*. Baltimore, MD: The Johns Hopkins University Press.

Wade, R. (2000). From a distance: Service learning and social justice. In C. O'Grady (Ed.), *Integrating service learning and multicultural education in colleges and universities* (pp. 93–111). Mahwah, NJ: Lawrence Erlbaum.

Waller, A. A. (2005). What is feminist pedagogy and how can it be used in Cset education? In *Proceedings of the 35th ASEE/IEEE Frontiers in Education Conference*, Indianapolis, IN. Piscataway, NJ: IEEE. Retrieved from fie-conference.org/fie2005/papers/1585.pdf

Waller, A. A., Riley, D., Cashman, E., Eschenbach, E., & Lord, S. (2006). Workshop – Classroom border crossings: Incorporating feminist and liberative pedagogies into uour CSET classroom. In *Proceedings of the 36th ASEE/IEEE Frontiers in Education Conference*, San Diego, CA. Piscataway, NJ: IEEE. Retrieved from fie-conference.org/fie2006/papers/1445.pdf

Ward, K., & Wolf-Wendell, L. (2000). Community-centered service learning: Moving from doing for to doing with. *American Behavioral Scientist, 43*, 767–780.

Wulf, W. A. (1998). Diversity in engineering. *The Bridge, 28*(4), 8–13.

Yin, R. K. (2006). *Case study research: Design and methods*. Thousand Oaks, CA: SAGE.

Zoltowski, C., Oakes, W., & Cardella, M. (2012). Students' ways of experiencing human-centered design. *Journal of Engineering Education, 101*(1), 1–32.

CHAPTER 18

Community Engagement in Engineering Education as a Way to Increase Inclusiveness

Christopher Swan, Kurt Paterson, and Angela R. Bielefeldt

Introduction

There has been a recent surge in community engagement (CE) efforts in engineering education. These efforts have involved a spectrum of academic avenues – from curricular to co-curricular to extracurricular – that cover community-based projects in local to global settings. For example, many CE experiences can be embedded within courses in the traditional pedagogical form of service-learning, although in many cases CE activities are implemented and/or facilitated by nonacademic organizations, such as Engineers Without Borders (EWB). These activities continue to undergo increasing levels of design, management, and assessment, the latter driven, in part, by the outcomes assessment requirements for ABET engineering program accreditation in the United States, but also because of apparent positive impacts to student participants. Previous studies indicate that the knowledge and skills gained by the students are at least on par with gains from traditional education models (e.g., see Bielefeldt, Paterson, & Swan, 2010). Additional attention also is

being focused increasingly on the potential impacts of CE on student attitudes and identity (Paterson, Swan, & Guzak, 2012) as well as long-term impacts on students as they enter the professional ranks (Canney & Bielefeldt, 2012). It is in these areas that differences in the influence of CE may appear more profound, yet small numbers of student participants in various programs and a lack of coordinated assessment efforts provide limited evidence that such results exist. This chapter highlights the development of CE in engineering education and possible research endeavors that can be taken to shed new light on its potential impact.

CE is an overarching "'umbrella'" term that captures various educational models such as service-learning, community service-learning, and problem- or project-based service-learning (PBL; for more on PBL see Chapter 8 by Kolmos & De Graaff in this volume). More recently, the term Learning Through Service (LTS) has entered into the lexicon of ways to describe these pedagogical methods that infuse academic learning with service. The more recognized of these pedagogies, service-learning, has a storied (and varied) history, mostly because its concepts/constructs date back to the mid-1800s. As a distinct term in higher education, service-learning dates back to 1964 when it was used by the Oak Ridge Associated Universities as a way to describe their community service programs (Wutzdorff & Giles, 1997). This application of the term grew as many institutions in the United States adopted the terminology, leading to a number of educational programs focused on service-learning, notably at the University of Vermont, Michigan State University, and the University of Southern California. Federal programs, such as the University Year for Action (1972) and the National Center for Service-learning (1970s), continued the growth in popularity of service-based educational efforts. In 1985, Campus Compact was formed as a consortium of college and university presidents who supported the value of service in education and on their campuses. Today, service-learning efforts have grown substantially, finding a place in elementary, secondary, and higher education.

Project-based learning, connected to a community's need, provides a socio-cultural context, stimulating the process of collaborative problem solving. When the complementary pedagogies (project-based and service-learning) merge, there is potential for student development on cognitive (Dewey, 1916, 1938; Harrisberger, Heydinger, Seeley, & Talburtt, 1976; Piaget, 1977; Siegler, 1991), social (Duckworth, 1996; Felder & Brent, 2004; Vygotsky, 1978, 1986), and moral (DeVries & Kohlberg, 1975; Jacoby, 1997; Kohlberg, 1987) levels – three developmental processes that are tightly entwined, inseparable, and often trigger each other or occur simultaneously. The constructs are based on the theories of Dewey (1916, 1938), Piaget (1977), Kohlberg (1975), Vygotsky (1978, 1986), and Kolb (1984). An experience may spawn development on multiple levels, ultimately leading to maturation, heightened self-awareness, and greater complexity in cognitive thinking.

All theories agree that learning is a continuing, life-long endeavor that is vital to human development. Kolb (1984) suggests that this process represents a learning cycle in which experience translates into concepts that are used as guides toward new experiences. Jacoby (1997) points out that there are three prevalent implications of Kolb's model that are central to service-learning. First, the course must be structured with continual opportunities and challenges to enable students to move "completely and frequently" through the learning cycle. Second, Kolb's model underscores how central and important reflection is to the learning cycle. Third, reflection must "follow direct and concrete experience and precede abstract conceptualization and generalization" (Kolb, 1984). Kolb further identified strategies to increasing retention of knowledge in students. According to his theory, learning must begin with motivation, upon which theory, application, and analysis are founded.

Within the broad context of CE, other constructs, represented in theories on self-efficacy, epistemological beliefs, and motivation, also play a role in the effectiveness and impact that CE can have in engineering education. For example, EWB-USA, like many other service programs, is completely voluntary; the motivation to help others and to learn is instilled within those who join (Paterson & Fuchs, 2008). The service aspect of CE initially motivates students to participate, but the cycle of overcoming problems and continual learning sustains them. Regardless of the construct, each suggests that CE should offer a rich learning environment for engineering students, one that not only fosters their cognitive development, but also provides strong opportunities for social and moral development.

Motivation

As noted earlier, CE can lead to students' development in cognitive, social, and moral domains. Motivation and self-determination theories also provide explanations behind why students, and faculty, might want to engage with CE projects. For example, prosocial motivation, which is the desire to expend effort to benefit other people (Batson, 1987, 1998), provides an initial motivational framework through which faculty can leverage students' (and other stakeholders') interests and make the CE experience stronger. Prosocially motivated students see their work as a means to the end goal of benefiting others. Intrinsic motivation taps into students' engagement with a task or project purely for the joy or pleasure they get from the task (Amabile, 1993). Grant (2008) notes that prosocial and intrinsic motivations differ analytically along several dimensions but often act together. Self-determination theory (SDT) states that three basic psychological needs – autonomy, competence, and relatedness – form the essential constituent of psychological development (Deci & Ryan, 1985; La Guardia, 2009; Ryan & Deci, 2000). Autonomy refers to actions that are self-initiated and self-regulated. Competence refers to experience of mastery and challenge and is evidenced in curiosity and exploration. Relatedness refers to the feeling of belonging and being significant in the eyes of others.

The student developmental processes – cognitive, social, and moral – previously noted, along with prosocial motivation, intrinsic motivation, and self-determination theory, form the backbone of CE programs. CE projects tend to provide students the opportunity to act on their prosocial motivation while also being intrinsically motivated by the task at hand. With the three elements of SDT acting as a guide, educators can design and implement CE programs that lead to students' development in multiple dimensions.

Student interest in curricular and extracurricular CE efforts has created institutional momentum for integrating the approach within engineering curricula. CE programs haves been incorporated into first-year project courses, core engineering science courses, and senior design courses (Bielefeldt, 2005; Fuchs, 2007; Paterson, 2008; Piket-May, Avery, & Carlson, 1995; Swan, Rachell, & Sakaguchi, 2000). Previous research has shown many beneficial student outcomes from well-designed CE efforts and programs.

Service-Learning

Over the last twenty years, at least two hundred different definitions of service-learning have been published (Furco, 2003), the more recent of which have focused on service-learning as a pedagogy (e.g., Bringle & Hatcher, 1996, 2009; Furco, 1996; Jacoby and Associates, 1996; Morgan & Streb, 2002; Rhoads & Howard, 1998; Shumer, 2003; Sigmon, 1974). The distinguishing factor between service-learning and community-service is that service-learning is intentionally designed to meet academic course objectives. In the numerous definitions of service-learning that can be found in the literature, service-learning has been described not only as a method of teaching, but also as a pedagogy, a philosophy, and an academic strategy. It has been characterized as a form of experiential learning in which the service experience deepens the education experience. Though small differences in defining service-learning exist, essentially all definitions state service-learning is a structured linkage of educational learning with service efforts. Other components common in service-learning definitions include the need for student reflection on the value of the effort, the connection and integration of the effort with the community, and the attainment by participants of the desirable outcome of civic or social responsibility.

One of the first uses of the term "service-learning" was in the Atlanta Service-Learning Conference report (1970), which defined service-learning as "the integration of the accomplishment of a needed task with education growth" (p. iii). The conference then proceeds to define the goals of service-learning as: " . . . to accomplish needed public services, to add breadth and depth and relevance to students' learning; to offer a productive avenue of communication and cooperation between public agencies and institutions of higher education; to give students exposure to, testing of, and experience

	service	SERVICE
learning	**service learning** Service and learning goals completely separate	**SERVICE-learning** Service outcomes primary, learning goals secondary
LEARNING	**service-LEARNING** Learning goals primary, service outcomes secondary	**SERVICE-LEARNING** Service and learning goals of equal weight and each enhances the other for all participants

Figure 18.1. A service and learning typology. (Adapted from Sigmon, 1994.)

in public service careers; to increase the number of well-qualified young people entering public service careers, and to give young people, whatever line of work they choose to enter, front-line experience with today's problems so they will be better equipped to solve them as adult citizens" (p. iii).

For example, the National Society for Experiential Education (NSEE, 1994) broadly defined service-learning as "any carefully monitored service experience in which a student has intentional learning goals and reflects actively on what he or she is learning throughout the experience" (p. 2). Sigmon (1979) described it as "reciprocal learning" in which those who serve and those who are served learn and benefit from the effort. He further developed a typology for describing the various types of service-learning, reproduced in Figure 18.1 (Sigmon, 1994).

Furco (1996) recognized a need to distinguish service-learning from other forms of service, such as volunteerism and internship. These distinctions, represented in Figure 18.2, show a continuum of "service" and "learning" and how benefits to the server and the served are equally "optimized" in service-learning efforts. It should be recognized that this graphic does not apply to all cases nor is constant for all time. For example, today there are many service-based internships offered by nonprofit or not-for-profit entities, for example, Americorps, Public Interest Research Groups (PIRGs), and others, which lead to internships that have strong learning and service outcomes.

Eyler and Giles (1999), one of the more oft-cited references in the service-learning literature, described service-learning as "a form of experiential education where learning occurs through a cycle of action and reflection as students work with others through a process of applying what they are learning to community problems and, at

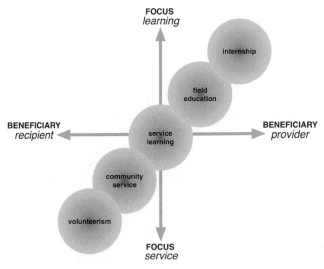

Figure 18.2. Continuum of service and learning. (Adapted from Furco, 1996.)

the same time, reflecting upon their experience as they seek to achieve real objectives for the community and deeper understanding and skills for themselves." Further, Butin (2005) also characterized service-learning as a postmodern pedagogy, one that "moving against the grain" (p. viii) of more traditional methods of teaching and learning.

The benefits of service-learning projects are well documented in the literature. For example, Eyler, Giles, Stenson, and Gray (2001) found that service-learning has a positive effect on students' personal, moral, and interpersonal development. Specific areas of development include students' sense of personal efficacy, personal identity, spiritual growth, ability to work well with others, and leadership and communication skills. In addition, service-learning appeared to have a positive effect on reducing stereotypes, facilitating cultural and racial understanding, giving one a sense of social responsibility, and strengthening citizenship skills and students' commitment to service. Learning outcomes were also positively impacted, including academic outcomes such as demonstrated complexity of understanding, problem analysis, critical thinking, and cognitive development.

Astin, Vogelgesang, Ikeda, and Yee (2000) performed an extensive, longitudinal study of students who participated in a variety of service efforts – from curricular-based service-learning to participation in various community service activities. Based on the study of more than 22,000 college undergraduates, the authors suggest that significant positive impacts occurred for a variety of outcomes including academic performance (e.g., GPA, writing skills, critical thinking skills), leadership, self-efficacy, values (e.g., activism and promotion of racial understanding), choice of a service career, and plans to participate in service after college. In addition, this study found that service-learning (i.e., course-based service) had a stronger impact on many of these outcomes than a student's participation solely in community service efforts. Furthermore, the study's results suggest that providing students with an opportunity to reflect on their service experience with each other is a powerful component of service-learning and other service-based efforts, with the primary forms of reflection being written reflection in the form of journals and papers and discussions among students and with instructors.

One of the more commonly accepted definitions is provided by Learn and Serve, a national organization that serves as a clearinghouse for implementing organized service efforts (Corporation for National Community Service [CNCS], 2012). Learn and Serve uses a definition of service-learning developed for the National Community and Service Trust Act (1993, p. 59) which describes service-learning as:

A method whereby participants learn and develop through active participation in thoughtfully organized service that:

- *is conducted in and meets the needs of a community;*
- *is coordinated with an elementary school, secondary school, institution of higher education, or community service program, and with the community;*
- *helps foster civic responsibility;*
- *is integrated into and enhances the [core] academic curriculum of the students, or the educational components of the community service program in which the participants are enrolled; and*
- *provides structured time for the students or participants to reflect on the service experience.*

Another common, more academically accepted definition of service-learning is provided by Bringle and Hatcher (1995, p. 112), who define service-learning as:

Course-based, credit bearing educational experience in which students (a) participate in an organized service activity that meets identified community needs, and (b) reflect on the service activity in such a way as to gain further understanding of course content, a broader appreciation of the discipline, and an enhanced sense of civic responsibility.

Though a precise definition would provide uniformity and conformity in how service-learning is described, it should be recognized that every service-learning effort is different, or idiosyncratic, not only in its development and implementation but also in what each stakeholder (i.e., student, instructor, institution, and community) has as an outcome. Therefore, it is difficult to develop a precise definition that encompasses all the variety of ways a community engagement effort can be developed, implemented, and assessed.

Community Engagement

The individualistic or idiosyncratic nature of service-based efforts has led the authors to create a broad definition for CE in engineering education; namely,

> Community engagement (CE) in engineering education is a form of active, experiential learning where students, instructors, and the community partners work collaboratively on projects that benefit a real community need and provide a rich learning experience for all participants. The CE must be properly planned, implemented, and assessed with expected outcomes, educational and others, in mind. Critical reflection is part of this assessment effort.

This definition provides an umbrella under which many previous definitions and descriptions of service-based learning activities can be housed, but is grounded in three interrelated elements that have been traditionally used to describe such learning activities – the effort being performed, the service effort being rendered, and the learning that occurs.

- Efforts must be authentic. The effort can be within a single discipline or multidisciplinary with components and participants spanning across engineering, math, technology, sciences, arts, and humanities. For example, a specific project can be curricular or extracurricular. The project can be initiated by the academy or by the affected community. Thus, CE can be practiced under a variety of project scenarios, BUT these projects require a connection of service and learning.

- Service must be intentional and appropriately developed. Service could be for a local industry, mobility-challenged individual, a developing rural community, or many other types of community partners. The service is grounded in the needs of the community but must be developed jointly by all stakeholders, most specifically the learners and the affected community. The service effort creates, nourishes, and sustains the overall effort; it deepens the learning experience of the project and serves as the basis for the partnership that is developed among the effort's stakeholders.

- Learning is the primary goal in carrying out the project and is enhanced by the service nature of the effort. Learning must be planned and assessed. Planning involves recognizing the educational objectives that need to be met and developing/implementing the assessment methodologies that will measure them. To deepen the learning, structured, critical reflection must be implemented in the assessment plan.

As defined, it is critical that both service and learning be considered in the CE effort's design, implementation, and assessment. There are situations in which a project is primarily service-oriented, such as volunteer or community service effort, and learning may have occurred. But unless this learning was planned and measured, the project is not optimizing this CE component. Similarly, projects may include technical evaluation or design of engineering components for a targeted community. However, unless a community connection is developed so that their input and direction are continually sought and valued, the project is not optimized for CE. For example, a service effort that installs a water distribution system for a rural community in India requires a structured learning process involving design, implementation, and assessment of desired educational objectives for it to be considered a CE project. Similarly, an engineering project that addresses the development of equipment for use by people with disabilities needs to engage and continually involve the targeted community in the product's development and use for it to be considered a CE project. As such, projects used in a course will not automatically qualify if a community's needs and input are not initially gathered and continually sought, or if the learning objectives are not appropriately measured and assessed.

In contrast to traditional engineering pedagogy in which the instructor develops a project and the learning path is fairly predictable, CE adds the community as a full partner and the outcomes are less clear. William Oakes notes: "the facilitation of the [service-learning] experience is more dependent upon capitalizing on teachable moments and learning opportunities than with traditional project-based learning. The service-learning therefore requires a more flexible curricular scaffolding to support the appropriate learning and presents additional assessment challenges since there is more uncertainty" (personal communication cited in Bielefeldt, Paterson, & Swan, 2009, p. 2).

In the United States, most engineering programs require undergraduate students to complete a one- or two-semester capstone senior design course. This capstone design experience has become almost universal given the ABET (formerly the Accreditation Board for Engineering and Technology) Engineering Criteria 2000 (EC 2000) requirements for program accreditation. This course is often considered a form of project-based learning, but the types of projects can vary tremendously. Some are fictitious "example" projects, some are based on historical real projects, while others are real projects conducted for industry partners or communities. The most common course model presented at the 2007 National Capstone Design Conference (Zable, 2007) was industry-sponsored projects, with two of the three keynote speakers discussing this topic and twenty-eight of ninety-two paper and poster abstracts highlighting industry-sponsored projects. In contrast, only six paper abstracts emphasized service-learning projects; four of these considered international service-based projects, including

those conducted in association with EWB-USA activities.

This difference between solely project-based learning and service-based project opportunities within capstone design is not surprising; numerous challenges with service-based projects have been identified (Aidoo, Hanson, Sutterer, Joughtalen, & Ahiamadi, 2007; Bielefeldt, Paterson, & Swan, 2009; Hanson et al., 2006):

- The project purpose needs to align with program outcomes, a challenge when communities are equal partners in the process.
- A meaningful relationship with the community is imperative, particularly an ongoing relationship to ensure that the community goals are served.
- A project planning phase before the beginning of the course is more critical to ensure a successful project.
- Site visits are very helpful so that students feel a connection; this can be difficult if class sizes are large or when working on international projects.
- A number of implementation challenges must be considered in project delivery including regulations, liability, local constraints, and sustainability.

Some infrastructure projects (e.g., civil or environmental engineering) for a community have a timeline to implementation longer than allowed in a single-course or academic year. This complicates student involvement, reflection, and assessment in CE; an individual student may not witness the impacts of his or her work to the community and thereby undervalue the service-learning opportunity. In addition, the strict regulatory requirements on these projects to be reviewed by a professional engineer (PE) and liability concerns can limit the use of CE in design projects. One solution to this problem is to conduct projects facilitated by EWB-USA, the International Center for Appropriate and Sustainable Technology (iCAST), or other nonprofit organizations. Perhaps partly owing to these curricular challenges with CE in engineering, EWB-USA has experienced tremendous popularity since its inception, growing from one to more than two hundred university chapters in six years (Paterson, 2009) and now hosting more than approximately three hundred chapters at U.S. institutions. Notably, this student interest in extracurricular CE, coupled with participant testimonials of the outcomes from their involvement, has created institutional momentum for integrating CE within engineering curricula.

Research, sampling, and feasibility studies are good targets for CE. CE can be incorporated into first-year project courses, core engineering science courses, and senior design courses (Bielefeldt, 2005, 2007; Fuchs, 2007; Paterson 2008; Piket-May et al., 1995; Swan et al., 2000). In addition, entire programs such as Engineering Projects in Community Service (EPICS) at Purdue University and Service-Learning Integrated in a College of Engineering (SLICE) program at the University of Massachusetts Lowell offer CE (service-learning) courses throughout the entire engineering discipline. Beyond formal courses, many students are active in conducting service projects for economically developing communities facilitated by university programs or extracurricular organizations. A student may work on the project for a number of semesters. Some students travel to partner communities, where they live and work on their projects for periods that typically range from one week to one month. Many community partners are often impoverished, rural areas in developing countries. This is a dramatically different setting than those in which most students in developed countries have previously worked. So although the project to provide clean drinking water has the same basic goal as municipal drinking water projects in the student's home country, the constraints, criteria, and demands for a successful project are vastly different. To ensure the long-term success of a project the community must feel ownership of the project and be engaged in determining its path. The world is littered with examples of engineering projects that

were implemented and then fell into disuse owing to cultural inappropriateness; insufficient local expertise, equipment, and financial resources; lack of interest by the community; or poor design and construction. To engage in successful service projects, domestically or abroad, requires a range of skills, attitudes, and non-technical attentiveness beyond those encountered in the classroom-based projects.

Inclusiveness of Community Engagement Efforts

Besides the impact of CE on engineering education noted previously, it also has been demonstrated that instructional strategies using holistic, real-world applications of science and technology tend to be more effective for attracting and retaining the interest of women and underrepresented minorities (Margolis & Fisher, 2002; Sadler, Coyle, & Schwartz, 2000). Therefore, programs that include a significant CE effort may be more attractive to women and underrepresented groups as well. For example, although women make up only 5% to 10% of practicing U.S. engineers, approximately 41% of professionals involved with EWB-USA are women (EWB, 2008), implying that women are disproportionately engaged in these sorts of programs after completing their formal education.

Romkey (2007) noted that women identify with and are more interested in science and technology when a variety of perspectives, including historical, philosophical, cultural, sociological, political, environmental, and ethical components, are included. Community engagement programs help to contextualize engineering, as well as attracting students, particularly women, by showing them how engineering helps others (Barrington & Duffy, 2007).

Because many CE programs tend to attract high percentages of women relative to the general population of engineering students, it is important to understand how CE, self-efficacy, motivation, and beliefs are interrelated. It can be hypothesized that CE increases confidence among women and leads to greater retention of women in engineering and that understanding this relationship will help institutions structure CE opportunities to help all students, and women in particular, overcome a lack of self-efficacy. In addition, linking a student's perceptions of the value and nature of engineering to real-life experiences that are important to communities will specifically enhance the views women hold of engineering.

Potential Outcomes from Community Engagement

The impacts of CE on the knowledge, skills, attitudes, and identity of student participants are of great interest to students, faculty, and employers alike. These outcomes will vary based on the learning goals for each CE opportunity and the context and intensity of the experience. Most of the research on CE uses a mixed methods approach of quantitative and qualitative research, and is typically summative. Little research on CE has combined formative and summative research methodology. Research conducted on course "artifacts" that students would otherwise produce as course assignments, test responses, or class projects may be the most effective. ABET has used this approach for accreditation in its site visits, where evaluators view examples of student work as direct evidence that various outcomes have been achieved. These artifacts provide direct evidence of what a student thinks, feels, and knows; however, the interpretations are often subjective. The development of rigorous rubrics and evaluative framework can standardize the assessment of student work. The scoring rubric for the Critical-thinking Assessment Test (CAT v. 5) is an excellent example of an interpretation guide for open-ended questions (Stein, Haynes, Redding, Ennis, & Cecil, 2007). Unique to service-learning is the opportunity to evaluate reflective essays, journals, or diaries that are a required element of the course to facilitate student learning. Some

design courses that are PBL rather than CE have also found the utility of the journaling process. Two key advantages of assessing student work that is submitted for course performance evaluation are that students do not feel that they are being "studied" or that they are burdened with additional activities.

Assessment of student knowledge and skill outcomes is now a fairly standard practice in U.S. engineering programs owing to the accreditation requirements of ABET (2008). Many universities outside the United States are also basing their program reviews on the ABET model. ABET requires all engineering students regardless of their specific discipline to have universal "A to K" knowledge and skills outcomes (see examples in Table 18.1). These encompass basic knowledge and skills that are fairly routine to assess and document across a curriculum. Curriculum-level assessment of knowledge often relies on standardized exams (such as the Fundamentals of Engineering exam). The ability to assess these outcomes from a single course is usually based on student performance demonstrated through exams, projects, and presentations; however, this generally reflects cumulative abilities rather than specific gains due to a single course as evaluated by pre- and post-assessments. Scoring rubrics created for various assignments can facilitate consistent evaluation of papers and projects for evidence of knowledge, skills, and attitudes. Research has generally found that CE is equally as effective or more so on teaching students knowledge and skills (i.e., cognitive development), with enhancements generally due to higher motivation associated with working on service projects (i.e., social and moral development). Table 18.1 highlights student learning outcomes where CE has been shown to deliver superior results over other traditional teaching methods.

Some ABET elements are more difficult to measure rigorously, such as the ability to engage in life-long learning. Some U.S.-based engineering professional societies have also defined further outcomes goals for engineering students in specific disciplines, such as civil engineering (ASCE, 2008) and

Table 18.1. Summary of Outcomes and Benefits for Engineering Students in Community Engagement

Desired Student Outcome	CE Benefits and Examples
Design a system or process within realistic constraints such as economic, environmental, social, political (ABET, 2008)	Greater complexity and range of constraints in CE settings deepened these abilities among students in capstone design courses (Bielefeldt, 2007a, 2007b).
Cultural competency	Developed as students work to understand the needs of communities with different cultural backgrounds from their own, both subtle or significant[9]; international community service experience beneficial in MTU D80 program (Paterson, 2008; Paterson & Fuchs, 2008).
Understand the impact of engineering solutions in a global and societal context (ABET, 2008)	Enhanced by working directly with a community (Duffy et al., 2007, 2008; McCormick et al., 2008; Pritchard & Tsang, 2000); >95% of students engaged in a CE capstone design experience self-reported high awareness of the social impact of engineering, significantly higher than non-CE project participants (Kremer & Burnette, 2008).
Understanding professional and ethical responsibility (ABET, 2008)	Enhanced on CE projects, even if not a central theme of the project (Bielefeldt et al., 2009; Duffy et al., 2008; McCormick et al., 2008; Pritchard & Tsang, 2000).
Attitudes toward community service (ACS)	Higher ACS scores for EWB-USA participants and high for students in Engineering for Developing World course (Bielefeldt, 2008).
Self-efficacy, self-confidence, self-esteem	Confidence in own abilities is enhanced, particularly as students achieve success and see the true benefits to a community (Gokhale & O'Dea, 2000).
Critical thinking/scientific reasoning	Critical thinking gains demonstrated for CE outside engineering (Astin et al., 2000; Sedlak et al., 2003).
Engineering identity	Redefine engineering as a helping profession particularly effective in first-year projects courses.
Ability to communicate effectively (ABET, 2008)	Students required to communicate with community members who are often nontechnical and across language and cultural differences (Bielefeldt, 2007a).
Function on multidisciplinary teams (ABET, 2008)	Greater stresses on CE projects may force students to learn better interaction skills; many CE projects are more multidisciplinary, including non-engineers (McCormick et al., 2008).
Recognize need for and ability to engage in lifelong learning (ABET, 2008)	Because CE projects are often less structured and can go in many directions, students commonly forced to a just-in-time learning model.
Sustainability; analyze systems of engineered works for sustainable performance	Length of time working with communities on service-learning projects directly influences usage and diversity of sustainability concepts (Paterson & Fuchs, 2008); evident in reflective essays from students in senior design who worked on CE projects (Bielefeldt, 2007a).
Leadership (ASCE BOK)	Students' have stronger understanding of leadership and skills to motivate others to achieve a common vision (Duffy et al., 2007, Duffy et al., 2008; McCormick et al., 2008).
Creativity; creative design	Open ended nature of many CE projects with vast array of non-technical and technical constraints forces students to be creative to find best solutions for communities (Christy and Lima, 2007).

environmental engineering (AAEE, 2009) in their body of knowledge (BOK) documents. For example, the Civil Engineering BOK includes "attitudes supportive of the professional practice of civil engineering" such as "commitment, confidence, consideration of others, curiosity, entrepreneurship, fairness, high expectations, honesty, integrity, intuition, judgment, optimism, persistence, positiveness, respect, self-esteem, sensitivity, thoughtfulness, thoroughness, and tolerance" (ASCE, 2008, p. 172). At the bachelor's level students are required only to explain these attitudes, although it is certainly desirable that they also model these attitudes. Items of this nature become much more challenging to demonstrate, leading to a greater complexity and proliferation of assessment instruments.

Faculty teaching these courses can evaluate student knowledge and skills directly via standard grading practices. Peer evaluations may also be an effective evaluation method. In addition, self-evaluation by the students on written "Likert-style" rating surveys can be used to gather information on students' perceptions of gains in knowledge and skills. However, such self-perceptions often inaccurately reflect actual levels of abilities. Therefore, Likert-based surveys are better suited to measure attitude and identity.

In general, findings have shown that when properly applied, CE is an equally effective method to develop core technical competencies in engineering. Community engagement also achieves a range of other, often unintended, benefits in an array of skills, attitude, and identity. A complete exploration of the methods that have been used to infer all of these outcomes from CE is beyond the scope of this chapter; however, some of the quantitative assessment instruments used to evaluate some of these other outcomes are listed in Table 18.1. It should be noted that developing assessment instruments that are fully validated and shown to be reliable is a challenging undertaking. As such, most engineering programs use instruments that are already available rather than developing their own. This may lead to a degree of compromise between what is desirable to

measure and what is easy to measure with an existing instrument.

There are a variety of models in how community engagement may be applied. A common curricular model, based on the Bringle and Hatcher (1995) definition of service-learning, integrates service into the classroom. In this case, the learning objectives and outcomes of the course are preeminent in designing the experience, if not already predetermined from previous non-service-learning–based course offerings.

Similar learning objectives/outcomes can be used in developing community engagement efforts that are extracurricular and/or not associated with a set of curricular, or other, educational objectives. Efforts by the student chapters of EWB-USA or Engineering for a Sustainable World represent such extracurricular CE efforts. In addition, various humanitarian engineering programs offer extracurricular efforts that are underpinned by curricular offerings that in themselves may not be service-learning courses.

These and other ways of incorporating community engagement in engineering education leads to the need to research the impacts of such efforts on specific learning outcomes – for what outcomes do various CE efforts provide the same benefit and impact? Provide different impacts? Are these impacts positive or negative to the overall development of future engineers?

Future research also needs to evaluate the impacts of CE on other stakeholders needs. Although anecdotal evidence has been presented, rigorous methods are needed to measure the levels of impacts to community partners who participate in the CE efforts. The impacts on faculty who participate in the inclusion of CE in engineering education is also needed. Such CE inclusion can have profound impact on the development of faculty who are not only technically competent, but also appropriately engaged in instructing the next generation of engineering students. Another area of research is how the use of CE impacts an institution of higher education. This research can focus on internal benefits/costs to institutions that deliver engineering programs that are centered on CE efforts (e.g., student attraction, retention, completion rates; potential employer interests in graduated students; use of faculty/staff time in delivering CE efforts; etc.) and the external benefits/costs of CE programs (e.g., institution status/ranking with other engineering programs; institution's positive/negative connection with community partners, institution alumni, the general public; and financial gain in/loss of support from external parties).

Research Area – Constraints on the Use of Community Engagement in Engineering Education

An example of the specific research topics that are worthy of future explorations is an examination of the constraints met in conceiving, designing, implementing, and assessing CE efforts. These constraints also relate to the model in which CE efforts can be incorporated into engineering education – for example, curricular or extracurricular CE efforts – as well as to the various stakeholders that are involved in CE efforts. For example, whereas the benefits to students engaged in CE efforts are relatively well documented, the constraints imposed on students by engaging in CE efforts are not as well explored. Similarly, it is well established that faculty who engage in CE activities find their efforts are constrained by a lack of available faculty time or resources to fully engage in CE. Similarly, faculty also see constraints with respect to other duties associated with their positions and the lack of value associated with doing CE – an extension of the traditional tension between "'teaching'" and "'research'" efforts. However, CE programs continue to grow as does the number of faculty who are involved with CE efforts. Why is this so?

Another set of research questions focus on the value of engineering education research associated with CE to higher education. Given the relative "youth" of engineering education research as a discipline, the overall value of engineering education research is still being assessed; that is, is education research as valued as traditional technical research by engineering faculty in existing programs and disciplines?

Future Direction of Research in CE in Engineering Education

The potential of CE as a viable pedagogy in engineering education has been implemented only recently and continues to attract students to engineering and excite faculty who want to engage students in the practice of engineering. The value of CE in education has a long history, dating back to the late 1800's but it is only recently that CE efforts have found common usage in engineering education with the rationales centered on the potentially positive benefits for students' learning engineering while also providing a common good to society. However, most of the previous evidence of these impacts in engineering education has rested on mostly anecdotal portrayals of CE impacts. Over the past fifteen years, research on the impacts of CE in engineering education, both negative and positive, have been more carefully documented, providing a richer and more rigorous evidence of CE's impact on students' learning of engineering.

However, research is still needed for evidence development, analysis, and synthesis of the impacts of CE on students. In addition to students, impacts also need to be documented for other stakeholders involved in these efforts, most notably engineering faculty, community partners, and the institutions of higher education who support such programs. Some of these efforts have begun only recently, but this work only begins to explore the multiple variables involved in the design, management, and assessment of CE, and more is needed to truly validate CE as an appropriate pedagogical strategy for engineering education.

References

ABET. (2008). Criteria for accrediting engineering programs effective for evaluations during the 2009–2010 accreditation cycle. ABET Engineering Accreditation Commission. Retrieved from www.abet.org

Aidoo, J., Hanson, J., Sutterer, K., Joughtalen, R., & Ahiamadi, S. (2007). International senior design projects – more lessons learned. Paper 11810 in *National Capstone Design Course Conference Proceedings*, Boulder, CO.

Amabile, T. M. (1993). Motivational synergy: Toward new conceptualizations of intrinsic and extrinsic motivation in the workplace. *Human Resource Management Review*, 3, 185–201.

American Academy of Environmental Engineers (AAEE). (2009). *Environmental engineering body of knowledge*. Retrieved from www.cecs.ucf.edu/bok/publications.htm

American Society for Civil Engineering (ASCE). (2008). *Civil engineering body of knowledge for the 21st century: Preparing the civil engineer for the future* (2nd ed.). Retrieved from ASCE. www.asce.org

Astin, A. W., Vogelgesang, L. J., Ikeda, E. K., & Yee, J. A. (2000). *How service learning affects students*. Higher Education Research Institute, University of California – Los Angeles.

Barrington, L., & Duffy, J. (2007). "Attracting underrepresented groups to engineering with service- learning." In *Proceedings of the ASEE 2007 Annual Conference*, Honolulu, HI.

Batson, C. D. (1987). Prosocial motivation: Is it ever truly altruistic? In L. Berkowitz (Ed.), *Advances in experimental social psychology* (Vol. 20, pp. 65–122). New York, NY: Academic Press.

Batson, C. D. (1998). Altruism and prosocial behavior. In D. T. Gilbert, S. T. Fiske, & G. Lindzey (Eds.), *The handbook of social psychology* (4th ed., Vol. 2, pp. 282–316). New York, NY: McGraw-Hill.

Bielefeldt, A. R. (2005). "Challenges and rewards of on-campus projects in capstone design." In *American Society for Engineering Education (ASEE) Conference and Exposition Proceedings*, Design in Engineering Education Division, Session 3625, Portland, OR. Retrieved from http://www.asee.org/search/proceedings

Bielefeldt, A. R. (2007a). Environmental engineering service learning projects for developing communities. Paper 12183. In *National Capstone Design Course Conference Proceedings*, University of Colorado – Boulder, CO.

Bielefeldt, A. R. (2007b). Engineering for the developing world course gives students international experience. Paper AC 2007–799. In *American Society for Engineering Education (ASEE) Conference and Exposition Proceedings*, Honolulu, HI.

Bielefeldt, A. R. (2008). Cultural competency assessment. Paper 2008–2313. In *American Society for Engineering Education (ASEE) Conference and Exposition Proceedings*, Pittsburgh, PA.

Bielefeldt, A., Paterson, K., & Swan, C. (2009). Measuring the impacts of project-based service learning. Paper AC 2009-1972. In *ASEE Annual Conference and Exposition*, Austin, TX.

Bielefeldt, A. R., Paterson, K., & Swan, C. (2010). Measuring the value added from service learning in project-based engineering education. *International Journal of Engineering Education*, 26(3), 1–12.

Bringle, R., & Hatcher, J. A. (1995). A service learning curriculum for faculty. *Michigan Journal of Community Service Learning*, 2, 112–122.

Bringle, R. G., & Hatcher, J. A. (1996). Implementing service-learning in higher education. *Journal of Higher Education*, 67, 221–239.

Bringle, R. G., & Hatcher, J. A. (2009). Innovative practices in service-learning and curricular engagement. *New Directions for Higher Education*, 147, 37–46.

Butin, D. W. (2005). "Service-learning as post-modern pedagogy." In D. W. Butin (Ed.), *Service-learning in higher education: Critical issues and directions*. Basingstoke, UK: Palgrave Macmillan.

Canney, N. & Bielefeldt, A. R. (2012). "Engineering students' views of the role of engineering in society." Paper AC 2012-3887. In *Proceedings of American Society of Engineering Education (ASEE) Conference and Exposition*, San Antonio, TX.

Christy, A. D., & Lima, M. (2007). Developing creativity and multidisciplinary approaches in teaching engineering problem solving. *International Journal of Engineering Education*, 23(4), 636–644.

Corporation for National Community Service (CNCS). (2012). Learn and Serve America. Retrieved from www.learnandserve.gov

Deci, E. L., & Ryan, R. (1985). *Intrinsic motivation and self-determination in human behavior*. New York, NY: Plenum Press.

DeVries, R., & Kohlberg, L. (1987). *Constructivist early education: Overview and comparison with other programs*. Washington, DC: National Association for the Education of Young Children.

Dewey, J. (1916). *Democracy and education*. New York, NY: Free Press.

Dewey, J. (1938). *Experience and education*. New York, NY: Collier Books.

Duckworth, E. (1996). Either we're too early and they can't learn it, or we're too late and they know it already: The dilemma of 'applying Piaget.' In *The having of wonderful ideas and other essays on teaching and learning* (2nd ed., pp. 31–49). New York, NY: Teachers College Press.

Duffy, J., Kazmer, D., Barrington, L., Ting, J., Barry, C. Zhang, X., . . . Rux, A. (2007). Service-learning integrated into existing core courses throughout a college of engineering. Paper 2007–2639. In *American Society for Engineering Education (ASEE) Conference and Exposition Proceedings*, Honolulu, HI.

Duffy, J., Moeller, W., Kazmer, D., Crespo, V., Barrington, L., Barry, C., & West, C. (2008). Service-learning projects in core undergraduate engineering courses. *International Journal for Service Learning in Engineering*, 3(2), 18–41.

Eyler, J. S., & Giles, D. E., Jr. (1999). *Where's the learning in service-learning?* San Francisco: Jossey-Bass.

Eyler, J. S., Giles, D. E. Jr., Stenson, C. M., & Gray, C. J. (2001). At a glance: What we know about the effects of service-learning on college students, faculty, institutions and communities, 1993–2000. Vanderbilt University. Retrieved from http://www.compact.org/wp-content/uploads/resources/downloads/aag.pdf.

Felder, R. M., & Brent, R. (2004). The intellectual development of science and engineering students. 2. Teaching to promote growth. *Journal of Engineering Education*, 93(4), 279–291.

Fuchs, V. J. (2007). *International engineering education assessed with the sustainable futures model* (Master of Science thesis). Michigan Technological University, Houghton, MI.

Furco, A. (1996). Service-learning and school to work: Making the connections. *Journal of Cooperative Education*, XXXII(1), 7–14.

Furco, A. (2003). Issues of definition and program diversity in the study of service-learning. In S. H. Billig (Ed.), *Studying service-learning* (pp. 13–34). Mahwah, NJ: Lawrence Erlbaum.

Gokhale, S., & O'Dea, M. (2000). Effectiveness of community service in enhancing student learning and development. Session 1621. In *American Society for Engineering Education (ASEE) Conference and Exposition Proceedings*, St. Louis, MO.

Grant, A. M. (2008). Does intrinsic motivation fuel the prosocial fire? Motivational synergy in predicting persistence, performance, and productivity. *Journal of Applied Psychology*, 93(1), 48–58.

Hanson, J. H., Houghtalen, R. J., Houghtalen, J., Johnson, Z., Lovell, M., & Van Houten, M. (2006). Our first experience with international senior design projects – lessons learned. In *Proceedings, American Society for Engineering Education (ASEE) Conference and Exposition*, Chicago, IL.

Harrisberger, L., Heydinger, R., Seeley, J., & Talburtt, M. (1976). *Experiential learning in engineering education*. Project Report, American Society for Engineering Education, Washington, DC.

Jaccoby, B. and Associates (1997). *Service-learning in higher education*. San Francisco, CA: Jossey-Boss.

Kohlberg, L. (1975). The cognitive development approach to moral education, *The Phi Delta Kappan*, 56(10), 670–677.

Kolb, D. A. (1984). *Experiential learning: Experience as the source of learning and development*. Englewood Cliffs, NJ: Prentice–Hall.

Kremer, G., & Burnette, D. (2008). Using performance reviews in capstone design courses for development and assessment of professional skills. Paper 2008–1041. In *American Society for Engineering Education (ASEE) Conference and Exposition Proceedings*, Pittsburgh, PA. Retrieved from http://www.ent.ohiou.edu/~me470/Resources/ASEE2008_presentation_PerformanceReviews.ppt

La Guardia, J. G. (2009). Developing who I am: A self-determination theory approach to the establishment of healthy identities. *Educational Psychologist*, 44(2), 90–104.

Margolis, J., & Fisher, A. (2002). *Unlocking the clubhouse: Women in computing*. Cambridge, MA: MIT Press.

McCormick, M., Swan, C., & Matson, D. (2008). Reading between the lines: Evaluating self-assessments of skills acquired during an international service-learning projects. In *American Society for Engineering Education (ASEE) Conference and Exposition Proceedings*, Pittsburgh, PA, Paper AC 2008-1937.

Morgan, W., & Streb, M. (2002). Promoting civic activism: Student leadership in service learning. *Politics and Policy*, 30(1), 161–188.

National Society for Experiential Education (NSEE). (1994). Partial list of experiential learning terms and their definitions. Raleigh, NC: Author.

National Community and Service Trust Act (1993). U.S. public law 103-82 [h.r. 2010], 108 pp.

Nieusma, D., & Riley, D. (2010). Designs on development: Engineering, globalization, and social justice. *Engineering Studies*, 2(1), 29–59.

Oakes, W.C. (2009). Creating effective and efficient learning experiences while addressing the needs of the poor: An overview of service-learning in engineering education. In *Proceedings of the ASEE Annual Conference and Exposition*, Austin, TX, Paper AC 2009-2091.

Paterson, K. G. (2008). Development for the other 80%: Assessing program outcomes. In *Global Colloquium on Engineering Education Proceedings*, Cape Town, South Africa.

Paterson, K. G. (2009, September). *A national summit on service learning. ASME/ASCE/EWB-USA*. Boulder, CO.

Paterson, K. G., & Fuchs, V. J. (2008). Development for the other 80%: Engineering hope. *Journal for Australasian Engineering Education*, 14(1), 1–12.

Paterson, K., Swan, C., & Guzak, K. L. (2012). Impacts of service on engineering students. In *Proceedings of American Society of Engineering Education (ASEE) Conference and Exposition*, San Antonio, TX, Paper AC 2012-4404.

Piaget, J. (1977). *The development of thought: Equilibration of cognitive structures*. New York, NY: Viking Press.

Piket-May, M. J., Avery, J. P., & Carlson, L. E. (1995). 1st year engineering projects: A multidisciplinary, hands-on introduction to engineering through a community/university collaboration in assistive technology. Session 3253. In *American Society for Engineering Education (ASEE) Conference and Exposition Proceedings*, Anaheim, CA (pp. 2363–2365).

Pritchard, M. S., & Tsang, E. (2000). Service learning: A positive approach to teaching

engineering ethics and social impact of technology. Session 3630. In *American Society for Engineering Education (ASEE) Conference and Exposition Proceedings*.

Rhoads, R. A., & Howard, J. (Eds.). (1998). *Academic service learning: A pedagogy of action and reflection*. San Francisco, CA: Jossey-Bass.

Romkey, L. (2007). Attracting and retaining females in engineering programs: Using an STSE approach. In *Proceedings of the ASEE 2007 Annual Conference*, Honolulu, HI, Paper AC 2007-2256.

Ryan, R. M., & Deci, E. L. (2000). Self-determination theory and the facilitation of intrinsic motivation, social development, and well-being. *American Psychologist*, 55, 68–78.

Sadler, P. M., Coyle, H. P., & Schwartz, M. (2000). Engineering competitions in the middle school classroom: Key elements in developing effective design challenges. *Journal of the Learning Sciences*, 9(3), 299–327.

Sedlak, A., Doheny, M. O., Panthofer, N., & Anaya, E. (2003). Critical thinking in students' service learning experiences. *College Teaching*, 51(3), 99–103.

Shumer, R. (1993). *Defining service learning: A delphi study*. St. Paul, MN: Center for Experiential Education and Service Learning, College of Education, University of Minnesota.

Siegler, R. (1991). Piaget's theory on development. In *Children's thinking* (pp. 21–61). Englewood Cliffs, NJ: Prentice–Hall.

Sigmon, R. (1974). Service-learning in North Carolina. *New Directions for Higher Education*, 6, 23–30.

Sigmon, R. L. (1979). Service-learning: Three principles. Synergist. National Center for Service-Learning. *ACTION*, 8(1), 9–11.

Sigmon, R. (1994). *Linking service with learning: A report from CIC*. Washington, DC: Council of Independent Colleges.

The Southern Regional Education Board. (1970). *Atlanta service-learning conference report*. Atlanta Service Learning Conference. Retrieved from http://www.eric.ed.gov

Stein, B., Haynes, A., Redding, M., Ennis, T., & Cecil, M. (2007). Assessing critical thinking in STEM and beyond. In M. Iskander (Ed.), *Innovations in e-learning, instruction technology, assessment, and engineering education* (pp. 79–82). Dordrecht, The Netherlands: Springer.

Swan, C., Rachell, T., & Sakaguchi, K. (2000). Community-based, service learning approach to teaching site remediation design. In *American Society for Engineering Education (ASEE) Conference and Exposition Proceedings*, St. Louis, MO.

Vygotsky, L. S. (1978). Interaction between learning and development. In *Mind and society: The development of higher psychological processes* (pp. 70–91). Cambridge, MA: Harvard University Press.

Vygotsky, L. S. (1986). The development of scientific concepts in childhood. In *Thought and language* (pp. 146–209). Cambridge, MA: MIT Press.

Wutzdorff, A., & Giles, D. (1997). Service learning in higher education. In J. Schine (Ed.), *Service learning: Ninety-sixth yearbook of the National Society for the Study of Education* (pp. 105–177). Chicago, IL: National Society for the Study of Education.

Zable, J. (2007). *National Capstone Design Course Conference Proceedings*, Sponsored by the ASEE and NSF. University of Colorado – Boulder, CO. Retrieved from http://www.capstoneconf.org/

Part 4

ENGINEERING EDUCATION AND INSTITUTIONAL PRACTICES

CHAPTER 19

Translating Research to Widespread Practice in Engineering Education

Thomas A. Litzinger and Lisa R. Lattuca

Introduction

Governmental, academic, and professional organizations around the world have pointed to the need for changes in engineering education to meet global and national challenges (see, e.g., Australian Council of Engineering Deans, 2008; National Academy of Engineering, 2004; Royal Academy of Engineering, 2007). Some of these organizations have specifically pointed to the need for the changes in engineering education to be based on educational research (Jamieson & Lohmann, 2009, 2012; National Research Council [NRC], 2011). In spite of these calls for change, researchers are finding that the rate of change and the nature of the change are not keeping pace with the calls for change.

Reidsema, Hadgraft, Cameron, and King (2011) ask "why has change (in engineering education in Australia) not proceeded more rapidly nor manifested itself more deeply within the curriculum" (p. 345) in spite of funding from the national government and continuing efforts of engineering professional societies and Australian Council of Engineering Deans? Reidsema et al. report that interviews of sixteen coordinators of engineering science units at four different universities in Australia revealed that traditional lecture combined with tutorials remained the dominant model of instruction. An in-depth study of the state of engineering education in the United States by Sheppard, Macatangay, Colby, and Sullivan (2009) makes the case that "in the midst of worldwide transformation, undergraduate engineering programs in the United States continue to approach problem-solving and knowledge acquisition in an outdated manner" (Schmidt, 2009, p. 1).

A study of the awareness and adoption of innovations within U.S. engineering programs found high awareness, but low adoption. Borrego, Froyd, and Hall (2010) surveyed engineering department heads in the United States on the use of seven innovations in engineering education, including student-active pedagogies and curriculum-based service learning. Awareness of these two research-based innovations was high, at approximately 80% of the 197 respondents. Just over 70% reported that student-active pedagogies were being used in their program, whereas only 28% indicated service learning was being used in their programs. The use of student-active pedagogies, at least, would seem to be quite common. However, when asked what fraction of their faculty members used student-active pedagogies, the department heads indicated that only about one third were using them.

This state of affairs is not unique to engineering educators or even to educators in general. As Henderson and Dancy (2009) have shown, slow adoption of research-based teaching practices exists in science education as well. In fact, workshops sponsored by the U.S. NRC suggest that these problems exist for science, technology, engineering, and mathematics (STEM) education throughout K–12[1] and higher education in the United States (NRC, 2011). Indeed, writing about K–12 education, Cohen and Ball (p. 31) note: "We expect innovative activity at every level of education, but typically sketchy implementation. . . . and even when there is broad adoption, to expect variable, and often weak, use in practice." Other fields, such as healthcare (Bero et al., 1998; Kreuter & Bernhardt, 2009) and social work (Dearing, 2009; Nutley, Walter, & Davies, 2009), also report that research-based practices are not readily taken up by practitioners.

Fortunately, the literature on change and diffusion of innovations, as well as on the use of research-based practices in education and other fields, provides insights into the causes of low rates and low quality of adoption as well as strategies for increasing the chances of successful transfer. Drawing on this literature, we have attempted to do the following:

- Identify likely causes for the slow adoption and low quality of the adoption of research-based practices.
- Provide summaries of strategies that have been found to be effective at promoting high-quality adoption of research-based practices.

- Discuss opportunities and challenges for further research into the processes of adoption of research-based practices in engineering education.
- Offer an overall summary, in the Final Thoughts section, of key messages for researchers who are developing research-based practices with the goal of widespread use and for leaders of educational change processes.

Before taking up our main discussion, however, we define what we mean by research-based practices. We also discuss the use of research-based practices in engineering education to set the context for the remainder of the discussion.

Research-Based Practices

So what is a "research-based practice?" Related terms that appear in the literature are "evidence-based practices" and "innovations." A recent report on STEM education published by the NRC of the U.S. National Academies (2011) uses the term "promising practices." We use the term research-based practice to encompass all of these elements. We take research-based practices to be those that have been studied in well-designed investigations that collect convincing evidence showing that the practice can be effective in promoting learning. Quantitative research studies supporting the development of research-based practices should provide reliable and valid evidence that the practice has a significant and substantial effect on learning. As we shall see later in the chapter, however, demonstrating that a new practice has a sizeable, statistically significant effect is not sufficient. High-quality adoption of a practice is more likely when those who adopt the new practice understand why it works. Therefore, a research-based practice must also be based on research that establishes why the practice is effective. Generally, this research will be qualitative and will not involve statistical analysis.

Limitations of this Review

Our approach to writing this chapter and the literature that we were able to access led to two limitations that are important to state explicitly. First, we focused the chapter on processes for bringing about large-scale change in faculty practice driven by education research. We do not address the factors that affect why individual educators decide to engage in a large-scale change effort nor do we address the experiences of those who undertake translation of research to practice as a personal journey. The other major limitation stems from the literature base that we were able to access, which is dominated by studies in the United States. We were able to locate some excellent work done outside of the United States, but still the majority of the references carry a U.S. perspective. Furthermore, most of the materials from outside the United States come from other Western countries. As discussed later in the chapter, adapting a practice to local context and culture is a critical part of successful transfer to widespread use. So, the dominance of a single country and cultural perspective (Western) in this review is a potentially significant limitation.

Research-based Practices in Engineering Education

Research-based practices enter engineering education primarily through two pathways. Until the last decade, the dominant pathway was through the adoption/adaptation of research-based educational practices developed outside of engineering. Over the last ten to fifteen years, however, educational research within engineering has grown dramatically and has begun to provide additional research-based practices for engineering educators. The scope of research-based practices in education and engineering education is very broad, spanning from recruitment of students to the performance of early career graduates in the workplace and everything in between. In this chapter, we focus on pedagogical practices, but much of what we discuss also applies to increasing the use of research-based practices independent of the specific type of practice.

We use team-based learning to illustrate the time scale of adoption of an innovation in engineering education. Team-based learning was recently identified as the most widely adopted research-based practice in engineering education in the United States by participants in a workshop on diffusion of innovations in engineering education (Center for the Advancement of Scholarship in Engineering Education, 2011). To create a the timeline of the adoption of team-based learning in engineering education, we used the American Society for Engineering Education (ASEE) proceedings database to search for the terms – teams, cooperative learning, and collaborative learning[2]. Two different searches were conducted: one for papers with any of these terms in the title and one with any of the terms appearing in the full paper, including references. The title search is taken as an indicator of scholarly use of team-based learning, whereas the full paper search is an indicator of awareness of team-based learning. Because of the number of papers involved, no attempt was made to judge the sophistication of the practice described in the papers.

Figure 19.1 presents the timelines for the number of papers that include teams or cooperative or collaborative learning in the title and anywhere in the paper, for the period from 1996 to 2011 (the full range of dates in the database). The curves show similar trends with a ratio of number of papers with any of the terms to the number with the terms in the title of roughly 20:1. To give a visual indication of the rate of change in the years prior to 1996, the time scale begins at 1980 because 1981 was the year when the first paper on cooperative learning was presented at an engineering conference in the U.S (Smith, Johnson, & Johnson, 1981; Smith, 1998, 2011). The dashed line connects the first paper with the term cooperative learning in the title to the data from the ASEE database. The figure shows that it took nearly twenty-five years for the number

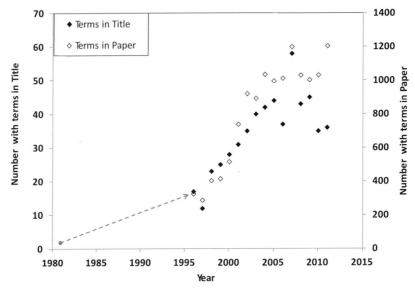

Figure 19.1. Number of papers containing terms related to cooperative learning; data from 1996 to 2011 were generated from the *Proceedings of the ASEE Annual Meeting*.

of papers on team-based learning to reach steady-state, which we take as indicator of the end of change process.

This time scale is consistent the work of Getz, Siegfried, and Anderson (1997), who studied the adoption of innovations in higher education in the United States. They conducted a survey study of the adoption of thirty innovations in six categories from curriculum to financial services at more than two hundred colleges and universities. The number of years between the first percentile adopters to the median percentile was twenty-six years. For the four curricular innovations in their study, women's studies, computer science major, interdisciplinary major, and formal study abroad, that difference was fifteen, seventeen, fifty-one, and fifty years, respectively. Thus, their work suggests a time scale measured in decades for change in higher education.

The time scale suggested by the publication data on team-based learning and the work of Getz, Siegfried, and Anderson is discouragingly long. The literature on change in educational systems and on translation of research to practice provides important insights into the factors that lead to such a slow pace of change and to the reasons

why such efforts often fail. We provide an overview of this literature in the next section.

Challenges to Successful Transfer from Research to Practice

In this discussion, we are not concerned here with what Cohen and Ball refer to as "agentless diffusion" through which a research-based practice is discovered and adopted without any direct action on the part of the developer, because such a process is highly unlikely to lead to widespread use of the research-based practice. Rather, we are concerned with the translation of research-based practices to widespread use through direct action on the part of the developers of the practice and/or other agents. The process by which the developers of a research-based practice seek to persuade others to adopt their research-based practice is often referred to as dissemination.

A common approach to dissemination is the "replication model" in which the instructor targeted as an adopter is expected to passively accept and apply the new practice just as it was developed (Bodilly,

Glennan, Kerr, & Galegher, 2004). In this model, the researcher identifies the need for a new practice, develops and assesses it, and then seeks to disseminate it to potential adopters. Trowler, Saunders, and Knight (2003) describe the change theory underpinning this approach as technical-rational; in this approach "experts plan and then manage faithful implementation" (p. 7). The underlying belief of the replication approach is that "well designed interventions will cause change" (p. 7). As we shall see, there are a number of issues with the replication model of dissemination.

According to Bodilly et al. (2004), the replication model was commonly used in the 1960s and 1970s in U.S. higher education. The model involved the development of an educational innovation along with associated training for educators that would lead to precise adoption of the innovation. The communication was essentially one-way, from the developers to the educators. Studies of the replication approach found "few new sites that had implemented the design with fidelity" (Bodilly et al., 2004, p. 12). In an article on the state of large-scale education reform around the world, Fullan (2009) confirms the assessment that the replication model failed to achieve widespread adoption of innovative practices in the United States. He writes that in spite of large expenditures of resources on major curriculum reforms, "by the early 1970s there was mounting evidence that the yield was miniscule, confined to isolated examples" (p. 103). Clearly, the replication model was a failure.

A major issue with the replication model is that it does not treat the educators as active participants who bring prior knowledge, experience, and beliefs about teaching and learning to the adoption process. The parallels between the replication model, which treats the potential adopter as a vessel to be filled, and the transmission model of teaching, which looks at students in a similar way, are somewhat disturbing. A related issue is that developers fail to meet the needs of potential adopters. Cohen and Ball note that the particular practice that the developer seeks to disseminate often does not address an "urgent" need of the potential adopters. In this situation, the developer is faced with creating a market for his or her research-based practice.

The nature of research-based practice that is being transferred to classroom practice can also have a significant impact on the likelihood of successful transfer to large numbers of educators. Regarding the process of reform in K–12 education in the United States, Elmore (1996) writes that:

> Innovations that require large changes in the core of educational practice seldom penetrate more than a small fraction of American schools and classrooms, and seldom last for very long when they do. By 'core of educational practice', I mean how the teachers understand the nature of knowledge and the student's role in learning, and how these ideas about knowledge and learning are manifested in teaching and classwork. (p. 1)

In a similar vein, Cohen and Ball (2007) note that "ambitious" pedagogical practices that seek to change significantly what an educator does in the classroom face the greatest challenges. They note that such practices are likely to lead to a feeling of "incompetence" on the part of potential adopters because familiar and conventional practices are being uprooted and challenged.

The points made by Elmore and Cohen and Ball are related to *compatibility* of an innovation as defined by Rogers (1995) within his book, *Diffusion of Innovations*. He describes diffusion of innovations as "the process through which an innovation is communicated through certain channels over time among members of a social system" (Rogers, 1995, p. 10). The innovation itself is one of the four main elements of the model of diffusion of innovations; the other elements are the social system within which potential adopters of the innovation live and/or work, the communication channels through which others learn about the innovation, and the temporal characteristics of the diffusion process. Rogers defines compatibility, one of five key attributes of an innovation, as "the degree to which an innovation is perceived as consistent with the values, past experiences, and needs of potential adopters" (p. 224). Research-based practices aimed at making substantial changes in the core of educational practice are likely to be perceived as incompatible with past experiences and possibly with the needs of potential adopters.

Dearing (2009) discusses research transfer to practice in the field of social work using the framework of diffusion of innovations. He provides a list of the "top ten dissemination mistakes"; a number of the mistakes are also relevant to transfer to practice in higher education. One of his top ten mistakes is that developers create and advocate only a single research-based practice, rather than offering a set of practices from which potential adopters can choose. Another mistake noted by Dearing is that developers assume that evidence of effectiveness will persuade potential adopters to implement the new practice. He suggests emphasizing other attributes of the practice, such as compatibility. On a similar note, Henderson and Dancy (2010) suggest emphasizing personal connections over presentation of data.

Dearing also considers using the developers as the leaders for dissemination as a mistake because the developers are often not the persons most likely to be able to engage and persuade potential adopters. Other researchers (e.g., Baker, 2007; Elmore, 1996; Horwitz, 2007; Schoenfeld, 2006) make a related point that the lack of organizations specifically focused on translating research to practice is a major barrier to widespread adoption of research-based practices. National governments have created such bodies, for example, the National Diffusion Network and the What Works Clearinghouse in the United States and Learning and Teaching Support Network in the United Kingdom. In the United States at least, the success at bringing about large-scale translation of research to practice has been limited (Fullan, 2009).

Challenges to the successful transfer of research-based practices can also arise as educators adapt them to meet personal and local needs. Coburn (2003) summarizes past work that relates to the nature and

quality of the implementation of new practices. She notes the following characteristics of the transfer process (p. 4):

- Even when educators adopt new practices, they do so in ways that show substantial variation in depth and substance.
- Educators' knowledge, beliefs, and experience influence how they choose, interpret, and implement new practices, making it likely that they "gravitate" to new practices that align with their prior experiences.
- Educators tend to prefer new practices that affect "surface features" such as new materials or classroom organizations, rather than practices involving deeper pedagogical principles.
- Finally, educators tend to "graft new approaches" onto normal classroom practices rather than changing those practices.

The findings of Henderson and Dancy (2009) on transfer of physics education research to practice in higher education are consistent with the trends noted by Coburn.

The sheer number of research-based practices available in the literature presents another challenge to widespread adoption. This situation is consistent with Cohen and Ball's observation that the present approach to creating research-based practices and translating them to practice will result in "innovative activity at every level of education but typically sketchy implementation" (p. 31). Their observation is consistent with Schoenfeld's (2006) observation that the process of research is more highly valued than the process of implementation. Within engineering education, the situation is complicated by a lack of a common vision on what needs to be changed and what research-based methods should be adopted.

Past work has also shown that ignoring the reality of the environment in which instructors find themselves, and the challenges that environment may present to the adoption of the new practice, also contribute to failure of transfer (e.g., see

Elmore, 1996). Environmental characteristics include instructional resources, disciplinary expectations, policies, and management. Lack of sufficient institutional resources and appropriate facilities can also hinder implementation of novel teaching practices. Disciplinary and institutional teaching norms can further impede or discourage experimentation with novel methods (Henderson & Dancy, 2010). Cohen and Ball (2007) note that many developers of research-based practices fail to consider the need for special equipment and spaces on the transferability of their innovative practice. Lack of incentives and recognition for the use of innovative pedagogies is widely noted (e.g., Cohen & Ball, 2007; Elmore, 1996; Fairweather, 2005) as a reason for the lack of use of innovative practices. Fairweather (2008) notes yet another challenge to widespread adoption of research-based practices: faculty and institutions bear the costs of implementing and sustaining new practices whereas the majority of the benefits accrue to the students and those who employ them.

A recent study of some of the most improved school systems around the world has demonstrated that cultural differences can have an impact on the adoption process and what is required for success (Mourshed, Chijioke, & Barber, 2010). One example of how culture can affect the implementation process relates to the use of evaluation data. Mourshed and colleagues make the point that evaluating the impact of the new practices is crucial to successful implementation, but that the results of those assessments must be used in a culturally sensitive manner. They report that it is common to make assessment data public in Anglo-American school systems, but that public release of such data would not be acceptable in many Asian and Eastern European school systems. A leader of an Asian system is quoted on this topic: "No good for our students could ever come from making school data public and embarrassing our educators" (p. 70).

Other work suggests that the culture of engineering education itself may contribute to failure, or at least increase the challenges to successful translation to widespread use. A study of more than 10,000 faculty at 517 colleges and universities by Nelson Laird, Shoup, Kuh, and Schwarz (2008) investigated the importance that faculty members in a variety of disciplines placed on deep approaches to learning.[3] In comparison to colleagues in other fields with less codified knowledge, for example, philosophy and literature, faculty members in engineering and science rated the importance of deep approaches to learning lower by nearly 0.75 standard deviations ($p < .001$). Thus, the culture of teaching in engineering seems to be a significant challenge to the use of many research-based pedagogies that are intended to increase student engagement. Student resistance to changing accepted practices in the classroom is also a potential challenge to the use of nontraditional teaching methods (Dancy & Henderson, 2004).

Another cultural tension common in engineering (as well as other fields) is the relative value placed on research and teaching in decisions regarding tenure and promotion (Fairweather, 2008). Fairweather's research, using data on approximately 17,000 faculty who responded to the National Survey on Postsecondary Faculty in 1992–3 and 1998–9, showed that the more time a faculty member spends in the classroom, the lower his or her salary, regardless of the type of four-year institution (Fairweather, 2005). His work also shows that the strongest predictor of faculty salary is the number of career publications. Comparing the differential cost/benefit of one hour teaching or publishing "in the mean" demonstrates that time spent teaching costs a faculty member money whereas time spent publishing is rewarded with higher pay. Fairweather (2008) concludes that:

> These findings strongly suggest that enhancing the value of teaching in STEM fields requires much more than empirical evidence of instructional effectiveness. It requires active intervention by academic leaders at the departmental, college, and institutional level. It requires efforts to encourage a culture within academic programs that values teaching. (p. 24)

Adopting research-based practices that lead to major shifts from traditional practices for teaching require a substantial investment of time to learn about and implement the new practices appropriately. The data from Fairweather indicate that investing effort in a process adopting new pedagogical practices is not the most productive use of time, at least when measured by salary compensation.

Schoenfeld (2006) makes a complementary point about the effect of values on the process of transfer to practice. He asserts that the academy places higher value on research, that is, the process that creates and evaluates innovative teaching methods, compared to development, that is, the process of transfer to practice. This difference in value would make it less likely that researchers would undertake studies of transfer to practice.

An additional set of influences, external to colleges and universities, that can affect the process of adoption of research-based practices are offered by Lattuca (2010). In the case of engineering education, these include accreditation agencies professional societies, and organizations, such as the National Academies in the United States, which attempt to influence educational practice. Ideally, external organizations should be drivers for change rather than barriers. Indeed the growth of interest in the use of teams in engineering education, evident in Figure 19.1, to some extent can be attributed to ABET's accreditation criterion 3, which includes the requirement that all engineering graduates develop team skills.

Fishman (2005) suggests a three-part framework for judging the "usability of innovations" that provides additional insights into reasons for failure to achieve widespread adoption. The three dimensions of his framework encompass many of the elements discussed in this section; they are Capability, Culture, and Policy and Management. The *capability* of potential adopters is an indication of the extent to which they have the conceptual and practical knowledge required to use the new practice.

Culture refers to the "norms, beliefs, values, and expectations for practice." *Policy and management* are organizational features such as faculty reward structures and support for professional development sets. He envisions these as coordinates of a three-dimensional space in which one can plot, at least conceptually, the characteristics of the adopters and the organization in which they work and the characteristics required of the adopters and organization for the research-based practice to be successfully transferred to practice. Gaps will exist that must be closed if the translation to large-scale practice is to be successful.

In sum, the literature on transfer of educational research to practice identifies a number of reasons that a dissemination approach is unlikely to succeed; these include:

- Failing to focus on the needs that potential adopters see as most important
- Offering only a single practice rather than a cluster of practices
- Failing to account for the desire of adopters to adapt, modify, and choose new practices to suit their teaching preferences
- Failing to assist adopters in understanding and incorporating the key elements of the new practice that ensure its effectiveness
- Failing to address potential barriers in the environment in which the potential adopters work, which include resource limitations, academic culture, and reward systems.

Increasing the Chances of Successful Transfer

In this section, we discuss strategies that address a number of the reasons for failure summarized in the preceding section. We also discuss an overall model that integrates many of the individual strategies. In addition, we have included summaries of two studies of successful implementations of new pedagogical practices around the world;

one study focuses on engineering programs and the other on K–12 school systems. Both provide insights into achieving and sustaining change in pedagogical practices.

Strategies

Consistent with the literature on diffusion of innovations (Rogers, 1995), several authors note the importance of addressing needs that educators see as important (see, e.g., Cohen & Ball, 2007; Glennan, Bodily, Galegher, & Kerr, 2004). To ensure that they are addressing important needs, the research team developing a new practice must understand the needs of potential users before beginning their research. Traditional needs assessment will not be adequate, however, because continuing dialogue among developers and users is needed as the research-based practice is developed. Therefore, strategies that involve continuing dialog from the beginning of a project, such as including potential adopters from the beginning of the project, should be utilized. Indeed, Fairweather (2008) recommends that every research study of pedagogical innovation should be conducted from the beginning as if the ultimate goal of the work were to take the innovation to widespread practice.

Dearing (2009) and also Cohen and Ball (2007) suggest that providing educators with more than one practice that will address an important pedagogical need will increase the chances of successful transfer to practice by allowing educators to choose the practice that best matches their teaching preferences and environment. This strategy is consistent with the use of "intervention clusters" that are composed of alternative practices to address the same need (Rogers, 1995). Chances of widespread adoption should also be increased if researchers design a practice that can be adapted to meet local needs and that supports local innovation (Baker, 2007; Henderson & Dancy, 2010).

Dearing (2009) suggests the use of "guided adaptation" of research-based practices through which educators come to understand which aspects of the practice are central to its success and why the prac-tice works. This approach would encourage effective adaptation of the practice, and it embraces the educator as an active participant in the implementation process. Cohen and Ball (2007) similarly argue that educators must understand the "underlying pedagogical principles" of the new practice if successful transfer is to occur. They describe two processes that are important to helping educators learn about and adopt new practices – *elaboration*, "the detail with which a reform is developed," and *scaffolding*, "the degree to which the innovation includes a design for and other means of learning to carry it out" (p. 24). Detailed elaboration allows potential users to understand the new practice more fully and should, Cohen and Ball contend, include the underlying pedagogical principles. Cohen and Ball point out, however, that a highly elaborated design could be seen as restrictive and conflict with the desire of educators to adapt the new practice to best suit their needs. Thus, a balance must be struck between the level to which a research-based practice is elaborated and the need to allow educators to adapt that practice to their needs, without losing the key elements that made it successful.

Goldman (2005) provides a list of design principles for educational improvement. She advocates inquiry-based approaches to allow educators to construct understanding of new practices and how they can be implemented. She further notes the potential for learning communities and practitioner networks to facilitate implementation and support educators as they learn new practices. McLaughlin and Mitra (2001) echo the potential of strong communities of practice to improve successful transfer of research to practice. Mourshed and colleagues (2010) note that peer led learning was particularly important for sustaining new practices and for creating a culture of innovation to drive continued improvement. Recent discussions of change in higher education have focused on the need for sociocognitive strategies that address the learning needs of instructors and instructional staff, suggesting a variety of learning experiences to promote the adoption or adaptation of curricular and instructional innovations. Reading groups, staff development, and ongoing professional development all provide opportunities for instructors to understand and learn new skills, roles, and educational beliefs associated with curricular change (Lattuca & Stark, 2009). Kezar (2001) notes that these strategies are well aligned with the academic culture of colleges and universities.

In "Change Thinking, Change Practices," Trowler and colleagues (2010) focus on the role of leaders of academic departments and programs in promoting and embedding good practices in higher education. They contrast a technical-rational model for change to a social practices model and conclude that the latter is a better approach for leaders in higher education. Some of the implications of this model for leaders of change in higher education include the following: expect that the people that you are trying to persuade to adopt a new practice will see that practice differently than you do; expect different faculty members to implement the practice in different ways; and be sensitive to the different histories of individual faculty members and departments, if you want to maximize the chances of successful adoption of the new practice (p. 19). Lattuca and Stark (2009) observe that changing academic programs requires knowledge of program norms and the social skills necessary to work with these norms. Those who study change note that practices and artifacts reflect values and commitments (e.g., Eckel, Hill, & Green, 1998). Understanding how changes in classroom practices affect deeply held beliefs is essential to understanding how to promote change, just as understanding a department's cultural norms will suggest strategies for building support for educational improvements.

Based on a review of 650 studies in education, healthcare, social care, and criminal justice, Walter, Nutley, and Davies (2003) identified eight mechanisms for translation of research to practice. In a later publication (Nutley et al., 2009), they grouped these into five strategies: *Dissemination*, characterized as a one-way flow of information;

Interaction, characterized as two-way flow of information; *Social Influence*, defined as using influential peers to persuade potential adopters; *Facilitation*, defined as giving technical, financial, organizational, and emotional support to potential adopters; and *Incentives and Reinforcement*, including financial incentives and feedback. An evaluation of the effectiveness of these strategies led to the conclusion that "interactive approaches currently seem to show most promise in improving use of research" (Nutley et al., 2009, p. 554). This observation is consistent with recommendation of a social practices model of change in higher education by Trowler et al. (2010) and with recommendations of Kezar (2001, 2012) that combining social cognition approaches to change with other strategies yields the greatest results in higher education settings.

An Overall Model for Translating Research to Practice

In *Extending the Reach of Education Reforms*, Glennan and colleagues (2004) offer a "mutual adaptation model for a translation of research to practice that relies on a nonsequential process of interaction, feedback and adaptation among groups of actors" (p. 27). Their model, which falls in the interaction category as defined by Walter and colleagues (2003), was developed for a K–12 context and advocates interaction among developers, educators, schools, and their district/state. Glennan et al. note three key elements of this model: (1) developing approaches and tools to enable multiple users to implement the new practice at a variety of sites; (2) ensuring high-quality implementation at each site; and (3) evaluating and improving the new practice. This interactive approach is intended to address the major reasons for failure of more traditional approaches through intensive interaction among all those involved, by focusing on adaptation, as opposed to adoption, and by attending to the context in which the research-based practices will be implemented. Goldman (2005) echoes

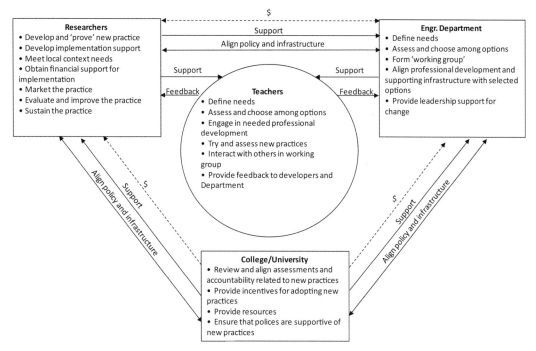

Figure 19.2. Mutual adaptation model for engineering education. (After Glennan et al., 2004, p. 649.)

the key role of ongoing interaction among all parties involved. The model of Glennan et al. also explicitly includes attention to processes required to sustain the practices. A variation on the mutual adaptation model for an engineering education context is presented in Figure 19.2. It is important to note that model is based on a single practice, which is not consistent with the need to provide adopters with multiples practices from which they can choose.

Case Studies

The Royal Academy of Engineering (RAE) and the Gordon Engineering Leadership Program at the Massachusetts Institute of Technology (MIT) funded a study on achieving sustainable change in engineering education (Graham, 2012). The final report summarizes common themes about change in engineering education based on interviews with more than seventy international experts from fifteen countries with significant experience in bringing about

change in engineering education. It also provides six case studies of successful change in engineering programs in Australia, Hong Kong, the United Kingdom, and the United States. The case studies provide important insights into how change is initiated, implemented, and sustained.

McKinsey & Company supported a study with a similar approach to the RAE–MIT study, but focused on K–12 school systems (Mourshed et al., 2010). In the McKinsey project, twenty highly successful school systems on five continents were studied. The schools fell into two broad categories: "sustained improvers" with five years or more of consistent increases on international assessments of student performance and "promising starts" who "have embarked on large-scale reform journeys employing innovative techniques that have shown significant improvements in national assessments in a short period of time" (p. 11). The report provides important results on starting, implementing, evaluating, and sustaining change in very different environments and cultural contexts.

Research Opportunities and Challenges

A number of authors point to the need to study the process of translation to widespread practice, for example, McLaughlin and Mitra (2001), Glennan et al. (2004), Goldman (2005), and Fairweather (2008). In this section, we take up this topic, highlighting major challenges to conducting such research and providing connections to related literature.

Numerous authors advocate the use of theory-based approaches in the design of research studies of transfer to practice. However, some among them question whether available theories are adequate to guide rigorous research on transfer to practice. Constas and Brown (2007) assert that the field is lacking true theories. They write about the need to design and conduct systematic studies that will yield generalizable findings about strategies for achieving widespread transfer to practice. Ideally, such studies are "built upon a set of disciplinary-based theoretical propositions and analytical models capable of guiding decisions about how best to collect, analyze and interpret data. Currently no well codified set of propositions or empirically anchored analytical frameworks exist" (p. 247). They also note that "little progress has been made in developing a comprehensive theory about how school improvement works and how such efforts might be scaled across schools, across programs, and across populations of students and teachers" (p. 245). Schoenfeld (2006) echoes this sentiment: "the theoretical state of the field . . . and the current state of theoretical disputation seriously undermine the R↔P (research to practice) process" (p. 22). It would appear that an important issue in studies of translation to widespread practice is development of an adequate theory to guide the research. Constas and Brown offer an example of a possible research design based on theories from other fields – implementation theory and developmental systems thinking.

Although not rising to the level of theory as defined by Constas and Brown, there are conceptual frameworks related to change at the individual and organizational level that can inform research in this area. The classic work of Rogers (1995) on diffusion of innovations synthesizes much of what is known about how novel practices propagate in a wide range of fields. Dearing (2009) provides a good summary of Rogers' work and describes how he has applied it in his studies of translation to practice in healthcare.

Senge's work on learning organizations (1990) provides another lens through which to view the actions that are needed within an organization to build a culture that values and invests in learning new practices. In her book, *Changing Academic Work: Developing the Learning University*, Martin (1999) applies the five disciplines from Senge's work – personal mastery, mental models, shared visions, team learning, and systems thinking – to academe. Her work provides insight into the organizational challenges involved in making substantive change based on a survey and interviews of academics in the United Kingdom and Australia.

The "Concern-Based Adoption Model," first described by Hall, Wallace, and Dorsett in 1973, is focused on the process by which individual educators adopt innovations and also provides a process for facilitating the adoption process. The current version of the Concern-Based Adoption Model (CBAM) is described in *Implementing Change* (Hall & Hord, 2011). A key aspect of the model is attending to the concerns of the potential adopters as they learn about and adapt the new practice for their use. The two scales within CBAM are the Stages of Concern and Levels of Use. The Stages of Concern range from unconcerned to refocusing. In the first stage, the potential user is unconcerned about the new practice; in the highest stage of the scale, the refocused user has substantial experience with the innovation and is exploring ways to improve it. The Levels of Use scale ranges from non-use to renewing. The highest stage on this scale is a user who is evaluating and improving the innovation. Focused very tightly on the individual educator, this model complements organizational change models.

Motivational factors are present in many of the models that we have discussed and are among the challenges to successful transfer to practice. For example, we earlier noted Rogers's focus on the compatibility of an innovation, which suggests that innovations will be more successful if they are "perceived as consistent with the values, past experiences, and needs of potential adopters" (1995, p. 224). Dearing (2009) similarly stressed compatibility and included among his top ten mistakes the assumption that evidence of effectiveness is sufficient to persuade individuals to implement new practices. In the Royal Academy of Engineering's report on successful change in engineering schools, Graham (2012) argued that although pedagogical evidence may influence course-level change,

> ... successful widespread changes are usually triggered by significant threats to the market position of the department/school. The issues faced are strongly apparent to faculty and, in some cases, university management have stipulated that a fundamental change is necessary for the long-term survival of the programme and/or department. (p. 2)

From the perspective of motivation theory, this statement highlights the role of external and internal influences on motivation for change. In general, motivation theories view motivation as potentially "intrinsic" to the individual or "situational," that is, stimulated by external factors (see, e.g., Renniger, 2000). In addition, motivation is influenced by an individual's expectations about the consequences of a particular behavior or activity as well as the value he or she places on that behavior or activity. "Expectancies" of success or failure and one's perceptions of whether adopting new practice will yield rewards or be personally satisfying affect the individual's motivation to learn and engage in new practices (see, e.g., Eccles & Wigfield, 2002). As noted in our earlier discussion of the work of Trowler et al. (2010), individuals in the same setting (a school or department) will often interpret the same events or information differently, which will lead to different levels of motivation. Social cognition

models (see Kezar, 2001) acknowledge these differences and suggest that change is more likely to succeed if individuals can come to a common understanding of the need for change and of the meaning of that change for themselves and for their organization. Clearly, theories of motivation are important to understanding how change can be successfully initiated and sustained.

Beyond identifying appropriate models, or perhaps creating them, researchers studying translation to practice must decide what constitutes successful translation to practice, how to measure it, and how to design and conduct appropriate experiments. In early research on translation to practice and the adoption of educational reform, the measure of success was simply the number of educators who were counted as using the new practice (Coburn, 2003). This simple counting approach proved to be unsatisfactory, so more complex measures have been proposed. Coburn's definition of success provides an example of a more rigorous set of measures. She recommends that the researchers studying the degree of success in the adoption of new practices consider four elements: Depth, Sustainability, Spread, and Shift in reform ownership. Successful transfer to widespread practice would correspond to

- Depth – the process of implementing the innovation leads to changes in "teachers' beliefs, norms of social interaction, and pedagogical principles as enacted in the curriculum" (p. 4).

- Sustainability – the innovation continues to be used widely even after the implementation process, and associated external resources, have ended.

- Spread – spread of the use of the innovation is accompanied by the spread of "underlying beliefs, norms, and principles" (p. 7).

- Shift in reform ownership – the ownership of the reform shifts from the external researchers who developed and spreads to the educators and schools who subsequently "sustain, spread, and deepen reform principles themselves" (p. 7).

Research built upon these four elements would examine the processes by which individual educators adopt the new practice, the impact of the process on educators' beliefs and conceptions of teaching and learning, the communities of practice that play a role in sustaining and continuing to develop the practice, and how different administrative levels within an organization support and sustain the new practice. Engaging such a large-scale study presents substantial challenges. Schoenfeld (2006) asserts that the effort to take research-based practices to widespread use is not valued highly in academia. He also notes that forming and sustaining the teams of researchers and users over the time period required to develop and take successful practices to widespread use is also very difficult.

Beyond these issues are those related to selecting the types of study and designing the complex experiments that would be required to execute them. Glennan et al. (2004) outline two different classes of research studies that can be undertaken: studies conducted during the process of development and spread of an innovation and studies of major scale-up efforts. They suggest that both successful and failed scale-up efforts are worthy of study. In "Designing Field Trials of Educational Innovations," Raudenbush (2007) proposes a conceptual model for studies of the transfer of research to practice similar to that used in clinical trials in medicine. Raudenbush (2007) also discusses issues related to the design such as randomization, generalization, and minimizing bias.

Raudenbush's conceptual model for studies of transfer to practice has two stages. In the first stage, the research-based practice is studied under ideal conditions, for example, use by highly motivated educators supported by generous resources, to establish its efficacy. In the second stage, which he describes as field trials, the research-based practice is tested under conditions that will exist when the practice is put in place under realistic conditions, for example, potential users are skeptical and they are not supported with generous resources. Such a two-stage study would uncover many challenges to the successful transfer to widespread practice.

Conducting research on transfer to widespread practice clearly presents formidable challenges. First, there are the issues of scale, the large number of educators and students who would be involved and the timescale over which the effort must be sustained. Then there is the complexity of the parameters involved in establishing success including effects on student learning, changes in classroom practice, and changes in educators' beliefs about teaching and learning. The early stages of the transfer to practice are much more amenable to study because the size and duration of the studies will be substantially reduced. However, the issues of establishing appropriate measurement methods and analyzing the data remain.

Final Thoughts

In writing this chapter, we had three groups in mind: researchers undertaking investigations of the process of translating research to practice, researchers developing innovative practices that they hope will achieve widespread use, and academic leaders who wish to increase the use of research-based practices in engineering education. In the section on research opportunities and challenges, we highlighted some of the research topics from the literature for those interested in studying the process of transfer to widespread practice. There are many exciting opportunities for research including further development of the theoretical foundations of this field of study. However, the scale, duration, and complexity of investigations of the process of transfer of research to practice are significant challenges to researchers, especially if they wish to study the entire process from the conception of the practice to large-scale implementation.

In the sections on challenges to successful transfer and strategies for increasing the chances of successful transfer, we summarized results from the literature that we hope will assist researchers who are developing new practices with the goal of widespread use. Some of the key messages for those researchers include: (1) align the practice with important needs of intended users; (2) begin planning for transfer to widespread practice from the very start of the development process; (3) engage the intended users as early as possible in the development of the research-based practice and of the transfer methodology; (4) incorporate research approaches that will determine why the practice is effective; and (5) plan for the fact that many users will want to adapt the practice to match their needs and work environment. Much of the literature cited in this chapter points to the importance of viewing the process of change as a learning process for participants; structuring ongoing interactions among those who seek to enact change with those who are being asked to implement that change is an overarching recommendation.

Finally, we believe that this chapter has salience for academic leaders who are attempting to bring about change in engineering education in response to calls for change by governments and professional organizations. These academic leaders face unique challenges. One of the major challenges is that answering the calls for change will require significant changes in how engineering instructors teach. The literature makes quite clear that such change is among the most difficult to achieve. Another major challenge is that the research-based practices that are best aligned with the calls for change are not likely to align with urgent needs of the intended users, that is, those who teach engineering. Many who teach engineering feel that they are doing just fine, with some justification, based on the success of their students in finding good jobs or spots at top graduate programs. Consequently, they see little need for change in their teaching approach. Even in the face of these challenges, however, there is hope for success as evidenced by engineering programs around the world that have achieved and sustained substantial changes in how engineering is taught and learned (Graham, 2012).

Acknowledgments

We thank the reviewers for their insightful comments and suggestions, which have significantly improved this chapter, and also Dr. Sarah Zappe for her input on an earlier version of this chapter.

Footnotes

1. K–12 refers to pre-elementary, elementary, and secondary education, i.e., kindergarten to Grade 12.

2. Cooperative learning compared to collaborative learning is "more structured, more prescriptive to teachers, more directive to students about how to work together, and more targeted (at least it was in its beginnings) to the public school population than to post-secondary or adult education" (Oxford, 1997). For a more in-depth comparison of the two see Matthews, Cooper, Davidson, and Hawkes (1995). Team-based learning may be either form, but is likely to describe students working together with little or no guidance on how they should conduct themselves within the team.

3. The construct of deep approaches to learning, a term related to the work of Marton and Säljö (1976), was originally used to describe students who read text with the intention of understanding and used strategies such as looking for main themes and underlying principles and examining arguments critically (Entwistle & Peterson, 2004).

References

Australian Council of Engineering Deans. (2008). *Engineers for the future: Addressing the supply and quality of Australian engineering graduates for the 21st Century.* New South Wales, Australia: Australian Council of Engineering Deans.

Baker, E. L. (2007). Principles for scaling up: Choosing, measuring effects, and promoting widespread use of educational innovation. In B. Schneider & S.-H. McDonald (Eds.), *Scale-up in education*, Vol. I: *Ideas in principle* (pp. 37–54). Plymouth, U.K.: Rowman & Littlefield.

Bero, L. A., Grilli, R., Grimshaw, J. M., Harvey, E., Oxman, A. D., & Thomson, M. A. (1998). Closing the gap between research and practice: An overview of systematic reviews of interventions to promote the implementation of research findings. *BMJ*, 317, 465. Retrieved from http://www.bmj.com/content/317/7156/465.full.

Bodilly, S. J., Glennan, T. K., Kerr, K. A., & Galegher, J. R. (2004). Introduction: Framing the problem. In T. K. Glennan, S. J. Bodilly, J. R. Galegher, & K. A. Kerr (Eds.), *Expanding the reach of education reforms: Perspectives from leaders in the scale-up of educational interventions* (pp. 1–39) Santa Monica, CA: RAND Corporation. Retrieved from http://www.rand.org/pubs/monographs/MG248.html

Borrego, M., Froyd, J. E., & Hall, T. S. (2010). Diffusion of engineering education innovations: A survey of awareness and adoption rates in U.S. engineering departments. *Journal of Engineering Education*, 99(3), 185–207.

Center for the Advancement of Scholarship in Engineering Education (2011, February). Forum on the impact and diffusion of transformative engineering education innovations. Workshop materials. Retrieved from http://www.nae.edu/File.aspx?id=42593

Coburn, C. E. (2003). Rethinking scale: Moving beyond numbers to deep and lasting change. *Educational Researcher*, 32(6), 3–12.

Cohen, D. K., & Ball, D. L. (2007). Educational innovation and the problem of scale. In B. Schneider & S.-H. McDonald (Eds.), *Scale-up in education*, Vol. I: *Ideas in principle* (pp. 19–36). Plymouth, U.K.: Rowman & Littlefield.

Constas, M. A., & Brown, K. L. (2007). Toward a program of research on scale-up: Analytical requirements and theoretical possibilities. In B. Schneider & S.-H. McDonald (Eds.), *Scale-up in education*, Vol. I: *Ideas in principle* (pp. 247–257). Plymouth, U.K.: Rowman & Littlefield.

Dancy, M. H., & Henderson, C. (2004). Beyond the individual instructor: systemic constraints in the implementation of research-informed Practices. In S. Franklin, J. Marx, & P. Heron (Eds.), *AAPT Physics Education Research Conference*, Sacramento, CA.

Dearing, J. W. (2009). Applying diffusion of innovation theory to intervention development. *Research on Social Work Practice*, 19, 503–518.

Eccles, J. S., & Wigfield, A. (2002). Motivational beliefs, values, and goals. *Annual Review of Psychology*, 53, 109–132.

Eckel, P., Hill, B., & Green, M. (1998). *En route to transformation*. (On Change: An Occasional Paper Series of the ACE Project on Leadership and Institutional Transformation). Washington, DC: American Council on Education. (ERIC Document Reproduction Service No. ED435293)

Elmore, R. F. (1996). Getting to scale with good educational practice. *Harvard Review*, 66(1), 1–26.

Entwistle, N. J., & Peterson, E. R. (2004). Conceptions of learning and knowledge in higher education: Relationships with study behavior and influences of learning environments. *International Journal of Educational Research*, 41(6), 407–428.

Fairweather, J. (2005). Beyond the rhetoric: Trends in the relative value of teaching and research in faculty salaries. *Journal of Higher Education*, 76, 401–422.

Fairweather, J. (2008). Linking evidence and promising practices in science, technology, engineering and mathematics (STEM) undergraduate education. Invited Paper for National Academies National Research Council Workshop, *Evidence on Promising Practices in Undergraduate Science, Technology, Engineering, and Mathematics Education*. Washington, DC.

Fishman, B. J. (2005). Adapting innovations to particular contexts of use: A collaborative framework. In C. Dede, J. P. Honan, & L. C. Peters (Eds.), *Scaling up success* (pp. 48–66). San Francisco, CA: Jossey-Bass.

Fullan, M. (2009). Large-scale reform comes of age. *Journal of Educational Change*, 10, 101–113.

Getz, M., Siegfried, J. J., & Anderson, K. H. (1997). Adoptions of innovations in higher education. *The Quarterly Review of Economics and Finance*, 37(3), 605–631.

Glennan, T. K., Bodilly, S. J., Galegher, J. R., & Kerr, K. A. (2004). Summary: Toward a more systematic approach to expanding the reach of educational interventions In T. K. Glennan, S. J. Bodilly, J. R. Galegher, & K. A. Kerr (Eds.), *Expanding the reach of education reforms: Perspectives from leaders in the scale-up of educational interventions* (pp. 647–685). Santa Monica, CA: RAND Corporation. Retrieved from http://www.rand.org/pubs/monographs/MG248.html

Goldman, S. (2005). Designing scalable educational improvements. In C. Dede, J. P. Honan, & L. C. Peters (Eds.), *Scaling up success* (pp. 67–96). San Francisco, CA: Jossey-Bass.

Graham, R. (2012). Achieving excellence in engineering education: The ingredients of successful change. Royal Academy of Engineering & MIT Press. Retrieved from http://www.raeng.org.uk/change.

Hall, G. E., & Hord, S. M. (2011). *Implementing change: Patterns, principles, and potholes.* Upper Saddle River, NJ: Pearson.

Hall, G. E., Wallace, R. C., & Dossett, W. F. (1973). A developmental conceptualization of the adoption process within educational institutions. Research and Development Center for Teacher Education. University of Texas at Austin. (ERIC Report ED 095 126).

Henderson, C., & Dancy, M. H. (2009). Impact of physics education research on the teaching of introductory quantitative physics in the United States. *Physical Review Special Topics – Physics Education Research, 5*, 1–9.

Henderson, C., & Dancy, M. H. (2010). Increasing the impact and diffusion of STEM education innovations. Invited paper for the National Academy of Engineering, Center for the Advancement of Engineering Education Forum, Impact and Diffusion of Transformative Engineering Education Innovations. Retrieved from http://www.nae.edu/File.aspx?id=36304

Horwitz, P. (2007). Successful scale-up in three stages: Insights and challenges for educational research and practice. In B. Schneider & S.-H. McDonald (Eds.), *Scale-up in education*, Vol. I: *Ideas in principle* (pp. 259–268). Plymouth, UK: Rowman & Littlefield.

Jamieson, L. H., & Lohmann, J. R. (2009). Creating a culture for scholarly and systematic innovation in engineering education. Washington, DC: American Society for Engineering Education. Retrieved from http://www.asee.org/about-us/the-organization/advisory-committees/CCSSIE

Jamieson, L. H., & Lohmann, J. R. (2012). Innovation with impact: Creating a culture for scholarly and systematic innovation in engineering education. Washington, DC: American Society for Engineering Education. Retrieved from http://www.asee.org/about-us/the-organization/advisory-committees/Innovation-with-Impact

Kezar, A. J. (2001). *Understanding and facilitating organizational change in the 21st century: Recent research and conceptualizations* (ASHE – ERIC Higher Education Reports, Vol. 28, No. 4). San Francisco, CA: Jossey-Bass.

Kezar, A. (2012). The path to pedagogical reform in the sciences: Engaging mutual adaptation and social movement models of change. *Liberal Education, 98*(1). Retrieved from http://www.aacu.org/liberaleducation/le-wi12/kezar.cfm

Kreuter, M. W., & Bernhardt, J. M. (2009). Reframing the dissemination challenge: A marketing and distribution perspective. *American Journal of Public Health, 99*(12), 2123–2127.

Lattuca, L. R. (2010). *Influences on engineering faculty members' decisions about educational innovations: A systems view of curricular and instructional change.* Invited paper for the National Academy of Engineering, Center for the Advancement of Engineering Education Forum, Impact and Diffusion of Transformative Engineering Education Innovations, New Orleans, LA. Retrieved from http://www.nae.edu/File.aspx?id=36674

Lattuca, L. R., & Stark, J. S. (2009). *Shaping the college curriculum: Academic plans in context.* San Francisco, CA: Jossey-Bass.

Martin, E. (1999). Changing academic work: Developing the learning university. Society for Research into Higher Education. Philadelphia, PA: Open University Press.

Marton, F., & Säljö, R. (1976). On qualitative differences in learning. I – Outcome and process. *British Journal of Educational Psychology, 46*(part 1), 4–11.

Matthews, R. S., Cooper, J. L., Davidson, N., & Hawkes, P. (1995). Building bridges between cooperative and collaborative learning. *Change, 27*(4), 34–40.

McLaughlin, M. W., & Mitra, D. (2001). Theory-based change and change-based theory: Going deeper, and going broader. *Journal of Educational Change, 2*, 301–323.

Mourshed, M., Chijioke, C. & Barber, M. (2010). *How the world's most improved school systems keep getting better.* McKinsey & Company. Retrieved from http://mckinseyonsociety.com/downloads/reports/Education/How-the-Worlds-Most-Improved-School-Systems-Keep-Getting-Better_Download-version_Final.pdf

National Academy of Engineering (NAE). (2004). *The engineer of 2020: Visions of engineering in the new century*. Washington, DC: The National Academies Press.

National Research Council (NRC). (2011). *Promising practices in undergraduate science, technology, engineering, and mathematics education: Summary of two workshops.* Washington, DC: The National Academies Press. Retrieved from http://www.nap.edu/catalog.php?record_id=13099

Nelson Laird, T. F., Shoup, R., Kuh, G. D., & Schwarz, M. J. (2008). The effects of discipline on deep approaches to student learning and college outcomes. *Research in Higher Education, 49*(6), 469–494.

Nutley, S., Walter, I., & Davies, T. O. (2009). Promoting evidence-based practice: Models and mechanisms from cross-sector review. *Research on Social Work Practice, 19,* 552–559.

Oxford, R. L. (1997). Cooperative learning, collaborative learning, and interaction: Three communicative stands in the language classroom. *The Modern Language Journal, 81*(4), 443–456.

Raudenbush, S. W. (2007). Designing field trials of educational innovation. In B. Schneider & S.-H. McDonald (Eds.), *Scale-up in education,* Vol. *II: Issues in practice* (pp. 23–40). Plymouth, UK: Rowman & Littlefield.

Reidsema, C., Hadgraft, R., Cameron, I., & King, R. (2011). Change strategies for educational transformation. In *Proceedings of the 2011 Australasian Engineering Education Conference,* Fremantle, Western Australia. Retrieved from http://www.aaee.com.au/conferences/2011/papers/AAEE2011/PDF/AUTHOR/AE110209.PDF

Renniger, K. A. (2000). Individual interest and its implication for understanding intrinsic motivation. In J. M. Harackiewicz & C. Sansone (Eds.), *Intrinsic and extrinsic motivation: The search for optimal motivation and performance* (pp. 373–404). San Diego: Academic Press.

Rogers, E. M. (1995). *Diffusion of innovations* (4th ed.). New York, NY: The Free Press.

The Royal Academy of Engineering (RAE). (2007). *Educating engineers for the 21st century.* London: The Royal Academy of Engineering.

Schmidt, P. (2009, January 15). Carnegie Foundation calls for overhaul of engineering education. *The Chronicle of Higher Education.* Retrieved from http://chronicle.com/article/Carnegie-Foundation-Calls-for/42254

Schoenfeld, A. H. (2006). Notes on the educational steeplechase: Hurdles and jumps in the development of research-based mathematics instruction. In M. A. Constas & R. J. Sternberg (Eds.), *Translating theory and research into educational practice: Developments in content domains, large-scale reforms, and intellectual capacity* (pp. 9–30). Mahwah, NJ: Lawrence Erlbaum.

Senge, P. M. (1990). *The fifth discipline: The art and practice of the learning organization.* New York, NY: Doubleday.

Sheppard, S., Macatangay, K., Colby, A., & Sullivan, W. M. (2009). *Educating engineers: Designing for the future of the field.* The Carnegie Foundation for the Advancement of Teaching. San Francisco, CA: Jossey-Bass.

Smith, K. A. (1998, June). *Cooperative learning.* Paper presented at American Society for Engineering Education Annual Meeting, Seattle, WA.

Smith, K. A. (2011, October). *Cooperative learning: Lessons and insights from thirty years of championing a research-based innovative practice.* Paper presented at the 41st Annual Frontiers in Education Conference, Rapid City, SD.

Smith, K. A., Johnson, D. W., & Johnson, R. T. (1981). *The use of cooperative learning groups in engineering education.* Paper presented at the Eleventh Annual Frontiers in Education Conference, Rapid City, SD.

Trowler, P., Saunders, M., & Knight, P. (2003). Change thinking, change practices: A guide to heads of departments, programme leaders and other change agents in higher education. Learning and Teaching Support Network Generic Centre, UK. Retrieved from http://www.heacademy.ac.uk/assets/documents/resources/database/id262_Change_Thinking_Change_Practices.pdf

Walter, I. C., Nutley, S. M., & Davies, H. T. O. (2003). Developing a taxonomy of interventions used to increase the impact of research. Research Unit for Research Utilisation, Discussion Paper 3. Retrieved from http://www.ruru.ac.uk/PDFs/LSDA%20literature%20review%20final.pdf

CHAPTER 20

Research-Guided Teaching Practices

Engineering Threshold Concepts as an Approach to Curriculum Renewal

Sally A. Male and Caroline A. Baillie

Introduction

In this chapter we introduce threshold concept theory and present the case for its use in curriculum, pedagogy, and assessment in engineering education, as an example of how we might consider using theory to develop practice. The process has been developed and tested in engineering at The University of Western Australia, with collaborators at the Universities of Oxford and Birmingham and The University of Queensland.

We begin with an introduction to threshold concept theory. We then step through the stages of curriculum design using threshold concepts. At each stage, the approach, critical issues that must be considered, and examples are discussed.

Threshold concept theory is a major new theory in higher education, first developed in the United Kingdom by Jan (Erik) Meyer, Ray Land, and others (Meyer & Land, 2003). The theory developed from a large research program in which it was noted that for many disciplines there are concepts, or ways of thinking, that are transformative, opening up new ways of thinking and understanding, yet troublesome for many students (Meyer & Land, 2003). These are "threshold concepts."

Since 2003, an interdisciplinary community of threshold concept researchers has developed. Four biennial threshold concepts symposia have been held and three seminal books published (Land, Meyer, & Smith, 2008; Meyer & Land, 2006; Meyer, Land, & Baillie, 2010). Threshold concept theory has been used to enhance curricula in fields as wide ranging as economics, computer science, and engineering mechanics, as demonstrated by the bibliography developed and maintained by Flanagan at University College London (2011). Approaches to identifying threshold concepts over the last eight years have included analysis of assessments, student observations, student and teacher interviews, analysis of concept maps, and student surveys. Most papers are discussions of the concepts themselves – whether they are or are not "threshold" – and increasingly there has been a focus on the translation to teaching practice. Very few studies to date have been concerned with transformation of assessment practices.

As is evident in this chapter, identifying threshold concepts is significantly different from other approaches to curriculum development. This approach allows us to focus teaching, learning, and assessment on concepts that are critical to students' progress and troublesome for students, rather than steadily working through a comprehensive syllabus. Using this approach we ensure that students have sufficient opportunities to learn the concepts that are transformative and for which they are likely to need the most time and help. The approach also helps us to identify concepts that are critical to students' learning yet frequently overlooked in syllabi, such as concepts that are intuitive to teachers. In addition, the approach helps us to design assessments that test understanding rather than ability to mimic understanding.

We are undertaking the first study to identify and investigate threshold concepts in the first and second years across all disciplines of engineering, for curriculum design. In this chapter we present our approach and demonstrate it using examples from our study.

Threshold Concept Theory

Threshold concept theory has three main aspects. First is the notion of the existence of threshold concepts and the value in recognizing this and identifying threshold concepts in a course. Second are the critical features of threshold concepts. The third relates to an educational theory about how students experience and overcome threshold concepts. A *liminal space* has been described representing the state of mind of a student while he or she experiences a concept as troublesome. We can use the information about this space to help students overcome threshold concepts.

The Existence of Threshold Concepts

First let us consider the significance of the existence of threshold concepts. This seems a simple and obvious idea. This attribute

of the theory helps it to be more readily adopted by academics than other learning theories (Cousin, 2010). In our study, several academics we interviewed already identified, tested, and emphasized critical and tricky concepts although the academics had no knowledge of threshold concept theory. For these thoughtful teachers, threshold concept theory provided a formalization, justification, and additional understanding to enhance their practice. Threshold concept theory provides an approach for academics more generally to focus their teaching, students' learning, and students' assessments on the concepts that can be transformative for students and troublesome for many students. As a minimum, conversations about the potential threshold concepts in a course draw academics' attention toward teaching and learning, potentially inspiring deviation from habitual teaching practices. As we describe in this chapter, threshold concept theory can be a central aspect of a complete curriculum development process. We use the term curriculum to include the complete learning experience of students including the syllabus, pedagogical, and assessment practices. By identifying and investigating threshold concepts we can focus otherwise crowded curricula on those aspects that we believe students need to understand but that often cause problems (Cousin, 2006).

Many readers will be aware of substantial previous research to develop concept inventories based on supporting the learning of core concepts (Kirk, 2007). Threshold concepts are not necessarily the same as core concepts. Many core concepts are not threshold concepts because they do not transform a students' way of thinking. Furthermore, some threshold concepts are not among the core concepts previously identified in courses. These threshold concepts are transformative for students yet tacit for academics and therefore not identified in syllabi. Scott and Harlow (2011) noted that threshold concepts they identified in a fundamental electronics course, namely visualizing a three-dimensional circuit from a two-dimensional circuit diagram and understanding the physical significance of graphed relationships, were only implicit in an existing electronics concept inventory. These are tacit for electronics engineering academics and therefore noticed only through the lens provided by threshold concept theory.

Features of Threshold Concepts

Identifying features of threshold concepts are their transformative and troublesome characteristics as discussed in the following subsections.

TRANSFORMATIVE

Threshold concepts are like gateways for students. They open new ways of thinking and practicing necessary to proceed in a course and, in the case of engineering courses, learn to think, speak, and identify as a professional engineer. Meyer and Land describe threshold concepts as *transformative epistemologically* and *ontologically*. These terms require explanation.

Epistemological transformation implies a shift in the student's conceptual understanding of valid knowledge and consequently how the world is understood, described, and explained. "Epistemes are manners of justifying, explaining, solving problems, conducting enquiries and designing and validating various kinds of products or outcomes" (Perkins, 2006, p. 42). In the field of engineering dynamics, we have identified vectors and changing coordinate systems as potential threshold concepts (Hesterman, Male, & Baillie, 2011). These fit Perkins's definition of episteme. Understanding coordinate systems that change with time can provide a student with a new perspective from which to perceive physical systems.

Ontological transformation refers to a shift in the student's identity. Threshold concepts can change a student's way of being. An academic and a student at our university independently described their understanding of the concept of stress, as used in engineering, in terms of a changing perspective. They both told us that their ability

to visualize stresses in buildings meant that they saw buildings differently – a building became a combination of stresses and strains and a shaped beam was recognized as elegant because of its design for the stresses it experienced. Having this insight and appreciation is likely to be accompanied by a sense of being an engineer and taking pride in this identity. The transformative feature of threshold concepts is the most important distinguishing feature of threshold concepts.

TROUBLESOME

Accompanying the transformative feature of threshold concepts is usually a *troublesome* feature. Threshold concepts can be troublesome for students in any one of a number of ways. Perkins (1999, 2006, pp. 37–41) identified and described five kinds of troublesome knowledge: *ritual knowledge, inert knowledge, conceptually difficult knowledge, tacit knowledge*, and *foreign* or *alien knowledge*. Baillie and Johnson (2008, pp. 137–8) also identified *fear of uncertainty* as a reason knowledge can be troublesome for students. Meyer and Land (2003, pp. 8–9) identified an additional type of troublesome knowledge: *knowledge with troublesome language*.

Ritual knowledge is learned and applied habitually without understanding. Academics at our university realized that students had adopted rituals in physics at school. They had learned the forces experienced by a body that is "simply supported" from below without understanding the sources of the forces. When presented with beams supported differently at university, they drew forces at points where no force was acting on the beam. Students were surprised that this was incorrect because they had not realized that the forces they had learned to draw in physics applied under limited conditions only.

Inert knowledge is learned without a context. Students often learn mathematics without an application. Students at our university felt that they would learn mathematics concepts better with awareness of their relevance to engineering (Male, 2011). To address this, engineering students in our new course that began in 2012 will study mathematics in tutorial groups with tutors from engineering. Similarly, students requested that engineering concepts be introduced initially without the mathematical abstraction. Academics at one of our regional workshops suggested that the abstract nature of electrical engineering is a reason first-year students often struggle with this discipline.

Conceptually difficult concepts are numerous in engineering. Perkins notes that conceptually difficult concepts are often counterintuitive. He provides motion as an example that can be conceptually difficult owing to commonly misinterpreted experiences, understandable misconceptions, and complexity of scientific explanations. We have identified transforms as a potential threshold concept. Transforms allow a variable to be changed from a representation as a function of one variable to a representation as a function of a different variable. For example, a voltage or current can be transformed from a representation as a function of time to a representation as a function of frequency, indicating how components of the voltage or current at different frequencies have different magnitude and phase. The concept of moving between domains, such as the time (t) domain, frequency (ω) domain, or Laplace ($s = j\omega$) domain, is not intuitive as it is inconsistent with our experience of life most obviously changing with time. An electrical engineering professor at our university noted in an interview:

> Intuition in one domain does not apply in another domain. Students don't have trouble taking the Laplace Transform. Trouble comes in the form of baggage they carry from one domain to another. For example, impedance is only defined in the frequency domain and not in the time domain. Impedance of an inductor is $j\omega L$. This does not apply in the time domain but they think it does. Similarly, $v = L\frac{di}{dt}$ cannot be written as a product of frequency. Students do not appreciate that abstraction puts you in a new world and baggage from other worlds must not be used.

The professor quoted here draws attention to one of the significant features of curriculum design using threshold concepts.

By designing curricula using threshold concepts we can focus on the troublesome and transformative aspects of a concept rather than the recipe-like steps in applying a concept. Although it is common to spend time teaching the mathematical steps involved in using transforms, performing the transforms mathematically is not especially troublesome. However, understanding that transforms take us into different domains, where different rules apply, is troublesome. Therefore we can leave students to read about the mathematical algorithm outside class and focus lesson time on the implications of using different domains.

Tacit knowledge is knowledge that is not explicitly recognized, taught, or learned. Engineering educators are aware that significant tacit knowledge is required by engineers. Regarding design, Schön (1983) noted the importance of tacit knowledge in engineering practice, particularly engineering design. Three other areas of engineering in which tacit knowledge can be seen are the practical application of theory, visualization of representations, and engineering processes.

Regarding practical application, there is much literature in the field of engineering education that refers to engineering graduates lacking practical competence, that is, ability to apply technical theory (Johnston, King, Bradley, & O'Kane, 2008; Male, 2010; Male, Bush, & Chapman, 2010a; Spinks, Silburn, & Birchall, 2006; World Chemical Engineering Council, 2004). Academics at one of our regional workshops noted the significance of experience as a way to learn. One of the reasons this is considered valuable is the possibility of developing tacit skills. In his investigation on curriculum innovations aligning engineering science and practice, Cameron (2009, p. 3) emphasized the "fundamental importance of learning spaces and places in building graduate capabilities" recognizing that students learn from experiences.

Visualization of the physical meaning of representations such as circuit diagrams and graphs used in engineering can be troublesome for students (Carstensen & Bernhard,

2008; Scott, Harlow, Peter, & Cowie, 2010). Part of students' difficulty could be the tacit nature of these concepts.

As a final example of tacit knowledge in engineering, "engineering process" is a term sometimes used by engineers to describe the processes or steps taken in an engineering project or program in an organization (Male, 2012). Parts of this are explicitly stipulated in many organizations. However, many parts of engineering process are tacit concepts learned on the job by graduate engineers (Trevelyan, 2010).

Perkins describes foreign or *alien knowledge* as founded in an unfamiliar perspective. One of the difficulties with understanding concepts that are foreign is lack of awareness of one's own perspective. Part of one's perspective comes from culture, and this is least visible to those central to the cultural group (Ihsen, 2005). Features consistent with masculine cultures have been observed in engineering faculties and workplaces (e.g., Fletcher, 1999; Godfrey & Parker, 2010; Hacker, 1981; Tonso, 2007). Within such cultures, men and women can subconsciously adopt values that reinforce a hierarchy privileging masculine practices. Fletcher found that men and women in a design engineering firm were unaware of actions taken by engineers to strengthen relationships – actions critical to the success of projects. For engineering students and faculty members, concepts such as sustainability, that require an understanding that engineering is not value-free, can be foreign because the values that have influenced engineering are part of the culture in which engineering students and faculty members are immersed.

Baillie and Johnson found that engineering students in their study experienced knowledge as troublesome due to *fear of uncertainty*. This could be related to the foreign or alien nature of open-ended questions for engineering students. Students who have focused on science and mathematics are often most familiar with questions with one correct answer and the idea that valid knowledge is objective.

Meyer and Land suggested that knowledge can be troublesome due to new or unfamiliar language. Students and academics in our study recalled confusion in use of terms such as "load" in mechanics and "potential difference" in electrical engineering. Academics at one of our regional workshops suggested that the language of mathematics can be troublesome for many parts of engineering.

OTHER COMMON FEATURES

In addition to transformative and troublesome features of threshold concepts, within threshold concept theory other common features of threshold concepts have been identified (Meyer & Land, 2003, pp. 6–7). Meyer and Land describe threshold concepts as commonly *irreversible, bounded, integrative, discursive*, and *reconstitutive*.

Within the threshold concept framework, threshold concepts are considered to be commonly *irreversible* because students generally do not lose understanding of a threshold concept once it has been fully learned. Scott and Harlow (2011) interviewed experienced engineers to ascertain whether they still understood threshold concepts in fundamental electronic engineering. A preliminary finding was that engineers who thought they still understood the concept were not always able to apply it when asked. It is possible that understanding of some engineering threshold concepts remains only if the concepts are used, especially as many engineering concepts are conceptually difficult. It is also possible that, if, as is often the case with ritual knowledge, a mimicry of the concept has been learned and not a full understanding, students will forget it again soon after passing their examinations.

Threshold concepts are commonly *bounded* in the sense that they apply to a specific field or discipline. In our study, we have found threshold concepts that are bounded to specific engineering disciplines. However, more commonly, we have found that these are examples of higher level concepts. For example, Thevenin's theorem of equivalent circuits was suggested as a threshold concept in electrical engineering. Electrical engineers use this theorem to replace linear circuits with a simplified model for analysis. However, other academics regarded this to be an example of abstraction which has broader applicability. Despite being a high level concept rather than firmly bounded within the discipline, the concept of abstraction holds part of the significance of the bounded feature of threshold concepts. The bounded nature of threshold concepts is linked to identity with a discipline. Understanding a threshold concept can contribute to part of the identity of members of the discipline. Fluent use of abstraction and modeling can be an identifying skill of engineers.

Threshold concepts are commonly *integrative* because they help students to see connections between concepts where they did not previously. One student in our study described threshold concepts as pieces in a jigsaw, or bringing an impressionist painting into focus. These analogies are slightly different from one another. One refers to a concept at the same level as those it integrates, and the other refers to a concept that overarches the concepts it integrates. We have found examples of both of these and also concepts that underlie other concepts (Male & Baillie, 2011b). Scott and Harlow (2011) and, independently, with our collaborators at the Universities of Oxford and Birmingham, we have used concept maps (Petrucci & Quinlan, 2007) to represent the relevance of individual threshold concepts (Quinlan et al., 2013). We found that threshold concepts are interrelated. For example, abstraction, visualization of representations, and vectors are interdependent. Concepts maps are a helpful tool for mapping the relationships between threshold concepts.

Threshold concepts are commonly *discursive*, meaning that a student enhances his or her use of language in a discipline by developing understanding of a threshold concept. As noted earlier, students in our study nominated terms that caused engineering concepts to be troublesome. Furthermore, Perkins (2006, p. 41) describes how concepts can be "double trouble," not only contributing ways to categorize the world but also as tools for solving problems. An electrical engineer in our study noted that he would like students to understand the concept of a proof, not to recognize a proof, but to be able to understand and use terms such as "necessary and sufficient" in order to be able to test the viability of an assumed operating mode for a transistor. In this case the enhanced use of language accompanying the understanding of proofs is significant.

Threshold concepts can be *reconstitutive*, meaning that their understanding is accompanied by reconstitution of the student's subjectivity (Meyer, Land, & Davies, 2006, pp. 20–1). Just as the discursive features of threshold concepts arise from troublesome knowledge related to unfamiliar or new use of language, reconstitutive features of threshold concepts arise from concepts that are at first foreign or alien, or counterintuitive for students.

How Students Experience and Overcome Threshold Concepts

An important feature of threshold concept theory is the *liminal space* or *state of liminality* (Meyer & Land, 2003, p. 10). This is the state experienced by a student when he or she has become aware of a concept but does not yet understand and accept the concept. Meyer and Land (2005, pp. 375–7) have noticed that a student in the liminal space is likely to *mimic* understanding by following learned steps rather than using understanding to solve problems. Given numeric examples, engineering students can often mimic solutions by inserting numbers into equations. This part of the theory becomes especially relevant when designing assessments.

For any threshold concept, some students pass through the liminal space relatively quickly and easily, while others require extended periods – possibly years, and some students never pass through the liminal space. This is relevant to curriculum design. One semester might provide insufficient opportunity for students to understand a threshold concept. Similarly, we might unintentionally discourage students from taking sufficient time. Engineering academics at one of our regional workshops noted that "asynchronous learning" experiences available for many students through online lectures and electronic communication could encourage students to believe they can learn a concept just before it is needed. In many cases this is not a realistic possibility, as it does not allow the student time to pass through the liminal space.

Meyer and Land (2006, pp. 22–3) suggest that the liminal space can resemble a rite of passage. Passing through the liminal space is a difficult experience, and by passing through the space a student takes a step toward gaining new status or privilege, such as membership of a profession. This raises a recurring issue in engineering. A mechanical engineering program in Linköping University was found to resemble a rite of passage, with many technical concepts that are not necessarily required for engineering practice (Dahlgren, Hult, Dahlgren, Hard, & Johansson, 2006). Yet technical prowess has been observed to be an important part of engineers' identities (Male, Bush, & Chapman, 2010b). The irony was noted in the report on an international survey of chemical engineering graduates:

> The two attributes which are rated as more important during education than for employment are **Appreciation of the potential of research and Ability to apply knowledge of basic science**. These are, in fact, the traditional priorities of a classical university education. For work, their relevance ranks 21st and 14th respectively. (WCEC, 2004, p. 58)

Threshold concepts are critical to ways of thinking and practicing in the relevant discipline (Meyer & Land, 2006, p. 15). This perhaps partly justifies the aforementioned apparent anomaly in engineering education. Many threshold concepts in engineering programs are similar to rites of passage. They are difficult and they might seem to have little relevance to engineering practice. However, passing through the liminal space can be transformative in a way that helps students think and practice like an engineer, for example, using abstraction even if they

do not remember the details of Thevenin's theorem.

The final aspect of threshold concept theory that we wish to introduce is the term *preliminal variation* (Meyer & Land, 2005, p. 384). Students come to university with diverse experiences, ways of thinking, and standpoints. Therefore they first experience concepts, or enter the liminal space for a concept, differently. This must be taken into account when teaching.

Terminology Within the Threshold Concept Framework

Table 20.1 summarizes terms introduced above and their relevance to the threshold concept framework.

Engineering Threshold Concepts as an Approach to Curriculum Renewal

We now proceed to six stages of curriculum renewal using threshold concept theory as a foundational framework. We have developed this approach and implemented it to design the first two years of the entirely new engineering science major (beginning 2012) at The University of Western Australia (UWA). These will be taken by students in all engineering disciplines. The six stages are (1) Identification of Three Big Ideas, (2) Identification of Learning Outcomes, (3) Identification of Potential Threshold Concepts, (4) Negotiation of Threshold Concepts, (5) Investigation of Threshold Concepts, and (6) Helping Students Overcome Threshold Concepts.

Stage 1. Identification of Three Big Ideas

In the first stage, engineering academics meet together to establish the three big ideas in their program of study. Academics answer: "what are the three big ideas students should understand after completing the program of study in their discipline?" Roger Hadgraft of the University of Melbourne and Carl Reidsema of University of

Table 20.1. Threshold Concept Terminology

Term	Relevance	Meaning in Brief
Threshold concept	Central idea of the threshold concept theory	A concept that is transformative and usually troublesome for many students in a specific discipline
Transformative	Compulsory feature of all threshold concepts	Opening up new ways of thinking and understanding necessary for future progress in a course or into a profession
Troublesome	Extremely common feature of threshold concepts due to transformative feature noted above	Difficult to understand and accept
Ritual knowledge	Some reasons a concept can be troublesome	Knowledge used habitually without understanding
Inert knowledge		Knowledge presented without context
Conceptually difficult knowledge		Knowledge that is complex or counter-intuitive
Tacit knowledge		Knowledge that is difficult to identify
Alien knowledge		Knowledge requiring a way of thinking that is foreign to many students
Fear of uncertainty		Fear associated with unknown possibilities such as arise in open-ended problems
Troublesome language		Language that is unfamiliar to the student or different from common usage
Irreversible	Other common features of threshold concepts	Not "unlearned" after understood
Bounded		Pertaining to a field or discipline
Integrative		Revealing otherwise unapparent connections between other concepts
Discursive		Enhancing a student's use of language
Reconstitutive		Repositioning a student's points of view
Liminal space	State experienced by a student	State a student is in when a threshold concept has come into view and is still troublesome for the student – in this state the transformation is not yet complete.

Queensland suggested this approach. This is an opportunity to think across units and topics. The ideas are often found to be similar. Previously, different units in the engineering degree had been taught by academics from different schools in the faculty. For example, dynamics was taught by the mechanical engineers, and electrical circuits and systems by electrical engineers. A vision of the current curriculum development was to overcome the disciplinary silos and teach concepts in ways that strengthen the key concepts across engineering disciplines by demonstrating the applications and variation in each concept in different engineering disciplines.

At UWA a foundation team of ten academics from all engineering disciplines led the discussion. After arriving at several different potential sets of three big ideas, members of the foundation team took these to their schools for comment, thereby involving approximately 150 academics in negotiation across the faculty.

Below is an example of an initial cluster of big ideas

1. Free body diagrams – separating the body of interest from all others and identifying the forces and moments (loads) acting on it.
2. Equilibrium of forces and moments in 2D and 3D. This underpins the topics of calculating reactions, analyzing trusses and shear force and bending moment diagrams.
3. Stress and strain, allowing bending and shear stresses to be calculated, again using the principle of equilibrium.

Stage 2. Identification of Learning Outcomes

The learning outcomes underpinning the three big ideas and the capabilities (Bowden, 2004) needed to understand them are identified in Stage 2. Learning outcomes are statements that stipulate desired capabilities of students, including knowledge, skills, and abilities. They describe what a learner is expected to be able to do after completing a program, and begin with verbs such as "appreciate," "understand," or "be able to define," "be able to select," or "be able to apply." The verb is followed by an object and usually a context (Moon, 2007). Verbs should be selected carefully. Curriculum designers should design for increasing cognitive levels within a program, such as described by Bloom's Taxonomy (Krathwohl, 2002). However, although it is sometimes suggested that different verbs indicate different cognitive levels, threshold concept theory reveals this simplification to be unrealistic. "Understanding" conceptually difficult knowledge does not necessarily involve a lower level of cognition than "applying" a simple concept. Using Bowden's capability theory (2004, pp. 42–4), we were able to note a developmental aspect to the learning of students as they moved from Bowden's most basic "'scoping'" and "'enabling'" levels to higher "'training'" and "'relating'" levels of capability as follows:

- *Scoping level* (*S*). Basic entry level – the student will be exploring what the capability means and what might be possible.

No demonstrable ability is necessarily developed. For example, a student begins to understand that communication skills are required for engineering but does not actually develop these skills in the unit.
- *Enabling level* (*E*). Some basic abilities are developed without reference to meaning or context.
- *Training level* (*T*). Specific abilities relating to the context are developed. For example, students develop the ability to write an engineering report and present engineering data.
- *Relating level* (*R*). Here the student will begin to develop a relation between meaning and context. For example, students will develop the skills to alter the way they present data when discussing with farmers, technicians, or other engineering professionals.

Members of the foundation team identified the learning outcomes in terms of these capabilities and then sub divided these into

1. Process learning outcomes:
   a. Guiding principles
   b. Personal professional skills
2. Technical fundamental learning outcomes

Individual academics identified comprehensive lists. Through team discussion the lists were then merged and refined to reduce redundancy. The list was then further refined through a shared document. As before, team members took this back to their schools for comment. The final list of learning outcomes was agreed by all schools.

Below are examples from the above clusters.

1. Process learning outcomes
   a. Guiding principles
      i. Be able to define a system (and articulate the difference between a model of a system and the system itself) to allow explorations of quantitative solutions to engineering problems
      ii. Understand the fundamental conservation laws (mass, material,

momentum, energy), be able to identify the property whose conservation allows solution to the problem, and be able to apply the conservation law to the control volume

b. Personal professional skills
    i. Be able to demonstrate sensitivity and inclusivity towards cultural and gender diversity, especially in relation to Indigenous knowledge, values, and culture

2. Technical fundamental learning outcomes
    a. Build simple digital and analogue circuits to perform various functions
    b. Calculate unknown forces in statically determinate systems
    c. Understand how the structure of the material affects the mechanical, electrical, chemical, optical, magnetic, thermal, and transport properties of material

Stage 3. Identification of Potential Threshold Concepts

At Stage 3, potential threshold concepts among the learning outcomes are identified by interviewing academics and asking about the concepts that are transformative and troublesome. Focus groups are held with students to discover the concepts they find troublesome. This is a large stage. It is preferable to allow at least a few months for this stage.

Stages 3 and 4 use a methodology we have developed to identify and negotiate potential threshold concepts (Male & Baillie, 2011b). Other researchers have used interviews, surveys, laboratory observations, grade distributions, and course feedback (Baillie & Johnson, 2008; Boustedt, 2010; Carstensen & Bernhard, 2008; Holloway, Alpay, & Bull, 2010; Kabo & Baillie, 2009; Scott et al., 2010; Zander et al., 2008). Threshold concepts are potentially experienced by students as transformative and usually troublesome. Therefore, for our methodology people who knew about

students' responses to engineering concepts were selected as participants. These included engineering academics, tutors, and students. In our case we interviewed academics individually or in pairs and tutors in pairs or focus groups. To be less intimidating, we interviewed students in a focus group.

During Stages 3 and 4 in our project, an inventory of potential threshold concepts was developed iteratively. Relevant items already identified were shown to current participants. The inventory included sufficient description of each concept for recognition by an engineer, how the concept was significant for future parts of the course or for engineering practice, and how the concept could be troublesome. If suggested by participants, ways to help students overcome the threshold concepts were also noted.

Academics, tutors, and students were asked to describe concepts they thought could be threshold concepts, and why they believed students found the concepts transformative and troublesome. Academics drew on questions asked by students, and assignment and examination scripts. Tutors were senior undergraduate or postgraduate students. Their responses during the research were based on extensive interaction with first- and second-year students during tutorials, examinations, and assignments they had marked, and their perspectives as recent students. Students based their responses on their own experiences of concepts and the experiences of peers they had helped.

There was overlap among different participants' responses. However, owing to their different perspectives, different groups of participants tended to provide different kinds of responses. Tutors spoke about specific concepts students failed to apply correctly. An example was the difference between acceleration and change in speed in curvilinear motion, and indicating frictional force in the wrong direction. Lecturers spoke about specific slightly higher level concepts such as isolation of the body in a free body diagram in order to draw a force due to friction in the correct direction, or using Thevenin's theorem to model part of a circuit with an equivalent circuit. One senior professor spoke about even higher levels of concepts. He described Thevenin's theorem as an example of modeling and abstraction. For him, evidence that students found abstraction troublesome was observed in students' pleads for numeric examples. The professor's perception is consistent with the expectation based on threshold concept theory, that students mimic solutions while in the liminal space. Given quantities, students can mimic the solution to a problem by inserting numbers into equations. The professor realized that deeper understanding, requiring understanding of relations between concepts, is required to solve problems without nominated quantities.

In this stage we interviewed sixteen academics, seven tutors, and thirteen students. Details of the interview questions and durations are provided in a separate paper (Male & Baillie, 2011a). Most participants gave permission to be recorded. In all interviews and focus groups, the interviewer also took notes.

With care, the process can help to secure support from academics who value the topics they teach. We found that, although academics frequently questioned the risk involved with curriculum design based on a research project, they warmed to the opportunity to discuss concepts in their fields. Unlike other threshold concept researchers, we introduced participants to threshold concept theory at the beginning of all interviews and focus groups. Where possible, we confirmed suggested potential threshold concepts with the participants.

Stage 4. Negotiation of Threshold Concepts

At Stage 4, the potential threshold concepts are negotiated across disciplines and multiple universities. Rather than interviews with individuals or small groups of participants from one discipline, this stage involves workshops. Stage 3 continued during Stage 4, when interviewees became available or

potential gaps requiring further investigation became apparent.

As part of Stage 4 we held a student workshop at which we asked similar questions to those in the interviews (Male, 2011; Male, MacNish, & Baillie, 2012). Thirteen students worked in small groups, first identifying potential threshold concepts individually and then negotiating them within the small groups and eventually with the whole group. We followed this with a student–staff workshop at which fifteen participants in small groups, and then the whole group, negotiated previously identified potential threshold concepts.

We also held four regional workshops in Perth, Adelaide, and Melbourne in Australia, and Auckland, New Zealand. At these workshops academics negotiated potential threshold concepts previously identified and new threshold concepts they identified themselves. Participants discussed whether the concepts were transformative and troublesome and therefore threshold concepts, why, and how concepts were troublesome. Participants' conversations were recorded with permission and groups also summarized their ideas on handouts. Altogether these were attended by ninety-nine participants. Most were engineering academics. Some were mathematics and physics academics and a few were postgraduate students studying engineering education.

For this process it is imperative that groups are well facilitated. It is suggested that each "table" of participants has one allocated facilitator and, where possible, one observer who can feed into strategic discussions. Feedback from our participants indicated that these facilitators were appreciated. The facilitators also increased the data collected.

We provided a Web discussion forum for follow-up discussion and further negotiation. This allowed us to provide discussion summaries to participants.

Our experience is that the inventories of potential threshold concepts can be reduced and refined considerably at this point. For example, workshop participants noticed that many potential threshold concepts relating to mechanics, which had arisen in interviews, depended on the concept of "vectors" (Hesterman et al., 2011). This concept involves troublesome language because vectors can be represented as arrows, using vector notation, and as arrays. This knowledge is also tacit, because teachers frequently move between the different representations without noticing or explaining. However, perhaps owing to its tacit nature, this concept was not recognized as a potential threshold concept until Stage 4. Commonalities between threshold concepts were revealed by the participants during the negotiations, and also afterwards through analysis of transcripts and notes on the handouts (Male, Guzzomi, & Baillie, 2012).

Stage 5. Investigation of Threshold Concepts

After negotiation of the potential thresholds, an inventory of the negotiated thresholds can be investigated further through interviews and analysis. It is necessary to investigate the transformative features of thresholds, and the reasons they are troublesome. Do the thresholds arise from the learning outcomes, how they are taught, or the students' backgrounds? This is the final stage of deciding which potential threshold concepts are indeed threshold concepts, structuring the final list, and ensuring that all technical details in the final list are correct.

It is easier to identify a concept that is troublesome than to uncover the specific barriers to students understanding the threshold concept and to describe the transformative aspect of the concept. We believe that the experience from working with threshold concepts in Stages 3 and 4 helps to prepare researchers for this stage. Workshop participants have noted that the notion of a threshold concept is itself a threshold concept.

Ensuring technical details are correct is critical. In our study, Stages 3 and 4 revealed diverse ways of seeing concepts among participants. The various constructs, especially those allowing a common concept to be understood in different disciplines of engineering, enriched the threshold study. However, it was critical that, for example, we did not confuse different concepts for two ways of seeing the same concept. Similarly, the method relies on comments made in interviews, focus groups, and workshops. People do not always communicate precisely when they speak and tend to use dot bullet points when they complete handouts in workshops. Therefore there is potential for inaccurate or incomplete details to appear. The confirmation with participants is one step to improve validity. This requires rigorous review of the inventory of threshold concepts among experts in relevant fields to ensure the concepts are clear and correct.

Creating a concept map for the program or unit could be useful. As previously noted, threshold concepts are interrelated and have various levels. Others have used concept maps to describe threshold concepts (e.g., Scott & Harlow, 2011). Concept maps can also indicate the relationships between threshold concepts (Quinlan et al., 2013).

Stage 6. Helping Students Overcome Threshold Concepts

The curriculum can be designed to focus on identified threshold concepts. Threshold concepts should not be clustered. Teaching time and learning experiences should be spent on threshold concepts, rather than simpler learning outcomes that can be picked up quickly by students alone. Teachers can alert students to threshold concepts as they are introduced. Assessments and learning experiences can be designed to encourage students to use deep learning approaches. Interactive learning such as problem-based learning, opportunities to teach peers, and hands-on or at least visual experiences are recommended. Students should be helped to understand the relationship between concepts and be provided with opportunities to use concepts in meaningful real-life problems that demonstrate relevance and significance for the students.

Many approaches are available to help students overcome identified thresholds. Recognizing how a threshold concept is troublesome can help to design ways to help students overcome the threshold. Perkins (1999, 2006) described ways to help improve students' understanding of each kind of troublesome knowledge he identified.

Following the use of capability theory, we also adopt *variation theory* for the design of pedagogical experiences for students (Bowden & Marton, 1998). Experiencing variation can be valuable to learning (Bowden & Marton, 1998). Variation theory is based on the notion that students learn through experiencing difference. A child learning to understand the concept "color" needs to be able to differentiate "red" from "blue" and know that this is color. Showing the child a red shoe and a blue book will not help. However, showing the child a red book, a red shoe, and a red coat to explain "red" and then the red book followed by a blue book and a green book to explain red as color, enables the student to "'experience'" the variation. They need to differentiate between object and color. In many engineering examples teachers vary the experience by giving many examples, but do not help students experience this variation. We recommend designing problem sets and projects that allow the students to experience the critical aspects that cause the threshold concepts to be troublesome.

A threshold concept approach can also be used to design assessments to focus students' learning on the threshold concepts rather than learning to mimic solutions to standard problems, and to assess understanding rather than ability to mimic (Land & Meyer, 2010). Open-ended problems are more likely to achieve this than simple problems with all required quantities provided. Oral examinations that can probe understanding are commonly used, for example, at MIT. At the least, examinations should be open book, thereby enabling the teacher to examine the way in which students use the basic facts and formulae, rather than testing simple recall.

Engineering lecturers can create questions that they believe really do test students' understanding of the threshold concept. Examiners need to check their papers or their assignments after they have developed them and ask themselves: "will I be able to know, after marking this paper, that the student has passed through the liminal space?"

To draw students' attention to these troublesome concepts, the examinations and assessments can be completely focused on the threshold capabilities or there could be an additional examination that assesses these critical threshold areas. At Imperial College in the United Kingdom a mastery examination is held for students each year, to test their knowledge of concepts considered to be critical. This is in addition to the standard tests and must be passed at a high level to move through to the next year.

Summary

This chapter has provided an example of how it is possible to use research on student learning in engineering to influence directly the creation of curricula and pedagogical and assessment strategies. It represents possibly the most direct form of such practice that exists in current Higher Education research. This is largely because of its discipline-specific nature but also because of its appeal. It is extremely attractive as an approach, not simply because it is based on serious educational theory and arguably one of the most important developments in student learning research to date, but also because of its potential to intrigue the most hard-bitten academic, normally not interested in pedagogy.

Acknowledgments

The success of this project is dependent on the generously shared insights of the numerous participants: students, tutors, and academics. We acknowledge the curriculum development by the foundation team members, originally: Caroline Baillie, Liang Cheng, Hui Tong Chua, Richard Durham, Adrian Keating, Tyrone Fernando, Marco Ghisalberti, Nick Spadaccini, Thomas Stemler, and Angus Tavner. We gratefully acknowledge the Engineering Thresholds Project Team: James Doherty, Andrew Guzzomi, Matt Hardin, Jasmine Henry, Dianne Hesterman, Kay Horn, Jeremy Leggoe, Cara MacNish, Carolyn Oldham, Gordon Royle, Nathan Scott, Angus Tavner, James Trevelyan; the project Reference Group: Charlotte Taylor, Denise Chalmers, Euan Lindsay, Erik Meyer, Robin King, Roger Hadgraft, Sally Sandover; collaborators: Johnny Fill, Zahira Jaffer, Kathleen Quinlan, Artemis Stamboulis, Chris Trevitt; John Bowden and Elizabeth Godfrey.

Support for this research has been provided from the Australian Learning and Teaching Council Ltd., an initiative of the Australian Government Department of Education, Employment and Workplace Relations. The views expressed in this chapter do not necessarily reflect the views of the Australian Learning and Teaching Council Ltd. Support for the regional workshops was provided by the Australian Council of Engineering Deans. Support for the collaboration with Birmingham and Oxford Universities has been provided by a UWA Research Collaboration Award.

References

Baillie, C., & Johnson, A. (2008). A threshold model for attitudes in first year engineering students. In R. Land, J. H. F. Meyer & J. Smith (Eds.), *Threshold concepts within the disciplines* (pp. 129–141). Rotterdam & Taipei: Sense.

Boustedt, J. (2010). *On the road to software profession* (Doctor of Philosophy). Uppsala University, Sweden.

Bowden, J. A. (2004). Capabilities-driven curriculum design. In C. Baillie & I. Moore (Eds.), *Effective learning & teaching in engineering* (pp. 36–47). Abingdon, Oxon: RoutledgeFalmer.

Bowden, J. A., & Marton, F. (1998). *The university of learning: Beyond quality and competence.* London: Kogan Page.

Cameron, I. (2009). *Engineering science and practice: Alignment and synergies in curriculum innovation.* Strawberry Hills, NSW, Australia: Australian Learning & Teaching Council.

Carstensen, A.-K., & Bernhard, J. (2008). Threshold concepts and keys to the portal of understanding: Some examples from electrical engineering. In R. Land, J. H. F. Meyer & J. Smith (Eds.), *Threshold Concepts within the Disciplines* (pp. 143–154). Rotterdam, The Netherlands: Sense.

Cousin, G. (2006). An introduction to threshold concepts. *Planet, 17,* 4–5.

Cousin, G. (2010). Neither teacher-centred nor student-centred: Threshold concepts and research partnerships. *Journal of Learning Development in Higher Education, 1*(2), 1–9.

Dahlgren, M. A., Hult, H., Dahlgren, L. O., Hard, H., & Johansson, K. (2006). From senior student to novice worker: Learning trajectories in political science, psychology and mechanical engineering. *Studies in Higher Education, 31*(5), 569–586.

Flanagan, M. (2011). Threshold concepts: Undergraduate teaching, postgraduate training and professional development: A short introduction and bibliography. Retrieved from http://www.ee.ucl.ac.uk/~mflanaga/thresholds.html

Fletcher, J. K. (1999). *Disappearing acts: Gender, power and relational practice at work.* Cambridge, MA: MIT Press.

Godfrey, E., & Parker, L. (2010). Mapping the cultural landscape in engineering education. *Journal of Engineering Education, 99*(1), 5–22.

Hacker, S. (1981). The culture of engineering: Woman, workplace and machine. *Women, Technology and Innovation, 4*(3), 341–353.

Hesterman, D. C., Male, S. A., & Baillie, C. A. (2011). *Some potential underlying threshold concepts in engineering dynamics.* Paper presented at the 22nd Annual Conference for the Australasian Association for Engineering Education, Fremantle, Western Australia. Retrieved from http://www.aaee.com.au/conferences/2011/papers/AAEE2011/PDF/AUTHOR/AE110021.PDF

Holloway, M., Alpay, E., & Bull, A. (2010). *A quantitative approach to identifying threshold concepts in engineering education.* Paper presented at the Engineering Education 2010:

Inspiring the Next Generation of Engineers, Aston University, UK.

Ihsen, S. (2005). Special gender studies for engineering? *European Journal of Engineering Education*, 30(4), 487–494.

Johnston, A., King, R., Bradley, A., & O'Kane, M. (2008). *Addressing the supply and quality of engineering graduates for the new century*. Sydney, Australia: The Carrick Institute for Learning and Teaching in Higher Education.

Kabo, J., & Baillie, C. (2009). Seeing through the lens of social justice: A threshold for engineering. *European Journal of Engineering Education*, 34(4), 317–325.

Kirk, A. (2007). Concept inventory central. Retrieved from http://engineering.purdue.edu/SCI/workshop/tools.html

Krathwohl, D. R. (2002). A revision of Bloom's Taxonomy: An overview. *Theory Into Practice*, 41(4), 212–218.

Land, R., & Meyer, J. H. F. (2010). Threshold concepts and troublesome knowledge (5): Dynamics of assessment. In J. H. F. Meyer, R. Land, & C. A. Baillie (Eds.), *Threshold concepts and transformational learning* (pp. 61–79). Rotterdam, The Netherlands: Sense.

Land, R., Meyer, J. H. F., & Smith, J. (Eds.). (2008). *Threshold concepts within the disciplines*. Rotterdam, The Netherlands: Sense.

Male, S. A. (2010). Generic engineering competencies: A review and modelling approach. *Education Research and Perspectives*, 37(1), 25–51.

Male, S. A. (2011). *Today's relevance of feminist theory and gender inclusive engineering curricula to help students overcome thresholds in engineering education*. Paper presented at the 15th International Conference of Women Engineers and Scientists, Adelaide, South Australia.

Male, S. A. (2012). Generic engineering competencies required by engineers graduating in Australia: The Competencies of Engineering Graduates (CEG) Project. In M. Rasul (Ed.), *Developments in engineering education standards: Advanced curriculum innovations* (pp. 41–63). Hershey, PA: IGI Global.

Male, S. A., & Baillie, C. A. (2011a). Engineering threshold concepts. Paper presented at the SEFI Annual Conference, Lisbon. Retrieved from http://www.sefi.be/?page_id=24

Male, S. A., & Baillie, C. A. (2011b). *Threshold concept methodology*. Paper presented at the Research in Engineering Education Symposium, Madrid, Spain.

Male, S. A., Bush, M. B., & Chapman, E. S. (2010a). Perceptions of competency deficiencies in engineering graduates. *Australasian Journal of Engineering Education*, 16(1), 55–68.

Male, S. A., Bush, M. B., & Chapman, E. S. (2010b). *Understanding generic engineering competencies*. Paper presented at the 21st Annual Conference of the Australasian Association for Engineering Education: Past, Present, Future – The 'Keys' to Engineering Education, Research and Practice, Sydney, Australia. Retrieved from http://www.ceg.ecm.uwa.edu.au/__data/page/67580/AaeE_2010_Male_Bush_Chapman_Understanding_generic_engineering_competencies.pdf

Male, S. A., Guzzomi, A. L., & Baillie, C. A. (2012). *Interdisciplinary threshold concepts in engineering*. Paper presented at the Research and Development in Higher Education: Connections in Higher Education, Hobart, Australia. Retrieved from http://www.herdsa.org.au/wp-content/uploads/conference/2012/HERDSA_2012_Male.pdf

Male, S. A., MacNish, C. K., & Baillie, C. A. (2012). *Engaging students in engineering curriculum renewal using threshold concepts*. Paper presented at the CDIO Conference, Brisbane, Australia.

Meyer, J. H. F., & Land, R. (2003). Enhancing teaching-learning environments in undergraduate courses. Occasional Report 4. Retrieved from http://www.etl.tla.ed.ac.uk/docs/ETLreport4.pdf

Meyer, J. H. F., & Land, R. (2005). Threshold concepts and troublesome knowledge (2): Epistemological considerations and a conceptual framework for teaching and learning. *Higher Education*, 49(3), 373–388.

Meyer, J. H. F., & Land, R. (Eds.). (2006). *Overcoming barriers to student understanding: Threshold concepts and troublesome knowledge*. London and New York: Routledge.

Meyer, J. H. F., Land, R., & Baillie, C. (2010). *Threshold concepts and transformational learning* (Vol. 42). Rotterdam, Boston, & Taipei: Sense.

Meyer, J. H. F., Land, R., & Davies, P. (2006). Implications of threshold concepts for course design and evaluation. In J. H. F. Meyer & R. Land (Eds.), *Overcoming barriers to student understanding: Threshold concepts and troublesome knowledge*. In J. H. F. Meyer & R. Land

(Eds.), *Overcoming barriers to student understanding: Threshold concepts and troublesome knowledge* (pp. 195–206). London and New York: Routledge.

Moon, J. (2007). Linking levels, Learning outcomes and assessment criteria – EHEA version. The Centre for Excellence in Media Practice, Bournemouth University, Poole, UK. Retrieved from http://www.cemp.ac.uk/people/jennymoon.php

Perkins, D. (1999). The many faces of constructivism. *Educational Leadership, 57*(3), 6–11.

Perkins, D. (2006). Constructivism and troublesome knowledge. In J. H. F. Meyer & R. Land (Eds.), *Overcoming barriers to student understanding: Threshold concepts and troublesome knowledge* (pp. 33–47). London and New York: Routledge.

Petrucci, C. J., & Quinlan, K. M. (2007). Bridging the research-practice gap: Concept mapping as a mixed methods strategy in practice-based research and evaluation. *Journal of Social Services Research, 34*(2), 25–42.

Quinlan, K. M., Male, S. A., Baillie, C. A., Stamboulis, A., Fill, J., & Jaffer, Z. (2013). Methodological challenges in researching threshold concepts: A comparative analysis of three projects. *Higher Education, Online First*. doi: 10.1007/s10734-013-9623-y.

Schön, D. (1983). *The reflective practitioner*. New York, NY: Harper Collins.

Scott, J., & Harlow, A. (2011). *Identification of threshold concepts involved in early electronics: Some methods and results*. Paper presented at the 22nd Annual Conference of the Australasian Association for Engineering Education Fremantle, Western Australia.

Scott, J., Harlow, A., Peter, M., & Cowie, B. (2010). *Threshold concepts and introductory electronics*. Paper presented at the 21st Annual Conference of the Australasian Association for Engineering Education, Sydney, Australia.

Spinks, N., Silburn, N., & Birchall, D. (2006, March). Educating engineers for the 21st century: The industry view. Retrieved from http://www.raeng.org.uk/news/releases/henley/pdf/henley_report.pdf

Tonso, K. L. (2007). *On the outskirts of engineering*. Rotterdam, The Netherlands: Sense.

Trevelyan, J. P. (2010). Reconstructing engineering from practice. *Engineering Studies, 2*(3), 21.

World Chemical Engineering Council. (2004, September). How does chemical engineering education meet the requirements of employment?, Retrieved from http://www.chemengworld.org/chemengworld_media/Downloads/survey.pdf

Zander, C., Boustedt, J., Eckerdal, A., McCartney, R., Mostrom, J. E., Ratcliffe, M., & Sanders, K. (2008). Threshold concepts in computer science: A multi-national empirical investigation. In R. Land, J. H. F. Meyer, & J. Smith (Eds.), *Threshold concepts within the disciplines* (pp. 105–18). Rotterdam & Tapei: Sense.

CHAPTER 21

Engineering Instructional Development

Programs, Best Practices, and Recommendations

Richard M. Felder, Rebecca Brent, and Michael J. Prince

Introduction

University Faculties: Unprepared Practitioners of a Highly Skilled Profession

University faculty* members face a broad range of challenges over the course of their careers. Laursen and Rocque (2009) identify career stages at which they need to acquire different skill sets to meet those challenges: *early career* (teaching, advising, research, negotiation, and time management skills); *mid-career* (leadership and administration, collaboration, and outreach skills), and *later career* (the skill to identify and evaluate possible changes in career direction).

This article is reprinted with permission (*Journal of Engineering Education*, 2011, 100(1), pp. 89–122). A number of common terms, such as *university* and *faculty* and *faculty development* and *instructional development*, mean different things in different countries. We follow common U.S. usage in this chapter. A glossary at the end of the chapter shows alternative definitions of these terms, and an asterisk is used after each term the first time it appears to call the reader's attention to its presence in the glossary.

For which of those challenges are new and experienced faculty members systematically prepared? Throughout most of the history of higher education, the answer has been "none." In the past half-century, *faculty development** programs have become available on many campuses, but unfortunately many faculty members are still expected to learn how to do everything their job requires by trial and error. Although there is much to be said for experiential learning, it is not terribly efficient. Studies by Boice (2000) show that for 95% of new faculty members it takes four to five years of trial and error to become fully productive in research and effective in teaching – and in teaching, the ones making the errors (the instructors) are not the ones paying for them (their students). Boice also found, however, that the other 5% – the "quick starters" – are effective in their first one to two years, and the actions that distinguish quick starters from their colleagues can be identified and taught. That is to say, a good faculty development program can cut several years off the normal faculty learning curve.

Given that finding, why is it not routine for engineering faculty members to participate in faculty development programs and for their administrators to encourage them to do it? The answer depends on which developmental areas are being discussed. Possible areas for faculty development include teaching, disciplinary research, educational research, learning the institutional culture, administration at the department and *college** levels, and changing activities and priorities at the mid- and late-career levels. For all of these areas *but* teaching, most faculty members do not participate in development programs because programs in those areas do not exist at their institutions.

On the other hand, faculty development programs that focus on improving teaching and learning (*instructional development** programs) can be found at many universities, but participation of faculty members in them is often low except in countries where it is mandatory (Groccia, 2010, p. 13), and many who do attend discount the relevance of the programs to them – sometimes unfairly, sometimes not. There are several reasons for this state of affairs. One is that many faculty members whose students perform inadequately do not acknowledge that the quality of their teaching may have anything to do with it. If their students get mediocre *grades** and/or give them low ratings, they argue that the students are incompetent or unmotivated, or that as instructors they maintain rigorous standards and high ratings only go to easy graders. Also, many instructors are unaware that alternatives exist to the traditional lecture-based approach with which they were taught. As long they believe they are teaching appropriately and poor student performance and low ratings only reflect deficiencies in the students, they have no incentive to get involved in instructional development.

Exacerbating the problem in engineering is that instructional development on most campuses is commonly provided by social scientists (generally education and psychology faculty members) to campus-wide audiences. In the absence of discipline-specific examples it is easy for engineers to dismiss program content as irrelevant to their

*courses**, subjects, students, and problems. Programs given specifically to engineering faculty members by teaching experts with engineering backgrounds are more likely to attract and influence larger audiences, but they require an investment that engineering administrators and campus centers may be unwilling or unable to make. In short, there is generally neither a meaningful incentive for engineering faculty members to participate in instructional development nor a meaningful reward for any improvements in teaching that may result from their participation.

The Case for Faculty Development in Engineering

Providing faculty development in engineering has always been a good idea, but the need for it has taken on new urgency in the past two decades (Adams & Felder, 2008). Here are some of the driving forces:

- *Outcomes-based program accreditation.* A large and growing number of engineering schools have adopted outcomes-based accreditation, including schools in the United States (ABET, http://www.abet.org) and countries that are signatories of the Bologna Process (http://www.ond.vlaanderen.be/hogeronderwijs/bologna) and the Washington Accord (http://www.washingtonaccord.org). Equipping engineering students with the skills for specified outcomes such as effective communication and teamwork requires teaching and assessment methods not traditionally found in engineering education and unfamiliar to most engineering educators (Felder & Brent, 2003).
- *Anticipated shortfalls in engineering graduation rates.* In many countries in the world, graduation rates in engineering fall well below anticipated demands for new engineers, for reasons having to do with both recruitment and retention (Jain, Shanahan, & Roe, 2009). A common myth among engineering faculties is that engineering dropouts are the weakest students, but research has shown that the

academic profile of those leaving engineering in the United States is indistinguishable from that of those staying, and many well-qualified students leave because they are dissatisfied with the quality of the teaching they have experienced (ASEE, 2009; Seymour & Hewitt, 1997). High-quality teaching is imperative to retain qualified engineering students in the numbers required to meet current and future needs for engineers.

- *Changing engineering student demographics.* For most of the twentieth century, students entering engineering school tended to come from the top echelon of their country's secondary school graduates. They were capable of learning even if they were taught using ineffective instructional methods. More recently, engineering schools have been admitting a much broader cross section of the student population, many of whom have the potential to be excellent engineers but who have to overcome deficiencies in their pre-college preparation. This is a positive development, but it means that ineffective teaching is no longer excusable.

- *Changing engineering student attributes.* Today's college students – the much written-about "Millennials" – tend to have certain characteristics that pose challenges to their instructors (Wilson & Gerber, 2008). They are idealistic and self-confident, and they gravitate to curricula and careers in which what they do can make a difference to society. They are used to getting information in short visual clips, and so they have less patience for lectures and textbooks than their counterparts in previous generations did. It has always been a mistake for faculty members to assume that the instructional methods that worked for them should work just as well for their students, but that assumption may be particularly bad when applied to the Millennials.

- *Changes in engineering practice in developed countries.* Many professional tasks for which the traditional engineering curriculum equips students are increasingly

performed by technicians on computers or by workers in developing countries where the jobs have migrated. Different skills, such as innovative, multidisciplinary, and entrepreneurial thinking, as well as the ability to work with people from different countries and cultures who speak different languages, will be needed by engineering graduates in the decades to come (Friedman, 2006; National Academy of Engineering, 2004). Most engineering faculty members were never trained in those skills, but they will nevertheless need to know how to help their students develop them.

- *Advances in instructional technology.* Teaching can be enhanced significantly by the full power of modern instructional technology, using such applications as interactive multimedia-based tutorials, system simulations, computer-aided design tools, course management software, personal response systems ("clickers"), and online communication tools. For engineering schools to continue to attract qualified applicants in an increasingly competitive market, faculty members must learn how to use the tools effectively, and their knowledge will have to be updated regularly.

- *Advances in cognitive science.* Thanks to modern cognitive science and a growing body of educational research (see next bullet point), we know a lot about how people learn, the instructional conditions that facilitate learning, and the conditions that hinder it (Adams, 2011; Ambrose, Bridges, DiPietro, Lovett, & Norman, 2010; Baars & Gage, 2007; Bransford, Brown, & Cocking, 2000; Fink, 2003; Heywood, 2005; Litzinger, Hadgraft, Lattuca, & Newstetter, 2011; Olds & Johri, 2011; Svinicki & McKeachie, 2009; Wankat, 2002; Weimer, 2010). Summarizing the research and demonstrating its applications to teaching is a vital function of campus faculty development programs.

- *The SOTL movement.* In the past two decades in the United States there have been growing incentives for engineering faculty members to engage in the *scholarship of teaching and learning* (SOTL) (Borrego & Bernhard, 2011; Case & Light, 2011; Jesiek, Newswander, & Borrego, 2009; Streveler & Smith, 2006). Not least of these incentives has been the National Science Foundation's major funding for educational research and its requirement that proposals for CAREER Awards and conventional disciplinary research grants must include educational components (NSF, http://www.nsf.gov). Educational research methods are in many ways different from the research methods engineers are accustomed to, and engineering faculty members who receive training in those methods will be better prepared to conduct classroom research and to secure research funding (Borrego & Bernhard, 2011).

For all these reasons, engineering schools in the future will feel mounting pressure to provide faculty development. Doing so will help them make their faculty more effective in teaching and more competitive for research grants; equip their students with the skills they will need to succeed in an increasingly globalized world; and enable their instructional programs to compete successfully with other engineering schools for growing numbers of students seeking degrees through distance education.

Focus, Intended Audience, and Structure

This article reviews practices in the design, delivery, and evaluation of engineering instructional development programs (i.e., programs designed to equip engineering instructors to improve their teaching and their students' learning) and offers recommendations for making such programs effective. It touches only briefly on other aspects of engineering faculty development such as improving faculty members' research and administrative skills and promoting their personal development, and improving courses, curricula, and organizational policies.

The emphasis of the article is on practical aspects of engineering instructional development rather than comprehensive faculty development or instructional development theory. The intended audience is faculty developers and teaching leaders who wish to address the instructional development needs of engineering faculty effectively, and engineering administrators who wish to understand those needs and build and nurture programs that address them. Readers interested in a conceptual framework for comprehensive faculty development program design and evaluation should consult Taylor and Colet (2010), who formulated such a model based on a comparative analysis of programs in Canada and four European countries. Scholars seeking information on the philosophy, theories, and political and social contexts of faculty development can find such information in the edited volumes of Eggins and Macdonald (2003) and Kahn and Baume (2003, 2004).

The structure of the chapter is as follows:

- Options are presented for the *content* of instructional development programs that target new, experienced, and future faculty members, followed by discussion of whether programs should focus on pedagogical strategies, learning theories, or human development issues.

- Possible program *structures* are then defined – workshops, courses, and seminar series; consulting, mentoring, and partnering arrangements' learning communities; and teaching certification programs for faculty members and *graduate students**. Pros and cons of campus-wide versus discipline-specific programs, external versus local program facilitators, and mandatory versus voluntary participation are outlined.

- Strategies for assessing and evaluating instructional development programs are surveyed.

- A research-based engineering instructional program design framework is formulated, and recommendations based on the framework are offered for developing and strengthening programs. Finally, questions for further study are suggested.

Table 21.1. Possible Content of Engineering Instructional Development Programs

Responsibility	Topics for Instructional Development Programs
Course design	Success strategies for new faculty members (Boice, 2000)[a]
	Cognitive science and modern theories of learning and teaching
	Theories of personal development[b]
	Outcomes-based education and ABET
	Writing and using learning objectives
	Taxonomies of objectives
	Constructing syllabi and defining course policies
	Getting a course off to a good start
	Making new course preparations manageable
Course instruction	Motivating and engaging students
	Effective lecturing and active learning
	Using instructional technology effectively
	Cooperative (team-based) learning[b]
	Inductive methods (inquiry, project-based and problem-based[b] learning)
Assessing learning	Basic concepts of assessment and evaluation
	Assessing quantitative skills, conceptual understanding, and professional skills (written and oral communication, critical and creative thinking, ethical awareness, etc.)
	Formative assessment: classroom assessment techniques
	Marking assignments and examinations and grading courses
Troubleshooting problems	Classroom management
	Academic misconduct
	Dealing with students' personal problems
Student diversity	Learning and teaching styles
	Approaches to learning (deep, surface, strategic)
	Levels of intellectual development
	Gender, racial, & ethnic diversity
Distance education	Techniques and challenges
	Engaging students interactively online

[a] Primarily for new faculty members.
[b] Primarily for experienced faculty members.

Instructional Development Program Content

In this section we review some of the topics that might profitably be covered in engineering instructional development programs. Just as with teaching at any level, the choice of content should be preceded by the definition of program learning objectives, and the content should then be selected to address the objectives.

Programs for Current Faculty Members

Table 21.1 lists the most common teaching responsibilities of faculty members and possible instructional development program content appropriate for each responsibility. Instructional development programs normally cover subsets of the topics shown in Table 21.1. The material may be presented in a concentrated workshop that lasts for one or more days, a series of seminars over the course of a semester, a summer institute, or a course for which academic credit is given.

Two topics in Table 21.1 – cooperative learning and problem-based learning – are marked as particularly suitable for more experienced faculty members. Those methods commonly evoke strong student resistance and interpersonal conflicts among students; before undertaking them, new faculty members are advised to gain experience and skill in more basic teaching methods.

This does not mean that new faculty members should never attempt cooperative or problem-based learning, but if they decide to use those methods they should do so with open eyes and good preliminary training.

In some countries such as Denmark, Norway, Sweden, and the United Kingdom, and at many individual institutions in other countries, new faculty members are required to participate in extensive instructional development programs to qualify for *tenure** and promotion (Schaefer & Utschig, 2008; Utschig & Schaefer, 2008). A more common practice in most parts of the world is to offer an optional new faculty orientation workshop that lasts between a day and a week. Brent, Felder, and Rajala (2006) describe an orientation workshop for science, technology, engineering, and mathematics (STEM) faculty members that covers most of the topics listed in Table 21.1 as well as topics related to research and balancing professional responsibilities.

If experienced program participants have previously been through the basic material in a workshop, brief reviews of that material might be provided in advanced programs and more emphasis can be placed on strategies such as cooperative and problem-based learning, modern cognitive science and learning theories, and human development. Romano, Hoesing, O'Donovan, and Weinsheimer (2004) describe an instructional development program for tenured faculty members. The program introduced participants to effective pedagogical strategies; supported them as they applied those strategies; gave them an opportunity to share ideas and experiences with peers; and provided a forum for them to discuss events that had an impact on their personal and professional lives. Another program designed specifically to address the developmental needs of mid-career faculty members is described by Baldwin, Dezure, Shaw, and Moretto (2008).

Programs for Future Faculty Members

Before they seek their first academic position, postdoctoral fellows and graduate stu-

Table 21.2. Possible Content of Future Faculty Instructional Development Programs

Workshops and seminars
- How students learn
- Learning styles and teaching styles
- Effective lecturing and active learning
- Grading tests, assignments, and project reports
- Facilitating process and computer laboratories[a]
- Tutoring in office hours
- Detecting and dealing with academic misconduct
- Dealing with student problems and problem students
- Designing and planning new courses
- Designing effective tests
- Using instructional technology effectively
- Applying and interviewing for a faculty position
- Success strategies for new faculty members

Preparing the Professoriate Programs
- Workshops and seminars
- Readings and discussions
- Class observations
- Co-teaching with a mentor
- Supervised teaching
- Classroom research
- Teaching portfolio development
- Certification

dents can benefit greatly from training on how to get a good faculty position and what they will need to know in their first one to two years. The training may culminate with some form of certification. Table 21.2 lists possible teaching-related elements of such programs.

Specific models and recommendations for future faculty programs are given by Pruitt-Logan, Gaff, and Jentoft (2002), Colbeck, O'Meara, and Austin (2008), and in other resources developed by the Preparing Future Faculty program (PFF, http://www.preparing-faculty.org/) and the Center for Integration of Research on Teaching and Learning (CIRTL, http://www.cirtl.net/). Brent and Felder (2008) describe a multifaceted graduate student training program that includes many of the elements shown in Table 21.2.

Critical Choices Regarding Program Content

Two important questions exist regarding the content of instructional development programs. First, should the main focus of the programs be pedagogical techniques and best practices, or fundamentals of how learning occurs (cognitive science), or explorations of the participants' personal experiences and their intellectual, emotional, and perhaps spiritual growth? Second, if a development program includes elements in all three categories, what is the optimal order of presenting them: cognitive and human development theories first followed by techniques and best practices (deductive presentation), or techniques and best practices first followed by interpretation in the light of theory (inductive presentation)?

PEDAGOGICAL PRACTICE VERSUS LEARNING THEORY VERSUS HUMAN DEVELOPMENT

Many faculty development experts caution against an overemphasis on teaching strategies. Susan Ambrose (2009, personal communication), a prominent faculty development authority who has worked extensively with engineers, observes, "Too many programs dispense tips and strategies, as opposed to educating faculty members about how learning works . . . and we do a great disservice when we do this because tips and strategies often do not transfer across contexts. The mission of all of these programs should be to try and bridge the gap between what we know about learning and how we design and teach courses." Ambrose and her colleagues provide a framework for that mission in the last chapter of their recent book *How Learning Works* (Ambrose et al., 2010).

In her preface to *Inspiring College Teaching*, Maryellen Weimer (2010, p. xii) similarly notes: "Although the book recognizes the value of techniques, it aspires to move faculty beyond them to conceptions of teach-ing that are more intriguing and intellectu-ally rich," and the book provides a powerful model of a developmental/experiential approach to faculty development. In *The Courage to Teach*, Parker Palmer (1998) refers to "the boredom many of us feel when teaching is approached as a question of 'how to do it'" (p. 11) and offers the premise that "Good teaching cannot be reduced to technique; good teaching comes from the identity and integrity of the teacher" (p. 10).

On the other hand, most engineers and scientists come to instructional development programs much more interested in finding out what they should do in their classes starting next Monday than in cognitive theories, educational research, and personal explorations and self-revelation. If they suspect that a program is likely to focus on any of the latter topics, many will choose not to attend it or not to sit through it if they do attend, and they will be unlikely to accept the ideas they hear and transfer them to their teaching practice. Svinicki and McKeachie (2011, p. 3) observe, "When you are just starting, discussions of philosophy of education and theories of learning and teaching can be helpful, but they are probably not as important as learning through techniques and simple skills to get through the first few weeks without great stress and with some satisfaction. Once some comfort has been achieved, you can think more deeply about the larger issues."

As is usually the case with questions of theory vs. practice, the correct answer is, both. As Kant (or Marx or Lenin or Einstein, depending on the attributor) said, "Theory without practice is sterile; practice without theory is blind." Both are needed, in science and in instructional development. Engineering educators are trained scientists, accustomed to critical thinking in their professional lives. If they are given a program that contains nothing but strategies and tips with no rigorous theoretical or empirical research support, few will see fit to change the way they teach. At the same time, if they get only general theories and educational research results or insights derived from reflection and sharing of personal experiences, and the translation of all of that into the context of teaching fluid dynamics or microprocessor design is left for them to work out, the desired changes are equally unlikely to occur. The right balance among practical teaching strategies, learning theories and research outcomes, and self-reflection depends on the needs, interests, and experience levels of the participants and the expertise and philosophy of the facilitators. The art of instructional development lies in finding that balance.

DEDUCTIVE VERSUS INDUCTIVE PRESENTATION

Pedagogical strategies, learning theories, and human developmental issues are each important components of instructional design, and a complete instructional development program should involve some balance of all three. The question then becomes, in which order should content in each area be presented? One choice is a *deductive* approach, starting with general theories of cognition and learning and human development and proceeding to apply the theories to deduce strategies for course and lesson planning, delivery, and assessment. The opposite is an *inductive* approach, which starts with specific observations and challenges, guides the participants to explanations and strategies, and then generalizes the results and provides the theoretical and research backing for the generalizations.

An argument for an inductive approach to instructional development is suggested by Fink, Ambrose, and Wheeler (2005), who observe that faculty members seeking to become better teachers tend to work through three increasingly sophisticated stages of development. Initially they try to learn about the nuts and bolts of teaching – making their lectures more interesting, writing better exams, and using technology effectively. They find that their teaching improves up to a point when they start using the strategies they learn, but they also recognize that there are still gaps between where their students are and where they want them to be. In the second stage,

they are ready to learn what cognitive science teaches about how people learn and the conditions that favor development of high-level thinking and problem-solving skills. Their students' learning often improves dramatically when they apply their new understanding to their teaching practices. They may be content to stop at that point, but some of them may move to the third stage and adopt the even deeper goal of helping to prepare their students to reach their full potential for achievement and personal fulfillment in life. To do that, they need to learn about students' intellectual and personal development and how they as instructors can serve as mentors and role models for the developmental process. They are then ready for instructional development of the type that Palmer (1998) advocates.

Once the content of an instructional development program has been selected, the question then arises of how the program might be structured.

Faculty Development Program Structures

Alternative Program Formats

Faculty development program elements generally fall into four categories: (1) workshops; (2) seminars and discussion sessions; (3) consulting, mentoring, and partnering arrangements; and (4) learning communities. Each of these formats has variations, advantages, and disadvantages. A campus program may involve any combination of them.

WORKSHOPS

The most common structure for instructional development is workshops that last from several hours to a week. The workshops may be organized by campus centers for teaching and learning or by university or individual college administrators. They may address multidisciplinary audiences or focus on individual disciplines, such as engineering, or groups of related disciplines such as STEM. Besides being held on individual campuses, workshops directed at engineering faculty members are often given at professional society conferences. In the United States workshops have been offered for many years at the annual conference of the American Society for Engineering Education (ASEE) and the Frontiers in Education conference sponsored by the ASEE and the Institute of Electrical and Electronics Engineers (IEEE), and offerings at conferences in other countries are becoming increasingly common.

The principal advantage of a single workshop is that if it is well promoted, many participants are likely to attend, especially if previous offerings or the facilitators have good reputations. A disadvantage is that one-time events are relatively unlikely to have a lasting impact on participants. The attendees may leave the workshop with good intentions and some may try implementing new practices, but in the absence of subsequent reinforcement their tendency is often to revert to familiar but less effective methods.

On the other hand, pedagogical experts sometimes go overboard in criticizing single workshops. Relatively short workshops (two days or less) attract faculty members who would be unwilling to commit to longer one-time programs or a series of events spread out over a semester or academic year. Those individuals are made aware that there are alternatives to the traditional teaching approach, which is likely to be the only way of teaching they know. When they try those alternatives and get good results, some will keep using them, and they will be much more likely to participate in future programs that require more of a time commitment. Also, in every workshop there are likely to be a few individuals who have been dissatisfied with their teaching and are ready to hear about alternatives. Single workshops have had a dramatic impact on the teaching of many of those instructors, some of whom have even been motivated to make faculty development a major focus of their subsequent academic careers.

COURSES AND SEMINAR SERIES

Another common structure is a series of sessions that take place over the course of a semester or academic year. The series may be a course open to graduate students and faculty members, regularly scheduled meetings of a faculty learning community with a fairly stable membership, or sessions on a variety of topics open to anyone interested. A session may involve a formal presentation by a speaker, a facilitated or unfacilitated discussion of a preassigned reading or topic, reporting and discussion of educational research projects or curriculum revisions, or open discussion.

A series of sessions attended by the same group of people can have a greater impact than a one-time seminar or workshop. Topics can be covered in greater depth in a series, and attendees can try suggested techniques following a session and report on the results and get feedback and guidance at the next session. The drawback of a series is that it can be difficult to persuade faculty members to commit to regular attendance, and dramatic drops in attendance after the first few sessions are common. The chances that a program will be successful are increased if incentives for participation are provided and if several colleagues from the same department participate.

CONSULTING, MENTORING, AND PARTNERING

In a third program format, a faculty member works with another individual on improving his or her teaching.

- *Individual Consulting.* The faculty member may have one or more sessions with a resident expert, such as a staff member in the campus teaching center. The consultant may observe or videotape the consultee's teaching, conduct student focus groups or individual interviews with students, review student evaluations, or simply talk with the consultee and suggest improvements in his or her teaching. Faculty members who have attended workshops or courses are most likely to avail themselves of consulting services

(Kolmos, Rump, Ingemarsson, Laloux, & Vinther, 2001). Lee (2000) discusses the difficulties that non-STEM consultants may encounter when dealing with STEM faculty members and suggests ways to overcome those difficulties, and Finelli et al. (2008) outline ways instructional consultations can be used to enhance the teaching performance of engineering faculty members.

- *Mentoring* (Felder, 1993). The faculty member either finds or is assigned (usually by the department head) an experienced colleague to function as a teaching mentor. Formal assignment generally works better than self-selection, for several reasons. Introversion keeps many new faculty members from asking more experienced colleagues to serve as mentors; it takes time for new faculty members to get to know their colleagues well enough to make a good choice; and not everyone who volunteers to be a mentor is qualified to be one. The mentor and mentee work together for a semester or academic year, perhaps co-teaching a course or periodically exchanging classroom visits and debriefing their observations afterwards. The key to a successful mentorship is meeting regularly: many so-called mentorships involve one or two initial meetings and invitations to the mentee to feel free to drop in and ask questions if problems arise, which mentees rarely do. Bland, Taylor, Shollen, Weber-Main, and Mulcahey (2009) provide many excellent ideas for initiating and maintaining successful mentoring programs, and Bullard and Felder (2003) offer an example of a successful engineering teaching mentorship.

- *Partnering.* The two previous arrangements involve expert–novice relationships. In the third one, peers in the same department or in different departments or disciplines support one another in an effort to improve their teaching. The arrangement may consist of two colleagues informally agreeing to periodic exchanges of classroom observations and

debriefings over the course of a semester (Sorcinelli & Yun, 2007), or it may involve three or four faculty members exchanging visits and subsequently engaging in discussion sessions. The latter arrangement is sometimes referred to as *teaching squares* (Wessely, 2002).

LEARNING COMMUNITIES

Another instructional development structure involves forming a community of faculty members who organize themselves around individual or communal activities intended to improve their teaching and to offer support and guidance to one another (Cox, 2004). The activities may include reading articles and books on education, viewing and discussing videos and webinars, observing one another's classes, implementing selected teaching methods in their own classes, and conducting informal classroom research or formal (and possibly funded) educational research. There may be a group facilitator who gives presentations on topics of interest to the group but whose primary function is to provide encouragement and consulting assistance when it is needed. A learning community may arise simply when faculty members from the same department attend an instructional development workshop together, an occurrence that increases the subsequent likelihood of change in the department's teaching practices (Kolmos et al., 2001).

Good examples of learning communities built around the scholarship of teaching and learning are found at the University of Lund in Sweden (Roxå, Olsson, & Mårtensson, 2008) and at Iowa State University in the United States (Jungst, Licklider, & Wiersema, 2003). The second of these programs is Project LEA/RN (Learning Enhancement Action/Resource Network), whose elements include seminars, discussions of readings, paired classroom observations, and classroom research projects. Project LEA/RN has had a particularly broad impact since its inception in 1991, reaching 1,200 educators at five universities and ten community colleges. Courter,

Freitag, and McEniry (2004) describe the pilot study of a distance education-based learning community organized under the auspices of CIRTL (http://www.cirtl.net/). In that approach, engineering and science faculty members participate in weekly online instructional development sessions and report on individual classroom research projects.

Teaching Certification Programs

When explanations are sought for the low levels of participation in instructional development programs at research universities, the first one offered is invariably the low status of teaching in the faculty recognition and reward system. Fairweather (2005), for example, found that even at the most teaching-oriented colleges and universities, the more time faculty members spent in the classroom, the lower their salaries. One way administrators can convey the importance of teaching to the university and their high expectations for their faculties in this area is through a certification program that recognizes and rewards faculty members who complete a specified course of training. Certification attests to the individual's qualification to teach, either at a basic or advanced level. Requiring certification as a prerequisite to teaching is universally required in pre-college education but not in higher education. The unstated assumption in the latter is that anyone who has studied a subject in college must be qualified to teach it. Anyone who has attended college knows how wrong that assumption can sometimes be.

In reviews of certification programs around the world, Schaefer and Utschig (2008) and Utschig and Schaefer (2008) conclude that to be successful, programs should (1) be supported by a national respected society or academy, (2) include qualifying criteria or standards at several levels of expertise, and (3) accommodate flexibility in implementation across various university cultures. They describe a national certification program in the United Kingdom administered by the Higher Education Academy (HEA, http://www.heacademy .ac.uk/) that meets all three criteria. As part of its accreditation function, the HEA supports Postgraduate Certificate in Higher Education Programs at institutions of higher education in England, Scotland, Wales, and Northern Ireland. At some of those institutions, completion of a certification program is a condition for being awarded tenure. One such program conducted at Imperial College in London (Imperial College, http://www8.imperial. ac.uk/content/dav/ad/workspaces/edudev/ CASLAT_Brochure_2010v3.pdf) includes face-to-face and online sessions on many aspects of teaching and assessment, research supervision, educational theories, technology, educational research, and teaching in the disciplines, plus two summative peer evaluations and preparation of a teaching portfolio.

Certification programs are in place in many countries. Kolmos, Vinther, Andersson, Malmi, and Fuglem (2004) note that the primary vehicle for instructional development in Denmark is compulsory courses for new faculty members that involve a total of 175 to 200 hours and culminate with the preparation of a teaching portfolio. Completion of the program is a condition for promotion. The Ministry of Higher Education in Malaysia is moving toward making teaching certification compulsory for academic staff, with new faculty members at several universities now being required to complete a series of modules to obtain the Certificate of Teaching in Higher Education (Yusof, 2009, personal communication). An international certification program specific to engineering education called ING-PAED (IGP, http://www.igip.org/pages/ aboutigip/ing-paed.html) is administered in seventy-two countries by the International Society of Engineering Education in Austria. It consists of a series of core modules (engineering pedagogy and laboratory methodology), theory modules (psychology, sociology, ethics, and intercultural competencies), and practice modules (oral communications, technical writing, project work, and instructional technology), and is open to

Table 21.3. Comparison of Engineering-Specific and Campus-Wide Faculty Development

	Engineering- or STEM-Specific	*Campus-Wide*
Suitable topics	• Single-discipline teaching and research • School-level services • department and school culture (including tenure and promotion requirements)	• Multidisciplinary teaching and research • Campus policies, services, and facilities (including computer resources), and safety • Employee benefits
Facilitators	• Top teachers and researchers in the discipline(s) • Deans, associate deans, and department heads	• Campus faculty development staff • University-level administrators
Pros	• Presenters understand participants' needs, interests, and problems related to teaching, research, and service • Presenters have credibility with participants	• Efficient and economical for discipline-independent topics • Get cross-fertilization of ideas, build community across disciplines
Cons	• May fall below critical mass of participants in some years • Possibly low cost-effectiveness	• Absence of engineering-relevant examples
Conditions for success	• Articulate facilitators with appropriate content knowledge and experience • Active engagement of participants • Practical, just-in-time content, with minimal emphasis on supporting research and theories (cite them but don't dwell on them) • Opportunities for interactions among participants • Good facilities, adequate staff support	

instructors with at least one year of teaching experience. Those who complete it are certified as "International Engineering Educators."

Critical Choices Regarding Program Structure

CAMPUS-WIDE VERSUS DISCIPLINE-SPECIFIC PROGRAMS

Roughly 70% of all U.S. research or doctorate-granting institutions have instructional development programs, most administered by campus centers for teaching and learning, and 40% of masters institutions have them (Kuhlenschmidt, 2009). The program elements generally include workshops for faculty members from all disciplines facilitated by campus administrators,

educators, or psychologists who address teaching-related issues, and computer systems administrators and programmers who deliver technology-related workshops. An alternative instructional development model provides workshops and seminars to faculty members in individual disciplines or groups of closely related disciplines, such as STEM. Table 21.3 summarizes the strengths of discipline-specific and campus-wide instructional development programs and suggests conditions for successful program implementation.

Baillie (2007), Healy and Jenkins (2003), and McAlpine et al. (2005) have noted the importance of a balance of discipline-based versus university-wide initiatives. Their arguments include that faculty members' primary allegiance is to their discipline (Jenkins, 1996); disciplinary thinking influences many teaching and learning tasks (Saroyan et al., 2004); and university-wide initiatives generally do not generally lead to faculty members applying what they have learned to their own teaching environment (Boud, 1999). Sometimes programs targeted specifically to engineering or STEM disciplines are organized by teaching and learning centers based in the institution's engineering school, such as the Leonhard Center for the Enhancement of Engineering Education at Penn State University, the EnVision program at Imperial College-London, and the Engineering Learning Unit at the University of Melbourne. In other cases, one or more pedagogical experts with STEM faculty positions give the programs; and in still others the programs are organized by a central campus administrative unit and presented by teams of STEM educators and pedagogical experts in other fields.

The question of whether instructional development programs should be campus-wide or discipline-specific has the same answer as the question about whether programs should focus on theory or practice – namely, both are needed. In a seminal paper, Shulman (1986) observed that a false dichotomy had arisen in education between *content knowledge* (the facts, methods, and theories of the discipline and subject being taught) and *pedagogical knowledge* (theories and methods of teaching and learning independent of the discipline and subject), and introduced the vitally important bridge category of *pedagogical content knowledge* (representations of subject topics that maximize their comprehensibility to students; understanding which topics are easy and difficult and what makes them that way; and knowing the misconceptions students bring into class with them and how to correct them). A good teacher should have a high level of mastery in all three of these domains. With rare exceptions, individual faculty members do not possess such broad knowledge, which suggests the wisdom of instructional development partnerships between content experts and pedagogical experts.

In engineering, the content experts usually reside in engineering departments and the pedagogical experts are more likely to be found in campus centers or on education or psychology faculties. Ideally, teams of facilitators from each domain can be formed to present in discipline-specific programs. If separate campus-wide and discipline-specific workshops are given, the key is to make sure that they are each done as effectively as they can be and that they function synergistically (Hicks, 1999).

EXTERNAL VERSUS LOCAL FACILITATORS

Bringing in an outside expert to present a workshop offers several attractive possibilities. External presenters enjoy credibility that derives simply from the fact that they come from somewhere else and are being brought in at the institution's expense, and they can be sources of fresh ideas and role models to campus faculty developers. They can also convey the message that good teaching is a goal at many universities and not just a concern of the local faculty developer. If the presenter has a strong reputation and the workshop is well promoted, faculty members who are not normally inclined to go to teaching programs might be induced to attend, and if the program is skillfully conducted, it can stimulate them to try some new strategies.

On the other hand, external experts cannot provide sustained support for improving teaching because they are normally limited to one or two visits to a particular campus per year. Local experts can give courses on teaching, facilitate regular seminars and discussion sessions, and observe faculty members teaching and provide subsequent feedback. They can also organize and conduct follow-up sessions after presentations by external experts to help faculty members apply what they learned. Such sessions can play a vital role in ensuring that the external presentations have the greatest possible impact.

For longer teaching workshops (a day and upwards), there is much to recommend team facilitation, with at least one facilitator from engineering or another STEM field and at least one with general pedagogical expertise. A good way to initiate such a program is to bring in an outside expert to give campus facilitators some training on effective instructional development for engineers and scientists. Without such training, the facilitators might have to work their way up a lengthy and steep learning curve, making their initial efforts less than effective, and the resulting word-of-mouth reporting could compromise continuation of the program. With the training, the likelihood of good initial workshops and long-term program sustainability both increase.

Sometimes teaching experts belong to the faculty of a specific department. They may have come from the department's discipline and made a decision to make teaching the focus of the remainder of their careers, or they may come from an education-related discipline and be "attached" to the department (as often occurs in the United Kingdom), or they may have been hired as a "teaching professor" (an increasingly common model in the United States).

Baillie (2007) suggests ways for an instructional expert within a department to have the greatest possible impact on the department's teaching program: (1) Give an initial seminar in the department to establish credibility; (2) try for some "quick wins" such as the establishment of a peer tutoring program; (3) work with individual colleagues in a consulting role; and (4) listen for the first months to understand key concerns, problems, and goals in the department. In all interactions with faculty colleagues, the pedagogical specialist should offer suggestions without being critical or prescriptive.

MANDATORY VERSUS VOLUNTARY PARTICIPATION WITH AND WITHOUT INCENTIVES

Most instructional development programs are voluntary. Exceptions include programs for new faculty members that are required for tenure and promotion in Denmark, Norway, Sweden, and the United Kingdom, and at individual institutions in other countries.

The potential benefits of requiring faculty members to attend instructional development programs are obvious, but there are also dangers in doing so. Anyone forced to attend is likely to resent it, and it is common for program facilitators to become proxy targets for the administrators responsible for the requirement. The resentful attendees may simply withdraw from active engagement, or they may use frequent and aggressive questioning in an attempt to discredit the program content or the facilitators' expertise. Skillful presenters know how to handle difficult participants, but the confrontations diminish the experience for the majority of participants who are there to learn. The benefits of required participation are most likely to outweigh the risks when teaching is clearly an important component of the faculty reward system (particularly if it counts toward tenure and promotion in a meaningful way) and the program has an excellent reputation among past participants. If both of these conditions are not in place, required participation may be counterproductive.

If an instructional development program is voluntary, the risk is that too few engineering faculty members will attend for it to have a measurable impact on the school's teaching program. To minimize that risk, sometimes incentives are offered for participation. Possible incentives are cash or other tangible resources (such as computers), travel funds, or release from teaching or service responsibilities. Even token stipends can indicate that administrators are serious enough about teaching to be willing to invest money to improve it, which may induce significant numbers of faculty members to participate.

The importance of recognition and reward to the success of instructional development programs cannot be overemphasized. "If quality teaching is not explicitly expected and rewarded as an institutional priority, faculty may feel that participation in such a program to strengthen teaching and improve student learning is not highly valued by administrators compared to other activities. Therefore, administrators may need to provide some form of external motivation for faculty participation" (Romano et al., 2004).

Evaluation of Instructional Development Programs

Although a considerable and growing body of work on instructional development program evaluation exists, most program directors simply administer participant satisfaction surveys, making little or no effort to determine how well their programs succeeded in meeting their objectives.

Chism and Szabó (1997) proposed that evaluation of an instructional development program can be performed at three levels:

- Level 1: How satisfied were the participants with the program?
- Level 2: What was the impact of the program on the participants' teaching knowledge, skills, attitudes, and practices? (To those measures might be added their evaluations by students and peers.)
- Level 3: What was the impact of the program on the participants' students' learning (knowledge, skills, and attitudes)?

An analogous system for evaluating corporate training programs is that of Kirkpatrick and Kirkpatrick (2006, 2007), who propose that evaluation can be performed at four different levels: *reactions* (which corresponds to Level 1), *learning* (Level 2 – change in knowledge, skills, and attitudes), *behavior* (Level 2 – change in teaching practices), and *results* (Level 3).

Table 21.4 summarizes some of the assessment measures that may be used to evaluate instructional development programs at each of the specified three levels.

Because the ultimate goal of teaching is learning, the true measure of the effectiveness of an instructional development program is the improvement in the participants' students' learning that can be attributed to the program (Level 3). Such improvements

cannot be assumed to follow from their teachers' satisfaction with a workshop (Level 1), and may only be inferred indirectly from changes in the teachers' instructional practices, attitudes, and evaluations following workshop attendance (Level 2). The Level 3 question is therefore the one that matters most, and if we could get an unequivocal answer to it there would be little need to ask the other two.

Unfortunately, it is difficult to obtain that answer, and next to impossible to obtain it such that observed improvements in learning can be unequivocally attributed to participation in the instructional development program. For that reason, evaluation generally consists of asking participants to rate the programs and the facilitators on some scale and perhaps to comment on things they liked and disliked (Level 1), or asking program alumni to retrospectively evaluate the effects of the program on their teaching (Level 2). Chism and Szabó (1997) found that 85% of the instructional development programs on the 200 campuses they surveyed assessed at Level 1. It is of course important to assess participant satisfaction to identify problems and obtain guidance on how to improve subsequent offerings, but satisfaction surveys provide no indication of the subsequent impact of the workshops on either teaching or learning. Fewer than 20% of Chism and Szabó's respondents indicated that they always or usually evaluate the impact of programs on the participants' teaching (Level 2), and none attempted to evaluate impact on students' learning.

The validity of using participants' self-assessments of their teaching as part of a Level 2 workshop evaluation has been examined by D'Eon, Sadownik, Harrison, and Nation (2008), who cite a number of studies that compared self-assessments of teaching with external evaluations by trained observers. Those studies support two conclusions:

- An individual's assessment of his or her teaching skill before or after a workshop cannot be taken at face value, but

Table 21.4. Instructional Development Program Evaluation

Assessment Instrument	Notes
Level 1 – Participant Satisfaction	
• End-of program satisfaction survey • End-of-program interviews • Retrospective survey and interviews	• Not a measure of program effectiveness (but probably a necessary condition)
Level 2 – Impact on Participants' Teaching Attitudes and Practices	
• Immediate post-program and retrospective surveys of attitudes and practices • Assessment of student-centeredness of teaching philosophy • Pre- and post-program student ratings of teaching • Pre- and post-program peer ratings of teaching	• Make sure rating form asks about targeted attitudes and practices, ideally from a professionally developed instrument with tested reliability and validity. • Use a reliable peer rating protocol, not just one class observation (see Brent & Felder, 2004).
Level 3 – Impact on Participants' Students' Learning	
• Performance on standardized or identical learning assessments of students taught before and after instructor's participation • Assessed program learning outcomes and course objectives for accreditation • Students' tendency to adopt a deep approach to learning	• Make sure both groups of students have statistically equivalent entering credentials, and the assessment instruments address targeted skills and are valid and reliable.

aggregated self-assessments from workshop participants generally match closely with external assessments and can provide the basis for a valid and reliable evaluation of workshop effectiveness.

• Individual gains in skill calculated from separate pre-workshop and post-workshop assessments are also suspect because before the workshop individuals often lack a legitimate basis for judging their skill levels. On the other hand, individuals' retrospective (post-workshop) self-assessments of pre–post workshop gains in skill levels correlate reasonably well with more objective external ratings. Skeff, Stratos, and Bergen (1992) reached the same conclusion in a study of a medical instructional development program.

Numerous publications report Level 2 evaluations of instructional development programs, most of which involved retrospective assessments of program participants'

teaching attitudes and practices and in some cases their student ratings before and after their participation (Brawner, Felder, Allen, & Brent, 2002; Camblin & Steger, 2000; Conley, Ressler, Lenox, & Samples, 2000; Estes et al., 2010; Felder & Brent, 2010; Fink et al., 2005; Gibbs & Coffey, 2004; Ho, Watkins, & Kelly, 2001; McAlpine et al., 2005; Postareff, Lindblom-Ylänne, & Nevgi, 2008; Strader, Ambrose, & Davidson, 2000). Gibbs & Coffey (2004) and Postareff et al. (2008) used the Approaches to Teaching Inventory (Trigwell & Prosser, 2004) to determine where the participants' conceptions of teaching fell on a continuum between teacher-centered (transmission of information) versus learner-centered (facilitation of learning). One study (Ho et al., 2001) used the Course Experience Questionnaire (Ramsden, 1991) to assess students' perceptions of the course and instruction, and another (Gibbs & Coffey, 2004) used a subset of the Student Evaluation of Educational Quality (Marsh, 1982) for the same purpose. Significant numbers of survey respondents indicated that their teaching had improved in their estimation and/or that of their students, and many reported adopting learner-centered teaching practices and conceptions of teaching that had been advocated in the programs they attended.

Level 3 evaluations are much scarcer. McShannon et al. (2006) examined the effectiveness of a program in which participants were trained in teaching methods designed to address different student learning styles and were then observed teaching classes and given individual feedback. Student cohorts taught by those instructors were compared with cohorts taught the same classes by the same instructors in the previous year. The grades earned by the experimental cohorts were on average 5.6% higher for freshmen and 6.7% higher for sophomores, and retentions in engineering were 7.8% higher for first-year students and 12.9% higher for sophomores, with all of those differences being statistically significant.

Ho et al. (2001) used the Approaches to Studying Inventory (Entwistle, 1992) to assess students' approaches to studying (surface vs. deep) at the beginning and end of courses taught by program participants, and Gibbs and Coffey (2004) did the same thing using the Module Experience Questionnaire (Lucas, Gibbs, Hughes, Jones, & Wisker, 1997). In the latter study, the questionnaire was also administered to a control group of non-participating teachers and their students. A matched-pair study showed that the students taught by the trained instructors were significantly less likely to take a surface approach to learning following the training and slightly more likely to take a deep approach. No change in approach occurred in the students taught by the untrained instructors. Assessing students' approaches to learning falls short of a direct Level 3 evaluation of learning, but numerous studies have shown that students who take a deep approach display a wide range of superior learning outcomes relative to students who take a surface approach (Meyer, Parsons, & Dunne, 1990; Ramsden, 2003).

Making Engineering Instructional Development Effective

Theoretical Foundations

Instructional development involves teaching adults, and the same theories, principles, and heuristics that have been validated for adult education by cognitive science and/or empirical educational research should be applicable to instructional development (King & Lawler, 2003). A critically important determinant of the effectiveness of teaching adults is the students' motivation to learn (Hofer, 2009; Wlodkowski, 1999, 2003). In a review of corporate training programs, Quiñones (1997) cites a number of studies demonstrating that trainees' motivation to learn has a significant effect on how effective a program is for them. Incorporating anything into the design and delivery of an instructional development program that increases the participants' motivation to learn the content should increase the program's effectiveness.

Wlodkowski (1999) suggests that five attributes of a learning environment have a motivational effect on adult learners (Table 21.5). We hypothesize that those five attributes should provide a good basis for the design of engineering instructional development programs.

Another well-validated model for effective instruction is the cognitive science-based How People Learn (HPL) framework (Bransford et al., 2000; VaNTH-ERC, http://repo.vanth.org/portal). The HPL criteria are compatible with Wlodkowski's motivational factors.

- The HPL framework calls for a *learner-centered* environment, which takes into account the knowledge, skills, and attitudes of the learners. This environment is promoted by establishing the *relevance* of course material and giving learners the freedom to make *choices* among alternatives. Learner-centeredness is also supported by having at least one workshop facilitator who comes from a similar discipline to that of the participants

Table 21.5. Factors that motivate adult learning

Factor	Rationale
1. Expertise of presenters	Adults expect their teachers to be experts in the material being taught, well-prepared to teach it, and knowledgeable about the interests, needs, and problems of their audience.
2. Relevance of content	Adults may quickly become impatient with material they cannot easily relate to their personal interests or professional needs.
3. Choice in application	Adults respond well when given options about whether, when, and how to apply recommended methods, and are skeptical of "one size fits all" prescriptions.
4. Praxis (action plus reflection)	Adults appreciate opportunities to see implementations of methods being taught and to try the methods themselves, and then to reflect on and generalize the outcomes.
5. Groupwork	Adults enjoy and benefit from sharing their knowledge and experiences with their colleagues.

Wlodkowski (1999).

and so shares their content knowledge (*expertise*). Moreover, active learning, which when done effectively involves both action and reflection and can almost be considered synonymous with *praxis*, is almost invariably included on lists of learner-centered teaching methods.

- HPL calls for a *knowledge-centered* environment, meaning that the content being taught should focus on the most important principles and methods associated with the subject of the presentation and should build on the learners' current knowledge and conceptions, and the presentation should utilize techniques known to promote skill development, conceptual understanding and metacognitive awareness rather than simple factual recall. Wlodkowski's *relevance* factor covers the first of those conditions, and *praxis* fosters high-level skills and metacognition if the activities require such skills and are followed by reflection and feedback on the outcomes.

- HPL calls for an *assessment-centered* environment, suggesting that feedback be regularly provided to help learners know where they stand in terms of meeting the learning objectives of the instructional program. Giving the participants opportunities to practice recommended techniques and providing immediate feedback

on their efforts (*praxis*) helps to establish such an environment.

- The final HPL requirement is a *community-centered* environment characterized by supportive interactions among learners and de-emphasis of individual competition. Both conditions are consistent with Wlodkowski's *groupwork* criterion.

It is consequently reasonable to suggest that an instructional program for adult learners that meets all five of Wlodkowski's criteria also complies with the HPL framework. As long as such a program is well organized and has skilled facilitators and a good classroom environment, it is likely to be effective and well received by the participants.

The ineffectiveness of many campus-wide workshops for engineering instructional development is understandable in the light of Wlodkowski's conditions for adult learner motivation (Table 21.5). Although educators and educational psychologists may be experts on pedagogical theories and good teaching practices, they usually lack the disciplinary content knowledge to construct examples that would make the workshop material clearly applicable to engineering courses. Even if the presenters had that knowledge, they would probably refrain from using it for fear of losing participants from other disciplines in campus-wide workshops. Many engineering faculty participants consequently cannot see the *relevance* of the workshop material to what they do, and they are also likely to conclude erroneously that the presenters lack the *expertise* to tell them anything useful about their teaching. Teaching workshops are sometimes prescriptive, giving the participants no *choice* in whether, when, and how to implement each recommendation. The participants get the message that they have been teaching wrong and must make the all of recommended changes to be acceptable teachers, a message most do not appreciate. Finally, some workshops consist almost entirely of lectures on educational theories and methods, with little opportunity for *praxis* (practice and reflection) in the methods and little or no content-related *groupwork* among participants.

Example: The ASEE National Effective Teaching Institute

The National Effective Teaching Institute (NETI, http://www.ncsu.edu/felderpublic/NETI.html) is a three-day teaching workshop given annually in conjunction with the ASEE Conference. Since 1991, it has been attended by 1047 engineering faculty members from 220 institutions. In the early spring of 2008, a web-based survey sent to 607 workshop alumni asked about the effects of the NETI on their teaching practices, their students' and their own ratings of their teaching, their involvement in educational research and instructional development, and their attitudes regarding various aspects of teaching and learning. Valid responses were received from 319 of the survey recipients, for a 53% rate of return. The complete survey and full analysis of the results is given by Felder and Brent (2009), and a synopsis of the results and their implications for engineering instructional development is given by Felder and Brent (2010).

Substantial percentages of the respondents incorporated learning styles, learning objectives, and active learning (the concepts most heavily emphasized in the workshop) into their teaching following their participation in the NETI. Fifty-two percent of the respondents felt that the NETI motivated them to get involved in instructional development; 44% had engaged in it (9% extensively and 35% occasionally), and 21% had not yet done so but planned to in the future. High percentages of the respondents reported engaging in practices that characterize scholarly teaching: 89% read education-related journal articles and 73% had participated in an education conference, with roughly half of each group having been motivated to do so by the NETI; and 69% belonged to the ASEE, roughly a third of whom were persuaded by the NETI to join. Three-quarters of the respondents had engaged in classroom research and/or formal educational research, with 50% having been stimulated to do so by the NETI. In their open-ended responses, many participants indicated that as a result of participating in the NETI, they had become more effective and/or more learner-centered and/or more reflective in their teaching. Felder and Brent (2010) proposed that the success of the NETI derives in large measure from the extent to which it has met Wlodkowski's criteria, giving specific examples of workshop features that directly address the criteria.

Increasing the Appeal and Relevance of Instructional Development to Engineers

There are two broad approaches to making engineering instructional development effective: (1) modifying programs to make them more appealing and relevant to both new and experienced engineering faculty members, and (2) changing the campus climate to make continuing professional development an expectation for all faculty members. This section offers recommendations in the first category and the next one deals with the second category.

Many faculty development experts and teaching leaders have presented suggestions for increasing the effectiveness of instructional development programs (Eble & McKeachie, 1985; Eggins & Macdonald, 2003; Felder & Brent, 2010; Fink, 2006; Garet, Porter, Desimone, Birman, & Yoon, 2001; Hendricson et al., 2007; Ho et al., 2001; Kahn & Baume, 2003, 2004; Lockhart, 2004; Sorcinelli, 2002; Sorcinelli, Austin, Eddy, & Beach, 2005; Sunal et al., 2001; Weimer, 2010; Wright & O'Neil, 1995). Most of the suggestions either directly or indirectly address one or more of the five factors of Wlodkowski's theory of adult motivation to learn (Table 21.5). The following list applies them specifically to engineering instructional development.

- *Be sure program facilitators have expertise in both engineering and pedagogy.* Except in those rare cases in which such expertise resides in a single individual, programs should be facilitated by teams that collectively provide these qualifications. Content expertise can be provided by engineering teaching leaders, while pedagogical expertise can be supplied either by campus teaching and learning center staff or by education or psychology faculty members. (*Expertise, relevance*)
- *Use external facilitators strategically.* External facilitators should be used to attract a wide audience, lend visibility and credibility to instructional development efforts, and provide expertise not available on campus. (*Expertise*)
- *Use engineering-related examples and demonstrations to the greatest possible extent.* The closer program content is to the sorts of things the participants teach, the less likely they will be to dismiss the content as irrelevant to their work. (*Relevance*)
- *Target program content to the needs and interests of the participants.* For new faculty members, emphasize basic instructional issues and strategies. Midcareer and senior faculty members can benefit from programs that introduce advanced pedagogical approaches such as cooperative and problem-based learning, infor-

mation on modern cognitive theories and models, and exploration of attitudes and values related to teaching and learning. (*Relevance*)
- *Provide choices in application of recommended methods.* Effective instructional development programs should not mandate the adoption of new strategies (as in "You can only be an effective teacher if you use active learning!"). The facilitators should instead outline strategies and invite participants to try two or three that look reasonable in their next course rather than setting out to adopt every program recommendation starting on Day 1. Above all, there should be no intimation that the teaching methods the participants have been using (such as lecturing) are wrong. The idea is not for them to stop using those methods but to gradually supplement them with new ones. (*Choice*)
- *Model recommended techniques and provide as many opportunities as possible for participants to practice them and formulate applications to their own courses.* (*Praxis*)
- *Actively engage the participants.* Adult learners like having the chance to try out recommended techniques and to offer their own ideas based on their experience. Programs that consist almost entirely of lectures without such opportunities are likely to be ineffective. (*Groupwork*)

Creating a Supportive Campus Culture

Weimer (2010, p. xiii) asks, "Why don't more faculty take advantage of professional development opportunities?" and responds "Because there are no consequences if they don't." Fink (2009, personal communication) articulated the crucial role of administrators in motivating faculty members to participate in instructional development programs and maximizing the impact of the programs on institutional teaching quality:

On many campuses, the faculty correctly perceives the view of the administration toward participation in instructional development to be: "If you want to participate in this, that is

*OK; if you don't want to, that is OK too."
So long as that is the faculty perception, you
will only get the 20% or so of faculty who
are "eager learners" to regularly participate.
What we need is for the organization to send
a clear message that in essence says: "Teaching at this institution is a major responsibility
and we view all faculty members as professional educators. As such, they should engage
in regular and continuous professional development directly related to their roles as educators." Then and only then will we get the rate
of participation up from 20% of all faculty to
where it ought to be, 80% to 100%.*

Following are recommendations for creating
a campus culture in which continuing professional development is an expectation for
all faculty members.

- *Make it clear in position advertisements,
  interviews, and offer letters that participation in faculty development is a job requirement, and add a section to the yearly faculty
  plan or activity report called "professional
  development"* (Fink, 2006; Fink et al.,
  2005).
- *Evaluate teaching meaningfully, taking into
  account assessments of course design and
  student learning, peer ratings, and alumni
  ratings along with student ratings. Then
  take the results into account meaningfully when making personnel decisions.* The
  more heavily the evaluation outcomes
  count in decisions regarding tenure, promotion, and merit raises, the more faculty
  members will be motivated to participate
  in instructional development and the better the institutional teaching quality will
  become (Felder & Brent, 2004; Fink, 2006;
  Fink et al., 2005).
- *Recognize faculty members' efforts to improve their teaching and reward their success in doing so.* Nominate excellent teachers for local, regional, national, and international awards, and publicize successes
  to the same extent research achievement
  is publicized.
- *Encourage direct administrator participation.* Personally taking part in instructional development programs helps deans
  and department heads understand the

developmental needs of their faculty
members. If the programs are good, they
may subsequently be inclined to increase
their moral and financial program support, and if they publicly announce their
intention to participate, faculty enrollment tends to rise.

Questions for Research

The given recommendations for making engineering instructional development
effective are based on evaluations of successful programs, suggestions from a diverse
group of faculty development authorities
and teaching leaders, and a hypothesis that
conditions known to motivate adult learners
and facilitate student learning should also
promote effective instructional development. As yet, however, there is no validated framework for instructional development program design, and most of the
recommendations are empirically grounded
but nonetheless speculative. Answering the
following questions should help confirm the
extent to which they are valid.

- What conditions are necessary to obtain
  a valid Level 3 evaluation (determining
  the impact of the program on the participants' students' learning)? What (if any)
  sufficient conditions can be formulated?
  Under what conditions can Level 2 evaluation (determining the program impact
  on the participants' teaching practices,
  attitudes, and student and peer ratings)
  provide a valid proxy for Level 3?
- How sound is the recommendation to
  focus on effective pedagogical practices
  in programs for relatively inexperienced
  faculty members and to place a greater
  emphasis on exploration of learning theories and teaching-related attitudes and
  values in programs for more experienced
  faculty members?
- How effective at promoting lasting
  change are different program structures (single workshops, seminars and
  discussion series, consulting, mentoring,
  partnering, learning communities, and
  teaching certification programs)?

- Under what conditions are discipline-specific instructional development programs more effective than campus-wide programs and vice versa?
- Under what conditions are program facilitators who come from fields outside engineering effective with engineering faculty members?
- Under what circumstances (if any) should participation in instructional development programs be mandatory?
- How effective are different incentives offered to engineering faculty members to attend optional instructional development programs? What differences in post-workshop changes exist between attendees who would go under any circumstances and those who go specifically because of incentives?
- How sound are the recommendations offered in this section for broadening the appeal and relevance of engineering instructional development programs? Can any of the recommendations be considered necessary conditions for program effectiveness? Can any subset of them be considered sufficient?
- How effective are on-line instructional development programs compared to their traditional counterparts? What conditions maximize their effectiveness?
- Which (if any) engineering schools have created cultures that support and reward effective teaching and instructional development? What actions and policies were responsible for the success of their efforts?

Answers to these and similar questions will be needed to help engineering school administrators and instructional development personnel gain a nuanced understanding of what makes engineering instructional development effective for specified target audiences and program outcomes. Our hope is that another review of engineering instructional development in five years will include a validated research-based framework for effective program design and delivery.

Acknowledgments

This chapter originally appeared in the *Journal of Engineering Education*, Vol. 100, Issue 1, and we gratefully acknowledge ASEE's permission to reprint the updated article.

Appendix Glossary of Selected U.S. Academic Terms and International Synonyms

Faculty development. "Some form of organized support to help faculty members develop as teachers, scholars, and citizens of their campuses, professions, and broader communities" (Sorcinelli et al. 2005, p. xiii). Faculty development programs may include efforts to enhance faculty members' teaching, research, and administrative skills and to facilitate their personal development, to modernize courses and curricula, to promote the scholarship of teaching and learning, and to redesign organizational structures and policies (including faculty reward systems) to better support the previous functions. The term *educational development* originated in Canada and is now used in several European countries as well to cover all aspects of faculty development related to teaching and learning (Saroyan & Frenay, 2010), and the terms *staff development*, *academic development*, and *professional development* (which takes into account the development of research skills) are also widely used to cover some or all faculty development functions.

College of engineering, school of engineering. The personnel (including faculty members, administrators, technicians, secretaries, other employees, and students) and other resources that comprise the academic unit devoted to teaching and research in engineering. In some countries this entity might be referred to as the *faculty of engineering.*

Course. A semester-long (or quarter-long) unit of instruction on a particular topic that may include lectures, recitations (also known as *tutorials* or *problem sessions*), laboratory sessions, regular assignments, projects, and examinations. In some countries the term *discipline* would have this meaning. The term *class* is sometimes used synonymously with course, although it might also mean the group of students taking a course.

Faculty, faculty members. The collection of individuals at a university who teach courses and/or conduct research and are not students. In most countries these people would collectively be called *staff* or (to distinguish them from support personnel such as secretaries and lab technicians) *academic staff*. They include tenured and untenured professors, associate professors, assistant professors, research professors, teaching professors, instructors, and lecturers.

Grades. Numbers or letters (A, B, C+, . . .) given by instructors to rate students' performance levels in courses, examinations, and assignments. The term *marks* is more common in many countries when applied to examinations and assignments.

Graduate students. Students who have received their initial university degrees and are enrolled for advanced study, possibly leading to masters or doctoral degrees. In other countries they would be referred to as *postgraduate students*, and students who are called graduate students in those countries would be called *undergraduates* in the United States.

Instructional development. A subset of faculty development (and educational development) specifically intended to help faculty members become better teachers.

Tenure. In the United States and Canada, after a period usually between four and six years, a faculty member may be granted tenure by a vote of their department faculty and approval of higher administrators. Tenure is tantamount to a guaranteed permanent faculty position, although a tenured faculty member can be fired for extreme incompetence or criminal conduct or if the faculty member's academic unit (department or program) is eliminated. In other countries, such as those in the United Kingdom, the term does not exist but after several years of competent performance faculty members automatically move into a permanent position.

University. Any institution of higher education. We will use this term to denote all institutions that have "university," "college," or "institute" in their names.

References

Adams, R., Evangelou, D., English, L., Figueiredo, A. D., Mousoulides, N., Pawley, A. L., . . . & Wilson, D. M. (2011). Multiple perspectives on engaging future engineers. *Journal of Engineering Education*, 100(1), 48–88.

Adams, R. S., & Felder, R. M. (2008). Reframing professional development: A systems approach to preparing engineering educators to educate tomorrow's engineers. *Journal of Engineering Education*, 97(3), 239–240.

Ambrose, S. A., Bridges, M. W., DiPietro, M., Lovett, M. C., & Norman, M. K. (2010). *How learning works: Seven research-based principles for smart teaching*. San Francisco, CA: Jossey-Bass.

American Society for Engineering Education (ASEE). (2009, October 28). High-achievers defect from STEM fields, study finds. *First Bell*.

Baars, B. J., & Gage, N. M. (Eds.) (2007). *Cognition, brain, and consciousness: Introduction to cognitive neuroscience*. London: Elsevier.

Baillie, C. (2007). Education development within engineering. *European Journal of Engineering Education*, 32(4), 421–428.

Baldwin, R. G., Dezure, D., Shaw, A., & Moretto, K. (2008). Mapping the terrain of mid-career faculty at a research university: Implications for faculty and academic leaders. *Change*, 40(5), 46–55.

Bland, C. J., Taylor, A. L., Shollen, S. L., Weber-Main, A. M., & Mulcahey, P. A. (2009). *Faculty success through mentoring: A guide for mentors, mentees, and leaders*. Lanham, MD: Rowman & Littlefield.

Boice, R. (2000). *Advice for new faculty members: Nihil Nimus*. Boston, MA: Allyn and Bacon.

Borrego, M., & Bernhard, J. (2011). The emergence of engineering education research as an internationally connected field of inquiry. *Journal of Engineering Education*, 100(1), 14–47.

Boud, D. (1999). Situating academic development in professional work: Using peer learning. *International Journal for Academic Development*, 4(1), 3–10.

Bransford, J., Brown, A. L., & Cocking, R. R. (Eds.). (2000). *How people learn: Brain, mind, experience, and school* (Expanded edition). Washington, DC: The National Academies Press. Retrieved from http://www.nap.edu/books/0309070368/html.

Brawner, C. E., Felder, R. M., Allen, R., & Brent, R. (2002). A survey of faculty teaching practices and involvement in faculty development activities. *Journal of Engineering Education*, 91(4), 393–396. Retrieved from http://www.ncsu.edu/felder-public/Papers/Survey_Teaching-Practices.pdf

Brent, R., & Felder, R. M. (2004). A protocol for peer review of teaching. In *Proceedings, 2004 ASEE Annual Conference*. Washington, DC: ASEE. Retrieved from http://www.ncsu.edu/felder-public/Papers/ASEE04(Peer-Review).pdf

Brent, R., & Felder, R. M. (2008). A professional development program for graduate students at N.C. State University. In *2008 ASEE Annual Conference Proceeding*. Washington, DC: ASEE. Retrieved from http://www.ncsu.edu/felder-public/Papers/ASEE08(GradStudentTraining).pdf

Brent, R., Felder, R. M., & Rajala, S. A. (2006). Preparing new faculty members to be successful: A no-brainer and yet a radical concept. In *2006 ASEE Annual Conference Proceedings*. Washington, DC: ASEE. Retrieved from http://www.ncsu.edu/felder-public/Papers/ASEE06(NewFaculty).pdf

Bullard, L. G., & Felder, R. M. (2003). Mentoring: A personal perspective. *College Teaching*, 51(2), 66–69. Retrieved from http://www.ncsu.edu/felder-public/Papers/Mentoring(Coll_Tch).html

Camblin, L. D., Jr., & Steger, J. (2000). Rethinking faculty development. *Higher Education*, 39, 1–18.

Case, J. M., & Light, G. (2011). Emerging research methodologies in engineering education research. *Journal of Engineering Education*, 100(1), 186–210.

Chism, N. V. N., & Szabó, B. S. (1997). How faculty development programs evaluate their services. *Journal of Staff, Program, and Organization Development*, 15(2), 55–62.

Colbeck, C. L., O'Meara, K. A., & Austin, A. E. (Eds.). (2008). *Educating integrated professionals: Theory and practice on preparation for the professoriate*. New Directions for Teaching and Learning, No. 113. San Francisco, CA: Jossey-Bass.

Conley, C. H., Ressler, S. J., Lenox, T. A., & Samples, J. W. (2000). Teaching teachers to teach engineering – T⁴E. *Journal of Engineering Education*, 89(1), 31–38.

Courter, S. S., Freitag, C., & McEniry, M. (2004). Professional development on-line: Ways of knowing and ways of practice. In *Proceedings of the Annual ASEE Conference*. Washington, DC: ASEE.

Cox, M. D. (2004). Introduction to faculty learning communities. In M. D. Cox & L. Richlin (Eds.), *Building faculty learning communities*. New Directions for Teaching and Learning, no. 97 (pp. 5–23). San Francisco, CA: Jossey-Bass.

D'Eon, M., Sadownik, L., Harrison, A., & Nation, J. (2008). Using self-assessments to detect workshop success: Do they work? *American Journal of Evaluation*, 29(1), 92–98.

Eble, K. E., & McKeachie, W. J. (1985). *Improving undergraduate education through faculty development*. San Francisco, CA: Jossey-Bass.

Eggins, H., & Macdonald, R. (Eds.) (2003). *The scholarship of academic development*. Milton Keynes, UK: Open University Press.

Entwistle, N. J. (1992). *Scales and items for revised approaches to studying inventory*. Personal communication cited by Ho (2000).

Estes, A. C., Welch, R. W., Ressler, S. J., Dennis, N., Larson, D., Considine, C., . . . Lenox,

T. (2010). Ten years of ExCEEd: Making a difference in the profession. *International Journal of Engineering Education, 26*(1), 141–154.

Fairweather, J. (2005). Beyond the rhetoric: Trends in the relative value of teaching and research in faculty salaries. *Journal of Higher Education, 76*, 401–422.

Felder, R. M. (1993). Teaching teachers to teach: The case for mentoring. *Chemical Engineering Education, 27*(3), 176–177. Retrieved from http://www.ncsu.edu/felder-public/Columns/Mentoring.html

Felder, R. M., & Brent, R. (2003). Designing and teaching courses to satisfy the ABET Engineering Criteria. *Journal of Engineering Education, 92*(1), 7–25. Retrieved from http://www.ncsu.edu/felder-public/Papers/ABET_Paper_(JEE).pdf

Felder, R. M., & Brent, R. (2004). How to evaluate teaching. *Chemical Engineering Education, 38*(3), 200–202. Retrieved from http://www.ncsu.edu/felder-public/Columns/Teacheval.pdf

Felder, R. M., & Brent, R. (2009). Analysis of fifteen years of the National Effective Teaching Institute. In 2009 *ASEE Annual Conference Proceedings*. Washington, DC: ASEE. Retrieved from http://www.ncsu.edu/felder-public/Papers/NETIpaper.pdf.

Felder, R. M., & Brent, R. (2010). The National Effective Teaching Institute: Assessment of impact and implications for faculty development. *Journal of Engineering Education, 99*(2), 121–134. Retrieved from http://www.ncsu.edu/felder-public/Papers/NETIpaper.pdf

Finelli, C. J., Ott, M., Gottfried, A. C., Hershock, C., O'Neal, C., & Kaplan, M. (2008). Utilizing instructional consultations to enhance the teaching performance of engineering faculty. *Journal of Engineering Education, 97*(4), 397–411.

Fink, D. (2003). *Creating significant learning experiences*. San Francisco, CA: Jossey-Bass.

Fink, D. (2006). Faculty development: A medicine for what ails academe today. *The Department Chair, 17*(1), 7–10.

Fink, D. L., Ambrose, S., & Wheeler, D. (2005). Becoming a professional engineering educator: A new role for a new era. *Journal of Engineering Education, 94*(1), 185–194.

Friedman, T. L. (2006). *The world is flat: A brief history of the twenty-first century*. New York: Farrar, Straus and Giroux.

Garet, M. S., Porter, A. C., Desimone, L., Birman, B. F., & Yoon, K. S. (2001). What makes professional development effective? Results from a national sample of teachers. *American Educational Research Journal, 38*(4), 915–945.

Gibbs, G., & Coffey, M. (2004). The impact of training of university teachers on their teaching skills, their approach to teaching and the approach to learning of their students. *Active Learning in Higher Education, 5*(1), 87–100.

Groccia, J. (2010). Why faculty development? Why now? In A. Saroyan and M. Frenay (Eds.), *Building teaching capacities in higher education: A comprehensive international model* (pp. 1–20). Sterling, VA: Stylus.

Healy, M., & Jenkins, A. (2003). Discipline-based educational development. In H. Eggins & R. Macdonald (Eds.), *The scholarship of academic development* (pp. 47–57). Milton Keynes, UK: Open University Press.

Hendricson, W. D., Anderson, E., Andrieu, S. C., Chadwick, D. G., Cole, J. R., George, M. C., . . . & Young, S. K. (2007). Does faculty development enhance teaching effectiveness? *Journal of Dental Education, 71*(12), 1513–1533.

Heywood, J. (2005). *Engineering education: Research and development in curriculum and instruction*. Hoboken, NJ: John Wiley & Sons.

Hicks, O. (1999). Integration of central and departmental development–reflections from Australian universities. *The International Journal for Academic Development, 4*(1), 43–51.

Ho, A. S. P. (2000). A conceptual change approach to staff development: A model for programme design. *International Journal for Academic Development, 5*(1), 30–41.

Ho, A., Watkins, D., & Kelly, M. (2001). The conceptual change approach to improving teaching and learning: An evaluation of a Hong Kong staff development programme. *Higher Education, 42*(2), 143–169.

Hofer, B. (2009). Motivation in the college classroom. In M. Svinicki & W. J. McKeachie (Eds.), *McKeachie's teaching tips: Strategies, research, and theory for college and university teachers* (13th ed., pp. 139–146). Florence, KY: Cengage Learning.

Jain, R., Shanahan, B., & Roe, C. (2009). Broadening the appeal of engineering – addressing factors contributing to low appeal and high attrition. *International Journal of Engineering Education*, 25(3), 405–418.

Jenkins, A. (1996). Discipline-based academic development. *International Journal for Academic Development*, 1, 50–62.

Jesiek, B. K., Newswander, L. K., & Borrego, M. (2009). Engineering education research: Field, community, or discipline? *Journal of Engineering Education*, 98(1), 39–52.

Jungst, J. E., Licklider, B. L., & Wiersema, J. A. (2003). Providing support for faculty who wish to shift to a learning-centered paradigm in their higher education classrooms. *Journal of Scholarship of Teaching and Learning*, 3(3), 69–81.

Kahn, P., & Baume, D. (Eds.). (2003). *A guide to staff and educational development*. London: Kogan Page.

Kahn, P., & Baume, D. (Eds.). (2004). *Enhancing staff and educational development*. Abingdon, U.K.: RoutledgeFalmer.

Lawler, P. A., & King, K. P. (2003). Changes, challenges, and the future. *New Directions for Adult and Continuing Education*, 2003(98), 83–92.

Kirkpatrick, D. L., & Kirkpatrick, J. D. (2006). *Evaluating training programs: The four levels* (3rd ed.). San Francisco, CA: Berrett-Koehler.

Kirkpatrick, D. L., & Kirkpatrick, J. D. (2007). *Implementing the four levels: A practical guide for effective evaluation of training programs*. San Francisco, CA: Berrett-Koehler.

Kolmos, A., Rump, C., Ingemarsson, I., Laloux, A., & Vinther, O. (2001). Organization of staff development – strategies and experiences. *European Journal of Engineering Education*, 26(4), 329–342.

Kolmos, A., Vinther, O., Andersson, O., Malmi, L., & Fuglem, M. (Eds). (2004). *Faculty development in Nordic engineering education*. Aalborg, Denmark: Aalborg University Press.

Kuhlenschmidt, S. (2009). *Where are CTLs? Implications for strategic planning and research*. Paper presented at the meeting of the Professional and Development Network in Higher Education, Houston, TX.

Laursen, S., & Rocque, B. (2009). Faculty development for institutional change: Lessons from an ADVANCE project. *Change*, 41(2), 18–26.

Lee, V. S. (2000). The influence of disciplinary differences on consultations with faculty. In M. Kaplan & D. Lieberman (Eds.), *To improve the academy* (Vol. 18, pp. 278–290). Bolton, MA: Anker.

Litzinger, T., Lattuca, L. R., Hadgraft, R., & Newstetter, W. (2011). Engineering education and the development of expertise. *Journal of Engineering Education*, 100(1), 123–150.

Lockhart, M. (2004). Using adult learning theory to create new friends, conversations, and connections in faculty development programs. *Journal of Faculty Development*, 19(3), 115–122.

Lucas, L., Gibbs, G., Hughes, S., Jones, O., & Wisker, G. (1997). A study of the effects of course design features on student learning in large classes at three institutions: A comparative study. In C. Rust & G. Gibbs (Eds.), *Improving student learning through course design* (pp. 10–24). Oxford: Oxford Centre for Staff and Learning Development.

Marsh, H. W. (1982). SEEQ: A reliable, valid, and useful instrument for collecting students' evaluations of university teaching. *British Journal of Educational Psychology*, 52, 77–95.

McAlpine, L., Gandell, T., Winer, L., Gruzleski, J., Mydlarski, L., Nicell, J., & Harris, R. (2005). A collective approach towards enhancing undergraduate engineering education. *European Journal of Engineering Education*, 30(3), 377–384.

McShannon, J. M., Hines, P., Nirmalakhandan, N., Venkataramana, G., Ricketts, C., Ulery, A., & Steiner, R. (2006). Gaining retention and achievement for students program: A faculty development program. *Journal of Professional Issues in Engineering Education and Practice*, 132(3), 204–208.

Meyer, J. H. F., Parsons, P., & Dunne, T. T. (1990). Individual study orchestrations and their association with learning outcome. *Higher Education*, 20, 67–89.

National Academy of Engineering. (2004). *The engineer of 2020: Visions of engineering in the new century*. Washington, DC: The National Academies Press.

Johri, A., & Olds, B. M. (2011). Situated engineering learning: Bridging engineering education research and the learning sciences. *Journal of Engineering Education*, 100(1), 151–185.

Palmer, P. (1998). *The courage to teach*. San Francisco, CA: Jossey-Bass.

Postareff, L., Lindblom-Ylänne, S., & Nevgi, A. (2008). A follow-up study of the effect of pedagogical training on teaching in higher education. *Higher Education*, 56(1), 29–43.

Pruitt-Logan, A. S., Gaff, J. G., & Jentoft, J. E. (2002). *Preparing future faculty in the sciences and mathematics: A guide for change.* Washington, DC: Association of American Colleges and Universities. Retrieved from http://www.preparing-faculty.org/PFFWeb.PFF3Manual.pdf

Quiñones, M. A. (1997). Contextual influences on training effectiveness. In M. A. Quiñones & A. Ehrenstein (Eds.), *Training for a rapidly changing workplace: Applications of psychological research* (pp. 177–199). Washington, DC: American Psychological Association.

Ramsden, P. (1991). A performance indicator of teaching quality in higher education: The Course Experience Questionnaire. *Studies in Higher Education, 16*(2), 129–150.

Ramsden, P. (2003). *Learning to teach in higher education* (2nd ed.). London: Taylor and Francis.

Romano, J. L., Hoesing, R., O'Donovan, K., & Weinsheimer, J. (2004). Faculty at mid-career: A program to enhance teaching and learning. *Innovative Higher Education, 29*(4), 21–48.

Roxå, T., Olsson, T., & Mårtensson, K. (2008). Appropriate use of theory in the scholarship of teaching and learning as a strategy for institutional development. *Arts and Humanities in Higher Education, 7*(3), 276–294.

Saroyan, A., Amundsen, C., McAlpine, L., Weston, C., Winer, L., & Gandell, T. (2004). Assumptions underlying workshop activities. In A. Saroyan & C. Amundsen (Eds.), *Rethinking teaching in higher education: From a course design workshop to a faculty development framework* (pp. 15–29). Sterling, VA: Stylus.

Saroyan, A., & Frenay, M. (Eds.). (2010). *Building teaching capacities in higher education: A comprehensive international model.* Sterling, VA: Stylus.

Schaefer, D., & Utschig, T. T. (2008). A review of professional qualification, development, and recognition of faculty teaching in higher education around the world. In *2008 ASEE Annual Conference Proceedings.* Washington, DC: ASEE.

Seymour, E., & Hewitt, N. M. (1997). *Talking about leaving: Why undergraduates leave the sciences.* Boulder, CO: Westview Press.

Shulman, L. S. (1986). Those who understand: Knowledge growth in teaching. *Educational Researcher, 15,* 4–14.

Skeff, K. M., Stratos, G. A., & Bergen, M. R. (1992). Evaluation of a medical faculty development program: A comparison of traditional pre-post and retrospective pre-post self-assessment ratings. *Evaluation and the Health Professions, 15*(3), 350–366.

Sorcinelli, M. D. (2002). Ten principles of good practice in creating and sustaining teaching and learning centers. In K. H. Gillespie, L. R. Hilsen, & E. C. Wadsworth (Eds.), *A guide to faculty development: Practical advice, examples, and resources* (pp. 9–23). Bolton, MA: Anker.

Sorcinelli, M. D., Austin, A. E., Eddy, P. L., & Beach, A. L. (2005). *Creating the future of faculty development: Learning from the past, understanding the present.* Bolton, MA: Anker.

Sorcinelli, M. D., & Yun, J. (2007). From mentor to mentoring networks: Mentoring in the new academy. *Change, 39*(6), 58–61.

Strader, R., Ambrose, S., & Davidson, C. (2000). An introduction to the community of professors: The engineering education scholars workshop. *Journal of Engineering Education, 89*(1), 7–11.

Streveler, R. A., & Smith, K. A. (2006). Guest editorial: Conducting rigorous research in engineering education. *Journal of Engineering Education, 95*(2), 103–105.

Sunal, D. W., Hodges, J., Sunal, C. S., Whitaker, K. W., Freeman, L. M., Edwards, L., . . . Odell, M. (2001). Teaching science in higher education: Faculty professional development and barriers to change. *School Science and Mathematics, 101,* 1–16.

Svinicki, M., & McKeachie, W. J. (2011). *McKeachie's teaching tips: Strategies, research, and theory for college and university teachers* (13th ed.). Florence, KY: Cengage Learning.

Taylor, K. L., & Rege Colet, N. (2010). Making the shift from faculty development to educational development: A conceptual framework grounded in practice. In Saroyan, A., & Frenay, M. (Eds.), *Building teaching capacities in higher education: A comprehensive international model* (pp. 139–167). Sterling, VA: Stylus.

Trigwell, K., & Prosser, M. (2004). Development and use of the Approaches to Teaching Inventory. *Educational Psychology Review, 16*(4), 409–424.

Utschig, T. T., & Schaefer, D. (2008). *Critical elements for future programs seeking to establish excellence in engineering education through professional qualification of faculty teaching in higher education.* Presented at the 8th World

Conference on Continuing Engineering Education, Atlanta, GA.

Wankat, P. C. (2002). *The effective, efficient professor.* Boston, MA: Allyn and Bacon.

Weimer, M. (2010). *Inspired college teaching.* San Francisco, CA: Jossey-Bass.

Wessely, A. (2002). *Teaching squares handbook for participants.* Retrieved from http://www .ntlf.com/html/lib/suppmat/ts/tsparticipant handbook.pdf.

Wilson, M., & Gerber, L. E. (2008). How generational theory can improve teaching: Strategies for working with the millennials. *Currents in Teaching and Learning, 1*(1), 29–44.

Wlodkowski, R. J. (1999). *Enhancing adult motivation to learn: A comprehensive guide for teaching all adults* (2nd ed.). New York, NY: John Wiley & Sons.

Wlodkowski, R. J. (2003). Fostering motivation in professional development programs. In K. P. King & P. A. Lawler (Eds.), (*New Directions for Adult and Continuing Education,* 98, 39–48.

Wright, W. A., & O'Neil, M. C. (1995). Teaching improvement practices: International perspectives. In W. A. Wright and Associates (Eds.), *Teaching improvement practices: Successful strategies for higher education* (pp. 1–57). Bolton, MA: Anker.

CHAPTER 22

Understanding Disciplinary Cultures

The First Step to Cultural Change

Elizabeth Godfrey

Introduction

The need to "know where we are" has been identified (Rover, 2008) as the first step in moving to "where we want to go." This chapter aims to demonstrate that understanding how our culture is formed and sustained, at a departmental, disciplinary, or institutional level, is the first step toward sustainable cultural change. The suggested conceptual framework of cultural dimensions has the ability to act as a practical tool for evaluating and positioning the culture of engineering education in a specific context, as a precursor to developing strategies for transformational change. These cultural dimensions are based on the work of Godfrey (2003a, 2007, 2009) and Godfrey and Parker (2010), who used ethnographic methods within an overarching interpretivist research paradigm to identify the shared assumptions and understandings that underpinned the lived experiences of staff and students as manifested in one institution as the basis for theory development.

References to "culture" and cultural change as key to systemic reforms have been plentiful in literature, including in engineering education (Bucciarelli, Einstein, Terenzini, & Walser, 2000, p. 141; Cordes, Evans, Frair, & Froyd, 1999; IEAust, 1996). The implicit assumption underlying these calls for cultural change – that a common, recognizable engineering education culture exists – has been questioned by scholars (Godfrey, 2007; Williams, 2002). Engineering educators undoubtedly recognize practices and behaviors that transcend differences in engineering specialization, institutions, and even national boundaries. Comments such as "[T]he predominant engineering school culture [is] based on compartmentalization of knowledge, individual specialization, and a wholly research-based reward structure" (Bucciarelli et al., 2000, p. 141), and "In engineering schools, it is generally assumed that propositional technical knowledge, discovered using a reductionist research paradigm, is the prime source of professional knowledge necessary for preparing students for the profession" (Radcliffe, 2006, p. 263), have been viewed as incontestable assumptions.

The calls for cultural change made by professional and government bodies are not necessarily matched by the understanding of engineering educators of how to change the culture at an operational level. The focus for engineering educators has predominantly been on the characteristics of behaviors and practices, "what is and what they should be" rather than the values, beliefs, and assumptions that underpin "how they came to be." Baba and Pawlowski (2001) and Godfrey (2003b) have both suggested that much of this discourse around cultural change has been incorrectly based on the assumption that engineering educators are familiar with theories and models of culture and cultural change, which have their origins in anthropology, sociology, and, in recent years, in business and organizational studies. In 2003, Godfrey offered a theoretical model, later developed in Godfrey (2009) and Godfrey and Parker (2010), that has been accepted as one avenue by which change leaders could understand the dimensions of their own school's engineering education culture (Bullard, Visco, Silverstein, & Keith, 2010; McGrath & Sheppard, 2007; Merton, Froyd, Clark, & Richardson, 2004).

Until very recently, research specifically investigating the culture of engineering education arose in the context of women's lack of participation (Dryburgh, 1999; Hacker, 1983; Lewis, McLean, Copeland, & Lintern, 1998; Tonso, 1996b) rather than from mainstream engineering education. Almost all of that research used qualitative research methods, predominantly ethnographic methods. Increasingly, and in accord with recommendations from the engineering education research agenda (Anon, 2006), multidisciplinary teams of engineering educators working with social scientists and educational researchers are producing a growing body of work that supports a better understanding of the complex interactions necessary to implement cultural change.

A close examination of the literature discussing the concept of culture in engineering

education identifies a range of studies, each offering a valuable perspective of the cultural landscape of engineering education. These perspectives have included: culture as gendered (Cronin & Roger, 1999; Lewis et al., 1998; Tonso, 1996b), culture as an agent in student attrition (Courter, Millar, & Lyons, 1998), student engagement and enculturation (Ambrose, 1998; Lattuca, Terenzini, & Volkwein, 2006), the development of engineering identity (Dryburgh, 1999; Stevens, O'Connor, Garrison, Jocuns, & Amos, 2008; Tonso, 2006a), faculty cultures (McKenna, Hutchinson, & Trautvetter, 2008), campus cultures (Tonso, 2006), sub-disciplinary cultures (Gilbert, 2009; Godfrey, 2007; Murphy et al., 2007), national cultures (Downey & Lucena, 2005), assessment cultures (Borrego, 2008; Yost, Roberts-Kirchhoff, & Zarkowski, 2011), the role of institutional culture in effecting change (Covington & Froyd, 2004; Kezar & Eckel, 2002; Merton et al., 2004), and measuring cultural change (Fromm & McGourty, 2001; Lattuca et al., 2006). Each of these perspectives offers a valuable yet partial view of the dimensions of the culture of engineering education.

The Study of Disciplinary Cultures

The calls for culture change in the 1990s were made in an environment when the culture of higher education was being studied at institutional (Bergquist, 1992; Tierney, 1988) and disciplinary (Becher, 1989) levels. Academic cultures were seen to reflect tacitly accepted theories of teaching and learning derived from long-accepted practices of course delivery and assessment. Academic disciplines were seen as involving a sense of identity and personal commitment, "a way of being in the world... a matter of taking on a cultural frame that defines a great part of one's life" (Geertz, 1973, p. 155) and as "academic tribes" (Becher, 1989). Becher's oft-cited work, based on earlier work by Biglan (1973), acknowledged that academic communities are both epistemological and social communities. His proposed four cell matrix of disciplines – hard-pure/soft-pure/hard-applied/soft-applied, and unsurprisingly placed engineering in the hard-applied quadrant. In a later work, in combination with Trowler (Becher & Trowler, 2001), the need for further theory development on how academic cultures (shaped around distinctive knowledge domains) affect teaching and learning was recommended. Another source of sustained investigation into disciplinary differences has been the work of Donald (1984, 2002). She focused on creating a framework for understanding student intellectual development and learning to think in different disciplines and studied the physics, engineering, and psychology disciplines using an ethnographic approach.

These discussions of academic institutional and disciplinary cultures, and indeed, the calls for culture change within engineering education all conceptualize culture as existing when a group shares both explicit and tacit knowledge, values and attitudes developed through a history of shared experience. In all cultural studies there is a sense that the object of the study is to uncover the deeply embedded, often unconscious, cultural knowledge and understandings that are used by participants to interpret experience and generate behavior. Inherent also in this conceptualization is the notion of a process of enculturation as newcomers join the group.

In recent years, this view of culture as a "relatively enduring, coherent and bounded way of living" (Eisenhart, 2001, p. 17) has been contested as limited, particularly in relation to the multiple influences impacting students and faculty in educational settings. Eisenhart suggests that postmodernist ideas from philosophy, feminism, and cultural studies have demonstrated that "we can no longer conceive of social groups of people with a culture that is bounded, and determined, internally coherent, and uniformly meaningful" (p. 19). She does, however, emphasize that "though untidy, culture is still useful" (p. 20), particularly in understanding how people act and make sense of their worlds. Eisenhart cites Ortner (1991) as suggesting that cultures remain sources of

value, meaning and ways of understanding for the people who live within them.

Disciplinary cultures exist within institutional contexts and cultures, which suggests that local context will modify the potential for change in a disciplinary culture. Kezar and Eckel (2002) built on the work of Bergquist (1992) and Tierney (1988) in examining the impact of institutional culture on the change process in colleges and universities. They confirmed that change strategies were likely to be successful if they were culturally coherent and aligned with the culture. The questions Tierney (1988) saw as needing answering for determining an institutional culture seem equally applicable at departmental level. Answering questions such as the following would provide valuable information before attempting to devise strategies for change implementation: How are decisions arrived at?: Who makes decisions? What do you need to know to excel or succeed in this environment?

It must be noted that the culture of engineering education is not only that of an academic discipline with a distinctive knowledge domain influenced by the institutional structures and traditions of higher education within which teaching and learning take place, but also incorporates cultural influences from the engineering profession. The engineering profession can readily be viewed as an occupational community with its own unique work culture and well established priorities and expectations (ABET, 1996; IPENZ, 2003). Considerations of the culture of engineering education culture must, therefore, of necessity move beyond concepts of epistemological and social communities, to take account of the range of dimensions necessary to provide a holistic view of the culture of engineering.

What Do We Know About the Engineering Culture?

Research in engineering education has been said to "almost exclusively depend on positivist methods of research using experimental evidence, usually quantitative, to support a defined hypothesis" (Tonso, 1996b, p. 143). As engineers stepping outside their perceived discipline boundaries, McLean, Lewis, Copeland, Lintern, and O'Neill (1997); Radcliffe, Crosthwaite, and Jolly (2002); Tonso (1996b); and Waller (2001) all recognized that qualitative research methodologies were the methods of choice in studies addressing culture and culture change where local knowledge and interpretation were needed to assign meaning to words and actions consistent with the meanings assigned by members of the group under study.

The theoretical foundation, or conceptualization of culture, for the study on which this chapter rests is the influential cultural framework proposed by Edgar Schein (1985, 1992). Originally promulgated for the study of organizational cultures, Schein's framework clearly delineates three levels when considering culture. Schein named the most accessible and visible elements of a culture, encompassing day-to-day behaviors and practices as well as physical objects as artifacts. Values and behavioral norms underlay the artifacts and, even deeper lay the core of shared beliefs and assumptions often unconsciously held and rarely articulated. Schein saw a cultural study as looking beneath observable practices, behaviors, and other visible cultural manifestations to the tacit knowledge, shared values, and understandings that guide and direct them. Kuh and Whitt (1988) appear to have adapted Schein's definition to the context of higher education using very similar language to define the culture of higher education as:

> ... mutually shaping patterns of norms, values, practices, beliefs, and assumptions that guide the behavior of individuals and groups in an institute of higher education and provide a frame of reference within which to interpret the meaning of events and actions on and off campus. (p. 28)

Godfrey and Parker's study (2010) used ethnographic methods within an overarching interpretivist research paradigm to investigate the culture of engineering education as manifested in one institution. Using a case study for the purpose of building theory has been recognized (Stake, 1994) as a common, and valid research design, particularly in education. In such cases, the case itself is said to be of secondary interest, "playing a supportive role as it facilitates our understanding of something else" (Stake, 1994, p. 237) in this instance, defining the dimensions of the culture of engineering education.

Adapting Schein's cultural framework, data were collected and analyzed to distil from observable behaviors and practices the essence of the culture in the form of tacitly known cultural norms, shared assumptions, and understandings that underpinned the lived experiences of staff and students. Godfrey and Parker worked with Schein's suggestion that the shared beliefs and assumptions that a group held as personal and collective answers to "issues of external adaptation and internal integration" lying at the very heart of a culture could be viewed as the dimensions of the culture. Issues such as what knowledge is valued, how people communicate, and how do people succeed in this culture are part of this picture, but following Schein's suggestion that a culture's beliefs and assumptions could be grouped in more abstract dimensions, Godfrey and Parker identified six cultural dimensions for engineering education. These were identified as:

- An Engineering Way of Thinking
- An Engineering Way of Doing
- Being an Engineer
- Acceptance of Difference
- Relationships
- Relationship to the Environment

The detailed findings from this study, combined with evidence from other studies, support the view that the proposed six dimensions have the potential to be transferred to other institutions as a practical tool for evaluating and positioning the culture of engineering education.

An Engineering Way of Thinking

It was suggested in the 2006 Research Agenda (Anon., 2006) for engineering education that an "implicit understanding of the essence of engineering thinking and knowing, was evidenced in current educational systems and in reports seeking to facilitate improvements in engineering education" but it was also acknowledged that "the profession needs research that will help characterize the nature of engineering knowledge (i.e., its technical, social, and ethical aspects) and ways of engineering thinking that are essential for identifying and solving technical problems within dynamic and multidisciplinary environments" (p. 259).

When considering the nature of engineering knowledge or way of thinking, it is beneficial to seek the understandings of a group around questions such as: What kinds of knowledge are valued? What is seen as truth? Is there a prevalent "way of thinking?"

The suggestion that there was an engineering way of thinking or knowing was readily acknowledged by faculty and students at the case study institution who articulated their understandings in Godfrey (2003a) with comments such as:

> The biggest single thing about engineering is that we are interested in things that work – as far as engineering is concerned, if it doesn't work, if it has no function, has no utility, then it is of but academic interest and engineers aren't interested in things for only academic interest.
>
> Engineering is all about achieving outcomes usually a physical thing or resulting from a physical thing – might even be delivered by a physical thing – applies even to software. Because of that relationship to physical things, you need to be able to represent them.... In some way – need some sort of abstraction to represent the problem and the desired outcome..... need a way of describing or modeling more effective than words

And from a first-year student:

> Thinking like an engineer, kind of being taught it I suppose. The whole course is directed at making you think differently, that is how I feel it.

Greater detail is provided in the original paper (Godfrey & Parker, 2010) but one of the most deeply ingrained assumptions associated with the Engineering Way of thinking was that engineering dealt with a tangible, definable, measurable, quantifiable reality. Valued knowledge was seen as knowledge that is relevant to real life. "What would we use this for?" was the justification for learning. The emphasis on contextual learning beginning in first year and across all courses had been an explicit goal of the 1996 restructuring of the degree at the case study institution. Abstract, philosophical concepts, such as ethics and sustainability were viewed as unacceptable to both faculty and students unless taught in a practical, relevant context.

Other values and beliefs that emerged as themes in the engineering way of thinking were: the pervasive role of mathematics, with truth and reality in engineering being proven and described by mathematics used as both a tool and a language; the prevalence of visual communication using diagrams and graphics rather than a reliance on words as integral modes of communication; and the focus on problem solving and design, dominated by reductionist and top-down methods. Design seemed to epitomize the essence of what faculty and students believed to be an "engineering way of thinking" including the acceptance that their focus was on "best" rather than "right" answers to predominantly open ended problem solving.

The evidence provided in Godfrey and Parker's study confirmed the duality and tension in conceptions referred to by Radcliffe (2006). The objectivity of engineering science courses and the subjectivity of engineering practice perspectives presented a disconnect or paradox in engineering thinking. One assumes certainty of knowledge and formulae underpinning analysis, and the other works with constraints and compromise, "best" rather than perfect answers.

Of note was an unquestioned assumption by faculty that the knowledge, the mathematical procedures and scientific processes, and the laws on which problem solutions were based were race and gender free. No recognition appeared to exist that the ethnocentricity and masculinity of engineering knowledge and procedures, identified by Lewis et al. (1998) and Stonyer (2001), might affect problem definition and accepted methods of problem solution, teaching and assessment. The problematic nature of this assumption and its prevalence in U.S. colleges, was noted by Bucciarelli et al. (2000), who acknowledged that ways of knowing and patterns within them are socially constructed.

An Engineering Way of Doing

The primary task of a School of Engineering is to educate and graduate students with a professional engineering degree. Shared beliefs and assumptions around how teaching and learning are accomplished can be grouped as An Engineering Way of Doing.

Although individual features of a curriculum might be influenced by pragmatism and compromise with resource issues, the curriculum content, teaching and assessment of an engineering program have evolved from beliefs and assumptions around the "right" way to teach and learn engineering. The international accreditation requirement of a sizeable core of technical content appear to be deeply entrenched in the beliefs of the faculty as defining essential content to be covered. Much of this "essential content" continues to be taught internationally through traditional, lecture-based courses and is seen as the fundamental knowledge that distinguishes engineers as experts in their field.

Engineering is distinguished from other disciplines such as those classified as the "pure" sciences by the inclusion of design subjects and project-based learning in the majority of degree curricula. Diversity, depth, and focus in implementation often reflect local value- priorities. In Godfrey and Parker's (2010) New Zealand study design courses had long been core and highly valued activities across all years/levels of the engineering degree, in contrast to engineering curricula in many other countries that appeared to restrict these courses largely to the first and final years of study.

One of the most basic assumptions at the case study institution was the belief that anything worthwhile was hard. The theme of "hardness" permeated the conversations of both faculty and students, conveying worth and status, with a devaluing of content or subject areas that were seen as "easy" or "soft." There were complex interwoven understandings around these words.

The heavy workload is one of the features of the engineering education culture that has been internationally recognized (Brainard, Staffin-Metz, & Gillmore, 1999; Seymour & Hewitt, 1997; Stevens, 2007; Stevens et al., 2008). Two factors appeared to contribute to the persistent perception of the heavy workload in engineering:

- The high proportion of the final grade gained from on-course assessment, often in the form of open-ended problems or projects. The nature of open-ended tasks was that there was always more that could be done to make the solution "better."
- The very prescriptive program structure, common to most professional degree programs, but in contrast to the structure of more flexible degree programs that enabled students to work to their strengths and avoid areas of weakness.

The frequent use of open-ended projects and assignments may contribute to perceptions of "hard" by their "challenge and stretch" teaching paradigm. The strength and ability to "take it" and succeed within this paradigm appeared to contribute to the pride and sense of achievement that students spoke of as an outcome of completing the degree. Similar beliefs, found in the multi-institutional Academic Pathways Study, have been described as a "meritocracy of difficulty" (Stevens, Amos, Jocuns, & Garrison, 2007).

A vast amount of literature covers topics that would be included as The Engineering Way of Doing. These include understandings about the appropriate ways to teach, and learn; curriculum design and delivery; course structures; assessment methods;

problem-based learning; how professional skills such as management, communication, ethics, and environmental responsibility should be incorporated into the curriculum; the balance between competitive and cooperative forms of student behavior; plagiarism; and time management. In particular, this dimension yielded several instances in which the values espoused in mission statements and goals were not necessarily matched in practice or in the shared understandings of staff and students. The most common example was the belief expressed by a majority of the faculty that time spent improving their teaching or developing course materials would not be rewarded in any way other than (hopefully) improved student satisfaction.

Being an Engineer

In engineering education, the development of identity has been cited (Seymour & Hewitt, 1997) as a key factor in retention of students in the discipline. The current focus on outcomes assessment and graduate attributes in accreditation procedures has focused attention further on the growth of a student's engineering identity. Research into enculturation (Tonso, 1996a), becoming an engineer (Stevens et al., 2008, Stonyer, 2002), and engineering identities (Chachra, Kilgore, Loshbaugh, McCain, & Chen, 2008; Pierrakos, Beam, Constantz, Johri, & Anderson, 2009; Tonso, 2007; Walker, 2001) has grown rapidly in the last ten years, especially with the acceptance of ethnographic research methods. Chapter 14 by Karen Tonso in this volume provides an in-depth discussion and synthesis of current research of this dimension.

The most notable feature of this dimension of the culture was the strength and pride with which both faculty and students commonly spoke of themselves as engineers rather than specifying their sub-discipline. This strong sense of pride, bordering on arrogance, was manifested in language, publications, dress, and the relationship to the rest of the university, as well as in the discourse that framed engineering as a "hard"

degree. "Engineering does have a special and valued identity. It is "I am an engineer" based on getting a job, working harder, doing something practical and useful . . . . . . The pride in being an engineer, combined with how much time you spend with classmates results in a family feeling" (Godfrey, 2003a, p. 257).

The rapid self-identification as engineers by students at the case study institution appeared to be a consequence of direct admission to the engineering program, with common first-year courses all taught by engineering faculty. In institutions where the norm is at least one year of fundamental science courses with only one "engineering" course, Stevens et al. (2008) saw students' identification as an engineer affected by when they were actually "admitted" to the engineering program. Once admitted, increased solidarity over time with other engineering students occurred.

Identity and a sense of belonging as an engineer have been suggested as closely linked to a male norm, or form of hegemonic masculinity (Stonyer, 2002; Tonso, 1999). Female participation at the time of Godfrey's study was approximately 20%. In the case study institution the norm was to be male; to be female was to be "other" – an accepted and often respected "other," but different nevertheless. This was an environment in which belonging continued to be associated with "being one of the guys."

Acceptance of Difference

Issues of homogeneity and diversity, with the implicit potential for shared or diverse values and norms, provide another dimension to be studied as part of a cultural analysis. Questions such as the following have rarely been addressed, other than for gender, in studies of the culture of engineering education: Is homogeneity desirable or necessary? How is difference accepted? Diverse educational or ethnic backgrounds have the potential to result in newcomers bringing diverse personal values and beliefs. In return, these would be likely to affect the speed and extent to which the individuals adapt and adopt the values and norms, and ultimately beliefs and assumptions of the disciplinary culture at a particular institution; indeed they have the potential to affect the culture itself.

In the case study institution, for example, a core of long-serving male faculty primarily educated in New Zealand had been a storehouse of cultural memory of "the way we do things round here" with a higher degree of homogeneity in educational values and background than might be encountered in other institutions. These assumptions regarding homogeneity of faculty values and cultural norms around teaching and learning have been increasingly challenged as older faculty retired and fewer new faculty were recruited locally. As one faculty member commented: "Some staff aren't assessing students the same way as other staff. It is not their fault really, they come from other universities with different expectations of students, and we have a set of unwritten guidelines."

When considering homogeneity and students, it is necessary to separate homogeneity in an academic sense from the social. The selection procedures in the case study institution, for example, made assumptions possible about homogeneity of academic background. High-achieving school leavers made up a very high proportion of new entrants, with considerably fewer numbers of "mature" age entrants. Assumptions regarding homogeneity in a social sense were increasingly challenged as changing immigration policies in the early 1990s saw marked changes in the academic and social backgrounds of the entry cohort. This diversity, coupled with increased female participation, as also noted by Bucciarelli et al. (2000), impact/act on academic experiences in the form of potentially diverse learning styles and different attitudes and abilities when working with others in team situations. Acceptance and inclusion of students, both academically and socially, regardless of race, gender, or age appeared to depend on fitting in with an engineering "norm," a situation seen by Tonso (2007) to have the potential to marginalize some students.

Relationships

The nature of relationships, the "right" way for people to relate to one another in a culture is an essential dimension in a cultural study. The unconsciously held beliefs and assumptions about appropriate levels of intimacy, and formality in a particular setting, determine responses and behaviors for members in the culture in both familiar and unfamiliar situations. Relationships particularly incorporate aspects of campus culture, institutional culture, national culture, and the values people bring from family, church, schooling, and community. It is challenging to discriminate and forefront those beliefs and assumptions that apply to the educational setting, while recognizing that this setting includes both in and out of class contexts.

As examples, two quite consciously held beliefs ran throughout the experience of both faculty and students at the case study institution. The first was the assumption that the school operated and would continue to operate in an integrated way, with strong interdepartmental cooperation a "given." This assumption was explicitly identified in strategic plans and conversation as well as manifested in administrative, financial, curriculum, and social interactions as evidenced by a department head's comment: " . . . manifests .for example in that we could run without a Dean for a year without major divisions, schisms or upsets." The integrated nature of the culture was also exemplified by the strong "family-like" relationships verbalized and enacted by both faculty and students: " . . . there is a feeling more of a big family and it is perhaps a little easier to relate to an engineer than someone with a BTech or another student."

In many engineering degrees, with prescribed curricula the opportunity afforded by prolonged close proximity, particularly in class groups, enables close relationships of varying degrees of intimacy to develop. Such cohesiveness cannot be assumed, however, and as with many aspects of culture is subject to external influences such as rapid growth of student numbers, increasing student diversity, the effects of flexible delivery modes, and increasing proportions of off-campus students.

The second belief about relationships was the importance of "mates" and the linking of friendship relationships to a sense of belonging in engineering. It was recognized that academic survival and success would be very difficult for any student who was marginalized or a "loner." Study teams that "stick together," studying, eating, and spending extraordinarily long hours together, were also identified by Stevens et al. (2008) as cultural norms in their study, particularly in relation to work on design or project assessments.

Other authors have written about a sense of "fit" (Foor, Walden, & Trytten, 2007; Lewis et al., 1998) and relationships (Seymour & Hewitt, 1997). In this study, the strength of the perceived need for these relationships emphasized the importance of what happened outside the classroom in the educative process, particularly an educative process characterized by the "challenge and stretch" paradigm or "hardness" identified earlier. The need to belong and have mates had the potential for students who might be viewed as different (by reason of age, ethnicity, educational background or gender) to be at risk in this environment.

Relationship to the Environment

Engineering education operates within the academic environment of higher education generally and more specifically its own institution, but also within an environment of oversight by the engineering profession. Taking a step even further back in the view of the cultural landscape, both higher education, and the professional environments themselves are also influenced by the political, economic, and social contexts of the nation.

The case study institution, for example, appeared to see itself as having a separate identity from the rest of the university, with the ability and desire to control its own destiny. A unified, "go it alone" mentality often prevailed. This may have been a legacy of the former time of physical separation from the main campus, but it also appeared to have its roots in a pride and sense of superiority in its ability to solve problems, whether the problems related to space usage, allocation of funding, timetabling, or parking. This pride extended to students as exemplified by the student comment: "It's almost like we're proud of not going to the rest of the university –across the road, we say "Ooh, I'm crossing the road" as if it is sort of shameful . . . " The symbolism of the often used term "across the road," epitomized the position of the engineering education culture within that of the main university, on the edge, rather than completely within the university as a whole. Similar symbolism was reported by Stevens et al. (2008), where intense habitation of engineering-specific spaces had led to the use of the term "down the hill" from the rest of the campus at one of the institutions they studied.

However, despite engineering's belief in its self-sufficiency, it was the university that conferred the qualifications, received government funding, and had regulatory control. Therefore, many of the behaviors and practices found in this case study reflected the beliefs and assumptions about teaching and learning found within the wider university. The university regulatory requirement for individual assessment, even in group projects, was one such example of engineering's belief in the valuing of teamwork being subsumed by the constraints of the overarching university.

A close relationship with the consulting and industrial engineering sectors was both valued and taken for granted. It was assumed that the curriculum would be not only monitored but also shaped by the needs of the profession.

Changing economic and political environments have the potential to affect changes in the prioritization of practices and behaviors. If these changes are sustained, shifts in values and cultural norms, which develop from changing behaviors and practices, have the potential to become embedded and ultimately become part of the culture as shared assumptions. Changing patterns of government education subsidies, for example, can lead to growth (or reduction) in undergraduate numbers, and the introduction of a performance-based research funding scheme noticeably increased pressure for faculty to focus on research at the expense of teaching. Demographic patterns and immigration policies are other examples of changing environments that may not affect the espoused values and ideals of engineering education, but do affect the reality of the enacted values and cultural norms by which the culture is sustained.

Institutional structures and culture are another aspect of this dimension. In countries such as Australia and America, with large numbers of engineering degree granting institutions, quite different institutional structures and cultures within engineering education are evident. Ranging from very traditional research-led universities concerned with their public rankings, universities that have grown from vocationally oriented technical colleges, to liberal arts colleges without a history of teaching engineering, the constraints imposed by structure and funding not only influence management and reward structures, but also permeate a variety of beliefs such as the importance of industrial experience and breadth versus depth of curriculum knowledge.

As mentioned earlier, the history and traditions embodied in national cultures also impact on how engineering education is enacted. Downey and Lucena (2005) have provided examples of the impact of national culture on values and cultural norms. In a study incorporating interviews with students from Australia, Germany, Thailand, and the United States, Walther, Boonchai, and Radcliffe (2008) graphically demonstrated the opportunities for truly international cross-cultural research provided the cultural assumptions on which research is based are recognized. Cultural research, with its implicitly interpretive nature, must therefore be context sensitive and adapt as appropriate.

Engineering Sub-disciplines

The majority of the literature pertaining to the culture of engineering education, and cultural change appears to assume that this culture is a unified and homogeneous culture with a significant lack of research into national, institutional, and discipline-specific subcultures within engineering education. Although little research is available specifically addressing the cultural differences between engineering sub-disciplines, members of multidisciplinary schools of engineering appear to recognize sub-disciplinary differences in aspects of the aforementioned cultural dimensions.

In the case study institution, civil, mechanical, and electrical engineering were the most traditional and well-established disciplines, with international as well as national professional associations. That background appeared to be linked to entrenched beliefs and assumptions about curriculum content, pedagogies, and professional issues appropriate to each discipline.

The discipline of civil engineering was linked to conservatism both by its traditions as the "oldest" engineering discipline bound by safety standards and regulations and by the dominance of its long-serving, male faculty members. As with any academic discipline, civil engineering had its own technical jargon, but the tangible, visible nature of its products and potential for "connection" with the subject matter appeared to be important for some students:

> Civil is more real in terms of what people experience every day – like you have pegs you can hang things on.

By way of contrast, the essence of mechanical engineering appeared to lie in the focus on design, the creative, innovative use of knowledge and theory. A practical "can do" attitude prevailed, with assumptions of prior knowledge closely aligned to male socialization norms.

Electrical engineering, compared to other disciplines that dealt with tangible outcomes, often seemed to be concerned with less visible or accessible concepts and products. A preponderance of terms such as impedance, resistance, bandwidths, and signal processing, which were accessible only to those with the same training, may have contributed to a sense of isolation and difficulty in communicating with nonelectrical engineers: "Most of us can have a reasonable discussion with a Civil about the new building being built next door but if we get into a discussion on electromagnetic radiation or interference or whatever because you can't actually point to it, it requires a language and a set of concepts – one can't make that kind of constructionist leap."

As another example, a senior faculty member from electrical engineering (rightly or wrongly) saw electrical engineering using high-order mathematics and a variety of visual images to analyze and solve complex but intangible systems requiring a high degree of abstraction, in contrast to disciplines such as civil engineering in which mathematics appeared to be used in more of a "plug in the formula" way. As a side issue, a sense of increased status appeared to be linked to subjects or research that incorporated a higher level of mathematics.

Space limits the opportunity at this time to develop sub-disciplinary differences further. A fruitful area for future research would be the investigation of the sub-disciplinary cultures within engineering, particularly for the dimensions Way of Thinking, Doing, and Being. Attempts to define a sub-disciplinary culture will, however, be complicated by the difficulty in separating epistemological features of the sub-discipline from the behaviors and practices emanating from the unique departmental or school culture in which that discipline is located. Bullard et al. (2010) suggested that each engineering department has a unique departmental culture based on its history, faculty makeup, geography, and a myriad of other factors. They recognized that the success of a department in creating and sustaining a desirable culture would have a significant impact on recruiting, retention, and general satisfaction of its undergraduate students. The comment (Godfrey & Parker, 2010) of a senior faculty member, "We are all engineers but the departments are like different tribes," will resonate with many.

Assumptions that emerged in the Engineering Way of Thinking dimension such as the quantifiable, definable, tangible nature of reality and truth are being contested by new, emerging, often interdisciplinary, fields of engineering such as software engineering, systems engineering, and biomedical engineering. The question "What is an engineer?" which presupposes common understandings of the term engineering, is challenged by these emerging disciplines. Although some of these fields may be considered too new to have developed epistemic traditions or disciplinary cultures, traditional beliefs and assumptions about "Engineering" Ways of Thinking, Doing, and Being do not match the values and practices already evident. As Williams (2002) has suggested, engineers cannot claim a unique professional identity or culture based on problem solving, and with many engineers now working in multidisciplinary teams, fuzzy boundaries are developing around what was a clear-cut professional identity.

The Impact of Disciplinary and Sub-disciplinary Cultures on Diversity and Inclusivity

As mentioned earlier, much of the work on engineering culture in the 1990s stemmed from researchers making a causal link between the engineering culture and the low participation of women. The masculine nature of the culture of engineering education has been named as a major inhibiting factor to the increased participation of women, with masculine values, norms, and assumptions identified not only at the level of social interaction and discourse, but also at the deeper levels of knowledge generation and transmission (Stonyer, 2002; Tonso, 1999). Godfrey (2003a) provided evidence that this masculinity pervaded each aspect of the cultural dimensions named earlier.

Over the last fifteen years, the plateauing of female participation at near 20% in the United States and other countries, has masked the continuing differential participation between the engineering disciplines and even between institutions (Godfrey, 2007). It is suggested that a connection exists between local versions of the engineering education culture and this differential participation.

A project report commissioned by the European Union (CuWaT, 1998) similarly commented that national, institutional and discipline specific differences appeared highly likely to provide a complex picture, where the engineering culture would be influenced by the nature and content of a discipline and how it was practiced in the "real world," as well as influenced and reinforced by how staff and students played their role.

The repetition of trends over time and location suggest that, if it is accepted that a relationship exists between the culture of engineering education and the underrepresentation of women, then it is reasonable to presume that studies of discipline-specific subcultures will provide insights into the persistence of the differences in female participation. Student perceptions in disciplinary differences have been investigated (Shivy & Sullivan, 2003; Trytten, Walden, & Reed-Rhoads, 2005), with Parikh, Chen, Donaldson, & Sheppard (2009) confirming significant differences in the motivation for choice by male and female students for different engineering majors.

Those subcultures (institutional or discipline-specific) with a high participation of women that appear to provide an inclusive cultural experience for women students and others of minority groups may provide the basis of appropriate cultural change strategies. It seems likely that they have developed norms, values, and beliefs that have "opened" up the culture of engineering education to value and welcome the participation of women. Similarly, an investigation of the features, both academic and social, of those disciplines that remain obdurately low in women's participation may provide revealing information about potential barriers.

Internationally the disciplines biomedical engineering, chemical and materials engineering, environmental engineering, and industrial engineering consistently have female participation figures of greater than 30%. Godfrey (2007) linked the attraction biomedical engineering and engineering science (a local near-equivalent to industrial engineering) had for female students to the level of connection with real-life identifiable problems despite the high level mathematics and very specialized computational mathematics techniques involved in such problems as modeling electrical activity around the heart. In electrical engineering, by contrast, the high degree of abstraction and intangibility of the outcomes appeared to mask connective links for many of the students. These observations confirmed the work of Murphy et al. (2007) investigating the culture within industrial engineering departments. They found strong evidence that both disciplinary and departmental cultural features were linked to high female participation. In particular, industrial engineering undergraduate majors viewed industrial engineering as more of "an approach to thinking" rather than a "focus on specific sets of problems," with approximately 80% of their respondents (both male and female) describing industrial engineering as "people oriented."

The hypothesis that disciplinary cultures in engineering are gendered and have a gendering effect of their own has also been supported by the work of Gilbert (2009) considering the culture of engineering research groups in Europe. Gilbert suggested that the group culture in materials science valued individuality and plurality, hence leaving more scope for gender diversity, whereas the group culture in mechanical engineering valued the subordination of individual needs to group norms and tended to reproduce features of homosocial male worlds.

Evidence from the studies mentioned appears to confirm that role models, perceptions of a sense of personal fit with the perceived social culture of a department or discipline, and retention in a discipline and on to postgraduate study is likely to be influenced by a sense of connection to the content and personal sense of confidence and belonging. The importance of relationships was emphasized in Murphy and colleagues' study (2007) suggesting that "most (but not all) of the students majoring in IE at OU described strong social networks which extended vertically to connections with the faculty and horizontally to a high level of camaraderie among the students themselves" and a comment from a student in Godfrey's study: "The atmosphere in Engineering Science was great – the faculty were all neat people who were really interested in us as people."

No one factor explains the persistence of sub-disciplinary participation rates. Rather, it appears to be the combination of several interrelated factors that become even more complex when other dimensions such as the influence of teaching and learning practices, the availability and type of relationships within the group, and the level of acceptance of diversity or difference are included.

The impact of sub-disciplinary cultures on other underrepresented groups has not been well explored. The author's experience in the Australian and New Zealand contexts suggests that a far greater proportion of indigenous students in both countries is attracted to civil and environmental engineering than to other sub-disciplines, but the reasons for, and validity of this perception, are yet to be investigated.

Sustained Cultural Change

The need to "know where we are" has been identified (Rover, 2008) as the first step in moving to "where we want to go." It has been the proposition of this chapter that an understanding of our culture can show us "where we are" and an understanding of how culture is formed and sustained "how we do things round here" can guide us in "how to get where we want to go."

Quick fixes, such as adding a module or course on ethics into the curriculum, or appointing a new faculty member from an underrepresented group, cannot in themselves change a culture. Sustained systemic cultural change, as suggested by professional and government bodies, requires shifts at the deepest level of shared beliefs and assumptions, followed by changed values and cultural norms.

Schein (1992) envisage change of this type initiated by strong and motivated leadership able to articulate beliefs and assumptions, and the values and cultural norms that would manifest them. The new vice chancellor at the case study institution, for example, had a vision and belief in the increased participation of underrepresented groups. He used his authority and persuasion to ensure that mission statements and strategic plans across all operating units included clear statements valuing the participation of all sections of the community. Annual reporting of operational practices in support of these strategies was then required. A variety of initiatives ranging from scholarships, outreach activities, mentoring schemes, and peer tutoring networks were funded manifesting these espoused values. After several years, these initiatives became accepted as cultural norms. It would be inaccurate to say that the beliefs of the leader were shared by all but pride developed in the university's role as a leader in the field of equal educational opportunity.

In practice, change emanating from shifts in shared beliefs and assumptions appears idealistic, and sometimes unlikely. Cultural change is more likely to come from strategic planning at the level of identifying the values and norms that are desired, and what behaviors and practices at an operational level would manifest those values and norms. Newcomers initially learn what is valued in a culture, from what they see as practices that lead to success and acceptance. If these behaviors and practices reflect desired values and norms, sustained cultural change can occur.

One of the calls for cultural change (IEAust, 1996) required engineering educators to produce graduates with an appreciation of the social, economic, and environmental consequences of their activities, and increased communication skills. If these generic skills were valued then it is clear that students must have opportunities not only to acquire them by their explicit inclusion in the curriculum, teaching, and assessment practices, but also to appreciate their value.

Similarly, if excellence in teaching is valued and part of the culture, rather than a goal aspired to in mission statements and publicity material, then that valuing must be embedded as a cultural norm in the reality of university practices. Reward systems such as public recognition and promotion could be viewed as indicators to faculty of those practices that were valued, just as good grades in assessment led students to understandings of what was valued in their learning.

For any department, discipline or institution considering cultural change, the process begins with stepping back, taking an objective view, first of the visible manifestations of the culture, then to identifying the taken-for-granted understandings and tacitly shared knowledge that underpin the everyday lived experience. Keeping in mind that shared understandings may not match espoused values or goals, a starting point could be for members of a group to reflect on "the way we do things round here" using questions such as:

What kinds of knowledge are valued? What is perceived as truth? Is there a prevalent way of thinking? What constitutes reality?

How is teaching and learning accomplished – what do our practices tell us about our assumptions of the "right" way to teach/learn?

Are there attributes and qualities inherent in being "an engineer"? Who fits in and is successful?

How is difference accepted and valued?

How do people relate to one another in this culture?

What is our relationship to the rest of the university and academia in general, the profession and the community?

Questions of this nature, which link well with the six cultural dimensions named in this chapter, are rarely raised as a topic for discussion in the normal day-to-day experience of engineering educators. By reflecting on questions such as these, individual departments, disciplines, and institutions would gain an understanding of the values, beliefs, and attitudes that dominate their own manifestation of the engineering education culture.

From the starting point of "knowing where we are" the process of implementing change could begin, constantly remembering that unless practices and behaviors unambiguously manifest the desired changes in values, beliefs, and attitudes they will not be sustained to become embedded in the culture. Explicitly articulating to both faculty and, where appropriate, students, the values and beliefs that underpin proposed changes and ensuring that sufficient support and resources are available to achieve effective implementation are essential in defining "where we want to get to."

Departmental and institutional cultures are fragile. Innovative and exemplary teaching and assessment practices, for example, can be interrupted by lack of funding for teaching assistant support, centrally controlled timetable scheduling, faculty changes, and other external drivers. An example of fragility is provided by the project initiated in an industrial engineering department (Murphy et al., 2007) that looked at the department culture and other factors as the source of the very high proportion of women (>50% at the start of the study). The decline of this proportion in subsequent years, albeit still to levels above the female participation in most other engineering disciplines, was identified as emanating from a variety of factors, including changes in course advising procedures, changes in faculty, and graduation of a particularly enthusiastic student cohort, all of which, as the authors suggest, demonstrate that seemingly small changes can have a large impact on a culture.

The need to be aware of the interaction of the culture of the institution with that of the discipline of engineering has been recognized. Efforts at curriculum change (Merton et al., 2004) and moving to an inclusive culture (Covington & Froyd, 2004) in the Foundation Coalition resulted in differing levels of success in implementing "best practice" curriculum change processes across diverse institutions. One of the factors identified as affecting the process of curricular change was the assumptions held by faculty (i.e., faculty culture) about how change occurs (Clark, Froyd, Merton, & Richardson, 2004).

Covington and Froyd (2004) suggested that creating pervasive, transformational change within engineering programs in higher education requires institutional change as well as change among the faculty, the principal population that develops and maintains the institutional culture. They cited Eckel and Kezar (2003), "A key part of transformation is changing mindsets, which, in turn, alters behaviors, appreciations, commitments, and priorities . . . people develop new beliefs and interpretations and adopt new ways of thinking and perceiving that help create the foundation of significant change."

Despite the apprehensions of culture theorists at attempts to quantify culture, or indeed to treat culture as an entity that can be manipulated to some desired end, there is often a need to evaluate or measure a desired cultural change, particularly following an intervention (Fromm & McGourty, 2001). Performance indicators of cultural change are the subject of a considerable body of literature, although rarely within higher education or engineering education. Considering inclusivity in engineering education, Mills, Ayre, and Gill (2011) provide an example of a set of performance indicators in their "Guidelines for faculty managers and academic staff to benchmark current and future performance in inclusivity," which built on

the publication "Benchmarks for Cultural Change" (Jost, 2004). The opportunity exists for further research in this area of evaluation of cultural change, particularly the identification of observable, measurable performance indicators that can demonstrate that espoused values, that is, the desired changes, have been developed and sustained to at least the level of becoming cultural norms.

Conclusions

National and international calls for cultural change, as mentioned at the beginning of this chapter, are not easily enforced. Driving forces such as changes in accreditation requirements to shift the focus of engineering programs from one of content to outcomes based assessment can provide incentives for change, but for many institutions change is slow. In Australia, for example, King (2008) reviewed progress since the 1996 report entitled "Changing the culture." He found that "curriculum change as proposed by the review has not been embedded nearly as far into the thinking of engineering schools as it should be. The content and methodology of many courses and programs has not changed substantially over the decade" (p. 29).

Considerable change and variation has become evident in engineering education institutions, indicating that generalizations about the culture must be approached with caution. It will continue to be important to locate cultures and cultural studies not only in the context of their physical or geographical space, but also in time, and with reference to their institutional, departmental, and social environment. Downey and Lucena's (2005) work demonstrated how important context is, commenting that "what counts as an engineer and engineering knowledge has varied over time and from place to place" (p. 252). As well as major curriculum changes such as the introduction of design- or project-based introductory courses in almost all first-year engineering programs, new disciplines discussed earlier have emerged in the last ten years that strongly challenge assumptions about the nature of engineering knowledge and practice.

The cultural dimensions suggested in this chapter provide a practical tool for evaluating and positioning the culture of engineering education in a specific context, particularly as a precursor to developing strategies for cultural change. The most essential "take home message" from this chapter is the recognition that change at the levels of curricula, structures, and behaviors is not sufficient for sustained cultural change. Cultural change requires transformation – forming new collective understandings and creating new beliefs about what is valued in engineering education.

It has become apparent that engineering education can be viewed as an overarching culture containing readily identifiable subcultures. As new engineering disciplines emerge and engineering education and the engineering profession respond to changing technologies and needs, particularly those that will be interdisciplinary in nature, it is likely that understandings of what counts as engineering knowledge, thinking, and practice will continue to evolve.

Exemplars of sustained cultural change are sorely needed so that one can "take from yesterday, to guide what is done today and move toward tomorrow."

References

ABET. (1996). *Engineering criteria 2000*. Engineering Accreditation Commission, Accrediting Board for Engineering and technology. Baltimore, MD.

Ambrose, S. (1998, June). Changing the culture: What's at the center of engineering education? In *Proceedings of Annual Conference and Exposition American Society for Engineering Education*, Seattle, WA.

Anon. (2006). The research agenda for the new discipline of engineering education. *Journal of Engineering Education*, 95(4), 257–261.

Baba, M. L., & Pawlowski, D. (2001, August). Creating culture change: An ethnographic approach to the transformation of Engineer-

ing Education. In *Proceedings of International Conference on Engineering Education*, Oslo, Norway.

Becher, T. (1989). *Academic tribes and territories: Intellectual enquiry and the cultures of disciplines*. Milton Keynes, UK: SRHE and Open University Press.

Becher, T., & Trowler, P. R. (2001). *Academic tribes and territories: Intellectual enquiry and the culture of disciplines*, (2nd ed.). Buckingham, UK: SRHE and Open University Press.

Bergquist, W. H. (1992). *The four cultures of the academy*. San Francisco, CA: Jossey-Bass.

Biglan, A. (1973). The characteristics of subject matter in different academic areas. *Journal of Applied Psychology*, 57(3), 195–203.

Borrego, M. (2008, October). Creating a culture of assessment within an engineering academic department. In *Proceedings of IEEE/ASEE Frontiers in Education Conference*, Saratoga Springs, NY.

Brainard, S., Staffin-Metz, S., & Gillmore, G. (1999). WEPAN Pilot Climate survey: Exploring the environment for undergraduate engineering students. Retrieved from http://www.wepan.org/associations/5413/files/Climate%20Survey.pdf

Bucciarelli, L, Einstein, H. H., Terenzini, P. T., & Walser, A. D. (2000). ECSEL/MIT Engineering Education Workshop '99: A report with recommendations. *Journal of Engineering Education*, 89(2), 141–150.

Bullard, L., Visco, D., Silverstein, D., & Keith, J. (2010, June). Strategies for creating and sustaining a departmental culture. In *Proceedings of the Annual Conference and Exposition of the American Society for Engineering Education*, Louisville, KY.

Chachra, D., Kilgore, D., Loshbaugh, H., McCain, J., & Chen, H. (2008, June). Being and becoming: Gender and identity formation of engineering students. In *Proceedings of the Annual Conference and Exposition of the American Society for Engineering Education*, Pittsburgh, PA.

Clark, C. M., Froyd, J., Merton, P., & Richardson, J. (2004). The evolution of curricular change models within the Foundation Coalition. *Journal of Engineering Education*, 93(1), 37–47.

Cordes, D., Evans, D., Frair, K., & Froyd, J. (1999). The NSF Foundation Coalition: The first five years. *Journal of Engineering Education*, 88(1), 73–77.

Courter, S. S., Millar, S. B., & Lyons, L. (1998). From the students' point of view: Experiences in a freshman engineering design course. *Journal of Engineering Education*, 87(3), 283–288.

Covington, K., & Froyd J. (2004, June). Challenges of changing faculty attitudes about the underlying nature of gender inequities. In *Proceedings of the Annual Conference and Exposition of the American Society for Engineering Education*. Salt Lake City, UT.

Cronin, C., & Roger, A. (1999). Theorizing progress: Women in science, engineering, and technology in higher education. *Journal of Research in Science Teaching*, 36(6), 637–661.

CuWaT (1998). *Changing the curriculum – changing the balance.* Uni of Central Lancashire, UK/European Union "Leonardo da Vinci Project.

Donald, J. G. (1984). Science students' learning: Ethnographic studies in three disciplines. In P. R. Pintrich, D. R. Brown, & C. E. Weinstein (Eds.), *Student motivation, cognition, and learning: Essays in Honor of Wilbert J. McKeachie* (pp. 79–112). Hillsdale, NJ: Lawrence Erlbaum.

Donald, J. G. (2002). *Learning to think: Disciplinary perspectives.* San Francisco, CA: Jossey-Bass.

Downey, G. L., & Lucena, J. C. (2005). National identities in multinational worlds: Engineers and "engineering cultures." *International Journal of Continuing Engineering Education and Lifelong Learning*, 15(3–6), 252–260.

Dryburgh, H. (1999). Work hard, play hard – Women and professionalization in engineering – Adapting to the culture. *Gender & Society*, 13(5), 664–682.

Eckel, P. D. & Kezar, A. (2003). Taking the Reins: Institutional Transformation in Higher Education, Westport, CT: Praeger.

Eisenhart, M. (2001). Educational ethnography past, present and future: Ideas to think with. *Educational Researcher*, 30(8), 16–27.

Foor, C. E., Walden, S. E., & Trytten, D. A. (2007). "I wish that I belonged more in this whole engineering group": Achieving individual diversity." *Journal of Engineering Education*, 96(2), 103–115.

Fromm, E., & McGourty, J. (2001, June). Measuring culture change in engineering education. In *Proceedings of the Annual Conference and Exposition American Society for Engineering Education*, Salt Lake City, UT.

Geertz, C. (1973). *The interpretation of cultures.* New York, NY: Basic Books.

Gilbert, A.-F. (2009). Disciplinary cultures in mechanical engineering and materials science. *Equal Opportunities International*, 28(1), 24–35.

Godfrey, E. (2003a). The culture of engineering education and its interaction with gender: A case study of a New Zealand university (Doctoral dissertation). Curtin University of Technology, Perth. Retrieved from http://espace.library.curtin.edu.au:80/R?func=dbin-jump-full&local_base=gen01-era02&object_id=14178

Godfrey, E. (2003b, June). A theoretical model of the engineering education culture: A tool for change. In *Proceedings of the Annual Conference and Exposition of the American Society for Engineering Education*. Nashville, TN.

Godfrey, E. (2007, June). Cultures within cultures: Welcoming or unwelcoming for women? In *Proceedings of 2007 Annual Conference and Exposition of American Society for Engineering Education*, Honolulu, HI.

Godfrey, E. (2009). Exploring the culture of engineering education: The journey. *Australasian Journal of Engineering Education*, 15(1), 1–12.

Godfrey, E., & Parker, L. (2010). Mapping the cultural landscape in engineering education. *Journal of Engineering Education*, 99(1), 5–22.

Hacker, S. (1983). Mathematization of engineering: Limits on women and the field. In J. Rothschild (Ed.), *Machine ex dea* (pp. 38–58). New York, NY: Pergamon Press.

IEAust. (1996). *Changing the culture: Engineering education into the future.* ACT: Institution of Engineers Australia.

IPENZ. (2003). *Requirements for initial academic education for professional engineers.* Wellington: Institution of Professional Engineers New Zealand.

Jamieson, L., & Lohmann, J. (2009). *Creating a culture for scholarly and systematic innovation in engineering education.* Phase 1 Report. Washington, DC: American Society for Engineering Education. Retrieved from http://www.asee.org/about-us/the-organization/advisory-committees/CCSSIE

Jost, R. (2004). Benchmarks for cultural change in engineering education. Newcastle: University of Newcastle.

Kezar, A., & Eckel, P. (2002). The effect of institutional culture on change strategies in higher education. *Journal of Higher Education*, 73(4), 435–461.

King, R. (2008). *Addressing the quality and supply of engineering graduates for the new century.* ALTC/ACED. Retrieved from http://www.olt.gov.au/project-ensuring-supply-quality-uts-2006

Kuh, G. D., & Whitt, E. J. (1988). *The invisible tapestry: Culture in American colleges and universities.* ASHE_ERIC Higher Education Report 17(1). Washington, DC: The George Washington University, Graduate School of Education and Human Development.

Lattuca, L. R., Terenzini, P. T., & Volkwein, J. F. (2006). *Engineering Change: A study of the impact of EC2000.* Baltimore, MD: ABET Inc.

Lewis, S., McLean, C., Copeland, J., & Lintern, S. (1998). Further explorations of masculinity and the culture of engineering. *Australasian Journal of Engineering Education*, 8(1), 59–78.

McGrath, E., & Sheppard, K. (2007, September). Growing a culture of intellectual inquiry in engineering education and research. In *Proceedings of International Conference on Engineering Education*, Coimbra, Portugal.

McKenna, A., Hutchison, M., & Trautvetter, L. (2008, July). *The engineer of 2020: Case studies of organizational features of effective engineering education.* Paper presented at the Research in Engineering Education Symposium, Davos, Switzerland.

McLean, C., Lewis, S., Copeland, J., Lintern, S., & O'Neill, B. (1997). Masculinity and the culture of engineering. *Australasian Journal of Engineering Education*, 7(2), 143–156.

Merton, P., Froyd, J. Clark, C., & Richardson, J. (2004, June). Challenging the norm in engineering education: Understanding organizational culture and curricular change. In *Proceedings of the Annual Conference and Exposition American Society for Engineering Education.* Salt Lake City, UT.

Mills, J., Ayre, M., & Gill, J. (2011). Guidelines for the design of inclusive engineering education programs. Australian Learning and Teaching Council. Retrieved from http://resource.unisa.edu.au/course/view.php?id=568

Murphy, T. J., Shehab, R. L., Reed-Rhoads, T., Foor, C. E., Harris, B. J., Trytten, D. A., Walden, S. E., Besterfield-Sacre, M., Hallbeck, M. S., & Moor, W. C. (2007). Achieving parity of the sexes at the undergraduate level: A

study of success. *Journal of Engineering Education*, 96(3), 241–252.

Parikh, S., Chen, H., Donaldson, K., & Sheppard, S. (2009, June). Does major matter? A look at what motivates engineering students in different majors. In *Proceedings of the Annual Conference and Exposition of the American Society for Engineering Education*, Austin, TX.

Pierrakos, O., Beam, T. K., Constantz, J., Johri, A., & Anderson, R. (2009, October). On the development of a professional identity: Engineering persisters vs engineering switchers. In *Proceedings 39th ASEE/IEEE Frontiers in Education conference*, San Antonio, TX.

Radcliffe, D. (2006). Shaping the discipline of engineering education. *Journal of Engineering Education*, 95(4), 263–264.

Radcliffe, D., Crosthwaite, C., & Jolly, L. (2002, June). Catalyzing cultural change in a research intensive university. In *Proceedings of the Annual Conference and Exposition American Society for Engineering Education*, Montreal, Canada.

Rover, D. T. (2008). Engineering Identity. *Journal of Engineering Education*, 97(3), 389.

Schein, E. H. (1985). *Organizational culture and leadership.* San Francisco, CA: Jossey-Bass.

Schein, E. H. (1992). *Organizational culture and leadership* (2nd ed.). San Francisco, CA: Jossey-Bass.

Seymour, E., & Hewitt, N. M. (1997). *Talking about leaving: Why undergraduates leave the sciences.* Boulder, CO: Westview Press.

Shivy, V., & Sullivan, T. (2003). Engineering students' perceptions of engineering specialties. *Journal of Vocational Behavior*, 67 (1), 87–101.

Stake, R. E. (1994). Case Studies. In L. K. Denzin & Y. S. Lincoln (Eds.), *Handbook of qualitative research* (pp. 236–247). Thousand Oaks, CA: SAGE.

Stevens, R., Amos, D. L., Jocuns, A., & Garrison, L. (2007, June). Engineering as lifestyle and a meritocracy of difficulty: Two pervasive beliefs among engineering students and their possible effects. In *Proceedings of the Annual Conference and Exposition American Society for Engineering Education.* Honolulu, HI.

Stevens, R., O'Connor, K., Garrison, L., Jocuns, A., & Amos, D. L. (2008). Becoming an engineer: Toward a three dimensional view of engineering learning. *Journal of Engineering Education*, 97(3), 355–368.

Stonyer, H. (2001). The problem of women in engineering – is it women, engineering academics, curriculum or engineering – where to act? *Australasian Journal of Engineering Education*, 9(2), 147–160.

Stonyer, H. (2002). Making engineering students – Making women: The discursive context of engineering education. *International Journal of Engineering education*, 18(4), 392–399.

Tierney, W. G. (1988). Organizational culture in higher education. *Journal of Higher Education*, 59(1), 2–21.

Tonso, K. L. (1996a). The impact of cultural norms on women. *Journal of Engineering Education*, 85(3), 217–225.

Tonso, K. L. (1996b). Student learning and gender. *Journal of Engineering Education*, 85(2), 143–150.

Tonso, K. (1999). Engineering gender-gendering engineering: A cultural model for belonging. *Journal of Women and Minorities in Science and Engineering*, 5, 365–405.

Tonso, K. L. (2006a). Student engineers and engineer identity: Campus engineer identities as figured world. *Cultural Studies of Science Education*, 1, 273–307.

Tonso, K. L. (2006b). Teams that work: Campus culture, engineer identity, and social interactions. *Journal of Engineering Education*, 95(1), 25–37.

Tonso, K. L. (2007). *On the outskirts of engineering; Learning identity, gender and power via engineering practice* (Vol. 6). Rotterdam, The Netherlands: Sense.

Trytten, D., Walden, S., & Rhoads, T. R. (2005, October). Industrial engineering student perceptions of computer science, computer engineering, and electrical engineering. In *Proceedings of ASEE/IEEE Frontiers in Education Conference*, Indianapolis, IN.

Walker, M. (2001). Engineering identities. *British Journal of Sociology of Education*, 22(1), 75–89.

Walther, J., Boonchai, C., & Radcliffe, D. F. (2008, July). *Understanding fundamental assumptions underlying educational research through the lens of a cultural 'Verfremdungseffekt.'* Paper presented at the Research in Engineering Education Symposium, Davos, Switzerland.

Williams, R. (2002). *Retooling: A historian confronts technological change*. Cambridge, MA: MIT Press.

Yost, S., Roberts-Kirchhoff, E., & Zarkowski, P. (2011, June). "We're all in the same boat": Promoting an institutional culture of assessment. In *Proceedings of the Annual Conference and Exposition of the American Society of Engineering Education*, Vancouver, BC, Canada.

CHAPTER 23

Preparing Engineering Educators for Engineering Education Research

Maura Borrego and Ruth A. Streveler

Introduction

The engineering profession is facing unprecedented challenges arising from globalization, poor public image, and low interest among students. To solve problems in sustainability, climate change, civil infrastructure, energy, and public health, the enterprise of engineering education must attract and retain a diverse group of students while preparing them to solve complex problems (Borri & Maffioli, 2007; Duderstadt, 2008; King, 2008). Clearly, this challenge requires effort from a wide range of stakeholders in industry, government, and both the teaching and research missions of academia.

In this chapter, we focus on research as just one position on a spectrum of inquiry activities that advances the collective goals of quality education of engineers. We seek not simply to distinguish engineering education research from other education-related activities, but to situate many teaching, assessment, evaluation, inquiry, and research activities with respect to each other and their complementary aims. We hope to convey to readers the benefits and limitations of each, as well as the necessity of efforts across the spectrum. Anyone with concern for the future of engineering education can contribute in a systematic way that will help to move collective efforts forward rather than continuously reinventing the wheel. We note that most of our experience, data, and theory are drawn from the U.S. context, so certain aspects (such as discussion of disciplines and departments of engineering education) may be less relevant to other countries. In addition, we limit our use of terms such as "scholarly" and "rigorous" that are popular in the U.S. context but may have less positive connotations elsewhere. Instead, we focus on quality inquiry through systematic, intentional, thoughtful efforts.

We begin by presenting an Inquiry Framework that compares and contrasts different types of engineering education work. Then, we present examples of programs for faculty members and graduate students that address each type of inquiry. After that, we focus more closely on engineering education research by describing the

Table 23.1. Engineering Education Inquiry Framework

Type of Inquiry	Attributes of that Type
Teaching	Teaching as taught.
Effective teaching	Involves the use of accepted teaching theories and practices.
Scholarly teaching	Good content and methods *and* classroom assessment and evidence gathering, informed by best practice and best knowledge, inviting of collaboration and review.
Scholarship of teaching and learning (SoTL)	Is public and open to critique and evaluation; is in a form that others can build on; involves question-asking, inquiry, and investigation, particularly about student learning.
Engineering education research	Also is public, open to critique, and involves asking questions about student learning, but it includes a few unique components. (1) Begins with a *research* question focusing on the "why" or "how" of learning (Paulsen, 2001). (2) Ties the question to learning, pedagogical, or social theory and interprets the results of the research in light of theory. This allows for the research to build theory and can increase the significance of the findings. (3) Pays careful attention to design of the study and the methods used. This will enable the study to hold up to scrutiny by a broad audience, again creating a potential for greater impact of results.

Adapted from (Streveler et al., 2007). The authors credit Hutchings & Shulman (1999) for types 1–3.

conceptual hurdles engineering academics face when learning to do engineering education research. We present the context for engineering education research as an emerging discipline. In the final sections, we move beyond our prior empirical work to offer suggestions for how engineering education inquiry and research in particular can be supported in higher education institutions.

The topics in this chapter are complemented by several others in this volume. In Chapter 21, Felder, Brent, and Prince delve much deeper into educational development (previously called "faculty development"). In Chapter 19, Litzinger and Lattuca provide strategies for using educational research to inform teaching. Several others describe specific aspects of engineering education research theories (Chapter 2 by Newstetter & Svinicki) and methodology (Chapters 27 by Case & Light; 25 by Kelly; 24 by Lattuca & Litzinger; and 26 by Moskal, Reed-Rhodes, & Strong).

Engineering Education Inquiry Framework

In this section, we present a framework that situates various engineering education inquiry activities with respect to each other. Readers are likely to be familiar with Scholarship of Teaching and Learning (SoTL), introduced by Boyer in 1990. Although this is a socially constructed concept that has gained the most traction in the United States where it was first published, literature from the United Kingdom (Yorke, 2003), Sweden (Roxa & Andersson, 2004), and Belgium (Clement & Frenay, 2010) also cites it as a useful way to think about the goals of educational development. In 1999, Hutchings and Shulman expanded the idea to include teaching and scholarly teaching to provide much-needed distinctions. Engineering education research was added in 2007 by Streveler, Borrego, and Smith.

The framework is presented in Table 23.1. Effective teaching is based on documented theory and practices of good teaching. Scholarly teaching is distinguished from effective teaching by the act of gathering assessment data and inviting collaboration and review by colleagues (Hutchings & Shulman, 1999). SoTL makes the work public, often by publishing results in a form others can use, such as a conference paper, thereby inviting public critique (Hutchings & Shulman, 1999). SoTL allows for the often very solitary act of teaching to be opened up to the community and for best practices to be more easily assimilated (Boyer, 1990; Hutchings & Shulman, 1999). However, the focus of SoTL usually is confined to very discipline-specific, even topic- or classroom-specific focus. To capture the broader work seeking to generalize or theorize across specific settings, engineering education research was added (Streveler, Borrego, & Smith, 2007). The difference between SoTL and engineering education research lies in the impact of the study – from individual topic-specific classroom to a larger population of learners. In engineering education research, educational and social science theories are used to shape the research design (Yorke, 2003), and the implications of the research are seen in terms of student learning as a whole – not just within one kind of classroom. Our other publications expand upon these differences (Borrego, 2007a; Borrego, Streveler, Miller, & Smith, 2008; Streveler et al., 2007).

The order of the types listed in Table 23.1 highlights the journey many engineering academics take in their own engineering education work. For example, initial interest may be motivated by very specific and localized problems, but as assessments are formalized and the work is made more public, opportunities for collaboration and engagement with broader questions present themselves. In his discussion of U.K. higher education, Yorke describes a very similar "spectrum of engagements" with pedagogical research: "there are stages through which new pedagogical researchers might progress until they reach the peaks of pedagogical research capability as represented by gaining major external funding" (Yorke, 2003,

Figure 23.1. Innovation cycle linking educational research and practice in engineering. (From ASEE [2009], who adapted the cycle from Booth, Colomb, & Williams [2008].)

p. 115). This is not to say that each step is mastered before moving to the next or that each individual is equally suited to each type of inquiry. In fact, individuals may be at different phases simultaneously for different projects or courses (Borrego, Streveler, et al., 2008). This is a major reason it is difficult to describe engineering education research without discussing other types of inquiry. In the Conceptual Hurdles section, we discuss the conceptual difficulties engineering faculty members encounter when moving through these phases.

In previous publications, this framework was presented as a table with numbered levels that brought with it the connotation that some kinds of inquiry are better than others. (For example, one might think that engineering education research is somehow more valuable than SoTL.) On the contrary, we are not proposing that this is the case. In describing the work that contributes to improving engineering education, we and others prefer a cyclic representation that better captures the iterative nature of this work. A five-component cycle had been used in a longstanding U.S. National Science Foundation (NSF) funding program: Course, Curriculum and Laboratory Improvement (reproduced in Jesiek, Borrego, & Beddoes, 2010a, 2010b). A more current and simplified version is presented in the report *Creating a Culture for Scholarly and Systematic Innovation in Engineering Education* (American Society for Engineering Education [ASEE], 2009). As depicted in Figure 23.1, research results do not simply "trickle down" into practice; real problems encountered in educational practice inspire research questions, a process that helps to ensure that the findings will be useful. This also reinforces the important value of all types of engineering education inquiry.

Before narrowing our focus specifically to engineering education research, we briefly review programs for faculty members and graduate students to develop their skills in these different types of engineering education inquiry. The list is not intended to be exhaustive; examples were selected on the basis of longevity, broad national or international scope, and availability of archival references.

Programs for Faculty Members

Programs Addressing Effective Teaching, Scholarly Teaching, and Scholarship of Teaching and Learning

In Chapter 21 in this volume, Felder, Brent, and Prince focus on "programs designed to help engineering instructors improve their teaching and their students' learning" while only briefly mentioning other faculty responsibilities such as research. (We complement this chapter by touching only briefly on the teaching improvement programs Felder, Brent, and Prince describe in detail.) As a specific case, they describe the National Effective Teaching Institute (Felder & Brent, 2010), which has been offered in conjunction with the American Society for Engineering Education conference since 1991. However, they also describe general considerations for designing effective international, national, and local engineering educational development programs.

Programs Addressing Engineering Education Research

Programs designed to assist engineering faculty to conduct engineering education research appear to have been first offered in 2004. One of the most well-known, Conducting Rigorous Research in Engineering

Education: Creating a Community of Practice (RREE), was funded by the U.S. National Science Foundation and presented week-long workshops for approximately 150 engineering academics from 2004 to 2006. The RREE introduced engineering faculty members to the theories and methods of engineering education research, while building a community of practice among engineering education researchers (Streveler, Smith, & Miller, 2005). The sense of belonging to a community of engineering education researchers was a powerful, long-term result of the RREE, and was most important to academics who felt isolated and unsupported in their interest in engineering education research (Allendoerfer, Streveler, Clarke Douglas, & Smith, unpublished data). At about same time as the RREE, a second NSF-funded project to build engineering education research capacity was also underway in the U.S. The Institute for Scholarship on Engineering Education, an element of the Center for the Advancement of Engineering Education, brought instruction on the theories and methods of engineering education research to engineering graduate students and faculty members (Atman et al., 2010). Along similar lines, the Australasian Association for Engineering Education's (AaeE) Educational Research Methods Special Interest Group (2008) has compiled online resources and presents workshops on engineering education research methods to engineering academics in Australia and New Zealand.

The aforementioned programs mentioned were face-to-face events. More recently, efforts have been made to reach a virtual audience. For example, the Collaboratory for Engineering Education Research (CLEERhub.org) is designed to act as a knowledge base, learning environment, and collaboration space for engineering education researchers spread throughout the globe (Malik et al., 2011). Just as the design of the RREE workshops was informed by Wenger's theoretical work on communities of practice (Wenger, 1998; Wenger, McDermott, & Snyder, 2002), CLEERhub was informed by Wenger's more recent *Digital Habitats: Stewarding Technology for Communities* (Wenger, White, & Smith, 2009). In this later work, Wenger et al. explain that three inherent polarities challenge virtual communities: togetherness and separateness, participation and reification, and individual and group (2009). Input from engineering education research experts familiar with technology emphasized well-organized content that is complete and easy to search, content contributions from the community, and the ability to see others' work and engage with authors. These findings were reflected in the literature, along with the need for a community manager to maintain quality and engagement with the virtual community (Streveler, Magana, Smith, & Clarke Douglas, 2010). This perspective allows program developers to look beyond the content that is needed to help engineers better understand engineering education research, to see how activities can be created which foster community members learning from each other.

One example of how a virtual learning community might evolve was created by a small group of officers in the United States Military. CompanyCommand.com is a website that provides a "virtual meeting place where [Army] company commanders could share with one another what they were learning about company-level command" (Dixon, Allen, Burgess, Kilner, & Schweitzer, 2005, p. v). First launched in the spring of 2000, and initially supported only by the volunteer efforts of the four core group members (authors Allen, Burgess, Kilner and Schweitzer), participation in CompanyCommand grew rapidly, and in 2002 the website was "gifted" by the core group to the U.S. military, which now hosts the servers and support staff. The philosophy of CompanyCommand is that connecting members of this community (past, present, and future company commanders) allows for conversations that have grown into content. This organic growth of knowledge via community conversations is aligned with Community of Practice models such as those underlying the development of CLEERhub.

Programs for Graduate Students as Future Faculty Members

Programs Addressing Effective Teaching, Scholarly Teaching, and Scholarship of Teaching and Learning

Given the constraints and challenges of educational development, a particularly promising practice is to train graduate students, as future faculty members, in the skills they need to be effective teachers and potential educational investigators (American Society for Engineering Education [ASEE], 2012). Felder, Brent, and Prince (2011) briefly highlight the major sources of information on how to do this. Interest is high; a special session on "Connecting Engineering Education Research Programs from Around the World" at the 2010 American Society for Engineering Education Annual Conference included eight U.S. universities describing engineering education-related certificate programs. Crede, Borrego, and McNair (2010) reviewed some of the engineering-specific programs presented in ASEE conference papers, while research on mentoring teaching assistants (Cox et al., 2011; Crede et al., 2010) provides new models and evidence for recommended practices.

Beyond but including engineering, Preparing Future Faculty (PFF) (Council of Graduate Schools, http://www.preparing-faculty.org/) has provided resources and support since 1993 to hundreds of programs at universities across the United States, including a guide written specifically for math and science disciplines (Pruitt-Logan, Gaff, & Jentoft, 2002). In addition to a focus on scholarly teaching, these programs may also include training in time management or balancing teaching and research responsibilities.

Programs Addressing Scholarship of Teaching and Learning

One particularly successful STEM-focused project emphasizing SoTL has been the Center for the Integration of Research on

Teaching and Learning (CIRTL, http://www.cirtl.net/), a project "committed to advancing the teaching of STEM disciplines in higher education." CIRTL has advanced the idea of "teaching-as-research" and concepts clearly tied to the SoTL. CIRTL provides a variety of courses, workshops, and materials to advance effective teaching practices. The Center also hosts the CIRTL Café, an online community to connect those interested in effective science, technology, engineering, and mathematics (STEM) teaching with each other.

Programs Addressing Engineering Education Research

Graduate programs focused on engineering education research are most often at the doctoral level. A number of other sources list the universities offering these degrees (Benson, Becker, Cooper, Griffin, & Smith, 2010; Borrego & Bernhard, 2011; Streveler & Smith, 2010) including institutions in Brazil, Denmark, India, Malaysia, Mexico, Sweden, and several in the United States. A few notable smaller-scale programs for training graduate students in engineering education research are the certificate offered by the University of Michigan (Streveler & Smith, 2010) and the Ph.D. course offered by the Engineering Education Working Group of Société Européenne pour la Formation des Ingénieurs (Société Européenne pour la Formation des Ingénieurs [SEFI; European Society for Engineering Education, http://www.it.uu.se/research/group/upcerg/EER?lang=en).

Fensham (2004) has argued that a new discipline creates its own theories and methods. In the field of engineering education, theories and methods are still largely being borrowed from education and the social sciences (Streveler & Smith, 2010). This situation leads to a lack of a common core curriculum in engineering education research, which is an ongoing challenge. However, recent efforts to create journal special issues and handbooks such as this one are beginning to define a core of knowledge for this emerging field.

Employment Prospects

The expanding availability of graduate-level engineering education certificates and degrees raises the question of employability. A 2005 survey (Borrego, 2006) of seventy engineering academic deans provides some empirical evidence, but it should be noted that this was conducted prior to the graduation of any Ph.D.s from U.S. departments of engineering education. Two-thirds of respondents thought that a certificate in engineering education (with a Ph.D. in traditional engineering) would increase desirability of faculty candidates in traditional engineering departments, provided that teaching-related training was not pursued at the expense of quality research training. Between 33% and 51% of deans considered master's and doctoral degrees in engineering education as desirable credentials for various university-level faculty positions (instructors and tenure-track). Many acknowledged the importance of pedagogy but noted that engineering academics conducting engineering education research would not be eligible for tenure at their institution. Responses were more promising regarding faculty hires for first-year engineering programs, in which many U.S. engineering education doctoral students gain their teaching experience (Borrego, 2006).

Candidates with master's and doctoral degrees in engineering education were also viewed as desirable for a variety of academic staff positions in advising, minority programs, and assessment (50–80% somewhat desirable + very desirable, depending on degree and type of position). In open-ended responses, many deans expressed concern that graduates would be overqualified, but a comparison set of job postings from the same time period did not reflect most of the objections raised in the survey responses, frequently requesting credentials including doctorates and a technical degree (Borrego, 2006).

More recent experience based on employment of new engineering education Ph.D.s is even more promising. For example, the first sixteen graduates of Purdue

University's engineering education Ph.D. program have all found employment: in faculty positions (two engineering education, one mathematics education, two civil and mechanical engineering, and one electrical and computer engineering and computer science), research, administration, or in a science museum (Benson et al., 2010; Streveler & Smith, 2010). Similarly, Ph.D.s earning Virginia Tech's certificate in engineering education are employed in a variety of engineering faculty and industry positions (Benson et al., 2010).

Conceptual Hurdles

Now that we have described the different types of engineering education inquiry and presented examples of programs for faculty members and graduate students addressing each type, we first review some of the challenges that trained engineers face when moving through the Inquiry Framework toward engineering education research. In 2007, Borrego published a study of engineering academics learning to conduct engineering education research that identified five conceptual difficulties. The first challenge is closely related to moving through the types of inquiry: framing questions with broad appeal. Engineering faculty members are often initially attracted to engineering education venues such as conferences and workshops because these events address problems specific to their classrooms. This local focus is reflected in initial attempts to frame research questions, which at first tend to be too focused on a specific course at a specific institution. By engaging in inquiry (perhaps initially types other than research), academics increase their awareness of related work, including the prevalence of similar problems among instructors of similar courses. One marker of the transition to research would be framing a question initially stated as, "What contributes to the low retention rate of students at my institution?" to "What factors affect the retention of engineering students at historically black colleges and universities?" (Borrego, 2007a).

The second difficulty is grounding research in a theoretical framework. Academics from a wide range of disciplines (including engineering) will tend to approach their teaching and educational research as they do their technical work (Huber, 2002; Lattuca & Stark, 1995), not even realizing that they assume the research process to remain essentially the same. However, educational research is different from engineering research in some very fundamental ways. Engineering academics who are used to researching inanimate objects may find that student variables cannot be controlled as easily[1] (Wankat, Felder, Smith, & Oreovicz, 2002). The complexity of human research subjects can make it much more difficult to identify what data and interventions to focus on. To concentrate on a specific phenomenon and the important factors influencing its outcomes, for example, cheating, a theoretical framework must be selected from among several viable options. Engineering research also uses theoretical frameworks, but they may be more straightforward, for example, encoded in equations. Engineering academics may be overwhelmed by the range of social science theories, which could result in more anxiety over selecting the "wrong" theory when conducting educational research (Borrego, 2007a).

The third challenge is that once student variables are identified, measurement is not trivial. In fact, it is more appropriate to refer to complex variables such as creativity, motivation, and intelligence as "constructs" because measuring them is not as straightforward as connecting a thermocouple or other sensors. Many years of research effort in cognitive psychology and related fields have developed theory and instruments (e.g., surveys) to measure them. Thus, considering the validity of measurement instruments is a potentially more significant step in educational research than in some other areas of engineering research, where the limitations of widely used measurement devices are relatively well documented and well understood (Borrego, 2007a).

When studying students, qualitative methods may also be more useful than in traditional engineering research, particularly when studying a phenomenon for which there is not much existing theory or prior research, or when seeking to understand the underlying reasons for a surprising quantitative result. Appreciating qualitative and mixed methods is the fourth conceptual difficulty, which can be overcome by deep thinking about how to answer a research question framed with broad appeal that focuses on "how" or "why" (Borrego, 2007a).

Fifth and finally, trained engineers wishing to conduct educational research should recognize that potential collaborators in education, psychology and similar departments are well versed in the methods, theories, and literature foundational to this kind of work (Borrego, 2007a). Borrego and Newswander (2008) note that collaboration, in addition to reading the literature and formal coursework, is a primary mechanism for engineering faculty members to learn how to do educational research. Participants in engineering education research sessions at international conferences on six continents identified all branches of engineering, social and cognitive psychology, sociology, education, ethnic studies, women's studies, international relations, and other STEM or STEM education disciplines as the expertise needed for engineering education research (Borrego, Beddoes, & Jesiek, 2009). In other words, beginning engineering education researchers are not expected to master a second discipline. Collaborators who are expert in relevant literature, theories, and methods are available at most higher education institutions and other research organizations, so the journey need not be undertaken alone.

Engineering Education Research as an Emerging Discipline

In the United States, there has been much debate about whether engineering education research is or should be a discipline and the role of various organizational structures (e.g., departments, centers) in supporting engineering education inquiry. We take the stance that engineering education is an emerging field, rather than a discipline, to emphasize its openness to interdisciplinary approaches and scholars. A summary of the arguments provides additional context for preparing and subsequently supporting engineering educators in their inquiry.

Despite claims about the establishment or "birth" of a new discipline of engineering education (Haghighi, 2005; Wankat, 2004), many engineering education researchers are uncomfortable with the implications of this assertion. Jesiek, Newswander, and Borrego analyzed discourse at the 2007 International Conference on Research in Engineering Education (ICREE) to find that many participants preferred the connotations of "field" or "community" to those of "discipline" (Jesiek, Newswander, & Borrego, 2009). This conference was attended by sixty-eight engineering education researchers from the United States, Europe, Australia, Africa, South America, Canada, and Asia, and outcomes included a special issue of *Journal of Engineering Education* (Jesiek et al., 2009). Identification as a discipline provides certain stability and resources, for example, departments, which are particularly important in the United States (Abbott, 2001), along with clarification of what "counts" as research. But the act of defining what counts as research necessarily also identifies what does not count. This exclusionary nature of disciplines is the aspect that many object to. Rather, engineering education researchers might identify with greater openness to researchers, theories, methods, and perspectives from other disciplines that characterizes "fields" or "communities." At the ICREE conference, the term "community" was used in reference to social networking, idea sharing, and collaboration. "Field" was used as an uncontroversial alternative to "discipline" that implied a more open and inclusive concept, more inviting and more compatible with an interdisciplinary identity. Some participants were also concerned that disciplinary engineering education research

Table 23.2. Typical Characteristics of Disciplines, Communities of Practice, Fields

Concept	Typical Characteristics
Discipline	• Mainly located in the academy (universities, colleges, etc.) • Structure often includes departments, degree programs, professional societies, journals, conferences, research centers • Claims to academic work and disciplinary knowledge are complex, shifting, and based on heterogeneous criteria (Abbott, 2001) • Do not pass through regular stages of development, but instead temporarily achieve consensus and recognition (Good, 2000)
Community of Practice	• Based on a *domain* of knowledge, *community* of people, and shared *practice(s)* (Wenger et al., 2002) • Organizations and institutions serve as backdrops (Wenger, 1998)
Field	• An area of activity, operation, or investigation ("field, n.," 1989) • May be broad (mechanical engineering) or narrow (kinematics)

Reproduced with permission from Jesiek et al. (2009).

could become too academic and therefore too far removed from practical application in engineering classrooms (Jesiek et al., 2009).

The literature on disciplines, communities of practice, and fields provides additional guidance. Table 23.2 is reproduced from Jesiek, Newswander, and Borrego (2009). Field is the least well defined, which explains its role as an uncontroversial label for engineering education. Communities of practice focus more on the self-organization of people around a practice than on the organizations and institutions that may support (or inhibit) them (Wenger, 1998; Wenger et al., 2002). Thus, discipline is the term most associated with structures.

Fensham's (2004) criteria for defining the discipline of science education research provide a simple framework for summarizing extensive literature on the development of disciplines. As listed in Table 23.3, structural criteria are separated from research criteria. The engineering education journals, associations, conferences, centers, and doctoral programs addressing the structural criteria are enumerated elsewhere (Borrego, 2007b; Borrego & Bernhard, 2011; Litzinger, 2010).

Research criteria are associated with the intellectual coherency of the discipline. Jesiek et al. (2009) cites Mahoney's explana-

tion that disciplinary unity is achieved when a group establishes its own agenda, or "what its practitioners agree ought to be done, a consensus concerning the problems of the field, their order of importance or priority, the means of solving them, and perhaps most importantly, what constitutes a solution" (Mahoney, 2004, p. 9). Similarly, Burrell and Morgan define a paradigm as a "commonality of perspective which binds the work of a group of theorists together" (1979, p. 23). In the terms used by Thomas Kuhn in his seminal book, *The Structure of Scientific Revolutions* (Kuhn, 1970), a paradigm is defined by consensus about such vital issues as standards of rigor in research, including important questions, accepted methods, and forms of convincing evidence (Guba, 1970).

We have argued that engineering education research is *pre-paradigmatic* (Borrego, 2007b; Borrego, Streveler, et al., 2008), as recent debate (Gabriele, 2005; Streveler & Smith, 2006) suggests that the guiding philosophies or paradigms that will later guide engineering education research appear to be still in formation. The implications for engineering education are many. Differences between high- and low-consensus disciplines are well documented (Braxton & Hargens, 1996), and these serve to explain

Table 23.3. Fensham's Criteria for Defining the Discipline of Science Education Research

Structural Criteria	Research Criteria	Outcome Criteria
Academic Recognition	Scientific Knowledge	Implications for Practice
Research Journals	Asking Questions	
Professional Associations	Conceptual and Theoretical	
Research Conferences	Development	
Research Centers	Research Methodologies	
Research Training	Progression	
	Model Publications	
	Seminal Publications	

Reproduced with permission from Jesiek et al. (2009).

some of the challenges that engineers face when moving from high-consensus traditional engineering disciplines to low-consensus engineering education. Hargens and Kelly-Wilson (1994) conducted a factor analysis to show statistically that a single underlying factor, whether titled paradigm development or consensus, appeared to account for the observed differences in co-authorship, publication length, and journal rejection rates. In short, it can be said that publishing in engineering education is objectively more challenging than in more traditional engineering disciplines because the criteria are less clear, and as a result, acceptance rates would be lower in engineering education journals. (As described in detail under Valuing Engineering Education Research and Other Inquiry in Institutional Contexts, journal acceptance rates can be powerful evidence to use when arguing that engineering education research should be valued as highly as traditional engineering research.)

This discourse makes it an exciting time to be involved in the engineering education community. However, it also emphasizes the need to support engineering educators as they learn to conduct inquiry. In the preceding arguments, community was advocated because it would be most open to new people and new perspectives. Welcoming newcomers is indeed important for building capacity to improve engineering education, as an increasing number of engineering education interest groups within professional societies are doing (ASEE ERM, n.d.; AaeE,

2008; SEFI, 2008; SEFI, n.d.). In addition, the collaborative aspects of community also provide opportunities for knowledge and skills development, along with a sense of identity and the support required to sustain researchers through challenging periods. This applies to both local and distributed communities. However, in the next section, we focus on local institutional contexts.

Valuing Engineering Education Research and Other Inquiry in Institutional Contexts

In this section, we move beyond our prior empirical work to offer suggestions for how engineering education inquiry generally, and research in particular, can be supported in higher education institutions.

As mentioned earlier, engineering education is an emerging discipline, and there are some conceptual hurdles to overcome when engineering academics first begin this work. Given this situation, it is not surprising that there are also important considerations for how the work is perceived by colleagues and administrators. This is where a strong understanding of the Inquiry Framework in Table 23.1, including the goals of each type and some representative activities, can help make the case for the value of engineering education. In terms of the traditional faculty responsibilities of research (or scholarly activity), teaching and service, the types of inquiry in Table 23.1 span teaching and research.[2]

This section is intended to address the widely held notion that promotion and tenure and other faculty reward systems are a disincentive to most engineering academics exerting too much effort on their teaching, and by extension, engineering education inquiry (Borrego, Jesiek, & Beddoes, 2008). It is widely lamented that despite efforts to integrate the knowledge building mission of higher education, the nature of faculty work has split into a dichotomy of teaching and research activities, with research typically receiving priority (Cooper, 2004; Gordon, 2004; Kolmos, Gynnild, & Roxa, 2004; Naukkarinen & Malmi, 2004; Roxa & Andersson, 2004). "Put simply, the argument is that the ambitious, even prudent, academic should maximize the allocation of their discretionary time to their research rather than to enhancing their teaching" (Gordon, 2004, p. 13). The situation is complicated further, as "Increasingly, engineering faculty who are interested in teaching are becoming 'education developers' for their department or faculty. In other words, they often take on the role of facilitating others to become good teachers" (Baillie, 2007, p. 421), which can reinforce their identity as more teaching-focused than research-focused.

This research–teaching dichotomy is particularly troubling given the evidence that engineering administrators tend to view all types of engineering education inquiry as teaching. Emerging results from a national survey of U.S. engineering deans and department chairs indicate that, for the most part, engineering education inquiry is viewed primarily as a teaching activity and not as a viable research area (ASEE, 2009, 2012; Olds, Borrego, Besterfield-Sacre, & Cox, 2012). (These administrators were asked to respond to the model in Figure 23.1 and accompanying text designed to convey a cyclic relationship between various types of engineering education inquiry.) Thus, more work needs to be done in helping administrators understand the role of various types of engineering education inquiry in the academy.

Beyond the tendency to view all engineering education inquiry as teaching, there are

structural organizational challenges to supporting and recognizing engineering education inquiry. If an institution places more value on teaching excellence (not necessarily at the expense of research), then this opens the door to engineering education research. Conversely, lack of attention to quality teaching can prevent engineering education research from ever taking hold (Yorke, 2003). Government and institutional policies vary greatly and can significantly influence attention to teaching (Kolmos, Vinther, Andersson, Malmi, & Fuglem, 2004; Saroyan & Frenay, 2010). For example, if some instruction in didactics or pedagogy is required for faculty members to earn tenure (as in Denmark), then more resources will be directed to educational development centers and programs. It is more efficient from an institutional standpoint to staff one centralized teaching and learning center, but engineering academics are less likely to access it owing to perceived (or actual) lack of engineering-specific expertise (Felder et al., 2011). Discipline-specific units, whether they are programs, staff members, centers, or departments, lower the barrier for engineering faculty members to seek assistance in developing their inquiry skills, and increasingly, these centers are expanding to include more resources for engineering education research (Litzinger, 2010).

Beyond these faculty development arguments, there is strong precedent for engineering education research to be conducted in departments, schools, and colleges of engineering instead of relying on education units. Across STEM disciplines, the label discipline-based educational research (DBER) collectively refers to educational research conducted by disciplinary experts such as engineering or physics faculty members (National Research Council, 2012). Particularly at the undergraduate and graduate levels, a high degree of disciplinary knowledge is required to investigate student learning and, for example, identify misconceptions (Streveler, Brown, Herman, & Montfort, 2012). Being embedded in the engineering organization facilitates access to research participants (e.g., students) and dissemination of findings to practitioners who can apply them.

As we noted earlier, engineering academics who were initially trained as engineers may take time to work up to full-scale engineering education research. Effective teaching and scholarly teaching correspond to teaching. Both of these are focused on raising the profile of teaching by keeping up-to-date with engineering content and pedagogical procedures. In *Scholarship Reconsidered*, Boyer (1990) argued that the Scholarship of Teaching (now called the Scholarship of Teaching and Learning [SoTL]) was a kind of scholarship and should be considered as such. However, because SoTL usually focuses on improving the learning experience of students, it could be perceived by promotion and tenure committees as falling into the category of teaching, not research.

We and many of our colleagues feel strongly that inquiry that genuinely qualifies as engineering education research should be categorized as research in terms of the traditional faculty responsibilities of teaching, research, and service. Several parallels can be drawn between traditional engineering research and engineering education research. Both ask significant questions of broad appeal. Both are informed by the findings of previous studies, particularly those that link to theory, which would predict cause and effect and help identify important variables.[3] Both involve careful consideration of data collection and analysis methods, including limitations and trade-offs. Both draw only conclusions that logically follow from the empirical data, and both consider alternative explanations. Both are situated with respect to prior work and seek to extend relevant theories. Finally, both are published in high-quality, peer-reviewed journals (Borrego, Streveler, et al., 2008; Shavelson & Towne, 2002). In addition to journal publications, other traditional measures of research productivity such as research grants and prestigious awards can signal that this type of inquiry has more in common with research than with teaching.

When arguing to administrators that engineering education research should be credited as research, it is important to draw these parallels. Acceptance rates are available from the relevant professional societies and should be used in these arguments. In addition, impact on practice is an important value in engineering education.[4] However, when drawing parallels to traditional engineering research, it may be important to note that, for example, thermodynamics researchers are not denied tenure if automobiles do not become more fuel-efficient as a direct and demonstrable result of their research.

Advocates of engineering education research sometimes make two arguments simultaneously, with different emphases and to different audiences. On one hand, they argue that engineering education research is as rigorous as traditional engineering research and should be regarded with the same respect. On the other hand, they argue that teaching should be more valued in the academy. But many in the intended audience hear only one message that may be an inaccurate combination of the two. For example, many administrators' primary engagement with engineering education is through venues such as the ASEE annual conference (ASEE, 2012). Even if they are not intimately familiar with criteria for high-quality engineering education research, they realize that the SoTL presented at these conferences might not be research. When a faculty member argues that "engineering education is (or can be) research," administrators may hear, "engineering education conferences are as research-focused as conferences in other engineering fields." This argument is quickly dismissed, and the considerations shift to whether engineering education conference papers should be categorized for the purposes of promotion and tenure as teaching, service or research contributions. Travel to these conferences and debriefing afterwards can be an important opportunity to explain to administrators that Engineering Education Research tracks (i.e., at SEFI) and Educational Research and Methods sessions (i.e., at ASEE, http://erm.asee.org) have higher standards for empirical data and impact. In fact, we recommend reporting the differential acceptance rates (as relevant) in promotion and tenure materials (i.e., ASEE conference acceptance rates are available from the different divisions).

We propose that as engineering education inquiry flourishes at an institution, there is a progression parallel to the individual engineering faculty member trajectory laid out in Table 23.1 and described earlier. Faculty members may experiment with innovative curricula or pedagogies in their own courses and departments. If supported, these programs become a source of pride to the engineering program, but one that is squarely categorized as teaching. Presentations at engineering education conferences contribute to the reputation of the program, institution, and individuals as engineering education leaders. If these faculty members have the time and inclination, they may move into engineering education research. If they want their primary research focus to be engineering education, then they must shift their arguments in ways that acknowledge the contribution of past efforts to the teaching mission, demonstrate the criteria for high-quality research, and clarify that by being more systematic, new work will contribute to the research mission. In other words, we are arguing, as Yorke (2003) also did, that teaching typically needs to be valued first for engineering education research to flourish in a college, school, or department of engineering. Both arguments are needed, but more careful attention to what various audiences should focus on at various points in time can assist the transition. Repeated conversations with administrators and fellow academics can help develop a more nuanced appreciation of the different contributions of various types of engineering education inquiry.

Future Directions and Opportunities

Many different models, curricula, and design principles exist for creating programs to prepare engineering educators for engineering education research. Programs may focus on graduate students or current faculty members, teaching, or research. The variety reflects and addresses the need to build capacity in all types of inquiry, from effective teaching through scholarly teaching and SoTL to engineering education research. Clear articulation of the goals of a specific program with respect to the Inquiry Framework will help communicate expectations to students, faculty members, and administrators. These efforts will ultimately contribute to sustainability of programs and of engineering education inquiry itself.

Periodic reflection on the aims and accomplishments of various efforts is important. What worked a decade ago may not be what is needed now, particularly in a specific engineering program or institution. We have presented progressions, types, stages, and multiple arguments to help proponents of engineering education identify strategies and arguments to garner support and ensure success. Specifically, proponents of engineering education should bring distinctions between inquiry types and their goals into the discourse of the importance of engineering education. In the absence of criteria to help identify and judge the quality of SoTL or engineering education research, important stakeholders such as administrators will draw their own (perhaps erroneous) conclusions. As a community, we must continue to demonstrate and publicize impact on engineering education, accompanied by more sophisticated and nuanced arguments for engineering education inquiry and infrastructure to support it.

Conclusion

Engineering education is not likely to ever become as large a branch of engineering as, say mechanical or electrical engineering, but it is still important to the future of the profession. Departments and graduate programs will not be needed at every institution. However, opportunities for people to work toward the collective goals of engineering education are expanding. Some academics will choose engineering education research as their primary research focus – this should be supported with appropriate benchmarks, arguments, measures, criteria, and champions. Others will simply want to spend more time improving their teaching, and again, they should have resources to do this in a systematic and thoughtful way. Openly discussing the different aims will help us all better understand how each type of engineering education inquiry contributes to the goals of the department, institution, and profession.

Acknowledgments

The work described in this chapter was funded by the U.S. National Science Foundation (NSF) through grants 0227558, 0517528, 0645736, 0810990, and 0817461. M. Borrego's work on this chapter was supported by the NSF, while working at the Foundation. Any opinions, findings, and conclusions or recommendations expressed in this material are those of the authors and do not necessarily reflect the views of the NSF. The authors also wish to thank numerous collaborators who have influenced their thinking (many of whom are cited throughout), as well as anonymous reviewers whose critique improved the quality of the final product.

Footnotes

1. Of course there are many exceptions, including human factors engineering and bioengineering, in which engineers are already dealing with human complexity and/or developing cutting-edge measurement techniques. This is one reason to frame engineering education research as one of many varieties of engineering research.
2. In some focal areas, such as recruiting and retaining students from underrepresented groups, it may be more appropriate to say that the types of inquiry span service and research.
3. We note that engineering education research, as a relatively new field, has less developed theories and less engagement with theories than many other disciplines (Borrego, 2007a; Yorke, 2003).
4. The nature of engineering higher education research as "directed inward towards the system in which it exists" may contribute to its low status (Yorke, 2003, p. 105).

References

Abbott, A. (2001). *Chaos of disciplines.* Chicago and London: The University of Chicago Press.

American Society for Engineering Education (ASEE). (2009). *Creating a culture for scholarly and systematic innovation in engineering education: Ensuring U.S. engineering has the right people with the right talent for a global society.* Washington, DC: American Society for Engineering Education.

American Society for Engineering Education (ASEE) (Ed.). (2012). *Innovation with impact: Creating a culture for scholarly and systematic innovation in engineering education.* Washington, DC: American Society for Engineering Education.

American Society for Engineering Education (ASEE) Educational Research and Methods Division (ERM). (n.d.). Homepage for the Educational Research and Methods Division. Retrieved from http://erm.asee.org

Atman, C. J., Sheppard, S. D., Turns, J., Adams, R. S., Fleming, L. F., Stevens, R., . . . Lund, D. (2010). *Enabling engineering student success: The final report for the Center for the Advancement of Engineering Education.* San Rafael, CA: Morgan & Claypool.

Australasian Association for Engineering Education (AaeE) Educational Research Methods Special Interest Group. (2008, March). ERM (Educational Research and Methods). *AAEE Newsletter.* Retrieved from http://www.aaee.com.au/newsletters/Newsletter%20March%202008_files/index.html

Baillie, C. (2007). Education development within engineering. *European Journal of Engineering Education, 32*(4), 421–428.

Benson, L. C., Becker, K., Cooper, M. M., Griffin, O. H., & Smith, K. A. (2010). Engineering education: Departments, degrees and directions. *International Journal of Engineering Education, 26*(5), 1042–1048.

Booth, W. C., Colomb, G. G., & Williams, J. M. (2008). *The craft of research.* Chicago, IL: University of Chicago Press.

Borrego, M. (2006). *The higher education job market for engineering education program graduates.* Paper presented at the American Society for Engineering Education Annual Conference, Chicago, IL.

Borrego, M. (2007a). Conceptual difficulties experienced by engineering faculty becoming engineering education researchers. *Journal of Engineering Education, 96*(2), 91–102.

Borrego, M. (2007b). Development of engineering education as a rigorous discipline: A study of the publication patterns of four coalitions *Journal of Engineering Education, 96*(1), 5–18.

Borrego, M., Beddoes, K., & Jesiek, B. K. (2009). *International perspectives on the need for interdisciplinary expertise in engineering education scholarship.* Paper presented at the Australasian Association for Engineering Education Conference, Adelaide, Australia.

Borrego, M., & Bernhard, J. (2011). The emergence of engineering education research as an internationally connected field of inquiry. *Journal of Engineering Education, 100*(1), 14–47.

Borrego, M., Jesiek, B. K., & Beddoes, K. (2008). *Advancing global capacity for engineering education research: Preliminary findings.* Paper presented at the ASEE/FIE Frontiers in Education Conference, Saratoga, NY. Retrieved from http://www.fie-conference.org/fie2008/papers/1365.pdf

Borrego, M., & Newswander, L. K. (2008). Characteristics of successful cross-disciplinary engineering education collaborations. *Journal of Engineering Education, 97*(2), 123–134.

Borrego, M., Streveler, R. A., Miller, R. L., & Smith, K. A. (2008). A new paradigm for a new field: Communicating representations of engineering education research. *Journal of Engineering Education, 97*(2), 147–162.

Borri, C., & Maffioli, F. (2007). *TREE: Teaching and research in engineering in Europe: Re-engineering engineering education in Europe.* Firenze, Italy: Firenze University Press.

Boyer, E. L. (1990). *Scholarship reconsidered: Priorities of the professoriate.* Princeton, NJ: Carnegie Foundation for the Advancement of Teaching.

Braxton, J. M., & Hargens, L. (1996). Variation among academic disciplines: Analytical frameworks and research. In J. C. Smart (Ed.), *Higher education handbook of theory and research* (Vol. XI, pp. 1–46). New York, NY: Agathon Press.

Burrell, G., & Morgan, G. (1979). *Sociological paradigms and organizational analysis.* London: Heinemann Books.

Clement, M., & Frenay, M. (2010). Faculty development in Belgian universities. In A. Saroyan & M. Frenay (Eds.), *Building teaching capacities in higher education* (pp. 82–103). Sterling, VA: Stylus.

Cooper, A. (2004). Leading programmes in learning and teaching. In D. Baume & P. Kahn (Eds.), *Enhancing staff & educational development* (pp. 56–80). Birmingham, UK: Staff and Educational Development Association.

Cox, M. F., Hahn, J., McNeill, N., Cekic, O., Zhu, J., & London, J. (2011). Enhancing the quality of engineering graduate teaching assistants through multidimensional feedback. *Advances in Engineering Education, 2*(3), 1–20.

Crede, E., Borrego, M., & McNair, L. D. (2010). Application of community of practice theory to the preparation of engineering graduate students for faculty careers. *Advances in Engineering Education, 2*(2), 1–22.

Dixon, N. M., Allen, N., Burgess, T., Kilner, P., & Schweitzer, S. (2005). *Company command: Unleashing the power of the army profession.* West Point, NY: Center for the Advancement of Leader Development and Organizational Learning.

Duderstadt, J. J. (2008). *Engineering for a changing world: A roadmap to the future of engineering practice, research, and education.* University of Michigan.

Felder, R. M., & Brent, R. (2010). The National Effective Teaching Institute: Assessment of impact and implications for faculty development. *Journal of Engineering Education, 99*(2), 121–134.

Felder, R. M., Brent, R., & Prince, M. (2011). Engineering faculty development. *Journal of Engineering Education, 100*(1), 89–122.

Fensham, P. J. (2004). *Defining an identity: The evolution of science education as a field of research.* Dordrecht, The Netherlands: Kluwer Academic.

Field, N. (1989). *The Oxford English dictionary.* Oxford: Oxford University Press.

Gabriele, G. (2005). Advancing engineering education in a flattened world. *Journal of Engineering Education*, 94(3), 285–286.

Good, G. A. (2000). The assembly of geophysics: Scientific disciplines as frameworks and consensus. *Studies in History and Philosophy of Modern Physics*, 31(3), 259–292.

Gordon, G. (2004). Locating educational development: Identifying and working with national contexts, policies and strategies. In D. Baume & P. Kahn (Eds.), *Enhancing staff & educational development* (pp. 1–17). Birmingham, UK: Staff and Educational Development Association.

Guba, E. G. (1970). *The alternative paradigm Dialog*. In E. G. Guba (Ed.), *The paradigm dialog* (pp. 17–27). Newbury Park, CA: SAGE

Haghighi, K. (2005). Quiet no longer: Birth of a new discipline (Guest Editorial). *Journal of Engineering Education*, 94(4), 351–353.

Hargens, L. L., & Kelly-Wilson, L. (1994). Determinants of disciplinary discontent. *Social Forces*, 72(4), 1177–1195.

Huber, M. T. (2002). Disciplinary styles in the scholarship of teaching. In M. T. Huber & S. P. Morrealle (Eds.), *Disciplinary styles in the scholarship of teaching and learning: Exploring common ground* (pp. 25–43). Sterling, VA: Stylus.

Hutchings, P., & Shulman, L. S. (1999). The scholarship of teaching: New elaborations, new developments. *Change*, 31(5), 10–15.

Jesiek, B. K., Borrego, M., & Beddoes, K. (2010a). Advancing global capacity for engineering education research (AGCEER): Relating research to practice, policy and industry. *Journal of Engineering Education*, 99(2), 107–119.

Jesiek, B. K., Borrego, M., & Beddoes, K. (2010b). Advancing global capacity for engineering education research: Relating research to practice, policy and industry. *European Journal of Engineering Education*, 35(2), 117–134.

Jesiek, B. K., Newswander, L. K., & Borrego, M. (2009). Engineering education research: Field, community, or discipline? *Journal of Engineering Education*, 98(1), 39–52.

King, R. (2008). Addressing the supply and quality of engineering graduates for the new century. Australian Council of Engineering Deans, Epping, New South Wales, Australia.

Kolmos, A., Gynnild, V., & Roxa, T. (2004). The organisational aspect of faculty development. In A. Kolmos, O. Vinther, P. Andersson, L. Malmi, & M. Fuglem (Eds.), *Faculty develop-*

ment in Nordic engineering education (pp. 67–88). Aalborg, Denmark: Aalborg University Press.

Kolmos, A., Vinther, O., Andersson, P., Malmi, L., & Fuglem, M. (Eds.). (2004). *Faculty development in Nordic engineering education*. Aalborg, Denmark: Aalborg University Press.

Kuhn, T. (1970). *The structure of scientific revolutions*. Chicago, IL: University of Chicago Press.

Lattuca, L. R., & Stark, J. S. (1995). Modifying the major – Discretionary thoughts from 10 disciplines. *Review of Higher Education*, 18(3), 315–344.

Litzinger, T. A. (2010). Engineering education centers and programs: A critical resource. *Journal of Engineering Education*, 99(1), 3–4.

Mahoney, M. (2004). Finding a history for software engineering. *IEEE Annals of the History of Computing*, 26(1), 8–19.

Malik, Q. H., Perova, N., Hacker, T. J., Streveler, R. A., Magana, A. J., Vogt, P. L., & Bessenbacher, A. M. (2011). Creating a virtual learning community with HUB architecture: CLEERhub as a case study. *Knowledge Management and E-Learning*, 3(4), 665–681.

National Research Council. (2012). *Discipline-based educational research: Understanding and improving learning in undergraduate science and engineering*. Washington, DC: The National Academies Press.

Naukkarinen, J., & Malmi, L. (2004). Faculty development in engineering education in Finland. In A. Kolmos, O. Vinther, P. Andersson, L. Malmi & M. Fuglem (Eds.), *Faculty development in Nordic engineering education* (pp. 97–110). Aalborg, Denmark: Aalborg University Press.

Olds, B. M., Borrego, M., Besterfield-Sacre, M., & Cox, M. F. (2012). Continuing the dialog: Possibilities for community action research in engineering education. *Journal of Engineering Education*, 101(3), 407–411.

Paulsen, M. P. (2001). The relationship between research and the scholarship of teaching. In C. Kreber (Ed.), *Scholarship revisited: Perspectives of the scholarship of teaching* (Vol. 86, pp. 19–29). San Francisco, CA: Jossey-Bass.

Pruitt-Logan, A. S., Gaff, J. G., & Jentoft, J. E. (2002). *Preparing future faculty in the sciences and mathematics: A guide for change*. Washington, DC: Association of American Colleges and Universities Retrieved from

http://www.preparing-faculty.org/PFFWeb
.PFF3Manual.pdf

Roxa, T., & Andersson, P. H. (2004). The break-through project – A large-scale project of pedagogical development. In A. Kolmos, O. Vinther, P. Andersson, L. Malmi, & M. Fuglem (Eds.), *Faculty development in Nordic engineering education* (pp. 25–48). Aalborg, Denmark: Aalborg University Press.

Saroyan, A., & Frenay, M. (Eds.). (2010). *Building teaching capacities in higher education*. Sterling, VA: Stylus.

Shavelson, R., & Towne, L. (2002). *Scientific research in education*. Washington, DC: The National Academies Press.

Société Européenne pour la Formation des Ingénieurs (SEFI; European Society for Engineering Education (2008). Engineering education research. Retrieved from http://www.sefi.be/index.php?page_id=1192

Société Européenne pour la Formation des Ingénieurs (SEFI; European Society for Engineering Education) Engineering Education Research Working Group (EER-WG). (n.d.) Retrieved from http://www.it.uu.se/research/group/upcerg/EER?lang=en

Streveler, R. A., Borrego, M., & Smith, K. A. (2007). Moving from the 'scholarship of teaching and learning' to 'educational research': An example from engineering. In D. R. Robertson & L. B. Nilson (Eds.), *To Improve the Academy* (Vol. 25, pp. 139–149). Boston, MA: Anker.

Streveler, R. A., Magana, A. J., Smith, K. A., & Clarke Douglas, T. (2010). *CLEERhub.org: Creating a digital habitat for engineering education researchers*. Paper presented at the American Society for Engineering Education Annual Conference, Louisville, KY.

Streveler, R. A., & Smith, K. A. (2006). Guest Editorial: Conducting rigorous research in engineering education. *Journal of Engineering Education*, 95(2), 103–105.

Streveler, R. A., & Smith, K. A. (2010). Guest Editorial: From the margins to the mainstream: The emerging landscape of engineering education research. *Journal of Engineering Education*, 99(4), 285–287.

Streveler, R. A., Smith, K. A., & Miller, R. (2005). *Rigorous research in engineering education: Developing a community of practice*. Paper presented at the American Society for Engineering Education Annual Conference, Portland, OR.

Wankat, P. C. (2004). *The emergence of engineering education as a scholarly discipline*. Paper presented at the 2004 ASEE Annual Conference and Exposition, Salt Lake City, UT.

Wankat, P. C., Felder, R. M., Smith, K. A., & Oreovicz, F. S. (2002). The scholarship of teaching and learning in engineering. In M. T. Huber & S. P. Morrealle (Eds.), *Disciplinary styles in the scholarship of teaching and learning: Exploring common ground* (pp. 217–237). Sterling, VA: Stylus.

Wenger, E. (1998). Communities of practice: Learning as a social system. *The Systems Thinker*, 9(5), 1–12.

Wenger, E., McDermott, R. A., & Snyder, W. (2002). *Cultivating communities of practice: A guide to managing knowledge*. Cambridge, MA: Harvard Business School Press.

Wenger, E., White, N., & Smith, J. D. (2009). *Digital habitats: Stewarding technology for communities*. Portland, OR: CPSquare Press.

Yorke, M. (2003). Pedagogical research in UK higher education: An emerging policy framework. In H. Eggins & R. Macdonald (Eds.), *The scholarship of academic development* (pp. 104–116). London: Society for Research into Higher Education and Open University Press.

Part 5

RESEARCH METHODS AND ASSESSMENT

CHAPTER 24

Studying Teaching and Learning in Undergraduate Engineering Programs

Conceptual Frameworks to Guide Research on Practice

Lisa R. Lattuca and Thomas A. Litzinger

Introduction

Teaching and learning have been studied from a variety of different theoretical perspectives over time. Collins, Greeno, and Resnick (2001; also, Greeno, Collins, & Resnick, 1996) suggest that each of the major perspectives on learning – behavioral, cognitive, and situative – provide us with useful, if partial, understandings of how people learn. Although these three perspectives are derived from different assumptions about what learning is and how it happens, they agree that learning is an interaction between a learner and his or her environment. A key distinction, however, is the extent to which each perspective emphasizes the learner or the environment in that interaction.

In the behavioral perspective, the individual learns when presented with stimuli from his or environment. In contrast, the cognitive perspective conceptualizes the learner as an active problem-solver, adapting to his or her environment rather than primarily reacting to it. Thus, although the environment is the catalyst for learning in the

behavioral perspective, it is the learner, rather than the setting he or she learns in, that is the primary focus of theories and studies that take a cognitive perspective. The situative perspective might appear to split the difference – if it were not for the very different assumptions about learning that are at its foundation. The situative – or what we call the sociocultural perspective – argues that separating learner from the environment is impossible; learning always occurs in context – or more precisely, in multiple, overlapping contexts. This is another way of saying that learning occurs in particular people with particular histories and characteristics (physical, mental, cultural, etc.) who exist in particular places in time (for discussion, see Greeno and the Middle School Mathematics Through Applications Project Group, 1998; Lave & Wenger, 1991; Wertsch, 1985; Wertsch, Del Rio, & Alvarez, 1995). Learning is thus strongly influenced by *who* we are, *where* and *when* we live and learn, and by *what* we seek to learn and *how* we go about learning it. (See Chapter 2 by Newstetter and Svinicki and Chapter 3 by Johri, Olds, & O'Connor in this volume for a more detailed discussion.)

Today we see the influence of these difference perspectives in research on learning in many academic domains – including engineering – but researchers increasingly agree that the environment in which one learns is a factor that must be taken into account in studies of learning. In the past two decades, researchers who study learning in K–12 and in undergraduate engineering settings have tended to focus on the classroom as the site for learning.[1] In such studies, classroom activities and interactions are assumed to be major influences on learners and learning. These micro-level studies are crucial to our understanding of how students learn engineering, but they often isolate the classroom from the larger educational institution, effectively conceptualizing learners and their learning as unaffected by other environmental influences that might affect their learning – for example, students' experiences in their other courses in their academic program.

Although researchers who study learning in colleges and universities have increasingly studied classroom interactions and activities, most research in the field of higher education has taken a more macro-level approach. In contrast to the micro-level approaches of researchers in K–12 and engineering education, higher education researchers have long relied on conceptual frameworks that view learning as an interaction of person (or persons) and a broader educational environment. These conceptual models do not ignore classroom-level experiences, but rather situate them in what we might call "systems models" that acknowledge (and sometimes try to capture empirically) the many influences that can affect student learning.

Alexander Astin's foundational work in this area offered a simplified systems model in which inputs (i.e., what students bring to the learning situation) and environment (the setting in which learning occurs) influence the kinds of outputs (e.g., learning, persistence, or degree attainment) that students

achieve. Astin introduced this basic model in the 1970s to allow researchers to correct or adjust for input differences and thus get a less biased estimate of the effects of different environments on student outcomes (Astin, 1993, p. 19). Learning theories support the argument that prior learning and experiences affect learning and thus must be accounted for in assessment of student learning. Astin's Input–Environment–Output (I–E–O) model has been modified in a variety of ways by higher education researchers, and in this chapter we review recent models that have built on this foundation, as well as on the insights of sociocultural perspectives on learning, to study learning in undergraduate engineering programs.

In recent research on undergraduate engineering education, we have been using and refining a greatly expanded I–E–O model to explore a range of potential program-level environmental influences – from faculty practices related to curriculum development and revision, to the kinds of instructional methods that are common in a program, to co-curricular programs such as design competitions and engineering professional societies – to understand how experiences in different educational environments influence students' knowledge and skill development.

In this chapter, we examine a number of conceptual frameworks that expand the concept of the educational environment well beyond the classroom to include a wide range influences on student learning and on curriculum planning and development. We begin by specifying what we mean by the term "educational environment," which has been used in the literature on higher education in a number of ways. In this section, we also briefly review research that supports the need to focus on the curriculum and the broad scope of students' experiences in order to understand what engineering students learn and why. In the next major section of the chapter, we explore "college impact" models that have been widely used to study the effects of college and university environments on students' learning outcomes. These models help researchers to think expansively about the variety of influences on undergraduate students as they learn both inside and outside the classroom. Although comprehensive in scope, college impact models do not provide much guidance for program-level studies that must carefully examine the impact of the particular curricular and instructional practices in an academic program or department. In the last section of this chapter, we identify three conceptual models of postsecondary curriculum development and planning that could remedy this and thus strengthen studies of the educational environment by identifying critical elements of postsecondary curricula as well as an array of influences on curriculum development. By presenting a range of models of the educational environment, we hope to assist engineering education researchers in thinking more broadly about possible influences on student learning as they develop conceptual models to guide data collection and analysis for studies of learning by engineering undergraduates.

Although researchers are increasingly exploring the influence of K–12 educational experiences on engineering knowledge and pathways into engineering schools and practice, this chapter focuses on postsecondary education as a site for the development of engineering knowledge and skills. Significant differences between the organizational and cultural features of schools and colleges/universities preclude a thorough discussion of conceptual models that can guide studies of these educational environments, although the need for such aids to research is clear.

Specifying and Studying the "Educational Environment"

Recent modifications of the Astin's I–E–O model have significantly expanded our understanding of the "environment" in which college students learn. For many years, higher education researchers conceptualized this environment rather broadly, in terms of institutional types (e.g., research universities, baccalaureate colleges) and control (whether an institution is public or private), as well as selectivity in admissions and size of enrollment. Researchers expected these institutional-level characteristics to influence levels of student engagement (and thus student learning) on a campus (Astin, 1993; Kuh et al., 2005; Pace, 1980), yet a number of studies suggested that the influential characteristics tend to be related to a college or university's culture rather than to what are typically referred to as structural features. Findings from studies by Kinzie, Schuh, and Kuh (2004), Pike (1999), and Pike, Schroeder, and Berry (1997), for example, contended that the key influence on student learning appears to be how an institution intentionally shapes its academic and co-curricular programs to encourage student involvement in the educational process.

After reviewing more than thirty years of research on student learning outcomes in higher education, educational researchers Ernest Pascarella and Patrick Terenzini, in fact, concluded that once studies of college learning took account of students' personal characteristics (such as SAT scores or socioeconomic status), none of these institutional dimensions were associated with important differences in student learning (Pascarella & Terenzini, 2005, p. 641). Researchers have argued that although these characteristics likely have some influence on the culture of the institution and its arrangements for learning, factors that are more proximal to the student experience – for example, the nature of the peer environment in a student's engineering major program or the professional society chapters in which she participates, have a more substantial (and measurable) impact on a student's learning (Pike, Smart, Kuh, & Hayek, 2006; Porter, 2006; Smart, Feldman, & Ethington, 2000). Such factors, which would be among the internal organizational features of an institution, might be more likely to influence learning directly. A study of the academic and co-curricular experiences of engineering students, however, indicated that both an institution's "structural-demographic" variables (e.g., size, control, institutional type, selectivity) and its internal organizational features (e.g., curricular emphases, faculty reward systems) were significantly, and sometimes substantially related, to the kind of academic and co-curricular experiences that engineering students had on their campuses (Ro, Terenzini, & Yin, 2013). The authors speculated, however, that institutional type might be a proxy for the more specific features of the institution's internal organizational context, and thus that the structural-demographic variables might yet be too distant from the student experience to explain with clarity "what institutional differences really make a difference" (p. 277).

The state of empirical evidence to date suggests that researchers should include both structural-demographic and organizational variables in conceptual frameworks guiding studies of student experiences and learning. In studies of learning in colleges and universities, this organizational context is particularly complex. In engineering, for example, the organizational context can be conceptualized at the institutional level (i.e., the university or campus), the unit level (i.e., the school of engineering), and the program level (i.e., the engineering department or program).

Experience, as well as large-scale studies, supports the claim that academic programs, in particular, provide distinctive learning environments for students. Smart et al. (2000) observed that as the number of disciplinary specialties grows, comprehensive categories such as "biology," "sociology," or "engineering" mask substantive variations in attitudes and behaviors of individuals who study and work within these broad disciplinary categories. Relying on John Holland's theory of occupations and environments (1966, 1973, 1985, 1997), they found "abundant evidence . . . that faculty in academic departments, classified according to the six academic environments proposed by Holland, differ in ways theoretically consistent with the postulates of Holland's theory" (Smart et al., 2000, p. 83). Their analyses of national data produced strong support for Holland's assumption that academic environments

are socializing mechanisms, reinforcing and rewarding different patterns of abilities, interests, and values while simultaneously discouraging others.

To explore the utility of Holland's typology of fields for understanding variations in engineering program environments, Lattuca, Terenzini, Harper, and Yin (2010) used data from 1,272 faculty members in 203 engineering programs on 39 U.S. campuses. Their analyses revealed variations in curricular and pedagogical patterns across engineering sub-disciplines consistent with the predictions of Holland's theory. The researchers concluded that the results of the study provided moderately strong evidence supporting the use of Holland's theory in research on engineering. Analyses using another national sample of 120 programs in 30 engineering schools are providing further and stronger support for significant differences in curricular and co-curricular experiences among students in different engineering fields (see, e.g., Knight et al., 2012; Knight, 2012).

Based on prior research, Lattuca et al. (2010) used institutional characteristics as control variables in their analyses, but found that two of these variables – an institution's status as a public or private institution and its Carnegie type (e.g., research, baccalaureate, etc.) defined one of the classification functions that distinguished among engineering programs. Although it is possible that this finding reflects the nature of engineering education in the United States, which is dominated by large, public institutions, it is also consistent prior research (e.g., Ro et al., 2013) and theory. Austin (1991), Smart and Ethington (1995), and Lattuca and Stark (1995) all argue that while the culture of academic fields is a strong influence on faculty work, faculty members also modify their teaching and research activities to align with local conditions. Put another way, institutional contexts may mediate disciplinary preferences.

These findings provide further support for the argument that researchers must take into account both institutional and program factors when exploring the experiences and learning of students in engineering. The curricular and instructional choices of engineering faculty, this research suggests, are influenced by both the type of institution in which they work as well as by the type of engineering program in which they teach. Variations in engineering faculty members' decisions about educational matters are likely the result of a complex array of factors, including institutional cultures that are strongly related to Carnegie type (as Kinzie et al. [2004] suggest), disciplinary predilections (as Holland's theory suggests), and external factors such as accreditation standards and industry needs (as the study of the impact of ABET's EC2000 criteria by Lattuca, Terenzini, and Volkwein, [2006] suggests). In studying engineering programs, researchers must thus seek to understand how an array of program and institutional characteristics interact with influences external to the institution to influence the design and implementation of educational experiences for students.

Finally, it is worth noting that program-level studies are necessary supplements to studies of individual classrooms because what happens in a classroom may or may not be consistent with the curricular and pedagogical characteristics of the larger program. For example, one engineering instructor may employ innovative learning strategies such as problem-based learning and simulations, but the engineering program in which the course is offered may be dominated by large lecture courses. The educational experience in this instructor's classroom cannot be considered in isolation because students do not arrive in this classroom unaffected by their previous learning experiences. Students' interest, motivation, and comfort with innovative approaches will be affected by their prior learning experiences (Alexander, Kulikowich, & Schulze, 1994; Dochy, 1994; Dochy, Moerkerke, & Segers, 1999) as well as those that are happening concurrently.

Not all studies of engineering education need to account for the broad range of factors that influence student learning. Yet,

studies that seek to understand, holistically, how engineering programs influence student learning experiences and learning must, by definition, take a broader perspective. This broader perspective, we argue, is well aligned with sociocultural theories of learning that view learning as embedded in – and thus affected by – a variety of overlapping social, cultural, and temporal contexts. The challenge is developing conceptual models and research designs that capture as much as possible of this complex set of educational interactions.

Broadening the Research Lens on Teaching and Learning in Engineering

In this section we present an overview of what are generally called "college impact models," focusing first on Astin's foundational I–E–O model and then describing models by Pascarella, Weidman, and Terenzini and Reason. Our purpose here is to identify similarities but also important departures that are relevant to the study of engineering programs. Next, we explore how curriculum models can help researchers further specify the educational environment by identifying key elements of the learning experience in college and university programs and classrooms.

Astin's I–E–O Conceptual Model for Assessment of Student Learning and Development

In an autobiographical discussion of how he came to develop his I–E–O model, Astin recalled his experiences as a Ph.D. trained clinical and counseling psychologist. As a psychologist, he viewed human behavior through a development framework that assumed that a judgment about the success of a treatment must always be measured in terms of how much it helped the individual improve. Because the initial condition of individuals varies, the efficacy of a treatment cannot be assessed in terms of its outcome; rather, the effectiveness of the treatment "has to be judged in terms of how much improvement takes place" (1993, p. 16). This early recognition translated into the first of a set of three principles that guided Astin's approach to assessment. The first principle is that the output (or the impact) of any educational program or institution must be evaluated in terms of inputs. In addition, no output measure can be determined using a single measure, such as student ability. Astin specifically mentions students' major field of study (as well as input variables such as gender) as important components in the study of outputs. Finally, he argues that no matter how good they are, input and output variables alone are of limited usefulness; information on the educational environment and student experiences are needed to understand how much students learn and why.

For Astin, outputs are educational outcomes – or "talents" – that instructors and others in colleges and universities want students to develop. Inputs are the personal qualities and characteristics that the student brings initially to the educational program. He used the term environment to refer to the student's actual experiences during his or her educational program. (Later in this chapter we will consider the differences between the academic plan that instructors devise for students and the actual experience that students have; the two are not necessarily the same.)

Translating outcomes, inputs, and environment into research terms, Astin wrote,

> . . . we could . . . refer to the outcome variables as dependent variables, criterion variables, posttests, outputs, consequents, ends, or endogenous variables. Environment and input variables are both types of independent variables, antecedent variables, or exogenous variables. Input could also be called control variables or pretests. Environmental variables might also be referred to as treatments, means, or educational experiences, practices, programs, or interventions. (p. 18)

Astin contended that his model was applicable to both qualitative and quantitative research designs that seek to identify causal connections between particular antecedent

events (inputs) or environments and any outcome(s) of interest. The I–E–O model is, Astin (1993) wrote, "specifically designed to produce information on how outcomes are affected by different educational policies and practices" (p. 37). He discussed a continuum of environmental variables, from those that are most distant from the student experience (such as the institutional characteristics discussed earlier, e.g., the size and location of the university, the characteristics of its faculty, and its endowment) to the most proximate (e.g., a student's instructors, classroom experiences, and peer interactions). This is a broad definition of the learning environment, as Astin acknowledged, "encompass[ing] everything that happens to a student during the course of an educational program that might conceivably influence the outcomes under consideration" (p. 81). He also noted, however, that environments are to a certain extent, "self-produced": students choose and create their own educational environments when they choose, for example, residential options, courses and course sections, majors, and peer groups (p. 83). This view is consistent with currently popular views of student engagement (or "involvement" as Astin would have called it), which contend that the amount of effort a student puts into his or her education will influence the kinds of outcomes he or she achieves.

We have discussed Astin's model in some detail because it lays the groundwork for other models of college impact, including those of Pascarella (1985), Weidman (1989), and Terenzini and Reason (2005). In the next sections, we summarize these models, giving greatest attention to our interpretations of Terenzini and Reasons' model for studies of engineering education.

Pascarella's General Model for Assessing Change

Causal models such as Pascarella's general model seek to make explicit the relationships among critical variables. Pascarella (1985) argues that "change" in students' knowledge, skills, and dispositions results from the direct and indirect effects of five sets of variables. The first set consists of the input variables discussed earlier. Depending on the outcome of interest, researchers are encouraged to consider a range of students' background experiences and pre-college characteristics. For example, in addition to students' engineering design knowledge, researchers might consider students' academic achievements in high school, their participation in summer engineering experiences before they attended college, and their parents' occupations, as well as their career and other aspirations. Ongoing work by P. K. Imbrie and his colleagues is also modeling the impacts of both cognitive and affective influences, such as goal orientations, self-efficacy, and self-worth, on students' decisions to persist in engineering (Lin, Reid, & Imbrie, 2009; Imbrie, Lin, & Malyscheff, 2008). Factors that influence persistence may also affect learning, so such studies are likely to contribute significantly to our understanding of the many influences on the development of engineering knowledge and skills.

An institution's structural characteristics comprise a second set of variables. Pascarella again identified a potential set of relevant characteristics – such as size, student enrollments, faculty–student ratios, selectivity, and percentage of students living on campus – but other variables might be considered depending on the student outcome of interest. Taking our design knowledge and skills example further, we might specify the size of the engineering school, the variety of majors offered, the emphasis on design competitions, and the provision of dedicated space to support student design teams.

Pascarella argued that these two sets of variables – the structural/organizational characteristics of the institution and a student's background/pre-college traits – shape a third set of variables that constitute the "institutional environment" that students experience. In turn, these three sets of variables affect a fourth set of variables, which Pascarella termed "interactions with agents of socialization." In short, this set of influences includes the frequency and content of

students' interactions with socializing agents on campus, primarily, other students and faculty. A fifth variable, "quality of student effort," is the result of the combination of student traits, the institutional environment, and student interactions with key socializing forces. What a student learns (what Pascarella called learning and cognitive development) is also a function of the interactions of these same variables.

Earlier we noted that Pascarella's model suggested both direct and indirect effects on student outcomes during college. The General Model for Assessing Change makes these direct and indirect paths of influence clear. The structure and organization of the institution have an indirect effect on development because their influence is "mediated" through the institutional environment (as Pascarella defined it), through the students' own quality of effort, and her interactions with peers and faculty. Only students' background and precollege traits are assumed to have a direct – that is, unmediated – influence on learning and cognitive develop-ment.

Weidman's Model of Undergraduate Socialization

John Weidman proposed a model of undergraduate socialization that seeks to understand student outcomes such as career choice, life style preferences, aspirations, and values – what some have called the noncognitive outcomes of college. Engineering education researchers are increasingly interested in such outcomes as educators and policymakers express concerns about pathways into engineering programs and work. A number of studies in recent years have focused particularly on major and career choices and/or the decision to remain or leave the field of engineering during college or afterward (see, e.g., Amelink & Creamer, 2010; Lord et al., 2009; Ro, 2011; Sheppard et al., 2010).

Like other models of college impact, Weidman's model posits an important role for students' background characteristics, Weidman's own research and sociological

grounding prompted the explicit identification of many more environmental variables
than Astin or Pascarella suggest. According
to Weidman, these environmental variables
reflect the formal and informal "normative
contexts" and socialization processes that
a student experiences as he or she attends
college. For example, Weidman's model
includes an academic normative context
and socialization process that emanates
from "formal" influences such as the institution's mission and quality and the major
department that students choose, as well
as "informal" contexts and processes from
the "hidden curriculum."[2] In addition,
the normative context of the institution is
shaped by formal "social" factors, such as the
institution's size, its residence facilities, and
organizations and informal social pressure
from peer groups. In Weidman's model,
however, parents continue to have socializing role, even if students live on campus
during college. "Parental socialization"
forces are defined in the model as family
socioeconomic status, life style, and parent–
child relationships. In addition to these
external normative and socializing forces,
Weidman adds non-college reference
groups – peers, employers, and members of
community organizations – who also shape
or try to shape student's career choices, aspirations, and values. Engineering researchers
have documented the influence of parents,
family members, and family friends who are
engineers on the choice of an engineering
major (e.g., Mannon & Schreuders, 2007;
Trenor, Yu, Waight, Zerda, & Sha, 2008),
but few have explored how the variety
of normative pressures and socialization
processes interact as students move through
their engineering programs (for an example,
see Trenor et al., 2008). Engineering education researchers studying career choice
should, in addition to exploring Weidman's
expansive model, consider the theory of
vocational choice offered and tested over
many years by John Holland as well as
the thorough examination of the utility of
this theory in U.S. college populations by
Smart, Feldman, and Ethington (2000).

Terenzini and Reason's College Impact Model

After years of conducting, reading, and synthesizing higher education literature in two
comprehensive co-authored volumes entitled *How College Affects Students: Findings
and Insights from Twenty Years of Research*
(Pascarella & Terenzini, 1991) and *How
College Affects Students: A Third Decade
of Research* (Pascarella & Terenzini, 2005),
Patrick Terenzini began to develop his own
conceptual model of college impacts in collaboration with his colleague, Robert Reason. The model not only builds on the work
of others who have proffered college impact
models, but it also attempts to incorporate
all the advances these models offer by identifying a broad array of influences that shape
college students' outcomes. It thus avoids
focusing on a narrow set of factors that may
present "only a partial picture of the forces
at work" (Pascarella & Terenzini, 2005, p.
630). Even the widely used National Survey of Student Engagement, Terenzini and
Reason (2010) argue, examines a relatively
restricted set of influences, neglecting the
potential influences of faculty cultures; peer
environments; and internal structural, programmatic, and policy considerations.

Terenzini and Reason (2010) acknowledge that like other college impact models, their model hypothesizes that students
come to college with personal, academic,
and social background characteristics and
experiences "that both prepare and dispose
them, to varying degrees, to engage with
the various formal and informal learning
opportunities their institution offers" (p.
10). Also, as in other models, these precollege characteristics are believed to shape
students' college experiences as students
interact with an institution's environment
and its major agents of socialization (as in
other models, presumed to be, primarily,
peers and faculty members). The Terenzini and Reason model departs from the
other models in its specification of three sets
of primary influences: (1) the institution's
internal organizational context and (2) peer

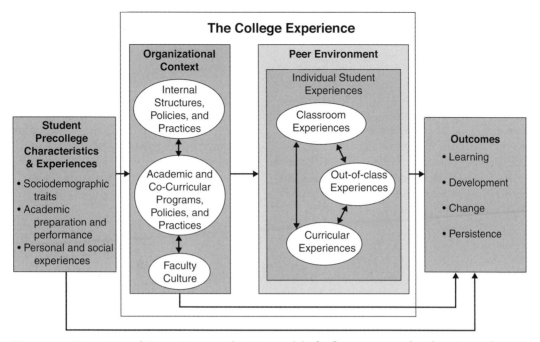

Figure 24.1. Terenzini and Reason's comprehensive model of influences on student learning and persistence.

environment, which the researchers argue exist separate from and "temporally prior" to (3) a student's individual experiences in various academic and non-academic settings (see Figure 24.1).

These three sections of the framework – which distinguish it from the others we have reviewed – focus researchers' attention on the three categories of organizational features: internal structures, policies, and practices; academic and co-curricular program, policies, and practices; and faculty culture. These categories ensure that researchers study what institutions "do" rather than what they "are" (Terenzini & Reason, 2011, p. 11). They are, Terenzini and Reason argued, more immediate to the student experience than the structural characteristics typically explored in studies of collegiate impact because they "are more likely to influence student outcomes (at least indirectly, if not also directly) through the kinds of student experiences and values they promote or inhibit" (p. 11). For this reason, the three categories of organizational influences are considered to be temporally prior to the student learning experience inside and outside the classroom.

The curriculum (including instructional approaches) is planned by faculty, but what students actually experience or perceive may be different than what is intended by faculty. Terenzini and Reason note a similar distinction between the programs, policies, and practices that constitute a student's out-of-class experiences. The student experience, they argue, is influenced heavily by the peer environment in which students learn and sometimes live. Accordingly, they categorize student's classroom experiences, curricular experiences, and out-of-class experiences as three sets of influences within a larger "peer environment," which they define as "the system of dominant and normative values, beliefs, attitudes, and expectations that characterize a campus' student body" (p. 16). They compare this environment to the "ethos" of the student body, which they assume is distinct from any "institutional ethos," because it will also reflect faculty and administrative values, beliefs, attitudes, and behaviors.

The model thus implies a chain of related influences that shape student outcomes; research by Ro et al. (2013) provides empirical support for this "causal chain," which is typically underspecified in studies of college effects on students (p. 277). The generality of Terenzini and Reason's model makes it useful to scholars studying a variety of post-secondary educational outcomes, such as persistence to degree, but also learning outcomes such as design skills or decisions such as whether to practice engineering after college graduation. In addition, its focus on organizational context factors makes it particularly useful for multi-institutional studies (which must compare these features of the environment in order to understand variations in student outcomes). The framework may also be adapted for studies of academic programs since these learning environments vary even within the same college or university.

Two studies of undergraduate engineering programs have utilized versions of the model (which has been in development for some time) with success. Both studies, funded by the National Science Foundation, studied engineering programs and student learning to understand whether and how undergraduate programs were responding to the National Academy of Engineering's (2004) report, *The Engineer of 2020: Visions of Engineering in the New Century*. One of these research efforts was a large-scale quantitative study of thirty engineering schools; the other was an in-depth qualitative investigation of the conditions that supported effective education at six schools empirically identified as outperforming their peers on a number of measures relevant to the attributes and vision of the *Engineer of 2020* report. In both studies, the model guided data collection in engineering programs and schools, specifically focusing the researchers' attention on three sets of organizational factors: (1) organizational structures, policies, and practices; (2) academic and co-curricular programs, policies, and practices; and (3) faculty culture.

A brief overview of some of these organizational factors will illustrate how they can be operationalized for studies of specific student outcomes. The two studies referenced earlier, Prototype to Production (NSF EEC 0550608) and Prototyping the Engineer of 2020 (NSF DUE-0618712), identified three "focal outcomes" for study: design and problem-solving, interdisciplinary competence, and contextual competence. The research teams working on the projects thus explored a variety of environmental factors, some of which were assumed to be particularly relevant to the development of one of these outcomes.

These program-level variables were embedded in a larger "organizational context" in acknowledgment of the fact that engineering programs are often embedded in larger schools or colleges of engineering that also influence the nature of the student experience in those programs. Researchers for the projects thus examined university- and school-level policies to understand how these might affect program-level practices and policies. Although institutional leadership was not specified as a key variable a priori, the qualitative case studies revealed how this influence shaped college- and program-level practices over time and thus suggested an additional variable that should be included in subsequent studies of the conditions that support the provision of high-quality engineering education.

College Impact Models: A Summary

The four models that we summarized in the preceding section represent a conceptual advance over research designs that sought to understand student learning and other outcomes of college without taking into account the prior learning and experiences of students that are likely to shape these outcomes. Astin's I–E–O model, in fact, was motivated by the need to correct analytical deficiencies in studies that assessed student learning and other outcomes. Pascarella, Weidman, and Terenzini and Reason built on Astin's important methodological insight to offer more explicit conceptual models that acknowledged the role of personal and environmental factors on student outcomes.

Another advantage of the foregoing conceptual models of college impacts is their ability to prompt researchers to move beyond purely psychological models of student learning and development that dominated early research on college students (see Pascarella & Terenzini, 2005). That research tended to view changes in students during college as the result of intraindividual developmental processes, neglecting the role of the environment in their growth and development. Dannefer (1984) contended that a psychological and developmental view of students conceptualizes environmental factors as instrumental but not as a critical influence on individual change. The college impact models we have explored begin to remedy this problem by further specifying the elements of the environment that affect student outcomes – but these models continue, as Feldman (1972) argued, to consider environmental influences important precisely because they "impinge" on cognitive development, attitudes, and other outcomes (p. xix). A sociocultural perspective significantly elevates the role of the environment in educational research, making it a strong focus of study. Such approaches to understanding learning may be accomplished more easily through qualitative and multiple-method research approaches that enable the researcher to study the learner and the context in which she or he learns as interacting and mutually constitutive (for discussions, see Minick, Stone, & Forman, 1993; Rogoff, Mosier, Mistry, & Goncu, 1993; Wertsch, 1985).

Curriculum Models: Exploring the Curricular Experience

Until recently, college impact models focused on instructional and curricular experiences in the classroom. Terenzini and Reason added conceptual depth when they recommended that researchers study not only the curriculum that faculty plan, and the one that students experience, but also the practices that influence the nature of that curriculum. In many studies of learning outcomes in higher education, however, the concept of the curriculum is under-theorized and underspecified. In this section, we examine potential contributions that curriculum models make to our conceptualizations of the educational "environment" and our understanding of the factors that shape the curriculum that becomes the foundation of the student learning experience.

The Academic Plan Model

The academic plan model posits an array of influences, both internal and external to the program and institution, that shape instructors' decisions about curriculum design at the college, program, and course levels (Lattuca & Stark, 2009). The model not only identifies the influences on curriculum planners (i.e., faculty members and instructors) as they create and revise academic plans for courses and programs in higher education programs, but also intends to inform research and practice in higher education by encouraging a thorough consideration of factors that affect the development of curricular activities. Therefore, it can guide inquiries into educational innovations as well as more routine curriculum planning and revision in engineering programs by helping researchers specify variables of interest.

Although the academic plan model is applicable at the course and institutional levels, our goal here is to consider how it can inform the design of studies of the impact of engineering program curricula on students' development of important knowledge and skills (such as ABET's a-k outcomes, the attributes of the "engineer of 2020" or the specific learning goals established by faculty in their engineering programs). Researchers interested in the studying the impact of innovations at the course level could, for the most part, substitute the word "course" for "program" in the following discussion.

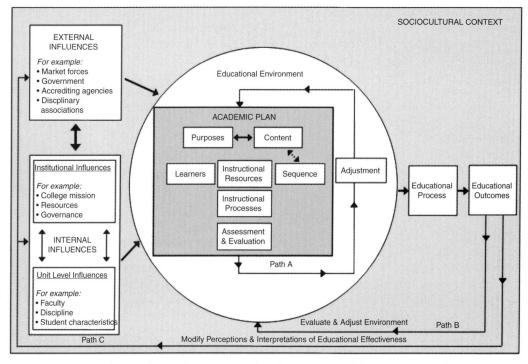

Figure 24.2. The Academic Plan in Sociocultural Context Model (Lattuca & Stark, 2009).

Figure 24.2 reveals a broad definition of the term *curriculum* with implications for how we design studies of student learning. The box entitled "Academic Plan" consists of a set of eight elements, or decision points, that are addressed in every academic plan – whether intentionally or not – by instructors as they develop programs. To understand why a program curriculum looks as it does, and how it affects student learning, researchers would examine each of these elements, examining the actual decisions that instructors have made, why they have made them, and how these elements are expected to interact as well as how they actually interact when the plan is enacted. Turns, Yellin, Huang, and Sattler (2008) have pursued such questions using interviews with engineering faculty to understand the kind of instructional decisions made in courses and the rationales behind these decisions.

The first element, purposes, concerns how instructors in an academic program individually and collectively define the *purposes* of the program curriculum. We can think of purposes as the views of education that inform decisions about the goals of program and of courses within a program. In addition, the academic plan model suggests researchers seeking to understand the impact of a program's curriculum must identify and interpret decisions about content; sequences of courses (and possibly, sequences of topics within courses); instructional processes used by program instructors[3]; available and utilized instructional resources; assessment strategies; and approaches to evaluation of courses and the overall program. An eighth element, adjustment (Path A in the diagram), represents the changes that instructors make to courses or programs after they are delivered and assessed, either formally or informally.

The academic plan model explicitly acknowledges the larger sociocultural factors in which curricula – and indeed the people who create them and the institutions they work in – are situated. Within this sociocultural context, two subsets of influences operate: (1) influences external to the institution (in engineering, we think about the palpable influence of employers and accreditation agencies on engineering faculty as they plan courses) and (2) influences internal to the institution. The internal influences are further divided into institutional-level influences (e.g., mission, leadership, resources) and unit-level influences (e.g., program goals, faculty characteristics and collective beliefs about their discipline and program, and student characteristics). In a large university, unit-level contexts are typically the school/college of engineering, as well as departments and programs that organize faculty work.

Internal and external influences interact to create the "educational environment" in which academic plans are created. Here we see additional connections to the college impact models discussed in the previous section. First, like Weidman's model, the academic plan model looks outside the college or university to understand the normative societal influences that shape what happens in colleges and universities and their programs. Our studies of undergraduate engineering programs, for example, assume that the National Academy of Engineering's *Engineer of 2020* report is one potential influence on the thinking of engineering administrators and faculty as they review and revise their programs. The report, of course, is itself a response to sociocultural factors such as the increasing globalization of markets for labor and goods, changes in how engineering is practiced as a result of technological advances, and changing demands on engineers in the workplace. Few studies of engineering programs take such external factors into account, in part because it is difficult to do so. The *Engineering Change* study (Lattuca, Terenzini, & Volkwein, 2006) attempted to assess the impact of at least some of these influences on engineering program faculty and chairs, asking them to report on the influence of (an admittedly limited) range of potential external (and internal) influences on changes they made on their program curricula, including ABET, industry, and even funding from the National Science Foundation.

Like all of the college impact models, the academic plan model assumes that an array of influences internal to the institution shape the educational environment. The variables suggested in the academic plan model overlap to some extent with those in the typical college impact model. For example, both identify college mission and resources as institutional influences on curriculum development. The academic plan model suggests, however, that these are only two possible institutional-level influences. Additional institutional characteristics, for example, the culture of a college or university, might also affect the nature or conduct of its educational programs. At the program level, we again see some of the same influences identified in college impact models, such as student characteristics. In addition, however, the academic plan highlights the role of faculty beliefs and attitudes, as well as faculty members' disciplinary socialization, as strong influences on the curriculum. This relationship is discussed further in the next section, which describes the contextual filters model of curriculum planning that influenced the development of the academic plan model.

It is important to note the distinction between academic plans and the "enacted" curriculum that we referred to earlier. The academic plan model describes how curricula are shaped and planned; it is not a model of how a curriculum works. As shown in Figure 24.2, educational processes and outcomes are placed outside the educational environment to recognize that many influences are beyond the control of those who plan (and teach) a curriculum. These include, for example, the academic preparation of students who enroll in a course and the normative pressures that influence students' reactions to curricula. The peer environment identified in the college impact models is presumably one of these normative pressures.

Finally, the academic plan model portrays several evaluation and adjustment paths for the plan. Path A depicts the adjustment path for a course or program. Path B suggests that the outcomes of educational plans may influence the educational environment itself. Path C reflects the potential influence of external and internal audiences that form perceptions and interpretations of educational outcomes. Engineering employers, for example, generalize about the preparation of new engineers for practice and these opinions shape their feedback to engineering schools and programs. Thus, once a curriculum is enacted, a new series of potential influences, in the form of results from formal or informal evaluations, emerges.

The academic plan model suggests a number of variables and relationships among variables that could be studied in a program evaluation or in studies that seek to understand which student outcomes are promoted by different curricula (and curricular elements) and why these outcomes vary across programs. Researchers who seek to explore faculty decision making about the curriculum might further explore the Contextual Filters Model, which informed the development of the academic plan model.

The Contextual Filters Model

The contextual filters model, which is based on the work of William Toombs (1977–8; Toombs & Tierney, 1993) and validated through interview and survey research, models the interaction of instructors' personal and professional experiences and their curriculum work in their program and institution. Stark and colleagues (1990) found that faculty decision making about college-level courses and programs is influenced first and foremost by what faculty members bring to the table – specifically, their (1) personal experiences and backgrounds, (2) views of their academic fields, and (3) beliefs about purposes of education. Following Toombs, Stark and her colleagues termed these "content" influences. Toombs argued for the role of contextual features – the constraints and affordances – presented by a given setting. These contextual filters mediate the "content" influences resulting in the "form" of a given course. Research by

Stark and her colleagues found support for the role of these features in the decisions related to curricula.

Stark et al. (1990) proposed a set of contextual "filters" based on research with a large sample of faculty teaching introductory courses in twelve liberal arts and professional fields in community and baccalaureate colleges and in comprehensive and doctoral universities. In addition to limiting their focus to introductory courses, they did not include faculty from research universities in their sample, which limits the generalizability of the study and suggests the need for future research. Still, the results of their study are helpful for informing research on engineering programs. Results of their interview and surveys supported Toombs' contention that most (but not all) instructors selected content as their first step in course planning. The studies also demonstrated a strong relationship between the subject-matter arrangements that instructors said they use in their courses, their beliefs about educational purposes, and their views of their academic fields. These findings are consistent with the body of research that established academic discipline as one of the strongest influences on faculty work (for a review see Smart et al., 2000). Interestingly, however, the study did not find that instructors' gender or race/ethnicity to be strong influences on their course planning decisions, but subsequent research has identified these as influences on the use of instructional methods (Lindholm, Szelényi, Hurtado, & Korn, 2005).

Seven specific contextual influences that affect course development, empirically derived from interview and survey data, are portrayed in the model: student characteristics; student goals; college and program goals; external influences; the literature on teaching and learning and advice available on campus (e.g., an instructional development center); pragmatic factors (such as the term calendar or the size of the course); and facilities, opportunities, and assistance. The model also acknowledges that other factors on a campus may be influential. Of course, the samples chosen and questions asked in such investigations always limit the researcher's vision. Relying on the contextual filters model for the conceptual grounding of her study of curriculum planning practices of research university faculty committed to the educational goal of civic engagement, Domagal-Goldman (2010) found that the faculty reward system on campus – a factor not identified in the study by Stark and her colleagues – also influenced junior faculty members' course planning. This influence was apparently not as salient to the faculty from the two- and four-year institutions Stark et al. (1990) studied. In addition, Domagal-Goldman found that community organizations outside the university influenced faculty decisions about course assignments and activities in courses in which civic engagement was a learning goal for students. She thus proposed modifying the contextual filters model to acknowledge these additional contextual and external influences, suggesting that the type of curriculum being planned influenced perceptions of relevant contextual filters.

Extrapolating from these findings on external influences community partners in service learning, we can imagine how engineering curricula might be influenced by external influences. The need to ensure that students receive instruction and develop the capabilities identified in the ABET EC2000 accreditation standards is clearly an external influence on college and program goals. The availability and degree of industry sponsorship also seems likely to influence the nature of the projects undertaken in undergraduate capstone design courses. Client sponsorship of design projects provides numerous opportunities for authentic assessment[4] of key engineering knowledge and skills. As researchers study engineering programs, they need to consider the particular array of "contextual filters," as well as external influences, that are likely to affect curriculum decision making and the curriculum as a whole.

Biggs's Constructive Alignment Model

John Biggs (1999) offers a process-oriented model of curriculum development that suggests key linkages among elements of an academic plan – for example, content and learning assessment. Grounded in constructivist learning theories that assume that students construct meaning when they engage in relevant learning activities, the model focuses on how instructors design learning environments that support the learning activities required to achieve desired learning outcomes. Biggs argues that the key components in the "teaching system" are the learning outcomes that guide selection of course content as well as suggest the level of understanding students should attain through course activities. Assessment, another key component of the teaching system, provides information on how well individual students have achieved the specified learning outcomes.

Biggs's model makes explicit what the academic plan model, which, as we have noted, is not a guide for course or program design – does not. Whereas the academic plan model identifies the curricular elements in a course or program (and suggests an array of influences on decisions about these elements), the constructive alignment model specifies an effective curriculum development process. It suggests, specifically, that instructors begin by establishing desired student learning outcomes, design-relevant course activities that will motivate students to engage in learning that will lead to these outcomes, and create assessments that provide detailed information on the extent to which students are learning what they are supposed to learn.

The admonition to begin with a set of clear learning outcomes and objectives is consistent across many different models of course design. In the United States, many faculty will be familiar with the "backwards design" model popularized by Grant Wiggins and Jay McTighe (2005), the course and curriculum development processes recommended by Robert Diamond (2008), who honed these processes through many years of instructional development work at Syracuse University, and others (e.g., Posner & Rudnitsky, 2006) who may be more familiar to K–12 educators than to college instructors but whose guidebooks take a similar approach. Each of these authors begins with the premise that an instructor cannot design a curriculum without first establishing what it is that students should know and be able to do at the end of a course or program. Wiggins and McTighe argue that focusing on outcomes first helps instructors avoid the "twin sins of design" – activity-focused teaching and coverage-focused teaching – both of which focus on teaching instead of learning. Diamond's model for course and program development is similarly goal-driven, requiring instructors to state general course and instructional goals.

These curriculum development models suggest that studies of program curricula (and their constituent courses) should investigate not only the structure of courses and course sequences, but the learning activities that instructors select and implement to understand if learning goals are supported by relevant readings, in-class activities, out-of-class assignments, and assessments that actually measure progress toward desired learning goals. Attention to such curricular variables is common in micro-level studies of classrooms, but much less apparent in studies of engineering programs. In our large-scale studies, we have struggled with how to collect the types and level of data that will permit in-depth understandings of the relationships between program curricula and student learning. Although survey data allow for easy comparison and statistical analysis, multiple-method approaches that combine interviews, observations, and surveys may be required to ensure systematic and comprehensive data collection (for guidance see, e.g., Johnson, Onwuegbuzie, & Turner, 2007; Plowright, 2011; Teddlie & Tashakkori, 2009). The curriculum models we have described here offer guidance to researchers as to the sources and types of data needed.

Curriculum Models as Research Heuristics: A Summary

The models discussed in the preceding section provide researchers with frameworks that encourage comprehensive explorations of course and program curricula. They suggest a variety of curricula elements (or decisions) that comprise a course or a set of courses in a program, as well as the influences on instructors as they make decisions about these curricular elements. Depending on the research goals, one or both of these categories of variables may be necessary, and the curriculum models discussed here can serve as heurisitics as researchers seek to think systematically and comprehensively about how to study the complexity of a program curriculum.

Curriculum and college impact models overlap to large degree in their identification of key components of the educational environment. However, the curriculum models begin the task of further specifying what researchers mean when they say curriculum. They look past surface features of courses and programs to understand the constituent parts of the curriculum as well as the factors that influence the decisions that faculty make about, for instance, content, instructional methods, and assessment strategies for their courses, course sequences, and overall program.

Conceptualizing Studies of Learning in Undergraduate Engineering Programs

Our goal in this chapter was to consider how models of college impact and curriculum development and planning can guide the conceptualization of research on engineering education programs. The models we selected and described are not the only models available to researchers. They provide, however, a useful overview of the type of variables that education researchers studying postsecondary experiences have considered important. Whatever model researchers choose to conceptually ground a study of engineering programs and their impacts on students should recognize that a complex set of sociocultural contexts shape student learning. These include influences external to the college or university and to the lives of students and faculty, internal influences that organize the work life of faculty and that create the peer environments in which students learn, and program-level structures and cultures that may or may not align with institutional and disciplinary norms and beliefs.

The addition of curriculum models to our discussion also makes clear the fact that curricula themselves are learning environments that are, to varying extents, purposefully designed and implemented. The study and evaluation of engineering programs is thus complicated by the nesting of these various environments – course- and classroom-level environments are situated in larger program environments, which are often in turn nested in engineering schools, which may in turn be located in multi-school universities. Researchers can benefit from an awareness of this complex context for learning and by reflecting on its implications for their studies of engineering programs.

Footnotes

1. Notable exceptions include the Academic Pathways of People Learning Engineering (APPLES) study (Sheppard et al., 2010); the Academic Pathways Study that included the APPLES study (Atman et al., 2010); and the *Engineering Change* study of the impact of the implementation of the EC2000 accreditation criteria on engineering programs and student outcomes (Lattuca et al., 2006).

2. Scholars have used the term "hidden curriculum" to describe the tacit rules of the game that students in schools must learn in order to succeed (Barnett & Coate, 2005; Snyder, 1973). Some suggest that the hidden curriculum teaches unintended lessons and reproduces social inequities (e.g., Giroux & Penna, 1983).

3. Although some individuals consider instruction to be separate from curriculum, this definition makes it clear that instruction is a critical element of every curriculum plan.

4. Authentic assessments ask students to apply their knowledge and skills in realistic simulations or actual situations. Authentic assessments involve complex tasks, criteria, and standards to provide evidence of a student's ability to succeed in expected roles.

References

Alexander, P. A., Kulikowich, J. M., & Schulze, S. K. (1994). How subject-matter knowledge affects recall and interest. *American Educational Research Journal, 31*(2), 313–337.

Amelink, C. T., & Creamer, E. G. (2010). Gender differences in elements of the undergraduate experience that influence satisfaction with the engineering major and the intent to pursue engineering as a career. *Journal of Engineering Education, 99*(1), 81–92.

Astin, A. W. (1991). *Assessment for excellence: The philosophy and practice of assessment and evaluation in higher education.* New York: Macmillan.

Astin, A. W. (1993). *What matters in college? Four critical years revisited.* San Francisco, CA: Jossey-Bass.

Atman, C. J., Sheppard, S. D., Turns, J., Adams, R. S., Fleming, L. N., Stevens, R., . . . Lund, D. (2010). *Enabling engineering student success: The final report for the Center for the Advancement of Engineering Education.* San Rafael, CA: Morgan & Claypool.

Barnett, R., & Coate, K. (2005). *Engaging the curriculum in higher education.* Berkshire, U.K.: Society for Research into Higher Education & Open University Press.

Biggs, J. (1999). *Teaching for quality learning at university.* Buckingham, UK: Open University Press.

Collins, A., Greeno, J. G., & Resnick, L. B. (2001). Educational learning theories. In N. Smelser & P. Baltes (Eds.), *International encyclopedia of the social and behavioral sciences* (pp. 4276–4279). Oxford: Elsevier Sciences.

Dannefer, D. (1984). Adult development and social theory: A paradigmatic reappraisal. *American Sociological Review, 49*(1), 100–116.

Diamond, R. M. (2008). *Designing and assessing courses and curricula: A practical guide* (3rd ed.). San Francisco, CA: Jossey-Bass.

Dochy, F. J. R. C. (1994). Prior knowledge and learning. In T. Husen & T. N. Postlewaite (Eds.), *International encyclopedia of education* (2nd ed., pp. 4698–4702). Oxford: Pergamon.

Dochy, F. J. R. C., Moerkerke, G., & Segers, M. (1999). The effect of prior knowledge on learning in educational practice: Studies using prior knowledge state assessment. *Evaluation and Research in Education*, 13(3), 114–131.

Domagal-Goldman, J. (2010). *Teaching for civic capacity & engagement: How faculty members align teaching with purpose* (Doctoral dissertation). The Pennsylvania State University.

Feldman, K. A. (1972). Some theoretical approaches to the study of change and stability of college students. *Review of Educational Research*, 42(1), 1–26.

Giroux, H., & Penna, A. (1983). Social education in the classroom: The dynamics of the hidden curriculum. In H. Giroux & D. Purpel (Eds.), *The hidden curriculum and moral education* (pp. 100–121). Berkeley, CA: McCutchan.

Greeno, J. G., Collins, A. M., & Resnick, L. B. (1996). Cognition and learning. In D. C. Berliner & R. C. Calfee (Eds.), *Handbook of educational psychology* (pp. 15–46). New York, NY: Macmillan.

Greeno, J. G. and the Middle School Mathematics through Applications Project Group (1998). The situativity of knowing, learning, and research. *American Psychologist*, 53(1), 5–26.

Holland, J. L. (1966). *The psychology of vocational choice: A theory of personality types and model environments*. Waltham, MA: Blaisdell.

Holland, J. L. (1973). *Making vocational choices: A theory of careers* (1st ed.). Englewood Cliffs, NJ: Prentice-Hall.

Holland, J. L. (1985). *Making vocational choices* (2nd ed.). Englewood Cliffs, NJ: Prentice-Hall.

Holland, J. L. (1997). *Making vocational choices: A theory of vocational personalities and work environments* (3rd ed.). Odessa, FL: Psychological Assessment Resources.

Imbrie, P. K., Lin, J. J., & Malyscheff, A. (2008). Artificial intelligence methods to forecast engineering students' retention based on cognitive and non-cognitive factors. In *Proceedings of the*

2008 *American Society for Engineering Education Annual Conference & Exposition*, Pittsburgh, PA.

Johnson, R. B., Onwuegbuzie, A. J., & Turner, L. A. (2007). Toward a definition of mixed methods research. *Journal of Mixed Methods Research*, 1(2), 112–133.

Kinzie, J., Schuh, J., & Kuh, G. D. (2004, November). *A deeper look at student engagement: An examination of institutional effectiveness*. Paper presented at the Annual Meeting of the Association for the Study of Higher Education, Kansas City, MO.

Knight, D. B. (2012). In search of the engineers of 2020: An outcomes-based typology of engineering undergraduates. In *Proceedings of the 119th Annual Conference of the American Society for Engineering Education*, San Antonio, TX.

Knight, D. B., Lattuca, L. R., Yin, A. C., Kremer, G., York, T., & Ro, H. K. (2012). An exploration of gender diversity in engineering programs: A curriculum and instruction-based perspective. *Journal of Women and Minorities in Science and Engineering*, 18(1), 55–78.

Kuh, G. D., Kinzie, J., Schuh, J. H., Whitt, E. J., & Associates. (2005). *Student success in college: Creating conditions that matter*. San Francisco, CA: Jossey-Bass.

Lattuca, L. R., & Stark, J. S. (1995). Modifying the major: Discretionary thoughts from ten disciplines. *Review of Higher Education*, 18(3), 315–344.

Lattuca, L. R., & Stark, J. S. (2009). *Shaping the college curriculum: Academic plans in context*. San Francisco, CA: Jossey-Bass.

Lattuca, L. R., Terenzini, P. T., Harper, B. J., & Yin, A. C. (2010). Academic environments in detail: Holland's theory at the subdiscipline level. *Research in Higher Education*, 51(1), 21–39.

Lattuca, L. R., Terenzini, P. T., & Volkwein, J. F. (2006). Engineering change: Findings from a study of the impact of EC2000, Final Report. Baltimore, MD: ABET, Inc.

Lave, J., & Wenger, E. (1991). *Situated learning: Legitimate peripheral participation*. New York, NY: Cambridge University Press.

Lin, J. J., Reid, K. J., & Imbrie, P. K. (2009). Work in progress – Predicting retention in engineering using an expanded scale of affective characteristics from incoming students. In *Proceedings of the 39th ASEE/IEEE Frontiers in Education Conference*, San Antonio, TX (pp. 616–617). Piscataway, NJ: IEEE.

Lindholm, J. A., Szelényi, K., Hurtado, S., & Korn, W. S. (2005). *The American college teacher: National norms for the 2004–2005 HERI faculty survey*. Los Angeles: Higher Education Research Institute, UCLA.

Lord, S. M., Brawner, D. E., Camacho, M. M., Layton, R. A., Ohland, M. W., & Wasburn, M. H. (2009). Work in progress – Engineering students' disciplinary choices: Do race and gender matter? In *Proceedings of the 39th ASEE/IEEE Frontiers in Education Conference*, San Antonio, TX (pp. 1233–1234). Piscataway, NJ: IEEE.

Mannon, S. E., & Schreuders, P. D. (2007). All in the (engineering) family? The family occupational background of men and women engineering students. *Journal of Women and Minorities in Science and Engineering, 13*(4), 333–351.

Minick, N., Stone, C. A., & Forman, E. A. (1993). Introduction: Integration of individual, social, and institutional processes in accounts of children's learning and development. In E A. Forman, N. Minick, & C. A. Stone (Eds.), *Contexts for learning: Sociocultural dynamics in children's development* (pp. 3–16). New York, NY: Oxford University Press.

National Academy of Engineering. (2004). *The engineer of 2020: Visions of engineering in the new century*. Washington, DC: The National Academies Press.

Pascarella, E. (1985). College environmental influences on learning and cognitive development: A critical review and synthesis. In J. Smart (Ed.), *Higher education: Handbook of theory and research* (Vol. I). New York, NY: Agathon Press.

Pascarella, E. T., & Terenzini, P. T. (1991). *How college affects students: Findings and insights from twenty years of research*. San Francisco, CA: Jossey-Bass.

Pascarella, E. T., & Terenzini, P. T. (2005). *How college affects students*, Vol. 2: *A third decade of research*. San Francisco, CA: Jossey-Bass.

Pike, G. R. (1999). The effects of residential learning communities and traditional residential living arrangements on educational gains during the first year of college. *Journal of College Student Development, 40*(3), 269–284.

Pike, G. R., Schroeder, C. C., & Berry, T. R. (1997). Enhancing the educational impact of residence halls, the relationship between residential learning communities and first-year college experiences and persistence. *Journal of College Student Development, 38*(6), 609–621.

Pike, G. R., Smart, J. C., Kuh, G. D., & Hayek, J. C. (2006). Educational expenditures and student engagement: When does money matter? *Research in Higher Education, 47*(7), 847–872.

Plowright, D. (2011). *Using mixed methods: Frameworks for an integrated methodology*. Los Angeles, CA: SAGE.

Porter, S. R. (2006). Institutional structures and student engagement. *Research in Higher Education, 47*(5), 521–558.

Posner, G. J., & Rudnitsky, A. N. (2006). *Course design: A guide to curriculum development for teachers* (7th ed.). Boston, MA: Pearson.

Reason, R. D., Terenzini, P. T., & Domingo, R. J. (2006). First things first: Developing academic competence in the first year of college. *Research in Higher Education, 47*(2), 149–175.

Ro, H. K. (2011). *An investigation of engineering students' post-graduation plans inside or outside of engineering* (Doctoral dissertation). The Pennsylvania State University.

Ro, H., Terenzini, P., & Yin, A. (2013). Between-college effects on students reconsidered. *Research in Higher Education, 54*(3), 253–282.

Rogoff, B., Mosier, C., Mistry, J., & Goncu, A. (1993). Toddlers' guided participation with their caregivers in cultural activity. In E. A. Forman, N. Minick, & C. A. Stone (Eds.), *Contexts for learning: Sociocultural dynamics in children's development* (pp. 230–253). New York, NY: Oxford University Press.

Sheppard, S., Gilmartin, S., Chen, H. L., Donaldson, K., Lichtenstein, G., Eris, Ö.,... Toye, G. (2010). *Exploring the engineering student experience: Findings from the Academic Pathways of People Learning Engineering Survey (APPLES)* (TR-10–01). Seattle, WA: Center for the Advancement for Engineering Education.

Smart, J. C., & Ethington, C. A. (1995). Disciplinary and institutional differences in undergraduate education goals. In N. Hativah (Ed.), *Disciplinary differences in teaching and learning: Implications for practice* (pp. 49–57). New Directions for Teaching and Learning, No. 64. San Francisco, CA: Jossey-Bass.

Smart, J., Feldman, K., & Ethington, C. (2000). *Academic disciplines: Holland's theory and the study of college students and faculty*. Nashville, TN: Vanderbilt University Press.

Snyder, B. R. (1973). *The hidden curriculum*. London: MIT Press.

Stark, J. S., Lowther, M. A., Bentley, R. J., Ryan, M. P., Martens, G. G., Genthon, M. L., Wren, P. A., & Shaw, K. M. (1990). *Planning introductory college courses: Influences on faculty*. Ann Arbor, MI: University of Michigan, National Center for Research to Improve Postsecondary Teaching and Learning. (ERIC Document Reproduction Service No. ED330277)

Teddlie, C., & Tashakkori, A. (2009). *Foundations of mixed methods research: Integrating quantitative and qualitative approaches in the social and behavioral sciences*. Los Angeles, CA: SAGE.

Terenzini, P. T., & Reason, R. D. (2005). *Parsing the first year of college: A conceptual framework for studying college impacts*. Paper presented at the Association for the Study of Higher Education, Philadelphia, PA.

Terenzini, P. T., & Reason, R. D. (2010). Toward a more comprehensive understanding of college effects on student learning. Paper presented at the Annual Conference of the Consortium of Higher Education Researchers (CHER), Oslo, Norway.

Toombs, W. (1977–1978). The application of design-based curriculum analysis to general education. *Review of Higher Education, 1*, 18–29.

Toombs, W., & Tierney, W. G. (1993). Curriculum definitions and reference points. *Journal of Curriculum and Supervision, 8*(3), 175–195.

Trenor, J. M., Yu, S. L, Waight, C. L., Zerda, K. S., & Sha, T-L. (2008). The relations of ethnicity to female engineering students' educational experiences and college and career plans in an ethnically diverse learning environment. *Journal of Engineering Education, 97*(2), 449–465.

Turns, J., Yellin, J. M. J., Huang, Y-M., & Sattler, B. (2008). We all take learners into account in our teaching decisions: Wait, do we? In *Proceedings of the American Society for Engineering Education Annual Conference*, Pittsburgh, PA, Session 2008-1199.

Weidman, J. C. (1989b). Undergraduate socialization: A conceptual approach. In J. Smart (Ed.), *Higher education: Handbook of theory and research*, Vol. V (pp. 289–322). New York, NY: Agathon Press.

Wertsch, J. V. (1985). *Vygotsky and the social formation of mind*. Cambridge, MA: Harvard University Press.

Wertsch, J. V., Del Rio, P., & Alvarez, A. (1995). *Sociocultural studies of mind*. New York, NY: Cambridge University Press.

Wiggins, G. T., & McTighe, J. (2005). *Understanding by design*. Alexandria, VA: Association for Supervision and Curriculum Development.

CHAPTER 25

Design-Based Research in Engineering Education

Current State and Next Steps

Anthony E. Kelly

Introduction

Engineering education research is looking to research methods in education and related social sciences as a source for new approaches (Adams et al., 2011; Borrego & Bernhard, 2011; Case & Light, 2011; Davison, 2010; Ganesh, 2011; Johri & Olds, 2011; Pears, Fincher, Adams, & Daniels, 2008; Streveler & Smith, 2010). Design-based research in education, the focus of this chapter, is a natural source for ideas because this emerging methodology draws on engineering practices for some of its key values and approaches (e.g., Brown, 1992; Hjalmarson & Lesh, 2008; Middleton, Gorard, Taylor, & Bannan-Ritland, 2008). Indeed, inspiration for one of the early stage models for design-based research in education (Bannan-Ritland, 2003) was proposed by Woodie Flowers, an MIT engineer, at a National Science Foundation (NSF)–funded workshop (Kelly & Lesh, 2001). Design-based research can contribute to engineering education research because it also draws on (1) a tradition of studies in mathematics and science education (e.g., Cobb, McClain,

& Gravemeijer, 2003; Kelly, Baek, Lesh, & Bannon-Ritland, 2008) and (2) frameworks from diffusion of innovations (Zaritsky et al., 2003) and more recently (3) from educational data mining (e.g., Baker & Yacef, 2009).

In this chapter, I use one model for design-based research in education, the Integrative Learning Design (ILD) framework (Bannan-Ritland, 2003), and illuminate its use with examples from education and an engineering education study by Hundhausen, Agarwal, Zollars, and Carter (2011). I suggest next steps in design-based research for engineering education research, including the creation of a Design Exchange for Scholars that would actively integrate insights from educational and engineering education research and place a greater emphasis on learning analytics and educational data mining.

Introduction

Engineering education in the United States has been the subject of calls for reform for more than 100 years (Seely, 2005) – calls that continue (e.g., Ambrose & Norman, 2006; Atman et al., 2010; Cheville & Bunting, 2011; National Academy of Engineering, 2005, 2013 Sullivan, 2006). This reform is an ongoing effort because modern engineering continues to multiply into new hybrids (Tadmore, 2006), and the aspirations set out for engineering education are quite challenging. As Charles Vest (2006), the president of the National Academy of Engineering, noted:

> ...engineering students prepared for 2020 and beyond must be excited by their freshman year; must have an understanding of what engineers actually do; must write and communicate well; must appreciate and draw on the richness of American diversity; must think clearly about ethics and social responsibility; must be adept at product development and high-quality manufacturing; must know how to merge the physical, life, and information sciences when working at the micro- and nanoscales; and must know how to conceive, design, and operate engineering systems of great complexity. They must also work within a framework of sustainable development, be creative and innovative, understand business and organizations, and be prepared to live and work as global citizens. That is a tall order...perhaps even an impossible order. (pp. 41–2)

The National Academy of Engineering's Committee on Engineering Education (Davison, 2010, p. 1) tied this complex goal to engineering curricular reform in a recent workshop:

> Topics addressed in the workshop included (a) the rationale for the scope and sequence of current engineering curricula, considering both the positive aspects as well as those aspects that have outlived their usefulness, (b) the potential to enhance engineering curricula through creative uses of instructional technologies, (c) the importance of inquiry-based activities as well as authentic learning experiences grounded in real world contexts, and (d) the opportunities provided by looking more deeply at what personal and professional outcomes result from studying engineering. General themes that appeared to underlie the workshop attendees' discussions included desires to (a) restructure engineering curricula to focus on inductive teaching and

> learning, (b) apply integrated, just-in-time learning of relevant topics across STEM fields, and (c) make more extensive use and implementation of learning technologies.

However, innovations aimed at reforming engineering education can be frustrating. For example, Froyd (2005), commenting on the Engineering Education Coalitions (EEC) programs, noted that even innovative practices, backed by research data, may not diffuse quickly or broadly (see also Borrego, Froyd & Hall, 2010; Rogers, 2003):

> The EEC Program demonstrated that engineering faculty members can construct out-of-the-box, effective models for curricular and systemic reform, and assessment data indicate that they lead to increased retention and improved student learning. However, the EEC Program also demonstrated that institutional and cultural barriers to change are more complex, intricate, and subtle than is often appreciated and that innovative models for reform are seldom enough to overcome the challenges to institutionalizing change. In addition, the program demonstrated that effective models, even when well supported by assessment data, do not catalyze systemic reform. To achieve that goal, resources matched to the extent, complexity, and dynamics of the system of engineering education must be assembled and deployed through intense, informed, and sustained conversation. (pp. 96–7)

These findings correspond with the history of general educational reform, which is a catalog of partially successful efforts (e.g., Cuban, 1986, 2001, 2009; Tyack & Cuban, 1995). Although scientific research in education can identify effective instructional strategies that apply under certain conditions (e.g., the What Works Clearinghouse, http://ies.ed.gov/ncee/wwc/), and in some cases it can point to more fundamental empirical underpinnings for these results (e.g., Dehaene, 2009; Kelly, 2011), it remains a clinical, designing, and prototyping endeavor. Research on education will not converge on immutable causal laws. Rather, it will describe and advise the enactment of solutions occurring within a context of diffusion of innovations (Rogers, 2003) shot through by the constraints of human agency (Cobb, Zhao, & Dean, 2009; Hattie, 2009; Howie & Plomp, 2006; Reeves, 2011). Education agents in clinical settings *reason with* scientific principles in a context of tacit knowledge and the wisdom of practice that cannot be comprehensively captured by laboratory studies (e.g., Nelson & Stolterman, 2003; Toulmin, 1998, 2001). As William James (1962, p. 3) noted in the late 1800s, "Psychology is a science and teaching is an art; and sciences never generate arts directly out of themselves."

Persistence in engineering education research remains critical despite its incremental impact. Engineering education will provide basic research insights and viable solutions that can be enacted within a complex system involving policymakers, funding agencies, university administrators, faculty, and students (e.g., Maroulis et al., 2010; Stephens & Richey, 2011; Weiss, Miller, Heck, & Cress, 2004). The impact will be felt both directly and indirectly over an extended period of time and will require not only basic research, but also an infrastructure to sustain efforts at scaling solutions (e.g., Bhide, 2008; McDonald et al., 2006; Penuel, Fishman, Cheng, & Sabelli, 2011).

Design-Based Research in Education

Design-based research in education is a field-driven, situated, collaborative, and interventionist research methodology (Shavelson et al., 2003), which represents a move from a laboratory- to a field-based science (Brown, 1992; Kelly, 2004, 2009). Please note that a number of different labels are used to describe this methodology: in addition to design-based research, one may encounter "design research" (e.g., Kelly et al., 2008; Yun Dai, 2012), "development research" (Van den Akker, 1999), or "design experiment" (Cobb, Confrey, diSessa, Lehrer, & Schauble, 2003). The literature on design-based research is rich and growing (e.g., Barab & Squire, 2004; Dede, 2005; Kelly, 2003; Plomp & Nieveen, 2008; Reinking & Bradley, 2008; Sandoval & Bell, 2004; Van den Akker, Gravemeijer, McKenney & Nieveen, 2006).

Although there is no methodological unanimity among design-based researchers in education (e.g., Kelly et al., 2008), there is a general commitment to designing and developing prototype solutions to problems of educational practice. The design, development, deployment, and refinement of prototypes generate data on learning and teaching within and across stages of research, from brainstorming to scaling. Iterative design cycles within and across development stages are used to understand what constructs are understandable and teachable, how these constructs can be assessed and modeled, and how data from these models can be mined for theoretical insights. Data modeling from any stage of research may inform modeling at some other stage. In other words, design-based research involves proactive and retroactive analyses (e.g., Cobb et al., 2003).

Design-based research has been applied at the level of the student (e.g., Lobato, 2008; Rasmussen & Stephan, 2008), the teacher (e.g., Bannan-Ritland, 2008; Zawojewski, Chamberlin, Hjalmarson, & Lewis, 2008), simultaneously at the teacher and student levels (Lesh, Kelly, & Yoon, 2008), the administrator level (e.g., Brazer & Keller, 2008; Wolf & Le Vasan, 2008), the research project level (e.g., Bannan-Ritland & Baek, 2008; Schwartz, Chang, & Martin, 2008), and at the level of programs of studies (e.g., Clements, 2008; Middleton et al., 2008; Roschelle, Tatar, & Kaput, 2008; Van den Akker et al., 2006). In fact, the Division of Research on Learning at the NSF explicitly describes multistage approaches to research that draw on design-based research models (see Figure 25.1, from the DR-K12 program announcement (http://www.nsf.gov/pubs/2011/nsf11588/nsf11588.htm).

The Integrative Learning Design Framework

Design-based research views the act of research as being distributed across and co-informing a number of distinct stages of data gathering and modeling. The Integrative Learning Design Framework (ILD)

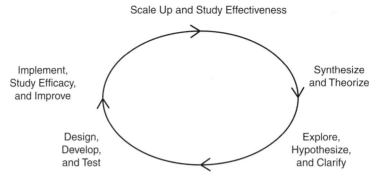

Figure 25.1. National Science Foundation, Division of Research on Learning, Cycle of Research and Development.

captures this complexity (Bannan-Ritland, 2003; see Figure 25.2), although other stage models exist (e.g., Bryk, 2009; Middleton et al., 2008; Sloane, 2008a). Researchers do not follow a strictly linear process across these stages, but may revisit "earlier" stages as emerging evidence warrants or may enter at a "later" stage if the maturity of the field allows (Kelly, 2009). More recently, online education studies can now move directly to "at scale" activity, once initial ideas are formulated, without touching each of the prior steps (e.g., the *Foldit*[1] and *Refraction*[2] programs at the University of Washington).

Bannan-Ritland (2003) situated design-based research among different approaches to innovation development. In Figure 25.2, she contrasts traditional educational research (bottom line) to each of: (1) Rogers's diffusion model, (2) usage-centered design, (3) instructional system design (ISD), (4) product design; combining elements of each in her comprehensive Integrative Learning Design (ILD) framework. More details on the methods and measures particular to each stage of ILD are available in Figure 1 in her paper (Bannan-Ritland, 2003, p. 22).

I apply the ILD stage model to two studies in education, and to the engineering education research study by Hundhausen et al. (2011), though other candidates, of course, exist (e.g., Dahm et al., 2009; Dalrymple, Sears, & Evangelou, 2011; Delale, Liaw, Jiji, Voiculescu, & Yu, 2011; Dinehart & Gross, 2010; Eggermont, Brennan, & Freiheit, 2010; Garcia, Varnasi, Acevedo, & Guturu, 2011;

Hayden et al., 2011; Impelluso, 2009; Mantri, Dutt, Gupta, & Chitkara, 2009).

Hundhausen and colleagues' two-part iterative study describes the development of a scaffolded software environment, called ChemProV. ChemProV is designed to address the problem that "students struggle because of misconceptions regarding the basic syntax and semantics of disciplinary diagrams and corresponding mathematical equations" (2011, p. 574).

ILD Stage 1-*Informed Exploration*

Informed exploration in Bannan-Ritland (2003) captures initial brainstorming, literature reviewing, and benchmarking against existing solutions. Research methods at this stage can include needs analyses, interviews of stakeholders, surveys of experts, observational studies, and case studies. The studies in a literature review are examined not just for a history of findings, but, like the other methods, are viewed as repositories of intervention designs, which are critiqued as design objects.

EDUCATION EXAMPLE: LITERACY ACCESS ONLINE

Literary Access Online (LAO; Bannan-Ritland & Baek, 2008) is a Web-based system to help young struggling readers. It provides support for teachers, tutors, and parents (acting as literacy facilitators). LAO was designed over multiple cycles with each of these stakeholders. Bannan-Ritland and Baek (2008) described how the LAO project

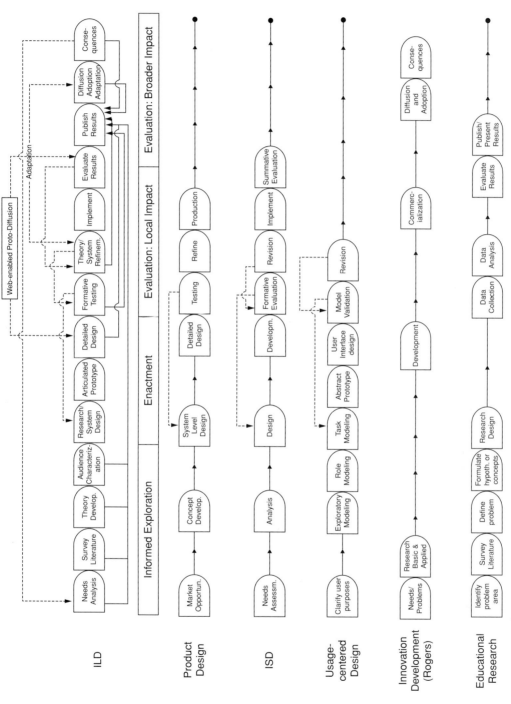

Figure 25.2. The Integrative Learning Design (ILD) framework and how it fits with related approaches. (Reproduced with permission from Bannan-Ritland, 2003.)

interpreted, framed, and re-framed research questions through design problems cycles, and shows the many factors, decisions, and judgments that impacted the process of design-based research project.

They conducted a needs analysis phase because the goal was user adoption of the software system, LAO. Thus, the team interviewed parents, children in the fourth through eighth grades with or without disabilities, teachers, reading specialists, special education personnel, and literacy tutors. These interviews used a usage-centered design process (Constantine & Lockwood, 1999) and led to the development of "personas" (Cooper, 1999) or abstracted characterization of users that, acting as role players, would help the team consider the design of LAO from the perspective of different users.

With these personas in mind, and analyzing the data from the interviews, the team significantly changed their original plan, which was to design a computer-to-child reading tutor, to designing *a support system* for *adults* helping struggling readers (i.e., from a computer–child dyad to a literacy facilitator–child dyad). As Bannan-Ritland and Baek (2008) noted, "These design moves resulted not from preconceived theory but from direct experience with individual and combinations of target audience members, analysis of the context of performance, and close investigation of the potential intersections of collaborative exchange among the participants" (p. 305).

We see here how design-based research is not following pre-scripted steps in a routinized attempt to establish and maintain methodological rigor. The LAO team recognized that any effectiveness metrics would depend on adoption and use, so the designed was swayed by the contexts and values of the adopters.

Following the user-centered interviews, the team engaged in extended brainstorming, which led to the creation of rapid prototypes of the early versions of the final design. Proposing and designing prototypes often involve both explicit and of implicit assumptions about learning and teaching.

During early stages, there are yet no clear hypotheses about models of learning; rather, there is a set of conjectures or heuristics that are expressed in some rough format, which may be as simple as storyboards or sketches of ideas. As the prototypes are made more detailed, and are subject to critique and analysis (by both the researchers and the users), the initial learning conjectures that they embody may develop into firmer hypotheses about learning.

A significant design question facing those who design for educational uses is to entertain one or more overarching perspectives about learning. It is beyond the scope of this chapter to describe learning theories, but each has different assumptions about learning and teaching and will influence content, practices, and assessment protocols. These perspectives include sociocultural theories (the one adopted by Bannan-Ritland and Baek), constructivist theories, social learning theories, behavioral theories, and so forth (e.g., National Research Council [NRC], 2000, 2005a, 2005b).

Whichever theoretical perspective is adopted, it is important to set learning targets (see Nitko & Brookhart, 2006) to ground both formative and summative assessments of progress. For Bannan-Ritland and Baek (2008, pp. 306–7), these learning targets were that:

- Literacy facilitators (represented by parents, teachers, or others) will acknowledge the importance of and demonstrate their ability to implement research-based reading strategies when provided technology-based support in a collaborative reading session with a child.

- Children with or without disabilities will demonstrate their abilities to access information, activities, and assistive technology support related to reading as well as to interact with literacy facilitators in a collaborative reading session.

- The facilitator–child dyad will be able to explore and select appropriate assistive technology integrated with Internet-based supports that can facilitate performance in reading and writing.

- Children, regardless of their disability, will be able to capitalize on technology-based supports and a collaborative process to improve their literacy skills.

In their chapter, the authors show how

> ...the task analysis of the learning targets isolated and explicated the interaction among the components such as a subject *(the facilitator–child dyad)*, an object *(the LAO system)*, the tools *(literacy strategies, assistive technologies)*, the division of labor *(the division of tasks between the facilitator and the child in the reading process)*, the community *(facilitator–child dyads are members of school, home, and tutoring communities)*, and the rules *(the social norms, relationships, or constraints that guide the collaborative literacy process)*. The identification of, and interaction among, these elements shaped the resulting evaluation procedures as well as informed the theoretical model of collaborative literacy that was embedded subsequently in the instructional innovation. *(Bannan-Ritland & Baek, 2008, p. 307, emphasis in the original)*

Hundhausen and colleagues similarly reviewed the literature in how people learn in general, and how people learn chemical processes. They also benchmarked their proposed design against existing tools. In contrast to Bannan-Ritland and Baek, however, they did not stress as strongly creating a persona of likely adopters, or the concerns of a comprehensive set of stakeholders (e.g., administrators; see Borrego et al., 2010).

ILD Stage 2-Enactment

The stage of enactment expresses conjectures as prototypes. The act of prototyping necessarily narrows the scope of future moves because some of the brainstormed options must be (at least temporarily) discarded or held in abeyance. This stage of the process will see reliance on storyboards, task analyses, and special attention to user feedback (Bannan-Ritland & Baek, 2008). Users can include students, teachers, other educational stakeholders, in addition to the educational researchers and expert advisors

(Kelly et al., 2008; Thomas, Barab, & Tuzun, 2009). Because the prototype will be field tested, it is important to document contextual data that not only may explain equivocal results, but that also illuminate design principles that later researchers may use (e.g., Kali, 2008; Kali, Levin-Peled, & Dori, 2009; Kali, Levin-Peled, Ronen-Fuhrmann, & Hans, 2009).

It is important at this stage of research to maintain design logs – written and multimedia records that reflect on what design choices were made, and why, and which paths were not pursued. These logs will prove valuable not only as a source of design principles for later use in other projects, but may also help illuminate why later stages of development fail and what design decisions were originally adopted (Bannan-Ritland & Baek, 2008; Clements, 2008; Kelly et al., 2008; Martinez et al., 2008; Schwartz et al., 2008).

LITERACY ACCESS ONLINE

For LAO, enactment of prototypes led to data on their effectiveness as judged by the researchers, the children, and their parents. The team ran two pilot studies on the developing system. The first pilot test involved observations of five parent–child interactions with the prototype and semistructured interviews. Having revised the design, they ran a second qualitative study with eight dyads, involving children with various disabilities in grades four through six who were reading at least two grades below grade level. Over a number of iterations, the emerging system was designed to function like an electronic performance support system (e.g., Gery, 2003), which provided up-to-date hints and tips (drawn from the research literature on reading) to the literacy facilitators. It also allowed children to write their own stories, and involved other features such as disability supports.[3]

For enactment, Hundhausen et al. (2011) were concerned with three key features: (1) diagram consistency, (2) equation correctness, and (3) equation solvability. They created a prototype system that provides a toolbox of process flow diagram components, a drag-and-drop equation building facility, and a dynamic feedback system of student alerts.

ILD Stage 3-Evaluation: Local Impact

Local evaluation asks if expected learning goals are being realized by the prototype (which could include a range of digital, human, and blended instructional strategies or products). Evaluations can include achievement measures or noncognitive outcomes such as engagement (e.g., Adams et al., 2011; Purzer, 2011). Although some research traditions privilege quasi-experimental or randomized controlled trials as the sine qua non to assess impact (e.g., Shadish, Cook, & Campbell, 2002), design-based research recognizes that the clinical and complex nature of education settings rarely allow laboratory controls (Collins, Joseph, & Bielaczyc, 2004). Thus, the focus is on evidence that tentatively supports a range of research claims drawn from usability testing, expert feedback, interviews, and case studies (Kelly & Yin, 2007; Yin, 2009). Design-based research seeks evidence to help *explicate* and *explain* changes in processes of learning that appear tied to aspects of learning interventions (Hjalmarson & Lesh, 2008; Maxwell, 2012). Where feasible, it focuses on growth models of individual learning (Sloane, 2008b; Sloane & Gorard, 2003; Sloane & Kelly, 2008), perhaps using interrupted time series (e.g., Sloane, Helding & Kelly, 2008) or single-case analyses (e.g., Shadish & Cook, 2009).

Design-based research does not separate the context of justification from the context of discovery (see Schickore & Steinle, 2010), but argues, instead, for a context of enacted innovation in which demonstrated regularities (i.e., the focus of justification studies) must engage dynamic clinical settings – where system variables external to laboratory controls and the caprice of human agency ultimately determine impact (Rogers, 2003).

SIMCALC MATHWORLDS

Since it has been funded longer than LAO, and as such has reached the "later" stages of ILD, I now discuss the SimCalc MathWorlds program, a design-based research project that has been funded for more than fifteen years. A design history of the project is available in Roschelle et al. (2008).[4]

A core early idea in this project was to reconceptualize mathematics curriculum by designing learner-centered interventions using the affordances of new representational technologies (e.g., computer simulations); see Roschelle et al. (2008). In its current expression, SimCalc encompasses both software and curricular support for learning the mathematics of change, and calculus (see project website).

In Figure 25.3, we see a screenshot from SimCalc in which learners can observe and act on the speeds of different fish and track their progress in different representations.

According to Roschelle et al. (2008), this multiyear initiative of iterative design cycles (some micro, within projects; and some macro, across projects), demonstrates some clear patterns. An initial concern with students' representations of mathematics was supplemented with a greater concern about teachers' interactions with the student and with the materials. Once teacher concerns were raised, then the larger systemic world of the teaching profession came into play: national and state legislation and standards, assessment practices, teacher professional development, and so forth. Over time, to generate more rigorous evidence, there was a push for larger samples of students and teachers, which, in turn, raised the complexity of the factors that required modeling. The project team size and the number of related disciplines it addressed grew over the life of the project.

At the same time, funding agencies (with an initial impulse from the Interagency Educational Research Initiative [IERI], which began around 2000[5]) began to raise concerns about whether novel technology interventions would scale to large numbers of students and teachers. SimCalc received IERI

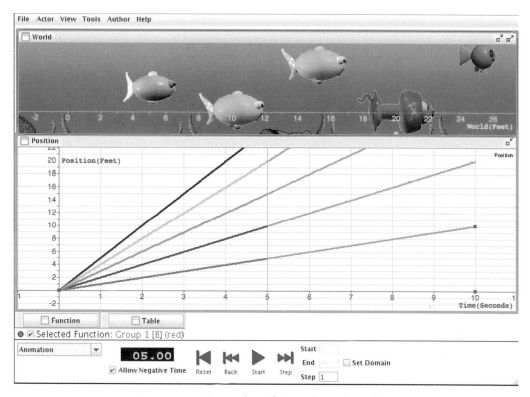

Figure 25.3. Screenshot of SimCalc MathWorlds.

funding in 2002 (NSF IERI Grant No. REC 0228515). This work on scaling an intervention continues (Roschelle et al., 2010).

Scaling a designed intervention such as SimCalc involved years of work in foundational activities. Roschelle et al. (2008) reported its phases:

> Work in Phase One had no empirical methodology; the basics of the design relied on historical, curricular, and literature analyses. Phase Two used primarily microanalysis of a small number of students. Phase Three involved design experiments, with pretest and posttest measures. Phase Four was a planning phase and not particularly methodological. Phases Five and Six focused on an experimental design but include embedded case study analyses. At each phase, there are still a huge number of design questions to resolve, more than can be settled using only benchmark methods. Research on scaling-up innovation is a complex enterprise.
>
> A particularly important and difficult transition occurred during Phase Four, when the team finally had to confront the question: "What is the essence of SimCalc and what is worth subjecting to a rigorous test?" We suspect this transition was difficult because of the lack of overlap between design researchers and experimental implementation researchers. Thus, although copious effort in the first three phases of SimCalc had addressed questions of scale, too little effort had been devoted to the eventual demands of scale-up research. In retrospect, we would have put more effort earlier into firming up measures and instruments and documenting better potential recruitment difficulties... Finally, the role of technology, the basic form of the SimCalc representational system, and the commitment to integrate technology with textual curricula have remained unchanged since 1995. (pp. 386–387)

For local impact, Hundhausen et al. (2011) compared performance on ChemProV with a paper-and-pencil approach using a matched-samples between subjects experimental study with two conditions

(ChemProV or paper-and-pencil first). They measured time on task and reviewed student solutions for: (1) process-flow diagram layout accuracy, (2) stream accuracy, and (3) material balance accuracy. The results were equivocal with no clear advantage for ChemProV.

Yet, the student log data showed how little the alerting system of ChemProV was used by students. The authors then created a refined prototype of ChemProV with an improved student alert system and conducted a second local impact study. The ChemProV system with feedback led to greater accuracy, more on-task learning, and better transfer of effect to unscaffolded problems than using ChemProV without the feedback feature. This local impact studies points to a potential broader impact study.

ILD Stage 4-Evaluation: Broader Impact

Broader impact evaluations are concerned with identifying implementable and mutable features of interventions for later scrutiny (and, perhaps revision) in implementation studies (Osmundson, Herman, Ringstaff, Dai, & Timms, 2012; Penuel et al., 2011).

Once implementation factors are better understood, the focus turns to the scaling of effective findings (e.g., McDonald, Keesler, Kauffman, & Schneider, 2006; Schneider & McDonald, 2006a, 2006b). Thus, the attention is on broad adoption (and adaptation) of principled interventions in which laboratory controls will not be possible (Collins et al., 2004), rather than on conducting large-scale studies to pursue causal claims that are assumed to apply nomothetically (e.g., Mosteller & Boruch, 2002; Shadish et al., 2002). That being said, the VaNTH learning modules have been subjected to sophisticated quasi-experimental and experimental trials (e.g., Barr et al., 2007; Cordray, Harris, & Klein, 2009). Over time, any solutions that work in well-structured situations must contend with the demands of ill-structured worlds (see Berliner, 2002; Bransford et al., 2009; Cobb & Smith, 2008; Cronbach, 1975),

where human agency, interpretation, and decision making (e.g., Brazer & Keller, 2008), not mechanical principles, apply.

Design-based research in education fosters innovations that are likely to be adopted, adapted by users, over time, and over wide geographic areas (e.g., Zaritsky, Kelly, Flowers, Rogers, & O'Neill, 2003). Hundhausen et al. (2011) give a clear exposition of the experimental limitations of the design they chose, and the resulting restrictions on generalizable claims. Yet, they have uncovered a key insight about student problem-solving behavior: the lack of attention to what experts might term superficial errors. Attention is a crucial aspect of learning (e.g., Barriga et al., 2002; Forster & Lavie, 2007). Thus, these apparently superficial errors have a "downstream" effect on and corrupt later calculations, which can retard the learning process. From a design-based research perspective, Hundhausen et al. show how to design a prototype system that can provide timely alerts to students to scaffold their growth from novice to expert problem solvers in this and in later studies.

By comparison, consider the engineering education research study by Taraban, Craig, and Anderson (2011). These authors used a mixed-methods approach (supplementing quantitative with video analyses) and showed that paper-and-pencil grading can incorrectly penalize stronger students (who make superficial algebraic errors, for example). Affirming the findings of Hundhausen et al. (2011), weaker students who incorrectly translated a problem into a free body diagram were highly likely to corrupt later forward inferences and frustrate their learning.

From a design-based research perspective, it would be welcome to see Taraban et al. (2011) emulate Hundhausen et al. (2011) and test a scalable alerting system (or provide the design principles for such an intervention). This comment is not meant as a criticism of Tarban et al. (2011). Rather, it is made to illustrate a difference between reasonable approaches to social science research: (1) *demonstrating* that an effect that supports an insight from cognitive learning theory and (2) the engineering step of designing, enacting, and testing scalable *implementations* of that insight. Incidentally, current work in educational data mining has begun to focus on automatic detection of student behaviors, which could take student alert systems to scale (Aleven, Roll, McLaren, & Koedinger, 2010; Baker, Goldstein, & Heffernan, 2011; Baker, Gowda, & Corbett, 2011a, 2011b; Reese et al., 2012; Roll, Aleven, & Koedinger, 2011; Roll, Aleven, McLaren, & Koedinger, 2011; Vendlinski, Chung, Binning, & Buschang, 2011). Note that automatic detector systems can track behaviors deeper than simple attention to errors (e.g., San Pedro, Baker, Gobert, Montalvo, & Nakama, 2011; San Pedro, Gobert, & Baker, 2012).

DIFFUSION OF INNOVATIONS

Ultimately, at the broadest level of the ILD model, the adoption of any research product for instructors and administrators is a process of action among agents working in complex social systems (e.g., Rogers, 2003), not a mechanical change to a simple machine. In an extended project in design-based research, Roschelle et al. (2008), who examined the effectiveness of a software program to support students' learning of the mathematics of change (SimCalc MathWorlds), reaffirmed the canonical concerns identified by Rogers (2003): (1) the importance of relative advantage (compared to current practices), that is, compatibility with existing practices; (2) simplicity (overly complex innovations tend not to be adopted); (3) trialability (can the innovation be tried out at little cost?); and (4) observability (can potential adopters see the innovation in action?).

Roschelle et al. (2008) thus called for design principles that ensure that innovations are: (1) specified in testable and scalable detail, (2) of large-scale benefit to the public, (3) adaptable and attractive to large number of users, and (4) and employed in valid experimental settings. Equally, in engineering education research, Borrego, Froyd, and Hall (2010) showed that awareness of innovative practices in engineering education research (which is widespread) does not easily translate into adoption and implementation. They provided a rich diffusion of innovations analyses of these findings, which will repay a close reading.

Next Steps in Design-Based Research

Design-based research in education emerged in the early 1990s at the early stages of the use of digital technologies (i.e., as e-mail was gaining popularity, but before the Mosaic and Netscape Internet browsers, and the Yahoo! Search engine). I can only sketch how design-based research must embrace the possibilities available through: (1) global social connectivity; (2) a pervasive cyber-infrastructure with massive digital libraries (e.g., Cyberlearning and Workforce Development Task Force, 2011); (3) the emergence of educational data mining (Baker & Yacef, 2009) and learning analytics (e.g., the data visualization for education key note by Borner, 2012); and (4) the growing national interest in addressing engineering and other grand challenges.[6]

Key features in these conversations are likely to include: (1) how to inform the design and redesign of iterated prototypes that apply not only to traditional K–12 education, but also to informal, after school, and life-long learning; (2) how to revisit constructs of validity and modernize psychometric theories (Kelly, 2012; Scardamalia, Bransford, Kozma, & Quellmalz, 2012); and (3) how to link generative approaches to assessment design, data collection, and theory building to real global challenges, not just the pursuit of higher test scores on standardized (i.e., context-insensitive) measures (e.g., Kelly et al., 2008).

Integrated and Cooperative Solutions via a Design Scholar Exchange

To foster progress toward these goals for design-based research in engineering and in education research, it may make sense to establish a Design Exchange for Scholars. This exchange could draw on similar efforts, for example, Peer Reviewed Research Offering Validation of Effective and Innovative Teaching (http://www.pr2ove-it.org/proveit/) and nanohub.com (to cite two examples). The Design Exchange would serve as a shared social space where researchers, intending to perform some intervention study, could seek advice from other researchers or developers. If one or more party shared a similar experimental goal (e.g., what is the impact of multimedia variability on learning?), then it should be possible to coordinate replications of the intervention in co-informing studies with different learners in different learning contexts. Such coordinated replication studies, with carefully mapped hypotheses, shared literature reviews, shared protocol, and outcome data, would also provide a rich source of secondary data analysis for educational data mining. Further, the Design Exchange would amplify the return of investment for the individual funding agencies, create new shared professional communities, and lead to the aggregation of results, the validation of instruments, and a welcome growth in methodological sophistication.

Consider, for example, the similarity of Hundhausen et al. (2011) and Taraban et al. (2011) to design-based research work in education by Davenport, Yaron, Klahr, and Koedinger (2008). Not unlike Hundhausen et al. (2011), Davenport et al. found that adding diagrams to instruction on chemical equilibrium, in this case using Mayer's principles on multimedia learning (e.g., Mayer, 2003), had little impact on learning.

Subsequent think-aloud protocol analyses, however, uncovered students' misunderstanding of chemical interactions (Davenport et al., 2008). Some students thought of chemistry equations as describing processes unfolding from left to right for which there was no expectations that values of variables could have different values before, during, and in achieving chemical equilibrium. These insights (as in Hundhausen et al., 2011) led to the design of a more

effective, and potentially scalable, intervention. Further, the ability to conduct A/B randomized testing (which is a feature of online education research; see publications of the Center for Advanced Technology in Schools: http://cats.cse.ucla.edu/index.php) could be augmented by ideas proposed by Sloane and colleagues, such as interrupted time series designs (e.g., Sloane et al., 2008; Sloane & Kelly, 2008).

Thus, each of these three teams of researchers could benefit from the synchronous sharing of intermediate results during the various stages (e.g., ILD) of research. Such stage-linked cross fertilization would work against "silo" research in each content area. It would also dramatically shorten the time that one community could learn from the other. Under current dissemination models, an engineering education researcher is highly unlikely to chance on a paper by a chemistry educator – one that would be published many months or even years after the actual experiments.

Backward-Chaining Diffusion of Innovations

For design-based research in education, some innovation is created and its spread and scaling is studied (e.g., the software program SimCalc of Roschelle et al., 2008). Of as much concern in educational research is the diffusion of *learning to learn*, the diffusion of a general process or capacity (for a review of transfer in learning studies, see Lobato, 2006). The "21st Century Skills" enterprise has a similar aspiration, except the skills are more general than learning mathematics of change (e.g., creativity or critical thinking[7]).

One of the goals of engineering education is to prepare students to apply what they learn to the workplace by exposing them to some sequence of college courses. This, effectively, is a *push* model: teach many concepts in mathematics, physics, and related fields and hope that some of these will be of value in the workplace. Engineering education research can also examine a simultaneous *pull* model by linking curricular strands to problems posed by industry: that is, draw on the perspectives of diffusion of innovations and transfer of learning that drive much of educational design-based research. Current work between SRI and Boeing Corporation (Sabelli, Lemke, Cheng, & Richey, 2012) could form the basis for a useful test case – for example, using the design of mutually valuable model-eliciting activities (MEAs) (Hamilton, Lesh, Lester, & Brilleslyper, 2008) to link modeling problems in industry with undergraduate projects. The use of a Design Exchange in this way could help ground engineering education projects more directly in the terminal contexts of use (transfer of learning) and build solutions more likely to adopted (diffusion of innovations). Equally, such mutual research activities could add to the ongoing professional development of engineers in industry (see Stephens & Richey, 2011).

Educational Data Mining

With appropriate funding, a Design Exchange could extend the reach of engineering education research to a shared data shop, such as Koedinger's LearnLab's DataShop (http://pslcdatashop.web.cmu.edu/index.jsp). Data alerts could be sent to networked researchers when new related data, measures (e.g., Borrego, Newswander, McNair, McGinnis, & Paretti, 2009; Hamilton et al., 2008), or findings arrive. Links to projects in K–12 such as ASSISTments (http://www.assistments.org/ and http://youtu.be/SEjB19BhCPk) would also be possible. Thus, Design Exchange participants could contribute to educational data mining, which is currently an area of growing strength with the capacity to increase the impact of design-based research (see Baker & Yacef, 2009; Baker et al., 2011; Rodrigo & Baker, 2011; Romero, Ventura, Pechenizkiy, & Baker, 2012).

Design-Principles Database

As projects leaders begin to understand the heuristics or design principles that informed

their work (e.g., precede multimedia manipulation with think-aloud protocol analyses), these principles could be added to a Design-Principles Database. Such a data base already exists for design-based research (see http://www.edu-design-principles.org/dp/aboutDPD.php). What is powerful about sharing use cases with large and dynamic data sets is that it may spur the next generation of multiteam, multidisciplinary design-based research. In this new model, different research teams would share materials, resources, literature reviews, assessments, outcome measures, practices, and insights on related projects in close-to real time.

Undergraduate Research and Graduate Training

The growing shared digital data resources sketched here could also improve apprenticeship to the ranks of researchers who can design and analyze big data resources in education. This need is pressing. A recent review of psychology doctoral programs decries the declining statistical and methodological preparation of doctoral students (see Aiken, West, & Millsap, 2008).

Summary

In this chapter, I reviewed the symbiotic relationship between design-based research in education and related efforts in engineering education research. The resonance between both efforts is high because they share metaphors of iterative design cycles and the prototyping and refinement of solutions to problems of practice. I look forward to even deeper connections between researchers in both fields supported, perhaps, by some mechanism akin to the proposed Design Exchange for Scholars. Such an Exchange could help accelerate the conduct of studies that have mutual benefit to academia and industry, and that add to theories of learning. Like educational practice, engineering education research grows out of the needs of practitioners (Stokes, 1997); thus, the continued dialog among researchers and practitioners from both fields (Johri & Olds, 2011) is crucial (McCandliss, Kalchman, & Bryant, 2003).

Footnotes

1. http://fold.it/portal/
2. http://games.cs.washington.edu/Refraction/
3. A website for the project shows the final prototype, and documents the various design strategies: http://www.literacyaccessonline.com/Site/siteindex.asp.
4. Work on the project is ongoing (http://www.kaputcenter.umassd.edu/projects/simcalc).
5. http://www.nsf.gov/pubs/2000/nsf0074/nsf0074.htm
6. http://www.whitehouse.gov/blog/2012/04/09/21st-century-grand-challenges
7. http://www7.nationalacademies.org/BOTA/Assessment_of_21st_Century_Skills_Homepage.html

References

Adams, R., Evangelou, D., English, L., De Figueiredo, A., Mousoulides, N., Pawley, A. L., ... Wilson, D. (2011). Multiple perspectives on engaging future engineers. *Journal of Engineering Education*, 100(1), 48–88.

Aiken, L. S., West, S. G., & Millsap, R. E. (2008). Doctoral training in statistics, measurement, and methodology in psychology: Replication and extension of the Aiken, West, Sechrest, and Reno (1990) Survey of Ph.D. Programs in North America. *American Psychologist*, 63(1), 32–50.

Aleven, V., Roll, I., McLaren, B. M., & Koedinger, K. R. (2010). Automated, unobtrusive, action-by-action assessment of self-regulation during learning with an intelligent tutoring system. *Educational Psychologist*, 45(4), 224–233.

Ambrose, S. A., & Norman, M. (2006). Preparing engineering faculty as educators. *The Bridge*, 36(2), 25–32.

Atman, C. J., Kilgore, D., & McKenna, A. (2008). Characterizing design learning: A mixed-methods study of engineering designers' use of language. *Journal of Engineering Education*, 97(3), 309–326.

Baker, R. S. J. D., Goldstein, A. B., & Heffernan, N. T. (2011). Detecting learning moment-by-moment. *International Journal of Artificial Intelligence in Education*.

Baker, R. S. J. D., Gowda, S. M., & Corbett, A. T. (2011a). Towards predicting future transfer of learning. In *Proceedings of 15th International Conference on Artificial Intelligence in Education*, Auckland, NZ (pp. 23–30).

Baker, R. S. J. D., Gowda, S. M., & Corbett, A. T. (2011b). Automatically detecting a student's preparation for future learning: Help use is key. In *Proceedings of the 4th International Conference on Educational Data Mining*, Eindhoven, The Netherlands (pp. 179–188).

Baker, R. S. J. D., & Yacef, K. (2009). The state of educational data mining in 2009: A review and future visions. *Journal of Educational Data Mining*, 1(1), 3–17.

Bannan-Ritland, B. (2003). The role of design in research: The integrative learning design framework. *Educational Researcher*, 32(1), 21–24.

Bannan-Ritland, B. (2008). Teacher design research: An emerging paradigm for teachers' professional development. In A. E. Kelly, R. A. Lesh, & J. Y. Baek (Eds.), *Handbook of design research methods in education: Innovations in science, technology, engineering, and mathematics learning and teaching* (pp. 246–262). New York, NY: Routledge.

Bannan-Ritland, B., & Baek, J. Y. (2008). Investigating the act of design in design research: The road taken. In A. E. Kelly, R. A. Lesh, & J. Y. Baek (Eds.), *Handbook of design research methods in education: Innovations in science, technology, engineering, and mathematics learning and teaching* (pp. 299–319). New York, NY: Routledge.

Barab, S., & Squire, K. (2004). Design-based research: Putting a stake in the ground. *Journal of the Learning Sciences*, 13(1), 1–14.

Barr, R. E., Pandy, M. G., Petrosino, A. J., Roselli, R. J., Brophy, S., & Freeman, R. A. (2007). Challenge-based instruction: The VaNTH biomechanics learning modules. *Advances in Engineering Education*, 1(1), 1–30.

Barriga, A. Q., Doran, J. W., Newell, S. B., Morrison, E. M., Barbetti, V., & Robbins, B. E. (2002). Relationships between problem behaviors and academic achievement in adolescents: The unique role of attention problems. *Journal of Emotional and Behavioral Disorders*, 10(4), 233–240.

Berliner, D. C. (2002). Educational research: The hardest science of all. *Educational Researcher, 31*(8), 18–20.

Bhide, A. (2008). *The venturesome economy: How innovation sustains prosperity in a more connected world*. Princeton, NJ: Princeton University Press.

Borner, K. (2012). *Visual analytics in support of education*. Keynote Presentation, LAK12, Vancouver. Retrieved from http://ivl.cns.iu.edu/km/pres/2012-borner-lak.pdf

Borrego, M., & Bernhard, J. (2011). The emergence of engineering education research as an internationally connected field of inquiry. *Journal of Engineering Education, 100*(1), 14–47.

Borrego, M., Froyd, J., & Hall, T. (2010). Diffusion of engineering education innovations: A survey or awareness and adoption rates in U.S. engineering departments. *Journal of Engineering Education, 99*(3), 185–207.

Borrego, M., Newswander, C. B., McNair, L. D., McGinnis, S., & Paretti, M. C. (2009). Using concept maps to assess interdisciplinary integration of green engineering knowledge. *Advances in Engineering Education, 2*(1), 1–26.

Bransford, J. D., Stipek, D. J., Vye, N. J., Gomez, L. M., & Lam, D. (Eds.). (2009). *The role of research in educational improvement*. Cambridge, MA: Harvard Education Press.

Brazer, D. S., & Keller, R. (2008). A design research approach to investigating educational decision making. In A. E. Kelly, R. A. Lesh, & J. Y. Baek (Eds.), *Handbook of design research methods in education: Innovations in science, technology, engineering, and mathematics learning and teaching* (pp. 284–296). New York, NY: Routledge.

Bryk, A. (2009). Support a science of performance improvement. *Phi Delta Kappan, 90*(8), 592–595.

Brown, A. L. (1992). Design experiments: Theoretical and methodological challenges in creating complex interventions in classroom settings. *The Journal of the Learning Sciences, 2*(2), 141–178.

Case, J., & Light, G. (2011). Emerging methodologies in engineering education research. *Journal of Engineering Education, 100*(1), 186–210.

Cheville, A., & Bunting, C. (2011). Engineering students for the 21st century: Student development through the curriculum. *Advances in Engineering Education, 2*(4), 1–37.

Clements, D. H. (2008). Design experiments in curriculum research. In A. E. Kelly, R. A. Lesh & J. Y. Baek (Eds.), *Handbook of design research methods in education: Innovations in science, technology, engineering, and mathematics learning and teaching* (pp. 410–422). New York, NY: Routledge.

Cobb, P., & Smith, T. (2008). The challenge of scale: Designing schools and districts as learning organizations for instructional improvement in mathematics. In K. Krainer & T. Wood (Eds.), *International handbook of mathematics teacher education*: Vol. 3. *Participants in mathematics teacher education: Individuals, teams, communities and networks* (pp. 231–254). Rotterdam, The Netherlands: Sense.

Cobb, P., Confrey, J., diSessa, A., Lehrer, R., & Schauble, L. (2003). Design experiments in educational research. *Educational Researcher, 32*, 9–13.

Cobb, P., McClain, K., & Gravemeijer, K. (2003). Learning about statistical covariation. *Cognition and Instruction, 21*, 1–78.

Collins, A., Joseph, D., & Bielaczyc, K. (2004). Design research: Theoretical and methodological issues. *The Journal of the Learning Sciences, 13*(1), 15–42.

Constantine, L. L., & Lockwood, L. A. (1999). *Software for use: A practical guide to the models and methods of usage-centered design*. Boston, MA: Addison Wesley Professional.

Cooper, A. (1999). *The inmates are running the asylum: Why high-technology products drive us crazy and how to restore the sanity*. Indianapolis, IN: SAMS.

Cordray, D. S., Harris, T. R., & Klein, S. (2009). A research synthesis of the effectiveness, replicability, and generality of the VaNTH challenge-based instructional modules in bioengineering. *Journal of Engineering Education, 98*(4), 335–348.

Cronbach, L. J. (1975, February). Beyond the two disciplines of scientific psychology. *American Psychologist, 30*(2), 116–127.

Cuban, L. (1986). *Teachers and machines: The classroom use of technology since 1920*. New York, NY: Teachers College Press.

Cuban, L. (2001). *Oversold and underused: Computers in the classroom*. Cambridge, MA: Harvard University Press.

Cuban, L. (2009). *Hugging the middle: how teachers teach in an era of testing and accountability*. New York, NY: Teachers College Press.

Cyberlearning and Workforce Development Task Force. (2011). *National Science Foundation advisory task force on cyberlearning and workforce development: Final report.* Arlington, VA: NSF.

Dahm, K., Riddell, W., Constans, E., Courtney, J., Harvey, R., & Von Lockette, P. (2009). Implementing and assessing the converging-diverging model of design in a sequence of sophomore projects. *Advances in Engineering Education*, 1(3), 1–16.

Dalrymple, O., Sears, D., & Evangelou, D. (2011). The motivational and transfer potential of disassemble/analyze/assemble activities. *Journal of Engineering Education*, 100(4), 741–759.

Davenport, J. L., Yaron, D., Klahr, D., & Koedinger, K. (2008). *When do diagrams enhance learning? A framework for designing relevant representations.* Paper presented at 2008 International Conference of the Learning Sciences, Utrecht, The Netherlands.

Davison, R. C. (2010). *Engineering curricula: Understanding the design space and exploiting the opportunities: Summary of a workshop.* Washington, DC: The National Academies Press. Retrieved from http://www.nap.edu/catalog.php?record_id=12824

Dede, C. (2005). Why design-based research is both important and difficult. *Educational Technology*, 45(1), 5–8.

Delale, F., Liaw, B. M., Jiji, L. M., Voiculescu, I., & Yu, H. (2011). Infusion of emerging technologies and new teaching methods into the mechanical engineering curriculum at the City College of New York. *Advances in Engineering Education*, 2(4), 1–36.

Dinehart, D. W., & Gross, S. P. (2010). A service learning structural engineering capstone course and the assessment of technical and non-technical objectives. *Advances in Engineering Education*, 2(1), 1–19.

Eggermont, M., Brennan, R., & Freiheit, T. (2010). Improving a capstone design course through mindmapping. *Advances in Engineering Education*, 2(1), 1–26.

Forster, S., & Lavie, N. (2007). High perceptual load makes everybody equal: Eliminating individual differences in distractibility with load. *Psychological Science*, 18(5), 377–381.

Froyd, J. (2005). The Engineering Education Coalitions program. Educating the engineer of 2020: Adapting engineering education to the new century. Washington, DC: National

Academies Press. Retrieved from http://www.nap.edu/openbook.php?record_id=11338%26;page=82

Ganesh, T. G. (2011). Design-based research: A framework for designing novel teaching and learning experiences in middle school engineering education. In *41st ASEE/IEEE Frontiers in Education (FIE) 2011* (Paper no. 1534, pp. 1–7). Rapid City, SD: ASEE/IEEE.

Garcia, O. N., Varnasi, M. R., Acevedo, M. R., & Guturu, P. (2011). An innovative project and design oriented electrical engineering curriculum at the University of North Texas. *Advances in Engineering Education*, 2(4), 1–34.

Gery, G. (2003). Ten years later: A new introduction to attributes & behaviors and the state of performance-centered systems. In G. J. Dickelman (Ed.), *EPSS Revisited: A Lifecycle for Developing Performance-Centered Systems* (pp. 1–3). Silver Spring, MD: Society for Performance Improvement.

Hamilton, E., Lesh, R., Lester, F., & Brilleslyper, M. (2008). Model-eliciting activities (MEAs) as a bridge between engineering education research and mathematics education research. *Advances in Engineering Education*, 1(2), 1–25.

Hattie, J. A. C. (2009). *Visible learning: A synthesis of over 800 meta-analyses related to achievement.* New York, NY: Routledge.

Hayden, N. J., Rizzo, D. M., Dewoolkar, M. M., Neumann, M. D., Lathem, S., & Sadek, A. (2011). Incorporating a systems approach into civil and environmental engineering curricula: Effect on course redesign, and student and faculty attitudes. *Advances in Engineering Education*, 2(4), 1–27.

Howie, S. J., & Plomp, T. (Eds.). (2006). *Contexts of learning mathematics and science: Lessons learned from TIMSS.* London: Routledge.

Hjalmarson, M. A., & Lesh, R. A. (2008). Engineering and design research: Intersections for education research and design. In A. E. Kelly, R. A. Lesh, & J. Y. Baek (Eds.), *Handbook of design research methods in education: Innovations in science, technology, engineering, and mathematics learning and teaching* (pp. 96–110). New York, NY: Routledge.

Hundhausen, C., Agarwal, P., Zollars, R., & Carter, A. (2011). The design and experimental evaluation of a scaffolded software environment to improve engineering students' disciplinary problem-solving skills.

Journal of Engineering Education, 100(3), 574–603.

Impelluso, T. J. (2009). Leveraging cognitive load theory, scaffolding, and distance technologies to enhance computer programming for non-majors. *Advances in Engineering Education*, 1(4), 1–19.

James, W. (1962). *Talks to teachers on psychology and to students on some of life's ideals*. New York, NY: Henry Holt 1899 reprint. Mineola, NY: Dover.

Johri, A., & Olds, B. (2011). Situated engineering learning: Bridging engineering education esearch and the learning sciences. *Journal of Engineering Education*, 100(1), 151–185.

Kali, Y. (2008). The Design Principles Database as means for promoting design-based research. In A. E. Kelly, R. A. Lesh & J. Y. Baek (Eds.), *Handbook of design research methods in education: Innovations in science, technology, engineering, and mathematics learning and teaching* (pp. 423–438). New York, NY: Routledge.

Kali, Y., Levin-Peled, R., & Dori, Y. (2009). The role of design-principles in designing courses that promote collaborative learning in higher-education. *Computers in Human Behavior*, 5, 1067–1078.

Kali, Y., Levin-Peled, R., Ronen-Fuhrmann, T., & Hans, M. (2009). The Design Principles Database: A multipurpose tool for the educational technology community. *Design Principles & Practices: An International Journal*, 3(1), 55–65.

Kelly, A. E. (2004). Design research in education: Yes, but is it methodological? *Journal of the Learning Sciences*, 13(1), 115–128.

Kelly, A. E. (2009). When is design research appropriate? In T. Plomb & N. Nieeven (Eds.), *Introduction to educational design research* (pp. 73–88), Retrieved from http://www.slo.nl/downloads/2009/Introduction_2oto_2oeducation_2odesign_2oresearch.pdf/download

Kelly, A. E. (2011). Can cognitive neuroscience ground a science of learning? *Theme issue on educational neuroscience. Educational Philosophy and Theory*, 43(1), 17–23.

Kelly, A. E. (2012). Developing validity and reliability criteria for assessments in innovation and design research studies. In D. Yun Dai (Ed.), *Design research on learning and thinking in educational settings: Enhancing intellectual growth and functioning*. New York, NY: Taylor & Francis.

Kelly, A. E., Baek, J. Y., Lesh, R. A., & Bannan-Ritland, B. (2008). Enabling innovations in education and systematizing their impact. In A. E. Kelly, R. A. Lesh, & J. Y. Baek (Eds.), *Handbook of design research methods in education: Innovations in science, technology, engineering, and mathematics learning and teaching* (pp. 3–18). New York, NY: Routledge.

Kelly, A. E. (Director), & Lesh, R. (Co-director). (2001). *Understanding and explicating the design experiment*. National Science Foundation Award No. 0107008.

Kelly, A. E., & Yin, R. (2007). Strengthening structured abstracts for educational research. *Educational Researcher*, 36(3), 133–138.

Lesh, R. A., Kelly, A. E., & Yoon, C. (2008). Multitiered design experiments in mathematics, science, and technology education. In A. E. Kelly, R. A. Lesh, & J. Y. Baek (Eds.), *Handbook of design research methods in education: Innovations in science, technology, engineering, and mathematics learning and teaching* (pp. 131–148). New York, NY: Routledge.

Lobato, J. (2006). Alternative perspectives on the transfer of learning: History, issues, and challenges for future research. *Journal of the Learning Sciences*, 15(4), 431–449.

Lobato, J. (2008). Research methods for alternative approaches to transfer: Implications for design experiments. In A. E. Kelly, R. A. Lesh, & J. Y. Baek (Eds.), *Handbook of design research methods in education: Innovations in science, technology, engineering, and mathematics learning and teaching* (pp. 167–194). New York, NY: Routledge.

Mantri, A., Dutt, S., Gupta, J. P., & Chitkara, M. (2009). Using PBL to deliver course in digital electronics. *Advances in Engineering Education*, 1(4), 1–17.

Maroulis, S., Guimera, R., Petry, H., Stringer, M. J., Gomez, L. M., Amaral, L. A. N., & Wilensky, U. (2010). Complex systems view of educational policy research. *Science*, 330(6000), 38–39.

Martinez, M. E., Peterson, M. W., Bodner, M., Coulson, A., Vuong, S., Hu, W., . . . Shaw, G. L. (2008). Music training and mathematics achievement: A multiyear iterative project designed to enhance students' learning. In A. E. Kelly, R. A. Lesh, & J. Y. Baek (Eds.), *Handbook of design research methods in* education: Innovations in science, technology, engineering, and mathematics learning and teaching (pp. 396–409). New York, NY: Routledge.

Maxwell, J. A. (2012). *A realist approach for qualitative research*. Thousand Oaks, CA: SAGE.

Mayer, R. E. (2003). The promise of multimedia learning: Using the same instructional design methods across different media. *Learning and Instruction*, 13(2), 125–139.

McCandliss, B. D., Kalchman, M., & Bryant, P. (2003). Design experiments and laboratory approaches to learning: Steps toward collaborative exchange. *Educational Researcher*, 32(1), 14–16.

McDonald, S.-K., Keesler, V., Kauffman, N., & Schneider, B. (2006). Scaling-up exemplary interventions. *Educational Researcher*, 35(3), 15–24.

Middleton, J., Gorard, S., Taylor, C., & Bannan-Ritland, B. (2008). The "compleat" design experiment: From soup to nuts. In A. E. Kelly, R. A. Lesh & J. Y. Baek (Eds.), *Handbook of design research methods in education: Innovations in science, technology, engineering, and mathematics learning and teaching* (pp. 21–46). New York, NY: Routledge.

Mosteller, F., & Boruch, R. F. (Eds.). (2002). *Evidence matters: Randomized trials in education research*. Washington, DC: Brookings Institution Press.

National Academy of Engineering, Committee on the Engineer of 2020, Phase II, Committee on Engineering Education (2005). *Educating the engineer of 2020: Adapting engineering education to the new century*. Washington, DC: The National Academies Press.

National Academy of Engineering (NAE). (2013). *Educating engineers: Preparing 21st Century leaders in the context of new modes of learning*. Washington, DC: The National Academies Press.

National Research Council (NRC). (2000). *How people learn: Brain, mind, experience, and school* (expanded edition). Washington, DC: The National Academies Press.

National Research Council (NRC). (2005a). *How students learn: Mathematics in the classroom*. M. S. Donovan & J. D. Bransford (Eds.). Washington, DC: The National Academies Press. Retrieved from http://www.nap.edu/openbook.php?record_id=10126%26;page=R1

National Research Council (NRC). (2005b). *How students learn: Science in the classroom*. M. S.

Donovan & J. D. Bransford (Eds.). Washington, DC: The National Academies Press.

Nelson, H. G., & Stolterman, E. (2003). *The design way: Intentional change in an unpredictable world: Foundations and fundamentals of design competence*. Englewood Cliffs, NJ: Education Technology Publications.

Nitko, A. J., & Brookhart, S. M. (2006). *Educational assessment of students* (5th ed.). New York, NY: Pearson.

Osmundson, E., Herman, J., Ringstaff, C., Dai, Y., & Timms, M. (2012). *Measuring fidelity of implementation – Methodological and conceptual issues and challenges*. (CRESST Report 811). Los Angeles, CA: University of California, National Center for Research on Evaluation, Standards, and Student Testing (CRESST).

Pears, A., Fincher, S. A., Adams, R., & Daniels, M. (2008). Stepping stones: Capacity building in engineering education. In *Proceedings of the Annual ASEE/IEEE Frontiers in Education Conference*, Saratoga, NY.

Penuel, W., Fishman, B., Cheng, B. H., & Sabelli, N. (2011). Organizing research and development at the intersection of learning, implementation, and design. *Educational Researcher*, 40, 331–337.

Plomb, T., & Nieveen, N. (Eds.). (2010). *Introduction to educational design research*. The Netherlands : Netzodruk, Enschede.

Purzer, S. (2011). The relationship between team discourse, self-efficacy, and individual achievement: A sequential mixed-methods study. *Journal of Engineering Education*, 100(4), 655–679.

Rasmussen, C., & Stephan, M. (2008). A methodology for documenting collective activity. In A. E. Kelly, R. A. Lesh, & J. Y. Baek (Eds.), *Handbook of design research methods in education: Innovations in science, technology, engineering, and mathematics learning and teaching* (pp. 195–216). New York, NY: Routledge.

Reese, D. D., Seward, R. J., Tabachnick, B. G., Hitt, B., Harrison, A., & McFarland, L. (2012). Timed report measures learning through game-based embedded assessment. In D. Ifenthaler, D. Eseryel & X. Ge (Eds.), *Assessment in game-based learning: Foundations, innovations, and perspectives*. New York, NY: Springer.

Reeves, T. C. (2011). Can educational research be both rigorous and relevant? *Educational*

Designer, 1(4). Retrieved from http://www.educationaldesigner.org/ed/volume1/issue4/article13

Reinking, D., & Bradley, B. A. (2008). *On formative and design experiments.* New York, NY: Teachers College Press.

Rodrigo, M. M. T., & Baker, R. S. J. D. (2011). Comparing learners' affect while using an intelligent tutor and an educational game. *Research and Practice in Technology Enhanced Learning, 6*(1), 43–66.

Rogers, E. M. (2003). *Diffusion of innovations* (5th ed.). New York, NY: Free Press.

Roll, I., Aleven, V., & Koedinger, K. R. (2011). Metacognitive practice makes perfect: Improving students' self-assessment skills with an intelligent tutoring system. In G. Biswas (Ed.), *Proceedings of the international conference on artificial intelligence in education* (pp. 288–295). Berlin, Germany: Springer-Verlag.

Roll, I., Aleven, V., McLaren, B. M., & Koedinger, K. R. (2011). Improving students' help-seeking skills using metacognitive feedback in an intelligent tutoring system. *Learning and Instruction, 21*, 267–280.

Romero, C., Ventura, S., Pechenizkiy, M., & Baker, R. S. J. D. (Eds.). (2012). *Handbook of educational data mining.* Boca Raton, FL: CRC Press.

Roschelle, J., Tatar, D., & Kaput, J. (2008). Getting to scale with innovations that deeply structure how students come to know mathematics. In A. E. Kelly, R. A. Lesh, & J. Y. Baek (Eds.), *Handbook of design research methods in education: Innovations in science, technology, engineering, and mathematics learning and teaching* (pp. 369–395). New York, NY: Routledge.

Roschelle, J., Shechtman, N., Tatar, D., Hegedus, S., Hopkins, B., Empson, S., Knudsen, J., & Gallagher, L. (2010). Integration of technology, curriculum, and professional development for advancing middle school mathematics: Three large-scale studies. *American Educational Research Journal, 47*(4), 833–878.

Sabelli, S., Lemke, J., Cheng, B., & Richey, M. (2012). *Education as a complex adaptive system: Report of progress.* Retrieved from http://msce.sri.com/publications/Education_As%20A_Complex_Adaptive%20System.pdf

Sao Pedro, M. A., Baker, R. S. J .D., Gobert, J., Montalvo, O., & Nakama, A. (2011). Leveraging machine-learned detectors of systematic inquiry behavior to estimate and predict transfer of inquiry skill. In *User modeling and user-adapted interaction.* Retrieved from http://users.wpi.edu/~rsbaker/UMUAI-MSP-2011.pdf

Sao Pedro, M. A., Gobert, J., & Baker, R. S. J. D. (2012). *The development and transfer of data collection inquiry skills across physical science microworlds.* Paper presented at the American Educational Research Association Conference. Retrieved from http://users.wpi.edu/~rsbaker/SaoPedroetal_AERA2012_FINAL.pdf

Sandoval, W. A., & Bell, P. L. (2004). Design-based research methods for studying learning in context: Introduction. *Educational Psychologist, 39*(4), 199–201.

Scardamalia, M., Bransford, J., Kozma, B., & Quellmalz, E. (2012). New assessments and environments for knowledge building. In P. Griffin, B. McGaw, & E. Care (Eds.), *Assessment and teaching of 21st century skills* (pp. 231–300). New York, NY: Springer.

Schickore, J., & Steinle, F. (Eds), (2010). *Revisiting discovery and justification: Historical and philosophical perspectives on the context distinction.* Dordrecht, The Netherlands: Springer.

Schneider, B., & McDonald, S.-K. (Eds.). (2006a). *Scale-up in practice.* Lanham, MD: Rowman & Littlefield.

Schneider, B., & McDonald, S.-K. (Eds.). (2006b). *Scale-up in principle.* Lanham, MD: Rowman & Littlefield.

Schwartz, D. L., Chang, J., & Martin, L. (2008). Instrumentation and innovation in design experiments: Taking the turn to efficiency. In A. E. Kelly, R. A. Lesh, & J. Y. Baek (Eds.), *Handbook of design research methods in education: Innovations in science, technology, engineering, and mathematics learning and teaching* (pp. 47–67). New York, NY: Routledge.

Seely, B. E. (2005). Patterns in the history of engineering education reform: A brief essay, for National Academy of Engineering. *Engineer of 2020: National Education Summit* (pp. 114–130). Washington, DC: The National Academies Press.

Shadish, W. R., & Cook, D. T. (2009). The renaissance of field experimentation in evaluating interventions. *Annual Review of Psychology, 60*, 607–629.

Shadish, W. R., Cook, T. D., & Campbell, D. T. (2002). *Experimental and quasi-experimental designs for generalized causal inference.* Boston: Houghton Mifflin.

Shavelson, R. J., Phillips, D. C., Towne, L. T., & Feuer, M. J. (2003). On the science of education design studies. *Educational Researcher*, 32(1), 25–28.

Sloane, F. C. (2008a). Randomized trials in mathematics education: Recalibrating the proposed high watermark. *Educational Researcher*, 37(9), 624–630.

Sloane, F. C. (2008b). Multilevel models in design research: A case from mathematics education. In A. E. Kelly, R. A. Lesh, & J. Y. Baek (Eds.), *Handbook of design research methods in education: Innovations in science, technology, mathematics and engineering* (pp. 459–476). New York, NY: Routledge.

Sloane, F. C., & Gorard, S. (2003). Exploring modeling aspects of design experiments. *Educational Researcher*, 32(1), 29–31.

Sloane, F. C., Helding, B., & Kelly, A. E. (2008). Longitudinal analysis and interrupted time series designs: Opportunities for the practice of design research: A case from mathematics education. In A. E. Kelly, R. A. Lesh, & J. Y. Baek (Eds.), *Handbook of design research methods in education: Innovations in science, technology, mathematics and engineering* (pp. 449–458). New York, NY: Routledge.

Sloane, F. C., & Kelly, A. E. (2008). Design research and the study of change: Conceptualizing individual growth in designed settings. In A. E. Kelly, R. A. Lesh, & J. Y. Baek (Eds.), *Handbook of design research methods in education: Innovations in science, technology, mathematics and engineering* (pp. 441–448). New York, NY: Routledge.

Stephens, R., & Richey, M. (2011). Accelerating STEM capacity: A complex adaptive system perspective. *Journal of Engineering Education*, 100(3), 417–423.

Stokes, D. (1997). *Pasteur's quadrant: Basic science and technological innovation*. Washington, DC: Brookings Institution Press.

Streveler, R., & Smith, K. (2010). From the margins to the mainstream: The emerging landscape of engineering education research. *Journal of Engineering Education*, 99(4), 285–287.

Sullivan, J. F. (2006). Broadening engineering's participation: A call for K-16 engineering education. *The Bridge*, 36(2), 17–24.

Tadmore, Z. (2006). Redefining engineering disciplines for the twenty-first century. *The Bridge*, 36(2), 33–37.

Taraban, R., Craig, C., & Anderson, E. (2011). Using paper-and-pencil solutions to assess problem solving skills. *Journal of Engineering Education*, 100(3), 498–519.

Thomas, M. K., Barab, S. A., & Tuzun, H. (2009). Developing critical implementations of technology-rich innovations: A cross-case study of the implementation of Quest Atlantis. *Journal of Educational Computing Research*, 41(2), 125–153.

Toulmin, S. (1998). The idol of stability. Presented at the *Tanner Lectures on Human Values*. Retrieved from http://tannerlectures.utah.edu/_documents/a-to-z/t/Toulmin99.pdf

Toulmin, S. (2001). *Return to reason*. Cambridge, MA: Harvard University Press.

Tyack, D., & Cuban, L. (1995). *Tinkering toward utopia*. Cambridge, MA: Harvard University Press.

van den Akker, J. (1999). Principles and methods of development research. In J. van den Akker, N. Nieveen, R. M. Branch, K. L. Gustafson, & T. Plomp (Eds.), *Design methodology and developmental research in education and training* (pp. 1–14). Dordrecht, The Netherlands: Kluwer Academic.

van den Akker, J., Gravemeijer, K., McKenney, S., & Nieveen, N. (2006). *Educational design research*. London: Routledge.

Vendlinski, T. P., Chung, G. K. W. K., Binning, K. R., & Buschang, R. E. (2011). *Teaching rational number addition using video games: The effects of instructional variation*. (CRESST Report 808). Los Angeles, CA: University of California, National Center for Research on Evaluation, Standards, and Student Testing (CRESST).

Vest, C. (2006). Educating engineers for 2020 and beyond. *The Bridge*, 36(2), 38–44.

Weiss, I. R., Miller, B. A., Heck, D. J., & Cress, K. (2004). *Handbook for enhancing strategic leadership in the math and science partnerships*. Chapel Hill, NC: Horizon Research Inc.

Wolf, J., & Le Vasan, M. (2008). Toward assessment of teachers' receptivity to change in Singapore: A case study. In A. E. Kelly, R. A. Lesh, & J. Y. Baek (Eds.), *Handbook of design research methods in education: Innovations in science, technology, engineering, and mathematics learning and teaching* (pp. 265–283). New York, NY: Routledge.

Yin, R. K. (2009). *Case study research: Design and methods* (4th ed.). Thousand Oaks, CA: SAGE.

Yun Dai, D. (Ed.) (2012). *Design research on learning and thinking in educational settings: Enhancing intellectual growth and functioning.* New York, NY: Taylor & Francis.

Zaritsky, R., Kelly, A. E., Flowers, W., Rogers, E., & O'Neill, P. (2003). Clinical design sciences: A view from sister design efforts. *Educational Researcher, 32*(1), 32–34.

Zawojewski, J., Chamberlin, M., Hjalmarson, M. A., & Lewis, C. (2008). Developing design studies in mathematics education professional development: Studying teachers' interpretive systems. In A. E. Kelly, R. A. Lesh, & J. Y. Baek (Eds.), *Handbook of design research methods in education: Innovations in science, technology, engineering, and mathematics learning and teaching* (pp. 219–245). New York, NY: Routledge.

CHAPTER 26

Quantitative and Mixed Methods Research

Approaches and Limitations

Barbara M. Moskal, Teri Reed, and Scott A. Strong

Introduction

A concern across the field of education, as well as within engineering education, is the identification of effective instructional approaches. "Effective" can be defined in many ways, including increased learning gains, improved attitudes, and changes in the general appeal of a subject or topic to students. To determine the effectiveness of an approach, it is often necessary to measure changes in student constructs over time or to acquire a snapshot of students' performances at a given point. In addition, teachers and researchers may be concerned with determining whether their approaches are equally effective across different student populations.

In engineering education, each of these assessment purposes receives increased emphasis at the program and student level owing to the existence of an accreditation board, ABET, Inc. (formerly known as the Accreditation Board for Engineering and Technology or ABET; see www.abet.org/history.shtml). ABET, Inc. requests that each accredited program demonstrate that

its graduating seniors have achieved a set of program outcomes that can be found at the referenced website.

ABET, Inc. does not require the measurement of changes in students' performances; rather, it requests that evidence be provided that students have acquired specific competencies by the completion of their degrees. In other words, some students may theoretically have entered the program with these competencies, while others develop these competencies over the course of study. In terms of being an engineer, it does not matter whether individuals develop their skills through their course of studies; it does matter that they have specific skills before entering the field. At many universities, however, tracking changes in these competencies over time is desirable. This allows universities to evaluate the impact that they have on their students' learning. Some universities also compare measurements across subpopulations, in an effort to demonstrate the effectiveness of their programs for diverse populations.

An appropriate methodology for demonstrating the attainment of student outcomes, as well as other competencies across a student population, is likely to be quantitative assessment (Borrego, Douglas, & Amelink, 2009). Quantitative assessment refers to a collection of assessment techniques that result in data sets or numbers that can be summarized or analyzed statistically (Moskal, Leydens, & Pavelich, 2002). The analysis methods can range from descriptive statistics to parametric and nonparametric analyses. Quantitative methods are different from qualitative methods, which are discussed in Chapter 27 by Case and Light and Chapter 28 by Johri, in that their purpose is to identify trends within data sets that will be used to draw conclusions with respect to populations of interest. Quantitative methods can support researchers as they seek to generalize beyond the participants to a defined, broader population. Qualitative approaches seek to provide rich and detailed descriptions of the participants in the investigation or classroom (Bernard, 2000; Leydens, Moskal, & Pavelich, 2004).

For many educational researchers and instructors, the most appropriate form of assessment is likely to be a mixed methods approach (Bernard, 2000; Olds, Moskal, & Miller, 2005). Mixed methods research refers to the combining of quantitative and qualitative methods such that broad conclusions can be drawn across a larger population, and thick, rich detail can be acquired with respect to the participating group. Often, mixed methods are also used as a validation technique or a method of acquiring additional data to support stronger conclusions than those that can be supported through quantitative methods. Because of these features, Johnson and Onwuegbuzie (2004) have argued that mixed methods research results in "superior research" when compared to research that is restricted to a single paradigm.

This chapter primarily concerns quantitative assessment methodologies, or designs that are used to collect quantitative data, from an educational psychology perspective. The purpose of this chapter is to illustrate that there are many diverse and effective assessment methodologies available to

support engineering education research. The intended audience for this chapter includes those who are novice to the field of educational assessment as well as those who are more seasoned but desire a review of the various educational measurement methods employed in educational psychology. The section that follows addresses validity, reliability, and triangulation, concepts that are important for informing the appropriateness of the conclusions that are drawn in a quantitative investigation. Next is a discussion of various quantitative research designs that are commonly used in educational psychology. The final section addresses the use of mixed methods, or the combining of quantitative and qualitative approaches in a single investigation. Examples are embedded throughout for illustrative purposes.

Validity and Reliability

This section describes two fundamental concepts within quantitative research or within research that requires educational measurement: validity and reliability. Owing to the length restrictions, these topics are briefly presented. The interested reader can find a more extensive treatment of these topics in Cohen, Manion, and Morrison (2007), Bernard (2000), and Johnson and Christensen (2008). The final two subsections address the connections between validity and reliability and the use of mixed methods and triangulation to enhance the validity argument.

Validity

Validity refers to the extent to which the interpretations that are drawn from the data reflect the true nature of the underlying phenomena. Evidence to support the validity of conclusions drawn from an investigation provides the strength that is needed to make research claims and, therefore, should be of central concern to educational researchers (Cohen et al., 2007). Researchers need to consider both the evidence to support the validity of the instruments that are used in a study as well as the broader assessment process.

There are three forms of validity evidence often discussed in reference to measurement instruments in educational psychology (Hopkins, 1998): Content-, Construct-, and Criterion-related validity evidence. Content validity refers to evidence that is collected to demonstrate that the material addressed within a given instrument is appropriate to and covers the intended content domain. Construct validity evidence refers to the extent to which the measurement instrument appropriately and effectively addresses the intended, underlying psychological construct. A psychological construct is a process that is internal to a person, such as an individual's reasoning or attitudes. Often, experts in the targeted field examine content and construct validity by reviewing the content that comprises the instrument.

Criterion validity refers to the extent a given instrument can be used to predict desired future performances, such as a student's future success in a field. This requires correlating the outcomes of the instrument with the achievement of the future performance. This is a difficult form of evidence to collect because it requires longitudinal follow-up. Another approach used to support criterion validity is correlating the results of the new instrument with the results from another, previously validated instrument. If a high correlation exists with the previous instrument and the expected outcome, *and* a high correlation exists with the previous instrument and the new instrument, then it can be argued that a high correlation will exist between the new instrument and the expected outcome. A shortcoming of this approach is that given the validity of the previous instrument, the argument justifying the need for the new instrument is often weak. Researchers typically seek to create instruments that measure something different.

In the late 1980s, Messick (1989, 1998) introduced another form of validity with respect to measurement instruments – consequential validity. Consequential validity is evidence that examines the "fairness" of an instrument for use with various subpopulations. For example, some instruments have better predictive or criterion validity with a given subpopulation than with another subpopulation. If the researcher has not examined the differing predictive value, some subpopulations' performances could be unintentionally and unknowingly under- or overestimated. This is of increased concern when an instrument has serious consequences, such as high-stakes testing.

Two additional forms of validity evidence are frequently discussed when considering the broader measurement process, internal and external validity evidence (Campbell & Stanley, 1963). Internal validity refers to the examination of the extent to which a causal relationship can be implied based on the results of the assessment process. This requires the examination of unintended, unplanned, or uncontrolled factors, which may impact the interpretation of the results. External validity refers to the extent to which the results may be appropriately generalized beyond the participating population. This form of validity evidence is strongly influenced by the selection process used when determining who will participate.

Although there are many forms of validity evidence that may be examined to support an assessment process, all of these forms share a common goal – to support that the assessment appropriately measures what it purports to measure. The type of validity evidence that is necessary to support a given assessment process is dependent on the purpose or intention of the assessment. For example, for instruments that are used with a defined homogeneous population, consequential validity may not be a concern because subpopulations do not exist.

Reliability

Reliability refers to the precision of a measurement (American Psychological Association, 1999). There are two primary forms of reliability: test and rater reliability. Each is briefly described here.

In the case of test reliability, the term refers to the repeatability of a measurement within or across the use of the instrument. Test–retest reliability examines the extent to which the same instrument produces similar results when used repeatedly, without intervention, on the same population. It also refers to the extent to which different versions or forms of a test result in similar conclusions when administered to the same population. Both types of test reliabilities are measured through correlations. Internal test reliability refers to the extent to which a given test measures a single, predefined construct or the extent to which a subset of questions measure a given construct. This is typically calculated using a statistical procedure referred to as Coefficient Alpha (Cortina, 1993; Miller, 1995) or through a Factor Analysis (Costello & Osboren, 2005; Gorsuch, 1983; Suhr, 2003). In all of these cases, reliability is estimated through a correlation, or series of correlations for various constructs, and the closer the correlation is to 1, the greater the reliability the instrument or subsection of the instrument is assumed to have.

Rater reliability (Moskal & Leydens, 2000) refers to consistency in the scoring of a test. A comparison between the scores of two independent raters who score the same test is referred to as interrater reliability. Measurements of deviations among the scores provided by the same person on the same set of responses at different times is referred to as intrarater reliability. Reliable scores are consistent regardless of when they are scored or who scored them. In all of the preceding cases, reliability refers to the repeatability of a measurement when various factors (time, rater, test form) are manipulated.

Connecting Validity and Reliability

An unreliable test or assessment procedure cannot be valid. Because the instrument is unreliable, a researcher cannot validly interpret or correctly draw accurate conclusions from the results of the assessment process. It is also possible for an assessment process to be very reliable, resulting in consistent scores, but still be invalid or lead to an incorrect conclusion. In other words, validity and reliability are closely linked concepts, each impacting the other, and need to be considered in combination.

Enhancing Validity Through Mixed Methods and Triangulation

Mixed methods refer to the concept of mixing quantitative and qualitative approaches in a single investigation (Borrego et al., 2009; Cohen et al., 2007; Creswell & Plano Clark, 2007; Johnson & Onweugbuzie, 2004). When quantitative and qualitative data support similar conclusions, the combined evidence is usually more compelling than the use of one method. Of course, combining across methods can be achieved by using two different quantitative measures or two different qualitative measures, and thus remaining within a given assessment paradigm. When multiple methods are used to increase the validity of the conclusions drawn, this is called "triangulation" (Denzin, 1978). The goal of triangulation is to provide multiple sources of data that support similar conclusions, increasing the likelihood that the conclusions are accurate, well understood, and compelling to a broader research population (Creswell & Plano Clark, 2007; Tashakkori & Teddlie, 2003).

Although mixed methods can be used as a triangulation technique in the evaluation process, it is also considered a design methodology with its own purposes. This is discussed in the section "Examples of Mixed Approaches." Before a researcher can combine information across designs, he or she needs to understand what some of the design options are. The next two sections overview various quantitative designs.

Fully Randomized Design and Its Limitations

As discussed previously, there are at least two levels a researcher should consider when designing a quantitative investigation: the appropriateness of the instruments used and the broader assessment process into which the instruments are embedded. When interpreting the results of an educational research investigation, both the instruments and the broader process need to be examined for validity and reliability concerns. This section discusses a theoretically "ideal" quantitative design for the broader assessment process.

There are many different types of quantitative research designs that can be used to define the broader assessment process. The fully randomized design with a treatment and control group has previously been referenced as the "gold standard" or the most preferred method of quantitative educational assessment (U.S. Department of Education, Institute of Education Science, and National Center for Education, Evaluation and Regional Assistance, 2003). Usually, this method includes "blind" placement or the randomized placement of subjects into the treatment and control groups without the distinction between these groups being known to participants. This design may employ a pre- and post-test structure or a post-test only design (both described later in this chapter). With respect to the internal validity of the conclusions drawn from such a study, this elite status as the "gold standard" is justifiable. The use of random assignment effectively controls for many external factors that complicate the interpretation of the results and strengthens the inferences that can be drawn from the results. This is one of the most common research methods used in the fields of medicine, welfare, employment policy, and psychology (U.S. Department of Education, Institute of Education Science & National Center for Education, Evaluation and Regional Assistance, 2003).

This design, however, often cannot be implemented in educational research (Olds et al., 2005). The fully randomized design requires that the researcher or instructor randomly assign participants to two separate groups, one that will participate in educational intervention (the treatment) and one that will not (the control). This is a significant shortcoming of the design for several reasons. First, in educational research, it is often not considered appropriate to withhold a promising educational approach from students for the sake of research. Second, the restrictions placed on human subjects research make it difficult for researchers to place participants into different groups without their consent. Once consent is acquired, the treatment and controls have been compromised because the participants know the purpose of the research. Human subjects' restrictions further require that participants be fully informed as to the nature of the research and that they have the option of opting out of the research. The self-selection option eliminates the possibility of random and blind placement. Given these challenges, fully randomized designs are infrequently implemented in educational research.

Other Quantitative Designs and Their Limitations

Although the fully randomized design is ideal from a statistical perspective, this design is often not feasible in educational research. Educational researchers may maintain many of the features of a randomized design but with adjustments as to how the treatment and control groups are established. This section addresses some of the most common quantitative approaches used in educational research: matching, baseline or historical data, pre- and post-testing, post-only designs, and longitudinal designs. When possible, the examples presented are drawn from the authors' own research, providing greater insight into the design selection process.

Matching

Matching is the process of identifying two groups, a treatment and a control group, which share similar characteristics. The treatment group receives the instructional intervention while the control group does not (Addison, Hipp, & Lyons, 2007; Merino & Abel, 2003). Individuals may be matched on degree program, gender, ethnicity, and other characteristics.

The College of Engineering at Purdue University uses matching to assess the effectiveness of a National Science Foundation (NSF)–funded program, Scholarships in Science, Technology, Engineering and Mathematics (S-STEM). The S-STEM program awards scholarships, through the oversight of universities, to students with high financial need who are majoring in science, technology, engineering, or mathematics.

To evaluate the impact of Purdue's S-STEM, a treatment group, those who received scholarships, was matched to a control group, a comparable group who did not receive the scholarship. Members of the treatment group were paired with members of the control group based on the following student factors: level of financial need as indicated by a "gap" (defined by the difference of a family's ability to contribute to tuition expenses as measured by the Free Application for Federal Student Aid to the overall cost of education), performance scores on the Scholastic Assessment Test (SAT) or the American College Test (ACT), high school core grade point average (GPA), gender, and ethnicity/race. The list of factors provided here is ordered based on priority as defined by the study. Students were initially paired on the first factor, then the second, and so forth, moving through the list in a systematic manner and then paired on the latter factors only when viable. The likelihood of pairing students with appropriate counterparts on the later factors was low owing to the limited size of the matching pool. Matching reduces bias in the interpretation process by reducing the impact of uncontrolled variables, such as individual student characteristics. Often researchers cannot control all factors of interest and must limit the matching process to the most important factors, as was done in the S-STEM project.

In the Purdue S-STEM project, the intervention extends beyond the availability of scholarship funds. All scholarship recipients were required to complete an experiential learning opportunity by the second semester of their junior year. S-STEM scholarship recipients also met with each other monthly throughout the academic year, building a community of scholars. Scholarship recipients worked with Purdue's Recruiting Office, inviting high school students to visit and eventually attend Purdue. Although the control group had access to many of these opportunities, they had to seek these resources rather than have them readily available and often required.

In this example, there were known variables (those later in the priority list) that may impact the outcomes but cannot be controlled owing to sample size. Additionally, non-scholarship recipients may select to participate in some of the S-STEM activities. These confounding variables and the open participation policy may interfere with the study's outcomes, threatening the internal validity of the study. Although traditional forms of quantitative analysis are often used to examine matched data, for example, Analysis of Variance or Analysis of Co-Variance, the interpretation of the results should be completed with caution. Researchers need to acknowledge, upfront, that unplanned, confounding variables may be influencing or skewing the results in an unexpected manner.

Matching designs can also be problematic owing to the requirements of Institutional Research Boards (IRB). For example, the comparison group of students or control may change their behavior based on the knowledge that they are being studied. This is referred to as the Hawthorne Effect, based on a study by Harvard researchers at the Hawthorne Works plant of Western Electric in 1924. In addition, control or treatment group participants may refuse consent. Also, some IRBs may allow only the aggregation of the control group data such that individuals cannot be identified, limiting the statistical analyses that can be completed.

Baseline or Historical Data

Another approach used in educational research is the collection of baseline data. Baseline data are collected before an instructional intervention occurs (Kashy, Albertelli, Kashy, & Thoennessen, 2001; Yadav, Subedu, Lundeber, & Bunting, 2011). This is related to "historical" data or data that are collected as part of policy over time. Despite the benefit of baseline or historical data, these samples have the potential of introducing unexpected, confounding variables into the investigation. The existence of confounding variables makes it difficult for researchers to tease out the impact of the intervention from the impact of other, unintended causes, threatening the internal validity of the investigation.

An example of using baseline data is provided by a study that was completed at the Colorado School of Mines (CSM). At CSM and through funding of Hewlett Packard, a tablet computer classroom was established in the spring of 2011. All sections of Probability and Statistics for Engineers were taught in this classroom. At the start of each class, every student picked up a tablet computer and logged into InkSurvey (Kowalski & Kowalski, 2008), a Web-based system that supports the anonymous interaction of the instructor and students. The instructor posted and released problems, which the students solved. The students responded to the posted questions and the instructor reviewed their anonymous submissions from his or her computer at the front of the room. This allowed the instructors to provide feedback to the students and adjust the pace of instruction based on submitted student responses in real time.

To evaluate the effectiveness of the tablet computer instructional approach, historical data collected in 2009 were used. One participating instructor had taught the course in the spring of 2009, two years prior to the study and, owing to school policy, had retained all finals. These finals provided historical data for the course. During the offering of the course in the spring of 2011, this instructor taught using the same lecture notes as in 2009. Even the day and time at which the course was offered were consistent. The primary difference between the two course offerings was the use of the

tablet and Inksurvey in the spring of 2011 and traditional lecture in the spring of 2009. A subset of questions from the earlier final administered to students in the spring of 2009 was administered in the spring of 2011. Statistical comparisons were made among student performances within this teacher's classrooms between the two administrations to determine the effectiveness of the tablet instructional approach. As anticipated, the treatment group performed statistically significantly better than the control group. It is possible that students who completed the course in 2009 are different in a manner that is important to the research from those who completed the course in 2011. Therefore, the outcomes may be a result of some unknown difference between the two groups rather than the impact of the intervention, resulting in a threat to the internal validity of the investigation. In addition, the external validity of the investigation is threatened by the completion of the study within a school of science and engineering. Generalizations should not be made beyond this restricted population.

Another example of the use of historical data is provided through the work of Holloway, Reed-Rhoads, and Groll (2011). They compared the likelihood of students dropping out of college during their first and second year of college using historical data (data collected through a natural process of the university) as a methodology to examine the "sophomore slump." This refers to a phenomenon in which sophomore students' behaviors change from their freshmen year in areas such as decreased retention rates and fewer hours studying (Hunter et al., 2010). Figure 26.1 identifies first- and second-year retention rates of men and women for cohort years 2000 to 2009. In general, the retention rates are between 74% and 87%. For the last four years, the first-year retention rates for both men and women have been increasing; the second-year retention rates for both men and women have been decreasing (except for women in the 2008 cohort, the most recent data available). Recognizing this phenomenon, researchers can now investigate the probable causes or influential factors.

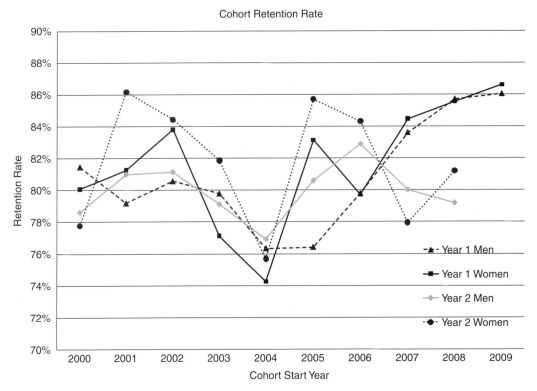

Figure 26.1. First- and second-year retention rates based on sex.

Statistical methods are often used for analysis purposes between treatment data and baseline data as well as historical data. However, the same concerns exist here that existed for matching. Because students were not randomly assigned, there may be confounding variables that are unknown to the researcher that impact the results. In other words, the researcher once again needs to acknowledge that there are limitations on the interpretations that can be drawn.

Pre–Post Designs

Pre- and post-test designs are also common within educational research. Pre-test scores are collected prior to the instructional intervention; post-test scores are collected immediately following the instructional intervention. These types of designs are most effective when a comparison group is available, preferably through matching or random assignment.

Occasionally, pre–post designs are used without a comparison group. In this case, the same set of students' pre-test scores provide baseline data to which the researcher compares the same students' post-test scores. In this case, the statistical analysis selected must account for the dependency of the data, such as the paired t-test for pre- and post-test student performances or the nonparametric equivalent, the Wilcoxon signed rank test (Corder & Foreman, 2009). When a comparison group is available and the sample is appropriately large, statistical comparisons can be made between treatment and control groups difference scores using an analysis of covariance (ANCOVA).

When using a pre–post design without a comparison group, it is highly desirable to use an instrument that already has some validity and reliability evidence that supports its use in pre–post format. Also, the researcher will want to be confident in the content and construct validity of the instrument and, if appropriate, the criterion and consequential validity of the instrument. Often the act of completing a test a second time results in increases in students' performance owing to practice taking the test rather than as a result of real change in student knowledge, reducing test–retest reliability. Tests that have been validated for pre–post use have research evidence that support their use in a pre–post format. In other words, there is already evidence to suggest that scores are unlikely to change from pre to post unless some type of intervention occurs.

Many such instruments and evidence of their validity are housed through the ciHUB.org, an online community of concept inventory developers, where instruments in science and engineering are housed for online access by faculty, instructors, and students. Concept Inventories (CIs) are used to assess students' understanding of fundamental concepts most commonly found in foundational courses, such as thermodynamics and statics in engineering and random processes and genetics in biology. Many of these instruments are further designed to track changes in students' knowledge on these fundamental skills from the beginning to end of a course. Because the purpose of a CI is to track change, test– retest reliability is an essential concern in establishing the appropriateness of the instrument.

One example of a pre–post assessment is included in the study that was previously discussed at CSM on the use of tablet computers. The Statistics Concept Inventory (Reed-Rhoads & Imbrie, 2008) is housed in the ciHUB.org and was administered in a pre–post format in the spring of 2011 in the tablet computer study. This was the first time that tablet computers were being used as part of the course. The purpose of this pre–post administration was to track whether the students gained knowledge in statistics while using the tablet in the classroom and whether these gains differed across instructional implementation methods. In this particular example, there was no control group but there were different levels in which the tablets were used in the classroom. One instructor used it daily, another weekly, and others only occasionally.

A major limitation of the pre–post format without a comparison group is that the only conclusion that can be drawn is whether change has occurred from pre- to post-administration. This change is not necessarily attributable to the instructional intervention itself (an internal validity concern). In other words, the pre–post methodology, without an associated control group, does not provide sufficient evidence to support the effectiveness of an instructional approach. Other causes for such an increase include maturation and the impact of repeated testing. Maturation refers to a natural process of students' knowledge, attitudes, and feelings changing, or advancing as a part of the aging process. Repeated testing refers to the impact that practice in the testing process can have on the outcome of the scores.

Post-Only Designs

Post-only designs are one of the weaker measures available to quantitative researchers (Ogot, Elliott, & Glumac, 2003; Sicker, Lookabaugh, Santos, & Barnes, 2005). In a post-only design, students complete a single test following an intervention and their knowledge or understanding is evaluated based on the results of that test. An example of such a design is the completion of a final at the end of a course or the completion of the Fundamentals of Engineering Exam at the end of a degree program. In these cases, the evaluator is concerned with whether the students have achieved a set of established learning goals by the end of a course or program.

Post-only designs are considered appropriate for accreditation purposes because the purpose is to verify that students possess a set of skills by the conclusion of a degree. Internal and external validity are not typically a concern. It does not matter whether a given student entered the degree program with the desired skills or whether the student developed these skills as a result of the program. It also does not matter whether the results can be generalized beyond the current population. The important component of such an exam is establishing that a given set of students possess the desired knowledge prior to entering the field.

Occasionally, post-only designs include treatment and control groups, which are established through matching, or random assignment. By comparing the performance of the treatment and control group, the researcher may imply the effectiveness of the intervention. However, any conclusions that are drawn cannot be strongly established because it is unknown how the two groups differed prior to the intervention.

An example of a post-only design is being used as part of the Bechtel K–5 Educational Excellence Initiative (Moskal & Skokan, 2011) at CSM. In this program, graduate students are providing classroom support to kindergarten through fifth grade teachers in mathematics and science for up to fifteen hours, each week, throughout the academic year. The participating teachers attend an intense two-week summer workshop immediately before receiving the graduate students' support. During the workshop, the teachers learn about mathematics and science and how these fields can be applied to understanding renewable energy concepts. To examine the impact of this effort on the participating students' attitudes with respect to mathematics and science, a post-only attitudes assessment is administered to treatment and control students. The treatment teachers attended the workshop and had the support of a graduate student in their classroom throughout the previous academic year; the control teachers will enter the program next year, and have not yet completed the workshop nor have they previously had the support of a graduate student in their classroom. In other words, both sets of teachers have self-selected into the study, but one group of teachers has received the treatment at the time of measurement and the other has not. This (potentially) controls for one confounding variable: teacher self-selection. There may, however, be many other unknown confounding variables at play, suggesting that the results should be interpreted with caution. In addition, during the analysis phase, the treatment and control classrooms are being matched based on grade level.

Longitudinal Designs

A longitudinal design refers to a methodology in which data are collected from the same students over an extended period (Felder, Felder, & Dietz, 1998; Sheppard et al., 2004; Matusovich, Streveler, & Miller, 2010). Sometimes the same type of data, such as an attitudes survey, is used repeatedly to capture changes within the student population. Other times a variety of different methods are used in combination throughout the course of the investigation, providing a comprehensive understanding of the participating student population and how key characteristics change overtime. A major concern in this type of investigation is the validity and reliability of the instruments used. Flawed instruments are likely to result in faulty conclusions.

The Student Attitudinal Success Instrument (SASI; Immekus, Imbrie, & Maller, 2004; Immekus, Maller, Imbrie, Wu. & McDermott, 2005; Reid & Imbrie, 2008) provides one example of a longitudinal assessment effort. The SASI consists of 161 items, which address 13 noncognitive constructs. The items and constructs that comprise this instrument are based on prior attitudinal instruments and research concerning the impact of attitudes on student success. The SASI is designed to provide data on noncognitive characteristics for incoming engineering students (1) *prior to* the onset of the first year and (2) for which higher education institutions have an influence during students' first year. The systematic collection of data helps universities make programmatic decisions that target recruitment, admission, retention, and ultimately the success, defined as graduation, of all students, with an emphasis on those who are underrepresented.

This study lies in the category of longitudinal because there are multiple points

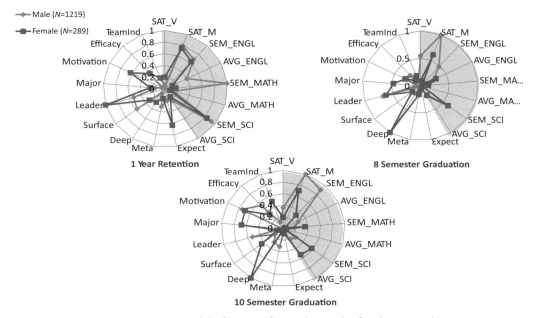

Figure 26.2. SASI model of success factors by gender for the 2004 cohort.

at which data are collected and compared across several years for the same students. The SASI is given as students enter their first year, at the end of their first year, and then again as they exit their capstone course. Data are also collected on their first year or two-semester retention rate, four-year or eight-semester graduation rate, and five-year or ten-semester graduation rate. The fifth year graduation rate is necessary here, because engineering degrees typically require more credit hours to complete than other degrees and/or students participate in alternating semester work opportunities. Data are also collected on eleven high school achievement factors when the students enter college, such as overall high school GPA; core GPA; standardized test results by sub-area on the SAT or ACT; average grades in mathematics, science, and English; and the number of semesters completed of mathematics, science, and English.

Based on the information that has been collected through this longitudinal investigation, the most important factors for predicting "success" for participating students

retained at the end of the first year and graduated at the end four and five years are determined and shown, by sex, in the radial plots of Figure 26.2. In these plots, the model's most important factors and those that have the most impact, or explain the most variability, extend from the center of the circle to its boundaries. For example, in the "1 Year Retention" radial plot it was found that for women, leadership is an important attribute for success; for men, the number of semesters of high school mathematics completed is an important predictor. The number of semesters of high school science was found to be an important predictor for both men and women's success in engineering.

To study the most significant factors in predicting success of students, a longitudinal design was necessary because these factors needed to be linked to eventual graduation with an engineering degree – a long-term outcome. Based on this research, a policy change was recommended that admission decisions be made using a set of priorities that include cognitive factors such as SAT verbal scores and the number of semesters of mathematics, science, and English completed, while also taking into account students' leadership potential, major decisions, and academic motivation (Holloway, Imbrie, & Reed-Rhoads, 2011). In addition, based on reports such as the National Academy of Engineering's report, *Changing the Conversation* (2008), social relevance of engineering as a discipline has been added to the decision process.

Analysis of Quantitative Designs

All of the designs discussed in the preceding sections result in numerical or quantitative data. The various quantitative analysis methods available, including descriptive statistics and parametric and nonparametric statistics, are far too vast to be addressed here. In fact, the study of these analysis methods comprises the field of Educational Measurement. In general, large samples are examined using parametric statistics; small samples are examined using descriptive or nonparametric statistics. In all cases, the assumptions that underlie the analysis method need to be considered and verified before implementing the analysis process. Major and minor violations of these assumptions and their impact should be considered and researched when a violation cannot be avoided through the use of an alternative technique. There are many textbooks available to guide the interested reader in selecting an appropriate analysis process (e.g., Box & Hunter, 2005; Corder & Foreman, 2009; Huck, 2011). There are also professionals in the field of educational measurement who can be consulted.

Software is also available to reduce the computation that is necessary in the analysis process. Examples of useful programs include: Minitab, M-Stats, and "R." The first two are purchased software whereas "R" is freely available for download online (see http://cran.r-project.org/bin/windows/base/).

Examples of Mixed Approaches

With the exception of the fully randomized design, the researcher is limited either by the lack of random assignment or the lack of an established control group. Missing either element in a quantitative research design results in challenges to the internal and external validity of the interpretations that can be drawn. One method that many researchers use to address these challenges is a mixed methods design. Mixed methods designs seek to provide additional evidence that supports a similar conclusion, strengthening the likelihood that the conclusion is correct.

In a mixed methods design, any of the quantitative methods described in this chapter could be combined with any of the qualitative methods described in Chapter 27 by Case and Light and Chapter 28 by Johri. The decision as to which methods to use is based on the purpose of the investigation. According to Greene, Caracelli, and Graham (1989), there are five purposes or motivations for using a mixed methods design (listed here from least to most flexible): triangulation, complementarity, development, initiation, and expansion. These purposes emerged from an empirical study of fifty-seven mixed methods evaluations. Although the design characteristics are similar, the motivation for selecting to use a mixed methods design is different. Formally, triangulation uses different methods, with different strengths and weaknesses, to assess the same phenomenon. The purpose of triangulation is to seek confirmation of results among the different approaches. Another approach is to implement a method that provides information that complements or provides further support for a result, but does not necessarily confirm the result. This is referred to as *complementary. Develop* refers to the use of assessment methods for the purpose of informing the design of another method, with the latter method being the primary method of the study. *Initiation* describes the use of mixed methods to resolve a paradox that emerges from the evaluation results, while *expansion* attempts to give breadth and scope to the evaluation. For greater details concerning the different purposes of a mixed methods approach, see Greene et al. (1989).

An example of a mixed methods approach can be found in a study completed by the Academic Pathways Study component of the Center for the Advancement of Engineering Education (Kilgore, Atman, Yasuhara, Barker, & Morozov, 2007). In this study, data were collected through surveys, interviews, and ethnographic observations and analyzed to determine the role motivation, skill, and abilities play in considering the contextual factors of a first-year engineering design process. When questioned in an interview setting, both the quantitative correctness of a student's response and the qualitative evaluation of the participant's motivations were examined. The two methods were combined in the analysis process to understand better the impact of motivation on the student's performance. This approach lies within the expansion category previously described. On a biannual basis, these same students responded to questions concerning their engagement in higher education. This was used to inform the interpretation and identification of factors that impact academic performance. In this study, it was found that gender played an important role in the ability of a first-year engineering student to contextualize a design problem. Specifically, it was determined that women emphasized the problem's context more than the men, which has implications on how educational paradigms affect recruitment and retention of engineering students. As this example illustrates, the evidence acquired through a mixed methods design is often more compelling than that which can be acquired through a quantitative method alone.

Chapter Summary

This chapter provides a brief overview of quantitative methods that are used in educational psychology. Because a fully randomized design is difficult to implement in an educational investigation, alternative designs are often used. When researchers implement these alternative designs, they need to be aware of the potential threats to the study's validity and openly acknowledge these threats during the data interpretation process.

Validity should be considered on two levels: instrument and assessment program. The validity of instruments is dependent on the concepts of content, construct, criterion, and consequential validity. Internal and external validity should be considered at the assessment program level. Reliability, with respect both to tests and scoring, should also be considered when establishing the validity of a given instrument. An unreliable instrument cannot be interpreted in an appropriate or valid manner.

A key concept that was introduced in this chapter is "triangulation" or the process of gathering evidence from multiple sources to more firmly support the conclusions that are drawn through a research investigation. Triangulation can be achieved by combining multiple quantitative methods, allowing the researcher to remain within a given assessment paradigm, or by combining quantitative and qualitative methods, crossing assessment paradigms using a mixed method approach. Mixed methods designs, however, are not limited to triangulation. Rather, there are four other potential purposes for employing mixed methods: complementarity, development, initiation, and expansion.

References

Addison, V., Hipp, C., & Lyons, J. (2007). A study on the effects of timing on engineering students' abilities to solve open-ended problems with computers. In *Proceedings of the 2007 American Society for Engineering Education Annual Conference & Exposition*. Honolulu, HI, Paper 2007-1383.

American Psychological Association. (1999). *Standards for educational and psychological testing*. Washington, DC: Author.

Bernard, R. (2000). *Social research methods: Qualitative and quantitative approaches*. Thousand Oaks, CA: SAGE.

Borrego, M., Douglas, E., & Amelink, C. (2009). Quantitative, qualitative, and mixed research methods in engineering education. *Journal of Engineering Education*, 98(1), 53–66.

Box, G. E. P., & Hunter, J. S. (2005). *Statistics for the experimenter: Design, innovation and discovery*. New York, NY: John Wiley & Sons.

Campbell, D. T., & Stanely, J. T. (1963). *Experimental and quasi-experimental designs for research*. Boston, MA: Houghton-Mifflin.

Cohen, L., Manion, L., & Morrison, K. (2007). *Research methods in education*. Milton Park, Abingdon, Oxon, UK: Routledge.

Corder, G., & Foreman, D. (2009). *Nonparametric statistics for non-statisticians: A step-by-step approach*. Hoboken, NJ: John Wiley & Sons.

Cortina, J. (1993). What is coefficient alpha? An examination of theory and applications. *Journal of Applied Psychology*, 73(1), 98–104.

Costello, A. B., & Osboren, J. W. (2005). Best practices in exploratory factor analysis: Four recommendations for getting the most from your analysis. *Practical Assessment, Research & Evaluation*, 10(7). Retrieved from http://pareonline.net/pdf/v10n7.pdf

Creswell, J. W., & Plano Clark, V. L. (2007). *Designing and conducting mixed methods research*. Thousand Oaks, CA: SAGE.

Denzin, N. K. (1978). *Sociological methods*. New York, NY: McGraw-Hill.

Felder, R., Felder, G., & Dietz, E. (1998). A longitudinal study of engineering student performance and retention. V. Comparisons with traditionally-taught students. *Journal of Engineering Education*, 87(4), 469–480.

Gorsuch, R. (1983). *Factor analysis*. New York, NY: Taylor & Francis.

Greene, J. C., Caracelli, V. J., & Graham, W. F. (1989). Toward a conceptual framework for mixed-method evaluation designs. *Educational Evaluation and Policy Analysis*, 11(3), 255–274.

Holloway, B. M., Imbrie, P. K., & Reed-Rhoads, T. (2011). A holistic review of gender differences in engineering admissions and early retention. In *15th International Conference for Women Engineers and Scientists*. Adelaide, Australia, Paper 0243. Retrieved from http://search.informit.com.au/documentSummary;dn=947873175834117;res=IELENG

Holloway, B. M., Reed-Rhoads, T., & Groll, L. E. (2011). Women as the miner's canary in undergraduate engineering education. In *Proceedings of the 2011 American Society for Engineering Education Annual Conference & Exposition*, Vancouver, BC, Canada, Paper 2011-1382.

Hopkins, K. D. (1998). *Educational and psychological measurement and evaluation* (8th ed.). Nedam Heights, MA: Allyn and Bacon.

Huck, S. (2011). *Reading statistics and research* (6th ed.). New York, NY: Pearson.

Hunter, S. H., Tobolowsky, B. F., Gardner, J. H., Evenbeck, S. E., Pattengale, J. A., Schaller, M. A., Schreiner, L. A., & Associates, (2010). *Helping sophomores succeed: Understanding and improving the second-year experience*. San Francisco, CA: Jossey-Bass.

Immekus, J. C., Imbrie, P. K., & Maller, S. J. (2004). The influence of pre-college factors on first-year engineering students' academic success and persistence. In *34th ASEE/IEEE Frontiers in Education Conference Proceedings*, Savannah, GA (pp. F3F-1–2).

Immekus, J. C., Maller, S. J., Imbrie, P. K., Wu, N., & McDermott, P. A. (2005). Work In Progress: An analysis of students' academic success and persistence using pre-college factors. In *35th ASEE/IEEE Frontiers in Education Conference Proceedings*, Indianapolis, IN (pp. S2C-3–4).

Johnson, B., & Christensen, L. (2008). *Educational research: Quantitative, qualitative and mixed methods approaches* (3rd ed.). Thousand Oaks, CA: SAGE.

Johnson, R. G., & Onweugbuzie, A. J. (2004). Mixed methods research: A research paradigm whose time has come. *Educational Researcher*, 33(7), 14–26.

Kashy, D., Albertelli, G., Kashy, E., & Thoennessen, M. (2001). Teaching with ALN technology: Benefits and costs. *Journal of Engineering Education*, 90(4), 499–505.

Kilgore, D., Atman, C., Yasuhara, K., Barker, T., & Morozov, A. (2007). Considering context: A study of first-year engineering students. *Journal of Engineering Education*, 96(4), 321–334.

Kowalski, F., & Kowalski, S. (2008). Exploring the role of tablet PCs in promoting active learning and real-time communication to enhance learning in the university setting. Retrieved from http://www.cfkeep.org/html/snapshot.php?id=79647250361412

Leydens, J., Moskal, B., & Pavelich, M. (2004). Qualitative methods used in the assessment of engineering education. *Journal of Engineering Education*, 93(1), 65–72.

Matusovich, H., Streveler, R., & Miller, R. (2010, October). Why do students choose engineering? A qualitative, longitudinal investigation of students' motivational values, *Journal of Engineering Education*, 289–303.

Merino, D., & Abel, K. (2003). Evaluating the effectiveness of computer tutorials versus traditional lecturing in accounting topics. *Journal of Engineering Education*, 92(1), 189–194.

Messick, S. (1989). *Educational measurement* (3rd ed.). R. L. Linn (Ed.), New York, NY: Macmillan.

Messick, S. (1998). Test validity: A matter of consequence. *Social Indicators Research*, 45(1), 35–44.

Miller, M. (1995). Coefficient Alpha: A basic introduction from the perspectives of classical test theory and structural equation modeling. *Structural Equation Modeling*, 2(3), 255–273.

Moskal, B. (2011). Bechtel K-5 educational excellence initiative. Retrieved from http://mcs.mines.edu/Research/bechtel/new/

Moskal, B., & Leydens, J. A. (2000). Scoring rubric development: Validity and reliability. *Practical Assessment, Research & Evaluation*, 7(10). Retrieved from http://pareonline.net/getvn.asp?v=7%26;n=10

Moskal, B., Leydens, J., & Pavelich, M. (2002). Validity, reliability and the assessment of engineering education. *Journal of Engineering Education*, 91(3), 351–354.

Moskal, B., & Skokan, C. (2011). Supporting the K-12 classroom through university outreach. *Journal of Higher Education Outreach and Engagement*, 15(1), 53–75. Retrieved from http://openjournals.libs.uga.edu/index.php/jheoe/issue/view/52

National Academy of Engineering, (2008). *Changing the conversation: Messages for improving public understanding of engineering.* Washington, DC: The National Academies Press.

Ogot, M., Elliott, G., &Glumac, N. (2003). An assessment of in-person and remotely operated laboratories. *Journal of Engineering Education*, 92(1).

Olds, B., Moskal, B., & Miller, R. (2005). Assessment in engineering education: Evolution, approaches and future collaborations. *Journal of Engineering Education*, 94(1), 13–25.

Reed-Rhoads, T., & Imbrie, P. K. (2008). Concept inventories in engineering education. *Evidence on Promising Practices in Undergraduate Science, Technology, Engineering, and Mathematics (STEM) Education Workshop 2 October 13–14*. Retrieved from http://sites.nationalacademies.org/dbasse/bose/dbasse_071087#.UdSNu5xsuMo

Reid, K. & Imbrie, P. K. (2008). Noncognitive characteristics of incoming engineering students compared to incoming engineering technology students: A preliminary examination. In *Proceedings of the 2008 American Society for Engineering Education Annual Conference & Exposition*, Pittsburgh, PA, Paper 2008-1995.

Sheppard, S., Atman, C., Stevens, R., Fleming, L., Streveler, R., Adams, R., & Barker, T. (2004). Studying the engineering student experience: Design of a longitudinal study. In *Proceedings of the 2004 American Society for Engineering Education Annual Conference & Exposition*, Salt Lake City, UT, Paper 2004-1736.

Sicker, D., Lookabaugh, T., Santos, J., & Barnes, F. (2005). Assessing the effectiveness of remote networking laboratories. In *35th ASEE/IEEE Frontiers in Education Conference Proceedings*, Indianapolis, IN (pp. S3F-7-12).

Suhr, D. D. (2003). Principal component analysis vs. exploratory factor analysis. In *SUGI 30 Proceedings*. Retrieved from www2.sas.com/proceedings/sugi30/203-30.pdf

Tashakkori, A., & Teddlie, C. (Eds.). (2003). *Handbook of mixed methods in social and behavioral research*. Thousand Oaks, CA: SAGE.

U.S. Department of Education, Institute of Education Science, and National Center for Education, Evaluation and Regional Assistance (2003). *Identifying and implementing educational practices Supported by rigorous evidence: A user friendly guide*. Retrieved from http://www2.ed.gov/rschstat/research/pubs/rigorousevid/guide.html

Yadav, A., Subedu, D., Lundeber, M., & Bunting, C. (2011). Problem-based learning: Influence on students' learning in an electrical engineering course. *Journal of Engineering Education*, 100(2), 254–280.

CHAPTER 27

Framing Qualitative Methods in Engineering Education Research

Established and Emerging Methodologies

Jennifer M. Case and Gregory Light

Introduction

In science and engineering research, methodologies based on quantitative methods of data collection are prominent, based on their power for building predictive models of the natural world. Research in the social world, of which engineering education is a subset, is only partially described by quantitative models. Much of the subtlety of human interaction rests in complex models of causality that require the use of qualitative data for building explanatory theory.

This chapter provides an introduction to the use of qualitative methods for engineering education researchers. A more substantial consideration than that of methods, however, is the way in which an argument is developed for the validity of the knowledge generated from the analysis of qualitative data. These arguments are encapsulated in a discussion of methodology, which can

This chapter is a revised and shortened version of the following journal article: Case, J. M., & Light, G. (2011). Emerging methodologies in engineering education research. *Journal of Engineering Education*, 100(1), 186–210.

be defined as referring to a theoretical justification for the methods used in a study (Burton, 2002; Clough & Nutbrown, 2002). This chapter focuses on methodology as a crucial area with which researchers need to grapple in order for the quality and scope of research to continue to develop. It is argued that to be able to answer the research questions at hand, methodological decisions need to be more explicitly represented in reports of research; and researchers need to consider a broad range of methodological options, in particular those methodologies that could be considered to be "emerging" in engineering education research.

In a recent article on data collection methods for engineering education research, Borrego, Douglas, and Amelink (2009) set out the terms of the debate focusing on the distinction among quantitative, qualitative, and mixed methods studies. They suggest that "engineering educators who have been trained primarily within the quantitative tradition may not be familiar with some of the norms of qualitative research" (p. 56) and they propose further that some journal reviewers may suffer from the same limitation. Borrego et al. (2009) make the crucial point that using a wider range of data collection methods would allow researchers to address a wider range of research questions. Their article thus gives a comprehensive overview of quantitative and qualitative methods, as well as a discussion of mixed methods studies that use both of these. A further useful contribution to this discussion has been the article by Koro-Ljungberg and Douglas (2008), who present an analysis of articles published in the *Journal of Engineering Education* in the period 2005–6 that use qualitative methods. In addition to noting a low frequency of such articles, they furthermore show that articles that use qualitative methods seldom provide an adequate justification of why these methods are used, or insufficiently integrate these justifications within their broader research design and theoretical orientation. The issues raised by these articles provide the departure point for the present chapter, which locates the debate on qualitative research methods within a broader discussion of methodology.

Defining the Domain of Methodology

> *Crudely, methods are best understood as the tools and procedures we use for our inquiries and methodology is about the framework within which they sit. (Cousin, 2009, p. 6)*

In building a methodological argument to support a study, there are a range of established methodological positions on which the researcher can draw. In this way one locates one's research design within a particular "tradition of enquiry" (Clough & Nutbrown, 2002, p. 31). Cousin (2009) makes the following important observation:

> *Although many methods are friendly to particular methodologies and to particular research contexts or questions, there is not always a straightforward association between method and methodology. This is because different people might use the same methods with quite different values and aims in mind. (p. 5)*

As mentioned previously, Koro-Ljungberg and Douglas (2008) have demonstrated that the issue of methodology has received limited explicit discussion in the engineering education literature. The same has been noted in other areas, for example, in the field of mathematics education research (Burton, 2002). The lack of an explicit discussion of methodology in these fields means that it is often hard to know what to make of particular research findings, how to interpret them, and what status to ascribe to them. More importantly, what often happens when methodological discussion is limited is that methodology is tacitly accepted as a given, meaning that one methodology is implicitly assumed to be the right one. As such, our focus on methodology in this chapter is, in the first instance, largely a call for making the whole question of methodology more explicit in engineering education. Second, although accepting, as Cousin (2009, p. 2) notes, that "randomized control trials remain a gold standard for some researchers," we also agree with her that educational research, including the growing field of engineering education research, is "a big playground where no one methodology needs to hog the best swing."

The choice of methodology (with its underlying theoretical perspective and its related set of methods) is determined by the kinds of research questions that one wishes to ask. The choice of methodology will constrain what questions can be addressed by the research, and conversely certain research questions can best be addressed with certain methodologies. It will thus be quickly seen that if a research community limits itself only to particular methodologies it will be likely that the research findings it is able to generate will also be limited. The relationship between research questions and methodology is usually not unidirectional but is rather two-way or what might be described as "dialectical." One might start out with some idea of what one wishes to research, identify an appropriate methodology, and then go back to the research questions and refine them, and so on. Cousin (2009) suggests that researchers would do well to move away from a traditional linear way of thinking about the research process where a researcher first produces a literature review, then formulates the research question, then collects data, then analyzes it, and finally writes it up. She notes that "Increasing numbers of researchers recognize that all of these activities need to be dynamically linked and continually enlivened by an engagement with a wide reading" (Cousin, 2009, p. 3). Research questions need to be informed by education theory, and thus another dialectical relationship exists between education theory and methodology.

Emerging Methodologies in Engineering Education Research

This chapter focuses largely on methodologies that can be used to support the incorporation of qualitative data in engineering education research. We have selected a number of methodologies that are promising but as yet not well represented in engineering education research, with uneven distribution across different global regions. In that sense they can be described as "emerging," although they might be well established in education research more generally. These are:

- Case Study
- Grounded Theory
- Ethnography
- Action Research
- Phenomenography
- Discourse Analysis
- Narrative Analysis

Each of the following sections consists of two parts. The first part offers a brief description of some of the defining features of each methodology, focusing in particular on the arguments that are offered to justify the use of the set of methods. The second part of each section presents an engineering education research article that exemplifies this methodology. Here the focus is on

the research questions and associated theory that were used in the study as well as the ways in which the methodology linked to these. Table 27.1 gives details of these exemplar articles, together with a brief outline of the data that were collected and analyzed in each study.

The selection of exemplar articles has been guided by the requirement that these are recent pieces of engineering education research published in the international literature that offer useful illustrations of how the methodology has been applied. They have also been selected as exemplars to demonstrate the wider range of research questions that the community might address if it broadens the range of methodologies with which it engages. It is worth noting that although some of these articles have been published in engineering education journals, others have been published in more generic education journals. There is much quality engineering education research published across a wide range of fields and the community would do well to keep alert to this broader literature.

Case Study

A case study involves a distinct, single instance of a class of phenomena. This could be an event, an individual, a group, an activity, or a community (Shepard & Greene, 2003). As a methodology, case study can be used to elucidate the validity of findings emerging either from an analysis of a single case or across multiple cases. Flyvbjerg (2001) identifies a range of different strategies that can be used in the selection of cases, depending on the nature of the research question involved. Options include choosing a set of cases with maximum variation in order to explore a range of different settings, or identifying unusual cases that allow the researcher to probe particularly problematic situations, or using critical cases that allow for logical deductions of the type "If this holds for this case, then it will hold for all other cases."

Case study as a methodological approach has frequently been critiqued for its assumed

Table 27.1. *Summary of Exemplar Articles and Research Methods for Each Methodology*

Methodology	Exemplar Article	Main Methods of Data Collection
Case study	Magin, D. J., & Churches, A. E. (1995). Peer tutoring in engineering design: A case study. *Studies in Higher Education*, 20(1), 73–85.	• Informal observation • Open-ended surveys of students and tutors • Interviews with tutor groups and teaching staff
Grounded theory	Jonassen, D., Strobel, J., & Lee, C. B. (2006). Everyday problem solving in engineering: Lessons for engineering educators. *Journal of Engineering Education*, 95(2), 139.	• Semistructured interviews with practicing engineers
Ethnography	Stevens, R., O'Connor, K., Garrison, L., Jocuns, A., & Amos, D. (2008). Becoming an engineer: Toward a three dimensional view of engineering learning. *Journal of Engineering Education*, 97(3), 355–368.	Over four-year longitudinal study: • Observations of students in daily life • Interviews with students
Action research	Jørgensen, F., & Kofoed, B-L. (2007). Integrating the development of continuous improvement and innovation capabilities into engineering education. *European Journal of Engineering Education*, 32(2), 181–191.	• Feedback from students in project steering committee meetings
Phenomenography	Booth, S. (2001). Learning computer science and engineering in context. *Computer Science Education*, 11(3), 169–188.	• Interviews with students and tutors
Discourse analysis	Kittleson, J. M., & Southerland, S. A. (2004). The role of discourse in group knowledge construction: A case study of engineering students. *Journal of Research in Science Teaching*, 41(3), 267–293.	• Audio and video recording of a group of students in laboratory sessions • Semistructured interviews with these students • Observations of the whole class
Narrative analysis	Walker, M. (2001). Engineering identities. *British Journal of Sociology of Education*, 22(1), 75–89.	• Interviews with students

limitations. Not surprisingly, these critiques are concerned with the issue of generalizability of the empirical results attained by case study. Flyvbjerg (2001) considers these critiques to rest on essential misunderstandings of case study methodology. The concrete, context-dependent nature of the knowledge that case studies unearth, on which many of these critiques focus, is precisely the source of its methodological strength. Case study can therefore be particularly appropriate to address research questions concerned with the specific application of initiatives or innovations to improve or enhance learning and teaching. The new knowledge here takes into consideration the particular idiosyncrasies of the institution, its resources, teachers, and students, as well as its overall culture. In this it can be contrasted with a more positivist kind of study aimed at evaluating the general applicability of an educational intervention – characterized, for example, by a more traditional randomized controlled study – primarily focused on "proving" the general effectiveness of the intervention.

Exemplar article:
Magin, D. J., & Churches, A. E. (1995). Peer tutoring in Engineering Design: A case study. *Studies in Higher Education*, 20(1), 73–85.

In 1992, the School of Mechanical and Manufacturing Engineering at the University of New South Wales commenced a program of conversion from traditional pencil-and-paper engineering design subjects to a program in which design teaching was to be based on computer graphics, employing a state-of the-art solid modeling designer's package. This provided a good opportunity to conduct a case study investigating the impact and success of this innovation.

In the first year of implementation, the number of students vastly outnumbered the number of workstations available for the students. The department therefore decided to run two separate courses in parallel, a traditional pencil-and-paper course for three quarters of the students and a workstation based course for the remaining quarter. With the arrival of new workstations the following year, all of the now second year students had access to work stations. The dilemma was that three quarters of them would need intensive training with the new software. It was decided the most efficient way to do to this training would be for the students who had workstation experience in the first year to peer tutor those who had not had the experience. During the first four weeks of the course, the students worked in groups of approximately one peer tutor for three tutored students.

A general research question drove the case study: What was the perceived enhancement in learning, for both the tutored students as well as the peer-tutors? The case study employed multiple methods to obtain data including informal observation of the program and tutoring groups, open-ended surveys of the tutored students and peer tutors, and interviews with tutorgroups, students, and teaching staff. The analysis of the data indicated that the conditions for learning facilitated by the program compared very favorably with the more traditional teacher-led approaches. For example, when compared with the alternative form of teaching that the students were offered – assistance from a teacher when required – about half the students preferred the peer tutoring compared to fewer than 5% who preferred the alternative of teacher assistance. Particular conditions that the students identified as helpful were enhanced interactions in terms of more individual assistance and immediate response to queries; better learning climate in terms of being more relaxed, easy, and friendly with a peer; and a more empathetic relationship with peers.

Grounded Theory

Grounded theory, originally developed by Glaser and Strauss (1967), was one of the first methodological positions put forward that supported the use of qualitative data in social research. It has since been described as "a general methodology for developing theory that is grounded in data systematically gathered and analyzed" (Strauss & Corbin, 1994, p. 273). At the heart of grounded theory is the idea that theory is generated from the data at hand, rather than already existing theory being used in the analysis as is generally common in education research.

The constant comparative method is a central data collection method in grounded theory methodology, even though it has also been subsequently utilized in many other areas of qualitative research. This method provides a clear step by step outline of a process for analyzing qualitative data. In the first stage of this procedure, also termed "open coding," initial categories are developed by grouping similar incidents together. During the coding of each incident it must be carefully compared with other incidents previously coded in the same category. In the next stage of the analysis, termed "axial coding," a further refinement is done on the categories and their properties by stepping back and testing all incidents coded in a category with the properties of that category. The categories are also compared for overlap and examined for possible relationships among categories. The endpoint of data collection is reached through a process termed "theoretical saturation," which occurs when additional data collection and analysis does not substantially change the findings. Another aspect of the grounded theory methodology is that data collection and analysis are tightly interwoven. The initial theories emerging from the data are used to direct further data collection. One possibility in this respect is to use theoretical sampling, where additional research subjects are selected as the study proceeds to explore issues that have arisen.

In its pure form grounded theory tends to find limited application in education research, where researchers often find it productive to use existing theoretical constructs in their analysis. However, as a mode of research for challenging preconceptions and allowing for alternative conceptualizations it has enormous strength.

Exemplar article:
Jonassen, D., Strobel, J., & Lee, C. B. (2006). Everyday problem solving in engineering: Lessons for engineering educators. *Journal of Engineering Education*, 95(2), 139.

This study aimed to identify the key attributes of workplace engineering problems, given the central role of problem solving in the engineering workplace, and current efforts in engineering curricula to prepare graduates better for the world of work. Semistructured interviews were conducted with over a hundred practicing engineers, and interviewees were required to focus on a single job or project that they had completed at some point in their career, and then to describe a typical problem that had needed to be solved.

The researchers did not want their research findings to be circumscribed by existing thinking on this topic, and thus they opted for a grounded theory methodology. However, they also needed the work to be informed by prior research and thus a particular grounded theory methodology termed "analytic induction" was employed. Each interview was transcribed and contributed towards a case library of engineering stories. The method of constant comparison was used to identify categories or themes, and stages of both open and axial coding were employed. Twelve themes were identified, which collectively describe important characteristics of workplace problems. These include that they are frequently ill structured, with constraints and unanticipated problems emerging only during the process. However, it was also found that these ill-structured problems often comprised aggregates of smaller, well-structured problems, which were important for the engineer to identify. It was further found that ill-structured problems frequently have multiple, conflicting goals and the task of the engineer was essentially to identify the goal with the highest priority.

A rather surprising theme in these findings is that the success of a solution is rarely judged using engineering standards that center largely on a technical analysis of risk and failure: the most significant criteria for judging success tend to involve satisfying the client, completing the job on time or under budget. In a similar vein it was noted that most constraints were not engineering related but most frequently related to time. A further theme is that problem solving tended to be distributed among team members, where different team members contribute their particular strengths to solving the problem. It was also found that engineers tend to rely primarily on experiential knowledge, especially in applying theoretical knowledge to real-world situations. Echoing the ill-structured problem characteristics mentioned initially, a further theme referred to the idea that unanticipated problems are often encountered along the way. The findings are used to support a call for engineering curricula that are more problem

based, and take into account these characteristics of real problems.

Ethnography

The genesis of ethnography as a research
method is generally attributed to early work
in anthropology, more than a century ago,
but one can also trace its ancestry to early
developments in sociology, where it was
argued that observation, later participant
observation, was critical for developing a
full understanding of an environment. Central to this early ethnographic research was
the idea of closely studying first-hand how
people live in particular social situations.
Drawing on this seminal work, ethnography
has, according to Hammersley and Atkinson (2007), more recently come to be understood as:

> ... a particular method or set of meth-
> ods ... (which) involves the ethnographer
> participating, overtly or covertly, in people's
> daily lives for an extended period of time,
> watching what happens, listening to what
> is said, asking questions – in fact, collecting
> whatever data are available to throw light on
> the issues that are the focus of the research.
> (p. 1)

The demands of coming to a rich understanding of people's day-to-day lives within
a social environment presents substantive
challenges – particularly with respect to
gathering and interpreting the data in terms
of the meanings that the members themselves attach to their own world (Bryman,
2001). In addition, the constraints of doing
educational research do not afford today's
ethnographer the luxury of living with the
people they are studying – and certainly not
for years. They are more likely only to be
able to follow their lives closely and carry
out their fieldwork over many months.

Ethnographers use many different methods to collect data during their fieldwork.
In addition to the field notes which may
include records of discussions, chance conversations, interviews, overheard remarks,
and observational notes, they may also
employ audio and video-recordings and
quantitative data gathered from surveys or
structured observation.

Exemplar article:
Stevens, R., O'Connor, K., Garrison, L.,
Jocuns, A., & Amos, D. (2008). Becoming an engineer: Toward a three dimensional view of engineering learning. *Journal
of Engineering Education*, 97(3), 355–368.

This study sought to identify critical dimensions of the experience of becoming an engineer over four years of undergraduate study,
using a longitudinal research design. The
authors aimed to identify the characteristics that describe both how a student makes
him- or herself into an engineer and how
he or she is made into an engineer with
respect to the formal and informal educational experiences related to academic engineering programs. To address this question,
the authors employ what they call a "person-
centered ethnography" (p. 355) that focuses
on the individual within the particular social
context. Indeed, their goal is described as
"recovering the person" within the wider
social context in which they are becoming
an engineer.

The ethnographic methodological approach taken in this study reveals qualitatively different ways of understanding this
process than might have been traditionally expected. Rather than, for example, a
focus on engineering knowledge in terms
of a stable curriculum of knowledge to be
acquired over time, this ethnographic study
disclosed the changing contextual nature
of what counts as disciplinary knowledge.
In the first years what counted was static
knowledge transmitted in traditional, prerequisite lecture courses in which students
were expected to give back the right answer
on exams. In later years, however, what
counted shifted more to open-ended problems, assessments of teamwork, and so on.
There were also changes in the expected
learner relationship to data, from data provided in laboratories to the expectation of
generating one's own data. Not surprisingly,
this was often accompanied by feelings of
frustration and anxiety.

Changes were also revealed in the ways in which students began to form personal identities with engineering. The article stresses the double-sided character of identification – both that students need to identify with engineering and that they need to feel accepted by engineering. Compared to universities in which students were admitted immediately into engineering programs, in those institutions where they were not admitted directly into engineering, students struggled to form an identity with the discipline. Likewise, concrete changes across the years had a substantive impact on identification: upper-level students began to be given the "keys to the clubhouse," so to speak, which included things like increased access to laboratories, online networks, disciplinary lounges, and so on.

Action Research

Action research, a term first used by Kurt Lewin in the 1940s, is a critical research methodology looking to foster change in social practices in the social situations in which they take place – "within everyday, natural contexts rather than within controlled settings" (Cousin, 2009, p. 150). The main aims and benefits of action research are strategic improvement of practice. This critical focus on continuous improvement raises the second defining feature that distinguishes action research from other educational methodologies: it is almost entirely determined and conducted by its various practitioners. Indeed, improvement occurs through the active engagement of the practitioners. As such, action research is research *with* subjects, not *on* them, an idea that, Cousins notes, "reverses the conventional scientific understanding of objectivity" (Cousin, 2009, p. 151).

In addition to being participative and focused on improvement, Kember (2000) also describes action research as being reflective, systematic, and cyclical. In its design, methods, and realization, it consciously and deliberately sets out to improve, enhance, and realize practice through actions informed, but not constrained, by

research and theory. Kemmis and McTaggart (1988, p. 7) describe the implementation of this strategic action as a continuous cycle of four moments:

- A plan of action to improve what is already happening
- Action to implement the plan
- Observation of the effects of action in the context in which it occurs
- Reflection on these effects as a basis for further planning, subsequent action and so on, through a succession of cycles

Action research of this kind can be a particularly effective methodology for engineering faculty who are interested not only in systematically researching their own educational practices but also in implementing substantial personal and social change in their practice.

Exemplar article:
Jørgensen, F., & Kofoed, B-L. (2007). Integrating the development of continuous improvement and innovation capabilities into engineering education. *European Journal of Engineering Education*, 32(2), 181–191.

This article reports on an action research project that is broadly situated in the context of ever-increasing demand for and proliferation of courses concerned with equipping a new generation of engineering students with knowledge about innovation. In this program, continuous improvement and innovation, rather than simply being topics on the curriculum, were used toward the building of student capabilities for continuous improvement and innovation. In light of the program's focus on improving student capabilities, it was felt that an action research approach was most appropriate. Moreover, bringing in students as active participants in the cyclical nature of the research, including planning and generating ideas and solutions, meant that the research began to engage them in continuous innovation as well as continuous improvement.

The students who participated in the research were all in the first year of a three-year Global Business Development (GBD) program at a Danish university. The program takes a problem-based and project-based approach with students working in six groups of five to seven participants. The projects are developed with a local company. During the research, two areas of improvement were identified: communication and projects. The communication issues centered on providing first year students with more practical information on the GBD program and on the feedback communicated by the steering committee coordinator. Critical comments about individual teachers tended to be censored, which the students felt eliminated their chance of communicating teaching problems. With respect to projects the students identified a mismatch between company goals of solving real problems and the learning outcomes of inexperienced first-year students.

Following the action research project the students asked if the study could be extended and a process of continuous improvement was enacted. This resulted in the development of a process of continuous innovation for the students that included establishing an online student network for planning and sharing ideas for improvement. Finally, the study reported important ongoing changes in the six participating students engaged in the action research. Four followed up their experience with further projects on continuous improvement and innovation and all six developed a collaborative plan to continue to work together to discover further ways to support improvement.

Phenomenography

Phenomenography was developed as a research methodology by researchers in Sweden in the 1970s. Its primary focus is the investigation of the different ways in which phenomena, or aspects of a phenomenon (such as specific concepts), are experienced or understood within particular educational and learning contexts; phenomenographic research is searching for a comprehensive record of the *variation* in the experiences of

people in such contexts. Marton and Booth (1997) describe this methodology as seeking "the totality of ways in which people experience . . . the object of interest and interpret it in terms of distinctly different categories that capture the essence of the variation . . . " (p. 121). This totality of different ways is often referred to as the "outcome space," a description, at the collective level, of different categories of experience.

Although phenomenography in some respects reflects another well-known methodology, phenomenology, it is unique in two respects. First, rather than looking for common shared experiences of a phenomenon, as is the case with phenomenology (Van Manen, 1990), phenomenography focuses on the ways in which learners *differ* (Marton, 1989). It is therefore important to maximize the potential variation of experience in the sample of individuals interviewed, ensuring the sample is fully representative of potential experience in respect of the phenomenon under consideration: not all the highest performing students, for instance, nor all the poorest performing students. The second key difference with phenomenology is that rather than making the phenomenon the subject of the research, as phenomenology does (Van Manen, 1990), phenomenography takes *experience* of the phenomenon as its unit of analysis. The data collected from individuals thus need to be pooled together and analyzed (in a careful iterative process) to identify a set of distinctive categories (and the critical dimensions of variation that differentiate these categories) by which the full collective experience can be described. The analysis aims at identifying the minimum number of logically related categories required to describe the totality of variation discerned in the pool of experience.

The identification of different conceptions makes phenomenography particularly well suited for the design of educational learning objectives, pedagogical strategies, assessments, and evaluations (Micari, Light, Calkins, & Streitwieser, 2007). Phenomenographic work provides program developers with a profile of the variation in experience across all of the participants in the program.

Exemplar article:
Booth, S. (2001). Learning computer science and engineering in context. *Computer Science Education*, 11(3), 169–188.

This article uses phenomenographic research to evaluate the impact of a new Computer Science and Engineering program. The phenomenographic approach to the evaluation of the new program consisted of interviewing a diverse selection of sixteen students from a range of different types of groups, including both successful and less successful groups, all male and mixed gender groups, and groups with members of similar ages and those with members of different ages. Results of the phenomenographic analysis focused on the variation in both student experiences of relevance structure, and on learning in groups. With respect to the former, three qualitatively distinct experience of the relevance of the course were identified: Those pointing "(A) nowhere at all, or (B) inwards to its parts in isolation, or (C) outwards to the coming education and/or professional field" (p. 179). Three qualitatively distinct categories of experience with respect to learning in groups were also identified: "(A) Learning in isolation within the group; (B) Learning as part of a distributed effort; and (C) Learning as part of a collaborative effort" (p. 182).

The results suggest that neither the students nor the tutors (teachers) were fully grasping the goals of the course and that further reform of the course was necessary. Indeed, an important insight learned was that student ability to gain richer experiences of relevance was not going to surpass the experience of relevance of the tutors themselves. Further training with tutors was conducted in the subsequent years. Similarly, the variation in student experiences of learning in groups suggested that the simple putting together of students with different perspectives in a groups was not sufficient for facilitating collaborative learning. They need to engage in meaningful collaborative tasks that require recognizing their peers as sources of knowledge and enlightenment.

Discourse Analysis

Discourse analysis is a methodology that emerged from the field of linguistics (cf. Fairclough, 2003; Gee, 2005) and that is now well established across a range of education research areas. The data that form the focus for this methodology are actual instances of language in use, for example, the transcript of a classroom discussion. Hicks (1995) emphasizes that the term discourse always refers to communication that is socially situated. The significance of an analysis of discourse is that it allows us to get insights into the beliefs, values, and worldviews that are held by participants because these are always reflected in the use of discourse.

In the context of engineering education, it is important to note that discourse comprises not only written text; it also includes mathematical equations, graphs, figures, verbal exchanges, and so on. The discourse of being an engineer will involve the practice of design to solve real-world problems, and this includes collecting and analyzing data, using empirical laws and correlations, doing mathematical calculations and modeling, as well as presenting one's results to a range of different audiences. Over time, this community has developed shared ways of talking about and understanding the issues and practices that matter to them. From this point of view, successful learning involves using a discourse in order to participate in this community (cf. Northedge, 2003).

From this description of discourse analysis it can be seen that this methodology is not limited to studies of student reading and writing, as might have previously been assumed, but has wide applicability to a range of research questions in engineering education. A central question concerns the role that language plays in student learning as well as the role it plays in social interaction. Importantly, a discourse perspective reminds us that the activities of academic discourse are never neutral and can pose particular difficulties when they clash with other discourses in which the student is engaged; learning a new discourse involves taking on a new identity (Gee, 2001).

Exemplar article:
Kittleson, J. M., & Southerland, S. A. (2004). The role of discourse in group knowledge construction: A case study of engineering students. *Journal of Research in Science Teaching*, 41(3), 267–293.

This article reports on a study of mechanical engineering students in a capstone design course at a large research university in the United States. This course had been recently reworked explicitly to combine aspects of numerical simulation and experimentation, in response to industry concerns. Engineering design requires an engagement with the underlying physical phenomena relevant to the project, and thus a key part of the group design process involved grappling collaboratively with these concepts. This study focused specifically on this aspect of the design project, termed here "concept negotiation." Another important dimension of this study was in understanding what really happens in project-based pedagogies, which are increasingly popular in engineering education.

For the study the researchers focused on one group of students in depth. The data collected for this study consisted of transcripts of the laboratory sessions where the group was working on its design project, semistructured interviews with the students, and field notes from participant observations of the class.

The research findings of this study demonstrate how this methodology revealed particular aspects that other methodologies might not have been able to surface. From a detailed analysis of transcripts of the group processes in action, it was found that instances of concept negotiation were, in fact, surprisingly rare. For example, in a typically laboratory session with a duration of 86 minutes and 41 seconds, only 7 minutes and 43 seconds was identified as concept negotiation. The majority of the time was

taken up with what they termed "off-task talk," "administrative talk," and "procedural talk," the latter dealing with the mechanics of the task, for example, on how to set up an experiment.

The study also identified the discourses at play that led to such a limited use of concept negotiation in the group. It was found that, despite the lecturers attempting to promote collaborative group work, students held a set of beliefs that seemed to focus on using group work for maximum efficiency and therefore dividing up work among the different group members and not working collaboratively. These discourses were directly related to students' views of what it was to be an engineer, standing in direct contradiction to the intentions of this teaching innovation and thus severely constraining the kind of learning that could take place in this course.

Narrative Analysis

Narrative analysis methodology originated in the field of literature studies, but has found applicability across a range of social science fields including education. Narrative methodology is focused on investigating the way people experience life. In fact, building on the work of John Dewey, which points to the deep interrelationships between experience, education, and life, some narrative researchers have argued that narrative methodology is particularly applicable to education research: "In its most general sense, when one asks what it means to study education, the answer is to study experience" (Clandinin & Connelly, 1998, p. 154).

What do we mean by the term "narrative"? Oliveira (2005) makes the important point that not all pieces of prose are automatically narratives. Hinchman and Hinchman (1997, p. xvi) define narratives as "discourses with a clear sequential order that connect events in a meaningful way for a definite audience, and thus offer insights about the world and/or people's experiences of it." An influential education scholar, Jerome Bruner (1986), has argued that "narrative cognition" is a particular form of human knowledge, distinct from what he terms "logical-scientific" knowledge. Telling stories is a fundamental human activity, a means by which we represent ourselves to others and make sense of our lives. Narrative methodology focuses on collecting and analyzing these stories to understand human experience.

Polkinghorne (1995) describes two modes of analysis that can be used within a narrative methodology. The more usual form of analysis he terms a "paradigmatic" analysis where the researcher attempts to identify common themes across the various narratives that have been collected as data. This has links to the method of constant comparison associated with grounded theory analysis mentioned previously. The second approach is termed "narrative analysis," where each narrative is considered on its own merits. A key activity in this approach is in organizing the various data elements present into a coherent account of that person's development.

Exemplar article:
Walker, M. (2001). Engineering identities. *British Journal of Sociology of Education*, 22(1), 75–89.

In this study, Walker seeks to address the question of why the participation of women in engineering programs remains relatively low, even though there have been dramatic shifts in society with regard to gender roles, including an equalizing of educational performance especially at the school level.

In order to address this question, Walker needed to explore the way students, both male and female, are experiencing engineering studies, and the identities that they develop in this regard. A narrative methodology was thus chosen, and she interviewed a small group of electrical engineering students at a research-intensive university in the United Kingdom. The interviewees included both male and female students, at a range of academic levels including undergraduate, postgraduate, and postdoctoral studies. The postdoctoral students were interviewed individually but the other students were interviewed in groups,

in order to allow for students to feel more comfortable with the interviewer, and also for the generation of richer material as they interacted with each other. In line with the narrative methodology the interviews were relatively unstructured and unrushed – mostly two hours in length – in order to obtain in-depth accounts of experience. In justifying her choice of a narrative form of methodology, Walker argues that "narrative interviews . . . expand and stretch complexity rather than reducing it, as generalizations are challenged by particular accounts" (p. 76).

A key finding of this study was that in narrating their experiences, female engineering students emphasized their difference from their other female peers, and projected identities that associated themselves with their male engineering peers. These can be termed "resistant identities," but Walker's analysis goes on to show that underneath the surface these in fact do not challenge the norm and that dominant stereotypes of gender roles remain intact. Thus, the female engineering students are seen by their male peers as more hard-working and organized rather than as more academically talented. Male students can be seen to make use of the nurturing roles taken on by women in the way group work is structured. This situation is played out further when students apply for work, with female students tending to underestimate their skills and male students doing the opposite. Walker also notes the way in which these dominant notions of masculinity and femininity within engineering education tend to constrain the opportunities for those male students who would prefer not to follow their traditional stereotype: A particular instance is given of a male postdoctoral student who took on a caring role with regard to his wife who had severe health problems; his supervisor was not able to accommodate this within his expectations of how a student's personal life should interact with his studies.

This brief outline of some of the key findings in this study demonstrates that the narrative methodology allowed for a research approach that was able to go beneath the surface of what is happening in engineering education: this analysis provides a compelling understanding for why the numbers of women in engineering program remains limited. For as long as the context requires that women need to take on these resistant identities it will only be a small group of women who will opt for this choice.

Discussion and Conclusion

Good research is not about good methods as much as it is about good thinking. (Stake, 1995, p. 19)

This chapter aimed to promote a more explicit engagement with methodological issues in engineering education research. Methodology refers not just to the methods of data collection and analysis that are used, but also to a theoretical justification for the use of these methods and the kinds of knowledge that they are able to generate. The chapter has outlined seven methodologies that were deemed to be "emerging" in the context of engineering education research. Case study is a methodology that argues for the significance of knowledge generated in particular contexts. It underpins much research in the interpretive and critical theoretical perspectives and is frequently combined with other methodologies. Grounded theory focuses on the generation of knowledge from empirical data without the use of a priori theory. Ethnography favors long-term engagement with the social context under investigation and aims for the generation of "rich descriptions" of the lives of the research participants. Although, in its pure form in education research, ethnography is employed in very specialized investigations, aspects of this methodology have been productively applied to a range of research contexts. Action research is aimed toward the improvement of practice, characterized by engaging the main participants in the research as co-researchers active in the research design, implementation, and analysis. Phenomenography seeks to uncover the different ways in which a phenomenon is experienced by people in a particular context and is well established in research on student learning in higher education. Discourse analysis and narrative analysis are both forms of linguistic analysis but focus their attention on different kinds of "texts." Discourse analysis is focused on instances of socially situated communication and seeks to link these to the underlying cultural ideas that they represent. Narrative analysis centers on the "stories" that people generate as they seek to make sense of their experiences.

The exemplar articles featured in this chapter have been used to provide concrete illustrations of how these methodologies have been used. Collectively the exemplar articles demonstrate a number of important points. First, as has been noted, a number of studies explicitly describe themselves as drawing on more than one methodology. Thus, the study by Jonassen et al. (2006) on the nature of engineering workplace problems, which was used to exemplify grounded theory, also made reference to narrative methodology to support the data collection of engineer's "stories." Moreover, it might be noted that in different ways, all of these exemplar studies draw on case study methodology, in that in-depth data is collected from particular contexts. Kittleson and Southerland (2004) focus on one group of engineering students working on a design problem. Walker (2001) interviews a small number of engineering students at one university. Both Booth (2001) and Jørgensen and Kofoed (2007) focus on a particular program intervention. Borrego et al. (2009) note the prominence of classroom-based studies in engineering education; the study by Magin and Churches (1995) featured as an exemplar of case study research is a useful example of how explicitly engaging with methodology can allow researchers to generate useful knowledge from classroom innovations, in this case the incorporation of peer tutoring in an engineering design course.

We have suggested that in expanding our methodological range we might be able to expand the kinds of research questions that can be addressed in engineering education research. The exemplar articles point prominently to some of these new directions.

For example, we need to understand in more depth what actually takes place in real engineering classrooms, particularly when we are implementing innovative pedagogies. Kittleson and Southerland (2004) used discourse analysis to investigate what happens when students work on a team assignment. Booth (2001), using phenomenography, is able to provide an explanation of why a program reform in computer engineering is not achieving its desired ends. Magin and Churches (1995) are able to look closely at peer tutoring and what it might achieve. With this broader range of methodologies we are also able to get deeper into some of the ongoing challenges in engineering education regarding student retention and diversity. Using narrative methodology, Walker (2001) is able to get closely into the reasons why the numbers of women in engineering remain stubbornly low. Stevens et al. (2008) provide a compelling understanding of the intertwined challenges that face engineering students and the ways in which institutions can facilitate or hinder their growing identification with the discipline. Jørgensen and Kofoed (2007) use action research to engage students in a continuous innovation and improvement process in their own program. Finally, we need to be able to look to the demands of the twenty-first century engineering workplace, and Jonassen et al. (2006) show how grounded theory methodology can be used to provide in-depth and contextual answers to these research questions.

The aim of this chapter has been to open further the debate about the range and kind of methodologies we use in engineering education research. There is an expectation that moving in this direction should yield significant achievements in the range of the research questions and findings that can be generated in the field.

References

Booth, S. (2001). Learning computer science and engineering in context. *Computer Science Education*, 11(3), 169–188.

Borrego, M., Douglas, E., & Amelink, C. (2009). Quantitative, qualitative, and mixed research methods in engineering education. *Journal of Engineering Education*, 98(1), 53.

Bruner, J. S. (1986). *Actual minds, possible worlds*. Cambridge, MA: Harvard University Press.

Bryman, A. (2001). Introduction. In A. Bryman (Ed.), *Ethnography*. London: SAGE.

Burton, L. (2002). Methodology and methods in mathematics education research: Where is "The Why"? In S. Goodchild & L. English (Eds.), *Researching mathematics classrooms: A critical examination of methodology* (pp. 1–10). Westport, CT: Praeger.

Clandinin, D. J., & Connelly, F. M. (1998). Personal experience methods. In N. K. Denzin & Y. S. Lincoln (Eds.), *Collecting and interpreting qualitative materials*. Thousand Oaks, CA: SAGE.

Clough, P., & Nutbrown, C. (2002). *A student's guide to methodology*. London: SAGE.

Cousin, G. (2009). *Researching learning in higher education: An introduction to contemporary methods and approaches*. New York, NY: Routledge.

Fairclough, N. (Ed.). (2003). *Analysing discourse: Textual analysis for social research*. London: Routledge.

Flyvbjerg, B. (2001). *Making social science matter*. Cambridge: Cambridge University Press.

Gee, J. P. (2001). Identity as an analytic lens for research in education. *Review of Research in Education*, 25, 99–125.

Gee, J. P. (2005). *An introduction to discourse analysis: Theory and method* (2nd ed.). London: Routledge.

Glaser, B., & Strauss, A. (1967). *The discovery of grounded theory: Strategies for qualitative research*. Chicago, IL: Aldine.

Hammersley, M., & Atkinson, P. (2007). *Ethnography: Principles in practice*. London: Routledge.

Hicks, D. (1995). Discourse, learning, and teaching. *Review of Research in Education*, 21, 49–95.

Hinchman, L. P., & Hinchman, S. (1997). *Memory, identity, community: The idea of narrative in the human sciences*. Albany, NY: State University of New York Press.

Jonassen, D., Strobel, J., & Lee, C. B. (2006). Everyday problem solving in engineering: Lessons for engineering educators. *Journal of Engineering Education*, 95(2), 139–151.

Jørgensen, F., & Kofoed, L. B. (2007). Integrating the development of continuous improvement and innovation capabilities into engineering education. *European Journal of Engineering Education*, 32(2), 181–191.

Kember, D. (2000). *Action learning and action research: Improving the quality of teaching and learning*. London: Routledge.

Kemmis, S., & McTaggart, R. (1988). *The action research planner* (3rd ed.). Geelong, Victoria.

Kittleson, J. M., & Southerland, S. A. (2004). The role of discourse in group knowledge construction: A case study of engineering students. *Journal of Research in Science Teaching*, 41(3), 267–293.

Koro-Ljungberg, M., & Douglas, E. (2008). State of qualitative research in engineering education: Meta-analysis of JEE articles, 2005–2006. *Journal of Engineering Education*, 97(2), 163.

Magin, D. J., & Churches, A. E. (1995). Peer tutoring in Engineering Design: a case study. *Studies in Higher Education*, 20(1), 73–85.

Marton, F. (1989). Towards a pedagogy of content. *Educational Psychologist*, 24(1), 1–23.

Marton, F., & Booth, S. (1997). *Learning and awareness*. Mahwah, NJ: Lawrence Erlbaum.

Micari, M., Light, G., Calkins, S., & Streitwieser, B. (2007). Assessment beyond performance: Phenomenography in educational evaluation. *American Journal of Evaluation*, 28(4), 458.

Northedge, A. (2003). Enabling participation in academic discourse. *Teaching in Higher Education*, 8(2), 169–180.

Oliveira, M. C. (2005). Review of "Using narrative in social research: Qualitative and quantitative approaches." *Narrative Inquiry*, 15(2), 421–429.

Polkinghorne, D. (1995). Narrative configuration in qualitative analysis. In J. A. Hatch & R. Wisniewski (Eds.), *Life history and narrative: Influences of feminism and culture* (pp. 5–23). London: Falmer.

Shepard, J., & Greene, R. W. (2003). *Sociology and you*. Columbus, OH: Glencoe/McGraw-Hill.

Stake, R. E. (1995). *The art of case study research*. Thousand Oaks, CA: SAGE.

Stevens, R., O'Connor, K., Garrison, L., Jocuns, A., & Amos, D. (2008). Becoming an engineer: Toward a three dimensional view of engineering learning. *Journal of Engineering Education*, 97(3), 355–368.

Strauss, A., & Corbin, J. (1994). Grounded theory methodology: An overview. In N. K. Denzin & Y. S. Lincoln (Eds.), *Handbook of qualitative research* (pp. 273–285). Thousand Oaks, CA: SAGE.

Van Manen, M. (1990). *Researching lived experience: Human science for an action sensitive pedagogy*. Albany, NY: State University of New York Press.

Walker, M. (2001). Engineering identities. *British Journal of Sociology of Education*, 22(1), 75–89.

CHAPTER 28

Conducting Interpretive Research in Engineering Education Using Qualitative and Ethnographic Methods

Aditya Johri

Introduction

This chapter provides an overview of conducting interpretive research using qualitative and ethnographic methods. The chapter complements Chapters 27 by Case and Light and 26 by Moskal, Reed-Rhoades, and Strong in this volume, which respectively provide an overview of the different qualitative and mixed method approaches to engineering education research. The primary purpose of this chapter is to provide entry-level researchers with background information on how to conduct interpretive research. Because every researcher has his or her own "tricks of the trade," this chapter necessarily presents methods and approaches found useful by this author, and represents an individual perspective on interpretive research. The number of references has been limited to around fifty[1]

I will like to thank students in my Ethnographic and Qualitative Research Methods course and students from Foundations of Engineering Education (Fall 2012) for their feedback, in particular John Sangster, Vaishali Nandy, Bushra Chowdhury, and Stephanie Kusano.

to avoid overwhelming the target audience; this chapter is not intended as a comprehensive review of interpretive research or of qualitative and ethnographic methods.

Interpretive Research in the Human Sciences

Social scientists currently distinguish interpretive research from other research traditions in that the researcher is taken to be a critical part of the process. The primary goal of the research is to understand participants' meaning-making, with less emphasis on the typical flow of the scientific method followed in other research methodologies. This type of research requires an investigator to include his or her own perceptions, becoming more of a sense-maker than an objective reporter: Interpretive research is an umbrella term used to describe studies that endeavor to understand a community in terms of the actions and interaction of the participants, from their own perspectives" (Tobin, 1999, p. 487).

Although interpretive methods are particularly well suited to sociocultural or situated approaches to learning and education (see Chapter 2 by Newstetter and Svinicki, or Chapter 3 by Johri, Olds, and O'Connor in this volume), they are appropriate in all cases when a researcher wishes to account for participants' perspectives in understanding an issue. The interpretive approach requires researchers to construct meaning from the data; there is a "perspective" aspect inherent to the process. Interpretive research sacrifices an element of objectivity in order to obtain situated meaning-making; the researchers and participants form a partnership in the development of understanding. Although interpretive research is by its nature subjective, there are standards of practice followed to ensure the quality of information within the field. An interpretive researcher "seeks to learn through systematic activity" (Tobin, 1999, p. 488) and builds an understanding through a skeptical search for evidence. Put another way, "the knowledge gained from research is put to test so that there is a close link between what is learned and the evidence used to support claims made from a study" (Tobin, 1999, p. 488).

This idea of an investigator's perspective in interpretive research is in direct conflict with the positivist approach toward research, in which settings are defined and described using external categories. There are three areas in which the interpretive approach deviates from a positivist approach. The first area of deviation involves the focus on large trends and central tendencies often used by positivist and other approaches that are primarily quantitative in nature; the interpretive approach deliberately seeks a sample including a high degree of diversity, or variability. In interpretive research, the decisions on who can or should participate in the study can be made while the study is ongoing, as new data are collected and new participants are approached. So long as the investigators' understanding of an issue is advanced, corroborated, or contradicted, participants may be added to the study. Snowball sampling is one example of an approach that lends itself well to interpretive research. The second area of deviation between the types of research comes in the design phase of a research project. Highly quantitative approaches devote significant effort to specifying and articulating the hypotheses and propositions of the study and to ensuring that the data collected are able to address the issues identified. In interpretive research initial questions are often open ended, requiring less engagement with and analysis of the data during the active portion of the study. Rather than a linear, or a strict cyclical scientific model of research, interpretive research processes tend to include more fluidity. The third area of deviation between interpretive and quantitative research approaches is that the results are dependent on the frame of reference of the researcher. It is important for the researcher to not only signal their worldview, but also consciously to use this viewpoint in interpreting, acknowledging, and documenting this activity. Researchers can be "invisible" in quantitative methodologies, but they are an integral part of interpretive research activities.

The effort to understand meaning-making and expressing or representing can be fraught with complexity. One of the earliest and strongest proponents of the interpretive tradition, Clifford Geertz (1973), argued that an intepretivist's task is to clarify the intersubjective meanings that form the foundation of social reality, as expressed in the everyday life experiences of people. Becker (1998) argues that just expressing the meanings is not enough, that we need to go beyond reproducing what we observe, to abstract the information that can help answer questions. Becker (1998) takes issue with Geertz's notion of "thick description" and related argument that the intepretivist's job is to provide or articulate meaning-making among participants. Nelson Goodman (1978) argues that researchers are bound to have a perspective, a frame, a way of seeing or worldmaking that is unique to them, raising questions about the ability of any researcher to faithfully represent what is being observed. Any researcher is bound to relate "affectively" or emotionally with experiences of data collection regardless of their actual disposition. As a consequence, the meaning-making in any social phenomenon being observed is inevitably linked with the meaning the researcher brings to the situation. The myth of objective observation is also questionable given that observation and meaning-making are interconnected. This process, according to Wagenaar (2011), is actually that of reconstructing the aggregated collective understanding of a social phenomenon by taking into account the individual subjective meanings of participants. This sense-making by researchers is dependent on our ability to situate a particular human action into a larger context; the act of interpretation involves an understanding at multiple levels – individuals, groups, and larger organizational or institutional levels. Wagenaar (2011) argues that there are three problems with this process creating epistemological and methodological issues. The first is that investigators may hold misconceptions about their own beliefs and intentions. The second is that investigators may or may not have access to objective meanings, rules, concepts, and norms that are implicit in practice (which may not be explicitly stated). The last is that investigators may lack an understanding of specific aspects of a social organization that are tacit and embodied, requiring actual participation to observe. In reviewing the work of other interpretive researchers, Wagenaar (2011) argues that:

> With important nuances and differences, they all agree that (1) explicit understanding can only be partial, (2) meaning resides not in the individual experiences of actors but in larger social configurations of which the actor is an integral part, (3) meaning emerges from acting upon concrete situations, and (4) meaning, in the sense of our actions exhibiting sensefulness, is created dynamically in an ongoing interaction among actions and between actors and larger social configurations. (p. 51)

Wagenaar further prescribes two ways of understanding meaning that go beyond or overcome the problematic aspects of interpretivist approach outlined in the preceding text. The discursive approach to meaning is thoroughly perspectivist, in the sense that it takes for granted that all human knowledge is a "conditioned point of view"; it implies that there can be no veracity of our knowledge of the social work. In the dialogic approach, meaning emerges only in relation to an interpreter and meanings may change depending on the interpretation. As a result it is essential for an interpretive researcher not only to capture and expression of intersubjective meaning-making, but also to articulate this insight in the service of a research question.

To make interpretive research more acceptable and mainstream, Yanow (2006) argues that researchers need to make their methods more explicit, as well as elaborating on the advantages of interpretive research. Researchers using an interpretive methodology should avoid using the word "qualitative" to describe their research; interpretive research extends beyond qualitative methods and is more precisely characterized as uncovering participant's meaning-making. Qualitative methods can be, and often are, used for positivist research without any interpretive component; any research with a hypothesis is well served by qualitative methods. Overall, Yanow (2006) argues that interpretive methodology has become an umbrella term subsuming several different schools of thought, including those drawing, explicitly or implicitly, on phenomenology, hermeneutics, some Frankfurt School critical theory, symbolic interaction, and ethnomethodology, among others. Many of these ideas dovetail with late-nineteenth to early-twentieth century pragmatism and later-twentieth century feminist epistemology research methods and science studies (p. 7).

In spite of the differences between interpretive researcher methodology and the traditional positivist framework, there are there any similarities. Yanow (2006) argues that:

Despite disagreements on ontological and epistemological matters, scientists working out of interpretive presuppositions, speaking broadly, share in common with those working out of positivist ones the two central attributes of scientific practice (what it means to "do" science or to "be" scientific) named above: an attitude of doubt, and a procedural systematicity. Where they differ is how these are enacted. Interpretive scientists share the appreciation for the possible fallibility of human judgment ... They also have a different understanding of what it means to prosecute "rigor" in research[...]; yet interpretive research, following its own canons of practice, is not less systematic than positivist-informed research, which renders the work "methodical" in different ways from that prescribed in the steps of the 'scientific method.' (p. 9)

Another important distinction Yanow makes is that the social world cannot be understood by interpretivists in the same fashion that the natural world can be understood and documented by positivists. This has specific consequences for engineering education researchers, particularly in disciplinary engineering research. Yanow (2006) argues that there is an essential between the natural and social worlds:

Unlike (to the best of our present knowledge) rocks, animals, and atoms, humans make, communicate, interpret, share and contest meaning. We act: we have intentions about our actions; we interpret others' actions; we (attempt to) make sense of the world: We are meaning-making creatures. Our institutions, our policies, our language, our ceremonies are human creations, not objects independent of us. And so a human (or social) science needs to be able to address what is meaningful to people in the social situation under study. It is this focus on meaning, the implications of that focus, that the various interpretive methods share. (p. 9)

Now that we have discussed interpretive research briefly and have some conception of what it is, let us explore why engineering educators should care about it and how they can go about conducting interpretive research.

Why Interpretive Engineering Education Research?

Empirical issues in furthering engineering education are similar to those in all other human or social sciences. Educators, students, and others who are involved with engineering education are embedded within the same socioeconomic webs as others. The problems they face in teaching or in learning often have the same root causes as barriers to furthering education, or other activity, in other settings. Therefore, positive change can be accomplished only by undertaking research that provides an interpretive understanding of issues at hand – by trying to understand matters from the perspective of those involved it might be possible to bring about change. This does not mean, of course, that structural change is not possible or desired and is often attempted. For instance, accreditation agencies such as ABET Inc. often play a significant role in forcing structural change within engineering education. An interpretive investigation, though, can alert us to the actual impact of these proposed and attempted changes as opposed to the intentioned change. For instance, what does it mean to implement a certain criterion? Does it result in a change in learning or are changes made in curriculum to align with a criterion without any meaningful learning gains? If change does happen over time, how does that take place? These questions are important questions and are raised frequently by authors of the handbook chapters. Although numeric and quantitative data is extremely helpful to document issues at a larger scale, an in-depth interpretive examination is equally important in meaningfully understanding the issue at hand.

Even though interpretive research is exceptionally useful, there are several barriers to conducting interpretive research and it is prudent to discuss these. First, interpretive research, particularly using qualitative and ethnographic methods, can be very time consuming. All research, when done rigorously, takes time interpretive research owing to the exhaustive data collection needs and the need for triangulation often needs more time. Sense-making is involved at every step of the research process as each step can be uncertain. Second, interpretive research often requires significant manpower resources and training for conducting research can often take longer than many other positivist approaches. Further, unlike other approaches that can utilize secondary data, interpretive research, owing to its localized settings and localized information needed for interpretation, has to rely overwhelmingly on primary data. Secondary data, even if they are available, have to be used sparingly and judiciously although it can often play a central role in triangulation. Finally, interpretive research requires significant post-processing effort in terms of outlining a convincing case. In many positivist traditions the basic structure of research output – how findings and results are to be presented – is standardized. An interpretive research project depends as much on its presentation for making a convincing argument as it does on data collection and analysis. Given these seemingly significant barriers to interpretive research for many researchers the benefits far outweigh its disadvantages. If the purpose of research is to truly understand an issue with the intent of effecting useful change, some form of interpretive research is essential.

In the rest of this chapter I provide a "methods" guide that can help researchers conduct interpretive research.[2] The guide is necessarily my own interpretation of how to do this research but can be useful to beginning researchers who are getting their feet wet. As with any research, researchers over time find their preferred data collection procedures, protocols, and coding procedures. The intent of the rest of the chapter is to show one way forward. Given the many excellent texts available on the design of research studies (see Table 28.1 for a selection) I do not discuss that aspect here. I assume that the researcher wants to conduct interpretive research and is looking for tools to help collect the data and analyze and write it up. Because analysis and writing depend significantly on the quality of data collected and in the interpretive skills of the researcher that often develop over time, I emphasize data collection above everything else. With a weak data set, interpretation is tricky if not impossible.

Starting Out – First Days in the Field

Every act of investigation begins with some uncertain moments. In interpretive research, especially when qualitative and ethnographic methods are being employed, the first few days in the field are often the most worrisome for researchers. It is in these initial moments that rapport has to be established, the purpose of research explained to participants, and "first impressions" formed and projected. This is the time period when the "learning" is the highest for any research. The lack of familiarity with a setting opens up our senses and we notice and absorb more information than at any other time during the research study. Therefore, it is crucial to take meticulous and detailed notes during this time period. The first few days also provide the researcher the opportunity to set the terms of the data collection. When and what kinds of data will be collected? What affordances does the setting actually provide as compared to the idealized vision in the research plan? Are there some participants who are likely to become key informants and can help interpret what is going on? Do I maintain a distance or do I create an informal relationship with all participants? All these are important questions that any researcher has to address during the initial days. Some other issues that come up are how to take observations notes. Although many young researchers are content to take mental notes, there is no substitute for written notes, and in any setting the note-taking procedure has to be established as soon as possible. Will you use a computer? Will you take short breaks and finish up your notes in another location? Finally, what kind of note-taking do you plan to do? Detailed, extremely details, cues to things you would remember. All these norms have to be established from inception and make the

Table 28.1. Overview of Introductory Texts on Design of Research Studies

Text	Brief Overview of the Text
Eisner: *Qualitative Inquiry in Education*	A text about different "ways of seeing" the world around us; the importance of having non-numerical representations to think about critical subjects; focused primarily on education; the critical use of a sample of one to make sense.
Glesne: *Becoming Qualitative Researchers*	A great introductory text for first time qualitative researchers; earlier editions were co-written with Alan "Buddy" Peshkin; Briefly covers all aspects of the research process; in the tradition of qualitative inquiry; covers different traditions of qualitative inquiry and differences with the positivist tradition.
Marshall & Rossman: *Designing Qualitative Research*	Classic text for understanding the design of qualitative research studies; covers the idea of having a "conceptual framework"; uses several examples from actual studies; great chapter on the "how" of field study.
Patton: *Qualitative Research and Evaluation Methods*	A "handbook" of sort; covers almost all topics related to qualitative research; a great reference; discusses the conceptual issues, design and data collection, and analysis and reporting; many examples from actual studies including codebook samples.
Taylor & Bogdan: *Introduction to Qualitative Research Methods*	The subtitle says it all: "The search of meanings"; this text goes to the core of qualitative inquiry – we want to understand how our informants make meaning and then we use the data to make meaning on our own; two comprehensive sections – How to conduct the research and How to write it up.
LeCompte & Schensul: *Designing & Conducting Ethnographic Research*	Several books and a whole series of seven books on how to conduct ethnographic research (Ethnographer's Toolkit); their field notes book in the toolkit is one of the few books that discuss it in depth; they talk a great deal about quantitative data as well. These are good reference books but keep in mind that they are not uniquely about interpretive research.
Rossman & Rallis: *Learning in the Field*	Does a great job of talking about the issues that arise in the field through actual studies of three of their students; lots of useful examples and samples; covers almost all topics.

first days in the field critical to the overall success of the project.[3]

Observing Others in Action

Observations are one of the most crucial components of any field study. Although observations form the backbone of any research, including positivist experimental research, they are critical in interpretive studies conducted in the field because the opportunity to observe any setting at any given time does not repeat. Therefore, the ability to capture action in the field accurately is indispensable. Of course, any interpretive researcher will take issue with the term "accurately" in the preceding sentence as any act of capturing data through observations is an interpretive act as well. What to look at and what to note down are all interpretive act; even the placing of a video camera to record a setting is an interpretive act. Therefore, when I say accurately I

Table 28.2. How to Conduct Observations

Text	Brief Overview
Lofland & Lofland: *Analyzing Social Settings*	Situation and strategies. Situations: encounters, roles, groups, organizations. Strategies: manner, identity, place timing; strategies of avoidance; pay attention to action within context by specific people.
Rossman and Rallis: *Learning in the Field*	Why observe?: Context, tacit patterns, see patterns not communicated in interviews, direct experience; What to observe? Social system – formal/informal, events, longitudinal changes, rituals, ceremonies, interaction patterns, unplanned activities, artifacts in use.
Brandt: *Studying Behavior in Natural Settings*	Different ways of studying behavior in varied settings through instruments such as observations, surveys, interviews; narrative data: anecdotes, specimen records (a specific person is chosen), field notes, ecological descriptions; examples of activity recording sheets and other devices to capture observational records systematically; timeline-based data collection.
Schatzman & Strauss: *Field Research*	A practical guide with a lot of strategies; strategy for entering, getting organized, watching, listening, recording, analyzing, communicating the research; angle of observation – it is a relative activity.

mean in the context of the research question and the purposes of the research study. The observers' perception has to be aligned with the goal of the observation and the observer has to ensure that his or her bias does not shape the observations. The texts reviewed in Table 28.2 provide further detailed guidance on conducting field observations. The key to good observations is being rigorous and steady – observe all the time and observe as much as you can.

Taking Observation Field Notes

One of the most critical skills one needs to develop as an interpretive researcher is the ability to take good field notes. The definition of what comprises good field notes is hard to pin down but there are a few characteristics that distinguish them from field notes that are inept. First, field notes have to be comprehensive. The data that need to be captured will vary based on the purpose of the research study but it is important to note different aspects of the site, the participants, and the activities that occur. Over time, and with experience, a researcher develops some tricks to keep track of all the happenings and is able to filter out information that might be

redundant[4] but it is good practice, initially as well as later in the study, to capture as much as possible. In addition to being comprehensive, field notes have to be retroactively useful and therefore they need to capture critical information such as time, place, and participants. It is a good habit to start each field note session by first making a note of the date and the time. There are many formats that can be used to facilitate better note taking, and the texts reviewed in Table 28.3 provide an assortment of such tools.

Conducting Interviews

Learning from participants about their worldview and their settings is one of the cornerstones of good interpretive work. But learning from others is much more difficult than it appears. It starts with developing good listening skills, which is often difficult for researchers, particularly academics, who are prone to advocating their knowledge. Yet, the importance of being a patient and sincere learner cannot be underestimated in interpretive research. Listening comes in many forms. Some of it has to do with avoiding interrupting your participants but

Table 28.3. How to Write Observations Field Notes

Text	Brief Overview
Emerson, Fretz, & Shaw: *Writing Ethnographic Fieldnotes*	Discusses the importance of field notes and then lays out the different issues involved when you are in the field, then writing up the fieldnotes afterward, and then processing them; mnemonic devices and jottings in the field; creating scenes from the notes, writing commentaries and memos; trying to understand informants' meanings from the write-up; scenes involve description and dialogues (and characterization); sketches, episodes, tales.
Lofland & Lofland: *Analyzing Social Settings*	Chapter 5: Logging data; words and action; mental notes, jotted notes, full fieldnotes – write promptly, content and style; start thinking about the unit.

it also has to do with giving them time to reflect and respecting silence. Often less experienced interviewers rush to fill in any empty space within a dialogue but that can be a mistake. Participants can take time to articulate an answer, or sometime hearing their own voice forces them to reflect and reform their response. These moments have to be respected within any interview. Table 28.4 lists some classic texts on interviewing, Table 28.5 presents a list of useful questions culled from a couple of texts, and Table 28.6 presents a sample interview protocol.

Interviews are often done on an individual basis but sometime the opportunity arises to interview more than one person, and this can change the dynamics of the interviewing process. In some cases, focus group interviews with more than two participants can be undertaken and these have their own dynamics. Although they were developed as a technique by mass media researchers to find out the opinion of participants, they can be used by interpretive researchers to delineate nuances and differences among participants. To do so, it is important to manage the group dynamics to ensure that each participant gets the opportunity to advance their viewpoint and that "loud" members do not take over and lead to groupthink. One technique for avoiding this, as noted in the sample protocol in Table 28.7, is to ask participants to first jot down ideas in writing on a piece of paper to ensure that their views are represented in some form before the group discussion. The participants can

then be asked to read this list or the interviewer can collect and use them as prompts.

Analyzing Field Study Data

Data analysis forms the most complicated but the most rewarding part of the interpretive research process as the fundamental argument for the approach relies on the ability to provide an analysis that accounts for the meaning-making of the study participants. Therefore, in addition to reporting the activities observed and the researcher's interpretation of them, an interpretive analysis provides a meta-analysis from the participants' perspective. The analysis process varies across researchers and research groups and many researchers prefer to immerse themselves in the data for a long period of time and emerge with an overall conception or argument. Many researchers follow the traditional approach of coding the data based on broad categories or themes and use these to triangulate the data. The codes in interpretive research are inductive; they emerge from the data. Coding techniques vary by researcher and project, but certain steps are common. The first step involves going through the data and documenting common themes that emerge. This process is usually called "free" coding as the intent is to capture all the major ideas that are present in the data. The codes generated through this process are termed *in vivo* codes as they emerge from the data.

Table 28.4. How to Conduct Interviews

Text	Brief Overview of the Text and Its Main Message
Spradley: *The Ethnographic Interview*	The "classic" text about ethnographic interviewing; "what do my informants know about their culture that I can discover?"; elements of the interview; kinds of descriptive questions (grand tour); structural questions (examples of something); contrast questions (alike, different); relationship between different kinds of questions and their placement within the interview is critical; domain analysis; cultural themes.
Merton, Fiske, & Kendall: *Focused Interview*	A contrast to the ethnographic interview; the style that led to focus groups; comes out of media effects research. The criteria: range (of situations and responses), specificity (highly specific aspects), depth (different meanings and degree of involvement), and personal context (informants' own distinctive meanings). This method is good for asking different informants about the same situation or event; can be carried out with a group of informants.
McCracken: *The Long Interview*	Deliberately more focused and efficient than the ethnographic interview; makes use of open-ended questions but in a focused and intensive interviewing process; good if interview is the sole data collection methods being used; literature review is considered part of the "long interview" process; examining the self is part of the process; questionnaire construction is an essential step; a scheme for data analysis based on the prior stages of the process.
Seidman: *Interviewing as Qualitative Research*	"Understanding the experience of other people and the meaning they make of that experience." Three-interview series: focused life history (everything in light of the current topic); details of experience (reconstructed); reflection on the meaning (what sense does this make to you?); each interview lasts around 90 minutes. Listen more, talk less; avoid leading questions, ask open-ended questions; keep participants focused and ask for concrete details; storytelling.
Kvale: *InterViews*	Phenomenological aspect of qualitative interviewing; informants' life-world; working through ambiguity; use of artifacts in the interviewing process – reflecting on artifacts; interaction with artifacts; hermeneutical aspects of interviewing (interpretation of texts).
Weiss: *Learning from Strangers*	Good introductory text with lots of examples and practical know-how of the interview process; The importance and design of an interview guide; analysis of transcriptions with comments; issues with interviewing; difference between survey and qualitative interviewing; also talks about analysis and writing.
Gubrium, J. F., & Holstein, J. A. (Eds.): *Handbook of Interview Research*	Contains chapters by a lot of experts that focus on specific issues related to interviewing; covers survey and journalistic interviews; chapters by Charmaz on grounded theory and by Adler & Alder on the "reluctant respondent" are particularly useful.

Table 28.5. Interviewing Tips

	Spradley (1979)
Tips for building rapport in early interviews (pp. 81–82)	
Make repeated explanations of why you're interviewing them.	"As I said earlier, I'm interested in . . . "
Restate what informants say – but don't *reinterpret* what they say.	Reinterpreting is when you state in different words what the interviewee has said; restating is using the same words – selecting key phrases and words and restating them.
Don't ask for meaning, ask for use – is a feature of ethnographic interviews.	Instead of "What do you mean by that?" or "Why would you do that" ask questions like "Could you tell me what you would say to the . . . " or "What kinds of things would I hear you saying?"
Tips for discovering questions (pp. 84–85)	
Record questions you hear people asking during the course of everyday life.	
Ask directly about the kinds of questions used by informants and others.	"What is an interesting question about . . . .?" "What is a question to which the answer is . . . ?"
Create a hypothetical situation and then ask for questions.	"If I listened to waitresses talking among themselves at the beginning of an evening, what questions would I hear them ask each other?"
Tips on descriptive questions/types of descriptive questions (pp. 85–90)	
Make the question longer – you will get a longer response.	Instead of "Could you tell me what jail is like?" try "I've never been inside the jail before, so I don't have much of an idea what it's like. Could you kind of take me through the jail and tell me what it's like, what I would see if I went into the jail . . . ?"
Phrasing questions personally or culturally	Might elicit different information or be differently sensitive for informant. Persona: "Can you describe a typical evening for you at Brady's Bar?" Cultural: "Can you describe a typical evening for most cocktail waitresses at Brady's Bar?" or "How would most tramps refer to the jail?"
Grand tour questions	"Can you describe all the different tools and other equipment you use in farming?" "What are all the things that you do during the initiation ceremony for new members who join the fraternity?"
Typical grand tour questions	"Could you describe a *typical* day at the office?"
Specific grand tour questions	Take the most recent event or space most familiar to the respondent to help him or her remember: "Tell me about the *last* time you made a . . . " "Could you describe what happened at Brady's Bar last night . . . ?"
Guided grand tour questions	Make them actually walk you around.
Task-related grand tour question	Get them to perform tasks and talk aloud while doing them. "Could you play a game of backgammon and explain what you are doing?" [Similar to think-aloud method]
Mini-tour questions	Same as grand tour, but about smaller aspects of experience.
Example questions	More specific. "Can you give me an example of someone giving you a hard time?"
Experience questions	Asks for experiences in actual settings – *tend to elicit atypical events*. "You've probably had some interesting experiences in jail; can you recall any of them?"
Native-language questions	"How would you refer to the jail?" "How would you talk about . . . ?"

	Spradley (1979)
Hypothetical interaction questions	*Makes informants less likely to translate for your sake.* "If I were to sit in the back of your classroom, what kinds of things would I hear kids saying to each other?"
Typical-sentence question	"What are some sentences that use the term . . . ?"

	Glesne and Peshkin (1992)
Don't say final goodbyes to respondents – leave door open for a return (p. 64).	
Keep questions free of words, syntax, or idioms respondents won't understand (p. 67).	
Don't make questions loaded or leading (p. 67).	No "Don't you suspect . . ." or "Isn't it the case that . . ."
Pilot interview questions with member of actual group you're studying (p. 68).	
Don't ask questions about "hot topics" before establishing rapport (p. 69).	
Keep questions from getting too vague. Provide mood and props to help interviewees recall previous times/events (p. 69).	"I'd like to have you go back to a time in your. . . . "
Use quotations from another source (p. 69).	Allows you to attribute to someone else ideas that are provocative, but that you don't want them to think reflect your opinion.
"Soliciting Advice" Questions – gets more idealized answers than direct "what do you do" questions (p. 70).	"I'd like you to put yourself in the position of my adviser. I'm a brand new teaching, never taught here before . . . What advice would you give me . . . ?"
Vary the voice or subject of questions; gets you different information (p. 70).	"Do you . . ." vs. "Do teachers like you . . ." vs. "Do teachers in your school . . ." vs. "Do teachers in general . . . ?"
Use language that respondents understand (pp. 70–71).	"Before we go on to other questions, I'd like to be sure what word or words you use to describe the different kinds of kids . . ."
Put questions that are easy to answer at the beginning (p. 71).	Gets interviewees talking and reassures them that your questions are manageable.
Watch for interaction effects of questions (p.71).	You'll need to keep some questions far apart, because one would influence the answer to another.
"Magic wand" questions (p. 72).	"If you could change anything at Riverview High School in any way you wanted, what, if anything, would you change?"
Get the interviewee to help you fill out the picture (p. 72).	"What else should we have asked that we haven't asked?" "What have we overlooked?" "Have we underemphasized important things?" "Have we overemphasized unimportant things?'
Try to promote regularity of interviews (p. 73).	"Same time and place next week?"
Give respondents an idea of how often you'll want to interview them (p. 73) [if you are conducting a multiple interview study]	"At least two times, and maybe more, certainly no more than is comfortable for you. And you may – without providing any explanation – stop and particular session or all further sessions."
Take notes on interviews, in addition to tapes [recording] (p. 75).	Account of an interview should include "old questions requiring elaboration; questions already covered; where to begin next time; special circumstances that you feel affected the quality of the interview; reminders about anything that might prepare you for subsequent interviews; and identification data."

(*continued*)

Table 28.5 (*continued*)

Glesne and Peshkin (1992)

Watch for body language (p. 77).	
Useful culminating statements (p. 78).	"Here is the ground we covered today. I was pleased to learn about . . . Would it be okay for next time if we went back to this and that point . . . ?"
Stick to the time frame you laid out going in (p. 78).	
Although being "naïve" is good, watch out for asking "What do you mean" too often; it can sound like you are second guessing (p. 80).	
Communicate that you are listening, without expressing your own opinion (p. 84).	"That must have hurt" and "Mm hmmm," *not* "I'm with you" or probably not "A-freakin'-men!" either
Don't rush (p. 85).	If you communicate your satisfaction with your respondents' short-shrift replies, then you teach them how little it takes to satisfy you. Say "Tell me more," and your interviewees will learn to respond accordingly.
Silence is one way to probe (pp. 85–86).	But don't stretch it too long so it gets awkward.
"Uh huh, uh huh" and "yes, yes" are also probes because they encourage talking (p. 86).	
Longer, more direct probes (p. 86).	"I'm not sure I got that straight. Would you please run by again what . . . ?"
	Summary statement followed by . . . "Did I understand you correctly?"
	Summary statement followed by . . . "Is there anything more you'd like to add?"
Take responsibility for all questions not understood; reframe/recast the questions, but if still not understood, move on instead of pushing it and making respondent feel stupid (p. 88).	
The too-talkative interviewee – set it up ahead of time that you might stop the talk from time to time, "in the interest of getting things straight" (p. 90).	
Do not point out contradictions during the interview (p. 90).	
To deal with contradicting statements, in a subsequent interview, you can take the two statements and put them in the mouths of other people and ask interviewee to respond (p. 90).	"I heard some people say . . . I've heard other people say . . . What is your thinking on these two positions? . . . Is it possible that both are right? Is there a third or even fourth position?"
Leave time at the end to just chat and thank the interviewee formally (p. 91).	

As a next step, codes are grouped together based on commonalities and tested extensively against the entire data or a specific corpus of data. This step is often called focused coding. In some cases, categories are related to subcategories and relationship between codes is examined. This step is termed axial coding and its usefulness lies in bringing together the disparate part of the study to provide coherence to the overall study. Initial coding often leads to viewing the data from too many different perspectives, and axial coding is a mechanism to provide an overarching analytical framework that groups things together. In interpretive research one has to be open to changing the

Table 28.6. Sample Interview Guide and Protocol

Date and Time: _____   Location: _____
Respondent: _____   Title: _____
Interviewer: _____

Guidelines for Introduction (5–10 minutes)
1. Thank informant for interview and explain project. The goal of this project is to understand how instructors teach, their thoughts about improving teaching, if any, and challenges instructors face in improving their teaching. We want to learn about *your* experiences. We'll be interviewing you and if possible conducting an observation of your teaching.
2. Reinforce confidentiality: the identity of the informants will be kept confidential. In addition, the findings of the study will be reported in a way that preserves the confidentiality of any private information provided.
3. The interview will be pretty open-ended and generally takes about an hour. (Ask interviewees if they have any "hard" constraints on time.)
4. Ask informant if he/she has any questions for you before starting.
5. Recording: "I usually transcribe all the interviews. Do you mind if I record our conversation? It will help me focus on the interview and keeps me from having to write everything down verbatim" (turn on the recorder).
6. Give them the consent form, explaining that it describes their rights as a study participant. As them to sign it as long as they feel comfortable participating. Also, ask them to initial that audio recording is okay.

Sample questions and probes from an interview protocol developed for understanding engineering work practices
1. Please describe your work.
   Alternatively or additionally, ask
   Tell me about your day last [Tuesday, Wednesday . . . choose a recent day].
Probes:
If teaching (or research) named predominantly, probe about the other ["you mentioned . . . , what about . . . "]:
- Can you tell me about the <teaching> or <research>?
- Can you tell me about the *relationship* between the two?
Additional probes:
- Tell me about *your* day last [Tuesday, Wednesday . . . choose a random, recent day].

2. Since you've been on this job/position, what kinds of changes have you seen?
Probes:
- How has your work changed?
- Has the relationship between research and teaching changed? How?
- Describe any *critical events*.

3. Can you describe a time when these differences between the X's and the Y's are most apparent?
- Can you give me an example?
- Do you think that the distinction affects how you work? In what way?
- If you could change something, what would you change?

4. One of the advantages of working in academia is the relative freedom to make changes. What have changes have you made in your work?
Probes:
- Changes in teaching
- Changes in research
- Frequency of changes
- Incentives for changes
- Describe any *critical events*.

(continued)

Table 28.6 (*continued*)

5. Did you teach before joining your present position? If yes, what, if anything, did you learn that helped you when you joined this position?

6. If you could make some changes in your work or its situation, what would they be? We could talk about this in two ways: changes that are feasible and changes that would require magic, that reflect how things would be in an ideal world.

Probe: For example, one of the goals of this study is to identify changes in technology that might more effectively facilitate teaching. Are there any technical tools or functionality that you think would be helpful?

Before the end of interview, ask once again if informants have any questions for you. Thank them. Encourage them to send you observations that they subsequently have about their teaching or anything they think is interesting about their teaching.

General probes:
- Can you give me an example of (or tell me a story about) that?
- What was your experience of that?

categories – either by eliminating them or collapsing them. Often a category will have to be divided into more subcategories to capture all the nuances. Ultimately, what an interpretive article does not do is imply enumerate the categories and their instances within the data and report them. Although the instances of occurrence of a category are important, their significance for the participants is a more important element in interpretive research. Therefore coding, categories, and their instances can be an important preliminary step; they have to be followed by meaning-making by the researcher. The purpose of this is to find not some hidden "truth" but the most illustrative reasoning. Table 28.8 lists some texts that contain information on how to analyze field study – observation and interview – data. As with other examples, these texts are from the qualitative or ethnographic traditions but their insights are equally applicable

Table 28.7. Focus Group Protocol

- Turn on the recorder.
- Pass informed consent forms.
- Take introductions as ice breakers.

Discussion starter question: Take a few minutes to write down what comes to your mind when you think of this course, your role, your design, collaboration, etc. (list some positives and some negatives).

Start group discussion.

Let's hear from each of you [give everyone the opportunity to participate and ask about their motivation for . . .]

[If you fail to hear a topic or if there is a common topic you want to explore]

One thing I heard several people mentions is . . . .

One thing I am surprised no one mentioned is . . . .

Other useful questions:

When you think about _____ what comes to mind?

Can you describe your project/artifact?

What kinds of things have made _____ easier/harder for you?

One of the things we are especially interested in is _____? What can you tell us about that?

That's something we're definitely interested in hearing more about. What can you tell us about it?

What will they improve? What can be removed or avoided? What did they learn?

Table 28.8. How to Analyze Field Study Data

	What Does it Entail?	Examples	When to Use
Charmaz: *Grounded Theory*	Coding: initial (line-by-line); focused, axial, theoretical; memo writing – raising codes to a conceptual level; theoretical sampling and saturation (earlier version: constant comparative method, sorting through data).		
Becker: *Tricks of the Trade*	The idea of a concept: What is it that we are talking about? We define them and the definitions are shaped by our collection of cases; tricks to figure out the concepts you are dealing with.		
Strauss: *Qualitative Analysis for Social Scientists*	A slightly different treatment of grounded theory; examples of coding – open, selective; memos and memo writing; illustrations and examples of working with data; integrative diagrams (frameworks); presentation and writing.		
Miles & Huberman: *Qualitative Data Analysis Sourcebook*	Great reference resource with chapters on doing the analysis and visual presentation of data; how to use tables, diagrams, figures to make meaning of the data and also to present it to the reader.		
Strauss & Corbin: *Basics of Qualitative Research*	Coding		

to interpretive research. Two texts that provide extensive and substantial discussions of the coding process are Charmaz (2006) and Strauss and Corbin (1998).

Writing Up Interpretive Research

As with any research process, the final writing and publication, or presentation, of the research is the usual culmination of the process. Although the publication and presentation of any research, even scientific research in the natural sciences, is a process of "social construction," in interpretive research it is often hard to separate the writing from the research – it is an integral part of the process. Whether it is to be consumed as a research article, a monograph, or a full-length book, writing up interpretive research is an often difficult and time-consuming process. One of the difficulties of writing is the lack of standard models. Interpretive researchers follow different styles and the final writing can take on many forms. Even if one eliminates such forms as "autoethnographic" texts from the canon of regular interpretive writing, the variation among research papers and books is immense. Yet, there are some commonalities underlying all interpretive writings and the most prominent

is the attempt to make sense through the eyes of the participants. Interpretive writing also balances researcher interpretation, data snippets, and often theoretical arguments in equal proportion. There are some common caveats in these writings where authors often make the leap from observing behavior to reporting it as participants' interpretation. The link between what is seen and what it means is hard to establish and often missing. This is a common criticism of interpretive research – do we ever really know what others think? This is also a challenge not only to establish but also to present in the form of written interpretation. Authors often take the support of other forms of representations such as analytical figures, illustrations, pictures, and if possible, video data. Table 28.9 provides some references to texts that can aid in writing and that describe different ways of writing that might be applicable to interpretive research.

Conclusion

This chapter provides a guide on conducting interpretive research using ethnographic and qualitative methods. This approach is becoming increasingly common in engineering education research and can benefit

Table 28.9. How to Write Up Interpretive Research

Text	Brief Description of the Text
Wolcott: *Writing Up Qualitative Research*	The most useful book on writing up qualitative research that I have encountered; Wolcott is a great writer himself and most of his books on qualitative research are worth reading (and some of them have been quite controversial).
Becker: *Writing for Social Scientists*	Not a book about qualitative writing per se but still a great book to think about how to write any social science topic; talks about the importance of editing; great complement to *Tricks of the Trade*.
Golden-Biddle & Locke:*Composing Qualitative Research*	They take examples of great qualitative journal articles, and through conversation with authors and documents of the reviewing process reconstruct how the arguments in the paper were constructed; the articles focus on organizational research but the process of constructing an argument, especially in the space offered by a journal article, is very well explained.
Van Mannen: *Tales of the Field*	A "classic" text about writing about ethnographic studies; the author outlines three different kinds of tales: realist (dispassionate, third-person voice), confessional (narrated through self), and impressionist (striking stories); the book starts with a great review of the "cultural" turn and where this piece fits in; gives examples from his own writing (and rewritings).
Goodall: *Writing the New Ethnography*	Represents the movement toward "auto-ethnography" and the engagement of self with the narrative; the trials and tribulations of an ethnographer who does not fit in the mainstream; issues of representation, voice, reflexivity.
Wolf: *A Thrice-Told Tale*	Anthropological critique of the "cultural" (and postmodern) turn; consists of three different texts that showcase the same set of events: a fictional short story, unanalyzed fieldnotes from observations and interviews, and a journal piece that appeared in *American Ethnologist*.

the coverall research agenda of the field by providing a more ecological valid look at why and how students persist in engineering, learning engineering, and institutional issues that shape engineering education.

Interpretive research can use many different approaches for data collection and analysis; Table 28.10 provides some examples of this diversity.

Finally, Table 28.11 recaps many of the issues that have to be considered in field study driven interpretive research. The list is broken down into four different sections: "Before the Field," "In the Field," "After the Field," and "Rest/All the Time." These are ideas based on my personal experiences, and the list can be useful if personalized by each researcher to meet his or her needs.

Table 28.10. Methods Appropriate for Interpretive Research

Linguistics	Heath: Language and social structure, linguistic analysis
Cognitive anthropology	Hutchins: Distributed cognition approach, learning in network of artifacts and people
Network analysis	Wasserman & Faust, Scott & Carrington: Network data, in combination with observations and interviews
Conversation analysis	Sacks & Silverman: Analysis of talk
Unobtrusive measures	Webb et al: Documents and other artifacts that are available in the setting
Interaction analysis	Jordon & Henderson: *Journal of Learning Sciences*
Ethnomethodology	Garfinkel, Sommerville et al.: Ethnography and system design

Table 28.11. Field Study Checklist

<hr>

Before entering the field

<hr>

Institutional Review Board application and consent form: Ensure you have the proper permissions and documentations to conduct the research and enough copies of the consent form for your participants.

Research questions: Ensure that the purpose of the study, in terms of research questions, is well articulated and is helpful in framing the data collection.

Access: Ensure you have access to the research site and have worked out details of entry and other logistics.

Site: Make sure you have done background research on the site and know its specifics – where it is, how to reach it, etc.

Logistics: Related to the site, prepare for logistical issues such as where to conduct the interview.

Proposals: If your research requires funding or support in some form and make sure you have done that background work.

Working hypothesis: Although interpretive research requires an open mind, it is often critical to have some working hypothesis to frame your work to support but more likely refute it.

Ethics: Think through the ethical implications of your work, especially how your presence in the field might impact them, particularly those who might be in a less powerful position.

Familiarity with site/Subjectivity: Do background research to build some form of familiarity with the site to be aid interpretation.

Technology (batteries): If you plan on using any technologies such as computers or recorders make sure they are in working condition and you have spare machines and batteries.

Interview/Observation protocol: Prepare the protocols and take copies of them with you to the field.

Make a checklist: Make your own checklist that you can use to quickly prepare before any trip to the field.

Format for filenames: Make a proper coding system/scheme for storing data.

In the field

Ethics/Subjectivity: Ethical issues become increasingly salient in the field given the numerous occasions that arise where ethics are tested. Any interaction with a participant or any opportunity to enter and observe an event or activity has potential ethical implications. Do you tell the people present who you are? How and when do you do that? How much information about your project do you divulge? If you tell them you are not going to record any information, do you completely stop data collection? Does your experience not even count towards your interpretation? Can it even be completely wiped out from your experience? These questions arise routinely and often the ethical implications come to individual subjectivity and doing what is right. When there is no IRB or another researcher to monitor, your own sense of ethics will take precedence and therefore to undertake ethical research more than rules, regulations, and training, one has to develop a self-sense of ethical doing.

First days: As discussed in this chapter in detail, navigating the first days in the field is often tricky and your impressions as a researcher can play a significant role in subsequent data collection. Therefore, it is important to map out the first days in the field as well as be extra vigilant in note taking. Introductions, including who you are and what you are doing at the site, are two important questions for which you must be prepared with credible and honest answers.

Memos: Memos are a useful mechanism to keep track of the field study as well as reflect on the data collection. Memos taken during the field study can also be important resources in data interpretation in later stages.

Observations: Obvious, but observing the right thing in the right amount is crucial for good quality data and even though "right," often can be assessed only post hoc; it helps to have a plan.

(*continued*)

Table 28.11 (*continued*)

Interviews: With interviews, it is important to remind yourself that the critical thing is using open-ended questions and respecting silences.

Unobtrusive methods/Archival data: Often, a research site will provide the opportunity to collect data through unobtrusive means, often archival data, and one has to be prepared to collect this data as well as make efforts to gain this data if possible. These data can be extremely useful in triangulating.

Rapport: Building rapport with informants is important and it requires building a safe environment and developing trust. The important thing to keep in mind is to not be devious but honest.

Field notes/Recording of data: Ensure data are being recorded in a reliable manner. Complete all field notes by the end of the day, especially if what you have from the field are rough jottings. Memory fails us often.

Discussion with colleagues: Sometimes, especially in large projects with multiple data collectors, it can be helpful to discuss your data collection with your colleagues. Of course, it is important to maintain participant privacy.

Rest/All of the time

Planning: Plans are helpful although things do and will change as the field study goes on. Before entering the field and even while you are in the field it is important to plan your data collection efforts with the goal of obtaining good and adequate data in mind.

Reading: Often, it helps to read other studies related to your current field study or research methods texts to keep your data collection fresh. You get new ideas and can often come with new ways to triangulating your data and interpretations.

Practice: Practicing your observation and interviewing skills is very important. Practice helps you smoothen out the rough edges but also lets you try out different ways of asking the same question.

Thinking: Thinking and reflecting about the data collection are important to ensure an in-depth analysis.

The politics: Every field sites has its politics and it is almost impossible to stay neutral but that should always be the first goal. Never pick sides. If in a difficult position of agreeing or disagreeing with someone, be honest, be consistent, and explain your stance.

After the field

Subjectivity: Interpretive research is subjective research and therefore your stance and framing of the problem and research should be laid out in your text.

Validity: The validity of your research, of any research, comes from doing a thorough job of the field study and being able to support your findings and arguments with proper evidence from your data.

Transcription: Using a professional transcriptionist or transcribing on your own often depends on the resources available and often on personal taste. The level of transcription also depends on the research question but verbatim transcriptions should be preferable to summaries.

Inductive/Deductive/Grounded theory/Cross-comparative: During analysis the process to follow depends on the objective but different options can include grounded theory or cross-comparative analysis and following an inductive approach overall.

Software: Which software to use for analysis, and whether or not to use software, is an important decision and the choice depends on several factors such as the team involved in analysis, the nature of data – interview transcripts and observations notes or other media, and cost considerations. Several popular options include NVivo™, InqScribe™, Atlasti™, and others.

Writing: Writing considerations involve how much to write, how to pace the writing, how to balance analysis and writing. During writing, tying data with theory is very important. Representing data in different ways to help with analysis and making findings accessible to others are other issues to consider.

Footnotes

1. One of the most comprehensive online resources for qualitative research methods is the bibliography maintained by the University of Georgia, Athens, GA, which contains list of references for almost all approaches to qualitative and ethnographic research: http://www.coe.uga.edu/leap/academic-programs/qualitative-research/qualitative-research-resources/bibliography/

2. I often use the term "ethnographically informed" to describe my data collection approach as it captures best the close involvement with the research site but is not an ethnography in the true sense of the word because the ethnographic tradition requires long-term engagement with the research site – extending several years. The ethnographically informed approach still follows many of principles and traditions outlined by ethnographers – in terms of the structure of interviews, the interpretive nature of the work, and so on – but deviates in other ways.

3. A useful reference is B. Geer, "First Days in the Field," in P. Hammond (Ed.), *Sociologists at Work* (New York, NY: Basic Books, 1964), pp. 322–344.

4. This is always problematic as researcher bias might be at play and what the researcher might not think is important may prove so later on.

References

Bailey, C. (1996/2006). *A guide to field research*. Pine Forge Press.

Becker, H. (1988). *Writing for social scientists*. Chicago, IL: University of Chicago Press.

Becker H. (1998). *Tricks of the trade: How to think about your research while you're doing it*. Chicago, IL: University of Chicago Press.

Brandt, R. M. (1972). *Studying behavior in natural settings*. New York: Holt, Rinehart and Winston.

Charmaz, K. (2006). *Constructing grounded theory*. London: SAGE.

Creswell, J. W. (1998). *Qualitative inquiry and research design: Choosing among five traditions*. Thousand Oaks, CA: SAGE.

Denzin, N., & Lincoln, Y. (2000). *Handbook of qualitative research*. Thousand Oaks, CA: SAGE.

Eisner, E. W. (1991). *The enlightened eye: Qualitative inquiry and the enhancement of educational practice*. New York, NY: Macmillan.

Eisner, E. W., & Peshkin, A. (1990). *Qualitative inquiry in education: The continuing debate*. New York, NY: Teachers College Press.

Emerson, R. M., Fretz, R. I., & Shaw, L. L. (1995). *Writing ethnographic fieldnotes*. Chicago, IL: University of Chicago Press.

Emerson, R. M., Fretz, R. I., & Shaw, L. L. (2001). Participant observation and fieldnotes. In P. Atkinson, A. Coffey, S. Delamont, J. Lofland, & L. Lofland (Eds.), *Handbook of ethnography* (pp. 352–368). London: SAGE.

Geertz, C. (1973). *The interpretation of cultures: Selected essays*. New York, NY: Basic Books.

Glaser, B., & Strauss, A. (1967). *The discovery of grounded theory: Strategies for qualitative research*. New York, NY: Aldine De Gruyter.

Glesne, C., & Peshkin, A. (1992). *Becoming qualitative researchers*. White Plains, NY: Longman.

Golden-Biddle, K & Locke, K. (2006). *Composing qualitative research: Crafting theoretical points from qualitative research* (2nd ed.). Thousand Oaks, CA: SAGE.

Goodall, H. (2000). *Writing the new ethnography*. Lanham, MD: AltaMira Press.

Goodman, N. (1978). *Ways of worldmaking*. Cambridge, MA: Hackett.

Garfinkel, H. (1967). *Studies in ethnomethodology*. NJ: Englewood Cliffs.

Gubrium, J. F., & Holstein, J. A. (Eds.). (2002). *Handbook of interview research: Context and method*. Sage Publications.

Heath, S. B. (1983). *Ways with words: Language, life and work in communities and classrooms*. Cambridge University Press.

Heath, S. B. (1986). *Sociocultural contexts of language development*. Beyond language: Social and cultural factors in schooling language minority students, 143–186.

Hutchins, E. (1995). *Cognition in the wild*. Cambridge, MA: MIT press.

Jordan, B., & Henderson, A. (1995). Interaction analysis: Foundations and practice. The Journal of the Learning Sciences, 4(1), 39–103.

Kvale, S. (1996). *InterViews: An introduction to qualitative research interviewing*. Thousand Oaks, CA: SAGE.

LeCompte, M. D., & Schensul, J. J. (2010). *Designing & conducting ethnographic research: An introduction* (2nd ed.). Lanham, MD: AltaMira Press.

Lofland, J., & Lofland, L. (2006). *Analyzing social settings*. Belmont, CA: Wadsworth.

Martin, D., & Sommerville, I. (2004). Patterns of cooperative interaction: Linking ethnomethodology and design. *ACM Transactions on Computer-Human Interaction (TOCHI)*, *11*(1), 59–89.

Marshall, C., & Rossman, G. B. (1999). *Designing qualitative research* (3rd ed.). Thousand Oaks, CA: SAGE.

McCracken, G. (1988). *The long interview*. Newbury Park, CA: SAGE.

Merton, R. K., Fiske, M., & Kendall, P. L. (1990). *Focused interview: A manual of problems and procedures* (2nd ed.). New York, NY: The Free Press.

Miles, M. B., & Huberman, A. M. (1994). *Qualitative data analysis: An expanded sourcebook* (2nd ed.). Thousand Oaks, CA: SAGE.

Morgan, D. L. (1997). *Focus group as qualitative research*. Newbury Park, CA: SAGE.

Patton, M. Q. (2001). *Qualitative research & evaluation methods* (3rd ed.). Thousand Oaks, CA: SAGE.

Rossman, G. B., & Rallis, S. F. (2003). *Learning in the field* (2nd ed.). Thousand Oaks, CA: SAGE.

Rubin, H. J., & Rubin, I. S. (1998). *Qualitative interviewing: The act of hearing data*. Thousand Oaks, CA: SAGE.

Sacks, H. (1984). Notes on methodology. *Structures of social action: Studies in conversation analysis*, 21–27.

Scott, J., & Carrington, P. J. (Eds.). (2011). *The SAGE handbook of social network analysis*. SAGE publications.

Schatzman, L., & Strauss, A. L. (1972). *Field research: Strategies for a natural sociology*. Upper Saddle River, NJ: Pearson.

Seidman, I. (2006). *Interviewing as qualitative research: A guide for researchers in education and the social sciences* (3rd ed.). New York, NY and London: Teachers College Press.

Silverman, D. (1998). *Harvey Sacks: Social science and conversation analysis*. Oxford University Press.

Spradley, J. (1979). *The ethnographic interview*. Fort Worth, TX: Harcourt Brace Jovanovich.

Spradley, J. P. (1980). *Participant observation*. New York, NY: Holt, Rinehart & Winston.

Stake, R. E. (1995). *The art of case study research: Perspectives on practice*. Thousand Oaks, CA: SAGE.

Strauss, A. (1987). *Qualitative analysis for social scientists*. New York, NY: Cambridge University Press.

Strauss, A., & Corbin, J. (1998). *Basics of qualitative research: Grounded theory procedures and techniques* (2nd ed.). London: SAGE.

Taylor, S. J., & Bogdan, R. (1984). *Introduction to qualitative research methods*. New York, NY: John Wiley & Sons.

Tobin, K. (1999). Interpretive research in science education. In A. E. Kelly & R. Lesh (Eds.), *Handbook of research design in mathematics and science education* (pp. 487–512). Mahwah, NJ: Lawrence Erlbaum.

Wasserman, S., & Faust, K. (1994). *Social network analysis: Methods and applications*. Cambridge University Press.

Webb, E. J., Campbell, D. T., Schwartz, R. D., & Sechrest, L. (2000). *Unobtrusive measures* (Rev. ed.). Thousand Oaks, CA: SAGE.

Wagenaar, H. (2011). *Meaning in action: Interpretation and dialogue in policy analysis*. Armonk, NY: M. E. Sharpe.

Weiss, R. S. (1994). *Learning from strangers: The art and method of qualitative interview studies*. New York: The Free Press.

Wolcott, H. F. (1995). *The art of fieldwork*. Walnut Creek, CA: Altamire Press.

Wolcott, H. (2001). *Writing up qualitative research* (2nd ed.). Thousand Oaks, CA: SAGE.

Wolf, M. (1992). *A thrice-told tale: Feminism, postmodernism, and ethnographic responsibility*. Stanford, CA: Stanford University Press.

Van Maanen, J. (1988). *Tales of the field: On writing ethnography*. Chicago, IL: University of Chicago Press.

Yanow, D. (2006). Thinking interpretively. In D. Yanow & P. Schwartz-Shea (Eds.), *Interpretation and method: Empirical research methods and the interpretive turn* (pp. 5–27). Armonk, NY: M. E. Sharpe.

Yin, R. K. (1994). *Case study research: Design and methods*. Thousand Oaks, CA: SAGE.

CHAPTER 29

The Science and Design of Assessment in Engineering Education

James W. Pellegrino, Louis V. DiBello, and Sean P. Brophy

Chapter Overview and Goals

In 2001, a report was issued by the National Research Council (NRC) entitled "*Knowing What Students Know: The Science and Design of Educational Assessment*" (Pellegrino, Chudowsky, & Glaser, 2001). The goal was to evaluate the state of research and theory on educational assessment and establish the scientific foundations for their design and use. As argued in that volume, many of the debates that surround educational assessment emanate from a failure to understand its fundamental nature, including the ways in which theories and models (1) of learning and knowing and (2) of measurement and statistical inference interact with and influence processes of assessment design, use, and interpretation. In this chapter we review some of the key issues regarding educational assessment raised in that report as well as examples from science, technology, engineering, and mathematics (STEM) education and educational research contexts. Our goal in explicating current understanding of the science and design of educational assessment and its applications to STEM

education is to provide background knowledge that supports the effective design and use of assessment in engineering education and sharpens the focus of engineering education R&D.

In the first section we briefly introduce some key ideas critical to understanding educational assessment. This includes consideration of formal and informal uses of assessment and some conceptual issues associated with assessment design, interpretation, and use. In the second section we discuss three related conceptual frameworks about assessment that should be considered by anyone using assessment for instructional or research purposes. These include (1) assessment as a process of reasoning from evidence, (2) the use of an evidence-centered design process to develop and interpret assessments, and (3) centrality of the concept of validity in the design, use, and interpretation of any assessment. In the third section we turn to a discussion of concepts of measurement as applied to assessment and the role of statistical inference. This includes the assumptions underlying different types of psychometric models used to estimate student proficiency. The fourth section then presents applications of key ideas from the preceding sections in the form of illustrative examples of assessments used in engineering education research. In the final section we close by briefly considering the significance of a careful and thoughtful approach to assessment design, use, and interpretation in engineering education research.

Educational Assessment in Context

Assessment Purposes and Contexts

From teachers' classroom quizzes, mid-term or final exams, to nationally and internationally administered standardized tests, assessments of students' knowledge and skills have become a ubiquitous part of the educational landscape. Assessments of school learning provide information to help educators, administrators, policymakers, students, parents, and researchers judge the state of student learning and make decisions about implications and actions. The specific purposes for which an assessment will be used are an important consideration in all phases of its design. For example, assessments used by instructors in classrooms to assist or monitor learning typically need to provide more detailed information than assessments whose results will be used by policymakers or accrediting agencies. One of the central points of the *Knowing What Students Know* report was that assessments are developed for specific purposes, and the nature of their design is very much constrained by their intended interpretive use.

ASSESSMENT TO ASSIST LEARNING

In the classroom context, instructors use various forms of assessment to inform day-to-day and month-to-month decisions about next steps for instruction, to give students feedback about their progress, and to motivate students. One familiar type of classroom assessment is a teacher-made quiz, but assessment also includes more informal methods for determining how students are progressing in their learning, such as class-room projects, feedback from computer-assisted instruction, classroom observation, written work, homework, and conversations with and among students – all interpreted by the teacher in light of additional information about the students, the schooling context, and the content being studied.

These situations are referred to as *assessments to assist learning,* or the *formative use of assessment.* These assessments provide specific information about students' strengths and difficulties with learning. For example, statistics teachers need to know more than the fact that a student does not understand probability; they need to know the details of this misunderstanding, such as the student's tendency to confuse conditional and compound probability. Teachers can use information from these types of assessments to adapt their instruction to meet students' needs, which may be difficult to anticipate and are likely to vary from one student to another. Students can use this information to determine which skills and knowledge they need to study further and what adjustments in their thinking they need to make.

ASSESSMENT OF INDIVIDUAL ACHIEVEMENT

Another type of assessment used to make decisions about individuals is that conducted to help determine whether a student has attained a certain level of competency after completing a particular phase of education, whether it be a two-week curricular unit, a semester long course, or twelve years of schooling. This is referred to as *assessment of individual achievement,* or the *summative use of assessment.* Some of the most familiar forms of summative assessment are those used by classroom instructors, such as end-of-unit or end-of-course tests, which often are used to assign letter grades when a course is finished. Large-scale assessments – which are administered at the direction of users external to the classroom – also provide information about the attainment of individual students, as well as comparative information about how one individual performs relative to others. Because large-scale assessments are typically given only once a year and involve a time lag between testing and availability of results, the results seldom provide information that can be used to help teachers or students make day-to-day or month-to-month decisions about teaching and learning.

ASSESSMENT TO EVALUATE PROGRAMS

Another common purpose of assessment is to help administrators, policymakers, or researchers formulate judgments about the quality and effectiveness of educational programs and institutions. Instructional evaluation can be considered formative in nature when used to improve the effectiveness of instruction. Summative uses of assessment for evaluation are incorporated increasingly in making high-stakes decisions not only about individuals but also about programs and institutions. For instance, public reporting of state assessment results by school and district can influence the judgments of parents and taxpayers about the quality and efficacy of their schools and affect decisions about resource allocations. Just as with individuals, the quality of the measure is of critical importance in the validity of these decisions.

Further Considerations of Purposes, Levels, and Timescales

As noted previously, assessment occurs in multiple contexts, has a variety of formal and informal uses, and is conducted to meet different purposes. The purpose of an assessment determines priorities, and the context of use imposes constraints on the design. Thus it is essential to recognize that one type of assessment does not fit all purposes or contexts of use. In general, the more purposes a single assessment aims to serve, the more each purpose will be compromised and the overall product will represent a suboptimal design for each intended use. A persistent mistake is to assume that an assessment is appropriate and interpretable for a particular context of use without determining if there is evidence regarding the validity

of such assumptions (see later discussions of validity) within that context. The one-size-fits-all fallacy is especially frequent and problematic since it produces inappropriate choices of assessments for instructional or research purposes that in turn can lead to invalid conclusions regarding persons, programs, and/or institutions.

Although assessments are currently used for many purposes in the educational system, a premise of the *Knowing What Students Know* report is that their effectiveness and utility must ultimately be judged by the extent to which they promote student learning. The aim of assessment should be *"to educate and improve* student performance, not merely to *audit* it" (Wiggins, 1998, p. 7). Because assessments are developed for specific purposes, the nature of their design is very much constrained by their intended use. Although it may seem reasonable to dichotomize between *internal* classroom assessments, administered by instructors, and *external* tests, administered by districts, states, or nations or other agencies, such a dichotomy is an oversimplification of a continuum that reflects the proximity of an assessment to the enactment of specific instructional and learning activities. Ruiz-Primo, Shavelson, Hamilton, and Klein (2002) defined five discrete points on a continuum of assessment distance: *immediate* (e.g., observations or artifacts from the enactment of a specific instructional activity), *close* (e.g., embedded assessments and semiformal quizzes of learning from one or more activities), *proximal* (e.g., formal classroom exams of learning from a specific curriculum), *distal* (e.g., criterion-referenced achievement tests such as required by the federal No Child Left Behind legislation), and *remote* (broader outcomes measured over time, including norm-referenced achievement tests and some national and international achievement measures). Different assessments should be understood as different points on this continuum if they are to be aligned effectively with each other and with curriculum and instruction. In essence, an assessment is a test of transfer and it can be near or far transfer depending on where the assessment falls along the continuum noted earlier.

The level at which an assessment is intended to function, which involves varying distance in "space and time" from the enactment of instruction and learning, has implications for how and how well it can fulfill various functions of assessment, be they formative, summative, or program evaluation (NRC, 2003). As argued elsewhere (Hickey & Pellegrino, 2005; Pellegrino & Hickey, 2006), it is also the case that the different levels and functions of assessment can have varying degrees of match with theoretical stances about the nature of knowing and learning (see also Chapter 2 Newstetter & Svinicki in this volume).

Although assessments used in various contexts and for differing purposes and at different timescales often look quite different, they share certain common principles. One such principle is that assessment is always a process of reasoning from evidence. By its very nature, moreover, assessment is imprecise to some degree. Assessment results are only estimates of what a person knows and can do. We elaborate on both of these issues in the following two sections.

Conceptual Frameworks

Assessment as a Process of Evidentiary Reasoning: The Assessment Triangle

Educators assess students to learn about what they know and can do, but assessments do not offer a direct pipeline into a student's mind. Assessing educational outcomes is not as straightforward as measuring height or weight; the attributes to be measured are mental representations and processes that are not outwardly visible. Thus, an assessment is a tool designed to observe students' behavior and produce data that can be used to draw reasonable inferences about what students know. Deciding what to assess and how to do so is not as simple as it might appear.

The process of collecting evidence to support inferences about what students know represents a chain of reasoning from evidence about student learning that characterizes all assessments, from classroom quizzes and standardized achievement tests, to computerized tutoring programs, to the conversation a student has with his or her teacher as they work through a math problem or discuss the meaning of a text. People reason from evidence every day about any number of decisions, small and large. When leaving the house in the morning, for example, one does not know with certainty that it is going to rain, but may reasonably decide to take an umbrella on the basis of such evidence as the morning weather report and the threatening clouds in the sky.

The first question in the assessment reasoning process is "evidence about what?" *Data* become *evidence* in an analytic problem only when one has established their relevance to a conjecture being considered (Schum, 1987, p. 16). Data do not provide their own meaning; their value as evidence can arise only through some interpretational framework. What a person perceives visually, for example, depends not only on the data she receives as photons of light striking her retinas, but also on what she thinks she might see. In the present context, educational assessments provide data such as written essays, marks on answer sheets, presentations of projects, or students' explanations of their problem solutions. These data become evidence only with respect to conjectures about how students acquire knowledge and skill.

In the *Knowing What Students Know* report the process of reasoning from evidence was portrayed as a triad of three interconnected elements: the *assessment triangle*. The vertices of the assessment triangle (see Figure 29.1) represent the three key elements underlying any assessment: a model of student *cognition* and learning in the domain of the assessment; a set of assumptions and principles about the kinds of *observations* that will provide evidence of students' competencies; and an *interpretation* process for making sense of the evidence in light of the assessment purpose and student understanding. These three elements may be explicit

observation   interpretation

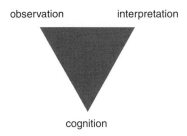

cognition

Figure 29.1. The assessment triangle.

or implicit, but an assessment cannot be designed and implemented, or evaluated, without consideration of each. The three are represented as vertices of a triangle because each is connected to and dependent on the other two. A major tenet of the *Knowing What Students Know* report is that for an assessment to be effective and valid, the three elements must be in synchrony. The assessment triangle provides a useful framework for analyzing the underpinnings of current assessments to determine how well they accomplish the goals we have in mind, as well as for designing future assessments and establishing validity (e.g., see Marion & Pellegrino, 2006).

The *cognition* corner of the triangle refers to theory, data, and a set of assumptions about how students represent knowledge and develop competence in a subject matter domain (e.g., fractions, Newton's laws, thermodynamics). In any particular assessment application, a theory of learning in the domain is needed to identify the set of knowledge and skills that is important to measure for the intended context of use, whether that be to characterize the competencies students have acquired at some point in time to make a summative judgment or to make formative judgments to guide subsequent instruction so as to maximize learning. A central premise is that the cognitive theory should represent the most scientifically credible understanding of typical ways in which learners represent knowledge and develop expertise in a domain.

Every assessment is also based on a set of assumptions and principles about the kinds of tasks or situations that will prompt students to say, do, or create something

that demonstrates important knowledge and skills. The tasks to which students are asked to respond on an assessment are not arbitrary. They must be carefully designed to provide evidence that is linked to the cognitive model of learning and to support the kinds of inferences and decisions that will be made on the basis of the assessment results. The *observation* vertex of the assessment triangle represents a description or set of specifications for assessment tasks that will elicit illuminating responses from students. In assessment, one has the opportunity to structure some small corner of the world to make observations. The assessment designer can use this capability to maximize the value of the data collected, as seen through the lens of the underlying assumptions about how students learn in the domain.

Every assessment is also based on certain assumptions and models for interpreting the evidence collected from observations. The *interpretation* vertex of the triangle encompasses all the methods and tools used to reason from fallible observations. It expresses how the observations derived from a set of assessment tasks constitute evidence about the knowledge and skills being assessed. In the context of large-scale assessment, the interpretation method is usually a statistical model, which is a characterization or summarization of patterns one would expect to see in the data given varying levels of student competency (see the section "Issues of Measurement: Interpretation and Statistical Inference"). In the context of classroom assessment, the interpretation is often made less formally by the teacher, and is often based on an intuitive or qualitative model rather than a formal statistical one. Even informally teachers make coordinated judgments about what aspects of students' understanding and learning are relevant, how a student has performed one or more tasks, and what the performances mean about the student's knowledge and understanding.

A crucial point is that each of the three elements of the assessment triangle not only must make sense on its own, but also must connect to each of the other two elements in a meaningful way to lead to an effective assessment and sound inferences. Thus, to have an effective assessment, all three vertices of the triangle must work together in synchrony. Central to this entire process, however, are theories and data on how students learn and what students know as they develop competence for important aspects of the curriculum.

Assessment Development: Evidence-Centered Design

Although it is especially useful to conceptualize assessment as a process of reasoning from evidence, the design of an actual assessment is a challenging endeavor that needs to be guided by theory and research about cognition as well as practical prescriptions regarding the processes that lead to a productive and potentially valid assessment for a particular context of use. As in any design activity, scientific knowledge provides direction and constrains the set of possibilities, but it does not prescribe the exact nature of the design, nor does it preclude ingenuity to achieve a final product. Design is always a complex process that applies theory and research to achieve near-optimal solutions under a series of multiple constraints, some of which are outside the realm of science. In the case of educational assessment, the design is influenced in important ways by variables such as its purpose (e.g., to assist learning, to measure individual attainment, or to evaluate a program), the context in which it will be used (classroom or large-scale), and practical constraints (e.g., resources and time).

The tendency in assessment design is to work from a somewhat "loose" description of what it is that students are supposed to know and be able to do (e.g., standards or a curriculum framework) to the development of tasks or problems for them to answer. Given the complexities of the assessment design process, it is unlikely that such a loose process can lead to generation of a quality assessment without a great deal of artistry, luck, and trial and error. As a consequence, many assessments are insufficient on a

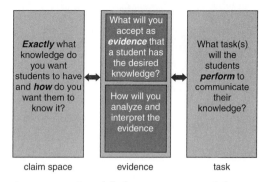

Figure 29.2. Simplified representation of three critical components of the evidence-centered design process and their reciprocal relationships.

number of dimensions including representation of the cognitive constructs and content to be covered and uncertainty about the scope of the inferences that can be drawn from task performance.

Recognizing that assessment is an evidentiary reasoning process, it has proven useful to be more systematic in framing the process of assessment design as an evidence-centered design process (e.g., Mislevy & Haertel, 2006; Mislevy & Riconscente, 2006). Figure 29.2 suffices to capture three essential components of the overall process. As shown in the figure, the process starts by defining as precisely as possible the claims that one wants to be able to make about student knowledge and the ways in which students are supposed to know and understand some particular aspect of a content domain. Examples might include aspects of algebraic thinking, ratio and proportion, force and motion, heat and temperature, and so forth. The most critical aspect of defining the claims one wants to make for purposes of assessment is to be as precise as possible about the elements that matter and express these in the form of verbs of cognition that are much more precise and less vague than high-level cognitive superordinate verbs such as *know* and *understand*. Example verbs might include compare, describe, analyze, compute, elaborate, explain, predict, justify, and so forth. Guiding this process of specifying the claims is theory and research on the nature of domain-specific knowing and learning (see Chapter 2 by Newstetter & Svinicki in this volume).

Although the claims one wishes to make or verify are about the student, they are linked to the forms of evidence that would provide support for those claims – the warrants in support of each claim. The evidence statements associated with given sets of claims capture the features of work products or performances that would give substance to the claims. This includes which features need to be present and how they are weighted in any evidentiary scheme – that is, what matters most and what matters least or not at all. For example, if the evidence in support of a claim about a student's knowledge of the laws of motion is that the student can analyze a physical situation in terms of the forces acting on all the bodies, then the evidence might be a free body diagram that is drawn with all the forces labeled including their magnitudes and directions.

The precision that comes from elaborating the claims and evidence statements associated with a domain of knowledge and skill pays off when one turns to the design of tasks or situations that can provide the requisite evidence. In essence, tasks are not designed or selected until it is clear what forms of evidence are needed to support the range of claims associated with a given assessment situation. The tasks need to provide all the necessary evidence and they should allow students to "show what they know" in a way that is as unambiguous as possible with respect to what the task performance implies about student knowledge and skill – that is, the inferences about student cognition that are permissible and sustainable from a given set of assessment tasks or items.

Assessment Validity: Argumentation and Evidence

The joint AERA/APA/NCME Standards (1999) frame validity largely in terms of "the concept or characteristic that a test is designed to measure" (1999, p. 5). In Messick's construct-centered view of validity, the theoretical construct the test score is purported to represent and score use are the foundations for interpreting the validity of any given test (Messick, 1994). For Messick, validity is "an integrated evaluative judgment of the degree to which empirical evidence and theoretical rationales support the adequacy and appropriateness of inferences and actions based on test scores" (1989, p. 13). Moving beyond a construct-centered view of validity, a more contemporary perspective calls for an interpretive validity argument, an argument that "specifies the proposed interpretations and uses of test results by laying out the network of inferences and assumptions leading from the observed performances to the conclusions and decisions based on the performances" (Kane, 2006, p. 23).

Kane (2006) and others (Haertel & Lorie, 2004; Mislevy, Steinberg, & Almond, 2003), distinguish between the interpretive argument, that is, the propositions that underpin test score interpretation, and the evidence and arguments that provide the necessary warrants for the propositions or claims of the interpretive argument. This view has been elaborated by Mislevy et al (2003) by asserting that validity evidence can be represented more formally in terms of a structured argument, an approach initially introduced and outlined in the context of legal reasoning by Stephen Toulmin (2003). In Toulmin's view a good argument is one that provides strong evidentiary support for a claim that, in turn, permits it to withstand criticism. Toulmin's practical approach begins by explicating the various "claims of interest" and then provides justification for those claims by identifying the evidence (data and/or expert opinion) to support those claims. The approach also calls for supplying a "warrant" that interprets the data and explicates how the data support the claims of interest. Toulmin's work, historically, focused on the formal role of argumentation and was used to evaluate the rational basis of arguments presented typically in the courtroom. Appropriating this approach, contemporary educational measurement theorists have more recently framed test validity as a reasoned argument backed by evidence (e.g., Kane, 2006). The particular forms of evidence are thus associated with the claims that one wishes to make about what a given assessment is and does and how its scores are to be interpreted.

Issues of Measurement: Interpretation and Statistical Inference

Measurement's Role in Evidentiary Reasoning

We have conceptualized assessments as evidentiary reasoning systems specifically constructed for making inferences about states of student knowledge (to be formally represented by latent variables) from observable evidence. The goal of an assessment can be construed in general terms as making particular inferences about examinee knowledge and thinking based on examinee performance on specific tasks. By design, the desired inferences can be expressed in the form of claims about what examinees know and can do under given circumstances. As part of assessment design, once the claims have been articulated explicitly, we can define a formal probability space called the student model as a unidimensional or multidimensional space of proficiency variables in such a way that each student can be thought of as possessing one set of values of the proficiency variables. Inferences about the values of student model variables can be interpreted as claims about students' knowledge and understanding.

Operationally, an examinee responds to a number of predesigned tasks. The examinee's performance results in some "work product" produced by the examinee and consisting of the observable trace left behind by the examinee, which hopefully reflects her thinking and understanding. A work product for a single task could be as simple as a multiple choice option selected – the examinee selected response option c on question 13 – or as complex and extended as a full-blown essay or a set of lab notes for a scientific experiment including drawings,

equations, and text. In either case the work product is whatever is produced by the examinee and left behind to be observed as her assessment performance.

An examinee's work product is then "scored" in some way according to a local evidence model. In the case of a multiple-choice option, the question response option c might be scored simply as right or wrong according to an answer key. For a concept inventory, a multiple-choice option may be linked to a specific misconception or common student error. In that case the "score" for the question response may be the misconception that was linked to the examinee's answer choice. For example, the examinee chose option c to question 13 and option c is linked to misconception 5, so we score the response to this question as misconception 5.

For questions with extended responses the examinee's work product for the responses may present multiple features that are observable and that are related to the knowledge intended to be measured by the item. In such a case, the "score" for the question response may consist of multiple feature scores of the work product, one such feature score for each of the observable features that has been identified for relevance to the construct measured and flagged for scoring. Alternatively, a more "holistic" rubric with multiple scoring categories (e.g., scores of 1–5) may be applied to the product (e.g., an essay or explanation) and a single score assigned to the work product.

The heart of the assessment system is the evidentiary reasoning step whereby we make inferences about examinee proficiencies (which can then be translated into claims about the examinee) based on evidence represented by the question scores as described earlier. The fundamental Bayes' theorem (Bayes, 1763; Stigler, 1982) translates $P(X_{ni} = x | \underline{\theta}_n = \underline{\theta})$ the conditional probability of a particular response to a question given an examinee's proficiency, into $P(\underline{\theta}_n = \underline{\theta} | X_{ni} = x)$ the conditional probability of the examinee's proficiency given his or her question responses. Bayes' theorem expresses the information about an examinee's proficiencies that is provided by question response $X_{ni} = x$ as follows:

$$P(\underline{\theta}_n = \underline{\theta}\,|\,X_{ni} = x)$$
$$= \frac{P(X_{ni} = x\,|\,\underline{\theta}_n = \underline{\theta})\,P(\underline{\theta}_n = \underline{\theta})}{P(X_{ni} = x)}.$$

For simplicity, in the equation above we have suppressed item parameter $\underline{\beta}_i$. The expression on the left $P(\underline{\theta}_n = \underline{\theta}\,|\,X_{ni} = x)$ is what we want to know, namely the probability that the examinee's proficiency value $\underline{\theta}_n = \underline{\theta}$ given his or her question response. In the numerator on the right, the first expression $P(X_{ni} = x\,|\,\underline{\theta}_n = \underline{\theta})$ is the item response model as we discuss in the section below on types of item response models. The second expression in the numerator on the right is called the prior probability of the proficiency variable $\underline{\theta}$, where "prior" means before taking into account the data $X_{ni} = x$. The denominator is a normalizing quantity that ensures that the expressions on the left add to 1.0 over all $\underline{\theta}$ (or integrate to 1.0 over the latent $\underline{\theta}$ space for continuous $\underline{\theta}$). That is required for the expression $P(\underline{\theta}_n = \underline{\theta}\,|\,X_{ni} = x)$ on the left to be a probability.

A similar expression can be written for $P(\underline{\theta}_n = \underline{\theta}\,|\,\underline{X}_n = \underline{x})$ where $\underline{X}_n = (X_1, \ldots, X_I)$ represents the vector of *all* scored item responses for a given student. In this more general case, the corresponding expression on the right hand side of Bayes' theorem can be expressed further in terms of the individual item response models $P(X_{ni} = x\,|\,\underline{\theta}_n = \underline{\theta})$ for each of the questions in the test. Thus Bayes' Theorem encapsulates the general evidence engine for an assessment and turns question performance $X_n = (x_1, \ldots, x_I)$ into inference about proficiencies $P(\underline{\theta}_n = \underline{\theta}\,|\,\underline{X}_n = \underline{x})$.

The Machinery of Measurement: Item Response Theory

Item response theory (IRT), in its broadest sense, serves as a general set of methods and approaches for quantitative modeling of assessment systems and analysis of assessment data (a number of classical references are available, including Baker, 2001; Crocker & Algina, 2006; Hambleton, Swaminathan, & Rogers, 1991). The general item response function is formulated as a probability model that expresses the probability of an examinee's response to a test question as a function of characteristics of the examinee and properties of the question:

$$P(X_{ni} = x\,|\,\underline{\theta}_n, \underline{\beta}_i).$$

Here $X_{ni} = x$ represents a particular scored response x to question i by examinee n; $\underline{\theta}_n$ represents one or more latent proficiencies of examinee n; and $\underline{\beta}_i$ consists of one or more parameters that represent properties of question i. For example, question i may be scored as $X_{ni} = 1$ for correct, and $X_{ni} = 0$ for incorrect. Underscores, as in $\underline{\theta}_n$ and $\underline{\beta}_i$, indicate that in general these variables can be vectors. Examinee proficiencies $\underline{\theta}_n$ can be continuous or discrete. The most common latent proficiency variable θ_n is a single real number, in which case we call the model a unidimensional, continuous model.

In expressing the probability of a question response as a function of examinee proficiency θ_n and question properties β_i, such a model postulates that examinee knowledge states and question parameters can be considered independently of any particular set of questions or fixed sample of students. This sample- and test-independence moves IRT beyond the more primitive ideas of classical test theory in which the basic unit of analysis is an administration of an intact test that consists of a fixed set of item responses from a particular sample of examinees. Classical test theory provides no theoretical base for inferring from results of a particular test to properties of a given question or subset of questions. If a subset of questions is taken from one test and added to a second test, a classical test theory analysis must be performed again from scratch. By contrast, IRT allows the model parameters of a given set of items to be estimated through a model calibration procedure from a representative set of data. Those items and their estimated model parameter values then can be used individually and in groups without the need for recalibrating the question parameters. A dual property holds for examinees. Once an examinee's proficiency variables have been estimated through one set of questions, IRT allows us to predict that examinee's performance on other calibrated items from the same domain (see Crocker & Algina, 2006).

Types of IRT Measurement Models

To illustrate the measurement process we provide examples of two types of item response models, one continuous and the other a discrete model suitable for diagnostic assessment. A unidimensional, continuous latent proficiency variable θ is appropriate when the knowledge being measured can reasonably be thought of as a "level" along a continuum. This can be justified for relatively broad measurement domains, as in the SAT Mathematics test or the Graduate Record Exam subject matter tests in fields such as physics, mathematics, and computer science. Modeling on a continuous scale may be appropriate for end-of-year or midterm tests in a science or engineering courses. In these cases the test outcome is a "scale score" that represents a level of achievement along a continuum.

The first example is called the 2-*parameter logistic (2PL) model:*

$$P(X_i = 1|\theta, a_i, b_i) = \frac{1}{1 + \exp(-Da_i(\theta - b_i))}.$$

Here θ is a continuous real number that represents *examinee ability level*, and there are two item parameters $\beta_i = (a_i, b_i)$ for item i, where b_i is a real number called the *item difficulty parameter*, and a_i is a positive real number called the *item discrimination parameter*. The parameter $D = 1.7$ is a scaling factor used for technical reasons and is held fixed over all items. Both θ and b_i can be any real numbers, positive, zero, or negative, and a_i can be any positive real number (see Baker, 2001; Crocker & Algina, 2006; Hambleton et al., 1991).

An important special case of the 2PL model is the *Rasch model* or 1-*parameter logistic (1PL) model*. This is defined by the same

function as the 2PL model with the restriction that the discrimination values a_i must be the same for all items on a given test, for example, restricting all $a_i = 1$. A typical structural assumption on the examinee proficiency space is that the set of all population θ values is normally distributed with mean 0.0 and variance 1.0.

For any fixed item parameters a_i and b_i we can consider the 2PL item response model as a function of the proficiency variable θ :

$$f(\theta) = f_{a_i, b_i}(\theta) = P(X_i = 1|\theta, a_i, b_i).$$

We can see from the graphs of 2PL models shown in Figure 29.3 that the probability of a correct response increases as θ increases from a probability of nearly 0.0 on the left to nearly 1.0 on the right. Thus the model shows an increased probability of answering the question correctly as ability θ increases. The two item response functions with $a_i = 1$ are "parallel" across the entire latent space – that is, the curves are exactly the same shape and they do not cross. The curve with $a_i = 2$ increases more steeply at $\theta = 0.0$, and that curve crosses both of the other curves.

The 2PL item difficulty parameter b_i identifies the point $\theta = b_i$ on the θ scale at which the probability of a correct answer is 0.5. That is $f(b_i) = P(X_i = 1|\theta = b_i, a_i, b_i) = 0.5$. We can see from Figure 29.3 that as b_i increases, the curve moves to the right and the item becomes more difficult for any given θ level. For a fixed item difficulty value b_i, as the discrimination parameter a_i increases, the slope of the curve at that center point $\theta = b_i$ increases, and the item becomes easier for ability levels $\theta < b_i$ and more difficult for θ levels $\theta > b_i$. Under reasonable assumptions it can be shown that a_i is a monotonic function of the biserial correlation between item response and total score, and so is a measure of how well an item discriminates between high- and low-scoring examinees.

EXAMPLES OF DIAGNOSTIC MODELS

Diagnostic classification models (DiBello & Stout, 2007) are a type of item response

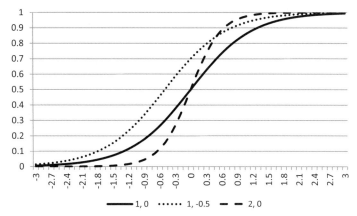

Figure 29.3. Graphs of three logistic item response functions. The legend gives the (a, b) values. The first two curves (solid and dotted lines) have parameters $(a, b) = (1, 0)$ and $(a, b) = (1, -0.5)$. The third curve (dashes) has $(a, b) = (2, 0)$. Since the first two curves have equal discrimination parameters (each has $a = 1$) they can be considered Rasch models with $b = 0$ and $b = -0.5$ respectively.

model in which the proficiency variable is a discrete profile of mastery levels for a set of aspects of knowledge and understanding:

$$P(X_i = x | \underline{\alpha}, \underline{\beta}_i).$$

Here the proficiency variable is a discrete vector $\underline{\alpha} = (\alpha_1, \ldots, \alpha_K)$ where each α_k represents an examinee's level of mastery of a particular facet k of knowledge or understanding. For example we could interpret $\alpha_k = 0, 1, 2$, as low, middle, or high proficiency in facet k knowledge; or, more commonly, α_k can be dichotomous, where $\alpha_k = 0$ or 1 represents nonmastery or mastery, respectively, of facet k knowledge. As part of its definition a diagnostic classification model includes an $I \times K$ Q-matrix of 0/1 elements, where $I =$ number of items, $K =$ number of identified facets of knowledge and understanding, and element $q_{ik} = 1$ or 0 indicates that item i either requires or does not require knowledge or understanding k.

An example of a diagnostic classification model is the fusion model (DiBello, Roussos, & Stout, 2007; DiBello & Stout, 2003, 2007) which is defined for item i as:

$$P(X_i = 1 | \underline{\alpha}, \pi_i^*, r_{ik}^*) = \pi_i^* \prod_{k=1}^{K} (r_{ik}^*)^{q_{ik}(1-\alpha_k)}.$$

The latent space for the fusion model is the set of all alpha vectors $\underline{\alpha} = (\alpha_1, \ldots, \alpha_K)$ where each $\alpha_K = 0$ (nonmastery) or 1 (mastery), or other ordered levels. The fusion model expression is easy to parse. A student who is a master $\alpha_k = 1$ of each facet of knowledge or understanding that is required for a given item will have every exponent $q_{ik}(1 - \alpha_k) = 0$, and the fusion model expression reduces to π_i^*. That is the maximum possible value for the model probability. By contrast, if the student is a nonmaster for some k required by the item, then the exponent $q_{ik}(1 - \alpha_k) = 1$, and the probability of a correct answer includes multiplication by r_{ik}^*. Each r_{ik}^* is a number less than 1.0 and greater than or equal to 0.0, and acts as a multiplicative penalty that diminishes the probability of a correct response. The more nonmastered skills k there are and the smaller the values of r_{ik}^* the harsher is the penalty and the smaller the probability of a correct answer.

Diagnostic classification models are designed for cases when items require one or more than one skill. The fusion model makes a *conjunctive* assumption that high probability of answering an item correctly requires that the examinee be a master of *all* skills required for that item. Other diagnostic models are available with alternative assumptions. The latent space of all possible $\underline{\alpha}$ vectors is finite. When each α_k is dichotomous, there are 2^K possible latent proficiency profiles $\underline{\alpha}$. The probability structure on the latent proficiency space can be any discrete probability distribution $p(\underline{\alpha})$ on the space of all 2^K $\underline{\alpha}$ vectors such that each $p(\underline{\alpha})$ is a nonnegative real number and the sum of $p(\underline{\alpha})$ over all $\underline{\alpha}$ is 1.0.

Reliability

TEST SCORE RELIABILITY

An important property of an assessment is the reliability of its score, traditionally discussed in the context of the observed total score $T_{observed}$. The reliability of measurement can be thought about from first principles in two different ways. First we can ask how closely correlated our observed total score $T_{observed}$ is to the unobservable true score T_{true}. For technical reasons we define the squared value of this correlation as a measure of score reliability. A different way of thinking about reliability is to imagine that a given examinee takes a test once, his brain is wiped clean, and then he takes the same test again. If his observed total score is $T_{observed}^{(1)}$ from the first performance and $T_{observed}^{(2)}$ from the second, the second idea of reliability is how closely $T_{observed}^{(1)}$ and $T_{observed}^{(2)}$ are correlated. Under the assumptions of classical test theory it can be shown that

$$\left(\rho_{T_{observed}, T_{true}}\right)^2 = \rho_{T_{observed}^{(1)}, T_{observed}^{(2)}},$$

using $\rho_{X,Y}$ for the correlation between random variables X and Y. Thus the two notions are equivalent (Lord & Novick, 1968). Of course neither correlation can actually be computed. The true score T_{true} is not known and not observable, and a person's brain cannot be wiped clean. A variety of reliability indices are available that approximate reliability for a given test score, including Kuder–Richardson 20 (KR20) and Cronbach's alpha (see Baker, 2001; Crocker & Algina, 2006; Hambleton et al., 1991; van der Linden & Hambleton, 1997). In the

framework of IRT the measurement outcome for a given examinee is his estimated proficiency level $\hat{\theta}$, and the analogous question is how reliable is the estimated proficiency level $\hat{\theta}$? We emphasize that the KR20 and Cronbach alpha indices for score reliability are appropriate in the context of IRT for the case when the proficiency variable is a total score or a unidimensional continuous latent ability $\hat{\theta}$.

RELIABILITY IN IRT MEASUREMENT MODELS

As discussed earlier, measurement plays a critical role in conceptualizing and formalizing aspects of inferential validity for an assessment. In particular, all model parameter estimates and proficiency level inferences produced by IRT are qualified by numerical measures of error or uncertainty. For example, for the unidimensional models 1PL or 2PL we are able to compute a conditional standard error of measurement for estimated proficiency $\hat{\theta}$ that effectively provides a 95% confidence interval around the estimate $\hat{\theta}$ of student latent ability. In the case of diagnostic classification models, discussed earlier, the proficiency variables are discrete and the reliability of diagnostic classifications is defined as an index of classification consistency. For example, in the case of a diagnostic assessment the mastery classification inference is about whether an examinee is a master or nonmaster of a particular skill, and we can ask the same two reliability questions from first principles as follows: What is the degree of consistency (1) between the true and inferred classifications, or (2) between the classifications inferred from administering the same test twice, wiping the brain between the two? The fundamental notions are the same for *any* IRT model, though the actual indices computed will differ from those used in unidimensional cases.

RELIABILITY CRITERIA

Educational measurement is imprecise to the extent that it yields best-possible inferences of unobservable latent proficiencies from observed student performance information. It is common for large-scale standardized assessments to set high minimum acceptable reliability criteria, such as, for example, Cronbach's alpha should be greater than 0.90. In fact the needed level of reliability for a given assessment depends on the purpose and context of use of the assessment. In particular the requirement of very high reliability is largely moot for a classroom test whose use is primarily formative in providing learners and instructors with information about levels of knowledge and understanding of individual students or groups of students for purposes of instructional decision making and actions. For formative classroom assessment use it is important to determine the value of information provided by an assessment for purposes of deciding what kind of activity or classroom discussion to have next.

Putting the Pieces Together: Model Selection and Finding the Sweet Spot

The item response model most appropriate for a given assessment must be selected based on substantive domain assumptions. In the framework of the assessment triangle and evidence-centered design, the claims about students must be translated into concrete latent variables, either continuous or discrete, with one or more dimensions. Definitions of the student variables and their levels of proficiency must be chosen in such a way as to be able to translate freely between proficiency levels and claims about what students know and can do.

   To apply the tools and methods of educational measurement to assessment means making the conceptual assessment framework concrete by (1) formally selecting and identifying the latent proficiency space as a probability space (the student model); (2) providing specific rules for turning observable evidence obtained from examinee performance on the selected assessment tasks into evidence that is usable by the item response model (which we will call the local evidence model); and (3) explicitly connecting the evidence thus obtained to updating the student model proficiency variables (called the global evidence model or

the assessment scoring model). In complex cases such as those found for assessments intended for classroom use, the selection of an appropriate model can present a significant challenge. A suitable set of student model variables together with an appropriate item response model must satisfy the following three properties simultaneously:

1. The proficiency variables and the model must reflect the purpose and intent of the assessment design;
2. The proficiency variables must be aligned with both curricular materials and instructional practice; and
3. The skills and model must be statistically supportable.

These three properties are in tension with one another. For example, satisfaction of properties 1 and 2 calls for many relatively small-grained variables. Property 3 requires fewer variables for a given test length. Consequently the grain size will be coarser for at least some of these fewer variables. Searching for a set of variables and item response model that satisfy all three properties simultaneously can be thought of as chasing after an elusive "sweet spot." The latent variables and the item response model must satisfy all three properties to be able to achieve high statistical quality and high assessment validity.

Summary of Measurement Applied to Assessments in Engineering Education

The general class of models that fall under the category of item response models provides a broad set of modeling capabilities that can be applied to a wide range of engineering education settings. Selecting model type and details of the modeling is a necessary creative act that must occur before satisfactory analysis of assessment data can be accomplished. It can be considered as the mathematical and statistical formalization of the assessment triangle and evidence-centered design foundations as discussed earlier. In general, the item response modeling approach provides a broad set of alternative models that can be fit to a wide range of assessment designs. Further, IRT comes equipped with a set of tools and methods for evaluating the performance of the model relative to the testing purpose by measuring errors of estimate and degree of data–model fit. Item response models help organize and interpret data, perform inferences about student proficiency relative to a particular student model and set of proficiency variables, and estimate the values of model parameters so that they can be used and interpreted.

Examples of Assessment and Measurement Applications in Engineering Education

There are many different types of assessment one might wish to develop and deploy for purposes of engineering education research. They can vary with respect to the constructs of interest (e.g., conceptual knowledge, problem solving, engineering design, interest, motivation) as well as the intended use (formative, summative, or program evaluation). It is beyond the scope of this chapter to consider each of these in depth. Rather, we illustrate how the ideas discussed in the first three main sections can be applied to an understanding and analysis of some particular examples drawn from research in engineering education.

To start our discussion of examples, it is useful to think about engineering knowledge, skills, and attributes in terms of two major claim spaces: (1) engineering knowledge and problem solving and (2) engineering professional skills. This distinction follows from a review of the engineering accreditation board's major criteria for evaluating engineering education programs. Five of the major ABET a–k criteria define knowledge and problem solving using descriptive action verbs fundamental to defining competency for the engineering tasks of design, troubleshooting and analysis (Brophy, Klein, Portsmore, & Rogers, 2008). ABET recommends engineering stu-

dents be able to (1) design systems to meet needs (criterion c), which requires abilities to (2) identify, formulate, and solve engineering problems (criterion e); (3) apply mathematics, science, and engineering principles (criterion a); (4) use the techniques, skills, and tools of modern engineering (criterion k); and (5) design and conduct experiments and analyze and interpret data (criterion b). This ordering of criteria provides a sequence of knowledge and skills consistent with various models of problem solving in engineering (Woods, 1997, mathematics (Polya, 1945), and creative problem solving (Bransford & Stein, 1993. The specificity of the language associated with these criteria is an important step in the direction of clarifying the claims one would want to make about student competence and it makes the definition of observable and measurable performances easier to specify in terms of specific actions and products expected of learners.

The other six ABET criteria (d, f, g, h, i, j) identify important knowledge, skills, and attributes that students need to develop to be productive participants in their engineering professional careers. These criteria highlight the need to *function* on multidisciplinary teams, *communicate* effectively, *understand* the impact of engineering on society, *understand* professional and ethical responsibility, demonstrate *knowledge* of contemporary issues, and *recognize* the need for life-long learning. These criteria are admittedly broad and require refinement to articulate better the nature of the competencies desired as well as what one would need to observe in students' behavior to provide evidence of proficiency. Although discussion of measures of professional skills is outside the scope of this chapter, the design, development, and validation of assessments of such competencies would necessarily follow the logic presented earlier in the first three main sections.

In the remainder of this section we consider assessments that have proven to be of significant interest in STEM education relative to the general category of knowledge and problem solving. We start with the case of concept inventories, then consider related examples of technology-based classroom formative assessment and end with a short discussion of problem-solving innovation. The theme throughout this section is how the assessment foundations provided by the assessment triangle and the evidence-centered design approach, combined with psychometric modeling and analysis, offer an interpretive framework for understanding the design and validity of an assessment relative to its intended purpose.

Concept Inventories in Engineering Education

A fundamental characteristic of expertise is the ability to reason with and apply disciplinary concepts to solve problems (e.g., Chi, Glaser, & Rees, 1982). Therefore it is not surprising that ABET lists this criterion first and engineering instructors make it a primary learning outcome for their courses. Engineers use their conceptual understanding of the sciences and the tools of engineering to comprehend and explain how systems work. These explanations are a critical step in troubleshooting existing systems, designing new systems, and constructing quantitative models to analyze these systems. Unfortunately, instructors in engineering programs regularly raise concerns about students' difficulties in developing deep conceptual understanding within their particular engineering disciplines and cite the challenges they face as instructors in assessing such understanding. Broad coverage of established sets of topics as commonly taught in engineering courses and a focus on developing practical problem-solving skill too often result in students able to pass course exams without achieving deep disciplinary understanding. Students are often unable to generalize their knowledge beyond the specific context of problems with which they are familiar. Concept inventories have been touted as one solution to this instructional and assessment dilemma and they are increasingly deployed in STEM courses as tools for both student and instructional evaluation.

Concept inventories provide good examples of the use of conceptual models of understanding in instructional domains to generate sets of test questions systematically (the cognition and observation vertices of the assessment triangle). In general, concept inventories are intended to measure students' deep conceptual understandings within a relatively narrow domain and they share a number of common features. Typically, they consist of multiple-choice questions whose distractor options were developed by first asking students open-ended questions and analyzing the student responses obtained. The questions are frequently based on science and engineering education research, including research on misconceptions and common student errors. Many inventories have developed explicit links between incorrect multiple-choice distractors and specific misconceptions. A wide array of inventories is now available in STEM education (http://ciHUB.org). For purposes of discussion, we use a particular example of a concept inventory to illustrate how issues of assessment design connect with issues of measurement, inference, and interpretation.

CONCEPT ASSESSMENT TOOL FOR STATICS

Concept Assessment Tool for Statics (CATS) is a concept inventory whose purpose is to diagnose statics concepts that engineering students typically have difficulty understanding and to identify persistent types of student errors (Steif & Dantzler, 2004; Steif & Hansen, 2006). The test consists of twenty-seven multiple-choice questions with three questions representing each of nine concepts: *Drawing Forces on Separated Bodies (Free Body Diagrams), Newton's Third Law, Static Equivalence, Roller Joint, Pin-in-Slot Joint, Loads at Surfaces with Negligible Friction, Representing Loads at Connections, Limits on Friction Force,* and *Equilibrium* (Steif & Hansen, 2007). Each question was designed to test one of the nine concepts in isolation and to require little or no calculation. Wrong answers were designed to reflect known student errors.

Table 29.1. Cronbach's α for CATS Subscales

Concept	α	N
Free body diagram	.72	3
Newton's third law	.71	3
Static equivalence	.61	3
Roller	.68	3
Slot	.68	3
Frictionless point contact	.47	3
Representation	.33	3
Friction	.64	3
Equilibrium	.37	3

CATS is available online through the website http://ciHUB.org for access by instructors and for administration to students and is used regularly by numerous instructors. Instructor reports give item right and wrong, total score, and student concept scores (0, 1, 2, or 3 correct of the three questions for that concept). How answers connect to the specific errors and misconceptions linked to wrong answer choices are not part of the typical reporting process. Paul Steif, the developer of CATS, was kind enough to provide us with 1,372 cases of CATS student response data that had been collected online during a one-year period. We were able to analyze those data to generate evidence about CATS given its purpose, design, and scores. We frame our analysis in terms of a set of questions one would want to ask about the reliability and validity of any given concept inventory or similar type of assessment.

Some Basic Item and Test Statistics. One question that is important to justify use of the CATS overall total score is its reliability (see definition of reliability in the section "Issues of Measurement: Interpretation and Statistical Inference"). Our analyses reveal that the CATS total score is highly reliable (twenty-seven items; $\alpha = 0.84$). Given that scores also are reported for the nine separate concept groups we also calculated reliabilities for the subscales. All but two demonstrated reasonable reliabilities for a three-item scale as shown in Table 29.1.

A second set of issues important for interpretation of score outcomes relates to how well the individual items function as probes of student proficiency in this domain of thinking. Individual item difficulties can be defined as the proportion of students answering a given item correctly. For CATS these ranged from 0.16 to 0.78, and, with the exception of question 26 at 0.16, most would consider a range of item difficulties from about 0.20 to 0.80 to be reasonable. A spread of item difficulties supports the process of discriminating among examinees who vary in their proficiencies with respect to this domain of knowledge and skill. A second characteristic of item functioning is item discrimination, which is the biserial correlation between each person's score on a given item and his or her total score on the rest of the instrument. For CATS the values ranged from 0.18 to 0.49. In essence, an item's discrimination value is a measure of its sensitivity to the student's overall ability, so that those who do well on the overall assessment also tend to do well on the individual item. Again, except for Q26 with discrimination 0.18, the values obtained for the CATS items constitute reasonable evidence that each item is generally positively related to the overall proficiency represented by the aggregate of performance on the rest of instrument.

Application of Unidimensional IRT to CATS. As noted earlier, IRT provides a powerful method for evaluating the functioning of an assessment. For CATS, the model-data fit indices of both 1PL and 2PL models were good. One interpretive advantage of an IRT analysis is that it allows us to place the items and examinees in the same space. An example is the "Wright map" shown in Figure 29.4. The vertical axis represents a scale for the "latent CATS ability" θ, which runs in the figure from -3.0 to $+3.0$. Indicated to the left of the axis are the locations of each student's θ and the distribution of those scores across the total examinee sample. The abilities are scaled to have a mean of 0.0 and a standard deviation of 1.0.

The twenty-seven questions numbered Q1 through Q27 are shown to the right of the central axis. Each question is located at the scale value of its b_i 1PL item difficulty

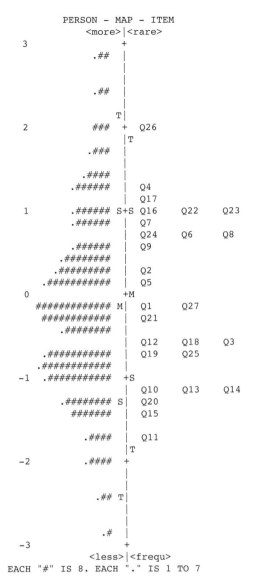

```
                PERSON - MAP - ITEM
                   <more>|<rare>
    3                     +
                   .##    |
                          |
                          |
                   .##    |
                          |
                        T |
    2                ###  +    Q26
                          |T
                  .###    |
                          |
                 .####    |
                .#####    |    Q4
                          |    Q17
    1           .##### S+S   Q16    Q22    Q23
                .#####    |    Q7
                          |    Q24    Q6     Q8
                .#####    |    Q9
               .#######   |
               .######## |
              .########## |    Q2
            .############ |    Q5
    0                     +M
         ############## M |    Q1     Q27
          ############    |    Q21
           .#######       |
                          |    Q12    Q18    Q3
            .##########    |   Q19    Q25
            .##########    |
   -1       .##########   +S
                          |    Q10    Q13    Q14
            .#######   S  |    Q20
             ######       |    Q15
                          |
                .####     |    Q11
                          |T
   -2           .####     +
                          |
                          |
                  .##  T  |
                          |
                          |
                   .#     |
   -3                     +
                   <less>|<frequ>
       EACH "#" IS 8. EACH "." IS 1 TO 7
```

Figure 29.4. Wright map of CATS data. The scale on the central axis goes from –3.0 at the bottom to +3.0 at the top. The symbols on the left stand for location of examinee abilities θ, with "#" representing eight examinees and "*" representing one to seven examinees. Mean of examinee θ is 0.0. The right side of the axis shows item numbers Q1 through Q27. Each item number is located at the 1PL difficulty parameter b_i.

parameter (see earlier discussion in the section "Issues of Measurement: Interpretation and Statistical Inference"). Recall that the b_i item difficulty parameters are monotonically related to the proportions of examinees

answering an item correctly. Thus items located at low points on the scale (such as Q11, Q16, and Q20, for example) are relatively easy, and items located at higher levels of the scale (like Q26 [the most difficult item], Q4, and Q17) are more difficult.

Placing the questions and examinees on the same scale allows us to see whether the question difficulties and examinee abilities are well matched. Notice that there is a set of about 5% of the students with estimated abilities –2.0 and below, and there are no comparable questions with difficulty parameters in such a low range. After the most difficult and problematic question (Q26), the next most difficult question was question Q4 with difficulty about 1.25, and a comparable set of about 10% of students have ability levels above 1.3. In general, the set of questions seems reasonably well matched to this range of student abilities.

Structural Analysis of the Data. An important empirical test of the developers' claims about questions clustered by concept is the structural consistency between the data and the developers' concepts. We performed a series of exploratory factor analyses (EFAs) to determine the best number of factors supported by the data, and the degree of match or mismatch between the factor loadings of questions and the concepts to which the question are assigned. An EFA with varimax rotation identified nine factors that collectively explained 57.6% of the variance. These results generally supported the developer's a priori nine clusters and indicated that the assessment is tapping important and differentiable aspects of student knowledge and understanding in the domain of statics.

We also performed confirmatory factor analyses (CFAs) to examine causal effects of the developers' concepts or combinations of concepts on question performance and associations among concepts. The best fitting model (see Figure 29.5 for a representation of the estimated model) was a seven-factor model that omitted problematic question 26 and combined two pairs of the hypothesized nine categories: Equivalence and Equilibrium and Frictionless and

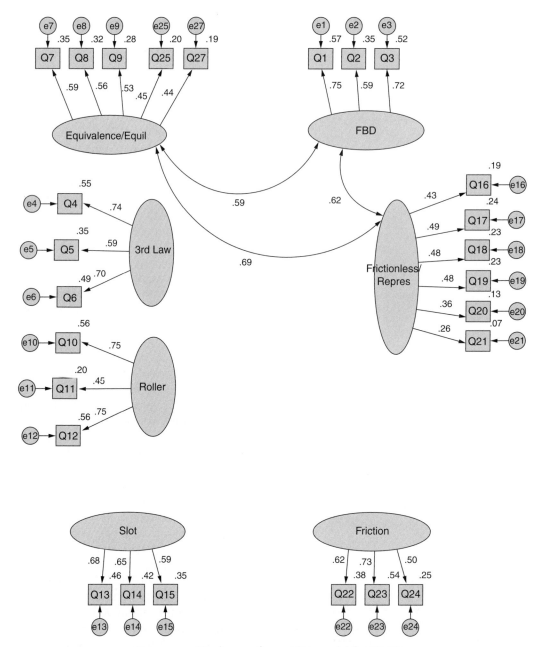

Figure 29.5. Final seven-factor CFA model for CATS.

Representation. This model included correlations among three latent constructs: Free Body Diagrams, Equivalence/Equilibrium, and Frictionless/Representation.

We also examined the diagnostic possibilities of the CATS through the complementary lens of a fusion model analysis (Santiago-Román, 2009). The CATS instrument showed strong diagnostic strength

relative to identified facets of knowledge and understanding. A set of ten skills was developed that consisted of some original concepts unchanged and some new skills made up of combinations of questions from several of the original concepts. Items were freely coded as requiring one or more than one of the defined skills. The ten skills were identified through an iterative sequence of substantive modifications to the current skills and Q matrix followed by diagnostic model calibration runs. Model–data fit measures strongly supported the set of ten skills. Details can be found in Santiago-Román (2009), Santiago-Román, Streveler, and DiBello (2010) and Santiago Román, Streveler, Steif, & DiBello (2010). The diagnostic analysis showed a striking concordance with the seven-factor best fitting confirmatory factor analysis model shown in Figure 29.5. All of the diagnostic skill definitions that were different from the original nine concepts were confined within the correlated portion of the model shown in Figure 29.5, namely the three latent variables identified as Equivalence/Equilibrium, Free Body Diagrams, and Frictionless/Representation.

The structural analyses taken together – the EFA, CFA, and diagnostic analyses – shed light on the functioning of CATS relative to its design and suggest possibilities for formative classroom use. The analyses offer an interpretive argument that links the CATS substantive foundations with the correlational structure found in the data as represented by Figure 29.5. Just as important they provide a reasonably strong warrant for the formative classroom use of diagnostic outcomes from CATS in addition to the existing and more typical use of summative scores.

APPLICABILITY OF THE APPROACH TO ALL CONCEPT INVENTORIES

In light of the rich conceptual and cognitive models guiding concept inventory (CI) item development, the analytic framework for assessments that is made up of the assessment triangle and the evidence-centered design approach, along with the application of a reasoned and principled combination of more powerful measurement models, makes possible a more extended understanding and appreciation of the inferential linkage from observation to interpretation and from interpretation back to cognition. The methods exemplified here constitute a good start at confirming the strong interpretive frameworks through more sophisticated psycho-metric techniques. They can help capture the possible diagnostic power of instruments such as CIs, and enhance their usefulness as both summative and formative assessment tools. Hopefully, this conversation about methods and approaches encourages engineering educators to raise the bar for the validity and quality of assessments for use in STEM classrooms and heightens an appreciation for what constitutes sufficient conceptual and empirical support for assessment claims.

Technology and Classroom Formative Assessment Processes

The issues highlighted in the preceding discussion regarding student difficulties with conceptual knowledge and understanding have resulted in many examples of attempts to design active learning environments in engineering education. The goal is to provide effective methods to engage learners in the classroom and overcome some of the conceptual challenges that students otherwise face (e.g., Prince, 2004; Roselli & Brophy, 2006; Smith, Johnson, Johnson, & Sheppard, 2005). Many of these designs rely on the use of classroom based formative assessment processes that begin by posing concept-based questions to students (e.g., Crouch & Mazur, 2001). Similar to questions found on concept inventories, these probes of student thinking are designed to require qualitative reasoning with key concepts to either explain or predict outcomes. Students are asked to generate a response (e.g., select a multiple-choice option; generate code, a diagram, written explanation, or a graph) and then share it with the instructor using various technologies. A very popular method uses Classroom Response System (CRS) technology to poll students' responses (e.g., Beatty & Gerace, 2009). Student responses are sent with wireless devices and aggregated into a summary chart at the instructor's computer. An instructor can use the results to determine how well students have comprehended key concepts. An instructional benefit occurs when a substantial portion of the students select

Figure 29.6. Athlete in an iron cross position.

responses that represent one or more common misunderstandings. The instructor then can use a number of pedagogical approaches to provide feedback to students. It must be noted, however, that engaging in these activities requires very thoughtful item design to ensure that the desired range of performance outcomes is likely to occur.

For example, both the Force Concept Inventory (Hestenes, Wells, & Swackhamer, 1992) and CATS (Steif & Hansen, 2007) target multiple concepts associated with analyzing a system in static equilibrium (i.e., all forces balance, resulting in a motionless system). A critical step in this analysis is representing the forces acting on the system using a free body diagram (FBD) and gener-

ating Newton's laws that define the behavior of these forces. A course in biomechanics included this kind of analysis as part of its learning outcomes (Roselli & Brophy, 2003, 2006). A CRS system was used in the course to support development of students' ability to generate appropriate FBD representations (Barr, Cone, Roselli, & Brophy, 2003; Roselli & Brophy, 2003, 2006). The course used challenge-based instruction to engage students in a design task to develop practice equipment for novice athletes to perform the iron cross position with rings shown in Figure 29.6. The instructor engaged learners in an inquiry process to investigate various sub-problems involved in this subtask. As part of a lecture session the instructor posed the subtask shown in Figure 29.7 that asked students to generate a free body diagram of the athlete's hand and ring assembly. After a few minutes the students were then shown possible solutions (Figure 29.8) and asked to pick the one that best matches their solution. Students used a small remote device to make their selection and a frequency plot of all the responses was displayed.

At this point an instructor has a range of pedagogical options for providing feedback. The first is simply to state the correct

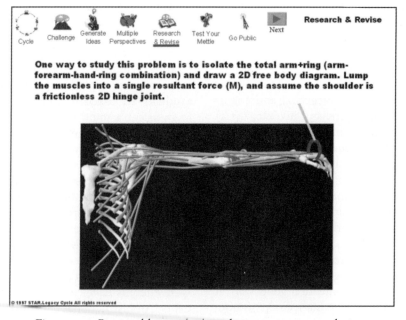

Figure 29.7. Pose problem and ask students to generate a solution.

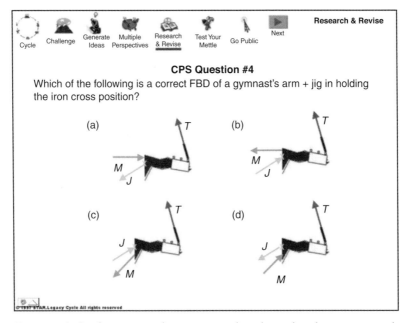

Figure 29.8. Students respond to a question based on what they constructed.

answer and move on. Students who answered correctly receive confirmation and will assume they reasoned correctly. Those who answered incorrectly know they need to refine their thinking, but have no additional information to help them do so. The instructor's second option is to work through the correct solution. This better assists students who chose the correct answer because it confirms the appropriateness of their reasoning about the problem. Students selecting one of the foils will need to recognize their errors on their own. Unfortunately, the worked out example may not address the learner's specific misconception or provide additional information to refine his or her thinking. The instructor's third choice would be to open a dialog with students to have them justify their responses. She could start with the most popular response first. If this response is not the correct answer, then the discussion can be quite productive because students will hear the different ways their peers conceive of the concepts underlying the problem. The instructor can repeat the process for each of the answer options as a way to review systematically all the ways in which members of the class may have

been thinking about the problem situation. This approach provides a level of individual feedback that addresses the needs of all students, and research has illustrated its positive impact on overall student learning (e.g., Roselli & Brophy, 2006).

Although using a CRS is efficient, answering multiple-choice questions does not necessarily require students to reflect on their own work. The earlier example asked students to generate a diagram first and then evaluate four other solutions. Learners could choose to ignore what they generated and try to work backwards from the solutions provided. Therefore, an alternative to multiple choice would be an interactive system in which students could generate a FBD and then receive feedback on what they generated. Figure 29.9 illustrates an interactive software program that presents learners with a sequence of statics problems requiring analysis (Roselli, Howard, & Brophy, 2006). Students can insert vectors and moments to represent their analysis of forces on the system. Feedback is generated once they submit their answer. They can refine their answer and resubmit for additional feedback. The feedback is designed to provide increasing levels of specificity.

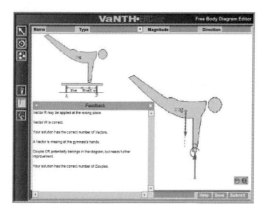

Figure 29.9. Free body diagram assistant.

The first round of feedback simply evaluates what is missing or potentially inappropriate. The second round of information is more specific about what is missing, such as the location or direction of a vector or moment. After three attempts students are provided the final solution for the diagram. Students using this system demonstrate an increase in their ability to generate diagrams for the specific joint types (e.g., contact, pin, or clamp). For example, with each new problem students required fewer submissions to achieve the correct solution. Systems like these have the potential of engaging students in problem-solving activities that replicate the professional experiences they need for engineering practice, and the systems provide mechanisms to help students develop the conceptual understanding they will need to be flexible and adaptive. Consistent with what we have argued in this chapter, designing such systems requires a very careful analysis of the claims one wishes to make about student competence and then designing ways to obtain relevant forms of evidence for either formative or summative assessment purposes.

Assessing Innovation and Adaptiveness in Problem Solving

Engineers continually generate new ideas designed to improve existing devices and systems (Dym et al., 2005). This kind of expertise involves skills associated with problem finding, identification, and formu-lation (Atman, Kilgore, & McKenna, 2008). As with other aspects of competence, assessment requires clarification of the knowledge, skills, and attributes associated with these various initial phases of problem solving. A related challenge is identifying the forms of evidence needed to support claims that students possess the relevant skills, as well as the tasks or environments that will elicit the relevant evidence. Given the importance of these competencies in engineering, it is not surprising that engineering education continues to expand its use of problem-based learning, which frequently relies on open-ended problems and projects that are presumed to tap the requisite knowledge and skills. Such problems may require weeks to solve and students often work in teams, making individual student evaluation a challenge.

Several approaches to measuring problem identification have focused on students' ability to evaluate novel problems using their domain knowledge and engineering tools to generate ideas, questions, and plans to solve a problem. This ability to adapt to novel problems illustrates how experts can transfer what they know into new contexts. This idea of adaptive expertise (Schwartz, Bransford, & Sears, 2005) highlights experts' ability both to solve familiar problems efficiently and systematically generate new knowledge to solve new problems. These two dimensions of expertise help instructors design differential learning objectives for their students. The first dimension is to rapidly and effortlessly solve common problems found in the target domain area. Experts' conceptual understanding of the domain can be used to solve routine problems with great speed and accuracy. The second dimension illustrates what occurs when an expert faces an unfamiliar problem. Though their speed and efficiency to approach the problem may be diminished, experts with the appropriate knowledge, skills, and attitudes engage in an inquiry process to generate the knowledge needed to identify possible solutions. This requires noticing what knowledge they don't know, generating questions they need to answer, and formulating a plan and a process to find information and synthesize it into new knowledge. This requires looking beyond their normal perspective and managing uncertainty. Therefore, experts also display abilities to persevere and a willingness to deal with ambiguity. Note that the display of efficiency for applying conceptual knowledge is not always an indicator of adaptiveness (Hatono & Inagaki, 1986), or of the ability to think innovatively. The design of instruction and assessment requires a framework that better illustrates the relationships between efficiency and innovation and how the latter develops in learners (Schwartz et al., 2005). New measures are being explored to better assess engineering students' development of adaptiveness.

For example, a scenario-based problem method was used to assess student adaptiveness. Researchers were interested in measuring the development of efficient use of domain knowledge and the growth in adaptiveness (innovation) over time. Also, they wanted to explore the development of these characteristics relative to students' confidence. The method consisted of a pre–post assessment using a design (analysis) scenario. The assessment method presented students with a cardiologist's design for an implantable defibrillator that he postulated would redefine the industry's approach to the device. To measure efficiency ability, or their ability to generate an appropriate response, the students were asked "What do you need to do to test the doctor's hypothesis?" Second, to measure students' adaptiveness they were asked "What questions do you have for the doctor?" Finally students were asked "How confident are you in your response?" Each question for the doctor posed by the student was scored with a rubric developed with the assistance of domain experts in cardiology and the design process. The responses to the first question about testing the hypothesis were coded as frequency of strategies generated and the quality of the responses (e.g., 1, novice; 2, expert). The difference in the quality of questions posed by students was based on whether students' strategies

were vague (e.g., gather data at sites) versus providing viable alternatives for testing the devices performance to prove feasibility (compare data at two sites to determine if both are necessary). The second question that asked students to generate questions for the doctor was coded by counting the number of questions asked and the quality of the questions (1, novice; 2, proficient; 3, expert). Novice questions focused more on factual information, like definitions. This included questions specific to the immediate problem (e.g., "Could I see tracing of the ECG?"). "Problem-specific knowledge" was coded as proficient. For example, expert questions focus more on understanding the rationale of the proposal and understanding of the range of alternatives already considered. These rubrics provided a quantitative score for the number and quality of students' responses.

This pre–post assessment was administered to both seniors in a capstone design class and first-year engineering students studying cardiac signal analysis in an introductory course. Seniors were better at defining an approach for testing the doctor's hypothesis and generated more questions to ask the doctor that were more like what an expert would ask. Seniors were also more confident than first-year students, but even first-year students showed an increase in these measures across their first year. This measure of adaptiveness illustrated that first-year students had a significant increase over the semester on both the efficiency and innovation scales. This kind of assessment could also be used to illustrate the growth of students' proficiency across the engineering curriculum (Walker, Cordray, King, & Brophy, 2006).

Assessing students' innovation skills (problem identification and formulation) can also be integrated into standard methods at the course level. Exams might be designed with various levels of questions. The first level of test items might focus on assessing students' efficiency defined in terms of basic taxonomic knowledge associated with facts and simple procedures. These items measure students' ability to apply fluently the concepts and procedures they have practiced throughout the semester. Innovation questions are designed to evaluate students' ability to formulate a plan of action for a larger open-ended problem. The goal is not to generate and execute a final solution, but to formulate the problem appropriately by defining problems to be explored and questions to ask to acquire needed information. The third level of questions would focus on transfer and would consist of questions that are conceptually similar to problems in the course, but within a different context. These are adaptive expertise question items. The approach is similar to the scenario-based instrument described earlier, but the addition of specific targeted questions at differing levels can make for a richer assessment of what students know and can do. A version of this approach was explored in a third year biotransport course using a challenge-based instructional approach. The course had a strong focus on developing students' problem solving abilities while learning the constitutive properties (taxonomic knowledge) that govern the transport behavior of biological systems. The course was organized around three major types of biotransport: fluid, heat, and mass transfer. An exam was given after instruction via a series of challenge-based modules. Analysis of performance on three exams distributed across the semester illustrated that students increased their abilities for all three areas of competency. The efficiency and innovation skills developed quickly from exam 1 to 2, but adaptiveness was relatively flat over that same period. However, by the end of the semester all three dimensions had increased significantly (Martin, Petrosino, Rivale, & Diller, 2006).

As new instructional approaches make their way into the engineering education classroom, additional assessment methods will need to be designed and validated to evaluate student learning outcomes. Some efforts have been made to measure dimensions of engineering thinking and the generalizability of such approaches (Atman et al., 2008; Cordray, Harris, & Klein, 2009; Prince & Felder, 2006). Assessment methods

must be sought that can demonstrate that students are developing both the problem-solving skills they need to solve routine problems and the executive skills needed for managing the process of investigating new problems. The previous examples illustrated text-based assessments of students' ability to articulate what they notice and what knowledge they need to generate. The next steps of measurement will involve students' ability to model systems to facilitate their comprehension and analysis of complex systems (Brophy & Li, 2011). A model-based reasoning approach (e.g., Nersessian, 2009) can involve qualitative reasoning with diagrams and graphs and analytic modeling to evaluate the behavior of systems quantitatively. These graphical images and how learners reason with them to generate new knowledge and insights makes engineering an interesting domain to rethink what and how we assess. Another dimension of engineering involves the decisions engineers make once they have evidence from their models and analysis. Therefore, other areas of investigation for assessment design can be in decision theory and choice (Schwartz & Arena, 2009). The challenge will be to create new measures that are valid, reliable, feasible, and useful to support students' learning and that can enhance the quality of an engineering education program.

Final Thoughts: The Role of Assessment in a Comprehensive R&D Program

High-quality evidence that permits instructors, researchers, and administrators to ask and answer critical questions about the outcomes of learning and instruction – what students know and are able to do – is critical to advancing an R&D agenda in engineering education. The first requirement for developing quality assessments is that the concepts and skills that signal progress toward mastery of a domain be understood and specified. These activities constitute fundamental components of a prospective R&D agenda as applied to multiple engineering courses and content domains. Assessment of the overall outcomes of instruction is important to the R&D agenda because it allows for tests of program effectiveness. But it is more broadly important because the content of such assessments can drive instructional practice for better or for worse. The deployment of the force concept inventory in physics illustrates a case in which an assessment tool based on cognitive and instructional research has had a powerful, positive impact on the redesign of instruction, serving as an evaluation tool to determine the effectiveness of new instructional methods (e.g., Hake, 1998). Assessment of the impact of long-term programs of R&D is also important. University faculty, administrators, and alumni need ways to determine the return on investing in research in engineering education. Such a use of assessment for program evaluation purposes needs measures that are sensitive to learning and the impact of quality teaching. Researchers, educators, and administrators must therefore concern themselves with supporting research and development on effective and appropriate assessment procedures that can serve multiple purposes and functions as part of a coordinated system of assessments.

Although we have distinguished among three purposes of assessment, quality instruments for each depend on the same three components: (1) theories and data about content-based cognition that indicate the knowledge and skills that should be assessed; (2) tasks and observations that can provide information on whether students have mastered the knowledge and skills of interest; and (3) qualitative and quantitative techniques for scoring student performance that capture fairly the differences in knowledge and skill among students being assessed. Research is clearly needed in all three areas of work, as well as their integration.

Research also needs to explore (1) how new forms of assessment can be made accessible for instructors, and practical for use in classrooms; (2) how they can be made efficient for use in university teaching contexts; and (3) how various new forms of assessment affect student learning, instructor practice, and educational decision making. Research must explore ways that instructors can be assisted in integrating new forms of assessment into their instructional practices and how they can best make use of information from such assessments. Also to be studied are ways that structural features of instructional delivery in higher education (e.g., length of class time, class size and organization, and opportunities for students and/or instructors to work together) impact the feasibility of implementing new types of assessments and their effectiveness.

As a field, engineering education continues to make progress. Hopefully this chapter's discussion of ways to think about the design and uses of assessment provides a useful set of ideas and approaches that can further advance the field of engineering education research.

Acknowledgments

Preparation of this chapter was supported by the following grants from the U.S. National Science Foundation to James W. Pellegrino and Louis V. DiBello: CCLI Project No. 0920242, REESE Project No. 0918552, and REESE Project No. 0815065 and to Sean Brophy: DUE-0817486.

References

American Educational Research Association, American Psychological Association, & National Council on Measurement in Education (AERA/APA/NCME). (1999). *Standards for educational and psychological testing.* Washington, DC: Author.

Atman, C. J., Kilgore, D., & McKenna, A. (2008). Characterizing design learning: A mixed-methods study of engineering designers' use of language. *Journal of Engineering Education*, 97(3), 309–326.

Baker, F. (2001). *The basics of item response theory.* ERIC Clearinghouse on Assessment and Evaluation, University of Maryland, College Park, MD. Retrieved from http://echo.edres.org:8080/irt/baker/http://edres.org/irt/baker/

Barr, R. W., Cone, J., Roselli, R. J., & Brophy, S. P. (2003). Initial experiences using an interactive classroom participation system (CPS) for presenting the iron cross biomechanics module. In *Proceedings of the 2003 ASEE Gulf-Southwest Annual Conference*, the University of Texas at Arlington. Lubbock, TX.

Bayes, T. P. (1763). An essay towards solving a problem in the doctrine of chances. *Philosophical Transactions of the Royal Society of London*, 53, 370–418.

Beatty, I. D., & Gerace, W. J. (2009). Technology-enhanced formative assessment: A research-based pedagogy for teaching science with classroom response technology. *Journal of Science Education and Technology*, 18(2), 1146–1162.

Bransford, J. D., & Stein, B. S. (1993). *The IDEAL problem solver* (2nd ed.). New York, NY: W. H. Freeman.

Brophy, S., Klein, S., Portsmore, M., & Rogers, C. (2008). Advancing engineering in P-12 classrooms. *Journal of Engineering Education*, 97(3), 369–387.

Brophy, S., & Li, S. (2011). Problem definition in design by first year engineering students. In *Proceedings of the American Society of Engineering Education Annual Conference*, Vancouver, BC, Canada.

Chi, M. T. H., Glaser, R., & Rees, E. (1982). Expertise in problem solving. In R. Sternberg (Ed.), *Advances in the psychology of human intelligence* (Vol. 1, pp. 7–76). Hillsdale, NJ: Lawrence Erlbaum.

Cordray, D. S., Harris, T. R., & Klein, S. (2009). A research synthesis of the effectiveness, replicability, and generality of the VaNTH challenge-based instructional models in bioengineering. *Journal of Engineering Education*, 98, 335–348.

Crocker, L., & Algina, J. (2006). *Introduction to classical and modern test theory*. New York, NY: Wadsworth.

Crouch, C. H., & Mazur, E. (2001). Peer instruction: Ten years of experience and results. *American Journal of Physics*, 69, 970–977.

DiBello, L. V., Roussos, L. A., & Stout, W. F. (2007). Review of cognitively diagnostic assessment and a summary of psychometric models. In C. R. Rao & S. Sinharay (Eds.), *Handbook of statistics*, Vol. 26: *Psychometrics* (pp. 979–1030). Amsterdam: Elsevier.

DiBello, L., & Stout, W. F. (2003). Student profile scoring methods for informative assessment. In H. Yanai, A. Okada, K. Shigemasu, & J. J. Meulmann (Eds.), *New developments in psychometrics* (pp. 81–92). Tokyo: Springer.

DiBello, L. V., & Stout, W. F. (Eds.). (2007) Special issue on IRT-based cognitive diagnostic models and related methods, *Journal of Educational Measurement*, 44(4), 285–292.

Dym, C. L., Agogino, A. M., Eris, V., Frey, D., & Leifer, L. J. (2005). Engineering design, thinking, teaching and learning. *Journal of Engineering Education*, 94(1), 103–120.

Haertel, E. H., & Lorie, W. A. (2004). Validating standards-based test score interpretations. *Measurement: Interdisciplinary Research and Perspectives*, 2(2), 61–103.

Hake, R. R. (1998). Interactive-engagement versus traditional methods: A six-thousand-student survey of mechanics test data for introductory physics courses. *American Journal of Physics*, 66(1), 64–74.

Hambleton, R. K., Swaminathan, H., & Rogers, H. J. (1991). *Fundamentals of item response theory*. Measurement Methods for the Social Science. Newbury Park, CA: SAGE.

Hatano, G., & Inagaki, K. (1986). Two courses of expertise. In H. Stevenson, H. Azuma, & K. Hakuta (Eds.), *Child development and education in Japan*. New York, NY: W. H. Freeman.

Hestenes, D., Wells, M., & Swackhamer, G. (1992, March). Force concept inventory. *The Physics Teacher*, 30, 159–166.

Hickey, D., & Pellegrino, J. W. (2005). Theory, level, and function: Three dimensions for understanding transfer and student assessment. In J. P. Mestre (Ed.), *Transfer of learning from a modern multidisciplinary perspective* (pp. 251–293). Greenwich, CT: Information Age.

Kane, M. T. (2006). Validation. In R. L. Brennan (Ed.), *Educational measurement* (4th ed., pp. 17–64). Westport, CT: Praeger.

Lord, R. M., & Novick, M. R. (1968). *Statistical theories of mental test scores*. Reading, MA: Addison-Wesley.

Marion, S., & Pellegrino, J. W. (2006, Winter). A validity framework for evaluating the technical quality of alternate assessments. *Educational Measurement: Issues and Practice*, 47–57.

Martin, T., Petrosino, A. J., Rivale, S., & Diller, K. R. (2006). The development of adaptive expertise in biotransport. *New Directions for Teaching and Learning*, 108, 35–47.

Messick, S. (1994). The interplay of evidence and consequences in the validation of performance assessments. *Educational Researcher*, 23(2), 13–23.

Mislevy, R. J., & Haertel, G. (2006). Implications of evidence-centered design for educational assessment. *Educational Measurement: Issues and Practice*, 25, 6–20.

Mislevy, R. J., & Riconscente, M. M. (2006). Evidence-centered assessment design: Layers, concepts, and terminology. In S. Downing & T. Haladyna (Eds.), *Handbook of test development* (pp. 61–90). Mahwah, NJ: Lawrence Erlbaum.

Mislevy, R. J., Steinberg, L., & Almond, R. (2003). On the structure of educational assessments. *Measurement: Interdisciplinary Research and Perspectives*, 1, 3–67.

National Research Council. (2003). *Assessment in support of learning and instruction: Bridging the gap between large-scale and classroom assessment*. Washington, DC: The National Academies Press.

Nersessian, N. J. (2009). How do engineering scientists think? Model-based simulation in biomedical engineering laboratories. *Topics in Cognitive Science*, 1, 730–757.

Pellegrino, J. W., Chudowsky, N., & Glaser, R. (Eds.). (2001). *Knowing what students know: The science and design of educational assessment*. Washington, DC: The National Academies Press.

Pellegrino, J. W., & Hickey, D. (2006). Educational assessment: Towards better alignment between theory and practice. In L. Verschaffel, F. Dochy, M. Boekaerts, & S. Vosniadou (Eds.), *Instructional psychology: Past, present and future trends. Sixteen essays in honour of Erik De Corte* (pp. 169–189). Oxford: Elsevier.

Polya, G. (1945). *How to solve it*. Princeton, NJ: Princeton University Press.

Prince, M. (2004). Does active learning work? A review of the research. *Journal of Engineering Education*, 93(3), 223–231.

Prince, M. J., & Felder, R. M. (2006). Inductive teaching and learning methods: Definitions, comparisons, and research bases. *Journal of Engineering Education*, 95, 123–138.

Roselli, R. J., & Brophy, S. P. (2003). Redesigning a biomechanics course using challenge-based instruction. *IEEE Engineering in Medicine and Biology Magazine*, 22(4), 66–70.

Roselli, R. J., & Brophy, S. P. (2006). Experiences with formative assessment in engineering classrooms. *Journal of Engineering Education*, 95(4), 325–333.

Roselli, R. J., Howard, L., & Brophy, S. P. (2006). A computer-based free body diagram assistant. *Computer Applications in Engineering Education*, 14(4), 281–290.

Ruiz-Primo, M. A., Shavelson, R. J., Hamilton, L., & Klein, S. (2002). On the evaluation of systemic science education reform: Searching for instructional sensitivity. *Journal of Research in Science Teaching*, 39, 369–393.

Santiago-Román, A. I. (2009). *Fitting cognitive diagnostic assessment to the concept assessment tool for statics (CATS)* (Doctoral dissertation). Purdue University, Lafayette, IN.

Santiago-Román, A. I., Streveler, R. A., & DiBello, L. (2010). *The development of estimated cognitive attribute profiles for the Concept Assessment Tool for Statics*. Paper presented at the 40th ASEE/IEEE Frontiers in Education Conference, Washington, DC.

Santiago-Román, A. I., Streveler, R. A., Steif, P., & DiBello, L. V. (2010). *The development of a Q-matrix for the Concept Assessment Tool for Statics*. Paper presented at the ERM Division of the ASEE Annual Conference and Exposition, Louisville, KY.

Schum, D. (1987). *Evidence and inference for the intelligence analyst*. Lantham, MD: University Press of America.

Schwartz, D. L., & Arena, D. (2009). *Choice-based assessments for the digital age*. Stanford University, School of Education. Retrieved from http://courses.ischool.berkeley.edu/i290-pm4e/f10/sites/default/files/assessment_schwartz.pdf

Schwartz, D. L., Bransford, J. D., & Sears, D. (2005). Efficiency and innovation in transfer. In J. P. Mestre (Ed.), *Transfer of learning from a modern multidisciplinary perspective* (pp. 1–51). Greenwich, CT: Information Age.

Smith, K. A., Johnson, D. W., Johnson, R. W., & Sheppard, S. D. (2005). Pedagogies of engagement: Classroom-based practices. *Journal of Engineering Education*, 94(1), 87–101.

Steif, P. S., & Dantzler, J. A. (2004). A statics concept inventory: Development and psychometric analysis. *Journal of Engineering Education*, 94(4), 363–371.

Steif, P. S., & Hansen, M. A. (2006). Comparisons between performances in a statics concept inventory and course examinations, *International Journal of Engineering Education*, 22, 1070–1076.

Steif, P. S., & Hansen, M. A. (2007). New practices for administering and analyzing the results of concept inventories. *Journal of Engineering Education*, 96, 205–212.

Stigler, S. M. (1982). Thomas Bayes' Bayesian inference. *Journal of the Royal Statistical Society, Series* A, 145, 250–258.

Toulmin, S. E. (2003). *The uses of argument*. Cambridge: Cambridge University Press.

van der Linden, W. J., & Hambleton, R. K. (Eds.) (1996). *Handbook of modern item response theory*. New York, NY: Springer.

Walker, J. T., Cordray, D. S., King, P. H., & Brophy, S. P. (2006). Design scenarios as an assessment of adaptive expertise. *International Journal of Engineering Education*, 22(1), 1–7.

Wiggins, G. (1998). *Educative assessment: Designing assessments to inform and improve student performance*. San Francisco, CA: Jossey-Bass.

Woods, D. R. (1997). Developing problem-solving skills: The McMaster problem solving program. *Journal of Engineering Education*, 86(2), 75–92.

Part 6

CROSS-CUTTING ISSUES AND PERSPECTIVES

CHAPTER 30

Engineering Communication

Marie C. Paretti, Lisa D. McNair, and Jon A. Leydens

Communication has long been considered a core professional skill and one clearly central to global engineering practice. In Sales' (2006) survey of U.K. engineers, more than 50% of her respondents spent more than 40% of their time on writing. In Australia, surveys by Male, Bush, and Chapman consistently identify communication as a critical (and deficient) quality in engineering graduates (Male, Bush, & Chapman, 2010, 2011). Surveys by Tenopir and King (2004) and Covington, Barksdale, Egan-Warren, Larsen, and Trunzo (2007) found that U.S. engineers spent upwards of 30% of their time writing and speaking, while Kreth's (2000) survey of U.S. graduates indicated that new engineers spent, on average, 38% of their time writing. In non–English-speaking regions and countries, the prevalence of English for Special Purposes (ESP), English for Academic Purposes (EAP), and second language (L2, sometimes referred to as English as a Foreign Language, or EFL) instruction for engineers attests to the ongoing need for communication competency in English as well as in engineers' native languages (e.g., Cismas, 2010; Orr et al., 1995).

Beyond its ubiquity in practice, engineering communication can profoundly affect both the development and the impact of technology. Most dramatically, as studies of the U.S. space shuttle *Challenger* explosion and the near-meltdown of the Three Mile Island nuclear facility demonstrate, failures in engineering communication are often central factors in engineering disasters (e.g., Dombrowski, 1992; Herndl, Fennell, & Miller, 1991; Winsor, 1988). Even a failure to appropriately identify units of measure can cost millions, as in the case of the NASA orbiter developed by Lockheed Martin (Isbell, Hardin, & Underwood, 1999). Effective communication in engineering environments, such cases remind us, can have exceptionally high stakes.

In this chapter, we provide an overview of relevant research to identify (1) historical and contemporary influences on practice and pedagogy, (2) prominent theoretical frameworks and research methodologies, (3) critical findings that impact practice and teaching, (4) approaches to assessment, and (5) emerging trends. Because communication encompasses a wide range of skills, concerns, and fields, we have necessarily limited our scope in several ways. First, although this overview presents work from a range of international settings, we focus on L1 contexts (individuals working and learning in their first/native language). Communication in L2 contexts is a separate, extensive field, and although we highlight some references for L2 in engineering, a full review of that literature is beyond our scope. Notably, however, L1 engineering communication research is set in the broader context of fields that include technical and professional communication, linguistics, and composition and rhetoric; relevant cross-disciplinary insights from those fields are included here even when they do not explicitly address engineering. Second, we focus on research that addresses formal writing and speaking – that is, documents and presentations, with some attention to electronic modes of communication. Studies that address multimodal communication have risen rapidly in the twenty-first century, often as a way to understand team communications in virtual and/or global contexts, but robust research in this area focused on professional contexts, including engineering, is still emerging. We thus address broader trends in multimodal communication, and point toward engineering-focused work where possible. Third, although we include a small number of studies on team communication, that, too, is a separate field with an extensive body of literature beyond the scope of this chapter. Interested readers can find studies of professional practices and effective pedagogies in organizational communication literature; the U.S.-based National Communication Association, for example, provides excellent resources for assessing both formal and conversational oral communication (http://www.natcom.org/Secondary.aspx?id=119).

Collectively, this chapter raises several important issues with salient implications for researchers and educators:

- Historical divides between learning engineering content and learning communication skills are blurring; communication learning increasingly occurs within engineering courses and contexts.
- Engineering communication research has evolved to encompass sociocultural conceptual frameworks and a broad range of established and emerging methodologies.
- Communication is a socially situated practice. As a result, abstract, universal standards for "good" communication do not exist; instead, "effective" communication depends on the social and professional context, the goals, the individuals involved, and related factors.
- Communication learning is also socially situated and occurs most effectively with assignments that reflect professional contexts and expectations.
- Learning to communicate within engineering is linked to engineering thinking and the development of a professional identity.
- Learning in L1 higher education contexts requires explicit feedback geared primarily toward issues of appropriate claims, evidence, argument, and understanding rather than format, grammar, and mechanics.
- Assessment practices are most often holistic and rubric- and/or portfolio-based, and they emphasize situated concerns of audience, purpose, and context.

History and Influences: Engineering Communication in the Twenty-first Century

A range of influences globally and locally have shaped the development of engineering communication research and pedagogy, including trends in communication instruction, pressures from industry and government groups, accreditation standards, globalization of work, and an expanding range of communication media.

The roots of engineering communication as a defined field date back at least to educational practices of the nineteenth century.

Although explicit writing instruction has not historically been prominent in European higher education (Björk, Bräuer, Rienecker, & Stray Jörgensen, 2003), engineering writing courses (initially called "engineering English") have been part of U.S. curricula since the late 1800s (Connors, 1982). In 1896, the Massachusetts Institute of Technology hired a communication specialist so that its communication program could be "explicitly designed to meet the writing needs of engineers" (Russell, 2002, p. 109). Communication within engineering programs emerged more formally elsewhere a few decades later, when specialists began directing engineering communication programs at Ohio State, Case Institute, and Rensselaer Institute in 1919 and at Purdue in 1920 (Gianniny, 2004). The twentieth century also saw the growth of standalone technical writing (later called technical communication) courses that serve a range of disciplines, including engineering, and an outpouring of associated textbooks. As Wolfe (2009) points out, however, these textbooks do not serve engineering students effectively.

The limits of standalone courses, coupled with recent developments in both global engineering practice and engineering accreditation standards, have accelerated a trend toward communication instruction within engineering programs. In the United States, the impetus for change is easily traced to a few major publications that identified global shifts in engineering practice that demand educational transformation, including *Rising Above the Gathering Storm* (Committee on Science, Engineering, and Public Policy, 2006) and *The Engineer of 2020* (National Academy of Engineering [NAE], 2004). Authored by educators, practitioners, and researchers from a variety of sectors and fields, such publications sound the same call: to compete in the global economy, engineers need to expand beyond technical skills to develop as innovative professionals. Industry echoed this call, adding that fresh engineers came to the workplace unable to write, present their work, function on teams, and manage interpersonal relationships – weaknesses that prevented them from rising to management positions and caused companies to fear for their own futures.

These calls set the tone for a decade in which engineering education, in the United States and internationally, has expanded beyond the narrow discourse of problem-solving grounded in engineering science to a more integrated approach that brings together technical and professional development, including communication. The publication of ABET, Inc.'s outcomes-based criteria for accreditation of engineering programs (ABET Engineering Accreditation Commission, 2007; Williams, 2001) played a major role in this shift in the United States as skills once derided as "soft" were reclassified as "professional" and foregrounded as central curricular outcomes. Similar trends appear internationally. ABET, according to its website, currently accredits engineering programs in twenty-three countries. Standards set by Engineers Australia include "effective oral and written communication in professional and lay domains" as one of six "professional and personal attributes" (Engineers Australia, 2011), while accreditation standards set forth by the United Kingdom's Engineering Council include communication as a core transferable skill (Engineering Council, 2011).

This inclusion of effective communication as a core outcome has opened space for partnerships between engineering and communication faculty (Williams, 2001). Such partnerships are supported by current efforts among language faculty to work with disciplinary faculty on the integration of communication and content learning. In the United States, this movement is embodied in Writing Across the Curriculum (WAC) and Writing in the Disciplines (WID) programs that began in the 1980s (Russell, 2002) and have recently shifted from "Writing..." to "Communication..." (i.e., Communication across the Curriculum [CAC], Communication in the Disciplines [CID]). A similar effort in Europe, referred to as Content and Language Integrated Learning (CLIL) or Integrated Content Learning (ICL), began about fifteen years ago

(Gustafsson et al., 2011), though the teaching of writing at the university level was not a common practice at many European universities for much of the twentieth century (see Bjork et al. [2003] for an excellent collection on the approaches to writing instruction in Europe). The emergence of Integrating Content and Language in Higher Education (ICLHE), formed in Europe in 2010 as an outgrowth of several conferences, points to the strength of such programs across the continent. Moreover, as the ongoing study of international WAC/CAC programs by Thaiss et al. suggests, this movement is not limited to the United States and Europe; the survey includes responses from at least fifty-one countries beyond the United States and Canada, including those in Australasia, South America, Asia, and Africa (Thaiss, 2010). Such programs reflect moves by experts in composition, rhetoric, literacy studies, professional communication, applied linguistics, and related fields to collaborate with faculty in other disciplines, including engineering, to develop assignments and approaches that support the simultaneous development of content expertise and communication skills. Thus even as those responsible for engineering curricula are turning more attention to the communication practices of students, communication specialists complement this attention by consistently seeking to support communication in the disciplines. Such partnerships emphasize overlapping goals: both engineering content and communication specialists are recognizing the strong connections between the process of learning to *think* like an engineer and learning to *communicate* like one (e.g., see Bean, 1996; Winsor, 1996).

Finally, technology is continually redefining how engineers communicate as global collaboration, mobile computing, and 24/7 product development permeate practice. As work by Madhavan and Lindsay (Chapter 31, this volume), Downey et al. (2006), McNair and Paretti (2010; Paretti & McNair, 2008a), and others make clear, today's engineers need to communicate across cultural and disciplinary boundaries amid ever-changing tools, from e-mail to Skype to virtual meetings and more. In a recent study of Indian and U.S. engineering firms, for example, Levine, Allard, and Tenopir (2011) found that engineers typically spend at least half the day in some type of communication activity, using a variety of media (e.g., phone, messaging, virtual or face-to-face meetings). Not surprisingly, they noted that engineers increasingly need multiple communication channels to be effective.

As such work makes clear, the emergence of new technologies has meant that engineering communication no longer involves simply writing technical reports; it means developing fluency on e-mail, synchronous chat technologies, virtual conferencing systems, mobile phones, Web-based collaborative tools, cloud data management, and whatever the next wave of development brings. The increase in visual and graphical communication, including tools for developing complex graphics, means that engineers need expertise not only in the written word, but also in the rhetoric of data presentation and visual communication (see Poe, Lerner, and Craig [2010] for an excellent discussion of effective pedagogies in this domain). Integrated product development teams accentuate engineers' need to communicate not only with one another, or even with managers, but with marketing specialists, sociologists, scientists, historians, industrial designers, mass media, the public, and other stakeholders. Moreover, global collaboration and 24/7 product development teams mean that communication isn't limited to one's own culture; any project may involve colleagues from the United States, Europe, India, China, Australia, and beyond, and all team members need skills in cross-cultural communication. While "the technical report" has not disappeared, it is now only one mode among many that engineers must employ to communicate their knowledge. The complexity of communication media, needs, and strategies means that engineering communication is a constantly evolving domain that, from a research perspective, sits at the intersection of a range of different fields internationally.

Theoretical Frameworks: The Contextual Nature of Engineering Communication

The need for and the increasingly complex nature of engineering communication has long offered a rich space for researchers from a range of fields. Engineering communication research is broadly framed by three general domains: writing and rhetoric, learning, and media. Although the theoretical frameworks described in this section may be of interest predominantly for researchers, they have critical implications for engineering faculty who wish to address communication in their courses. Table 30.1 summarizes these theories and their implications for practice, while the sections that follow provide more detail for practitioners and researchers seeking a richer understanding of these frameworks.

Frameworks for Understanding Engineering Communication

Although the early 2000s have seen a sea change in the role of communication in engineering, engineering communication as an identifiable realm of teaching and inquiry dates back to those early "engineering English" course. Those courses were built on a transmission model of communication that treats writing as an encoding process – writers simply encode existing ideas or meanings into texts, which readers then decode, much like data moving through a network. As Slack, Miller, and Doak (2003) point out in their history of technical communication, "Engineering English courses were designed, among other things, to teach students to encode the special meanings of engineering" (p. 174) so that the meanings are clear, accurate, and easily understood.

This view of writing began to shift, however, as "engineering writing" broadened and "technical communication" emerged as an academic discipline that included engineering writing, but also technical, scientific, and professional writing in other domains, as well as other modes of communication. As the discipline developed, it drew on fields such as rhetoric, sociolinguistics, composition, and literacy studies; at the same time, the communication practices of engineers both at school and at work became objects of study across these and related fields internationally. As a result, contemporary theoretical frameworks now reflect more complex views of communication and of the relationships between communication and engineering practice. A full history of this development is outside the scope of this chapter, but the work of Connors (1982), as well as Volume 1 of Kynell-Hunt and Savage's *Power and Legitimacy in Technical Communication* (2003) provide excellent starting points.

As this interdisciplinary history suggests, several theoretical frameworks inform studies of engineering communication. The most prominent are those aligned with sociocultural perspectives, including discourse communities, rhetorical genre theory, academic literacy, and activity theory. These frameworks treat communication not as an isolated process of encoding, transmitting, and decoding, but as a complex negotiation in which meaning is constructed by individuals interacting through language that shapes and is shaped by particular contexts. That is, "meaning" does not exist in abstract form; rather, it is created as individuals communicate (in words, images, mathematical symbols) with one another in particular contexts. Importantly, this perspective frames writing instruction not only in the United States, but, as Gustafsson et al. (2011) note, in Europe, South Africa, and elsewhere. Moreover, these theories are not unique to engineering communication; they shape and are shaped by a range of language-related fields that examine how communication operates in context and how individuals develop as communicators in social and professional domains.

Table 30.1. Theoretical Frameworks and Their Implications for Practice

Theory	Brief Definition	Implications
Discourse communities	Individuals in different groups have different conventions for communicating, including not only language but also ways of making and supporting arguments.	"Good" communication does not exist as an abstract, decontextualized set of features but instead varies across contexts and fields. Students need instruction in the accepted language patterns and modes of argument and reasoning within their fields.
Academic literacy	The language of a discipline is central to disciplinary identity; as students learn to write and speak, they are learning how to think and act like professionals in their fields.	
Rhetorical genre theory	The format of a given text (its organization and structure) is closely linked to how it is used; documents are designed to serve particular functions.	Students must go beyond learning "formats" for specific documents to understand how those documents are used in practice, and how they can be adapted to different contexts and reader needs.
Activity theory	Communication serves as a tool that mediates work among a group of people seeking to accomplish a goal.	
Situated learning	Students learn effectively when they are engaged in the authentic (or "real world") practices of their fields.	Assignments should be designed to reflect workplace communication practices, and where possible they should provide a basis for meaningful exchange of information.
Media richness theory	Different communication media (e-mail, voice, paper) are capable of providing different kinds and levels of information.	Students need to learn to make informed choices not only about what to say, but also about which communication media are most likely to be effective.
Social presence theory	Different communication media create different kinds of relationships among participants.	

To understand recent evolutions in engineering communication research, then, we present a brief discussion of salient frameworks. We begin with the term *rhetoric* because it is used widely across several frameworks. Although in popular use *rhetoric* often refers to tricky, empty, or idle talk, particularly when used to sway public opinion, the term has ancient roots. Among communication researchers, it refers broadly to the ways in which arguments are structured, claims are made, evidence is used, and knowledge is constructed in a given domain. Rhetoric also emphasizes how texts (writ large to include figures, reports, PowerPoint slides, and more) affect and in particular persuade audiences.

Specific rhetorical patterns are often aligned with different discourse communities. The term *discourse community* (Nystrand, 1982) describes a group of individuals who share sets of conventions – specific terms, but also accepted forms of evidence, accepted kinds of arguments, accepted patterns of writing – that facilitate and regulate communication within the group. Thus mechanical engineers form one discourse community while historians form another, just as the nuclear engineers at Three Mile Island formed one discourse community while the physicists formed another (Herndl et al., 1991). Importantly, the differences are not simply in what we might call "jargon" – specialized terms unique to each field. Instead, they encompass differences in values and epistemologies as discourse communities create and share meaning in the process of doing knowledge work.

Where discourse communities describe how groups of people share communication norms, *rhetorical genre theory* (sometimes called North American genre theory) helps describe how texts operate both within and beyond those communities. *Genre* here refers not simply to literary conventions – fiction, poetry, drama – but rather to any type of document with a readily identifiable set of features. Thus lab reports are a genre, as are résumés, proposals, design specifications, and so on. *Rhetorical* genre theory considers not just the surface features of a text (e.g., the IMRAD structure of lab reports – Introduction, Methods, Results, and Discussion), but also how those texts function within specific contexts – their social uses. This social (rather than structural) understanding of genre stems from Carolyn Miller's seminal article, "Genre as Social Action," which defines genre as "typified rhetorical actions based in recurrent situations" (1984, p. 159). Most simply, Miller argues that the patterns in a text (e.g., the order of the sections, the types of information or evidence presented) arise from and are integrated with the functions that text serves within a given context. Thus research articles in engineering follow the IMRAD structure because that structure both reflects the nature of engineering research and enables researchers to understand, evaluate, and expand on one another's work; articles in disciplines such as English or history do not apply this structure because it is neither reflective of the nature of inquiry nor useful for building new knowledge.

Academic literacy, which grounds much of European CLIL, adopts a similar sociocultural approach. Emerging out of New Literacy Studies (Barton, Hamilton, & Ivanič, 2000; Zamel & Spack, 1998), academic literacy positions language as central to disciplinary identity; it explicitly attends to differences across disciplines and it addresses issues of power as personal and institutional identities collide (Lea & Street, 1998). As a result, learning the language of a discipline involves not simply a mechanical mastery of language (i.e., grammar, syntax), but also an understanding of the social practices of the discipline in terms of what constitutes knowledge, how it is created and tested, and how meaning is socially constructed within the domain. Importantly, as Gustafsson notes, this approach moves beyond a simple model of disciplinary socialization, in which communication skills can be taught before or apart from the content work of the field, and instead positions "learning the field" and "learning to communicate in the field" as commensurate, even inseparable (Gustafsson et al., 2011). Language and content are both part of disciplinary identity and the process of constructing disciplinary knowledge and meaning.

Although rhetorical genre theory and academic literacy provide two means to situate communication in context, in recent years scholars have also begun employing *activity theory*, a framework that grows out of the work of Vygotsky (1962/1986, 1978) and Leont'v, and has been developed extensively by Engeström et al. (Cole, Engeström, & Vasquez, 1997; Engeström & Middleton, 1998; Engeström, Miettinen, & Punämaki, 1999). Traditional rhetorical approaches generally consider the "rhetorical triangle" of audience, purpose, and context. Activity theory, in contrast, considers the entire scope of the activity, including the individuals involved, the activity itself, the desired outcomes, the rules governing the system, the community(ies) to which the actors belong, the division of labor among individuals, and the mediating artifacts – including texts – that circulate through the activity. As David Russell (1997) argues, an activity theory approach to studies of communication, particularly in educational contexts, "allow[s] us to theorize and trace the interactions among people and the inscriptions called texts (and other material tools) without separating either from collective, ongoing, motivated action over time" (p. 509). Activity theory has proved particularly salient in engineering as a way to understand the relationship between classroom and workplace communication and to design assignments and classroom interactions that help students effectively transfer skills from school to professional practice (Dannels, 2003; Paretti, 2008). It helps identify and explain the need for meaningful communication assignments that function in the context of engineering work, rather than decontextualized reports with no audience and no purpose.

Frameworks for Understanding Learning

The preceding theoretical frameworks reflect a sociocultural view of language, in which meaning is socially constructed in specific contexts and negotiated by individuals who continually enact and contest linguistic norms. Collectively, these frameworks emphasize the relationships among communication, content, and context, reinforcing for researchers and educators alike that "good writing" does not exist in the abstract, apart from the knowledge and meaning-making practices of the discipline and the context in which communication occurs. These same frameworks inform studies of engineering students learning to communicate – for example, both Dannels (2003) and Paretti (2008) draw on activity theory to analyze capstone design courses, while Artemeva (2007) and Artemeva,

Logie, and St-Martin (1999) examine rhetorical genre theory as pedagogical framework.

Such studies, however, often also rely heavily on social theories of learning, and particularly on situated learning, described in more detail in Chapters 2 by Newstetter and Svinicki and 3 by Johri, Olds, and O'Connor in this volume. *Situated learning* accentuates the importance of field and context by advocating that learning should incorporate authentic, "real world" situations. As Johri et al. (Chapter 3) explain, situated learning draws on Lave and Wenger's work on legitimate peripheral participation, which examines how individuals move from novice to expert by engaging in the work of the community under the guidance of mentors (1991), and on Vygotsky's (1962/1986, 1978) work on social cognition, which suggests that individuals learn not in isolation but through their interactions with others. Situated learning thus emphasizes engagement in activities that reflect the actual practices of a domain, are embedded in social contexts and meaningful actions, and involve not only acquiring information but also, as social constructivist frameworks stress, developing ways of thinking, perceiving, and solving problems. That is, engineering students learn to communicate not by completing practice exercises, style drills, or decontextualized research papers, but by communicating about their engineering work to individuals in meaningful ways that reflect workplace practices (Artemeva et al., 1999; Dias, Freedman, Medway, & Pare, 1999; Freedman & Adam, 2000; Paretti, 2006, 2008; Spinuzzi, 1996). The 2010 book-length study by Poe, Lerner, and Craig provides one of the most comprehensive analyses of situated learning in engineering communication to date. In addition, situated learning disrupts the notion that engineering work is exclusively or almost exclusively technical, divorced from social dimensions that render it rhetorical and sociotechnical. That disruption is evident in a large body of twentieth and twenty-first century scholarship in both science and technology studies and engineering studies (e.g., Bijker, Hughes,

& Pinch, 1987; Downey, 2013; Latour, 1987; Latour & Woolgar, 1986).

Frameworks for Understanding Communication Media

In addition to the sociocultural frameworks described in the previous sections, researchers who examine multimodal communication, often in the context of virtual collaboration, also draw on theories surrounding communication media. As with L2 learning, computer-mediated communication is an extensive field and a full review of the literature is outside our scope. Two theories, however, bear mention because they are invoked in several studies of engineering communication in virtual environments: media richness and social presence. *Media richness theory* addresses the capacity of a given communication medium to carry multiple kinds of information (Kock, 2005; Robert & Dennis, 2005). E-mail, for example, is a lean medium in that it provides asynchronous communication of primarily textual information. Virtual conferencing systems that include video, shared writing or drawing space, and ability to exchange documents are rich media because they provide synchronous, interactive access to verbal, visual, and textual information. The degree of richness is considered to affect the effectiveness of communication, though as Robert and Dennis (2005) note, it is not always true that richer media are inherently better; in some instances, for example, e-mail may be more effective because it allows for sufficient time to review and process complex documents, including those produced during engineering design work.

Where media richness focuses on the kinds of information available through a given technology, *social presence theory* (Short, Williams, & Christie, 1976) focuses on the experience of the collaborators and the degree to which they experience connection and a sense of one another's presence. Along with media richness theory, social presence theory has frequently been referenced in studies of communication in virtual teams, including engineering teams (Farrell & Holkner, 2004; Kiesler & Cummings, 2002; Robert & Dennis, 2005; Roberts, Lowry, & Sweeney, 2006; Walther & Bunz, 2005). Collectively, the results of such studies make clear that engineers and engineering students need to develop fluency not only with asynchronous written texts designed for paper, but also with a full range of communication media and the ways those media support and hinder effective workplace communication. Work by McNair, Paretti et al., for example, shows that engineering students' social fluency with a variety of electronic media does not inherently transfer to professional fluency (McNair & Paretti, 2010, Paretti, McNair, & Holloway-Attaway, 2007); however, targeted interventions with strategies that support virtual collaboration can improve effective multimodal communication (McNair, Paretti, & Davitt, 2010).

Research Methods: Conceptual Frameworks for Studying Engineering Communication

Given the diversity of fields concerned with engineering communication, it is not surprising that research methods vary widely. However, as Case and Light (Chapter 27, this volume) and others (Borrego, 2007; Borrego, Douglas, & Amelink, 2009; Leydens, Moskal, & Poavelich, 2004) note, the research questions themselves should inform the choice of methods. That recommendation accentuates the importance of multidisciplinary collaboration among researchers to develop the most appropriate qualitative, quantitative, or mixed methods approach for a given set of research questions. It also destabilizes any claim that a particular, privileged method exists for engineering communication research.

Instead, when we consider the broad range of research questions that fall within "engineering communication," the need for methodological diversity and cross-disciplinary collaboration is clearly apparent. Engineering communication scholars hail from backgrounds such as technical communication, composition and rhetoric, speech communication, applied linguistics, literacy studies, and engineering. Research questions thus vary considerably; they include studies of events such as the *Challenger* explosion (Dombrowski, 1992; Herndl et al., 1991; Winsor, 1988), studies of practices such as global collaboration (McNair & Paretti, 2010), studies of organizations (Spinuzzi, 2003; Tenopir & King, 2004; Winsor, 2003, 2006), studies of communication beliefs (Leydens, 2008; Matusovich, Paretti, Motto, & Cross, 2012), and studies of student learning (Dannels, 2000, 2003; Dannels & Martin, 2008; Herrington, 1985; McNair et al., 2010; Paretti, 2008; Paretti & McNair, 2008a; Paretti, McNair, & Holloway-Attaway, 2007; Winsor, 1996).

To provide researchers with a map of methods and questions, and practitioners with a clearer sense of the types of findings that emerge from research, Table 30.2 offers an overview of research methods, focus areas, example studies, and representative findings.

Although many of these methods are likely familiar to engineering education researchers, and are discussed in more detail in Chapters 24 to 28 of this volume, a few are more localized to communication-related fields. A full discussion of these methods is beyond the scope of this chapter, but we offer short definitions and references for interested readers:

- Discourse analysis is a broad term that encompasses a range of strategies for analyzing language use to explore the ways in which language structures and is structured by social, cultural, and political relationships. It is often used to study verbal language, and thus is useful in studies of team communication as well as faculty/student dynamics. A number of different approaches to discourse analysis exist; interested researchers may want to begin with Gee (2005) and Fairclough (2003).
- Corpus linguistics employs statistical as well as qualitative analysis to identify semantic patterns across a large collection of written or oral texts (or "corpus"); Biber, Conrad, and Reppen (1998), Kennedy (1998), and McEnery and Wilson (2001) all provide helpful introductions to this methodology.
- Rhetorical analysis also focuses on the study of documents, but where corpus linguistics attends primarily to grammatical and semantic patterns, rhetorical analysis attends more to patterns in the ways arguments are made, evidence is deployed, and persuasive appeals are constructed, along with issues of style and tone. Useful starting points include the collections by Bazerman et al. (2010) and Bazerman and Prior (2004) on writing research.

Perhaps most importantly, as noted at the beginning of this section, methods for studying engineering communication are as varied as the research questions themselves. The approaches described here reflect a critical level of interdisciplinary flexibility as scholars from a range of disciplinary traditions embrace and adapt data collection and analysis methods that help illuminate both what engineering communication practices look like and how engineering communication expertise develops.

Engineering Communication Practices

As suggested in the preceding section, one broad strand of engineering communication research examines engineers' communication patterns in both academic and workplace settings. Though methodologically and epistemologically diverse, collectively such studies point to three critical findings:

- Engineering communication shapes and is shaped by engineering identity.
- Engineering communication is always situated, and always rhetorical.

Table 30.2. Summary of Research Methods Used in Engineering Communication Research

Method	Research Focus	Example Studies	Representative Findings
Surveys	Generalizable descriptions of practice	Kreth, 2000; Lattuca, Terenzini, & Volkwein, 2006; Paretti, McNair, Belanger, & George, 2009; Sageev & Romanowski, 2001; Tenopir & King, 2004	Engineers spend more than 30% of their time in written and oral communication. The ability to communicate effectively is one of, and often the, top-ranked skill valued by employers.
Outcomes measures (e.g., concept inventories)	Impact of communication assignments on content learning	Burrows, McNeill, Hubele, & Bellamy, 2001; Venters, McNair, & Paretti, 2012.	Well-designed writing assignments can enhance content learning.
Discourse analysis	How language patterns shape and are shaped by academic and workplace identities	Paretti & McNair, 2012.	Language use reflects and shapes available professional identities.
		McNair et al., 2010; Wolfe & Powell, 2009.	Language patterns (e.g., pronoun use, commands versus requests, gendered patterns) can foster or inhibit effective collaboration.
Corpus linguistics	Language patterns in academic and workplace settings	Conrad, Dusicka, Pfeiffer, & Evans, 2009; Conrad & Pfeiffer, 2011; Flowerdew, 2011; Krishnamurthy & Kosem, 2007; Orr, 2006.	Engineering-specific language patterns can be used to enhance L2 learning and to differentiate expert and novice communicators.
Rhetorical analysis	Common patterns and argument structures in engineering communication	Dombrowski, 1992; Herndl, et al., 1991; Winsor, 1988.	Differences in argument patterns (uses of evidence, validity of claims, organization of ideas) can create barriers to effective project work that, in extreme cases such as the *Challenger*, have disastrous consequences.
Interviews	Perceptions about communication and/or communication learning	Leydens, 2008; Matusovich, et al., 2012.	Engineers' beliefs about the nature of communication and about communication learning vary substantially, and do not always align with situated, contextual frameworks for learning and practice.

(continued)

Table 30.2 (*continued*)

Method	Research Focus	Example Studies	Representative Findings
Think-aloud protocols	In-situ understanding of communication beliefs and practices	Smith Taylor, 2011; Winsor, 1996. Smith, 2003.	Like faculty, students often do not understand the situated nature of writing and need concrete feedback with explicit reasoning to support learning. Engineering and technical writing faculty share some standards with respect to defining good writing, but differences emerge around issues such as evaluation of content validity and articulation of organizational or structural problems.
Ethnographies and case studies	The ways in which communication practices function, shape, and are shaped by larger academic and workplace contexts	McNair & Paretti, 2010; Winsor, 2003. Dannels, 2000, 2002, 2003; Dannels & Martin, 2008; Herrington, 1985. Harran, 2011.	Communication not only provides ways to exchange or transmit information, but also structures both the work and the relationships among the workers. Expectations around communication differ from context to context. Students are aware of the differences across classroom contexts as well as between academic and workplace contexts, but may need explicit instruction to help them both articulate and adapt to those differences. Collaborations between language and technical faculty can yield fruitful pedagogies, but require intentional, engaged dialogue from both sides and may face institutional barriers.

- Engineers' perspectives about communication practices vary along a rhetorical continuum.

The following sections address each of these findings in turn.

Engineering Communication Shapes and Is Shaped by Engineering Identity

The link between discourse and identity is central to several of the theoretical frameworks addressed in the section "Theoretical Frameworks: The Contextual Nature of Engineering Communication," and evidence of that link appears across a range of workplace and school settings. The issue is critical for researchers and educators in three ways. First, it emphasizes the fact that writing a report or delivering a presentation is not simply something done after the "real" engineering work is over, but is instead always part of the work itself, and failures within a given document can reflect failures in the work. Second, it reminds educators in particular that learning to "write like an engineer" (to use Winsor's term) is closely aligned with learning to think like an engineer. Engineering communication is not a skill students bring fully formed with them to the engineering classroom; rather, it is a practice they develop as they learn the evidence, the chains of reasoning, the language, and the values of the field. And the more explicit faculty are about those issues, the more effective feedback on communication assignments becomes. Finally, the link between communication and identity points to the challenge of expecting students to engage in professional discourse while positioning them in school-based assignment and evaluation contexts.

Several studies have explored the link between communication and identity in engineering. Winsor's book-length ethnography, *Writing Power* (2003), provides perhaps the most detailed exploration of the ways documents that circulate through an engineering center shape the kinds of work engineers do and the kinds of identities (and

power relations) available to them. As her study reveals, texts such as reports, work orders, budgets, graphs, and charts become tools not only for representing knowledge, but also for allocating resources, awarding power, and making critical decisions. Understanding the ways in which these texts operate is part of what new engineers learn as they enter the workplace, and part of what enables individuals to exert access control over their own work. Paretti and McNair's (2012) discourse analysis of design teams points to a similar link on a smaller scale. Their study demonstrates that the ways in which engineers and other members of the team (including managers or facilitators) use language can expand or constrain the kinds of roles engineers can assume – from narrow instrumentalists who function only to "make it work" (where "it" is someone else's idea), to participatory designers who can bring a rich and complex set of insights not only to the "mechanics" of a product, but to its core design and development.

Analyses such as those by Perelman (1999), Geisler (1993, 1998, 2007), and Herndl et al. (1991) focus on the ways in which engineering texts illuminate the kind of logical patterns that underpin the field. Perelman (1999), for example, points to the deliberative logic found in design reports, and the ways in which engineering communication reflects engineers' focus on the question, "Should we do X, and if so, what is the best way to do it?" (p. 69). Similarly, Geisler looks at the ways in which the talk and the texts that circulate through a design project shape the kinds of issues, debates, and decisions within the team. The terms of these arguments and discussions, as the next section suggests, become particularly crucial when engineers engage in dialogue with colleagues outside their field.

Finally, work by Dannels (2000) points to the ways in which students construct professional identities as they work through the communication assignments associated with design projects. As her study shows, students' construction of design reports and presentations embodies their understanding of their own identities and the identities of their customers, clients, and users. Importantly, both the texts and the students' talk about the texts reflect disconnects between academic and professional expectations. As Dannels (2000) points out, the students ultimately worked, and communicated, in ways that were dominated by their student identities: "they did not learn to design through designing a product for a professional client; they designed a product for the classroom in order to learn" (p. 26). The ways in which faculty structure and evaluate communication assignments, that is, influences the kinds of identities students are able to assume, the kinds of questions they are able to ask, and the kinds of values they are able to enact in their work.

Engineering Communication Is Always Situated and Always Rhetorical

Understanding the situated nature of engineering communication is central to helping engineers communicate effectively with those outside their field to share knowledge and inform decision making. One of the greatest struggles for engineers reaches back to the transmission model of communication that undergirds the "engineering writing" courses of the late 1800s. Work by Leydens (2008), Winsor (1996), and others consistently points to the ways in which some engineers (students and professionals) embrace that transmission model and see communication simply as a process of reducing "noise" in the system so that objective messages can be encoded and decoded clearly and efficiently. Most commonly, this perception manifests itself in the belief that engineering communication is a context-free, neutral presentation of data that need no spokesperson. Yet rhetorical analyses of disasters such as the *Challenger* and Three Mile Island point to the ways in which differences in argument patterns such as those described by Perelman (1999) can lead to catastrophic decisions even when "the data" seem clear to engineers (Dombrowski, 1992; Herndl et al., 1991; Winsor, 1988). Herndl et al., for example, identified a series of key

communication breakdowns, rooted in different ways of making an argument, from both incidents that help explain why available corrective actions were not taken. Winsor's rhetorical analysis of publicly available documents from the *Challenger* program identify significant unresolved differences in perception and interpretation between managers and engineers, as well as engineers' inability to communicate bad news; both factors help explain the decision to launch in the face of near-certain technical failure. The engineers who designed the solid-rocket boosters believed, based on data, that the joint that sealed the O-ring (which eventually failed) was unsafe when operating under cold conditions. Those engineers sent that test data to their own managers and NASA personnel, assuming that both would arrive at the same conclusion. However, the managers at NASA placed more stock in the twenty-four previous successful launches as evidence of the joint's safety. What was missing from the engineers' communication was not data, but a clear rationale for the significance of that data framed in terms of the values, concerns, and expertise of those making the launch decision in ways that accounted for both history and the public context of the launch. In assuming that data speak for themselves, and ignoring the situated rhetorical context, the engineers failed to communicate effectively (Winsor, 1996).

The corollary to the belief that data speak for themselves is a persistent belief in engineering communication as neutral, objective, and essentially (to use Winsor's term) *arhetorical*: divorced from any need to interpret and persuade specific audiences in specific contexts. This blind spot is the subject of research involving engineers (or future engineers) in diverse contexts. For instance, studies of students in engineering design courses reveal that written prose is devalued (Geisler, 1993) and that the rhetorical nature of engineering communication – especially the idea that persuading audiences is necessary to gain approval or consent – is rendered virtually invisible in the document composition process (Mathison, 2000). Winsor's landmark longitudinal study of student development over five years poses evidence-based propositions for this rhetorical blind spot in engineering students (Winsor, 1996). First, Winsor contends that the denial of rhetoric is reinforced by engineers' commitment to objectivity. That is, the notion that data speak for themselves emanates from a positivist value on objective knowledge, which either negates, calls into question, or ignores the notions that (1) data can be interpreted in diverse ways, and (2) engineers need to convince others that their designs or services are efficient, useful, or otherwise sound. Second, engineers, especially when newcomers in any given organizational culture, have not yet learned culture- and context-specific ways of thinking, communicating, and interrelating and thus fail to account for context in their communication (Winsor, 1996). And the failure to learn can lead not only to misunderstanding, but to disaster.

Engineers' Perspectives About Communication Practices Vary Along a Rhetorical Continuum

Despite the importance of understanding the rhetorical nature of engineering communication, multiple studies suggest that engineers themselves have widely differing perspectives. Winsor's study not only provided clarity on the neutrality myth; it also opened new questions: What kinds of experiences can destabilize the myth that data speak for themselves? At what junctures in one's engineering career does the myth seem to dissipate? Leydens' phenomenological research (2008) extended Winsor's study by exploring engineers' perspectives on rhetoric and communication. Study participants ranged from students at the end of an undergraduate engineering program to engineers at various stages beyond graduation, including three, fifteen, twenty-five, and more years out. From that study a continuum emerged in which rhetorical awareness increases after the acquisition of industry experience, and often especially after leadership experience in industry. Albeit confounded by contradictory factors, undergraduate engineering students within a semester or less of graduation still held remnants of the neutrality myth. By contrast, engineers who had graduated from the same program and then gained three years of industry experience conveyed no traces of that myth, but instead exhibited a richer, more nuanced understanding of the role of rhetoric in engineering. Some (but not all) engineers working in leadership positions seemed to have the most sophisticated understanding of the role of rhetoric. Engineers at the more sophisticated end of the rhetorical awareness continuum advocated that data, rather than speaking for themselves, require ethical, persuasive spokespersons and interpretive guides (Leydens, 2008). Moreover, recent work suggests that a rhetorically informed understanding of communication is associated with the shift from envisioning engineering communication as purely or almost exclusively technical to seeing it as sociotechnical (Leydens, 2012a) – that is, informed not only by technical dimensions but also social ones, such as economic, environmental, ethical, political, organizational, and interpersonal.

Such findings are consistent with both academic literacy approaches that explore how individual disciplines make meaning from accepted forms of evidence (in this case, data) and rhetorical genre approaches that explore how texts function in different contexts. Given the high stakes associated with engineering communication, however, and the rapid technology development cycles at play in global engineering, it may no longer be feasible to allow students the luxury of time on the job to "figure it out for themselves." The continuum, instead, points to both the possibility and the need for engineering faculty to create classrooms and assignments that accelerate students' development on the rhetorical continuum.

Engineering Communication Pedagogy

Because engineering communication as a field of inquiry emerged in the context of teaching engineers to "write better" (Connors, 1982), it has a rich tradition of pedagogical research. This research has been the subject of two special journal issues – *Language and Learning Across the Disciplines* (now *Across the Disciplines*) (Youra, 1999) and *IEEE Transactions on Professional Communication* (Paretti & McNair, 2008b). In addition, special journal issues and collections examining integrated content and language learning often include discussions of engineering-specific programs (e.g., Bazerman & Russell, 1994; Gonzalez, Weiser, & Fehler, 2009; Gustafsson, 2011; Russell, 2002). Book-length studies of engineering communication pedagogy are more rare, but the discussion of pedagogies at MIT by Poe et al. (2010) provides an excellent overview of a range of curricular and course approaches. Overall, research on engineering communication practice, described earlier, coupled with research on pedagogy, described in this section, point to three key findings:

- When possible, communication instruction should be situated.
- Ideal learning contexts include well-structured communication-intensive courses in the major.
- Communication assignments can enhance content learning.

When Possible, Communication Instruction Should Be Situated

The rhetorical nature of engineering communication, and particularly the need to actively and intentionally persuade and address specific audiences, complicates the transition from school – often rife with decontextualized research papers and laboratory reports – to work (e.g., Beaufort, 2007; Dias et al., 1999; Herrington, 1985). As Winsor (1996) noted in her landmark study, "it may be hard for students to see writing as rhetorical and contextualized in a school setting. Perhaps for most people, the rhetorical nature of writing becomes most obvious when they engage in authentic language tasks, such as those required by the practice of a profession"

(p. 8). Fortunately, more recent studies suggest that although academic courses cannot perfectly mimic industrial contexts, such courses, when well designed, can effectively, if imperfectly, simulate workplace expectations (Artemeva, 2005, 2007, 2008; Artemeva et al., 1999; Dannels, 2000, 2003; Paretti, 2006, 2008, 2009; Paretti & Burgoyne, 2005a; Poe et al., 2010). Such courses are successful when they situate writing assignments within the context of engineering work; as both Dannels and Paretti note, although students readily understand the distinctions between school and work, they are able to engage successfully in workplace communication around school design tasks, particularly when the assignments are intentionally structured to foster that engagement.

Ideal Learning Contexts Include Well-Structured Communication-Intensive Courses in the Majors

Curricular approaches for engineering communication include standalone technical communication courses, individual writing- or communication-intensive technical courses, and curricula designed to integrate communication across multiple years. Although each can support student learning, research strongly suggests that integrating communication into engineering courses across the curriculum is critical.

Effective standalone courses can ground students' meta-knowledge of communication, and both Jarratt, Mack, Sartor, and Watson's (2009) university-wide study and Ford's (2004) study of engineering students demonstrate that students do transfer some skills from writing courses to technical courses, particularly when supported by further in-discipline work that helps make the connections explicit. To support such transfer, Boiarsky (2004), echoing current trends in student learning (Bransford, Brown, & Cocking, 2000), argues for an explicit metacognitive approach to such courses that focuses on understanding audience, purpose, and context (rather than format). Similarly, Artemeva's use of rhetorical genre theory to inform assignment design familiarizes engineering students with the common reporting structures (1999, 2007) in ways that show how "format" links to audience, purpose, and context. Standalone courses designed specifically for engineering students can also integrate multiple professional practices, as in Ballentine's (2008) incorporation of "ethics, intellectual property, design, and globalization" (p. 328). First-year composition programs also provide possibilities for partnerships to more closely align such general courses with workplace communication practices (Leydens & Schneider, 2009).

However, standalone courses are limited in their ability to serve engineering students, in part because, as noted earlier, textbooks for such courses often lack critical discussions about representing technical data and fail to address the key forms of evidence and argument central to engineering discourse (Wolfe, 2009). Moreover, when communication education occurs exclusively outside a discipline-enriched context, compartmentalization occurs and writing becomes simply a post–problem-solving chore rather than an integral work task (Poe et al., 2010). The view is reflected, as well, in engineering faculty who perceive communication as something taught "elsewhere" – often in an English course or at best in a single capstone design course (Matusovich et al., 2012) – and communicate that split to their students.

Such limitations mean that the literature is also rich with descriptions of approaches to integrating communication into engineering courses, often through collaboration. Collaborations between language and engineering faculty have been described in detail by several researchers, along with guidelines for establishing such collaborations at the classroom level (e.g., Harran, 2011; Jacobs, 2010). The embedded model supported by such collaborations offers a richer definition of communication skills and recognizes that writing occurs throughout the engineering or scientific process, even when it is not seen as "writing" (Locke, 1992; Poe et al., 2010). For interested faculty, a number of useful case studies

describe successful assignments. For example, Craig, Learner, and Poe describe ways to address laboratory research skills, teamwork, and visual data-driven arguments for disciplines from biology to aeronautics to biomedical engineering (Craig, Lerner, & Poe, 2008; Poe et al., 2010). Paretti (2006) provides guidelines for design and assessment of assignments grounded in situated learning to mirror meaningful workplace contexts. Design education researchers have described an array of approaches to establishing workplace communication practices in design courses at both first-year and capstone courses (Dannels, 2002, 2003; Dannels et al., 2003; Davis, Beyerlein, Harrison, Thompson, & Trevisan, 2007; Hirsch et al., 2001; Paretti, 2008; Shwom, Hirsch, Yarnoff, & Anderson, 1999). The Transferable Integrated Design Engineering Education group (http://www.tidee.wsu.edu/) has provided a number of online resources for teaching and evaluating communication, and the emerging Capstone Design Hub promises communication resources specifically for capstone courses in engineering (Howe, 2012).

But even individual disciplinary courses are limited when they occur in isolation. Numerous studies, both within engineering and in the broader language-related fields noted earlier, describe the complexity of transferring knowledge from one context to another and reinforce the need for cross-contextual learning. Within engineering, for example, Herrington identified significant differences in expectations for "good writing" even between two chemical engineering courses at the same university in the same department (1985). Similarly, Thaiss and Zawacki's 2006 study of writing across a range of disciplines points to the variations across and within disciplines as faculty articulate different expectations and norms. Such work suggests that students need to acquire facility with disciplinary communication not in one course once, but iteratively as they move through multiple courses in their field. Studies in other disciplines (Berkenkotter & Huckin, 1995; Berkenkotter, Huckin,

& Ackerman, 1994; Jarratt et al., 2009; McCarthy, 1994) as well as studies of school-to-work transition (Dias et al., 1999; Freedman & Adam, 2000) point to similar conclusions and stress students' need to experience "frequent writing for a variety of teachers and courses" (Thaiss & Zawacki, 2006, p. 121). Faculty interested in developing integrated curricula can benefit from the case study presented by Patton (2008), which describes the processes by which one civil engineering department developed an integrated curriculum that addressed, in concrete terms, the communication needs and norms of the profession. A similar project describes innovations in a materials science and engineering department (Hendricks & Pappas, 1996), an approach later expanded by Paretti and Burgoyne (2005b) to intentionally vary genre knowledge and leverage situated learning. Airey (2011) provides an alternate framework that embraces multimodal communication as well as academic, workplace, and societal contexts to provide a framework for language and content teachers to engage in productive dialogue around curriculum.

Because of the challenge of integrating writing into large courses, researchers have also begun to examine approaches that facilitate student learning without overburdening instructors. One such approach is Calibrated Peer Review (CPR), a computer-based system that provides formative and summative assessment through computer-mediated peer review (Carlson & Berry, 2003, 2008). The system provides tools for instructors to tailor expectations for various assignments and provide models, and then enables students to "train" themselves through interacting with the system to identify strengths and weaknesses of the specified document types. The process-writing approach developed by Hanson and Williams and further enhanced by Venters et al. similarly focuses on small-scale assignments that, even without computer-mediated grading, can still be managed within the constraints of engineering service courses such as statics (Hanson & Williams, 2008; Venters et al., 2012). Although such

systems provide manageable approaches to incorporating small writing tasks into high-enrollment courses, their effectiveness depends heavily on the assignment design, and they are limited in their ability to address the broader rhetorical concerns that can overcome the myths surrounding neutrality and decontextualization. They are thus most useful in conjunction with integrated curricular-based approaches.

Communication Assignments Can Enhance Content Learning

One of the most frequent concerns faculty raise about integrating communication assignments into engineering is that such assignments take time away from "content" (Matusovich et al., 2012). Yet multiple studies suggest that writing assignments can support content learning. Klein's (1999) overview of cognitive theories and empirical studies provides a useful summary of the ways in which writing practices can support learning: (1) the initial act of writing itself brings forth new knowledge as writers allow ideas to flow; (2) learning occurs as writers revise to organize, link, evaluate, and refine their ideas; (3) different genre structures help students move through different logical processes; and (4) learning occurs as students plan their writing and set goals for problem solving. The systematic review of studies on writing-to-learn conducted by Bangert-Downs, Hurley, and Wilkinson (2004) found that in a majority of cases, statistically significant content learning gains were associated with writing exercises. More importantly, these gains were most often correlated with writing that included metacognitive reflection that allowed students to examine their level of understanding and their learning processes.

Recent research has continued to explore the ways in which writing supports learning. For example, recent work has examined the ways in which writing in college science courses helped socialize students into the practices and norms of the discipline, acting as a means of enculturation (Carter, Ferzli, & Wiebe, 2007). Their study shows how writing discipline-specific lab reports supported students' understanding of how to "do" biology as well as how to communicate experimental results. Within engineering, Venters et al. demonstrate that the short "process problem" assignments developed by Hanson and Williams (2008) to support communication skills, in which students write brief explanations of problems in statics, can be employed in large sections of several hundred students to enhance conceptual understanding, as measured by a validated concept inventory (Venters et al., 2012). These and similar studies (see Bean, 1996, 2011) can help overcome faculty concerns about losing "content" when they incorporate communication assignments.

Summary: Implications for Practitioners

The findings described in the preceding section, together with the findings regarding communication practice described in the previous section, lead to the guidelines in Table 30.3. These guidelines synthesize core principles in the teaching and learning of communication across written, oral, and multimodal forms, in formal and informal settings. The listed citations provide more detail for each guideline, but are by no means comprehensive. Perhaps most importantly, what characterizes many of the most effective approaches is interdisciplinary engagement between engineering and communication faculty that fosters mutual understanding in which engineering faculty begin to learn the frameworks, approaches, and values of communication specialists and communication specialists learn the frameworks, approaches, and values of engineers. Such collaborations can be complicated by power, disciplinary biases, time, and other issues, but when effective, they can result in effective learning contexts for engineering students.

Assessment

Approaches to assessing engineering communication reflect larger trends in communication assessment across fields. As noted in the introduction to *Assessment of Writing* (part of the *Assessment in the Disciplines* series from Association of Institutional Research), writing assessment has a rich history that, like the history of communication research, has moved from formulaic evaluation of grammar and format to rubric-based holistic assessment of the effectiveness of a given document, presentation, poster, Web page, or other communication product (Paretti & Powell, 2009a). Several books and edited collections are available to help guide faculty through designing and implementing classroom and program writing assessment in particular (e.g., Elliot & Perelman, 2012; Huot, 2002; Huot & O'Neill, 2008; Paretti & Powell, 2009b; Yancey & Huot, 1997).

The principles for assessing writing and other forms of communication follow principles common to all assessment (Schneider, Leydens, Olds, & Miller, 2009): attention to stakeholders, clear connections between values and method, support for ongoing learning from the process and results, local grassroots involvement, and awareness of the situated nature of writing. In enacting these principles, however, two trends are critical for engineering communication: the development of meaningful rubrics to guide students and faculty (e.g., House et al., 2009; Paretti, 2006; Swarts & Odell, 2001), and the implementation of portfolios to assess student performance across a wide array of disciplinary genres and contexts (McNair, Paretti, Knott, & Wolfe, 2006; Ostheimer & White, 2005; Paretti, 2005; Scott & Plumb, 1999; Turns & Lappenbusch, 2006; Turns, Sattler, Eliot, Kilgore, & Mobrand, 2012; Williams, 2002). Rubrics describe performance criteria and thus provide critical tools to help faculty and students develop a common language and a shared set of expectations about communication in specific contexts. In doing so, they align well with the theoretical frameworks described at the outset of this chapter, and they provide a means to define the "effective communication" called for in ABET, Inc.'s Criterion 3-g in concrete

Table 30.3. Guidelines for Engineering Communication Assignments

- Design of communication assignments and curricula are often most effective when they result from partnerships between communication and engineering experts (Harran, 2011; Jacobs, 2010; Paretti, 2011).
  - Effective partnerships involve active mutual learning – communication experts learning about engineering and engineering faculty learning about communication – not only to "speak one another's language" but to design assignments and curricula that best support students' sociocognitive development as professionals within their field. Engaged dialogue is critical to successful collaboration.
  - Such partnerships do not always have to be team teaching arrangements; they can include communication experts helping engineering faculty design manageable assignments within engineering courses, as well as engineering experts helping communication faculty better understand the norms and expectations of different engineering fields.
- Communication learning should occur across a curriculum, embedded within engineering courses, to reflect students' developing professional identities, their need to engage with different audiences (e.g., peers, experts, managers, public), and their need to practice a range of genres (e.g., lab reports, proposals, journal articles, elevator pitches, design briefs) for different contexts and purposes (Paretti & Burgoyne, 2005b; Patton, 2008; Reave, 2004).
- When communication skills are taught in standalone "technical writing" or "technical communication" courses, cross-curricular dialogue is needed to help ensure transfer (Ford, 2004; Jarratt et al., 2009).
  - Engineering faculty can actively support transfer by learning and referencing the language and core concepts taught in technical communication courses.
  - Communication faculty can help engineering faculty better understand how concepts taught in a general communication course can transfer to or serve as the foundation for discipline-specific communication assignments.
- Assignments should reflect discipline-appropriate workplace genres to the extent possible (Artemeva, 2007; Artemeva et al., 1999; Patton, 2008); where the constraints of academic courses require different approaches, instructors should make those differences explicit to students.
  - Common genres such as design or laboratory reports should not be taught as arbitrary formats to follow, but rather as structures that are useful for achieving discipline-specific purposes (Artemeva et al., 1999).
  - Assignments should engage real audiences with explicit, meaningful information needs to help students learn how audience, purpose, and context shape communication (Paretti, 2006, 2008; Spinuzzi, 1996).
- Effective feedback, best understood as coaching rather than correcting (Smith Taylor, 2011), is critical to student development, particularly when opportunities for revision are available.
  - Marking that focuses on basic mechanics or formatting is generally less helpful to students; form-focused comments in particular may appear as simply arbitrary rules to follow.
  - Instructors should provide explanations and rationales for suggested changes to help students understand how or why the changes will help the document better achieve its purpose.
  - Suggestions are most helpful when they identify the reader's perspective and link changes to reader needs or experiences (e.g., confusion, inability to follow logic).
- Effective evaluation approaches generally employ rubrics that highlight the skills targeted by the assignment, including both technical competence (e.g., ability to interpret experimental results in light of theory) and communication competence (e.g., ability to effectively target the intended audience) (House et al., 2009).

terms – a central assessment requirement (Driskill, 2000). Such rubrics reflect the kind of holistic evaluation model described by Pappas et al. (2000, 2004) for engineering communication.

As a starting point, the Association of American Colleges and Universities (AAC&U) developed rubrics for liberal education outcomes, including communication (The VALUE Project Overview, 2009); the communication rubric (available at http://www.aacu.org/value/rubrics/) provides a framework adaptable to a range of programs and assignments. House et al. (2009) provide case studies of assignments and rubrics in both civil and chemical engineering, at the program and classroom levels respectively, that serve as exemplars for other programs. Brinkman and van der Geest (2003) provide a similar approach, identifying competencies for engineering communication at four layers (text, genre, strategy, and feedback), with criteria at each layer. Like House et al., they provide approaches to both student and program assessment that should prove useful to engineering educators. Critical, however, is the attention to local context; rubrics should align with the overall goals and design of the assignment as well as with the specific technical and communication outcomes targeted; they cannot simply be imported uncritically from one case to another. Finally, as the case studies by House et al. (2009) and Poe et al. (2010) make clear, discussions of the effectiveness of any communication artifact are always embedded in discussions around the technical work at stake; the data, as we noted earlier, do not speak for themselves.

Rubrics can be developed to evaluate both individual assignments and portfolios. Portfolios allow programs to conduct assessment across a range of skills and contexts, and they are seeing increasing use in engineering programs not only to assess communication (Williams, 2002), but also to provide both engineering students and faculty a means to understand engineering students' development holistically (McNair & Garrison, 2012; Turns & Lappenbusch, 2006; Turns et al., 2012). From an engineering communication perspective, portfolios align particularly well with the theoretical frameworks discussed earlier. They provide a mechanism to assess students' work across a variety of genres for different audiences in the context of different kinds of engineering work (Paretti & Burgoyne, 2009), and they provide opportunities for developing metacognitive awareness through reflective writing (Thaiss & Zawacki, 2006).

One additional component of assessment that has received general treatment in the writing pedagogy literature and some treatment in the engineering communication literature is formative assessment – that is, faculty providing meaningful and actionable feedback. Work by both Swarts and Odell (2001) and Smith Taylor (2011) emphasizes the need to provide meaningful feedback to students as they work on drafts. Smith Taylor's work is particularly useful for educators because it identifies characteristics of effective communication (including asking meaningful questions and providing explanations and rationales for suggestions).

Finally, although assessment of communication itself is one critical issue, it is also important to note that communication assignments are themselves often effective tools for assessing other core engineering learning outcomes. For example, in addition to effective communication (Criterion 3-g), ABET, Inc. includes five other professional skills that programs need to teach and measure via direct methods that "judge student work, projects, and portfolios developed as a result of the learning experiences" (Spurlin, Rajala, & Lavelle, 2008, p. 32). Most of these outcomes can be demonstrated through artifacts such as papers, reports, and portfolios. Evidence such as graded laboratory reports and student presentations were recommended for even some of the more technical skills, such as conducting experiments and analyzing data (Criterion 3-b) and designing products or process (Criterion 3-c) (Spurlin et al., 2008). Using communication assignments to provide measurable outcomes means, notably, that programs need to take responsibility for teaching the skills to produce those outcomes; for example, students need to have the ability not only to complete a report within a course, but also to write clear, accurate, and effective reports of their research beyond the confines of the classroom.

Emerging Research

The preceding descriptions depict engineering communication as a vibrant, growing, and dynamic area of study. As engineering communication continues to evolve, several important new directions merit mention. First, new research has recently appeared on the horizon that promises a broader reimagining of the concept of *communication*. For instance, in a review of emerging trends in several engineering communication programs, one study showed evidence of increasing interest in communication reconceptualized as WOVE (Leydens & Schneider, 2009). That acronym stands for writing, oral, visual, and electronic communication, and recognizes that writing and speaking do not occur in a vacuum but are often embedded in dynamic, multimodal contexts. As others have noted, teaching students writing using pre-digital age methods fails to acknowledge the realities and complexities of twenty-first century composition processes (e.g., Selfe, 2007; Sheridan & Inman, 2010). Other research has suggested possible additions to WOVE, including adding N for nonverbal communication (e.g., Burnett, 2008) and L for listening. Research on listening has indicated that listening is highly valued by prominent engineering education stakeholders yet virtually absent from the engineering education curriculum (Leydens & Lucena, 2009).

Second, the prevalence of electronic communication, particularly as necessitated by global, virtual design collaborations, has led to new research on the role of such technologies in engineering (Bøhn & Anderl, 2005; Coppola, Hiltz, & Rotter, 2004; Fruchter & Lewis, 2001; Gosavi, 2010; Kock & Nosek, 2005; McNair & Paretti, 2010; McNair et al., 2010; Paretti et al., 2007; Robey, Khoo, & Powers, 2004; Rutkowski, Vogel, van Genuchten, Bemelmans, & Favier, 2002). But in general, engineering curricula have been slower to embrace multimodal communication, relying instead on traditional forms of reports and presentations. Few studies to date have explored strategies for embedding multimodal communication within engineering courses in ways that meaningfully reflect workplace needs and practices, though such approaches are increasingly part of technical communication courses more broadly. Clearly, as communication technologies continue to shift and work becomes increasingly distributed, more work needs to be done by engineering communication researchers to identify ways to integrate these communication practices into the curriculum.

A final strand of emerging research appears in a turn toward engineering communication as it interfaces with engineering and social justice. For instance, some research proposes a theory of contextual listening for engineers working in sustainable community development contexts (Leydens & Lucena, 2009; Lucena, Schneider, & Leydens, 2010). Contextual listening is a multidimensional, integrated understanding of the listening process wherein listening facilitates meaning making, enhances human potential, and helps foster community-supported change (Leydens & Lucena, 2009x; Lucena et al., 2010). Other research has sought to understand communication gaps that emerge between engineers who advocate for integrating social justice into engineering education and engineers who resist such integration (Leydens, Lucena, & Schneider, 2012). Similarly, one study examined the reasons why social justice may be an appropriate – yet resisted – fit for much professional communication research (Leydens 2012b) As with research on the *Challenger* disaster, applications of communication theory to such real engineering contexts bring that theory alive, providing rich opportunities for student learning.

This chapter has also raised several opportunities for future investigation. Future research should continue to expand our understanding of effective context specific and generalizable practices that foster deep learning of both professional communication and engineering concepts simultaneously. More research is also welcome on how communication practices shape and are shaped by larger sociocultural and sociotechnical contexts, especially within cross-cultural settings and with implications for engineering communication pedagogy, and how such shaping impacts the identities of engineers. Engineering communication research that astutely builds on existing conceptual frameworks and methodologies is important, yet so is research that shows the limitations of these frameworks and methodologies. Even better, in this dynamic research area, the day will soon come when we forge connections to new, relevant frameworks and methodologies. Future inquiry is also needed on how local, institutional interpretations of accreditation and other guiding forces position engineering communication to range from being integrated with to separated from (and points in between) engineering courses and program assessments. As an area of research, engineering communication is poised to reach exciting new levels of development.

Conclusion

We began this chapter by considering the centrality of communication in engineering work, as reflected in both its pervasiveness and its potential consequences. As the research described in this chapter makes clear, engineering knowledge is continuously created and enacted in communicative practices that define and are defined by the nature of engineering work itself. Scholars continue to employ a variety of theoretical and methodological approaches to understand the arguments, evidence, values, and identities embodied in engineers' ways of speaking and writing. In doing so, these scholars help identify the kinds of barriers engineers face when trying to communicate effectively in the digitally mediated global workplace.

This research into communication practices in turn helps define communication goals for engineering education. As engineers seek to make their knowledge accessible to colleagues, managers, technicians, and the public, they need tools and approaches that help them move beyond grammatical correctness, adherence to formats, and abstract concepts of "clarity," and instead enable them to analyze rhetorical situations to design texts that meet audience needs and accomplish critical goals, and to contextualize their work in larger social and cultural contexts. Learning such tools and approaches, as the pedagogical research outlined here demonstrates, involves situated learning across a range of contexts that helps students understand the rhetorical nature of communication and the sociotechnical nature of engineering work so that they can enter into meaningful cross-disciplinary, cross-cultural collaborations. Such learning emerges most effectively, studies suggest, when engineering and communication faculty come together in dialogue to learn one another's language; understand one another's work; and collaborate to design assignments, feedback systems, and assessment tools that help students learn to think and write like globally competent engineers.

Acknowledgments

This material is based in part upon work supported by the National Science Foundation under Grant Nos. 1025189 and 1129338. Any opinions, findings, and conclusions or recommendations expressed in this material are those of the authors and do not necessarily reflect the views of the National Science Foundation.

References

ABET Engineering Accreditation Commission (2007). *Criteria for accrediting engineering programs.* Baltimore, MD: ABET, Inc.

Airey, J. (2011). The disciplinary literacy discussion matrix: A heuristic tool for initiating collaboration in higher education. *Across the Disciplines*, 8(3). Retrieved from http://wac .colostate.edu/atd/clil/airey.cfm

Artemeva, N. (2005). A time to speak, a time to act: A rhetorical genre analysis of a novice engineer's calculated risk taking. *Journal of Business and Technical Communication*, 19(4), 389–421.

Artemeva, N. (2007, August). *Becoming an engineering communicator: Novices learning engineering genres*. In A. Bonini, D. de Carvalho Figueiredo, & F. J. Rauen (Eds.), *Proceedings of the 4th International Symposium on Genre Studies*, Tubarão, Brazil (pp. 253–265).

Artemeva, N. (2008). Toward a unified social theory of genre learning. *Journal of Business and Technical Communication*, 22(2), 160–185.

Artemeva, N., Logie, S., & St-Martin, J. (1999). From page to stage: How theories of genre and situated learning help introduce engineering students to discipline-specific communication. *Technical Communication Quarterly*, 8(3), 301–316.

Ballentine, B. D. (2008). Professional communication and a 'whole new mind': Engaging with ethics, intellectual property design, and globalization. *IEEE Transactions on Professional Communication*, 51(3), 328–340.

Bangert-Drowns, R. L., Hurley, M. M., & Wilkinson, B. (2004). The effects of school-based writing-to-learn interventions on academic achievement: A meta-analysis. *Review of Educational Research*, 74(1), 29–58.

Barton, D., Hamilton, M., & Ivanič, R. (Eds.). (2000). *Situated literacies: Reading and writing in context*. London: Routledge.

Bazerman, C., Krut, R., Lunsford, K., McLeod, S., Rogers, P., & Stansell, A. (Eds.). (2010). *Traditions of writing research*. New York, NY: Routledge.

Bazerman, C., & Prior, P. (Eds.). (2004). *What writing does and how it does it: An introduction to analyzing texts and textual practices*. Mahwah, NJ: Lawrence Erlbaum.

Bazerman, C., & Russell, D. R. (Eds.). (1994). *Landmark essays on writing across the curriculum*. Davis, CA: Hemagoras Press.

Bean, J. (1996). *Engaging ideas: The professor's guide to integrating writing, critical thinking, and active learning in the classroom* (1st ed.). San Francisco, CA: Jossey-Bass.

Bean, J. (2011). *Engaging ideas: The professor's guide to integrating writing, critical thinking, and*

active learning in the classroom (2nd ed.). Hoboken, NJ: John Wiley & Sons.

Beaufort, A. (2007). *College writing and beyond: A new framework for university writing instruction*. Logan, UT: Utah State University Press.

Berkenkotter, C., & Huckin, T. (1995). *Genre knowledge in disciplinary communication: Cognition/culture/power*. Hillsdale, NJ: Lawrence Erlbaum.

Berkenkotter, C., Huckin, T., & Ackerman, J. (1994). Social context and socially constructed texts: The initiation of a graduate student into a writing research community. In C. Bazerman & D. R. Russell (Eds.), *Landmark essays: Writing across the curriculum* (Vol. 6, pp. 211–232). Davis, CA: Hermagoras Press.

Biber, D., Conrad, S., & Reppen, R. (1998). *Corpus linguistics: Investigating language structure and use*. Cambridge: Cambridge University Press.

Bijker, W. E., Hughes, T. P., & Pinch, T. J. (1987). *The social construction of technological systems: New directions in the sociology and history of technology*. Cambridge, MA: MIT Press.

Björk, L., Bräuer, G., Rienecker, L., & Stray Jörgensen, P. (Eds.). (2003). *Teaching academic writing in European higher education* (Vol. 12). New York, NY: Springer.

Bøhn, J. H., & Anderl, R. (2005). Transatlantic curse on 24/7 collaborative engineering and product data management. In *Proceedings of the American Society for Engineering Education Southeast Section Conference*. Retrieved from http://se.asee.org/proceedings/ asee-proceedings_old.htm

Boiarsky, C. (2004). Teaching engineering students to communicate effectively: A metacognitive approach. *International Journal of Engineering Education*, 20(2), 251–260.

Borrego, M. (2007). Conceptual difficulties experienced by trained engineers learning educational research methods. *Journal of Engineering Education*, 96(2), 91–102.

Borrego, M., Douglas, E., & Amelink, C. T. (2009). Quantitative, qualitative, and mixed research methods in engineering education. *Journal of Engineering Education*, 98(1), 53–66.

Bransford, J. D., Brown, A. L., & Cocking, R. R. (Eds.). (2000). *How people learn: Brain, mind, experience, and school*. Washington, DC: The National Academies Press.

Brinkman, G. W., & van der Geest, T. M. (2003). Assessment of communication competencies in engineering design projects. *Technical Communication Quarterly*, 12(1), 67–81.

Burnett, R. (2008, September). *Refiguring technologies in multimodal communication: Ways to improve engagement and learning.* Paper presented at the Bedford/St. Martin's Composition Symposium, Denver, CO.

Burrows, V. A., McNeill, B., Hubele, N. F., & Bellamy, L. (2001). Statistical evidence for enhanced learning of content through reflective journal writing. *Journal of Engineering Education*, 90(4), 661–667.

Carlson, P. A., & Berry, F. C. (2003). CPR: A tool for addressing EC2000, Item 'g' – Ability to Communicate Effectively (WIP). In *Proceedings of the ASEE/IEEE Frontiers in Education Conference.* Retrieved from http://ieeexplore.ieee.org/xpl/conhome.jsp?punumber=1000297

Carlson, P. A., & Berry, F. C. (2008). Using computer-mediated peer review in an engineering design course. *IEEE Transactions on Professional Communication*, 51(3), 264–279.

Carter, M., Ferzli, M., & Wiebe, E. (2007). Writing to learn by learning to write in the disciplines. *Journal of Business and Technical Communication*, 21(3), 278–302.

Cismas, S. C. (2010, July). Foreign language technical writing abilities for power engineering students in the polytechnic university of Bucharest. Paper presented at the Proceedings of the 7th WSEAS International Conference on Engineering Education, Corfu Island, Greece.

Cole, M., Engeström, Y., & Vasquez, O. (Eds.). (1997). *Mind, culture, and activity: Seminal papers from the laboratory of comparative human cognition.* Cambridge: Cambridge University Press.

Committee on Science, Engineering, and Public Policy. (2006). *Rising above the gathering storm: Energizing and employing America for a brighter economic future.* Washington, DC: The National Academies Press.

Connors, R. J. (1982). The rise of technical writing instruction in America. *Journal of Technical Writing and Communication*, 12(4), 329–352.

Conrad, S., Dusicka, P., Pfeiffer, T., & Evans, R. (2009). Work in progress – A new approach for understanding student and workplace writing in engineering. In *Proceedings of the ASEE/IEEE Frontiers in Education Conference.* Retrieved from http://ieeexplore.ieee.org/xpl/conhome.jsp?punumber=1000297

Conrad, S., & Pfeiffer, T. J. (2011). Preliminary analysis of student and workplace writing in civil engineering. In *Proceedings of the American Society for Engineering Education Annual Conference and Exposition.* Retrieved from http://www.asee.org/search/proceedings

Coppola, N. W., Hiltz, S. R., & Rotter, N. G. (2004). Building trust in virtual teams. *IEEE Transactions on Professional Communication*, 47(2), 95–104.

Covington, D., Barksdale, E., Egan-Warren, S., Larsen, J., & Trunzo, S. (2007). *Communication in the workplace: What can NC State students expect?* Raleigh: North Carolina State University. Retrieved from http://courses.ncsu.edu/eng331/common/resources/ciw2007/

Craig, J. L., Lerner, N., & Poe, M. (2008). Innovation across the curriculum: Three case studies in teaching science and engineering communication. *IEEE Transactions on Professional Communication*, 51(3), 280–301.

Dannels, D. P. (2000). Learning to be professional: Technical classroom discourse, practice, and professional identity construction. *Journal of Business and Technical Communication*, 14(1), 5–37.

Dannels, D. P. (2002). Communication across the curriculum and in the disciplines: Speaking in engineering. *Communication Education*, 51(3), 254–268.

Dannels, D. P. (2003). Teaching and learning design presentations in engineering: Contradictions between academic and workplace activity systems. *Journal of Business and Technical Communication*, 17(2), 139–169.

Dannels, D. P., Berardinelle, P., Anson, C. M., Bullard, L., Kleid, N., Kmeic, D., & Peretti, S. (2003). Integrating teaming, writing, and speaking in CHE Unit Operations Lab. In *Proceedings of the American Society for Engineering Education Annual Conference and Exposition.* Retrieved from http://www.asee.org/search/proceedings

Dannels, D. P., & Martin, K. N. (2008). Critiquing critiques: A genre analysis of feedback across novice to expert design studios. *Journal of Business and Technical Communication*, 22(2), 135–159.

Davis, D., Beyerlein, S., Harrison, O., Thompson, P., & Trevisan, M. (2007). Assessments for three performance areas in capstone engineering design. In *Proceedings of the American Society for Engineering Education Annual*

Conference and Exposition. Retrieved from http://www.asee.org/search/proceedings

Dias, P., Freedman, A., Medway, P., & Pare, A. (1999). *Worlds apart: Acting and writing in academic and workplace contexts*. Mahwah, NJ: Lawrence Erlbaum.

Dombrowski, P. (1992). *Challenger* and the social contingency of meaning: Two lesson for the technical communication classroom. *Technical Communication Quarterly, 1*(1), 73–85.

Downey, G. L., Lucena, J. C., Moskal, B., Parkhurst, R., Bigley, T., Hays, C., . . . Nichols-Belo, A. (2006). The globally competent engineer: Working effectively with people who define problems differently. *Journal of Engineering Education, 95*(2), 107–122.

Driskill, L. (2000). Linking industry best practices and EC3(g) Assessment in Engineering Communication. In *Proceedings of the American Society for Engineering Education Annual Conference and Exposition*. Retrieved from http://www.asee.org/search/proceedings

Elliot, N., & Perelman, L. (Eds.). (2012). *Writing assessment in the 21st Century: Essays in honor of Edward M. White*. Cresskill, NJ: Hampton Press.

Engeström, Y., & Middleton, D. (Eds.). (1998). *Cognition and communication at work*. Cambridge: Cambridge University Press.

Engeström, Y., Miettinen, R., & Punämaki, R.-L. (Eds.). (1999). *Perspectives on activity theory*. Cambridge: Cambridge University Press.

Engineering Council. (2011). *The accreditation of higher education programmes: UK standard for professional engineering competence*. London: Engineering Council.

Engineers Australia. (2011). *Stage 1 competency standard for the professional engineer*. Barton ACT: Engineers Australia.

Fairclough, N. (2003). *Analysing discourse*. New York, NY: Routledge.

Farrell, L., & Holkner, B. A. M. U. A. (2004). Points of vulnerability and presence: Knowing and learning in globally networked communities. *Discourse: Studies in the Cultural Politics of Education, 25*(2), 133–144.

Flowerdew, L. (2011). ESP and corpus studies. In D. Belcher, A. M. Johns & B. Paltridge (Eds.), *New directions for ESP research* (pp. 222–251). Ann Arbor: University of Michigan Press.

Ford, J. D. (2004). Knowledge transfer across disciplines: Tracking rhetorical strategies from a technical communication classroom to an engineering classroom. *IEEE Transactions on Professional Communication, 47*(4), 1–15.

Freedman, A., & Adam, C. (2000). Write where you are: Situating learning to write in university and workplace settings. In P. Dias & A. Pare (Eds.), *Transitions: Writing in academic and workplace settings* (pp. 31–60). Cresskill, NJ: Hampton Press.

Fruchter, R., & Lewis, S. (2001). Mentoring models in an A/E/C global teamwork e-learning environment. In *Proceedings of the American Society for Engineering Education Annual Conference and Exposition*. Retrieved from http://www.asee.org/search/proceedings

Gee, J. P. (2005). *An introduction to discourse analysis: Theory and method* (2nd ed.). New York, NY: Routledge.

Geisler, C. (1993). The relationship between language and design in mechanical engineering: Some preliminary observations. *Technical Communication, 40*(1), 173–175.

Geisler, C., & Lewis, B. (2007). Remaking the world through talk and text: What we should learn from how engineers use language to design. In R. Horowitz (Ed.), *Talking texts: How speech and writing interact in school learning*. Mahwah, NJ: Lawrence Erlbaum.

Geisler, C., Rogers, E., & Haller, C. R. (1998). Disciplining discourse: Discourse practices in the affiliated professions of software engineering design. *Written Communication, 15*(1), 3–24.

Gianniny, O. A., Jr. (2004). A century of ASEE and liberal education (or how did we get here from there, and where does it all lead?). In D. F. Ollis, K. A. Neeley, & H. C. Luegenbiehl (Eds.), *Liberal education in twenty-first century engineering: Responses to ABET/EC 2000 criteria* (pp. 320–346). New York, NY: Peter Lang.

Gonzalez, A., Weiser, E., & Fehler, B. (Eds.). (2009). *Engaging audience: Writing in an age of new literacies*. Urbana, IL: NCTE Press.

Gosavi, A. (2010). Building communication skills through virtual teaming. In *Proceedings of the American Society for Engineering Education Annual Conference and Exposition*. Retrieved from http://www.asee.org/search/proceedings

Gustafsson, M. (Ed.). (2011). Collaborating for content and language integrated learning [Special Issue]. *Across the Disciplines 8*(3). Retrieved from http://wac.colostate.edu/atd/clil/index.cfm

Gustafsson, M., Eriksson, A., Räisänen, C., Stenberg, A.-C., Jacobs, C., Wright, J., . . . Winberg, C. (2011). Collaborating for content and language integrated learning: The situated character of faculty collaboration and student learning. *Across the Disciplines*, 8(3). Retrieved from http://wac.colostate.edu/atd/clil/index.cfm

Hanson, J. H., & Williams, J. M. (2008). Using writing assignments to improve self-assessment and communication skills in an engineering statics course. *Journal of Engineering Education*, 97(4), 515–530.

Harran, M. (2011). Engineering and language discourse collaboration: Practice realities. *Across the Disciplines*, 8(3). Retrieved from http://wac.colostate.edu/atd/clil/index.cfm

Hendricks, R. W., & Pappas, E. C. (1996). Advanced engineering communication: An integrated writing and communication program for materials engineers. *Journal of Engineering Education*, 85(4), 343–352.

Herndl, C. G., Fennell, B., & Miller, C. (1991). Understanding failures in organizational discourse: The accident at Three Mile Island and the shuttle *Challenger* disaster. In C. Bazerman & J. Paradis (Eds.), *Textual dynamics of the professions* (pp. 279–305). Madison: University of Wisconsin Press.

Herrington, A. J. (1985). Writing in academic settings: A study of the contexts for writing in two college chemical engineering courses. In C. Bazerman & D. R. Russell (Eds.), *Landmark essays on writing across the curriculum* (Vol. 6, pp. 97–124). Davis, CA: Hermagoras Press.

Hirsch, P., Shwom, B., Yarnoff, C., Anderson, J. C., Kelso, D. M., Olson, G. B., & Colgate, J. E. (2001). Engineering design and communication: The case for interdisciplinary collaboration. *International Journal of Engineering Education*, 17(4–5), 342–348.

House, R., Livingston, J., Minster, M., Taylor, C., Watt, A., & Williams, J. (2009). Assessing engineering communication in the technical classroom: The case of Rose-Hulman Institute of Technology. In M. C. Paretti & K. Powell (Eds.), *Assessment of writing* (pp. 127–158). Tallahassee, FL: Association of Institutional Research.

Howe, S. (2012). Capstone design hub: Building the capstone design community. In *Proceedings of the American Society for Engineering Education Annual Conference and Exposition*. Retrieved from http://www.asee.org/search/proceedings

Huot, B. (2002). *(Re)Articulating writing assessment for teaching and learning*. Logan: Utah State University Press.

Huot, B., & O'Neill, P. (2008). *Assessing writing: A critical sourcebook*. Boston: Bedford/St. Martins.

Isbell, D., Hardin, M., & Underwood, J. (1999). Mars climate orbiter team finds likely cause of loss. Retrieved from mars.jpl.nasa.gov/msp98/news/mco990930.html

Jacobs, C. (2010). Collaboration as pedagogy: Consequences and implications for partnerships between communication and disciplinary specialists. *Southern African Linguistics & Applied Language Studies*, 28(3), 227–237.

Jarratt, S. C., Mack, K., Sartor, A., & Watson, S. E. (2009). Pedagogical memory: Writing, mapping, translating. *WPA: Writing Program Administration*, 33(1–2), 46–73.

Kennedy, G. (1998). *An introduction to corpus linguistics*. London: Longman.

Kiesler, S., & Cummings, J. N. (2002). What do we know about proximity and distance in work groups? A legacy of research. In P. Hinds & S. Kiesler (Eds.), *Distributed work* (pp. 57–80). Cambridge: MIT Press.

Klein, P. D. (1999). Reopening inquiry into cognitive processes in writing-to-learn. *Educational Psychology Review*, 11(3), 203–270.

Kock, N. (2005). Media richness or media naturalness? The evolution of our biological communication apparatus and its influence on our behavior toward E-communication tools. *IEEE Transactions on Professional Communication*, 48(2), 117–130.

Kock, N., & Nosek, J. (2005). Expanding the boundaries of E-collaboration. *IEEE Transactions on Professional Communication*, 48(1), 1–9.

Kreth, M. L. (2000). A survey of the co-op writing experiences of recent engineering graduates. *IEEE Transactions on Professional Communication*, 43(2), 137–151.

Krishnamurthy, R., & Kosem, I. (2007). Issues in creating a corpus for EAP pedagogy and research. *Journal of English for Academic Purposes*, 6(4), 356–373.

Kynell-Hunt, T., & Savage, G. J. (Eds.). (2003). *Power and legitimacy in technical communication*, Vol. 1: *The historical and contemporary struggle for professional status*. Amityville, NY: Baywood.

Latour, B. (1987). *Science in action: How to follow scientists and engineers through society.* Cambridge, MA: Harvard University Press.

Latour, B., & Woolgar, S. (1986). *Laboratory life: The construction of scientific facts.* Princeton, NJ: Princeton University Press.

Lattuca, L. R., Terenzini, P. T., & Volkwein, J. F. (2006). *Engineering change: A study of the impact of EC 2000.* Baltimore, MD: ABET, Inc.

Lave, J., & Wegner, E. (1991). *Situated learning: Legitimate peripheral participation.* Cambridge: Cambridge University Press.

Lea, M., & Street, B. (1998). Student writing in higher education: An academic literacies approach. *Studies in Higher Education, 23*(2), 157–172.

Levine, K. J., Allard, S., & Tenopir, C. (2011). The changing communication patterns of engineers [Point of View]. *Proceedings of the IEEE, 99*(7), 1155–1157.

Leydens, J. A. (2008). Novice and insider perspectives on academic and workplace writing: Toward a continuum of rhetorical awareness. *IEEE Transactions on Professional Communication, 51*(3), 242–263.

Leydens, J. A. (2012a). Sociotechnical communication in engineering: An exploration and unveiling of common myths. *Engineering Studies, 4*(1), 1–9.

Leydens, J. A. (2012b). What does professional communication research have to do with social justice? Intersections and sources of resistance. In *Proceedings of the IEEE International Professional Communication Conference.* Retrieved from http://ieeexplore.ieee.org/xpl/conhome.jsp?punumber=1000591

Leydens, J. A., & Lucena, J. C. (2009). Listening as a missing dimension in engineering education: Implications for sustainable community development efforts. *IEEE Transactions on Professional Communication, 52*(4), 359–376.

Leydens, J. A., Lucena, J. C., & Schneider, J. (2012). Are engineering and social justice (in)commensurable? A theoretical exploration of macro-sociological frameworks. *International Journal of Engineering, Social Justice and Peace, 1*(1), 63–82.

Leydens, J. A., Moskal, B. M., & Pavelich, M. (2004). Qualitative methods used in the assessment of engineering education. *Journal of Engineering Education, 93*(1), 65–72.

Leydens, J. A., & Schneider, J. (2009). Innovations in composition programs that educate engineers: Drivers, opportunities, and challenges. *Journal of Engineering Education, 98*(3), 255–271.

Lucena, J. C., Schneider, J., & Leydens, J. A. (2010). *Engineering and sustainable community development.* San Rafael, CA: Morgan & Claypool.

Locke, D. M. (1992). *Science as writing.* New Haven, CT: Yale University Press.

Male, S., Bush, M., & Chapman, E. (2010). Perceptions of competency deficiencies of engineering graduates. *Australasian Journal of Engineering Education, 16*(1), 55–67.

Male, S., Bush, M., & Chapman, E. (2011). Understanding generic engineering competencies. *Australasian Journal of Engineering Education, 17*(3), 147–156.

Mathison, M. (2000). "I don't have to argue my design–the visual speaks for itself": A case study of mediated activity in an introductory mechanical engineering course. In S. Michell & R. Andrews (Eds.), *Learning to argue in higher education* (pp. 74–84). Portsmouth, NH: Boynton/Cook.

Matusovich, H., Paretti, M. C., Motto, A., & Cross, K. J. (2012). Understanding faculty and student beliefs about teamwork & communication skills. In *Proceedings of the American Society for Engineering Education Annual Conference and Exposition.* Retrieved from http://www.asee.org/search/proceedings

McCarthy, L. P. (1994). A stranger in strange lands: A college student writing across the curriculum. In C. Bazerman & D. R. Russell (Eds.), *Landmark essays: Writing across the curriculum* (Vol. 6, pp. 125–155). Davis, CA: Hermagoras Press.

McEnery, T., & Wilson, A. (2001). *Corpus linguistics* (2nd ed.). Edinburgh, UK: University of Edinburgh Press.

McNair, L. D., & Garrison, W. (2012). Portfolios to professoriate: Helping students integrate professional identities through ePortfolios. In *Proceedings of the American Society for Engineering Education Annual Conference and Exposition.* Retrieved from http://www.asee.org/search/proceedings

McNair, L. D., & Paretti, M. C. (2010). Activity theory, speech acts, and the "doctrine of infelicity": Connecting language and technology in globally networked learning environments. *Journal of Business and Technical Communication, 24*(3), 323–357.

McNair, L. D., Paretti, M. C., & Davitt, M. (2010). Towards a pedagogy of relational space and trust: Analyzing distributed collaboration using discourse and speech act analysis. *IEEE Transactions on Professional Communication*, 53(3), 233–248.

McNair, L. D., Paretti, M. C., Knott, M., & Wolfe, M. L. (2006). Work in progress: Using e-Portfolio to define, teach, and assess ABET professional skills. In *Proceedings of the ASEE/IEEE Frontiers in Education Conference*. Retrieved from http://ieeexplore.ieee.org/xpl/conhome. jsp?punumber=1000297

Miller, C. (1984). Genre as social action. *Quarterly Journal of Speech*, 70(2), 151–167.

National Academy of Engineering (NAE). (2004). *The engineer of 2020: Visions of engineering in the new century*. Washington, DC: The National Academies Press.

Nystrand, M. (1982). *What writers know: The language, process, and structure of written discourse*. New York, NY: Academic Press.

Orr, T. (2006). Introduction to the special issue: Insights from corpus linguistics for professional communication. *IEEE Transactions on Professional Communication*, 49(3), 213–216.

Orr, T., Bowers, R., Busch, D., Kushner, S., Mueller, E., & de Prospero, A. (1995). *Serving science and technology: Five programs around the globe* (Tech. Rep. No. 95-5–001). Aizuwakamatsu, Japan: University of Aizu; reproduced by ERIC Document Reproduction Service No. ED 389 173.

Ostheimer, M. W., & White, E. M. (2005). Reliable data verify that curriculum reform made in response to portfolio assessment findings improves student writing: Successful writing portfolio assessment at the University of Arizona's Electrical and Computer Engineering Department improves student writing. In *Proceedings of the American Society for Engineering Education Annual Conference and Exposition*. Retrieved from http://www.asee.org/search/proceedings

Pappas, E. C., & Hendricks, R. W. (2000). Holistic grading in science and engineering. *Journal of Engineering Education*, 89(2), 403–408.

Pappas, E. C., Kampe, S. L., & Hendricks, R. W. (2004). An assessment methodology and its application to an advanced engineering communications program. *Journal of Engineering Education*, 93(3), 233–246.

Paretti, M. C. (2005, June 14). Using project portfolios to assess design in materials science and engineering. In *Proceedings of the American Society for Engineering Education Annual Conference and Exposition*. Retrieved from http://www.asee.org/search/proceedings

Paretti, M. C. (2006). Audience awareness: Leveraging problem-based learning to teach workplace communication practice. *IEEE Transactions on Professional Communication*, 49(6), 189–198.

Paretti, M. C. (2008). Teaching communication in capstone design: The role of the instructor in situated learning. *Journal of Engineering Education*, 97(4), 491–503.

Paretti, M. C. (2009). When the teacher is the audience: Assignment design and assessment in the absence of "real" readers. In A. Gonzalez, E. Weiser, & B. Fehler (Eds.), *Engaging audience: Writing in an age of new literacies* (pp. 165–185). Urbana, IL: NCTE Press.

Paretti, M. C. (2011, January). *Theories of language and content together: The case for interdisciplinarity*. Paper presented at the Dynamic content and language collaboration in higher education: theory, research, and reflections, Cape Town, South Africa.

Paretti, M. C., & Burgoyne, C. B. (2005a). Integrating engineering and communication: A study of capstone design courses. In *Proceedings of the ASEE/IEEE Frontiers in Education Conference*. Retrieved from http://ieeexplore. ieee.org/xpl/conhome.jsp?punumber=1000297

Paretti, M. C., & Burgoyne, C. B. (2005b). Work-in-progress: An integrated engineering communications curriculum for the 21st century. In *Proceedings of the ASEE/IEEE Frontiers in Education Conference*. Retrieved from http://ieeexplore.ieee.org/xpl/conferences. jsp?queryText=frontiers%20in%20education

Paretti, M. C., & Burgoyne, C. B. (2009). Assessing excellence: Using activity theory to understand assessment practices in engineering communication. In *Proceedings of the ASEE/IEEE Frontiers in Education Conference*. Retrieved from http://ieeexplore.ieee.org/xpl/conhome. jsp?punumber=1000297

Paretti, M. C., & McNair, L. D. (2008a). Communicating in global virtual teams: Managing complex activity systems. In P. Zemliansky & K. St. Amant (Eds.), *Handbook of research on virtual workplaces and the new nature of business practices* (pp. 24–38). Hershey, PA: Idea Group.

Paretti, M. C., & McNair, L. D. (2012). Analyzing the intersections of institutional and discourse identities in engineering work at the local level. *Engineering Studies*, 4(1), 55–78.

Paretti, M. C., & McNair, L. D. (Eds.). (2008b). Communication in engineering curricula [Special issue]. *IEEE Transactions on Professional Communication*, 51(3).

Paretti, M. C., McNair, L. D., Belanger, K., & George, D. (2009). Reformist possibilities? Exploring cross-campus writing partnerships. *WPA: Writing Program Administration*, 33(1–2), 74–113.

Paretti, M. C., McNair, L. D., & Holloway-Attaway, L. (2007). Teaching technical communication in an era of distributed work: A case study of collaboration between U.S. and Swedish students. *Technical Communication Quarterly*, 16(3), 327–352.

Paretti, M. C., & Powell, K. (2009a). Introduction: Bringing voices together: Partnerships for assessing writing across contexts. In M. C. Paretti & K. Powell (Eds.), *Assessment of writing* (pp. 1–10). Tallahassee, FL: Association of Institutional Research.

Paretti, M. C., & Powell, K. M. (Eds.). (2009b). *Assessment of writing*. Tallahassee, FL: Association for Institutional Research.

Patton, M. D. (2008). Beyond WI: Building an integrated communication curriculum in one department of civil engineering. *IEEE Transactions on Professional Communication*, 51(3), 313–327.

Perelman, L. C. (1999). The two rhetorics: Design and interpretation in engineering and humanistic discourse. *Language and Learning Across the Disciplines*, 3(2), 64–82.

Poe, M., Lerner, N., & Craig, J. (2010). *Learning to communicate in science and engineering: Case studies from MIT*. Cambridge, MA: The MIT Press.

Reave, L. (2004). Technical communication instruction in engineering schools: A survey of top-ranked U.S. and Canadian programs. *Journal of Business and Technical Communication*, 18(4), 452–490.

Robert, L. P., & Dennis, A. R. (2005). Paradox of richness: A cognitive model of media choice. *IEEE Transactions on Professional Communication*, 48(1), 10–21.

Roberts, T. L., Lowry, P. B., & Sweeney, P. D. (2006). An evaluation of the impact of social presence through group size and the use of collaborative software on group member "voice" in face-to-face and computer-mediated task groups. *IEEE Transactions on Professional Communication*, 49(1), 28–43.

Robey, D., Khoo, H. M., & Powers, C. (2004). Situated learning in cross-functional virtual teams. In J. Dubinsky (Ed.), *Teaching technical communication: Critical issues for the classroom* (pp. 541–567). Boston: Bedford St. Martins.

Russell, D. R. (1997). Rethinking genre in school and society: An activity theory analysis. *Written Communication*, 14(4), 504–554.

Russell, D. R. (2002). *Writing in the academic disciplines: A curricular history* (2nd ed.). Carbondale, IL: Southern Illinois University Press.

Rutkowski, A. F., Vogel, D. R., van Genuchten, M., Bemelmans, T. M. A., & Favier, M. (2002). E-Collaboration: The reality of virtuality. *IEEE Transactions on Professional Communication*, 45(4), 219–230.

Sageev, P., & Romanowski, C. J. (2001). A message from recent engineering graduates in the marketplace: Results of a survey on technical communication skills. *Journal of Engineering Education*, 90(4), 685–697.

Sales, H. E. (2006). *Professional communication in engineering*. Basingstoke, UK and New York, NY: Palgrave Macmillan.

Schneider, J., Leydens, J. A., Olds, B., & Miller, R. (2009). Guiding principles in engineering writing assessment: Context, collaboration, and ownership. In M. C. Paretti & K. Powell (Eds.), *Assessment of writing* (pp. 65–81). Tallahassee, FL: Association of Institutional Research.

Scott, C., & Plumb, C. (1999). Using portfolios to evaluate service courses as part of an engineering writing program. *Technical Communication Quarterly*, 8(3), 337–350.

Selfe, C. L. (2007). *Multimodal composition: Resources for teachers*. Cresskill, NJ: Hampton Press.

Sheridan, D. M., & Inman, J. A. (2010). *Multiliteracy centers: Writing center work, new media, and multimodal rhetoric*. Cresskill, NJ: Hampton Press.

Short, J., Williams, E., & Christie, B. (1976). *The social psychology of telecommunications*. New York, NY: John Wiley & Sons.

Shwom, B., Hirsch, P., Yarnoff, C., & Anderson, J. C. (1999). Engineering design and communication: A foundational course for freshmen.

Language and Learning Across the Disciplines, 3(2), 107–112.

Slack, J. D., Miller, D. J., & Doak, J. (2003). The technical communicator as author: Meaning, power, authority. In T. Kynell-Hunt & G. J. Savage (Eds.), *Power and legitimacy in technical communication*, Vol. 1: *The historical and contemporary struggle for professional status* (pp. 169–192). Amityville, NY: Baywood.

Smith, S. (2003). What is 'good' technical communication? A comparison of the standards of writing and engineering instructors. *Technical Communication Quarterly*, 12(1), 7–24.

Smith Taylor, S. (2011). "I really don't know what he meant by that": How well do engineering students understand teachers' comments on their writing? *Technical Communication Quarterly*, 20(2), 139–166.

Spinuzzi, C. (1996). Pseudotransactionality, activity theory, and professional writing instruction. *Technical Communication Quarterly*, 5(3), 295–308.

Spinuzzi, C. (2003). *Tracing genres through organizations: A sociocultural approach to information design*. Cambridge, MA: MIT Press.

Spurlin, J. E., Rajala, S. A., & Lavelle, J. P. (Eds.). (2008). *Designing better engineering education through assessment: A practical resource for faculty and department chairs on using assessment and ABET criteria to improve student learning*. Sterling, VA: Stylus.

Swarts, J., & Odell, L. (2001). *Rethinking the evaluation of writing in engineering courses*. In *Proceedings of the ASEE/IEEE Frontiers in Education Conference*. Retrieved from http://ieeexplore. ieee.org/xpl/conhome.jsp?punumber=1000297

Tenopir, C., & King, D. W. (2004). *Communications patterns of engineerings*. Piscataway, NJ: Wiley-Interscience: IEEE Press.

Thaiss, C. (2010). The international WAC/WID mapping project: Objectives, methods, and early results. In C. Bazerman, R. Krut, K. Lundsford, S. McLeod, S. Null, P. Rogers, & A. Stansell (Eds.), *Traditions of writing research* (pp. 251–264). New York, NY: Routledge.

Thaiss, C., & Zawacki, T. M. (2006). *Engaged writers and dynamic disciplines: Research on the academic writing life*. Portsmouth, NH: Heinemann.

Turns, J., & Lappenbusch, S. (2006). Tracing student development during construction of engineering professional portfolios. In *Proceedings of the American Society for Engineering Education Annual Conference and Exposition*. Retrieved from http://www.asee.org/search/proceedings

Turns, J., Sattler, B., Eliot, M., Kilgore, D., & Mobrand, K. (2012). Preparedness portfolios and portfolio studios. *International Journal of ePortfolio*, 2(1). Retrieved from http://www.theijep.com/current.cfm

The VALUE Project Overview. (2009). *Peer Review*, 11(1), 4–7.

Venters, C., McNair, L. D., & Paretti, M. C. (2012). Using writing assignments to improve conceptual understanding in statics: Results from a pilot study. In *Proceedings of the American Society for Engineering Education Annual Conference and Exposition*. Retrieved from http://www.asee.org/search/proceedings

Vygotsky, L. S. (1962/1986). *Thought and language* (Rev. ed.). Cambridge, MA: MIT Press.

Vygotsky, L. S. (1978). *Mind in society*. Cambridge, MA: Harvard University Press.

Walther, J. B., & Bunz, U. (2005). The rules of virtual groups: Trust, liking, and performance in computer-mediated communication. *Journal of Communication*, 55(4), 828–846.

Williams, J. M. (2001). Transformations in technical communication pedagogy: Engineering, writing, and the ABET Engineering Criteria 2000. *Technical Communication Quarterly*, 10(2), 149–167.

Williams, J. M. (2002). The engineering portfolio: Communication, reflection, and student learning outcomes assessment. *International Journal of Engineering Education*, 18(2), 199–207.

Winsor, D. A. (1988). Communication failures contributing to the *Challenger* accident: An example for technical communicators. *IEEE Transactions on Professional Communication*, 31(3), 101–108.

Winsor, D. A. (1996). *Writing like an engineer: A rhetorical education*. Mahwah, NJ: Lawrence Erlbaum.

Winsor, D. A. (2003). *Writing power: Communication in an engineering center*. Albany, NY: State University of New York Press.

Winsor, D. A. (2006). Using writing to structure agency: An examination of engineers' practice. *Technical Communication Quarterly*, 15(4), 411–430.

Wolfe, J. (2009). How technical communication textbooks fail engineering students.

Technical Communication Quarterly, 18(4), 351–375.

Wolfe, J., & Powell, E. (2009). Biases in interpersonal communication: How engineering students perceive gender typical speech acts in teamwork. *Journal of Engineering Education, 98*(1), 5–16.

Yancey, K. B., & Huot, B. (Eds.). (1997). *Assessing writing across the curriculum: Diverse approaches and practices*. Greenwich, CT: Ablex.

Youra, S. (1999). Letter from the guest editor. *Language and Learning Across the Disciplines, 3*(2), 1–12.

Zamel, V., & Spack, R. (Eds.). (1998). *Negotiating academic literacies: Teaching and learning across languages and cultures*. Mahwah, NJ: Lawrence Erlbaum.

CHAPTER 31

Use of Information Technology in Engineering Education

Krishna Madhavan and Euan D. Lindsay

Introduction: The Opportunity of Information Technology in Engineering Education

Engineering as a profession is changing at a rapid pace in large part owing to an ever-evolving technology landscape. The technological changes are wrapped increasingly in societal transformations that allow for better and rapid exchange of information and practices. Distance, time, and space are no longer boundaries for the practice of engineering. Essentially, the transition of computing infrastructure from an individual "sitting at the desk in isolation" paradigm to a "network-based" paradigm has resulted in knowledge production and sharing at a pace previously never considered possible. More importantly, the transformation of engineering practice through information technology is seen in the convergence of theory and experimentation (which are traditional approaches to engineering problem solving) with modeling and simulation, where design and problem solving are primarily *in silico* (meaning performed on a computer [chip] or using computer simulation). "Advances in computing and simulation coupled with technologies that mimic rudimentary attributes in analysis, may radically redefine common practices in engineering" (National Academy of Engineering [NAE], 2004, p. 15). Information technology (IT) impacts our daily lives in fundamental ways. Nowhere is this impact more obvious than on a university campus.

A new generation of students – digital natives – armed with a dizzying array of gadgets and gizmos roam the hallways of academic institutions struggling to keep pace with speed of the digital world. Learning happens everywhere and at any given time. A recently published U.S. National Academy of Sciences report entitled *Learning Science in Informal Environments: People. Places, and Pursuits* states, "all learning environments, including school and non-school settings, can be said to fall on a continuum of educational design or structure" (National Research Council [NRC], 2009, p. 47). Information technology allows this continuum to be traversed and possibly controlled in very useful and effective ways.

Engineering at its very core is a highly creative endeavor, as the former president of the U.S. NAE William Wulf argued, ". . . down to my toes I believe that engineering is profoundly creative . . . what comes out is a function of the life experiences of the people who do it" (Wulf, 1998). The impact of information technology on creativity is so profound that it has inspired a new generation of applications such as YouTube, Flickr, Twitter, Facebook, and so many others resulting in radical changes to the delivery of materials and training to engineering and science students. These applications allow non-IT experts to create content in a nonthreatening environment, thus sparking the era of Web 2.0. Further, the effect of IT is so significant that *TIME* magazine celebrated this ability to create and contribute by naming "You" the 2006 Person of the Year (*TIME*, 2006). *TIME* went further by stating "Yes You. You control the Information Age. Welcome to your world" (*TIME*, 2006, cover page). We argue that IT can allow educators to put students in charge of their own learning process. In this chapter, we discuss a variety of technologies used for engineering education that can enable real-world problem solving in the engineering classroom (and beyond) in ways that were not possible before. We, however, acknowledge the vast nature of information technologies that would fall within the purview of this chapter. "The scope of IT-enabled creative practices is suggested (but by no means exhausted) by a host of coinages that have recently entered common language – computer graphics, computer-aided design, computer music, computer games, digital photography, digital video, digital media, new media, hypertext, virtual environments, interaction design, and electronic publishing, to name just a few" (NRC, 2003, p. 15).

In this chapter, we present a slightly different perspective on the role of technology, in particular information and communication technologies, in the engineering education context. Our argument focuses on how the fundamental characteristics of problem solving in engineering have changed. We present the case that despite the rapidly shifting technology base, engineering education must remain aggressive in using and understanding the impact of information technology in enabling engineering learning. Further, we argue that engineering has played an interesting dual role in the evolution of information technology. On the one hand, the fundamental nature and approach to problem solving in engineering has undergone radical change in the last two decades due to the availability of advanced information technology tools and services. We provide an in-depth discussion of how this transformation has occurred and provide evidence for the same. On the other hand, these very advances in information and communication technologies have spawned a virtuous cycle of research and development in engineering that is feeding and accelerating the development of extremely fast computation, data, and network infrastructures.

IT-Enabled Changes in Engineering Problem Solving

IT tools are already starting to democratize the creative process allowing experts and non-experts alike to contribute and participate meaningfully in scientific discourse. Further, IT tools blur the boundaries of space and time that have traditionally plagued teaching and learning. Learning anytime, anywhere has been the rallying cry behind numerous innovations in technology-enabled learning. Given this major push in IT, many researchers question if and how to use technology appropriately in learning contexts. Before we dive deeper into a discussion on the use of information technology in the engineering learning context, we acknowledge these concerns in a brief discussion.

Although on the one hand IT is being used heavily in the engineering curricula, prominent researchers such as Todd Oppenheimer (2003) have raised the question of whether computers and indeed the Internet are just another blip in the rapidly evolving technology landscape. They question if the educational system, by being reactionary and over-focused on the latest and greatest technologies, has moved far from the core tenets of enabling learning. Cuban (2001) raises the very fundamental question of whether the notion of e-learning in schools (particularly in the American public school context) has "turned out to be word processing and Internet searches" (p. 178). Many researchers have raised the issue of whether the use of technology is coming at the cost of evidence-based approaches.

Bennett, Maton, and Kervin (2008) question the long-held belief that because we are dealing with students who are immersed in modern information and communication technology, the pedagogical paradigm should reflect this technology sophistication. Empirical evidence examined as part of their work leads them to the conclusion that "the claim that there is a distinctive new generation of students in possession of sophisticated technology skills and with learning preferences for which education is not equipped to support has excited much recent attention. Proponents arguing that education must change dramatically to cater for the needs of these digital natives have sparked an academic form of a 'moral panic' [quotes in original] using extreme arguments that have lacked empirical evidence" (Bennett et al., 2008, p. 783). Another study by Kennedy et al. (2008) concluded that "[. . .] first year students are highly tech-savvy. However, when one moves beyond entrenched technologies and tools (e.g. computers, mobile phones, email), the patterns of access and use of a range of other technologies show considerable variation" (p. 108).

Although these are valid concerns, there are larger global forces at work that need to be considered carefully. We understand that "chasing technologies" that become viable for adoption is not a scalable or pedagogically sound strategy for engineering education. More fundamentally, in this chapter, we argue that changes on the technology front are dictated by economic and political forces well outside the purview of the global engineering education system as a whole. These forces require educators and researchers to adopt evidence-based strategies to guide the use of information technology in the curricula. Further, in this chapter, we argue that the engineering education as a system has played more of reactionary role in the adoption of technology than strategic.

One of the most important transformations that have fueled the importance of IT use and adoption in the engineering education system is that engineering practice, research, and discovery processes as a whole are moving toward a more collaborative socially immersive model. Figure 31.1 shows how engineering and science practice has evolved over time. Before the dawn of the Internet, workflows in engineering practice were focused largely on the computing infrastructure available to a single researcher. Usually, the resource was a powerful (for its time) computer system that was owned either by the organization or by the engineer. Using these systems required the engineer to interact physically with the computer system. The networking infrastructure was not the focus of the engineer's workflow at that time. Therefore, the flow of ideas and problem solving strategies were largely constrained by the distances involved, time, and ability to interact face to face. As the engineering methodology at that time was largely focused on theory and experimentation, the deluge of data was not a huge methodological problem.

With the advent of powerful networking infrastructures, the engineering problem-solving paradigm became one of organization-wide teams that were geographically highly distributed, taking on design and problem-solving that had never been seen before. For example, the design of the Boeing 777 by engineers who were distributed in multiple parts of the world using modeling and simulations as the primary problem-solving modality is a watershed transformation in engineering practice. More recently, the design, testing, and indeed the production of major engineering products are based on services leveraged from throughout the world. Cloud computing environments (Armbrust et al., 2010)

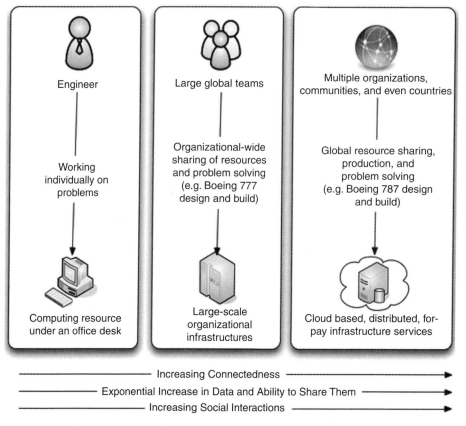

Figure 31.1. The increasing social nature of engineering problem solving. As a result of a combination of cheap storage, powerful networks, and extremely fast computers, data generation and sharing have increased exponentially over time.

provide the computing infrastructure needed to scale and develop complex products and services. In essence, the network has become the fabric for intelligent computing. The combination of extremely cheap storage, fast multicore computer processors, and powerful networking infrastructures has fueled this transformation in engineering practice. Not only is the quantity of data growing exponentially[1], but the cost of producing and transporting these data on our networks is also plummeting. Simple and powerful software makes data sharing easier than ever before.

As engineering problem solving became more scaffolded by social structures, one other major transformation occurred that eliminated the gap between the architectures of IT systems supporting engineering and science problem solving and the

services that students see in their daily lives. Figure 31.2 shows the narrowing of this systems architecture gap. One side of the figure represents the engineering and science workflow. Engineers and scientists use powerful online gateways such as nanoHUB.org[2] that provide a rich user interface that hides the complexity of the infrastructure. However, the user interface (which is also essentially software) interacts with software components, which in turn connect to the data and compute resources needed to provide the users with insights. This generalized system architecture is very closely related to the type of infrastructures that power many social networking sites and services and video gaming infrastructures. The networking infrastructure allows these types of distributed, n-tier applications. Although Figure 31.2 presents a largely simplified view

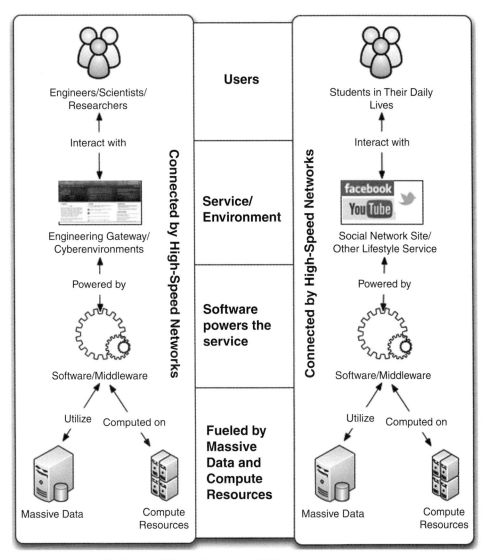

Figure 31.2. Simplified view of the architecture of IT systems used to support engineering gateways (or cyberenvironments) and the systems that support day-to-day activities. Fundamentally there is a convergence and more standardization of IT architectures that support a wide range of applications.

of the infrastructure stack, the point here is that there is a convergence and standardization of architectures to a very great extent. This convergence leads to a large number of similarities in how simple workflows in our day-to-day living start to resemble more advanced workflows once the domain of high-end engineering problem solving.

In his book *The World Is Flat 3.0: A Brief History of the Twenty-first Century*, Thomas Friedman talks about the notion of "triple convergence." He states that "thanks to the triple convergence, this new flat-world platform is, in effect, blowing away our walls, ceilings, and floors – all at the same time. That is, the wiring of the world with fiber-optic cable, the Internet, and work flow software has blown down many of the walls that prevented collaboration" (Friedman, 2007, p. 231). Engineering has a direct role in enabling the type of complementary convergence that Friedman outlines in his book.

However, engineering is also the beneficiary of this convergence. Essentially solving major engineering problems increasingly involves a global team. Richards (1998) points out that "many of the major engineering achievements are the result of the efforts of large teams and organizations."

One of the other significant convergences that has significantly impacted engineering problem-solving is the notion of "progressive convergence" (Bainbridge & Roco, 2006) – which refers to the convergence of nanotechnology, biotechnology, information technology, and cognitive science. In describing the impact of progressive convergence Bainbridge and Roco (2006) argue strongly that

> [. . .] the great convergence that is taking place today should not be mistaken for the mundane growth of interdisciplinary or multidisciplinary fields. For many decades, small-scale convergence has taken place in areas such as astrophysics, biochemistry, and social psychology. However significant these local convergences have seemed for the scientists involved in them, they pale in comparison to the global convergence that is poised to occur in the next decade. It will constitute a major phase change in the nature of science and technology, with the greatest possible implications for the economy, society, and culture. (p. 2)

The notion of complementary convergence as proposed by Friedman (2007) is facilitated in great part due to the progressive convergence as elaborated in Bainbridge and Roco (2006). Although the intersection of these two types of convergences points to fundamental engineering and science progress, one transformation that has changed our daily lives is the innovations in software technology. If the Internet provides the infrastructure for the flow of information and data across the globe, the new generation of software is the lifeline that makes this infrastructure transparent to the end-user, thereby increasing human productivity to levels not previously known in the history of modern science and engineering.

Web 2.0 is a popular term that is used in day-to-day conversations. Many websites and services use this term to describe themselves. Web 2.0 essentially calls a new paradigm of software using "the web as a platform" (O'Reilly, 2007, p. 19). Web 2.0 also marks the movement of the individual user from the role of an individual working in isolation to that of a participant in a larger network where knowledge is truly reflective of the larger community in which the user participates. O'Reilly (2007) terms this transformation "harnessing the collective intelligence" (p. 22). Engineers not only work in an environment where they are actively participating in the creation of new knowledge, but they also actively utilize the collective intelligence of crowds offered by the Web 2.0 paradigm as part of their problem-solving strategy.

Engineers routinely deal with a combination of what Suroweicki (2004) terms "cognition problems," "coordination problems," and "cooperation problems." In essence cognition problems require engineers to have tremendous levels of expertise in their chosen technical space. Coordination problems require them to work with others to coordinate their problem-solving behaviors in effective ways, assuming that a cooperative mindset is forthcoming from their collaborators. Finally, cooperation problems require engineers to use what has been traditionally termed "soft skills" to communicate with others – sometimes groups that disagree with their approach – to come up with solutions to major global engineering problems. In 2008, the U.S. NAE released a set of fourteen grand challenges for the twenty-first century (NAE, 2008). Every one of the fourteen grand challenges identified by NAE is a complex problem that is composed of multiple cognition, coordination, and cooperation problems. The only viable way to even attempting to propose solutions to these engineering grand challenges is to rely heavily on technologies that allow us tap into the rich information that is derived from the "wisdom of the crowds" (Suroweicki, 2004).

Engineering educators have long understood that "engineering lies at the interface between science, on the one hand, and society on the other" (Grimson, 2002, p. 31). In the early part of the twentieth century, engineering education focused primarily on the underlying scientific and technical content rather than on the societal aspects of engineering. However, there is now a new call for more holistic approaches to engineering education. The notion of technology mediation is a necessary condition for training future engineers. We expand on this notion in the next section. Further, we discuss the impact of technology mediation on notions of space and time in enabling engineering learning.

Technology as the Fabric for Engineering Learning Innovation

The increasing saturation of technology in the everyday environment means that students are living Internet-based lives. As students spend more of their non-study lives online, they now expect that their study environment will similarly have an online presence. To remain relevant, it becomes essential that engineering education incorporate technology-mediated interfaces.

Computers are the dominant technology in engineering education, and as the nature of computing evolves so does the way in which computers are employed in the educational context. The earliest computers were artifacts in the purest sense – filling entire rooms and requiring teams of technicians to keep them functioning correctly. Over time, Moore's law has continued to hold true, with computer power doubling every two years, and computers themselves shrinking until they can fit onto a fingertip.

Parallel to this physical evolution there has been an evolution of functionality. The advent of the Internet allowed computers to communicate and for users to share the data and information stored on each machine. In the early days, it was the computers that were smart, and the network that connected them was dumb – it was simply a medium for conveying information.

In the early part of the twenty-first century, however, the situation has reversed. Computers are no longer smart artifacts – they are simply the dumb interfaces to the smart network of the Internet. Videos are uploaded to YouTube, not to a particular computer – the actual physical location of the data no longer matters. The World Wide Web has gone from being an information system to an interaction system. It is the interactions themselves, rather than the machines on which they occur, that are the future of the Internet – the hardware is now merely the vehicle. As this cloud grows, the concept of technology as artifacts will grow increasingly weaker, leaving behind instead the services and functionality that the technology can provide.

Further, the ways in which people interact with these smart networks are converging. The same content is available in multiple devices and the networks act as a vehicle for content synchronization across multiple devices with a range of form factors. Whereas in the past an engineer would engage with technology in one way in his or her professional capacity and in a different way for his or her personal entertainment needs, the paradigms that underpin both are converging.

This convergence offers great opportunities, because it means that students are in many cases already familiar with the workflows they need to operate as engineers. More importantly, it shifts the focus from the technology per se to the services and affordances that the technology provides.

Clarifying Technology versus Technologies

In discussing how technology forms a critical base for engineering learning, it is critical to distinguish between *technology* as a collection versus *individual technologies*. There is a danger in writing a chapter on the topic of "Technology in Engineering Education" that one instead ends up writing the chapter

"Technologies in Engineering Education." The former focuses on the transformative nature of technology in the education process; the latter becomes a catalogue of the seemingly infinite range of technology that can be or has been employed in an engineering classroom at some time. Specific technologies can be found all throughout the literature; in this chapter we focus instead on the issues surrounding technology.

Perhaps the clearest indication of why the catalogue approach is inappropriate is provided by the Gartner Hype Cycle (Gartner Research, 2012). The hype cycle models the flow of technology from the initial idea, through early expectations, disillusionment, and then through to mainstream adoption. Every year the Gartner Institute presents an analysis of current and emerging technologies, locating a selection of them on the hype cycle curve and assessing a time horizon for when these technologies will be adopted into the mainstream. Figure 31.3 shows the 2010 (*a*) and 2011 (*b*) curves.

For every technology that has dropped off the curve, there is a new technology that has emerged to take its place. These technologies are not developed for educational purposes; educators are not engaged in the requirements analysis phase. Yet these technologies find their way into the lives of our students and then bleed into their learning experiences. Trying to chase technologies is futile. The focus instead must be on things that technology can do. Technology is enabling new ways of learning by removing many of the barriers inherent in "traditional" engineering education:

- The barrier of synchronization: technology allows learners to schedule their own learning, rather than having to fit in with the university or other organizational imposed structures.
- The barrier of co-location: technology allows learners to learn from anywhere, rather than having to attend a university physically.
- The barrier of class size: technology allows teachers to teach thousands of students at once.

- The barrier of one-size-fits-all: technology allows learners to customize their learning environment to their preferences and their needs to a great extent.

Certainly there are traditional approaches that allow some of these barriers to be overcome separately; however, twenty-first century technology is allowing for these barriers to be overcome in tandem.

Technology Itself versus the Affordances It Provides

Debate regarding technology versus what can be done with it is not new; indeed it was the core of what has become known as the Clark–Kozma debate in the early 1980s. The debate has been reported in a number of places in the literature – for example, Lindsay, Naidu, and Good (2007). It centered on Robert Kozma and Richard Clark's opposing views on whether media could influence learning. Clark's viewpoint was that "media are *mere vehicles* that deliver instructions but do not influence student achievement any more than the truck that delivers our groceries causes changes in nutrition" (Clark, 1983, p. 445). Kozma, however, argued that the medium is not an inert vehicle for transferring information – that there is interaction between the medium and the message (Kozma, 1994). The learner can construct his or her learning only from the building blocks that the instructional method provides – and the range of these blocks is constrained further by the selection of the medium involved.

Ultimately, the analogy of a delivery truck breaks down, but not before it provides one further useful insight into the difference between technology itself, and the affordances it provides: the *refrigerated* delivery truck. A refrigerated delivery truck can deliver products that an unrefrigerated truck cannot. However, ultimately it is these new products, and not the truck that brings them, that are important. New technologies often allow new teaching methods that would otherwise be impossible without the technology. However, it is

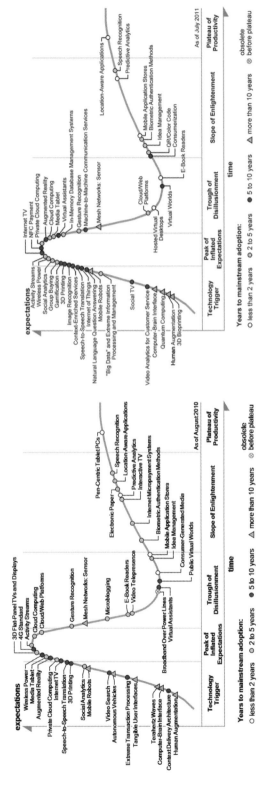

Figure 31.3. The Gartner technology hype cycle and a broad range of technologies; what is perhaps most telling is the inconsistency between the two curves, even though they are only a year apart. Gartner is a technology research company, and even their experts struggle to keep up; how is a busy engineering professor to even try?

these new teaching methods, and not the technology, that matters.

A good example of a technology being confounded with its use is that of "clickers." The term "clickers" covers a wide range of devices that allow students to respond to questions from a lecturer and for these responses to be aggregated and presented back to the class in real time. Clickers were pioneered by Eric Mazur at Harvard University, who used them to implement his paradigm of peer instruction (Mazur, 1997). By including in his lectures conceptual questions (called ConcepTests) that addressed key points of understanding, Mazur was able to determine quickly whether his students had in fact grasped the concepts he was teaching. These multiple-choice questions, with alternatives addressing key misconceptions, were presented to the students. The learners then worked together for a short period of time, and their answers were collected instantly and automatically through the clickers. If the clicker responses showed that very few students understood the concept, Mazur was able to cover the material a second time, providing an alternative delivery perspective, and allowing the students a second chance to learn. On the other hand, if the clicker responses showed that most of the students understood the concept, Mazur was able instead to focus simply on the remaining misconceptions held by the students. If the responses showed that the class understood entirely, then the learning goals had been met.

The peer instruction approach showed significant improvements in student learning, but to ascribe these results to the technology of clickers is misleading. Similar results have been achieved elsewhere using colored cards rather than clickers (Lasry, 2008). It is ultimately the teaching approach, and only partly the technology, that is important. At the start of the twenty-first century, this conclusion holds truer than ever. New technology has opened up new ways of learning, but the technology is still the vehicle. Certainly the vehicle may be leading the context rather than the other

way around – for example, people adapting iPads to learning, rather than developing iPads to meet a learning need – but the reality is that it is the new ways of learning that the technology enables that matter. Technology is less about artifacts and increasingly more about the services – in this case learning services – it provides.

Asynchronous Learning as an Information Technology Effect

One effect of information technology has been to time-shift the learning to take on more of an asynchronous mode. Just as digital video recorders (DVRs) allow people to reschedule their television viewing, a range of technologies have allowed students to schedule their learning activities and added time flexibility as an affordance. Recording of lectures is now almost ubiquitous, with most universities operating some system that allows students to access recordings of lectures. In some institutions, recording of lectures is now the default setting – any lecturer wishing to not be recorded must opt out, rather than opting in to be recorded. Most modern systems also allow for the projected displays to be recorded, and these images are synchronized with the audio recordings.

Many academics have concerns that recording lectures will have a negative impact on lecture attendance. This concern is founded on the assumption that students can learn only if they are physically present in the lecture theatre or location. It largely neglects the paradigm that alternative access to lectures may be equally valid. However, research in fact shows that recordings do not have an impact on lecture attendance (Brotherton & Abowd, 2004; Copley, 2007). Further, the literature in fact shows that students do not use lecture recordings just as a simple "time-shift" of the lecture.

Analysis of access patterns shows that students' time is not evenly spread throughout the duration of the recording – rather, some areas are viewed multiple times, while others are skipped completely (Pinder-Grover, Millunchick, Bierwert, & Shuller, 2009). Therefore, it seems that students value recordings not for the time shift functionality (or asynchronicity) but for the rewind functionality. "Rewinding" a live lecture is very difficult, as few students have the courage to ask their professors/instructors to stop and repeat themselves, and further not all instructors will do so. More importantly, in learning various engineering concepts, each student will have his or her own areas of confusion. A lecture recording allows students to focus their revision on the areas that are most valuable to themselves, thereby allowing some level of instructional personalization. The aforementioned paradigm has also been implemented for laboratory-based feedback at Sheffield Hallam University (Nortcliffe & Middleton, 2008). By recording the "walkthrough feedback" given to students when they demonstrate their work in the laboratory, students are able to later revisit the advice they have received. This revisiting allows the students to "reactivate their train of thought" and prevent the erosion of the value of the feedback over time.

Virtual (or Remote) Presence as a New Technology Affordance

With lecture recordings the focus is on time shifting (asynchronicity). However, implicit in the process of time flexibility is also the phenomenon of location shifting. Students do not access lecture recordings in the same place that they experience the live synchronous lecture. In fact, the power of technology allows them to access course materials (including lectures) from libraries, or from their homes, or from wherever is most convenient. In the context of accessing learning materials, this relocation goes unremarked. But when it comes to remotely or virtually accessing laboratory equipment the affordance that information technology offers for virtual presence becomes a critical feature.

Using Internet-based tele-control to operate equipment is an increasingly common reality in industrial practice (Engineers Australia, 2012). Over the past two decades it has also become increasingly commonplace in undergraduate teaching. Remote or virtual access to laboratory equipment certainly offers the ability to time shift the learning experience. However, it is the ability to location shift it that was the driving force for development. Essentially bringing laboratories into spaces that were not previously open to such learning experiences allows broader inclusivity and is one of the foundational tenets for democratizing learning and participation.

The earliest remote laboratories were developed to allow for flexibility of access to equipment (Aktan, Bohus, Crowl, & Shor, 1996; Murray & Lasky, 2006). The experimental rigs that were converted were already instrumented through computer connections. All that was required was to extend this computer interface into an Internet or networked interface. Students were already interacting with the hardware through a computer interface. Now, rather than being compelled to be in a specific room at a specific time to use that interface, they were able to access that interface from anywhere at any time. Students certainly embrace the flexibility of access, with most remote laboratory experiments showing access times around the clock. Figure 31.4 shows the pattern of student access to an MIT iLabs rig between the release of an assignment and its submission deadline a week later.

This flexibility of access was the initial driver for the development of remote laboratories, and early literature in this problem space focused on the technical feasibility of the approach. The remote laboratory field has since evolved to the point where there are now consortia of universities sharing resources. Further, new remote laboratory equipment is now being developed with the express purpose of being shared across institutions (Harward et al., 2008; Lowe et al., 2011). Feasibility of implementation

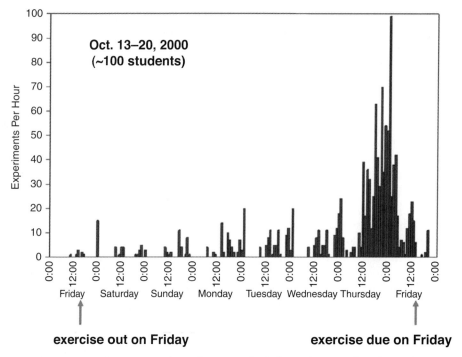

Figure 31.4. Accesses per hour for a sample iLabs rig (Harward et al., 2008).

has given way to sustainability of implementation and the standardization of interfaces to allow interoperability. The educational support material has also evolved, with guides to the selection of suitable experiments for remote access (Murray, Lindsay, Lowe, & Tuttle, 2010), and how remote laboratories can be used to address accreditation outcomes (Stumpers & Lindsay, 2011).

The evaluations of remote laboratories have evolved in parallel with the educational and technical development. The earliest attempts to evaluate learning in remote access modes focused on overall assessment outcomes (Ogot, Elliot, & Glumac, 2003), showing that the average performance achieved by students in different modes was the same. More fine-grained analysis, however, shows that there are differences in the learning outcomes, but that gains on some outcomes are offset by losses on others (Lindsay & Good, 2005). For instance, students who experienced remote or virtual access were more likely to identify ways in which experimental results differed from

theory-based expectations. Further, of all the students who identified these differences, students in the remote and virtual modes were more likely to demonstrate an understanding of the consequences of these differences.

Students in the simulation mode showed a poorer understanding of the limitations of the accuracy of their results. Also in the remote mode, learners showed a more reflective approach to their laboratory work. Although it is unclear whether they were made more reflective by the remote presence nature of the learning, or whether the virtuality better supported a pre-existing reflective tendency.

Lindsay and Good (2005) also observed statistically significant differences in the students' perceptions of the objectives of the laboratory experience as a function of which access mode they had experienced. Simulation mode students reported less emphasis on the hardware-specific outcomes, instead showing a bias toward abstract concepts such as principles of calibration. Conversely, a greater proportion of students in the remote mode felt they were supposed to learn about the hardware than those in the face-to-face mode.

More recent work has shown that these differences in outcomes can depend on whether students adopt an individual approach or a group approach to data collection (Corter, Esche, Chassapis, Ma, & Nickerson, 2011).

It is increasingly clear that remote laboratories are not an equivalent substitute for the in-person experience. They support the achievement of different learning outcomes, and as such they cannot be simply cut-and-pasted into a curriculum. They can, however, be an invaluable tool to supplement existing laboratory work. Indeed, the remote access infrastructure can be used to provide real-time data to students in a face-to-face laboratory (Mason, Shih, & Dragovich, 2007). Remote laboratories are also extremely effective in providing laboratory experiences where there would otherwise not be the possibility.

As remote laboratories have matured, the focus has moved from a simple location shift from an in-person laboratory to a time, location, and pedagogical shift. Just as the broader trend for computers is away from the computer itself toward the network it can access, so too remote laboratories shift from using a specific piece of equipment to accessing a network of shared facilities.

Technology Leads to Mediation and Mediation Affects Presence

Ideally, students must be able to anchor what they are learning into their prior knowledge (Ausubel, 2000), much of which will have been learned in a "real" context. If the students have new experiences in which to anchor new knowledge, the logical presumption is that it will be easier if these new experiences are themselves perceived as real. There are certainly good grounds for this presumption. However, the desire for the learning experience to be realistic is in fact a simplification of a number of other factors that correlate well with realism.

Social presence is "the degree to which a person is perceived as 'real' in a mediated communication" (Gunawardena & Zittle, 1997, p. 8). Social presence is strongly linked to academic outcomes, both as a predictor for perceived achievement (Hackman & Walker, 1990; Richardson & Swan, 2003) and as a strong predictor for student satisfaction (Aragon, 2003; Gunawardena & Zittle, 1997; Moore, Masterton, Christophel, & Shea, 1996). Social presence can be established through a number of immediacy behaviors – actions that promote a sense of presence and interaction. The choice of expressions and vocabulary of the teacher is one such behavior. Learning the students' names is another. Punctuality demonstrates a sense of perceived value in the learning process on the part of the instructor and serves to enhance social presence. These immediacy behaviors have an impact on student outcomes – greater instructor immediacy results in higher cognitive and affective achievement (Messman & Jones-Corley, 2001).

Establishing social presence through a technology-mediated interface is more challenging, but it is not impossible. Some immediacy behaviors do not depend on proximity. For example, learning students' names is possible regardless of any physical or psychological distance. Many of the behaviors, however, are not as robust to the separation of the instructor and the learner. Rather, they are dependent on the nature of the communication medium used to close the distance. The medium moderates the interactions, and in doing so potentially can impact the social presence of the parties involved (Jonassen, Campbell, & Davidson, 1994; Reiser, 1994). One of the critical challenges to overcome is the question of how to establish presence – and how much presence to establish – through the mediation of technology.

Fidelity – the extent to which a mediated experience faithfully represents the unmediated experience – is an important aspect of establishing social presence. Fidelity impacts learners' abilities to transfer knowledge they have learned. As practicing engineers, the contexts in which they will need the knowledge will be real, rather than educational simulations. Learners transfer better in high-fidelity situations. But if they do not learn in the first place, then there can be no transfer (Alessi & Trollip, 1991). An inexperienced learner in an aircraft flight deck gets overwhelmed rather than supported. "Increasing fidelity, which theoretically should increase transfer, may inhibit initial learning which in turn would inhibit transfer" (Alessi & Trollip, 1991, p. 234). Another example in which higher levels of fidelity are inappropriate was raised by Aldrich, who observed that the best selling bird-watching guides use illustrations of birds rather than photographs (Aldrich, 2004).

One of the determinants of presence is *media form* (Ijsselsteijn, de Ridder, Freeman, & Avons, 2000), which depends heavily on the transparency of the medium. The information provided by the medium must be sufficiently abundant to provide the user with a sense of presence. Further, this information must correspond to the information that the user would receive in an equivalent unmediated environment. The second factor deals with how the user interacts with the mediated environment. The environment should respond to changes in the same way the unmediated environment would (such as views changing when the user changes his or her viewpoint). The first two factors are consistent with Lombard and Ditton's definition of presence as "the perceptual illusion of non-mediation" (Lombard & Ditton, 1997) – that presence occurs when the users of media are oblivious to the media itself.

There can be a danger in students becoming oblivious to the interface – the goal is learning, rather than fidelity. There is evidence to show that students using simulations of hardware can lose sight of the real hardware being simulated, and instead get caught up in the "computer game" attitude toward the software (Schofield, Lester, & Wilson, 2004). This issue – whether the students focus on the equipment being simulated or the interface of the simulation – is known as transparency.

Lombard and Ditton provide an extensive review of the literature dealing with media factors that affect presence, with the recurring theme being that richer feedback leads to a greater sense of presence (Lombard & Ditton, 1997). Bigger screens promote a greater sense of *actually being there*. Better image resolution promotes engagement. Color images are better than black and white. Multimodal feedback is superior to single-mode feedback. The richer the experience, and the more information provided – assuming that this information remains consistent with itself and the environment – the greater the sense of presence reported by the users. Further, a transparent interface increases the sense of presence.

A more realistic simulation offers many advantages in light of these desirable outcomes. The mental distance between the learning experience and the students' previous (and indeed future) experiences is lower. In these instances reality is good, but it must not be reality for reality's sake. The "realness" of the interface must be focused to achieve the desired outcomes. Without an adequate sense of realness, the option of a technology-mediated interface is inferior to that of the in-person traditional experience. Once this realness has been achieved, however, the introduction of information technology allows for the removal of many of the constraints of the traditional approach. This can be the removal of the constraint of time through the use of asynchronous approaches; the removal of the constraint of place through remote or virtual access; or the removal of the constraint of homogeneity through the personalization of the learning process.

Simulations in Engineering Education

Modeling and simulation tools are finding extensive use in engineering education as a way for introducing abstract, difficult to observe phenomena into the curricula. More importantly, simulations "(a) allow transfer of knowledge, (b) skill development, and (c) the application of both knowledge and skills. During a simulation students typically acquire broad discipline-specific knowledge that they are able to later transfer into a professional setting" (Hertel & Millis, 2002, p. 1). Indeed in the engineering context, the use of simulations in the curriculum is also a reflection of the broad acceptance of modeling and simulations as a methodology for conducting day-to-day engineering work. We have pointed out earlier in this chapter that theory, experimentation, and modeling and simulations have converged, marking a synergistic blending of scientific methodologies. But this convergence brings about a larger issue that we elaborate next.

De Jong and Van Joolingen (1998) point out that "learning with simulations is closely related to a specific form of constructivistic learning, namely, *scientific discovery learning*" [italics in original] (p. 179). Many of the engineering cyber-environments (or science gateways) such as nanoHUB.org attempt explicitly to bridge processes of discovery with processes of learning. De Jong and Van Joolingen (1998) "identify a number of characteristic problems that learners may encounter in discovery learning, and classify them according to the main discovery learning processes: hypothesis generation, design of experiments, interpretation of data, and regulation of learning" (p. 183). Simulations allow students to fail gracefully and develop an intuition for the underlying engineering artifact. Lunetta and Hofstein (1981) argue that

> ... *simulations in education can provide participants with opportunities to become actively involved in:*
>
> (1) *experiencing phenomena not normally accessible because of administrative, technical, or economic constraints;*
>
> (2) *developing certain kinds of inquiry and problem solving skills (interpreting data, hypothesizing, communicating, decision-making); and*
>
> (3) *developing understanding of certain concepts and models."* (p. 250)

In modern engineering disciplines such as nanotechnology, bioengineering, and sustainability engineering, simulations are the primary methodology for even experts to experience and understand phenomena at scales that either are too small or too large to observe physically. Essentially, simulations allow learners to be placed in authentic contexts within which engineering and science problem solving occurs. "Authentic settings have the capability to motivate and encourage learner participation by facilitating students' willing suspension of disbelief" (Oliver, Harrington, & Reeves, 2002). Brown, Collins, and Duguid (1989) underscore the importance of using authentic settings when they state that "the activities of a domain are framed by its culture. [...] Authentic activities then, are most simply defined as the ordinary practices of the culture" (p. 34).

Many modern engineering problem-solving environments allow researchers to contribute simulation tools and mathematical models that are directly applicable in engineering and science courses. For example, most of the tools available on nanoHUB.org have been developed as a result of cutting-edge research. These tools are used directly in the classroom. Such use essentially provides students with a direct conduit to the results and indeed the culture of problem solving occurring within a specific engineering problem space. Lave and Wenger (1991) introduced the notion of "legitimate peripheral participation" wherein learning is positioned as an artifact of participating in a community of scientific practices. They argue that learners initially participate as just passive consumers of the knowledge artifacts but slowly increase their participation and contribution to the community. The use of simulations allows learners initially to use the models and knowledge tacit in the design of the tools.

By allowing learners to participate in a "community of practice" (Wenger, 1998) the use of simulation tools essentially promotes the notion of "cognitive apprenticeship" (Collins, Brown, & Newman, 1989) wherein the learner is following the experts' workflow and problem-solving patterns designed by expert engineers and scientists. Not only does the use of simulation tools allow

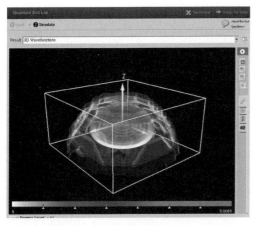

Figure 31.5. nanoHUB simulation tool Quantum Dot Lab is used primarily in engineering classes to help students understand that quantum dots can be produced in a variety of material systems and geometries. The core engine that powers this learning tool is the same engine that is used by researchers to publish advanced scientific papers in top journals such as *Science*. Simulation tools allow students to learn the workflow of expert engineers and scientists, allowing for a "cognitive apprenticeship model."

learners to gain knowledge about the culture and problem-solving strategies of the content domain, but it also allows learners to get access to resources that go beyond the boundaries of any physical space. Advanced simulation tools allow learners to solve problems designed for learning in exactly the same way that expert engineers solve problems. Figure 31.5 shows a tool from an engineering cyber-environment called Quantum Dot Lab (Klimeck et al., 2005) that is used extensively to help students understand how simple quantum dots behave. However, the very same engine and core workflows used in this particular simulation tools were also used to produce a highly visible paper in *Science* (Weber et al., 2012).

It is in this context that the role of information technology in general and simulation tools, specifically, comes into clear view. Modern simulation tools by virtue of their ability to access large-scale computation and data resources in transparent ways fundamentally lower barriers of entry into advanced engineering and science. The ability of simulation tools to democratize the

learning process is clearly visible not only in the preceding example with nanoHUB, but also is generally true of other major efforts. For example, the Physics Education Technology Project (PhET)[3] allows access resources that help students learn fundamental physics concepts. These resources are free and are broadly accessible. Numerous other examples such as the MIT OpenCourseWare,[4] [US] National STEM Digital Library,[5] Khan Academy,[6] as well as many others, make resources and learning content freely available online. The combination of asynchronous delivery and remoteness allows for the possibility of students learning anywhere, anytime. One key consequence of this combination is that it removes one of the key constraints on teaching – student capacity. With the flexibility to deliver learning through technology, the marginal cost per student falls, and it is possible to service audiences that are orders of magnitude larger than a traditional classroom. In essence, through the widespread use of simulation tools in combination with the exponential growth in networking bandwidths that deliver audio, video, and other multimedia, technology has allowed true democratization of learning. These transformations in the global engineering learning landscape are so massive that technology is now being considered in many circles as the primary tool for free expression and transforming learning within communities that have traditionally endured harsh censorship and social injustice. Such a discussion is well beyond the scope of this chapter.

Technology Allows Personalization

Engineering problem solving and practice are increasingly relying on cloud-like[7] infrastructures that allow engineers to select and tailor the type of resources to their problem-solving style. The focus on data within scientific systems has opened avenues for a new breed of informatics that allow for the customization of research and learning experiences. Although on the one hand the engineering problem-solving infrastructure is able to support a global scientific enterprise, the underlying infrastructure resembles the power grid – where it plays merely a utility role. Scientists have termed this phenomenon "service-oriented science" (Foster, 2005).

The excitement about "service-oriented science" stems from the fact that on the one hand infrastructure has turned into a service, and on the other the very same "service-orientedness" allows for large-scale personalization of the scientific enterprise. West-Burnham and Coates (2007) point out that "personalizing learning is emerging as a dominant theme in the reconceptualization of the way in which education is provided. [. . .] At the heart of the personalization debate is recognition of the concept of a *service* [italics in original] provided to *individuals* [italics in original] to meet their specific and personal needs" (p. 9). Asynchronous learning frees students from the constraints of time; remote access frees students from the constraints of space. The next step is to free students from the constraint of homogeneity of teaching approaches. In 2010 the NAE identified "Advancing Personalized Learning" as one of the 14 Grand Challenges for Engineering (NAE, 2008). There is a clear expectation that technology will play an important role in addressing these challenges with "computer programs designed to match the way it presents content with a learner's personality," (NAE, 2008) a key example of how learning can be tailored to suit an individual's needs.

The impetus to personalize learning is by no means new; in many ways it has been preceded by a push to personalize assessment. With the advent of large-scale assessment – such as the Graduate Records Examination (GRE©)[8] – there had always been a major need for theories and approaches that would allow assessment to overcome some of the basic "mass" nature of classical testing – namely, one test fits all paradigm. Highly efficient database systems could organize test data in a form that can be utilized by assessment experts for post hoc analyses. The key term borrowed into the field of testing from computer science is "data

leverage." Item response theory (Hambleton, Swaminathan, & Rogers, 1991), based on what was at that time a new psychometric model, allowed assessment developers to peel away from assessing the characteristics of a group performance on a test to more of an individual model of assessment.

At the simpler level are individually tailored assessment tasks, in which each student is given his or her own "version" of a standard structured problem. An example of this approach is the Weekly Assessed Tutorial Sheets (WATS) implemented in a freshman fluid mechanics class at Hertfordshire University (Russell, 2005). Each week the students are required to complete an assessable tutorial question. The structure of the question is the same for all students, but the WATS system randomizes the parameters (flow rate, pressure, etc.) so that each student receives a unique version of the tutorial question to complete. In this way "students could only discuss their methodology and not the answers" (Russell, 2005, p. 30) – a benefit for promoting deep learning approaches in students as well as deterring plagiarism.

The necessity for assessing an individually tailored assignment offered a number of learning opportunities beyond the deep learning approach. Assessment was by necessity automated. The alternative was for the lecturer to solve the assignment question individually for each student and then check his or her answers. The WATS system automatically determines the correct answer for each student's combination of parameters, and an acceptable error margin is employed to determine whether the student's submitted answers are close enough to be correct.

The automation of the assessment process allows the WATS system to go further and to provide helpful feedback to the students. Just as the correct answer can be automatically determined by the system, so can the most likely incorrect answers – such as from using a *sine* instead of a *cosine*, or through forgetting a particular minus sign. In this way the WATS system is able to detect when a student submits an answer most likely caused by a specific error and to provide automated feedback to the student as to the likely error he or she has made. Seventy-two percent of students felt that this feedback was helpful to them, and 95% of students felt that the WATS system would help them in their performance on the final examination (Russell, 2006).

This approach ensures that the assessment tasks for students are personalized, which offers some learning and anti-collusion benefits. But it does not fully harness the opportunities for personalization that modern technology can provide. To fully personalize the entire learning process, it is now abundantly clear that there is a need to treat all of the data emanating from the entire learning process as a single ecosystem that leads to deep insights about the learner. Further, it is not sufficient to just derive these insights, but the system needs to translate this insight into some action that can in turn scaffold the learning process.

Traditionally, when we spoke of a personalized learning system that extends beyond assessment, the discussion always revolved around intelligent tutors (Bursilovsky & Peylo, 2003; Woolf, 2009) that are tightly coupled with approaches that utilize artificial intelligence techniques. There are numerous implementations of intelligent tutors. However, we just point to a trend toward these types of systems. At the core of any effort to personalize learning in technology-mediated environments is the issue of data, the need for better instrumentation (Borgman et al., 2008), and availability of better algorithms to understand in real time what the learner is doing within the environment.

Adams et al. (2010) introduce the notion of user flow informatics that is geared toward real-time analysis and decision making within engineering cyber-environments such as nanoHUB.org. The notion of flow moves away from a model where student performance is treated as a static measure in time. In this method, they treat learning data as dynamic streams that can lead to more insights. They also introduce the notion of

"software as sensor" – meaning software within engineering cyber-environments can not only gather data about what learners are doing at any given time, but also change artifacts in the environment to accommodate learner-specific pathways of content and progressions.

Conclusions

In this chapter, we have attempted to capture the zeitgeist of technology in engineering education. Many times an attempt to write in depth about the role of technology descends into a laundry list discussion of the individual technologies themselves. We have attempted to avoid this. Our argument is that engineering education as a system cannot afford to chase specific technologies without a clear grand strategy and argument. As a system, engineering education stakeholders have very little to no impact on the progress of technology as a whole. For the most part, we remain passive consumers of technology and our analyses are post hoc. In the chapter, we present the case that this passiveness is no longer affordable. We have argued that the fundamental nature of problem solving in engineering has undergone a technology-fueled change. By adopting a strategy that closely aligns with how technology can bring the engineering education system in sync with the demands of the engineering industry, we believe we become active in directing the path of technological progress. Throughout this chapter, we have drawn parallels between how highly impactful theories of learning are not very far from the tenets that govern real-world engineering problem solving.

Networked technologies play a dual role in the engineering education arena. In its first role networked technologies allow content delivery and tie this content to pedagogical frameworks. In its second role, networked technologies fundamentally change how research is conducted in engineering education. For example, there are major efforts such as the Interactive Knowledge Networks for Engineering Education Research[9] (currently known as Deep Insights Anytime, Anywhere – DIA2), which attempt to utilize ultra-large-scale data in combination with data mining and visual informatics to characterize better the engineering education problem space. We have chosen not to purse a discussion on this second role of networked technologies.

We close with the statement that networked technologies are an unavoidable part of the landscape that students today bring into their learning contexts. By harnessing the power of networked technologies to open the window of our learning environments to real-world problems uninhibited by the false constraints of the "classroom," by embracing the inherent complexity of the engineering grand challenges we deal with, and by using technology to help derive insights into the learning process that we otherwise could not, we transform the learning itself for the better.

Footnotes

1. EMC, a company focused on data storage, estimates that our digital universe is approximately 1.8 trillion gigabytes. This amount doubles every year and the cost for data storage is plummeting. Source: http://www.emc.com/collateral/about/news/idc-emc-digital-universe-2011-infographic.pdf

2. http://www.nanohub.org

3. Available online at http://phet.colorado.edu/. Retrieved on on April 5, 2012.

4. Available at http://ocw.mit.edu/index.htm. Retrieved on April 5, 2012.

5. Available at http://nsdl.org/. Retrieved on on April 5, 2012.

6. Available at http://www.khanacademy.org/. Retrieved on on April 4, 2012.

7. Meaning users are able to get a service and not worry about where the infrastructure is located or how it is managed.

8. http://www.ets.org/gre/

9. http://www.ikneer.org. Retrieved on on April 6, 2012.

References

Adams, G. B. III, Madhavan, K. P. C., Zentner, M. G., Denny, N., Shivarajapura, S., & Klimeck, G. (2010). *User flow informatics to personalize learning in engineering cyber-environments – nanoHUB.org: A case study.* Paper presented at the ASEE 9th Global Colloquium on Engineering Education, Singapore.

Aktan, B., Bohus, C. A., Crowl, L. A., & Shor, M. H. (1996). Distance learning applied to control engineering laboratories. *IEEE Transactions on Education, 39*(3), 320–326.

Aldrich, C. (2004). *Simulations and the future of learning.* San Francisco, CA: Pfeiffer.

Alessi, S. M., & Trollip, S. R. (1991). *Computer-based instruction: Methods and development* (2nd ed.). Englewood Cliffs, NJ: Prentice-Hall.

Aragon, S. R. (2003). Creating social presence in online environments. *New Directions for Adult and Continuing Education, 2003*, 100, 57–68.

Armbrust, M., Fox, A., Griffith, R., Joseph, A. D., Katz, R., Konwinski, A., . . . Zaharia, M. (2010). A view of cloud computing. *Communications of the ACM, 53*(4), 50–58.

Ausubel, D. P. (2000). *The acquisition and retention of knowledge: A cognitive view.* Dordrecht, The Netherlands and Boston, MA: Kluwer Academic.

Bainbridge, W. S., & Roco, M. C. (2006). Progressive convergence. In W. S. Bainbridge & M. C. Roco (Eds.), *Managing nano-bio-info-cogno innovations: Converging technologies in society* (pp. 1–8). Dordrecht, The Netherlands: Springer.

Bennett, S., Maton, K., & Kervin, L. (2008). The 'digital natives' debate: A critical review of the evidence. *British Journal of Educational Technology, 39*(5), 775–786.

Borgman, C. L., Abelson, H., Dirks, L., Johnson, R., Koedinger, K. R., Linn, M. C., . . . Szalay, A. (2008). *Fostering learning in a networked world: The cyberlearning opportunity and challenge* (N. T. F. o. Cyberlearning, Trans.). Washington, DC: National Science Foundation.

Brotherton, J. A., & Abowd, G. D. (2004). Lessons learned from eClass: Assessing automated capture and access in the classroom. *Transactions on Computer-Human Interaction, 11*(2), 121–155.

Brown, J. S., Collins, A., & Duguid, P. (1989). Situated cognition and the culture of learning. *Educational Researcher*, 18(1), 32–42.

Bursilovsky, P., & Peylo, C. (2003). Adaptive and intelligent web-based eduational systems. *International Journal of Artificial in Higher Education*, 13(2–4), 159–172.

Clark, R. (1983). Reconsidering research on learning from media. *Review of Educational Research*, 53, 445–459.

Collins, A., Brown, J. S., & Newman, S. E. (1989). Cognitive apprenticeship: Teaching the craft of reading, writing, and mathematics. In L. A. Resnick (Ed.), *Knowing, learning, and instruction: Essays in honor of Robert Glaser* (pp. 453–494). Hillsdale, NJ: Lawrence Erlbaum.

Copley, J. (2007). Audio and video podcasts of lectures for campus-based students: Production and evaluation of student use. *Innovations in Education and Teaching International*, 44(4), 387–399.

Corter, J. E., Esche, S., Chassapis, C., Ma, J., & Nickerson, J. V. (2011). Process and learning outcomes from remotely operated, simulated, and hands-on student laboratories. *Computers and Education*, 57(3), 2054–2067.

Cuban, L. (2001). *Oversold and underused: Computers in the classroom*. Cambridge, MA: Harvard University Press.

De Jong, T., & Van Joolingen, W. R. (1998). Scientific discovery learning with computer simulations of conceptual domains. *Review of Educational Research*, 68(2), 179–201.

Engineers Australia. (2012). Autonomous trucks roll out in the Pilbara. *Mining 2012 – Special Insert of the Engineers Australia* magazine. Retrieved from http://www.engaust.com.au/magazines/mining/pdf/Mining_Iss1_2012.pdf?xml=Infrastructure_Australia

Foster, I. (2005). Service-oriented science. *Science*, 308(5723), 814–817.

Friedman, T. L. (2007). *The world is flat 3.0: A brief history of the twenty-first century*. New York, NY: Picador Reading Group.

Gartner Research. (2012). Hype cycles. Retrieved from http://www.gartner.com/technology/research/methodologies/hype-cycle.jsp

Grimson, J. (2002). Re-engineering the curriculum for the 21st century. *European Journal of Engineering Education*, 27(1), 31–37.

Gunawardena, C. N., & Zittle, F. J. (1997). Social presence as a predictor of satisfaction within a computer-mediated conferencing environment. *American Journal of Distance Education*, 11(3), 8–26.

Hackman, M. Z., & Walker, K. B. (1990). Instructional communication in the televised classroom: The effects of system design and teacher immediacy on student learning and satisfaction. *Communications Education*, 39(3), 196–209.

Hambleton, R. K., Swaminathan, H., & Rogers, H. J. (1991). *Fundamentals of item response theory*. Newbury Park, CA: SAGE.

Harward, V. J., del Alamo, J. A., Lerman, S. R., Bailey, P. H., Carpenter, J., DeLong, K., . . . Zych, D. (2008). The iLab shared architecture: A Web services infrastructure to build communities of Internet accessible laboratories. *Proceedings of the IEEE*, 96(6), 931–950.

Hertel, J. P., & Millis, B. J. (2002). *Using simulations to promote learning in higher education: An introduction* (1st ed.). Sterling, VA: Stylus.

Ijsselsteijn, W. A., de Ridder, H., Freeman, J., & Avons, S. E. (2000). Presence: Concept, determinants and measurement. In *Proceedings of the SPIE*, 3959.

Jonassen, D. H., Campbell, J. P., & Davidson, M. E. (1994). Learning with media: Restructuring the debate. *Educational Technology Research & Development*, 42(4), 31–39.

Kennedy, G. E., Judd, T. S., Churchward, A., Gray, K., & Krause, K.-L. (2008). First year students' experiences with technology: Are they really digital natives? *Australasian Journal of Educational Technology*, 24(1), 108–122.

Klimeck, G., Bjaalie, L., Steiger, S., Ebert, D., Kubis, T. C., Mannino, M., . . . Povolotskyi, M. (2005). Quantum Dot Lab. Retrieved from https://nanohub.org/resources/qdot. (DOI: 10.4231/D3B27PR0Q).

Kozma, R. B. (1994). Will media influence learning? Reframing the debate. *Educational Technology Research & Development*, 42(2), 7–19.

Lasry, N. (2008). Clickers or flashcards: Is there really a difference? *The Physics Teacher*, 46, 242–244.

Lave, J., & Wenger, E. (1991). *Situated learning: Legitimate peripheral participation*. New York, NY: Cambridge University Press.

Lindsay, E. D., & Good, M. C. (2005). Effects of laboratory access modes upon learning outcomes. *IEEE Transactions on Education*, 48(4), 619–631.

Lindsay, E. D., Naidu, S., & Good, M. (2007). A different kind of difference: Theoretical implications of using technology to overcome separation in remote laboratories. *International Journal of Engineering Education*, 23(4), 772–779.

Lombard, M., & Ditton, T. (1997). At the heart of it all: The concept of presence. *Journal of Computer Mediated Communication*, 3(2). Retrieved from http://onlinelibrary.wiley.com/doi/10.1111/j.1083-6101.1997.tb00072.x/full

Lowe, D. B., Conlon, S., Murray, S., Weber, L., de la Villefromoy, M., Lindsay, E. D., . . . Tang, T. (2011). Labshare: Towards cross-institutional laboratory sharing. In A. K. M. Azad, M. E. Auer, & J. Harward (Eds.), *Internet accessible remote laboratories: Scalable E-learning tools for engineering and science disciplines*. IGI Global.

Lunetta, V. N., & Hofstein, A. (1981). Simulations in science education. *Science Education*, 65(3), 243–252.

Mason, G., Shih, F., & Dragovich, J. (2007, June 24–27). *Real-time access to experimental data using tablet PCs*. Paper presented at the American Society for Engineering Education Annual Conference, Honolulu, HI.

Mazur, E. (1997). *Peer instruction: A user's manual*. Upper Saddle River, NJ: Prentice Hall.

Messman, S. J., & Jones-Corley, J. (2001). Effects of communication environment, immediacy, and communication apprehension on cognitive and affective learning. *Communications Monographs*, 68(2), 184–200.

Moore, A., Masterton, J. T., Christophel, D. M., & Shea, K. A. (1996). College teacher immediacy and student ratings of instruction. *Communications Education*, 45, 29–39.

Murray, S. J., & Lasky, V. L. (2006). A remotely accessible embedded systems laboratory. In N. Sarkar (Ed.), *Tools for teaching computer networking and hardware concepts*. London: Information Science.

Murray, S., Lindsay, E. D., Lowe, D. B., & Tuttle, S. W. (2010). *Derivation of suitability metrics for remote access mode experiments*. Paper presented at the Remote Engineering and Virtual Instrumentation, Stockholm, Sweden.

National Academy of Engineering (NAE). (2004). *The engineer of 2020: Visions of engineering in the new century*. Washington, DC: The National Academies Press.

National Academy of Engineering (NAE). (2008). Grand challenges for engineering in the 21st century. Retrieved from http://www.engineeringchallenges.org/

National Research Council (NRC). (2003). Beyond productivity: Information technology, innovation, and creativity (D. o. E. a. P. Sciences, Trans.). Retrieved from http://www.nap.edu/catalog.php?record_id=10671

National Research Council (NRC). (2009). Learning science in informal environments: People, places, and pursuits (D. o. B. a. S. S. a. Education, Trans.). Retrieved from http://www.nap.edu/catalog.php?record_id=12190

Nortcliffe, A., & Middleton, A. (2008). A three year case study of using audio to blend the engineers learning environment. *Engineering Education*, 3(2), 45–57.

Ogot, M., Elliot, G., & Glumac, N. (2003). An assessment of in-person and remotely operated laboratories. *Journal of Engineering Education*, 92(1), 57–64.

Oliver, R. G., Harrington, J. A., & Reeves, T. C. (2003). Patterns of engagement in authentic online learning environments. *Australasian Journal of Educational Technology*, 19(1), 59–71. Retrieved from http://www.ascilite.org.au/ajet/ajet19/herrington.html

Oppenheimer, O. (2003). *The flickering mind: The false promise of technology in the classroom and how learning can be saved*. New York, NY: Random House.

O'Reilly, T. (2007). What is web 2.0: Design patterns and business models for the next generation of software. *Communications and Strategies*, 65(1), 17–38. Retrieved from http://papers.ssrn.com/sol3/papers.cfm?abstract_id=1008839

Pinder-Grover, T., Millunchick, J. M., Bierwert, C., & Shuller, L. (2009). *The efficacy of screencasts on diverse students in a large lecture course*. Paper presented at the American Society for Engineering Education, Austin, TX.

Reiser, R. A. (1994). Clark's invitation to the dance: An instructional designer's response. *Educational Technology Research & Development*, 42(4), 45–48.

Richards, L. G. (1998). *Teaching engineers to be creative*. Paper presented at the Frontiers in Education, Tempe, AZ. Retrieved from https://docs.google.com/viewer?url=http%3A%2F%2Ffie-conference.org%2Ffie98%2Fpapers%2F1243.pdf

Richardson, J. C., & Swan, K. (2003). Examining social presence in online courses in relation to students' perceived learning and satisfaction. *Journal of Asynchronous Learning Networks*, 7(1), 68–88.

Russell, M. (2005). Evaluating the weekly-assessed tutorial sheet approach to assessment: Background, pedagogy and impact. *Journal for the Enhancement of Learning and Teaching*, 2(1), 26–35.

Russell, M. (2006). Evaluating the weekly-assessed tutorial sheet approach to assessment: The students' experience. *Journal for the Enhancement of Learning and Teaching*, 3(1), 37–47.

Schofield, D., Lester, E., & Wilson, J. A. (2004). *Virtual reality interactive learning environments.* Paper presented at the EE2004, Wolverhampton, UK.

Stumpers, B. D., & Lindsay, E. D. (2011, December 4–7). *Remote laboratories: Enhancing accredited engineering degree programs.* Paper presented at the 22nd Annual Australasian Association for Engineering Education Conference, Fremantle, Australia.

Suroweicki, J. (2004). *The wisdom of crowds.* New York, NY: Anchor.

TIME magazine. (2006). *Time.* Retrieved from http://www.time.com/time/covers/0,16641,20061225,00.html

Weber, B., Mahapatra, S., Ryu, H., Lee, S., Fuhrer, A., Reusch, T. C. G., . . . Simmons, M. Y. (2012). Ohm's law survives to the atomic scale. *Science*, 335(6064), 64–67.

Wenger, E. (1998). *Communities of practice: Learning, meaning, and identity.* New York, NY: Cambridge University Press.

West-Burnham, J., & Coates, M. (2007). *Transforming education for every child.* London: Continuum.

Woolf, B. P. (2009). *Building intelligent interactive tutors: Student centered strategies for revolutionizing e-learning.* Burlington, MA: Morgan Kaufman.

Wulf, W. A. (1998). Diversity in engineering. *The Bridge*, 28(4). Retrieved from http://www.nae.edu/Publications/Bridge/CompetitiveMaterialsandSolutions/DiversityinEngineering.aspx#about_author7488

CHAPTER 32

Global and International Issues in Engineering Education

Aditya Johri and Brent K. Jesiek

"There is an urgent need for research on engineering in a global context. The phenomenon of global engineering is still emerging. There is a need for a theoretical foundation on learning behaviors and models as well as on organizational processes and management methods focused on instilling global competence in engineers."
Global Engineering Excellence Initiative (GEEI) Report (Italics in original, 2006, p. 2)

Introduction

Although globalization has been a political and economic concern for centuries (O'Leary, Orlikowski, & Yates, 2002), its impact on the engineering education community has become particularly salient since the publication of the *Engineer of 2020* (National Academy of Engineering [NAE], 2003) and *The World Is Flat* (Friedman, 2005). These and subsequent publications (Committee on Science Engineering and Public Policy [CSEPP], 2006; Duderstadt, 2008,

Engineers Against Poverty (EAP)/Institute of Education (IOE), 2008; GEEI, 2006; Grandin & Hirleman, 2009; NAE, 2005) have emphasized the critical need to train globally competent engineers. Academics, practitioners, and policymakers alike have realized that solving the critical problems facing the world today, including addressing the NAE Grand Challenges (NAE, http://www .engineeringchallenges.org/) and achieving a "better quality of life," will require close collaboration among engineers from around the world. Not surprisingly, educational institutions are investing millions of dollars to train students for success in the globalized economy through initiatives such as study abroad. Students, in turn, are spending significant amounts of money to participate in these programs. Recent efforts to pass a study abroad bill provide still another indication of the importance of this issue at the national level (Study Abroad Bill, 2009). And from a research perspective, the National Science Foundation (NSF) itself spends upwards of $50 million each year funding international programs (Office of International Science and Engineering [OISE], 2008).

The greatest investment by far, however, has been by industry. In 2011 alone, U.S. firms spent more than $156 billion on workforce training, an average of $1,182/employee (American Society for Training & Development [ASTD], 2012). Almost 21% of that amount was spent on training people to work more effectively with others, and global issues are becoming a bigger slice of that pie amidst growing recognition that global work experiences give a boost to industry careers. It is within this context that engineering educators are trying to make an impact to better prepare future engineers for the global world of work.

In addition to these recent trends, in many ways engineering has long been a field of comingled local and global activities. Early approaches to educating engineers – such as apprenticeships in Great Britain or formal instruction in France – gradually diffused, leaving a lasting impact on engineering education and practice in countries across the globe (Meiksins & Smith, 1996). At the same time, a growing body of comparative research has highlighted similarities and differences among engineers in different countries, including as related to their educational backgrounds, social status, preferred career pathways, and typical ways of defining and solving problems (Downey & Lucena, 2004; Meiksins & Smith, 1996). These kinds of differences become ever more salient as engineers find themselves working cross-nationally to carry out academic research, industry projects, humanitarian initiatives, or international standards work. Paradoxically, most engineering graduates have limited awareness of the global dynamics that increasingly define their fields and subfields, and many engineering educators see global education as largely outside their sphere of activity. But as highlighted in the sections that follow, an expanding body of scholarly literature on global engineering has started to emerge and outlines numerous opportunities for further research as well as ways to innovate engineering education to address global issues.

Overview of Current Literature on Global Engineering Education and Practice

To categorize contemporary understandings and approaches to global engineering among engineering educators and researchers, we first reviewed all studies related to global and international education published in the proceedings of the annual meeting of the American Society for Engineering Education (ASEE) (144 papers) and the Frontiers in Education Conference (66 papers) between the years 2003–2007. The proceedings were selected for their broad representation of the issues, at least from the U.S. academic and practitioner community, and for the short publication cycle that accurately reflects current concerns. This five-year range was chosen as it coincided with the publication of *Engineer of 2020* (NAE, 2003) and *The World Is Flat* (Friedman, 2005). A content analysis of the titles and abstracts was carried out and all papers relevant to the proposal were read in full. Table 32.1 provides an overview of the major issues that emerged.

In summary, our analysis reveals that almost 40% of the papers describe or discuss some program or project that an institution, college, or department has implemented to make its offerings more global. In addition, almost 15% of the articles are about international activities such as study abroad, student exchange, or international research experience, with study abroad being the most common among these. Another 12% of the articles report virtual or online classes offered between institutions in two different geographic locations. These classes often aim to provide first-hand experiences with virtual work, giving students knowledge of issues such as effective use of technology, working across time zones, and forming common ground (McNair, Paretti, & Kakar, 2008).

Articles discussing the introduction of content related to international issues and globalization within existing curricula or new courses represent 15% of the papers in the data set. One successful form taken

Table 32.1. Distribution of ASEE and FIE Papers, 2003–2007

Category	Number (Total = 210)	Percentage
Program description (institutional/departmental/college level)	80	38
International activities (study abroad, exchange programs)	30	15
Virtual class (online courses/labs)	25	12
Course content (full or largely global component)	30	15
Others (accreditation, opinion, panel, workshop)	40	20
Workplace related studies (out of the total papers)	10	5

Journal of Engineering Education (JEE) and *International Journal of Engineering Education (IJEE)* were initially included in our sample but owing to significantly low number of papers related to global and international issues in the journals we decided to exclude them from the final analysis. JEE had two articles and IJEE had nineteen articles, all part of a single special issue. We also acknowledge the U.S.-centric view we present by not including other international outlets.

by such efforts is to teach students about other cultures and cultural differences, for example, the "Engineering Cultures" course and learning modules developed by Downey and colleagues (2006), which teach students about differences that exist among engineers from different nations, including as related to their educational backgrounds, career pathways, social status, and roles in national development. Other relevant issues, such as accreditation guidelines or the opinions of individuals, are discussed in 20% of all papers, while just 5% involved workplace-related studies. As this overview reveals, there remain significant opportunities to grow this area of scholarly literature, including in emerging domains such as studies of global engineering work. Another emerging area of international activities within engineering education that is not covered in this review is international service learning (Johri & Sharma, 2013). Informal and extracurricular activities such as Engineers without Borders (EWB) and Engineers for a Sustainable World (ESW) have existed for years, but widespread formal integration of such activities within engineering curriculum are relatively new but increasing at a significant pace (see Chapter 18 by Swan, Paterson, & Bielefeldt in this volume).

As can be seen from the review in the preceding text, many areas of interest exist within the umbrella of global and international issues in engineering education. In this chapter we focus primarily on two areas. First, we review global and international engineering programs to look closely at the kinds of globally oriented educational experiences now available to engineering students and their impacts. Second, we review studies of global engineering work, which has significant implications for the professional future of engineering students. Given the lack of work within engineering education on global engineering work (for some background see Chapter 7 by Stevens, Johri, and O'Connor in this volume), we look primarily at implications of organizational studies for engineering work and outline an agenda for future research.

Global and International Engineering Programs: History, Formats, and Trends

The history of global engineering education in the United States can be traced back to at least the 1940s and 1950s (Jesiek & Beddoes, 2010). From the immediate postwar period into the 1970s, shifting concerns about international diplomacy, Cold War sociopolitical tensions, and growing demand for technical assistance in developing countries each in turn helped encourage both the internationalization of many universities and major expansions in study abroad. Against this backdrop, and starting in the late 1950s, a small but growing number of programs were also established to give American engineering students study and work opportunities abroad. In more recent decades, the number of such programs has continued to rise in response to new drivers. During the post–Cold War context of the 1980s, for instance, arguments about enhancing America's economic competitiveness emerged as an important new justification for internationalizing engineering education. Even more recently, globalization discourses and dynamics have surfaced as a prominent rationale for continuing to expand global engineering education.

As a number of commentators have observed, this more than decade long history has spawned many different types of global engineering programs and experiences for engineering students (Downey et al., 2006; Grandin & Hirleman, 2009; Parkinson, 2007). Most generally, these can be divided into learning experiences that students have abroad versus those occurring "at home." For engineering students who go abroad, some relevant program variables include the type of experience (e.g., taking courses, completing an internship, doing research in a lab, or participating in a guided field trip) and its duration and intensity (e.g., short vs. long term, one experience or a series of experiences). In addition, many programs involve project-based learning (PBL) and/or service learning activities, such as when students work on engineering projects in developing contexts or partner with peers abroad on senior/capstone design projects (e.g., Bloomgarden & Riley, 2006; Mohtar & Dare, 2012). Global engineering education "at home" can also take many forms, with coursework and extracurricular activities two typical experiences in this category (Downey et al., 2006).

Despite the diversity of programs that have emerged over the years, commentators continue to lament the relatively small number of students and schools seriously engaging in global engineering. Shuman, for instance, has estimated that just 10% to 15% of U.S. engineering schools are taking international education seriously (Bremer, 2008), while another group of leading stakeholders noted that "there has been no tradition of sending engineers to study or work abroad" and went on to identify sixteen major obstacles to scaling up global engineering education (Grandin & Hirleman, 2009). A recent edited volume by Downey and Beddoes (2010) more specifically highlights the career risks often associated with building new global engineering programs and initiatives, including through a series of "personal geographies" that richly illustrate the struggles and challenges faced by sixteen contributing engineering educators.

Other developments suggest a somewhat more positive outlook. Organizations such as the Global Engineering Education Exchange (Global E3, http://www.iie.org/programs/globale3), conferences like the annual Colloquia on International Engineering Education, and the online cybercommunity GlobalHUB (http://globalhub.org) support a growing network of instructors, administrators, and researchers engaged in global engineering education. A number of leading engineering schools have also seriously embraced global education. In 2007, Georgia Tech claimed that 34% of its students were studying abroad (Connell, 2007), and in 2008 Rensselaer Polytechnic Institute (RPI) announced a plan to have 25% of its engineering students studying or working abroad by 2009 and 100% by 2015 (Gerhardt & Smith, 2008).

More generally, the Institute of International Education reported that 8,330 engineering students studied abroad for academic credit in 2008–9, representing about 3.2% of all U.S. students studying abroad, and just under 9% of the total number of students earning undergraduate bachelor's degrees that year (Gibbons, 2009; Institute of International Education [IIE], 2010). Such statistics suggest ample room for future growth, but are also part of a longer and largely upward historical trend. Over five decades, the proportion of engineering students among all study abroad participants has gradually but steadily risen from about 1% in the 1950s and 1960s to above 3% in recent years (Jesiek & Beddoes, 2010). Further, these numbers do not include students who have participated in other relevant global experiences, such as short-term trips, elective coursework, cross-national design projects, extracurricular activities, and so forth, although this wide variability in formats raises questions about what specific kinds of learning outcomes these diverse programs are supporting.

Defining Global Competency in Engineering

As suggested above, workforce diversification trends, the increasingly international character of industry and higher education, and growing concerns about national competitiveness and global "grand challenges" have led to widespread recognition that global competency is imperative for a new generation of engineers who must be ready to work in a diverse, interconnected, and rapidly changing world (Boeing Company & Rensselaer Polytechnic Institute [RPI], 1997; Duderstadt, 2008; Grandin & Hirleman, 2009; Jamieson & Lohmann, 2009; Katehi, 2005; Sigma Xi, 2007; Wulf, 2003). These realities are also reflected in contemporary accreditation guidelines and curriculum reports, and in the rise of global engineering education as a burgeoning domain of activity. As a result, ever more engineering students, faculty, and professionals are "going global."

As global engineering gains prominence as a distinct field of research and practice, there remains a lack of consensus about what counts as global competency for engineers, much less how it might best be developed and assessed. Posing further challenges, there is the question of how to relate instruction focused on the traditional "core" of engineering education – including mathematical foundations, scientific principles,

and engineering "object worlds" (Bucciarelli, 1996) – to the wider array of "non-technical" attributes viewed as increasingly essential for practicing professionals (Downey, 2010). In short, global engineering initiatives may evoke skepticism and resistance when perceived to impinge on, displace, and/or transform the kinds of technical curricula that have been at the heart of engineering education for more than a century.

Nonetheless, many recent accreditation and curricular guidelines provide some tentative insights about the perceived importance and definitional status of global engineering. ABET EC2000 Criterion 3.h, for instance, states that engineering graduates must "understand the impact of engineering solutions in a global, economic, environmental, and societal context" (ABET, 2008). The American Society for Civil Engineers' (ASCE's) *Civil Engineering Body of Knowledge for the 21st Century* is more explicit, listing "Globalization" as a category of outcomes for civil engineering graduates who can "[o]rganize, formulate, and solve engineering problems within a global context" and "[a]nalyze engineering works and services in order to function at a basic level in a global context" (ASCE, 2008, p. 144). Recognition of such attributes has also occurred in other countries and regions, as reflected in both the EU-based EUR-ACE and Engineers Australia (EA) accreditation guidelines (European Network for Accreditation of Engineering Education [ENAEE], 2008; EA, 2011).

Although these kinds of criteria are helpful, they fail to provide clear, operational definitions regarding the specific capabilities of the global engineer. In fact, a broader literature review reveals a lack of agreement about what *specific* competencies are required for global engineering practice, although many frameworks, definitions, and lists have been proposed (Allert, Atkinson, Groll, & Hirleman, 2007; Ball et al., 2012; Downey et al., 2006; Jesiek, Sangam, Thompson, Chang, & Evangelou, 2010; Lohmann, Rollins, & Hoey, 2006; Parkinson,

2009; Patil, 2005). For example, Downey et al. (2006) use historical research and ethnographic observations to argue that globally competent engineering students should have the "knowledge, ability, and predisposition to work effectively with people who define problems differently than they do" (p. 110). Parkinson (2009), on the other hand, uses various sources – including prior reports, his own experiences, and a review of learning outcomes from relevant programs and courses – to identify thirteen key "dimensions" for globally competent engineering graduates.

The extant literature therefore provides some useful insights about what kinds of attributes scholars think are important for global engineers. For clarity, these can be placed into a number of major groupings. First, commentators such as Patil (2005) and Allert et al. (2007) argue that global engineering professionals should have solid foundational abilities in traditional technical areas (e.g., engineering fundamentals, problem solving, design) and the professional domain (e.g., leadership, teamwork, communication, etc.). Although these are certainly important attributes, they represent a broad swath of scholarly activity in engineering education, so are beyond the scope of this review.

More relevant are the various "global attributes" that these and other authors have identified as important for current and future engineers, and these can in turn be grouped in two major categories. The first includes those aspects of global competency that are perceived as important for most any global professional, and not just engineers, including: cross-cultural attitudes (awareness, respect, adaptability, etc.); foreign language proficiency; familiarity with the history, economy, and society of foreign locales; and knowledge of globalization trends and dynamics. Many global educational programs for engineers (and others) focus on learning outcomes in one or more of these areas.

In addition, there is the more specific domain we call *global engineering competency*, which consists of *those attributes uniquely or*

especially relevant for cross-national/cultural engineering practice. A review of the literature reveals a number of attributes in this domain:

- *Engineering ethics:* "Can effectively deal with ethical issues arising from cultural or national differences" (Parkinson, 2009, p. 10); "The challenge of practicing ethically in a global environment" (ASCE, 2008, p. 144); "[M]oral responsibility to improve conditions and take action in diverse engineering settings" (Ragusa, 2011, p. 3).

- *Engineering cultures:* "Understand implications of cultural differences on how engineering tasks might be approached" (Parkinson, 2009, p. 11); "Socio/political impact on problem definition" (Allert et al., 2007, p. 3); "[A]nalyze how national differences are important in engineering work" (Downey et al., 2006, p. 114); "Applying engineering solutions and applications with a global context" (Patil, 2005, p. 51).

- *Regulations and standards:* "Understand cultural differences relating to product design, manufacture and use" (Parkinson, 2009, p. 11); "Have some exposure to international aspects of topics such as supply chain management, intellectual property, liability and risk, and business practices" (Parkinson, 2009, p. 11); "Global product platforms" (Allert et al., 2007, p. 3); "International labour market and workplace imperatives" (Patil, 2005, p. 51).

Although these themes are rising in prominence and are being addressed in some programs and courses, many other attributes also seem to fall in this category. For instance, it has been recognized that cross-cultural/national dynamics may require engineers and other technical professionals to develop special competence in areas such as teamwork and leadership. This theme is explicit in current standards for a professional engineer in Australia, who is expected to function "as an effective member or leader of diverse engineering teams, including those with multi-level, multi-disciplinary and multi-cultural dimensions" (EA, 2011, p. 6). Ethnographic research by Johri (2008a), on the other hand, highlights still other related characteristics and abilities, such as the importance of "knowledge brokering" activities for engineers who work on global teams. As this overview suggests, the technical dimensions of engineering practice are often deeply intertwined with global and professional considerations. Nonetheless, additional empirical research is needed to define better the full range of attributes required for successful global engineering practice and develop a better understanding for how these kinds of competence develop.

Assessing Global Competency in Engineering

A variety of strategies have been used to assess the global competency of engineering students. Many of these have focused on attributes viewed as generally important for global professionals, often assessed through a combination of off-the-shelf and tailor-made assessment instruments. The comprehensive evaluation framework for Georgia Tech's International Plan (IP), for example, involves multiple quantitative and qualitative measures of second language proficiency, comparative global knowledge, intercultural assimilation, and intercultural sensitivity (Lohmann et al., 2006). In the specific area of intercultural assimilation, surveys are collected from employers of student interns, and students complete pre/post-experience questionnaires, reflective essays, and the Intercultural Development Inventory (IDI). Georgia Tech's efforts to systematically collect and analyze IDI data are now starting to generate promising results, including evidence of variations in cross-cultural competence related to demographic variables and prior experiences, and significant pre/post gains in competence among students in the school's study abroad and IP programs (Georgia Institute of Technology, 2011).

Jesiek et al. (2011) used a similar mix of strategies to evaluate IREE 2010 China, a research abroad program for U.S. engineering students that was in part designed to enhance global competency. For purposes of assessment, evaluation, and reflective learning, participants were asked to self-report levels of language proficiency, complete a Sojourn Readiness Assessment survey, submit a series of reflective writing exercises, and participate in interview and focus groups. In addition, the fifteen-question Miville–Guzman Universality-Diversity Scale short form (MGUDS-S) was used to evaluate their cross-cultural competence, with pre/post results showing statistically significant gains in their scores (Jesiek, Shen, & Haller, 2012). Still others have performed in-depth qualitative research on student experiences and learning outcomes in global engineering programs, such as McNeill's (2010) comparative case study of three such programs. Also important, albeit less research oriented, is a volume by Grandin (2011) that tells the stories of fifteen graduates of the University of Rhode Island's International Engineering Program (IEP).

Only a handful of studies have focused on global engineering competency. As Lohmann et al. (2006) acknowledge, "Largely absent are studies featuring rigorous methods for assessing . . . competencies specifically related to professional practice within the academic discipline" (p. 125). In response, Georgia Tech has been assessing "global disciplinary practice" through employer surveys and rubric-based evaluations of senior design projects. Other relevant work in this area includes Ragusa's Engineering Global Preparedness Index (EGPI), a thirty-question survey with four subscales: engineering ethics, engineering efficacy, engineering global-centrism, and engineering community connectedness (Ragusa, 2011). Preliminary results have been reported for a multi-institutional pilot of EGPI with nearly 500 respondents. Given its basis in global citizenry theory, the instrument appears most useful and relevant for certain types of programs and organizations (e.g., those focused on global service learning or development).

As another example of work in this area, Downey et al. have developed and used a number of strategies to assess students enrolled in an undergraduate elective course ($N = 146$) that supports development of their global competency, including an open-ended scenario-based instrument that evaluates the ability of students to "explain how national differences among engineers are important in engineering work" (Downey et al., 2006, p. 117). A pre/post-course multiple-choice quiz has also been used to measure attainment of course content knowledge, and a final course survey asked students to self-report gains related to specific course learning outcomes. This research showed positive pre/post-course results in attainment of various aspects of global competency. Jesiek et al. (2011) have continued to build on this work, including by developing new scenario-based instruments. Nonetheless, there remains a lack of evidence regarding the validity of these assessment tools, and their portability and scalability is limited by their open-ended format and alignment with learning outcomes specific to certain courses and programs.

In summary, current assessment efforts around global engineering competency have mostly relied on a mix of quantitative, self-report measures (e.g., self-efficacy questions) or qualitative measures (e.g., scenario-based assessments, or evaluations of student work) with limited portability, validity, and scalability. There remain many untapped opportunities to perform research to determine what constitutes global competency in engineering, including through studies of global work practices as described in the next section. In addition, there is a lack of both theory to account for how this competence develops and robust assessment tools to measure its attainment. To begin filling these gaps, Jesiek and Woo (2011) have proposed creation and use of situational assessment strategies, including situational judgment tests (SJTs), to assess global engineering competency

Studies of Global Engineering Work

The second core area of research that we review in this chapter is "global engineering work." Scholars have expressed their concern regarding the effectiveness of study abroad programs given their lack of scalability and argue that we need other formats and strategies for providing global education (Miller, 2007). One proposed avenue is a closer alignment of student training with the work of practitioners. Miller (2007) argues that there is a need to move beyond short and simple lists of attributes (e.g., ABET, 2008) to develop in-depth theoretical understanding of global engineering work that can be translated into teaching. This finding is echoed in the Global Engineering Education Initiative report (2006), which states that "there is a significant lack of knowledge about proven theories and effective practices for instilling global competence" (p. 41). This argument is well aligned with the recent Carnegie report on "Educating Engineers" (Sheppard, Macatangay, Colby, & Sullivan, 2008) and work of many other scholars (Aldridge, 1994; Bucciarelli & Kuhn, 1997; Jonassen, Strobel, & Lee, 2006; Lucena, 2006; Raju & Sankar, 1999) who argue for a better fit between engineering practice and pedagogy. But a major intellectual gap in our efforts to prepare an engineering workforce ready to work in a globalized world is that we have not yet sufficiently examined engineering work practices or how engineers work with each other in firms (see Chapter 7 by Stevens, Johri, & O'Connor in this volume for a review of previous research).

With increasing digitization of engineering work, significant changes have occurred in how engineers work globally (Schmiede & Will-Zocholl, 2011). As Sheppard et al. (2008) state in their book *Educating Engineers*, the products of engineering work are reshaping how engineering is practiced:

From the environment to medicine, transportation to communication, household appliances to space exploration, engineers affect the world. Yet just as the technology born of

engineering has transformed much about our world, so has it transformed the work of engineers. (p. 3)

The importance of understanding the constantly changing nature of work (National Research Council [NRC], 1999) cannot be underestimated for those who prepare future engineers, and engineering faculty have a significant role to play in this process. Yet, the study of professional engineering work is a complex task as fruitful understanding can only be built by examining day-to-day practices of engineers in an in-depth manner (Barley, 1996; Barley & Kunda, 2001; Luff, Hundmarsh, & Heath, 2000; Suchman, 2000). Several scholars have adopted this approach and prior research on professional engineering work is reviewed extensively in Chapter 7 by Stevens, Johri, and O'Connor in this handbook. In the next section, we look at what we know and what we need to know about professionals working in global contexts.

What do we know about the work practices of technical professionals working in global contexts? In a recent internal survey of IT staff at a large multinational firm (made available to the first author by the firm), 253 of 256 total respondents indicated that they worked as members of a team, with 40% of them working in teams of two to five and 35% in teams of six to ten. Furthermore, 24% of them worked with people across two time zones, and another 23% worked with people in three or more time zones. They were dispersed across ten nations in fifteen different locations, and together used a combination of twenty different kinds of technological devices or software to communicate; e-mail, phone, and remote desktop were the most popular. The survey also showed that most informants had no training in how to work on virtual or remote teams and/or manage offshore projects. The respondents listed "knowledge transfer" (41%) and "team building" (48%) as the biggest challenges they faced. This survey is very telling of the realities of working in a global firm. It highlights the prevalence of teams, use of

technology, diversity that comes from geographic dispersion, and problems associated with learning and sharing knowledge. These characteristics echo major findings of the NRC report on the Changing Nature of Work (1999) and other recent, related literature (Axtell, Fleck, & Turner, 2004; Gibson & Gibbs, 2006; Hinds & Bailey, 2003). Teams spread across locations, extensive use of technology, and diversity of workforce generate temporal, disciplinary, occupational, and organizational boundaries (Johri, 2008a). These boundaries in turn create severe interpersonal communication and coordination problems such as lack of mutual knowledge and common ground (Cramton, 2001), misattributions (Cramton, 2002), interpersonal conflict (Hinds & Mortensen, 2005), lack of trust (Jarvenpaa & Leidner, 1999), and difficulty in sharing knowledge and expertise (Majchrzak, Rice, Malhotra, King, & Ba, 2000). For workers on global teams, the world is not all that flat.

Open Issues in Global Engineering Education and Work

In this section we review prior work and outline three issues central to global teamwork – diversity, technology, and on-the-job learning – as well as another central concern of engineering work and education identified earlier: ethics.

Effect of Team Diversity on Alignment of Perspectives Among Global Engineering Team Members

The first factor affecting work on global teams is the increased diversity of team membership that creates boundaries not just of race, gender, and nationality, but also of discipline, age, experience and tenure, educational background, and other factors (Gibson & Gibbs, 2006). On global teams boundaries are often fuzzy (Mortensen & Hinds, 2002), making it hard to ascertain team composition and the expertise of members (Johri, 2012a). This increases work uncertainty and leads to the development of faultlines among workers and potentially subgroup formation (Fiol & O'Connor, 2005). Subgroups – an "us" versus "them" mentality – are historically affected by problems of trust and conflict (Cramton & Hinds, 2005). Within engineering, as Downey and colleagues have shown, differences often arise among engineers owing to the types of training they received in their home country (Downey et al., 2006; Lucena, Downey, Jesiek, & Elber, 2008; Pankhurst et al., 2008). Engineers in different nations may define and solve problems differently, and these differences create problems that should be taken into account while working on global teams. Yet practitioners are often able to move beyond an awareness of differences to successfully bridge differences and work across boundaries. Bucciarelli (1984) even argues that this is one of the main roles of design: "The task of design is then as much a matter of getting different people to share a common perspective . . . as it is a matter of concept formation, evaluation of alternatives, costing and sizing – all the things we teach" (p. 187). What this suggests is that working on a team requires alignment of diverse perspectives held by team members regarding the problem, solution, task, roles, and so on, and this alignment is harder to achieve in global teams owing to greater diversity. A recent study presents preliminary evidence that global team members were able to align their perspectives under certain conditions and complete their work tasks successfully (Johri, 2011, 2010). By following this line of inquiry further and studying the different mechanisms that allow global engineering team members to align diverse perspectives, we can build theoretical understanding that can be used as a basis to design case studies and other educational experiences that can impart these skills to students. In future work, the answer to this question can also help develop technology-based learning environments (e.g., Ellis, Luther, Bessiere, & Kellogg, 2008) where students can be trained to work more effectively on global teams.

Technology Pervasiveness and Work Coordination in Global Engineering Teams

The second aspect of globalization that has affected engineering work is the exponential increase in the use of information technology (Fulk & Collins-Jarvis, 2001; Hinds & Kiesler, 2002). This dynamic was well explored by Friedman (2005) from an international perspective but is equally applicable at the level of actual work coordination (Cummings, 2004; Malone & Crowston, 1994). Technology creates immense opportunities for collaboration across time and space but, as discussed earlier, it also generates boundaries that inhibit interpersonal interaction (Gibson & Gibbs, 2006; Hinds & Bailey, 2003). Because distributed workers are usually unable to share direct experiential knowledge, they must rely on interactional dynamics and category membership that are mediated by technology. Communication in these mediated environments is a "leaky process" (Cramton, 2001, p. 364) and can contribute to bias, partial information, a lack of trust, misunderstandings, and conflict, especially between people lacking shared knowledge (p. 349). The consequences of a failure to establish mutual knowledge are harsh and include poor decision quality, lower productivity, and less effective conflict resolution. They require constant maintenance of awareness (Weisband, 2002). These problems are particularly salient in coordinating work across locations. Coordination is critical for engineering work – especially large engineering projects – as different elements of a project being worked on at different places need to be aligned for the final product.

Dispersion across location increases the need for and complexity of coordination among different team members. Engineers in a single location share common norms, have easier access to expertise, and can engage in informal, spontaneous communication (see review in Kiesler & Cummings, 2002). Distance reduces collaborative opportunities and effectiveness (Galeghar, Kraut, & Egido, 1990). The interaction of distance with technology increases the complexity of work and affects professional engineers (Crowston & Howison, 2004; Espinosa et al., 2002) because digitization of engineering work has resulted in the use of systems that affect coordination directly. For instance, in software development coordination is achieved not just through increased use of communication channels, but also the use of the software code itself as a coordinating mechanism (Aneesh, 2006; Metiu, 2006). Recent studies suggest that technical artifacts often play the role of boundary objects – helping different stakeholders find common ground and bridge perspectives (Boland, Lyytinen, & Yoo, 2007; Carlile, 2002; Kellogg, Orlikowski, & Yates, 2006). As yet, we do not have in-depth understanding of work coordination through technology from the perspective of practicing engineers and these research themes will help us establish and understand the connections.

Informal Learning Practices in Global Engineering Teams

On-the-job learning forms the third dimension of global engineering work reviewed here. According to Katehi (2005) U.S. engineers wishing to become global "will have to know how to replenish their knowledge by self-motivated, self-initiated learning . . . The engineer of 2020 and beyond will need skills to be globally competitive over the length of her or his career" (pp. 152–153). Although learning can be categorized along many dimensions, formal and informal learning are two key dimensions that appear in the literature (NRC, 2003). Formal learning involves environments that are intentionally designed and institutionalized – classrooms and courses, for instance. Informal learning, which is a more common form of learning over the lifetime, involves interactions and knowledge building that occur as people engage in activity in the course of daily life or work (Greeno, 2006).

The workplace has been a site of investigation for research on learning for many decades, and scholars have characterized its nature as cognitive apprenticeship

(Collins, 2006; Lave & Wenger, 1990). Further, on-the-job learning is strongly dependent on informal interaction among workers (Orr, 1996). For engineers, and others in technical professions, learning involves acquiring both social and disciplinary knowledge that can allow workers to participate in a community of practice (Johri & Olds, 2011; Johri, 2012b; Wenger, 1998). Within the context of global work the ability to draw on expertise and knowledge of fellow workers is one of the primary advantages of working with people around the world. Studies show that global teams can produce better output and are more innovative as a result of their diversity and the diverse knowledge that team members bring to the table (Cummings, 2004).

Yet, this process is problematic owing to difficulties in communication and coordination – as discussed previously – and also owing to the situated nature of knowledge in the workplace (Brown & Duguid, 2001 Elsbach et al., 2005). Learning occurs in specific context and "what people learn" and "how they learn" are both embedded in particular space and time, making on-the-job informal learning across locations – without shared context and practices – hard to accomplish (Sole & Edmondson, 2002). For instance, a recent study (Johri, 2009b) found that newcomers often had a preference for face-to-face interaction during their initial time in the firm and then found it easier to work virtually as they were better able to learn on the job. Yet these same newcomers also reported successfully working on open source software projects without ever meeting other team members face to face. This indicates that one aspect of learning – learning about organizational context – might require being physically embedded in the work context and when that context is digital (such as in an open source project), a negligible amount of physical interaction might be sufficient (Johri, 2009b). This preliminary work suggests that on-the-job learning in global engineering teams is critical for the success of teams and of team members. It also suggests that it is a complex process that requires further in-depth investigation to go beyond identification of skills to build a detailed understanding of what is learned and how it is learned, with these findings in turn leveraged to design pedagogy for training future engineers.

Personal Ethical Dilemmas Confronted by Global Engineers

Finally, an important aspect of global work that has not received much consideration from the perspective of engineers on global teams is ethics. As Colby and Sullivan (2008) note: "Although the formal engineering codes of ethics do not generally spell out the complexities that globalization of engineering work has added to these issues, every one of these keys topics delineated by the ethics code is affected by the pervasive and growing influence of globalization" (p. 328). Within the context of engineering education, ethical issues are increasingly becoming part of the curriculum (ABET, 2008) but the coverage is still limited in both quantity and quality (Colby & Sullivan, 2008, pp. 332–333). In particular, Colby and Sullivan (2008) found that "though teamwork was used extensively in these programs, faculty seldom highlighted the key dimensions of professionalism entailed in successful teamwork, including a sense of personal responsibility, fairness and honesty within the group, and a climate of respect and trust" (p. 333). In global teams, members have to make decisions on a day-to-day basis that are not clear cut and require them to make ethical trade-offs. Standards of practices often differ across locations and issues such as honesty, fairness, power, and status are more convoluted in co-located teams. For instance, Johri found that workers blamed co-workers in locations with more resources – usually Western countries – for not doing enough work, whereas workers in resource-intensive locations argued that co-workers in other locations have it easier because they are not under close scrutiny all the time (Johri, 2007a). This study also found that team members in non-headquarter locations – usually Asia – made significant adjustments to their lifestyles to accommodate teleconference meetings with co-workers in the central locations – staying up until 1 or 2 AM or waking up at 5 or 6 AM. It was clear to them, and to their co-workers, that they were not on equal footing, which created an ethical interpersonal dilemma. These preliminary observations establish that ethics is a critical issue in global engineering teamwork and professional practice and needs to be investigated further to develop better theoretical understandings and a body of work that can be used in undergraduate education and professional development.

Conclusion

Overall, current research on global and international issues in engineering captures three fundamental ideas: (1) engineers design their own lived world and the world of engineers has changed considerably owing to technological advancement and globalization; (2) to better prepare future engineers for the global world we need to develop a thorough understanding of what global engineering entails and how global competency develops; and (3) this process needs to begin with much better understandings of engineering work and professional practice. Together, they underscore the critical need to build a comprehensive theoretical understanding of engineering work that can be leveraged to design effective pedagogical innovations and assessment strategies to prepare globally competent engineers.

Acknowledgements

Aditya Johri was partially supported by the U.S. National Science Foundation (NSF) Award #0954034 during the completion of the work reported here. Brent Jesiek was partially supported by NSF Awards #1160455 & 1254323. Any opinions, findings, and conclusions or recommendations expressed in this material are those of the author(s) and do not necessarily reflect the views of the National Science Foundation.

References

Accreditation Board for Engineering and Technology (ABET). (2008). Criteria for accrediting engineering programs, effective for evaluations During 2008–2009 cycle. ABET, Engineering Accreditation Commission, 2007.

Aldridge, M. D. (1994). Professional practice: A topic for engineering research and instruction. *Journal of Engineering Education*, 83(3), 231–236.

Allert, B., Atkinson, D., Groll, E., & Hirleman, E. D. (2007). Making the case for global engineering: Building foreign language collaborations for designing, implementing, and assessing programs. *Online Journal for Global Engineering Education*, 2(2), 1–14. Retrieved from http://digitalcommons.uri.edu/ojgee/vol2/iss2/1/

American Society of Civil Engineers (ASCE), Body of Knowledge Committee of the Committee on Academic Prerequisites for Professional Practice. (2008). *Civil engineering body of knowledge for the 21st century: Preparing the civil engineer of the future* (2nd ed.). Reston, VA: American Society of Civil Engineers.

Aneesh, A. (2006). *Virtual migration: The programming of globalization*. Durham, NC: Duke University Press.

American Society for Training & Development (ASTD). (2007). 2012 state of the industry. Alexandria, VA: ASTD. Retrieved from http://www.astd.org/Publications/Research-Reports/2012/2012-State-of-the-Industry

Axtell, C., Fleck, S., & Turner, N. (2004). Virtual teams: Collaborating across distance. *International Review of Industrial and Organizational Psychology*, 19, 205–248.

Ball, A. G., Zaugg, H., Davies, R., Tateishi, I., Parkinson, A. R., Jensen, C. G., & Magleby, S. R. (2012). Identification and validation of a set of global competencies for engineering students. *International Journal of Engineering Education*, 28(1), 156–168.

Barley, S. (1996). Technicians in the workplace: Ethnographic evidence for bringing work into organization studies. *Administrative Science Quarterly*, 41, 404–441.

Barley, S., & Kunda, G. (2001). Bringing work back in. *Organization Science*, 12(1), 76–95.

Bloomgarden, A. H., & Riley, D. (2006). Learning and service in engineering and global development. *International Journal for Service Learning in Engineering*, 1(2), 47–59.

Boeing Company, The, & Rensselaer Polytechnic Institute (RPI). (1997). *A manifesto for global engineering education. Summary report of the Engineering Futures Conference.*

Boland, R., Lyytinen, K., & Yoo, Y. (2007). Wakes of innovation in project networks: The case of digital 3-D representations in architecture, engineering, and construction. *Organization Science*, 18(4), 631–647.

Bremer, D. (2008). Engineering the world. *Online Journal of Global Engineering Education*, 3(2), Article 2. Retrieved from http://digitalcommons.uri.edu/ojgee/vol3/iss2/2

Brown, J. S., & Duguid, P. (2001). Knowledge and organization: A social-practice perspective. *Organization Science*, 12(2), 198–215.

Bucciarelli, L. (1984). Reflective practices in engineering design. *Design Studies*, 5(3), 185–190.

Bucciarelli, L. (1996). *Designing engineers*. Cambridge, MA: MIT Press.

Bucciarelli, L., & Kuhn, S. (1997). Engineering education and engineering practice: Improving the fit. In *Between craft and science: Technical work in U.S. settings* (pp. 210–229). Ithaca, NY: Cornell University Press.

Callahan, R. (1962). *Education and the cult of efficiency*. Chicago, IL: University of Chicago Press.

Carlile, P. (2002). A pragmatic view of knowledge and boundaries: Boundary objects in new product development. *Organization Science*, 13(4), 442–455.

Christensen, C., & Carlile, P. (2009). Course research: Using the case method to build and teach management theory. *Academy of Management Learning and Education*, 8(2), 240–251.

Colby, A., & Sullivan, W. (2008). Ethics teaching in undergraduate engineering education. *Journal of Engineering Education*, 97(3), 327–338.

Collins, A. (2006). Cognitive apprenticeship. In K. Sawyer (Ed.), *The Cambridge handbook of the learning sciences* (pp. 47–60). New York, NY: Cambridge University Press.

Committee on Science Engineering and Public Policy. (CSEPP) (2006). *Rising above the gathering storm: Energizing and employing America for a brighter economic future*. Washington, DC: The National Academies Press.

Connell, C. (2007). Georgia Tech's well-engineered engagement with the world. *International Educator*, XVI(6), 38–46, 66.

Cramton, C. D. (2001). The mutual knowledge problem and its consequences for dispersed collaboration. *Organization Science*, 12(3), 346–371.

Cramton, C. D. (2002). Attribution in distributed work groups. In P. J. Hinds & S. Kiesler (Eds.), *Distributed work: New ways of working across distance using technology* (pp. 191–212). Cambridge, MA: MIT Press.

Cramton, C., & Hinds, P. (2005). Subgroup dynamics in internationally distributed teams: Ethnocentrism or cross-national learning? *Research in Organizational Behavior, 26*, 231–263

Crowston, K., & Howison, J. (2004). The social structure of free and Open Source Software Development. *First Monday, 10*(2). Retrieved from http://firstmonday.org/ojs/index.php/fm/article/viewArticle/1207

Cummings, J. N. (2004). Work groups, structural diversity, and knowledge sharing in a global organization. *Management Science, 50*(3), 352–364.

Daft, R., & Weick, K. (1984). Toward a model of organizations as interpretation systems. *Academy of Management Review, 9*(2), 284–295.

Downey, G. (2010). Epilogue – Beyond global competence: Implications for engineering pedagogy. In In G. L. Downey & K. Beddoes (Eds.), *What is global engineering education for?: The making of international educators* (pp. 45–76). San Rafael, CA: Morgan and Claypool.

Downey, G., & Beddoes, K. (2010). *What is global engineering education for?: The making of international educators*. San Rafael, CA: Morgan and Claypool.

Downey, G., & Lucena, J. (2004). Knowledge and professional identity in engineering: Code-switching and metrics of progress. *History and Technology, 20*(4), 393–420.

Downey, G., Lucena, J. C., Moskal, B., Bigley, T., Hays, C., Jesiek, B., . . . Parkhurst, R. (2006). The globally competent engineer: Working effectively with people who define problems differently. *Journal of Engineering Education, 105*(2), 107–122.

Duderstadt, J. (2008). *Engineering for a changing world: A roadmap to the future of engineering practice, research, and education*. Ann Arbor, MI: The Millennium Project, University of Michigan.

Eisenhardt, K. (1989). Building theories from case study research. *Academy of Management Review, 14*(1), 532–550.

Ellis, J. B., Luther, K., Bessiere, K., & Kellogg, W. A. (2008, February). Games for virtual team building. In *Proceedings of the 7th ACM Conference on Designing Interactive Systems* (pp. 295–304). ACM Press.

Elsbach, K., Barr, P., & Hargadon, A. (2005). Identifying situated cognition in organizations. *Organization Science, 16*(4), 422–433.

Engineers Against Poverty (EAP)/Institute of Education (IOE). (2008). *The global engineer: Incorporating global skills within UK higher education of engineers*. London: EAP and Development Education Research Center.

Engineers Australia (EA). (2011). Stage 1 competency standard for professional engineer. Barton, ACT, Australia: Engineers Australia. Retrieved from http://www.engineersaustralia.org.au/sites/default/files/shado/Education/Program%20Accreditation/110318%20Stage%201%20Professional%20Engineer.pdf

Espinosa, J., Kraut, R. E., Lerch, J. F., Slaughter, S. A., Herbsleb, J. D., & Mockus, A. (2002). Shared mental models and coordination in large-scale, distributed software development. In ICIS 2002 Proceedings. Paper 39. Retrieved from http://aisel.aisnet.org/icis2002/39

European Network for Accreditation of Engineering Education (ENAEE). (2008). *EUR-ACE framework standards for the accreditation of engineering programmes*. Brussels, Belgium: ENAEE.

Fiol, C. M., & O'Connor, E. (2005). Identification in face-to-face, hybrid, and virtual teams. *Organization Science, 16*(1), 19–32.

Friedman, T. (2005). *The world is flat: A brief history of the twenty-first century*. New York, NY: Farrar, Straus and Giroux.

Fulk, J., & Collins-Jarvis, L. (2001). Wired meetings: Technological mediation of organizational gatherings. In L. Putnam & F. Jablin (Eds.), *The new handbook of organizational communication* (2nd ed., pp. 624–703). Newbury Park, CA: SAGE.

Galegher, J., Kraut, R. E., & Egido, C. (Eds.) (1990). *Intellectual teamwork: Social and technological bases for cooperative work*. Hillsdale, NJ: Lawrence Erlbaum.

Georgia Institute of Technology. (2011). *Georgia Institute of Technology's Quality Enhancement Plan: Impact report*. Atlanta, GA: Georgia Institute of Technology. Retrieved from http://www.accreditation.gatech.edu/wp-content/uploads/2011/03/QEP-Impact-Report SACSCOC_March-25-2011.pdf

Gerhardt, L., & Smith, R. (2008). Development of a required international experience for undergraduate engineering students. In *Proceedings of the 38th ASEE/IEEE Frontiers in Education Conference*, Saratoga Springs, NY (pp. S4E-12–16).

Gibbons, M. (2009). *Engineering by the numbers*. Washington, DC: American Society for Engineering Education (ASEE).

Gibson, C. B., & Gibbs, J. L. (2006). Unpacking the concept of virtuality: The effects of geographic dispersion, electronic dependence, dynamic structure, and national diversity on team innovation. *Administrative Science Quarterly, 51*(3), 451–495.

Global Engineering Excellence Initiative (GEEI) (2006). In search of global engineering excellence: Educating the next generation of engineers for the global workplace. Hanover, Germany: Continental AG. Retrieved from http://www.global-engineering-excellence.org/

Grandin, J. (2011). *Going the extra mile: University of Rhode Island engineers in the global workplace*. Newport, RI: John M. Grandin.

Grandin, J., & Hirleman, E. D. (2009). Educating engineers as global citizens: A call for action / A report of the National Summit Meeting on the Globalization of Engineering Education. *Online Journal of Global Engineering Education, 4*(1), 1–28. Retrieved from http://digitalcommons.uri.edu/ojgee/vol4/iss1/1/

Greeno, J. G. (2006). Learning in activity. In K. Sawyer (Ed.), *The Cambridge handbook of the learning sciences* (pp. 79–96). New York, NY: Cambridge University Press.

Herbsleb, J. D., Mockus, A., Finholt, T. A., & Grinter, R. E. (2001). An empirical study of global software development: Distance and speed. In *Proceedings of the 23rd International Conference on Software Engineering* Toronto, ON, Canada (pp. 81–90).

Hinds, P. J., & Bailey, D. E. (2003). Out of sight, Out of synch: Understanding conflict in distributed teams. *Organization Science, 14*(6), 615–632.

Hinds, P. J., & Kiesler, S. (2002.) Preface. In P. Hinds & S. Kiesler (Eds.), *Distributed work: New ways of working across distance using technology* (pp. xi–xviii). Cambridge, MA: MIT Press.

Hinds, P. J., & Mortensen, M. (2005). Understanding conflict in geographically distributed teams: The moderating effects of shared identity, shared context, and spontaneous communication. *Organization Science, 16*(3), 290–307.

Institute of International Education (IIE). (2010). *Open Doors 2010 fast facts*. New York, NY: Institute of International Education.

Jamieson, L., & Lohmann, J. (2009). *Creating a culture for scholarly and systematic innovation in engineering education: Ensuring U.S. engineering has the right people with the right talent for a global society*. Washington, DC: American Society for Engineering Education.

Jarvenpaa, S., & Leidner, D. (1999, Winter). Communication and trust in global virtual teams. *Organization Science*, 791–815.

Jesiek, B. K., & Beddoes, K. (2010). From diplomacy and development to competitiveness and globalization: Historical perspectives on the internationalization of engineering education. In G. L. Downey & K. Beddoes (Eds.), *What is global engineering education for? The making of international educators* (pp. 45–76). San Rafael, CA: Morgan and Claypool.

Jesiek, B. K., Chang, Y., Shen, Y., Lin, J. J., Hirleman, E. D., & Groll, E. (2011). International Research and Education in Engineering (IREE) 2010 China: Developing globally competent engineering researchers. In *Proceedings of the 2011 ASEE Annual Conference and Exposition*, Vancouver, BC, Canada.

Jesiek, B. K., Sangam, D., Thompson, J., Chang, Y., & Evangelou, D. (2010). Global engineering attributes and attainment pathways: A study of student perceptions. In *Proceedings of the 2010 ASEE Annual Conference and Exposition*, Louisville, KY.

Jesiek, B. K., Shen, Y., & Haller, Y. (2012). Cross-cultural competence: A comparative assessment of engineering students. *International Journal of Engineering Education, 28*(1), 144–155.

Jesiek, B. K., & Woo, S. E. (2011). Realistic assessment for realistic instruction: Situational assessment strategies for engineering education and practice. In *Proceedings of the 2011 SEFI Annual Conference*, Lisbon, Portugal.

Johri, A. (2008a). Boundary spanning knowledge broker: An emerging role in global engineering firms. In *Proceedings of 38th Annual Frontiers in Education Conference*, Saratoga Springs, NY.

Johri, A. (2008b). *Why we see coworkers differently: Situational and institutional shaping of impression*. Paper presented at Organizational

Communication and Information Systems Division Session on Individuals and Distributed Work, Academy of Management, Anaheim, CA.

Johri, A. (2009a). Preparing engineers for a global world: Identifying and teaching sensemaking and creating new practices strategies. In *Proceedings of 39th Annual Frontiers in Education Conference*, San Antonio, TX.

Johri, A. (2009b). Global software practices: Comparing open source and traditional development. Technical Report Submitted to Sun Microsystems, February 2009.

Johri, A. (2010). Open organizing: Designing sustainable work practices for the engineering workforce. *International Journal of Engineering Education*, 26(2), 278–286.

Johri. A. (2011). Sociomaterial bricolage: The creation of location-spanning work practices by global software developers. *Information and Software Technology*, 53(9), 955–968.

Johri, A. (2012a). From a distance: Impression formation and impression accuracy among geographically distributed coworkers. *Computers in Human Behavior*, 28(6), 1997–2006.

Johri, A. (2012b). Learning to demo: The sociomateriality of newcomer participation in engineering research practices. *Engineering Studies*, 4(3), 249–269.

Johri, A., & Olds, B. (2011). Situated engineering learning: Bridging engineering education research and the learning sciences. *Journal of Engineering Education*, 100(1), 151–185.

Johri, A., & Sharma, A. (2013). *Designing development: Case study of an international education and outreach program*. Morgan and Claypool Synthesis Lectures on Global Engineering. San Rafael, CA: Morgan and Claypool.

Jonassen, D., Strobel, J., & Lee, C. (2006, April). Everyday problem solving in engineering: Lessons for engineering educators. *Journal of Engineering Education*, 95(2), 139–151.

Katehi, L. (2005). The global engineer. In *Educating the engineer of 2020* (pp. 151–155). Washington, DC: National Academies Press.

Kellogg, K., Orlikowski, W., & Yates, J. (2006). Life in the trading zone: Structuring coordination across boundaries in post-bureaucratic organizations. *Organization Science*, 17(1), 22–44.

Kiesler, S., & Cummings, J. N. (2002). What do we know about proximity in work groups?

A legacy of research on physical distance. In P. Hinds & S. Kiesler (Eds.), *Distributed work: New ways of working across distance using technology* (pp. 57–80). Cambridge, MA: MIT Press.

Kolodner, J. L., Camp, P. J., Crismond, D., Fasse, B., Gray, J., Holbrook, J., . . . Ryan, M. (2003). Problem-based learning meets case-based reasoning in the middle-school science classroom: Putting Learning by Design™ into practice. The *Journal of the Learning Sciences*, 12(4), 495–547.

Kunda, G. (1992). *Engineering culture: Control and commitment in a high-tech organization*. Philadelphia, PA: Temple University Press.

Lambert, S. (2009). *Sustainable design throughout the curriculum using case studies*. Paper presented at the Mudd Design Workshop, Harvey Mudd, Claremont, CA. Retrieved from http://design.uwaterloo.ca/wcde/3.CASES/index.html

Lave, J., & Wenger, E. (1990). *Situated learning: Legitimate peripheral participation*. New York, NY: Cambridge University Press.

Lohmann, J., Rollins, H., & Hoey, J. (2006). Defining, developing, and assessing global competence in engineers. *European Journal of Engineering Education*, 31(1), 119–131.

Lucena, J. (2006). Globalization and organizational change: Engineers' experiences and their implications for engineering education. *European Journal of Engineering Education*, 31(3), 321–338.

Lucena, J., Downey, G., Jesiek, B., & Elber, S. (2008). Competencies beyond countries: The re-organization of engineering education in the United States, Europe, and Latin America. *Journal of Engineering Education*, 97(4), 433–447.

Luff, P., Hindmarsh, J., & Heath, C. (2000). *Workplace studies: Recovering work practice and informing systems design*. Cambridge: Cambridge University Press.

Majchrzak, A., Rice, R., Malhotra, A, King, N., & Ba, S. (2000). Technology adaptation: The case of a computer-supported inter-organizational virtual team. *MIS Quarterly*, 24(4), 569–600.

Malone, Y., & Crowston, K. (1994). The interdisciplinary study of coordination. *ACM Computing Reviews*, 26(1), 87–119.

Marshall, C., & Rossman, G. (1999). *Designing qualitative research*. Thousand Oaks, CA: SAGE.

Martin, T., Petrosino, A. J., Rivale, S., & Diller, K. (2007). The development of adaptive expertise in biotransport. *New Directions in Teaching and Learning*, 108, 35–49.

Martin, T., Rayne, K., Kemp, N. J., Hart, J., & Diller, K. R. (2005). Teaching for adaptive expertise in biomedical engineering ethics. *Science and Engineering Ethics*, 11(2), 257–276.

McNair, L., Paretti, M., & Kakar, A. (2008). A case study of prior knowledge: Expectations and identity constructions in interdisciplinary, cross-cultural, virtual collaboration. *International Journal of Engineering Education*, 24(2), 386–399.

McNeill, N. (2010). *Global engineering education programs: More than just international experiences* (Doctoral dissertation). Purdue University, West Lafayette, IN.

Meiksins, P., & Smith, C. (1996). *Engineering labour: Technical workers in comparative perspective*. London: Verso.

Metiu, A. (2006). Owning the code: Status closure in distributed groups. *Organization Science*, 17(4), 418–435.

Miller, R. (2007). Beyond study abroad: Preparing students for the new global economy. In *Proceedings of the ABET Annual Meeting*, Incline Village, NV.

Mohtar, R., & Dare, A. (2012). Global design team: A global service-learning experience. *International Journal of Engineering Education*, 28(1), 169–182.

Mortensen, M., & Hinds, P. (2002). Fuzzy teams: Boundary disagreement in distributed and collocated teams. In P. Hinds & S. Kiesler (Eds.), *Distributed work: New ways of working across distance using technology* (pp. 283–308). Cambridge, MA: MIT Press.

National Academy of Engineering (NAE). (2004). *The engineer of 2020: Visions of engineering in the new century*. Washington, DC: The National Academies Press. Retrieved from http://www.nap.edu/catalog.php?record_id=10999

National Academy of Engineering (NAE). (2005). *Educating the engineer of 2020: Adapting engineering education to the new century*. Washington, DC: The National Academies Press. Retrieved from http://www.nap.edu/catalog.php?record_id=11338

National Academy of Engineering (NAE). *Grand challenges for engineering*. Washington, DC:

National Academy of Engineering. Retrieved from http://www.engineeringchallenges.org/

National Research Council (NRC). (1999). *The changing nature of work: Implications for occupational analysis*. Washington, DC: The National Academies Press.

National Research Council (NRC), Committee on the Foundations of Assessment. (2001). *Knowing what students know: The science and design of educational assessment*. Pellegrino, J., Chudowsky, N., & Glaser, R. (Eds.). Washington, DC: The National Academies Press.

National Research Council (NRC). (2003). *How people learn: Brain, mind, experience, and school*. Washington, DC: National Academies Press.

National Science Foundation (NSF). (2006). *NSF's cyberinfrastructure vision for 21st century discovery*. Washington: DC: NSF Cyberinfrastructure Council.

National Science Foundation (NSF). (2008). *Fostering learning in a networked world: The cyberlearning opportunity and challenge. Report of the NSF Taskforce on Cyber-enabled Learning*. Washington, DC: NSF.

O'Leary, M., & Cummings, J. N. (2007). The spatial, temporal, and configurational characteristics of geographic dispersion in work teams. *MIS Quarterly*, 31(3), 433–452.

O'Leary, M., Orlikowski, W., & Yates, J. (2002). Distributed work over the centuries: Trust and control in the Hudson's Bay Company, 1670–1826. In P. J. Hinds & S. Kiesler (Eds.), *Distributed work: New ways of working across distance using technology* (pp. 27–54). Cambridge, MA: MIT Press.

Office of International Science and Engineering (OISE). (2008). Advisory committee meeting summary. Washington, DC: National Science Foundation. Retrieved from http://www.nsf.gov/attachments/110090/public/OISEAC-5-07-Notes.doc

Orr, J. E. (1996). *Talking about machines: An ethnography of a modern job*. Ithaca, NY: Cornell University Press.

Parkhurst, R., Moskal, B., Lucena, J., Downey, G., Bigley, T., & Elber, S. (2008). Engineering cultures: Comparing student learning in online and classroom based implementations. *International Journal of Engineering Education*, 24(5), 955–964.

Parkinson, A. (2007). Engineering study abroad programs: Formats, challenges, best practices.

Online Journal of Global Engineering Education, 2(2), 1–15. Retrieved from http://digitalcommons.uri.edu/ojgee/vol2/iss2/2/

Parkinson, A. (2009). The rationale for developing global competence. *Online Journal of Global Engineering Education*, 4(2), 1–15. Retrieved from http://digitalcommons.uri.edu/ojgee/vol4/iss2/2/

Patil, A. (2005). Global engineering criteria for the development of the global engineering profession. *World Transactions on Engineering and Technology Education*, 4(1), 49–52.

Ragusa, G. (2011). Engineering preparedness for global workforces: Curricular connections and experiential impacts. In *Proceedings of the 2011 ASEE Annual Conference and Exposition*, Vancouver, BC, Canada.

Raju, P., & Sankar, C. (1999). Teaching real-world issues through case studies. *Journal of Engineering Education*, 88(4), 501–508.

Scardamalia, M., & Bereiter, C. (2006). Knowledge building: Theory, pedagogy, and technology. In K. Sawyer (Ed.), *Cambridge handbook of the learning sciences* (pp. 97–118). New York: Cambridge University Press.

Schmiede, R., & Will-Zocholl, M. C. (2011). Engineers' work on the move: challenges in automobile engineering in a globalized world. *Engineering Studies*, 3(2), 101–121.

Sheppard, S., Macatangay, K., Colby, A., & Sullivan, W. (2008). *Educating engineers: Designing for the future of the field.* San Francisco, CA: Jossey-Bass.

Shuman, L., Besterfield-Sacre, M., & Olds, B. (2005). Ethics assessment rubrics. In C. Mitcham, L. Arnhart, D. Johnson, & R. Spiers (Eds.), Encyclopedia of science, technology, and ethics (Vol. 2, pp. 693–695). New York, NY: Macmillan.

Shuman, L., Sindelar, M., Besterfield-Sacre, M. Wolfe, H., Pinkus, R., Miller, R. Old, B., & Mitcham, C. (2004). Can our students recognize and resolve ethical dilemmas? In *Proceedings of the ASEE Annual Conference and Exposition*, Salt Lake City, UT.

Sigma Xi. (2007). Embracing globalization: Assuring a globally engaged science and engineering workforce. Workshop report. Washington, DC: Sigma Xi. Retrieved from www.sigmaxi.org/programs/global/FinalReport.pdf

Sole, D., & Edmondson, A. (2002). Situated knowledge and learning in dispersed teams. *British Journal of Management*, 13(S2), S17–S34.

Study Abroad Bill (2009). S. 473 (111th): Senator Paul Simon Study Abroad Foundation Act of 2009. Retrieved from http://www.govtrack.us/congress/bills/111/s473

Suchman, L. (2000). Embodied practices of engineering work. *Mind Culture and Activity*, 7(1&2), 4–18.

Weisband, S. (2002). Maintaining awareness in distributed team collaboration: Implications for leadership and performance. In P. Hinds & S. Kiesler (Eds.), *Distributed work: New ways of working across distance using technology* (pp. 311–333). Cambridge, MA: MIT Press.

Wenger, E. (1998). *Communities of practice: Learning, meaning, and identity.* Cambridge: Cambridge University Press.

Wulf, W. (2003). 2003 Annual meeting – President's remarks. Washington, DC: National Academy of Engineering. Retrieved from http://www.nae.edu/News/Speechesand Remarks/page2003AnnualMeeting-Presidents Remarks.aspx

CHAPTER 33

Engineering Ethics

Brock E. Barry and Joseph R. Herkert

Introduction

Instruction and research related to engineering ethics is by no means a new field of practice. However, as the field of engineering education has been formalized and seen significant growth, the field of engineering ethics has naturally benefited. This chapter is divided into four subsections. The first section is a relatively brief overview of what engineering ethics is and how is it defined. The second section is a review of the historical development of engineering ethics in professional practice and in higher education. The third section is focused entirely on engineering ethics in education and addresses issues of curriculum content, pedagogical methods, resources, and instructor qualifications, as well as providing an overview of assessment of moral development. Finally, the fourth section focuses on engineering ethics in practice and covers such topics as the environment and sustainability, research ethics, application of ethics in international context, academic dishonesty, macroethics, and other emerging issues.

What Is Engineering Ethics?

Two of the most popular textbooks in engineering ethics define engineering ethics in similar yet different ways. The definition offered by Martin and Schinzinger (1996) in their classic text is descriptive:

> *Engineering ethics is (1) the study of moral issues and decisions confronting individuals and organizations engaged in engineering and (2) the study of related questions about the moral ideals, character, policies and relationships of people and corporations involved in technological activity. (pp. 2–3)*

Harris, Pritchard, and Rabins (2000), on the other hand, offer a more normative definition:

> *Engineering ethics is concerned with the question of what the standards in engineering ethics should be and how to apply these standards to particular situations.* One of the values of studying engineering ethics is that it can serve the function of helping to promote responsible engineering practice. *[emphasis added] (p. 26)*

In both cases, the authors focus on the moral issues and standards of the engineering profession, but Martin and Schinzinger also include consideration of technological activity more generally. This broader focus is important because moral problems in engineering are often influenced by factors outside of engineering. On the other hand, the applied emphasis of Harris and colleagues, and their concern for promoting responsible engineering practice, is also important, especially in engineering education.

The study and teaching of engineering ethics are often closely tied to the notion of professional responsibility, where a profession can be defined as "[a] learned occupation requiring systematic knowledge and training, and commitment to a social good" (Wujek & Johnson, 1992, slide 15). Under this common notion of a profession, society grants the profession privileges such as prestige, financial rewards, and professional autonomy in return for the members of the profession committing themselves to a social good.

Philosopher Michael Davis points out (1999b) that professional ethics is integral to professional practice, asserting that "[p]rofessional ethics is as much a part of what members of a profession know – and others do not – as their 'technical' knowledge. Engineering ethics is part of thinking like an engineer." Further, unlike some philosophers, Davis (1999b) emphasizes the primary role of the members of the profession in prescribing the ethical standards of the profession:

> Professional ethics . . . belongs neither to common sense nor to philosophy but to the profession in question. Knowing engineering ethics is as much a part of knowing how to engineer as knowing how to calculate stress or design a circuit is. Indeed, insofar as engineering is a profession, knowing how to calculate stress or design a circuit is in part knowing what the profession allows, forbids, or requires.

History of Engineering Ethics

With the exception of the Code of Hammurabi, which largely covers medical regulations (British Medical Association, 1984), little documentation of engineering ethics can be found prior to the early 1900s. Petroski (2008) speculates that this may be due to the lack of an all-encompassing professional authority prior to that time. The first engineering society in the United States to adopt a code of ethics was the American Institute of Consulting Engineers (AICE) in 1911 (Luegenbiehl, 1991). The codes of engineering ethics for the Institute of Electrical and Electronics Engineers (IEEE) and the National Society of Professional Engineers date to 1912 and 1946, respectively (Baura, 2006). Fleddermann (2008) has suggested that engineering codes during the early part of the twentieth century were focused primarily on issues of how to conduct business. For example, many early engineering codes referenced service as a faithful agent or trustee. Although these early codes fail to mention public health, safety, and welfare, many authors (Davis, 2001; Pfatteicher, 2003) believe such items are implied. The Engineers' Council for Professional Development, a precursor to ABET, Inc. (formerly known as the Accreditation Board for Engineering and Technology), made the first explicit reference to the engineer's responsibility to the public in 1947 (Harris, Pritchard, & Rabins, 2005). The balance between business and professionalism is referred to as "one of the most important forces in the formation and evolution of engineering societies in America" in Layton's *The Revolt of the Engineers* (1971, p. 25). The commonly recognized phrase "[engineers] shall hold paramount the safety, health, and welfare of the public" first appears in the 1974 Engineer's Council for Professional Development code (Harris et al., 2005).

In the early days of American colleges, ethics instruction was considered a cornerstone of the curriculum. In fact, ethics was held in such high regard that teaching the subject was often an honor reserved for the college's president (Rosen & Caplan, 1980). However, around 1880, ethics content slowly began to be replaced as a result of pressures to increase specialization in undergraduate programs (Baum, 1980; Davis, 1999a; Hastings Center, 1980; Rosen & Caplan, 1980). As early as the 1930s, ethics content was relegated to elective courses (Hastings Center, 1980; Kelly, 1980). Then in the 1970s, engineering and medical professional societies began raising concerns that colleges and universities were not adequately preparing graduates to address personal and professional moral dilemmas (Hastings Center, 1980).

During the past several decades, there has been a growing awareness of the importance of ethics in engineering practice. Shuman, Besterfield-Sacre, & McGourty (2005) suggest that the single largest change agent bringing professional skills, including ethics, back into the curriculum has been ABET's Engineering Criteria 2000 (EC2000). Criterion 3 of EC2000 includes a series of 11 outcomes (a–k) that students are expected to embody on graduation from accredited engineering programs. Specifically, Criterion 3.f states that graduates must demonstrate an "understanding of professional and ethical responsibility" (ABET, 2007).

Several notable engineering-related disasters during the latter half of the 1900s, including the Ford Pinto design issues (1970s), Three Mile Island accident (1979), Kansas City Hyatt Regency walkway collapse (1981), Chernobyl (1986), and the space shuttle *Challenger* accident (1986), have focused greater attention on engineering ethics, both in academics and in practice. In addition, these events have increased awareness of the role of the engineer in society. In turn, this has led to an increase in the quantity and quality of academic research in the area of engineering ethics, and ultimately recognition of engineering ethics as a respected field of study. The National Science Foundation (NSF) has supported broad research in the areas of engineering ethics and engineering ethics education. The NSF's most recent funding initiative through its Ethics Education in Science and Engineering (EESE) program has focused primarily on graduate-level ethics education and research ethics.

Engineering Ethics in Education

Goals and Content of Engineering Ethics Education

The desired outcomes of engineering ethics education are aptly described by Davis (1999b):

> Teaching engineering ethics . . . can achieve at least four desirable outcomes: a) increased ethical sensitivity; b) increased knowledge of relevant standards of conduct; c) improved ethical judgment; and d) improved ethical will-power (that is, a greater ability to act ethically when one wants to).

Framed in this manner, the goals of engineering ethics instruction, as well as the ability to evaluate student performance, are similar to other subjects in the engineering curriculum, despite the fact that engineers sometimes question whether ethics can be taught and evaluated.

Engineering ethics is often conceptualized in a framework of "professional responsibility," characterized by many ethicists as moral responsibility arising from specialized knowledge possessed by an individual, where, according to Whitbeck (1998), "for someone to have a moral responsibility for some matter means that the person must exercise judgment and care to achieve or maintain a desirable state of affairs" (p. 37). The "desired state of affairs" can vary from profession to profession. Martin and Schinzinger argue (1996) that responsible engineers are dedicated to "the creation of useful and safe technological products while respecting the autonomy of clients and the public, especially in matters of risk-taking" (p. 42). Professional responsibility is also reflected in the issues traditionally considered in engineering ethics instruction, including (Rabins, 1998; Wujek & Johnson, 1992) public health, safety and welfare; risk (including the principle of informed consent); environmental quality; conflict of interest; truthfulness; integrity of data; whistle blowing; job choice; loyalty and accountability to clients and customers; plagiarism and giving due credit; quality control; confidentiality; industrial espionage and trade secrets; gifts and bribes; employer/employee relations; and discrimination. As seen in the fourth section of this chapter, in recent years the focus of engineering ethics has expanded beyond such traditional issues.

Ethics in the Engineering Curriculum

Any discussion of ethics in the engineering curriculum raises several questions that must be addressed: What methods of curriculum integration are available? How much curriculum content is enough to satisfy ABET Criterion 3.f? Is quantity or quality more important? The primary methods of incorporating ethics within curricula include required courses within the discipline, required courses outside the discipline, ethics across-the-curriculum, and linking ethics with societal implications of technology (Herkert, 2000a). The "required course within the discipline" approach is typically performed as a full-semester, multiple-credit class, which all students within a given discipline are required to complete. This approach has been successfully used for engineering ethics at Texas A&M (Rabins, 1998) and a few institutions with much smaller engineering programs; however, the high staffing costs and tightly packed engineering curriculum make the required-course model difficult if not impossible to adopt in most engineering programs. Even if it is possible, a required course needs to be supplemented by further ethics instruction in "real" engineering courses; otherwise students may be left with the impression that ethics is not an integral part of their engineering education. This is particularly true when the stand-alone course is not taught or co-taught by regular engineering faculty.

The "within the discipline" method focuses class content on discipline-specific issues, such as engineering ethics. The "course outside the discipline" method relies on course offerings outside of engineering, typically within philosophy or religion departments. This method often presents students with a more general ethics background, while sacrificing much of the disciplinary context captured in the "within the discipline" method. For engineering students, marginalization of ethics is even greater than in the "within the discipline" method.

The "ethics across-the-curriculum" approach, patterned after writing-across-the curriculum programs, frequently presents students with ethical considerations, in multiple courses, during a progression toward their degree. This method requires a collective commitment among department faculty to capitalize on ethics discussions within courses that have a non-ethics topic as their principal focus (Weil, 2003).

Finally, the "linking of ethics with the societal implications of technology" approach uses a curriculum model with a series of required core courses that strongly emphasizes professional ethics and the role of the discipline within society. This type of model is found in programs that design their complete curriculum around the principles of professionalism and society. The Science, Technology and Society Program (formerly the Program on Technology, Culture and Communication) at the University of Virginia's School of Engineering and Applied Science is an example. In that program, "all engineering students take a four-course core" and integration with the engineering curriculum occurs "through a required senior thesis on the social impacts of a technical project" (Herkert, 2000a, p. 310).

There is a wide variation in the amount of ethics content presently in use in the engineering curricula (Barry & Ohland, 2012; Herkert, 2000a, 2002; Rabins, 1998). Curriculum content refers to both the number of courses and number of course credits. Engineering education literature is full of opinions regarding the most effective amount of curriculum content (Baum, 1980; Drake, Griffin, Kirkman, & Swann, 2005; Hastings Center, 1980; Newberry, 2004; Rosen & Caplan, 1980). A study by Barry and Ohland (2012) was the first of its kind to make a quantitative and qualitative simultaneous analysis of multiple amounts of ethics-based curriculum content. Qualitative aspects of the study showed that the represented engineering programs were highly uncertain of how much professional and ethical content is sufficient to satisfy ABET during accreditation review and that the typical program reaction is to increase the quantity of applicable content in their curricula. When amount of curriculum content was statistically evaluated against student performance in the engineering ethics content on a nationally administered, engineering-specific examination, there was a lack of reported hierarchal structure between curriculum content and examination performance. This would appear to be in conflict with literature (Chickering & Gamson, 1991) that identifies a link between time-on-task and student achievement. One implication of these findings is that there are variables that have a greater influence on examination performance than the amount of curriculum content. A study by Vogt (2008) underscores the influence that faculty members have on academic achievement. Specifically, Vogt found that students had "greater engagement in course material if they felt positive toward faculty and their classroom environments" (p. 34). Similarly, Lambert, Terenzini, and Lattuca (2007) found significant, although indirect, effects of program characteristics and faculty behavior on student learning. Both quality and quantity are unquestionably important. Thus, although the engineering education literature is replete with arguments over the optimum quantity of curriculum content, the suggested conclusion is that the discourse would be better focused on the quality of instruction.

Pedagogical Methods

Curriculum-level decisions are largely made at the administrative level. However, classroom pedagogical methods are an instructor-level decision and have great bearing on the delivery of engineering ethics education.

Harris, Davis, Pritchard, and Rabins (1996) suggest that "there is widespread agreement that the best way to teach professional ethics is by using cases." The widely used engineering ethics text by Harris et al. (2005) makes extensive use of cases to illustrate ethical concepts. Fleddermann (2008) also utilizes cases to illustrate ethical concepts within his text. Loui (2005) discusses the use of case-based ethics instruction in developing student self-efficacy in dealing with moral dilemmas. Based on the common reference to case-based instruction related to engineering ethics, it would be difficult to argue that this pedagogical method is not the most common. However, the engineering literature is devoid of research that definitively identifies a most effective pedagogical method for introducing students to engineering ethics. It is important to recognize that other instructional methods are valid and in use. For example, Bird (2003) makes a strong argument for the use of team-based projects and workshop-type instruction as an alternative to case-based instruction. Although literature advocating for purely lecture-based ethics instruction was not identified, there is evidence that this method has support (Baum, 1980; Self & Ellison, 1998). An emphasis on instruction using codes of ethics and theoretical grounding was noted in a meta-analysis of conference papers conducted by Haws (2001). Similarly, codes and ethical theories are also discussed extensively in the texts of Harris et al. (2005, 2009) and Fleddermann (2008).

As the most widely used pedagogical method, case-based instruction provides students with an opportunity to connect ethical questions and theoretical concepts in context. Multiple learning theories advocate the idea of learning in context (see Chapter by Newstetter & Scinicki of this handbook for more information related to learning theories). Engineering ethics cases can be found in a variety of forms: long or short, real or fictional, technical or nontechnical (Yadav & Barry, 2009). They may be available in print, but are increasingly available in online, multimedia, or video formats. Cases are self-contained, or include documentation, such as book chapters (or even entire books), journal and magazine articles, news stories, and primary source data. In spite of such variety, Davis (1999a) observes several common characteristics of effective cases, including the encouragement of students to express ethical opinions, identify ethical issues, formulate and justify decisions, and "develop[ment] in students [of] a sense of the practical context of ethics" (p. 174).

Well-known cases used in engineering ethics instruction include the 1979 DC-10 crash in Paris that killed 346 people (Fielder & Birsch, 1992), the collapse of suspended atrium walkways at the Kansas City Hyatt Regency Hotel in 1981 that killed 114 and injured dozens (Pfatteicher, 2000), and the 1986 explosion of the space shuttle *Challenger* (Pinkus, Shuman, Hummon, & Wolfe, 1997).

Such high-profile cases are useful for gaining the attention of engineering students, but typically the ethical dilemmas encountered by most practicing engineers are far less spectacular. Case studies of more commonplace events, such as fictionalized reviews of actual cases considered by the National Society of Professional Engineers (NSPE) Board of Ethical Review, are also used in classrooms. NSPE cases focus on such varied topics as conflicts of interest, trade secrets, and gift giving (Smith, Harper, & Burgess, 2008).

Pritchard (1998) has called for the development and use of cases that focus on "good works;" that is, cases that demonstrate that making sound ethical judgments need not end in disaster or scandal. One such case is that of William LeMessurier, the civil engineer who designed New York's Citi-Corp Building. After discovering, when the building was already in use, that it had not been constructed to withstand hurricane-force winds, he informed his partners and CitiCorp of the problem and insisted that immediate action be taken to correct it. Another example cited by Pritchard is that of Fred Cuny, an engineer who dedicated his professional career to disaster relief, who disappeared in 1995 when he was attempting to negotiate a cease-fire in war-torn Chechnya. (See Chapter 9 by Davis and Yadav in this handbook for more content related to case studies in engineering.)

Educational Resources

Although there is no clear "best" instruction method for delivery of engineering ethics education, there are plenty of resources available to draw from. In addition to the large number and variety of engineering ethics textbooks (e.g., see Baura, 2006; Gunn & Vesilind, 2003; Fleddermann, 2008; Harris, Pritchard, & Rabins, 2009; Herkert, 2000b; Martin & Schinzinger, 1996; Seebauer & Barry, 2001), there is an extensive amount of electronic content that can be utilized.

The Online Ethics Center (OEC) for Engineering and Research (www.online ethics.org), initially funded through a series of NSF grants, is currently maintained by the Center for Engineering, Ethics and Society of the National Academy of Engineering. The mission of the OEC is "to provide engineers and engineering students with resources for understanding and addressing ethically significant problems that arise in their work, and to serve those who are promoting learning and advancing the understanding of responsible research and practice of engineering" (National Academy of Engineering, 2011b). The website contains an extensive number of cases and scenarios, as well as related discussion points. Guidelines and codes of ethics for both scientific and engineering societies from multiple countries are assembled in a single location. The website also provides a detailed listing of biographies of individuals working in the engineering ethics community.

The National Institute for Engineering Ethics (NIEE) is a not-for-profit educational corporation with the mission of promoting the study and application of ethics in engineering education and throughout the profession of engineering (www.niee.org). In 2001, the NIEE was absorbed into the Murdough Center for Engineering Professionalism at Texas Tech University. NIEE provides professional ethics workshops, seminars, presentations, and distance learning opportunities. In addition, NIEE obtained financing for, produced, and markets a series of highly successful engineering ethics videos: *Gilbane Gold* (1989; produced by the

National Society of Professional Engineers), *Incident at Morales* (2003), and *Henry's Daughters* (2010). A study guide, presentation handout, full script, and suggested assignments are freely available for use with each video. NIEE has also generated an excellent text titled *Engineering Ethics: Concepts, Viewpoints, Cases, and Codes* (Smith et al., 2008). Finally, the NIEE website also contains a large number of cases and modules designed for classroom use.

The University of Illinois has developed an NSF funded National Center for Professional and Research Ethics (NCPRE) (www.nationalethicsresourcecenter.net). The mission of the NCPRE is to "develop, gather, preserve, and provide comprehensive access to resources related to ethics for teachers, students, researchers, administrators, and other audiences" (Gudeman, 2011). The NCPRE website, Ethics CORE, contains a wealth of teaching resources including course lectures, course syllabi, role-playing scenarios, and various discussion scenarios.

A large number of professional journals and magazines are dedicated to engineering ethics research and instruction, including: *Science and Engineering Ethics* (published by Springer), *IEEE Technology and Society Magazine*, and *Teaching Ethics* (published by the Society for Ethics Across the Curriculum). In addition, several other journals routinely publish articles on engineering ethics research and instruction, including the *Journal of Engineering Education* (published by the American Society for Engineering Education), the *International Journal of Engineering Education* (published by TEMPUS), and the *Journal of Professional Issues in Engineering Education and Professional Practice* (published by the American Society of Civil Engineers).

Qualifications to Teach Engineering Ethics

The code of ethics for the National Society of Professional Engineers states that engineers shall "perform services only in their area of competence" (National Society of Professional Engineers, 2011). What then would qualify an engineering educator to be "competent" to teach engineering ethics? Without question, there is a certain minimum level of qualification required to teach any particular subject matter. Typically, institutions with a Carnegie Classification of "RU," or equivalent, require their faculty to hold terminal degrees in their field and to demonstrate competence in their area of specialization (Carnegie Foundation for the Advancement of Teaching, 2007). Thus, an individual who teaches courses in classical ethics would typically be expected to hold a Ph.D. in philosophy or religion. Likewise, an individual who teaches courses within an engineering discipline would be expected to have a Ph.D. in that subject matter. However, an apparent gray area is created when the fields of ethics and the engineering disciplines are combined. In addition, it is not difficult to find examples of individuals without an engineering, philosophy, or religion background who are successfully teaching courses in engineering ethics.

A series of reports by the Hastings Center identified a hesitation among faculty in various disciplines to teach professional ethics (Baum, 1980; Hastings Center, 1980; Kelly, 1980; Powers & Vogel, 1980; Rosen & Caplan, 1980). Preparation of faculty to comfortably engage in engineering ethics instruction remains one of the biggest challenges facing engineering ethics education. The Hastings Center reports indicate that as of 1980 the qualifications to teach professional ethics within various disciplines were unclear (Baum, 1980; Hastings Center, 1980; Powers & Vogel, 1980; Rosen & Caplan, 1980). In addition, the Hastings Center studies revealed that most individuals teaching professional ethics had little or no prior training in the subject area. The Hastings Center recommends that an individual qualified to teach professional ethics should have an advanced degree in their home discipline (e.g., engineering), as well as a solid background in ethics. Notably, the Hastings Center does not believe that an advanced degree in moral philosophy or moral theology is required (Rosen & Caplan,

1980). In Baum's report (1980) for the Hastings Center, he states that individuals with first-hand field experience are well-suited to teach professional ethics and that qualified individuals should be familiar with the history of their own discipline, including the development of professional societies and codes.

As Newberry (2004) notes, most "current engineering faculty members are products of the admittedly ethics-deficient undergraduate engineering educational system" (p. 349). In an effort to overcome this, an individual can develop an expertise in professional ethics in the course of self-education through reading, use of online resources, discussions with colleagues and, where available, faculty development seminars. Weil (2003) provides a detailed discussion of how faculty seminars and workshops can increase the awareness and comfort of faculty preparing to teach professional ethics. Thus, although a background and experience in philosophy and engineering might make an individual well prepared to teach engineering ethics, a well prepared instructor from history of science or technology, technical communications, science and technology studies, and so forth could be equally qualified. First and foremost, faculty must be enthusiastic about and comfortable with discussing ethical issues and the social implications of engineering.

Assessment of Understanding

Most forms of assessment in engineering ethics specifically evaluate the notion of moral reasoning. Although ethics and morals are often regarded as distinct concepts, earlier in this chapter the authors provided several definitions and interpretations of engineering ethics in terms of morality. Modern efforts to assess moral reasoning can trace their lineage back to Lawrence Kohlberg's theory of cognitive moral development. Kohlberg's work was initiated during his dissertation work at the University of Chicago and continued to develop over thirty years of renowned progressive research (Palmer, Cooper, &

Bresler, 2001). Underlying Kohlberg's theory is the belief that individuals progress through a series of six stages of moral reasoning. Those six stages can subsequently be grouped into three levels of morality: preconventional, conventional, and postconventional (Crain, 2005). Kohlberg's theory built upon the works of Socrates, Jean Piaget, and John Dewey (Palmer et al., 2001). Piaget's stage theory of cognitive development forms the basis for Kohlberg's work (Colby & Kohlberg, 1987; Ginsburg & Opper, 1988). Piaget's stage theory is based on five primary tenets: stages are qualitatively different, each stage is a structured whole, the sequence is invariant, hierarchical progression, and universalization (Ginsburg & Opper, 1988). Space constraints don't allow us to provide greater detail into the complexities of Kohlberg's theory. The interested reader is referred to Colby and Kohlberg (1987); Crain (2005); Ginsburg and Opper (1988); Killen and Smetana (2006); Kohlberg, Levine, and Hewer (1983); Kohlberg (1981); or Kohlberg et al. (1983) for further background related to Kohlberg's theory.

The moral reasoning assessment tool developed by Kohlberg is known as the Moral Judgment Interview (MJI). In using this tool, participants read three prepared moral dilemmas and then provide oral responses to a series of standardized questions that the test administrator uses to probe the participant's reasons for their statements. Justification for the participant's reasoning, rather than merely right versus wrong, is the focus of the assessment. Participants are scored based on the relation between their responses to the dilemmas and Kohlberg's six predefined stages. The MJI has been shown to be a valid and reliable assessment of moral reasoning (Colby & Kohlberg, 1987; Kohlberg, 1981). Although other tools for the general assessment of moral reasoning have been developed by various researchers, they each grounded their studies in the works of Kohlberg and in turn Piaget.

Although the MJI was the first well known tool for assessment of moral reason, perhaps the most widely employed moral reasoning assessment tool is the Defining Issue Test (DIT) (Killen & Smetana, 2006) as developed by James Rest (Self & Ellison, 1998). The DIT owes its popularity among researchers to its format and ease of scoring. Unlike the MJI, the DIT is a paper-and-pencil–based multiple-choice examination that can be group administered and computer scored. Thus, it eliminates the need for a trained administrator and rater familiar with the 800+ page scoring guide used in the MJI (Rest, Narvaez, Bebeau, & Thoma, 1999). The most recent version of the Defining Issues Test is known as the DIT-2. The DIT-2 maintains the framework of the original, while streamlining the instructions and updating the dilemmas (Rest, 1999). After reviewing each of five moral dilemmas, participants make a selection from a three-point scale relative to what they believe the protagonist in the dilemmas should do. Subsequent to the action choice, participants evaluate twelve prewritten items to identify which items they believe to be the most important in addressing the dilemma. In turn, each of the twelve items is ranked using a five-point Likert-type scale. Participants must recognize and select issue statements that best reflect their understanding of the moral dilemma (Rest, 1994; Rest et al., 1999). Scoring of the DIT results in several numerical values, the most commonly discussed being the P-score. A P-score is defined as the relative importance that a subject gives to items representing moral thinking (Duckett et al., 1997; Rest et al., 1999). Like the MJI, both the DIT and the DIT-2 have been shown to be valid and reliable measures of moral reasoning (Duckett et al., 1997; Duckett & Ryden, 1994; Self & Ellison, 1998; Sutton, 1992).

Another common general moral reasoning assessment tool, based on the work of Kohlberg, is known as the Sociomoral Reflection Measure (SRM). The SRM was developed by John Gibbs and, unlike the MJI or DIT, the SRM does not utilize a dilemma-based questionnaire. Instead, the SRM uses eleven open-ended lead-in statements and participants are asked to consider each and then respond to a series of evaluation questions. The SRM short form (SRM-SF) can be group administered and computer scored (Basinger, Gibbs, & Fuller, 1995). Scores are converted to a four-stage system of moral reasoning that is similar to Kohlberg's six-stage system and is also based on Piagetian stage theory. The SRM and SRM-SF have tested well for validity and reliability, but have not found widespread use to the extent that the DIT has (Self, Wolinsky, Baldwin, & DeWitt, 1989).

Other forms of general assessment, based on the work of Kohlberg, are in use, such as the Moral Judgment Test developed by Lind (1999). However, the literature would suggest that the MJI, DIT, and SRM have found the most wide spread use and recognition. The MJI, DIT, and SRM have been used extensively in various forms of research and have specifically been applied to the study of applied ethics within various professions (Bebeau, 1994, 2002; Drake et al., 2005; Duckett et al., 1997; Duckett & Ryden, 1994; Rest, 1994; Self & Ellison, 1998; Self et al., 1989). Although these assessment tools have proven to be valid for the evaluation of general moral reasoning, they lack the sensitivity and context to capture the unique aspects of professional ethics within specific disciplines. Accordingly, a variety of discipline-specific moral reasoning assessment tools have been developed. Several such assessment tools for the health professions, business, and law are discussed in Barry and Ohland (2009).

Although many professions, including health, business, and law, have advanced discipline-specific assessment tools to target professional ethics, engineering has only recently fully developed such an assessment tool. Borenstein, Drake, Kirkman, and Swann presented a conference paper in 2008 that detailed encouraging results toward the development of what they had termed the Test of Ethical Sensitivity in Science and Engineering (TESSE). Subsequently this same group published a journal article showing the results of research done using a new tool known as the Engineering and Science Issues Test (ESIT) (Borenstein et al., 2010). This assessment tool is patterned after the DIT-2, but uses scenarios in the context of engineering. In a recent NSF-funded project, Canary, Herkert, Ellison, and Wetmore (2012) have used the ESIT to evaluate three instructional models (traditional stand alone course, hybrid online/face-to face course, ethics material embedded in a required course) for delivering ethics instruction to graduate students in engineering and science. Although the ESIT is still in the early stages of application, it has the potential to be a highly valued engineering ethics research tool.

Engineering Ethics in Practice

Academic Dishonesty

Although this section might have been included in Part III of this chapter, the authors view academic dishonesty as an all too common reality of the practice and action of students, who are in effect apprentices in the engineering profession. More specifically, this includes the existence of cheating among the students we teach. In his book *The Cheating Culture*, David Callahan (2004) provides details related to a large number of academic cheating scandals. Individuals working in academics typically don't need to look beyond their own classes or departments to cite such unfortunate examples of academic dishonesty.

Some of the most extensive and well structured research in the area of academic dishonesty has been conducted by a group of researchers known as the E[3] Team. This group of engineering educators and educational researchers is led by Drs. Donald D. Carpenter, Cynthia J. Finelli, and Trevor S. Harding. The long-term goals of this research team are to "quantify the frequency of cheating among engineering undergraduate students and to clarify their perceptions and attitudes about cheating" (E3, 2007). The team draws their motivation, at least in part, from prior research that identified engineering students are more likely to

self-report frequent cheating than their peers in other disciplines.

Several successful studies related to these goals have been completed to date. The PACES-1 Study included development and testing of a 139-item survey to evaluate engineering students' definition of and frequency of cheating. Among the findings of this study was that context (homework, examination, etc.) influenced a student's decision to cheat (Passow, Mayhew, Finelli, Harding, & Carpenter, 2006). Further, prior experience with cheating (such as during high school) becomes a strong predictor of the likelihood to cheat in college (Carpenter, Harding, Finelli, & Passow, 2004). This study also determined that students tend to rationalize their cheating by suggesting that instructor-based actions (e.g., too much assigned homework) justifies their actions (Carpenter, Harding, Finelli, Montgomery, & Passow, 2006). The follow-on study, PACES-2, focused on the development and testing of a theoretical decision-making process model, known as the modified Theory of Planned Behavior. Model validation was performed using a self-developed, multiquestion survey, as well as application of Rest's DIT-2 (discussed previously in this chapter). This study made a specific comparison between engineering students and humanities students. Findings of the PACES-2 study include the determination that engineering students reported cheating at higher frequencies than their humanities counterparts and those differences exist only in college (not in high school). Further, the modified Theory of Planned Behavior provided an accurate prediction of an individual's intention to cheat (Harding, Mayhew, Finelli, & Carpenter, 2007).

More recently, the E[3] team has advanced the Student Engineering Ethical Development (SEED) study in an effort to identify "the factors which positively affect the ethical development of engineering undergraduates" (E3, 2007). This study has included collection of qualitative and quantitative data from nineteen institutions, representing a range of geographic locations and Carnegie Foundation classifications. After performing a large number of focus group and individual interviews, the team created the Survey of Engineering Ethical Development. This detailed survey was administered to nearly 4,000 engineering undergraduate students (Holsapple et al., 2011). Although the data from this study are just now being evaluated, one of the early reported findings suggest that students who possess a higher level of ethical reasoning are more likely to report dissatisfaction with ethics education (Holsapple et al., 2011). Likewise, the amount of content and methods used to deliver ethics education affects student satisfaction (Holsapple et al., 2011). It is anticipated that the rich data associated with this study will generate many additional insightful conclusions.

Engineering Design

Although engineering design is implicit in most treatments of engineering ethics, more and more it has come to be treated explicitly. In an oft-cited chapter of their textbook, Martin and Schinzinger (1996) discuss engineering design in the context of "engineering as social experimentation." van Gorp and van de Poel (2001) highlight ethical issues in the design process using the sinking of the ferry Herald of Free Enterprise as a case study. Roeser (2010) argues not only that engineering design is not value neutral, but also that emotional reflection offers valuable ethical insight in engineering design. Herkert (2003) discusses the relationship between engineering ethics and product liability arising from engineering design. Busby and Coeckelbergh (2003) focus on the public's expectations of engineers' responsibilities for the products they design. In the aftermath of the Oklahoma City and 9/11 terrorist attacks, Kemper (2004) argues that design responsibility extends to anticipating evil intent. Whitbeck (1998) observes that the problem-solving approach employed in engineering design can serve as a paradigm for addressing ethical problems, an argument that is useful in showing engineering students how applied ethics and engineering are mutually reinforcing. Engineering design

is addressed further in Chapter 10 by Lord and Chen, as well as Chapter 11 by Atman, Borgford-Parnell, McDonnell, Eris, Delft, and Cardella, in this handbook.

Macroethics

As engineering ethics grew as an academic field, several authors, notably the ethicist John Ladd (1991), began to observe that the content of engineering ethics could include multiple perspectives. Drawing on the work of Ladd and others, Herkert argued (2001) that engineering ethics has three frames of reference – individual, professional, and social – that can be divided into "microethics," that is, ethical decision making by individual engineers and the internal relationships of the engineering profession, and "macroethics," that is, the profession's collective social responsibility and the role of engineers in societal decisions about technology.

Engineering ethics research and teaching traditionally focused on microanalysis of individual ethical dilemmas in such areas as health and safety implications of engineering design, conflicts of interest, integrity of test data, and trade secrets (Herkert, 2001), with little or no attention being paid to macroethical issues in engineering and still less to attempts at integrating microethical and macroethical approaches to engineering ethics. Although the strong identification of engineering ethics with professionalism has contributed significantly to the intellectual underpinnings of the field, by emphasizing issues internal to the profession, it has also historically deflected the attention of engineering ethicists from macroethical issues (O'Connell & Herkert, 2004).

Engineering ethicists and engineering leaders have recently been turning their attention to macroethical issues. In 2003, for example, the National Academy of Engineering (NAE) convened a workshop on emerging technologies and ethical issues in engineering at the initiative of then NAE President Bill Wulf. Wulf argues for the importance of macroethics largely on the basis of the complexity of emerging technologies:

> *Several things have changed, and are changing, in engineering that raise macroethical questions. I'm going to talk only about the one that is closest to my professional experience – complexity. The level of complexity of the systems we are engineering today, specifically systems involving information technology, biotechnology, and increasingly nanotechnology, is simply astonishing. When systems reach a sufficiently high level of complexity, it becomes impossible to predict their behavior. It's not just hard to predict their behavior, it's impossible to predict their behavior. (Wulf, 2004, p. 4)*

The NAE subsequently established the Center for Engineering, Ethics and Society (CEES), which in addition to assuming responsibility for the Online Ethics Center in Engineering and Research (discussed earlier), has embarked on an ambitious program of addressing macroethical issues in engineering. The CEES's first major project, for example, was a workshop on Engineering Ethics, Social Justice and Sustainable Community Development held in 2008 that focused on both technical and nontechnical perspectives on social and environmental justice and sustainable development. Recent examples of case studies that feature integration of microethical and macroethical concepts are Newberry's (2009) analysis of Hurricane Katrina and Pfatteicher's (2010) book on the attack against and collapse of the World Trade Center towers.

Environment and Sustainability

Especially in the United States, codes of engineering ethics are seen as expressions of the profession's ethical commitments (Davis, 1988). Not until the 1990s did provisions regarding the environment and sustainable development begin appearing in the codes.

All modern engineering ethics codes contain what is known as the "paramountcy clause" which stipulates concern for the public health, safety, and welfare as the "paramount" ethical obligation of an engineer. For example, consider the codes of the four major engineering disciplines:

> *ASME Code of Ethics of Engineers – "Engineers shall hold paramount the safety, health and welfare of the public in the performance of their professional duties." (2009)*

> *American Institute of Chemical Engineers (AIChE) Code of Ethics – "Hold paramount the safety, health and welfare of the public and protect the environment in performance of their professional duties." (2003)*

> *American Society of Civil Engineers (ASCE) Code of Ethics – "Engineers shall hold paramount the safety, health and welfare of the public and shall strive to comply with the principles of sustainable development in the performance of their professional duties." (2011)*

> *Institute of Electrical and Electronics Engineers (IEEE) Code of Ethics – "... to accept responsibility in making engineering decisions consistent with the safety, health and welfare of the public, and to disclose promptly factors that might endanger the public or the environment..." (2011)*

Beyond the general obligation to uphold the public safety, health, and welfare, the current versions of these four codes also acknowledge the engineer's responsibilities for the environment and/or sustainable development. The IEEE and the AIChE specifically list environmental protection in their paramountcy clause. In addition to including sustainable development in the paramountcy clause, the ASCE code's first "Fundamental Principle" pledges engineers to "using their knowledge and skill for the enhancement of human welfare and the environment." The ASME Code contains a separate canon providing that "Engineers shall consider environmental impact and sustainable development in the performance of their professional duties."

In addition to the codes, engineering societies have widely promoted the concept of sustainable development and the prominent role of engineering. For example, a document prepared by several of the U.S.

engineering societies for the Johannesburg Earth Summit 2002 states:

> *Creating a sustainable world that provides a safe, secure, healthy life for all peoples is a priority for the US engineering community. It is evident that US engineering must increase its focus on sharing and disseminating information, knowledge and technology that provides access to minerals, materials, energy, water, food and public health while addressing basic human needs. Engineers must deliver solutions that are technically viable, commercially feasible and, environmentally and socially sustainable. (American Association of Engineering Societies, American Institute of Chemical Engineers, ASME – Environmental Engineering Division, National Academy of Engineering, & National Society of Professional Engineers, 2001)*

Engineering educators have risen to the challenge imposed by changes in the codes by incorporating concern for the environment and sustainable development in the engineering curriculum, including the ethics curriculum. Notable contributions include a textbook by Vesilind and Gunn (1998), chapters in some of the better known engineering texts (e.g., Harris et al., 2009), and a number of journal articles (Allenby, 2004; Beder, 1994; Herkert, 1998; Woodhouse, 2001).

International Issues

Contributions to engineering ethics research and education in the area of international issues have been many and varied (see chapter 32 by Johri and Jesiek in this handbook for additional information). For example, Harris (2004) called for internationalizing engineering codes of ethics and suggested nine "culture transcending guidelines" and methods for applying the codes in specific contexts. Downey, Lucena, and Mitcham (2007) conducted a comparative study of the varying interest in engineering ethics in three countries based on the concept of "engineering identities." Several authors have focused on engineering ethics in the context of technological disasters in developing countries

such as the Bhopal, India chemical leak (Unger, 1994).

An area of great concern in internationalizing engineering ethics is the different expectations from country to country regarding bribery and corruption (Pritchard, 1998).

Ethicana (http://www.ethicana.org/), a video and training module developed under the sponsorship of several corporate and professional society sponsors, is a useful tool for discussing these issues with students and practitioners alike.

Although this chapter has primarily focused on engineering ethics education in the United States, the field has also flourished in several other countries. Notable examples include the Netherlands (van de Poel, Zandvoort, & Brumsen, 2001) and Japan (Fudano, 2009).

Research Ethics

With few exceptions (e.g., Whitbeck 1998), conventional treatments of engineering ethics have not included consideration of research ethics. This picture changed, however, with passage of The America COMPETES Act of 2007 that requires funding proposals to the NSF to include plans for "appropriate training and oversight in the responsible and ethical conduct of research" for all students and post-docs working on the projects.

The NSF's Ethics Education for Engineering and Science (EESE) program is focused on research and education in the area of graduate studies in engineering and science (Hollander, 2005). EESE-funded projects have addressed a number of issues and pedagogical methods. For example, Newberry et al. (2009) focus on introducing engineering ethics to international graduate students through the use of online instructional materials. Canary, Herkert, Ellison, and Wetmore (2012) have developed and assessed four instructional models for combining micro-and macroethics in graduate education for engineers and scientists. The models considered include a stand-alone course, a hybrid face-to-face/online course, ethics material embedded in a required course, and engagement of lab groups. Moore, Hart, Randall, and Nichols (2006) have built on their experience designing online material to supplement discussion of engineering ethics in undergraduate engineering classes to design similar materials for use with graduate engineering students.

Emerging Technologies

As engineering ethics advances as a field, one area demanding more attention is the ethical challenges of "emerging technologies" (Herkert, 2011), which generally refers to developments in nanotechnology, neurotechnology (and cognitive science), biotechnology, and robotics, as well as advanced information and communication technology.

An example of the many hundreds of emerging technologies under development is "pervasive computing," which expands the popular conception of the "smart house" to the entire built environment (perhaps even the natural environment). The Steven Spielberg film *Minority Report* includes a fictional representation of pervasive computing but pervasive computing is hardly science fiction. Its outlines are already taking shape in today's world of smart phones (with built-in geographical positioning systems), microprocessors embedded in everyday objects, smart cards, radiofrequency identification tags and implants, and face recognition technology, all potentially interconnected in faster and faster wireless broadband networks. Technical possibilities such as these pose daunting ethical challenges, especially in protecting personal privacy in a system designed to know who you are, where you are, and all of your personal preferences.

A key question being asked by ethicists is whether such emerging technologies have unique characteristics that distinguish them from previous technologies. Many observers who believe that to be the case point to such novel characteristics as accelerating pace of development, mind-boggling systems complexity, seemingly unlimited reach, embeddedness, specificity, and malleability of form. Another factor often highlighted is that these technologies are not being developed in a vacuum but rather tend to converge with one another in both processes and products (Khushf, 2006; Moor, 2005; Nordmann, 2004; Wulf, 2004).

The engineers and computer scientists behind these technologies often seem quite convinced of the ethical imperative of their work, from battle field robots that can be programmed to follow the rules of war (Arkin, 2009) to autonomous machines that transcend human moral character in addition to human intelligence (Hall, 2007). Ethicist James Moor (2005) and others have taken a more cautious view, arguing that emerging technologies require more than "ethics as usual," including ethical thinking that is better informed, more proactive, and characterized by more and better interdisciplinary collaboration among scientists, engineers, ethicists, and others.

In addressing emerging technologies in research and teaching, we can sometimes draw on concepts that have been applied successfully to previous technologies (Herkert, 2011). Moral imagination (Berne & Schummer, 2005), for example, is useful in addressing the complexity and malleability of emerging technologies; preventive ethics (Harris, 1995) provides useful lessons on the need for more proactive ethical analysis. New ethical tools are also needed. Deborah Johnson (2011), for example, and others, drawing on concepts from science and technology studies, have been arguing for an anticipatory ethics geared to the pace, complexity, and embeddedness of emerging technologies.

Social Issues

The past decade has seen an explosion of efforts aimed at expanding the focus of engineering ethics to include many topics heretofore not considered in traditional engineering ethics education. Many of these efforts fall into the general category of social issues. In addition to providing a window into important areas of ethical concern, such endeavors also provide linkages to ABET EC 2000 criteria calling for "the broad education necessary to understand the impact of engineering solutions in a global, economic, environmental, and societal context" and "a knowledge of contemporary issues" (ABET, 2007). Especially notable has been work in the area of gender issues (Adam, 2001), peace studies (Vesilind, 2009), social justice (Riley, 2008) and humanitarian engineering (Moskal, Skokan, Munoz, & Gosink, 2008). Chapter 17 by Riley, Slayton, & Pawley in this handbook provides an in-depth examination of social justice, women, and minorities in engineering.

Conclusion

During much of the early twenty-first century there has been a general movement to reflect on the conditions and challenges that engineers will face in the future. For example, the National Academy of Engineering's Engineer of 2020 (National Academy of Engineering, 2004) attempts to predict the roles that engineers will play in the future, while the follow-on report, Educating the Engineer of 2020 (National Academy of Engineering, 2005) discusses how to educate those future engineers. Similarly, the National Academy of Engineering developed a list of fourteen grand challenges and opportunities for engineering that must be addressed in the century ahead and published their findings in an internet based report (National Academy of Engineering, 2011a). Although each of these reports mentions professional ethics, by no means do they dwell on the topic extensively. Perhaps the reason is that in reality the ethical challenges that modern engineers face have not changed significantly from those that several prior generations of engineers have been asked to consider. What has changed, in relatively recent times, is the potential modern engineers have for broader and more significant impacts on society (locally, nationally, and globally).

With the ever increasing potential for impact in mind, it should be evident that ethics remains a critical component of engineering education and practice. The application of engineering knowledge will hold little value if not performed in an ethical manner. Accordingly, the performance of research related to the instruction, retention, and application of engineering ethics is a field that deserves and requires continued funding, sustained exploration, and persistent dissemination of findings.

References

ABET. (2007). *Criteria for accrediting engineering programs.* Baltimore, MD: ABET, Inc.

Adam, A. (2001). Heroes or sibyls? Gender and engineering ethics. *IEEE Technology and Society Magazine,* 20(3), 39–46.

Allenby, B. (2004). Engineering and ethics for an anthropogenic planet. In National Academy of Engineering., National Academies (U.S.) & National Academies Press (U.S.) (Eds.), *Emerging technologies and ethical issues in engineering: Papers from a workshop, October 14–15, 2003* (pp. 9–28). Washington, DC: The National Academies Press.

American Association of Engineering Societies, American Institute of Chemical Engineers, ASME – Environmental Engineering Division, National Academy of Engineering, & National Society of Professional Engineers. (2001). A declaration by the US engineering community to the world summit on sustainable development. Retrieved from http://www.ieeeusa .org/policy/POLICY/2002/02June24.html

American Institute of Chemical Engineers (AICE). (2003). Code of ethics. Retrieved from http://www.aiche.org/About/Code.aspx

American Society of Civil Engineers (ASCE). (2011). Code of ethics. Retrieved from http:// www.asce.org/Leadership-and-Management/ Ethics/Code-of-Ethics/

American Society of Mechanical Engineers (ASME). (2009). Code of ethics of engineers. Retrieved from http://files.asme.org/ ASMEORG/Governance/3675.pdf

Arkin, R. C. (2009). Ethical robots in warfare. *IEEE Technology and Society Magazine,* 28(1), 30–33.

Barry, B. E., & Ohland, M. W. (2009). Applied ethics in the engineering, health, business, and law professions: A comparison. *Journal of Engineering Education*, 98(4), 377–388.

Barry, B. E., & Ohland, M. (2012). ABET Criterion 3.f: How much curriculum content is enough? *Science and Engineering Ethics*, 18(2), 369–392.

Basinger, K. S., Gibbs, J. C., & Fuller, D. (1995). Context and the measurement of moral judgment. *International Journal of Behavioral Development*, 18(3), 537–556.

Baum, R. J. (1980). *Ethics and engineering curricula*. Hastings-on-Hudson, NY: The Hastings Center, Institute of Society, Ethics, and the Life Sciences.

Baura, G. D. (2006). *Engineering ethics: An industrial perspective*. Boston, MA: Elsevier.

Bebeau, M. J. (1994). Influencing the moral dimensions of dental practice. In J. R. Rest & D. Narvâaez (Eds.), *Moral development in the professions: Psychology and applied ethics* (pp. 121–146). Hillsdale, NJ: Lawrence Erlbaum.

Bebeau, M. J. (2002). The defining issues test and the four component model: Contributions to professional education. *Journal of Moral Education*, 31(3), 271–295.

Beder, S. (1994). Role of technology in sustainable development. *IEEE Technology and Society Magazine*, 13(4), 14–19.

Berne, R. W., & Schummer, J. (2005). Teaching societal and ethical implications of nanotechnology to engineering students through science fiction. *Bulletin of Science, Technology and Society*, 25(6), 459–468.

Bird, S. J. (2003). Integrating ethics education at all levels. In W. Wulf (Ed.), *Emerging technologies and ethical issues in engineering: Papers from a workshop, October 14–15, 2003* (pp. 79–93). Washington, DC: The National Academies Press.

Borenstein, J., Drake, M., Kirkman, R., & Swann, J. (2008). *The test of ethical sensitivity in science and engineering (TESSE): A discipline-specific assessment tool for awareness of ethical issues*. Paper presented at the American Society for Engineering Education Annual Conference & Exposition, Pittsburgh, PA.

Borenstein, J., Drake, M., Kirkman, R., & Swann, J. (2010). The engineering and science issues test (ESIT): A discipline-specific approach to assessing moral judgment. *Science and Engineering Ethics*, 16(2), 387–407.

British Medical Association. (1984). *The handbook of medical ethics*. London: British Medical Association.

Busby, J., & Coeckelbergh, M. (2003). The social ascription of obligations to engineers. *Science and Engineering Ethics*, 9(3), 363–376.

Callahan, D. (2004). *The cheating culture: Why more Americans are doing wrong to get ahead* (1st ed.). Orlando, FL: Harcourt.

Canary, H., Herkert, J. R., Ellison, K., & Wetmore, J. (2012). *Microethics and macroethics in graduate education for scientists and engineers: Developing and assessing instructional models*. Paper presented at the American Society for Engineering Education Annual Conference & Exposition.

Carnegie Foundation for the Advancement of Teaching. (2007). The Carnegie classification of institutions of higher education. Retrieved from http://www.carnegiefoundation.org/classifications/

Carpenter, D., Harding, T. S., Finelli, C. J., Montgomery, S. M., & Passow, H. J. (2006). Engineering students' perceptions of and attitudes towards cheating. *Journal of Engineering Education*, 95(3), 181–194.

Carpenter, D., Harding, T., Finelli, C., & Passow, H. (2004). Does academic dishonesty relate to unethical behavior in professional practice? An exploratory study. *Science and Engineering Ethics*, 10(2), 311–324.

Chickering, A. W., & Gamson, Z. F. (1991). *Applying the seven principles for good practice in undergraduate education*. San Francisco, CA: Jossey-Bass.

Colby, A., & Kohlberg, L. (1987). *The measurement of moral judgment*. Cambridge: Cambridge University Press.

Crain, W. C. (2005). *Theories of development: Concepts and applications* (5th ed.). Upper Saddle River, NJ: Pearson Prentice Hall.

Davis, M. (1988). The special role of professionals in business ethics. *Business and Professional Ethics Journal*, 7(2), 51–62.

Davis, M. (1999a). *Ethics and the university*. London: Routledge.

Davis, M. (1999b). Teaching ethics across the engineering curriculum. In *Online Proceedings of International Conference on Ethics in Engineering and Computer Science*. Cleveland, OH. Retrieved from http://www.onlineethics.org/Education/instructessays/curriculum.aspx

Davis, M. (2001). Three myths about codes of engineering ethics. *Technology and Society Magazine, IEEE*, 20(3), 8–14.

Downey, G., Lucena, J., & Mitcham, C. (2007). Engineering ethics and identity: Emerging initiatives in comparative perspective. *Science and Engineering Ethics*, 13(4), 463–487.

Drake, M. J., Griffin, P. M., Kirkman, R., & Swann, J. L. (2005). Engineering ethical curricula: Assessment and comparison of two approaches. *Journal of Engineering Education*, 94(2), 223–231.

Duckett, L. J., Rowan, M., Ryden, M., Krichbaum, K., Miller, M., Wainwright, H., & Savik, K. (1997). Progress in the moral reasoning of baccalaureate nursing students between program entry and exit. *Nursing Research*, 46(4), 222–229.

Duckett, L. J., & Ryden, M. B. (1994). Education for ethical nursing practice. In *Moral development in the professions: Psychology and applied ethics* (pp. 51–70). Hillsdale, NJ: Lawrence Erlbaum.

E3. (2007). Exploring ethical decision-making in engineering. Retrieved from http://www.engin.umich.edu/research/e3/index.html

Fielder, J. H., & Birsch, D. (1992). *The DC-10 case: A study in applied ethics, technology, and society*. Albany, NY: State University of New York Press.

Fleddermann, C. B. (2008). *Engineering ethics* (2nd ed.). Upper Saddle River, NJ: Pearson Education.

Fudano, J. (2009). Engineering ethics education and the role of engineering professional societies: Toward the construction of value-sharing type ethics program. *Journal of Japan Society of Civil Engineers*, 1(0), 1–5.

Ginsburg, H., & Opper, S. (1988). *Piaget's theory of intellectual development* (3rd ed.). Englewood Cliffs, NJ: Prentice-Hall.

Gudeman, K. (2011). Illinois to develop a national center for ethics in science, mathematics, and engineering Retrieved from http://www.ece.illinois.edu/mediacenter/article.asp?id=1162

Gunn, A. S., & Vesilind, P. A. (2003). *Hold paramount: The engineer's responsibility to society*. Pacific Grove, CA: Thomson-Brooks/Cole.

Hall, J. S. (2007). *Beyond AI: Creating the conscience of the machine*. Amherst, NY: Prometheus Books.

Harding, T. S., Mayhew, M. J., Finelli, C. J., & Carpenter, D. D. (2007). The theory of planned behavior as a model of academic dishonesty in engineering and humanities undergraduates. *Ethics & Behavior*, 17(3), 255–279.

Harris, C. (1995). Explaining disasters: The case for preventive ethics. *IEEE Technology and Society Magazine*, 14(2), 22–27.

Harris, C. (2004). Internationalizing professional codes in engineering. *Science and Engineering Ethics*, 10(3), 503–521.

Harris, C., Davis, M., Pritchard, M., & Rabins, M. (1996). Engineering ethics: What? why? how? and when? *Journal of Engineering Education*, 85(2), 93–96.

Harris, C., Pritchard, M., & Rabins, M. (2000). *Engineering ethics: Concepts and cases* (2nd ed.). Belmont, CA: Wadsworth.

Harris, C., Pritchard, M., & Rabins, M. (2005). *Engineering ethics: Concepts and cases* (3rd ed.). Belmont, CA: Thomson/Wadsworth.

Harris, C., Pritchard, M., & Rabins, M. (2009). *Engineering ethics: Concepts and cases* (4th ed.). Belmont, CA: Wadsworth Cengage Learning.

Hastings Center. (1980). *The teaching of ethics in higher education: A report*. Hastings-on-Hudson, NY: Hastings Center, Institute for Society, Ethics, and the Life Sciences.

Haws, D. R. (2001). Ethics instruction in engineering education: A (mini) meta-analysis. *Journal of Engineering Education*, 90(2), 223–229.

Herkert, J. R. (1998). Sustainable development, engineering and multinational corporations: Ethical and public policy implications. *Science and Engineering Ethics*, 4(3), 333–346.

Herkert, J. R. (2000a). Engineering education in the USA: Content, pedagogy and curriculum. *European Journal of Engineering Education*, 25(4), 303–313.

Herkert, J. R. (2000b). *Social, ethical, and policy implications of engineering: Selected readings*. New York: IEEE Press.

Herkert, J. R. (2001). Future directions in engineering ethics research: Microethics, macroethics and the role of professional societies. *Science and Engineering Ethics*, 7(3), 403–414.

Herkert, J. R. (2002). Continuing and emerging issues in engineering ethics education. *The Bridge*, 32(3), 8–13.

Herkert, J. R. (2003). Professional societies, microethics, and macroethics: Product liability as an ethical issue in engineering design.

International Journal of Engineering Education, 19(1), 163–167.

Herkert, J. R. (2011). Ethical challenges of emerging technologies. In G. E. Marchant, A. B. R. & J. R. Herkert (Eds.), *The growing gap between emerging technologies and legal-ethical oversight: The pacing problem* (Vol. 7, pp. 35–44). New York, NY: Springer.

Hollander, R. (2005). Ethics education at NSF. *Science and Engineering Ethics,* 11(3), 509–511.

Holsapple, M., Sutkus, J., Carpenter, D., Finelli, C., Burt, B., Ra, E., Harding, T., & Bielby, R. (2011, June). *We can't get no satisfaction!: The relationship between students' ethical reasoning and their satisfaction with engineering ethics education.* Paper presented at the American Society for Engineering Education Annual Conference & Exposition, Vancouver, BC, Canada.

Institute or Electrical and Electronics Engineers. (2011). IEEE code of ethics. Retrieved from http://ieee.org/portal/pages/iportals/aboutus/ethics/code.html

Johnson, D. G. (2011). Software agents, anticipatory ethics and accountability. In G. E. Marchant, Allenby, B. R., & J. R. Herkert (Eds.), *The growing gap between emerging technologies and legal-ethical oversight: The pacing problem* (1st ed., Vol. 7, pp. 61–76). New York, NY: Springer.

Kelly, M. J. (1980). *Legal ethics and legal education.* Hastings-on-Hudson, NY: Hastings Center, Institute of Society, Ethics and the Life Sciences.

Kemper, B. (2004). Evil intent and design responsibility. *Science and Engineering Ethics,* 10(2), 303–309.

Khushf, G. (2006). An ethic for enhancing human performance through integrative technologies. In W. S. Bainbridge & M. C. Roco (Eds.), *Managing nano-bio-info-cogno innovations: Converging technologies in society* (pp. 255–278). Dordrecht, The Netherlands: Springer.

Killen, M., & Smetana, J. G. (2006). *Handbook of moral development.* Mahwah, NJ: Lawrence Erlbaum.

Kohlberg, L. (1981). *Essays on moral development* (1st ed.). San Francisco, CA: Harper & Row.

Kohlberg, L., Levine, C., & Hewer, A. (1983). *Moral stages: A current formulation and a response to critics.* Basel: Karger.

Ladd, J. (1991). The quest for a code of professional ethics: An intellectual and moral confusion. In D. G. Johnson (Ed.), *Ethical issues in engineering* (pp. 130–136). Englewood Cliffs, NJ: Prentice Hall.

Lambert, A. D., Terenzini, P. T., & Lattuca, L. R. (2007). More than meets the eye: Curricular and programmatic effects on student learning. *Research in Higher Education,* 48(2), 141–168.

Layton, E. T. (1971). *The revolt of the engineers; Social responsibility and the American engineering profession.* Cleveland, OH: Press of Case Western Reserve University.

Lind, G. (1999). An introduction to the Moral Judgment Test. University of Konstanz. Retrieved from http://www.uni-konstanz.de/FuF/SozWiss/fg-psy/ag-moral/pdf/Lind-1999-MJT-Introduction-E.pdf

Loui, M. C. (2005). Ethics and the development of professional identities of engineering students. *Journal of Engineering Education,* 94(4), 383–390.

Luegehbiehl, H. C. (1991). Codes of ethics and the moral education of engineers. In D. G. Johnson (Ed.), *Ethical issues in engineering* (pp. 137–154). Englewood Cliffs, NJ: Prentice Hall.

Martin, M. W., & Schinzinger, R. (1996). *Ethics in engineering* (3rd ed.). New York, NY: McGraw-Hill.

Moor, J. (2005). Why we need better ethics for emerging technologies. *Ethics and Information Technology,* 7(3), 111–119.

Moore, C., Hart, H., Randall, D. A., & Nichols, S. (2006). PRiME: Integrating professional responsibility into the engineering curriculum. *Science and Engineering Ethics,* 12(2), 273–289.

Moskal, B., Skokan, C., Munoz, D., & Gosink, J. (2008). Humanitarian engineering: Global impacts and sustainability of a curricular effort. *International Journal of Engineering Education,* 24(1), 162–174.

National Academy of Engineering (NAE). (2004). *The engineer of 2020: Visions of engineering in the new century.* Washington, DC: The National Academies Press.

National Academy of Engineering (NAE). (2005). *Educating the engineer of 2020: Adapting engineering education to the new century.* Washington, DC: The National Academies Press.

National Academy of Engineering (NAE). (2011a). Grand challenges for engineering. Retrieved from http://www.engineeringchallenges.org/

National Academy of Engineering (NAE). (2011b). Online ethics center: Mission. Retrieved from http://www.onlineethics.org/about.aspx

National Society of Professional Engineers (NSPE). (2011). NSPE code of ethics for engineers. Retrieved from http://www.nspe.org/Ethics/CodeofEthics/index.html

Newberry, B. (2004). The dilemma of ethics in engineering education. *Science and Engineering Ethics*, 10(2), 343–351.

Newberry, B. (2009). Katrina: Macro-ethical issues for engineers. *Science and Engineering Ethics*, 16(3), 535–571.

Newberry, B., Austin, K., Lawson, W., Gorsuch, G., & Darwin, T. (2009). Acclimating international graduate students to professional engineering ethics. *Science and Engineering Ethics*, 17(1), 171–194.

Nordmann, A. (2004). *Converging technologies: Shaping the future of European societies*. Luxembourg: European Commission and the Directorate-General for Research. Retrieved from http://www.philosophie.tu-darmstadt.de/media/philosophie_nanobu-ero/pdf_2/nordmannalfredconvergingtechnol-ogiesshapingthefutureofeuropeansocieties.pdf

O'Connell, B. M., & Herkert, J. R. (2004). Engineering ethics and computer ethics: Twins separated at birth? *Techne: Research in Ethics and Engineering*, 8(1), 35–56.

Palmer, J., Cooper, D. E., & Bresler, L. (2001). *Fifty modern thinkers on education: From Piaget to the present*. London: Routledge.

Passow, H., Mayhew, M., Finelli, C., Harding, T., & Carpenter, D. (2006). Factors influencing engineering students' decisions to cheat by type of assessment. *Research in Higher Education*, 47(6), 643–684.

Petroski, H. (2008, October). Founding societies. *Prism*, 18, 31.

Pfatteicher, S. K. A. (2000). The Hyatt horror: Failure and responsibility in American engineering. *Journal of Performance of Constructed Facilities*, 14(2), 62–66.

Pfatteicher, S. K. A. (2003). Depending on character: ASCE shapes its first code of ethics. *Journal of Professional Issues in Engineering Education and Practice*, 129(1), 21.

Pfatteicher, S. K. A. (2010). *Lessons amid the rubble: An introduction to post-disaster engineering and ethics*. Baltimore, MD: The Johns Hopkins University Press.

Pinkus, R. L. B., Shuman, L. J., Hummon, N. P., & Wolfe, H. (1997). *Engineering ethics: Balancing cost, schedule, and risk–lessons learned from the space shuttle*. Cambridge: Cambridge University Press.

Powers, C. W., & Vogel, D. (1980). *Ethics in the education of business managers*. Hastings-on-Hudson, NY: Hastings Center.

Pritchard, M. (1998). Professional responsibility: Focusing on the exemplary. *Science and Engineering Ethics*, 4(2), 215–233.

Rabins, M. (1998). Teaching engineering ethics to undergraduates: Why? What? How? *Science and Engineering Ethics*, 4(3), 291–302.

Rest, J. R. (1994). Background: Theory and research. In J. R. Rest & D. Narvâaez (Eds.), *Moral development in the professions: Psychology and applied ethics* (pp. 1–26). Hillsdale, NJ: Lawrence Erlbaum.

Rest, J. R. (1999). *Postconventional moral thinking: A Neo-Kohlbergian approach*. Mahwah, NJ: Lawrence Erlbaum.

Rest, J. R., Narvaez, D., Bebeau, M., & Thoma, S. (1999). A Neo-Kohlbergian approach: The DIT and schema theory. *Educational Psychology Review*, 11(4), 291–324.

Riley, D. (2008). Engineering and social justice. *Synthesis Lectures on Engineers, Technology, and Society*, 7, 1–163.

Roeser, S. (2010). Emotional engineers: Toward morally responsible design. *Science and Engineering Ethics*, 1–13.

Rosen, B., & Caplan, A. L. (1980). *Ethics in the undergraduate curriculum*. Hastings-on-Hudson, NY: The Hastings Center.

Seebauer, E. G., & Barry, R. L. (2001). *Fundamentals of ethics for scientists and engineers*. New York, NY: Oxford University Press.

Self, D. J., & Ellison, E. M. (1998). Teaching engineering ethics: Assessment of its influence on moral reasoning skills. *Journal of Engineering Education*, 87(1), 29–34.

Self, D. J., Wolinsky, F. D., Baldwin, J., & DeWitt C. (1989). The effect of teaching medical ethics on medical students' moral reasoning. *Academic Medicine*, 62(12), 755–759.

Shuman, L. J., Besterfield-Sacre, M., & McGourty, J. (2005). The ABET "Professional Skills" – Can they be taught? Can they be assessed? *Journal of Engineering Education,* 94(1), 41–55.

Smith, J. H., Harper, P. M., & Burgess, R. A. (Eds.). (2008). *Engineering ethics: Concepts, viewpoints, cases, and codes* (2nd ed.). Lubbock, TX: National Institute for Engineering Ethics.

Sutton, R. E. (1992). Review of the defining issues test. In J. J. Kramer & J. C. Conoley (Eds.), *The mental measurements yearbook.* Lincoln, NE: Buros Institute of Mental Measurements.

Unger, S. H. (1994). *Controlling technology: Ethics and the responsible engineer* (2nd ed.). New York, NY: John Wiley & Sons.

van de Poel, I., Zandvoort, H., & Brumsen, M. (2001). Ethics and engineering courses at Delft University of Technology: Contents, educational setup and experiences. *Science and Engineering Ethics,* 7(2), 267–282.

van Gorp, A., & van de Poel, I. (2001). Ethical considerations in engineering design processes. *IEEE Technology and Society Magazine,* 20(3), 15–22.

Vesilind, P. A. (2009). *Engineering peace and justice: The responsibility of engineers to society* (1st ed.). New York, NY: Springer.

Vesilind, P. A., & Gunn, A. S. (1998). *Engineering, ethics, and the environment.* New York, NY: Cambridge University Press.

Vogt, C. M. (2008). Faculty as a critical juncture in student retention and performance in engineering programs. *Journal of Engineering Education,* 97(1), 27–36.

Weil, V. (2003). Ethics across the curriculum: Preparing engineering and science faculty to introduce ethics in their teaching. In W. Wulf (Ed.), *Emerging technologies and ethical issues in engineering: Papers from a workshop, October 14–15, 2003* (pp. 79–93). Washington, DC: The National Academies Press.

Whitbeck, C. (1998). *Ethics in engineering practice and research.* Cambridge: Cambridge University Press.

Woodhouse, E. J. (2001). Curbing overconsumption: Challenge for ethically responsible engineering. *IEEE Technology and Society Magazine,* 20(3), 23–30.

Wujek, J. W., & Johnson, D. G. (1992). *How to be a good engineer.* Washington, DC: Institute of Electrical and Electronics Engineers.

Wulf, W. A. (2004). Keynote Address. In W. Wulf (Ed.), *Emerging technologies and ethical issues in engineering: Papers from a workshop, October 14–15, 2003, National Academy of Engineering* (pp. 1–6). Washington, DC: The National Academies Press.

Yadav, A., & Barry, B. E. (2009). Using case-based instruction to increase ethical understanding in engineering: What do we know? What do we need? *International Journal of Engineering Education* 25(1), 138–143.

CHAPTER 34

The Normative Contents of Engineering Formation

Engineering Studies

Gary Lee Downey

Introduction

Engineering educators will be persuaded to alter pedagogy and curricula only when presented with rigorous research demonstrating better learning outcomes through alternative practices. So goes one justification for the recent rise of engineering education research (EER) as an arena of scholarship (Borrego, 2007; Lohmann, 2005). It builds on the view that engineering educators will be influenced most by educational research that best emulates the high value they place on rational, rigorous, especially quantitative, approaches to their own research. Rigorous EER will increase the frequency through which engineering educators adopt curricular innovations. Although engineering educators of the 1980s and 1990s showed themselves to not be influenced by the many calls for change that lacked supporting evidence, and no one advocates bad educational research, overemphasizing the

This contribution draws insights and some text from (Downey, 2011).

assumption that engineering educators are rational actors who are most responsive to rational argument may prove, however, to be self-limiting. It fails in particular to place sufficient emphasis on the normative contents of engineering practice and, hence, the normative commitments in engineering teaching and learning. EER must critically examine not only what engineering educators do and how they do it but also why they do it.

What is engineering for? What are engineers for? Engineering knowledge and practices always have normative contents. That is, engineers work in relation to or on behalf of a multitude of broader sociomaterial projects, that is, projects whose participants and implications extend beyond engineers and engineering. Because the education and training of would-be engineers seeks to prepare them for engineering work, a key issue for engineering education researchers to consider is the normative contents of engineering formation. I use the European term "formation" because it helpfully refers and calls attention to both formal education and informal training.

This contribution introduces readers to research on the normative contents of engineering formation undertaken by scholars in and around the scholarly field of engineering studies. Its distinctive lens is to focus on the question of "critical participation" in practices of engineering formation. By virtue of its theoretical and methodological commitments, every study of engineering education and training delineates potential pathways for participating critically in practices of engineering formation, whether explicitly or implicitly (Downey, 2009). This contribution identifies and reviews pathways for critical participation enacted, recommended, or implied by engineering studies research on the normative contents of engineering.

A key question permeating this review (addressed explicitly in the conclusion) concerns ways in which different approaches to engineering studies research might be self-limiting as well, in the sense that they inhibit or throw up significant barriers to critical participation in engineering formation, whether purposely or not. In recently reviewing possibilities for bringing engineering education and liberal education closer together, for example, Catherine Koshland, Vice Provost at University of California–Berkeley, argued, "Engineering and the liberal arts need to engage the other in a more compatible relationship" (Koshland, 2010, p. 58). Such statements and associated initiatives have a long history, especially in the United States. In 1893, for example, William Burr, Professor of Civil Engineering at the Columbia College School of Mines, argued for placing "as the first and fundamental requisite in the ideal education of young engineers, a broad, liberal education in philosophy and arts" whose main purpose "should be such a cultivation of human qualities as will subsequently enable engineers to meet men as well as matter" (Burr, 1893, p. 20).[1] It is worth reflecting on ways in which modes of thought and analysis originating in the liberal arts might limit the depths in engineering education to which pedagogies that draw on such modes might achieve.

The next section briefly introduces engineering studies by recounting initiatives over

the past decade to build it into a scholarly field of research, teaching, and outreach. The discussion takes care to mention other parallel efforts that frequently overlap. The following section sharpens the central theme of this review by identifying the contents and practices of "normative holism" in engineering formation as a key challenge for research and critical participation by engineering studies scholarship. It shows how the common tendency among engineers to distinguish technical from nontechnical dimensions of engineering work performs a normative judgment with variable implications. This section also frames what follows by calling readers' attention to four other contributions in this volume that review research on the normative contents of engineering formation. The present review is by no means comprehensive.

The main section provides brief overviews of research on the issues of engineering formation and (1) engineering as a profession; (2) private industry, (3) technology through design; and (4) the nation, the state, and the country. The final section examines how what engineering studies work selects as its object of study and how it goes about conducting that study may affect the range of possibilities it affords for critical participation in engineering formation.

Overall, this review argues that, to achieve effective critical participation in engineering formation, engineering studies research must ground practices of critical self-reflection that the makers of engineers can successfully scale up and integrate with other practices of engineering teaching and learning.

Engineering Studies

At a breakfast gathering during the 2002 joint meeting of the Society for Social Studies of Science (4S) and Society for History of Technology (SHOT), nineteen attendees expressed interest in establishing a scholarly network devoted to engineering studies. A 1989 review essay titled "The Invisible

Engineer" had documented how researchers in sociology, philosophy, and history had come to find engineering and engineers intellectually uninteresting, albeit for different reasons (Downey, Donovan, & Elliott, 1989). A 1994 review essay titled "Engineering Studies" hopefully traced expanding work by social scientists and humanists on questions of engineering knowledge, engineering as technical work, and gender in engineering, as well as recent scholarly reflections by engineers on engineering (Downey & Lucena, 1994).

Nineteen was a good number for a busy meeting. By 2002, many researchers were finding engineering and engineers to offer a significant collection of research sites for retheorizing the traditional model of knowledge as creation, diffusion, and utilization, especially by highlighting the importance of materializing effective, directional knowledge practices. One could feel growing momentum. Yet many scholars were finding it difficult to advance research projects because grant agencies, academic departments, publication outlets, and even professional societies continued to place primary emphasis on the terms "science" and "technology." In many arenas in and around interdisciplinary science and technology studies (STS), including the societies holding the joint meeting, engineers and engineering were of interest to the extent that they shed light on practices of science and/or technology but generally not in and of themselves.[2]

The work of building a networked arena of research in engineering studies started with the premise that engineers and engineering can and should be analyzed as central rather than peripheral actors and practices in worlds of science and technology. The International Network for Engineering Studies (INES) was established in Paris in 2004 at the joint meeting of 4S and the European Association for Studies of Science and Technology (EASST).[3] Imagining a network whose participants had primary allegiances in many different organizations, INES would help advance research, teaching, and outreach in historical, social, cultural,

political, philosophical, rhetorical, and organizational studies of engineers and engineering. INES-related scholarship would be built around the core question: What are the relationships among the technical and the nontechnical dimensions of engineering practices, and how do these relationships change over time and from place to place? The presence of INES as a formal organization would also encourage engineering studies researchers to become critical participants in the practices they study, including, for example, engineering formation, engineering work, engineering design, equity in engineering (gender, racial, ethnic, class, geopolitical), and engineering service to society.

Beginning in 2006, INES held workshops in Blacksburg, Virginia, USA; Lisbon, Portugal; Grafton, New York, USA; Cleveland, Ohio, USA; and Copenhagen, Denmark. Also in 2006, INES members began organizing INES-affiliated paper sessions at 4S and SHOT. In 2008, Taylor & Francis/Routledge began publishing *Engineering Studies: Journal of the International Network for Engineering Studies*. Morgan & Claypool began publishing the engineering studies series *Global Engineering* in 2011. In 2012, the MIT Press began publishing the capstone book series *Engineering Studies*.

INES emerged in the company of five related initiatives in the humanistic and social science study of engineers and engineering. In Europe, an international collection of historians began in 2002 organizing meetings and publications rethinking the history and current status of engineering formation and practice (Chatzis, 2007; de Matos, Diogo, Gouzévitch, & Grelon, 2009; Gouzevitch & Inkster, 2007). In the Society for the History of Technology (SHOT), historians of engineering interested in reestablishing linkages to engineering education and training established in 2005 the special interest group Prometheans (Brown, Downey, & Diogo, 2009).[4] In 2006, philosophers affiliated with the Society for Philosophy of Technology began mapping out a philosophy of engineering and its relations to the philosophy of technology in a series of publications and workshops (Christensen,

Delahousse, & Meganck, 2007, 2009; van de Poel & Goldberg, 2010). Scholar-activists interested in engineering and social justice established a series of paper sessions, workshops, and conferences beginning in 2004[5]; the Morgan & Claypool book series *Engineers, Society, and Technology* in 2006[6]; the Engineering, Social Justice, and Peace network in 2010[7]; and the *International Journal of Engineering, Social Justice, and Peace* in 2011.[8] Finally, members of the Liberal Education Division of the American Society of Engineering Education responded to new accreditation criteria for engineering programs in 2000 with research and learning practices that would go beyond bringing engineering students to liberal education and highlight the critical participation of liberal education practices within engineering education (Neeley, 2003; Ollis, Neeley, & Luegenbiehl, 2004; Steneck, Olds, & Neeley, 2002). Acknowledging a significant expansion of participation by members interested in critical participation within engineering education, division members in 2011 changed its name to the Liberal Education/Engineering & Society Division.[9]

Although many participants in these latter initiatives also contributed to the formal development of INES and engineering studies, many also do not identify their scholarship in the first instance with the label "engineering studies." For the benefit of readers, this review takes the risk of blurring some of these boundaries in order to highlight the question of normativity in engineering formation.

Normative Holism in Engineering Formation

Leaders in engineering formation around the world have long portrayed the making of engineers as contributing directly to human progress.[10] They have tended to assert an equivalence between the technical contents of engineering practices and material advancements throughout the world for human benefit. Official reports and vision statements for engineering formation

around the world regularly invoke such connections as "benefit to humankind" (National Academy of Engineering, 2004, p. 1), "human development" (International Federation of Engineering Education Societies, 2010), "development of society" (Japan Accreditation Board for Engineering Education, 2010), "economic and social development," (Engineering for the Americas, 2010), "match the social, economic, social, technological needs of the today society [sic]" (European Society for Engineering Education (SEFI), 2005, p. 5), "service of mankind and the advancement of general welfare" (Indian Society for Technical Education, 2010), "collective well-being [*bien-être collectif*]" (*Comité d'études sur les formations d'ingénieurs* [CEFI], 2010), and "complex and interdependent global challenges" (Anderl et al., 2006, p. 1).

The most prominent feature of these normative projects is their holism. Engineering work contributes to human advancement as a whole. The making of engineers produces as its outcomes people who contribute to human advancement as a whole. As an inherent feature of technical engineering practice, normative holism is a key dimension of engineering learning.[11]

The normative commitment to holism in engineering formation has a crucial, far-ranging implication. It grounds what Wendy Faulkner (2007) has called "technical-social dualism," the sharp division between the technical contents of engineering work and the social practices engineers participate in as people. Engineering curricula routinely portray the core of engineering work as wholly technical in content, especially the engineering sciences. Both as individuals and as a collectivity, engineers learn to encounter and engage flows of experience as endless sources of technical problems to solve, seeking optimal gain (Alder, 1997, p. 60) on behalf of humanity as a whole.

Technical-social dualism has a number of correlates. One is that distinctions among different fields and disciplines of engineering become functional technical differences that complement one another, for example, among civil, mechanical, electrical, and chemical engineering and among mechanics, circuit theory, thermodynamics, and vibrations (Gilbert, 2009). Although quantitative material practices on the job may range from design to manufacturing to sales, technical learning provides the core preparation for all of them. Also, engineering work becomes technological design. Until fairly recently, engineers have been relatively free to claim jurisdiction over the design and development of new technologies, without competition from other knowledge workers (Wisnioski, 2012). This has made it easy for both engineers and non-engineers to equate engineering with technology, frequently to the point of confusion. Engineers design technologies, and the making of engineers is always about the making of technologists. Finally, because quality engineering work necessarily produces technological outcomes that benefit humanity, engineers must be careful only to make sure nothing interferes with the high-quality technical practices they learned in school and in subsequent training.

Engineering Studies and Normative Holism

Much work in engineering studies consists of critical reactions to normative holism in engineering formation and practice, both explicitly and implicitly. Most work calls attention to its limitations by demonstrating inconsistencies with actual features of engineering formation and practice. It makes visible complexes of socio-technical relationships that normative holism hides.

Also, because the makers of engineers tend to rely on technical-social dualism in understanding and characterizing their work, research that examines relationships *between* the technical and the nontechnical dimensions of engineering practices necessarily gains normative content. It gains directionality in relation to the image and practices of normative holism. Even simple descriptions of relationships between technical and nontechnical agencies in engineering formation raise the possibility that

practices of engineering teaching and learning might have been otherwise, could become otherwise, and, hence, perhaps should be otherwise. Indeed, when one diligently pursues hard the questions – What is engineering for? and What are engineers for? (Downey, 2009) – a resolute and narrowly defined commitment to normative holism can appear at best delusional, at worst conspiratorial, hiding other normativities. But the key question remains, at least in this review: What then?

At least three valuable areas of EER research that overlap with engineering studies work on normativity in engineering formation are covered in depth elsewhere in this volume. One makes a case for making social justice issues, especially involving women and minorities, a pervasive consideration in engineering teaching and learning and engineering education research (see Chapter 17 by Riley, Slaton, & Pawley).[12] A second reviews the history of and possibilities for engineering ethics in engineering curricula (Herkert & Barry, 2012). And a third calls attention to global and international issues facing engineering students and working engineers today (Johri, 2012). To minimize duplication, I do not consider these further here.[13]

For a Profession?

A persistent and vexed issue has been the tendency of engineers to claim the status of professionals. The claim itself makes sense. Embracing human progress as a whole grants the work of engineers a special kind of autonomy that seems analogous to holistic service in the classical professions of law, medicine, and divinity. Professionals serve the community as a whole. The difficulty with working engineers has been that they largely do not fit the image, raising difficult questions about building engineering formation around normative holism.

During the 1960s and 1970s, American sociologists struggled unsuccessfully to theorize the professions in a way that would include engineers. The big problem was that engineers tended to work within large organizations (Gerstl & Perrucci, 1969). The unpopularity of separate career ladders for engineers in industry during the 1950s and 1960s, for example, affirmed that success for more elite engineers entailed success within the organization (Rothstein, 1968). And as recently as the mid-1950s, nearly one-tenth of the engineering labor force of 500,000 engineers was unionized, with organizational concerns emphasizing salary levels and work conditions. For American sociologists, examining engineers more closely actually contributed to a more general disciplinary shift to organization studies (Downey et al., 1989, pp. 191–198). In 1985, Robert Zussman (1985) sought unsuccessfully to revive sociological interest in engineers by examining them through the lenses of career and class rather than profession. The normative commitment of engineers to career mobility by changing jobs, he argued, could be traced to their middle-class status. But this move did not overcome the continuing problem of engineers claiming autonomy and normative holism in engineering education yet pursuing advancement within organizations.

Some historians of engineering have followed the assertion that engineers serve humanity as a whole by examining the organization of professional societies and professional institutions and their recruiting initiatives among students. Bruce Sinclair and J. P. Hull (1980) explored how the American Society for Mechanical Engineering achieved organizational independence worthy of a biography. Terry Reynolds (1983) showed how the American Institute of Chemical Engineers played a key role in defining and maintaining the discipline of chemical engineering. And A. Michal McMahon (1984) combined an organizational biography of the Institute of Electrical and Electronics Engineers with an account of its role in shaping the development of electrical engineering, a discipline born within corporate structures. In the United Kingdom, Angus Buchanan (1989) has both asserted the normative holism of engineering formation and recounted the persistent struggles of engineers to achieve the status of gentlemen. Exploring the role of professional institutions in structuring formal engineering education, W. J. Reader (1987) examined the Institution for Electrical Engineers and Colin Divall and Sean Johnston (2000) the Institution of Chemical Engineers.

Historians have long questioned normative holism in engineering formation precisely by calling attention to the positioning of engineers within organizations. In the late nineteenth century United States, the most elite civil engineers did achieve a public reputation for independent craft genius bordering on heroic status (McCullough, 1972). Yet Daniel Calhoun (1960, p. 199) found that even civil engineers on commission had almost always been supervised within and served the interests of governing organizations. Indeed, Buchanan's (1989, p. 11) extensive research on the formation of professional engineers in Britain forced him to admit, however, that they "have always co-existed with a much larger number of non-professional engineers."

Edwin Layton's *Revolt of the Engineers* (1986) has become a classic study of a tension between normative holism and localized commitments to the projects of employers. *Revolt* traces the failed attempts of early twentieth-century civil engineers to position themselves, their professional organizations, and by extension practices of education based on an image of autonomy and social responsibility. The study identifies a transitional process wherein the leaders of professional engineering societies tended to also be leaders in industry. Peter Meiksins (1988, p. 402) later showed that Layton's account of "patrician reformers" masked struggles of the far more numerous rank-and-file engineers whose concerns were less about professional recognition and autonomy than a "more humane, comfortable form of employment and a better chance as individuals to rise through the ranks."

Taking a lead from the engineers' commitment to normative holism, applied philosophers have devoted enormous effort to interrogating ethical tensions between normative holism and organizational loyalty.

Early contributions explored engineers' rights and responsibilities as employees, examined the role of ethics training in engineering curricula, and assessed formal codes of ethics. See Chapter 33 by Barry and Herkert in this volume for elaboration and update.

The bottom line is that engineering studies scholarship on the whole has not supported the implication of normative holism that engineers are or should be treated as autonomous professionals. Most recently, Carroll Seron and Susan Silbey (2009) describe how despite engineering claims to professional status, the accreditation process for engineering curricula enacts an "instrumental logic" that restricts the discretionary judgment characteristic of professions. The focus becomes making sure engineering students achieve mastery in applying the instrumental logic.

For Private Industry?

That engineers in, especially, Anglo-American contexts have accepted private industry in particular as the dominant venue for their work has been a primary source of interest and concern for engineering studies work on engineering formation.[14] Monte Calvert (1967) famously attributed the bonding of mechanical engineers to industrial organizations to a contingent shift in educational practices from "shop culture" to "school culture." This shift greatly expanded attention to formal education, especially at land-grant institutions. While mechanical engineers had been elite apprentices preparing on the job for leadership positions, they became tertiary-school graduates prepared for entry-level positions and armed with a desire to move up.

A 1960s survey of working engineers concluded that recruits to engineering schools generally came from lower social origins than recruits into medicine, law, and the clergy, and that upward social mobility was a primary motivation for selecting an engineering career (LeBold, Perrucci, & Howland, 1966). A "consequence of this mobility experience," argued Robert Perrucci (1971,

p. 500), was that engineers direct their "particular loyalties . . . to their employers and to organizational careers, rather than to their colleagues." In other words, an employee orientation and organizational identity were part of the novice's original conception of an engineering career and were both reinforced and solidified by the socialization process.

If engineering service is routed through industrial organizations, then are engineering students preparing to be servants of industrial corporations? David Noble (1977, pp. 47, 170, 322, 324) put the issue most starkly in a classic Marxian study loaded with evidence about the emergence of engineering formation in the United States. The normative holism of engineering service was essentially false consciousness, he argued, for engineering education served as "a major channel of corporate power" by providing the "immediate manpower needs of industry and the long-range requirements of continued corporate development." Its output was a "domesticated breed" that "convinced themselves that they served the interests of society as a whole . . . [but] in reality served only the dominant class in society."

The influence, if not control, of normative projects from private industry in engineering formation remains a given in perhaps most engineering studies work today. Among initial reactions that complicated the claim of dominance, Bernard Carlson's (1988, p. 396) study of academic entrepreneurship at MIT argued the corporation "could not simply order entry-level engineers from engineering schools" and linkages to industry were "marked by a clash of values and expectations." Peter Meiksins and Chris Smith (1996, p. 253) advanced a theory of "structural contingency," which starts with capitalist relations of production and then adds contingency, holding that "under capitalism engineers are shaped by and organized around the central contradictions of capitalism, but that this in no way points to any eventual convergence on a single way of organizing technical labor." Also Gary Downey's (1998) study of education and research in computer-aided design documented normative tensions among academic, governmental, and corporate projects.

Studies of engineering formation in Britain have had to confront the issues of employer scorn for formal engineering education and relative disinterest by government. A whole body of literature debates the extent to which British industrial decline in the late nineteenth century can be attributed to an overemphasis on craft training for engineers and their attendant low status (Edgerton, 1996; Sanderson, 1999; Wiener, 1981). Colin Divall (1990) showed how industrial employers during the early decades of the twentieth century found higher education in engineering forced upon them, while Chris Smith (1987) documented tensions between engineering staff and senior management. In contrast with the assertion of corporate control, Peter Whalley (1986) characterized engineering formation as the preparation of "trusted workers" who manage dual commitments inside and outside the organization.

The many official assertions in recent years that practices of engineering formation have actually fallen out of step with the expectations and desires of industrial employers has produced a surge of engineering studies scholarship on engineering work, with implications for engineering formation. Kevin Anderson and colleagues (Anderson, Courter, McGlamery, Nathans-Kelly, & Nicometo (2010) document how engineers carry technical-social dualism onto the job, expressing a primary interest in solving technical problems and frustrated by responsibilities for what they consider to be non-engineering work. Gian Marco Campagnolo and Giolo Fele (2010) describe how vagueness can be crucial to the distribution of engineering software. Julie Gainsburg, Carlos Rodriguez-Lluesma, and Diane Bailey (2010) describe how workplace knowledge in an engineering occupation derived more from on-the-job experience than from prior formal education. And both Bernard Delahousse (2007) and Rudi Schmiede and Mascha Will-Zocholl (2011) describe how the globalization of engineering work and rise of information and communication technologies is marginalizing purely technical work.

For Technology Through Design?

An integral component of the engineer's commitment to normative holism is that engineering education consists primarily of students learning the engineering sciences in order to be able to design technologies. The flow is linear and the emphasis is technical. Much engineering studies scholarship has problematized this flow by complicating each step.

Philosophers of technology have wrestled with the relationship between epistemic content and normative holism in the engineering sciences at least since the 1966 debate among Mario Bunge, Joseph Agassi, and Henryk Skolimowski over how to analyze the relationship between applied science and technology. Searching for epistemic categories freed from context, Bunge (1966) drew a sharp line between pure and applied science and Agassi (1966) between applied science and technology. Skolimowski (1966) began to introduce the question of normativity by drawing his line between science and social practice. If engineering theory is linked to social practice, then engineering work is always linked in some way to broader material projects.

One outcome was the rise of a social philosophy of engineering, which has tended to analyze engineering formation work from epistemics to normativity, and under the auspices of the Society for Philosophy of Technology. Early contributions theorized preparation for a "meaning community" with "inner subdivisions, structures and functions" (Durbin, 1985); the development of professional competence (and incompetence) (Sinclair & Tilston, 1982); production of a "technological intelligentsia" whose success depended on a wide range of social, ethical, and epistemological criteria (Lenk, 1984); propagation of "captive knowledge" organized in pursuit of managerial goals" (Goldman, 1984); and the development of curricula to serve an "occupation" (Davis, 1996).

Questioning the linear thread led researchers to also examine normativities in engineering design. Walter Vincenti (1990) laid down a challenge by seeking to identify universal, context-free categories of design knowledge that fit an overarching commitment to normative holism. These included fundamental design concepts, criteria and specifications, theoretical tools, quantitative data, practical considerations, and design instrumentalities. Larry Bucciarelli (1994) took a step toward highlighting normative differences by showing that contrasts in engineering design work exceed pure functional differentiation. Designers work, he argued, in different "object worlds," each of which comes with distinct epistemological and material commitments. Ibo van de Poel and A.C. van Gorp (2006) and Wade Robison (2010) extend this work by showing that ambiguity in design judgments means that ethical judgment is necessarily an integral part.

Michel Callon (1980) and other actor-network theorists have examined design practices as themselves the sources of broader material projects. He joined with John Law (Law & Callon, 1988) to portray engineering formation as the making of "engineer-sociologists" who are "not just people who sit in drawing offices and design machines" but are "social activists" who design structures or social institutions to fit the machines." Dominique Vinck and colleagues (Vinck, 2011; Vinck & Blanco, 2003) dive deeply into engineering design practices to show how they effectively organize things and people at the same time.

Recent work has tended to demonstrate the multiplicity or complexity of considerations in engineering judgments involving engineering science, design, and technology. Maarten Ottens (2010) shows that even supposedly comprehensive systems engineering depends on limiting assumptions that exclude key social considerations, e.g., the perspectives of users and significance of variable legislation. Arto Mutanen (2007) describes the engineering methodology that students learn as simultaneously epistemic, ethical, and aesthetic in content. Michael Davis (2010) points out that engineers do many things beyond design that have significant technical content, for example, inspecting, writing regulations, and evaluating patents. Eugene Coyle, Mike Murphy, and William Grimson (2007) characterize engineering design as poly-paradigmatic in the sense that it is often a balancing act among financial, environmental, aesthetical, and sociological commitments.

Finally, engineering studies researchers have examined some unforeseen implications of committing engineering formation to normative holism. Matthew Wisnioski (2009b, 2012) shows how identifying engineering formation with technology produced a crisis in elite engineering circles during the 1960s when technology itself lost its assumed link with progress. A frequent outcome was awkward attempts to appropriate modes of thought and analysis from the humanities and social sciences. Gary Downey (1998) follows the struggles of engineering students learning practices of computer-aided design as they come to realize they cannot control the technology but must also yield to it. Indeed, Wilhelm Bomke (2009) shows that some instructors and employers still resist its use. Jose Avila and Manuel Arias (2007) and Alessandro Gasparetto (2007) maintain that the technical-social dualism in normative holism has led to a decline in the public perception of engineers across European countries. And Rosalind Williams (2002) has argued that attempting to produce increasingly diverse technologies is multiplying subfields in engineering and producing an "expansive disintegration" of engineering research and teaching.

For the Nation, the State, the Country?

In 1990, German historian Peter Lundgreen greatly complicated studies of the normative projects in engineering formation by resisting the image of the contents of engineering education responding to the needs or demands of private industry. Comparing the emergence of formal engineering education in France and Germany with that in Britain and the United States, Lundgreen (1990, p. 45) highlighted the fact that French engineers since the mid-eighteenth century had been committed to formation through schools and to work in government. Also, the "driving forces" behind the emergence of engineering education across Germany could "scarcely be attributed to the rising needs of the job market in the private sector." Emphasizing the paramount role played by the state in continental technical education, Lundgreen argued for a focus on the practices of supply, taking care to distinguish between the qualification of engineers and their commitments at work.

Enormous bodies of work now exist on the histories of engineering formation across different countries, often written in English or French as well as in languages of origin. Detailing what Antoine Picon (2007) calls "local specificities," these works map contrasting emphases on formal schoolwork and on-the-job training and in degrees of participation by governments at various levels. One outcome has been to integrate histories of engineering formation with histories of the development of nations, states, and/or countries.

One approach has been to describe territorial patterns in engineering formation and engineering work as engineering cultures. Eda Kranakis (1997), for example, compares the emergence of what she described as engineering cultures in France and the United States, including both educational properties and design practices. She characterizes these as the contingent product of distinct "social and institutional factors" that happened to appear in the two countries, which then gained the force of influence through their persistence. John Brown (2000) details differences in the nineteenth century drawing practices of British and American engineers as, again, examples of distinct engineering cultures." In this case, both the "applications of plans (and the drawings themselves) came to reflect and reinforce their host cultures." In a related approach, Mikael Hård and

Andreas Knie describe German and French diesel engineers in the early twentieth century as developing distinct engineering "styles," with Germans actively codifying an engineering grammar of design and the French actively avoiding doing so, concentrating their attention on copying and preparing designs from other locations. Comparing U.S. and Japanese approaches to teaching and practicing design, Joel Moses (2010) describes design practices as deeply embedded in national cultures.

Perhaps the most common form of analysis is to recount the relation between engineering formation and the state or the relative role of the state in both institutional and curricular developments in engineering education. Providing an overview would warrant at least another complete review, if not more. Consider some starting points in France, for example. Antoine Picon (1992, 2007, 2009) describes how formation of engineers became distinct from that of architects in the eighteenth century, acquiring quasi-judicial authority as they pursue technocratic ideals. Bruno Belhoste and Konstantinos Chatzis (2007) review the preparation of state engineers in the nineteenth century and Chatzis (2009) examines the multiplication of small schools. And André Grelon (1986, 1993, 2007) has focused his attention on mapping relations and differences between state engineers and preparation of the much larger population of engineers who have worked in the private sector. Recent work by Continental historians has examined the travels of "models" of formation, the role of key "reference schools," and the distinctive normativities of schools on the European "periphery" in efforts to theorize "appropriation," "circulation," and "transnationality" (de Matos et al., 2009). Such concerns have extended worldwide.

This review has itself been informed by an approach that critiques normative holism in engineering formation not by mapping what it hides but by demonstrating its multiplicity. That is, what has counted as progress or human advancement has varied dramatically across territories and over time. This approach began with an analysis of struggles in teaching and learning computer-aided design in the United States in response to economic competitiveness as a newly dominant image of progress (Downey, 1998). It was further developed in a comparative overview of engineering formation and "metrics of progress" (Downey & Lucena, 2004); a history of U.S. governmental policies for science and engineering education in relation to images of a "nation under threat" (J. Lucena, 2005); case examples of the United States (Downey, 2007), Mexico (J. C. Lucena, 2007), Brazil (J. Lucena, 2009), and Colombia (Valderrama et al., 2009); and a theoretical overview (Downey, 2012b).

Scaling Up Engineering Studies into Engineering Education

In parallel with what I have analyzed here as normative holism in engineering formation, Byron Newberry (2009, p. 40) points to "an axiomatic presumption, perhaps instinctive to many engineers, that their work is organically beneficent, at least in a utilitarian sense." Many engineers, he continues, tend to judge engineering work to operate "like a Smithian invisible hand that inexorably promotes the collective benefit." If normative holism remains invisible as a given in the making of engineers, then engineers will have no reason to reflect critically either on their commitments or on the outcomes of their work. Yet an enormous body of engineering studies work calls attention to multiple normativities in both engineering formation and engineering work.

This review raises the possibility of scaling up findings from engineering studies into pathways for critical participation in the making of engineers. It calls attention to flows of images and practices originating from one or both of two distinct categories of sources.

Scaling Up Findings

One source is engineering studies scholars who draw on their findings to help scale up new or altered images and practices in the making of engineers. Likely a common feature of different approaches is to help make critical reflection on normative commitments an integral part of engineering education and training. A key implication of this essay is that likely an essential component in any such effort is to integrate practices for making visible and critically analyzing the dominant commitment to normative holism – the equivalence drawn between technical engineering work and the production of benefits for humanity as a whole.

What possibilities for grounding this initial step lie in different approaches to examining normative projects in engineering formation? And how deeply into formal engineering education and informal engineering training are potential alternative practices of critical reflection likely to reach? The opportunities and limitations vary with what studies take as their objects of study and how they frame their questions.

ENGINEERING PROFESSION

Research taking the engineering profession as its object of analysis may find it difficult to achieve deep inroads in engineering formation. On the one hand, such work allies itself with normative holism by challenging practitioners to rise above localized considerations, especially as employees, and focus attention and energy on problems and issues facing humanity as a whole. It points researchers in the direction of supporting or helping to create courses, curricula, and institutions devoted to addressing and solving planetary problems, such as poverty, water supply, energy supply and usage, climate change, and so forth. When done effectively, critical participation that emphasizes responsibilities to humanity as a whole can help students recognize and reflect on normative commitments implicitly built into existing curricula, for example, that a key future responsibility will be to support the private interests of an industrial employer.

On the other hand, critically assessing engineering practice in relation to an image of the profession, whether explicitly or implicitly, may limit the ability of curricular interventions to move beyond the periphery of the new course, specialized track, or voluntary membership organization. The image of a profession serving humanity as a whole does not fit too much of what engineers actually do.

PRIVATE INDUSTRY

Taking the relationship between engineering formation and private industry as one's primary object of study has different implications depending upon how one theorizes the relationship. Much of the resistance that emerged over the years to David Noble's argument that engineers are a "domesticated breed" designed to serve only the dominant class lies in his analysis that capitalism is a singular, pervasive structure with dominating agencies that invariably produce the exploitation of labor by capital. His interpretation renders spurious any efforts to build practices of critical reflection into engineering formation except through a revolutionary movement that would transform engineers into an entirely different breed altogether, not to mention all of social life as well. It effectively accepts resolute pessimism.

Most alternative approaches to the study of engineering formation in relation to private industry maintain that responsibilities to employers constitute one source of normative projects, or one collection of sources, among many. Carlson's (1988) analysis of academic entrepreneurship, Anderson at al.'s (2010) analysis of tension between authentic engineering and non-engineering projects, and Downey's (1998) analysis of tensions among industrial, academic, and governmental projects in developing and teaching computer-aided design point in this direction. Adopting an alternative to Noble's analysis ultimately requires questioning the Marxian labor theory of value that grounds it.

One approach is that articulated by Lundgreen (1990), holding that the variable role of governments in engineering

education and training make clear that serving industrial needs is but one normative project. Another possible approach might be to hold that the humanist assumption in the labor theory of value insufficiently considers technological agencies in industrial production. Such would make the alienation of labor a variable phenomenon the extent of which would depend upon actual wage rates. In any case, Noble's thorough and consistent analysis continues to throw down a gauntlet that any scholarly inquiry into the making of engineers or any effort to build practices of critical reflection into engineering formation ignores at its peril.

NATION, STATE, COUNTRY

Approaches to engineering formation and the nation, state, or country introduce distinct opportunities with distinct limitations depending on how the analysis conceives the relationship. A significant strength in identifying patterns in engineering education and training as distinct cultures, for example, is that such locates the technical contents of engineering formation in the same analytical (and ontological) category as practices of governmental action, political economy, educational innovation, activities, of work, and so forth. They all become cultural phenomena.

A key limitation lies in the same point. The project of normative holism already has a place for culture in engineering practices. It serves as a cover label for everything that is not technical, that is, not authentically engineering. Cultural phenomena that live on the margins may be of great interest to the makers of engineers, especially in an era of globalization (Downey, 2011), but it is therefore difficult for that very reason to persuade engineers that the technical contents of engineering work are cultural phenomena as well. Such could appear remove the technical rigor from engineering judgment that normative holism authorizes and values. Focusing on culture could prove to throw up a barrier to significant cultural participation.

The burgeoning of studies of the role of the state in engineering formation, often in relation to the role of the private sector, introduces opportunities for critically participating in engineering formation through the image of engineering "in context." This approach is becoming increasingly popular among engineering studies scholars. It is relatively straightforward, for example, to demonstrate to would-be engineers that they will have to participate in a variety of social contexts, including relations with employers, fellow employees, clients, professional associations, standards organizations, regulatory bodies, and so on. And making visible such contexts can help students recognize that technical problem solving will be only part of what they do and that they would do well to prepare for participation in a variety of contexts.

A limitation in this approach parallels that of analyzing engineering as a cultural phenomenon. It leaves relatively untouched the technical-social dualism that normative holism defines and supports. The word "context" comes to refer to everything that does not have to do with the linear flow of technical engineering practices from engineering science to technology through design. It does not take account of the fact that in taking normative holism and its associated technical-social dualism for granted, engineers already routinely "contextualize themselves," as Wisnioski (2009a) puts it.

MULTIPLYING NORMATIVE HOLISMS

The scholarly approach demonstrating that normative holism in engineering formation is not singular but multiple has the strength of telling the makers of engineers not that they are wrong in some way but that what they take for granted is in an important sense highly localized. It encourages engineering educators, in the first instance, to re-characterize as localized knowledge practices they currently teach accepted by all good engineers everywhere. A longer-term possibility is that, akin to Bucciarelli's (1994)

delineation of the distinct object worlds that different fields of engineering postulate, the makers of engineers could teach every engineering science and/or every engineering practice as a normative project that makes some things visible and doable while hiding others (Downey, 2005).

A limitation in this approach is that it is not obvious how critical pedagogy of this sort would actually make it to the heart of instruction in the engineering sciences. Although it does at least envision a pathway to this core of practices of engineering formation, persuading engineering instructors to actually travel that pathway to revising what they do may also require something akin to a mass social movement in engineering instruction.

Appropriating Findings

A second category of sources for scaling up findings from engineering studies into pathways for critical participation in the making of engineers is other engineering education researchers. EER scholars who read engineering studies could appropriate and integrate it into the practices that emerge from their own research. Such could be accomplished in one of two ways.

The first is for EER scholars to make visible and examine how they might be building normative contents into the process of constructing proposals. A multitude of possibilities lie here. Might it be the case, for example, that particular efforts to expand design education in engineering assume that the desired goal is consumer products from the private sector? What are the normative implications of such commitments? What might be the normative implications across engineering curricula of integrating new attention to issues such as interdisciplinarity, nanoscale technologies, or climate change research? Might all makers of engineers, including contemporary innovators, be attempting what the STS scholar Sheila Jasanoff (1996, 2004) calls "co-producing" the technical and normative contents of engineering formation at the same time?

The other way of appropriating findings from engineering studies research is to study the normative struggles of students as they confront engineering curricula. Every existing practice and proposed innovation in the making of engineers raises normative questions around student identities. The greatest amount of existing research in this respect has focused on how the sex or gender of students affects the types of challenges they experience in becoming engineers and how they might respond to those challenges. But in principle any dimension of student identity could be engaged by any effort to scale up new practices in engineering education and training. EER scholars must be sensitive to the normative dimensions of students' multiple identities as those students pursue pathways toward becoming engineers.

In his 1954 novel *The New Men*, C. P. Snow blurred the two cultures of science and the humanities when he characterized engineers. "[E]ngineers," he wrote, " . . . were in nine cases out of ten . . . acceptant of any regime in which they found themselves" (p. 176).[15] Albeit fiction, this characterization of engineers still stands out because it likely strikes the vast majority of readers as true. The key question it raises is: Why?

If engineers are bred to be wholly domesticated agents of powerful external forces, engineers are weak and their acceptance of the regime in which they find themselves is a kind of surrender. From this point of view, it makes little sense for critical scholars to go beyond critique to seek critical participation. They will be co-opted. But the situation changes, however, if engineers are in a regime *because* they accept it, because they, as Snow continues, are mainly "interested in making their machine work" and perhaps find regimes that facilitate that end. In that case, reason does exist to interrogate the normative projects in that acceptance and their implications for the making of engineers.

Acknowledgments

The research reported here is based on work supported by National Science Foundation under award no. DUE-1022898. Any opinions, findings, and conclusions or recommendations expressed in this publication are those of the author and do not necessarily reflect the views of the National Science Foundation. I greatly appreciate the three thorough and extremely helpful anonymous reviews I received on the original manuscript and a first revision. Thank you for reading so closely and thoughtfully. I also appreciate the initial invitation and helpful guidance from the handbook's editors, Aditya Johri and Barbara Olds.

Footnotes

1. Thanks to Larry Bucciarelli, emeritus MIT professor, for calling my attention to this quotation.

2. My own account of this tendency is that STS work has gained and retained broader legitimacy by highlighting and calling attention to limitations in dominant everyday images of science and technology. Attempts to scale up new images of science and technology tended not to include highlighting the relatively devalued arenas of engineering and the applied sciences. See Downey (2009) for elaboration. The contemporary shift to what Gibbons et al. (1994) called "Mode 2" science is an important factor changing this.

3. www.inesweb.org. The founding INES coordinators include Gary Downey, Alumni Distinguished Professor, Department of Science and Technology in Society, Virginia Tech, USA; Maria Paula Diogo, Associate Professor, Department of History, New University of Lisbon, Portugal; and Chyuan Yuan Wu, Director, Institute of Sociology and Program in Science, Technology and Society, National Tsing Hua University, Taiwan. Atsushi Akera, Associate Professor, Department of Science and Technology Studies, Rensselaer Polytechnic Institute, USA became a coordinator in 2011. The founding INES Web Editor is Brent Jesiek, Assistant Professor, School of Engineering Education and School of Electrical and Computer Engineering, Purdue University.

4. www.historyoftechnology.org/sigs.html.

5. http://esjp.org/esjp-conference.

6. www.morganclaypool.com/toc/ets/1/1.

7. http://esjp.org.
8. http://esjp.org/journal.
9. www.asee.org/member-resources/groups/divisions.
10. This section draws on (Downey, 2012a, 2012b).
11. Thanks to the reviewer who argued that the term "holism" typically "evokes a kind of completeness in which nothing has been glossed over or left out, that is simultaneously attentive to the particular and the whole, to the local and the global." I am using the term in a commonly used sense that the whole is greater than or has an existence other than the sum of its parts. I considered using the term "(w)holism" to label this usage but ultimately decided against it after receiving comments from philosophers of engineering who find the existing usage not only acceptable but also accurately evocative.
12. I would like to call particular attention to Cech & Waidzunas (2011), Faulkner (2007, 2009), Fujino (1963), Slaton (2010), and Tonso (2007).
13. Thanks to the reviewer who noted that this chapter focuses primarily on North American and European perspectives. I do want to acknowledge the rapidly growing literatures in engineering studies in other parts of the world, especially Latin America and East and South Asia, which have bearing on the making of engineers in and beyond those regions. Complex colonial and postcolonial histories, as well as engagements with Marxist/Leninist/Maoist political philosophies and Eastern political and social philosophies, just for starters, all present the makers of engineers with normative challenges. Although undertaking some of this work myself, I am not yet competent to offer anything approaching a reasonable review of the normativities in such experiences. Sorry.
14. Some text in this section draws from (Downey, 2012b).
15. Thanks to Matthew Wisnioski for calling this quotation to my attention.

References

Agassi, J. (1966, Summer). The confusion between science and technology in the standard philosophies of science. *Technology and Culture, 7,* 348–366.

Alder, K. (1997). *Engineering the revolution: Arms and enlightenment in France, 1763–1815.* Princeton, NJ: Princeton University Press.

Anderl, R., Gong, K., Cai Li, N, Kaminski, P., Netto, M., Kimura, F., . . . Widdig, B. (2006). *In search of global engineering excellence: Educating the next generation of engineers for the global workplace.* Hanover, Germany: Continental AG.

Anderson, K., Courter, S. S., McGlamery, T., Nathans-Kelly, T., & Nicometo, C. G. (2010). Understanding engineering work and identity: A cross-case analysis of engineers within six firms. *Engineering Studies, 2*(3), 153–174.

Avila, J. C., & Arias, M. J. (2007). Current problems in engineering historically rooted in the search for status as a profession. In S. H. Christensen, B. Delahousse, & M. Meganck (Eds.), *Philosophy in engineering* (pp. 369–390). Copenhagen: Academica.

Belhoste, B., & Chatzis, K. (2007). From technical corps to technocratic power: French state engineers and their professional and cultural universe in the first-half of the 19th century. *History and Technology, 23*(3), 209–225.

Bomke, W. (2009). A historical perspective of engineering design. In S. Hyldgaard Christensen, B. Delahousse, & M. Meganck (Eds.), *Engineering in context* (pp. 245–262). Aarhus, Denmark: Academica.

Borrego, M. (2007). Development of engineering education as a rigorous discipline: A study of the publication patterns of four coalitions. *Journal of Engineering Education, 96*(1), 5–18.

Brown, J. K. (2000). Design plans, working drawing, national styles. *Technology and Culture, 41,* 195–238.

Brown, J. K., Downey, G. L., & Diogo, M. P. (2009). The normativities of engineers: Engineering education and history of technology. *Technology and Culture, 50*(4), 737–752.

Bucciarelli, L. L. (1994). *Designing engineers.* Cambridge, MA: MIT Press.

Buchanan, R. A. (1989). *The engineers: A history of the engineering profession in Britain 1750–1914.* London: Jessica Kingsley.

Bunge, M. (1966, Summer). Technology as applied science. *Technology and Culture, 7,* 329–347.

Burr, W. H. (1893, July 31). *The ideal engineering education.* Paper presented at the Society for the Promotion of Engineering Education, Chicago, IL.

Calhoun, D. H. (1960). *The American civil engineer: Origins and conflict.* Cambridge, MA: Harvard University Press.

Callon, M. (1980). The state and technical innovation: A case study of the electric vehicle in France. *Research Policy, 9,* 358–376.

Calvert, M. A. (1967). *The mechanical engineer in America: Professional cultures in conflict.* Baltimore, MD: The Johns Hopkins University Press.

Campagnolo, G. M., & Fele, G. (2010). From software specifics to software-specific vagueness: System engineering and the enterprise architecture software market. *Engineering Studies, 2*(3), 221–243.

Carlson, B. (1988). Academic entrepreneurship and engineering education: Dugald C. Jackson and the MIT-GE Cooperative Engineering course, 1907–1932. *Technology and Culture, 29,* 536–567.

Cech, E., & Waidzunas, T. (2011). Navigating the heteronormativity of engineering: The experiences of lesbian, gay, and bisexual students. *Engineering Studies, 3*(1), 1–24.

Chatzis, K. (2007). Introduction: National identities of engineers. *History and Technology, 23*(3), 193–196.

Chatzis, K. (2009). Coping with the Second Industrial Revolution: Fragmentation of the French engineering education system, 1870s to the present. *Engineering Studies, 1*(2), 77–99.

Christensen, S. H., Delahousse, B., & Meganck, M. (2007). *Philosophy in engineering* (1st ed.). Copenhagen: Academica.

Christensen, S. H., Delahousse, B., & Meganck, M. (2009). *Engineering in context.* Aarhus, Denmark: Academica.

Comité d'Études sur les Formations d'Ingénieurs (CEFI). (2010). Etre ingénieur. Retrieved from http://www.cefi.org/EMPLOIS/EC_JOBS.HTM

Coyle, E., Murphy, M., & Grimson, W. (2007). Engineering science as opposed to applied science and natural science. In S. Hyldgaard Christensen, M. Meganck, & B. Delahousse (Eds.), *Philosophy in engineering* (1st ed., pp. 139–160). Aarhus, Denmark: Academica.

Davis, M. (1996). Defining "engineer": How to do it and why it matters. *Journal of Engineering Education, 85*(2), 97–101.

Davis, M. (2010). Distinguishing architects from engineers: A pilot study in differences between engineers and other technologists. In I. van de

Poel & D. E. Goldberg (Eds.), *Philosophy and engineering: An emerging agenda* (pp. 15–30). New York, NY: Springer.

Delahousse, B. (2007). Industry requirements for the new engineer. In S. H. Christensen, B. Delahousse, & M. Meganck (Eds.), *Engineering in context* (pp. 315–338). Copenhagen: Academica.

de Matos, A. C., Diogo, M. P., Gouzévitch, I., & Grelon, A. (Eds.). (2009). *Les enjeux identitaires des ingénieurs: Entre la formation et l'action/The quest for a professional identity: engineers between training and action.* Lisbon: Colibri.

Divall, C. (1990). A measure of agreement: Employers and engineering studies in the universities of England and Wales, 1897–1939. *Social Studies of Science, 20*(1), 65–112.

Divall, C., & Johnston, S. F. (2000). *Scaling up: The institution of chemical engineers and the rise of a new profession.* Dordrecht, The Netherlands: Kluwer Academic.

Downey, G. L. (1998). *The machine in me: An anthropologist sits among computer engineers.* New York, NY: Routledge.

Downey, G. L. (2005). Keynote address: Are engineers losing control of technology?: From "problem solving" to "problem definition and solution" in engineering education. *Chemical Engineering Research and Design, 83*(A8), 1–12.

Downey, G. L. (2007). Low cost, mass use: American engineers and the metrics of progress. *History and Technology, 22*(3), 289–308.

Downey, G. L. (2009). What is engineering studies for? Dominant practices and scalable scholarship. *Engineering Studies: Journal of the International Network for Engineering Studies, 1*(1), 55–76.

Downey, G. L. (2011). Epilogue: Beyond global competence: Implications for engineering pedagogy. In G. L. Downey & K. Beddoes (Eds.), *What is global engineering education for?: The making of international educators* (pp. 415–432). San Rafael, CA: Morgan & Claypool.

Downey, G. L. (2012a). *Technicians of progress: Dominant images of engineers* (unpublished manuscript).

Downey, G. L. (2012b). Normative holism in engineering formation. In S. H. Christensen, C. Mitcham, I. Bocong, & Y. An (Eds.), *Engineering, development and philosophy: Chinese, American, and European perspectives.* New York, NY: Springer.

Downey, G. L., Donovan, A., & Elliott, T. J. (1989). The invisible engineer: How engineering ceased to be a problem in science and technology studies. *Knowledge and Society* 8, 189–216.

Downey, G. L., & Lucena, J. C. (1994). Engineering studies. In S. Jasanoff, G. Markle, J. Petersen & T. Pinch (Eds.), *Handbook of science, technology, and society*. Newbury Park, CA: SAGE.

Downey, G. L., & Lucena, J. (2004). Knowledge and professional identity in engineering: Code-switching and the metrics of progress. *History and Technology*, 20(4), 393–420.

Durbin, P. T. (1978). Toward a social philosophy of technology. In P. T. Durbin (Ed.), *Research in philosophy and technology: Annual compilation of research* (Vol. 1, pp. 67–97). Greenwich, CT: JAI Press.

Edgerton, D. (1996). *Science, technology and the British industrial 'decline,' 1870–1970*. Cambridge: Cambridge University Press.

Engineering for the Americas. (2010). About EFTA. Retrieved from http://www.efta.oas.org/english/cpo_sobre.asp

European Society for Engineering Education (SEFI). (2005). SEFI Mission Statement 16.09.05 Retrieved from http://www.sefi.be

Faulkner, W. (2007). 'Nuts and bolts and people': Gender-troubled engineering identities. *Social Studies of Science*, 37(3), 331–356.

Faulkner, W. (2009). Doing gender in engineering workplace cultures II: Gender in/authenticity and the in/visibility paradox *Engineering Studies*, 1(3), 169–189.

Fujino, T. (1963). Edo Bakufu. In S. Ienaga (Ed.), *Kinsei 2 [Early Modern 2]* (Vol. 10, pp. 1–55). Tokyo: Iwanami Shoten.

Gainsburg, J., Rodriguez-Lluesma, C., & Bailey, D. E. (2010). A "knowledge profile" of an engineering occupation: Temporal patterns in the use of engineering knowledge. *Engineering Studies*, 2(3), 197–219.

Gasparetto, A. (2007). The status of the engineer in Europe in view of social changes. In S. H. Christensen, B. Delahousse, & M. Meganck (Eds.), *Philosophy in engineering* (pp. 339–352). Copenhagen: Academica.

Gerstl, J. E., & Perrucci, R. (1969). *Profession without community*. New York, NY: Random House.

Gibbons, M., Scott, P., Nowotny, H., Limoges, C., Schwartzmann, S., & Trow, M. (1994). *The new production of knowledge: The dynamics of science and research in contemporary societies*. Newbury Park, CA: SAGE.

Gilbert, A.-F. (2009). Disciplinary cultures in mechanical engineering and materials science: Gendered/gendering practices? *Equal Opportunities International*, 28, 24–35.

Goldman, S. L. (1984). The *Techne* of philosophy and the philosophy of technology. In P. T. Durbin (Ed.), *Research in philosophy and technology: Annual compilation of research* (Vol. 7, pp. 115–144). Greenwich, CT: JAI Press.

Gouzevitch, I., & Inkster, I. (2007). Introduction: Identifying engineers in history. *History of Technology*, 27, 101–106.

Grelon, A. (Ed.). (1986). *Les ingénieurs de la crise: Titre et profession entre les deux guerres*. Paris: L'École des Hautes Études en Sciences Sociales.

Grelon, A. (1993). The training and career structures of engineers in France, 1880–1939. In R. Fox & A. Guagnini (Eds.), *Education, technology and industrial performance in Europe, 1850–1939*. Cambridge: Cambridge University Press.

Grelon, A. (2007). French engineers: Between unity and heterogeneity. *History of Technology*, 27, 107–123.

Indian Society for Technical Education. (2010). President's message. Retrieved from http://www.isteonline.in/index.php

International Federation of Engineering Education Societies (Producer). (2010). IFEES members value propositions. Retrieved from http://www.sefi.be/ifees/wp-content/uploads/IFEES_Members_Value_Propositions.pdf

Japan Accreditation Board for Engineering Education. (2010). About JABEE Retrieved from http://www.jabee.org/english/OpenHomePage/e_about_jabee.htm

Jasanoff, S. (1996). Beyond epistemology: Relativism and engagement in the politics of science. Special issue: The politics of SSK: Neutrality, commitment and beyond. *Social Studies of Science*, 26(2), 393–418.

Jasanoff, S. (Ed.). (2004). *States of knowledge: The co-production of science and social order*. London: Routledge.

Koshland, C. P. (2010). Liberal arts and engineering. In D. Grasso & M. Brown Burkins (Eds.), *Holistic engineering education: Beyond technology* (pp. 53–67). New York, NY: Springer.

Kranakis, E. (1997). *Constructing a bridge: An exploration of engineering culture, design, and research in nineteenth-century France and America*. Cambridge, MA: MIT Press.

Law, J., & Callon, M. (1988). Engineering and sociology in a military aircraft project: A network analysis of technological change. *Social Problems*, 35, 284–297.

Layton, E. T. (1986). *The revolt of the engineers: Social responsibility and the American engineering profession*. Baltimore, MD: The Johns Hopkins University Press.

LeBold, W. K., Perrucci, R., & Howland, W. (1966, March). The engineer in industry and government. *Journal of Engineering Education*, 56, 237–273.

Lenk, H. (1984). Toward a pragmatic social philosophy of technology and the technological intelligentsia. In P. T. Durbin (Ed.), *Research in philosophy and technology: Annual compilation of research* (Vol. 7, pp. 23–58). Greenwich, CT: JAI Press.

Lohmann, J. R. (2005). Building a community of scholars: The role of the *Journal of Engineering Education* as a research journal. *Journal of Engineering Education*, 94(1), 1–4.

Lucena, J. (2005). *Defending the nation: U.S. policymaking to create scientists and engineers from Sputnik to the 'War Against Terrorism'*. Lanham, MD: University Press of America.

Lucena, J. C. (2007). *De criollos a Mexicanos*: Engineers' identity and the construction of Mexico. *History and Technology*, 23(3), 275–288.

Lucena, J. (2009). Imagining nation, envisioning progress: Emperor, agricultural elites, and imperial ministers in search of engineers in 19th century Brazil. *Engineering Studies*, 1(3), 24–50

Lundgreen, P. (1990). Engineering education in Europe and the U.S.A., 1750–1930: The rise to dominance of school culture and the engineering profession. *Annals of Science*, 47, 33–75.

McCullough, D. G. (1972). *The great bridge*. New York, NY: Simon and Schuster.

McMahon, A. M. (1984). *The making of a profession: A century of electrical engineering in America*. New York, NY: IEEE Press.

Meiksins, P. (1988). "The revolt of the engineers" reconsidered. *Technology and Culture*, 29(2), 219–246.

Meiksins, P., & Smith, C. (1996). Engineers and convergence. In P. Meiksins & C. Smith (Eds.), *Engineering labour: Technical workers in comparative perspective* (pp. 256–285). London: Verso.

Moses, J. (2010). Architecting engineering systems. In I. van de Poel & D. E. Goldberg (Eds.), *Philosophy and engineering: An emerging agenda* (pp. 275–284). New York, NY: Springer.

Mutanen, A. (2007). Methodology of engineering science as a combination of epistemic, ethical and aesthetic aspects. In S. Hyldgaard Christensen, M. Meganck, & B. Delahousse (Eds.), *Philosophy in engineering* (1st ed., pp. 123–138). Aarhus, Denmark: Academica.

National Academy of Engineering. (2004). *The engineer of 2020: Visions of engineering in the new century*. Washington, DC: The National Academies Press.

Neeley, K. (2003). Liberal studies and the integrated engineering education of ABET 2000. In *Reports from a planning conference at the University of Virginia, April 4–6, 2002*. Charlottesville, VA: University of Virginia.

Newberry, B. (2009). The dialectics of engineering. In S. Hyldgaard Christensen, B. Delahousse, & M. Meganck (Eds.), *Engineering in context* (1st ed., pp. 33–47). Aarhus, Denmark: Academica.

Noble, D. (1977). *America by design: Science, technology, and the rise of corporate capitalism*. New York, NY: Alfred A. Knopf.

Ollis, D. F., Neeley, K. A., & Luegenbiehl, H. (Eds.). (2004). *Liberal education in twenty-first century engineering: Responses to ABET/EC 2000 criteria*. New York, NY: Peter Lang.

Ottens, M. M. (2010). Limits to systems engineering. In I. van de Poel & D. E. Goldberg (Eds.), *Philosophy and engineering: An emerging agenda* (pp. 109–122). New York, NY: Springer.

Perrucci, R. (1971, March/April). Engineering: Professional servant of power. *American Behavioral Scientist*, 14, 492–506.

Picon, A. (1992). *French architects and engineers in the age of enlightenment*. Cambridge: Cambridge University Press.

Picon, A. (2007). French engineers and social thought, 18th–20th centuries. *History and Technology*, 23(3), 197–208.

Picon, A. (2009). The engineer as judge: Engineering analysis and political economy in 18th century France. *Engineering Studies*, 1(1), 19–34.

Reader, W. J. (1987). *A history of the institution of electrical engineers, 1871–1971*. London: Peregrinus on behalf of the Institution of Electrical Engineers.

Reynolds, T. S. (1983). *75 Years of progress: A history of the American Institute of Chemical*

Engineers, 1908–1983. New York, NY: American Institute of Chemical Engineers.

Robison, W. (2010). Design problems and ethics. In I. van de Poel & D. E. Goldberg (Eds.), *Philosophy and engineering: An emerging agenda* (pp. 205–214). New York, NY: Springer.

Rothstein, W. (1968, October). The American Association of Engineers. *Industrial and Labor Relations Review, 22,* 48–72.

Sanderson, M. (1999). *Education and economic decline in Britain, 1870s to the 1990s.* Cambridge: Cambridge University Press.

Schmiede, R., & Will-Zocholl, M. C. (2011). Engineers' work on the move: Challenges in automobile engineering in a globalized world. *Engineering Studies, 3*(2), 101–121.

Seron, C., & Silbey, S. (2009). The dialectic between expertise knowledge and professional discretion: Accreditation, social control, and the limits of instrumental logic. *Engineering Studies, 1*(2), 101–127.

Sinclair, B., & Hull, J. P. (1980). *A centennial history of the American Society of Mechanical Engineers, 1880–1980.* Toronto: University of Toronto Press.

Sinclair, G., & Tilston, W. V. (1982). The relationship of technology to engineering. In P. T. Durbin (Ed.), *Research in philosophy and technology: Annual compilation of research* (Vol. 5, pp. 87–97). Greenwich, CT: JAI Press.

Skolimowski, H. (1966, Summer). The structure of thinking in technology. *Technology and Culture, 7,* 371–383.

Slaton, A. (2010). *Race, rigor and selectivity in U.S. engineering: The History of an occupational color line.* Cambridge, MA: Harvard University Press.

Smith, C. (1987). *Technical workers: Class, labour and trade unionism.* Basingstoke, U.K.: Macmillan Education.

Steneck, N., Olds, B., & Neeley, K. A. (2002). Recommendations for liberal education in engineering: A white paper from the Liberal Education Division of the American Society for Engineering Education (endorsed by the Liberal Education Division of ASEE). In K. A. Neeley (Ed.), *Liberal studies and the integrated engineering education of ABET 2000.* Charlottesville, VA: Liberal Education Division of the American Society for Engineering Education.

Tonso, K. (2007). *On the outskirts of engineering: Learning identity, gender and power via engineering practice.* Rotterdam, The Netherlands: Sense.

Valderrama, A., Mejía, I., Mejía, A., Lleras, E., Garcia, A., & Camargo, J. (2009). Engineers' identity and engineering education in Colombia, 1887–1972. *Technology and Culture, 50*(4), 811–838.

van de Poel, I., & van Gorp, A. C. (2006). The need for ethical reflection in engineering design: The relevance of type of design and design hierarchy. *Science, Technology, & Human Values, 31*(3), 333–360.

van de Poel, I., & Goldberg, D. E. (Eds.). (2010). *Philosophy and engineering: An emerging agenda.* New York, NY: Springer.

Vincenti, W. G. (1990). *What engineers know and how they know it: Analytical studies from aeronautical history.* Baltimore, MD: The Johns Hopkins University Press.

Vinck, D. (2011). Taking intermediary objects and equipping work into account in the study of engineering practices. *Engineering Studies, 3*(1), 25–44.

Vinck, D., & Blanco, E. (2003). *Everyday engineering: An ethnography of design and innovation.* Cambridge, MA: MIT Press.

Whalley, P. (1986). *The social production of technical work.* Albany, NY: The State University of New York Press.

Wiener, M. J. (1981). *English culture and the decline of the industrial spirit, 1850–1950.* Cambridge: Cambridge University Press.

Williams, R. H. (2002). *Retooling: A historian confronts technological change.* Cambridge, MA: MIT Press.

Wisnioski, M. (2009a). How engineers contextualize themselves. In Steen H. Christensen, B. Delahousse, & M. Meganck (Eds.), *Engineering in context* (pp. 403–416). Aarhus, Denmark: Academica.

Wisnioski, M. (2009b). "Liberal education has failed": Reading like an engineer in 1960s America. *Technology and Culture, 50*(4), 753–782.

Wisnioski, M. (2012). *Engineers for change: Competing vision of technology in 1960s America.* Cambridge, MA: MIT Press.

Zussman, R. (1985). *Mechanics of the middle class.* Berkeley: University of California Press.

CHAPTER 35

Interdisciplinarity in Engineering Research and Learning

Nancy J. Nersessian and Wendy C. Newstetter

Disciplines are distinguished partly for historical reasons and reasons of administrative convenience (such as the organization of teaching and appointments) and partly because the theories which we construct to solve our problems have a tendency to grow into unified systems. But all this classification and distinction is a comparatively unimportant and superficial affair. We are not students of the same subject matter but students of problems. And problems may cut right across the border of any subject matter or discipline.

Sir Karl Popper Conjectures and Refutations (Popper, 1962, p. 67)

Introduction

Moves beyond disciplinary thought and practice abound today. National and international funding agencies are creating, facilitating, fostering boundary crossing and cross-disciplinary synergy and integration as a focal point of their agendas across the sciences, medicine, engineering, humanities, and arts. Research on interdisciplinarity (ID)

as it is practiced in humanities and the sciences is also abundant, ranging from rich case studies of specific instances to bibliometric analyses that aim to map such things as patterns of interaction in scientific fields. To date, however, the research on ID as practiced in engineering fields is scant, both with respect to practice and to education.[1] Yet, as noted in the National Academy of Engineering (NAE) report on the engineer of 2020, the demands of twenty-first century engineering are such that education needs to be redefined starting at the undergraduate level:

The dissolution of boundaries between disciplines such that 'imagination, diversity and capacity to adapt quickly have become essential qualities for both institutions and individuals, not only to facilitate research, but also to ensure immediate and broad-based application of research results related to the environment. To meet these complex challenges as well as urgent human needs, we need to . . . frame integrated interdisciplinary research questions and activities to merge data, approaches, and ideas across spatial, temporal and societal scales. (NAE, 2005, p. 36, quoting AC-ERE, 2003)

In this chapter we focus primarily on ID as it is enacted in engineering research laboratories. Our focus on practice stems from what we call a *translational approach* to transforming engineering education, by which education researchers first investigate the cognitive strategies and learning ecologies as they occur in the practices of a specific field and then translate findings from these investigations into instructional environments using design-based research. This approach infuses the actuality of the cognitive and learning practices found in the engineering workplace into the classroom setting. The goal is to achieve greater parity between the synthetic environment of the classroom (*in vitro*) and the authentic environment outside the classroom (*in vivo*) (Newstetter, Behravesh, Nersessian, & Fasse, 2010). As with Popper's claim earlier, our research supports the position that ID in engineering is *problem-driven*, and, further, that the nature of the problems and the variety of approaches to them require that we differentiate among different kinds of ID practices. A recent National Academies report defines ID as "a mode of research by teams or individuals that integrates information, data, techniques, tools, perspectives, concepts and/or theories from two or more disciplines or bodies of specialized knowledge to advance fundamental understanding or to solve problems whose solutions are beyond the scope of a single discipline or field of research practice" (NAS, NAE, & IM, 2005, p. 26). What is powerful about this statement is that it goes beyond the usual focus on language and communication to highlight the numerous dimensions across which ID integration needs to take place. However, although there is recognition in the literature on ID that there are several forms, policy statements such as this tend to treat it as though it were all of one kind. As developed in later sections, our research supports the need for a more nuanced understanding of the varieties of ID fields, which can be characterized as *multidiscipline*, *interdiscipline*, and *transdiscipline*. In this chapter, each is exemplified with specific cases from engineering.

The structure of the chapter is as follows. Before beginning the analysis that derives from our *in situ* investigations of ID, we begin with a brief survey of the landscape of the main conceptual understandings of ID to provide readers with entre to the literature on ID broadly construed. We then draw on analyses of ID processes in the science studies literature and in our own research to provide some analytic tools for thinking about ID in practice. We then focus on what ID looks like in action by considering the variety of ID in engineering practice, and conclude with implications for learning in engineering education.

Tools for Analyzing Interdisciplinarity

The characterization of ID by the leading contemporary scholar in the field, Julie Klein, resonates with the sentiment expressed nearly thirty years earlier by Popper (see earlier):

> *Interdisciplinarity has been variously defined in this century: as a methodology, a concept, a process, a way of thinking, a philosophy, and a reflexive ideology. It has been linked with attempts to expose the dangers of fragmentation, to reestablish old connections, to explore emerging relations, and to create new subjects adequate to handle our practical and conceptual needs. Cutting across all these theories is a recurring idea, interdisciplinarity is a means of solving problems and answering questions that cannot be satisfactorily addressed using single methods or approaches. (Klein, 1990, p. 196)*

Thus, ID is best understood as a *process* (Klein & Newell, 1996) of problem solving, and there is widespread agreement that the hallmark of ID processes is *integration* (Klein, 1990, 1996; Lattuca, 2001; NAS, NAE, & IM, 2005). Much of the focus of research on ID has been on collaboration in ID teams comprising members coming from different disciplines (see Derry, Schunn, & Gernsbacher, 2005 for a number of case studies), a practice for which, according to Klein (Klein, 2005) World War II was a "watershed" (see also Galison, 1997). Given the focus on integration in most research on ID, *multidisciplinarity* is most often contrasted with *interdisciplinarity* because it is argued to fail to achieve lasting integration of disciplinary components. Individuals come together from different disciplines, work together on a problem, and then return to their disciplinary habits and abodes largely unchanged. The other less widely recognized form of ID is *transdisciplinarity*, which is variously construed in the literature as *trans*cending disciplinary boundaries through a kind of overarching synthesis toward the pursuit of applications (Klein, 2010). Some have also suggested that transdisciplinarity invites a broader range of stakeholders from the public or practitioners recruited to solve an authentic problem (Borrego & Cutler, 2010). In the discussion that follows, we address all three kinds of ID while trying to distinguish to the extent possible among them.

Disciplinary research is often characterized as taking place in "silos" and interdisciplinarity, as moves out of these. How are these moves made? What facilitates interaction and integration? What are the characteristics of the interactions and integrations? Here we introduce some metaphors in the science studies literature on interdisciplinarity that we have found to be useful for articulating and analyzing these dimensions.

Trading Zones

In characterizing the development of microphysics by experimentalists, theorists, engineers, and mathematicians, spurred by various problems that fueled research and development during World War II, historian Peter Galison sought a metaphor that would capture the movement and interactions across boundaries that occurred within these cultures. He called the process he was trying to capture "intercalation": coordination without homogenization, and used the metaphor of "trading zone" to colorfully capture this concept. He found the trading zone metaphor in the thinking of anthropologists and linguists about how communication and exchange of goods of value can take place among radically different communities with no cultural point of reference or language:

> *Two groups can agree on the rules of exchange even if they ascribe utterly different significance to the objects being exchanged; they may even disagree on the meaning of the exchange process itself. Nonetheless, the trading practices can hammer out a local coordination, despite vast global differences. In an even more sophisticated way, cultures in action frequently establish contact languages, systems of discourse that can vary from the most function-specific jargons, through semi-specific pidgins, to full-fledged creoles. (Galison, 1997, pp. 782–783)*

The notion of a *trading zone* designates a bounded, delimited space in between disciplines where trading processes can occur because each participant group needs something from the other to address problems that lead to shared projects and goals. In his analysis of the trading zones, "language" is expanded to mean any structured symbolic system, which can include graphical and mathematical representations. The central metaphor is *exchange*: researchers come together for a period and exchanges take place, and then everyone goes back to where they came from. It is possible that fundamentally new concepts and techniques emerge in the zones that then impact the original disciplines as in the case of Julian Schwinger discussed later. Schwinger took back some of the "pidgin" of the MIT Radiation Lab and used it to great effect. However, emergent disciplines are not the focus of Galison's interpretations, and as we discuss later, trading is not the best metaphor for these. The main point is that transactions take place within the trading zone and everyone goes back to their disciplinary silos, with the disciplines largely unchanged although occasionally individuals have significant impact on their disciplines from this cross-cultural encounter.

Boundary Objects

In their study of how workers at the Museum of Vertebrate Zoology (curators, amateur collectors, professional biologists, occasional field hands, science club members) managed both diversity and cooperation, sociologist Susan Leigh Star and philosopher James Griesemer introduced a notion that has had wide impact on studies of interdisciplinarity: *boundary object*. The example they use is specimens of dead birds which had differing meanings for the intersecting worlds of amateur bird watchers and professional biologists in the context of various problems involved in museum work. They designated as boundary objects:

> . . . *those scientific objects which both inhabit several intersecting social worlds . . . and satisfy the informational needs of each of them. Boundary objects are both plastic enough to adapt to local needs and the constraints of the several parties employing them, yet robust enough to maintain a common identity across sites. They are weakly structured in common use, and become strongly structured in individual use. These objects may be abstract or concrete. They have different meanings in different social worlds but their structure is common enough to more than one world to make them recognizable as a means of translation.* (Star & Griesemer, 1989, p. 391)

A boundary object is an entity (concrete or abstract) that has a complex structure such that it is compatible with more than one interpretation. Parts of that structure meaningfully intersect across the communities concerned with it. The notion of a *boundary object* has been extended to a wide range of instances in the subsequent literature, such as the everglades (conservation science: scientists of various kinds, environmentalists, government agents), soil (soil science: geologists and botanists), hand-drawn sketches (engineers and architects), and cloud chamber traces (particle physics: experimentalists, theorists, instrument makers).

An object or entity is a static notion but one thing boundary objects can do is to lead to the construction of spaces of dynamic interaction between disciplines, which is

how Galison characterizes the trading zone. Thinking about how new spaces can lead to the emergence of ID engineering fields has led our research group to introduce the notion of *adaptive problem spaces* where disciplines intersect and hybridization and other forms of emergence occur. Further, we introduce the notion of *boundary agents* to capture agency of participants in constructing these ID spaces.

Adaptive Spaces and Boundary Agents

Whereas zones and objects are bounded in that they are delimited or constrained spatially and temporally, we think of adaptive problem spaces as finite but unbounded spaces (on analogy with the Einsteinian conception of the physical space of the universe) where problem-driven adaptation takes place in a complex system. In our work, we have come to formulate the notion of adaptive spaces as follows:

> *Adaptation of complex systems is a process of continually revising and reconfiguring the components from which these are built, as these gain experience. Research in adaptive spaces is driven by complex interdisciplinary problems, and these require that the individuals themselves achieve a measure of interdisciplinary integration in methods, concepts, models, materials – in how they think and how they act. Adaptive spaces are distributed in space and time. They are dynamic and diachronic and span mental and material worlds.* (Nersessian, 2006)

Unlike the inhabitants of trading zones who return to their disciplines after working on a problem, researchers and artifacts within the adaptive space become to varying extents hybrid systems and inhabit regions that can themselves give rise to new hybrid disciplines (interdisciplines such a biomedical engineering) or can be more varied (transdisciplines such as integrative systems biology). Although the central metaphor of a trading zone is exchange, the central metaphor of an adaptive space is *emergence*. The people who are forging the adaptive space to advance the processes of interdisciplinary emergence through their activities we designate as *boundary agents*.

The literature on ID has tended to focus more on integration of language, methods, theories, and so forth, with less attention directed toward the individuals who do the integration and the ways ID impacts them as researchers. Understanding the kinds of adaptations and transformations researchers need to undergo to become boundary agents raises issues of cognitive development through learning, identity, and the development of interactional skills suited to the variety of ID practice. The latter skills have been called *interactional expertise* by the sociologist Harry Collins (Collins & Evans, 2002). Although his use of the term focuses on the idea that participants in ID work need, to some extent, to learn the languages of the other discipline(s), we expand the notion to comprise other facets of ID interaction.

Interdisciplinary Engineering in Action

Our research has led us to classify varieties of ID practice in terms of the kinds of engineering these have been producing. In this section we provide examples of ID engineering in research laboratories for each of the varieties: *multidiscipline, interdiscipline,* and *transdiscipline*. In a multidiscipline, participants from disciplines come together in response to a problem, create a local integration to solve that problem, and go back to their respective disciplines, with these largely unchanged by the *transient* interaction. In contrast to the current ID literature surveyed in the preceding text, we cast such multidisciplinary interactions as falling within the category of ID research because problem solutions do require and achieve integration, even if the disciplines themselves are largely not impacted. We have not investigated this form of ID ourselves, but use an interesting historical case as illustration: microwave engineering research within the MIT Rad Lab in which engineers and physicists collaborated on the problem of radar development in World War II.

In our investigations of engineering research laboratories, we have found that an interdiscipline might be thought of as a *hybrid discipline* – one that emerges when the integrative activities of participants move beyond collaboration to create a new hybrid field in which there is *stable and sustainable integration* in concepts, methods, technologies, and materials in the service of addressing an ongoing range of problems. We use biomedical engineering as an exemplar.

The notion of a transdiscipline is harder to articulate, but the basic idea is that researchers draw largely on the knowledge, methods, etc. of a discipline, but address problems that require *penetration* by one or more other disciplines. That is, interactions are likely to mutually effect changes in understanding, methods, and other practices in regions of the participating disciplines that seep into the adaptive space. The case we look at here is integrative systems biology, which involves *interdependence* among researchers in engineering, computing, and biosciences. In this context, the prefix "trans" signifies that this enterprise seeps into, penetrates, specific prior practices of the mother fields and a further emergent problem space opens with multiple possibilities for interaction and integration.

Multidiscipline: Microwave Engineering in the MIT Rad Lab in World War II

This exemplar provides an instance of ID collaboration that was driven by a problem external to the communities and the collaboration was pragmatic and expedient. The problem of creating radar systems for the war effort brought electrical engineers and theoretical physicists together in what came to be known as the MIT Rad Lab. We provide only a brief account because our discussion relies on secondary sources. As Galison analyzes the interactions in the trading zones between physics and engineering, the main interdisciplinary problem was one of translation of the complex mathematics of electromagnetic field theory into a form electrical engineers, accustomed to algebra and circuit language, could understand and use to analyze wave guides, which are long, hollow metal boxes with discontinuities (Galison, 1997). Electrical engineers were by-and-large not familiar with the mathematics of field theory and even for the physicists, the usual method of solving the electromagnetic field equations for all points in the field proved an intractable problem. The physicist Julian Schwinger developed a mathematical notion of "equivalent circuits." Reducing complex electromagnetic field representations to circuit representations with which electrical engineers were familiar greatly simplified the calculations required and enabled the engineers to predict various aspects significant to radar design in advance of constructing the artifact. Thus, exchange was affected by means of simplified diagrammatic representations and equations, which performed as boundary objects, equally meaningful to the physicists and the engineers.

The exchange led to the development of radar, and with success in the trading zone everyone then went back to their disciplinary silos. However, the exchange process also had a significant impact on the thinking of the person we would call the boundary agent, Julian Schwinger. As Galison details, not only did Schwinger's methodological and conceptual innovation figure centrally in the development of radar, but when he left the zone and went back to particle physics he had a new way of thinking that ultimately led to his notion of renormalization in quantum electrodynamics. That is, in developing interactional expertise in electrical engineering, Schwinger also developed a mode of thinking about complex phenomena in terms of minimal structural aspects. Schwinger himself linked the seemingly unrelated domains of radar and QED in a memorial lecture for the Japanese physicist Tomonaga: "The waveguide investigations showed the utility of organizing a theory to isolate those inner structural aspects that are not probed under the given experimental circumstances.... And it is this viewpoint that [led me] to the quantum

electrodynamics concept of self-consistent subtraction or renormalization" (Galison, 1997, p. 826).

Interdiscipline: Biomedical Engineering Research Labs

This exemplar is drawn from our eight-year study of cognitive and learning practices in research laboratories in biomedical engineering (BME). In this case the interdisciplinarity is explicit, reflective, and intentional, with the ultimate aim of stabilizing into an "interdisciplinary discipline" or interdiscipline. Pioneering engineers in the field wanted to move beyond multidisciplinary collaborations to creating the *integrative individual* biomedical engineer. Researchers believe the challenge of biomedical engineering now and in the future to be that the research problems are inherently interdisciplinary, calling for the integration of concepts, methods, materials, models, and so forth into *emergent hybrid systems* within the adaptive problem space of BME. The cases we have examined in some depth come from tissue engineering and neural engineering. Here we discuss salient interdiscipline features these labs have in common and then provide some brief details of a hybrid researcher in tissue engineering.

The BME labs are hybrid engineering and biological science environments. The hybrid nature of these laboratories is reflected in the bioengineered physical simulation model-systems designed, built, and experimented with by the labs. This hybridity is also found in the characteristics of the researcher-students who are part of an educational program designed explicitly to produce individuals who are interdisciplinary, integrative biomedical engineers who can also act in industry and in academia as boundary agents in interaction with collaborators from any of the three disciplines. Research in biomedical engineering often confronts the problem that it is both impractical and unethical to carry out experiments directly on animals or human subjects. In our studies of two pioneering biomedical engineering research laboratories we have found a common investigative practice is to design, build, and experiment by means of *in vitro* systems, which parallel certain features of *in vivo* systems. When biological and engineering components are brought together in an investigation, researchers refer to this as a "model-system." As one respondent stated: "when everything comes together I would call it a 'model-system' [...] I think you would be very safe to use that [notion] as the integrated nature, the biological aspect coming together with an engineering aspect ... " These physical models are hybrid artifacts engineered to capture what researchers deem to be salient properties and behaviors of biological systems (Nersessian & Patton, 2009). They are structural, behavioral, or functional analogs of *in vivo* biological phenomena of interest with engineering constraints that impose simplifications and idealizations unrelated to the biological systems they model. These emergent hybrid objects are not boundary objects, but are integrated artifacts understood in the same way by the community of researchers.

Lab A, in tissue engineering, seeks to design off-the-shelf vascular tissue replacements for the human cardiovascular system. Some intermediate problems that drive the research are: producing "constructs" (blood vessel wall models composed of living tissue that mimic properties of natural vessels); examining and enhancing their mechanical properties; and creating endothelial cell sources through mechanical manipulation of stem cells. Lab D, in neural engineering, seeks to understand the ways neurons learn in the brain and, potentially, to create aids for neurological disabilities. Its intermediate investigations center on finding evidence of plasticity in a "dish" of multi-electrode neuron arrays, and producing controlled "muscle" activity in robots or in simulated agents, all of which constitute their model-systems. Given space constraints we can consider only a brief example from one lab, tissue engineering, which illustrates its nature as an adaptive space in which hybrid BME researchers and artifacts emerge.

Research in Lab A stems from the insight its director had in the early 1970s: "characteristics of blood flow [mechanical forces] actually were influencing the biology of the wall of a blood vessel. And even more than that.... it made sense to me that, if there was this influence of flow on the underlying biology of the vessel wall, that somehow that cell type [endothelial] had to be involved." The central problem became that of understanding the nature of these influences of mechanical force on the vascular biology, codified in the hybrid concept arterial shear: frictional force of blood flow parallel to the plane of flow through the lumen. Lab research is directed toward both fundamental problems, such as of endothelial cell biology, and potential application problems, such as engineering a viable artery substitute. Tackling these problems had led to the development of a number of hybrid model-systems. The processes of creating and using models drive researchers to be integrative interdisciplinary individuals. The design of model-systems incorporates engineering and biological constraints, making them hybrid objects used to simulate the in vivo phenomena of interest and provide sites of experimentation.

The construct model (see earlier) is now the central focus of research in Lab A. The nature of the model changes along various dimensions depending on the constraints of the experiment in which it will be used. For instance, it can be seeded with smooth muscle cells and endothelial cells, or simply the latter, and the components of the collagen scaffolding can vary. At any given time, its design is based on what is currently understood of the biological environment of endothelial cells in cell and vascular biology, the kinds of materials available, and bioengineering techniques thus far developed. Building the construct has led to new hybrid methods for the engineering of living tissue. Once built, a specific construct model can be manipulated by various means as part of an engineered model-system. One form of manipulation is by the flow channel device ("flow loop"), an engineered model of the *in vivo* force of blood flow over the lumen.

The flow loop design is based on the fluid mechanics of a long channel with a rectangular cross-section. Exposing the endothelial cells lining the construct to shear stresses "conditions" the cells, and can be the locus of experiment itself (e.g., relating to cell morphology or gene expression), or just one step in a multi-model process.

A diagram, drawn by the Lab A director in response to our request that he "draw a picture" of the research in his lab, provides a glimpse into the dimensions of the adaptive space of Lab A (Figure 35.1). He mapped not only the problems ("major barriers"), but also the technologies (at the bottom), the relations of researchers to both of these and to one another in that space. This map begins to articulate an adaptive space distributed across problems, methods, technologies, and members; as well as connecting the lab to resources and communities external to it. For example, for Lab A "gene profiling" requires using technology at a nearby medical school. The investigative practice of *in vitro* simulation is deeply implicated in these mappings. To address the "barrier" of "mechanical properties" of endothelial cells *in vivo*, for instance, requires designing and using flow chambers and collagen gel constructs.

Given the hybrid nature of the *in vitro* models, a major learning challenge for these researchers is to develop selective, integrated understandings of biological concepts, methods, and materials and engineering concepts, methods, and materials. By "selective," we mean that a researcher-learner needs to integrate, in thinking and experimenting, only those dimensions of biology and engineering relevant to his or her research goals and problems. For example, in Lab A, researchers need to develop an integrated understanding of the endothelial cell in terms of the stresses of fluid dynamics of blood flow in an artery. Further, in designing and conducting experiments with devices, researchers need to understand what engineering constraints they possess deriving from their design and construction, and what limitations these impose on the simulation and subsequent interpretation

and inferences. That is, the device needs to be understood both as device *qua* model of *in vivo* phenomena and device *qua* engineered model.

Building activities centered on the *in vitro* models are pervasive in the lab and serve several functions. The artifact models connect the cognitive practice of *in vitro* simulation with social practices; for instance, much initial mentoring and learning of laboratory ethos takes place in the context of cell culturing – something all newcomers must master. The processes of building hybrid physical models also provide opportunities for the researcher to build integrated mental representations (Nersessian, 2009). For instance, building a physical construct to condition with the flow loop facilitates building a mental representation that selectively integrates concepts from cell biology and fluid dynamics – one that represents, for example, biological aspects of the endothelial cells with respect to mechanical forces in terms of the integrated concept of arterial shear rate (force of blood as it flows over these cells, causing elongation, proliferation, and so forth).

In experimental situations models tend to be put into interlocking configurations, that is, models stand in particular relations to other models. A brief look at the design and execution of a significant Lab A experiment will provide a means of articulating this dimension of bioengineering integration. Soon after one graduate student (designated A7 in Figure 35.1) arrived (with a background in chemical engineering) she was designated the "*person who would take the construct in vivo,*" meaning that her research was directed toward conducting experiments with an animal that serves as a model for the human body in the context of the experiment. This objective immediately required that she would (1) need to design and build a construct that would both more closely mimic the functional characteristics of an in vivo artery than was used in most other experiments and would have sufficient strength to withstand the force of in vivo blood flow; (2) modify the flow loop so that it would work with constructs in

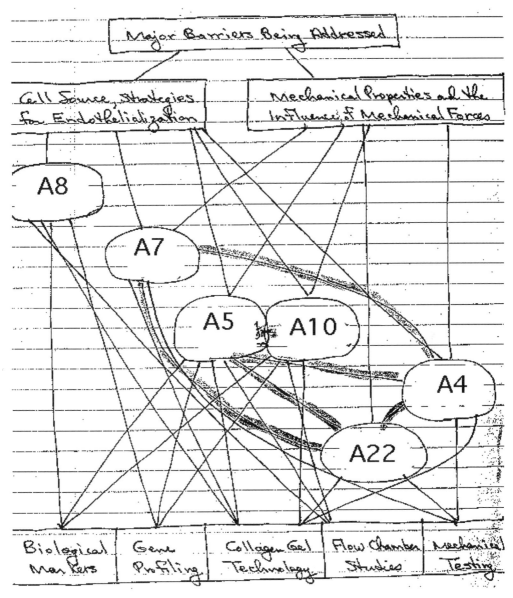

Figure 35.1. The adaptive problem space of the tissue engineering lab at a specific period in time as drawn by the lab director.

tubular form, and (3) arrange for an animal (baboon) to be surgically altered so as to experiment with the construct outside of its body and in a minimally invasive way. It also required that she bring together the strands of research being conducted by nearly all the other lab members, as represented by the lab director in Figure 35.1. As she expressed it, "to go to an in vivo model we have to have all, well most of the aspects that people have studied."

When we started, she had been in the laboratory about a year, but was still in the process of defining the specific goals and problems of her research. Her final overarching formulation of the problem was to determine whether it would be possible to use circulating endothelial cells ("progenitor cells") derived from a patient's peripheral blood to line the vascular graft. The endothelial cells that line the artery are among the most immune sensitive cells in the body. If the patient's own endothelial progenitor cells could be harvested and used that would greatly enhance the potential of a vascular graft. However, the progenitor cells do not modulate thrombosis, which is a function of the mature cells. She hypothesized that shear stress conditioning (by means of the flow loop) the construct before implantation would solve the problem of platelet formation and the resulting thrombosis.

It is instructive to examine her own succinct summary statement as an example of dimensions of hybridization.

> We used the **shunt to evaluate platelet deposition** and that would be – in other words – were **the cells, as a function of the treatment** that they were given before they were seeded onto **the engineered tissue**, able to prevent blood clotting? And so we specifically measured the number of platelets that would sit down on the surface. More platelets equals a clot. So, it ended up being that we were able to look at **the effects of shear stress preconditioning on the cells ability to prevent platelets** and found that it was actually necessary to **shear precondition** these blood derived cells at **an arterial shear rate**, which I used 15 dynes per square centimeter compared to a low shear rate, which in my case I used like 1 dyne per square centimeter, so, a pretty big difference. But I found that **the arterial shear** was necessary **to enhance their expression of anti-coagulant proteins** and therefore prevent clotting. So in other words, **the shear that they were exposed to before going into the shunt** was critical in terms of magnitude, for sure.

The bold terms mark reference both to hybrid interdisciplinary models as they function in her understanding and reasoning and to the various hybrid physical models. To unpack a few of her expressions "the cells" are the endothelial progenitor cells she extracted from baboon blood and seeded onto the "engineered tissue" (vascular construct model). The "treatment" they received was "shear stress preconditioning" conducted by using the modified flow loop model. The objective of her research was to determine if, and at what level, the preconditioning ("arterial shear" simulation)

of constructs would "enhance their [cells] expression of anti-coagulant proteins" ("prevent platelets"). She found, through several iterations of the entire model-system ("used the shunt [animal model] to evaluate platelet deposition"), that the in vivo human arterial shear rate ("15 dynes/cm²") was required for sufficient protein expression ("was critical in terms of magnitude"). Likewise, her designing and experimenting by means of the hybrid baboon model-system led to a revision of her conceptual understanding of the vascular construct so as to reflect the necessity of using arterial shear in order to prevent thrombosis. From this articulate summary of research project and findings, we can detect that this researcher has developed into both an integrative biomedical engineer and a potential boundary agent with interactional expertise for the relevant disciplinary communities, such as medicine.

Transdiscipline: Integrative Systems Biology Research Labs

This exemplar draws from our ongoing investigation of an emerging transdiscipline, integrative systems biology (ISB), which focuses on two labs that self-identify as conducting research in ISB; one that does only computational modeling and the other that does both modeling and experimentation. They are both largely populated by student researchers with engineering backgrounds. The modeling lab has various bioscience external collaborators. The overarching problem of the labs – and the field of ISB in general – is: How to develop a non-reductionist understanding of how multilevel biological systems function? As the modeling lab director stated, "[systems modeling] allows us to merge diverse data and contextual pieces of information into quantitative conceptual structures; analyze these structures with the rigor of mathematics; yield novel insight into biological systems; suggest new means of manipulation and optimization." Our findings are preliminary in this exemplar since we have only been conducting this research for two years.

What is striking is that the various possible configurations for research in this adaptive space are numerous and continue to emerge, and our cases provide only a subset. Still, we have gained some important insights into transdisciplinary engineering research in ISB.

Although there are many different kinds of researchers in this space at present, the aspiration of this field, still in its infancy, might be characterized as addressing the research problems that lie at the intersections of computing/applied mathematics, biosciences, and engineering to create an emergent transdisciplinary space that allows for multiple kinds of adaptations and boundary agents. The goal is integration of novel high throughput technologies, modeling, and experimentation to address biological problems, but the way the field looks at present, researchers in ISB will largely remain in disciplinary fields while working in collaboration with other disciplinary partners. But unlike the multidiscipline, in this case each field in the adaptive ISB problem space will likely penetrate and change significant practices in regions of the collaborating fields. For instance, for the aspirations of the field to succeed, modeling needs will lead to changes in biological practices, for example, with respect to the kinds of data collected; high-throughput technologies that generate reams of data have and will continue to change the practices of both bioscientists and modelers.

Initially, we, along with much that is written about this emerging field, cast the participants as computer scientists/applied mathematicians, biologists, and engineers. So, it was quite interesting to note early on that they tend to identify themselves and other members of the community functionally as "modelers" (those who apply mathematics and develop computational models/simulations) and "experimentalists" (those who conduct bench top experimentation) which we see as already a move into a *transdisciplinary* adaptive space. In our study nearly all the researchers have engineering backgrounds, though many identify themselves as having had a *"generalist"* education. As one participant from Europe stated of her electrical engineering degree: "they learned us to learn, not to learn something." As discussed in the Conclusion section such a generalist background might be important for developing the cognitive flexibility required to become integrative researchers in this field.

Lab G comprises only modelers and is led by a senior pioneer in biological systems modeling. The overarching problem of the lab is to develop rigorous computational models of biological phenomena at the systems level. Most have never done this kind of modeling before entering the lab. The biological problems researchers work on are provided by experimentalists external to the lab. Experimentalists usually contact the lab director asking him to model some data they consider to have potential to benefit from such analysis in areas as varied as biofuels, Parkinson's disease, atherosclerosis, and heat shock in yeast. The fact that they largely depend on bioscience problems that are generated external to the lab has the implication that researchers (who are engineers) have to develop the facility to go deeply into the experimental literature that changes with each modeling project with little course work or bench top experience in biology. The lab does formulate its own research problems in methods development, such as new methods of parameter estimation.

Lab C comprises both researchers who do only modeling and those who do both modeling and experimentation. The director is a young assistant professor who is fully conversant in both modeling and experimental methods. The overarching problem of the lab is to understand cell signaling dynamics in a reduction–oxidation (redox) environment in immunological contexts; in effect, to integrate redox cell biology and biochemistry research through systems modeling. The specific biological problems, which thus far have been chosen by the lab, include immunosenescence and drug resistance in acute lymphoblastic anemia. Their methodological research has largely been in the experimental area, such as the design and development of microfluidic devices to generate high-throughput data for their models. Experimentation in Lab C is directed toward getting parameters needed to develop and validate models that they develop initially from the same kind of literature searching as lab G. Experimentation is conducted either by the student who is building the model, who is developing into a hybrid researcher, or, for the pure modelers, by the lab director or the lab technician, who has an MS in molecular biology.

Unlike BME, which has a relatively unified vision of how research and training should proceed, ISB is experiencing what the Lab G director calls a "philosophical divide." First, with respect to lab structure, Lab G is an instance of what we call a *unimodal lab*. Such a lab can comprise all modelers or all experimentalists, such that the transdisciplinarity manifests as two separate research partners undertaking complimentary but different activities. The philosophy that underlies this research modality is that a lab does the best research if its members are deeply engaged in only one kind of activity. A potential disadvantage of this modality is that each research partner is dependent on the sustained interest and engagement of the other for successful biosystems modeling, despite there being little interaction between them. From the modeler's perspective, there is often a significant phase lag between model building and generation of the needed experimental data, as the Lab director put it, "you need 10 experimentalists for every modeler" and everyone needs "ongoing technical problems to work on so that time is not wasted [waiting for experimental data]." As for advantages of unimodality, our research provides insight into only the modeler's perspective. One advantage is that the development of operating principles and novel theoretical approaches are driven by researchers developing perspectives across different domains and also by the poverty of data.

Lab C is what we characterize as a bimodal lab were bimodality can manifest as either a within-lab collaboration between an experimentalist (or hybrid researcher) and a modeler or by a hybrid researcher who carries out his or her own modeling and experimentation. This approach has the advantage of the lab being able to focus deeply on biological problems of its own choosing and of being able to design its own experiments and collect data as needed in a more timely manner. A potential disadvantage is that the within lab collaborations could prove not to be able to meet the needs of someone who does only modeling (unless the lab has a large number of people engaged in experimentation); another is that it is an open question as to whether the hybrid researchers will be able to develop the requisite level of expertise in both modeling and experimentation. This latter question points to another philosophical divide in the field: how best to train those wishing to become hybrid researchers.

The main divide in training hybrid bimodal researchers is over whether such training should be sequential or parallel. All three possible configurations of training are represented in the labs we have studied. A postdoctoral researcher, who collaborates with Lab G was first trained as an experimentalist and then migrated to learning and doing only modeling. The Lab C director first did only modeling and then nearly five years into her Ph.D. started on experimental work. She believes her student researchers who want to become hybrids need to be trained simultaneously in both. As she said, "I tell my students never to do this [sequential]. You should always do these things in parallel. I ran into the learning curve early graduate students face – only here I was 4.5 years in and starting from scratch on some of these things." The Lab G director believes that students should only be trained in one or the other, because otherwise there is "the problem of diluting both sides" with "modeling lite and experimenting lite." If a person wants to become bimodal, then he or she should train sequentially through a post-doc in the other area.

Although the jury will be out on this divide for some time, it is interesting to compare the perspectives of the two who have recently completed their training. The post-doc in Lab G when asked how he would run his own lab and student training stated, "I lose a lot of time going from one side to the other . . . its more efficient to have a student doing lab work and another dealing with the problems of modeling . . . They should be in the same lab, they need to see each other working." On the other hand, the Lab C student who recently graduated with a dissertation project that combined both said she was concerned about having someone else do her experimental work because they might not "really understand the modeling project can't accurately come up with a good enough experimental protocol to get what it is I need." Of her own experience, she said "I like the idea that as I'm building my model things are popping up in my head on wow this would be a good experiment. I plan out the experiment and then do it. I like the idea I'm being trained to do both so I have enough tools in my toolbox."

General Discussion

In the Rad Lab case we see multidisciplinary interaction within a trading zone where Schwinger was a central boundary agent. The equivalent circuits are the boundary objects. There was integration of concepts and methods specific to the wave guide calculations. We would argue in some cases trading zones can also be adaptive spaces. For instance, in this example one emergent phenomenon was a new way of thinking for Schwinger which he used in a highly productive manner to resolve problems in his discipline. On the engineering side, a new field of microwave engineering emerged. The interaction was driven by problems stemming from a specific situation, World War II, and collaboration was not driven by internal problems originating in the disciplinary fields. Multidisciplinary interactions are often serendipitous, related to a specific problem, and transient. Once the problem has a satisfactory solution (or proves insoluble) the collaboration ends and the practices of the participating disciplines are largely unaltered, even when a new field might have been spun off.

In the BME lab cases there is emergent hybridization in an adaptive space initiated by pioneering engineers who participated in collaborations and thus acted as boundary agents in the early days of the field. To create the emergent hybrid systems of thought, methods and materials they believed would move the field forward required a different model of research than that of two researchers from different disciplines collaborating. BME's answer has been to design a different kind of researcher – individuals, who might be considered themselves as "hybrid systems." The integrative biomedical engineer is both a self-sufficient ID researcher and prepared to be a boundary agent able to collaborate with researchers in other disciplines. Over time this ID field is transforming into an interdiscipline that integrates elements of all three original disciplines, and creates individuals who identify as hybrid biomedical engineers.

In the ISB lab cases there is interpenetration of disciplines that are mutually effecting changes as well as various kinds of emergent adaptations at their intersections. This is a newly configuring adaptive space and how its research practices and researchers will evolve is quite open at present. Although there are fully hybrid individuals emerging in this space, the current aims of the participants seem likely to make this exception, rather than the rule. The implication of this is that developing interactional expertise is crucial to functioning as boundary agents in this adaptive space. From our research thus far, it appears that the requisite interactional expertise is not sufficiently developed for what one researcher called "synergistic" collaboration in at least three ways. First, model building begins with the development of a pathway for the biological phenomenon under investigation. The pathway performs as a boundary object in that it is a meaningful representation for both experimentalists and modelers. At present it is the primary means of communication of research results and ideas between collaborators. However, our investigation indicates that the model needs to become a boundary object because it is the vehicle of integration and the engine that is pushing systems understanding forward. Experimentalists need to understand how the data will be used in order to conduct experiments that will provide sufficiently informative data. Second, modelers need to develop experimental understanding at the bench top level in order to know enough about experimental design and execution to have a realistic sense of such things as what is experimentally feasible, the reliability of the data, and the costly and time consuming nature of experimentation. Finally, all participants need a basic systems understanding as provided by engineering fields.

Conclusion: Implications for Learning

Given the varieties of interdisciplinarity illustrated in these case studies, how can we best prepare engineering students to participate in these interdisciplinary configurations? Are there certain pedagogical configurations that best support the development of interactional expertise making it possible for teams to find common ground, to identify and leverage boundary objects and to more smoothly exchange information and intent toward reaching a commonly valued goal? And where do we situate these learning experiences in the overall curriculum?

We contend that while each variety of interdisciplinarity implies a particularized pedagogic approach, in all cases, students need to develop what has been called *cognitive flexibility* (Spiro, Feltovich, Jacobson, & Coulson, 1992) or the ability and knowledge to engage a problem domain, an object or a representation from more than one perspective. We see this as the ability to adapt in interdisciplinary problem spaces, becoming boundary agents in problem situations that require them, while also leveraging boundary objects with their intersecting social worlds/meanings. Developing this ability requires students to work in complex, ill-structured knowledge domains not simple, well-structured ones. Problem-driven learning experiences are particularly appropriate in developing cognitive flexibility because such experiences situate students in real-world, complex situations that require teams to work together to achieve a goal and possibly to pursue and evaluate multiple routes and solutions. To date there has not been much research conducted on designing learning environments targeted specifically toward creating interdisciplinary engineers (Richter & Paretti, 2009). Here, we offer an example from our own institution of three different learning environments where students can practice a specific form of interdisciplinarity.

Bio-inspired Design: A Multidisciplinary Experience

In recent years, engineers have started to look to biology as a source and inspiration for design solutions, on the assumption that evolutionary adaptation has produced simple but elegant solutions to complex problems in the natural world. However, translating between the descriptive world of biology and the quantitative systems world of the engineer is challenging. Like the trading zone on an African river, these "tribes' do not share a common language even though they may share common concepts. Developing the kind of interactional expertise that makes it possible to span the boundaries of these disciplines is of paramount importance. Working with this concept, the bio-inspired design course seeks to develop individuals who are able to translate biological solutions into engineered designs. To do this, students from biology, engineering and industrial design work on two team projects over the term towards developing the interdependent skills of interactional expertise and analogical reasoning (Vattam & Goel, 2011; Vattam, Helms, & Goel, 2010) that will enable them to find commonalities across the disciplines. Analogical reasoning entails looking beyond surface features or application of a given object in one domain for a deeper structure than can be mapped or translated into another domain or application.

Such reasoning facilitates exchange across disciplinary borders.

Leveraging the idea of a boundary object, teams first are invited to identify a biological solution in the natural world and then to translate that into a solution in the engineering world (solution-driven design). In this first case, the design in nature can be viewed from the biological perspective (evolutionary adaptation) but also the engineering perspective (function and form). The second project works in the opposite direction where an engineering problem is identified and the team seeks a biological solution that can be applied. The multidisciplinary teams work together on both projects learning to translate from one perspective/world to the others while practicing looking at the same object from another perspective. Whereas, the first project brings the engineer and design student deeply into the world of biology, the second brings the biologist into the world of the engineer and designer. Learning to parse a problem solution from biological, engineering, and industrial design perspectives promotes the kind of cognitive flexibility indicative of boundary agents and those able to barter in trading zones. This educational model for promoting a multidisciplinary practice can also be found in a number of capstone design experiences (Adams, Beltz, Mann, & Wilson, 2010; Adams, Mann, Forin, & Jordan, 2009; Adams, Mann, Jordan, & Daly, 2009; McNair, Newswandera, & Borrego, 2011) as well as community-based service projects.

Problems in Biomedical Engineering: An Interdiscipline Experience

Biomedical engineers need to be true integrative, hybrid thinkers and problem-solvers if they are to utilize engineering analysis and methods to design healthcare solutions. The *model-based reasoning* exemplified in the development of in vitro device discussed in the preceding text depends on the ability to simultaneously view an object or application or representation from multiple dimensions and perspectives (Nersessian, 2002, 2008; Nersessian & Patton,

2009). This integrative ability needs to be practiced repeatedly over time and in a variety of circumstances. Because learners need multiple opportunities to practice this integration, developing a biomedical engineer is not about a single course, but a total curriculum systematically designed to foster flexible, responsive model-based problem solving. The *Problems in Biomedical Engineering* course was developed using a translational approach whereby the design principles for the course were derived from our ethnographic investigations of learning in bioengineering research laboratories (Newstetter, 2006; Newstetter et al., 2010). The goal was to create an adaptive space by replicating some of the kinds of (authentic) activities undergraduates and graduate students undertake in the research labs to the extent possible in a (synthetic) classroom.

Learning in this class is driven by the need to solve three complex, interdisciplinary problems over the term. The problem sequence is set but each problem can be changed sufficiently on the surface so that every term is different. The problems all require teams of eight students to integrate knowledge and skills from biosciences and engineering in arriving at a problem solution related to a medical context. As an example, the first problem focuses on the challenges of screening for disease. The team needs to evaluate current screening technologies for a given cancer and then make recommendations for future screening protocols based on research using peer-reviewed science and medical and engineering articles. Overall, the problem requires the integration of cancer biology, probability statistics, and screening technologies across the molecular to the whole body scale. Often cost–benefit analyses and social issues become part of the problem solution. To support this integration and complex problem solving, teams work in specially designed 10×10 classrooms with writable walls which they use to represent, explain, and speculate individually and as a group. They also benefit from interacting with a faculty or post doc facilitator who makes his/her reasoning and problem solving strategies more "visible" through asking probing questions just as a lab director would do with lab members. The goal of this course is for students to start the process of learning to integrate skills, knowledge, methods, and representations from the sciences and engineering toward solving real-world problems, practices that define what it means to be interdisciplinary.

Three Possible Models: A Transdisciplinary Experience

Preparing students for transdisciplinary practice implies a very different educational scenario from the two above. First, our studies suggest that graduate school is where this needs to happen. We have found that the researchers who inhabit this world and who claim identities as either modelers or experimenters have sophisticated skills and knowledge already, which they bring to bear on the lab problems. This deep disciplinary training gives them what they need to begin their graduate work. At the same time, they commonly have blind spots to the needs, values, or constraints of the other camp. Modelers need a certain kind of data, which the experimenters may not value and so they will not take the time to perform the experiments. At the same time, the experimentalist may see the modeler as just reproducing her study *in silico*, which is not particularly interesting or relevant. These misalignments can lead to a certain stereotyping one of the other, which is counterproductive. We offer three different models for creating adaptive spaces to bring greater alignment and understanding to the modeler/experimenter configuration.

A first remedy followed by Lab G was a temporary summer excursion into the other camp. Two graduate student modelers from engineering backgrounds spent two months learning experimental procedure, conducting their own experiments and collecting data. In doing so, they have begun to develop an appreciation for the challenge of gathering data both from the time and expense perspective. They also developed the ability to read papers with enhanced understanding of techniques, equipment, and procedures used in lab work. Another benefit was the confidence the modelers developed in talking directly with the experimenters (interactional expertise) from spending some intense time at the bench top. On the flip side, if experimentalists were to spend time with modelers, they could better see the possibilities of modeling for prediction, speculation, and experimentation. They could also better understand why the modeler needs certain kinds of data. Further, the experimentalist could become a better consumer of modeling papers for his or her own work.

A second model would be the design of a collaborative laboratory-based graduate course that paired modelers with experimenters. The task would be to address a problem that could benefit from both approaches working in tandem. The need for interactional expertise would be obvious as the team traversed and engaged the same task from two different methodological and epistemological perspectives.

A final model would be an integrative modeling course where experimenters and modelers were again paired to address a problem area, for example, a disease like cystic fibrosis. The experimenter could be very helpful in finding resources for the model and translating them for the modeler. Likewise, the modeler would need to articulate her needs in a way that would allow the experimenter to be a resource. These educational models do not advocate for developing fully hybrid (bimodal) researchers, but rather for individual adaptation of the kind that creates symbiosis or mutualism, where both come to appreciate and see the value of the practices of the other. Such educational experiences can promote the kind of cognitive flexibility and interactional expertise for the transdisciplinary space where values, practices, and epistemologies differ.

Acknowledgments

We gratefully acknowledge the support of the U.S. National Science Foundation grants

REC0106733, DRL0411825, and DRL0909971 in conducting this research. Our analysis derives from research conducted with Elke Kurz-Milcke, Lisa Osbeck, Ellie Harmon, Christopher Patton, Vrishali Subramanian, and Sanjay Chandrasekharan. We thank the members of the research labs for allowing us into their work environment, letting us observe them, and granting numerous interviews. Finally we appreciate the comments of the anonymous reviewers of this chapter and the editors for their helpful recommendations.

Footnote

1. An exception on practice is the chapter on civil engineering by Culligan and Pena-Mara in the recent *Oxford handbook of interdisciplinarity* (Frodeman, Klein, & Mitcham, 2010, pp. 161–174), which provides a valuable resource for developing a broad understanding of the current "terrain" of ID research. Notable exceptions on education are a survey of the current landscape with respect to undergraduate education (Lattuca, Trautvetter, Codd, Knight, & Cortes, 2011) and research on undergraduate design teams (see, e.g., Adams, Beltz, Mann, & Wilson, 2010; Adams, Mann, Forin, & Jordan, 2009; Adams, Mann, Jordan, & Daly, 2009; McNair, Newswandera, Chad, & Borrego, 2011).

References

Adams, R. S., Beltz, N., Mann, L., & Wilson, D. (2010). Exploring student differences in formulating cross-disciplinary sustainability problems. *International Journal of Engineering Education*, 26(2), 324–338.

Adams, R. S., Mann, L., Forin, T., & Jordan, S. (2009). *Cross-disciplinary practice in engineering contexts*. Paper presented at the 17th International Conference on Engineering Design (ICED'09), Stanford University.

Adams, R. S., Mann, L., Jordan, S., & Daly, S. (2009). Exploring the boundaries: Language, roles, and structures in cross-disciplinary design teams. In J. McDonnell & P. Lloyd (Eds.), *About designing: Analysing design meetings*. London: Taylor and Francis.

Borrego, M., & Cutler, S. (2010). Constructive alignment of interdisciplinary graduate curriculum in engineering and science: An analysis of successful IGERT proposals. *Journal of Engineering Education*, 99(3), 355–369.

Collins, H. M., & Evans, R. J. (2002). The third wave of science studies: Studies of expertise and experience. *Social Studies of Sciences*, 32(2), 235–296.

Derry, S. J., Schunn, C. D., & Gernsbacher, M. A. (Eds.). (2005). *Interdisciplinary collaboration: An emerging cognitive science*. Mahwah, NJ: Lawrence Erlbaum.

Frodeman, R., Klein, J. T., & Mitcham, C. (Eds.). (2010). *The Oxford handbook of interdisciplinarity*. New York, NY: Oxford University Press.

Galison, P. (1997). *Image and logic: A material culture of microphysics*. Chicago, IL: University of Chicago Press.

Klein, J. T. (1990). *Interdisciplinarity: History, theory, and practice*. Detroit: Wayne State University Press.

Klein, J. T. (1996). *Crossing boundaries: Knoweldge, disciplinarities, and interdisciplinarities*. Charlottesville: University Press of Virginia.

Klein, J. T. (2005). Interdisciplinary teamwork: The dynamics of collaboration and integration. In S. J. Derry, C. D. Schunn, & M. A. Gernsbacher (Eds.), *Interdisciplinary collaboration: An emerging cognitive science* (pp. 51–84). Mahwah, NJ: Lawrence Erlbaum.

Klein, J. T. (2010). A taxonomy of interdisciplinarity. In R. Frodeman, J. T. Klein, & C. Mitcham (Eds.), *The Oxford handbook of interdisciplinarity* (pp. 15–30). New York, NY: Oxford University Press.

Klein, J. T., & Newell, W. (1996). Interdisciplinary studies. In J. Graff & J. Ratcliffe (Eds.), *Handbook for the undergraduate curriculum* (pp. 393–415). San Francisco, CA: Jossey-Bass.

Lattuca, L. R. (2001). *Creating interdisciplinarity: Interdisciplinary research and teaching among college and university faculty*. Nashville, TN: Vanderbilt University Press.

Lattuca, L. R., Trautvetter, L. C., Codd, S. L., Knight, D. B., & Cortes, C. M. (2011). *Promoting interdisciplinary competence in the engineers of 2020*. Paper presented at the Annual Meeting of the American Society for Engineering Education, New Orleans, LA.

McNair, L. D., Newswander, C., Boden, D., & Borrego, M. (2011). Student and Faculty Interdisciplinary Identities in Self-Managed

Teams. *Journal of Engineering Education*, 100(2), 374–396.

National Academy of Engineering (NAE). (2005). *Educating the engineer of 2020: Adapting engineering education to the new century.* Washington, DC: The National Academies Press.

National Academy of Sciences (NAS), National Academy of Engineering (NAE), & Institute of Medicine (IM). (2005). *Facilitating interdisciplinary research.* Washington, DC: The National Academies Press.

Nersessian, N. J. (2002). The cognitive basis of model-based reasoning in science. In P. Carruthers, S. Stich, & M. Siegal (Eds.), *The cognitive basis of science* (pp. 133–153). Cambridge: Cambridge University Press.

Nersessian, N. J. (2008). *Creating scientific concepts.* Cambridge, MA: MIT Press.

Nersessian, N. J. (2006, October). *Boundary objects, trading zones and adaptive spaces: How to create interdisciplinary emergence.* NSF Science of Learning Centers Address.

Nersessian, N. J. (2009). How do engineering scientists think? Model-based simulation in biomedical engineering research laboratories. *Topics in Cognitive Science*, 1, 730–757.

Nersessian, N. J., & Patton, C. (2009). Model-based reasoning in interdisciplinary engineering: Two case studies from biomedical engineering research laboratories. In A. Meijers (Ed.), *Philosophy of technology and engineering sciences* (pp. 678–718). Amsterdam: Elsevier.

Newstetter, W. C. (2006). Fostering integrative problem solving in biomedical engineering: The PBL approach. *Annals of Biomedical Engineering*, 34(2), 217–225.

Newstetter, W. C., Behravesh, E., Nersessian, N. J., & Fasse, B. B. (2010). Design principles for problem-driven learning laboratories in biomedical engineering education. *Annals of Biomedical Engineering*, 38(10), 3257–3267.

NSF Advisory Committee for Environmental Research and Education. (2003). *Complex environmental systems: Synthesis for earth life and society in the 21st century.* Washington, DC: National Science Foundation.

Popper, K. R. (1962). *Conjectures and refutations.* New York, NY: Basic Books.

Richter, D. M., & Paretti, M. C. (2009). Identifying barriers to and outcomes of interdisciplinarity in the engineering classroom. *European Journal of Engineering Education*, 34(1), 29–45.

Spiro, R. J., Feltovich, P. L., Jacobson, M. J., & Coulson, R. L. (1992). Cognitive flexibility, constructivism, and hypertext: Random access for advanced knowledge acquisition in ill-structured domains. In D. T. M. & D. Jonassen (Eds.), *Constructivism and the technology of instruction: A conversation.* Hillsdale, NJ: Lawrence Erlbaum.

Star, S. L., & Griesemer, J. G. (1989). Institutional ecology, 'translations' and boundary objects: Amateurs and professionals in Berkeley's Museum of Vertebrate Zoology, 1907–39. *Social Studies of Science*, 19, 387–420.

Vattam, S., & Goel, A. (2011). *Model-based tagging: Promoting access to online texts on complex systems for interdisciplinary learning.* Paper presented at the 11th International Conference on Advanced Learning Technologies, Athens, GA.

Vattam, S., Helms, M., & Goel, A. (2010). A content account of creative analogies in biologically inspired design. *Artificial Intelligence for Engineering Design, Analysis and Manufacturing*, 24, 467–481.

Engineering at the Crossroads

Implications for Educational Policy Makers

John Heywood

Introduction

In a recent report Sparks and Waits (2011) for the U.S. National Governors Association (NGA) faced the association with a crossroads in the development path of higher education. They suggested that the governors should make a radical change in direction. They recommended that the states should redirect their support away from traditional four-year programs of general education toward the provision of courses that would "better prepare students for high paying, high demand jobs" (p. 22). At the same time they should also consider the provision made for shorter programs. To determine the requirements for these courses the states should take much more notice of the comments of employers than they had in the past. Although these proposals result from a review of labor market trends in the United States there is no reason to believe that they are not representative of trends among the labor markets of the industrialized nations. In any case it is clear that higher education is facing a crisis of enormous proportions that some believe will be solved only by

revolutionary changes in the structure of higher education and the curriculum in particular, irrespective of developments in educational technology.

It seems that the cliché that individuals will have several if not many changes of job during their lifetimes is becoming a truism. If that is the case, then the relationships among employers, employees, and society will have to change.

Whether the substantial paradigm shift proposed by Sparks and Waits is evolutionary or revolutionary, it has reignited a debate between two opposing philosophies about the relative merits of liberal/ general education on the one hand, and on the other hand, job-oriented (vocational) education.[1] Such discussion provides an international opportunity to open a fundamental debate about the aims of education in a technological age as well as the curriculum structures for achieving those aims. Hence it is intended to examine the role that engineering and technological literacy can play in reconciling these two philosophies, and therefore contribute to the liberal and vocational education of the non-engineering workforce, and to outline a curriculum model to achieve the aims that this requires. It is argued that although the activity of engineering is inherently professional (vocational), preparation for that activity requires that students be able to perceive the relationships between many areas of knowledge that the activity of engineering forces them to perceive.

Engineering is the art and science of making things that meet the needs of self and society. It is both an activity and a system that serves both individuals and society and creates new problems for both. Therefore, engineering literacy is necessarily interdisciplinary and a liberal study. Engineering literacy is about the process of engineering whereas technological literacy is about the products of engineering and their impact on society (Krupczak et al., 2012). Through the production of a design a student is brought face to face with the social purposes and consequences of engineering through the technologies it creates, the practice of manufacturing, the management of people, and the personal transferable skills required by individuals that are demanded by continually changing patterns of work.

Utilitarian Education for the Creation of Economic Capital

The model presented in the NGA report derives from a utilitarian view of education. For example, the systems of education in the British Isles are largely based on this philosophy. Universities offer a large range of specialisms. Depending on the system, most students study one of these specialisms for three (or four) years without reference to any other subject.[2] There is no concept of the unity of knowledge. In contrast to the United States, there is no such thing as a liberal arts curriculum. STEM (science, technology, engineering, and mathematics) subjects are accorded more status than the humanities, at least in terms of government expenditure. Universities have no unity of purpose other than that which is economic.

One consequence of this view is that an in-depth debate about the aims of education is considered unnecessary. Another and equally important consequence of specialist study in Western society is that it has become isolated from the very subjects that are causing substantial change in social structure, beliefs and values, namely engineering and technology. A specialist education deprives individuals of the unified view of knowledge they need for living and working in an increasingly complex society, a point recognized by many engineering educators. Equally, a general education intended to help a person to become a "rounded" individual cannot be said to be truly liberal and achieve that goal if it does not embrace engineering/technological literacy. It is argued that such liberal education is, at the same time, more likely to meet the work goals of society as the patterns of employment and living change.

Changing Patterns of Employment

Certain assumptions have accompanied this utilitarian view of education.[3] One is that there is a shortage of highly qualified technological personnel. Another is that the schools do not do enough to attract students into STEM courses (e.g., *First Bell*, 2012). At the same time the view that there is no shortage of highly qualified scientific and technological personnel has was put forcibly to a committee of the U.S. House of Representatives by M. S. Teitelbaum, a program director at the Alfred P. Sloan Foundation. He said at a conference on the U.S. Scientific and Technical Workforce "the supposed causes are weaknesses in elementary, secondary, or higher education, inadequate financing of the fields, declining interests in science and engineering among American students, or some combination of these. Thus it is said that the United States must import students, scientists, and engineers from abroad to fill universities and work in the private sector-though even this talent pool may dry up eventually as more foreign nationals find attractive opportunities elsewhere"

(Teitelbaum, 2007 in Kelly, p. 12. See also Kourdi, 2010; Lowell & Saltzman, 2007). But Teitelbaum went on to argue that such data that were available were weak and often misinterpreted (Lowell & Saltzman, 2007). There was no evidence for a shortage of qualified personnel, and in a submission to a subcommittee of the House of Representatives Teitelbaum argued that "despite lawmakers being told by corporate lobbyists that R & D is being globalized in part due to shortages of scientists in the US no one who has studied the matter with an open mind has been able to find any objective data of such general shortages." He concluded with the view that, "Federal policy encourages an over production of science professionals" (cited in *First Bell*, 2011a). It has created its own system of vested interests. Such views are highly contentious, especially when they take on the establishment, and cognitive dissonance can work on both side.[4] Nevertheless the federal government in the United States (and other governments) consider there is a shortage of science, technology, engineering, and mathematics (STEM) people and so continue to pour funds into STEM education (*First Bell*, 2011b, 2012). Those who take that view would have to explain why there is a rise in the unemployment of engineers in their middle age (Wadhwa, 2011). Not withstanding either view, there is clearly a need for very refined information about what engineers do and what knowledge they use, especially in small organizations.

Support for the position taken by Sparks and Waits is also to be found in a report published at the time of writing by the Boston Consulting Group (2012). Their investigators found that the so-called skills gap does exist but it is much less severe than many believe. They say it is less than 1% of the nation's manufacturing workers, and less than 8% of its 1.4 million highly skilled manufacturing workers. Although the term industry engineer is used, it is not defined and there are no data on them; the Boston Consulting Group considers that most of the high-skill manufacturing jobs require only a high school education in conjunction with on-the-job training. No information is given on the survival times of the jobs in question or of the need for training development. The average age of the manufacturing workforce is fifty-six and will soon have to be replaced. The same problem of training for life, which is a principal consideration of this chapter, applies as much to this workforce as it does to those seeking higher qualifications. Thus the comments on the NGA report about shorter programs have relevance because they show the possibilities of career development and give some indication of what an industry engineer might do (see the next paragraph). Unfortunately, this survey is based on firms with more than $1 billion annual sales, so we continue to know little about the small and medium sized firms on which most economies depend. Whatever else is said about the manpower controversy, it seems undeniable that in the future workers will require an education that helps them to be more flexible and adaptable than they have been in the past.

Some current thinking in the United States is of the view that if the drive to develop advanced manufacturing is to succeed there will be a demand for "technicians who have a grounding in math and science" (Matthews, 2011, p. 6). Support for this view will be found in Washington State's Assessment of Education Credentials and Employer Needs. Eleven Centers of Excellence have been established by the state in two-year colleges. The occupations for which skills standards have been developed are all for varying grades of technician and craftsman (Sparks & Waits, 2011, pp. 20–23). And, in respect of manufacturing, the State of Minnesota has established a career and education pathways for manufacturing and applied engineering workers that can bring them as far as middle management on the one hand and on the other hand an M.S. degree (Sparks & Waits, 2011, pp. 22–27). More generally, the U.S. Department of Commerce has a new program that will invest $2 billion in community colleges, and in the U.K. technical college universities have been established that specialize in

Table 1. Top areas of job growth in the next six years simplified but ranked in order of demand

Taking care of people
Making computers work
Taking care of business
Building and maintaining our Infrastructure
Teaching children
Designing things: solving problems
Keeping businesses running
Selling goods and providing basic services

Adapted from Waldock, C. Closing America's job gap. US Bureau of Labor Statistics. Cited in Sparks, E., & Waits, M. J. (2011). *Degrees for what jobs? Raising expectations for universities and colleges in a global economy.* Washington, DC: National Governors Association.

technical studies for fourteen- to nineteen-year-olds.

The Organisation for Economic Co-operation and Development (OECD) suggests that 80% of the decline in so-called "labour share," which measures wages as a proportion of total income generated in an economy, is due to automation and computerization. According to the president of the Illinois Community College Trustees Association Barbara Oilschlager, 41% of jobs will be at the middle level, requiring more education than high school but less than a bachelor's degree (*First Bell*, 2010). In the United Kingdom this would be called technician level education applied across business and technology, with distinctions being made between two levels of technician in engineering: those requiring one or two years beyond high school (engineering technicians) and those requiring a basic degree, for example, engineering technology (Incorporated Engineers in the United Kingdom[5]).

In the United States, at the top of the list of areas of employment growth shown in Table 1, jobs in the area of "taking care of people" have the greatest potential for growth. At least half of the jobs in "making computers work" will not require degree-level qualifications. It is evident that patterns of employment are changing radically, a view that is supported by analyses of the likely areas of maximum growth in the United States during the next six years.

Of course it may not be true of other countries at their particular stage of development. Nevertheless, changing technology is a major driver of social and occupational change from which few nations are likely to be free.

Engineering Employment and Employer Responsibilities

Related to employment in the software industry is the "E mail" column of the November 2011 issue of *ASEE Prism*. It contains an exchange of letters between Professor Allen Plotkin and columnist Vivak Wadwha about an article that Wadwha had written in the September issue of the magazine. He had asked: Why should a company pay a 40-year-old engineer a considerable salary if it can get the same job done much more cheaply by an entry level employee? He said that this was happening in the software industry. "After all the graduate is likely to have more up-to-date skills and will work harder." He went on to say that "if you listen to the heart-wrenching stories of older engineers" (who have become unemployed), "you learn that they have a great many skills, but no one wants to hire them" (p. 32). Professor Plotkin questions whether or not anyone would want to work in an industry that treats its workers in the ways described in the article. Nevertheless it seems there is a serious unemployment problem among middle aged and older engineers in some sectors of the United States. In the same vein G. Paschal Zachary (2011), writing in *IEEE Spectrum*, said that often emerging technologies require far fewer workers. The new titans of Silicon Valley employ far fewer workers than the older titans and this is likely to apply equally to their offshore establishments. At the same time, some emerging technologies destroy jobs. He also draws attention to the phenomenon of "jobless" innovation. This occurs when an innovation is off-shored to countries where qualified manpower is much cheaper to employ. Zachary goes on to ask "How can Americans capture more of the employment associated with job expanding innovations? They can start by examining their faith in the traditional equation of technological innovation with healthy markets." None of this is to say that there will not be specific shortages of professional workers or that many will have to be replaced as the experience of Germany shows (Blau, 2011; Schneiderman, 2010).

Wadwha's (2011) response to the data is to cite the metaphor of a roller-coaster and suggest that the universities need to prepare students for that ride so that when the need arises they are able and interested to change jobs. Hence the need to take the concept of life-long learning more seriously and to design courses of continuing professional development that support engineers on that roller-coaster. Such programs are likely to be as much about personal development as they are about specific topics in engineering. But they will have to be as much the responsibility of the employer as they are the employee. More importantly, higher education will have to provide the basis for that flexibility.

Industrial Criticisms of Higher Education in the United Kingdom and United States

The one similarity between the NGA report and U.K. policy is that for twenty or more years the U.K. government has accepted employer complaints about graduates that irrespective of their field of study they are not suitably prepared for work in industry. But, and it is an important "but," what they were said to lack were what came to be called "personal transferable skills." Under the heading of "enterprise learning" the Employment Department listed four categories. These were:

1. Cognitive knowledge. Key concepts of "enterprise learning" (accounting, economics, organizational behavior, inter and intrapersonal behavior) and skills (i.e., the ability to handle information,

evaluate evidence, think critically, think systematically [in terms of systems], solve problems, argue rationally, and think creatively).

2. Social skills (e.g., working with others in varied roles including as leaders and as team members, and communicating with others).

3. Managing one's self (includes initiatives, independence, risk-taking, achieving, willingness to change, adaptability, and knowing oneself and one's values).

4. Learning to learn (includes knowing how one learns in different contexts and being able to deploy a range of appropriate styles of learning) (cited in Heywood, 1994).

These personal transferable skills have many similarities with the needs listed by the State of Minnesota Office of Higher Education and summarized in Sparks and Waits (2011). A rearranged and simplified list of the top skills required by Minnesota employers is shown in Table 2. It can be inferred from these lists that the affective domain and personal qualities are likely to be as important as the cognitive. They are not skills that separate the academic from the vocational, but skills that any educated person would require and that liberal educators profess to develop (Heywood, 1994). The Employment Department believed that these could be developed within an undergraduate's specialist study provided that learning was designed for that purpose. But it was also argued by others that (1) some special provision should be made for separate learning in what might be called organizational behavior and behavior in organizations and (2) that university programs should be designed to meet the developmental needs of students, a proposal that had the theories of Perry and King and Kitchener in mind (Heywood, 1994; King and Kitchener, 1994; Perry, 1970).

However, it is clear that these developments in higher education cannot be separated from development in school systems. Many governments are concerned that

Table 2. A re-arranged summary of the top skill needs of Minnesota Employers

Attributes (most frequent ratings of "very important" by employers
Professionalism (punctuality, time management, attitude)
Self-direction, ability to take initiative
Adaptability, willingness to learn
Professional ethics, integrity
Verbal communication skills

Most frequent ratings of "not at all" or "not very important," last 5
Advanced mathematical reasoning (linear algebra, statistics, calculus)
Technical communications
Fluency in a language other than English
Knowledge of specific computer applications required for the job
Application of knowledge from a particular field of study

Other
Capability for promotion and advancement
Creativity
Ability to work in a culturally diverse environment
Ability to work in teams
Written communication skills
Basic mathematical reasoning (arithmetic, basic algebra)
Critical thinking and analysis
Problem solving, application of theory
General computer skills (word processing, spread sheets)
Knowledge of technology/equipment required for job

Minnesota Office of Higher Education. Minnesota measure. 2009 Report on Higher Education Performance. Cited in full in Sparks, E., & Waits, M. J. (2011) *Degrees for what jobs? Raising expectations for universities and colleges in a global economy* (p. 27). Washington, DC: National Governors Association.

teaching should lead to more deep learning than it has in the past, and that what are called "21st century skills" should be acquired by their pupils (Aniandou & Claro, 2009). In the United States the National Research Council established a committee, whose report has now been published, to "define the skills that are referenced by the labels "deeper learning," twenty first century skills," "college and career readiness," "Student–centered learning," "next generation learning," "new basic skills" and "higher order thing," and to describe how "they relate to each other and more traditional academic learning and content in these areas." From the perspective of the present discussion suffice it to say that they too look like the earlier reports that distinguish between three domains of competence, interpersonal, and intrapersonal (Pellegrino & Hilton, 2012). They point out that

research that demonstrates a causal relationship between these competencies and adult outcomes is limited although there is a positive relationship between these outcomes and early academic competencies. Although this has a bearing on the work of those who are developing courses in technological literacy, it highlights the failure of policymakers, educators, and researchers to consider the whole system of education. Clearly there is need for developmental models that cover both early and adult development.

Both directly and indirectly the Minnesota and U.K. Employment Department lists draw attention to the need for employees to be adaptable and flexible, by which is meant the ability to learn a new job when the one they are in becomes redundant. Because it is anticipated that employees will have to change their jobs more than in the past what was a cliché has become a reality for employees but it brings with it responsibilities for employers. Continuing personal (professional) development will have to become a sine qua non and a responsibility of both employee and employer so that employees have the skills necessary for transfer to a wide range of occupations. For this they will require a wider and more unified form of knowledge than in the past.

In Britain in the past, employers have been unwilling to accept entrants whose qualifications and experience are not perceived to be directly related to the jobs they have on offer. They have been unwilling to accept that individuals have transferable skills that can be of value in any job (Thomas & Madigan, 1974). Coupled with the need to acquire a new work identity, this can lead to a self-fulfilling hypothesis that employees come to believe that they are suitable for work only in the areas for which they have been specifically trained as indicated by their job titles. Is this "occupational transfer gap" real or imagined? It may be argued strongly that this is much more likely to be a perceptual problem and therefore imagined. Most employees are likely to have skills that cross the divide of job perceptions, for which reason there needs to be a skills approach to defining jobs that is accompanied by educational programs that encourage the possession of wide-ranging interests and knowledge that fit an "arena" of jobs. Youngman et al. (1978) described a "labour arena" as a group of skills that is already possessed or that may be readily acquired. It has a bearing on the design of records of achievement.

The implications for the institutional structures of higher education of changing patterns in employment are profound. If individuals are to be able to undertake new jobs that currently they or their potential employers perceive to be beyond their capabilities, they will require a liberal education that enlarges the mind. By that is meant something that is far more than the study of a range of disparate subjects that apparently have no connection with one another.

Enlargement of the Mind and Adaptability

The requirement that we should become more adaptable makes its own case for a broad education. However, it does not dictate what that breadth should be or how it should be structured. It does mean that however the curriculum is structured it should focus on what John Henry Newman called "enlargement or expansion of mind" (Newman, 1947, p. 118). For Newman enlargement, like knowledge, was a process through which a person obtained a philosophical disposition or wisdom. Therefore, one of the aims of higher education is to help students acquire the knowledge and skills necessary to enlarge their minds. Often when he is writing about what today are called the outcomes of university education, Newman uses terms and phrases from which it can only be concluded that he embraced all those domains of the person that are noncognitional. For example, "such an intellect [. . .] cannot be impetuous, cannot be at a loss, cannot but be patient, collected, and majestically calm" (Newman, 1947, p. 122). It is to be found in his idea of a university tutor and the value placed on residence (Culler, 1955, p. 46). It is also a reminder that the skills of the affective domain are as important as those in the cognitive domain, which is the point that industrialists are making when they demand personal transferable skills.

Newman argued that "knowledge itself, though a condition of the mind's enlargement, yet whatever be its range, is not the very thing that enlarges it" (Newman, 1890, p. 287). Rather, it is the ability to perceive the relationships between subjects. "Enlargement consists in the comparisons of the subjects of knowledge one with another. We feel ourselves to be ranging freely when we not only learn something but when we also refer it to what we know before" (Newman, 1890, p. 287; see also Culler, 1955, p. 182 ff). This would seem to be consistent with that present-day view of learning that it is the process by which experience develops new and reorganizes old responses (Heywood, 1984; Saupé, 1960). This is clearly what happens or should happen within courses. Without it there could be no development or movement within a course. But demonstrating that knowledge has been acquired is no guarantee that there has been enlargement.

Adaptability arises from a person's ability to think and reason. It is these abilities applied across a range of subjects that enlarge the mind. The one thing in which we are all engaged is reasoning. We are all engaged in "deducing well or ill, conclusions from premises, each concerning the subject of his own particular business" – "The man who has learnt to think, and to reason and to compare, and to discriminate and analyse... will not at once be a lawyer [...], or a physician, or a good landlord, or a man of business, or soldier, or engineer, or chemist [...] but he will be placed in that state of intellect in which he can take up any one of the sciences, or callings I have referred to, or any other for which he has a taste or special talent [...]" (Newman, 1947, p. 146). That is the essence of an educated person.

But in today's understanding "transfer" will not take place if these subjects are taught independently of each other (Saupé, 1960). Because transfer will occur only to the extent we expect it to occur, the curriculum has to show how it can occur in what might be best described as interdisciplinary or transdisciplinary situations. From Fischer's skill theory it is clear that if transferable skills are to be developed they must be practiced throughout a person's educational program, irrespective of the structure of that program (Fischer, Kenny, & Pipp, 1990). The failure to approach study in this way is the reason why a general education that comprises the study of a number of independently organized subjects is not liberal. It is the reason why in subject specialisms such as engineering so many students are unable to combine knowledge from the sub-disciplines to solve complex problems. Taught in a way that overcomes this problem, that is, in a spirit of uni-versality, engineering is as much a liberal study as any other (Heywood, 2010). Any detailed analysis of the activity (process) of engineering will demonstrate that this is so.

The Activity (Process) of Engineering

Engineering and its core activity design, as Louis Bucciarelli (2003) affirms, is an intensely social activity (process). Consider, for example the runaway impact of developments in the social media and the loss of control over the direction of their lives that some people already experience. Clearly if people are to control their lives then the skills of enterprise learning already referred to earlier become all important. Individuals will have to become good decision makers and will also have to become thoroughly good at evaluating information and making appropriate judgments, moral and otherwise, to exert the control they would wish over their jobs and their lives. Without a liberal education and the skills it helps to develop when it is designed for such learning they are unlikely to be able to exert this control.

Figure 1 shows a model of the interrelationships between the areas of knowledge and the achievement of a technological artefact for society and the economy (Heywood, 1986). Engineering is the art and science of making things that meet the needs of self and society. It has to function within the constraints, legal and otherwise, imposed by society and the environment.

The base represents the person. The mind that supports the whole activity is the source of our values, beliefs, and technical understanding: it is the source of our attitudes and opinions in the different social systems in which we find ourselves; it is the driver of our actions. It is the source of our ideas and creativity. Understanding how our beliefs and values (moral and otherwise) are formed is important to our conduct as engineers and individuals. It is at the core of any program of liberal education. It belongs

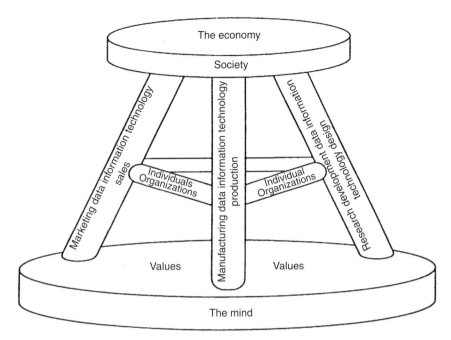

Figure 1. A simplified socio-technical model of the engineering process and its technological outputs in relation to the individual and society.

primarily to the domains of philosophy and theology.

The three legs of the stool represent the technological aspects of engineering: research, development, data acquisition, information technology, and design; manufacturing data and production; and marketing data and sales. The first two legs are the domains of engineering science, design, and manufacturing. The third leg is the knowledge domain of business, legal, and economic understanding. Supporting the legs are the trusses that represent individuals and the way the organization is structured. These are the domains of organizational behavior and behavior in organizations.

It follows that engineering is both an activity and a system that is a service to both individuals and society that continually creates new problems for both. It is used to create change and at the same time creates the need for change. It requires individuals to develop the skill of adaptability, which includes the ability to judge the merits or otherwise of a particular innovation. Engineering literacy requires that we

understand how individuals, organizations, and society interact with technology, and this requires an appreciation of the values that are brought to that understanding. The components of engineering literacy are: (1) the engineering of artefacts or the art and science of designing and making things; the term artefact is to be interpreted broadly to include such things as a computer program; (2) the engineering (technology) of organizations or the art and science of making organizations work for employees as well as employers; (3) information technology or the art and science of acquiring data in an appropriate form for problem solving; and (4) the art and science of understanding self and others and the factors that contribute to our aspirations and behavior.

Engineering literacy has as its objective the appreciation of engineering through an understanding of the relationships as represented by the model. In the model it is conflated with technological literacy because a major aim of such programs should be to address that major misunderstanding of Western society that assumes that technology has a "life of its own." It is as Bucciarelli says "romantic nonsense to think and talk this way out here in the big world. So too to imagine we can perfect a missile defense shield, that we can profit from the genetic manipulation of life at all levels without occasioning significant collateral damage, or that we can convince every scientist that global warming is upon us before it is too late to do anything about it – all of this is wishful thinking. It follows from a seriously flawed vision of technology, one that sets it apart and aloof, distant and seemingly out of reach of ordinary people. As citizens we ought to know and do better" (Bucciarelli, 2003, p. 101). It also follows that a person who has not an acquaintance with engineering is not a liberally educated person. Nevertheless, from the perspective of liberal education the model is incomplete. For example, there is no requirement for history, the fine arts and music, literature, or the learning of a language other than one's own. It is generally agreed that a liberal education would not be complete without attention to these. Engineering is as much a liberal study as it is vocational: it is the task of engineering literacy to demonstrate that this is so, and this requires attention to how it should be constructed for learning and consequently how it should be taught.

Learning and Teaching

Just as Newman's epistemology justifies the curriculum model presented so there is an epistemology that indicates the most appropriate approach to teaching and learning that integrated or trans-disciplinary studies require in the first instance. It comes from the work of the Scottish philosopher John Macmurray, which has been considered in more detail elsewhere (Heywood, 2012b). For the purpose of this chapter it is enough to say that Macmurray substituted "I do" for "I think" in Descarte's dictum *cogito ergo sum*. As Macmurray put it in one of his lectures *cogito non ergo sum* (Costello, 2002). All our activities begin with the practical. Finding out how to do what we want to

do leads to our theories. So it is with learning we learn that which we do. So the first stage of this curriculum should necessarily be problem or project based in which the problems or projects are arranged to ensure that the need for the other dimensions of knowledge and behavior become apparent and worthy of exploration. Key to the choice of activities is the centrality of engineering design for which engineering science is a necessary support. This first stage in any thought process is discovery, but even more so it is one of "romance," to cite the mathematician/philosopher Alfred North Whitehead (Whitehead, 1932).

Whitehead's idea of romance comes from his view that "life is essentially periodic" (p. 27). Intellectual progress begins with novelty. "Knowledge is not dominated by systematic procedure." [. . .] "Such a system as there must be is created piecemeal ad hoc" [. . .] "Romantic emotion is essentially the excitement consequent on the transition from bare facts to the first realisations of the import of unexplored relationships" (p. 28). The next stage of learning he calls "precision." "It is the stage of grammar, the grammar of language and the grammar of science. It proceeds by forcing on the students' acceptance of a given way of analysing the facts, bit by bit. New facts are added, but they are the facts which fit into the analysis" (p. 29). This stage, he argues, is "barren" without the stage of romance. The final stage is "generalisation." "It is a return to romanticism with added advantage of classified ideas and relevant technique" (p. 30). It is a stage of synthesis that requires romance. These stages may be used to characterize the differences between primary (elementary), post-primary (secondary) and university education. But some commentators have argued that universities have never got beyond the stage of precision and that is why some university educators including engineers advocate learning from the activities that take place in primary (elementary) school classrooms (Crynes & Crynes, 1997). In terms of curriculum structure it can also be seen that primary (elementary) education is a stage of interdisciplinarity, post-primary a stage for the subject disciplines and third level a return to interdisciplinarity.

But these cycles are going on all the time – short ones, long ones, and even for a single event in the classroom. It would be a mistake to align them with the Piagetian stages of cognitive development because the evidence from children in philosophy curriculum schemes suggests that young children are capable of abstract thought (Matthews, 1980). If that is so it would seem they are capable of precision and generalization. Much of Whitehead's theory of education is based on his experience, but it seems that our everyday experience of tackling new fields is consistent with his theory. Learning is a process of discovery: to begin a new study we have to be in a stage of romantic emotion, that is "the excitement consequent on the transition from the bare facts to the first realisations of the import of their unexplored relationships" (Whitehead, p. 28). This is the challenge of general education and more particularly teaching to provide for romance in a range of subject areas for, if it can do that, and if the curriculum can be designed to show the relations ships between subjects, then skill in transfer should be developed. As Saupé (1960, p. 68) noted, "transfer will only occur when there is a recognized similarity between the learning and the transfer situation" and "transfer will only occur to the extent that students expect it to" (p. 69). To develop this skill so that it takes in knowledge distant from the core is the challenge that Newman issues.

Practicalities

The argument that has been presented is open to the objection that might be made by supporters of the Sparks and Waits (2011) proposals that it takes little or no account of content. The list of aims and objectives in Table 3 is intended to remedy this defect and to stimulate discussion. At first sight, the achievement of these objectives would appear to be a tall order for what is likely to be a relatively short course, so the question of approach looms large.

It certainly could not be achieved without some integration. However, Newman's idea of recombination coupled with Whitehead's concept of "romance" and Bruner's (Bruner, 1963, 1974) view of discovery suggest there are three methods that are most likely to achieve the integration that is required. These are case studies, mini and major projects, and debates coupled with reading in advance following the learning technique of the advanced organizer. Atman and Bursic found in a controlled study of engineering students that those who read a textbook about engineering design subsequently exhibited more complex design processes than those who had not (Atman & Bursic, 1996). But such books have to be chosen carefully and might have to have specifically designed materials, as was done by Heywood and Montagu Pollock in the radio astronomy section of a course in physics for arts (humanities) students (Heywood & Montagu Pollock, 1977). Case studies are often used in the teaching of engineering ethics and carefully chosen ones can be used to illustrate the role of science in design. In respect of the former the well-documented failure of the *Challenger* space shuttle disaster that exploded and killed its crew is still in use even though the event occurred more than twenty years ago (Davis, 1998). It shows quite clearly the need to understand how other people think through illustrations of the thinking of managers and engineers as they tried to determine if the shuttle should be launched. It also deals with the problems in the relationships between professionals (engineers) and managers, and it is a reminder of need for managers in these circumstances to be engineering literate. There are, of course, more recent cases as a paper *Learning from Disaster* shows (van de Voort & de Berijn, 2009). There are other articles that can be used as case studies to both develop and assess basic engineering science. Other case studies may be provided to show how engineering science is used in engineering design. But there is nothing like practice, and this can be achieved through different types of projects such as design or make projects undertaken by an individual, design

Table 3. Aims and objectives of engineering literacy

Some aims and objectives for engineering literacy

- The prime aim of a course in engineering literacy would be to show the relevance of engineering to the solution of societal problems and to develop capabilities in the judgment of the merits of solutions that are proposed. It would involve showing the students what engineers do when they are doing engineering, an aim that implies giving them practical experience of designing and making things.
- For engineering this aim may also be expressed in terms of the ability to comprehend papers of the type that appear in IEEE Technology and Society and IEEE Spectrum but especially articles and comments in quality newspapers.

The materials and activities used should

- Illustrate the structure of thought and methods of inquiry in engineering and how it differs from science, especially physics
- Illustrate the centrality of design in engineering
- Provide an insight into the processes of engineering to such a point that students would be able to increase their understanding unaided, when circumstances required it
- Enable them to compare the structure of thought and methods of inquiry in engineering to those of their own subjects and interests
- Consider the implications of the view that engineering design is a social process
- Help them gain an insight into their own moral purpose

From inspection of these objectives it would appear that a course in engineering literacy would consider the following:

- The sources of engineering ideas and problems (creativity, invention, and innovation)
- The design and manufacturing cycle; modeling
- The potential and limitations of materials and processes
- The potential and limitations of machines for the manufacturing process
- The determinants of cost, quality, and morality in engineering and manufacturing processes (optimization)
- The determination of risk and safety in engineering taking into account moral purpose
- The significance of product design in engineering, and design in styling
- The potential of electronic devices in information technology, and control technology
- The assistance that the application of scientific principles can give to each stage of the engineering activity
- The significance of industry in wealth creation and the roles of the entrepreneur and innovator
- The significance of the market in product design and manufacture
- The voting contribution of the individual in a democratic society as a determinant of prevailing and future socioeconomic models
- The understanding of human behavior and factors that contribute to personal and interpersonal competence

---

and make projects undertaken in teams, and projects (investigations) for the purpose of undertaking a technical investigation (Kelly & Heywood, 1996). A well-known example of the latter is the construction of bridges with balsa-wood for the purpose of investigating the parameters of different types of structures. Projects as conceived here are a form of problem based learning. But none of these methods will contribute to liberal knowledge unless there is reflection on the process and outcomes achieved. At the same time project work in teams provides a sound base for the development of personal transferable skills that are so necessary for satisfactory performance at work.

Macmurray's Other Challenge

The focus placed on Macmurray's epistemology in the earlier section does little justice to his philosophy as expressed in the *Self
as Agent* and makes no reference to the other
volume of his Gifford lectures – *Persons in
Relation* (Macmurray, 1961). Linked together
they have much to say about the problems
of present society and the way forward. In
the *Self as Agent*, writing about its major
theme Macmurray says, "consider now the
Self in relation to the world. When I act I
modify the world. Action is causally effective, even if it fails of the particular effect
that is intended. This implies that the Self
is part of the world in which it acts, and in
dynamic relation with the rest of the world.
On the other hand, as subject the Self stands
'over against' the world which is its object.
The Self as subject then is not part of the
world it knows, but withdrawn from it, and
so, in conception, outside it or other than
its object. But to be part of the world is to
exist, while to be excluded from the world
is to be non-existent" (Macmurray, 1957,
p. 91). In these terms we can choose to exist
or non-exist as a person or a professional.
If we choose the former then there is an
obligation to understand the world beyond
that of the technicalities of our chosen profession. That is the case for liberal education, and since the "Self is a person" and
"persons only develop as persons in relation
to other persons. We come to be who we
are as personal individuals only in personal
relationships" (Costello, 2002, p. 326). The
outcome of Macmurray's philosophy is that
it challenges the "primacy placed on individuality and therefore self interest, creates
an inevitable primacy for competition over
cooperation, to say nothing of communal
trust and affection, in public relationships"
(Costello, 2002, p. 328). For Macmurray all
competition is for the sake of cooperation,
and all cooperation is for the sake of commu-
nion (friendship). These are issues that are
at the very heart of current debates about
capitalism in the Western world, and the
relations between employers and employees
as the latter strive to continually develop.

Macmurray would say that a new contract
has to be made between them and society.
"While such a philosophy may seem distant from the practical realities of what to
do in the classroom, this cannot be further
from the truth. What students are taught
to relate to-other people or the world of
technology-influences the types of contracts
students will forge throughout their careers.
Whether they see intrinsic value in learning
engineering or whether value is determined
extrinsically by their relationship with an
employer" (Alan Cheville, personal communication, 2013).

Discussion

However good, however poor the data several inferences may be drawn. They are
(1) that policymaking should be undertaken
from a systems perspective that embraces
elementary education at one end of the
spectrum and lifelong (permanent) education at the other; (2) that educators
together with industrialists should pay much
more attention to lifelong education, and in
respect of engineers continuing professional
development in both technical and personal dimensions; (3) that educators more
generally and engineering educators specifically should better prepare students with the
skills of flexibility and adaptability required
to cope with ever changing knowledge, that
is "personal transferable skills"; and (4) that
industrialists have as much responsibility as
their employees for continuing professional
development.

The NGA report provides an international opportunity to open a fundamental
debate about the aims of education in a technological age particularly at a time when
higher education is in a state of crisis. The
proposals in the report are more revolutionary than they are evolutionary. They highlight the need to bring about reconciliation
between liberal and vocational education.

A utilitarian view of education that sees
its purpose as economic is one sided, and
ultimately corrosive of the individual. Education has both social and personal goals for

as Whitehead (p. 1) said "the valuable intellectual development is self-development." All our activities depend on our belief systems for which reason every course of study has to be grounded in the central issues we face in affirming or denying these systems. No major activity in life is a "closed" area of knowledge: most major problems require knowledge from other sources for their solution. Closed systems die; open systems adapt and live. An equally valid exemplar of this argument as applied to the curriculum would be nursing. It would, like engineering, have to be completed by other studies that make up this conception of liberal education.

From the practical perspective the curriculum paradigm has been related to that stage of learning that Whitehead calls romance, and this has consequences for learning and teaching. It brings us to another crossroads. In one direction curriculum structure persists. In the other direction curriculum structure is changed. It is difficult to see how an integrated curriculum of the kind proposed can be achieved within a traditional unit (per lecture) structure. For example, at the level of romance, it is easy to demonstrate that the key skills of making things are best learnt during a concentrated period of time (Parlett & King, 1971). It can also be argued that this is true when learning the skills of technical investigation (Kelly & Heywood, 1996). It is also evident that at this level different approaches will have to be sought to the teaching of engineering science in the service of engineering design (Heywood, 2010).

It has been argued that there is a necessary link between liberal and vocational education. The one cannot do without the other. The activity of engineering as defined here contributes a dimension of knowledge without which a liberal education could not be said to be liberal or unified. Engineering is at one and the same time liberal and vocational. A vocational education that does not embrace liberal knowledge is not an education but training. It defeats the objectives of helping individuals to become more adaptable and flexible on the one hand, and on

the other hand of preparing them for life. These arguments apply generally to thinking about higher education irrespective of culture, even though how they are resolved will be a function of the particular socioeconomic culture to which they are addressed.

The crossroads that finally brought this writer to a halt pointed engineering along to one route labeled modified tradition while another road pointed to a general education that had embraced engineering literacy as the base for professional study. Together with every other student engineers would begin with a liberal education focused on engineering and technological literacy. Only when this was completed would they proceed to professional studies.

Research

There is a need to continue and refine studies of changing patterns in the workforce, particularly with respect to changes to individual careers over their lifetimes. There is also a need to investigate in fine detail the jobs that engineers do and the knowledge they use in fulfilling these tasks. In particular, there is a need to know what engineers do in small organizations because the models educators have of what the process of engineering is are derived from large organizations. Without such knowledge it is difficult to see how an adequate statement of curriculum goals can be reached.

Given the proposition that the system of higher education is in a state of crisis there is a need to investigate and experiment with alternative curriculum structures, both horizontally and vertically by year. Engineering educators have not engaged in much research and development on the curriculum and its structures. There is a need to engage the engineering community in bottom-up development rather than top-down, and accrediting organizations need to recognize the value of a "many approaches" model.

Such investigations and experiments should be extended to teaching methods and assessment and how they relate to the syllabus (list of content). Lists of content are often too long and pile pressure on students. For example, very little is known about how long it takes students of varying abilities to learn a fundamental concept or principle. In regard to assessment and its relation to learning it would be useful to know how many engineering teachers practice those classroom assessment procedures recommended by Angelo and Cross (1993) that bear on this issue, and to undertake case studies to evaluate their impact on student learning and teaching. Time line studies of learning (of the kind used by Atman in her studies of how students learn design; see Atman et al. [2010] for a summary) could be developed for more general use in the evaluation of the curriculum from this perspective. The assessment of student learning has received little attention until recently. There is need for substantive research in this area (Heywood, 2000).

It is clear that relatively little is known about the effectiveness of the teaching of transferable skills and that particular attention to be paid to evaluation research in this area. Case studies can make an important contribution to understanding in this area.

Finally higher education should begin to appreciate that it is a subsystem within a more general system of education. Such thinking would be helped by a more general theory of human development that does not separate out the school from the university, and the school and university from adulthood. There are many dimensions to human behavior and learning that have to be considered when determining the goals that dictate a curriculum.

Acknowledgments

This chapter is a revised and expanded version of the 2012 distinguished lecture sponsored by the Technological Literacy, Liberal Education and Engineer in Society, Ethics and Engineering, and Public Policy divisions of the American Society for Engineering Education at the Society's annual conference. I am grateful to Dr. Alan Cheville and Dr. John Krupczak for their advice in the preparation of this version and to the editors for their insights. Sections of this paper are based on Heywood (2012b).

Footnotes

1. The term vocational is used a number of ways throughout the world. It is used to relate to particular careers that are said to be a "calling," e.g., nursing, the priesthood. It is also used as the opposite of academic to describe courses that rely on specialist techniques, e.g., technicians, clerical officers. It is used in the second sense here.

2. Broadly speaking, in England and Wales most courses are of three years' duration. The exceptions include engineering, which is four years for a professional qualification. Entry is via the Advanced level of the General Certificate of Education. A similar system is used in Northern Ireland. Scottish degrees are of four years' duration following entry by a broad-based school Higher Certificate. In the Republic of Ireland there are both three- and four-year courses following, as in Scotland, a broadly based school Leaving Certificate, where students enter university at 17. Most courses in the University of Dublin (Trinity College are of four years' duration.

3. Much of this section is based on Heywood (2012a).

4. A 1961 report from the U.K. manpower committee (Cmnd 1490) suggested that by 1965 supply and demand of manpower should not be much out of balance and that a surplus may exist after that date brought forth considerable criticism from the establishment (Report of a debate in the House of Lords. *Guardian*, 16, 11: 1961), and the committee were more or less forced to back track (Heywood, 1969). Similarly, Heywood ran into trouble with the establishment when as part of a study of the system of technological education in the United Kingdom he was required to establish industrial demand for the output of graduates from university and college programs. No evidence of a shortage was found. It was confidently expected by his steering committee that he would find to the contrary. When he put the case at the 1963 conference of the British Association for Commercial and

Industrial Education, the chairman, who was also chairman of the government subcommittee on scientific and technological manpower, with others severely castigated him and considered the data to be wrong. Several years later he wrote to Heywood to say that he had been wrong and that the real shortage was of technicians. This was a view promoted by Cotgrove (1958, 1963) from research that he first reported in 1958. Another frequently made criticism of industry during the late 1950s and 1960s is that many qualified personnel were underutilized. Heywood eventually published his data in 1974.

5. In the United Kingdom the registered body for engineers distinguishes between Chartered Engineers, Incorporated Engineers, and Engineering Technicians. See http://www.engc.org.uk/. Engineering Council UK.

References

Ananiadou, K., & Claro, M. (2009). *21st century skills and competencies for new millenium learners in OECD countries*. Paris: Organization for Economic Cooperation and Development. Retrieved from http://www.oecd-ilibrary.org/education/21st-century-skills-and-competencies-for-new-millenium-learnerners-in-oecd countries_218525261

Angelo, T., & Cross, P. K. (1993). *Classroom assessment techniques*. San Francisco: Jossey Bass.

Atman, C. J., & Bursic, K. M. (1996). Teaching engineering design. Can reading a textbook make a difference? *Research in Engineering Design*, 8, 240–250.

Atman, C. J., Sheppard, S. D., Turns, J., Adams, R. S., Fleming, L. N., Stevens, R., . . . Lund, D. (2010). *Enabling engineering student success. The final report for the Center for the Advancement of Engineering Education*. San Rafael, CA: Morgan and Claypool.

Blau, J. (2011) (updated 19:08:2011). Germany faces shortage of engineers. *IEEE Spectrum*. Retrieved from http://spectrum.ieee.org/at-work/tech-careers/germany-faces-a-shortage-of-engineers

Boston Consulting Group (2012). Skills gap in US manufacturing is less pervasive than many believe. Summarized by Sims, D. in IMT industry market trends. ThomasNet. Com news. Retrieved from http://news.thomasnet.com/IMT/2012/10/18/skilled-worker-shortage-may be -exaggerated/?

Bruner, J. (1974). *The relevance of education*. Harmondsworth, UK: Penguin.

Bucciarelli, L. L. (2003). *Engineering philosophy*. Delft, The Netherlands: Delft University Press.

Carter, G, Heywood, J., & Kelly, D. T. (1986). *A case study in curriculum assessment: GCE engineering science (advanced)*. Manchester, UK: Roundthorn.

Cmnd 1490 (1961). Advisory Council on Scientific Policy. Committee on Scientific Manpower. *The long term demand for scientific manpower*. London: Her Majesty's Stationery Office.

Costello, J. E. (2002). *John Macmurray: A biography*. Edinburgh: Floris Books.

Cotgrove, S. F. (1958). *Technical education and social change*. London: Allen and Unwin.

Cotgrove, S. F. (1960). The education and training of technicians. *B.A.C.I.E Journal*, 14, 144. London: British Advisory Council for Commercial and Industrial Education.

Crynes, B. I., & Crynes, D. A. (1997). They already do it: Common practices in primary education that engineering should use. In *Proceedings of Frontiers in Education Conference*, 3, 12–19. Piscataway, NJ: IEEE.

Culler, A. D. (1955). The *imperial intellect. A study of Newman's educational ideal*. New Haven, CT: Yale University Press.

Culver, R. S., & Hackos, J. T. (1982). Perry's model of intellectual development. *Engineering Education*, 73(2), 221–226.

Davis, M. (1998). *Thinking like an engineer: Studies in the ethics of a profession*. New York: Oxford University Press.

First Bell. (2010). Programs touted as helping students for middle skill level jobs. (First Bell is a digest of the most important news selected from thousands of sources by the editors of *Custom Briefings* for the American Society for Engineering Education [ASEE], which circulates it to its members at regular short intervals).

First Bell. (2011a). *Labor researchers tell Congress U.S not lacking in scientists, engineers*. Washington, DC: ASEE.

First Bell. (2011b). *Demand for STEM skills increasing*. Washington, DC: ASEE.

First Bell. (2012). *US failing to produce enough STEM workers*. Washington, DC: ASEE.

First Bell. (ASEE). (2012, April 13). *p1. Report.* US failing to produce enough STEM workers. (*First Bell* is a more or less weekly summary of press cuttings relevant to engineering compiled for members of the American Society for Engineering Education).

Fischer, K. W., Kenny, S. L., & Pipp, S. L. (1990). How cognitive processes and environmental conditions organize discontinuities in the development of abstractions. In C. N. Alexander & E. J. Langer (Eds.), *Higher stages of human development.* Oxford: Oxford University Press.

Heywood, J. (1969). The manpower debate. In *An evaluation of certain post-war developments in technological education* (pp. 470–499). M. Litt Thesis, Lancaster, UK: University of Lancaster.

Heywood, J. (1974). Trends in the supply and demand for qualified manpower in the sixties and seventies. *The Vocational Aspect of Education, 26*(64), 65–72.

Heywood, J. (1986). Toward technological literacy in Ireland: An opportunity for an inclusive approach. In J. Heywood & P. Matthews (Eds.), *Technology, society and the school curriculum: Practice and theory in Europe.* Manchester: Roundthorn.

Heywood, J. (1994). *Enterprise learning and its assessment in higher education.* Technical Report No. 20. Sheffield, UK: Learning Methods Branch, Employment Department.

Heywood, J. (2000). *Assessment in higher education: Student learning, teaching, programmes and institutions.* London: Jessica Kingsley.

Heywood, J. (2005). Aims and objectives. In *Engineering education: Research and development in curriculum and instruction* (pp. 19–52). Hoboken, NJ: IEEE/John Wiley & Sons.

Heywood, J. (2010). Engineering literacy: A component of liberal education. In *Proceedings of Annual Conference of the American Society for Engineering Education.* CD Paper 1505. Washington, DC: American Society for Engineering.

Heywood, J. (2012a). The response of higher and technological education to changing patterns of employment. In *Proceedings of Annual Conference of the American Society for Engineering Education.* CD Paper 3626. Washington, DC: American Society for Engineering Education.

Heywood, J. (2012b). Philosophy and undergraduate teaching and learning. Perspectives for engineering education. In *Proceedings of Annual Conference of the American Society for Engineering Education.* CD Paper 5738. Washington, DC: American Society for Engineering Education.

Heywood, J., & Montagu Pollock, H. (1977). *Science for arts students: A case study in curriculum development.* Guildford, UK: Society for Research into Higher Education.

Kelly, D. T., & Heywood, J. (1996). Alternative approaches to K-12 school technology illustrated by an experimental course in technical investigations. In *Proceedings of Frontiers in Education Conference.* Piscataway, NJ: IEEE (pp. 388–393).

King, P. M., & Kitchener, K. S. (1994). *Developing reflective practice.* San Francisco: Jossey-Bass.

Kourdi, J. (2010). The future is not what it used to be. *The Professional Manager 19*(2), 27–28.

Krupczak, J., Blake, J. W., Disney, K. A., Hilgarth, C. O., Libros, R., Mina, M., & Walk, S. R. (2012). Defining engineering and technological literacy. In *Proceedings of Annual Conference of the American Society for Engineering Education.* CD. Paper 5100. Washington, DC: American Society for Engineering Education.

Lowell, L. B., & Salzman, H. (2007). *Into the eye of the storm: Assessing the evidence on science and engineering education, quality and workforce demand.* Washington, DC: Urban Institute.

Macmurray, J. (1957). *The self agent.* London: Faber and Faber.

Macmurray, J. (1961). *Persons in relation.* London: Faber and Faber.

Matthews, M. (2011, November). Editorial. *ASEE Prism, 21*(3), 6.

Newman, J. H. (1890). *Fifteen sermons preached before the University of Oxford* (3rd ed.). London: Rivingtons.

Newman, J. H. (1852/1947). *The idea of a university* (with additional lectures added in 1873. (1947 edition edited by C. F. Harrold). London: Longmans, Green.

Organisation for Economic Co-operation and Development (OECD). (2012, July 11). *Employment outlook 2012.* Paris OECD. Cited in *The Times,* p. 32.

Owen, S., & Heywood, J. (1990). Transition technology in Ireland: An experimental course. *International Journal of Technology and Design Education, 1*(1), 21–32.

Parlett, M. R., & King, J. G. (1971). *Concentrated study.* London: Society for Research into Higher Education.

Pellegrino, J. W., & Hilton, M. L. (Eds.) (2012). *Education for life and work: Developing transferable skills in the 21st century.* Washington, DC: The National Academies Press. (Prepublication copy: uncorrected proofs).

Perry, W. B. (1970). *Forms of intellectual and ethical development in the college years.* New York: Holt, Reinhart and Winston.

Saupé, J. L (1961). Learning and evaluation processes. In P. Dressel (Ed.), *Evaluation in higher education* (pp. 55–78). Boston, MA: Houghton Mifflin.

Schneiderman, R. (2010, March). Economy and shortages affect European job outlook. *IEEE Spectrum.* Retrieved from http://spectrum.ieee/at-work/tech-careers-and-shortages-affect-the-europ

Sparks, E., & Waits, M. J. (2011). *Degrees for what jobs? Raising expectations for universities and colleges in a global economy.* Washington, DC: National Governors Association.

Teitelbaum, M. S. (2003). Do we need more scientists? *The Public Interest* No 153. Washington, DC: National Affairs Inc. Also in T. K. Kelly, W. Butz, S. J. Carroll, D. N. Adamson, & G. Bloom (Eds.) (2007). *The US scientific and technical workforce: Improving data for decision making.* Report of Conference Proceedings. Rand Corporation. Retrieved from WWW.rand.org

Thomas, B., & Madigan, C. (1974). Strategy and job choice after redundancy: A case study in the aircraft industry. *Sociological Review, 22,* 83–102.

van de Voort, H., & de Berijn, H. (2009). Learning from disasters. Competing perspectives on tragedy. *IEEE Technology and Society, 28*(3), 28–36.

Wadwha, V. (2011). Leading edge. Over the hill at 40. *ASEE Prism, 21*(1), 32.

Whitehead, A. N. (1932–1959, 9th impression). *The aims of education and other essays.* London: Benn.

Youngman, M. B., Oxtoby, R., Monk, J. D., & Heywood, J. (1978). Implications. *Analysing jobs* (pp. 106 and 107). Aldershot, UK: Gower Press.

Zachary, G. P (2011, April). Jobless innovation. *IEEE Spectrum,* 8.

Index